Lecture Notes in Computer Science 7073

Commenced Publication in 1973
Founding and Former Series Editors:
Gerhard Goos, Juris Hartmanis, and Jan v

Dong Hoon Lee   Xiaoyun Wang (Eds.)

Advances in Cryptology – ASIACRYPT 2011

17th International Conference on the Theory
and Application of Cryptology and Information Security
Seoul, South Korea, December 4-8, 2011
Proceedings

 Springer

Volume Editors

Dong Hoon Lee
Korea University
Center for Information Security Technologies
Anam Dong 5-ga, Seungbuk-gu, Seoul, South Korea
E-mail: donghlee@korea.ac.kr

Xiaoyun Wang
Tsinghua University
Institute for Advanced Study
Beijing 100084, China
E-mail: xiaoyunwang@tsinghua.edu.cn

ISSN 0302-9743   e-ISSN 1611-3349
ISBN 978-3-642-25384-3   e-ISBN 978-3-642-25385-0
DOI 10.1007/978-3-642-25385-0
Springer Heidelberg Dordrecht London New York

Library of Congress Control Number: 2011940813

CR Subject Classification (1998): E.3, D.4.6, F.2, K.6.5, G.2, I.1, J.1

LNCS Sublibrary: SL 4 – Security and Cryptology

Typesetting: Camera-ready by author, data conversion by Scientific Publishing Services, Chennai, India

Printed on acid-free paper

Springer is part of Springer Science+Business Media (www.springer.com)

Preface

ASIACRYPT 2011, the 17th International Conference on Theory and Application of Cryptology and Information Security, was held during December 4–8 in the Silla Hotel, Seoul, Republic of Korea. The conference was sponsored by the International Association for Cryptologic Research (IACR) in cooperation with Korea Institute of Information Security and Cryptology (KIISC), Digital Contents Society (DCS), Korea Internet Security Agency (KISA), and National Security Research Institute (NSRI). It was also co-sponsored by the Center for Information Security Technologies of Korea University (CIST), the Korean Federation of Science and Technology Societies (KOFST), Seoul National University, Electronics and Telecommunications Research Institute (ETRI), and Seoul Metropolitan Government.

We received 266 valid submissions, of which 42 were accepted for publication. With two pairs of papers merged, these proceedings contain the revised versions of 40 papers. The Program Committee (PC) was aided by 243 external reviewers. Every paper received at least three independent reviews, and papers with PC contributions got five or more. Several questions from PC members to authors were relayed in order to increase the quality of submissions. ASIACRYPT 2011 used a rolling Co-chair model and we made all decisions by consensus by sharing a great deal of e-mails.

For the Best Paper Award, the PC selected "A Framework for Practical Universally Composable Zero-Knowledge Protocols" by Jan Camenisch, Stephan Krenn, and Victor Shoup and "Counting Points on Genus 2 Curves with Real Multiplication" by Pierrick Gaudry, David Kohel, and Benjamin Smith. There were two invited talks; Joan Daemen delivered "15 Years of Rijndael" on December 6 and Úlfar Erlingsson spoke on "Securing Cloud Computing Services" on December 7.

We would like to thank the authors of all submissions regardless of whether their papers were accepted or not. Their work made this conference possible. We are extremely grateful to the PC members for their enormous investment of time and effort in the difficult and delicate process of review and selection. A list of PC members and external reviewers can be found on succeeding pages of this volume. We would like to thank Hyoung Joong Kim, who was the General Chair in charge of the local organization and finances. Special thanks go to Shai Halevi for providing and setting up the splendid review software. We are most grateful to Kwangsu Lee and Jong Hwan Park, who provided support for the entire ASIACRYPT 2011 process. We are also grateful to Masayuki Abe, the ASIACRYPT 2010 Program Chair, for his timely information and replies to the host of questions we posed during the process.

September 2011

Dong Hoon Lee
Xiaoyun Wang

ASIACRYPT 2011

The 17th Annual International Conference on the Theory and Application of Cryptology and Information Security

December 4–8, 2011, Seoul, Korea

Sponsored by
the International Association of Cryptologic Research (IACR)

in cooperation with
Korea Institute of Information Security and Cryptology (KIISC),
Digital Contents Society (DCS),
Korea Internet Security Agency (KISA),
and
National Security Research Institute (NSRI)

General Chair

Hyoung Joong Kim   Korea University, Korea

Program Chairs

Dong Hoon Lee   Korea University, Korea
Xiaoyun Wang   Tsinghua University, China

Program Committee

Michel Abdalla   ENS and CNRS, France
Masayuki Abe   NTT, Japan
Kazumaro Aoki   NTT, Japan
Jung Hee Cheon   Seoul National University, Korea
Carlos Cid   Royal Holloway University of London, UK
Craig Gentry   IBM Research, USA
Vipul Goyal   Microsoft Research, India
Jens Groth   University College London, UK
Iftach Haitner   Tel Aviv University, Israel
Dennis Hofheinz   Karlsruhe Institute of Technology, Germany

Antoine Joux	DGA and Universite de Versailles, PRISM, France
Aggelos Kiayias	University of Connecticut, USA
Eike Kiltz	Ruhr University Bochum, Germany
Jongsung Kim	Kyungnam University, Korea
Lars R. Knudsen	Technical University of Denmark, Denmark
Dong Hoon Lee	Korea University, Korea
Arjen K. Lenstra	EPFL, Switzerland
Stefan Lucks	Bauhaus-University Weimar, Germany
Willi Meier	FHNW, Switzerland
Alfred Menezes	University of Waterloo, Canada
Payman Mohassel	University of Calgary, Canada
Phong Q. Nguyen	INRIA and ENS, France
Jesper Buus Nielsen	Aarhus University, Denmark
Chris Peikert	Georgia Tech, USA
Thomas Peyrin	NTU, Singapore
Christian Rechberger	ENS, France
Palash Sarkar	Indian Statistical Institute, India
Nigel P. Smart	University of Bristol, UK
Willy Susilo	University of Wollongong, Australia
Xiaoyun Wang	Tsinghua University, China
Hoeteck Wee	George Washington University, USA
Hongbo Yu	Tsinghua University, China

External Reviewers

Hadi Ahmadi
Martin Albrecht
Mohsen Alimomeni
Jacob Alperin-Sheriff
Tadashi Araragi
Frederik Armknecht
Man Ho Au
Jean-Philippe Aumasson
Chung Hun Baek
Joonsang Baek
Endre Bangerter
Masoud Barati
Paulo S.L.M. Barreto
Stephanie Bayer
Amos Beimel
Mihir Bellare
David Bernhard
Rishiraj Bhattacharyya
Sanjay Bhattacherjee

Simon Blackburn
Bruno Blanchet
Andrey Bogdanov
Julia Borghoff
Joppe Bos
Wieb Bosma
Charles Bouillaguet
Elette Boyle
Christina Brzuska
Florian Böhl
Jan Camenisch
Angelo De Caro
David Cash
Dario Catalano
Debrup Chakraborty
Sanjit Chatterjee
Céline Chevalier
Kyu Young Choi
Seung Geol Choi

Ashish Choudhury
Sherman S.M. Chow
Cheng-Kang Chu
Ji Young Chun
Kai-Min Chung
Iwen Coisel
Véronique Cortier
Joan Daemen
Ivan Damgård
M. Prem Laxman Das
Yi Deng
Yvo Desmedt
Claus Diem
Léo Ducas
Nico Döttling
Pooya Farshim
Sebastian Faust
Serge Fehr
Matthieu Finiasz

Dario Fiore
Ewan Fleischmann
Christian Forler
Pierre-Alain Fouque
Georg Fuchsbauer
Atsushi Fujioka
Eiichiro Fujisaki
Jakob Funder
Steven Galbraith
Nicolas Gama
Praveen Gauravaram
Ran Gelles
Michael Gorski
Rob Granger
Fuchun Guo
Jian Guo
Kishan Chand Gupta
Shai Halevi
Mike Hamburg
Dong-Guk Han
Jinguang Han
Carmit Hazay
Jens Hermans
Shoichi Hirose
Hyunsook Hong
Qiong Huang
Xinyi Huang
Pavel Hubacek
Jung Yeon Hwang
Yuval Ishay
Kouichi Itoh
Tetsu Iwata
Abhishek Jain
David Jao
Jeremy Jean
Ik Rae Jeong
Dimitar Jetchev
Nam-Su Jho
Saqib Kakvi
Seny Kamara
Koray Karabina
Jonathan Katz
Shahram Khazaei
Jihye Kim
Kitak Kim

Minkyu Kim
Myungsun Kim
Sungwook Kim
TaeHyun Kim
Thorsten Kleinjung
Edward Knapp
Simon Knellwolf
Woo Kwon Koo
Daniel Kraschewski
Mathias Krause
Stephan Krenn
Ranjit Kumaresan
Hidenori Kuwakado
Soonhak Kwon
Junzuo Lai
Gregor Leander
Hyung Tae Lee
Kwangsu Lee
Mun-Kyu Lee
Yuseop Lee
Anja Lehmann
Allison Lewko
Jin Li
Benoît Libert
Seongan Lim
Huijia Lin
Yehuda Lindell
Richard Lindner
Jake Loftus
Jiqiang Lu
Vadim Lyubashevsky
Daegun Ma
Subhamoy Maitra
Hemanta Maji
Mark Manulis
Alexander May
James McKee
Sigurd Meldgaard
Florian Mendel
Alexander Meurer
Andrea Miele
Marine Minier
Tal Moran
Ciaran Mullan
Sean Murphy

Mridul Nandi
Kris Narayan
María Naya-Plasencia
Salman Niksefat
Ryo Nishimaki
Peter Sebastian Nordholt
Miyako Ohkubo
Tatsuaki Okamoto
Eran Omri
Claudio Orlandi
Onur Ozen
Dan Page
Periklis
    Papakonstantinou
Hyun-A Park
Je Hong Park
Jong Hwan Park
Jung Youl Park
Anat
    Paskin-Cherniavsky
Valerio Pastro
Kenny Paterson
Arpita Patra
Serdar Pehlivanoglu
Ludovic Perret
Christiane Peters
Viet Pham
Duong Hieu Phan
Krzysztof Pietrzak
Benny Pinkas
David Pointcheval
Tal Rabin
Somindu C Ramanna
Vanishree Rao
Mariana Raykova
Mohammad Reza
    Reyhanitabar
Thomas Ristenpart
Aaron Roth
Ron Rothblum
Carla Ràfols
S. Sharmila Deva Selvi
Subhabrata Samajder
Yu Sasaki
Takakazu Satoh

Christian Schaffner
Martin Schläffer
Dominique Schröder
Jacob Schuldt
Jörg Schwenk
Sven Schäge
Mike Scott
Jae Hong Seo
Hakan Seyalioglu
Abhi Shelat
Shashank Singh
Adam Smith
Martijn Stam
Paul Stankovski
Douglas Stebila
John Steinberger
Rainer Steinwandt
Mario Strefler
Christoph Striecks

Koutarou Suzuki
Tsuyoshi Takagi
Qiang Tang
Tamir Tassa
Aris Tentes
Stefano Tessaro
Abhradeep
   Guha Thakurta
Nicolas Theriault
Enrico Thomae
Søren S Thomsen
Mehdi Tibouchi
Tomas Toft
Berkant Ustaoglu
Yevgeniy Vahlis
Frederik Vercauteren
Damien Vergnaud
Ivan Visconti
Martin Vuagnoux

Lei Wang
Bogdan Warinschi
Gaven Watson
Lei Wei
Daniel Wichs
Christopher Wolf
Hongjun Wu
Qianhong Wu
Keita Xagawa
Guomin Yang
Kan Yasuda
Kazuki Yoneyama
Tsz Hon Yuen
Greg Zaverucha
Erik Zenner
Hong-Sheng Zhou
Angela Zottarel

Sponsoring Institutions

Center for Information Security Technologies of Korea University (CIST)
Korean Federation of Science and Technology Societies (KOFST)
Seoul National University
Electronics and Telecommunications Research Institute (ETRI)
Seoul Metropolitan Government

Table of Contents

Database Privacy

Hash Function

Symmetric Key Encryption

Zero Knowledge Proof

Universal Composability

Foundation

Secure Computation and Secret Sharing

Public Key Signature

Leakage Resilient Cryptography

BKZ 2.0: Better Lattice Security Estimates

Yuanmi Chen[1] and Phong Q. Nguyen[2]

[1] ENS, Dept. Informatique, 45 rue d'Ulm, 75005 Paris, France
http://www.eleves.ens.fr/home/ychen/
[2] INRIA and ENS, Dept. Informatique, 45 rue d'Ulm, 75005 Paris, France
http://www.di.ens.fr/~pnguyen/

Abstract. The best lattice reduction algorithm known in practice for high dimension is Schnorr-Euchner's BKZ: all security estimates of lattice cryptosystems are based on NTL's old implementation of BKZ. However, recent progress on lattice enumeration suggests that BKZ and its NTL implementation are no longer optimal, but the precise impact on security estimates was unclear. We assess this impact thanks to extensive experiments with BKZ 2.0, the first state-of-the-art implementation of BKZ incorporating recent improvements, such as Gama-Nguyen-Regev pruning. We propose an efficient simulation algorithm to model the behaviour of BKZ in high dimension with high blocksize ≥ 50, which can predict approximately both the output quality and the running time, thereby revising lattice security estimates. For instance, our simulation suggests that the smallest NTRUSign parameter set, which was claimed to provide at least 93-bit security against key-recovery lattice attacks, actually offers at most 65-bit security.

1 Introduction

Lattices are discrete subgroups of $\mathbb{R}^m$. A lattice L is represented by a *basis*, *i.e.* a set of linearly independent vectors $\mathbf{b}_1, \ldots, \mathbf{b}_n$ in $\mathbb{R}^m$ such that L is equal to the set $L(\mathbf{b}_1, \ldots, \mathbf{b}_n) = \{\sum_{i=1}^n x_i \mathbf{b}_i, x_i \in \mathbb{Z}\}$ of all integer linear combinations of the $\mathbf{b}_i$'s. The integer n is the dimension of L. The goal of *lattice reduction* is to find bases consisting of reasonably short and nearly orthogonal vectors. Lattice reduction algorithms have many applications (see [35]), notably public-key cryptanalysis where they have been used to break special cases of RSA and DSA, among others (see [32] and references therein). There are roughly two types of lattice reduction algorithms:

- *Approximation* algorithms like the celebrated LLL algorithm [22,35], and its blockwise generalizations [41,42,7,8]. Such algorithms find relatively short vectors, but usually not shortest vectors in high dimension.
- *Exact* algorithms to output shortest or nearly shortest vectors. There are space-efficient enumeration algorithms [38,20,6,42,43,10] and exponential-space algorithms [3,36,30,29], the latter being outperformed in practice by the former despite their better asymptotic running time $2^{O(n)}$.

D.H. Lee and X. Wang (Eds.): ASIACRYPT 2011, LNCS 7073, pp. 1–20, 2011.

In high dimension, only approximation algorithms can be run, but both types are complementary: approximation algorithms use exact algorithms as subroutines, and exact algorithms use approximation algorithms as preprocessing. In theory, the best approximation algorithm is Gama-Nguyen's reduction [8]. But experiments (such as that of [9], or the cryptanalyses [31,21] of GGH challenges [12]) suggest that the best approximation algorithm known in practice for high dimension is BKZ, published by Schnorr and Euchner in 1994 [42], and implemented in NTL [44]. Like all blockwise algorithms [41,7,8], BKZ has an additional input parameter – the blocksize β – which impacts both the running time and the output quality: BKZ calls many times an enumeration subroutine [38,20,6,42], which looks for nearly-shortest vectors in projected lattices of dimension $\leq \beta$. As β increases, the output basis becomes more and more reduced, but the cost increases significantly: the cost of the enumeration subroutine is typically superexponential in β, namely $2^{O(\beta^2)}$ polynomial-time operations (see [10]); and experiments [9] show that the number of calls increases sharply with both β and the lattice dimension n: for fixed $\beta \geq 30$, the number of calls looks superpolynomial if not exponential in n. This leads to two typical uses of BKZ:

1. A small blocksize β around 20 in any dimension n, or a medium blocksize β around 30-40 in medium dimension n (say, around 100 at most). Here, BKZ terminates in a reasonable time, and is routinely used to improve the quality of an LLL-reduced basis.
2. A high blocksize $\beta \geq 40$ in high dimension n, to find shorter and shorter lattice vectors. Here, BKZ does not terminate in a reasonable time, and the computation is typically aborted after say, a few hours or days, with the hope that the current basis is good enough for the application: we note that Hanrot et al. [14] recently proved worst-case bounds for the output quality of aborted-BKZ, which are only slightly worse than full-BKZ. And one usually speeds up the enumeration subroutine by a pruning technique [42,43,10]: for instance, the implementation of BKZ in NTL proposes Schnorr-Hörner (SH) pruning [43], which adds another input parameter p, whose impact was only clarified in [10]. The largest GGH cryptographic challenges [12] were solved [31,21] using an aborted BKZ of blocksize $\beta = 60$ and SH factor $p = 14$.

One major issue is to assess the output quality of BKZ, especially since lattice algorithms tend to perform better than theoretically expected. The quality is measured by the so-called Hermite factor, as popularized by Gama and Nguyen [9]. In practice, the Hermite factor of all lattice algorithms known is typically exponential in the dimension, namely c^n where c depends on the parameters of the algorithm. The experiments of [9] show that in practice, the Hermite factor of BKZ is typically $c(\beta, n)^n$ where $c(\beta, n)$ quickly converges as n grows to infinity for fixed β. However, the limit values of $c(\beta, n)$ are only known for small values of β (roughly ≤ 30), and theoretical upper bounds [9,14] on $c(\beta, n)$ are significantly higher than experimental values.

All security estimates and proposed parameters (such as recent ones [28,39,23] and NTRU's [18]) of lattice cryptosystems are based on benchmarks of NTL's old implementation of BKZ, but the significance of these estimates is rather debatable. First, these benchmarks were all computed with only usage 1: NTRU [18] *"never observed a noticeable improvement from the pruning procedure, so the pruning procedure was not called"* and used $\beta \leq 25$, while [39,23] use $\beta \leq 30$. This means that such security estimates either assume that BKZ cannot be run with $\beta \geq 30$, or they extrapolate $c(\beta, n)$ for high values of β from low values $\beta \leq 30$. Second, recent progress [10] in enumeration shows that enumeration can now be performed in much higher dimension (*e.g.* $\beta \approx 110$) than previously imagined, but no approximate value of $c(\beta, n)$ is known for large $\beta \geq 50$. And NTL's implementation does not include these recent improvements, and is therefore suboptimal.

Our results. We report the first extensive experiments with high-blocksize BKZ ($\beta \geq 40$) in high dimension. This is made possible by implementing BKZ 2.0, an updated version of BKZ taking into account recent algorithmic improvements. The main modification is the incorporation of the sound pruning technique developed by Gama, Nguyen and Regev [10] at EUROCRYPT '10. The modifications significantly decrease the running time of the enumeration subroutine, without degrading its output quality for appropriate parameters, which allow much bigger blocksizes. BKZ 2.0 outperforms NTL's implementation of BKZ, even with SH pruning [43], which we checked by breaking lattice records such as Darmstadt's lattice challenges [24] or the SVP-challenges [40]: for instance, we find the shortest vector in NTRU [18]'s historical 214-dimensional lattices within $2^{42.62}$ clock cycles, at least 70 times less computation than previously reported [25].

More importantly, our experiments allow us to propose an efficient simulation algorithm to model the execution of BKZ with (arbitrarily) high blocksize ≥ 50, to guess the approximate length of the output vector and the time required: in particular, this algorithm provides the first ever predictions for $c(\beta, n)$ for arbitrarily high values of $\beta \geq 50$. For a given target length, the simulation predicts what is the approximate blocksize β required to obtain such short lattice vectors, and how many enumeration calls will be required approximately. This can be converted into an approximate running time, once we know a good approximation of the cost of enumeration. And we provide such approximations for the best enumeration subroutines known.

Our simulation refines the Gama-Nguyen security estimates [9] on the concrete hardness of lattice problems, which did not take into account pruning, like the security estimates of NTRU [19,16] and those of [23,39]. We illustrate the usefulness of our simulation by revising security estimates. For instance, our simulation suggests that the smallest NTRUSign parameter set, which was claimed to provide at least 93-bit security against key-recovery lattice attacks, actually offers at most 65-bit security. And we use our simulation to provide the first concrete security assessment of the fully-homomorphic encryption challenges [11] recently proposed by Gentry and Halevi. It seems that none of these challenges offers a very high security level, except the largest one, which seems to offer at most a 100-bit security level.

Roadmap. We start in Sect. 2 with background and notation on lattices. In Sect. 3, we recall the BKZ algorithm. In Sect. 4, we present BKZ 2.0 by describing our modifications to BKZ. In Sect. 5, we briefly report on new lattice records obtained. We present in Sect. 6 a simulation algorithm to predict the performances of BKZ 2.0 with (arbitrarily) high blocksize, which we apply to revise security estimates in Sect. 7. More information can be found in the full version.

2 Preliminaries

We use row representations of matrices (to match lattice software), and use bold fonts to denote vectors: if $B = (\mathbf{b}_1, \ldots, \mathbf{b}_n)$ is a matrix, its row vectors are the $\mathbf{b}_i$'s. The Euclidean norm of a vector $\mathbf{v} \in \mathbb{R}^m$ is $\|\mathbf{v}\|$. We denote by $\mathrm{Ball}_n(R)$ the n-dim Euclidean ball of radius R, and by $V_n(R) = R^n \cdot \frac{\pi^{n/2}}{\Gamma(n/2+1)}$ its volume. The n-dim unit sphere is denoted by S^{n-1}. Let L be an n-dim lattice in $\mathbb{R}^m$. Its *volume* $\mathrm{vol}(L)$ is the n-dim volume of the parallelepiped generated by any basis of L.

Orthogonalization. An $n \times m$ basis $B = (\mathbf{b}_1, \ldots, \mathbf{b}_n)$ can be written uniquely as $B = \mu \cdot D \cdot Q$ where $\mu = (\mu_{i,j})$ is $n \times n$ lower-triangular with unit diagonal, D is $n \times n$ positive diagonal, and Q is $n \times m$ with orthonormal row vectors. Then μD is a lower triangular representation of B (with respect to Q), $B^* = DQ = (\mathbf{b}_1^*, \ldots, \mathbf{b}_n^*)$ is the Gram-Schmidt orthogonalization of the basis, and D is the diagonal matrix formed by the $\|\mathbf{b}_i^*\|$'s. For $1 \leq i \leq n+1$, we denote by π_i the orthogonal projection over $(\mathbf{b}_1, \ldots, \mathbf{b}_{i-1})^{\perp}$. For $1 \leq j \leq k \leq n$, we denote by $B_{[j,k]}$ the local projected block $(\pi_j(\mathbf{b}_j), \pi_j(\mathbf{b}_{j+1}), \ldots, \pi_j(\mathbf{b}_k))$, and by $L_{[j,k]}$ the lattice spanned by $B_{[j,k]}$, whose dimension is $k - j + 1$.

Random Lattices. There is a natural notion of random (real) lattices of given volume, based on Haar measures of classical groups (see [1]). And there is a simple notion of random integer lattices, used in recent experiments: For any integer V, a random n-dim integer lattice of volume V is one chosen uniformly at random among the finitely many n-dim integer lattices of volume V. It was shown in [13] that, as V grows to infinity, the uniform distribution over integer lattices of volume V converges towards the distribution of random (real) lattices of unit volume, once the integer lattice is scaled by $V^{1/n}$. In experiments with random lattices, we mean an n-dim integer lattice chosen uniformly at random with volume a random prime number of bit-length $10n$: for prime volumes, it is trivial to sample from the uniform distribution, using the Hermite normal form. A bit-length $\Theta(n^2)$ would be preferable in theory (in order to apply the result of [13]), but it significantly increases running times, without affecting noticeably experimental results.

Gaussian Heuristic. Given a lattice L and a "nice" set S, the Gaussian Heuristic predicts that the number of points in $S \cap L$ is $\approx \mathrm{vol}(S)/\mathrm{vol}(L)$. In some cases, this heuristic can be proved [1] or refuted [27].

Shortest vector. A *shortest vector* of L has norm $\lambda_1(L) = \min_{\mathbf{v} \in L, \mathbf{v} \neq 0} \|\mathbf{v}\|$, the *first minimum* of L. If the Gaussian heuristic was true for any ball S, we would expect $\lambda_1(L) \approx GH(L)$ where $GH(L) = \mathrm{vol}(L)^{1/n} \cdot V_n(1)^{-1/n}$. Minkowski's theorem shows that $\lambda_1(L) \leq 2GH(L)$ for any lattice L. For random real lattices, $\lambda_1(L)$ is asymptotically equivalent to $GH(L)$ with overwhelming probability (see [1]).

Reduced bases. We recall a few classical reductions. A basis $B = (\mathbf{b}_1, \dots, \mathbf{b}_n)$ is:

- *size-reduced* if its Gram-Schmidt matrix μ satisfies $|\mu_{i,j}| \leq 1/2$ for $1 \leq j < i \leq n$.
- *LLL-reduced* [22] with factor ε such that $0 < \varepsilon < 1$ if it is size-reduced and its Gram-Schmidt orthogonalization satisfies $\|\mathbf{b}_{i+1}^* + \mu_{i+1,i}\mathbf{b}_i^*\|^2 \geq (1-\varepsilon)\|\mathbf{b}_i^*\|^2$ for $1 \leq i < n$. If we omit the factor ε, we mean the factor $\varepsilon = 0.01$, which is the usual choice in practice.
- *BKZ-reduced* [41] with blocksize $\beta \geq 2$ and factor ε such that $0 < \varepsilon < 1$ if it is LLL-reduced with factor ε and for each $1 \leq j \leq n$: $\|\mathbf{b}_j^*\| = \lambda_1(L_{[j,k]})$ where $k = \min(j + \beta - 1, n)$.

One is usually interested in minimizing the *Hermite factor* $\|\mathbf{b}_1\|/\mathrm{vol}(L)^{1/n}$ (see [9]), which is completely determined by the sequence $\|\mathbf{b}_1^*\|, \dots, \|\mathbf{b}_n^*\|$. This is because the Hermite factor dictates the performance of the algorithm at solving the most useful lattice problems: see [9] for approx-SVP and unique-SVP, and [28,39,23] for SIS and LWE. It turns out that the Gram-Schmidt coefficients of bases produced by the main reduction algorithms (such as LLL or BKZ) have a certain "typical shape" [9,34], provided that the input basis is sufficiently randomized. To give an idea, the shape is roughly such that $\|\mathbf{b}_i^*\|/\|\mathbf{b}_{i+1}^*\| \approx q$ where q depends on the reduction algorithm, except for the first indexes i. This means that the Hermite factor will typically be of the form c^n where $c \approx \sqrt{q}$.

3   The Blockwise Korkine-Zolotarev (BKZ) Algorithm

3.1   Description

The Blockwise-Korkine-Zolotarev (BKZ) algorithm [42] outputs a BKZ-reduced basis with blocksize $\beta \geq 2$ and reduction factor $\varepsilon > 0$, from an input basis $B = (\mathbf{b}_1, \dots, \mathbf{b}_n)$ of a lattice L. It starts by LLL-reducing the basis B, then iteratively reduces each local block $B_{[j,min(j+\beta-1,n)]}$ for $j = 1$ to n, to make sure that the first vector of each such block is the shortest in the projected lattice. This gives rise to Algorithm 1, which proceeds in such a way that each block is already LLL-reduced before being enumerated: there is an index j, initially set to 1. At each iteration, BKZ performs an enumeration of the local projected lattice $L_{[j,k]}$ where $k = \min(j + \beta - 1, n)$ to find $\mathbf{v} = (v_1, \ldots, v_n) \in \mathbb{Z}^n$ such that $\|\pi_j(\sum_{i=j}^{k} v_i \mathbf{b}_i)\| = \lambda_1(L_{[j,k]})$. We let $h = \min(k + 1, n)$ be the ending index of the new block in the next iteration:

- If $\|\mathbf{b}_j^*\| > \lambda_1(L_{[j,k]})$, then $\mathbf{b}^{new} = \sum_{i=j}^{k} v_i \mathbf{b}_i$ is inserted between $\mathbf{b}_{j-1}$ and $\mathbf{b}_j$. This means that we no longer have a basis, so LLL is called on the generating set $(\mathbf{b}_1, \ldots, \mathbf{b}_{j-1}, \mathbf{b}^{new}, \mathbf{b}_j, \ldots, \mathbf{b}_h)$, to give rise to a new LLL-reduced basis $(\mathbf{b}_1, \ldots, \mathbf{b}_h)$.
- Otherwise, LLL is called on the truncated basis $(\mathbf{b}_1, \ldots, \mathbf{b}_h)$.

Thus, at the end of each iteration, the basis $B = (\mathbf{b}_1, \ldots, \mathbf{b}_n)$ is such that $(\mathbf{b}_1, \ldots, \mathbf{b}_h)$ is LLL-reduced. When j reaches n, it is reset to 1, unless no enumeration was successful, in which case the algorithm terminates: the goal of z in Alg. 1 is to count the number of consecutive failed enumerations, to check termination.

Algorithm 1. The Block Korkin-Zolotarev (BKZ) algorithm

Input: A basis $B = (\mathbf{b}_1, \ldots, \mathbf{b}_n)$, a blocksize $\beta \in \{2, \ldots, n\}$, the Gram-Schmidt triangular matrix μ and $\|\mathbf{b}_1^*\|^2, \ldots, \|\mathbf{b}_n^*\|^2$.
Output: The basis $(\mathbf{b}_1, \ldots, \mathbf{b}_n)$ is BKZ-β reduced
1. $z \leftarrow 0$; $j \leftarrow 0$; LLL$(\mathbf{b}_1, \ldots, \mathbf{b}_n, \mu)$;// *LLL-reduce the basis, and update μ*
2. **while** $z < n - 1$ **do**
3. $j \leftarrow (j \mod (n - 1)) + 1$; $k \leftarrow \min(j + \beta - 1, n)$; $h \leftarrow \min(k + 1, n)$; // *define the local block*
4. $\mathbf{v} \leftarrow$ Enum$(\mu_{[j,k]}, \|\mathbf{b}_j^*\|^2, \ldots, \|\mathbf{b}_k^*\|^2)$; // *find $\mathbf{v} = (v_j, \ldots, v_k) \in \mathbb{Z}^{k-j+1} - \mathbf{0}$ s.t.* $\|\pi_j(\sum_{i=j}^{k} v_i \mathbf{b}_i)\| = \lambda_1(L_{[j,k]})$
5.   **if** $\mathbf{v} \neq (1, 0, \ldots, 0)$ **then**
6. $z \leftarrow 0$; LLL$(\mathbf{b}_1, \ldots, \sum_{i=j}^{k} v_i \mathbf{b}_i, \mathbf{b}_j, \ldots, \mathbf{b}_h, \mu)$ at stage j; //*insert the new vector in the lattice at the start of the current block, then remove the dependency in the current block, update μ.*
7.   **else**
8. $z \leftarrow z + 1$; LLL$(\mathbf{b}_1, \ldots, \mathbf{b}_h, \mu)$ at stage $h - 1$; // *LLL-reduce the next block before enumeration.*
9.   **end if**
10. **end while**

3.2 Enumeration Subroutine

BKZ requires a subroutine to find a shortest vector in a local projected lattice $L_{[j,k]}$: given as input two integers j and k such that $1 \leq j \leq k \leq n$, output $\mathbf{v} = (v_j, \ldots, v_k) \in \mathbb{Z}^{k-j+1}$ such that $\|\pi_j(\sum_{i=j}^{k} v_i \mathbf{b}_i)\| = \lambda_1(L_{[j,k]})$. In practice, as well as in the BKZ article [42], this is implemented by enumeration. One sets $R = \|\mathbf{b}_j^*\|$ as an initial upper bound of $\lambda_1(L_{[j,k]})$. Enumeration goes through the *enumeration tree* formed by "half" of the vectors in the local projected lattices $L_{[k,k]}, L_{[k-1,k]}, \ldots, L_{[j,k]}$ of norm at most R. The tree has depth

$k - j + 1$, and for each $d \in \{0, \ldots, k - j + 1\}$, the nodes at depth d are 0 and all $\pi_{k-d+1}(\mathbf{u}) \in L_{[k-d+1,k]}$ where $\mathbf{u} = \sum_{i=j}^{k'} u_i \mathbf{b}_i$ with $j \leq k' \leq k$, $u_{k'} > 0$ and $\|\pi_{k-d+1}(\mathbf{u})\| \leq R$. The parent of a node $\mathbf{u} \in L_{[k-d+1,k]}$ at depth d is $\pi_{k+2-d}(\mathbf{u})$ at depth $d - 1$. Child nodes are ordered by increasing Euclidean norm. The Schnorr-Euchner algorithm [42] performs a Depth First Search of the tree to output a nonzero leaf of minimal norm, with the following modification: everytime a new (nonzero) leaf is found, one updates the enumeration radius R as the norm of the leaf. The more reduced the basis is, the less nodes in the tree, and the cheaper the enumeration. The running time of the enumeration algorithm is N polynomial-time operations where N is the total number of tree nodes. If the algorithm did not update R, Hanrot and Stehlé [15] noticed that the number of nodes at depth d could be estimated from the Gaussian heuristic as:

$$H_d(R) = \frac{1}{2} \cdot \frac{V_d(R)}{\prod_{i=k-d+1}^{k} \|\mathbf{b}_i^*\|} = \frac{1}{2} \cdot \frac{R^d V_d(1)}{\prod_{i=k-d+1}^{k} \|\mathbf{b}_i^*\|}. \tag{1}$$

Gama *et al.* [10] showed that this heuristic estimate is experimentally very accurate, at least for sufficiently large $k - j + 1$ and typical reduced bases. We can therefore heuristically bound the number of nodes at depth d in the actual Schnorr-Euchner algorithm (with update of R) by setting $R = \lambda_1(L_{[j,k]})$ and $R = \|\mathbf{b}_j^*\|$ in Eq. (1). It is shown in [10] that for typical reduced bases, $H_d(R)$ is maximal around the middle depth $d \approx (k - j)/2$, and the remaining $H_d(R)$'s are significantly smaller.

3.3   Analysis

No good upper bound on the complexity of BKZ is known. The best upper bound known for the number of calls (to the enumeration subroutine) is exponential (see [14]). In practice (see [9]), BKZ with $\beta = 20$ is very practical, but the running time significantly increases for $\beta \geq 25$, making any $\beta \geq 40$ too expensive for high-dimensional lattices. In practice, the quality of bases output by BKZ is better than the best theoretical worst-case bounds: according to [9], the Hermite factor for high-dimensional lattices is typically $c(\beta, n)^n$ where $c(\beta, n)$ seems to quickly converge as n grows to infinity, whereas theoretical upper bounds are $c'(\beta)^n$ with $c'(\beta)$ significantly larger than $c(\beta, n)$. For instance, $c(20, n) \approx 1.0128$ for large n. Furthermore, [14] recently showed that if one aborts BKZ after a suitable polynomial number of calls, one can obtain theoretical upper bounds which are only slightly worse than $c'(\beta)^n$.

4   BKZ 2.0

When the blocksize is sufficiently high, namely ≥ 30, it is known [9] that the overall running time of BKZ is dominated by the enumeration subroutine, which finds a shortest vector in the m-dimensional local projected lattice $L_{[j,k]}$, using a radius R initially set to $\|\mathbf{b}_j^*\|$, where $1 \leq j \leq k \leq n$ and $m = k - j + 1$.

In this section, we describe BKZ 2.0, an updated version of BKZ with four improvements, which we implemented by modifying NTL [44]'s implementation of BKZ [42]. The first improvement is simply an early-abort, which is common practice in cryptanalysis, and is partially supported by the recent theoretical result of [14]: we add a parameter that specifies how many iterations should be performed, *i.e.* we choose the number of oracle calls; this already provides an exponential speedup over BKZ, because the number of calls seems to grow exponentially for fixed $\beta \geq 30$ according to the experiments of [9]. The other three improvements aim at decreasing the running time of the enumeration subroutine: sound pruning [10], preprocessing of local bases, and shorter enumeration radius. Though these improvements may be considered as folklore, we stress that none had been incorporated in BKZ (except that a weaker form of pruning had been designed by Schnorr and Hörner [43], and implemented in NTL [44]), and that implementing them is not trivial.

4.1   Sound Pruning

Pruning speedups enumeration by discarding certain branches, but may not return any vector, or maybe not the shortest one. The idea of pruned enumeration goes back to Schnorr and Euchner [42], and was first analyzed by Schnorr and Hörner [43] in 1995. It was recently revisited by Gama *et al.* [10], who noticed that the analysis of [43] was flawed and that the pruning was not optimal. They showed that a well-chosen high-probability pruning leads to an asymptotical speedup of $2^{m/4}$ over full enumeration, and introduced an extreme pruning technique which gives an asymptotical speedup of $2^{m/2}$ over full enumeration. We incorporated both pruning with non-negligible probability, and extreme pruning using randomization. Formally, pruning replaces each of the $k - j + 1$ inequalities $\|\pi_{k+1-d}(\mathbf{u})\| \leq R$ for $1 \leq d \leq k - j + 1$ by $\|\pi_{k+1-d}(\mathbf{u})\| \leq R_d \cdot R$ where $0 \leq R_1 \leq \cdots \leq R_{k-j+1} = 1$ are $k - j + 1$ real numbers defined by the pruning strategy. For any bounding function $(R_1, \ldots, R_{k-j+1})$, [10] consider the quantities N' and p_{succ} defined by:

- $N' = \sum_{d=1}^{k-j+1} H_d'$ is a heuristic estimate of the total number of nodes in the pruned enumeration tree, where $H_d' = \frac{1}{2} \frac{R^d V_{R_1,\ldots,R_d}}{\prod_{i=k+1-d}^{k} \|\mathbf{b}_i^*\|}$ and $V_{R_1,\ldots,R_d}$ denotes the volume of $C_{R_1,\ldots,R_d} = \left\{ (x_1, \ldots, x_d) \in \mathbb{R}^d, \ \forall 1 \leq i \leq d, \ \sum_{l=1}^{i} x_l^2 \leq R_i^2 \right\}$.

- $p_{\mathrm{succ}} = p_{\mathrm{succ}}(R_1, \ldots, R_m) = \Pr_{\mathbf{u} \sim S^{m-1}} \left(\forall i \in [1,m], \ \sum_{l=1}^{i} u_l^2 \leq R_i^2 \right)$. Let $\mathbf{t} \in L_{[j,k]}$ be a target vector such that $\|\pi_j(\mathbf{t})\| = R$. If the local basis $B_{[j,k]}$ is assumed to be randomized, then p_{succ} is the probability that $\pi_j(\mathbf{t})$ is a leaf of the pruned enumeration tree, under the (idealized) assumption that the distribution of the coordinates of $\pi_j(\mathbf{t})$, when written in the normalized Gram-Schmidt basis $(\mathbf{b}_j^*/\|\mathbf{b}_j^*\|, \ldots, \mathbf{b}_k^*/\|\mathbf{b}_k^*\|)$ of the local basis $B_{[j,k]}$, look like those of a uniformly distributed vector of norm $\|\pi_j(\mathbf{t})\|$.

We stress that the assumption is only an idealization: in practice, when m is small, for a non-negligible fraction of the local blocks $B_{[j,k]}$, one of the vectors of $B_{[j,k]}$ is a shortest vector of $L_{[j,k]}$, which should have had zero probability. For the application to BKZ, it makes sense to consider various bounding functions of various p_{succ}, say ranging from 1% to 95%, but with a cost N' as small as possible. Based on the methodology of [10], we performed an automated search to generate such bounding functions, for blocksizes β ranging from 35 to 90 by steps of 5, and p_{succ} ranging from 1% to 95%.

It should be noted that BKZ calls the enumeration subroutine on lattices $L_{[j,k]}$ whose dimension $m = k - j + 1$ is not necessarily equal to β. When $j \leq n - \beta + 1$, the dimension m of the block is equal to β, but when $j \geq n - \beta$, the dimension m of the block is strictly less than β. To avoid generating bounding functions for every dimension, we decided in this case to interpolate the bounding function found for β, and checked that interpolating does not affect much p_{succ}. Finally, in order to boost p_{succ}, we added an optional parameter ν, so that BKZ actually performs ν pruned enumerations, each starting with a different random basis of the same local block. This corresponds to the extreme pruning of [10].

4.2   Preprocessing of Local Blocks

The cost of enumeration is strongly influenced by the quality of the local basis, especially as the blocksize increases: the more reduced the local basis, the bigger the volumes of the local projected lattices $L_{[k-d+1,k]}$, and therefore the less nodes in the most populated depths of the enumeration tree. This is folklore, but since BKZ improves regularly the quality of the basis, one might think there is no need to change the local basis before enumeration. However:

– For each enumeration, the local basis is only guaranteed to be LLL -reduced, even though the whole basis may be more than LLL-reduced.
– In high blocksizes, most enumerations are successful: they find a shorter vector than the first block vector. This implies that a local LLL-reduction will be performed to get a basis from a generating set: see Line 1 in Alg. 1. At the next iteration, the enumeration will proceed on a typical LLL-reduced basis, and not something likely to be better reduced.

This suggests that for most enumerations, the local basis is only LLL-reduced, and nothing more, even though other local bases may be better reduced: this was confirmed by experiments.

Hence, we implemented a simple speedup: ensure that the local basis is significantly more reduced than LLL-reduced before each enumeration, but without spending too much time. We used a recursive aborted-BKZ preprocessing to the local basis before enumeration: we performed an automated search to find good parameters depending on β.

4.3   Optimizing the Enumeration Radius

It is folklore that the enumeration cost is also influenced by the choice of the initial radius R, even though this radius is updated during enumeration. Initially, the radius is $R = \|\mathbf{b}_j^*\|$, but if we knew before hand how short would be the output vector, we would choose a lower initial radius R, decreasing the enumeration time. Indeed, the number of nodes at depth d of the enumeration tree (pruned or not) is proportional to R^d. Unfortunately, not much is known (from a theoretical point of view) on how small should be $\lambda_1(L_{[j,k]})$, except general bounds. So we performed experiments to see what was the final norm found by enumeration in practice: Fig. 1 compares the final norm (found by enumeration) to $GH(L_{[j,k]})$, depending on the starting index j of the local block, for one round of BKZ. For the lowest indices j, one sees that the final norm is significantly lower than $GH(L_{[j,k]})$, whereas for the largest indices, it is significantly larger. In the middle, which accounts for most of the enumerations, the ratio between the final norm and the Gaussian heuristic prediction is mostly within 0.95 and 1.05, whereas the ratio between the norm of the first local basis vector and $GH(L_{[j,k]})$ is typically slightly below 1.1. We therefore used the following optimization: for all indexes j except the last 30 ones, we let $R = \min(\sqrt{\gamma}GH(L_{[j,k]}), \|\mathbf{b}_j^*\|)$ instead of $R = \|\mathbf{b}_j^*\|$, where γ is a radius parameter. In practice, we selected $\sqrt{\gamma} = \sqrt{1.1} \approx 1.05$.

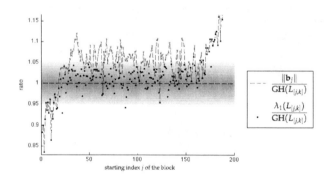

Fig. 1. Comparing $\|\mathbf{b}_j^*\|$, $\lambda_1(L_{[j,k]})$ and $GH(L_{[j,k]})$, for each local block $B_{[j,k]}$

5   New Lattice Records

Here, we briefly report on experiments using 64-bit Xeon processors to break some lattice records, which suggest that BKZ 2.0 is currently the best lattice reduction algorithm in practice.

5.1   Darmstadt's Lattice Challenge

Darmstadt's lattice challenge [24] started in 2008. For each dimension, the challenge is to find a vector of norm $< q$ in an Ajtai lattice [2], where q depends on the dimension; and try to minimize the norm. Until now, the highest challenge solved was 725: the first solutions to all challenges in dimension 575 to 725 were found by Gama and Nguyen in 2008, using NTL's implementation of BKZ with SH pruning. Shorter solutions have been found since (see the full list [24]), but no challenge of higher dimension had been solved. All solutions were found by reducing appropriate sublattices of much smaller dimension (typically around 150-200), whose existence follows from the structure of Ajtai lattices: we followed the same strategy.

BKZ 2.0 with blocksize 90 (18 pruned-enumerations at 5%) found the first ever solution to challenges 750, 775 and 800, and significantly shorter vectors in all challenges 525 to 725, using in total about 3 core-years, as summarized in Table 1: the first column is the dimension of the challenge, the second one is the dimension of the sublattice we used to find the solution, the third one is the best norm found by BKZ 2.0, the fourth one is the previous best norm found by former algorithms, the fifth one is the ratio between norms, and the sixth one is the Hermite factor of the reduced basis of the sublattice, which turns out to be slightly below $1.01^{\dim}$. The factor $1.01^{\dim}$ was considered to be the state-of-the-art limit in 2008 by Gama and Nguyen [9], which shows the improvement.

Table 1. New Solutions for Darmstadt's lattice challenge [24]

Dim(lattice)	Dim(sublattice)	New norm	Previous norm	Ratio	Hermite factor
800	230	120.054	Unsolved		1.00978^{230}
775	230	112.539	Unsolved		1.00994^{230}
750	220	95.995	Unsolved		1.0976^{220}
725	210	85.726	100.90	0.85	1.00978^{210}
700	200	78.537	86.02	0.91	1.00993^{200}
675	190	72.243	74.78	0.97	1.00997^{190}
650	190	61.935	66.72	0.93	1.00993^{190}
625	180	53.953	59.41	0.91	1.00987^{180}
600	180	45.420	52.01	0.87	1.00976^{180}
575	180	39.153	42.71	0.92	1.00977^{180}
550	180	32.481	38.29	0.85	1.00955^{180}
525	180	29.866	30.74	0.97	1.00990^{180}

5.2 SVP Challenges

The SVP challenge [40] opened in May 2010. The lattices L are random integer lattices of large volume, so that $\lambda_1(L) \approx GH(L)$ with high probability. The challenge is to find a nearly-shortest vector, namely a nonzero lattice vector of norm $\leq 1.05GH(L)$. Using BKZ 2.0 with blocksize 75, 20%-pruning, we were able to solve all challenges from dimension 90 to 112.

6 Predicting BKZ 2.0 by Simulation

We now present an efficient simulation algorithm to predict the performances of BKZ 2.0 with high blocksize $\beta \geq 50$ in high dimension, in terms of running time and output quality. Our simulation is fairly consistent with experiments using several core-years on 64-bit Xeon processors, on random lattices and Darmstadt's lattice challenges. Accordingly, we believe that our simulation can be used to predict approximately what can be achieved using much larger computational power than used in our experiments, thereby leading to more convincing security estimates.

6.1   Description

The goal of our simulation algorithm is to predict the Gram-Schmidt sequence $(\|\mathbf{b}_1^*\|, \|\mathbf{b}_2^*\|, \ldots, \|\mathbf{b}_n^*\|)$ during the execution of BKZ, more precisely at the beginning of every round: a *round* occurs whenever $j = 0$ in Step 1 of Alg. 1, so one round of BKZ costs essentially $n - 1$ enumeration calls. We assume that the input basis is a "random" reduced basis, without special property.

The starting point of our simulation is the intuition, based on Sect. 4.3, that the first minimum of most local blocks looks like that of a random lattice of dimension the blocksize: this phenomenon does not hold in small blocksize ≤ 30 (as noted by Gama and Nguyen [9]), but it becomes more and more true as the blocksize increases, as shown in Fig. 2, where we see that the expectation and the standard deviation of $\frac{\lambda_1(L)}{GH(L)}$ seem to converge to that of a random lattice. Intuitively, this may be explained by a concentration phenomenon: as

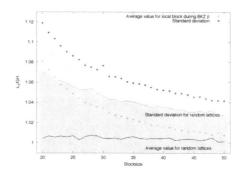

Fig. 2. Comparing $\frac{\lambda_1(L)}{GH(L)}$ for a non-extreme local block during BKZ-β reduction, with a random lattice of dimension β. Expectations with and without standard deviation are given.

the dimension increases, random lattices dominate in the set of lattices, so unless there is a strong reason why a given lattice cannot be random, we may assume that it behaves like a random lattice.

Once we can predict the value of $\lambda_1(L_{[j,k]})$ for each local block, we know that this will be the new value of $\|\mathbf{b}_j^*\|$ by definition of the enumeration subroutine, which allows to deduce the volume of the next local block, and therefore iterate the process until the end of the round. This gives rise to our simulation algorithm (see Alg. 2).

Algorithm 2. Simulation of BKZ reduction

Input: The Gram-Schmidt norms, given as $\ell_i = \log(\|\mathbf{b}_i^*\|)$, for $i = 1, \ldots, n$,
    a blocksize $\beta \in \{45, \ldots, n\}$, and a number N of rounds.

Output: A prediction for the Gram-Schmidt norms $\ell_i' = \log(\|\mathbf{b}_i^*\|)$, $i = 1, \ldots, n$, after
    N rounds of BKZ reduction.

1. **for** $k = 1, \ldots, 45$ **do**
2. $r_k \leftarrow$ average $\log(\|\mathbf{b}_k^*\|)$ of an HKZ-reduced random unit-volume 45-dim lattice
3. **end for**
4. **for** $d = 46, \ldots, \beta$, **do** $c_d \leftarrow \log(GH(\mathbb{Z}^d)) = \log(\frac{\Gamma(d/2+1)^{1/d}}{\pi^{1/2}})$ **end for**
5. **for** $j = 1, \ldots, N$ **do**
6. $\phi \leftarrow$ true //flag to store whether $L_{[k,n]}$ has changed
7.   **for** $k = 1$ to $n - 45$ **do**
8. $d \leftarrow \min(\beta, n - k + 1)$ // Dimension of local block
9. $f \leftarrow \min(k + \beta, n)$ //End index of local block
10. $\log V \leftarrow \sum_{i=1}^{f} \ell_i - \sum_{i=1}^{k-1} \ell_i'$
11.   **if** $\phi =$ true **then**
12.   **if** $\log V/d + c_d < \ell_k$ **then**
13. $\ell_k' \leftarrow \log V/d + c_d$;
14. $\phi \leftarrow$ false
15.   **end if**
16.   **else**
17. $\ell_k' \leftarrow \log V/d + c_d$
18.   **end if**
19.   **end for**
20. $\log V \leftarrow \sum_{i=1}^{n} \ell_i - \sum_{i=1}^{n-45} \ell_i'$
21.   **for** $k = n - 44$ to n **do**
22. $\ell_k' \leftarrow \frac{\log V}{45} + r_{k+45-n}$
23.   **end for**
24. $\ell_{1,\ldots,n} \leftarrow \ell_{1,\ldots,n}'$
25. **end for**

We predict this first minimum $\lambda_1(L_{[j,k]})$ as follows:

- For most indexes j, we choose $GH(L_{[j,k]})$, unless $\|\mathbf{b}_j^*\|$ was already better.
- However, for the last indexes j, namely those inside the last β-dimensional block $L_{[n-\beta+1,n]}$, we do something different: since this last block will be HKZ-reduced at the end of the round, we assume that it behaves like an HKZ-reduced basis of a random lattice of the same volume. Since these averages may be expensive to compute for large β, we apply a simplified rule: we determine the last 45 Gram-Schmidt norms from the average Gram-Schmidt norms (computed experimentally) of an HKZ-reduced basis of a random 45-dim lattice of unit volume, and we compute the first $\beta - 45$ Gram-Schmidt norms using the Gaussian heuristic. But this model may not work with bases of special structure such as partial reductions of the NTRU Hermite normal form, which is why we only consider random reduced bases as input.

This simulation algorithms allows us to guess the approximate Hermite factor achieved by BKZ 2.0, given an arbitrary blocksize, as summarized in Table 2: for a given dimension n, one should run the simulation algorithm, because the actual blocksize also depends on the dimension. As mentioned in Sect. 2, the

Table 2. Approximate required blocksize for high-dimensional BKZ, as predicted by the simulation

Target Hermite Factor	1.01^n	1.009^n	1.008^n	1.007^n	1.006^n	1.005^n
Approximate Blocksize	85	106	133	168	216	286

Hermite factor dictates the performances at solving lattice problems relevant to cryptography: see [9] for approx-SVP and unique-SVP, and [28,39,23] for SIS and LWE. Obviously, we can only hope for an approximation, since there are well-known variations in the Hermite factor when the input basis is randomized.

The simulation algorithm also gives us an approximate running time, using the number of rounds, provided that we know the cost of the enumeration subroutine: we will discuss these points more precisely later on.

6.2   Consistency with Experiments

It turns out that our simulation matches well with experiments using random lattices and Darmstadt's lattice challenges. First, the prediction of the Gram-Schmidt sequence $(\|\mathbf{b}_1^*\|, \|\mathbf{b}_2^*\|, \ldots, \|\mathbf{b}_n^*\|)$ by our simulation algorithm is fairly accurate for random reduced bases, as shown in Fig. 3 This implies that our simulation algorithm can give a good prediction of the Hermite factor of BKZ at any given number of rounds, which is confirmed by Fig. 4. Furthermore, Fig. 4 suggests that a polynomial number of calls seems sufficient to obtain a Hermite factor not very far from that of a full reduction: the main progress seems to occur in the early rounds of BKZ, which justifies the use of aborted-BKZ, which complements the theoretical results of [14].

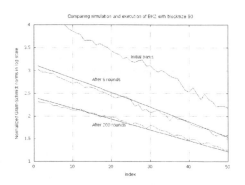

Fig. 3. Predicted vs. actual values of Gram-Schmidt norms during BKZ-50 reduction of a 200-dim random lattice

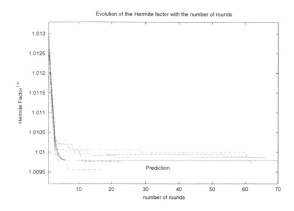

Fig. 4. Evolution and prediction of $(\|\mathbf{b}_1\|/\mathrm{vol}(L)^{1/n})^{1/n}$ during BKZ-90 reduction in dim 180 for Darmstadt's lattice challenges 500–625

6.3 Enumeration Subroutine

It remains to estimate the cost of the enumeration subroutine, with a radius equal to the Gaussian heuristic. First, we computed upper bounds, by applying extreme pruning on bases reduced with BKZ 2.0, following the search method of [10]: Table 3 gives the approximate cost (in terms of logarithmic number of nodes) of extreme pruning for blocksizes 100-250, using BKZ-75-20% as preprocessing, and radius equal to the Gaussian heuristic. Numbers of nodes can

Table 3. Upper bound on the cost of the enumeration subroutine, using extreme pruning with aborted-BKZ preprocessing. Cost is given as $\log_2$(number of nodes).

Blocksize	100	110	120	130	140	150	160	170	180	190	200	250
BKZ-75-20%	41.4	47.1	53.1	59.8	66.8	75.2	84.7	94.7	105.8	117.6	129.4	204.1
Simulation of BKZ-90/100/110/120	40.8	45.3	50.3	56.3	63.3	69.4	79.9	89.1	99.1	103.3	111.1	175.2

be approximately converted into clock cycles as follows: in the implementation of [10], one node requires about 200 clock cycles for double-precision enumeration, but this figure depends on the dimension, and for high blocksize, we may need higher precision than double precision. For instance, Table 3 says that applying extreme pruning in blocksize 120 would cost at most approximately 2^{53} nodes, which is less than 30 core-years on a 1.86-GHz Xeon, assuming double precision. This is useful to determine parameters for feasible attacks. However, these upper bounds should not be considered as tight: the performances of enumeration techniques depend on preprocessing, and it is likely that better figures (than Table 3) can be obtained with better preprocessing, including BKZ 2.0 with different parameters. In fact, Table 3 also provides a better upper bound, based on our simulation of BKZ with higher blocksizes 90–120 as a preprocessing. In order to provide security estimates with a good security margin, we need to estimate how much progress can be made. Interestingly, there are limits to enumeration techniques. Nguyen [33] established a lower bound on the number of nodes at each depth of the enumeration tree, assuming that the Gaussian heuristic estimates well the number of nodes (as is usual in analyzing the complexity of enumeration techniques). The lower bounds are based on the Rankin invariants $\gamma_{n,m}(L)$ of a lattice:

$$\gamma_{n,m}(L) = \min_{\substack{S \text{ sublattice of } L \\ \dim S = m}} \left(\frac{\text{vol}(S)}{\text{vol}(L)^{m/n}} \right)^2.$$

In particular, [33] shows that the number of nodes in the middle depth of a full enumeration of a d-dim lattice L with radius $GH(L)$ is $\geq V_{d/2}(1)$ $\sqrt{\gamma_{d,d/2}(L)/V_d(1)}$. For typical lattices L, the Rankin invariant $\gamma_{n,m}(L)$ is heuristically close to the following lower bound on Rankin's constant $\gamma_{n,m}$ (see [7]):

$$\gamma_{n,m} \geq \left(n \frac{\prod_{j=n-m+1}^{n} Z(j)}{\prod_{j=2}^{m} Z(j)} \right)^{\frac{2}{n}} \tag{2}$$

where $Z(j) = \zeta(j)\Gamma(\frac{j}{2})/\pi^{\frac{j}{2}}$ and ζ is Riemann's zeta function: $\zeta(j) = \sum_{p=1}^{\infty} p^{-j}$. These lower bounds are for full enumeration, but they can be adapted to pruning by taking into account the actual speedup of pruning (as analyzed in [10]), which is asymptotically $2^{n/4}$ for high-probability pruning and $2^{n/2}$ for extreme pruning. Table 4 gives the figures obtained with respectively the actual speedup of the so-called linear pruning, and the asymptotical speedup $2^{n/2}$ of extreme pruning. Compared to the upper bounds of Table 3, there is a significant gap: the lower

Table 4. Lower bounds on the cost (in log-nodes) of the enumeration subroutine using linear pruning or extreme pruning, following [33,10]

Blocksize	100	120	140	160	180	200	220	240	280	380
Linear pruning	33.6	44.5	56.1	68.2	80.7	93.7	107.0	120.6	148.8	223.5
Extreme pruning	9	15	21.7	28.8	36.4	44.4	52.8	61.5	79.8	129.9

bound of linear pruning tells us how much progress could be made if a stronger preprocessing was found for enumeration.

Finally, we note that asymptotically, heuristic variants [36,30,45] of sieve algorithms [3] are faster than pruned enumeration. However, it is unclear how meaningful it is for security estimates, since these variants require exponential space and are outperformed in practice. And more experiments than [36,30] would be required to evaluate precisely their practical running time. But our model can easily adapt to new progress in the enumeration subroutine, due to Table 2.

7   Revising Security Estimates

Here, we illustrate how our simulation algorithm can be used to obtain arguably better security estimates than previously known.

7.1 NTRU Lattices

In the NTRU cryptosystem [18], recovering the secret key from the public key amounts to finding a shortest vector in high-dimensional lattices of special structure. Because NTRU security estimates are based on benchmarks with BKZ, it is interesting to see the limits of this methodology.

In the original article [18], the smallest parameter set NTRU-107 corresponds to lattices of dimension 214, and it was estimated that key recovery would cost at least 2^{50} elementary operations. The best experimental result to recover the secret key for NTRU-107 by direct lattice reduction (without ad-hoc techniques like [25,26,9] which exploit the special structure of NTRU lattices) is due to May in 1999 [25], who reported one successful experiment using BKZ with SH pruning [43], after 663 hours on a 200-MHz processor, that is $2^{48.76}$ clock cycles. We performed experiments with BKZ 2.0 on 10 random NTRU-107 lattices: We applied LLL and BKZ-20, which takes a few minutes at most; We applied BKZ -65 with 5%-pruning, and checked every 5 minutes if the first basis vector was the shortest vector corresponding to the secret key, in which case we aborted. BKZ 2.0 was successful for each lattice, and the aborted BKZ-65 reduction took less than 2000s on the average, on a 2.83Mhz single core. So the overall running time is less than 40 minutes, that is $2^{42.62}$ clock cycles, which gives a speedup of at least 70, compared to May's experiment, and is significantly lower than 2^{50} elementary operations. Hence, there is an order of magnitude between the initial security estimate of 2^{50} and the actual security level, which is approximately at most 40-bit.

Now, we revisit recent parameters for NTRUSign. In the recent article by Hoffstein *et al.* [17], a summary of the latest parameters for NTRU encryption and signature is given. In particular, the smallest parameter for NTRUsign is $(N, q) = (157, 256)$, which is claimed to provide 80-bit security against all attacks knowns, and 93-bit security against key-recovery lattice attacks. Similarly to [9], we estimate that finding the secret key is essentially as hard as recovering a vector of norm $< q$ in a lattice of dimension $2N = 314$ and volume q^N, which corresponds to a Hermite factor of 1.00886^{2N}. We ran our simulation algorithm for these parameters to guess how many rounds would be required, depending on the blocksize, starting from a BKZ-20 reduced basis (whose cost is negligible here): about six rounds of BKZ-110 should be sufficient to break NTRUSign-157, which corresponds to roughly 2^{11} enumerations. And according to Table 3, extreme pruning enumeration in blocksize 110 can be done by searching through at most 2^{47} nodes, which corresponds to roughly 2^{54} clock cycles on a typical processor. This suggests that the security level of the smallest NTRUSign parameter against state-of-the-art lattice attacks is at most 65-bit, rather than 93-bit, which is a significant gap.

7.2 Gentry-Halevi's Fully-Homomorphic Encryption Challenges

We now turn to Gentry-Halevi's main Fully-Homomorphic Encryption Challenges [11], for which no concrete security estimate was given. Decrypting a ciphertext amounts to solve a BDD instance, which can be done up to the distance $\min_i \|\mathbf{b}_i\|^*/2$ using Babai's nearest plane algorithm. Targetting a given value of $\min_i \|\mathbf{b}_i\|^*$ can be transformed into a target Hermite factor in the dual lattice. This allows us to estimate the required Hermite factor to solve the BDD instance, based on the approximate distance of the BDD instance and the lattice volume, which is summarized in Table 5.

Table 5. Security Assessment of Gentry-Halevi's main challenges [11]

Dimension n	512	2048	8192	32768
Name	Toy	Small	Medium	Large
Target Hermite factor [9]	1.67^n	1.14^n	1.03^n	1.0081^n
Algorithm expected to decrypt a fresh ciphertext	LLL	LLL	LLL	BKZ with blocksize ≈ 130
Time estimate	30 core-days	≤ 45 core-years	≤ 68582 core-years	$\approx 2^{100}$ clock-cycles

Accordingly, we speculate that decryption for the toy, small and medium challenge can be solved by LLL reduction, which is not straightforward due to the lattice dimension and the gigantic bit-size of the basis (note that there is new theoretical progress [37] on LLL-reduction for large entries). We checked that this was indeed the case for the toy challenge, by performing an actual reduction using a modification of fplll [4]. For the small and medium challenges, we extrapolated running times from truncated challenges, using the fact that our modification of fplll has heuristic running time $O(n^3 d^2)$ where d is the bit-size of the lattice volume, where the $\mathcal{O}$ constant depends on the floating-point precision (which increases with the dimension). According to our simulation, breaking the large challenge would require a blocksize ≈ 130 and approximately 60000 rounds (starting from an LLL basis), that is, 2^{31} enumeration calls. Based on Table 3, this enumeration routine would cost at most 2^{60} nodes, so the security offered by the large challenge is at most roughly 100-bit. On the other hand, if ever a stronger preprocessing for enumeration is found, Table 4 suggests that the security level could potentially drop by a factor in the range $2^{10} - 2^{40}$.

Acknowledgements. Part of this work is supported by the Commission of the European Communities through the ICT program under contract ICT-2007-216676 ECRYPT II.

References

1. Ajtai, M.: Generating random lattices according to the invariant distribution (draft of March 2006)
2. Ajtai, M.: Generating hard instances of lattice problems. In: Proc. STOC 1996, pp. 99–108. ACM (1996)
3. Ajtai, M., Kumar, R., Sivakumar, D.: A sieve algorithm for the shortest lattice vector problem. In: Proc. 33rd STOC 2001, pp. 601–610. ACM (2001)
4. Cadé, D., Pujol, X., Stehlé, D.: FPLLL library, version 3.0 (September 2008)
5. Devroye, L.: Non-uniform random variate generation (1986),
   http://cg.scs.carleton.ca/~luc/rnbookindex.html

6. Fincke, U., Pohst, M.: Improved methods for calculating vectors of short length in a lattice, including a complexity analysis. Mathematics of Computation 44(170), 463–471 (1985)
7. Gama, N., Howgrave-Graham, N., Koy, H., Nguyên, P.Q.: Rankin's Constant and Blockwise Lattice Reduction. In: Dwork, C. (ed.) CRYPTO 2006. LNCS, vol. 4117, pp. 112–130. Springer, Heidelberg (2006)
8. Gama, N., Nguyen, P.Q.: Finding short lattice vectors within Mordell's inequality. In: Proc. STOC 2008. ACM (2008)
9. Gama, N., Nguyen, P.Q.: Predicting Lattice Reduction. In: Smart, N.P. (ed.) EUROCRYPT 2008. LNCS, vol. 4965, pp. 31–51. Springer, Heidelberg (2008)
10. Gama, N., Nguyen, P.Q., Regev, O.: Lattice Enumeration Using Extreme Pruning. In: Gilbert, H. (ed.) EUROCRYPT 2010. LNCS, vol. 6110, pp. 257–278. Springer, Heidelberg (2010)
11. Gentry, C., Halevi, S.: Public challenges for fully-homomorphic encryption (2010), http://researcher.ibm.com/researcher/view_project.php?id=1548
12. Goldreich, O., Goldwasser, S., Halevi, S.: Challenges for the GGH cryptosystem (1997), http://theory.lcs.mit.edu/~shaih/challenge.html
13. Goldstein, D., Mayer, A.: On the equidistribution of Hecke points. Forum Math. 15(2), 165–189 (2003)
14. Hanrot, G., Pujol, X., Stehlé, D.: Analyzing Blockwise Lattice Algorithms Using Dynamical Systems. In: Rogaway, P. (ed.) CRYPTO 2011. LNCS, vol. 6841, pp. 447–464. Springer, Heidelberg (2011)
15. Hanrot, G., Stehlé, D.: Improved Analysis of Kannan's Shortest Lattice Vector Algorithm. In: Menezes, A. (ed.) CRYPTO 2007. LNCS, vol. 4622, pp. 170–186. Springer, Heidelberg (2007)
16. Hirschhorn, P.S., Hoffstein, J., Howgrave-Graham, N., Whyte, W.: Choosing NTRUEncrypt Parameters in Light of Combined Lattice Reduction and MITM Approaches. In: Abdalla, M., Pointcheval, D., Fouque, P.-A., Vergnaud, D. (eds.) ACNS 2009. LNCS, vol. 5536, pp. 437–455. Springer, Heidelberg (2009)
17. Hoffstein, J., Howgrave-Graham, N., Pipher, J., Whyte, W.: Practical lattice-based cryptography: NTRUEncrypt and NTRUSign. In: [35] (2010)
18. Hoffstein, J., Pipher, J., Silverman, J.H.: NTRU: A Ring-Based Public Key Cryptosystem. In: Buhler, J.P. (ed.) ANTS 1998. LNCS, vol. 1423, pp. 267–288. Springer, Heidelberg (1998)
19. Hoffstein, J., Silverman, J.H., Whyte, W.: Estimated breaking times for ntru lattices. Technical report, NTRU Cryptosystems, Report #012, v2 (October 2003)
20. Kannan, R.: Improved algorithms for integer programming and related lattice problems. In: STOC 1983, pp. 193–206. ACM (1983)
21. Lee, M.S., Hahn, S.G.: Cryptanalysis of the GGH cryptosystem. Mathematics in Computer Science 3, 201–208 (2010)
22. Lenstra, A.K., Lenstra Jr., H.W., Lovász, L.: Factoring polynomials with rational coefficients. Mathematische Ann. 261, 513–534 (1982)
23. Lindner, R., Peikert, C.: Better key sizes (and attacks) for lwe-based encryption. Cryptology ePrint Archive, Report 2010/613, Full version of the CT-RSA 2011
24. Lindner, R., Rückert, M.: TU Darmstadt lattice challenge, http://www.latticechallenge.org/
25. May, A.: Cryptanalysis of NTRU-107. Draft of 1999, available on May's webpage (1999)
26. May, A., Silverman, J.H.: Dimension Reduction Methods for Convolution Modular Lattices. In: Silverman, J.H. (ed.) CaLC 2001. LNCS, vol. 2146, pp. 110–125. Springer, Heidelberg (2001)

27. Mazo, J.E., Odlyzko, A.M.: Lattice points in high dimensional spheres. Monatsheft Mathematik 17, 47–61 (1990)
28. Micciancio, D., Regev, O.: Lattice-based cryptography. In: Post-Quantum Cryptography, pp. 147–191. Springer, Berlin (2009)
29. Micciancio, D., Voulgaris, P.: A deterministic single exponential time algorithm for most lattice problems based on Voronoi cell computations. In: STOC 2010. ACM (2010)
30. Micciancio, D., Voulgaris, P.: Faster exponential time algorithms for the shortest vector problem. In: SODA 2010, pp. 1468–1480. ACM-SIAM (2010)
31. Nguyên, P.Q.: Cryptanalysis of the Goldreich-Goldwasser-Halevi Cryptosystem from Crypto'97. In: Wiener, M. (ed.) CRYPTO 1999. LNCS, vol. 1666, pp. 288–304. Springer, Heidelberg (1999)
32. Nguyen, P.Q.: Public-key cryptanalysis. In: Luengo, I. (ed.) Recent Trends in Cryptography. Contemporary Mathematics, vol. 477. AMS–RSME (2009)
33. Nguyen, P.Q.: Hermite's constant and lattice algorithms. In: [35] (2010)
34. Nguyên, P.Q., Stehlé, D.: LLL on the Average. In: Hess, F., Pauli, S., Pohst, M. (eds.) ANTS 2006. LNCS, vol. 4076, pp. 238–256. Springer, Heidelberg (2006)
35. Nguyen, P.Q., Vallée, B. (eds.): The LLL Algorithm: Survey and Applications. Information Security and Cryptography. Springer, Heidelberg (2010)
36. Nguyen, P.Q., Vidick, T.: Sieve algorithms for the shortest vector problem are practical. J. of Mathematical Cryptology 2(2), 181–207 (2008)
37. Novocin, A., Stehlé, D., Villard, G.: An LLL-reduction algorithm with quasi-linear time complexity. In: Proc. STOC 2011. ACM (2011)
38. Pohst, M.: On the computation of lattice vectors of minimal length, successive minima and reduced bases with applications. SIGSAM Bull. 15(1), 37–44 (1981)
39. Rückert, M., Schneider, M.: Estimating the security of lattice-based cryptosystems. Cryptology ePrint Archive, Report 2010/137 (2010)
40. Schneider, M., Gama, N.: SVP challenge,
    http://www.latticechallenge.org/svp-challenge/
41. Schnorr, C.-P.: A hierarchy of polynomial lattice basis reduction algorithms. Theoretical Computer Science 53(2-3), 201–224 (1987)
42. Schnorr, C.-P., Euchner, M.: Lattice basis reduction: improved practical algorithms and solving subset sum problems. Math. Programming 66, 181–199 (1994)
43. Schnorr, C.-P., Hörner, H.H.: Attacking the Chor-Rivest Cryptosystem by Improved Lattice Reduction. In: Guillou, L.C., Quisquater, J.-J. (eds.) EUROCRYPT 1995. LNCS, vol. 921, pp. 1–12. Springer, Heidelberg (1995)
44. Shoup, V.: Number Theory C++ Library (NTL) version 5.4.1,
    http://www.shoup.net/ntl/
45. Wang, X., Liu, M., Tian, C., Bi, J.: Improved Nguyen-Vidick heuristic sieve algorithm for shortest vector problem. In: Cryptology ePrint Archive, Report 2010/647 (2010)

Functional Encryption for Inner Product Predicates from Learning with Errors

Shweta Agrawal[1,*], David Mandell Freeman[2,**], and Vinod Vaikuntanathan[3,***]

[1] University of California, Los Angeles, USA
shweta@cs.ucla.edu
[2] Stanford University, USA
dfreeman@cs.stanford.edu
[3] University of Toronto, Canada
vinodv@cs.toronto.edu

Abstract. We propose a lattice-based functional encryption scheme for inner product predicates whose security follows from the difficulty of the *learning with errors* (LWE) problem. This construction allows us to achieve applications such as range and subset queries, polynomial evaluation, and CNF/DNF formulas on encrypted data. Our scheme supports inner products over small fields, in contrast to earlier works based on bilinear maps.

Our construction is the first functional encryption scheme based on lattice techniques that goes beyond basic identity-based encryption. The main technique in our scheme is a novel twist to the identity-based encryption scheme of Agrawal, Boneh and Boyen (Eurocrypt 2010). Our scheme is weakly attribute hiding in the standard model.

Keywords: Functional encryption, predicate encryption, lattices, learning with errors.

1 Introduction

Traditional public-key encryption is "coarse," in the sense that any user in the system can decrypt only messages encrypted with that user's public key. In a line of research beginning with the work of Sahai and Waters [39], a number of researchers have asked how to make encryption more fine-grained. The result is the notion of *functional encryption* [16], in which secret keys allow users to learn functions of encrypted data. Two important examples of functional encryption are *attribute-based encryption*

---

* Part of this work done while at Microsoft Research Redmond. Research supported in part from a DARPA/ONR PROCEED award, and NSF grants 1118096, 1065276, 0916574 and 0830803.
** Research supported by NSF and DARPA.
*** Part of this work done while at Microsoft Research Redmond. Supported by an NSERC Discovery grant and by DARPA under Agreement number FA8750-11-2-0225. The U.S. Government is authorized to reproduce and distribute reprints for Governmental purposes notwithstanding any copyright notation thereon. The views and conclusions contained herein are those of the authors and should not be interpreted as necessarily representing the official policies or endorsements, either expressed or implied, of DARPA or the U.S. Government.

D.H. Lee and X. Wang (Eds.): ASIACRYPT 2011, LNCS 7073, pp. 21–40, 2011.

(ABE) [39, 27] and *predicate encryption* (PE) [17, 29]. In (key-policy) ABE and PE systems, each ciphertext c is associated with an attribute a and each secret key s is associated with a predicate f. A user holding the key s can decrypt c if and only if $f(a) = 1$. The difference between the two types of systems is in the amount of information revealed: an ABE system reveals the attribute associated with each ciphertext, while a PE system keeps the attribute hidden. (Formal definitions of these properties appear in Section 2.)

This hiding requirement has made predicate encryption systems much more difficult to construct than attribute-based encryption systems: while there exist ABE schemes that allow *any* access formula over attributes [35, 46], the most expressive PE scheme is that of Katz, Sahai, and Waters [29], who construct a PE scheme for *inner product predicates*. In such a scheme, attributes a and predicates f are expressed as vectors $\vec{v}_a$ and $\vec{w}_f$ respectively, and we say $f(a) = 1$ if and only if $\langle \vec{v}_a, \vec{w}_f \rangle = 0$. Despite this apparently restrictive structure, inner product predicates can support conjunction, subset and range queries on encrypted data [17] as well as disjunctions, polynomial evaluation, and CNF and DNF formulas [29].

All known constructions of attribute-based encryption [39, 27, 10, 21, 35, 26, 46, 8, 30, 34, 9] and predicate encryption [15, 1, 41, 17, 29, 42, 40, 33, 11, 30] make use of groups with bilinear maps, and the security of these schemes is based on many different, and often complex, assumptions. In particular, there is at present no known construction of predicate encryption for inner products based on a "standard" assumption in bilinear groups.[1] As an example of a "nonstandard" assumption used in previous constructions, Katz, Sahai, and Waters present an assumption [29, Assumption 1] where the challenge consists of ten elements chosen in a specified way from a group whose order is the product of three large primes p, q, r, and the problem is to determine whether one of these elements has an order-q component. While assumptions such as this one can often be shown to hold in a suitable "generic group model" (e.g., [29, Appendix A]), to obtain more confidence in security we would like to build ABE and PE schemes based on computational problems whose complexity is better understood.

Our Contribution. In this work we construct a lattice-based predicate encryption scheme for inner product predicates whose security follows from the difficulty of the *learning with errors* (LWE) problem. The LWE problem, in turn, is at least as hard as approximating the standard lattice problems GapSVP and SIVP in the *worst case* [38, 36] and is also conjectured to be difficult even for quantum adversaries. Our construction is the first functional encryption scheme based on lattice techniques that goes beyond basic identity-based encryption (which can be viewed as predicate encryption that tests equality on strings). Our construction is capable of instantiating all of the applications of predicate encryption proposed by Boneh and Waters [17] and Katz, Sahai, and Waters [29].[2] While our construction does not satisfy the strong notion of privacy defined by Katz, Sahai, and Waters [29], it does satisfy the slightly weaker notion considered by Okamoto and Takashima [33, 34] and Lewko *et al.* [30].

---

[1] Okamoto and Takashima [34] claim a PE construction from the decision linear assumption, but their paper only indicates how this is achieved for ABE.

[2] A detailed discussion of these applications can be found in the full version of this paper [2, §5].

1.1 Overview of the Construction

Our Approach. Just as functional encryption in bilinear groups builds on the ideas and techniques introduced in constructions of identity-based encryption (IBE) in bilinear groups [14, 25, 12, 13, 44, 22], our construction builds on the ideas and techniques used to achieve identity-based encryption from the LWE assumption [24, 5, 20, 3, 4]. However, there is a key difference between lattice IBE constructions (without random oracles) and bilinear-group constructions that makes this kind of generalization more difficult in the lattice setting. Namely, in the bilinear-group IBE constructions the groups remain fixed, while the ciphertexts and keys are manipulated so that group elements "cancel out" when a ciphertext matches a key. In the lattice IBE constructions, each key and ciphertext is constructed using a *different* lattice, and decryption only works when the key lattice and ciphertext lattice match. This structure does not easily generalize to the functional encryption setting, where each key may match many ciphertexts and each ciphertext may match many keys.

We solve this "lattice matching" problem using a new algebraic technique that builds on the IBE scheme of Agrawal, Boneh, and Boyen [3]. In our construction, we generate keys using a lattice Λ_f that depends only on the predicate f, and we generate ciphertexts c using a lattice Λ_a that depends only on the attribute a. Given a ciphertext c generated in this way and predicate f, we apply a suitable linear transformation that moves c into the lattice Λ_f if and only if $f(a) = 1$. Once this transformation is applied, we can decrypt using a key associated with Λ_f.

The details of our scheme and security proof are in Section 4. To prove security, we use a simulation technique that draws on ideas introduced in [3]. In particular, we construct our simulation using a "punctured" trapdoor that allows the simulator to generate secret keys for any predicate f such that $f(a) = 0$, where a is the "challenge" attribute. In the simulation we can use an LWE challenge to construct a ciphertext that either decrypts correctly or decrypts to a random message. While this technique suffices to prove that the system hides the message contents ("payload hiding"), it only allows us to prove a weak form of anonymity ("attribute hiding"). Specifically, given a ciphertext c and a number of keys that *do not* decrypt c, the user cannot determine the attribute associated with c. In the strong form of attribute hiding, the user cannot determine the attribute associated with c even when given keys that *do* decrypt c. (Formal definitions of these concepts appear in Section 2.) The weakened form of attribute hiding we do achieve is nonetheless more than is required for ABE and should be sufficient for many applications of PE.

Key Technical Ideas. Our encryption scheme is at its core based on the LWE scheme of Gentry, Peikert, and Vaikuntanathan [24, §7], which is itself a "dual" of the original Regev LWE scheme [38, §5]. From a geometric perspective, the public key in the GPV scheme describes a lattice Λ used to construct ciphertexts, and the secret key is derived from the dual lattice $\Lambda^{\perp}$. Existing constructions of lattice-based IBE in the standard model [5, 20, 3, 4] use the GPV encryption scheme but replace the fixed lattice Λ with a lattice Λ_{id} that depends on the user's identity id. Decryption only works when the ciphertext lattice Λ_{id} and secret key lattice $\Lambda_{id'}$ are duals of each other, and there are several methods of ensuring that this is the case if and only if $id = id'$.

In trying to adapt these constructions to the predicate encryption setting, we run into the problem that each ciphertext can be decrypted by many secret keys and each secret key can decrypt many ciphertexts. Thus we cannot require that key lattices match ciphertext lattices in the same way as above.

Before explaining our solution to this problem, let us recall the IBE scheme of Agrawal, Boneh, and Boyen [3]. In the ABB IBE scheme, the encryption lattice is constructed as

$$\Lambda_{\mathsf{id}} = \Lambda_q(\mathbf{A}_0 \parallel \mathbf{A}_1 + H(\mathsf{id})\mathbf{B}),$$

where $\mathbf{A}_0, \mathbf{A}_1, \mathbf{B}$ are $n \times m$ matrices over $\mathbb{Z}_q$ and $H(\mathsf{id})$ is a "full-rank difference" hash function. One can generate secret keys for $\Lambda_{\mathsf{id}}^{\perp}$ using a short basis of $\Lambda_q^{\perp}(\mathbf{A}_0)$ and the basis extension technique of [5, 20]. In the (selective-)security proof, the LWE challenge is embedded as the matrix $\mathbf{A}_0$, and the matrix $\mathbf{A}_1 + H(\mathsf{id})\mathbf{B}$ is equipped with a "punctured" trapdoor that allows the simulator to respond to secret key queries for all identities id not equal to the challenge identity id^*.

The algebraic structure of the ABB IBE scheme gives us the tools we need to solve the "lattice matching" problem described above. Specifically, in our predicate encryption scheme we encode an attribute vector $\vec{w} = (w_1, \ldots, w_\ell) \in \mathbb{Z}_q^\ell$ as the $n \times \ell m$ matrix

$$\mathbf{B}_{\vec{w}} := (w_1\mathbf{B} \parallel \cdots \parallel w_\ell \mathbf{B}).$$

where $\mathbf{B} \in \mathbb{Z}_q^{n \times m}$ is a uniformly random matrix chosen by the encryptor. We generate the ciphertext as a GPV encryption relative to the matrix

$$\Lambda_{\vec{w}} := \Lambda_q(\mathbf{A}_0 \parallel \mathbf{A}_1 + w_1\mathbf{B} \parallel \cdots \parallel \mathbf{A}_\ell + w_\ell \mathbf{B})$$

where the $\mathbf{A}_i$ are all $n \times m$ matrices. We view the ciphertext component that is close to $\Lambda_{\vec{w}}$ as a tuple $(\mathbf{c}_0, \ldots, \mathbf{c}_\ell) \in (\mathbb{Z}_q^m)^{\ell+1}$.

Since the recipient of a ciphertext does not know *a priori* which lattice was used to encrypt (indeed, this is exactly the anonymity property of predicate encryption), we cannot expect the recipient to possess a secret key derived from the dual of the ciphertext lattice as in the IBE case. Instead, we derive the key for a predicate vector $\vec{v}$ from the dual of a certain lattice $\Lambda_{\vec{v}}$ and apply a linear transformation $T_{\vec{v}}$ that moves the ciphertext into $\Lambda_{\vec{v}}$ exactly when $\langle \vec{v}, \vec{w} \rangle = 0$. If this linear transformation is "short" (in the sense of not increasing the length of vectors too much), then a GPV secret key derived from $\Lambda_{\vec{v}}^{\perp}$ can decrypt the ciphertext $T_{\vec{v}}(c)$.

Concretely, this transformation works as follows. For a predicate vector $\vec{v} = (v_1, \ldots, v_\ell) \in \mathbb{Z}_q^\ell$, we define the linear transformation $T_{\vec{v}} : (\mathbb{Z}_q^m)^{\ell+1} \to \mathbb{Z}_q^{2m}$ by

$$T_{\vec{v}}(\mathbf{c}_0, \ldots, \mathbf{c}_\ell) = (\mathbf{c}_0, \textstyle\sum_{i=1}^\ell v_i\mathbf{c}_i).$$

Some algebraic manipulation (detailed in Section 4) shows that applying this transformation to a ciphertext encrypted using $\Lambda_{\vec{w}}$ is equivalent to computing a GPV ciphertext using the lattice

$$\Lambda_{\vec{v},\vec{w}} := \Lambda_q\Big(\mathbf{A}_0 \parallel \sum_{i=1}^\ell v_i\mathbf{A}_i + \langle \vec{v}, \vec{w} \rangle\mathbf{B}\Big),$$

Letting the secret key for $\vec{v}$ be the GPV secret key associated to $\Lambda_q^{\perp}(\mathbf{A}_0 \parallel \sum_{i=1}^{\ell} v_i \mathbf{A}_i)$ allows the holder of a key for predicate $\vec{v}$ to decrypt a ciphertext associated with attribute $\vec{w}$ exactly when $\langle \vec{v}, \vec{w} \rangle = 0$. In this aspect our construction is inspired by that of Katz, Sahai, and Waters [29]: the matrix $\mathbf{B}$ corresponds to the "masking terms" in a KSW ciphertext that "cancel out" exactly when $\langle \vec{v}, \vec{w} \rangle = 0$.

The reader may have observed that in the above formulation, the requirement that the transformation $T_{\vec{v}}$ be "short" implies that we cannot use all vectors $\vec{v} \in \mathbb{Z}_q^{\ell}$ as predicates, but only ones whose entries have small absolute value (when viewed as integers in $(-q/2, q/2]$). In Section 4 we will show that decomposing the vector $\vec{v}$ into its binary representation enables our construction to use arbitrary vectors in $\mathbb{Z}_q^{\ell}$, at the expense of expanding the ciphertext by a factor of $\lg q$.

2   Predicate Encryption

We use the definition of predicate encryption proposed by Katz, Sahai, and Waters [29], which is based on the definition of *searchable encryption* proposed by Boneh and Waters [17]. We will let n denote the security parameter throughout this paper.

Definition 2.1 ([29, Definition 2.1]). A (key-policy) *predicate encryption scheme* for the class of predicates $\mathcal{F}$ over the set of attributes Σ consists of four probabilistic polynomial-time algorithms Setup, KeyGen, Enc, Dec such that:

- Setup takes as input a security parameter n and outputs a set of public parameters PP and a master secret key MK.
- KeyGen takes as input the master secret key MK and a (description of a) predicate $f \in \mathcal{F}$. It outputs a key sk_f.
- Enc takes as input the public parameters PP, an attribute $I \in \Sigma$, and a message M in some associated message space $\mathcal{M}$. It returns a ciphertext C.
- Dec takes as input a secret key sk_f and a ciphertext C. It outputs either a message M or the distinguished symbol $\perp$.

For correctness, we require that for all n, all $(\mathsf{PP}, \mathsf{MK})$ generated by $\mathsf{Setup}(1^n)$, all $f \in \mathcal{F}$, any key $\mathsf{sk}_f \leftarrow \mathsf{KeyGen}(\mathsf{MK}, f)$, all $I \in \Sigma$, and any ciphertext $C \leftarrow \mathsf{Enc}(\mathsf{PP}, I, M)$:

- If $f(I) = 1$, then $\mathsf{Dec}(\mathsf{sk}_f, C) = M$.
- If $f(I) = 0$, then $\mathsf{Dec}(\mathsf{sk}_f, C) = \perp$ with all but negligible probability.

In a *ciphertext-policy* scheme keys are associated with attributes and ciphertexts are associated with predicates; the syntax is otherwise the same.

Our construction in Section 4 satisfies a different correctness condition: If $f(I) = 1$ and $C = \mathsf{Enc}(\mathsf{PP}, I, M)$, then $\mathsf{Dec}(\mathsf{sk}_f, C) = M$, but if $f(I) = 0$ then $\mathsf{Dec}(\mathsf{sk}_f, C)$ is computationally indistinguishable from a uniformly random element in the message space $\mathcal{M}$. However, if $\mathcal{M}$ is exponentially large then we can easily transform our system into one satisfying Definition 2.1 by restricting the message space to some subset $\mathcal{M}' \subset \mathcal{M}$ with $|\mathcal{M}'|/|\mathcal{M}| = \mathsf{negl}(n)$.

2.1 Security

There are several notions of security for predicate encryption schemes. The most basic is *payload hiding*, which guarantees that no efficient adversary can obtain any information about the encrypted message, but allows information about attributes to be revealed. A stronger notion is *attribute hiding*, which guarantees in addition that no efficient adversary can obtain any information about the attribute associated with a ciphertext. Following Lewko *et al.* [30, Definition 17], we also define an intermediate notion, *weak attribute hiding*, which makes the same guarantee only in the case that the adversary cannot decrypt the ciphertext. Our definition of security is "selective," in the sense that the adversary must commit to its challenge attributes before seeing any secret keys.

Definition 2.2 ([29, Definition 2.2]). A predicate encryption scheme with respect to $\mathcal{F}$ and Σ is *attribute hiding* if for all probabilistic polynomial-time adversaries $\mathcal{A}$, the advantage of $\mathcal{A}$ in the following experiment is negligible in the security parameter n:

1. $\mathcal{A}(1^n)$ outputs $I_0, I_1 \in \Sigma$.
2. Setup(1^n) is run to generate PP and MK, and the adversary is given PP.
3. $\mathcal{A}$ may adaptively request keys for any predicates $f_1, \ldots, f_\ell \in \mathcal{F}$ subject to the restriction that $f_i(I_0) = f_i(I_1)$ for all i. In response, $\mathcal{A}$ is given the corresponding keys $\mathsf{sk}_{f_i} \leftarrow \mathsf{KeyGen}(\mathsf{MK}, f_i)$.
4. $\mathcal{A}$ outputs two equal-length messages M_0, M_1. If there is an i for which $f_i(I_0) = f_i(I_1) = 1$, then it is required that $M_0 = M_1$. A random bit b is chosen, and $\mathcal{A}$ is given the ciphertext $C \leftarrow \mathsf{Enc}(\mathsf{PP}, I_b, M_b)$.
5. The adversary may continue to request keys for additional predicates, subject to the same restrictions as before.
6. $\mathcal{A}$ outputs a bit b', and succeeds if $b' = b$. The *advantage* of $\mathcal{A}$ is the absolute value of the difference between its success probability and $1/2$.

We say the scheme is *weakly attribute hiding* if the same condition holds for adversaries $\mathcal{A}$ that are only allowed to request keys for predicates f_i with $f_i(I_0) = f_i(I_1) = 0$. We say the scheme is *payload hiding* if we require $I_0 = I_1$.

We observe that any scheme that is attribute hiding is weakly attribute hiding, and any scheme that is weakly attribute hiding is payload hiding. (In the payload hiding game no adversary can achieve nonzero advantage when requesting a key for a predicate f with $f(I_0) = f(I_1) = 1$, so we may assume without loss of generality that the adversary does not request such a key.)

Remark 2.3. In our construction the spaces $\mathcal{F}$ of predicates and Σ of attributes depend on the public parameters PP output by Setup. We thus modify the security game so as to give the adversary descriptions of $\mathcal{F}$ and Σ before Step 1 and run the remainder of the game (including any remaining steps in the Setup algorithm) as described.

3 Lattice Preliminaries

In this section we collect the results from the literature that we will need for our construction and the proof of security.

Notation. For any integer $q \geq 2$, we let $\mathbb{Z}_q$ denote the ring of integers modulo q and we represent $\mathbb{Z}_q$ as integers in $(-q/2, q/2]$. We let $\mathbb{Z}_q^{n \times m}$ denote the set of $n \times m$ matrices with entries in $\mathbb{Z}_q$. We use bold capital letters (e.g. $\mathbf{A}$) to denote matrices, bold lowercase letters (e.g. $\mathbf{x}$) to denote vectors that are components of our encryption scheme, and arrows (e.g. $\vec{v}$) to denote vectors that represent attributes or predicates. The notation $\mathbf{A}^\mathsf{T}$ denotes the transpose of the matrix $\mathbf{A}$. When we say a matrix defined over $\mathbb{Z}_q$ has *full rank*, we mean that it has full rank modulo each prime factor of q. The notation $\lfloor x \rceil$ denotes the nearest integer to x, rounding towards 0 for half-integers.

3.1 Lattices

An m-*dimensional lattice* Λ is a full-rank discrete subgroup of $\mathbb{R}^m$. A *basis* of Λ is a linearly independent set of vectors whose span is Λ. We will usually be concerned with *integer lattices*, i.e., those whose points have coordinates in $\mathbb{Z}^m$. Among these lattices are the "q-ary" lattices defined as follows: for any integer $q \geq 2$ and any $\mathbf{A} \in \mathbb{Z}_q^{n \times m}$, we define

$$\Lambda_q^\perp(\mathbf{A}) := \left\{ \mathbf{e} \in \mathbb{Z}^m : \mathbf{A} \cdot \mathbf{e} = \mathbf{0} \bmod q \right\}$$
$$\Lambda_q^\mathbf{u}(\mathbf{A}) := \left\{ \mathbf{e} \in \mathbb{Z}^m : \mathbf{A} \cdot \mathbf{e} = \mathbf{u} \bmod q \right\}.$$

The lattice $\Lambda_q^\mathbf{u}(\mathbf{A})$ is a coset of $\Lambda_q^\perp(\mathbf{A})$; namely, $\Lambda_q^\mathbf{u}(\mathbf{A}) = \Lambda_q^\perp(\mathbf{A}) + \mathbf{t}$ for any $\mathbf{t}$ such that $\mathbf{A} \cdot \mathbf{t} = \mathbf{u} \bmod q$.

The Gram-Schmidt norm of a basis. Let $\mathbf{S} = \{\mathbf{s}_1, \ldots, \mathbf{s}_k\}$ be a set of vectors in $\mathbb{R}^m$. We use the following standard notation:

- $\|\mathbf{S}\|$ denotes the length of the longest vector in $\mathbf{S}$, i.e., $\max_{1 \leq i \leq k} \|\mathbf{s}_i\|$.
- $\widetilde{\mathbf{S}} := \{\tilde{\mathbf{s}}_1, \ldots, \tilde{\mathbf{s}}_k\} \subset \mathbb{R}^m$ denotes the Gram-Schmidt orthogonalization of the vectors $\mathbf{s}_1, \ldots, \mathbf{s}_k$.

We refer to $\|\widetilde{\mathbf{S}}\|$ as the *Gram-Schmidt norm* of $\mathbf{S}$.

Ajtai [6] and later Alwen and Peikert [7] showed how to sample an almost uniform matrix $\mathbf{A} \in \mathbb{Z}_q^{n \times m}$ along with a basis $\mathbf{S}$ of $\Lambda_q^\perp(\mathbf{A})$ with low Gram-Schmidt norm.

Theorem 3.1 ([7, Theorem 3.2] with $\delta = 1/3$). *Let q, n, m be positive integers with $q \geq 2$ and $m \geq 6n \lg q$. There is a probabilistic polynomial-time algorithm* $\mathsf{TrapGen}(q, n, m)$ *that with overwhelming probability (in n) outputs a pair $(\mathbf{A} \in \mathbb{Z}_q^{n \times m}, \mathbf{S} \in \mathbb{Z}^{m \times m})$ such that $\mathbf{A}$ is statistically close to uniform in $\mathbb{Z}_q^{n \times m}$ and $\mathbf{S}$ is a basis for $\Lambda_q^\perp(\mathbf{A})$ satisfying*

$$\|\widetilde{\mathbf{S}}\| \leq O(\sqrt{n \log q}) \quad \text{and} \quad \|\mathbf{S}\| \leq O(n \log q).$$

Gaussian Distributions. Let L be a discrete subset of $\mathbb{Z}^n$. For any vector $\mathbf{c} \in \mathbb{R}^n$ and any positive parameter $\sigma \in \mathbb{R}_{>0}$, let $\rho_{\sigma,\mathbf{c}}(\mathbf{x}) := \exp\left(-\pi \|\mathbf{x} - \mathbf{c}\|^2 / \sigma^2\right)$ be the Gaussian function on $\mathbb{R}^n$ with center $\mathbf{c}$ and parameter σ. Let $\rho_{\sigma,\mathbf{c}}(L) := \sum_{\mathbf{x} \in L} \rho_{\sigma,\mathbf{c}}(\mathbf{x})$ be the discrete integral of $\rho_{\sigma,\mathbf{c}}$ over L (which always converges), and let $\mathcal{D}_{L,\sigma,\mathbf{c}}$ be the discrete Gaussian distribution over L with center $\mathbf{c}$ and parameter σ. Specifically, for all $\mathbf{y} \in L$, we have $\mathcal{D}_{L,\sigma,\mathbf{c}}(\mathbf{y}) = \frac{\rho_{\sigma,\mathbf{c}}(\mathbf{y})}{\rho_{\sigma,\mathbf{c}}(L)}$. For notational convenience, $\rho_{\sigma,0}$ and $\mathcal{D}_{L,\sigma,0}$ are abbreviated as ρ_σ and $\mathcal{D}_{L,\sigma}$, respectively.

The following lemma gives a bound on the length of vectors sampled from a discrete Gaussian. The result follows from [32, Lemma 4.4], using [24, Lemma 5.3] to bound the smoothing parameter.

Lemma 3.2. *Let Λ be an n-dimensional lattice, let $\mathbf{T}$ be a basis for Λ, and suppose $\sigma \geq \|\widetilde{\mathbf{T}}\| \cdot \omega(\sqrt{\log n})$. Then for any $\mathbf{c} \in \mathbb{R}^n$ we have*

$$\Pr\left[\|\mathbf{x} - \mathbf{c}\| > \sigma\sqrt{n} : \mathbf{x} \xleftarrow{\mathrm{R}} \mathcal{D}_{\Lambda,\sigma,\mathbf{c}}\right] \leq \mathrm{negl}(n)$$

3.2   Sampling Algorithms

We will use the following algorithms to sample short vectors from specific lattices. Looking ahead, the algorithm SampleLeft [3, 20] will be used to sample keys in the real system, while the algorithm SampleRight [3] will be used to sample keys in the simulation.

Algorithm SampleLeft($\mathbf{A}, \mathbf{B}, \mathbf{T_A}, \mathbf{u}, \sigma$):

> *Inputs:* a full rank matrix $\mathbf{A}$ in $\mathbb{Z}_q^{n \times m}$, a "short" basis $\mathbf{T_A}$ of $\Lambda_q^\perp(\mathbf{A})$, a matrix $\mathbf{B}$ in $\mathbb{Z}_q^{n \times m_1}$, a vector $\mathbf{u} \in \mathbb{Z}_q^n$, and a Gaussian parameter σ.   (3.1)
>
> *Output:* Let $\mathbf{F} := (\mathbf{A} \| \mathbf{B})$. The algorithm outputs a vector $\mathbf{e} \in \mathbb{Z}^{m+m_1}$ in the coset $\Lambda_q^\mathbf{u}(\mathbf{F})$.

Theorem 3.3 ([3, Theorem 17], [20, Lemma 3.2]). *Let $q > 2$, $m > n$ and $\sigma > \|\widetilde{\mathbf{T_A}}\| \cdot \omega(\sqrt{\log(m + m_1)})$. Then SampleLeft($\mathbf{A}, \mathbf{B}, \mathbf{T_A}, \mathbf{u}, \sigma$) taking inputs as in (3.1) outputs a vector $\mathbf{e} \in \mathbb{Z}^{m+m_1}$ distributed statistically close to $\mathcal{D}_{\Lambda_q^\mathbf{u}(\mathbf{F}),\sigma}$, where $\mathbf{F} := (\mathbf{A} \| \mathbf{B})$.*

Algorithm SampleRight($\mathbf{A}, \mathbf{B}, \mathbf{R}, \mathbf{T_B}, \mathbf{u}, \sigma$):

> *Inputs:* matrices $\mathbf{A}$ in $\mathbb{Z}_q^{n \times k}$ and $\mathbf{R}$ in $\mathbb{Z}^{k \times m}$, a full rank matrix $\mathbf{B}$ in $\mathbb{Z}_q^{n \times m}$, a "short" basis $\mathbf{T_B}$ of $\Lambda_q^\perp(\mathbf{B})$, a vector $\mathbf{u} \in \mathbb{Z}_q^n$, and a Gaussian parameter σ.   (3.2)
>
> *Output:* Let $\mathbf{F} := (\mathbf{A} \| \mathbf{AR} + \mathbf{B})$. The algorithm outputs a vector $\mathbf{e} \in \mathbb{Z}^{m+k}$ in the coset $\Lambda_q^\mathbf{u}(\mathbf{F})$.

Often the matrix $\mathbf{R}$ given to the algorithm as input will be a random matrix in $\{1, -1\}^{m \times m}$. Let S^m be the m-sphere $\{\mathbf{x} \in \mathbb{R}^{m+1} : \|\mathbf{x}\| = 1\}$. We define $s_R := \|\mathbf{R}\| := \sup_{\mathbf{x} \in S^{m-1}} \|\mathbf{R} \cdot \mathbf{x}\|$.

Theorem 3.4 ([3, Theorem 19]). *Let $q > 2, m > n$ and $\sigma > \|\widetilde{\mathbf{T_B}}\| \cdot s_R \cdot \omega(\sqrt{\log m})$. Then SampleRight($\mathbf{A}, \mathbf{B}, \mathbf{R}, \mathbf{T_B}, \mathbf{u}, \sigma$) taking inputs as in (3.2) outputs a vector $\mathbf{e} \in \mathbb{Z}^{m+k}$ distributed statistically close to $\mathcal{D}_{\Lambda_q^\mathbf{u}(\mathbf{F}),\sigma}$, where $\mathbf{F} := (\mathbf{A} \| \mathbf{AR} + \mathbf{B})$.*

3.3   The LWE Problem

The *learning with errors* problem, or LWE, is the problem of determining a secret vector over $\mathbb{Z}_q$ given an arbitrary number of "noisy" inner products. The decision variant is to distinguish such samples from random. More formally, we define the (average-case) problem as follows:

Definition 3.5 ([38]). *Let* $n \geq 1$ *and* $q \geq 2$ *be integers, and let* χ *be a probability distribution on* $\mathbb{Z}_q$. *For* $\mathbf{s} \in \mathbb{Z}_q^n$, *let* $A_{\mathbf{s},\chi}$ *be the probability distribution on* $\mathbb{Z}_q^n \times \mathbb{Z}_q$ *obtained by choosing a vector* $\mathbf{a} \in \mathbb{Z}_q^n$ *uniformly at random, choosing* $e \in \mathbb{Z}_q$ *according to* χ, *and outputting* $(\mathbf{a}, \langle \mathbf{a}, \mathbf{s} \rangle + e)$.

(a) *The* search-LWE$_{q,n,\chi}$ *problem is: for uniformly random* $\mathbf{s} \in \mathbb{Z}_q^n$, *given a* $\mathrm{poly}(n)$ *number of samples from* $A_{\mathbf{s},\chi}$, *output* $\mathbf{s}$.

(b) *The* decision-LWE$_{q,n,\chi}$ *problem is: for uniformly random* $\mathbf{s} \in \mathbb{Z}_q^n$, *given a* $\mathrm{poly}(n)$ *number of samples that are either (all) from* $A_{\mathbf{s},\chi}$ *or (all) uniformly random in* $\mathbb{Z}_q^n \times \mathbb{Z}_q$, *output 0 if the former holds and 1 if the latter holds.*

We say the decision-LWE$_{q,n,\chi}$ problem is *infeasible* if for all polynomial-time algorithms $\mathcal{A}$, the probability that $\mathcal{A}$ solves the decision-LWE problem (over $\mathbf{s}$ and $\mathcal{A}$'s random coins) is negligibly close to $1/2$ as a function of n.

The power of the LWE problem comes from the fact that for certain noise distributions χ, solving the search-LWE problem is as hard as finding approximate solutions to the shortest independent vectors problem (SIVP) and the decision version of the shortest vector problem (GapSVP) in the worst case. For polynomial size q there is a quantum reduction due to Regev, while for exponential size q there is a classical reduction due to Peikert. Furthermore, the search and decision versions of the problem are equivalent whenever q is a product of small primes. These results are summarized in the following:

Definition 3.6. *For* $\alpha \in (0,1)$ *and an integer* $q > 2$, *let* $\overline{\Psi}_\alpha$ *denote the probability distribution over* $\mathbb{Z}_q$ *obtained by choosing* $x \in \mathbb{R}$ *according to the normal distribution with mean 0 and standard deviation* $\alpha/\sqrt{2\pi}$ *and outputting* $\lfloor qx \rceil$.

Theorem 3.7 ([38]). *Let* n, q *be integers and* $\alpha \in (0,1)$ *such that* $q = \mathrm{poly}(n)$ *and* $\alpha q > 2\sqrt{n}$. *If there exists an efficient (possibly quantum) algorithm that solves decision-*LWE$_{q,n,\overline{\Psi}_\alpha}$, *then there exists an efficient quantum algorithm that approximates* SIVP *and* GapSVP *to within* $\tilde{O}(n/\alpha)$ *in the worst case.*

Theorem 3.8 ([36]). *Let* n, q *be integers and* $\alpha \in (0,1)$, *and* $q = \prod_i q_i \geq 2^{n/2}$, *where the* q_i *are distinct primes satisfying* $\omega(\sqrt{\log n})/\alpha \leq q_i \leq \mathrm{poly}(n)$. *If there exists an efficient (classical) algorithm that solves decision-*LWE$_{q,n,\overline{\Psi}_\alpha}$, *then there exists an efficient (classical) algorithm that approximates* GapSVP *to within* $\tilde{O}(n/\alpha)$ *in the worst case.*

The following lemma will be used to show correctness of decryption.

Lemma 3.9 ([3, Lemma 12]). *Let* $\mathbf{e}$ *be some vector in* $\mathbb{Z}^m$ *and let* $\mathbf{y} \leftarrow \overline{\Psi}_\alpha^m$. *Then the quantity* $|\langle \mathbf{e}, \mathbf{y} \rangle|$ *when treated as an integer in* $(-q/2, q/2]$ *satisfies*

$$|\langle \mathbf{e}, \mathbf{y} \rangle| \leq \|\mathbf{e}\| q \alpha \cdot \omega(\sqrt{\log m}) + \|\mathbf{e}\| \sqrt{m}/2$$

with overwhelming probability (in m).

4 A Functional Encryption Scheme for Inner Product Predicates

In our system, each secret key will be associated with a predicate vector $\vec{v} \in \mathbb{Z}_q^\ell$ (for some fixed $\ell \geq 2$) and each ciphertext will be associated with an attribute vector $\vec{w} \in \mathbb{Z}_q^\ell$. Decryption should succeed if and only if $\langle \vec{v}, \vec{w} \rangle = 0 \pmod{q}$. Hence the predicate associated with the secret key is defined as $f_{\vec{v}}(\vec{w}) = 1$ if $\langle \vec{v}, \vec{w} \rangle = 0 \pmod{q}$, and $f_{\vec{v}}(\vec{w}) = 0$ otherwise.

4.1 The Construction

Let $n \in \mathbb{Z}^+$ be a security parameter and ℓ be the dimension of predicate and attribute vectors. Let $q = q(n, \ell)$ and $m = m(n, \ell)$ be positive integers. Let $\sigma = \sigma(n, \ell)$ and $\alpha = \alpha(n, \ell)$ be positive real Gaussian parameters. Define $k = k(n, \ell) := \lfloor \lg q \rfloor$. The encryption scheme described below encrypts a single bit; we show how to encrypt multiple bits in the full version of this paper [2, §4.5].

LinFE.Setup$(1^n, 1^\ell)$: On input a security parameter n and a parameter ℓ denoting the dimension of predicate and attribute vectors, do:
1. Use the algorithm $\mathsf{TrapGen}(q, n, m)$ (from Theorem 3.1) to select a matrix $\mathbf{A} \in \mathbb{Z}_q^{n \times m}$ together with a full-rank set of vectors $\mathbf{T_A} \subseteq \Lambda_q^\perp(\mathbf{A})$ such that $\|\widetilde{\mathbf{T_A}}\| \leq m \cdot \omega(\sqrt{\log m})$.
2. Choose $\ell \cdot (1 + k)$ uniformly random matrices $\mathbf{A}_{i,\gamma} \in \mathbb{Z}_q^{n \times m}$ for $i = 1, \ldots, \ell$ and $\gamma = 0, \ldots, k$.
3. Select a uniformly random vector $\mathbf{u} \in \mathbb{Z}_q^n$.
   Output $\mathsf{PP} = (\mathbf{A}, \{\mathbf{A}_{i,\gamma}\}_{i \in \{1,\ldots,\ell\}, \gamma \in \{0,\ldots,k\}}, \mathbf{u})$ and $\mathsf{MK} = \mathbf{T_A}$.

LinFE.KeyGen$(\mathsf{PP}, \mathsf{MK}, \vec{v})$: On input public parameters PP, a master secret key MK, and a predicate vector $\vec{v} = (v_1, \ldots, v_\ell) \in \mathbb{Z}_q^\ell$, do:
1. For $i = 1, \ldots, \ell$, let $\hat{v}_i$ be the integer in $[0, q-1]$ congruent to $v_i \bmod q$. Write the binary decomposition of $\hat{v}_i$ as

$$\hat{v}_i = \sum_{\gamma=0}^k v_{i,\gamma} \cdot 2^\gamma, \tag{4.1}$$

where $v_{i,\gamma}$ are in $\{0, 1\}$.
2. Define the matrices

$$\mathbf{C}_{\vec{v}} := \sum_{i=1}^\ell \sum_{\gamma=0}^k v_{i,\gamma} \mathbf{A}_{i,\gamma} \qquad \in \mathbb{Z}_q^{n \times m},$$

$$\mathbf{A}_{\vec{v}} := [\mathbf{A} \parallel \mathbf{C}_{\vec{v}}] \qquad \in \mathbb{Z}_q^{n \times 2m}.$$

3. Using the master secret key $\mathsf{MK} = (\mathbf{T_A}, \sigma)$, compute $\mathbf{e} \leftarrow \mathsf{SampleLeft}(\mathbf{A}, \mathbf{C}_{\vec{v}}, \mathbf{T_A}, \mathbf{u}, \sigma)$.
   Then $\mathbf{e}$ is a vector in $\mathbb{Z}^{2m}$ satisfying $\mathbf{A}_{\vec{v}} \cdot \mathbf{e} = \mathbf{u} \mod q$.
   Output the secret key $\mathsf{sk}_{\vec{v}} = \mathbf{e}$.

LinFE.Enc(PP, $\vec{w}, M$): On input public parameters PP, an attribute vector $\vec{w}$, and a message $M \in \{0, 1\}$, do:

1. Choose a uniformly random matrix $\mathbf{B} \xleftarrow{\mathrm{R}} \mathbb{Z}_q^{n \times m}$.
2. Choose a uniformly random $\mathbf{s} \xleftarrow{\mathrm{R}} \mathbb{Z}_q^n$.
3. Choose a noise vector $\mathbf{x} \leftarrow \overline{\Psi}_\alpha^m$ and a noise term $x \leftarrow \overline{\Psi}_\alpha$.
4. Compute $\mathbf{c}_0 \leftarrow \mathbf{A}^\mathsf{T}\mathbf{s} + \mathbf{x} \quad \in \mathbb{Z}_q^m$.
5. For $i = 1, \ldots, \ell$ and $\gamma = 0, \ldots, k$, do the following:
   (a) Pick a random matrix $\mathbf{R}_{i,\gamma} \in \{-1, 1\}^{m \times m}$.
   (b) Compute $\mathbf{c}_{i,\gamma} \leftarrow (\mathbf{A}_{i,\gamma} + 2^\gamma w_i \mathbf{B})^\mathsf{T}\mathbf{s} + \mathbf{R}_{i,\gamma}^\mathsf{T}\mathbf{x} \quad \in \mathbb{Z}_q^m$.
6. Compute $c' \leftarrow \mathbf{u}^\mathsf{T}\mathbf{s} + x + M \cdot \lfloor q/2 \rfloor \in \mathbb{Z}_q$.

Output the ciphertext $\mathsf{CT} := (\mathbf{c}_0, \{\mathbf{c}_{i,\gamma}\}_{i \in \{1,\ldots,\ell\}, \gamma \in \{0,\ldots,k\}}, c')$.

LinFE.Dec(PP, $\mathsf{sk}_{\vec{v}}$, CT): On input public parameters PP, a secret key $\mathsf{sk}_{\vec{v}}$ for predicate vector $\vec{v}$, and a ciphertext $\mathsf{CT} = (\mathbf{c}_0, \{\mathbf{c}_{i,\gamma}\}_{i \in \{1,\ldots,\ell\}, \gamma \in \{0,\ldots,k\}}, c')$, do:

1. Define the binary expansion of the vector $\vec{v}$ as in (4.1) and compute

$$\mathbf{c}_{\vec{v}} := \sum_{i=1}^{\ell} \sum_{\gamma=0}^{k} v_{i,\gamma} \mathbf{c}_{i,\gamma}.$$

2. Let $\mathbf{c} := [\mathbf{c}_0 | \mathbf{c}_{\vec{v}}]$.
3. Compute $z \leftarrow c' - \mathbf{e}^\mathsf{T}\mathbf{c} \pmod{q}$.

Output 0 if $|z| < q/4$ (when interpreted as in integer in $(-q/2, q/2]$) and 1 otherwise.

For consistency with prior work, we choose the noise in Step 3 of Enc from the rounded continuous Gaussian $\overline{\Psi}_\alpha$. It was pointed out to us by a referee that one can instead use the discrete Gaussian $\mathcal{D}_{\mathbb{Z}, \alpha q}$ and obtain a system with the same security guarantee (up to a factor of $\sqrt{2}$); this result follows from [28, Lemma 2], using the work of Peikert [37].

4.2 Correctness

We now show that for certain parameter choices, if a bit M is encrypted to the attribute vector $\vec{w}$, the secret key $\mathbf{s}_{\vec{v}}$ corresponds to a predicate vector $\vec{v}$, and $\langle \vec{v}, \vec{w} \rangle = 0 \pmod{q}$, then the LinFE.Dec algorithm recovers M.

Lemma 4.1. *Suppose the parameters q and α are such that*

$$q/\lg q = \Omega\left(\sigma \cdot \ell \cdot m^{3/2}\right) \qquad \text{and} \qquad \alpha \leq \left(\log q \cdot \sigma \cdot \ell \cdot m \cdot \omega \sqrt{\log m}\right)^{-1}.$$

Let $\mathbf{e} \leftarrow$ KeyGen(PP, MK, $\vec{v}$), CT $\leftarrow$ Enc(PP, $\vec{w}$, M), and $\tilde{M} \leftarrow$ Dec(PP, e, CT). If $\langle \vec{v}, \vec{w} \rangle = 0 \pmod{q}$, then with overwhelming probability we have $M' = M$.

Proof. During the first step of LinFE.Dec we compute $\mathbf{c}_{\vec{v}}$, which is by definition:

$$\mathbf{c}_{\vec{v}} = \sum_{i=1}^{\ell} \sum_{\gamma=0}^{k} v_{i,\gamma} \mathbf{c}_{i,\gamma}.$$

This can be expanded as

$$\mathbf{c}_{\vec{v}} = \sum_{i=1}^{\ell} \sum_{\gamma=0}^{k} v_{i,\gamma} \left[(\mathbf{A}_{i,\gamma} + 2^{\gamma} w_i \mathbf{B})^{\mathsf{T}} \mathbf{s} + \mathbf{R}_{i,\gamma}^{\mathsf{T}} \mathbf{x} \right] \tag{4.2}$$

$$= \left(\sum_{i=1}^{\ell} \sum_{\gamma=0}^{k} v_{i,\gamma} \mathbf{A}_{i,\gamma} \right)^{\mathsf{T}} \mathbf{s} + \underbrace{\left(\sum_{i=1}^{\ell} \sum_{\gamma=0}^{k} 2^{\gamma} v_{i,\gamma} w_i \right)}_{\langle \vec{v}, \vec{w} \rangle \pmod{q}} \mathbf{B}^{\mathsf{T}} \mathbf{s} + \sum_{i=1}^{\ell} \sum_{\gamma=0}^{k} v_{i,\gamma} \mathbf{R}_{i,\gamma}^{\mathsf{T}} \mathbf{x}.$$

If $\langle \vec{v}, \vec{w} \rangle = 0 \pmod{q}$, then the middle term of (4.2) disappears, leaving

$$\mathbf{c}_{\vec{v}} = \left(\sum_{i=1}^{\ell} \sum_{\gamma=0}^{k} v_{i,\gamma} \mathbf{A}_{i,\gamma} \right)^{\mathsf{T}} \mathbf{s} + \sum_{i=1}^{\ell} \sum_{\gamma=0}^{k} v_{i,\gamma} \mathbf{R}_{i,\gamma}^{\mathsf{T}} \mathbf{x} \pmod{q}.$$

In the second step of LinFE.Dec we have:

$$\mathbf{c} = [\mathbf{c}_0 | \mathbf{c}_{\vec{v}}] = \left[\mathbf{A} \middle\| \sum_{i=1}^{\ell} \sum_{\gamma=0}^{k} v_{i,\gamma} \mathbf{A}_{i,\gamma} \right]^{\mathsf{T}} \mathbf{s} + \left[\mathbf{x} \middle| \sum_{i=1}^{\ell} \sum_{\gamma=0}^{k} v_{i,\gamma} \mathbf{R}_{i,\gamma}^{\mathsf{T}} \mathbf{x} \right] \pmod{q}$$

$$= \mathbf{A}_{\vec{v}}^{\mathsf{T}} \cdot \mathbf{s} + \left[\mathbf{x} \middle| \sum_{i=1}^{\ell} \sum_{\gamma=0}^{k} v_{i,\gamma} \mathbf{R}_{i,\gamma}^{\mathsf{T}} \mathbf{x} \right] \pmod{q}$$

In the third step of LinFE.Dec we multiply $\mathbf{c}$ with the key $\mathbf{e}$. Recall that by Theorem 3.3 we have $\mathbf{A}_{\vec{v}} \cdot \mathbf{e} = \mathbf{u} \pmod{q}$. It follows that

$$\mathbf{e}^{\mathsf{T}} \mathbf{c} = \mathbf{u}^{\mathsf{T}} \mathbf{s} + \mathbf{e}^{\mathsf{T}} \left[\mathbf{x} \middle| \sum_{i=1}^{\ell} \sum_{\gamma=0}^{k} v_{i,\gamma} \mathbf{R}_{i,\gamma}^{\mathsf{T}} \mathbf{x} \right] \pmod{q}.$$

Finally, we compute:

$$z = c' - \mathbf{e}^{\mathsf{T}} \mathbf{c} \pmod{q}$$

$$= (\mathbf{u}^{\mathsf{T}} \mathbf{s} + x + M \cdot \lfloor q/2 \rfloor) - \mathbf{u}^{\mathsf{T}} \mathbf{s} - \mathbf{e}^{\mathsf{T}} \left[\mathbf{x} \middle| \sum_{i=1}^{\ell} \sum_{\gamma=0}^{k} v_{i,\gamma} \mathbf{R}_{i,\gamma}^{\mathsf{T}} \mathbf{x} \right] \pmod{q}$$

$$= M \cdot \lfloor q/2 \rfloor + \underbrace{\left(x - \mathbf{e}^{\mathsf{T}} \left[\mathbf{x} \middle| \sum_{i=1}^{\ell} \sum_{\gamma=0}^{k} v_{i,\gamma} \mathbf{R}_{i,\gamma}^{\mathsf{T}} \mathbf{x} \right] \right)}_{\text{low-norm noise}} \pmod{q}$$

To obtain $\tilde{M} = M$, it suffices to set the parameters so that with overwhelming probability,

$$\left| x - \mathbf{e}^{\mathsf{T}} \left[\mathbf{x} \middle| \sum_{i=1}^{\ell} \sum_{\gamma=0}^{k} v_{i,\gamma} \mathbf{R}_{i,\gamma}^{\mathsf{T}} \mathbf{x} \right] \right| < q/4. \tag{4.3}$$

Writing $\mathbf{e} = [\mathbf{e}_1 | \mathbf{e}_2]$ with $\mathbf{e}_i \in \mathbb{Z}^m$ allows us to rewrite this "noise" term as

$$x - \left(\mathbf{e}_1 + \sum_{i=1}^{\ell} \sum_{\gamma=0}^{k} v_{i,\gamma} \mathbf{R}_{i,\gamma} \mathbf{e}_2\right)^{\mathsf{T}} \mathbf{x}.$$

By Theorem 3.3 and Lemma 3.2, we have $\|\mathbf{e}\| < \sigma\sqrt{2m}$ with overwhelming probability. By [3, Lemma 15], we have $\|\mathbf{R}_{i,\gamma} \cdot \mathbf{e}_2\| \le 12\sqrt{2m} \cdot \|\mathbf{e}_2\|$ with overwhelming probability. Since $v_{i,\gamma} \in \{0, 1\}$ it follows that

$$\left\|\mathbf{e}_1 + \sum_{i=1}^{\ell} \sum_{\gamma=0}^{k} v_{i,\gamma} \mathbf{R}_{i,\gamma} \mathbf{e}_2\right\| < \left(1 + 12 \cdot \ell \cdot (1 + k) \cdot \sqrt{2m}\right) \cdot \sigma\sqrt{2m} = O(\ell \cdot k \cdot \sigma \cdot m).$$

It now follows from Lemma 3.9 that the error term (4.2) has absolute value at most

$$\left(q\alpha \cdot \omega(\sqrt{\log m}) + \sqrt{m}/2\right) \cdot O\left(\ell \cdot \sigma \cdot m \cdot \lg q\right). \tag{4.4}$$

(Recall that $k = \lfloor \lg q \rfloor$.) For the quantity (4.4) to have absolute value less than $q/4$, it suffices to choose q and α as in the statement of the Lemma. $\square$

4.3 Security

We use the simulation technique of Agrawal, Boneh, and Boyen [3] to reduce the security of our system to the hardness of the decision-LWE problem.

Theorem 4.2. *Suppose $m \ge 6n \log q$. If the decision-$\mathsf{LWE}_{q,\alpha}$ problem is infeasible, then the predicate encryption scheme described above is weakly attribute hiding.*

To prove the theorem we define a series of three games against an adversary $\mathcal{A}$ that plays the weak attribute hiding game (subject to the modification described in Remark 2.3). The adversary $\mathcal{A}$ outputs two attribute vectors $\vec{w}_0$ and $\vec{w}_1$ at the beginning of each game, and at some point outputs two messages M_0 and M_1. Each game comes in two variants, reflecting the choice of attribute/message pair used to create the challenge ciphertext. The first game corresponds to the real security game. In the other two games we use "alternative" setup, key generation, and encryption algorithms Sim.Setup, Sim.KeyGen, and Sim.Enc. The algorithm Sim.Setup takes as additional input an attribute vector $\vec{w}^*$, and Sim.Enc takes as additional input the master key output by Sim.Setup. Recall that during the course of the game the adversary can only request keys for predicate vectors $\vec{v}$ such that $\langle \vec{v}, \vec{w}_0 \rangle \neq 0$ and $\langle \vec{v}, \vec{w}_1 \rangle \neq 0$.

$\mathsf{Game}_{0,b}$: For $b \in \{0, 1\}$, the challenger runs the LinFE.Setup algorithm, answers the adversary's secret key queries using the LinFE.KeyGen algorithm, and generates the challenge ciphertext using the LinFE.Enc algorithm with attribute $\vec{w}_b$ and message M_b.

$\mathsf{Game}_{1,b}$: For $b \in \{0, 1\}$, the challenger runs the $\mathsf{Sim.Setup}$ algorithm with $\vec{w}^* = \vec{w}_b$ and answers the adversary's secret key queries using the $\mathsf{Sim.KeyGen}$ algorithm. The challenger generates the challenge ciphertext using the $\mathsf{Sim.Enc}$ algorithm with attribute $\vec{w}_b$ and message M_b.

$\mathsf{Game}_{2,b}$: This game is the same as $\mathsf{Game}_{1,b}$, except the challenger generates the challenge ciphertext by choosing a uniformly random element of the ciphertext space.

We now define the alternative setup, key generation, and encryption algorithms.

$\mathsf{Sim.Setup}(1^n, 1^\ell, \vec{w}^*)$: On input a security parameter n, a parameter ℓ denoting the dimension of predicate and attribute vectors, and an attribute vector $\vec{w}^* \in \mathbb{Z}_q^\ell$, do the following:

1. Choose a random matrix $\mathbf{A} \stackrel{\mathrm{R}}{\leftarrow} \mathbb{Z}_q^{n \times m}$ and a random vector $\mathbf{u} \stackrel{\mathrm{R}}{\leftarrow} \mathbb{Z}_q^n$.
2. Use $\mathsf{TrapGen}(q, n, m)$ to generate a matrix $\mathbf{B}^* \in \mathbb{Z}_q^{n \times m}$ along with a basis $\mathbf{T}_{\mathbf{B}^*}$ of $\Lambda_q^\perp(\mathbf{B}^*)$.
3. For $i = 1, \ldots, \ell$ and $\gamma = 0, \ldots, k$, pick random matrices $\mathbf{R}_{i,\gamma}^* \stackrel{\mathrm{R}}{\leftarrow} \{-1, 1\}^{m \times m}$ and set

$$\mathbf{A}_{i,\gamma} \leftarrow \mathbf{A}\,\mathbf{R}_{i,\gamma}^* - 2^\gamma w_i^*\,\mathbf{B}^*.$$

Output the public parameters and master key

$$\mathsf{PP} = \left(\mathbf{A}, \{\mathbf{A}_{i,\gamma}\}_{i \in \{1, \ldots, \ell\}, \gamma \in \{0, \ldots, k\}}, \mathbf{u}\right), \quad \mathsf{MK} = \left(\vec{w}^*, \{\mathbf{R}_{i,\gamma}^*\}_{i \in \{1, \ldots, \ell\}, \gamma \in \{0, \ldots, k\}}, \mathbf{B}^*, \mathbf{T}_{\mathbf{B}^*}\right)$$

$\mathsf{Sim.KeyGen}(\mathsf{PP}, \mathsf{MK}, \vec{v})$: On input public parameters PP, a master key MK, and a vector $\vec{v} \in \mathbb{Z}_q^\ell$, do the following:

1. If $\langle \vec{v}, \vec{w}^* \rangle = 0$, output $\perp$.
2. Define the binary decomposition of v_i as in (4.1).
3. Define the matrices

$$\mathbf{C}_{\vec{v}} := \sum_{i=1}^\ell \sum_{\gamma=0}^k v_{i,\gamma} \mathbf{A}_{i,\gamma} \in \mathbb{Z}_q^{n \times m}, \qquad \mathbf{A}_{\vec{v}} := [\mathbf{A} \,\|\, \mathbf{C}_{\vec{v}}] \in \mathbb{Z}_q^{n \times 2m}.$$

Observe that

$$\mathbf{A}_{\vec{v}} = \left[\mathbf{A} \,\|\, \mathbf{A}\left(\sum_{i=1}^\ell \sum_{\gamma=0}^k v_{i,\gamma} \mathbf{R}_{i,\gamma}^*\right) - \underbrace{\left(\sum_{i=1}^\ell \sum_{\gamma=0}^k 2^\gamma v_{i,\gamma} w_i^*\right)}_{\langle \vec{v}, \vec{w}^* \rangle \pmod q} \mathbf{B}^*\right].$$

4. Let $\mathbf{e} \leftarrow \mathsf{SampleRight}\left(\mathbf{A}, \ -\langle \vec{v}, \vec{w}^* \rangle \mathbf{B}^*, \ \sum_{i=1}^\ell \sum_{\gamma=0}^k v_{i,\gamma} \mathbf{R}_{i,\gamma}^*, \ \mathbf{T}_{\mathbf{B}^*}, \ \mathbf{u}, \ \sigma\right) \in \mathbb{Z}_q^{2m}$.

Output the secret key $\mathsf{sk}_{\vec{v}} = \mathbf{e}$.

Sim.Enc($PP, \vec{w}, M, MK$)**:** This algorithm is the same as the LinFE.Enc algorithm, except:

1. In Step 1, matrix $\mathbf{B}^* \in MK$ is used instead of a random matrix $\mathbf{B}$.
2. In Step 5a, the matrices $\mathbf{R}_{i,\gamma}^* \in MK$ for are used instead of random matrices $\mathbf{R}_{i,\gamma}$ for $i = 1, \ldots, \ell$ and $\gamma = 0, \ldots, k$.

To prove security of our system, we show that the two games in each of the pairs ($\text{Game}_{0,b}, \text{Game}_{1,b}$), ($\text{Game}_{1,b}, \text{Game}_{2,b}$) and ($\text{Game}_{2,0}, \text{Game}_{2,1}$) are either statistically or computationally indistinguishable (under the decision-LWE assumption) from the point of view of the adversary. Theorem 4.2 then follows from a simple hybrid argument; details are in the full version of this paper [2].

Lemma 4.3. *For a given* $b \in \{0,1\}$, *the view of the adversary* $\mathcal{A}$ *in* $\text{Game}_{0,b}$ *is statistically close to the view of* $\mathcal{A}$ *in* $\text{Game}_{1,b}$.

The proof of Lemma 4.3 can be found in the full version of this paper [2].

Lemma 4.4. *For a given* $b \in \{0,1\}$, *if the decision-LWE assumption holds, then the view of the adversary* $\mathcal{A}$ *in* $\text{Game}_{1,b}$ *is computationally indistinguishable from the view of* $\mathcal{A}$ *in* $\text{Game}_{2,b}$.

Proof. Suppose we are given $m + 1$ LWE challenges $(\mathbf{a}_i, y_i) \in \mathbb{Z}_q^n \times \mathbb{Z}_q$ for $j = 0, \ldots, m$, where either $y_j = \langle \mathbf{a}_j, \mathbf{s} \rangle + x_j$ for some (fixed) random secret $\mathbf{s} \overset{R}{\leftarrow} \mathbb{Z}_q^n$ and Gaussian noise $x_j \leftarrow \overline{\Psi}_\alpha$, or y_j is uniformly random in $\mathbb{Z}_q$ (and this choice is the same for each challenge). We define the following variables:

$$\mathbf{A} := \begin{pmatrix} | & & | \\ \mathbf{a}_1 & \cdots & \mathbf{a}_m \\ | & & | \end{pmatrix} \in \mathbb{Z}_q^{n \times m} \quad \mathbf{u} := \mathbf{a}_0$$
$$\mathbf{c}_0 := (y_1, \ldots, y_m) \in \mathbb{Z}_q^m \quad c' := y_0 + M_b \cdot \lfloor \tfrac{q}{2} \rceil \tag{4.5}$$

We simulate the challenger as follows:

- **Setup:** Run Sim.Setup with $\vec{w}^* = \vec{w}_b$, and let $\mathbf{A}$ and $\mathbf{u}$ be as in (4.5).
- **Private key queries:** Run the Sim.KeyGen algorithm.
- **Challenge ciphertext:** For $i = 1, \ldots, \ell$ and $\gamma = 0, \ldots, k$, let $\mathbf{c}_{i,\gamma} = \mathbf{R}_{i,\gamma}^{*\mathsf{T}} \mathbf{c}_0$ (using $\mathbf{R}_{i,\gamma}^* \in MK$). Output $(\mathbf{c}_0, \{\mathbf{c}_{i,\gamma}\}_{i \in \{1,\ldots,\ell\}, \gamma \in \{0,\ldots,k\}}, c')$.

Now observe that for $i = 1, \ldots, \ell$ and $\gamma = 0, \ldots, k$, the Sim.Enc algorithm sets

$$\mathbf{c}_{i,\gamma} = \left(\mathbf{A}\mathbf{R}_{i,\gamma} - 2^\gamma w_i^* \mathbf{B}^* + 2^\gamma w_i^* \mathbf{B}^* \right)^\mathsf{T} \mathbf{s} + \mathbf{R}_{i,\gamma}^{*\mathsf{T}} \mathbf{x} = \mathbf{R}_{i,\gamma}^{*\mathsf{T}} (\mathbf{A}^\mathsf{T} \mathbf{s} + \mathbf{x}).$$

It follows that if $y_j = \langle \mathbf{a}_j, \mathbf{s} \rangle + x_j$, then $\mathbf{c}_{i,\gamma} = \mathbf{R}_{i,\gamma}^{*\mathsf{T}} \mathbf{c}_0$ and the simulator described above is identical to the challenger in $\text{Game}_{1,b}$.

On the other hand, if y_j is random in $\mathbb{Z}_q$, then the simulated ciphertext is $(\mathbf{c}_0, \overline{\mathbf{R}^*}^\mathsf{T} \mathbf{c}_0, c')$, where $\overline{\mathbf{R}^*}$ is the concatenation of the matrices $\mathbf{R}_{i,\gamma}^*$. By the standard leftover hash lemma (e.g. [43, Theorem 8.37]), the quantities $\mathbf{A}\overline{\mathbf{R}^*}$ and $\overline{\mathbf{R}^*}^\mathsf{T} \mathbf{c}_0$ are independent uniformly random samples. Thus in this case the ciphertext is uniformly random and the simulator described above is identical the challenger in $\text{Game}_{2,b}$.

We conclude that any efficient adversary that can distinguish $\text{Game}_{1,b}$ from $\text{Game}_{2,b}$ can solve the decision-LWE problem.   □

Lemma 4.5. *The view of the adversary* $\mathcal{A}$ *in* $\text{Game}_{2,0}$ *is statistically indistinguishable from the view of* $\mathcal{A}$ *in* $\text{Game}_{2,1}$.

Proof. Note that the only place where $\vec{w}^*$ appears in $\text{Game}_{2,b}$ is in the public parameter $\mathbf{A}_{i,\gamma} := \mathbf{A}\mathbf{R}_{i,\gamma}^* - 2^\gamma w_i^* \mathbf{B}^*$. Let $\overline{\mathbf{A}} \in \mathbb{Z}_q^{n \times m\ell(k+1)}$ and $\overline{\mathbf{R}^*} \in \mathbb{Z}_q^{m \times m\ell(k+1)}$ be the concatenations of the $\mathbf{A}_{i,\gamma}$ and the $\mathbf{R}_{i,\gamma}^*$, respectively. Then we have $\overline{\mathbf{A}} = \mathbf{A}\overline{\mathbf{R}^*}$. By [3, Lemma 13] the pair $(\mathbf{A}, \mathbf{A}\overline{\mathbf{R}}^*)$ is statistically indistinguishable from $(\mathbf{A}, \mathbf{C})$ where $\mathbf{C}$ is uniformly random. Since for any fixed value of $\mathbf{X}$ and uniformly random $\mathbf{C}$, the variable $\mathbf{C} - \mathbf{X}$ is also uniformly random, it follows that the distributions of $\mathbf{A}_{i,\gamma}$ in the two games are statistically indistinguishable.

4.4   Parameter Selection

We can extract from the above description the parameters required for correctness and security of the system. For correctness of decryption, by Lemma 4.1 we require

$$q/\lg q = \Omega\left(\sigma \cdot \ell \cdot m^{3/2}\right) \quad \text{and} \quad \alpha \le \left(\log q \cdot \sigma \cdot \ell \cdot m \cdot \omega\sqrt{\log m}\right)^{-1}. \quad (4.6)$$

In our security theorem (Theorem 4.2), we require $m > 6n \lg q$ in order for the output of TrapGen to be statistically random. The additional constraints imposed by our security reduction are the following:

- From the description of LinFE.Setup and LinFE.KeyGen, we have $\|\widetilde{\mathbf{T}_\mathbf{A}}\| = O(\sqrt{n \log q})$ (by Theorem 3.1) and $\mathbf{e} \leftarrow \mathcal{D}_{\Lambda_q^\mathbf{u}(\mathbf{A}_{\bar{v}}),\sigma}$ (by Theorem 3.3), subject to the requirement that

$$\sigma \ge \|\widetilde{\mathbf{T}_\mathbf{A}}\| \cdot \omega(\sqrt{\log m}) = O(\sqrt{n \log q}) \cdot \omega(\sqrt{\log m}).$$

- From the description of Sim.Setup and Sim.KeyGen, we have $\|\widetilde{\mathbf{T}_{\mathbf{B}^*}}\| = O(\sqrt{n \log q})$ (by Theorem 3.1), and $\mathbf{e} \leftarrow \mathcal{D}_{\Lambda_q^\mathbf{u}(\mathbf{A}_{\bar{v}}),\sigma}$ (by Theorem 3.4), subject to the requirement that

$$\sigma \ge \|\widetilde{\mathbf{T}_{\mathbf{B}^*}}\| \cdot s_R \cdot \omega(\sqrt{\log m}) \quad (4.7)$$

Since $\mathbf{R}$ is a sum of $\ell \cdot (\lg q + 1)$ random matrices with $\{1, -1\}$ entries, it follows from [3, Lemma 15] that $s_R = \sup_{\{\mathbf{x}:\|\mathbf{x}\|=1\}} \|\mathbf{R}\mathbf{x}\| = O(\ell \cdot (\lg q + 1) \cdot \sqrt{m})$ with overwhelming probability. Plugging this value into (4.7), we see that it suffices to choose

$$\sigma \ge O(\sqrt{n \log q}) \cdot O(\ell \cdot (\lg q + 1) \cdot \sqrt{m}) \cdot \omega(\sqrt{\log m}).$$

Thus to satisfy the more stringent of the above two conditions (i.e., the latter), we set

$$\sigma = \omega(m \cdot \ell \cdot \log q \cdot \sqrt{\log m}), \tag{4.8}$$

using the fact (noted above) that $m \geq 6n \log q$.

In order to reduce decision-LWE to approximating worst-case lattice problems to within $\mathrm{poly}(n)$ factors we have two options: for polynomial-size q we can use Regev's quantum reduction (Theorem 3.7) with $q\alpha > 2\sqrt{n}$ and $\alpha \geq 1/\mathrm{poly}(n)$, while for exponential-size q we can use Peikert's classical reduction (Theorem 3.8) with each prime factor q_i of q satisfying $\omega(\sqrt{\log n})/\alpha < q_i < \mathrm{poly}(n)$. (Note that a large value of q may be required for certain applications; see the full version of this paper [2, §5] for details.)

The following selection of parameters satisfies all of these constraints. For a given ℓ, pick a small constant $\delta > 0$, and set

$$
\begin{aligned}
m &= \lceil n^{1+\delta} \rceil, & &\text{to satisfy } m > 6n \lg q \\
\sigma &= \lceil n^{2+2\delta} \cdot \ell \rceil, & &\text{to satisfy (4.8)} \\
q_i &= \text{the } i\text{th prime larger than } (\ell \log \ell)^2 \cdot n^{7/2+5\delta} \\
\alpha &= \Omega\left((\ell \log \ell)^2 \cdot n^{3+5\delta}\right)^{-1} & &\text{to satisfy (4.6)}
\end{aligned}
$$

Observe that the above setting of parameters satisfies the conditions for applying Theorems 3.7 and 3.8. To obtain polynomial size q we use $q = q_1$, while to obtain exponential size q we use $q = \prod_{i=1}^{\tau} q_i$, where τ is chosen so that $q > 2^{n/2}$. In either case we can choose δ large enough so that $n^{1+\delta} > 6n \lg q$. In the former case, the security of the scheme can be based on the hardness of approximating SIVP and GapSVP to within a factor of $\tilde{O}(n/\alpha) = \tilde{O}((\ell \log \ell)^2 \cdot n^{4+5\delta})$ in the worst case (by quantum algorithms). In the latter case, security is based on the hardness of approximating GapSVP to within a factor of $\tilde{O}(n/\alpha) = \tilde{O}((\ell \log \ell)^2 \cdot n^{4+5\delta})$ in the worst case (by classical algorithms).

Note that since $m > n \lg q$ and $q_i > n$, the matrices $\mathbf{A}$ and $\mathbf{B}$ have full rank modulo each prime divisor of q with overwhelming probability, as required for successful execution of the SampleLeft and SampleRight algorithms.

Finally, we note that these parameter choices are not necessarily optimal, and one might be able to set the parameters to have somewhat smaller values while maintaining correctness and security. In particular, one might be able to reduce the ciphertext size by using the r-ary expansion of the vector $\vec{v}$ for some $r > 2$ instead of the binary expansion as described above.

5 Conclusion and Open Questions

We have presented a lattice-based predicate encryption scheme for inner product predicates whose security follows from the difficulty of the   learning with errors problem. Our construction can instantiate applications such as range and subset queries, polynomial evaluation, and CNF/DNF formulas on encrypted data. (A more detailed discussion of these applications appears in the full version of this paper [2].) Our construction is the first functional encryption scheme based on lattice techniques that goes beyond basic identity-based encryption.

Many open questions still remain in this field. One direction of research is to improve the security of our construction. Our scheme is weakly attribute hiding in the selective security model, but for stronger security guarantees we would like to construct a scheme that is fully secure and/or fully attribute hiding. Achieving either task will require new simulation techniques; a natural question is whether the "dual-system" approach introduced by Waters [45] and used to prove full security of attribute-based encryption and predicate encryption constructions using bilinear groups [30, 9, 34] can be adapted to lattice-based constructions.

Another direction of research is to improve the efficiency of our scheme. If $q = 2^{O(n)}$ is exponential size, as is needed for several of our applications, then setting the parameters as recommended in Section 4.4 gives public parameters of size $\Theta(\ell n m \lg^2(q)) = \Omega(\ell n^5)$ and ciphertexts of size $\Theta(\ell m \lg^2(q)) = \Omega(\ell n^4)$, which may be too large for practical purposes. A construction that achieved the same functionality with polynomial-size q would be a significant step forward. The *ring*-LWE problem introduced by Lyubashevsky, Peikert, and Regev [31] seems to be a natural candidate for such a construction.

Finally, it is a open question to construct predicate encryption schemes (via any technique) that support a greater range of functionality than inner product predicates. Ideally we would like a system that could support any polynomial-size predicate on encrypted data. Now that predicate encryption has moved into the world of lattices, perhaps techniques used to construct fully homomorphic encryption from lattices [23, 19, 18] could be used to help us move towards this goal.

Acknowledgments. The authors thank Dan Boneh, Brent Waters, Hoeteck Wee, and the anonymous referees for helpful discussions and comments.

References

1. Abdalla, M., Bellare, M., Catalano, D., Kiltz, E., Kohno, T., Lange, T., Malone-Lee, J., Neven, G., Paillier, P., Shi, H.: Searchable encryption revisited: Consistency properties, relation to anonymous IBE, and extensions. J. Cryptology 21(3), 350–391 (2008)
2. Agrawal, S., Freeman, D.M., Vaikuntanathan, V.: Functional encryption for inner product predicates from learning with errors. Cryptology ePrint Report 2011/410 (2011) (full version of this paper), http://eprint.iacr.org/2011/410
3. Agrawal, S., Boneh, D., Boyen, X.: Efficient Lattice (H)IBE in the Standard Model. In: Gilbert, H. (ed.) EUROCRYPT 2010. LNCS, vol. 6110, pp. 553–572. Springer, Heidelberg (2010), full version at,
   http://crypto.stanford.edu/~dabo/pubs/papers/latticebb.pdf
4. Agrawal, S., Boneh, D., Boyen, X.: Lattice Basis Delegation in Fixed Dimension and Shorter-Ciphertext Hierarchical IBE. In: Rabin, T. (ed.) CRYPTO 2010. LNCS, vol. 6223, pp. 98–115. Springer, Heidelberg (2010)
5. Agrawal, S., Boyen, X.: Identity-based encryption from lattices in the standard model (July 2009) (manuscript), http://www.cs.stanford.edu/~xb/ab09/
6. Ajtai, M.: Generating Hard Instances of the Short Basis Problem. In: Wiedermann, J., Van Emde Boas, P., Nielsen, M. (eds.) ICALP 1999. LNCS, vol. 1644, pp. 1–9. Springer, Heidelberg (1999)
7. Alwen, J., Peikert, C.: Generating shorter bases for hard random lattices. In: STACS, pp. 75–86 (2009), full version available at
   http://www.cc.gatech.edu/~cpeikert/pubs/shorter.pdf

8. Attrapadung, N., Imai, H.: Conjunctive Broadcast and Attribute-Based Encryption. In: Shacham, H., Waters, B. (eds.) Pairing 2009. LNCS, vol. 5671, pp. 248–265. Springer, Heidelberg (2009)
9. Attrapadung, N., Libert, B.: Functional Encryption for Inner Product: Achieving Constant-Size Ciphertexts with Adaptive Security or Support for Negation. In: Nguyen, P.Q., Pointcheval, D. (eds.) PKC 2010. LNCS, vol. 6056, pp. 384–402. Springer, Heidelberg (2010)
10. Bethencourt, J., Sahai, A., Waters, B.: Ciphertext-policy attribute-based encryption. In: IEEE Symposium on Security and Privacy, pp. 321–334 (2007)
11. Blundo, C., Iovino, V., Persiano, G.: Predicate Encryption with Partial Public Keys. In: Heng, S.H., Wright, R.N., Goi, B.M. (eds.) CANS 2010. LNCS, vol. 6467, pp. 298–313. Springer, Heidelberg (2010)
12. Boneh, D., Boyen, X.: Efficient Selective-ID Secure Identity-Based Encryption Without Random Oracles. In: Cachin, C., Camenisch, J. (eds.) EUROCRYPT 2004. LNCS, vol. 3027, pp. 223–238. Springer, Heidelberg (2004)
13. Boneh, D., Boyen, X.: Secure Identity Based Encryption Without Random Oracles. In: Franklin, M. (ed.) CRYPTO 2004. LNCS, vol. 3152, pp. 443–459. Springer, Heidelberg (2004)
14. Boneh, D., Franklin, M.: Identity-based encryption from the Weil pairing. SIAM J. Comput. 32(3), 586–615 (2003); extended abstract in CRYPTO 2001
15. Boneh, D., Di Crescenzo, G., Ostrovsky, R., Persiano, G.: Public Key Encryption with Keyword Search. In: Cachin, C., Camenisch, J. (eds.) EUROCRYPT 2004. LNCS, vol. 3027, pp. 506–522. Springer, Heidelberg (2004)
16. Boneh, D., Sahai, A., Waters, B.: Functional Encryption: Definitions and Challenges. In: Ishai, Y. (ed.) TCC 2011. LNCS, vol. 6597, pp. 253–273. Springer, Heidelberg (2011)
17. Boneh, D., Waters, B.: Conjunctive, Subset, and Range Queries on Encrypted Data. In: Vadhan, S.P. (ed.) TCC 2007. LNCS, vol. 4392, pp. 535–554. Springer, Heidelberg (2007)
18. Brakerski, Z., Vaikuntanathan, V.: Efficient fully homomorphic encryption from (standard) LWE. To appear in FOCS 2011 (2011), preprint available at http://eprint.iacr.org/2011/344
19. Brakerski, Z., Vaikuntanathan, V.: Fully Homomorphic Encryption from Ring-LWE and Security for Key Dependent Messages. In: Rogaway, P. (ed.) CRYPTO 2011. LNCS, vol. 6841, pp. 505–524. Springer, Heidelberg (2011)
20. Cash, D., Hofheinz, D., Kiltz, E., Peikert, C.: Bonsai Trees, or How to Delegate a Lattice Basis. In: Gilbert, H. (ed.) EUROCRYPT 2010. LNCS, vol. 6110, pp. 523–552. Springer, Heidelberg (2010)
21. Chase, M.: Multi-Authority Attribute Based Encryption. In: Vadhan, S.P. (ed.) TCC 2007. LNCS, vol. 4392, pp. 515–534. Springer, Heidelberg (2007)
22. Gentry, C.: Practical Identity-Based Encryption Without Random Oracles. In: Vaudenay, S. (ed.) EUROCRYPT 2006. LNCS, vol. 4004, pp. 445–464. Springer, Heidelberg (2006)
23. Gentry, C.: A fully homomorphic encryption scheme. Ph.D. thesis, Stanford University (2009), http://crypto.stanford.edu/craig
24. Gentry, C., Peikert, C., Vaikuntanathan, V.: Trapdoors for hard lattices and new cryptographic constructions. In: 40th ACM Symposium on Theory of Computing — STOC 2008, pp. 197–206. ACM (2008)
25. Gentry, C., Silverberg, A.: Hierarchical ID-Based Cryptography. In: Zheng, Y. (ed.) ASIACRYPT 2002. LNCS, vol. 2501, pp. 548–566. Springer, Heidelberg (2002)
26. Goyal, V., Jain, A., Pandey, O., Sahai, A.: Bounded Ciphertext Policy Attribute Based Encryption. In: Aceto, L., Damgård, I., Goldberg, L.A., Halldórsson, M.M., Ingólfsdóttir, A., Walukiewicz, I. (eds.) ICALP 2008, Part II. LNCS, vol. 5126, pp. 579–591. Springer, Heidelberg (2008)
27. Goyal, V., Pandey, O., Sahai, A., Waters, B.: Attribute-based encryption for fine-grained access control of encrypted data. In: ACM Conference on Computer and Communications Security, pp. 89–98 (2006)

28. Gordon, S.D., Katz, J., Vaikuntanathan, V.: A Group Signature Scheme from Lattice Assumptions. In: Abe, M. (ed.) ASIACRYPT 2010. LNCS, vol. 6477, pp. 395–412. Springer, Heidelberg (2010)

29. Katz, J., Sahai, A., Waters, B.: Predicate Encryption Supporting Disjunctions, Polynomial Equations, and Inner Products. In: Smart, N. (ed.) EUROCRYPT 2008. LNCS, vol. 4965, pp. 146–162. Springer, Heidelberg (2008), http://eprint.iacr.org/2007/404

30. Lewko, A., Okamoto, T., Sahai, A., Takashima, K., Waters, B.: Fully Secure Functional Encryption: Attribute-Based Encryption and (Hierarchical) Inner Product Encryption. In: Gilbert, H. (ed.) EUROCRYPT 2010. LNCS, vol. 6110, pp. 62–91. Springer, Heidelberg (2010), full version available at http://eprint.iacr.org/2010/110

31. Lyubashevsky, V., Peikert, C., Regev, O.: On Ideal Lattices and Learning with Errors Over Rings. In: Gilbert, H. (ed.) EUROCRYPT 2010. LNCS, vol. 6110, pp. 1–23. Springer, Heidelberg (2010)

32. Micciancio, D., Regev, O.: Worst-case to average-case reductions based on Gaussian measures. In: 45th IEEE Symposium on Foundations of Computer Science — FOCS 2004, pp. 372–381 (2004)

33. Okamoto, T., Takashima, K.: Hierarchical Predicate Encryption for Inner-Products. In: Matsui, M. (ed.) ASIACRYPT 2009. LNCS, vol. 5912, pp. 214–231. Springer, Heidelberg (2009)

34. Okamoto, T., Takashima, K.: Fully Secure Functional Encryption with General Relations from the Decisional Linear Assumption. In: Rabin, T. (ed.) CRYPTO 2010. LNCS, vol. 6223, pp. 191–208. Springer, Heidelberg (2010)

35. Ostrovsky, R., Sahai, A., Waters, B.: Attribute-based encryption with non-monotonic access structures. In: ACM Conference on Computer and Communications Security, pp. 195–203 (2007)

36. Peikert, C.: Public-key cryptosystems from the worst-case shortest vector problem. In: 41st Annual ACM Symposium on Theory of Computing — STOC 2009, pp. 333–342 (2009)

37. Peikert, C.: An Efficient and Parallel Gaussian Sampler for Lattices. In: Rabin, T. (ed.) CRYPTO 2010. LNCS, vol. 6223, pp. 80–97. Springer, Heidelberg (2010)

38. Regev, O.: On lattices, learning with errors, random linear codes, and cryptography. In: 37th Annual ACM Symposium on Theory of Computing — STOC 2005, pp. 84–93 (2005)

39. Sahai, A., Waters, B.: Fuzzy Identity-Based Encryption. In: Cramer, R. (ed.) EUROCRYPT 2005. LNCS, vol. 3494, pp. 457–473. Springer, Heidelberg (2005)

40. Shen, E., Shi, E., Waters, B.: Predicate Privacy in Encryption Systems. In: Reingold, O. (ed.) TCC 2009. LNCS, vol. 5444, pp. 457–473. Springer, Heidelberg (2009)

41. Shi, E., Bethencourt, J., Chan, H.T.H., Song, D.X., Perrig, A.: Multi-dimensional range query over encrypted data. In: IEEE Symposium on Security and Privacy, pp. 350–364 (2007)

42. Shi, E., Waters, B.: Delegating Capabilities in Predicate Encryption Systems. In: Aceto, L., Damgård, I., Goldberg, L.A., Halldórsson, M.M., Ingólfsdóttir, A., Walukiewicz, I. (eds.) ICALP 2008, Part II. LNCS, vol. 5126, pp. 560–578. Springer, Heidelberg (2008)

43. Shoup, V.: A Computational Introduction to Number Theory and Algebra, 2nd edn. Cambridge University Press (2008)

44. Waters, B.: Efficient Identity-Based Encryption Without Random Oracles. In: Cramer, R. (ed.) EUROCRYPT 2005. LNCS, vol. 3494, pp. 114–127. Springer, Heidelberg (2005)

45. Waters, B.: Dual System Encryption: Realizing Fully Secure IBE and HIBE Under Simple Assumptions. In: Halevi, S. (ed.) CRYPTO 2009. LNCS, vol. 5677, pp. 619–636. Springer, Heidelberg (2009)

46. Waters, B.: Ciphertext-Policy Attribute-Based Encryption: An Expressive, Efficient, and Provably Secure Realization. In: Catalano, D., Fazio, N., Gennaro, R., Nicolosi, A. (eds.) PKC 2011. LNCS, vol. 6571, pp. 53–70. Springer, Heidelberg (2011)

Random Oracles in a Quantum World

Dan Boneh[1], Özgür Dagdelen[2], Marc Fischlin[2],
Anja Lehmann[3], Christian Schaffner[4], and Mark Zhandry[1]

[1] Stanford University, USA
[2] CASED & Darmstadt University of Technology, Germany
[3] IBM Research Zurich, Switzerland
[4] University of Amsterdam and CWI, The Netherlands

Abstract. The interest in post-quantum cryptography — classical systems that remain secure in the presence of a quantum adversary — has generated elegant proposals for new cryptosystems. Some of these systems are set in the random oracle model and are proven secure relative to adversaries that have classical access to the random oracle. We argue that to prove post-quantum security one needs to prove security in the *quantum-accessible* random oracle model where the adversary can query the random oracle with quantum state.

We begin by separating the classical and quantum-accessible random oracle models by presenting a scheme that is secure when the adversary is given classical access to the random oracle, but is insecure when the adversary can make quantum oracle queries. We then set out to develop generic conditions under which a *classical* random oracle proof implies security in the *quantum-accessible* random oracle model. We introduce the concept of a *history-free reduction* which is a category of classical random oracle reductions that basically determine oracle answers independently of the history of previous queries, and we prove that such reductions imply security in the quantum model. We then show that certain post-quantum proposals, including ones based on lattices, can be proven secure using history-free reductions and are therefore post-quantum secure. We conclude with a rich set of open problems in this area.

Keywords: Quantum, Random Oracle, Signatures, Encryption.

1 Introduction

The threat to existing public-key systems posed by quantum computation [Sho97] has generated considerable interest in *post-quantum* cryptosystems, namely systems that remain secure in the presence of a quantum adversary. A promising direction is lattice-based cryptography, where the underlying problems are related to finding short vectors in high dimensional lattices. These problems have so far remained immune to quantum attacks and some evidence suggests that they may be hard for quantum computers [Reg02].

As it is often the case, the most efficient constructions in lattice-based cryptography are set in the random oracle (RO) model [BR93]. For example, Gentry,

D.H. Lee and X. Wang (Eds.): ASIACRYPT 2011, LNCS 7073, pp. 41–69, 2011.

Peikert, and Vaikuntanathan [GPV08] give elegant random oracle model constructions for existentially unforgeable signatures and for identity-based encryption. Gordon, Katz, and Vaikuntanathan [GKV10] construct a random oracle model group signature scheme. Boneh and Freeman [BF11] give a random oracle homomorphic signature scheme and Cayrel et al. [CLRS10] give a lattice-based signature scheme using the Fiat-Shamir random oracle heuristic. Some of these lattice constructions can now be realized without random oracles, but at a significant cost in performance [CHKP10,ABB10a,Boy10].

Modeling Random Oracles for Quantum Attackers. While quantum resistance is good motivation for lattice-based constructions, most random oracle systems to date are only proven secure relative to an adversary with *classical* access to the random oracle. In this model the adversary is given oracle access to a random hash function $O : \{0,1\}^* \to \{0,1\}^*$ and it can only "learn" a value $O(x)$ by querying the oracle O at the classical state x. However, to obtain a concrete system, the random oracle is eventually replaced by a concrete hash function thereby enabling a quantum attacker to evaluate this hash function on *quantum states*. To capture this issue in the model, we allow the adversary to evaluate the random oracle "in superposition", that is, the adversary can submit quantum states $|\varphi\rangle = \sum \alpha_x |x\rangle$ to the oracle O and receives back the evaluated state $\sum \alpha_x |O(x)\rangle$ (appropriately encoded to make the transformation unitary). We call this the *quantum(-accessible) random oracle model*. It complies with similar efforts from learning theory [BJ99,SG04] and computational complexity [BBBV97] where oracles are quantum-accessible, and from lower bounds for quantum collision finders [AS04]. Still, since we are only interested in classical cryptosystems, *honest* parties and the scheme's algorithms can access O only via classical bit strings.

Proving security in the quantum-accessible RO model is considerably harder than in the classical model. As a simple example, consider the case of digital signatures. A standard proof strategy in the classical settings is to choose randomly one of the adversary's RO queries and embed in the response a given instance of a challenge problem. One then hopes that the adversary uses this response in his signature forgery. If the adversary makes q random oracle queries, then this happens with probability $1/q$ and since q is polynomial this success probability is sufficiently high for the proof of security in the classical setting. Unfortunately, this strategy fails completely in the quantum-accessible random oracle model since *every* random oracle query potentially evaluates the random oracle at exponentially many points. Therefore, embedding the challenge in one response will be of no use to the reduction algorithm. This simple example shows that proving security in the classical RO model does not necessarily prove post-quantum security.

More abstractly, the following common classical proof techniques are not known to carry over to the quantum settings offhand:

- Adaptive Programmability: The classical random oracle model allows a simulator to program the answers of the random oracle for an adversary, often adaptively. Since the quantum adversary can query the random oracle with a state in superposition, the adversary may get some information about all exponentially many values right at the beginning, thereby making it difficult to program the oracle adaptively.

- Extractability/Preimage Awareness: Another application of the random oracle model for classical adversaries is that the simulator learns the pre-images the adversary is interested in. This is, for example, crucial to simulate decryption queries in the security proof for OAEP [FOPS01]. For quantum-accessible oracles the actual query may be hidden in a superposition of exponentially many states, and it is unclear how to extract the right query.
- Efficient Simulation: In the classical world, we can simulate an exponential-size random oracle efficiently via lazy sampling: simply pick random but consistent answers "on the fly". With quantum-accessible random oracles the adversary can evaluate the random oracle on all inputs simultaneously, making it harder to apply the on-demand strategy for classical oracles.
- Rewinding/Partial Consistency: Certain random oracle proofs [PS00] require rewinding the adversary, replaying some hash values but changing at least a single value. Beyond the usual problems of rewinding quantum adversaries, we again encounter the fact that we may not be able to change hash values unnoticed. We note that some form of rewinding is possible for quantum zero-knowledge [Wat09].

We do not claim that these problems are insurmountable. In fact, we show how to resolve the issue of efficient simulation by using (quantum-accessible) pseudo-random functions. These are pseudorandom functions where the quantum distinguisher can submit quantum states to the pseudorandom or random oracle. By this technique, we can efficiently simulate the quantum-accessible random oracle through the (efficient) pseudorandom function. While pseudorandom functions where the distinguisher may use quantum power but only gets classical access to the function can be derived from quantum-immune pseudorandom generators [GGM86], it is an open problem if the stronger quantum-accessible pseudorandom functions exist.

Note, too, that we do not seek to solve the problems related to the random oracle model which appear already in the classical settings [CGH98]. Instead we show that for post-quantum security one should allow for quantum access to the random oracle in order to capture attacks that are available when the hash function is eventually instantiated.

1.1   Our Contributions

Separation. We begin with a separation between the classical and quantum-accessible RO models by presenting a two-party protocol which is:

- secure in the classical random oracle model,
- secure against quantum attackers with classical access to the random oracle model, but insecure under *any* implementation of the hash function, and
- insecure in the quantum-accessible random oracle model.

The protocol itself assumes that (asymptotically) quantum computers are faster than classical (parallel) machines and uses the quadratic gap due to Grover's algorithms and its application to collision search [BHT98] to separate secure from insecure executions.

Constructions. Next, we set out to give general conditions under which a *classical* RO proof implies security for a *quantum* RO. Our goal is to provide generic tools by which authors can simply state that their classical proof has the "right" structure and therefore their proof implies quantum security. We give two flavors of results:

- For signatures, we define a proof structure we call a *history-free reduction* which roughly says that the reduction answers oracle queries independently of the history of queries. We prove that any classical proof that happens to be a history-free reduction implies quantum existential unforgeability for the signature scheme. We then show that the GPV random oracle signature scheme [GPV08] has a history-free reduction and is therefore secure in the quantum settings.

  Next, we consider signature schemes built from claw-free permutations. The first is the Full Domain Hash (FDH) signature system of Bellare and Rogaway [BR93], for which we show that the classical proof technique due to Coron [Cor00] is history-free. We also prove the quantum security of a variant of FDH due to Katz and Wang [KW03] which has a tight security reduction. Lastly, we note that, as observed in [GPV08], claw-free permutations give rise to preimage sampleable trapdoor functions, which gives another FDH-like signature scheme with a tight security reduction. In all three cases the reductions in the quantum-accessible random oracle model achieve essentially the same tightness as their classical analogs.

  Interestingly, we do not know of a history-free reduction for the generic Full Domain Hash of Bellare and Rogaway [BR93]. One reason is that proofs for generic FDH must somehow program the random oracle, as shown in [FLR+10]. We leave the quantum security of generic FDH as an interesting open problem. It is worth noting that at this time the quantum security of FDH is somewhat theoretical since we have no candidate quantum-secure trapdoor permutation to instantiate the FDH scheme, though this may change once a candidate is proposed.

- For encryption we prove the quantum CPA security of an encryption scheme due to Bellare and Rogaway [BR93] and the quantum CCA security of a hybrid encryption variant of [BR93].

Many open problems remain in this space. For signatures, it is still open to prove the quantum security of signatures that result from applying the Fiat-Shamir heuristic to a Σ identification protocol, for example, as suggested in [CLRS10]. Similarly, proving security of generic FDH is still open. For CCA-secure encryption, it is unknown if generic CPA to CCA transformations, such as [FO99], are secure in the quantum settings. Similarly, it is not known if lattice-based identity-based encryption systems secure in the classical RO model (e.g. as in [GPV08,ABB10b]) are also secure in the quantum random oracle model.

Related Work. The quantum random oracle model has been used in a few previous constructions. Aaronson [Aar09] uses quantum random oracles to construct unclonable public-key quantum money. Brassard and Salvail [BS08] give a modified version of Merkle's Puzzles, and show that any quantum attacker must query the random (permutation) oracle asymptotically more times than honest parties. Recently, a modified version was proposed that restores some level of security even in the presence of a quantum adversary [BHK+11]. Quantum random oracles have also been used to prove impossibility results for quantum computation. For example, Bennett et al. [BBBV97] show that relative to a random oracle, a quantum computer cannot solve all of NP.

Some progress toward identifying sufficient conditions under which classical protocols are also quantum immune has been made by Unruh [Unr10] and Hallgren et al. [HSS11]. These results show that, if a cryptographic protocol can be shown to be (computationally [HSS11] resp. statistically [Unr10]) secure in Canetti's universal composition (UC) framework [Can01] against classical adversaries, then the protocol is also resistant against (computationally bounded resp. unbounded) quantum adversaries. This, however, means that the underlying protocol must already provide strong security guarantees in the first place, namely, universal composition security, which is typically more than the aforementioned schemes in the literature satisfy. This also applies to similar results by Hallgren et al. [HSS11] for so-called simulation-based security notions for the starting protocol. Furthermore, all these results do not seem to be applicable immediately to the random oracle model where the quantum adversary now has *quantum* access to the random function (but where the ideal functionality for the random oracle in the UC framework would have only been defined for classical access according to the classical protocol specification), and where the question of instantiation is an integral step which needs to be considered.

2   Preliminaries

A non-negative function $\epsilon = \epsilon(n)$ is negligible if, for all polynomials $p(n)$ we have that $\epsilon(n) < p(n)^{-1}$ for all sufficiently large n. The variational distance between two distributions D_1 and D_2 over Ω is given by

$$|D_1 - D_2| = \sum_{x \in \Omega} |\Pr[x|D_1] - \Pr[x|D_2]|.$$

If the distance between two output distributions is ϵ, the difference in probability of the output satisfying a certain property is at most ϵ.

A classical randomized algorithm A can be thought of in two ways. In the first, A is given an input x, A makes some coin tosses during its computation, and ultimately outputs some value y. We denote this action by $A(x)$ where $A(x)$ is a random variable. Alternatively, we can give A both its input x and randomness r

in which case we denote this action as $A(x; r)$. For a classical algorithm, $A(x; r)$ is deterministic. An algorithm A runs is probabilistic polynomial-time (PPT) if it runs in polynomial time in the security parameter (which we often omit from the input for sake of simplicity).

2.1 Quantum Computation

We briefly give some background on quantum computation and refer to [NC00] for a more complete discussion. A quantum system A is associated to a (finite-dimensional) complex Hilbert space $\mathcal{H}_A$ with an inner product $\langle \cdot | \cdot \rangle$. The state of the system is described by a vector $|\varphi\rangle \in \mathcal{H}_A$ such that the Euclidean norm $\| |\varphi\rangle \| = \sqrt{\langle \varphi | \varphi \rangle}$ is 1. Given quantum systems A and B over spaces $\mathcal{H}_A$ and $\mathcal{H}_B$, respectively, we define the joint or composite quantum system through the tensor product $\mathcal{H}_A \otimes \mathcal{H}_B$. The product state of $|\varphi_A\rangle \in \mathcal{H}_A$ and $|\varphi_B\rangle \in \mathcal{H}_B$ is denoted by $|\varphi_A\rangle \otimes |\varphi_B\rangle$ or simply $|\varphi_A\rangle |\varphi_B\rangle$. An n-qubit system lives in the joint quantum system of n two-dimensional Hilbert spaces. The standard orthonormal computational basis $|x\rangle$ for such a system is given by $|x_1\rangle \otimes \cdots \otimes |x_n\rangle$ for $x = x_1 \ldots x_n$. Any (classical) bit string x is encoded into a quantum state as $|x\rangle$. An arbitrary pure n-qubit state $|\varphi\rangle$ can be expressed in the computational basis as $|\varphi\rangle = \sum_{x \in \{0,1\}^n} \alpha_x |x\rangle$ where α_x are complex amplitudes obeying $\sum_{x \in \{0,1\}^n} |\alpha_x|^2 = 1$.

Transformations. Evolutions of quantum systems are described by unitary transformations with $\mathbb{I}_A$ being the identity transformation on register A. Given a joint quantum system over $\mathcal{H}_A \otimes \mathcal{H}_B$ and a transformation U_A acting only on $\mathcal{H}_A$, it is understood that $U_A |\varphi_A\rangle |\varphi_B\rangle$ refers to $(U_A \otimes \mathbb{I}_B) |\varphi_A\rangle |\varphi_B\rangle$.

Information can be extracted from a quantum state $|\varphi\rangle$ by performing a positive-operator valued measurement (POVM) $M = \{M_i\}$ with positive semidefinite measurement operators M_i that sum to the identity $\sum_i M_i = \mathbb{I}$. Outcome i is obtained with probability $p_i = \langle \varphi | M_i | \varphi \rangle$. A special case are projective measurements such as the measurement in the computational basis of the state $|\varphi\rangle = \sum_x \alpha_x |x\rangle$ which yields outcome x with probability $|\alpha_x|^2$. We can also do a partial measurement on some of the qubits. The probability of the partial measurement resulting in a string x is the same as if we measured the whole state, and ignored the rest of the qubits. In this case, the resulting state will be the same as $|\phi\rangle$, except that all the strings inconsistent with x are removed. This new state will not have a norm of 1, so the actual superposition is obtained by dividing by the norm. For example, if we measure the first n bits of $|\phi\rangle = \sum_{x,y} \alpha_{x,y} |x, y\rangle$, we will obtain the measurement x with probability $\sum_{y'} |\alpha_{x,y'}|^2$, and in this case the resulting state will be

$$|x\rangle \sum_y \frac{\alpha_{x,y}}{\sqrt{\sum_{y'} |\alpha_{x,y'}|^2}} |y\rangle.$$

Following [BBC$^+$98], we model a quantum attacker $\mathcal{A}_Q$ with access to (possibly identical) oracles $O_1, O_2, \ldots$ by a sequence of unitary transformations $U_1, O_1, U_2, \ldots,$
O_{T-1}, U_T over $k = \text{poly}(n)$ qubits. Here, oracle $O_i : \{0,1\}^n \to \{0,1\}^m$ maps the first $n + m$ qubits from basis state $|x\rangle |y\rangle$ to basis state $|x\rangle |y \oplus O_i(x)\rangle$ for $x \in \{0,1\}^n$ and $y \in \{0,1\}^m$. If we require the access to O_i to be classical instead of quantum, the first n bits of the state are measured before applying the unitary transformation corresponding to O_i. Notice that any quantum-accessible oracle can also be used as a classical oracle. Note that the algorithm $\mathcal{A}_Q$ may also receive some input $|\psi\rangle$. Given an algorithm $\mathcal{A}_Q$ as above, with access to oracles O_i, we sometimes write $\mathcal{A}_Q^{|O_1(\cdot)\rangle, |O_2(\cdot)\rangle, \ldots}$ to indicate that the oracle is quantum-accessible (contrary to oracles which can only process classical bits).

To introduce asymptotics we assume that $\mathcal{A}_Q$ is actually a sequence of such transformation sequences, indexed by parameter n, and that each transformation sequence is composed out of quantum systems for input, output, oracle calls, and work space (of sufficiently many qubits). To measure polynomial running time, we assume that each U_i is approximated (to sufficient precision) by members of a set of universal gates (say, Hadamard, phase, CNOT and $\pi/8$; for sake of concreteness [NC00]), where at most polynomially many gates are used. Furthermore, $T = T(n)$ is assumed to be polynomial, too. Note that T also bounds the number of oracle queries.

We define the Euclidean distance $\||\phi\rangle + |\psi\rangle|$ between two states as the value $\left(\sum_x |\alpha_x - \beta_x|^2\right)^{\frac{1}{2}}$ where $|\phi\rangle = \sum_x \alpha_x |x\rangle$ and $|\psi\rangle = \sum_x \beta_x |x\rangle$.

Define $q_r(|\phi_t\rangle)$ to be the magnitude squared of r in the superposition of query t. We call this the query probability of r in query t. If we sum over all t, we get the total query probability of r.

We will be using the following lemmas:

Lemma 1 ([BBBV97] Theorem 3.1). *Let $|\varphi\rangle$ and $|\psi\rangle$ be quantum states with Euclidean distance at most ϵ. Then, performing the same measurement on $|\varphi\rangle$ and $|\psi\rangle$ yields distributions with statistical distance at most 4ϵ.*

Lemma 2 ([BBBV97] Theorem 3.3). *Let A_Q be a quantum algorithm running in time T with oracle access to O. Let $\epsilon > 0$ and let $S \subseteq [1,T] \times \{0,1\}^n$ be a set of time-string pairs such that $\sum_{(t,r) \in S} q_r(|\phi_t\rangle) \leq \epsilon$. If we modify O into an oracle O' which answers each query r at time t by providing the same string R (which has been independently sampled at random), then the Euclidean distance between the final states of A_Q when invoking O and O' is at most $\sqrt{T}\epsilon$.*

2.2 Quantum-Accessible Random Oracles

In the classical random oracle model [BR93] all algorithms used in the system are given access to the same random oracle. In the proof of security, the reduction algorithm answers the adversary's queries with consistent random answers.

In the quantum settings, a quantum attacker issues a random oracle query which is itself a superposition of exponentially many states. The reduction algorithm must evaluate the random oracle at all points in the superposition. To ensure that random oracle queries are answered consistently across queries, it is convenient to assume that quantum-resistant pseudorandom functions exist, and to implement this auxiliary random oracle with such a PRF.

Definition 1 (Pseudorandom Function). *A quantum-accessible pseudorandom function is an efficiently computable function* PRF *where, for all efficient quantum algorithms* D,

$$\left| \Pr[D^{\mathsf{PRF}(k,\cdot)}(1^n) = 1] - \Pr[D^{O(\cdot)}(1^n) = 1] \right| < \epsilon$$

where $\epsilon = \epsilon(n)$ *is negligible in* n, *and where* O *is a random oracle, the first probability is over the keys* k *of length* n, *and the second probability is over all random oracles and the sampling of the result of* D.

We note that, following Watrous [Wat09], indistinguishability as above should still hold for any auxiliary quantum state σ given as additional input to D (akin to non-uniformity for classical algorithms). We do not include such auxiliary information in our definition in order to simplify.

We say that an oracle O' is computationally indistinguishable from a random oracle if, for all polynomial time quantum algorithms with oracle access, the variational distance of the output distributions when the oracle is O' and when the oracle is a truly random oracle O is negligible. Thus, simulating a random oracle with a quantum-accessible pseudorandom function is computationally indistinguishable from a true random oracle.

We remark that, instead of assuming that quantum-accessible PRFs exist, we can often carry out security reductions relative to a random oracle. Consider, for example, a signature scheme (in the quantum-accessible random oracle model) which we prove to be unforgeable for quantum adversaries, via a reduction to the one-wayness of a trapdoor permutation against quantum inverters. We can then formally first claim that the scheme is unforgeable as long as inverting the trapdoor permutation is infeasible even when having the additional power of a quantum-accessible random oracle; only in the next step we can then conclude that this remains true in the standard model, if we assume that quantum-accessible pseudorandom functions exist and let the inverter simulate the random oracle with such a PRF. We thus still get a potentially reasonable security claim even if such PRFs do not exist. This technique works whenever we can determine the success of the adversary (as in case of inverting a one-way function).

2.3   Hard Problems for Quantum Computers

We will use the following general notion of a hard problem.

Definition 2 (Problem). *A problem is a pair* $P = (Game_P, \alpha_P)$ *where* $Game_P$ *specifies a game that a (possibly quantum) adversary plays with a classical challenger. The game works as follows:*

- *On input 1^n, the challenger computes a value x, which it sends to the adversary as its input*

- *The adversary is then run on x, and is allowed to make classical queries to the challenger.*

- *The adversary then outputs a value y, which it sends to the challenger.*

- *The challenger then looks at x, y, and the classical queries made by the adversary, and outputs 1 or 0.*

The value α_P is a real number between 0 (inclusive) and 1 (exclusive). It may also be a function of n, but for this paper, we only need constant α_P, specifically α_P is always 0 or $\frac{1}{2}$.

We say that an adversary A wins the game Game_P if the challenger outputs 1. We define the advantage $\text{Adv}_{A,P}$ of A in problem P as

$$Adv_{A,P} = |\Pr[A \text{ wins in Game}_P] - \alpha_P|$$

Definition 3 (Hard Problem). *A problem $P = (Game_P, \alpha_P)$ is hard for quantum computers if, for all polynomial time quantum adversaries A, $Adv_{A,P}$ is negligible.*

2.4   Cryptographic Primitives

For this paper, we define the security of standard cryptographic primitives in terms of certain problems being hard for quantum computers. We give a brief sketch here and refer to the full version [BDF+10] for supplementary details.

A trapdoor function $\mathcal{F}$ is secure if $\text{Inv}(\mathcal{F}) = (\text{Game}_{\text{INV}}(\mathcal{F}), 0)$ is a hard problem for quantum computers, where in Game_{INV}, an adversary is given a random element y and public key, and succeeds if it can output an inverse for y relative to the public key. A preimage sampleable trapdoor function, $\mathcal{F}$, is secure if $\text{Inv}(\mathcal{F})$ as described above is hard, and if $\text{Col}(\mathcal{F}) = (\text{Game}_{\text{Col}}(\mathcal{F}), 0)$ is hard for quantum computers, where in Game_{Col}, an adversary is given a public key, succeeds if it can output a collision relative to that public key. A signature scheme $\mathcal{S}$ is secure if the game $\text{Sig-Forge}(\mathcal{S}) = (\text{Game}_{\text{Sig}}(\mathcal{S}), 0)$ is hard, where Game_{Sig} is the standard existential unforgeability under a chosen message attack game. Lastly, a private (resp. public) key encryption scheme $\mathcal{E}$ is secure if $\text{Sym-CCA}(\mathcal{E}) = (\text{Game}_{\text{Sym}}(\mathcal{E}), \frac{1}{2})$ (resp. $\text{Asym-CCA}(\mathcal{E}) = (\text{Game}_{\text{Asym}}(\mathcal{E}), \frac{1}{2})$), where Game_{Sym} is the standard private key CCA attack game, and $\text{Game}_{\text{Asym}}$ is the standard public key attack game.

3   Separation Result

In this section, we discuss a two-party protocol that is provably secure in the random oracle model against both classical and quantum adversaries with classical access to the random oracle (and when using quantum-immune primitives). We then use the polynomial gap between the birthday attack and a collision finder based on Grover's algorithm to show that the protocol remains secure for certain hash functions when only classical adversaries are considered, but becomes insecure for any hash function if quantum adversaries are allowed. Analyzing the protocol in the stronger quantum random oracle model, where we grant the adversary quantum access to the random oracle, yields the same negative result.

Note that, due to the page limit, we discuss only the high-level idea of our protocol, for the full description and the formal security analysis we refer to the full version [BDF+10]. We start by briefly presenting the necessary definitions and assumptions for our construction.

Building Blocks. For sake of simplicity, we start with a quantum-immune identification scheme to derive our protocol; any other primitive or protocol can be used in a similar fashion. An *identification scheme* IS consists of three efficient algorithms (IS.KGen, $\mathcal{P}, \mathcal{V}$) where IS.KGen on input 1^n returns a key pair (sk, pk). The joint execution of $\mathcal{P}(\text{sk}, \text{pk})$ and $\mathcal{V}(\text{pk})$ then defines an interactive protocol between the prover $\mathcal{P}$ and the verifier $\mathcal{V}$. At the end of the protocol $\mathcal{V}$ outputs a decision bit $b \in \{0, 1\}$, indicating whether he accepts the identification of $\mathcal{P}$ or not. We say that IS is secure if an adversary after interacting with an honest prover $\mathcal{P}$ cannot impersonate $\mathcal{P}$ such that a verifier accepts the interaction.

A *hash function* H = (H.KGen, H.Eval) is a pair of efficient algorithms such that H.KGen for input 1^n returns a key k (which contains 1^n), and H.Eval for input k and $M \in \{0, 1\}^*$ deterministically outputs a digest H.Eval(k, M). For a random oracle H we use k as a "salt" and consider the random function $H(k, \cdot)$. The hash function is called *near-collision-resistant* if for any efficient algorithm $\mathcal{A}$ the probability that for $k \leftarrow$ H.KGen(1^n), some constant $1 \leq \ell \leq n$ and $(M, M') \leftarrow \mathcal{A}(k, \ell)$ we have $M \neq M'$ but H.Eval(k, M)$|_\ell =$ H.Eval(k, M')$|_\ell$, is negligible (as a function of n). Here we denote by $x|_\ell$ the leading ℓ bits of the string x. Note that for $\ell = n$ the above definition yields the standard notion of collision-resistance.

Classical vs. Quantum Collision-Resistance. In the classical setting, (near-) collision-resistance for any hash function is upper bounded by the *birthday attack*. This generic attack states that for any hash function with n bits output, an attacker can find a collision with probability roughly $1/2$ by probing $2^{n/2}$ distinct and random inputs. For (classical) random oracles this attack is optimal.

In the quantum setting, one can gain a polynomial speed-up on the collision search by using *Grover's algorithm* [Gro96,Gro98], which performs a search on an unstructured database with N elements in time $O(\sqrt{N})$. Roughly, this is achieved by using superpositions to examine all entries "at the same time". Brassard et al. [BHT98] use Grover's algorithm to obtain an algorithm for solving the collision problem for a hash function $H : \{0, 1\}^* \rightarrow \{0, 1\}^n$ with probability at least $1/2$, using only $O(\sqrt[3]{2^n})$ evaluations of H.

Computational and Timing Assumptions. To allow reasonable statements about the security of our protocol we need to formalize assumptions concerning the computational power of the adversary and the time that elapses on quantum and classical computers. In particular, we assume the following:

1. The speed-up one can gain by using a parallel machine with many processors, is bounded by a fixed term.
2. The time that is required to evaluate a hash function is independent of the input and the computational environment.
3. Any computation or action that does not require the evaluation of a hash function, costs zero time.

The first assumption basically resembles the fact that in the real world there is only a concrete and finite amount of equipment available that can contribute to a performance gain of a parallel system. Assumptions (2)+(3) are regarding the time that is needed to evaluate a hash function or to send a message between two parties and are merely for the sake of convenience, as one could patch the idea by relating the timings more rigorously. The latter assumption implicitly states that the computational overhead that quantum algorithms may create to obtain a speed-up is negligible when compared to the costs of a hash evaluation. This might be too optimistic in the near future, as indicated by Bernstein [Ber09]. That is, Bernstein discussed that the overall costs of a quantum computation can be higher than of massive parallel computation. However, as our work addresses conceptional issues that arise when *efficient* quantum computers exist, this assumption is somewhat inherent in our scenario.

3.1   Construction

We now present our identification scheme between a prover $\mathcal{P}$ and a verifier $\mathcal{V}$. The main idea is to augment a secure identification scheme IS by a collision-finding stage for some hash function H. In this first stage, the verifier checks if the prover is able to produce collisions on a hash function in a particular time. More precisely, the verifier starts for timekeeping to evaluate the hash function H.Eval$(k, \cdot)$ on the messages $\langle c \rangle$ for $c = 1, 2, \ldots, \left\lceil \sqrt[3]{2^\ell} \right\rceil$ for a key k chosen by the verifier and where $\langle c \rangle$ stands for the binary representation of c with $\log \left\lceil \sqrt[3]{2^\ell} \right\rceil$ bits. The prover has now to respond with a near-collision $M \neq M'$ such that H.Eval$(k, M) = $ H.Eval(k, M') holds for the first ℓ bits. One round of the collision-stage ends if the verifier either receives such a collision or finishes its $\sqrt[3]{2^\ell}$ hash evaluations. The verifier and the receiver then repeat such a round $r = \text{poly}(n)$ times, sending a fresh key k in each round.

Subsequently, both parties run the standard identification scheme. At the end, the verifier accepts if the prover was able to find enough collisions in the first stage or identifies correctly in the second stage. Thus, as long as the prover is not able to produce collisions in the required time, the protocol mainly resembles the IS protocol.

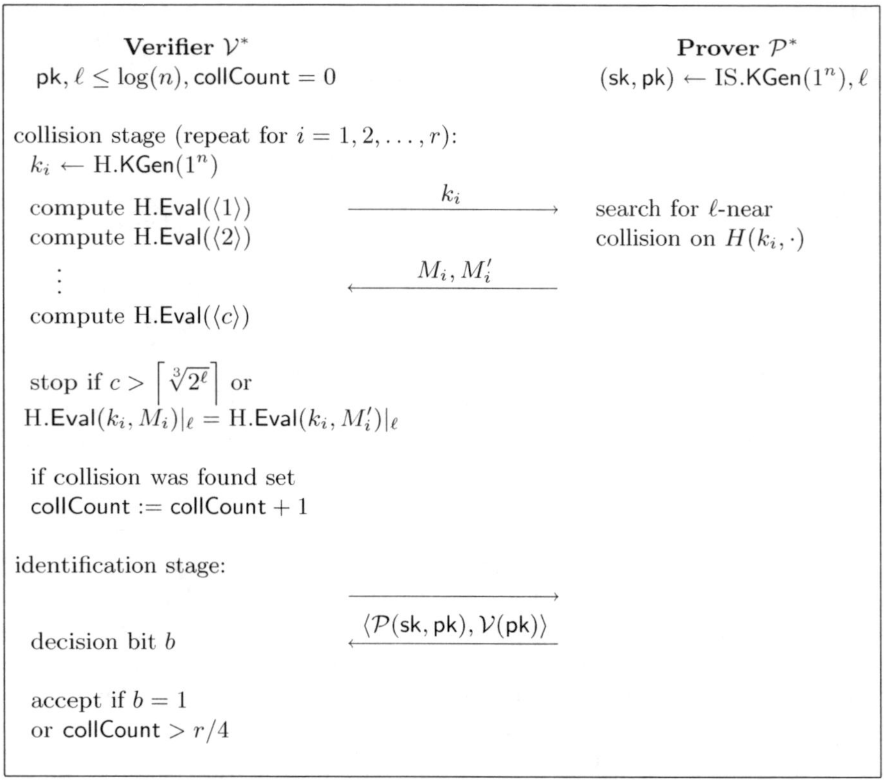

Fig. 1. The IS*-Identification Protocol

Completeness of the IS* protocol follows easily from the completeness of the underlying IS scheme.

Security against Classical and Quantum Adversaries. To prove security of our protocol, we need to show that an adversary $\mathcal{A}$ after interacting with an honest prover $\mathcal{P}^*$, can subsequently not impersonate $\mathcal{P}^*$ such that $\mathcal{V}^*$ will accept the identification. Let ℓ be such that $\ell > 6\log(\alpha)$ where α is the constant reflecting the bounded speed-up in parallel computing from Assumption (1). By assuming that $IS = (IS.KGen, \mathcal{P}, \mathcal{V})$ is a quantum-immune identification scheme, we can show that IS* is secure in the standard random oracle model against classical and quantum adversaries.

The main idea is that for the standard random oracle model, the ability of finding collisions is bounded by the birthday attack. Due to the constraint of granting only time $O(\sqrt[3]{2^\ell})$ for the collision search and setting $\ell > 6\log(\alpha)$, even an adversary with quantum or parallel power is not able to make at least $\sqrt{2^\ell}$ random oracle queries. Thus, $\mathcal{A}$ has only negligible probability to respond in more than $1/4$ of r rounds with a collision.

When considering only classical adversaries, we can also securely instantiate the random oracle by a hash function H that provides near-collision-resistance close to the birthday bound. Note that this property is particularly required from the SHA-3 candidates [NIS07].

However, for adversaries $\mathcal{A}_Q$ with quantum power, such an instantiation is not possible for *any* hash function. This stems from the fact that $\mathcal{A}_Q$ can locally evaluate a hash function on quantum states which in turns allows it to apply Grover's search algorithm. Then an adversary will find a collision in time $\sqrt[3]{2^\ell}$ with probability at least $1/2$, and thus will be able to provide $r/4$ collisions with noticeable probability. The same result holds in the quantum-accessible random oracle model, since Grover's algorithm only requires (quantum) black-box access to the hash function.

4   Signature Schemes in the Quantum-Accessible Random Oracle Model

We now turn to proving security in the quantum-accessible random oracle model. We present general conditions for when a proof of security in the classical random oracle model implies security in the quantum-accessible random oracle model. The result in this section applies to signatures whose classical proof of security is a *history-free reduction* as defined next. Roughly speaking, history-freeness means that the classical proof of security simulates the random oracle and signature oracle in a history-free fashion. That is, its responses to queries do not depend on responses to previous queries or the query number. We then show that a number of classical signature schemes have a history-free reduction thereby proving their security in the quantum-accessible random oracle model.

Definition 4 (History-free Reduction). *A random oracle model signature scheme $\mathcal{S} = (G, S^O, V^O)$ has a history-free reduction from a hard problem $P = (Game_P, 0)$ if there is a proof of security that uses a classical PPT adversary A for $\mathcal{S}$ to construct a classical PPT algorithm B for problem P such that:*

- *Algorithm B for P contains four explicit classical algorithms: START, RAND^{O_c}, SIGN^{O_c}, and FINISH^{O_c}. The latter three algorithms have access to a shared classical random oracle O_c. These algorithms, except for RAND^{O_c}, may also make queries to the challenger for problem P. The algorithms are used as follows:*

  (1) Given an instance x for problem P as input, algorithm B first runs START(x) to obtain (pk, z) where pk is a signature public key and z is private state to be used by B. Algorithm B sends pk to A and plays the role of challenger to A.

  (2) When A makes a classical random oracle query to $O(r)$, algorithm B responds with $\mathrm{RAND}^{O_c}(r, z)$. Note that RAND is given the current query as input, but is unaware of previous queries and responses.

  (3) When A makes a classical signature query $S(\mathsf{sk}, m)$, algorithm B responds with $\mathrm{SIGN}^{O_c}(m, z)$.

(4) When A outputs a signature forgery candidate (m, σ), algorithm B outputs $\text{FINISH}^{O_c}(m, \sigma, z)$.

- *There is an efficiently computable function* $\text{INSTANCE}(\mathsf{pk})$ *which produces an instance x of problem P such that* $\text{START}(x) = (\mathsf{pk}, z)$ *for some z. Consider the process of first generating* $(\mathsf{sk}, \mathsf{pk})$ *from $G(1^n)$, and then computing* $x = \text{INSTANCE}(\mathsf{pk})$. *The distribution of x generated in this way is negligibly close to the distribution of x generated in* Game_P.

- *For fixed z, consider the classical random oracle* $O(r) = \text{RAND}^{O_c}(r, z)$. *Define a quantum oracle* O_{quant}, *which transforms a basis element* $|x, y\rangle$ *into* $|x, y \oplus O(x)\rangle$. *We require that* O_{quant} *is quantum computationally indistinguishable from a random oracle.*

- SIGN^{O_c} *either aborts (and hence B aborts) or it generates a valid signature relative to the oracle* $O(r) = \text{RAND}^{O_c}(r, z)$ *with a distribution negligibly close to the correct signing algorithm. The probability that none of the signature queries abort is non-negligible.*

- *If (m, σ) is a valid signature forgery relative to the public key pk and oracle* $O(r) = \text{RAND}^{O_c}(r, z)$ *then the output of B (i.e.* $\text{FINISH}^{O_c}(m, \sigma, z)$*) causes the challenger for problem P to output 1 with non-negligible probability.* □

We now show that history-free reductions imply security in the quantum settings.

Theorem 1. *Let* $\mathcal{S} = (G, S, V)$ *be a signature scheme. Suppose that there is a history-free reduction that uses a classical PPT adversary A for $\mathcal{S}$ to construct a PPT algorithm B for a problem P. Further, assume that P is hard for polynomial-time quantum computers, and that quantum-accessible pseudorandom functions exist. Then $\mathcal{S}$ is secure in the quantum-accessible random oracle model.*

Proof. The history-free reduction includes five (classical) algorithms START, RAND, SIGN, FINISH, and INSTANCE, as in Definition 4. We prove the quantum security of $\mathcal{S}$ using a sequence of games, where the first game is the standard quantum signature game with respect to $\mathcal{S}$.

Game 0. Define Game_0 as the game a quantum adversary A_Q plays for problem Sig-Forge($\mathcal{S}$). Assume towards contradiction that A_Q has a non-negligible advantage.

Game 1. Define Game_1 as the following modification to Game_0: after the challenger generates $(\mathsf{sk}, \mathsf{pk})$, it computes $x \leftarrow \text{INSTANCE}(\mathsf{pk})$ as well as $(\mathsf{pk}, z) \leftarrow \text{START}(x)$. Further, instead of answering A_Q's quantum random oracle queries with a truly random oracle, the challenger simulates for A_Q a quantum-accessible random oracle O_{quant} as an oracle that maps a basis element $|x, y\rangle$ into the element $|x, y \oplus \text{RAND}^{O_q}(x, z)\rangle$, where O_q is a truly random quantum-accessible oracle. The history-free guarantee on RAND ensures that O_{quant} is computationally indistinguishable from random for quantum adversaries. Therefore, the success probability of A_Q in Game_1 is negligibly close to its success probability in Game_0, and hence is non-negligible.

Game 2. Modify the challenger from Game$_1$ as follows: instead of generating (sk, pk) and computing $x = \text{INSTANCE}(\text{pk})$, start off by running the challenger for problem P. When that challenger sends x, then start the challenger from Game$_1$ using this x. Also, when A_Q asks for a signature on m, answer with $\text{SIGN}^{O_q}(m, z)$. First, since INSTANCE is part of a history-free reduction, this change in how we compute x only negligibly affects the distribution of x, and hence the behavior of A_Q. Second, as long as all signing algorithms succeed, changing how we answer signing queries only negligibly affects the behavior of A_Q. Thus, the probability that A_Q succeeds is the product of the following two probabilities:

- The probability that all of the signing queries are answered without aborting.

- The probability that A_Q produces a valid forgery given that the signing queries were answered successfully.

The first probability is non-negligible by assumption, and the second is negligibly close to the success probability of A_Q in Game$_1$, which is also non-negligible. This means that the success probability of A_Q in Game$_3$ is non-negligible.

Game 3. Define Game$_3$ as in Game$_2$, except that for two modifications to the challenger: First, it generates a key k for the quantum-accessible PRF. Then, to answer a random oracle query $O_q(|\phi\rangle)$, the challenger applies the unitary transformation that takes a basis element $|x, y\rangle$ into $|x, y \oplus \text{PRF}(k, x)\rangle$. If the success probability in Game$_3$ was non-negligibly different from that of Game$_2$, we could construct a distinguisher for PRF which plays both the role of A_Q and the challenger. Hence, the success probability in Game$_3$ is negligibly close to that of Game$_2$, and hence is also non-negligible.

Given a quantum adversary that has non-negligible advantage in Game 3 we construct a quantum algorithm B_Q that breaks problem P. When B_Q receives instance x from the challenger for problem P, it computes $(\text{pk}, z) \leftarrow \text{START}(x)$ and generates a key k for PRF. Then, it simulates A_Q on pk. B_Q answers random oracle queries using a quantum-accessible function built from $\text{RAND}^{\text{PRF}(k,\cdot)}(\cdot, z)$ as in Game 1. It answers signing queries using $\text{SIGN}^{\text{PRF}(k,\cdot)}(\cdot, z)$. Then, when A_Q outputs a forgery candidate (m, σ), B_Q computes $\text{FINISH}^{\text{PRF}(k,\cdot)}(m, \sigma, z)$, and returns the result to the challenger for problem P.

Observe that the behavior of A_Q in Game$_3$ is identical to that as a subroutine of B_Q. Hence, A_Q as a subroutine of B_Q will output a valid forgery (m, σ) with non-negligible probability. If (m, σ) is a valid forgery, then since FINISH is part of a history-free reduction, $\text{FINISH}^{\text{PRF}(k,\cdot)}(m, \sigma, z)$ will cause the challenger for problem P to accept with non-negligible probability. Thus, the probability that P accepts is also non-negligible, contradicting our assumption that P is hard for quantum computers.

Hence we have shown that any polynomial quantum algorithm has negligible advantage against problem Sig-Forge($\mathcal{S}$) which completes the proof.   □

We note that, in every step of the algorithm, the adversary A_Q remains in a pure state. This is because, in each game, A_Q's state is initially pure (since it is classical), and every step of the game either involves a unitary transformation, a partial measurement, or classical communication. In all three cases, if the state is pure before, it is also pure after.

We also note that we could have stopped at Game$_2$ and assumed that the cryptographic problem P is hard relative to a (quantum-accessible) random oracle. Assuming the existence of quantum-accessible pseudorandom functions allows us to draw the same conclusion in the standard (i.e., non-relativized) model at the expense of an extra assumption.

4.1   Secure Signatures from Preimage Sampleable Trapdoor Functions (PSF)

We now use Theorem 1 to prove the security of the Full Domain Hash signature scheme when instantiated with a preimage sampleable trapdoor function (PSF), such as the one proposed in [GPV08]. Loosely speaking, a PSF $\mathcal{F}$ is a tuple of PPT algorithms $(G, \mathrm{Sample}, f, f^{-1})$ where $G(\cdot)$ generates a key pair $(\mathsf{pk}, \mathsf{sk})$, $f(\mathsf{pk}, \cdot)$ defines an efficiently computable function, $f^{-1}(\mathsf{sk}, y)$ samples from the set of pre-images of y, and $\mathrm{Sample}(\mathsf{pk})$ samples x from the domain of $f(\mathsf{pk}, \cdot)$ such that $f(\mathsf{pk}, x)$ is statistically close to uniform in the range of $f(\mathsf{pk}, \cdot)$. The PSF of [GPV08] is not only one-way, but is also collision resistant.

Recall that the full domain hash (FDH) signature scheme [BR93] is defined as follows:

Definition 5 (Full Domain Hash). *Let $\mathcal{F} = (G, f, f^{-1})$ be a trapdoor permutation, and O a hash function whose range is the same as the range of f. The full domain hash signature scheme is $\mathcal{S} = (G, T, V)$ where:*

- $G = G_0$
- $S^O(\mathsf{sk}, m) = f^{-1}(\mathsf{sk}, O(m))$
- $V^O(\mathsf{pk}, m, \sigma) = \begin{cases} 1 & \text{if } O(m) = f(\mathsf{pk}, \sigma) \\ 0 & \text{otherwise} \end{cases}$

Gentry et al. [GPV08] show that the FDH signature scheme can be instantiated with a PSF $\mathcal{F} = (G, \mathrm{Sample}, f, f^{-1})$ instead of a trapdoor permutation. Call the resulting system FDH-PSF. They prove that FDH-PSF is secure against classical adversaries, provided that the pre-image sampling algorithm used during signing is derandomized (e.g. by using a classical PRF to generate its random bits). Their reduction is not quite history-free, but we show that it can be made history-free.

Consider the following reduction from a classical adversary A for the FDH-PSF scheme $\mathcal{S}$ to a classical collision finder B for $\mathcal{F}$:

- On input pk, B computes $\mathrm{START}(\mathsf{pk}) := (\mathsf{pk}, \mathsf{pk})$, and simulates A on pk.
- When A queries $O(r)$, B responds with
$$\mathrm{RAND}^{O_c}(r, \mathsf{pk}) := f(\mathsf{pk}, \mathrm{Sample}(1^n; O_c(r))).$$

- When A queries $S(\mathsf{sk}, m)$, B responds with
$$\mathrm{SIGN}^{O_c}(m, \mathsf{pk}) := Sample(1^n; O_c(m)).$$
- When A outputs (m, σ), B outputs
$$\mathrm{FINISH}^{O_c}(m, \sigma, \mathsf{pk}) := \big(Sample(1^n; O_c(m)), \sigma\big).$$

In addition, we define $\mathrm{INSTANCE}(\mathsf{pk}) := \mathsf{pk}$. Algorithms INSTANCE and START trivially satisfy the requirements of history-freeness (Definition 4). Before showing that the above reduction is in history-free form, we need the following technical lemma whose proof is given in the full version [BDF+10].

Lemma 3. *Say A is a quantum algorithm that makes q quantum oracle queries. Suppose further that we draw the oracle O from two distributions. The first is the random oracle distribution. The second is the distribution of oracles where the value of the oracle at each input x is identically and independently distributed by some distribution D whose variational distance is within ϵ from uniform. Then the variational distance between the distributions of outputs of A with each oracle is at most $4q^2\sqrt{\epsilon}$.*

Proof Sketch. We show that there is a way of moving from O to O_D such that the oracle is only changed on inputs in a set K where the sum of the amplitudes squared of all $k \in K$, over all queries made by A, is small. Thus, we can use Lemma 2 to show that the expected behavior of any algorithm making polynomially many quantum queries to O is only changed by a small amount. $\square$

Lemma 3 shows that we can replace a truly random oracle O with an oracle O_D distributed according to distribution D without impacting A, provided D is close to uniform. Note, however, that while this change only affects the output of A negligibly, the effects are larger than in the classical setting. If A only made classical queries to O, a simple hybrid argument shows that changing to O_D affects the distribution of the output of A by at most $q\epsilon$, as opposed to $4q^2\sqrt{\epsilon}$ in the quantum case. Thus, quantum security reductions that use Lemma 3 will not be as tight as their classical counterparts.

We now show that the reduction above is history-free.

Theorem 2. *The reduction above applied to FDH-PSF is history-free.*

Proof. The definition of a PSF implies that the distribution of $f(\mathsf{pk}, Sample(1^n))$ is within $\epsilon_{\mathsf{sample}}$ of uniform, for some negligible $\epsilon_{\mathsf{sample}}$. Now, since $O(r) = \mathrm{RAND}^{O_c}(r, \mathsf{pk}) = f(\mathsf{pk}, Sample(1^n; O_c(r)))$ and O_c is a true random oracle, the quantity $O(r)$ is distributed independently according to a distribution that is $\epsilon_{\mathsf{sample}}$ away from uniform. Define a quantum oracle O_{quant} which transforms the basis state $|x, y\rangle$ into $|x, y \oplus O(x)\rangle$. Using Lemma 3, for any algorithm B making q random oracle queries, the variational distance between the probability distributions of the outputs of B using a truly random oracle and the "not-quite" random oracle O_{quant} is at most $4q^2\sqrt{\epsilon_{\mathsf{sample}}}$, which is still negligible. Hence, O_q is computationally indistinguishable from random.

Gentry et al. [GPV08] also show that $\text{SIGN}^{O_c}(m, \mathsf{pk})$ is consistent with $\text{RAND}^{O_c}(\cdot, \mathsf{pk})$ for all queries, and that if A outputs a valid forgery (m, σ), $\text{FINISH}^{O_c}(m, \sigma, \mathsf{pk})$ produces a collision for $\mathcal{F}$ with probability $1 - 2^{-E}$, where E is the minimum over all y in the range of $f(\mathsf{pk}, \cdot)$ of the min-entropy of the distribution on σ given $f(\mathsf{pk}, \sigma) = y$. The PSF of Gentry et al. [GPV08] has super-logarithmic min-entropy, so $1 - 2^{-E}$ is negligibly close to 1, though any constant non-zero min-entropy will suffice to make the quantity a non-negligible fraction of 1. □

We note that the security proof of Gentry et al. [GPV08] is a tight reduction in the following sense: if the advantage of an adversary A for $\mathcal{S}$ is ϵ, the reduction gives a collision finding adversary B for $\mathcal{F}$ with advantage negligibly close to ϵ, provided that the lower bound over y in the range of $f(\mathsf{pk}, \cdot)$ of the min-entropy of σ given $f(\mathsf{pk}, \sigma) = y$ is super-logarithmic. If the PSF has a min-entropy of 1, the advantage of B is still $\epsilon/2$.

The following corollary, which is the main result of this section, follows from Theorems (1) and (2).

Corollary 1. *If quantum-accessible pseudorandom functions exist, and $\mathcal{F}$ is a secure PSF against quantum adversaries, then the FDH-PSF signature scheme is secure in the quantum-accessible random oracle model.*

4.2   Secure Signatures from Claw-Free Permutations

In this section, we show how to use claw-free permutations to construct three signature schemes that have history-free reductions and are therefore secure in the quantum-accessible random oracle model. The first is the standard FDH from Definition 5, but when the underlying permutation is a claw-free permutation. We adapt the proof of Coron [Cor00] to give a history-free reduction. The second is the Katz and Wang [KW03] signature scheme, and we also modify their proof to get a history-free reduction. Lastly, following Gentry et al. [GPV08], we note that claw-free permutations give rise to a pre-image sampleable trapdoor function (PSF), which can then be used in FDH to get a secure signature scheme as in Section 4.1. The Katz-Wang and FDH-PSF schemes from claw-free permutations give a tight reduction, whereas the Coron-based proof loses a factor of q_s in the security reduction, where q_s is the number of signing queries.

Recall that a claw-free pair of permutations [GMR88] is a pair of trapdoor permutations $(\mathcal{F}_1, \mathcal{F}_2)$, where $\mathcal{F}_i = (G_i, f_i, f_i^{-1})$, with the following properties:

- $G_1 = G_2$. Define $G = G_1 = G_2$.

- For any key pk, $f_1(\mathsf{pk}, \cdot)$ and $f_2(\mathsf{pk}, \cdot)$ have the same domain and range.

- Given only pk, the probability that any PPT adversary can find a pair (x_1, x_2) such that $f_1(\mathsf{pk}, x_1) = f_2(\mathsf{pk}, x_2)$ is negligible. Such a pair is called a claw.

Dodis and Reyzin [DR03] note that claw-free permutations are a generalization of trapdoor permutations with a random self-reduction. A random self-reduction is a way of taking a worst-case instance x of a problem, and converting it into a random instance y of the same problem, such that a solution to y gives a solution to x. Dodis and Reyzin [DR03] show that any trapdoor permutation with a random self reduction (e.g. RSA) gives a claw-free pair of permutations.

We note that currently there are no candidate pairs of claw-free permutations that are secure against quantum adversaries, but this may change in time.

FDH Signatures from Claw-Free Permutations. Coron [Cor00] shows that the Full Domain Hash signature scheme, when instantiated with the RSA trapdoor permutation, has a tighter security reduction than the general Full Domain Hash scheme, in the classical world. That is, Coron's reduction loses a factor of approximately q_s, the number of signing queries, as apposed to q_h, the number of hash queries. Of course, the RSA trapdoor permutation is not secure against quantum adversaries, but his reduction can be applied to any claw-free permutation and is equivalent to a history-free reduction with similar tightness.

To construct a FDH signature scheme from a pair of claw-free permutations $(\mathcal{F}_1, \mathcal{F}_2)$, we simply instantiate FDH with $\mathcal{F}_1$, and ignore the second permutation $\mathcal{F}_2$, to yield the following signature scheme

- G is the generator for the pair of claw-free permutations.
- $S^O(\mathsf{sk}, m) = f_1^{-1}(\mathsf{sk}, O(m))$
- $V^O(\mathsf{pk}, m, \sigma) = 1$ if and only if $f_1(\mathsf{pk}, \sigma) = O(m)$.

We now present a history-free reduction for this scheme. The random oracle for this reduction, $O_c(r)$, returns a random pair (a, b), where a is a random element from the domain of $\mathcal{F}_1$ and $\mathcal{F}_2$, and b is a random element from $\{1, ..., p\}$ for some p to be chosen later.

We construct history-free reduction from a classical adversary A for S to a classical adversary B for $(\mathcal{F}_1, \mathcal{F}_2)$. Algorithm B, on input pk, works as follows:

- Compute $\mathrm{START}(\mathsf{pk}, y) = (\mathsf{pk}, \mathsf{pk})$, and simulate A on pk. Notice that $z = \mathsf{pk}$ is the state saved by B.
- When A queries $O(r)$, compute $\mathrm{RAND}^{O_c}(r, \mathsf{pk})$. For each string r, RAND works as follows: compute $(a, b) \leftarrow O_c(r)$. If $b = 1$, return $f_2(\mathsf{pk}, a)$. Otherwise, return $f_1(\mathsf{pk}, a)$
- When A queries $S(\mathsf{sk}, m)$, compute $\mathrm{SIGN}^{O_c}(m, \mathsf{pk})$. SIGN works as follows: compute $(a, b) \leftarrow O_c(m)$ and return a if $b \neq 1$. Otherwise, fail.
- When A returns (m, σ), compute $\mathrm{FINISH}^{O_c}(m, \sigma, \mathsf{pk})$. FINISH works as follows: compute $(a, b) \leftarrow O_c(m)$ and output (σ, a).

In addition, we have $\mathrm{INSTANCE}(\mathsf{pk}) = \mathsf{pk}$ and $\mathrm{START}(\mathrm{INSTANCE}(\mathsf{pk})) = (\mathsf{pk}, \mathsf{pk})$, so INSTANCE and START satisfy the required properties.

Theorem 3. *The reduction above is in history-free form.*

Proof. $\mathrm{RAND}^{O_c}(r, \mathsf{pk})$ is completely random and independently distributed, as $f_1(\mathsf{pk}, a)$ and $f_2(\mathsf{pk}, a)$ are both random ($f_b(\mathsf{pk}, \cdot)$ is a permutation and a is truly random). As long as $b \neq 1$, where $(a, b) = O_c(m)$, $\text{SIGN}^{O_c}(m, \text{pk})$ will be consistent with RAND. This is because because $V^{\text{RAND}^{O_c}(\cdot, \text{pk})}(\text{pk}, m, \text{SIGN}^{O_c}(m, \text{pk}))$ outputs 1 if $\text{RAND}^{O_c}(m, \text{pk}) = f_1(\text{pk}, \text{SIGN}^{O_c}(m, \text{pk}))$. But $\text{RAND}^{O_c}(m, \text{pk}) = f_1(\text{pk}, a)$ (since $b \neq 1$), and $\text{SIGN}^{O_c}(m, \text{pk})) = a$. Thus, the equality holds. The probability over all signature queries of no failure is $(1 - 1/p)^{q_{\text{SIGN}}}$. If we chose $p = q_{\text{SIGN}}$, this quantity is at least $e^{-1} - o(1)$, which is non-negligible.

Suppose A returns a valid forgery (m, σ), meaning A never asked for a forgery on m and $f_1(\text{sk}, \sigma) = \text{RAND}^{O_c}(m, \text{pk})$. If $b = 1$ (where $(a, b) = O_c(m)$), then we have $f_1(\text{sk}, \sigma) = \text{RAND}^{O_c}(m, \text{pk}) = f_2(\text{pk}, a)$, meaning that (σ, a) is a claw. Since A never asked for a signature on m, there is no way A could have figured out a, so the case where $b = 1$ and a is the preimage of $O(m)$ under f_2, and the case where $b \neq 1$ and a is the preimage of $O(m)$ under f_1 are indistinguishable. Thus, $b = 1$ with probability $1/p$. Thus, B converts a valid signature into a claw with non-negligible probability.   □

Corollary 2. *If quantum-accessible pseudorandom functions exists, and $(\mathcal{F}_1, \mathcal{F}_2)$ is a pair claw-free trapdoor permutations, then the FDH scheme instantiated with $\mathcal{F}_1$ is secure against quantum adversaries.*

Note that in this reduction, our simulated random oracle is truly random, so we do not need to rely on Lemma 3. Hence, the tightness of the reduction will be the same as the classical setting. Namely, if the quantum adversary A has advantage ϵ when making q_{SIGN} signature queries, B will have advantage approximately ϵ/q_{SIGN}.

The Katz-Wang Signature Scheme In this section, we consider a variant of FDH due to Katz and Wang [KW03]. This scheme admits an almost tight security reduction in the classical world. That is, if an adversary has advantage ϵ, the reduction gives a claw finder with advantage $\epsilon/2$. Their proof of security is not in history-free form, but it can be modified so that it is in history-free form. Given a pair of trapdoor permutation $(\mathcal{F}_1, \mathcal{F}_2)$, the construction is as follows:

- G is the key generator for $\mathcal{F}$.

- $S^O(\text{sk}, m) = f_1^{-1}(\text{sk}, O(b, m))$ for a random bit b.

- $V^O(\text{pk}, m, \sigma)$ is 1 if either $f_1(\text{pk}, \sigma) = O(0, m)$ or $f_1(\text{pk}, \sigma) = O(1, m)$

We construct a history-free reduction from an adversary A for S to an adversary B for $(\mathcal{F}_1, \mathcal{F}_2)$. The random oracle for this reduction, $O_c(r)$, generates a random pair (a, b), where a is a random element from the domain of $\mathcal{F}_1$ and $\mathcal{F}_2$, and b is a random bit. On input pk, B works as follows:

- Compute $\text{START}(\text{pk}, y) = (\text{pk}, \text{pk})$, and simulate A on pk. Notice that $z = \text{pk}$ is the state saved by B.

- When A queries $O(b, r)$, compute $\text{RAND}^{O_c}(b, r, \text{pk})$. For each string (b, r), RAND works as follows: compute $(a, b') = O_c(r)$. If $b = b'$, return $f_1(\text{pk}, a)$. Otherwise, return $f_2(\text{pk}, a)$.

- When A queries $S(\mathsf{sk}, m)$, compute $\mathrm{SIGN}^{O_c}(m, \mathsf{pk})$. SIGN works as follows: compute $(a, b) = O_c(m)$ and return a.

- When A returns (m, σ), compute $\mathrm{FINISH}^{O_c}(m, \sigma, \mathsf{pk})$. FINISH works as follows: compute $(a, b) = O_c(m)$. If $\sigma = a$, abort. Otherwise, output (σ, a).

In addition, we have $\mathrm{INSTANCE}(\mathsf{pk}) = \mathsf{pk}$ and $\mathrm{START}(\mathrm{INSTANCE}(\mathsf{pk})) = (\mathsf{pk}, \mathsf{pk})$, so INSTANCE and START satisfy the required properties.

Theorem 4. *The reduction above is in history-free form.*

Proof. $\mathrm{RAND}^{O_c}(b, r, \mathsf{pk})$ is completely random and independently distributed, as $f_1(\mathsf{pk}, a)$ and $f_2(\mathsf{pk}, a)$ are both random (f_b is a permutation and a is truly random). Observe that $f_1(\mathsf{pk}, \mathrm{SIGN}^{O_c}(m, \mathsf{pk})) = f_1(\mathsf{pk}, a) = O(b, m)$ where $(a, b) = O_c(m)$. Thus, signing queries are always answered with a valid signature, and the distribution of signatures is identical to that of the correct signing algorithm since b is chosen uniformly.

Suppose A returns a valid forgery (m, σ). Let $(a, b) = O_c(m)$. There are two cases, corresponding to whether σ corresponds to a signature using b or $1 - b$. In the first case, we have $f_1(\mathsf{pk}, \sigma) = O(b, m) = f_1(\mathsf{pk}, a)$, meaning $\sigma = a$, so we abort. Otherwise, $f_1(\mathsf{pk}, \sigma) = O(1 - b, m) = f_2(\mathsf{pk}, a)$, so (σ, a) form a claw. Since the adversary never asked for a signing query on m, these two cases are indistinguishable by the same logic as the proof for FDH. Thus, the probability of failure is at most a half, which is non-negligible. $\qquad\square$

Corollary 3. *If quantum-accessible pseudorandom functions exists, and $(\mathcal{F}_1, \mathcal{F}_2)$ is a pair claw-free trapdoor permutations, then the Katz-Wang signature scheme instantiated with $\mathcal{F}_1$ is secure against quantum adversaries.*

As in the case of FDH, our simulated quantum-accessible random oracle is truly random, so we do not need to rely on Lemma 3. Thus, the tightness of our reduction is the same as the classical case. In particular, if the quantum adversary A_Q has advantage ϵ then B will have advantage $\epsilon/2$.

PSF Signatures from Claw-Free Permutations. Gentry et al. [GPV08] note that Claw-Free Permutations give rise to pre-image sampleable trapdoor functions (PSFs). These PSFs can then be used to construct an FDH signature scheme as in Section 4.1.

Given a pair of claw-free permutations $(\mathcal{F}_1, \mathcal{F}_2)$, define the following PSF: G is just the generator for the pair of permutations. Sample(pk) generates a random bit b and random x in the domain of f_b, and returns (x, b). $f(\mathsf{pk}, x, b) = f_b(\mathsf{pk}, x)$, and $f^{-1}(\mathsf{sk}, y) = (f_b^{-1}(\mathsf{sk}, y), b)$ for a random b. Suppose we have a collision $((x_1, b_1), (x_2, b_2))$ for this PSF. Then

$$f_{b_1}(\mathsf{pk}, x_1) = f(\mathsf{pk}, x_1, b_1) = f(\mathsf{pk}, x_2, b_2) = f_{b_2}(\mathsf{pk}, x_2)$$

If $b_1 = b_2$, then $x_1 = x_2$ since f_{b_1} is a permutation. But this is impossible since $(x_1, b_1) \neq (x_2, b_2)$. Thus, $b_1 \neq b_2$, so one of (x_1, x_2) or (x_2, x_1) is a claw for $(\mathcal{F}_1, \mathcal{F}_2)$.

Hence, we can instantiate FDH with this PSF to get the following signature scheme:

- G is the generator for the permutations.
- $S^O(\mathsf{sk}, m) = (f_b^{-1}(\mathsf{sk}, O(m)), b)$ for a random bit b.
- $V^O(\mathsf{pk}, m, (\sigma, b)) = 1$ if and only if $f_b(\mathsf{pk}, \sigma) = O(m)$.

The security of this scheme follows from Corollary 1, with a similar tightness guarantee (this PSF has only a pre-image min-entropy of 1, which results in a loss of a factor of two in the tightness of the reduction). In particular, if we have a quantum adversary A_Q for $\mathcal{E}$ with advantage ϵ, we get a quantum algorithm B_Q for the PSF with advantage $\epsilon/2$, which gives us a quantum algorithm C_Q that finds claws of $(\mathcal{F}_1, \mathcal{F}_2)$ with probability $\epsilon/2$.

5   Encryption Schemes in the Quantum-Accessible Random Oracle Model

In this section, we prove the security of two encryption schemes. The first is the BR encryption scheme due to Bellare and Rogaway [BR93], which we show is CPA secure. The second is a hybrid generalization of the BR scheme, which we show is CCA secure.

Ideally, we could define a general type of classical reduction like we did for signatures, and show that such a reduction implies quantum security. Unfortunately, defining a history-free reduction for encryption is considerably more complicated than for signatures. We therefore directly prove the security of two random oracle schemes in the quantum setting.

5.1   CPA Security of BR Encryption

In this section, we prove the security of the BR encryption scheme [BR93] against quantum adversaries:

Definition 6 (BR Encryption Scheme). *Let* $\mathcal{F} = (G_0, f, f^{-1})$ *be an injective trapdoor function, and O a hash function with the same domain as $f(\mathsf{pk}, \cdot)$. We define the following encryption scheme, $\mathcal{E} = (G, E, D)$ where:*

- $G = G_0$
- $E^O(\mathsf{pk}, m) = (f(\mathsf{pk}, r), O(r) \oplus m)$ *for a randomly chosen r.*
- $D^O(\mathsf{sk}, (y, c)) = c \oplus f^{-1}(\mathsf{sk}, y)$

A candidate quantum-immune injective trapdoor function can be built from hard problems on lattices [PW08].

Theorem 5. *If quantum-accessible pseudorandom functions exists and $\mathcal{F}$ is a quantum-immune injective trapdoor function, then $\mathcal{E}$ is quantum CPA secure.*

We omit the proof of Theorem 5 because the CPA security of the BR encryption scheme is a special case of the CCA security of the hybrid encryption scheme in the next section.

5.2   CCA Security of Hybrid Encryption

We now prove the CCA security of the following standard hybrid encryption, a generalization of the BR encryption scheme scheme [BR93], built from an injective trapdoor function and symmetric key encryption scheme.

Definition 7 (Hybrid Encryption Scheme). *Let $\mathcal{F} = (G_0, f, f^{-1})$ be an injective trapdoor function, and $\mathcal{E}_S = (E_S, D_S)$ be a CCA secure symmetric key encryption scheme, and O a hash function. We define the following encryption scheme, $\mathcal{E} = (G, E, D)$ where:*

- $G = G_0$

- $E^O(\mathsf{pk}, m) = (f(\mathsf{pk}, r), E_S(O(r), m))$ *for a randomly chosen r.*

- $D^O(\mathsf{sk}, (y, c)) = D_S(O(r'), c)$ *where $r' = f^{-1}(\mathsf{sk}, y)$*

We note that the BR encryption scheme from the previous section is a special case of this hybrid encryption scheme where $\mathcal{E}_S$ is the one-time pad. That is, $E_S(k, m) = k \oplus m$ and $D_S(k, c) = k \oplus c$.

Theorem 6. *If quantum-accessible pseudorandom functions exists, $\mathcal{F}$ is a quantum-immune injective trapdoor function, and $\mathcal{E}_S$ is a quantum CCA secure symmetric key encryption scheme, then $\mathcal{E}$ is quantum CCA secure.*

Proof. Suppose we have an adversary A_Q that breaks $\mathcal{E}$. We start with the standard security game for CCA secure encryption:

Game 0. Define Game_0 as the game a quantum adversary A_Q plays for problem Asym-CCA($\mathcal{E}$).

Game 1. Define Game_1 as the following game: the challenger generates $(\mathsf{sk}, \mathsf{pk}) \leftarrow G(1^n)$, a random r in the domain of $\mathcal{F}$, a random k in the key space of $\mathcal{E}_S$, and computes $y = f(\mathsf{pk}, r)$. The challenger has access to a quantum-accessible random oracle O_q whose range is the key space of $\mathcal{E}_S$. It then sends pk to A_Q. The challenger answers queries as follows:

- Random oracle queries are answered with the random oracle O_{quant}, which takes a basis element $|x, y\rangle$ into $|x, y \oplus O_q(f(\mathsf{pk}, x))\rangle$.

- Decryption queries on (y', c') are answered as follows:

    Case 1: If $y = y'$, respond with $D_S(k, c')$.

    Case 2: If $y \neq y'$, respond with $D_S(O_q(y'), c')$.

- The challenge query on (m_0, m_1) is answered as follows: choose a random b. Then, respond with $(y, E_S(k, m_b))$.

When A_Q responds with b', we say that A_Q won if $b = b'$.

   Observe that, because f is injective and O_q is random, the oracle O_{quant} is a truly random oracle with the same range as O_q. The challenge ciphertext (y, c) seen by A_Q is distributed identically to that of Game_0. Further, it is a valid encryption of m_b relative to the random oracle being O_{quant} if $O_q(y) = k$. For $y' \neq y$, the decryption of (y', c') is

$$D_S(O_q(y'), c') = D_S(O_{\text{quant}}(f^{-1}(\text{sk}, y')), c') = D^{O_{\text{quant}}}(\text{sk}, (y', c'))$$

Which is correct. Likewise, if $O_q(y) = k$, the decryption of (y, c') is also correct. Thus, the view of A_Q in Game$_1$ is identical to that in Game$_0$ if $O_q(y) = k$. We now make the following observations:

- The challenge query and decryption query answering algorithms never query O_q on y.

- Each quantum random oracle query from the adversary to O_{quant} leads to a quantum random oracle query from the challenger to O_q. The query magnitude of y in the challenger's query to O_q is the same as the query magnitude of r in the adversary's query O_{quant}.

Let ϵ be the sum of the square magnitudes of y over all queries made to O_q (i.e. the total query probability of y). This is identical to the total query probability of r over all queries A_Q makes to O_{quant}.

We now construct a quantum algorithm $B_{\mathcal{F}}^{O_q}$ that uses a quantum-accessible random oracle O_q, and inverts f with probability ϵ/q, where q is the number of random oracle queries made by A_Q. $B_{\mathcal{F}}^{O_q}$ takes as input (pk, y), and its goal is to output $r = f^{-1}(\text{sk}, y)$. $B_{\mathcal{F}}^{O_q}$ works as follows:

- Generate a random k in the key space of $\mathcal{E}_S$. Also, generate a random $i \in \{1, ..., q\}$. Now, send pk to A_Q and play the role of challenger to A_Q.

- Answer random oracle queries with the random oracle O_{quant}, which takes a basis element $|x, y\rangle$ into $|x, y \oplus O_q(f(\text{pk}, x))\rangle$.

- Answer decryption queries on (y', c') as follows:

  Case 1: If $y = y'$, respond with $D_S(k, c')$.

  Case 2: If $y \neq y'$, respond with $D_S(O_q(y'), c')$.

- Answer the challenge query on (m_0, m_1) as follows: choose a random b. Then, respond with $(y, E_S(k, m_b))$.

- At the ith random oracle query, sample the query to get r', and output r' and terminate.

Comparing our definition of $B_{\mathcal{F}}^{O_q}$ to Game$_1$, we can conclude that the view seen by A_Q in both cases is identical. Thus, the total query probability that A_Q makes to O_{quant} at the point r is ϵ. Hence, the probability that $B_{\mathcal{F}}^{O_q}$ outputs r is ϵ/q. If we assume that $\mathcal{F}$ is secure against quantum adversaries that use a quantum-accessible random oracle, then this quantity, and hence ϵ, must be negligible. As in the case of signatures (Section 4), we can replace this assumption with the assumption that $\mathcal{F}$ is secure against quantum adversaries (i.e. with no access to a quantum random oracle) and that pseudorandom functions exists to reach the same conclusion.

Since ϵ is negligible, we can change $O_q(y) = k$ in Game$_1$, thus getting a game identical to Game$_0$ from the adversary's point of view. Notice that in Game$_0$

and Game_1, A_Q is in a pure state because we are only applying unitary transformations, performing measurements, or performing classical communication. We are only changing the oracle at a point with negligible total query probability, so Lemma 2 tells us that making this change only affects the distribution of the outcome of Game_1 negligibly. This allows us to conclude that the success probability of A_Q in Game_1 is negligibly close to that in Game_0.

Now, assume that the success probability of A_Q in Game_1 is non-negligible. We now define a quantum algorithm $B_{\mathcal{E}_S}^{O_q}$ that uses a quantum-accessible random oracle O_q to break the CCA security of $\mathcal{E}_S$. $B_{\mathcal{E}_S}^{O_q}$ works as follows:

- On input 1^n, generate $(\mathsf{sk}, \mathsf{pk}) \leftarrow G(1^n)$. Also, generate a random r, and compute $y = f(\mathsf{pk}, r)$. Now send pk to A_Q and play the role of challenger to A_Q.

- Answer random oracle queries with the random oracle O_{quant}, which takes a basis element $|x, y\rangle$ into $|x, y \oplus O_q(f(\mathsf{pk}, x))\rangle$.

- Answer decryption queries on (y', c') as follows:

  Case 1: If $y = y'$, ask the $\mathcal{E}_S$ challenger for a decryption $D_S(k, c')$ to obtain m'. Return m' to A_Q.

  Case 2: If $y \neq y'$, respond with $D_S(O_q(y'), c')$.

- Answer the challenge query on (m_0, m_1) by forwarding the pair $\mathcal{E}_S$. When the challenger responds with c (which equals $E_S(k, m_b)$ for some b), return (y, c) to A_Q.

- When A_Q outputs b', output b' and halt.

Comparing our definition of $B_{\mathcal{E}_S}^{O_q}$ to that of Game_1, we can conclude that the view of A_Q in both cases is identical. Thus, A_Q succeeds with non-negligible probability. If A_Q succeeds, it means it returned b, meaning $B_{\mathcal{E}_S}^{O_q}$ also succeeded. Thus, we have an algorithm with a quantum random oracle that breaks $\mathcal{E}_S$. This is a contradiction if $\mathcal{E}_S$ is CCA secure against quantum adversaries with access to a quantum random oracle, which holds since $\mathcal{E}_S$ is CCA secure against quantum adversaries and quantum-accessible pseudorandom functions exist, by assumption.

Thus, the success probability of A_Q in Game_1 is negligible, so the success probability of A_Q in Game_0 is also negligible. Hence, we have shown that all polynomial time quantum adversaries have negligible advantage in breaking in breaking the CCA security of $\mathcal{E}$, so $\mathcal{E}$ is CCA secure. □

We briefly explain why Theorem 5 is a special case of Theorem 6. Notice that, in the above proof, $B_{\mathcal{E}_S}$ only queries its decryption oracle when answering decryption queries made by A_Q, and that it never makes encryption queries. Hence, if A_Q makes no decryption queries, $B_{\mathcal{E}_S}$ makes no queries at all except the challenge query. If we are only concerned with the CPA security of $\mathcal{E}$, we then only need E_S to be secure against adversaries that can only make the challenge query. Further, if we only let A_Q make a challenge query with messages of length n, then E_S only has to be secure against adversaries making challenges of a specific length. But this is exactly the model in which the one-time pad is unconditionally secure. Hence, the BR encryption scheme is secure, and we have proved Theorem 5.

6 Conclusion

We have shown that great care must be taken if using the random oracle model when arguing security against quantum attackers. Proofs in the classical case should be reconsidered, especially in case the quantum adversary can access the random oracle with quantum states. We also developed conditions for translating security proofs in the classical random oracle model to the quantum random oracle model. We applied these tools to certain signature and encryption schemes.

The foremost question raised by our results is in how far techniques for "classical random oracles" can be applied in the quantum case. This stems from the fact that manipulating or even observing the interaction with the quantum-accessible random oracle would require measurements of the quantum states. That, however, prevents further processing of the query in a quantum manner. We gave several examples of schemes that remain secure in the quantum setting, provided quantum-accessible pseudorandom functions exist. The latter primitive seems to be fundamental to simulate random oracles in the quantum world. Showing or disproving the existence of such pseudorandom functions is thus an important step.

Many classical random oracle results remain open in the quantum random oracle settings. It is not known how to prove security of generic FDH signatures as well as signatures derived from the Fiat-Shamir heuristic in the quantum random oracle model. Similarly, a secure generic transformation from CPA to CCA security in the quantum RO model is still open.

Acknowledgments. We thank Chris Peikert for helpful comments. Dan Boneh was supported by NSF, the Air Force Office of Scientific Research (AFO SR) under a MURI award, and by the Packard Foundation. Marc Fischlin and Anja Lehmann were supported by grants Fi 940/2-1 and Fi 940/3-1 of the German Research Foundation (DFG). Özgür Dagdelen and Marc Fischlin were also supported by CASED (www.cased.de). Christian Schaffner is supported by a NWO VENI grant.

References

Aar09.   Aaronson, S.: Quantum copy-protection and quantum money. In: Structure in Complexity Theory Conference, pp. 229–242 (2009)

ABB10a.   Agrawal, S., Boneh, D., Boyen, X.: Efficient Lattice (H)IBE in the Standard Model. In: Gilbert, H. (ed.) EUROCRYPT 2010. LNCS, vol. 6110, pp. 553–572. Springer, Heidelberg (2010)

ABB10b. Agrawal, S., Boneh, D., Boyen, X.: Lattice Basis Delegation in Fixed Dimension and Shorter-Ciphertext Hierarchical IBE. In: Rabin, T. (ed.) CRYPTO 2010. LNCS, vol. 6223, pp. 98–115. Springer, Heidelberg (2010)

AS04. Aaronson, S., Shi, Y.: Quantum lower bounds for the collision and the element distinctness problems. Journal of the ACM 51(4), 595–605 (2004)

BBBV97. Bennett, C.H., Bernstein, E., Brassard, G., Vazirani, U.V.: Strengths and weaknesses of quantum computing. SIAM J. Comput. 26(5), 1510–1523 (1997)

BBC⁺98. Beals, R., Buhrman, H., Cleve, R., Mosca, M., de Wolf, R.: Quantum lower bounds by polynomials. In: Proceedings of the Annual Symposium on Foundations of Computer Science (FOCS) 1998, pp. 352–361. IEEE Computer Society Press (1998)

BDF⁺10. Boneh, D., Dagdelen, Ö., Fischlin, M., Lehmann, A., Schaffner, C., Zhandry, M.: Random oracles in a quantum world. Cryptology ePrint Archive, Report 2010/428 (2010), http://eprint.iacr.org/

Ber09. Bernstein, D.J.: Cost analysis of hash collisions: Will quantum computers make SHARCS obsolete? In: SHARCS 2009: Special-Purpose Hardware for Attacking Cryptographic Systems (2009)

BF11. Boneh, D., Freeman, D.M.: Homomorphic Signatures for Polynomial Functions. In: Paterson, K.G. (ed.) EUROCRYPT 2011. LNCS, vol. 6632, pp. 149–168. Springer, Heidelberg (2011)

BHK⁺11. Brassard, G., Høyer, P., Kalach, K., Kaplan, M., Laplante, S., Salvail, L.: Merkle Puzzles in a Quantum World. In: Rogaway, P. (ed.) CRYPTO 2011. LNCS, vol. 6841, pp. 391–410. Springer, Heidelberg (2011)

BHT98. Brassard, G., Høyer, P., Tapp, A.: Quantum Cryptanalysis of Hash and Claw-Free Functions. In: Lucchesi, C.L., Moura, A.V. (eds.) LATIN 1998. LNCS, vol. 1380, pp. 163–169. Springer, Heidelberg (1998)

BJ99. Bshouty, N.H., Jackson, J.C.: Learning DNF over the uniform distribution using a quantum example oracle. SIAM Journal on Computing 28(3), 1136–1153 (1999)

Boy10. Boyen, X.: Lattice Mixing and Vanishing Trapdoors: A Framework for Fully Secure Short Signatures and More. In: Nguyen, P.Q., Pointcheval, D. (eds.) PKC 2010. LNCS, vol. 6056, pp. 499–517. Springer, Heidelberg (2010)

BR93. Bellare, M., Rogaway, P.: Random oracles are practical: A paradigm for designing efficient protocols. In: Proc. of ACM Conference on Computers and Communication Security, pp. 62–73 (1993)

BS08. Brassard, G., Salvail, L.: Quantum Merkle Puzzles. In: Second International Conference on Quantum, Nano and Micro Technologies (ICQNM 2008), pp. 76–79 (February 2008)

Can01. Canetti, R.: Universally composable security: A new paradigm for cryptographic protocols. In: Proceedings of the Annual Symposium on Foundations of Computer Science (FOCS) 2001. IEEE Computer Society Press (2001), for an updated version see http://eprint.iacr.org

CGH98. Canetti, R., Goldreich, O., Halevi, S.: The random oracle methodology, revisited. In: Proceedings of the Annual Symposium on the Theory of Computing (STOC) 1998, pp. 209–218. ACM Press (1998)

CHKP10. Cash, D., Hofheinz, D., Kiltz, E., Peikert, C.: Bonsai Trees, or How to Delegate a Lattice Basis. In: Gilbert, H. (ed.) EUROCRYPT 2010. LNCS, vol. 6110, pp. 523–552. Springer, Heidelberg (2010)

CLRS10. Cayrel, P.-L., Lindner, R., Rückert, M., Silva, R.: Improved Zero-Knowledge Identification with Lattices. In: Heng, S.-H., Kurosawa, K. (eds.) ProvSec 2010. LNCS, vol. 6402, pp. 1–17. Springer, Heidelberg (2010)

Cor00.   Coron, J.-S.: On the Exact Security of Full Domain Hash. In: Bellare, M. (ed.) CRYPTO 2000. LNCS, vol. 1880, pp. 229–235. Springer, Heidelberg (2000)

DR03.   Dodis, Y., Reyzin, L.: On the Power of Claw-Free Permutations. In: Cimato, S., Galdi, C., Persiano, G. (eds.) SCN 2002. LNCS, vol. 2576, pp. 55–73. Springer, Heidelberg (2003)

FLR+10. Fischlin, M., Lehmann, A., Ristenpart, T., Shrimpton, T., Stam, M., Tessaro, S.: Random Oracles with(out) Programmability. In: Abe, M. (ed.) ASIACRYPT 2010. LNCS, vol. 6477, pp. 303–320. Springer, Heidelberg (2010)

FO99.   Fujisaki, E., Okamoto, T.: Secure Integration of Asymmetric and Symmetric Encryption Schemes. In: Wiener, M. (ed.) CRYPTO 1999. LNCS, vol. 1666, pp. 537–554. Springer, Heidelberg (1999)

FOPS01. Fujisaki, E., Okamoto, T., Pointcheval, D., Stern, J.: RSA-OAEP is Secure Under the RSA Assumption. In: Kilian, J. (ed.) CRYPTO 2001. LNCS, vol. 2139, pp. 260–274. Springer, Heidelberg (2001)

GGM86.   Goldreich, O., Goldwasser, S., Micali, S.: How to construct random functions. Journal of the ACM 33, 792–807 (1986)

GKV10.   Gordon, S.D., Katz, J., Vaikuntanathan, V.: A Group Signature Scheme from Lattice Assumptions. In: Abe, M. (ed.) ASIACRYPT 2010. LNCS, vol. 6477, pp. 395–412. Springer, Heidelberg (2010)

GMR88.   Goldwasser, S., Micali, S., Rivest, R.L.: A Digital Signature Scheme Secure Against Adaptive Chosen-Message Attacks. SIAM Journal on Computing 17(2), 281 (1988)

GPV08.   Gentry, C., Peikert, C., Vaikuntanathan, V.: Trapdoors for hard lattices and new cryptographic constructions. In: Proceedings of the Fourtieth Annual ACM Symposium on Theory of Computing - STOC 2008, p. 197 (2008)

Gro96.   Grover, L.K.: A fast quantum mechanical algorithm for database search. In: Proceedings of the Annual Symposium on the Theory of Computing (STOC) 1996, pp. 212–219. ACM (1996)

Gro98.   Grover, L.K.: Quantum Search on Structured Problems. In: Williams, C.P. (ed.) QCQC 1998. LNCS, vol. 1509, pp. 126–139. Springer, Heidelberg (1999)

HSS11.   Hallgren, S., Smith, A., Song, F.: Classical Cryptographic Protocols in a Quantum World. In: Rogaway, P. (ed.) CRYPTO 2011. LNCS, vol. 6841, pp. 411–428. Springer, Heidelberg (2011)

KW03.   Katz, J., Wang, N.: Efficiency improvements for signature schemes with tight security reductions. In: Proceedings of the 10th ACM Conference on Computer and Communication Security - CCS 2003, p. 155 (2003)

NC00.   Nielsen, M.A., Chuang, I.L.: Quantum Computation and Quantum Information. Cambridge University Press (2000)

NIS07.   NIST. National institute of standards and technology: Sha-3 competition (2007), http://csrc.nist.gov/groups/ST/hash/sha-3/

PS00.   Pointcheval, D., Stern, J.: Security arguments for digital signatures and blind signatures. Journal of Cryptology 13(3), 361–396 (2000)

PW08.   Peikert, C., Waters, B.: Lossy trapdoor functions and their applications. In: Proceedings of the 14th Annual ACM Symposium on Theory of Computing - STOC 2008, p. 187 (2008)

Reg02.   Regev, O.: Quantum computation and lattice problems. In: FOCS, pp. 520–529 (2002)

SG04.   Servedio, R.A., Gortler, S.J.: Equivalences and separations between quantum and classical learnability. SIAM Journal on Computing 33(5), 1067–1092 (2004)

Sho97.   Shor, P.: Polynomial-time algorithms for prime factorization and discrete logarithms on a quantum computer. SIAM J. Comput. 26(5), 1484–1509 (1997)

Unr10.   Unruh, D.: Universally Composable Quantum Multi-Party Computation. In: Gilbert, H. (ed.) EUROCRYPT 2010. LNCS, vol. 6110, pp. 486–505. Springer, Heidelberg (2010)

Wat09.   Watrous, J.: Zero-knowledge against quantum attacks. SIAM Journal on Computing 39(1), 25–58 (2009)

Lossy Encryption: Constructions from General Assumptions and Efficient Selective Opening Chosen Ciphertext Security

Brett Hemenway[1], Benoît Libert[2], Rafail Ostrovsky[3], and Damien Vergnaud[4]

[1] University of Michigan, USA
[2] Université catholique de Louvain, Belgium
[3] University of California, Los Angeles, USA
[4] École Normale Supérieure – C.N.R.S. - INRIA, France

Abstract. Lossy encryption was originally studied as a means of achieving efficient and composable oblivious transfer. Bellare, Hofheinz and Yilek showed that lossy encryption is also selective opening secure. We present new and general constructions of lossy encryption schemes and of cryptosystems secure against selective opening adversaries.

We show that *every* re-randomizable encryption scheme gives rise to efficient encryptions secure against a selective opening adversary. We show that statistically-hiding 2-round Oblivious Transfer implies Lossy Encryption and so do smooth hash proof systems. This shows that private information retrieval and homomorphic encryption *both* imply Lossy Encryption, and thus Selective Opening Secure Public Key Encryption.

Applying our constructions to well-known cryptosystems, we obtain selective opening secure commitments and encryptions from the *Decisional Diffie-Hellman*, *Decisional Composite Residuosity* and *Quadratic Residuosity* assumptions.

In an indistinguishability-based model of chosen-ciphertext selective opening security, we obtain secure schemes featuring short ciphertexts under standard number theoretic assumptions. In a simulation-based definition of chosen-ciphertext selective opening security, we also handle non-adaptive adversaries by adapting the Naor-Yung paradigm and using the perfect zero-knowledge proofs of Groth, Ostrovsky and Sahai.

Keywords: Public key encryption, commitment, lossy encryption, homomorphic encryption, selective opening, chosen-ciphertext security.

1 Introduction

In Byzantine agreement, and more generally in secure multiparty computation, it is often assumed that all parties are connected to each other via private channels. In practice, these private channels are implemented using a public-key cryptosystem. An adaptive adversary in a MPC setting, however, has very different powers than an adversary in an IND-CPA or IND-CCA game. In particular, an adaptive MPC adversary may view all the encryptions sent in a given round,

D.H. Lee and X. Wang (Eds.): ASIACRYPT 2011, LNCS 7073, pp. 70–88, 2011.

and then choose to corrupt a certain fraction of the players, thus revealing the decryptions of those players' messages *and the randomness used to encrypt them.* A natural question is whether the messages sent from the uncorrupted players remain secure. If the messages (and randomness) of all the players are chosen independently, then security in this setting follows from the IND-CPA security of the underlying encryption. If, however, the messages are not independent, the security does not immediately follow from the IND-CPA (or even IND-CCA) security of the underlying scheme. Although this problem was first investigated over twenty years ago, it remains an open question whether IND-CPA security implies this *selective opening* security.

Previous Work. There have been many attempts to design encryption protocols that can be used to implement secure multiparty computation against an adaptive adversary. The first protocols by Beaver and Haber [4] required interaction between the sender and receiver, required erasure and were fairly inefficient. The first non-interactive protocol was given by Canetti, Feige, Goldreich and Naor in [10]. In [10] the authors defined a new primitive called Non-Committing Encryption, and gave an example of such a scheme based on the RSA assumption. In [2], Beaver extended the work of [10], and created adaptively secure key exchange under the Diffie-Hellman assumption. In subsequent work, Damgård and Nielsen improved the efficiency of the schemes of Canetti *et al.* and Beaver, they were also able to obtain Non-Committing Encryption based on one-way trapdoor functions with invertible sampling. In [12], Canetti, Halevi and Katz presented a Non-Committing encryption protocols with evolving keys.

In [9], Canetti, Dwork, Naor and Ostrovsky extended the notion of Non-Committing Encryption to a new protocol which they called Deniable Encryption. In Non-Committing Encryption schemes there is a simulator, which can generate non-committing ciphertexts, and later open them to any desired message, while in Deniable Encryption, valid encryptions generated by the sender and receiver can later be opened to any desired message. The power of this primitive made it relatively difficult to realize, and Canetti *et al.* were only able to obtain modest examples of Deniable Encryption and left it as an open question whether fully deniable schemes could be created.

The notions of security against an adaptive adversary can also be applied to commitments. According to [21], the necessity of adaptively-secure commitments was realized by 1985. Despite its utility, until recently, relatively few papers directly addressed the question of commitments secure against a selective opening adversary (SOA). The work of Dwork, Naor, Reingold and Stockmeyer [21] was the first to explicitly address the problem. In [21], Dwork *et al.* showed that non-interactive SOA-secure commitments can be used to create a 3-round zero-knowledge proof systems for NP with negligible soundness error, and they gave constructions of a weak form of SOA-secure commitments, but left as an open question the existence of whether general SOA-secure commitments.

The question of SOA-secure commitments was put on firm foundations by Hofheinz [27] and Bellare, Hofheinz and Yilek in [5]. In [5], Bellare *et al.* provided simulation-based and indistinguishability-based definitions of security (these will be given the prefixes IND and SEM respectively) and gave a number of constructions and strong black-box separations, which indicated the difficulty of constructing selective opening secure commitments. Our results in the selective opening setting build on the breakthrough results of [5].

The independent work of Fehr, Hofheinz and Kiltz and Wee [23] also examines the case of CCA2 cryptosystems that are selective opening secure. In their work, they show how to adapt the universal hash proof systems of [17], to provide CCA2 security in the selective opening setting. Their constructions are general, and offer the first SEM-SO-CCA secure cryptosystem whose parameters are completely independent of n, the number of messages. Their work also considers selective opening security against chosen-plaintext attacks, and using techniques from Non-Committing Encryption [10] they construct SEM-SO-CPA secure systems from enhanced one-way trapdoor permutations.

Bellare, Waters and Yilek [7] show how to construct Identity-Based Encryption (IBE) schemes secure under selective-opening attacks. Our results are orthogonal to theirs. Their work constructs IBE schemes secure under selective-opening attacks, while our work starts with a tag-based encryption scheme, and uses it to construct encryption schemes that are secure against a selective-opening chosen-ciphertext attack, but are not identity-based.

Our Contributions. We primarily consider encryptions secure against a selective opening adversary. First we consider a selective-opening adversary who can mount a chosen-plaintext attack, and a the second part, we consider a selective-opening adversary who can mount a chosen-ciphertext attack.

Selective Opening Security Against Chosen-Plaintext Attacks. We formalize the notion of re-randomizable Public-Key Encryption and show that it implies Lossy Encryption [41,32,5]. Combining this with the observation (due to Bellare *et al.* [5]) that Lossy Encryption is IND-SO-CPA secure, we obtain an efficient construction of IND-SO-CPA secure encryption from any re-randomizable encryption (which generalizes and extends previous results). Moreover, these constructions retain the efficiency of the underlying re-randomizable cryptosystem.

Applying our results to the Paillier cryptosystem [39], we obtain an encryption scheme attaining a strong, simulation-based form of semantic security under selective openings (SEM-SO-CPA security). This is the first such construction from the Composite Residuosity (DCR) assumption. As far as bandwidth goes, it is also the most efficient SEM-SO-CPA secure encryption scheme to date. The possible use of Paillier as a lossy encryption scheme implicitly appears in [45]. To the best of our knowledge, its SEM-SO-CPA security was not reported earlier.

Next, we show that Lossy Encryption is also implied by (honest-receiver) statistically-hiding $\binom{2}{1}$-Oblivious Transfer and hash proof systems [17]. Combining this with the results of [42,41], we recognize that the relatively new Lossy Encryption primitive is essentially a different way to view the well-known statistically-hiding $\binom{2}{1}$-OT primitive. Applying the reductions in [5] to this result, yields constructions of SOA secure encryption from both private information retrieval (PIR) and homomorphic encryption.

These results show that the Lossy and Selective Opening Secure Encryption primitives (at least according to the latter's indistinguishability-based security definition), which have not been extensively studied until recently, are actually implied by several well-known primitives: $i.e.$, re-randomizable encryption, PIR, homomorphic encryption, hash proof systems and statistically-hiding $\binom{2}{1}$-OT. So far, the only known general constructions of lossy encryption were from lossy trapdoor functions. Our results show that they can be obtained from many seemingly weaker primitives (see figure 1).

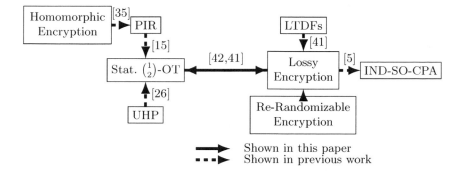

Fig. 1. Constructing Lossy Encryption

Selective Opening Security Against Chosen-Ciphertext Attacks: Continuing the study of selective-opening security, we present definitions chosen-ciphertext security (CCA2) in the selective opening setting (in both the indistinguishability and simulation-based models) and describe encryption schemes that provably satisfy these enhanced forms of security. Despite recent progress, relatively few methods are known for constructing IND-CCA2 cryptosystems in the standard model. The problem is even more complex with selective openings, where some known approaches for CCA2 security do not seem to apply. We note how the Naor-Yung paradigm, even when applied with statistical zero knowledge proofs fails to prove CCA2 security in the selective opening setting. Essentially, this is because the selective opening adversary learns the randomness used in the signature scheme, which allows him to forge signatures, and thus create ciphertexts that cannot be handled by the simulated decryption oracle.

The results of Fehr, Hofheinz, Kiltz and Wee [23] show how to modify universal hash proof systems [17] to achieve security under selective openings. We take a different approach and follow (a variant of) the Canetti-Halevi-Katz paradigm [11]. This too encounters many obstacles in the selective opening setting. Nevertheless, under standard assumptions (such as DDH or the Composite Residuosity assumption), we construct schemes featuring compact ciphertexts while resisting adaptive ($i.e.$, CCA2) chosen-ciphertext attacks according to our indistinguishability-based definition. When comparing our schemes to those of [23], we note that our public key size depends on n, the number of senders that can be possibly corrupted, while the systems of [23] are independent of n. On the other hand, to encrypt m-bit messages with security parameter λ, our ciphertexts are of length $\mathcal{O}(\lambda + m)$, while theirs are of length $\mathcal{O}(\lambda m)$. Our public-keys are longer than in [23] because our construction relies on All-But-N Lossy Trapdoor Functions (defined below), which have long description. The recent complementary work of Hofheinz [28] shows how to create All-But-Many Trapdoor Functions with short keys. Using his results in our construction eliminates the dependence of the public-key size on n. Regarding security definitions, our constructions satisfy an indistinguishability-based definition (IND-SO-CCA), whereas theirs fit a simulation-based definition (SEM-SO-CCA) which avoids the restriction on the efficient conditional re-sampleability of the message distribution.

The scheme of [23] is very different from ours and we found it interesting to investigate the extent to which well-known paradigms like [11] can be applied in the present context. Moreover, by adapting the Naor-Yung paradigm [38], under more general assumptions, we give a CCA1 construction that also satisfies a strong simulation-based notion of adaptive selective opening security.

One advantage of our IND-SO-CCA scheme is the ability to natively encrypt multi-bit messages. It is natural to consider whether our approach applies to the scheme of Bellare, Waters and Yilek [7] to achieve multi-bit IND-SO-CCA encryption. The scheme of [7], like [23], encrypts multi-bit messages in a bitwise manner. Applying a Canetti-Halevi-Katz-like transformation to the construction of [7] does not immediately yield IND-SO-CCA encryption schemes for multi-bit messages: the reason is that it is not clear how to prevent the adversary from reordering the bit encryptions without employing a one-time signature scheme.

2   Background

If $f : X \to Y$ is a function, for any subset $Z \subset X$, we let $f(Z) = \{f(x) : x \in Z\}$. If A is a PPT machine, then $a \overset{\$}{\leftarrow} A$ denotes the action of running A and obtaining an output a, which is distributed according to the internal randomness of A. Also, $\mathsf{coins}(A)$ denotes the distribution of A's internal randomness, so that the distribution $\{a \overset{\$}{\leftarrow} A\}$ is actually $\{r \overset{\$}{\leftarrow} \mathsf{coins}(A) : a = A(r)\}$. If R is a set, we use $r \overset{\$}{\leftarrow} R$ to denote sampling uniformly from R.

When λ is a security parameter, $\mathsf{negl}(\lambda)$ denotes the set of negligible functions (*i.e.*, which decrease faster than the inverse of any polynomial in λ). If X and Y are families of distributions indexed by λ, their statistical indistinguishability is written as $X \approx_s Y$. We write $X \approx_c Y$ to express that X and Y are computationally indistinguishable, *i.e.*, for all PPT adversaries A, for all polynomials p, then for all sufficiently large λ, we have $|\Pr[A^X = 1] - \Pr[A^Y = 1]| \in \mathsf{negl}(\lambda)$.

2.1   Selective Opening Secure Encryption

We recall the indistinguishability-based definition of encryption secure against a selective opening adversary, originally formalized in [5]. We define a real game and an ideal game which should be indistinguishable to any efficient adversary. The adversary receives *both* the messages and the randomness for his selection. This mirrors the fact that an adaptive MPC adversary learns the entire history of corrupted players (*i.e.*, there are no secure erasures). If the adversary receives only the messages this would reduce to standard CPA security.

As in the notations of [5], $\mathcal{M}$ denotes an n-message sampler outputting a n-vector $\mathbf{m} = (m_1, \ldots, m_n)$ of messages whereas $\mathcal{M}_{|I, \mathbf{m}[I]}$ denotes an algorithm that conditionally resamples another random n-vector $\mathbf{m}' = (m'_1, \ldots, m'_n)$ such that $m'_i = m_i$ for each $i \in I \subset \{1, \ldots, n\}$. If such a resampling can be done efficiently for all $I, \mathbf{m}$, then $\mathcal{M}$ is said to support efficient conditional resampling.

Definition 1. (*Indistinguishability under selective openings*). *A public key cryptosystem* (G, E, D) *is indistinguishable under selective openings (or IND-SO-CPA secure) if, for any message sampler* $\mathcal{M}$ *supporting efficient conditional resampling and any PPT adversary* $\mathcal{A} = (\mathcal{A}_1, \mathcal{A}_2)$, *we have*

$$\left| \Pr\left[\mathcal{A}^{\textit{ind-so-real}} = 1 \right] - \Pr\left[\mathcal{A}^{\textit{ind-so-ideal}} = 1 \right] \right| \in \mathsf{negl}(\lambda)$$

where the games ind-so-real *and* ind-so-ideal *are defined as follows.*

IND-SO-CPA (Real)	IND-SO-CPA (Ideal)
$\mathbf{m} = (m_1, \ldots, m_n) \overset{\$}{\leftarrow} \mathcal{M}$	$\mathbf{m} = (m_1, \ldots, m_n) \overset{\$}{\leftarrow} \mathcal{M}$
$r_1, \ldots, r_n \overset{\$}{\leftarrow} \mathsf{coins}(E)$	$r_1, \ldots, r_n \overset{\$}{\leftarrow} \mathsf{coins}(E)$
$(I, st) \overset{\$}{\leftarrow} \mathcal{A}_1\big(pk, E(m_1, r_1), \ldots$	$(I, st) \overset{\$}{\leftarrow} \mathcal{A}_1\big(pk, E(m_1, r_1), \ldots, E(m_n, r_n)\big)$
$\ldots, E(m_n, r_n)\big)$	$\mathbf{m}' = (m'_1, \ldots, m'_n) \overset{\$}{\leftarrow} \mathcal{M}_{|I, \mathbf{m}[I]}$
$b \overset{\$}{\leftarrow} \mathcal{A}_2\big(st, (m_i, r_i)_{i \in I}, \mathbf{m}\big)$	$b \overset{\$}{\leftarrow} \mathcal{A}_2\big(st, (m_i, r_i)_{i \in I}, \mathbf{m}'\big)$

In the real game, the challenger samples $\mathbf{m} = (m_1, \ldots, m_n) \overset{\$}{\leftarrow} \mathcal{M}$, chooses $r_1, \ldots, r_n \overset{\$}{\leftarrow} \mathsf{coins}(E)$ and sends $(E(m_1, r_1), \ldots, E(m_n, r_n))$ to $\mathcal{A}$ who responds with a subset $I \subset \{1, \ldots, n\}$ and obtains $\{r_i\}_{i \in I}$ as well as the entire vector $\mathbf{m} = (m_1, \ldots, m_n)$. Finally, $\mathcal{A}$ outputs a bit $b \in \{0, 1\}$.

In the ideal game, the challenger also samples $\mathbf{m} = (m_1, \ldots, m_n) \overset{\$}{\leftarrow} \mathcal{M}$, chooses $r_1, \ldots, r_n \overset{\$}{\leftarrow} \mathsf{coins}(E)$ and sends $(E(m_1, r_1), \ldots, E(m_n, r_n))$ to $\mathcal{A}$. The latter chooses a subset $I \subset \{1, \ldots, n\}$ and obtains $\{r_i\}_{i \in I}$. The only difference w.r.t. the real game is that, instead of revealing $\mathbf{m}$, the challenger samples a new vector $\mathbf{m}' \overset{\$}{\leftarrow} \mathcal{M}_{|I, \mathbf{m}[I]}$ and sends $\mathbf{m}'$ to $\mathcal{A}$. Eventually, $\mathcal{A}$ outputs a bit $b \in \{0, 1\}$.

This definition of IND-SO-CPA security (taken from [5]) does not allow the message distribution $\mathcal{M}$ to depend on the public key. However, all our proofs (as well as the proof that Lossy Encryption is IND-SO-CPA secure in [5]) go through essentially unchanged if $\mathcal{M}$ is allowed to depend on the public-key of the scheme. For consistency, we continue to use the definition of [5].

2.2   Lossy Encryption

Bellare *et al.* [5] define *Lossy Encryption*, expanding on the definitions of Dual-Mode Encryption [41] and Meaningful/Meaningless Encryption [32]. A 'lossy' (or 'messy' in the terminology of [41]) cryptosystem has two types of public keys which specify two different modes of operation. In the normal mode, encryption is injective, while in the lossy (or 'messy') mode, the ciphertexts generated by the encryption algorithm are independent of the plaintext. We also require that no efficient adversary can distinguish normal keys from lossy keys. Bellare *et al.* [5] introduce a property called *openability*, which allows a possibly inefficient algorithm to open a ciphertext generated under a lossy key to *any* plaintext.

Definition 2. *A* lossy public-key cryptosystem *is a tuple* (G, E, D) *such that*

- $G(1^\lambda, \mathsf{inj})$ *outputs keys* (pk, sk) *which are called* injective keys.
- $G(1^\lambda, \mathsf{lossy})$ *outputs keys* $(pk_{\mathsf{lossy}}, sk_{\mathsf{lossy}})$ *which are called* lossy keys.

Additionally, (G, E, D) *are efficient algorithms satisfying these properties:*

1. *We have* $\Pr[(pk, sk) \xleftarrow{\$} G(1^\lambda, \mathsf{inj}); \ r \xleftarrow{\$} \mathsf{coins}(E) : D(sk, E(pk, x, r)) = x] = 1$ *for all plaintexts* $x \in X$. *This property is called* correctness on injective keys.

2. Indistinguishability of keys. *In lossy mode, public keys are computationally indistinguishable from those in the injective mode. If* $\mathrm{proj} : (pk, sk) \mapsto pk$ *is the projection map, then* $\{\mathrm{proj}(G(1^\lambda), \mathsf{inj})\} \approx_c \{\mathrm{proj}(G(1^\lambda, \mathsf{lossy}))\}$.

3. Lossiness of lossy keys. *If* $(pk_{\mathsf{lossy}}, sk_{\mathsf{lossy}}) \xleftarrow{\$} G(1^\lambda, \mathsf{lossy})$, *for all* $x_0, x_1 \in X$, *the distributions* $E(pk_{\mathsf{lossy}}, x_0, R)$ *and* $E(pk_{\mathsf{lossy}}, x_1, R)$ *are statistically close.*

4. Openability. *If* $(pk_{\mathsf{lossy}}, sk_{\mathsf{lossy}}) \xleftarrow{\$} G(1^\lambda, \mathsf{lossy})$, *and* $r \xleftarrow{\$} \mathsf{coins}(E)$, *then for all* $x_0, x_1 \in X$ *with overwhelming probability, there exists* $r' \in \mathsf{coins}(E)$ *such that* $E(pk_{\mathsf{lossy}}, x_0, r) = E(pk_{\mathsf{lossy}}, x_1, r')$. *Hence, there is an unbounded algorithm* opener *that can open a lossy ciphertext to* any *plaintext.*

Although openability is implied by property (3), it is convenient to state it explicitly in terms of an algorithm. In [5], it was shown that, if the algorithm opener is efficient, then the encryption scheme is actually SEM-SO-CPA secure. We do not explicitly require schemes to be IND-CPA secure since semantic security follows from the indistinguishability of keys and lossiness of the lossy keys. In [5], it was shown that the IND-CPA secure cryptosystem based on Lossy Trapdoor Functions given in [42], is in fact a Lossy Encryption. Next, they proved that any Lossy Encryption scheme where the plaintext space admits a n-message sampler with efficient resampling is IND-SO-CPA secure.

3   Constructing Lossy Encryption Schemes

3.1   Re-Randomizable Encryption Implies Lossy Encryption

In many cryptosystems, given a ciphertext c and a public-key, it is possible to re-randomize c to a new ciphertext c' such that c and c' encrypt the same plaintext but are statistically independent. We call a public key cryptosystem given by algorithms (G, E, D) *statistically re-randomizable*[1] if

- (G, E, D) is semantically-secure in the standard sense (IND-CPA).
- There is negligible function ν, and an efficient function ReRand such that for all λ, pk, m, r_1 we have $\Delta(\{r_0 \xleftarrow{\$} \text{coins}(E) : E(pk, m, r_0)\}, \{r' \xleftarrow{\$} \text{coins}(\text{ReRand}) : \text{ReRand}(E(pk, m, r_1), r')\}) < \nu(\lambda)$.

Since re-randomization does not require any kind of group structure on the plain-text space or any method for combining ciphertexts, re-randomizable encryption appears to be a weaker primitive than homomorphic encryption, and all known homomorphic cryptosystems are re-randomizable.

Our first result is a simple lossy encryption system $(\bar{G}_{\text{inj}}, \bar{G}_{\text{lossy}}, \bar{E}, \bar{D})$ obtained from a statistically re-randomizable public-key cryptosystem (G, E, D).

- **Key Generation:** first, $\bar{G}(1^\lambda, \text{inj})$ generates $(pk, sk) \leftarrow G(1^\lambda)$. Then, it picks $r_0, r_1 \xleftarrow{\$} \text{coins}(E)$, computes $e_0 = E(pk, 0, r_0)$, $e_1 = E(pk, 1, r_1)$ and returns $(\bar{pk}, \bar{sk}) = ((pk, e_0, e_1), sk)$. Algorithm $\bar{G}(1^\lambda, \text{lossy})$ runs $G(1^\lambda)$, generating a pair (pk, sk). Then, it picks $r_0, r_1 \xleftarrow{\$} \text{coins}(E)$ and generates $e_0 = E(pk, 0, r_0)$, $e_1 = E(pk, 0, r_1)$. It returns $(\bar{pk}, \bar{sk}) = ((pk, e_0, e_1), sk)$.
- **Encryption:** $\bar{E}(\bar{pk}, b, r') = \text{ReRand}(pk, e_b, r')$ for $b \in \{0, 1\}$.
- **Decryption** $\bar{D}(\bar{sk}, c)$, simply outputs $D(sk, c)$.

It is not hard to show that this construction is a lossy encryption scheme, as formally proved in the full version of the paper. Although it only allows encrypting single bits, it can be easily modified to encrypt longer messages if the underlying cryptosystem is homomorphic and if the set of encryptions of zero can be almost uniformly sampled (the details are available in the full paper).

We also note that specific homomorphic cryptosystems such as Paillier [39] or Damgård-Jurik [20] provide more efficient constructions where multi-bit messages can be encrypted. In addition, as shown in the full version of the paper, the factorization of the modulus N provides a means for efficiently opening a lossy ciphertext to any plaintext. Thus this scheme is actually SEM-SO-CPA secure when instantiated with these cryptosystems. This provides the most efficient known examples of SEM-SO-CPA secure cryptosystems. Previously, the most efficient known SEM-SO-CPA secure construction was the Goldwasser-Micali cryptosystem [5] which can only encrypt single bits.

---

[1] This definition of re-randomizable encryption requires statistical re-randomization. It is possible to define re-randomizable encryption which satisfies perfect re-randomization (stronger) or computational re-randomization (weaker). Such definitions already exist in the literature (see for example [40,25,29,14]). Our constructions require statistical re-randomization, and do not go through under a computational re-randomization assumption.

3.2 Statistically-Hiding $\binom{2}{1}$-OT Implies Lossy Encryption

Honest-receiver two-round statistically-hiding $\binom{2}{1}$-oblivious transfer is a protocol between a sender Sen and a receiver $\mathsf{Rec} = (\mathsf{Rec}_q, \mathsf{Rec}_r)$. The former has two strings s_0, s_1 and the latter has a bit b. The receiver Rec_q generates a query q, which is sent to Sen, along with some state information sk. The sender evaluates $\mathsf{q}(s_0, s_1)$ and sends the result $\mathsf{rsp} = \mathsf{Sen}(\mathsf{q}, s_0, s_1)$ to Rec_r who uses sk to get s_b.

- **Correctness:** For all $s_0, s_1 \in \{0,1\}^k$, $b \in \{0,1\}$, there exists $\nu \in \mathsf{negl}(\lambda)$ s.t.

$$\Pr[(\mathsf{q}, sk) \xleftarrow{\$} \mathsf{Rec}_q(1^\lambda, b); \mathsf{rsp} \xleftarrow{\$} \mathsf{Sen}(\mathsf{q}, s_0, s_1) : \mathsf{Rec}_r(sk, \mathsf{rsp}) = s_b] \geq 1 - \nu(\lambda).$$

- **Receiver Privacy:** b remains computationally hidden from Sen's view. That is, we must have $\{(\mathsf{q}, sk) \xleftarrow{\$} \mathsf{Rec}_q(1^\lambda, 0) : \mathsf{q}\} \approx_c \{(\mathsf{q}, sk) \xleftarrow{\$} \mathsf{Rec}_q(1^\lambda, 1) : \mathsf{q}\}$, where the distributions are taken over the internal randomness of Rec_q.

- **Sender Privacy:** for any $b \in \{0,1\}$, for any strings s_0, s_1, s_0', s_1' such that $s_b = s_b'$ and any honest receiver's query $\mathsf{q} = \mathsf{Rec}_q(1^\lambda, b)$, it must hold that

$$\{(\mathsf{q}, sk) \xleftarrow{\$} \mathsf{Rec}_q(1^\lambda, b); \mathsf{rsp} \xleftarrow{\$} \mathsf{Sen}(\mathsf{q}, s_0, s_1) : \mathsf{rsp}\}$$

$$\approx_s \{(\mathsf{q}, sk) \xleftarrow{\$} \mathsf{Rec}_q(1^\lambda, b); \mathsf{rsp} \xleftarrow{\$} \mathsf{Sen}(\mathsf{q}, s_0', s_1') : \mathsf{rsp}\},$$

the distributions being taken over the internal randomness of Rec_q and Sen.

A two-round honest-receiver statistically-hiding $\binom{2}{1}$-OT $(\mathsf{Sen}, \mathsf{Rec})$ gives a lossy encryption as follows:

- **Key Generation:** Define $G(1^\lambda, \mathsf{inj}) = \mathsf{Rec}_q(1^\lambda, 0)$. Set $pk = \mathsf{q}$, and $sk = sk$. Define $G(1^\lambda, \mathsf{lossy}) = \mathsf{Rec}_q(1^\lambda, 1)$. Set $pk = \mathsf{q}$, and $sk = \bot$.

- **Encryption:** Define $E(pk, m, (r, r^*)) = \mathsf{Sen}(\mathsf{q}, m, r; r^*)$, where r^* is the randomness used in $\mathsf{Sen}(\mathsf{q}, m, r)$ and $r \xleftarrow{\$} \{0,1\}^{|m|}$ is a random string.

- **Decryption:** given $c = \mathsf{rsp}$ in injective mode, define $D(sk, \mathsf{rsp}) = \mathsf{Rec}_r(sk, \mathsf{rsp})$.

Lemma 1. *The scheme (G, E, D) forms a lossy encryption scheme.*

The (straightforward) proof of Lemma 1 can be found in the full version of this paper. Since single-server Private Information Retrieval (PIR) implies statistically-hiding OT [15], we find the following corollary.

Corollary 1. *One-round Single-Server PIR implies Lossy Encryption.*

Since homomorphic encryption implies PIR [33,35], the following result follows.

Corollary 2. *Homomorphic encryption implies Lossy Encryption.*

In the half simulation model, statistically hiding $\binom{2}{1}$-OT can rely [30,26] on smooth hash proof systems that fit a slight modification of the original definition [17] with suitable verifiability properties. In the honest-but-curious receiver setting (which suffices here), it was already noted in [26][Section 1.3] that ordinary hash proof systems are sufficient to realize $\binom{2}{1}$-OT. In the full version of the paper, we describe a simplification of the construction of lossy encryption from hash proof systems and obtain the next result.

Corollary 3. *Smooth projective hash functions imply Lossy Encryption.*

To summarize this section, since lossy encryption is selective-opening secure, we obtain the following theorem.

Theorem 1. *Statistically-hiding 2-round honest-receiver $\binom{2}{1}$-OT, single server PIR, smooth projective hash proof systems and homomorphic encryption all imply IND-SO-CPA secure encryption.*

4   Chosen-Ciphertext Security

When an adversary has access to a decryption oracle, many cryptosystems become insecure. The notion of chosen-ciphertext security [38,43,19] was created to address this issue and, since then, many schemes have achieved this security level. The attacks of Bleichenbacher on RSA PKCS#1 [6] emphasized the importance of security against chosen-ciphertext attacks (CCA).

The need for selective opening security was first recognized in the context of Multi-Party Computation (MPC), where an active MPC adversary can view all ciphertexts sent in a current round and then choose a subset of senders to corrupt. It is natural to imagine an adversary who, in addition to corrupting a subset of senders, can also mount a chosen-ciphertext attack against the receiver. Schemes proposed so far (based on re-randomizable encryption or described in [5]) are obviously insecure in this scenario.

In this section, we extend the notion of chosen-ciphertext security to the selective opening setting. As in the standard selective-opening setting, we can define security either by indistinguishability, or by simulatability. We will give definitions of security as well as constructions for both settings.

Classical techniques to acquire chosen-ciphertext security are delicate to use here. Handling decryption queries using the Naor-Yung paradigm [38] and non-interactive zero-knowledge proofs [44] is not straightforward as, when the adversary makes her corruption query, it should obtain the random coins that were used to produce NIZK proofs. Fehr, Hofheinz, Kiltz and Wee [23] showed how to use non-committing encryption [10] along with a modified hash proof system [17] to achieve chosen-ciphertext security in the selective opening setting in the simulation-based model (SEM-SO-CCA). Our work takes a different approach and seeks to apply the Canetti-Halevi-Katz paradigm [11]. As we shall see, adapting this methodology to the selective opening setting encounters a number of technical obstacles that need to be overcome.

4.1   Chosen-Ciphertext Security: Indistinguishability

We begin with the indistinguishability-based definition. We define a real game (ind-cca2-real) and an ideal game (ind-cca2-ideal). In both games, the challenger generates a key pair $(sk, pk) \leftarrow G(1^\lambda)$ and sends pk to $\mathcal{A}$. The adversary is then allowed to adaptively make the following types of queries.

- **Challenge Query:** let $\mathcal{M}$ be a message sampler. The latter samples a vector $\mathbf{m} = (m_1, \ldots, m_n) \overset{\$}{\leftarrow} \mathcal{M}$ and returns a vector containing n "target" ciphertexts $\mathbf{C} = (\mathbf{C}[1], \ldots, \mathbf{C}[n]) \leftarrow (E(pk, m_1, r_1), \ldots, E(pk, m_n, r_n))$.
- **Corrupt Query:** $\mathcal{A}$ chooses $I \subset \{1, \ldots, n\}$ and receives $\{(m_i, r_i)\}_{i \in I}$.
  - In ind-cca2-real, the challenger then sends $\{m_j\}_{j \notin I}$ to the adversary.
  - In ind-cca2-ideal, the challenger re-samples $\mathbf{m}' = (m_1', \ldots, m_n') \overset{\$}{\leftarrow} \mathcal{M}_{|I, \mathbf{m}[I]}$ (*i.e.*, so that $m_j' = m_j$ for each $j \in I$) and sends $\{m_j'\}_{j \notin I}$ to $\mathcal{A}$.
- **Decryption Queries:** $\mathcal{A}$ chooses a ciphertext C such that $C \neq \mathbf{C}[i]$ for each $i \in \{1, \ldots, n\}$ and sends C to the challenger which responds with $D(sk, C)$.

After polynomially-many queries, one of which is a challenge query and precedes the corrupt query (which is unique as well), the adversary outputs $b \in \{0, 1\}$.

Definition 3. *A public key cryptosystem is IND-SO-CCA2 secure if, for any polynomial n and any n-message sampler $\mathcal{M}$ supporting efficient conditional resampling, any PPT adversary $\mathcal{A}$ has negligibly different outputs in the real game and in the ideal game: for some negligible function ν, we must have*

$$\left| \Pr[\mathcal{A}^{ind\text{-}cca2\text{-}real} = 1] - \Pr[\mathcal{A}^{ind\text{-}cca2\text{-}ideal} = 1] \right| < \nu.$$

4.2   Chameleon Hash Functions

A chameleon hash function [34] $\mathcal{CMH} = (\mathsf{CMKg}, \mathsf{CMhash}, \mathsf{CMswitch})$ consists of an algorithm CMKg that, given a security parameter λ, outputs a key pair $(hk, tk) \overset{\$}{\leftarrow} \mathcal{G}(\lambda)$. The hashing algorithm outputs $y = \mathsf{CMhash}(hk, m, r)$ given the public key hk, a message m and random coins $r \in \mathcal{R}_{hash}$. On input of m, r, m' and the trapdoor key tk, the switching algorithm $r' \leftarrow \mathsf{CMswitch}(tk, m, r, m')$ outputs $r' \in \mathcal{R}_{hash}$ such that $\mathsf{CMhash}(hk, m, r) = \mathsf{CMhash}(hk, m', r')$. Collision-resistance mandates that it be infeasible to find pairs $(m', r') \neq (m, r)$ such that $\mathsf{CMhash}(hk, m, r) = \mathsf{CMhash}(hk, m', r')$ without knowing tk. Uniformity guarantees that the distribution of hashes is independent of the message m, in particular, for all hk, and m, m', the distributions $\{r \leftarrow \mathcal{R}_{hash} : \mathsf{CMHash}(hk, m, r)\}$ and $\{r \leftarrow \mathcal{R}_{hash} : \mathsf{CMHash}(hk, m', r)\}$ are identical. It is well-known that chameleon hashing can be based on standard number theoretic assumptions.

4.3   A Special Use of the Canetti-Halevi-Katz Paradigm

The Canetti-Halevi-Katz technique [11] allows building chosen-ciphertext secure cryptosystems from weakly secure identity-based or tag-based encryption scheme. A tag-based encryption scheme (TBE) [36,31] is a cryptosystem where the encryption and decryption algorithms take an additional input, named the *tag*, which is a binary string of appropriate length with no particular structure. A TBE scheme consists of a triple $\mathcal{TBE} = (\mathsf{TBEKg}, \mathsf{TBEEnc}, \mathsf{TBEDec})$ of efficient algorithms where, on input of a security parameter λ, TBEKg outputs a private/public key pair (pk, sk); TBEEnc is a randomized algorithm that outputs a ciphertext C on input of a public key pk, a string θ – called *tag* – and a message $m \in \mathsf{MsgSp}(\lambda)$; TBEDec$(sk, \theta, C)$ is the decryption algorithm that takes as input a secret key sk, a tag θ and a ciphertext C and returns a plaintext m or $\perp$. Associated with $\mathcal{TBE}$ is a plaintext space MsgSp. Correctness requires that for all $\lambda \in \mathbb{N}$, all key pairs $(pk, sk) \leftarrow$ TBEKg(1^λ), all tags θ and any plaintext $m \in \mathsf{MsgSp}(\lambda)$, it holds that TBEDec$(sk, \theta, \mathsf{TBEEnc}(pk, \theta, M)) = m$.

SELECTIVE OPENING SECURITY FOR TBE SCHEMES. In the selective opening setting, the weak CCA2 security definition of [31] can be extended as follows.

Definition 4. *A TBE scheme* $\mathcal{TBE} = (\mathsf{TBEKg}, \mathsf{TBEEnc}, \mathsf{TBEDec})$ *is selective-tag weakly* IND-SO-CCA2 *secure (or IND-SO-stag-wCCA2 secure) if, for any polynomial* n *and any* n-message sampler $\mathcal{M}$ *supporting efficient conditional re-sampling, any PPT adversary* $\mathcal{A}$ *produces negligibly different outputs in the real and ideal games, which are defined as follows.*

1. *The adversary* $\mathcal{A}$ *chooses* n *tags* $\theta_1^\star, \ldots, \theta_n^\star$ *and sends them to the challenger.*
2. *The challenger generates a key pair* $(sk, pk) \leftarrow$ TKEKg(1^λ) *and hands* pk *to* $\mathcal{A}$. *The latter then adaptively makes the following kinds of queries:*

   - ***Challenge Query:*** *let* $\mathcal{M}$ *be a message sampler for* $\mathsf{MsgSp}(\lambda)$. *The challenger samples* $(m_1, \ldots, m_n) \overset{\$}{\leftarrow} \mathcal{M}$ *and returns* $\mathbf{C} = (\mathbf{C}[1], \ldots, \mathbf{C}[n])$, *where* $\mathbf{C}[i] = \mathsf{TBEEnc}(pk, \theta_i^\star, m_i, r_i)$
   - ***Corrupt Query:*** $\mathcal{A}$ *chooses* $I \subset \{1, \ldots, n\}$ *and obtains* $\{(m_i, r_i)\}_{i \in I}$.
     - *In the real game, the challenger then sends* $\{m_j\}_{j \notin I}$ *to the adversary.*
     - *In the ideal game, the challenger re-samples* $(m_1', \ldots, m_n') \overset{\$}{\leftarrow} \mathcal{M}_{|I, \mathbf{m}[I]}$ *and reveals* $\{m_j'\}_{j \notin I}$.
   - ***Decryption Queries:*** $\mathcal{A}$ *sends a pair* (C, θ) *such that* $\theta \notin \{\theta_1^\star, \ldots, \theta_n^\star\}$. *The challenger replies with* TBEDec$(sk, \theta, C) \in \mathsf{MsgSp}(\lambda) \cup \{\perp\}$.

After polynomially-many queries, one of which is a challenge query, $\mathcal{A}$ *outputs* $b \in \{0, 1\}$. *Its advantage* $\mathbf{Adv}_{\mathcal{A}}^{\text{IND-SO-stag-wCCA2}}(\lambda)$ *is defined as in definition 3.*

At first, one may hope to obtain IND-SO-CCA2 security by applying the CHK method [11] to any IBE/TBE scheme satisfying some weaker level of selective opening security. Let $\mathcal{TBE} = (\mathsf{TBEKg}, \mathsf{TBEEnc}, \mathsf{TBEDec})$ be a secure TBE scheme in the sense of definition 4 and let $\Sigma = (\mathcal{G}, \mathcal{S}, \mathcal{V})$ be a strong one-time signature. The CHK technique turns $\mathcal{TBE}$ into a cryptosystem $\mathcal{PKE} = (G, E, D)$ which is obtained by letting $G(1^\lambda)$ output $(sk', (\Sigma, pk'))$ where $(sk', pk') \leftarrow$ TBEKg(1^λ). To encrypt a message m, E generates a one-time signature key pair $(\mathsf{SK}, \mathsf{VK}) \leftarrow \mathcal{G}(1^\lambda)$, computes $C_{tbe} = \mathsf{TBEEnc}(pk, \mathsf{VK}, m)$ under the tag VK and sets the $\mathcal{PKE}$ ciphertext as $(\mathsf{VK}, C_{tbe}, \sigma)$, where $\sigma = \mathcal{S}(\mathsf{SK}, C_{tbe})$.

In the selective opening setting, when the adversary makes its corruption query in the reduction, it must obtain the random coins that were used to generate one-time signature keys appearing target ciphertexts. Then, it is able to re-compute the corresponding private keys and make decryption queries for ciphertexts involving the same verification keys as target ciphertexts, which causes the reduction to fail. Although schemes using one-time signatures do not appear to become trivially insecure, the reduction of [11,31] ceases to go through.

It was showed in [46] that chameleon hash functions [34] can be used to turn certain TBE schemes, termed *separable*, into full-fledged IND-CCA2 cryptosytems and supersede one-time signatures in the CHK transform. A TBE scheme is said *separable* if, on input of pk, m, θ, algorithm $\mathsf{TBEEnc}(pk, t, m)$ uses randomness $r \in \mathcal{R}_{tbe}$ and returns $C_{tbe} = (f_1(pk, m, r), f_2(pk, r), f_3(pk, \theta, r))$, where functions f_1, f_2 and f_3 are computed independently of each other and are all deterministic (so that they give the same outputs when queried twice on the same (m, r), r and (θ, r)). In addition, f_2 must be injective.

The construction of $[46]^2$ uses chameleon hashing instead of one-time signatures. Key generation requires to create a TBE key pair (pk', sk') and a chameleon hashing public key hk. The private key of $\mathcal{PKE}$ is the TBE private key sk'. Encryption and decryption procedures are depicted hereafter.

$E(m, pk)$	$D(sk, C)$
Parse pk as (pk', hk)	Parse C as (u, v, w, r_2) and sk as sk'
$r_1 \leftarrow \mathcal{R}_{tbe}$; $r_2 \leftarrow \mathcal{R}_{hash}$	$\theta = \mathsf{CMhash}(hk, u\|v, r_2)$
$u = f_1(pk', m, r_1)$; $v = f_2(pk', r_1)$	Return $m \leftarrow \mathsf{TBEDec}(sk', \theta, (u, v, w))$
$\theta = \mathsf{CMhash}(hk, u\|v, r_2)$	
$w = f_3(pk', \theta, r_1)$	
Return $C = (u, v, w, r_2)$	

Unlike the CHK transform, this construction computes C without using any other secret random coins than those of the underlying TBE ciphertext. The tag is derived from a ciphertext component u and some independent randomness r_2 that *publicly* appears in C. For this reason, we can hope to avoid the difficulty that appears with the CHK transform. Indeed, we prove that any separable TBE that satisfies definition 4 yields an IND-SO-CCA2 cryptosystem.

Theorem 2. *If* $\mathcal{TBE} = (\mathsf{TBEKg}, \mathsf{TBEEnc}, \mathsf{TBEDec})$ *is a separable TBE scheme with IND-SO-stag-wCCA2 security, the transformation of figure* ?? *gives an IND-SO-CCA2 PKE scheme.* (The proof is given in the full version of the paper).

4.4   Lossy and All-But-n Trapdoor Functions

A tuple $(S_{\mathrm{ltdf}}, F_{\mathrm{ltdf}}, F_{\mathrm{ltdf}}^{-1})$ of PPT algorithms is called a family of (d, k)-lossy trapdoor functions [42] if the following properties hold:

---

[2] As described in [46], the construction uses a single function F instead of f_1 and f_2 (*i.e.*, we are re-writing it in the particular case $F(m, r) = (f_1(pk, m, r), f_2(pk, r))$). The security proof of [46] implicitly requires F to be such that no two pairs $(m, r) \neq (m', r')$ give $F(m, r) = F(m', r')$. Using functions f_1, f_2 is a way to enforce this.

Sampling injective functions: $S_{\mathrm{ltdf}}(1^\lambda, 1)$ outputs (s, t), where s is a function index and t its trapdoor. It is required that $F_{\mathrm{ltdf}}(s, \cdot)$ be injective on $\{0, 1\}^d$ and $F^{-1}_{\mathrm{ltdf}}(t, F_{\mathrm{ltdf}}(s, x)) = x$ for all x.

Sampling lossy functions: $S_{\mathrm{ltdf}}(1^\lambda, 0)$ outputs $(s, \perp)$ where s is a function index and $F_{\mathrm{ltdf}}(s, \cdot)$ is a function on $\{0, 1\}^d$ with image size at most 2^{d-k}.

Indistinguishability: $\{(s, t) \overset{\$}{\leftarrow} S_{\mathrm{ltdf}}(1^\lambda, 1) : s\} \approx_c \{(s, \perp) \overset{\$}{\leftarrow} S_{\mathrm{ltdf}}(1^\lambda, 0) : s\}$.

Along with lossy trapdoor functions, Peikert and Waters [42] defined all-but-one (ABO) functions. These are lossy trapdoor functions, except instead of having two branches (a lossy branch and an injective branch) they have many branches coming from a branch set $\mathcal{B}$, all but one of which are injective.

The Peikert-Waters system only requires ABO functions to have one lossy branch because the IND-CCA2 game involves a single challenge ciphertext and a single ABO function must be evaluated on a lossy branch. Since the IND-SO-CCA security game involves $n > 1$ challenge ciphertexts, we need to generalize ABO functions into all-but-n (ABN) functions that have multiple lossy branches and where all branches except the specified ones are injective. A tuple $(S_{\mathrm{abn}}, G_{\mathrm{abn}}, G^{-1}_{\mathrm{abn}})$ is a family of ABN functions if these conditions are satisfied.

- **Sampling with a given lossy set:** For any n-subset $I \subset \mathcal{B}$, $S_{\mathrm{abn}}(1^\lambda, I)$ outputs s, t where s is a function index, and t its trapdoor. We require that for any $b \in \mathcal{B} \setminus I$, $G_{\mathrm{abn}}(s, b, \cdot)$ is an injective deterministic function on $\{0, 1\}^d$, and $G^{-1}_{\mathrm{abn}}(t, b, G_{\mathrm{abn}}(s, b, x)) = x$ for all x. Additionally, for each $b \in I$, the image $G_{\mathrm{abn}}(s, b, \cdot)$ has size at most 2^{d-k}.

- **Hidden lossy sets:** For any distinct n-subsets $I^\star_0, I^\star_1 \subset \mathcal{B}$, the first outputs of $S_{\mathrm{abn}}(1^\lambda, I^\star_0)$ and $S_{\mathrm{abn}}(1^\lambda, I^\star_1)$ are computationally indistinguishable.

Just as ABO functions can be obtained from lossy trapdoor functions [42], ABN functions can also be constructed from LTDFs and a general construction is provided in the full version of the paper. The recent results of Hofheinz [28], show how to create All-But-Many Lossy Functions, which are Lossy Trapdoor Functions with a super-polynomial number of lossy branches. The advantage of his construction is that the description of the function is independent of N. Hofheinz's All-But-Many functions can be plugged into our constructions to shrink the size of the public-key in our constructions (see [28] for details).

4.5   An IND-SO-stag-wCCA2 TBE Construction

We construct IND-SO-stag-wCCA2 tag-based cryptosystems from lossy trapdoor functions. Let (CMKg, CMhash, CMswitch) be a chameleon hash function where CMhash ranges over the set of branches $\mathcal{B}$ of the ABN family. We eventually obtain an IND-SO-CCA2 public key encryption scheme as a LTDF-based construction that mimics the one [42] (in its IND-CCA1 variant).

Let $(S_{\mathrm{ltdf}}, F_{\mathrm{ltdf}}, F^{-1}_{\mathrm{ltdf}})$ be a family of (d, k)-lossy-trapdoor functions, and let $(S_{\mathrm{abn}}, G_{\mathrm{abn}}, G^{-1}_{\mathrm{abn}})$ be a family of (d, k') all-but-n functions with branch set $\{0, 1\}^v$ where v is the length of a verification key for a one-time signature. We require that $2d - k - k' \le t - \kappa$, for $\kappa = \kappa(t) = \omega(\log t)$. Let $\mathcal{H}$ be a pairwise independent hash family from $\{0,1\}^d \to \{0,1\}^\ell$, with $0 < \ell < \kappa - 2\log(1/\nu)$, for some negligible $\nu = \nu(\lambda)$. The message space will be $\mathsf{MsgSp} = \{0,1\}^\ell$.

- $\mathsf{TBEKg}(1^\lambda)$: choose $h \xleftarrow{\$} \mathcal{H}$ in the pairwise independent hash family and generate $(s,t) \leftarrow S_{\mathrm{ltdf}}(1^\lambda, inj)$, $(s',t') \leftarrow S_{\mathrm{abn}}(1^\lambda, \{0,1,\dots,n-1\})$. The public key will be $pk = (s,s',h)$ and the secret key will be $sk = (t,t')$.

- $\mathsf{TBEEnc}(m, pk, \theta)$: to encrypt $m \in \{0,1\}^\ell$ under the tag $\theta \in \mathcal{B}$, choose $x \xleftarrow{\$} \{0,1\}^d$. Compute $c_0 = h(x) \oplus m$, $c_1 = F_{\mathrm{ltdf}}(s,x)$ and $c_2 = G_{\mathrm{abn}}(s, \theta, x)$ and the TBE ciphertext is $C = (c_0, c_1, c_2) = (h(x) \oplus m,\ F_{\mathrm{ltdf}}(s,x),\ G_{\mathrm{abn}}(s', \theta, x))$.

- $\mathsf{TBEDec}(C, sk, \theta)$: given $C = (c_0, c_1, c_2)$ and $sk = t$, compute $x = F_{\mathrm{ltdf}}^{-1}(t, c_1)$ and $m = c_0 \oplus h(x)$ if $G_{\mathrm{abn}}(s, \theta, x) = c_2$. Otherwise, output $\bot$.

The scheme is separable since C is obtained as $c_0 = f_1(pk, m, x) = m \oplus h(x)$, $c_1 = f_2(pk, x) = F_{\mathrm{ltdf}}(s,x)$ and $c_2 = f_3(pk, \theta, x) = G_{\mathrm{abn}}(s', \theta, x)$.

Theorem 3. *The algorithms described above form an IND-SO-stag-wCCA2 secure tag-based cryptosystem assuming the security of the lossy and all-but-n families.* (The proof is given in the full version of the paper).

4.6 An All-But-n Function with Short Outputs

While generic, the all-but-n function described in the full version of the paper has the disadvantage of long outputs, the size of which is proportional to nk. Efficient all-but-one functions can be based on the Composite Residuosity assumption [22,3]. We show that the all-but-one function of [22,3] extends into an ABN function that retains short (*i.e.*, independent of n or k) outputs. Multiple lossy branches can be obtained using a technique that traces back to the work of Chatterjee and Sarkar [18] who used it in the context of identity-based encryption.

- **Sampling with a given lossy set:** given a security parameter $\lambda \in \mathbb{N}$ and the desired lossy set $I = \{\theta_1^\star, \dots, \theta_n^\star\}$, where $\theta_i^\star \in \{0,1\}^\lambda$ for each $i \in \{1, \dots, n\}$, let $\gamma \ge 4$ be a polynomial in λ.

  1. Choose random primes p, q s.t. $N = pq > 2^\lambda$.
  2. Generate a vector $\vec{U} \in (\mathbb{Z}_{N^{\gamma+1}}^*)^{n+1}$ as follows. Let $\alpha_{n-1}, \dots, \alpha_0 \in \mathbb{Z}_{N^\gamma}$ be coefficients of $P[T] = \prod_{i=1}^n (T - \theta_i^\star) = T^n + \alpha_{n-1}T^{n-1} + \dots + \alpha_1 T + \alpha_0$ in $\mathbb{Z}_{N^\gamma}[T]$ (note that $P[T]$ is expanded in $\mathbb{Z}_{N^\gamma}$ but its roots are all in $\mathbb{Z}_N^*$). Then, for each $i \in \{0, \dots, n\}$, set $U_i = (1+N)^{\alpha_i} a_i^{N^\gamma} \bmod N^{\gamma+1}$, where $(a_0, \dots, a_n) \xleftarrow{\$} (\mathbb{Z}_N^*)^{n+1}$ and with $\alpha_n = 1$.
  3. The evaluation key is $s' = \{N, \vec{U} = (U_0, \dots, U_n)\}$ and the domain of the function is $\{0, \dots, 2^{\gamma\lambda/2} - 1\}$. The trapdoor is $t' = \mathrm{lcm}(p-1, q-1)$.

- **Evaluation:** to evaluate $G_{\mathrm{abn}}(s', \theta, x)$, where $x \in \{0, \dots, 2^{\gamma\lambda/2} - 1\}$ and $\theta \in \{0,1\}^\lambda$, compute $c = \left(\prod_{j=0}^n U_i^{(\theta^i \bmod N^\gamma)} \right)^x \bmod N^{\gamma+1}$.

- **Inversion:** for a branch θ, $c = G_{\mathrm{abn}}(s', \theta, x)$ is a Damgård-Jurik encryption of $y = P(\theta)x \bmod N^\gamma$. Using $t' = \mathrm{lcm}(p-1, q-1)$, we apply the decryption algorithm of [20] to obtain $y \in \mathbb{Z}_{N^\gamma}$ and return $x = yP(\theta)^{-1} \bmod N^\gamma$.

As in [22,3], $G_{\mathrm{abn}}(s', \theta, \cdot)$ has image size smaller than N when $\theta \in I$ and it can be shown that $\tilde{H}_\infty\big(x|(G_{\mathrm{abn}}(s', \theta, x), N, \vec{U})\big) \geq \gamma\lambda/2 - \log(N)$.

We note that the ABN function $G_{\mathrm{abn}}(s', \theta, \cdot)$ is not injective for each branch $\theta \notin I$, but only for those such that $\gcd(P(\theta), N^\gamma) = 1$. However, the fraction of branches $\theta \in \{0,1\}^\lambda$ such that $\gcd(P(\theta), N^\gamma) \neq 1$ is bounded by $2/\min(p,q)$, which is negligible. Moreover, the proof of theorem 3 is not affected if the TBE scheme is instantiated with this ABN function and the LTDF of [22,3]. As explained in the full version of the paper, as long as factoring is hard (which is implied by the Composite Residuosity assumption), the adversary has negligible chance of making decryption queries w.r.t. to such a problematic tag θ.

Lemma 2. *The above ABN function is lossy set hiding under the Composite Residuosity assumption.* (The proof is given in the full version of the paper).

The above ABN function yields an IND-SO-CCA2 secure encryption scheme with ciphertexts of constant (*i.e.*, independent of n) size but a public key of size $\mathcal{O}(n)$. Encryption and decryption require $\mathcal{O}(n)$ exponentiations as they entail an ABN evaluation. On the other hand, the private key has $\mathcal{O}(1)$ size, which keeps the private storage very cheap. At the expense of sacrificing the short private key size, we can optimize the decryption algorithm by computing $x = G_{\mathrm{abn}}^{-1}(t', \theta, c_2)$ (instead of $x = F_{\mathrm{ltdf}}^{-1}(t, c_1)$) so as to avoid computing $G_{\mathrm{abn}}(s', \theta, x)$ in the forward direction to check the validity of ciphertexts. In this case, the receiver has to store $\alpha_0, \ldots, \alpha_{n-1}$ to evaluate $P(\theta)$ when inverting G_{abn}.

It is also possible to extend the DDH-based ABO function described in [42] into an ABN function. However, in the full version of the paper, we describe a more efficient lossy TBE scheme based on the DDH assumption.

Acknowledgements. We thank Yuval Ishai for suggesting a connection between Oblivious Transfer and Lossy Encryption.

Brett Hemenway was supported in part by NSF VIGRE Fellowship and NSF grants 0716835, 0716389, 0830803 and 0916574. Rafail Ostrovsky's research is supported in part by NSF grants 0830803, 09165174, 106527 and 1118126, US-Israel BSF grant 2008411, grants from OKAWA Foundation, IBM, Lockheed-Martin Corporation and the Defense Advanced Research Projects Agency through the U.S. Office of Naval Research under Contract N00014-11-1-0392. The views expressed are those of the author and do not reflect the official policy or position of the Department of Defense or the U.S. Government. Benoît Libert acknowledges the Belgian Fund for Scientific Research (F.R.S.-F.N.R.S.) for his "Chargé de recherches" fellowship and the BCRYPT Interuniversity Attraction Pole. Damien Vergnaud was supported in part by the European Commission through the ICT Program under contract ICT-2007-216676 ECRYPT II.

References

1. Boneh, D., Boyen, X.: Efficient Selective-ID Secure Identity-Based Encryption Without Random Oracles. In: Cachin, C., Camenisch, J.L. (eds.) EUROCRYPT 2004. LNCS, vol. 3027, pp. 223–238. Springer, Heidelberg (2004)
2. Beaver, D.: Plug and Play Encryption. In: Kaliski Jr., B.S. (ed.) CRYPTO 1997. LNCS, vol. 1294, pp. 75–89. Springer, Heidelberg (1997)
3. Boldyreva, A., Fehr, S., O'Neill, A.: On Notions of Security for Deterministic Encryption, and Efficient Constructions without Random Oracles. In: Wagner, D. (ed.) CRYPTO 2008. LNCS, vol. 5157, pp. 335–359. Springer, Heidelberg (2008)
4. Beaver, D., Haber, S.: Cryptographic Protocols Provably Secure Against Dynamic Adversaries. In: Rueppel, R.A. (ed.) EUROCRYPT 1992. LNCS, vol. 658, pp. 307–323. Springer, Heidelberg (1993)
5. Bellare, M., Hofheinz, D., Yilek, S.: Possibility and Impossibility Results for Encryption and Commitment Secure Under Selective Opening. In: Joux, A. (ed.) EUROCRYPT 2009. LNCS, vol. 5479, pp. 1–35. Springer, Heidelberg (2009)
6. Bleichenbacher, D.: Chosen Ciphertext Attacks Against Protocols Based on the RSA Encryption Standard PKCS #1. In: Krawczyk, H. (ed.) CRYPTO 1998. LNCS, vol. 1462, pp. 1–12. Springer, Heidelberg (1998)
7. Bellare, M., Waters, B., Yilek, S.: Identity-Based Encryption Secure Against Selective Opening Attack. In: Ishai, Y. (ed.) TCC 2011. LNCS, vol. 6597, pp. 235–252. Springer, Heidelberg (2011)
8. Bellare, M., Yilek, S.: Encryption schemes secure under selective opening attack. Cryptology ePrint Archive: Report 2009/101 (2009)
9. Canetti, R., Dwork, C., Naor, M., Ostrovsky, R.: Deniable Encryption. In: Kaliski Jr., B.S. (ed.) CRYPTO 1997. LNCS, vol. 1294, pp. 90–104. Springer, Heidelberg (1997)
10. Canetti, R., Feige, U., Goldreich, O., Naor, M.: Adaptively secure multi-party computation. In: STOC 1996, pp. 639–648. ACM Press (1996)
11. Canetti, R., Halevi, S., Katz, J.: Chosen-Ciphertext Security from Identity-Based Encryption. In: Cachin, C., Camenisch, J.L. (eds.) EUROCRYPT 2004. LNCS, vol. 3027, pp. 207–222. Springer, Heidelberg (2004)
12. Canetti, R., Halevi, S., Katz, J.: Adaptively-Secure, Non-Interactive Public-Key Encryption. In: Kilian, J. (ed.) TCC 2005. LNCS, vol. 3378, pp. 150–168. Springer, Heidelberg (2005)
13. Di Crescenzo, G., Ishai, Y., Ostrovsky, R.: Non-interactive and non-malleable commitment. In: STOC 1998. ACM (1998)
14. Canetti, R., Krawczyk, H., Nielsen, J.B.: Relaxing Chosen-Ciphertext Security. In: Boneh, D. (ed.) CRYPTO 2003. LNCS, vol. 2729, pp. 565–582. Springer, Heidelberg (2003)
15. Di Crescenzo, G., Malkin, T., Ostrovsky, R.: Single Database Private Information Retrieval Implies Oblivious Transfer. In: Preneel, B. (ed.) EUROCRYPT 2000. LNCS, vol. 1807, pp. 122–138. Springer, Heidelberg (2000)
16. Cramer, R., Shoup, V.: A Practical Public Key Cryptosystem Provably Secure Against Adaptive Chosen Ciphertext Attack. In: Krawczyk, H. (ed.) CRYPTO 1998. LNCS, vol. 1462, pp. 13–25. Springer, Heidelberg (1998)
17. Cramer, R., Shoup, V.: Universal Hash Proofs and a Paradigm for Adaptive Chosen Ciphertext Secure Public-Key Encryption. In: Knudsen, L.R. (ed.) EUROCRYPT 2002. LNCS, vol. 2332, pp. 45–64. Springer, Heidelberg (2002)

18. Chatterjee, S., Sarkar, P.: Generalization of the Selective-ID Security Model for HIBE Protocols. In: Yung, M., Dodis, Y., Kiayias, A., Malkin, T. (eds.) PKC 2006. LNCS, vol. 3958, pp. 241–256. Springer, Heidelberg (2006)
19. Dolev, D., Dwork, C., Naor, M.: Non-malleable cryptography. In: STOC 1991, pp. 542–552 (1991)
20. Damgård, I., Jurik, M.: A Generalisation, a Simplification and Some Applications of Paillier's Probabilistic Public-Key System. In: Kim, K.-c. (ed.) PKC 2001. LNCS, vol. 1992, pp. 119–136. Springer, Heidelberg (2001)
21. Dwork, C., Naor, M., Reingold, O., Stockmeyer, L.: Magic functions. J. of the ACM 50(6), 852–921 (2003)
22. Freeman, D.M., Goldreich, O., Kiltz, E., Rosen, A., Segev, G.: More Constructions of Lossy and Correlation-Secure Trapdoor Functions. In: Nguyen, P.Q., Pointcheval, D. (eds.) PKC 2010. LNCS, vol. 6056, pp. 279–295. Springer, Heidelberg (2010)
23. Fehr, S., Hofheinz, D., Kiltz, E., Wee, H.: Encryption Schemes Secure Against Chosen-Ciphertext Selective Opening Attacks. In: Gilbert, H. (ed.) EUROCRYPT 2010. LNCS, vol. 6110, pp. 381–402. Springer, Heidelberg (2010)
24. Groth, J., Ostrovsky, R., Sahai, A.: Perfect Non-Interactive Zero Knowledge for NP. In: Vaudenay, S. (ed.) EUROCRYPT 2006. LNCS, vol. 4004, pp. 339–358. Springer, Heidelberg (2006)
25. Groth, J.: Rerandomizable and Replayable Adaptive Chosen Ciphertext Attack Secure Cryptosystems. In: Naor, M. (ed.) TCC 2004. LNCS, vol. 2951, pp. 152–170. Springer, Heidelberg (2004)
26. Halevi, S., Tauman-Kalai, Y.: Smooth projective hashing and two-message oblivious transfer. Cryptology ePrint Archive, Report 2007/118 (2007)
27. Hofheinz, D.: Possibility and impossibility results for selective decommitments. Cryptology ePrint Archive, Report 2008/168 (2008)
28. Hofheinz, D.: All-but-many lossy trapdoor functions. Cryptology ePrint Archive: Report 2011/230 (2011)
29. Golle, P., Jakobsson, M., Juels, A., Syverson, P.: Universal Re-Encryption for Mixnets. In: Okamoto, T. (ed.) CT-RSA 2004. LNCS, vol. 2964, pp. 163–178. Springer, Heidelberg (2004)
30. Kalai, Y.T.: Smooth Projective Hashing and Two-Message Oblivious Transfer. In: Cramer, R. (ed.) EUROCRYPT 2005. LNCS, vol. 3494, pp. 78–95. Springer, Heidelberg (2005)
31. Kiltz, E.: Chosen-Ciphertext Security from Tag-Based Encryption. In: Halevi, S., Rabin, T. (eds.) TCC 2006. LNCS, vol. 3876, pp. 581–600. Springer, Heidelberg (2006)
32. Kol, G., Naor, M.: Cryptography and Game Theory: Designing Protocols for Exchanging Information. In: Canetti, R. (ed.) TCC 2008. LNCS, vol. 4948, pp. 320–339. Springer, Heidelberg (2008)
33. Kushilevitz, E., Ostrovsky, R.: Replication is not needed: Single database, computationally-private information retrieval. In: FOCS 1997, pp. 364–373 (1997)
34. Krawczyk, H., Rabin, T.: Chameleon signatures. In: Network and Distributed System Security Symposium, NDSS 2000 (2000)
35. Mann, E.: Private access to distributed information. Master's thesis, Technion - Israel Institute of Technology (1998)
36. MacKenzie, P., Reiter, M.K., Yang, K.: Alternatives to Non-Malleability: Definitions, Constructions, and Applications. In: Naor, M. (ed.) TCC 2004. LNCS, vol. 2951, pp. 171–190. Springer, Heidelberg (2004)

37. Naor, M., Pinkas, B.: Efficient oblivious transfer protocols. In: SODA 2001, pp. 448–457. ACM-SIAM (2001)
38. Naor, M., Yung, M.: Public-key cryptosystems provably secure against chosen ciphertext attacks. In: STOC 1990, pp. 427–437 (1990)
39. Paillier, P.: Public-Key Cryptosystems Based on Composite Degree Residuosity Classes. In: Stern, J. (ed.) EUROCRYPT 1999. LNCS, vol. 1592, pp. 223–238. Springer, Heidelberg (1999)
40. Prabhakaran, M., Rosulek, M.: Rerandomizable RCCA Encryption. In: Menezes, A. (ed.) CRYPTO 2007. LNCS, vol. 4622, pp. 517–534. Springer, Heidelberg (2007)
41. Peikert, C., Vaikuntanathan, V., Waters, B.: A Framework for Efficient and Composable Oblivious Transfer. In: Wagner, D. (ed.) CRYPTO 2008. LNCS, vol. 5157, pp. 554–571. Springer, Heidelberg (2008)
42. Peikert, C., Waters, B.: Lossy trapdoor functions and their applications. In: STOC 2008, pp. 187–196. ACM Press (2008)
43. Rackoff, C., Simon, D.R.: Non-Interactive Zero-Knowledge Proof of Knowledge and Chosen Ciphertext Attack. In: Feigenbaum, J. (ed.) CRYPTO 1991. LNCS, vol. 576, pp. 433–444. Springer, Heidelberg (1992)
44. Sahai, A.: Non-malleable non-interactive zero-knowledge, and adaptive chosen-ciphertext security. In: FOCS 1999, pp. 543–553 (1999)
45. Young, A., Yung, M.: Questionable Encryption And Its Applications. In: Dawson, E., Vaudenay, S. (eds.) Mycrypt 2005. LNCS, vol. 3715, pp. 210–221. Springer, Heidelberg (2005)
46. Zhang, R.: Tweaking TBE/IBE to PKE Transforms with Chameleon Hash Functions. In: Katz, J., Yung, M. (eds.) ACNS 2007. LNCS, vol. 4521, pp. 323–339. Springer, Heidelberg (2007)

Structure Preserving CCA Secure Encryption and Applications

Jan Camenisch[1], Kristiyan Haralambiev[2], Markulf Kohlweiss[3], Jorn Lapon[4], and Vincent Naessens[4]

[1] IBM Research Zürich, Switzerland
jca@zurich.ibm.com
[2] Computer Science Department, New York University, USA
kkh@cs.nyu.edu
[3] Microsoft Research Cambridge, UK
markulf@microsoft.com
[4] Katholieke Hogeschool Sint-Lieven, Ghent, Belgium
{jorn.lapon,vincent.naessens}@kahosl.be

Abstract. In this paper we present the first CCA-secure public key encryption scheme that is structure preserving, i.e., our encryption scheme uses only algebraic operations. In particular, it does not use hash-functions or interpret group elements as bit-strings. This makes our scheme a perfect building block for cryptographic protocols where parties for instance want to prove properties about ciphertexts to each other or to jointly compute ciphertexts. Our scheme is very efficient and is secure against adaptive chosen ciphertext attacks.

We also provide a few example protocols for which our scheme is useful. For instance, we present an efficient protocol for two parties, Alice and Bob, that allows them to jointly encrypt a given function of their respective secret inputs such that only Bob learns the resulting ciphertext, yet they are both ensured of the computation's correctness. This protocol serves as a building block for our second contribution which is a set of protocols that implement the concept of so-called oblivious trusted third parties. This concept has been proposed before, but no concrete realization was known.

Keywords: public-key encryption, structure preserving, oblivious trusted third party.

1 Introduction

Public key encryption and signature schemes have become indispensable building blocks for cryptographic protocols such as anonymous credential schemes, group signatures, anonymous voting schemes, and e-cash systems. In the design of such protocols, it is often necessary that one party be able to prove to another that it has correctly signed or encrypted a message without revealing the message and its signature or encryption. An efficient implementation of such proofs is possible if the signature and encryption schemes allows one to employ

D.H. Lee and X. Wang (Eds.): ASIACRYPT 2011, LNCS 7073, pp. 89–106, 2011.

generalized Schnorr [12] or Groth-Sahai proofs [21]. In the design of suitable signature and encryption schemes one should therefore stay within the realm of algebraic groups and not break the algebraic structures, for instance, by using hash-functions in an essential way.

When it comes to signature schemes, a designer can pick from a number of schemes that are suitable (e.g., [15,5,1]). For encryption schemes secure against adaptive chosen ciphertext attack (CCA) the situation is quite different. Two schemes that are somewhat suitable are the Camenisch-Shoup and the Cramer-Shoup encryption schemes [7,18], allowing for the verifiable encryption (and decryption) of discrete logarithms and group elements, respectively. Both these schemes make use of a cryptographic hash function to achieve security against chosen ciphertext attacks. These hash functions, unfortunately, prevent one from efficiently proving relations between the input and output of the encryption procedure. Such proofs, however, are an important feature in many advanced protocols. They are for instance required when two parties are to jointly encrypt (a function of) their respective inputs without revealing them or when a user is to prove knowledge of a ciphertext, e.g., as a part of a proof of knowledge of a leakage-resilient signature [22,20] (proving knowledge of a signature is a central tool in privacy-preserving protocols which so far is not possible for leakage-resilient signatures).

In this paper we present the first efficient *structure preserving* CCA secure encryption scheme. The term "structure-preserving" is borrowed from the notion of structure-preserving digital signatures [1]. An encryption scheme is called structure-preserving if its public keys, messages (plaintexts), and ciphertexts are group elements and the encryption and decryption algorithm consists only of group and pairing operations. We achieve structure preserving encryption by a novel implementation of the consistency check that ensures security against chosen ciphertext attacks. More precisely, we implement the consistency checks using a bilinear map between algebraic groups and embed all other ciphertext components in the pre-image group of that map. Our ciphertext consistency element(s) could be either one element in the target group or several group elements in the pre-image group. The former gives better efficiency, whereas the latter can be used in more scenarios, in particular those making use of Groth-Sahai proofs [21]. We prove our encryption scheme secure against chosen ciphertext attacks under the decisional linear assumption [6]. Our encryption scheme and protocols also support so-called labels [7] which are public messages attached to a ciphertext and are important in the scenario we consider in this paper to bind a decryption policy to the ciphertext.

Our new encryption scheme is well suited to build a variety of protocols. For instance, with our scheme the following protocol problems can be addressed which are common stumbling stones when designing advanced cryptographic protocols:

- Our scheme can be used in the construction of leakage-resilient signatures [20] which will then enable, for the first time, a user to efficiently prove knowledge of a leakage-resilient signature.

- A user, who is given a ciphertext and a Groth-Sahai proof that the ciphertext was correctly computed, is able to prove to a third party that it is in possession of such a ciphertext without revealing it.
- Two users can jointly compute a ciphertext (of a function) of two plaintexts such that neither party learns the plain text of the other party and only one of the parties learns the ciphertext.

The last problem typically appears in protocols that do some kind of conflict resolution via a trusted third party. Examples include anonymity lifting (revocation) in group signatures and in anonymous credential systems [14] and optimistic fair exchange [3]. In these scenarios, there are typically two parties, say Alice and Bob, who run a protocol with each other and then provide each other with ciphertexts that can in case of a mishap (such as abuse of anonymity, conflict, unfair abortion of the protocol, etc.) be presented to a third party for resolution by decryption. Hereby, it is of course important that (1) the trusted third party be involved in case of mishap only and (2) the parties can convince each other that the ciphertexts indeed contain the right information. Note that CCA security is crucial here, as the trusted third party effectively acts as a decryption oracle. So far, protocol designers have used verifiable encryption, which unfortunately has the disadvantage that both parties learn the ciphertext of the other party. Hence, Alice could for instance take Bob's ciphertext and bribe the TTP so that it would act normally for all decryption requests except when Bob's ciphertext is presented in which case the TTP would just ignore the request.

To address this problem Camenisch, Gross, and Heydt-Benjamin [10] propose the concept of *oblivious trusted third parties (OTP)*: here, such conflict resolution protocols are designed in such a way that the trusted third party is kept oblivious of the concrete instance of the conflict resolution protocol. This means if Bob goes to the TTP for resolution, he cannot possibly be discriminated as the TTP cannot tell whether it is contacted by Bob or some other person. Therefore, if the TTP would deny such requests too often, that would be known and so there is no reason for Bob to believe that the TTP will not resolve the conflict for him if need be. Unfortunately, Camenisch et al. only provide a high-level construction for such a protocol but do not present a concrete instantiation. Based on our new encryption scheme, we present the first concrete protocols that implement OTP.

We prove all our protocols secure under composable simulation-based security definitions [16,4,23].

Related Work. There is of course a lot of related work on encryption schemes, but our scheme is the first one that is structure preserving. Considering our second contribution, the protocols for oblivious trusted parties, the only related work is by Camenisch, Gross, and Heydt-Benjamin [10]. They introduced the concept of oblivious trusted third parties but, as we mentioned, do not provide any concrete protocol.

2   Structure Preserving Encryption

In this section, we define the notion of structure-preserving encryption and present the first instantiation of such a scheme. The term "structure-preserving" is borrowed from the notion of structure-preserving digital signatures [1], and, for encryption, represents the idea that ciphertexts are constructed purely using (bilinear) group operations.

Note that the well known Cramer-Shoup [17,18] and Camenisch-Shoup [7] encryption schemes are not structure preserving as they make use of a cryptographic hash function. Even the hash-free variant of Cramer-Shoup is not structure preserving; that is because its consistency check requires group elements to be interpreted as exponents, which is not a group operation. The details of a proof of knowledge of a hash-free ciphertext would depend on the group's internal structure, e.g., it might be based on so called double-discrete logarithm proofs [8], which are bit-wise and thus much less efficient than standard discrete logarithm representation proofs.

Definition 1. Structure Preserving Encryption. *An encryption scheme is said to be structure-preserving if (1) its public keys, messages, and ciphertexts consist entirely of elements of a bilinear group, (2) its encryption and decryption algorithm perform only group and bilinear map operations, and (3) it is provably secure against chosen-ciphertext attacks.*

2.1   Basic Notation

We work in a group $\mathbb{G}$ of prime order q generated by g and equipped with a non-degenerate efficiently computable bilinear map $\hat{e} : \mathbb{G} \times \mathbb{G} \to \mathbb{G}_T$. Also, recall the well-known DLIN assumption [6]:

Definition 2. *Decisional Linear Assumption (DLIN). Let $\mathbb{G}$ be a group of prime order q. For randomly chosen $g_1, g_2, g_3 \leftarrow \mathbb{G}$ and $r, s, t \leftarrow \mathbb{Z}_q$, the following two distributions are computationally indistinguishable:*

$$(\mathbb{G}, g_1, g_2, g_3, g_1^r, g_2^s, g_3^t) \approx (\mathbb{G}, g_1, g_2, g_3, g_1^r, g_2^s, g_3^{r+s}) .$$

2.2   Construction

We construct a structure-preserving encryption scheme secure under DLIN. The scheme shares some similarities with the Cramer-Shoup encryption and with the Linear Cramer-Shoup encryption described by Shacham [24], neither of which is structure-preserving (even for their hash-free variants).

For simplicity, we describe the scheme when encrypting a message that is a single group element in $\mathbb{G}$, but it is easily extended to encrypt vectors of group elements. The extension is presented in the full version of the paper. Also, our scheme supports labels. We consider the case when a label L is a single group element, but the scheme extends trivially for the case of a label which is a vector of group elements. Labels from the space $\{0, 1\}^*$ could be hashed to one or several group elements, though in such cases they have to be part of the statement rather than the witness for any NIZK proof.

- KeyGen(1^λ): Choose random group generators $g_1, g_2, g_3 \leftarrow \mathbb{G}^*$. For randomly chosen $\boldsymbol{\alpha} \leftarrow \mathbb{Z}_q^3$, set $h_1 = g_1^{\alpha_1} g_3^{\alpha_3}$ and $h_2 = g_2^{\alpha_2} g_3^{\alpha_3}$. Then, select $\boldsymbol{\beta}_0, \ldots, \boldsymbol{\beta}_5 \leftarrow \mathbb{Z}_q^3$, and compute $f_{i,1} = g_1^{\beta_{i,1}} g_3^{\beta_{i,3}}$, $f_{i,2} = g_2^{\beta_{i,2}} g_3^{\beta_{i,3}}$, for $i = 0, \ldots, 5$. Output $\mathsf{pk} = (g_1, g_2, g_3, h_1, h_2, \{f_{i,1}, f_{i,2}\}_{i=0}^5)$ and $\mathsf{sk} = (\boldsymbol{\alpha}, \{\boldsymbol{\beta}_i\}_{i=0}^5)$.
- Enc(pk, L, m): To encrypt a message m with a label L, choose random $r, s \leftarrow \mathbb{Z}_q$ and set

$$u_1 = g_1^r, \ u_2 = g_2^s, \ u_3 = g_3^{r+s}, \quad c = m \cdot h_1^r h_2^s,$$

$$v = \prod_{i=0}^3 \hat{e}(f_{i,1}^r f_{i,2}^s, u_i) \cdot \hat{e}(f_{4,1}^r f_{4,2}^s, c) \cdot \hat{e}(f_{5,1}^r f_{5,2}^s, L),$$

where $u_0 = g$. Output $\mathfrak{c} = (u_1, u_2, u_3, c, v)$.
- Dec$(sk, L, \mathfrak{c})$: Parse $\mathfrak{c}$ as (u_1, u_2, u_3, c, v). Then check whether

$$v \stackrel{?}{=} \prod_{i=0}^3 \hat{e}(u_1^{\beta_{i,1}} u_2^{\beta_{i,2}} u_3^{\beta_{i,3}}, u_i) \cdot \hat{e}(u_1^{\beta_{4,1}} u_2^{\beta_{4,2}} u_3^{\beta_{4,3}}, c) \cdot \hat{e}(u_1^{\beta_{5,1}} u_2^{\beta_{5,2}} u_3^{\beta_{5,3}}, L),$$

where $u_0 = g$. If the latter is unsuccessful, reject the ciphertext as invalid. Otherwise, output $m = c \cdot (u_1^{\alpha_1} u_2^{\alpha_2} u_3^{\alpha_3})^{-1}$.

Note that the ciphertext $\mathfrak{c} \in \mathbb{G}^4 \times \mathbb{G}_T$. Using the pairing randomization techniques of [2], $v \in \mathbb{G}_T$ can be replaced by six random group elements $v_0, \ldots, v_5 \in \mathbb{G}$ for which the following equation holds: $v = \prod_{i=0}^3 \hat{e}(v_i, u_i) \cdot \hat{e}(v_4, c) \cdot \hat{e}(v_5, L)$. This way, the ciphertext would consist only of elements in $\mathbb{G}$. The modification is straightforward and is described in the full version of this paper [11].

2.3 Correctness and Security

To observe the correctness of the decryption, note that

$$
\begin{aligned}
c \cdot (u_1^{\alpha_1} u_2^{\alpha_2} u_3^{\alpha_3})^{-1} &= m \cdot h_1^r h_2^s \cdot \left((g_1^r)^{\alpha_1} (g_2^s)^{\alpha_2} (g_3^{r+s})^{\alpha_3}\right)^{-1} \\
&= m \cdot (g_1^{\alpha_1} g_3^{\alpha_3})^r (g_2^{\alpha_2} g_3^{\alpha_3})^s \cdot \left((g_1^r)^{\alpha_1} (g_2^s)^{\alpha_2} (g_3^{r+s})^{\alpha_3}\right)^{-1} = m.
\end{aligned}
$$

The correctness of the validity element v can be verified similarly.

Next, we show the CCA security of the encryption scheme. Our security proof follows the high level idea of the Hash Proof System (HPS) paradigm [19]. Essentially, Lemma 1 says the "proof" π, which is used as a one-time pad for the encryption of the message, has a corresponding HPS which is 1-universal, whereas Lemma 2 shows that the "proof" φ, which constitutes the consistency check element, has a corresponding HPS that is 2-universal. To make the proof below more accessible to readers unfamiliar with the HPS paradigm, we opt for a self-contained proof which can be easily translated into the HPS framework.

Theorem 1. *If DLIN holds, the above public key encryption scheme is secure against chosen-ciphertext attacks (CCA).*

Proof sketch of Theorem 1: We proceed in a sequence of games. We start with a game where the challenger behaves like in the standard IND-CCA game (i.e., the challenge ciphertext is an encryption of m_b, for a randomly chosen bit b, where m_0, m_1 are messages given by the adversary), and end up with a game where the challenge ciphertext is an encryption of a message chosen uniformly at random from the message space. Then we show that all those games are computationally indistinguishable. Let W_i denote the event that the adversary $\mathcal{A}$ outputs b' such that $b = b'$ in Game i.

Game 0. This is the standard IND-CCA game. $Pr[W_0] = \frac{1}{2} + \mathsf{Adv}^{\mathcal{A}}(\lambda)$.

Game 1. For (m_0, m_1, L) chosen by the adversary, the challenge ciphertext $\mathfrak{c} = (\boldsymbol{u}, c, v)$ is computed using the "decryption procedure", i.e., $u_1 = g_1^r$, $u_2 = g_2^s$, $u_3 = g_3^{r+s}$, $c = m_b \cdot u_1^{\alpha_1} u_2^{\alpha_2} u_3^{\alpha_3}$ and $v = \prod_{i=0}^{3} \hat{e}(u_1^{\beta_{i,1}} u_2^{\beta_{i,2}} u_3^{\beta_{i,3}}, u_i) \cdot \hat{e}(u_1^{\beta_{4,1}} u_2^{\beta_{4,2}} u_3^{\beta_{4,3}}, c) \cdot \hat{e}(u_1^{\beta_{5,1}} u_2^{\beta_{5,2}} u_3^{\beta_{5,3}}, L)$. The change is only syntactical, so the two games produce the same distributions. $Pr[W_1] = Pr[W_0]$.

Game 2. The randomness vector $\boldsymbol{u} = (u_1, u_2, u_3)$ of the challenge ciphertext is computed as non-DLIN tuple, i.e., $u_1 = g_1^r$, $u_2 = g_2^s$, $u_3 = g_3^t$ where $r, s, t \leftarrow \mathbb{Z}_q$ and $r + s \neq t$. Game 1 and Game 2 are indistinguishable by DLIN. Therefore, $|\, Pr[W_2] - Pr[W_1]\,| = \mathsf{negl}(\lambda)$.

Game 3. First note that in the previous game, as well as in this one, any decryption query with "correct" ciphertext, i.e., which has a randomness vector a DLIN tuple, yields a unique plaintext. That is, regardless of the concrete choice of sk which matches pk seen by the adversary, such queries do not reveal any information about the secret key.

In this game, unlike the previous one, any decryption query with "malformed" ciphertext, i.e, which has a non-DLIN randomness vector $\widehat{\boldsymbol{u}}$, is rejected. Let's consider two cases:

- $(\widehat{\boldsymbol{u}}, \widehat{c}, \widehat{L}) = (\boldsymbol{u}, c, L)$. Such decryption query is rejected because it is either the challenge ciphertext (when $\widehat{v} = v$) or the verification predicate fails trivially (when $\widehat{v} \neq v$). So, this case is the same in Game 2 and Game 3.
- $(\widehat{\boldsymbol{u}}, \widehat{c}, \widehat{L}) \neq (\boldsymbol{u}, c, L)$. By Lemma 2, such decryption query is rejected in Game 2 with overwhelming probability, whereas in Game 3 it is always rejected.

As the number of decryption queries is polynomial, $|\, Pr[W_3] - Pr[W_2]\,| = \mathsf{negl}(\lambda)$.

Game 4. The challenge ciphertext encrypts a random message from the message space. Game 3 and Game 4 are (information theoretically) indistinguishable by Lemma 1. $Pr[W_4] = Pr[W_3]$.

In the last game, the challenger's choice b is independent from the ciphertext, so $Pr[W_4] = \frac{1}{2}$. Then, by the indistinguishability of the consecutive games $Pr[W_0] = \frac{1}{2} + \mathsf{negl}(\lambda)$, hence $\mathsf{Adv}^{\mathcal{A}}(\lambda) = \mathsf{negl}(\lambda)$. $\qquad\square$

Lemma 1 which we used in the above proof says that the one-time pad of the message, when computing the challenge ciphertext in Game 4, can be replaced by a random element. Whereas Lemma 2 shows that any decryption query with "malformed" ciphertext $\hat{c}$ is rejected with overwhelming probability because the adversary $\mathcal{A}$ can hardly do better than guess the correct validity element.

For the formulation and proof of the lemmas, let $g_1, g_2, g_3 \leftarrow \mathbb{G}^*$ and $u_1 = g_1^r$, $u_2 = g_2^s$, $u_3 = g_3^t$, where r, s, t are randomly chosen from $\mathbb{Z}_q$ and $r + s \neq t$. And for convenience, denote $z_1 = \mathrm{dlog}_g(g_1)$, $z_2 = \mathrm{dlog}_g(g_2)$, and $z_3 = \mathrm{dlog}_g(g_3)$.

Lemma 1. *For randomly chosen $\boldsymbol{\alpha} \leftarrow \mathbb{Z}_q^3$, let $h_1 = g_1^{\alpha_1} g_3^{\alpha_3}$, $h_2 = g_2^{\alpha_2} g_3^{\alpha_3}$, and $\pi = u_1^{\alpha_1} u_2^{\alpha_2} u_3^{\alpha_3}$. Then, for a randomly chosen $\psi \leftarrow \mathbb{G}$ it is true that the following distributions are equivalent: $(h_1, h_2, \pi) \equiv (h_1, h_2, \psi)$.*

Proof sketch of Lemma 1: Note that $h_1 = g^{\alpha_1 z_1 + \alpha_3 z_3}$ and $h_2 = g^{\alpha_2 z_2 + \alpha_3 z_3}$. Then, for the tuple (h_1, h_2, π) the following equation holds:

$$
\begin{pmatrix} z_1 & 0 & z_3 \\ 0 & z_2 & z_3 \\ r z_1 & s z_2 & t z_3 \end{pmatrix} \cdot \begin{pmatrix} \alpha_1 \\ \alpha_2 \\ \alpha_3 \end{pmatrix} = \begin{pmatrix} \mathrm{dlog}_g(h_1) \\ \mathrm{dlog}_g(h_2) \\ \mathrm{dlog}_g(\pi) \end{pmatrix}
$$

Denote the matrix with M. It has a determinant $\det(M) = z_1 z_2 z_3 (t - r - s)$ which is not equal to 0 due to the choice of the parameters. Therefore the matrix is invertible, and for any $\pi \in \mathbb{G}$, and fixed h_1, h_2, there exists a unique $\boldsymbol{x}$ which yields the tuple (h_1, h_2, π). □

Lemma 2. *Let $\hat{\boldsymbol{u}} = (\hat{u}_1, \hat{u}_2, \hat{u}_3)$ be any tuple such that $\hat{u}_1 = g_1^{\hat{r}}$, $\hat{u}_2 = g_2^{\hat{s}}$, and $\hat{u}_3 = g_3^{\hat{t}}$, for $\hat{r} + \hat{s} \neq \hat{t}$. And for randomly chosen $\boldsymbol{\beta}_0, \boldsymbol{\beta}_1, \ldots, \boldsymbol{\beta}_5 \leftarrow \mathbb{Z}_q^3$, let $f_{i,1} = g_1^{\beta_{i,1}} g_3^{\beta_{i,3}}$, $f_{i,2} = g_2^{\beta_{i,2}} g_3^{\beta_{i,3}}$, for $i = 0, \ldots, 5$. For any $\boldsymbol{m}$ and $\widehat{\boldsymbol{m}}$ in $\mathbb{G}^5$, let*

$$
\varphi = \prod_{i=0}^{5} \hat{e}(u_1^{\beta_{i,1}} u_2^{\beta_{i,2}} u_3^{\beta_{i,3}}, m_i) \quad and \quad \widehat{\varphi} = \prod_{i=0}^{5} \hat{e}((\hat{u}_1)^{\beta_{i,1}} (\hat{u}_2)^{\beta_{i,2}} (\hat{u}_3)^{\beta_{i,3}}, \widehat{m}_i),
$$

where $m_0 = \widehat{m}_0 = g$. Then, for any $\boldsymbol{m}$ and $\widehat{\boldsymbol{m}}$, $\boldsymbol{m} \neq \widehat{\boldsymbol{m}}$, it is true that the following two distributions are equivalent: $(\{f_{i,1} f_{i,2}\}_{i=0}^5, \varphi, \widehat{\varphi}) \equiv (\{f_{i,1} f_{i,2}\}_{i=0}^5, \varphi, \psi)$, where $\psi \leftarrow \mathbb{G}_T$ is randomly chosen.

Proof sketch of Lemma 2: Similarly to the proof of the previous lemma, let's define all variables which depend on $\{\boldsymbol{\beta}_i\}_{i=0}^5$ as the result of a constant matrix M multiplied by the vector $(\boldsymbol{\beta}_0^\top || \boldsymbol{\beta}_1^\top || \ldots || \boldsymbol{\beta}_5^\top)^\top$. For convenience, denote with $w_i = \mathrm{dlog}_g(m_i)$ and $\widehat{w}_i = \mathrm{dlog}_g(\widehat{m}_i)$, for $i = 1, \ldots, 5$. Then, we have:

$$
\begin{pmatrix}
z_1 & 0 & z_3 & - & - & - & \cdots & - & - & - \\
0 & z_2 & z_3 & - & - & - & \cdots & - & - & - \\
- & - & - & z_1 & 0 & z_3 & \cdots & - & - & - \\
- & - & - & 0 & z_2 & z_3 & \cdots & - & - & - \\
\vdots & \vdots & \vdots & \vdots & \vdots & \vdots & \ddots & \vdots & \vdots & \vdots \\
- & - & - & - & - & - & \cdots & z_1 & 0 & z_3 \\
- & - & - & - & - & - & \cdots & 0 & z_2 & z_3 \\
rz_1 & sz_2 & tz_3 & w_1rz_1 & w_1sz_2 & w_1tz_3 & \cdots & w_5rz_1 & w_5sz_2 & w_5tz_3 \\
\widehat{r}z_1 & \widehat{s}z_2 & \widehat{t}z_3 & \widehat{w}_1\widehat{r}z_1 & \widehat{w}_1\widehat{s}z_2 & \widehat{w}_1\widehat{t}z_3 & \cdots & \widehat{w}_5\widehat{r}z_1 & \widehat{w}_5\widehat{s}z_2 & \widehat{w}_5\widehat{t}z_3
\end{pmatrix}
\cdot
\begin{pmatrix}
| \\
\beta_0 \\
| \\
\vdots \\
| \\
\beta_5 \\
|
\end{pmatrix}
=
\begin{pmatrix}
\mathrm{dlog}_g(f_{0,1}) \\
\mathrm{dlog}_g(f_{0,2}) \\
\mathrm{dlog}_g(f_{1,1}) \\
\mathrm{dlog}_g(f_{2,2}) \\
\vdots \\
\mathrm{dlog}_g(f_{5,1}) \\
\mathrm{dlog}_g(f_{5,2}) \\
\mathrm{dlog}(\varphi) \\
\mathrm{dlog}(\widehat{\varphi})
\end{pmatrix}.
$$

We would like to argue that the rows of the matrix M are linearly independent. As there exists i, $i \geq 1$, such that $m_i \neq \widehat{m}_i$, if we choose the sub-matrix M' consisting of the intersection of the last two rows and rows 1, 2, $2i+1$, $2i+2$ with columns 1, 2, 3, $3i+1$, $3i+2$, $3i+3$, we get:

$$
M' =
\begin{pmatrix}
z_1 & 0 & z_3 & 0 & 0 & 0 \\
0 & z_2 & z_3 & 0 & 0 & 0 \\
0 & 0 & 0 & z_1 & 0 & z_3 \\
0 & 0 & 0 & 0 & z_2 & z_3 \\
rz_1 & sz_2 & tz_3 & w_irz_1 & w_isz_2 & w_itz_3 \\
\widehat{r}z_1 & \widehat{s}z_2 & \widehat{t}z_3 & \widehat{w}_i\widehat{r}z_1 & \widehat{w}_i\widehat{s}z_2 & \widehat{w}_i\widehat{t}z_3
\end{pmatrix}.
$$

If the rows of M are not linearly independent, so are the rows of M'. However, M' has a determinant $\det(M') = \pm z_1^2 z_2^2 z_3^2 (w_i - \widehat{w}_i)(t - r - s)(\widehat{t} - \widehat{r} - \widehat{s})$ which is not equal to 0 due to choice of the parameters. Therefore, the rows of M are linearly independent. □

3 Secure Joint Ciphertext Computation

The CCA secure structure preserving encryptions scheme is well suited to build a variety of protocols. More specifically, it facilitates the construction of protocols that make use of practical ZK protocols to prove properties about partial ciphertexts. We consider a two-party protocol for the joint computation of a ciphertext under a third-party public key pk. The encrypted value is a function of two secrets, each of which remains secret from the other protocol participant. Moreover, only one participant gets to know the ciphertext. We study the case where only the first party learns the ciphertext whereas the second one has no output.

3.1 Preliminaries

Simulatability Model. We use strong simulation-based definitions that guarantee security under composition in the flavor of [16,4,23]. In particular we base our exposition on [23]. In [23] both ideal systems $\mathcal{I}$ and their realizations as cryptographic protocols $\mathcal{P}$ are configurations of multi-tape interactive Turing machines (ITMs). An ITM is triggered by another ITM if the latter writes a message on an output tape that corresponds to an input tape of the former. As a convention we bundle communication tapes into interfaces inf where an interface consists of named input/output tape pairs. An input/output tape pair is named inf.R after a combination of the interface name inf and a role name R. We refer to the set of all roles of an interface as inf.$\mathcal{R}$.

For simulation-based security definitions, the ideal system $\mathcal{I}$ and the protocol $\mathcal{P}$ that emulates this ideal system, have to present the same interface inf towards their environment, i.e., they must be *environment compatible*. We refer to an ideal system and a protocol that is environment compatible with respect to interface inf as $\mathcal{I}_{\mathsf{inf}}$ and $\mathcal{P}_{\mathsf{inf}}$, respectively. In addition $\mathcal{I}_{\mathsf{inf}}$ and $\mathcal{P}_{\mathsf{inf}}$ expose different network interfaces, the simulator interface $\mathsf{inf}_{\mathsf{Sim}}$ and the adversary interface $\mathsf{inf}_{\mathsf{Adv}}$, respectively.

Strong simulatability. A proof that $\mathcal{P}_{\mathsf{inf}}$ emulates $\mathcal{I}_{\mathsf{inf}}$, short $\mathcal{P}_{\mathsf{inf}} \leq^{SS} \mathcal{I}_{\mathsf{inf}}$ will need to prove existence of a simulator Sim that translates between the interfaces $\mathsf{inf}_{\mathsf{Sim}}$ and $\mathsf{inf}_{\mathsf{Adv}}$ such that for all p.p.t. Env: $\mathsf{Env}|\mathcal{P}_{\mathsf{inf}} \approx \mathsf{Env}|\mathsf{Sim}|\mathcal{I}_{\mathsf{inf}}$. This is formalized as *strong simulatability* which implies other simulatability notions such as universal composability with dummy adversaries and blackbox simulatability.

Corruption. We consider only static corruption. A corrupted role in the ideal and in the real world is controlled through $\mathsf{inf}_{\mathsf{Sim}}$.R and $\mathsf{inf}_{\mathsf{Adv}}$.R respectively, and acts as a proxy that allows the simulator, respectively, the environment to send messages to any of its other connected tapes. We consider ideal systems $\mathcal{I}_{\mathsf{inf}}$ that are fully described by a virtual incorruptible party $\mathcal{F}_{\mathsf{inf}}$. As the functionality $\mathcal{F}_{\mathsf{inf}}$ implements the security critical parts of an ideal system, the ITM's representing the different roles of the interface only need to implement forwarding and corruption. We refer to the dummy party of role R as $\mathcal{D}_{\mathsf{R}}$. When operating over an adversarially controlled network, even an ideal cryptographic system cannot prevent denial of service attacks. We therefor give the adversary the possibility to delay messages from the ideal functionality to dummies.

Practical Zero-Knowledge Proof of Knowledge Protocols. For the types of relations required in our protocols, there exist practical ZK protocols. We refer to Camenisch et al. [9,13] for details. We will be proving statements of the form $\exists\, w_1, \ldots, w_n : \phi(w_1, \ldots, w_n, bases)$ where w_i are exponents and ϕ is a predicate defining discrete logarithm representations. For a more detailed description, we refer to the full version of this paper.

We use a zero-knowledge ideal functionality as defined by Listing 1 that is a simplification of the $\mathcal{F}_{ZK}^{R,R'}$ functionality of [9] for which we consider only static corruption. This allows us to reuse their ZK protocol compiler to obtain efficient multi-session instantiations $\mathcal{P}_{\mathsf{zk}}$ of $\mathcal{I}_{\mathsf{zk}}(\mathfrak{R})$ in the hybrid secure channel and joint-state common reference string model.

Listing 1: Functionality $\mathcal{F}_{zk}(\mathfrak{R})$:

$\mathcal{F}_{zk}$ *receives input from* $\mathcal{D}_{Pv}$ *over* $\mathcal{F}_{zk}$.Pv *and provides output to* $\mathcal{D}_{Vf}$ *through the delayed communication tape* $\mathcal{F}_{zk}$.Vf. *Variable state is initialized to "ready".*

On (**Prove**, *inst*, *wit*) *from* $\mathcal{F}_{zk}$.Pv *where state = "ready" and* (*inst*, *wit*) $\in \mathfrak{R}$

- *let state = "final"; send* (**Prove**, *inst*) *to* $\mathcal{F}_{zk}$.Vf

---

Two-party computation. In conformance with the simulatability model discussed above, Listing 2 defines the ideal functionality for the joint computation of any function f on verifiable inputs inp_1 and inp_2. When performing such a two-party computation, party P_{1+i} is guaranteed that P_{2-i} knows a witness wit_{2-i} for its input inp_{2-i} such that $(inst, (wit_{2-i}, inp_{2-i})) \in \mathfrak{R}_{2-i}$. We restrict ourselves to tractable relations $\mathfrak{R}_i$ for which we can give efficient universally composable proofs of knowledge as described in the full paper.

Listing 2: Functionality $\mathcal{F}_{tpc}(f, \mathfrak{R}_1, \mathfrak{R}_2)$

$\mathcal{F}_{tpc}$ *communicates with* $\mathcal{D}_{P_1}$ *and* $\mathcal{D}_{P_2}$ *through delayed communication tapes* $\mathcal{F}_{tpc}$.P_1 *and* $\mathcal{F}_{tpc}$.P_2. *Variables inst, pub, inp_1 store the input of the first party; variable state is initialized to "ready".*

On (**Input**$_1$, *inst'*, *pub'*, *wit'$_1$*, *inp'$_1$*) *from* $\mathcal{F}_{tpc}$.P_1 *where state = "ready" and* (*inst'*, (*wit'$_1$*, *inp'$_1$*)) $\in \mathfrak{R}_1$

- *let inp_1 = inp'_1, inst = inst', pub = pub', and state = "input1"; send* (**Input**$_1$, *inst*, *pub*) *to* $\mathcal{F}_{tpc}$.P_2

On (**Input**$_2$, *wit$_2$*, *inp$_2$*) *from* $\mathcal{F}_{tpc}$.P_2 *where state = "input1" and* (*inst*, (*wit$_2$*, *inp$_2$*)) $\in \mathfrak{R}_2$

- *let state = "final"; send* (**Result**, $f(pub, inp_1, inp_2)$) *to* $\mathcal{F}_{tpc}$.P_1

We model an ideal secure two-party computation system $\mathcal{I}_{tpc}(f, \mathfrak{R}_1, \mathfrak{R}_2)$ with interface tpc as the combination of two dummy Parties $\mathcal{D}_{P_1}$ and $\mathcal{D}_{P_2}$ and an ideal two party computation functionality $\mathcal{F}_{tpc}$.

3.2   Construction

Model. The model of our joint ciphertext computation, is fully described by a secure two party computation as in Listing 2, where $inp_i = (l_i, \boldsymbol{x_i})$, $pub = \mathsf{pk}$, and f is $f_{JC}(\mathsf{pk}, (l_1, \boldsymbol{x}_1), (l_2, \boldsymbol{x}_2)) = \mathsf{Enc}(\mathsf{pk}, g^{l_1+l_2}, (g^{x_{1,1}+x_{1,2}}, \ldots, g^{x_{n,1}+x_{n,2}}))$.

Implementation. We present the protocol for the special case where the jointly computed ciphertext encrypts a single message (i.e., $n = 1$). This extends trivially in the multi-message case.

The idea of the protocol is as follows. The first party computes a partial and blinded encryption of her secret, she proves that the computation is carried out correctly, and sends the partial encryption to the other party. The second party takes the values from the first flow of the protocol and, using its secret and some randomness, computes a blinded full encryption of the agreed function of the two plaintext contributions. Then, the second party sends these values and proves that they are computed correctly. Finally, the first party unblinds the ciphertext and updates the consistency element to obtain a valid encryption of the function of the two secrets under jointly chosen randomness. The function can be a constant to the power of any polynomial of the two secrets; for simplicity, we consider the function $g^{x_1 + x_2}$ where g is a fixed group element and x_1, x_2 are the two secrets.

Listing 3: Protocol $\mathcal{P}_{\mathsf{jcc}}(\mathfrak{R}_1, \mathfrak{R}_2)$

Party P_1 and P_2 receive input from $\mathsf{jcc.P}_1$ *and* $\mathsf{jcc.P}_2$ *respectively and communicate over* $\mathcal{I}_{\mathsf{zk}_1}$ *and* $\mathcal{I}_{\mathsf{zk}_2}$.

On $(\mathtt{Input}_1, \mathit{inst}, \mathsf{pk}, \mathit{wit}_1, (l_1, x_1))$ *from* $\mathsf{jcc.P}_1$

- *if* $(\mathit{inst}, (\mathit{wit}_1, l_1, x_1)) \notin \mathfrak{R}_1$, P_1 *aborts*
- P_1 *computes* $(\mathit{msg}_1, \mathit{aux}_1) \leftarrow \mathsf{BlindEnc}_1(\mathsf{pk}, l_1, x_1)$ *and proves* $((\mathit{msg}_1, \mathsf{pk}, \mathit{inst}),$ $(\mathit{wit}_1, l_1, x_1, \mathit{aux}_1)) \in \mathfrak{R}_{\mathsf{P}_1}(\mathfrak{R}_1)$ *to* P_2 *using* $\mathcal{I}_{\mathsf{zk}_1}(\mathfrak{R}_{\mathsf{P}_1}(\mathfrak{R}_1))$
- P_2 *learns* $(\mathit{msg}_1, \mathsf{pk}, \mathit{inst})$ *from* $\mathcal{I}_{\mathsf{zk}_1}$ *and outputs* $(\mathtt{Input}_1, \mathit{inst}, \mathsf{pk})$ *to* $\mathsf{jcc.P}_2$

On $(\mathtt{Input}_2, \mathit{wit}_2, (l_2, x_2))$ *from* $\mathsf{jcc.P}_2$

- *if* $(\mathit{inst}, (\mathit{wit}_2, l_2, x_2)) \notin \mathfrak{R}_2$, P_2 *aborts*
- P_2 *runs* $(\mathit{msg}_2, \mathit{aux}_2) \leftarrow \mathsf{BlindEnc}_2(\mathsf{pk}, l_2, x_2, \mathit{msg}_1)$
- P_2 *proves* $((\mathit{msg}_2, \mathsf{pk}, \mathit{inst}), (\mathit{wit}_2, l_2, x_2, \mathit{aux}_2)) \in \mathfrak{R}_{\mathsf{P}_2}(\mathfrak{R}_2)$ *to* P_1 *using* $\mathcal{I}_{\mathsf{zk}_2}$ $(\mathfrak{R}_{\mathsf{P}_2}(\mathfrak{R}_2))$
- P_1 *learns* $(\mathit{msg}_2, \mathsf{pk}, \mathit{inst})$ *from* $\mathcal{I}_{\mathsf{zk}_2}$, *computes* $\mathfrak{c} \leftarrow \mathsf{UnblindEnc}(\mathsf{pk}, \mathit{msg}_2, \mathit{aux}_1)$, *and outputs* $(\mathtt{Result}, \mathfrak{c})$ *to* $\mathsf{jcc.P}_1$

Where abstractly, relations $\mathfrak{R}_{\mathsf{P}_1}(\mathfrak{R}_1)$ and $\mathfrak{R}_{\mathsf{P}_2}(\mathfrak{R}_2)$ are defined as

$$\mathfrak{R}_{\mathsf{P}_1}(\mathfrak{R}_1) = \{(\mathit{msg}_1, \mathsf{pk}, \mathit{inst}), (\mathit{wit}_1, l_1, x_1, \mathit{aux}_1)) \mid$$
$$\exists r : (\mathit{msg}_1, \mathit{aux}_1) = \mathsf{BlindEnc}_1(\mathsf{pk}, l_1, x_1; r) \wedge (\mathit{inst}, (\mathit{wit}_1, l_1, x_1)) \in \mathfrak{R}_1\}$$
$$\mathfrak{R}_{\mathsf{P}_2}(\mathfrak{R}_2) = \{((\mathit{msg}_2, \mathsf{pk}, \mathit{inst}), (\mathit{wit}_2, l_2, x_2, \mathit{aux}_2)) \mid$$
$$\exists r : (\mathit{msg}_2, \mathit{aux}_2) = \mathsf{BlindEnc}_2(\mathsf{pk}, l_2, x_2, \mathit{msg}_1; r) \wedge (\mathit{wit}_2, l_2, x_2)) \in \mathfrak{R}_2\} \,.$$

In the full paper, we show how to efficiently prove the relations $\mathfrak{R}_{\mathsf{P}_1}(\mathfrak{R}_1))$ and $\mathfrak{R}_{\mathsf{P}_2}(\mathfrak{R}_2))$ by giving a λ language statement.

We now give the details for the $\mathsf{BlindEnc}_1$, $\mathsf{BlindEnc}_2$, and $\mathsf{UnblindEnc}$ algorithms.

Listing 4: Algorithms of $\mathcal{P}_{\mathsf{jcc}}$

$(msg_1, aux_1) \leftarrow \mathsf{BlindEnc}_1(\mathsf{pk}, l_1, x_1)$
- parse pk as $(g_1, g_2, g_3, h_1, h_2, \{f_{i,1}, f_{i,2}\}_{i=0}^5)$.
- pick $\{\gamma_i\}_{i=1}^5$, $\{\delta_i\}_{i=1}^2$, r_1, and s_1 at random and compute

$$\bar{u}_1' = g^{\gamma_1} \cdot g_1^{r_1}, \qquad \bar{u}_2' = g^{\gamma_2} \cdot g_2^{s_1}, \qquad \bar{u}_3' = g^{\gamma_3} \cdot g_3^{r_1+s_1},$$
$$\bar{u}_4' = g^{\gamma_4} \cdot g^{x_1} \cdot h_1^{r_1} h_2^{s_1}, \qquad \bar{u}_5' = g^{\gamma_5} \cdot g^{l_1},$$
$$\bar{v}_1' = \hat{e}(g_1, g^{\delta_1}) \cdot \prod_{i=1} \hat{e}(f_{i,1}, g^{\gamma_i}), \qquad \bar{v}_2' = \hat{e}(g_2, g^{\delta_2}) \cdot \prod_{i=1} \hat{e}(f_{i,2}, g^{\gamma_i}).$$

- output $msg_1 = (\bar{u}_1', \bar{u}_2', \bar{u}_3', \bar{u}_4', \bar{u}_5', \bar{v}_1', \bar{v}_2')$
  and $aux_1 = (\{\gamma_i\}_{i=1}^5, \{\delta_i\}_{i=1}^2, r_1, s_1)$.

$(msg_2, aux_2) \leftarrow \mathsf{BlindEnc}_2(\mathsf{pk}, l_2, x_2, msg_1)$
- parse pk as $(g_1, g_2, g_3, h_1, h_2, \{f_{i,1}, f_{i,2}\}_{i=0}^5)$ and msg_1 as $(\bar{u}_1', \bar{u}_2', \bar{u}_3', \bar{u}_4', \bar{u}_5', \bar{v}_1', \bar{v}_2')$.
- pick r_2 and s_2 at random and compute

$$\bar{u}_1 = \bar{u}_1' \cdot g_1^{r_2}, \qquad \bar{u}_2 = \bar{u}_2' \cdot g_2^{s_2}, \qquad \bar{u}_3 = \bar{u}_3' \cdot g_3^{r_2+s_2},$$
$$\bar{u}_4 = \bar{u}_4' \cdot g^{x_2} \cdot h_1^{r_2} h_2^{s_2}, \qquad \bar{u}_5 = \bar{u}_5' \cdot g^{l_2},$$
$$\bar{v} = \left(\prod_{i=0} \hat{e}(f_{i,1}, \bar{u}_i)/\bar{v}_1'\right)^{r_2} \cdot \left(\prod_{i=0} \hat{e}(f_{i,2}, \bar{u}_i)/\bar{v}_2'\right)^{s_2},$$

where $\bar{u}_0 = g$.
- output $msg_2 = (\bar{u}_1, \bar{u}_2, \bar{u}_3, \bar{u}_4, \bar{u}_5, \bar{v})$ and $aux_2 = (r_2, s_2)$.

$\mathfrak{c} \leftarrow \mathsf{UnblindEnc}(\mathsf{pk}, msg_2, aux_1)$
- parse pk as $(g_1, g_2, g_3, h_1, h_2, \{f_{i,1}, f_{i,2}\}_{i=0}^5)$, msg_2 as $(\bar{u}_1, \bar{u}_2, \bar{u}_3, \bar{u}_4, \bar{u}_5, \bar{v})$ and $aux_1 = (\{\gamma_i\}_{i=1}^5, \{\delta_i\}_{i=1}^2, r_1, s_1)$.
- compute

$$u_1 = \bar{u}_1/g^{\gamma_1} = g_1^r, \qquad u_2 = \bar{u}_2/g^{\gamma_2} = g_2^s, \qquad u_3 = \bar{u}_3/g^{\gamma_3} = g_3^{r+s},$$
$$u_4 = \bar{u}_4/g^{\gamma_4} = g^{x_1+x_2} \cdot h_1^r h_2^s, \qquad u_5 = \bar{u}_5/g^{\gamma_5} = g^{l_1+l_2},$$
$$v = \bar{v} \cdot \hat{e}(u_1 g_1^{-r_1}, g^{\delta_1}) \cdot \hat{e}(u_2 g_2^{-s_1}, g^{\delta_2}) \cdot \prod_{i=0} \hat{e}(f_{i,1}^{r_1} f_{i,2}^{s_1}, u_i),$$

where $u_0 = g$.
- output $\mathfrak{c} = (u_1, u_2, u_3, u_4, v)$ encrypted with label u_5.

Correctness. Recall the structure of the ciphertext of the public-key encryption scheme described in Section 2: for a public key $\mathsf{pk} = (g_1, g_2, g_3, h_1, h_2, \{f_{i,1}, f_{i,2}\}_{i=0})$, label u_5, and randomly chosen $r, s \leftarrow \mathbb{Z}_q$, the ciphertext is computed as

$$(u_1, u_2, u_3, u_4, v) = \left(g_1^r, \; g_2^s, \; g_3^{r+s}, \; m \cdot h_1^r h_2^s, \; \prod_{i=0}^5 \hat{e}(f_{i,1}^r f_{i,2}^s, u_i)\right), \text{ where } u_0 = g.$$

Note that the protocol in Listing 3 computes a valid ciphertext because $u_1 = g_1^r$ for $r = r_1 + r_2$, $u_2 = g_2^s$ for $s = s_1 + s_2$, $u_3 = g_3^{r+s}$, $u_4 = m \cdot h_1^r h_2^s$ for $m = g^{x_1+x_2}$, and $v = \prod_{i=0} \hat{e}(f_{i,1}^r f_{i,2}^s, u_i)$. To see v is indeed computed this way, note that:

$$\bar{v} = \left(\prod_{i=0} \hat{e}(f_{i,1}, \bar{u}_i)/\bar{v}_1' \right)^{r_2} \cdot \left(\prod_{i=0} \hat{e}(f_{i,2}, \bar{u}_i)/\bar{v}_2' \right)^{s_2} = \frac{\prod_{i=0} \hat{e}(f_{i,1}^{r_2} f_{i,2}^{s_2}, u_i)}{\hat{e}(g_1, g^{\delta_1})^{r_2} \cdot \hat{e}(g_2, g^{\delta_2})^{s_2}}$$

and

$$\bar{v} \cdot \hat{e}\left(\frac{u_1}{g_1^{r_1}}, g^{\delta_1} \right) \cdot \hat{e}\left(\frac{u_2}{g_2^{s_1}}, g^{\delta_2} \right) = \bar{v} \cdot \hat{e}(g_1^{r_2}, g^{\delta_1}) \cdot \hat{e}(g_2^{s_2}, g^{\delta_2}) = \prod_{i=0} \hat{e}(f_{i,1}^{r_2} f_{i,2}^{s_2}, u_i).$$

Theorem 2. *The joint ciphertext computation protocol (Listing 3) strongly emulates the ideal two-party computation protocol (Listing 2) for function f_{JC}: $\mathcal{P}_{\mathsf{jcc}}(\mathfrak{R}_1, \mathfrak{R}_2) \leq^{SS} \mathcal{I}_{\mathsf{tpc}}(f_{\mathsf{JC}}, \mathfrak{R}_1, \mathfrak{R}_2)$. We refer to the full paper for details.*

4   Oblivious Third Parties

Modeling oblivious third parties. Transactions in the real world can be intricately related. They may depend on many conditions, of which the verification can be deferred to a number of (as oblivious as possible) third parties. For the sake of concreteness, we now formally model a system that involves two oblivious third parties: a satisfaction authority and a revocation authority. In our example scenario, after a service enrollment between a user U and a service provider SP, the user ought to make a payment for the service before t_{due}. Upon request, the satisfaction authority SA checks that the user indeed made the payment and provides the user with a blinded transaction token. The user unblinds the token and publishes it to prove the satisfaction of the payment. Finally, the revocation authority RA reveals the user's identity to the service provider if no payment has been made before the payment deadline (i.e. no token corresponding to the enrollment was published).

We model the security and privacy requirements of such a system with the help of an ideal functionality $\mathcal{F}_{\mathsf{otp}}$. As usual, corruption is modeled via dummies D_U, D_SP, D_SA, D_RA that allow to access the functionality both over the environment interface (before corruption) and the network interface (after corruption).

The ideal system $\mathcal{I}_{\mathsf{otp}}$ is depicted in Figure 1(a) and consists of the ideal functionality connected to the dummy parties over delayed communication tapes. Listing 5 specifies the reactive behavior of $\mathcal{F}_{\mathsf{otp}}$. A user that can prove his identity with the help of a witness such that $(inst, (id, wit)) \in \mathfrak{R}$, is allowed to enroll. In particular, this interface supports the case where wit and $inst$ are the secrets and the public key of a CL-signature [15] on the user's identity, i.e., an anonymous credential [14,5], or the opening and a commitment to the user's identity, i.e., a pseudonym [14]. For all these cases, the relation $\mathfrak{R}$ is tractable.

Enrollment consists of three rounds. The first round commits the user to her identity. The second round provides the user with a random satisfaction label

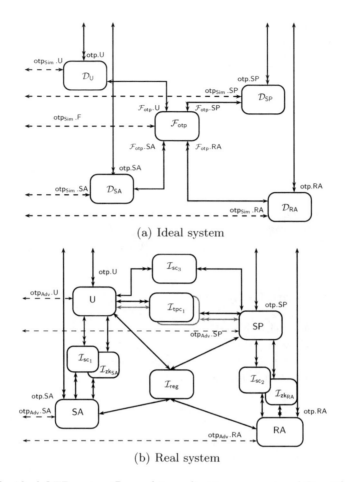

(a) Ideal system

(b) Real system

Fig. 1. The ideal OTP system $\mathcal{I}_{otp}$ and its realization as a protocol $\mathcal{P}_{otp}$: The realization makes use of ideal resources $\mathcal{I}_{sc_i}$, $\mathcal{I}_{zk_R}$, $\mathcal{I}_{reg}$, $\mathcal{I}_{jcc_i}$ for secure communication, proofs of knowledge, key registration, and joint ciphertext computation respectively.

with respect to which she can satisfy the condition, e.g., make the necessary payment. In this round the user is also made aware of the due date t_{due} for the payment. Note that the user has to check that t_{due} fulfills reasonable uniformity constraints to protect her privacy. The last round gives the service provider the possibility to ask the identity revocation authority for the user's identity. As a common limitation with other escrow mechanisms for anonymous credentials, we cannot extract the identity itself, but only the image of a bijection of it. We model this by giving the simulator the possibility to choose the bijection. As the identity space of realistic systems is small enough to allow for exhaustive search, this is not a serious limitation.

The client interface towards the ideal oblivious parties, i.e., the interface of the user and the service provider respectively, consists of two messages ReqAction

and `TestAction`, with `Action` $\in$ {`Satisfy`, `Open`}. The obliviousness requirement guarantees that oblivious parties do not learn anything about the transactions of their clients. Indeed the decision of an oblivious party cannot be
influenced in a transaction specific way, even if the other transaction participant
colludes with the oblivious party. This is modeled with the help of test requests
that are not related to any transaction. As these requests are indistinguishable
from real requests, they allow the user to check whether the oblivious party
indeed operates as required.[1]

Consequently, the decision of an oblivious party can only depend on explicit
and relevant information. For *satisfaction*, this is the user known satisfaction
label L with respect to which she makes her payment. For the *opening*, it is the
transaction token T that is secret until after satisfaction, when it is learned by
the user. We abstract from the way through which users make T available to
the revocation authority, but envision some kind of anonymous publicly available
bulletin board. It is in the responsibility of the user to make the token, learned
during satisfaction, available to RA, and in the responsibility of RA to check its
existence. All the protocol guarantees is that RA learns the same T value during
opening as the user learned during *satisfaction*.

Listing 5: Functionality $\mathcal{F}_{\text{otp}}$

Upon initialization, let state = "ready", $L = T = id = \widehat{T} = \widehat{id} = F = \mathbb{T} = \mathbb{L} = \epsilon$.
On (SetF, F', $\mathbb{T}'$, $\mathbb{L}'$) from otp$_{\text{Sim}}$.F where state = "ready":

  − *abort if F' is not an efficient bijection or $\mathbb{T}'$ or $\mathbb{L}'$ are not of sufficient size;
    set $F = F'$, $\mathbb{T} = \mathbb{T}'$, and $\mathbb{L} = \mathbb{L}'$.*

On (EnrollU, inst, (id', wit')) from $\mathcal{F}_{\text{otp}}$.U where state = "ready":

  − *if (inst, (id', wit')) $\notin \mathfrak{R}$) abort;*
  − *set state = "enrollu"; set id = id'; send (EnrollU, inst) to $\mathcal{F}_{\mathsf{T}}$.SP.*

On (DeliverEnrollU, t_{due}') from $\mathcal{F}_{\mathsf{T}}$.SP where state = "enrollu":

  − *set $t_{due} = t_{due}'$; set T, L to random values from $\mathbb{T}$ and $\mathbb{L}$ respectively;*
  − *set state = "deliverenrollu"; send (DeliverEnrollU, L, t_{due}) to $\mathcal{F}_{\text{otp}}$.U.*

On (DeliverEnrollSP) from $\mathcal{F}_{\text{otp}}$.U where state = "deliverenrollu":

  − *set state = "enrolled"; send (DeliverEnrollSP) to $\mathcal{F}_{\text{otp}}$.SP.*

On (ReqSatisfy) from $\mathcal{F}_{\text{otp}}$.U where $L \neq \epsilon$ and $\widehat{T} = \epsilon$:

  − *set $\widehat{T} = T$; send (ReqSatisfy, L) to $\mathcal{F}_{\text{otp}}$.SA.*

On (TestSatisfy, L', T') from $\mathcal{F}_{\text{otp}}$.U where $\widehat{T} = \epsilon$:

  − *set $\widehat{T} = T'$; send (ReqSatisfy, L') to $\mathcal{F}_{\text{otp}}$.SA.*

On (Satisfy, satisfied) from $\mathcal{F}_{\text{otp}}$.SA where $\widehat{T} \neq \epsilon$:

---

[1] An extension that allows not only the requester, but arbitrary external parties, e.g.
an auditor, to make test requests is a useful and cryptographically straightforward
extension to this interface.

- *if satisfied, set* $m = (\texttt{Satisfy}, \widehat{T})$, *otherwise set* $m = (\texttt{Satisfy}, \perp)$; *set* $\widehat{T} = \epsilon$; *send* m *to* $\mathcal{F}_{\mathsf{otp}}.\mathsf{U}$.

On $(\texttt{ReqOpen})$ *from* $\mathcal{F}_{\mathsf{otp}}.\mathsf{SP}$ *where state* $=$ *"enrolled" and* $\widehat{id} = \epsilon$:
- *set* $\widehat{id} = id$; *send* $(\texttt{ReqOpen}, T, t_{due})$ *to* $\mathcal{F}_{\mathsf{otp}}.\mathsf{RA}$.

On $(\texttt{TestOpen}, T', id', t_{due}')$ *from* $\mathcal{F}_{\mathsf{otp}}.\mathsf{SP}$ *where* $\widehat{id} = \epsilon$:
- *set* $\widehat{id} = id'$; *send* $(\texttt{ReqOpen}, T', t_{due}')$ *to* $\mathcal{F}_{\mathsf{otp}}.\mathsf{RA}$.

On $(\texttt{Open}, open)$ *from* $\mathcal{F}_{\mathsf{otp}}.\mathsf{RA}$ *where* $\widehat{id} \neq \epsilon$:
- *if open, set* $m = (\texttt{Open}, F(\widehat{id}))$, *otherwise set* $m = (\texttt{Open}, \perp)$; *set* $\widehat{id} = \epsilon$; *send* m *to* $\mathcal{F}_{\mathsf{otp}}.\mathsf{SP}$.

Implementing oblivious third parties. To construct a protocol that securely emulates the above functionality we make essential use of (adaptive chosen-ciphertext attack secure) encryption. As depicted in Figure 1(b) the protocol makes use of several cryptographic building blocks. But at the core of the protocol are two joint-ciphertext computations that, as described in Section 3, can be efficiently realized thanks to structure preserving encryption.

The enrollment protocol has a few more communication rounds, because of the zero-knowledge proofs, but otherwise closely follows the three phases of the ideal system. In the first phase the user commits to and proves her identity. Both the user and the service provider commit to randomness that they will use to jointly compute the transaction token T. The user proves knowledge of the opening of her commitment as part of the joint computation of the satisfaction ciphertext $\mathfrak{c}_1 = \mathsf{Enc}(pk_{\mathsf{SA}}, L, T \cdot g^r)$. In the second phase, the service provider transfers t_{due}, completes the joint ciphertext computation, and starts the computation of the revocation ciphertext $\mathfrak{c}_2 = \mathsf{Enc}(pk_{\mathsf{RA}}, g^{t_{due}}, (g^{id+r'}, T))$. In both cases, he proves knowledge of the opening to his commitment to guarantee that the transaction token is embedded correctly into both ciphertexts. The user outputs the label of $\mathfrak{c}_1$ as the random satisfaction label L. In the last phase the user again proves knowledge of openings for her commitments in the computation of $\mathfrak{c}_2$ to guarantee that it contains the transaction token T and a blinded user identity g^{id} under label $g^{t_{due}}$.

To satisfy her financial obligations, the user makes a payment with respect to label L and then asks the satisfaction authority to decrypt $\mathfrak{c}_1$. The user receives the blinded transaction token, that she unblinds using her locally stored randomness to learn T. She makes T available to the revocation authority, through some out-of-band anonymous bulletin board mechanism. Test satisfaction requests are just encryptions of blinded T' under label L'. To request the opening of a user identity, the service provider sends the ciphertext $\mathfrak{c}_2$ to the revocation authority, which checks the label t_{due}, decrypts the ciphertext to learn T and verifies whether T was posted by the user. If not, the revocation authority returns the blinded identity $g^{id+r'}$ to the service provider, which can unblind the identity. Test opening requests are just encryptions of T' and blinded $g^{id'}$ under label t_{due}'.

The Real System $\mathcal{P}_{\text{otp}}$. We omit the details of the protocol and refer to the full version for the description of $\mathcal{P}_{\text{otp}}$ and the proof that it securely emulates $\mathcal{F}_{\text{otp}}$.

5   Conclusion

We propose the first public key encryption scheme that is structure preserving and secure against adaptive chosen ciphertext attacks. We demonstrate the usefulness of this new primitive by the joint ciphertext computation protocol and our proposal for instantiating oblivious third parties. We conjecture, however, that the combination of the structure preserving encryption scheme and efficient zero-knowledge proofs facilitate a much larger set of efficient protocol constructions. All protocols are proven secure in the universal composability model.

Acknowledgements. Jorn Lapon is supported by the Interuniversity Attraction Poles Programme Belgian State, Belgian Science Policy and the Research Fund K.U.Leuven, and the IWT-SBO projects DiCoMas and MobCom.

References

1. Abe, M., Fuchsbauer, G., Groth, J., Haralambiev, K., Ohkubo, M.: Structure-Preserving Signatures and Commitments to Group Elements. In: Rabin, T. (ed.) CRYPTO 2010. LNCS, vol. 6223, pp. 209–236. Springer, Heidelberg (2010)
2. Abe, M., Haralambiev, K., Ohkubo, M.: Signing on group elements for modular protocol designs. Cryptology ePrint Archive, Report 2010/133 (2010), http://eprint.iacr.org
3. Asokan, N., Shoup, V., Waidner, M.: Optimistic fair exchange of digital signatures. IEEE Journal on Selected Areas in Communications 18(4), 591–610 (2000)
4. Backes, M., Pfitzmann, B., Waidner, M.: The reactive simulatability (rsim) framework for asynchronous systems. Inf. Comput. 205(12), 1685–1720 (2007)
5. Belenkiy, M., Chase, M., Kohlweiss, M., Lysyanskaya, A.: P-signatures and Noninteractive Anonymous Credentials. In: Canetti, R. (ed.) TCC 2008. LNCS, vol. 4948, pp. 356–374. Springer, Heidelberg (2008)
6. Boneh, D., Boyen, X., Shacham, H.: Short Group Signatures. In: Franklin, M. (ed.) CRYPTO 2004. LNCS, vol. 3152, pp. 41–55. Springer, Heidelberg (2004)
7. Camenisch, J., Shoup, V.: Practical Verifiable Encryption and Decryption of Discrete Logarithms. In: Boneh, D. (ed.) CRYPTO 2003. LNCS, vol. 2729, pp. 126–144. Springer, Heidelberg (2003)
8. Camenisch, J.: Group Signature Schemes and Payment Systems Based on the Discrete Logarithm Problem. Ph.D. thesis, ETH Zürich (1998)
9. Camenisch, J., Casati, N., Groß, T., Shoup, V.: Credential Authenticated Identification and Key Exchange. In: Rabin, T. (ed.) CRYPTO 2010. LNCS, vol. 6223, pp. 255–276. Springer, Heidelberg (2010)
10. Camenisch, J., Groß, T., Heydt-Benjamin, T.S.: Rethinking accountable privacy supporting services: extended abstract. In: ACM DIM – Digital Identity Management, pp. 1–8 (2008)

11. Camenisch, J., Haralambiev, K., Kohlweiss, M., Lapon, J., Naessens, V.: Structure preserving CCA secure encryption and its application to oblivious third parties. Cryptology ePrint Archive, Report 2011/319 (2011), http://eprint.iacr.org/
12. Camenisch, J., Kiayias, A., Yung, M.: On the Portability of Generalized Schnorr Proofs. In: Joux, A. (ed.) EUROCRYPT 2009. LNCS, vol. 5479, pp. 425–442. Springer, Heidelberg (2009)
13. Camenisch, J., Krenn, S., Shoup, V.: A framework for practical universally composable zero-knowledge protocols. Cryptology ePrint Archive, Report 2011/228 (2011), http://eprint.iacr.org/
14. Camenisch, J., Lysyanskaya, A.: An Efficient System for Non-Transferable Anonymous Credentials with Optional Anonymity Revocation. In: Pfitzmann, B. (ed.) EUROCRYPT 2001. LNCS, vol. 2045, pp. 93–118. Springer, Heidelberg (2001)
15. Camenisch, J., Lysyanskaya, A.: A Signature Scheme with Efficient Protocols. In: Cimato, S., Galdi, C., Persiano, G. (eds.) SCN 2002. LNCS, vol. 2576, pp. 268–289. Springer, Heidelberg (2003)
16. Canetti, R.: Universally composable security: A new paradigm for cryptographic protocols. In: FOCS, pp. 136–145 (2001)
17. Cramer, R., Shoup, V.: A Practical Public Key Cryptosystem Provably Secure Against Adaptive Chosen Ciphertext Attack. In: Krawczyk, H. (ed.) CRYPTO 1998. LNCS, vol. 1462, pp. 13–25. Springer, Heidelberg (1998)
18. Cramer, R., Shoup, V.: Design and analysis of practical public-key encryption schemes secure against adaptive chosen ciphertext attack. SIAM Journal on Computing 33, 167–226 (2001)
19. Cramer, R., Shoup, V.: Universal Hash Rroofs and a Paradigm for Adaptive Chosen Ciphertext Secure Public-Key Encryption. In: Knudsen, L.R. (ed.) EUROCRYPT 2002. LNCS, vol. 2332, pp. 45–64. Springer, Heidelberg (2002)
20. Dodis, Y., Haralambiev, K., López-Alt, A., Wichs, D.: Efficient Public-Key Cryptography in the Presence of Key Leakage. In: Abe, M. (ed.) ASIACRYPT 2010. LNCS, vol. 6477, pp. 613–631. Springer, Heidelberg (2010)
21. Groth, J., Sahai, A.: Efficient Non-Interactive Proof Systems for Bilinear Groups. In: Smart, N.P. (ed.) EUROCRYPT 2008. LNCS, vol. 4965, pp. 415–432. Springer, Heidelberg (2008)
22. Katz, J., Vaikuntanathan, V.: Signature Schemes with Bounded Leakage Resilience. In: Matsui, M. (ed.) ASIACRYPT 2009. LNCS, vol. 5912, pp. 703–720. Springer, Heidelberg (2009)
23. Küsters, R.: Simulation-based security with inexhaustible interactive turing machines. In: 19th IEEE Computer Security Foundations Workshop, pp. 309–320. IEEE (2006)
24. Shacham, H.: A cramer-shoup encryption scheme from the linear assumption and from progressively weaker linear variants. Cryptology ePrint Archive, Report 2007/074 (2007), http://eprint.iacr.org

Decoding Random Linear Codes in $\tilde{\mathcal{O}}(2^{0.054n})$

Alexander May[*], Alexander Meurer[**], and Enrico Thomae[***]

Faculty of Mathematics
Horst Görtz Institute for IT-Security
Ruhr-University Bochum, Germany
{alex.may,alexander.meurer,enrico.thomae}@rub.de

Abstract. Decoding random linear codes is a fundamental problem in complexity theory and lies at the heart of almost all code-based cryptography. The best attacks on the most prominent code-based cryptosystems such as McEliece directly use decoding algorithms for linear codes. The asymptotically best decoding algorithm for random linear codes of length n was for a long time Stern's variant of information-set decoding running in time $\tilde{\mathcal{O}}\left(2^{0.05563n}\right)$. Recently, Bernstein, Lange and Peters proposed a new technique called *Ball-collision decoding* which offers a speed-up over Stern's algorithm by improving the running time to $\tilde{\mathcal{O}}\left(2^{0.05558n}\right)$.

In this paper, we present a new algorithm for decoding linear codes that is inspired by a representation technique due to Howgrave-Graham and Joux in the context of subset sum algorithms. Our decoding algorithm offers a rigorous complexity analysis for random linear codes and brings the time complexity down to $\tilde{\mathcal{O}}\left(2^{0.05363n}\right)$.

Keywords: Information set decoding, representation technique.

1 Introduction

Linear codes have various applications in information theory and in cryptography. Many problems for random linear codes such as the so-called *syndrome decoding* are known to be NP-hard [2] and thus coding-based cryptography hopes to transfer this hardness to an average case hardness for cryptographic constructions. Since it is unlikely that hard coding problems are efficiently solvable on quantum computers, coding-based constructions are also one of the most prominent candidates for quantum-resistant cryptography.

Even many of today's lattice-based constructions like Regev's cryptosystem [12] or the HB protocol [7] inherently rely on the hardness of syndrome decoding via a variant called Learning Parity with Noise (LPN) problem. Given the importance of the syndrome decoding problem, it is a major task to understand its complexity in order to properly define cryptographic parameters that offer a

[*] Supported by DFG project MA 2536/7-1 and by ICT-2007-216676 ECRYPT II.
[**] Ruhr-University Research School Germany Excellence Initiative [DFG GSC 98/1].
[***] and by DFG through an Emmy Noether grant.

D.H. Lee and X. Wang (Eds.): ASIACRYPT 2011, LNCS 7073, pp. 107–124, 2011.

sufficient security level. Let us introduce some notion that helps to investigate the syndrome decoding problem for linear codes.

A binary linear $[n, k, d]$-code $\mathcal{C}$ of length n is a linear subspace of the vector space $\mathbb{F}_2^n$. The dimension k of $\mathcal{C}$ is the dimension of the subspace. The distance d of $\mathcal{C}$ is defined as the minimal Hamming distance between two codewords.

An $[n, k, d]$-code $\mathcal{C}$ can be defined via some basis matrix $\mathbf{G} \in \mathbb{F}_2^{k \times n}$ for the subspace, called a generator matrix, i.e. $\mathcal{C} = \{\mathbf{xG} : \mathbf{x} \in \mathbb{F}_2^k\}$. Alternatively, we can define $\mathcal{C}$ via a parity check matrix $\mathbf{H} \in \mathbb{F}_2^{(n-k) \times n}$ whose kernel equals $\mathcal{C}$, i.e. we have $\mathcal{C} = \{\mathbf{x} \in \mathbb{F}_2^n : \mathbf{Hx}^t = \mathbf{0}\}$. Moreover, let $\mathcal{C}$ have distance d and let $\mathbf{c} \in \mathcal{C}$ be a codeword. Assume that we transmit $\mathbf{x} = \mathbf{c} + \mathbf{e}$ for some error vector with Hamming weight $w := \mathrm{wt}(\mathbf{e}) \leq \lfloor \frac{d-1}{2} \rfloor$. Then $\mathbf{c}$ is the unique closest codeword in $\mathcal{C}$ to $\mathbf{x}$.

The term $\mathbf{s}(\mathbf{x}) := \mathbf{Hx}^t = \mathbf{H}(\mathbf{c}^t + \mathbf{e}^t) = \mathbf{He}^t$ is called the *syndrome* of $\mathbf{x}$. Notice that $\mathbf{e}$ defines the unique linear combination of exactly w columns of $\mathbf{H}$ that sum to $\mathbf{He}^t$ over $\mathbb{F}_2^t$. Finding this linear combination allows to recover the closest codeword $\mathbf{c} = \mathbf{x} + \mathbf{e}$. Hence, the so-called *syndrome decoding* of linear codes amounts to finding a subset I of ω out of n vectors from $\mathbb{F}_2^{n-k}$ such that the vectors in I sum to a fixed target value $\mathbf{s}(\mathbf{x})$.

A naive linear decoding algorithm is thus to search over all $\binom{n}{w}$ linear combinations of columns in $\mathbf{H}$. Obviously $w < \frac{n}{2}$, therefore the search space $\binom{n}{w}$ is maximal for w as large as possible. Thus, in coding based cryptosystems like McEliece [11] one usually fixes the weight of the error vector $\mathbf{e}$ to $w := \lfloor \frac{d-1}{2} \rfloor$. Throughout the paper, we assume for simplicity that we know w. We would like to stress that our decoding algorithm also works with the same asymptotical running time for unknown w, if we incorporate a loop over all possible values of w within the interval $(0, \lfloor \frac{d-1}{2} \rfloor]$, since our asymptotical running time is dominated by the largest value of w.

The running time of a decoding algorithm is a function of the three code parameters $[n, k, d]$. A random $[n, k, d]$-code is defined via a random parity check matrix $\mathbf{H} \in_R \mathbb{F}_2^{(n-k) \times n}$. It is well-known that for sufficiently large n random linear codes reach the so-called Gilbert-Varshamov bound (see [6], Chapter 2 for an introduction). More precisely, the code rate $\frac{k}{n}$ of a random linear code asymptotically reaches $1 - H(\frac{d}{n})$, where H is the binary entropy function. Solving for d allows us to express the asymptotical running time for random linear codes as a function of $[n, k]$ only. We obtain a worst case running time as a function of n if we take the maximum over all values of $0 \leq k \leq n$. For all decoding algorithms in this work the worst case appears for codes of rate $\frac{k}{n} \approx 0.47$.

Related Work. Let $\mathbf{s}(\mathbf{x}) = \mathbf{Hx}^t$ be the syndrome of some erroneous codeword $\mathbf{x} = \mathbf{c} + \mathbf{e}$ with $\mathbf{c} \in \mathcal{C}$ and weight-w error $\mathbf{e}$. We briefly show how to extract $\mathbf{e}$ from $\mathbf{s}(\mathbf{x})$ by an algorithm called *information set decoding*, that was already mentioned in the initial security analysis of McEliece [11] and further explored by Lee and Brickell [9].

The idea of information set decoding is to reduce the search space by linear algebra. The first step is to randomly permute the columns of $\mathbf{H}$, which basically permutes the coordinates of the error vector $\mathbf{e}$. Then, one transforms the permuted $\mathbf{H} \in \mathbb{F}_2^{(n-k) \times n}$ into systematic form $(\mathbf{Q} | \mathbf{I}_{n-k})$ with $\mathbf{Q} \in \mathbb{F}_2^{(n-k) \times k}$ and $\mathbf{I}_{n-k}$ the $(n - k)$-dimensional identity matrix. Next, one fixes a weight p and computes for all linear combinations of p columns in $\mathbf{Q}$ the sum with the given syndrome $\mathbf{s}(\mathbf{x})$. If this sum has Hamming weight exactly $w - p$, then we can simply choose another $w - p$ columns from the identity matrix $\mathbf{I}_{n-k}$ in order to obtain a weight-w linear combination of columns that sum to $\mathbf{s}(\mathbf{x})$.

Obviously, information set decoding succeeds if we permute $\mathbf{e}$ such that exactly p out of its w 1-entries are in the first k coordinates, and the remaining $(w - p)$ 1-entries fall into the last $n - k$ coordinates. Optimization of p yields a running time of $\tilde{\mathcal{O}}\left(2^{0.05751n}\right)$.

In 1988, Leon [10] and Stern [13] further improved information set decoding by enforcing a window of 0-entries of size ℓ in the last $n - k$ entries of $\mathbf{e}$. Assume that this length-ℓ window is e.g. in positions $k + 1, \ldots, k + \ell$ of $\mathbf{e}$. Then the weight-p linear combination of $\mathbf{Q}$ has to exactly match the syndrome $\mathbf{s}(\mathbf{x})$ in the ℓ positions $1, \ldots, \ell$, since we are no longer allowed to use the first ℓ columns from $\mathbf{I}_{n-k}$. Stern [13] proposed to compute those weight-p linear combinations of $\mathbf{Q}$ by a birthday technique via the sum of two disjoint weight-$\frac{p}{2}$ sums of columns in $\mathbf{Q}$. This algorithm lowers the time complexity to $\tilde{\mathcal{O}}\left(2^{0.05563n}\right)$ by increasing the memory complexity to $\tilde{\mathcal{O}}\left(2^{0.013n}\right)$.

In this work, we study a variant of Stern's information set decoding algorithm which is an instantiation of an algorithm by Finiasz and Sendrier from 2009 [5]. We call this instantiation FS-ISD. In FS-ISD, the 0-window is removed by simply removing the corresponding ℓ columns, i.e., by adjusting the systematic form to $\left(\mathbf{Q} \mid \begin{smallmatrix} \mathbf{0} \\ \mathbf{I}_{n-k-\ell} \end{smallmatrix}\right)$ with $\mathbf{Q} \in \mathbb{F}_2^{(n-k) \times (k+\ell)}$.

A different approach for removing the length-ℓ 0-window restriction in Stern's algorithm was recently proposed by Bernstein, Lange and Peters [4], called *Ball-collision decoding* by the authors. In Ball-collision decoding, one allows to have a small non-zero weight q in the length-ℓ window. Both algorithms, FS-ISD and Ball-collision decoding, share the same time complexity $\tilde{\mathcal{O}}\left(2^{0.05558n}\right)$ and memory complexity $\tilde{\mathcal{O}}\left(2^{0.014n}\right)$.

As a sideline of our work we show that any parameter choice (p, q, ℓ) for Ball-collision decoding can be transformed into parameters (p', ℓ') for the FS-ISD algorithm with the *same* asymptotic time complexity. That is, FS-ISD is asymptotically at least as efficient as Ball-collision decoding. We conjecture that both algorithms actually behave asymptotically equivalent. Since FS-ISD offers a simpler description than Ball-collision, we focus on improving the FS-ISD variant in this work.

Our Contribution.   We provide a new information set decoding algorithm based on FS-ISD. The major subproblem in FS-ISD is to find exactly p columns of an ℓ-row submatrix $\mathbf{Q}'$ of the $(n - k) \times (k + \ell)$ matrix $\mathbf{Q}$ that sum to the corresponding ℓ coordinates of the syndrome $\mathbf{s}(\mathbf{x})$.

More precisely, let $\mathbf{Q}' = [\mathbf{q}'_1 \ldots \mathbf{q}'_{k+\ell}]$ and $\mathbf{s}'(\mathbf{x})$ be the projections of $\mathbf{Q}$ and $\mathbf{s}(\mathbf{x})$ on the desired ℓ coordinates. Then we have to find an index set $I \subseteq \{1, \ldots, k + \ell\}$ with $|I| = p$ and $\sum_{i \in I} \mathbf{q}'_i = \mathbf{s}'(\mathbf{x})$. We call this problem the *submatrix matching problem*. Our improvement of information set decoding comes from a more efficient algorithm for the submatrix matching problem than the birthday algorithm of Stern. Our algorithm for the submatrix matching problem might be of independent interest as this problem is again a parametrized version of syndrome decoding.

In FS-ISD, the submatrix matching problem is solved by splitting the interval $[1, k + \ell]$ into the two disjoint intervals $[1, \frac{k+\ell}{2}]$ and $[\frac{k+\ell}{2} + 1, k + \ell]$. Then one searches in a birthday-type manner for two index sets $I_1 \subset [1, \frac{k+\ell}{2}]$ and $I_2 \subset [\frac{k+\ell}{2} + 1, k + \ell]$ of cardinality $\frac{p}{2}$ each, such that $\sum_{i \in I_1} \mathbf{q}'_i = \sum_{i \in I_2} \mathbf{q}'_i + \mathbf{s}'(\mathbf{x})$.

Our approach is inspired by a clever *representation technique* used in a recent subset sum algorithm of Howgrave-Graham and Joux from Eurocrypt 2010 [8]. We choose I_1 and I_2 in the submatrix matching problem both from the whole interval $[1, k + \ell]$ instead of taking two disjoint intervals of size $\frac{k+\ell}{2}$. Let I be a solution with $\sum_{i \in I} \mathbf{q}'_i = \mathbf{s}'(\mathbf{x})$ and $|I| = p$.

Then the major observation is that I has $\binom{p}{p/2}$ different representations of the form $I = I_1 \cup I_2$ with $|I_1| = |I_2| = \frac{p}{2}$. Thus, we also have $\binom{p}{p/2}$ identities of the form

$$\sum_{i \in I_1} \mathbf{q}'_i = \sum_{i \in I_2} \mathbf{q}'_i + \mathbf{s}'(\mathbf{x}), \tag{1}$$

instead of just one unique representation as in FS-ISD.

Interestingly, Finiasz and Sendrier also allow for non-disjoint splittings in [5]. However, their framework does not make use of different representations. It is precisely the representation technique that allow us to bypass their lower bound argument and to asymptotically beat the lower bound for information set decoding given in [5]. Our algorithms achieves an asymptotic running time of $\tilde{\mathcal{O}}\left(2^{0.05363n}\right)$ using memory $\tilde{\mathcal{O}}\left(2^{0.021n}\right)$.

The correctness of our algorithm is rigorously proven under the assumption that $\mathbf{H}$ is a uniformly random $\{0, 1\}$-matrix. This assumption is plausible in the cryptographic setting, since it is actually the goal of crypto designers to hide the structure of the underlying code, e.g. the Goppa code in McEliece, by linear transformations.

Table 1. Comparison of exponents in the asymptotic worst-case complexities

	time	space
Lee-Brickell	$0.05751n$	-
Stern	$0.05563n$	$0.013n$
FS-ISD / Ball-collision	$0.05558n$	$0.014n$
Lower bound from [5]	$0.05556n$	$0.014n$
Our algorithm with FS-ISD space	$0.05402n$	$0.014n$
Our algorithm	$0.05363n$	$0.021n$

Table 1 summarizes the worst-case complexity of decoding algorithms. Notice that Stern's algorithm, FS-ISD and Ball-collision are typical time-memory tradeoffs that decrease the running time complexity at the cost of an increased memory complexity. In contrast, our algorithm does not only benefit from a mere time-memory tradeoff. For example, if we restrict our memory complexity to $\tilde{\mathcal{O}}\left(2^{0.014n}\right)$ as in FS-ISD we still obtain an improved running time.

Roadmap. Our paper is organized as follows. We first introduce some useful notation in Section 2. In Section 3, we briefly recall the state of the art in information set decoding, including Stern's algorithm, FS-ISD and Ball-Collision decoding. In Section 4, we provide an algorithm for the submatrix matching problem. This leads to our new information set decoding algorithm in Section 5, for which we provide some experimental results in Section 6.

2 Notation

By $[k]$ we define the set of natural numbers between 1 and k, i.e. $[k] = \{1, \ldots, k\}$. The cardinality of a finite set I is denoted by $|I|$. For a better readability we represent matrices $\mathbf{Q}$ and vectors $\mathbf{e}$ by bold letters. For index sets $I \subset [n]$, $J \subset [k]$ and an $n \times k$ matrix $\mathbf{Q} = (q_{i,j})_{i \in [n], j \in [k]} \in \mathbb{F}_2^{n \times k}$, we denote by $\mathbf{Q}_J^I := (q_{i,j})_{i \in I, j \in J}$ the submatrix containing the $|I|$ rows and $|J|$ columns defined by I and J, respectively. When we consider submatrices of $\mathbf{Q}$ where *either* columns *or* rows are chosen, we simply write $\mathbf{Q}_J$ or $\mathbf{Q}^I$ meaning $\mathbf{Q}_J = \mathbf{Q}_J^{[n]}$ and $\mathbf{Q}^I = \mathbf{Q}_{[k]}^I$.

We extend this notion to vectors $\mathbf{s} \in \mathbb{F}_2^n$ and write $\mathbf{s}_L \in \mathbb{F}_2^{|L|}$ for the projection of $\mathbf{s}$ onto the coordinates defined by L. Further, for a matrix $\mathbf{Q} = (q_{i,j})_{i \in [n], k \in [k]} \in \mathbb{F}_2^{n \times k}$ and index sets $L \subseteq [n]$ with $|L| = \ell$, we define a mapping $\pi_L : \mathbb{F}_2^{n \times k} \to \mathbb{F}_2^{\ell}$ where

$$\pi_L(\mathbf{Q}) := \sum_{i=1}^{k} \mathbf{Q}_{\{i\}}^L \in \mathbb{F}_2^{\ell}$$

is the projection of the sum of $\mathbf{Q}'s$ columns onto the ℓ rows defined by L. As before, we sometimes omit the index set L which means that we consider the sum of $\mathbf{Q}$'s columns *without* projecting it to a certain number of rows, i.e. $\pi(\mathbf{Q}) = \pi_{[n]}(\mathbf{Q}) \in \mathbb{F}_2^n$.

By $\mathrm{wt}(\mathbf{x})$ we denote the Hamming weight of a vector $\mathbf{x} \in \mathbb{F}_2^n$, i.e., $\mathrm{wt}(\mathbf{x})$ counts the number of non-zero entries of $\mathbf{x}$. By $\mathrm{supp}(\mathbf{x}) := \{i \in [n] : x_i = 1\}$ we denote the support of a vector $\mathbf{x}$, i.e., the set of indices corresponding to non-zero coordinates of $\mathbf{x} \subset \mathbb{F}_2^n$. We represent the n-dimensional identity matrix by $\mathbf{I}_n$ and the i-th unit vector by $\mathbf{u}_i$. Observe that $\sum_{i \in \mathrm{supp}(\mathbf{x})} \mathbf{u}_i = \mathbf{x}$ for every $\mathbf{x} \in \mathbb{F}_2^n$. For a set of natural number $I \subset \mathbb{N}$, we introduce the shifted set $k + I := \{k + i : i \in I\}$ for arbitrary $k \in \mathbb{N}$.

Throughout the asymptotic complexity analysis of our *exponential* algorithms we make use of the soft Landau notation $\tilde{\mathcal{O}}$ which suppresses arbitrary *polynomial* factors, i.e., $p(n)2^n = \tilde{\mathcal{O}}(2^n)$ for every polynomial $p(n)$. We often need to estimate binomial coefficients of the form $\binom{\alpha n}{\beta n}$ asymptotically. Stirling's formula yields

$$\binom{\alpha n}{\beta n} = \tilde{\mathcal{O}}(2^{\alpha H(\beta/\alpha)n}), \tag{2}$$

where $H(x) = -x \log_2(x) - (1 - x) \log_2(1 - x)$ is the binary entropy function.

3   Information Set Decoding Algorithms

3.1   Information Set Decoding

Let $\mathcal{C}$ be an $[n, k, d]$-code with parity check matrix $\mathbf{H}$. Furthermore, let $\mathbf{x} = \mathbf{c} + \mathbf{e}$, $\mathbf{c} \in \mathcal{C}$ be an erroneous codeword with error $\mathbf{e}$, $\mathrm{wt}(\mathbf{e}) = \lfloor \frac{d-1}{2} \rfloor$. In order to find $\mathbf{e}$, information set decoding proceeds as follows.

Initially, we apply a random permutation to the columns of $\mathbf{H}$, resulting in a permuted matrix $\tilde{\mathbf{H}}$. Then we apply Gaussian Elimination on the right-hand square submatrix $\tilde{\mathbf{H}}_I$, $I = \{k + 1, \ldots, n\}$. If $\tilde{\mathbf{H}}_I$ is invertible, Gaussian Elimination will succeed and we obtain a systematic form[1] $(\mathbf{Q}|\mathbf{I_{n-k}})$ of $\tilde{\mathbf{H}}$, see Figure 1.

After the first step all the work can be done within the k columns of submatrix $\mathbf{Q}$. In the Lee-Brickell algorithm [9] one checks for every $I \subseteq [k]$ with cardinality $|I| = p$ whether $\mathrm{wt}(\pi(\mathbf{Q}_I) + \mathbf{s}(\mathbf{x})) = \omega - p$. If so, we can easily choose $\omega - p$ columns in the $\mathbf{I_{n-k}}$ part of $\tilde{\mathbf{H}}$ indexed by $J = k + \mathrm{supp}(\pi(\mathbf{Q}_I) + \mathbf{s}(\mathbf{x})) \subseteq [k + \ell + 1, n]$ which eliminate the remaining 1-entries. This in turn implies that $\sum_{i \in I} \mathbf{Q}_{\{i\}} + \sum_{j \in J} \mathbf{u}_{j-k} = \mathbf{s}(\mathbf{x})$.

Therefore, I and J determine the support of the permuted error vector $\tilde{\mathbf{e}} = \mathbf{e}\mathbf{U}_P$, i.e., we can set $\mathrm{supp}(\tilde{\mathbf{e}}) := I \cup J$ which finally reveals the error $\mathbf{e}$.

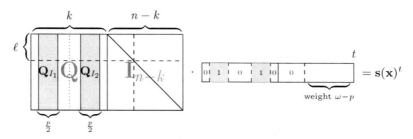

Fig. 1. Collision Decoding by Stern - $\mathbf{H}\mathbf{e}^t = \mathbf{s}(\mathbf{x})^t$. The error vector $\mathbf{e}$ contains two blocks each of $\frac{p}{2}$ 1's in its upper half corresponding to the columns of $\mathbf{Q}_{I_1}$ and $\mathbf{Q}_{I_2}$. Since $\mathbf{Q}_{I_1}$ and $\mathbf{Q}_{I_2}$ sum up to $\mathbf{s}(\mathbf{x})$ on the rows defined by $[\ell]$ we have to fix a corresponding zero-block in coordinates $\{k + 1, \ldots, k + \ell\}$ of $\mathbf{e}$. The remaining $(\omega - p)$ 1's are then distributed over the remaining coordinates $\{k + \ell + 1, \ldots, n\}$ of $\mathbf{e}$.

3.2   Stern's Algorithm

In the late 80s, Leon and Stern [13] introduced the idea of forcing the first ℓ coordinates of $\pi(\mathbf{Q}_I)$ already to the coordinates of $\mathbf{s}(\mathbf{x})$. Let $\mathbf{s}_{[\ell]}(\mathbf{x})$ be the projection of $\mathbf{s}(\mathbf{x})$ onto the coordinates in $[\ell]$.

We enumerate for all $I_1 \subseteq [1, \frac{k}{2}]$, $I_2 \subseteq [\frac{k}{2} + 1, k]$ the projected vectors $\pi_{[\ell]}(\mathbf{Q}_{I_1})$ and $\pi_{[\ell]}(\mathbf{Q}_{I_2}) + \mathbf{s}_{[\ell]}(\mathbf{x})$ in two lists. Then we search for collisions in these lists,

---

[1] In more detail, we transform $\mathbf{H}$ by multiplying it by two invertible matrices $\mathbf{U}_P \in \mathbb{F}_2^{n \times n}$, $\mathbf{U}_G \in \mathbb{F}_2^{n-k \times n-k}$ corresponding to the initial column permutation and the Gaussian Elimination, respectively. Then $(\mathbf{Q}|\mathbf{I}) = \mathbf{U}_G(\mathbf{H}\mathbf{U}_P)$. Notice, that the transformation $\mathbf{U}_G$ also needs to be applied to the syndrome $\mathbf{s}(\mathbf{x})$, which we omit for simplicity of exposition.

meaning that we look for two weight-$\frac{p}{2}$ sums of columns that are equal to the syndrome $\mathbf{s}(\mathbf{x})$ within the coordinates of $[\ell]$.

If $\text{wt}(\pi(\mathbf{Q}_{I_1}) + \pi(\mathbf{Q}_{I_2}) + \mathbf{s}(\mathbf{x})) = \omega - p$ holds for one of these collisions, we again set the corresponding $\omega - p$ coordinates in the second half of the permuted error vector $\tilde{\mathbf{e}}$ to 1, see Fig. 1 for an illustration.

To analyze Stern's algorithm we have to consider both the complexity of each iteration and the probability of success. The complexity of each iteration is dominated by the collision finding step in two lists. This can be done by a simple sort-and-match technique. Neglecting log factors, we obtain complexity

$$C_{\text{Stern}}(p, \ell) := \max \left\{ \binom{k/2}{p/2}, \frac{\binom{k/2}{p/2}^2}{2^{\ell}} \right\}. \tag{3}$$

In order to analyze the success probability, we need to compute the probability that a random permutation of the error $\mathbf{e} \in \mathbb{F}_2^n$ of weight $\text{wt}(\mathbf{e}) = \omega$ has a good weight distribution, i.e., $\tilde{\mathbf{e}}$ needs to have weight $p/2$ both on its coordinates in $[1, k/2]$ and $[k/2 + 1, k]$ *and* zero-weight on all coordinates with indices in the set $\{k+1, \ldots, k+\ell\}$ as illustrated in Fig. 1. Thus, we obtain success probability

$$P_{\text{Stern}}(p, \ell) := \frac{\binom{k/2}{p/2}^2 \binom{n-k-\ell}{\omega-p}}{\binom{n}{\omega}}. \tag{4}$$

The overall running time of Stern's algorithm is hence given by $C_{\text{Stern}} \cdot P_{\text{Stern}}^{-1}$. Optimizing this expression for p and ℓ under the natural constraints $0 \le p \le \omega$ and $0 \le \ell \le n - k - \omega + p$ we obtain time complexity $\tilde{\mathcal{O}}\left(2^{0.05563n}\right)$ and space complexity $\tilde{\mathcal{O}}\left(2^{0.013n}\right)$. The optimal parameter choice is given by $p = 0.003n$ and $\ell = 0.013n$.

3.3   The Finiasz-Sendrier ISD Algorithm

The idea of the FS-ISD algorithm is to increase the success probability for having a permuted error vector $\tilde{\mathbf{e}}$ of the desired form by allowing $\tilde{\mathbf{e}}$ to spread it's 1's over all coordinates, instead of fixing a certain ℓ-width 0-window. This is realized by changing the systematic form during the Gaussian Elimination process.

As before, we first randomly permute the columns of $\mathbf{H}$, which results in a permuted matrix $\tilde{\mathbf{H}} = \mathbf{H}\mathbf{U}_P$. Then we carry out a *partial Gaussian Elimination* on the right-hand lower square submatrix $\tilde{\mathbf{H}}_J^I \in \mathbb{F}_2^{(n-k-\ell) \times (n-k-\ell)}$ with index sets $I = \{\ell+1, \ldots, n-k\}$ and $J = \{k+\ell+1, \ldots, n\}$.

Next, we force an $\ell \times (n - k - \ell)$ zero block in the remaining ℓ rows of the submatrix $\tilde{\mathbf{H}}_J$ by adding rows of the identity matrix. Mathematically, we represent the partial Gaussian Elimination plus row elimination by a multiplication with an $(n - k) \times (n - k)$ invertible matrix $\mathbf{U}_G$. Therefore, the initial step in FS-ISD, which we denote $\text{Init}(\mathbf{H})$, yields a modified systematic form

$$\left(\mathbf{Q} \,\middle|\, \begin{matrix} \mathbf{0} \\ \mathbf{I}_{n-k-\ell} \end{matrix} \right) = \mathbf{U}_G \mathbf{H} \mathbf{U}_P.$$

In Fig. 2, we illustrate the Birthday collision step of FS-ISD which is the same as in Stern's algorithm but for a submatrix $\mathbf{Q}^{[\ell]}$ which now has $k + \ell$ columns instead of k columns.

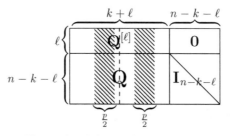

Fig. 2. Birthday collision search in FS-ISD

A straight-forward modification of the analysis of Stern's algorithm from Section 3 yields a complexity of

$$T_{\text{FS-ISD}}(p, \ell) := \max \left\{ S_{\text{FS-ISD}}(p, \ell), \frac{S_{\text{FS-ISD}}(p, \ell)^2}{2^\ell} \right\} \qquad (5)$$

per iteration, where $S_{\text{FS-ISD}}(p, \ell) = \binom{(k+\ell)/2}{p/2}$ denotes the size of the initial lists and thus represents also the space complexity. Furthermore, the success probability of getting an error vector $\mathbf{e}$ of the desired form is now given by

$$P_{\text{FS-ISD}}(p, \ell) := \frac{\binom{(k+\ell)/2}{p/2}^2 \binom{n-k-\ell}{\omega-p}}{\binom{n}{\omega}} \qquad (6)$$

Thus, we obtain a total complexity of $C_{\text{FS-ISD}}(p, \ell) = T_{\text{FS-ISD}}(p, \ell) \cdot P_{\text{FS-ISD}}(p, \ell)^{-1}$. Optimizing this expression yields a worst-case running time of $\tilde{\mathcal{O}}\left(2^{0.05558n}\right)$ within space complexity $\tilde{\mathcal{O}}\left(2^{0.014n}\right)$. The optimal parameter choice is given by $p = 0.003n$ and $\ell = 0.014n$.

3.4 Ball-collision Decoding

In 2011, Bernstein, Lange and Peters [4] presented another information set decoding algorithm, which they called *Ball-collision decoding* (BCD for shorthand). The general idea of BCD is very similar to the idea of the FS-ISD algorithm, namely the authors increase the success probability of one iteration in Stern's algorithm by allowing an additional number of ones within the fixed width-ℓ 0-window.

Therefore, BCD allows for q additional 1's within the 0-window, or in other words for a Hamming ball of radius q within the 0-window. More precisely, let I be an index set with $|I| = \frac{p}{2}$ chosen from the intervals $[1, k/2]$ or $[k/2+1, k]$. Each entry $(I, \pi_{[\ell]}(\mathbf{Q}_I))$ in the initial lists of Stern's algorithm has to be expanded by all possible projected weight-$q/2$ column sums $\pi_{[\ell]}(\mathbf{I}_J)$ of the identity matrix

I – for index sets J of size $|J| = q/2$ contained either in $[k+1, k+\ell/2]$ or $[k+\ell/2+1, k+\ell]$.

Analogously to the analysis of Stern's algorithm in Sect. 3, we obtain an asymptotic time complexity for one iteration of BCD of

$$T_{\text{BCD}}(p, \ell, q) := \max\left\{S_{\text{BCD}}(p, \ell, q), \frac{S_{\text{BCD}}(p, \ell, q)^2}{2^\ell}\right\}. \qquad (7)$$

The space consumption is $S_{\text{BCD}}(p, \ell, q) = \binom{k/2}{p/2}\binom{\ell/2}{q/2}$. Similarly one obtains a success probability of

$$P_{\text{BCD}}(p, \ell, q) := \frac{\left(\binom{k/2}{p/2}\binom{\ell/2}{q/2}\right)^2 \binom{n-k-\ell}{w-p-q}}{\binom{n}{w}}. \qquad (8)$$

Eventually, the overall complexity of BCD is given by $C_{\text{BCD}}(p, \ell, q) = T_{\text{BCD}}(p, \ell, q) \cdot P_{\text{BCD}}(p, \ell, q)^{-1}$.

Intuitively, FS-ISD and BCD proceed in a similar fashion by allowing $\tilde{\mathbf{e}}$ to spread its 1's in a more flexible way at the cost of slightly increasing the workload and space complexity per iteration. Indeed, the following theorem shows that FS-ISD is asymptotically at least as efficient as BCD.

Theorem 1. *Let (p, q, ℓ) be a parameter set for the BCD algorithm. Then $(p + q, \ell)$ is a parameter set for FS-ISD satisfying*

$$C_{\text{FS-ISD}}(p + q, \ell) \leq C_{\text{BCD}}(p, \ell, q) .$$

Proof. See full version, available from the authors.

Due to Theorem 1, we take the FS-ISD algorithm as a starting point for our new construction, in which we improve on the birthday-collision step.

4   How to Solve the Submatrix Problem

Recall that in each iteration of the FS-ISD algorithm one has to find in a projected $\ell \times (k + \ell)$ - submatrix a weight-p sum of columns that sums to a target syndrome. We call this problem the *submatrix matching problem*.

Definition 1. *The* submatrix matching problem *with parameters ℓ, k and $p \leq k + \ell$ is defined as follows. Given a random matrix $\mathbf{Q} = [\mathbf{q}_1 \dots \mathbf{q}_{k+\ell}] \in_R \mathbb{F}_2^{\ell \times (k+\ell)}$ and a target vector $\mathbf{s} \in \mathbb{F}_2^\ell$, find an index set I of size at most p such that the corresponding columns of $\mathbf{Q}$ sum to $\mathbf{s}$, i.e., find $I \subset [k + \ell], |I| \leq p$ with*

$$\pi(\mathbf{Q}_I) = \sum_{i \in I} \mathbf{q}_i = \mathbf{s} \in \mathbb{F}_2^\ell .$$

The submatrix matching problem is a vectorial variant of the well-known subset sum problem. In the following, we propose an algorithm COLUMNMATCH for the problem, based on a recently introduced *representation technique* for the subset sum problem by Howgrave-Graham and Joux [8].

When we use COLUMNMATCH in information set decoding, the input parameters p, ℓ are optimization parameters that guarantee that some solution I exists with a certain probability $P(p, \ell)$, compare e.g. with Eq.(6).

4.1   The COLUMNMATCH Algorithm

Let us briefly explain our COLUMNMATCH algorithm. We recommend the reader to follow our algorithm's description via the illustration given in Fig. 3 and the pseudocode description in Algorithm 1.

Let $\mathbf{Q} = [\mathbf{q}_1 \ldots \mathbf{q}_{k+\ell}] \in_R \mathbb{F}_2^{\ell \times (k+\ell)}$ and $\mathbf{s} \in \mathbb{F}_2^\ell$ be an instance of the submatrix matching problem. Assume that I is a solution to the problem of size exactly p. Similar to FS-ISD we construct I from two sets I_1, I_2 of size $\frac{p}{2}$ each.

As opposed to FS-ISD, we do not choose I_1 and I_2 from disjoint sets of size $\frac{k+\ell}{2}$. Rather we choose both I_1, I_2 from the full set $[k+\ell]$. This choice of the index sets is similar to what we call the *representation technique* due to Howgrave-Graham and Joux [8]. The effect of the choice is that we obtain $\binom{p}{p/2} \approx 2^p$ different partitions $I = I_1 \dot\cup I_2$ and therefore the same number of identities

$$\sum_{i \in I_1} \mathbf{q}_i = \sum_{i \in I_2} \mathbf{q}_i + \mathbf{s} \text{ in } \mathbb{F}_2^\ell. \tag{9}$$

Our goal is to find one of these identities with constant success probability, where the probability is taken over the random choice of $\mathbf{Q}$. Therefore we do not construct all possible sums of elements in I_1, I_2 but only those that satisfy additional constraints. To establish the constraints, we introduce shortening parameters ℓ_1, ℓ_2 with $\ell_1 + \ell_2 = \ell$ that correspond to disjoint subsets $L_1, L_2 \subset [l]$ of size ℓ_1, ℓ_2, respectively.

Our construction now proceeds in two steps. In the first step, we construct partial solutions that already sum to the target value $\mathbf{s}$ on the ℓ_2 positions of L_2. More precisely, we construct two lists

$$\mathcal{L}_1 := \left\{ (I_1, \pi_{L_1}(\mathbf{Q}_{I_1})) : I_1 \subset [k+\ell], |I_1| = \tfrac{p}{2} \text{ and } \pi_{L_2}(\mathbf{Q}_{I_1}) = \mathbf{0} \in \mathbb{F}_2^{\ell_2} \right\} \text{ and}$$

$$\mathcal{L}_2 := \left\{ (I_2, \pi_{L_1}(\mathbf{Q}_{I_2}) + \mathbf{s}_{L_1}) : I_2 \subset [k+\ell], |I_2| = \tfrac{p}{2} \text{ and } \pi_{L_2}(\mathbf{Q}_{I_2}) = \mathbf{s}_{L_2} \in \mathbb{F}_2^{\ell_2} \right\}.$$

Notice that out of the 2^p possible identities that satisfy Eq. (9), we consider only those identities where $\sum_{i \in I_1} \mathbf{q}_i$ is equal to $\mathbf{0} \in \mathbb{F}_2^{l_2}$ on the bits of L_2. Thus we expect that we already remove a $2^{-\ell_2}$-fraction of all solutions, which lets an expected number of $2^{p-\ell_2}$ solutions survive.

Once we have constructed the lists $\mathcal{L}_1, \mathcal{L}_2$ in the first step, we sort $\mathcal{L}_2$ according to the labels $\pi_{L_1}(\mathbf{Q}_{I_2}) + \mathbf{s}_{L_1}$ and search for all elements $\pi_{L_1}(\mathbf{Q}_{I_1})$ in $\mathcal{L}_1$ for a matching element in $\mathcal{L}_2$. Notice that every matching (I_1, I_2) fulfills Eq. (9) and hence is a solution to the submatrix matching problem.

Since we constructed I_1, I_2 in a non-disjoint way, their intersection $J = I_1 \cap I_2$ might be non-empty. In this case, all vectors in J appear on both sides of Eq. (9) and thus cancel out when we compute $\sum_{i \in I_1} \mathbf{q}_i + \sum_{i \in I_2} \mathbf{q}_i$ over $\mathbb{F}_2^\ell$. This means that we have found a solution $I' = I_1 \Delta I_2 = (I_1 \cup I_2) \setminus (I_1 \cap I_2)$ to the submatrix matching problem with size $|I'| = p - 2|I_1 \cap I_2|$.

How to construct $\mathcal{L}_1$ and $\mathcal{L}_2$. The initial lists $\mathcal{L}_1$ and $\mathcal{L}_2$ can be easily constructed by a classical sort-and-match step. Let us show how to construct

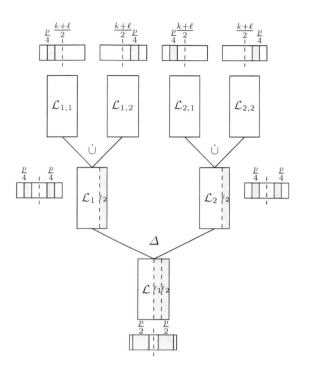

Fig. 3. Illustration of the COLUMNMATCH algorithm. The flat rectangles above, beside or below the lists represent the structure of the index sets $I_{i,j}$ contained in distinct lists, e.g., the level-2 list $\mathcal{L}_{1,1}$ contains index sets $I_{1,1}$ whose $\frac{p}{4}$ ones are spread over the first half of $[k + \ell]$ (as illustrated by the gray region).

$\mathcal{L}_1$, the construction of $\mathcal{L}_2$ is analogous. We partition $I_1 = I_{1,1} \dot{\cup} I_{1,2}$ with $|I_{1,1}| = |I_{1,2}| = \frac{p}{4}$ where $I_{1,1} \subset [1, \frac{k+\ell}{2}]$ and $I_{1,2} \subset [\frac{k+\ell}{2} + 1, k + \ell]$. More precisely, we compute two lists

$$\mathcal{L}_{1,1} := \left\{ \left(I_{1,1}, \pi_{L_2}(\mathbf{Q}_{I_{1,1}})\right) : I_{1,1} \subset [1, \tfrac{k+\ell}{2}], |I_{1,1}| = \tfrac{p}{4} \right\} \text{ and}$$

$$\mathcal{L}_{1,2} := \left\{ \left(I_{1,2}, \pi_{L_2}(\mathbf{Q}_{I_{1,2}})\right) : I_{1,2} \subset [\tfrac{k+\ell}{2} + 1, k + \ell], |I_{1,2}| = \tfrac{p}{4} \right\} \; .$$

We then sort $\mathcal{L}_{1,2}$ with respect to the second component and search for all second components in $\mathcal{L}_{1,1}$ for matching elements in $\mathcal{L}_{1,2}$.

Remark 1. Notice that the construction of $\mathcal{L}_1$ and $\mathcal{L}_2$ via disjoint splittings $I_1 = I_{1,1} \dot{\cup} I_{1,2}$ and $I_2 = I_{2,1} \dot{\cup} I_{2,2}$ lowers the number of representations $\mathcal{R}(p)$. Instead of considering *every* subset $I_1 \subset I$ of size $\frac{p}{2}$ we take every I_1 with an equal number of $\frac{p}{4}$ indices coming from $[1, (k + \ell)/2]$ and $[(k + \ell)/2 + 1, k + \ell]$, respectively. Hence, we only have $\binom{p/2}{p/4}^2$ instead of $\binom{p}{p/2}$ many different representations per solution in Eq. (9). Asymptotically, this can be neglected since both terms equal $2^{p(1-o(1))}$.

Algorithm 1. ColumnMatch

Input: $\mathbf{Q} \in \mathbb{F}_2^{\ell \times (k+\ell)}$, $\mathbf{s} \in \mathbb{F}_2^{\ell}$, $p \leq k + \ell$
Output: I with $\pi(\mathbf{Q}_I) = \mathbf{s}$ or $\perp$ if no solution is found
Parameters: L_1, L_2 with $[l] = L_1 \dot{\cup} L_2$ and $|L_i| = \ell_i$ for $i = 1, 2$.

01 Construct $\mathcal{L}_{1,1}, \mathcal{L}_{1,2}, \mathcal{L}_{2,1}, \mathcal{L}_{2,2}$.
02 Sort $\mathcal{L}_{1,2}, \mathcal{L}_{2,2}$ according to their labels $\pi_{L_2}(\mathbf{Q}_{I_{1,2}})$, $\pi_{L_2}(\mathbf{Q}_{I_{2,2}}) + \mathbf{s}_{L_2}$.
03 Join $\mathcal{L}_{1,1}$ and $\mathcal{L}_{1,2}$ to $\mathcal{L}_1$, i.e., **for all** $(I_{1,1}, \pi_{L_2}(\mathbf{Q}_{I_{1,1}})) \in \mathcal{L}_{1,1}$ **do**
04   **for all** $(I_{1,2}, \pi_{L_2}(\mathbf{Q}_{I_{1,2}})) \in \mathcal{L}_{1,2}$ with $\pi_{L_2}(\mathbf{Q}_{I_{1,1}}) = \pi_{L_2}(\mathbf{Q}_{I_{1,2}})$ **do**
05                $I_1 = I_{1,1} \cup I_{1,2}$. Insert $(I_1, \pi_{L_1}(\mathbf{Q}_{I_1}))$ into $\mathcal{L}_1$.
06 Join $\mathcal{L}_{2,1}$ and $\mathcal{L}_{2,2}$ to $\mathcal{L}_2$, i.e., **for all** $(I_{2,1}, \pi_{L_2}(\mathbf{Q}_{I_{2,1}})) \in \mathcal{L}_{2,1}$ **do**
07   **for all** $(I_{2,2}, \pi_{L_2}(\mathbf{Q}_{I_{2,2}}) + \mathbf{s}_{L_2}) \in \mathcal{L}_{2,2}$ with $\pi_{L_2}(\mathbf{Q}_{I_{2,1}}) = \pi_{L_2}(\mathbf{Q}_{I_{2,2}}) + \mathbf{s}_{L_2}$ **do**
08                $I_2 = I_{2,1} \cup I_{2,2}$. Insert $(I_2, \pi_{L_1}(\mathbf{Q}_{I_2}) + \mathbf{s}_{L_1})$ into $\mathcal{L}_2$.
09 Sort $\mathcal{L}_2$ according to the label $\pi_{L_1}(\mathbf{Q}_{I_2}) + \mathbf{s}_{L_1}$.
10 Join $\mathcal{L}_1$ and $\mathcal{L}_2$ to $\mathcal{L}$, i.e., **for all** $(I_1, \pi_{L_1}(\mathbf{Q}_{I_1})) \in \mathcal{L}_1$ **do**
11   **for all** $(I_2, \pi_{L_1}(\mathbf{Q}_{I_2}) + \mathbf{s}_{L_1}) \in \mathcal{L}_2$ with $\pi_{L_1}(\mathbf{Q}_{I_1}) = \pi_{L_1}(\mathbf{Q}_{I_2}) + \mathbf{s}_{L_1}$ **do**
12   Output $I_1 \triangle I_2 = (I_1 \cup I_2) \setminus (I_1 \cap I_2)$.
13 Output $\perp$.

Time and space complexity. Throughout the analysis, we will again ignore low-order terms that are polynomial in the parameters p, ℓ. The space complexity of constructing the four level-2 lists $\mathcal{L}_{1,1}, \mathcal{L}_{1,2}, \mathcal{L}_{2,1}, \mathcal{L}_{2,2}$ is bounded by the length $\binom{(k+\ell)/2}{p/4}$ of these lists. The sort-and-match step of these lists can be done in time

$$\max \left\{ \binom{(k+\ell)/2}{p/4}, \binom{(k+\ell)/2}{p/4}^2 \cdot 2^{-\ell_2} \right\} .$$

Joining lists $\mathcal{L}_{1,1}$ and $\mathcal{L}_{1,2}$ to list $\mathcal{L}_1$ produces a list of expected size

$$\mathbb{E}[|\mathcal{L}_1|] = \binom{(k+\ell)/2}{p/4}^2 \cdot 2^{-\ell_2} = \tilde{\mathcal{O}}(2^{(k+\ell)H(\frac{p}{2(k+\ell)})-\ell_2}).$$

The final sort-and-match step of $\mathcal{L}_1$ and $\mathcal{L}_2$ on level 1 then takes expected time

$$\max \left\{ \mathbb{E}[|\mathcal{L}_1|], \frac{\mathbb{E}[|\mathcal{L}_1|] \cdot \mathbb{E}[|\mathcal{L}_2|]}{2^{\ell_1}} \right\} = \max \left\{ \binom{(k+\ell)/2}{p/4}^2 \cdot 2^{-\ell_2}, \binom{(k+\ell)/2}{p/4}^4 \cdot 2^{-2\ell_2-\ell_1} \right\} .$$

The following table summarizes the exponents in the complexities for both levels of our algorithm ColumnMatch. This means that e.g. on level 2, we have space complexity $\tilde{\mathcal{O}}(2^{S_2(k,p,\ell)})$. All binomial coefficients are estimated via Eq.(2).

The total time and space complexity for ColumnMatch is hence given by

$$S(k, p, \ell, \ell_2) = \max\{S_2(k, p, \ell), S_1(k, p, \ell, \ell_2)\} \text{ and}$$

$$T(k, p, \ell, \ell_1, \ell_2) = \max\{S_2(k, p, \ell), S_1(k, p, \ell, \ell_2), 2S_1(k, p, \ell, \ell_2) - \ell_1\} .$$

Table 2. Exponents of time and space complexities

level	space	time
2	$S_2(k,p,\ell) := \frac{k+\ell}{2} H(\frac{p}{2(k+\ell)})$	$\max\{S_2(k,p,\ell), 2S_2(k,p,\ell) - \ell_2\}$
1	$S_1(k,p,\ell,\ell_2) := 2S_2(k,p,\ell) - \ell_2$	$\max\{S_1(k,p,\ell,\ell_2), 2S_1(k,p,\ell,\ell_2) - \ell_1\}$

Theorem 2. *Let* $\mathbf{Q} \in_R \mathbb{F}_2^{\ell \times (k+\ell)}$, $\mathbf{s} \in \mathbb{F}_2^\ell$ *and* $p \leq k + \ell$. *Let* $\hat{I}$ *be a solution of the submatrix matching problem for* $\mathbf{Q}, \mathbf{s}$. *For sufficiently large* p COLUMN-MATCH *finds* $\hat{I}$ *with probability at least* $\frac{1}{2}$ *in time* $\tilde{\mathcal{O}}(2^{T(k,p,\ell,\ell_1,\ell_2)})$ *and space* $\tilde{\mathcal{O}}(2^{S(k,p,\ell,\ell_2)})$ *as long as* $\ell_2 \leq p - 2$.

Proof. We already proved the claim about the time and space complexity. It remains to show that COLUMNMATCH succeeds with probability at least $\frac{1}{2}$.

To analyze the success probability of COLUMNMATCH we introduce a random variable X that counts the number of representations $I = I_1 \dot\cup I_2$ of the solution $\hat{I}$ in lists $\mathcal{L}_1$ and $\mathcal{L}_2$. Our goal is to show that at least one representation survives in our algorithm with probability at least $\frac{1}{2}$.

Notice that we have a total number of $\mathcal{R}(p) := \binom{p/2}{p/4}^2$ representations on level 1. To analyze X we introduce $\mathcal{R}(p)$ indicator variables X_I where $X_I = 1$ iff representation $I = I_1 \dot\cup I_2$ of $\hat{I}$ is contained in $\mathcal{L}_1$, i.e.,

$$X_I = \begin{cases} 1 & \text{if } \pi_{L_2}(\mathbf{Q}_{I_1}) = \mathbf{0} \\ 0 & \text{otherwise} \end{cases}.$$

Note that $X = \sum X_I$. The Second Moment Method [1] now lower bounds the success probability $\mathbf{Pr}[X \geq 1]$ by upper bounding $\mathbf{Pr}[X = 0] = 1 - \mathbf{Pr}[X \geq 1]$ using Chebyshev's inequality

$$\mathbf{Pr}[X = 0] \leq \frac{\text{Var}[X]}{\mathbb{E}[X]^2} = \frac{\sum_I \text{Var}[X_I] + \sum_{I \neq J} \text{Cov}[X_I, X_J]}{\mathbb{E}[X]^2}. \tag{10}$$

Here the covariance has to be computed over all different representations $I \neq J$ of the solution $\hat{I}$. Essentially, for every representation I there is exactly one different representation J for which X_I and X_J are dependent, otherwise they are pairwise independent and hence $\text{Cov}[X_I, X_J] = 0$.

We write $I = I_1 \dot\cup I_2$ with $|I_1|, |I_2| = \frac{p}{2}$ and analogously $J = J_1 \dot\cup J_2$. Notice that for all choices $J_1 \neq I \setminus I_1$, the random variables X_I and X_J are pairwise independent because $\mathbf{Q}$ contains randomly distributed columns.

Let $J_1 = I \setminus I_1$. Since $\pi(\mathbf{Q}_I) = \mathbf{s}$, we have

$$\pi_{L_2}(\mathbf{Q}_{I_1}) = \pi_{L_2}(\mathbf{Q}_{J_1}) + \mathbf{s}_{L_2}.$$

If $\mathbf{s}_{L_2} \neq \mathbf{0}$ then $\pi_{L_2}(\mathbf{Q}_{I_1}) \neq \pi_{L_2}(\mathbf{Q}_{J_1})$ which implies that $X_I X_J = 0$. Therefore $\text{Cov}[X_I, X_J] = \mathbb{E}[X_I X_J] - \mathbb{E}[X_I]\mathbb{E}[X_J] = -\mathbb{E}[X_I]\mathbb{E}[X_J] < 0$. Hence we can bound Eq.(10) as $\mathbf{Pr}[X = 0] \leq \frac{\sum_I \text{Var}[X_I]}{\mathbb{E}[X]^2}$.

If $\mathbf{s}_{L_2} = \mathbf{0}$ then $\pi_{L_2}(\mathbf{Q}_{I_1}) = \pi_{L_2}(\mathbf{Q}_{J_1})$ which implies $X_I = X_J$. This means that for every I there is exactly one $J \neq I$ such that $\mathrm{Cov}[X_I, X_J] = \mathrm{Cov}[X_I, X_I] = \mathrm{Var}[X_I]$. In this case, we can bound Eq.(10) as

$$\mathbf{Pr}\,[X = 0] \leq \frac{2\sum_I \mathrm{Var}[X_I]}{\mathbb{E}[X]^2}.$$

Example 1. Consider the case $k = 8$ and $p = 4$ with $\mathbf{Q} = (\mathbf{q}_1, \ldots, \mathbf{q}_8)$, $\mathbf{s} = \mathbf{0}$ and $\hat{I} = \{1, 2, 5, 6\}$. The representations $I = I_1 \,\dot\cup\, I_2 = \{1, 5\}\,\dot\cup\,\{2, 6\}$ and $J = J_1 \,\dot\cup\, J_2 = \{2, 6\}\,\dot\cup\,\{1, 5\}$ have identical indicator variables X_I, X_J. However I and $K = \{2, 5\}\,\dot\cup\,\{1, 6\}$ have independent indicator variables since $\mathbf{Pr}\,[X_K = 1 | X_I = 1] = \mathbf{Pr}\,[\mathbf{q}_2 + \mathbf{q}_5 = 0 | \mathbf{q}_1 + \mathbf{q}_5 = 0] = \mathbf{Pr}\,[\mathbf{q}_2 = \mathbf{q}_1] = 2^{-\ell_2} = \mathbf{Pr}\,[X_K = 1]$.

We further observe that

$$\mathrm{Var}[X_I] = \mathbb{E}[X_I^2] - (\mathbb{E}[X_I])^2 = \mathbb{E}[X_I] - (\mathbb{E}[X_I])^2 \leq \mathbb{E}[X_I].$$

Therefore, we obtain

$$\mathbf{Pr}\,[X = 0] \leq \frac{2\sum_I \mathrm{Var}[X_I]}{\mathbb{E}[X]^2} \leq \frac{2\sum_I \mathbb{E}[X_I]}{\mathbb{E}[X]^2} \leq \frac{2\,\mathbb{E}[\sum_I X_I]}{\mathbb{E}[X]^2} = \frac{2\,\mathbb{E}[X]}{\mathbb{E}[X]^2} = \frac{2}{\mathbb{E}[X]}.$$

Since $\mathbb{E}[X] = \mathcal{R}(p)2^{-\ell_2} \geq 2^{p(1-o(1))-\ell_2}$, putting the restriction $\ell_2 \leq p - 2$ on the choice of the parameter ℓ_2 yields for large enough p

$$\mathbf{Pr}\,[X = 0] \leq 2^{1-p(1-o(1))+\ell_2} \rightarrow 2^{\ell_2-(p-1)} \leq \frac{1}{2}.$$

This in turn implies that our algorithm COLUMNMATCH succeeds in constructing at least one representation of the solution with probability at least $\frac{1}{2}$. $\quad\square$

5   Our New Decoding Algorithm

Let us start by giving a high-level description of our new information set decoding algorithm which we call DECODE. Let $\mathbf{H} \in \mathbb{F}_2^{(n-k)\times n}$ be a parity check matrix of an $[n, k, d]$-code $\mathcal{C}$. Assume that we want to decode $\mathbf{x} = \mathbf{c} + \mathbf{e}$ with $\mathbf{c} \in \mathcal{C}$, $\omega := \mathrm{wt}(\mathbf{e}) = \lfloor \frac{d-1}{2} \rfloor$. That means we want to find ω columns in $\mathbf{H}$ that sum to the syndrome $\mathbf{s}(\mathbf{x}) = \mathbf{H}\mathbf{x}^t$. As described in Sect. 3.3, we start with the initial transformation on the parity check matrix $\mathbf{H}$ and obtain the modified systematic form

$$\tilde{\mathbf{H}} = \mathsf{Init}(\mathbf{H}) = \mathbf{U}_G \mathbf{H} \mathbf{U}_P = \left(\mathbf{Q} \,\middle|\, \begin{matrix} \mathbf{0} \\ \mathbf{I}_{n-k-\ell} \end{matrix} \right).$$

This process also permutes $\mathbf{e}$ to $\tilde{\mathbf{e}} = \mathbf{U}_P\mathbf{e}$. Let $p \leq \omega$ be an optimization parameter. We need that the ω ones in $\tilde{\mathbf{e}}$ are distributed as $\frac{p}{2}, \frac{p}{2}, \omega - p$ in the coordinate intervals $[1, (k + \ell)/2], [(k + \ell)/2 + 1, k + \ell], [k + \ell + 1, n]$ of $\tilde{\mathbf{e}}$, respectively.

Recall from Section 3.3 that $\tilde{\mathbf{e}}$ happens to have the correct form with probability

$$P_{ColumnMatch}(p, \ell) := \frac{\binom{(k+\ell)/2}{p/2}^2 \binom{n-k-\ell}{\omega-p}}{\binom{n}{\omega}} . \tag{11}$$

We now look within the submatrix $\mathbf{Q}^{[\ell]}$ of $\mathbf{Q}$ for a weight-p sum of the columns that exactly matches the projection of the syndrome to the first ℓ rows.

In the DECODE algorithm, we now apply our COLUMNMATCH algorithm to $\mathbf{Q}^{[\ell]} \in \mathbb{F}_2^{\ell \times (k+\ell)}$ with the projected syndrome as target vector and a solution weight of p.

In each iteration of DECODE, our COLUMNMATCH algorithm yields with probability at least $\frac{1}{2} \cdot P_{ColumnMatch}(p, \ell)$ at least one index set I, $|I| \leq p$ such that $\pi_{[\ell]}(\mathbf{Q}_I)$ exactly matches the projected syndrome. Thus we already match the syndrome on ℓ coordinates using a weight-$|I|$ linear combination of columns from $\mathbf{Q}$. If the remaining coordinates of $\pi(\mathbf{Q}_I)$ differ from the syndrome only by $w - |I|$ 1-entries, then we can correct these entries by choosing $w - |I|$ unit vectors from $\mathbf{I}_{n-k-\ell}$. Let us summarize our decoding algorithm by giving a pseudo-code description in Algorithm 2.

Algorithm 2. DECODE

Input: Parity check matrix $\mathbf{H} \in \mathbb{F}_2^{(n-k) \times n}$, syndrome $\mathbf{s(x)} = \mathbf{H}\mathbf{e}^t$ with wt$(\mathbf{e}) = \omega$.
Output: Error $\mathbf{e} \in \mathbb{F}_2^n$
Parameters: p, ℓ, ℓ_1, ℓ_2 with $\ell = \ell_1 + \ell_2$

00 **Repeat**
01   Compute $\tilde{\mathbf{H}} \leftarrow \mathsf{Init}(\mathbf{H})$ where $\tilde{\mathbf{H}} = \mathbf{U}_G \mathbf{H} \mathbf{U}_P$.
02   **For all** (solutions I found by COLUMNMATCH$(\mathbf{Q}^{[\ell]}, (\mathbf{U}_G \mathbf{s}^t(\mathbf{x}))_{[\ell]}, p, \ell_1, \ell_2))$ **do**
03   **If** wt$(\pi(\mathbf{Q}_I) + \mathbf{U}_G \mathbf{s}^t(\mathbf{x})) = \omega - |I|$ **then**
04   Compute $\tilde{\mathbf{e}} \in \mathbb{F}_2^n$ by setting
05                    $\tilde{e}_i = 1 \; \forall i \in I$
06                    $\tilde{e}_{k+\ell+j} = 1 \; \forall j \in \mathrm{supp}(\pi_{[n-k]\setminus[\ell]}(\mathbf{Q}_I + \mathbf{U}_G \mathbf{s}^t(\mathbf{x})))$
07   **Output** $\mathbf{e} = \tilde{\mathbf{e}} \mathbf{U}_P^t$.

The correctness of DECODE is implied by correctness of the COLUMNMATCH algorithm as we show in the following lemma.

Lemma 1. DECODE *is correct, i.e., if* DECODE *outputs error* $\mathbf{e}$ *then* $\mathbf{H}\mathbf{e}^t = \mathbf{s(x)}$ *and* wt $(\mathbf{e}) = \omega$.

Proof. Let I be an output of COLUMNMATCH, i.e., $\pi_{[\ell]}(\mathbf{Q}_I) = (\mathbf{U}_G \mathbf{s}^t(\mathbf{x}))_{[\ell]}$ and $0 < |I| \leq p$. Furthermore, we have

$$\mathbf{U}_G \mathbf{H}\mathbf{e}^t = \mathbf{U}_G \mathbf{H} \mathbf{U}_P \tilde{\mathbf{e}}^t = \tilde{\mathbf{H}}\tilde{\mathbf{e}}^t = \left(\mathbf{Q} \,\middle|\, \begin{matrix} \mathbf{0} \\ \mathbf{I}_{n-k-\ell} \end{matrix} \right) \tilde{\mathbf{e}}^t = \mathbf{Q}\tilde{\mathbf{e}}_{[k+\ell]}^t + \begin{pmatrix} \mathbf{0} \\ \mathbf{I} \end{pmatrix} \tilde{\mathbf{e}}_{[n]\setminus[k+\ell]}^t$$

$$= \pi(\mathbf{Q}_I) + \begin{pmatrix} \mathbf{0} \\ \pi_{[n-k]\setminus[\ell]}(\mathbf{Q}_I + \mathbf{U}_G \mathbf{s}^t(\mathbf{x})) \end{pmatrix} = \begin{pmatrix} \mathbf{U}_G \mathbf{s}_{[\ell]}^t(\mathbf{x}) \\ \mathbf{U}_G \mathbf{s}_{[n-k]\setminus[\ell]}^t(\mathbf{x}) \end{pmatrix} = \mathbf{U}_G \mathbf{s}^t(\mathbf{x}) .$$

Since $\mathbf{U}_G$ is invertible, it follows that $\mathbf{He}^t = \mathbf{s}(\mathbf{x})$. Moreover, from line 03 of DECODE we obtain that

$$\mathrm{wt}\,(\mathbf{e}) = \mathrm{wt}(\tilde{\mathbf{e}}) = |I| + \mathrm{wt}(\pi(\mathbf{Q}_I) + \mathbf{U}_G\mathbf{s}^t(\mathbf{x})) = |I| + \omega - |I| = \omega. \qquad \square$$

In the remaining part of this section we explain how to derive optimal parameter choices for the DECODE algorithm. We parametrize our code by $k = c_k n$, $\omega = c_\omega n$. We also parametrize the algorithm's optimization parameters as $\ell_1 = c_{\ell_1} n$, $\ell_2 = c_{\ell_2} n$ and $p = c_p n$.

Optimal parameters for the DECODE algorithm. Recall that a randomly permuted error $\tilde{\mathbf{e}} \in \mathbb{F}_2^n$ of weight $\mathrm{wt}\,(\tilde{\mathbf{e}}) = \omega$ has the desired weight distribution of 1-entries with probability $P_{ColumnMatch}(p, \ell)$ from Eq. (11) as in the FS-ISD algorithm. Thus the inverse success probability is asymptotically $P_{ColumnMatch}^{-1}(p, \ell) = \tilde{\mathcal{O}}(2^{\alpha n})$ with

$$\alpha(c_p, c_\ell) = H(c_\omega) - \left((c_k + c_\ell)H\left(\frac{c_p}{c_k + c_\ell}\right) + (1 - c_k - c_\ell)H\left(\frac{c_\omega - c_p}{1 - c_k - c_\ell}\right) \right).$$

For a fixed choice of the parameters ℓ_1, ℓ_2 and p, the asymptotic time and space complexities of one iteration of DECODE are given by $2^{T(k,p,\ell,\ell_1,\ell_2)n}$ and $2^{S(k,p,\ell,\ell_2)n}$ from Theorem 2. In order to apply Theorem 2 we need to further ensure that $\ell_2 \leq p - 2$, which asymptotically simplifies to $c_{\ell_2} \leq c_p + \frac{2}{n} \to c_p$.

In total, we have to solve the following optimization problem

$$\min\{ T(c_k, c_p, c_\ell, c_{\ell_1}, c_{\ell_2}) + \alpha(c_p, c_{\ell_1} + c_{\ell_2}) \} \qquad \text{(OPT)}$$
$$\text{s.t.} \quad 0 \leq c_p \leq c_\omega$$
$$0 \leq c_{\ell_1} + c_{\ell_2} \leq 1 - c_k - c_\omega + c_p$$
$$0 \leq c_{\ell_2} \leq c_p$$
$$0 \leq c_{\ell_1}.$$

We solve (OPT) numerically for various code rates $0 \leq c_k \leq 1$. Since random linear codes attain the Gilbert-Varshamov bound [6], we related the value c_w for the maximal error-correction capability to c_k by the identity $c_k = 1 - H(2c_\omega)$.

For every code rate $0 \leq c_k \leq 1$ on the x-axis we plotted the complexity of DECODE in comparison with the FS-ISD algorithm, see Fig. 4 and Fig. 5. This shows that our DECODE algorithm yields for all rates c_k an exponential improvement over the best-known decoding algorithms FS-ISD and Ball-collision decoding. If we additionally plot the lower bound curve from [5] in its asymptotical form, then this curve lies strictly below the FS-ISD curve and strictly above our new curve. This shows that the representation technique in our DECODE algorithm allows to bypass the lower bound framework from [5].

We obtained the worst-case complexity for $c_k \approx 0.47n$ with the parameter choice as stated in the following main result.

Theorem 3. DECODE *recovers* $\mathbf{e}$ *in time* $\tilde{\mathcal{O}}(2^{0.05363n})$ *and space* $\tilde{\mathcal{O}}(2^{0.021n})$, *where the optimal parameter choice is* $c_p = c_{\ell_2} = 0.006$ *and* $c_{\ell_1} = 0.028$.

Our formulation as an optimization problem (OPT) easily allows to specify additional space constraints. E.g. adding the restriction $S(c_k, c_p, c_\ell, c_{\ell_2}) \leq 0.014$ gives us a running time of $\tilde{\mathcal{O}}\left(2^{0.05402n}\right)$ using the same space $\tilde{\mathcal{O}}\left(2^{0.014n}\right)$ as in FS-ISD/Ball-collision decoding.

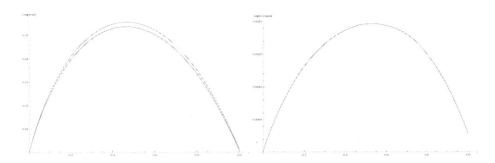

Fig. 4. Run time comparison with FS-ISD   **Fig. 5.** Improvement over FS-ISD

6   Experiments

We implemented our DECODE and COLUMNMATCH algorithms in C++ and tested them on three small McEliece instances with underlying $[n, k, \omega]$-Goppa codes. For each instance we computed optimal parameters p, ℓ_1, ℓ_2 (see second column of Table 3) using the *exact* formulas for the time and space complexities from Sect. 4 as well as for the respective probabilities from Eq. (11). We then carried out 10.000 experiments per McEliece instance with varying $\mathbf{Q}$. We computed the target syndrome $\mathbf{s} = \mathbf{Q}\mathbf{e}^t$ for an error vector $\mathbf{e}$ fulfilling the required weight distribution, i.e., we fixed $p/2$ coordinates to 1 in both intervals $[1, \frac{k+l}{2}]$ and $[\frac{k+l}{2} + 1, k + \ell]$.

Recall that our sole heuristic assumption was that $\mathbf{Q}$ behaves as a uniformly random matrix, implying that the projected partial sums $\pi_{L_j}(\mathbf{Q}_I)$ are distributed uniformly at random as well. To verify this assumption experimentally, we determined the average list size of $\mathcal{L}_1$ on level 1 and compared it to the theoretically expected size (see columns three and four of Table 3).

Furthermore, we counted the number of successful iterations where the error vector $\mathbf{e}$ was found (see column five of table 3). The results approximately match the theoretically predicted success probability of at least $\frac{1}{2}$ for COLUMNMATCH. The slight discrepancy is due to the small value of p.

For the sake of completeness, we also give the time per iteration as well as the number of repetitions P^{-1} that would be needed for the complete DECODE algorithm (see columns six and seven).

We would like to stress that the main goal of our implementation was to test the validity of the heuristic assumption, that $\mathbf{Q}$ behaves as a random matrix.

Table 3. Experimental results for the COLUMNMATCH algorithm

$[n, k, \omega]$	$[p, \ell_1, \ell_2]$	$|\mathcal{L}_1|$ theo.	$|\mathcal{L}_1|$ exp.	success prob.	time (ms)	P^{-1}
$[255, 135, 15]$	$[4, 11, 2]$	1369	1369.1	43.6%	11	$2^{8.12}$
$[511, 259, 28]$	$[4, 13, 2]$	4692.25	4692.08	44.2%	44	$2^{17.96}$
$[1024, 524, 50]$	$[4, 16, 2]$	18360	18360.4	43.3%	207	$2^{38.74}$

We did not put effort in optimizing our code for speed by e.g. using clever data structures or hash tables as it was done in [3]. We leave it has an open problem to implement an efficient version of our algorithm for determining the cut-off point with other variants of information set decoding, such as Stern, FS-ISD or Ball-collision decoding.

Acknowledgements. The authors would like to thank Antoine Joux for useful discussions and Jannik Pewny for carrying out the experiments.

References

1. Alon, N., Spencer, J.: The Probabilistic Method. Wiley (2008)
2. Berlekamp, E., McEliece, R., van Tilborg, H.: On the inherent intractability of certain coding problems (Coresp.). IEEE Transactions on Information Theory 24(3), 384–386 (1978)
3. Bernstein, D.J., Lange, T., Peters, C.: Attacking and Defending the McEliece Cryptosystem. In: Buchmann, J., Ding, J. (eds.) PQCrypto 2008. LNCS, vol. 5299, pp. 31–46. Springer, Heidelberg (2008)
4. Bernstein, D.J., Lange, T., Peters, C.: Smaller Decoding Exponents: Ball-Collision Decoding. In: Rogaway, P. (ed.) CRYPTO 2011. LNCS, vol. 6841, pp. 743–760. Springer, Heidelberg (2011)
5. Finiasz, M., Sendrier, N.: Security Bounds for the Design of Code-Based Cryptosystems. In: Matsui, M. (ed.) ASIACRYPT 2009. LNCS, vol. 5912, pp. 88–105. Springer, Heidelberg (2009)
6. Guruswami, V.: Introduction to Coding Theory. Lecture Notes (2010)
7. Hopper, N.J., Blum, M.: Secure Human Identification Protocols. In: Boyd, C. (ed.) ASIACRYPT 2001. LNCS, vol. 2248, pp. 52–66. Springer, Heidelberg (2001)
8. Howgrave-Graham, N., Joux, A.: New Generic Algorithms for Hard Knapsacks. In: Gilbert, H. (ed.) EUROCRYPT 2010. LNCS, vol. 6110, pp. 235–256. Springer, Heidelberg (2010)
9. Lee, P.J., Brickell, E.F.: An Observation on the Security of McEliece's Public-Key Cryptosystem. In: Günther, C.G. (ed.) EUROCRYPT 1988. LNCS, vol. 330, pp. 275–280. Springer, Heidelberg (1988)
10. Leon, J.S.: A probabilistic algorithm for computing minimum weights of large error-correcting codes. IEEE Transactions on Information Theory 34(5), 1354 (1988)
11. McEliece, R.J.: A public-key cryptosystem based on algebraic coding theory. In: DSN Progress Report 42–44 (1978)
12. Regev, O.: On lattices, learning with errors, random linear codes, and cryptography. In: Gabow, H.N., Fagin, R. (eds.) STOC, pp. 84–93. ACM (2005)
13. Stern, J.: A method for finding codewords of small weight. In: Wolfmann, J., Cohen, G.D. (eds.) Coding Theory 1988. LNCS, vol. 388, pp. 106–113. Springer, Heidelberg (1989)

Lower and Upper Bounds for Deniable Public-Key Encryption

Rikke Bendlin[1,*], Jesper Buus Nielsen[1,*,**], Peter Sebastian Nordholt[1,*],
and Claudio Orlandi[2,***]

[1] Aarhus University, Denmark
{rikkeb,jbn,psn}@cs.au.dk
[2] Bar-Ilan University, Israel
claudio.orlandi@biu.ac.il

Abstract. A deniable cryptosystem allows a sender and a receiver to communicate over an insecure channel in such a way that the communication is still secure even if the adversary can threaten the parties into revealing their internal states after the execution of the protocol. This is done by allowing the parties to change their internal state to make it look like a given ciphertext decrypts to a message different from what it really decrypts to. Deniable encryption was in this way introduced to allow to deny a message exchange and hence combat coercion.

Depending on which parties can be coerced, the security level, the flavor and the number of rounds of the cryptosystem, it is possible to define a number of notions of deniable encryption.

In this paper we prove that there does not exist any non-interactive receiver-deniable cryptosystem with better than polynomial security. This also shows that it is impossible to construct a non-interactive bi-deniable public-key encryption scheme with better than polynomial security. Specifically, we give an explicit bound relating the security of the scheme to how efficient the scheme is in terms of key size. Our impossibility result establishes a lower bound on the security.

As a final contribution we give constructions of deniable public-key encryption schemes which establishes upper bounds on the security in terms of key length. There is a gap between our lower and upper bounds, which leaves the interesting open problem of finding the tight bounds.

1 Introduction

Alice and Bob live in a country ruled by an evil dictator, Eve. If Alice wants to communicate with Bob, standard public-key cryptography can be used by

* Supported by the Danish National Research Foundation and The National Science Foundation of China (under the grant 61061130540) for the Sino-Danish Center for the Theory of Interactive Computation.
** Supported by a Sapere Aude grant from The Danish Council for Independent Research.
*** Supported by the European Research Council as part of the ERC project LAST.

D.H. Lee and X. Wang (Eds.): ASIACRYPT 2011, LNCS 7073, pp. 125–142, 2011.

Alice if she wants to keep Eve from learning the subject of her communication with Bob. However, if Eve controls the network she will be able to observe that a ciphertext is traveling from Alice to Bob. Once the evil Eve knows that a conversation took place, she might get suspicious and force Bob to reveal the content of the conversation. Can cryptography offer any help to Alice and Bob against such a powerful adversary? To solve this problem Canetti, Dwork, Naor and Ostrovsky [CDNO97] introduced the notion of deniable encryption as a tool to combat coercion.

Using a deniable cryptosystem Alice and Bob can communicate over an insecure channel in a way such that even if Eve records the transcript of the communication and later coerces Alice (resp. Bob, or both) to reveal their internal state (secret keys, randomness, ...), then Alice (resp. Bob, or both) has an efficient strategy to produce an alternative internal state that is consistent with the transcript and with a message different than the original one.

Threat model: First note that deniable encryption does not help if Eve has physical access to Alice and Bob's computers. In this case nothing can prevent Eve from seeing everything that Bob sees and therefore learn the encrypted message—since we want Alice and Bob to actually communicate information between them, this is unavoidable. On the other hand, if Alice and Bob can *erase* their secret information, they could simply lie about the content of a ciphertext: the standard indistinguishability security requirement implies that Eve cannot check whether the ciphertext is really an encryption of the message that Alice and Bob claim it to be. Therefore, as in [CDNO97], we consider the case where the parties hand their private keys and randomness to Eve, who can then check that the revealed message is in fact consistent with the ciphertext she observed earlier. If the parties are able to produce a reasonable explanation for the ciphertext that Eve observes, this is enough to fight this kind of coercion.

Sender/Receiver/Bi-Deniability: We distinguish between three kinds of deniability, according to which parties can be coerced by Eve. Note that, up to the number of rounds required by the protocol, sender and receiver deniability are equivalent: Bob can use a sender-deniable scheme to send a random key K to Alice, who can use it to encrypt the message M using a one-time pad and send back $C = M \oplus K$. Now if Bob is coerced he can claim to have received a different message M' by using the sender-deniable property and explain the transcript as if it contained a different K'.

When we consider bi-deniability, the case where Eve can coerce both Alice and Bob, the only coordination that we allow between Alice and Bob is to agree on which message to fake the ciphertext to. In particular this means that the parties cannot communicate to each other their internal states, when they have to produce a fake explanation. This seems to be the only meaningful definition: if Alice and Bob could communicate this information through a channel not controlled by Eve, why would they not use this channel to communicate the original message in the first place?

Fully-Deniable vs. Multi-Distributional: In a multi-distributional deniable cryptosystem a ciphertext produced with a "fake" encryption algorithm E_F can be later explained as an encryption of any message under the "standard" encryption algorithm E. In other words, for any m, m' it is possible to find appropriate randomness for E, E_F such that $E(m') = E_F(m)$. Note however, that Eve might not believe that the ciphertext was produced using E and ask to see the internal state for E_F and in this case the parties have no efficient strategy to lie about the content of the ciphertext. A fully-deniable scheme is a scheme where $E = E_F$ and therefore does not present this issue.

Public-key vs. Interactive Cryptosystems: A (receiver/sender/bi)-deniable *public-key* cryptosystem is a public-key cryptosystem that is (receiver/sender/bi)-deniable. I.e., the cryptosystem consist of a public key known by the sender and the communication protocol consists of sending a ciphertext to the receiver. A generic, or interactive, cryptosystem might involve arbitrary interaction.

Security Level: All notions of deniability can be quantified by $\varepsilon : \mathbb{N} \to \mathbb{R}_+$ which measures how indistinguishable the faked states are from the honest states. As an example, an ε-receiver-deniable public-key cryptosystem is one in which the faked secret key is ε-indistinguishable from the honest secret key to a computationally bounded distinguisher. We will distinguish between schemes where ε is a negligible function and where ε is of the form $1/p$, for some polynomial p. We will idiosyncratically say that the former kind has negligible security and the latter polynomial security.

Prior Work, Our Contributions and Open Questions: Deniable encryption was first introduced and defined in [CDNO97]. They constructed a sender-deniable public-key cryptosystem with polynomial security, and therefore a receiver-deniable interactive cryptosystem. In [OPW11] O'Neill, Peikert and Waters showed how to construct multi-distributional bi-deniable public-key encryption with negligible security. This is the first scheme that achieves any kind of deniability when both parties are corrupted. Recently, Dürmuth and Freeman announced a fully-deniable (receiver/sender)-deniable interactive cryptosystem with negligible security [DF11]. However their result was later showed to be incorrect by Peikert and Waters.

Our contribution to the state of the art on deniable-encryption is to derive upper and lower bounds on how secure a deniable public-key encryption scheme can be as a function of the key-size.

Lower bounds: As for lower bounds, we have the following results.
  Receiver: We show that any public-key cryptosystem with σ-bit keys can be at most $\frac{1}{2}(\sigma + 1)^{-1}$-receiver-deniable.
  Sender: We do not know of a non-trivial lower bound for sender-deniable public-key encryption.
  Bi: Since bi-deniable public-key encryption with σ-bit keys implies receiver-deniable public-key encryption with σ-bit keys, any public-key cryptosystem with σ-bit keys can be at most $\frac{1}{2}(\sigma + 1)^{-1}$-bi-deniable.
Upper bounds: We show three upper bounds.

Receiver: If we let κ denote the length of the secret key of the best multi-distributional receiver-deniable public-key encryption scheme, then there exists a $1/n$-receiver-deniable public-key encryption scheme with key length $\sigma = O(n^2 \kappa)$.

Sender: If we let κ denote the length of the sender randomness in the best multi-distributional sender-deniable public-key encryption scheme, then there exists a $1/n$-sender-deniable public-key encryption scheme where the sender randomness has length $\sigma = O(n\kappa)$.

Bi: If we let κ denote the length of the secret key of the best multi-distributional bi-deniable public-key encryption scheme, then there exists a $1/n$-bi-deniable public-key encryption scheme with key length $\sigma = O(n^4 \kappa)$.

We phrase the upper bounds in terms of the upper bounds for multi-distributional schemes. The reason for this is that we do not know of any assumption which allows to construct deniable public-key encryption with polynomial security, which does not also allow to construct multi-distributional deniable encryption. And, we do not know of any direct construction of deniable public-key encryption with polynomial security which is more efficient than going via a multi-distributional scheme. It therefore seems that multi-distributional schemes are the natural building block for deniable public-key encryption with polynomial security.

Our upper bounds for receiver-deniability and sender-deniability are similar to bounds which can be derived from constructions in [OPW11]. Our upper bound for bi-deniability is new. In [OPW11] a construction of a bi-deniable public-key encryption scheme is hinted, but no explicit construction is given which makes it impossible to estimate the complexity. The hinted construction is, however, different from the one we give here.

Our lower bound for receiver-deniability is a generalization of a result in [CDNO97], where a similar bound was proven for any so-called *separable* public-key encryption scheme. An encryption scheme being separable is, however, a very strong structural requirement, so it was unclear if the bound in [CDNO97] should hold for any scheme. In fact, we have not been able to find even a conjecture in the more than a decade of literature between [CDNO97] and the present result that polynomial security should be optimal in general. Our proof technique is completely different from the one in [CDNO97], as we cannot make any structural assumption about the encryption scheme in question.

Our work leaves a number of interesting open problems.

1. Our proof of the upper bounds are via black-box constructions of deniable public-key encryption with polynomial security from multi-distributional deniable public-key encryption. This shows that multi-distributional deniable public-key encryption is stronger than deniable public-key encryption with polynomial security. Is it *strictly* stronger, or does there exist a black-box construction of multi-distributional deniable public-key encryption from deniable public-key encryption with polynomial security?

Table 1. The current state of the art for deniable encryption. The first column distinguishes between fully-deniable schemes and schemes with multi-distributional deniability. The Sender/Receiver/Bi columns contains "✓" if any construction is known; a "✗" indicates an impossibility result; a "?" marks a question that is still open.

Notion	Security	Interaction	Sender	Receiver	Bi
Full-Deniability	Negligible	Interactive	?		?
		Public-key	?	✗	✗
	Polynomial	Public-key	✓	✓	✓
Multi-Distributional	Negligible	Public-key	✓	✓	✓

2. Our lower bounds do not apply to sender-deniable public-key encryption. Is it possible to construct sender-deniable public-key encryption with better than polynomial security?
3. Our lower bounds do not apply to interactive encryption schemes. Is it possible to construct deniable encryption schemes with better than polynomial security when arbitrary interaction is allowed?
4. There is a gap between our upper and lower bounds of at least a factor κ. Since κ itself is typically, for practical purposes, a rather large number (multi-distributional schemes are not simple objects on themselves), this gap is important in practice. What are the tight bounds on the security of a deniable public-key encryption scheme? We conjecture that the bound is in the order of σ^{-1}.

Non-committing encryption: Canetti, Feige, Goldreich and Naor introduced the notion of a non-committing cryptosystem, which is similar to the notion of a bi-deniable cryptosystem, but it is only required that the faking can be done by a simulator. This simulator is allowed to use public keys with a different distribution than those in the protocol. This is needed when showing adaptive security in simulation-based models. It is known [CFGN96] how to implement non-committing encryption with negligible security. Several improvements over the original scheme (both in terms of efficiency and assumptions) have been published in [Bea97, DN00, KO04, GWZ09, CDSMW09].

In [Nie02] it was shown that non-interactive non-committing encryption is impossible. This does not imply the negative result we are proving here, as receiver-deniable public-key encryption does not imply non-committing encryption. In non-committing encryption both sides have to be faked. In receiver-deniable encryption, only the receiver has to be faked. In this sense non-committing encryption is a stronger notion than receiver-deniable encryption. But, in fact, the notions are incomparable, as receiver-deniable encryption on other axes is stronger than non-committing encryption. As an example, it can be shown that if a public-key encryption scheme is receiver-deniable, then the parallel composition of the scheme where the same public key is used to encrypt many massages is also receiver-deniable. This is a property which non-committing encryption provably does not have. And, in fact, this self composition property is crucial in the proof of our lower bound. Also, the result in [Nie02] addresses the case of perfect non-committing encryption (the real-world and the simulated world must be indistinguishable). We are interested in the exact level of security which can be obtained i.e., given a public-key encryption scheme with a certain secret-key length, how deniable can the scheme be?

Structure: In Section 2 we formally define the different flavors of deniable public-key encryption. In Section 3 we show that receiver-deniability is maintained under parallel self-composition with at most a linear security loss. We use that fact to derive our lower bounds giving us the impossibility result of fully-receiver deniable encryption. Finally, section 4 contains our results on poly-deniable encryption schemes.

2   Deniable Public-Key Encryption

In this section we define three different notions of deniable public-key encryption schemes. These notions correspond respectively to an adversary with the ability to coerce the receiver, the sender or both parties simultaneously. We model coercion by letting the adversary request the secret information used in the encryption scheme by the coerceable parties. Deniability is obtained by letting the coerceable parties supply fake secret information.

Basic Scheme. All schemes are defined based on the following definition of a standard public-key encryption scheme consisting of three probabilistic polynomial-time algorithms $(\mathsf{G}, \mathsf{E}, \mathsf{D})$:

- $\mathsf{G}(1^\kappa)$ generates a key-pair (pk, sk), where pk is the public key, sk is the secret key and κ is the security parameter. Note that we consider sk to be the randomness used in $\mathsf{G}(1^\kappa)$.
- $\mathsf{E}_{pk}(m; r)$ generates a ciphertext c which is an encryption under the public key pk of message $m \in \{0, 1\}^\ell$ using randomness r. We sometimes write $\mathsf{E}_{pk}(m)$ to make the randomness be implicit.
- $\mathsf{D}_{sk}(c)$ outputs the message $m \in \{0, 1\}^\ell$ contained in the ciphertext c.

Let negl $: \mathbb{N} \to \mathbb{R}_+$ be a negligible function. For all notions defined below we require correctness, i.e., we require that $\Pr[\mathsf{D}_{sk}(\mathsf{E}_{pk}(m)) = m] > 1 - \mathrm{negl}(\kappa)$, and IND-CPA security i.e., we require that $\forall$ PPT $(A_1, A_2), \exists \, \mathrm{negl}(\cdot)$:

$$\Pr[(pk, sk) \leftarrow \mathsf{G}(1^\kappa), (m_0, m_1, \mathsf{st}) \leftarrow A_1(pk),$$
$$c = \mathsf{E}_{pk}(m_b), b' \leftarrow A_2(c, \mathsf{st}) : b = b'] < 1/2 + \mathrm{negl}(\kappa) .$$

Multi-distributional Encryption. We define a general form of deniable public-key encryption called multi-distributional deniable public-key encryption. Such a scheme essentially consists of two standard public-key schemes sharing a common decryption algorithm.

- The honest scheme $(\mathsf{G}, \mathsf{E}, \mathsf{D})$ does not provide deniability in itself.
- The fakeable scheme $(\mathsf{G_F}, \mathsf{E_F}, \mathsf{D})$ provides deniability in the sense that, for a ciphertext c fake secret information can be generated. The faked secret information will make c appear as an encryption of any chosen message m' in the honest scheme. How this is done depends on the notion of deniability as defined below.

For a multi-distributional deniable public-key encryption scheme to be correct we require standard correctness of all public-key schemes $(\mathsf{G'}, \mathsf{E'}, \mathsf{D})$ where $\mathsf{G'} \in \{\mathsf{G}, \mathsf{G_F}\}$ and $\mathsf{E'} \in \{\mathsf{E}, \mathsf{E_F}\}$.

The idea behind having two different schemes is to use the fakeable scheme to encrypt a message m on which the parties would like to have deniability. When coerced the parties simply claim that they used the honest scheme to encrypt the fake message m'. This approach has two disadvantages. First, the parties must decide beforehand whether they later want to deny. Secondly, is the question of why a coercer should believe the parties, when they claim to have used the honest scheme. Note that we cannot guarantee deniability, if the coercer insists on getting the secret information used in the faking process.

Fully-deniable Encryption. An important special case of multi-distributional deniable public-key encryption is fully-deniable public-key encryption (or just deniable public-key encryption). This notion addresses the disadvantages of multi-distributional encryption mentioned above. For a fully-deniable public-key encryption scheme we have that $(\mathsf{G}, \mathsf{E}, \mathsf{D}) = (\mathsf{G_F}, \mathsf{E_F}, \mathsf{D})$, that is there are no special faking key generation and encryption algorithms. We will often omit the prefix 'fully' for simplicity.

Receiver-Deniability. A multi-distributional receiver-deniable public-key encryption scheme consists of five probabilistic polynomial-time algorithms $(\mathsf{G}, \mathsf{G_F}, \mathsf{E}, \mathsf{D}, \mathsf{F_R})$. Here $(\mathsf{G}, \mathsf{E}, \mathsf{D})$ is the honest scheme and $(\mathsf{G_F}, \mathsf{E}, \mathsf{D})$ is the fakeable scheme. Notice that the honest and fakeable encryption algorithm are the same since faking is only done on the receiver's side. The faking algorithm $\mathsf{F_R}$ is defined as follows:

- For $(pk, sk) \leftarrow \mathsf{G_F}(1^\kappa)$ and $c \leftarrow \mathsf{E}_{pk}(m)$, $\mathsf{F_R}(sk, c, m')$ generates an alternative secret key sk' such that $\mathsf{D}_{sk'}(c) = m'$.

Sender-Deniability. A multi-distributional sender-deniable public-key encryption scheme consists of five probabilistic polynomial-time algorithms $(\mathsf{G}, \mathsf{E}, \mathsf{E_F}, \mathsf{D}, \mathsf{F_S})$. Here $(\mathsf{G}, \mathsf{E}, \mathsf{D})$ is the honest scheme and $(\mathsf{G}, \mathsf{E_F}, \mathsf{D})$ is the fakeable scheme. The faking algorithm $\mathsf{F_S}$ is defined as follows:

- $\mathsf{F_S}(pk, m, r, m')$ generates alternative randomness r' such that $\mathsf{E_F}_{pk}(m; r) = \mathsf{E}_{pk}(m'; r')$.

Bi-Deniability. We assume here to be in a setting where receiver and sender have individual faking algorithms. This models the fact that, after an initial stage where the parties can agree on which message to fake to, the sender and the receiver cannot communicate over a channel that is not controlled by the adversary—otherwise they could be using this channel to communicate the message m in the first place.

A multi-distributional bi-deniable public-key encryption scheme consists of seven probabilistic polynomial-time algorithms $(\mathsf{G}, \mathsf{G_F}, \mathsf{E}, \mathsf{E_F}, \mathsf{D}, \mathsf{F_R}, \mathsf{F_S})$. The faking algorithms $\mathsf{F_R}$ and $\mathsf{F_S}$ are defined similar to the receiver-deniable and sender-deniable notions respectively, that is:

- For $(pk, sk) \leftarrow \mathsf{G_F}(1^\kappa)$ and $c \leftarrow \mathsf{E_{F}}_{pk}(m)$, $\mathsf{F_R}(sk, c, m')$ generates an alternative secret key sk' such that $\mathsf{D}_{sk'}(c) = m'$.
- $\mathsf{F_S}(pk, m, r, m')$ generates alternative randomness r' such that $\mathsf{E_{F}}_{pk}(m; r) = \mathsf{E}_{pk}(m'; r')$.

2.1   Security Notions

The security notions of the three schemes above, are defined in terms of the following experiments performed with an adversary $A = (A_1, A_2)$, where $m, m' \in \{0, 1\}^\ell$.

Honest Game (Receiver)	Faking Game (Receiver)
$(pk, sk) \leftarrow \mathsf{G}(1^\kappa)$	$(pk, sk) \leftarrow \mathsf{G_F}(1^\kappa)$
$(m, m', \mathsf{st}) \leftarrow A_1(pk)$	$(m, m', \mathsf{st}) \leftarrow A_1(pk)$
$c \leftarrow \mathsf{E}_{pk}(m'; r)$	$c \leftarrow \mathsf{E}_{pk}(m; r)$
	$sk' \leftarrow \mathsf{F_R}(sk, c, m')$
$b \leftarrow A_2(\mathsf{st}, c, sk)$	$b \leftarrow A_2(\mathsf{st}, c, sk')$

Honest Game (Sender)	Faking Game (Sender)
$(pk, sk) \leftarrow \mathsf{G}(1^\kappa)$	$(pk, sk) \leftarrow \mathsf{G}(1^\kappa)$
$(m, m', \mathsf{st}) \leftarrow A_1(pk)$	$(m, m', \mathsf{st}) \leftarrow A_1(pk)$
$c \leftarrow \mathsf{E}_{pk}(m'; r)$	$c \leftarrow \mathsf{E_{F}}_{pk}(m; r)$
	$r' \leftarrow \mathsf{F_S}(pk, m, r, m')$
$b \leftarrow A_2(\mathsf{st}, c, r)$	$b \leftarrow A_2(\mathsf{st}, c, r')$

Honest Game (Bi)	Faking Game (Bi)
$(pk, sk) \leftarrow \mathsf{G}(1^\kappa)$	$(pk, sk) \leftarrow \mathsf{G_F}(1^\kappa)$
$(m, m', \mathsf{st}) \leftarrow A_1(pk)$	$(m, m', \mathsf{st}) \leftarrow A_1(pk)$
$c \leftarrow \mathsf{E}_{pk}(m'; r)$	$c \leftarrow \mathsf{E_{F}}_{pk}(m; r)$
	$sk' \leftarrow \mathsf{F_R}(sk, c, m')$
	$r' \leftarrow \mathsf{F_S}(pk, m, r, m')$
$b \leftarrow A_2(\mathsf{st}, c, sk, r)$	$b \leftarrow A_2(\mathsf{st}, c, sk', r')$

Let $h_A(\kappa)$ and $f_A(\kappa)$ be the random variables describing b when running the honest game and faking game respectively with security parameter κ. The advantage of A is

$$\mathrm{Adv}_A(\kappa) = |h_A(\kappa) - f_A(\kappa)| \ .$$

We say that a scheme is (receiver/sender/bi)-deniable if Adv_A is negligible in κ for any efficient A. Let $\varepsilon : \mathbb{N} \to \mathbb{R}_+$. We say that a scheme is ε-(receiver/sender/bi)-deniable if $\mathrm{Adv}_A(\kappa) \le \varepsilon(\kappa) + \mathrm{negl}(\kappa)$.

2.2   Full Bi-deniablity Implies Full Sender/Receiver-Deniability

Any fully bi-deniable scheme can trivially be turned into both a receiver-deniable and a sender-deniable scheme. On the surface this seems obvious, if both parties can fake then they should be able to fake individually as well. Surprisingly, however, this conclusion cannot be drawn in the multi-distributional setting—in [OPW11] the authors show that in this setting bi-deniability does imply sender deniability but not receiver deniability. As stated in Lemma 1 similar subtleties do not arise in the fully-deniable case. A proof of this can be found in the full version.

Lemma 1. *If* $(\mathsf{G}, \mathsf{E}, \mathsf{D}, \mathsf{F_R}, \mathsf{F_S})$ *is a fully ε-bi-deniable encryption scheme, then* $(\mathsf{G}, \mathsf{E}, \mathsf{D}, \mathsf{F_S})$ *is a fully ε-sender-deniable encryption scheme and* $(\mathsf{G}, \mathsf{E}, \mathsf{D}, \mathsf{F_R})$ *is a fully ε-receiver-deniable encryption scheme.*

3   Impossibility of Fully Receiver/Bi-deniable Encryption

In this section we prove the impossibility of fully receiver-deniable and fully bi-deniable public-key encryption with better than inverse polynomial security. Since, by Lemma 1, any fully bi-deniable public-key encryption scheme is also a fully receiver-deniable public-key encryption scheme, it is sufficient to prove impossibility of fully receiver-deniable public-key encryption. It turns out that the impossibility follows readily from the fact that full receiver-deniability is preserved under parallel self-composition with only a linear security loss.

We will use a slightly modified definition of receiver-deniability. Recall that in the definition from section 2 the faking algorithm $\mathsf{F_R}$ is invoked as $\mathsf{F_R}(sk, c, m')$, especially it is not given the sender's randomness r. In this section we will allow $\mathsf{F_R}$ to have access to r, that is $\mathsf{F_R}$ is invoked as $\mathsf{F_R}(sk, m, r, m')$. Since we are proving an impossibility result, this does not weaken the result.

3.1   Security of Parallel Self-composition

Let $(\mathsf{G}, \mathsf{E}, \mathsf{D}, \mathsf{F_R})$ be any receiver-deniable public-key cryptosystem. Let $n : \mathbb{N} \to \mathbb{N}$ be a polynomial in the security parameter κ. We define the parallel self-composition $(\mathsf{G}^n, \mathsf{E}^n, \mathsf{D}^n, \mathsf{F_R}^n)$ as follows:

$$\mathsf{G}^n(1^\kappa) = \mathsf{G}(1^\kappa)$$
$$\mathsf{E}^n_{pk}(m_1,\dots,m_n;r_1,\dots,r_n) = (\mathsf{E}_{pk}(m_1;r_1),\dots,\mathsf{E}_{pk}(m_n;r_n))$$
$$\mathsf{D}^n_{sk}(c_1,\dots,c_n) = (\mathsf{D}_{sk}(c_1),\dots,\mathsf{D}_{sk}(c_n))$$
$$\mathsf{F_R}^n(sk,(m_1,\dots,m_n),(r_1,\dots,r_n),(m'_1,\dots,m'_n)) = sk'\ ,$$

where $sk_0 = sk$, $sk_i \leftarrow \mathsf{F_R}(sk_{i-1}, m_i, r_i, m'_i)$ for $i = 1,\dots,n$ and $sk_n = sk'$.

Lemma 2. *If* $(\mathsf{G}, \mathsf{E}, \mathsf{D}, \mathsf{F_R})$ *is ε-receiver-deniable, then* $(\mathsf{G}^n, \mathsf{E}^n, \mathsf{D}^n, \mathsf{F_R}^n)$ *is $n\varepsilon$-receiver-deniable.*

Proof. Let $A^n = (A^n_1, A^n_2)$ be any probabilistic polynomial-time attacker against $(\mathsf{G}^n, \mathsf{E}^n, \mathsf{D}^n, \mathsf{F_R}^n)$. For $h = 1,\dots,n$ we construct from A^n a probabilistic polynomial-time attacker $A_h = (A_{h,1}, A_{h,2})$ against $(\mathsf{G}, \mathsf{E}, \mathsf{D}, \mathsf{F_R})$. We can then describe the advantage of A^n in terms of the advantages of A_h for $h = 1,\dots,n$. Since, by assumption on $(\mathsf{G}, \mathsf{E}, \mathsf{D}, \mathsf{F_R})$, we have a bound on the advantage of each A_h, this gives us the bound on the advantage of A^n. The attacker A_h runs as follows:

1. $A_{h,1}$: Receives pk.
2. $A_{h,1}$: Input pk to A^n_1 and run A^n_1 to obtain $(m_1,\dots,m_n)$, $(m'_1,\dots,m'_n)$ and state st_{A^n}.
3. $A_{h,1}$: For $i = 1,\dots,h-1$, sample $c_i \leftarrow \mathsf{E}_{pk}(m'_i)$.
4. $A_{h,1}$: Output (m_h, m'_h, st_{A_h}) where $st_{A_h} = ((m_1,\dots,m_n),\ (m'_1,\dots,m'_n),\ st_{A^n}, (c_1,\dots,c_{h-1}))$.
5. $A_{h,2}$: Receive (st_{A_h}, c, sk). Let $c_h = c$ and $sk_h = sk$.
6. $A_{h,2}$: For $i = h+1,\dots,n$, sample $c_i \leftarrow \mathsf{E}_{pk}(m_i; r_i)$ and $sk_i \leftarrow \mathsf{F_R}(sk_{i-1}, m_i, r_i, m'_i)$.
7. $A_{h,2}$: Input $(st_{A^n}, (c_1,\dots,c_n), sk_n)$ to A^n and run it to obtain a bit $b \in \{0,1\}$.
8. $A_{h,2}$: Output b.

Let b^0_h be the distribution of the bit b output by A_h when run in the honest game and let b^1_h be the distribution of the bit b output by A_h when run in the faking game.

When A_h is run in the honest game, then sk_n is computed from an honest secret key sk_h as $sk_i \leftarrow \mathsf{F_R}(sk_{i-1}, m_i, r_i, m'_i)$ for $i = h+1,\dots,n$. When A_h is run in the faking game, then sk_n is computed from an honest secret key sk_{h-1} as $sk_i \leftarrow \mathsf{F_R}(sk_{i-1}, m_i, r_i, m'_i)$ for $i = h,\dots,n$, where the first computation $sk_h \leftarrow \mathsf{F_R}(sk_{h-1}, m_h, r_h, m'_h)$ is performed by the faking game before sk_h is input to A_h. It follows that when A_h is run in the honest game and A_{h+1} is run in the faking game, the values input to A^n have identical distributions, so $b^1_h = b^0_{h-1}$. Let Adv_{A_h} denote the advantage of A_h against $(\mathsf{G}, \mathsf{E}, \mathsf{D}, \mathsf{F_R})$ and Adv_{A^n} be the advantage of A^n against $(\mathsf{G}^n, \mathsf{E}^n, \mathsf{D}^n, \mathsf{F_R}^n)$. We then have by definition $\mathrm{Adv}_{A_h}(\kappa) = |b^0_h - b^1_h|$ and by construction $\mathrm{Adv}_{A^n}(\kappa) = |b^0_n - b^1_1|$, where κ is the security parameter. It then follows using telescoping and the triangle inequality that $\mathrm{Adv}_{A^n}(\kappa) \le n\varepsilon(\kappa) + \sum_{h=1}^n \mathrm{negl}_h(\kappa)$, where all negl_h are negligible in κ. The lemma then follows from the fact that the sum of polynomially many negligible functions is negligible. $\qquad\square$

Notice that Lemma 2 means that a faked secret key sk_n, resulting from $\mathsf{F_R}^n$, must somehow *remember* the faking of each ciphertext involved in the process. In other words sk_n must not only fake a single ciphertext, it must ensure that every ciphertext c_i decrypts to the faked message m_i' with high probability. To see why consider the efficient adversary A of the receiver-deniable game against $(\mathsf{G}^n, \mathsf{E}^n, \mathsf{D}^n, \mathsf{F_R}^n)$ that simply outputs $b = 1$ if $m_i' = \mathsf{D}_{sk}(c_i)$ for all $i = 1, \ldots, n$ and $b = 0$ otherwise. By correctness of the encryption scheme and by Lemma 2 the above property of sk_n becomes clear.

Let s be a bit string of length n. In the proof of the following theorem we use this property to show how to associate each bit of s with a faking of a ciphertext and thus how to store s in the *memory* of the faked secret key sk_n. The impossibility result arises from the fact that this can be done even for random s longer than sk_n.

3.2   Lower Bound

We here show a lower bound on ε in an ε-receiver-deniable encryption scheme. This bound immediately gives that one cannot obtain better than polynomial security. The bound is stated formally in the following theorem:

Theorem 1. *Let* $(\mathsf{G}, \mathsf{E}, \mathsf{D}, \mathsf{F_R})$ *be ε-receiver deniable, and let σ be an upper bound on the length of the secret keys of* $(\mathsf{G}, \mathsf{E}, \mathsf{D}, \mathsf{F_R})$, *including the faked ones. Then* $\varepsilon \geq \frac{1}{2}(\sigma + 1)^{-1} - \mathrm{negl}(\kappa)$.

Proof. We reach our bound via impossibility of compressing uniformly random data. Let $n = \sigma + 1$. We can assume that $(\mathsf{G}, \mathsf{E}, \mathsf{D}, \mathsf{F_R})$ can encrypt at least one bit, so $(\mathsf{G}^n, \mathsf{E}^n, \mathsf{D}^n, \mathsf{F_R}^n)$ can encrypt n-bit messages. Furthermore $(\mathsf{G}^n, \mathsf{E}^n, \mathsf{D}^n, \mathsf{F_R}^n)$ is $n\varepsilon$-receiver-deniable.

Consider the following communication protocol parametrized by κ. Here is how the sender works:

1. Sample $(pk, sk) \leftarrow \mathsf{G}^n(1^\kappa)$.
2. Sample uniformly random $m' \leftarrow \{0, 1\}^n$ and let $m = 0^n$.
3. Sample $c \leftarrow \mathsf{E}_{pk}^n(m; r)$.
4. Let $sk' \leftarrow \mathsf{F_R}^n(sk, m, r, m')$.
5. Send (c, sk').

On receiving (c, sk') the receiver outputs $m'' = \mathsf{D}_{sk'}^n(c)$.

To bound the probability that this protocol fails i.e., that $m'' \neq m'$, consider the following adversary $A = (A_1, A_2)$ for the receiver-deniable security games against $(\mathsf{G}^n, \mathsf{E}^n, \mathsf{D}^n, \mathsf{F_R}^n)$. On input pk A_1 outputs (m, m', st), where the messages m and m' are sampled as in step 2 of the sender algorithm above. The state st is set to be m'. On input (st, c, sk') A_2 computes $\mathsf{D}_{sk'}^n(c) = m''$ and outputs 1 if $m'' = m' = \mathsf{st}$ and 0 otherwise. Now notice that steps 1-4 of the sender algorithm above correspond to the first four steps of the receiver-deniable faking game against A. That is the probability that the communication protocol fails i.e., that $m'' \neq m'$, is exactly the same as A_2 outputting 0 in the faking game. In the honest game we have by correctness of $(\mathsf{G}^n, \mathsf{E}^n, \mathsf{D}^n, \mathsf{F_R}^n)$ that A_2 only outputs 0 with negligible probability. Thus by $n\varepsilon$-receiver deniability we have $\Pr[m'' \neq m'] \leq n\varepsilon(\kappa) + \mathrm{negl}(\kappa)$. We later use this bound on the correctness of the communication protocol to derive our bound, but first we transform the protocol a bit.

For each κ, let r_κ be the value which minimizes the probability that $m'' \neq m'$ when $c_\kappa = \mathsf{E}_{pk}(0^n; r_\kappa)$. Consider then the following non-uniform communication protocol parametrized by κ. Here is how the sender works:

1. Sample $(pk, sk) \leftarrow \mathsf{G}(1^\kappa)$.
2. Sample $m' \leftarrow \{0, 1\}^n$.
3. Let $sk' \leftarrow \mathsf{F_R}(sk, 0^n, r_\kappa, m')$.
4. Send sk'.

The receiver outputs $m'' = \mathsf{D}_{sk'}(c_\kappa)$, where $c_\kappa = \mathsf{E}_{pk}(0; r_\kappa)$. Note that r_κ and c_κ are hardwired into the protocol and is therefore not communicated as part of the protocol. We still have that $\Pr[m'' \neq m'] \leq n\varepsilon(\kappa) + \mathrm{negl}(\kappa)$. Using that $n = \sigma + 1$ we get that $(\sigma + 1)\varepsilon(\kappa) \geq 1 - \Pr[m'' = m'] - \mathrm{negl}(\kappa)$. From incompressibility of uniformly random data it follows that $\Pr[m'' = m'] \leq 2^{\sigma - n} = 2^{-1}$, as the protocol sends only sk', which is at most σ bits long and because m' is uniformly random and $n = \sigma + 1$ bits long. Combining these bounds we get that $\varepsilon(\kappa) \geq \frac{1}{2}(\sigma + 1)^{-1} - \mathrm{negl}(\kappa)$. $\qquad\square$

In words, this bound says that any public-key cryptosystem with σ-bit keys can be be at most $\frac{1}{2}(\sigma + 1)^{-1}$-receiver-deniable. Thus to get negligible receiver-deniability keys must be superpolynomial in size. This however would contradict the key generation algorithm being polynomial-time as required by our definition of a public-key cryptosystem.

4   From Multi-distributional to Poly Deniability

We now give explicit constructions of poly-(sender/receiver/bi)-deniable public-key encryption schemes from any multi-distributional (sender/reciever/bi)-deniable public-key encryption scheme respectively. As in [CDNO97, OPW11], the basic idea in all these constructions is to encrypt a message bit b by first writing it as $b = \bigoplus_{i=1}^{n} b_i$ for random b_i's, and then encrypting each b_i independently using randomly either the honest or the fakeable encryption scheme. To fake we just have to identify an index j where the fakeable scheme was used and use the corresponding faking algorithm. This is no problem for sender and receiver deniablility since in those cases whoever is running the faking algorithm knows exactly on which indices the fakeable scheme was used. The bi-deniable case however is more challenging because sender and receiver must agree on an index j where they both used the fakeable scheme. As discussed in the introduction, a different solution for this problem was hinted in [OPW11]. All the constructions are for bit encryption: for longer plaintext space one can simply run the scheme in parallel.

In the following subsections we will need two technical lemmas which we state here. Let a *randomized encoding* E be a randomized function from $\{0,1\}$ to $\{0,1\}^n$. Consider the following game $\partial(A, E)$ between a randomized encoding E and an adversary A (an interactive Turing machine):

1. Run A to make it output a bit $b \in \{0,1\}$.
2. Sample $(b_1, \ldots, b_n) \leftarrow E(b)$.
3. Input $(b_1, \ldots, b_n)$ to A and run it to produce a guess $g \in \{0,1\}$.
4. Output g.

We define the *advantage* of A in distinguishing two randomized encodings E_0 and E_1 to be $\mathrm{Adv}_A(E_0, E_1) = |\Pr[\partial(A, E_0) = 0] - \Pr[\partial(A, E_1) = 0]|$. Notice that if we fix b, then $E_0(b)$ and $E_1(b)$ are random variables, making the statistical distance between them well-defined. Let σ_b denote the statistical distance between $E_0(b)$ and $E_1(b)$ and let $\sigma(E_0, E_1) = \max(\sigma_0, \sigma_1)$.

Lemma 3. *It holds for all adversaries A and all randomized encodings E_0 and E_1 that $\mathrm{Adv}_A(E_0, E_1) \leq \sigma(E_0, E_1)$.*

Lemma 4. *Let $s = 1, 2, \ldots$ be a parameter. Let $N : \mathbb{N} \to \mathbb{N}$, where $N_s = N(s)$ is the number of samples at setting s. For each s, let*

$$
D_s = \begin{cases} -p & \text{with probability } q \\ q & \text{with probability } p \\ 0 & \text{with probability } 1 - p - q \end{cases},
$$

where p and q might be functions of s. Let $X_{s,1}, \ldots, X_{s,N_s}$ be N_s i.i.d. variables, distributed according to D_s. Let $X_s = \sum_{i=1}^{N_s} X_{s,i}$ and let $S_s = \Pr[X_s \in [0, \frac{1}{2})]$. Then

$$
S_s \leq \frac{1}{\sqrt{pq(p+q)N_s}} \left(\frac{p^2 + q^2}{p+q} + \frac{1}{2\sqrt{2\pi}} \right).
$$

The first lemma is trivial to prove, and the second follows directly from the Berry-Esseen inequality [KS10]. Full proofs can be found in the full version.

4.1 Poly-Sender-Deniability

As a warm up we show that a multi-distributional sender-deniable scheme implies a poly-sender-deniable scheme. From a scheme (G, E, E_F, D, F_S) we produce a scheme (G', E', D', F_S') which encrypts a single bit b. The produced scheme is basically the *Parity Scheme* of [CDNO97] only whereas our scheme is based on a multi-distributional sender-deniable scheme, the scheme in [CDNO97] is based on a so-called *translucent set*.

Key Generation $G'(1^\kappa)$: Output $(pk, sk) \leftarrow G(1^\kappa)$.
Encryption $E'_{pk}(b)$: Sample a uniformly random index $j \in \{0, \ldots, n\}$ so that j is even for $b = 0$ and odd for $b = 1$. For $i = 1, \ldots n$ do the following.
    1. For $i \leq j$ sample $c_i \leftarrow E_{Fpk}(1; r_i)$.

2. For $i > j$ sample $c_i \leftarrow \mathsf{E}_{pk}(0; r_i)$.

Output $C = (c_i)_{i=1}^n$.

Decryption $\mathsf{D}'_{sk}(C)$**:** Parse C as $(c_i)_{i=1}^n$. Compute $b_i = \mathsf{D}_{sk}(c_i)$ for $i = 1, \ldots, n$ and output $b = \bigoplus_{i=1}^n b_i$.

Fake $\mathsf{F}_\mathsf{S}'(pk, b, (j, (r_i)_{i=1}^n), b')$**:** If $b = b'$ output $(j, (r_i)_{i=1}^n)$. Otherwise let $r'_j = \mathsf{F}_\mathsf{S}(pk, 1, r_j, 0)$ and $j' = j - 1$. Let all $r'_i = r_i$ for $i \neq j$ and output $(j', (r'_i)_{i=1}^n)$.

Theorem 2. *If* $(\mathsf{G}, \mathsf{E}, \mathsf{E}_\mathsf{F}, \mathsf{D}, \mathsf{F}_\mathsf{S})$ *is multi-distributional sender-deniable, then* $(\mathsf{G}', \mathsf{E}', \mathsf{D}', \mathsf{F}_\mathsf{S}')$ *is* $4/n$-*sender-deniable.*

Proof. Correctness and semantic security is obvious. To prove poly-sender-deniability we first consider the following hybrid game H_1.

H_1 proceeds exactly as the faking game for sender-deniability only it modifies the faking algorithm F_S' by simply sampling r'_j as randomness for the honest encryption algorithm E, and replaces the ciphertext $C = (c_i)_{i=1}^n$ with $C' = (c'_i)_{i=1}^n$ where $c'_j = \mathsf{E}_{pk}(0; r'_j)$ and $c'_i = c_i$ for all $i \neq j$. Notice that the H_1 only changes the distribution of r'_j and c'_j, the distribution of all other inputs to the adversary remains the same. In other words distinguishing the two games comes down to distinguishing an honest encryption of 0 from an encryption faked to an honest encryption of 0. Thus by the multi-distributional sender-deniability of $(\mathsf{G}, \mathsf{E}, \mathsf{E}_\mathsf{F}, \mathsf{D}, \mathsf{F}_\mathsf{S})$ the advantage of any adversary in distinguishing the two games will be negligible in κ.

Now consider another hybrid game H_2. H_2 proceeds exactly as the honest game for sender-deniability except that it modifies the encryption algorithm E' by picking j in the following way: first it picks a uniformly random index $i \in \{0, \ldots, n\}$ such that i is odd for $b = 0$ and even for $b = 1$ (i.e., the opposite of how E' picks j) and then sets $j = i - 1$. Notice now that H_2 outputs exactly the same as H_1 to the adversary only the output is generated in a slightly different order. I.e., H_1 and H_2 are perfectly indistinguishable. However since H_2 proceeds exactly as the honest game, except that it picks j from a different distribution, distinguishing H_2 from the honest game comes down to distinguishing the two different distributions of j.

In order to utilize Lemma 3 we can view these distributions as randomized encodings. Let us denote by E_0 and E_1 the encodings that encodes a bit b as j 1's followed by $n - j$ 0's. For E_0 j is sampled as in the honest game where the adversary outputs b and for E_1 j is sampled as in the hybrid game H_2 where the adversary outputs b. If $j = -1$ in the hybrid game E_1 will encode this as a special string, say a 0 followed by $n - 1$ 1's. First notice that for $b = 0$ both games sample j uniformly random in $\{0, 2, 4, \ldots, n-1\}$, i.e., $\sigma_0 = 0$. However for $b = 1$ the honest game samples j uniformly random in $\{1, 3, 5, \ldots, n\}$ whereas H_2 samples uniformly random in $\{-1, 1, 3, \ldots, n-2\}$. Thus clearly $\sigma_1 = 4/n$.

Now by Lemma 3 we have that any adversary has advantage at most $4/n$ in distinguishing the honest game from H_2. By the above hybrid argument it follows that any adversary has advantage at most $4/n + \mathrm{negl}(\kappa)$ in distinguishing the honest game from the faking game. I.e., $(\mathsf{G}, \mathsf{E}, \mathsf{E}_\mathsf{F}, \mathsf{D}, \mathsf{F}_\mathsf{S})$ is $4/n$-deniable.   □

4.2 Poly-Receiver-Deniability

We show that a multi-distributional receiver-deniable scheme implies a poly-receiver-deniable scheme. From a scheme (G, G_F, E, D, F_R) we produce a scheme (G', E', D', F_R') which encrypts a single bit b.

Key generation $G'(1^\kappa)$: For $i = 1, \ldots, n$ sample uniformly random bits $a_i \in \{0, 1\}$ and then sample $(pk_i, sk_i) \leftarrow G^{a_i}$, where $G^0 = G$ and $G^1 = G_F$. Output $(PK, SK) = ((pk_i)_{i=1}^n, (sk_i, a_i)_{i=1}^n)$.

Encryption $E'_{PK}(b)$: Parse PK as $(pk_i)_{i=1}^n$. For $i = 1, \ldots, n-1$, sample b_i uniformly at random and let $b_n = b \oplus \bigoplus_i^{n-1} b_i$, compute $c_i \leftarrow E_{pk_i}(b_i)$ and output $C = (c_i)_{i=1}^n$.

Decryption $D'_{SK}(C)$: Parse SK as $(sk_i, a_i)_{i=1}^n$ and C as $(c_i)_{i=1}^n$. Compute $b_i = D_{sk_i}(c_i)$ for $i = 1, \ldots, n$ and output $b = \bigoplus_{i=1}^n b_i$.

Fake $F_R'(SK, C, b')$: If $b' = D'_{SK}(C)$ output SK. Otherwise parse SK as $(sk_i, a_i)_{i=1}^n$ and C as $(c_i)_{i=1}^n$. Pick a uniformly random index i for which $a_i = 1$, compute $b_i = D_{sk_i}(c_i)$ and let $sk_i' = F_R(sk_i, c_i, 1 - b_i)$ and $a_i' = 0$. For all $j \neq i$, let $sk_j' = sk_j$ and $a_j' = a_j$. Output $SK' = (sk_j', a_j')_{j=1}^n$.

If κ is they key length of the underlying scheme then the above scheme has keys of length $n\kappa$. The following result then implies that one can build a $1/n$-receiver deniable scheme with keys of size $\sigma = O(n^2\kappa)$.

Theorem 3. *If* (G, G_F, E, D, F_R) *is multi-distributional receiver-deniable, then* (G', E', D', F_R') *is* $(n-1)^{-1/2}$*-receiver-deniable.*

Proof. In the following we assume for simplicity that n is odd, a similar analysis can be made in the case of n even. Correctness and semantic security is obvious. Using a hybrid argument, the distinguishing probability of any poly-time adversary against the above scheme is negligible close to the best distinguishing advantage between the two randomized encoding E_0 and E_1 defined as follows:

1. $E_0(b) = (b_1, \ldots, b_n)$, where the $b_i \in \{0, 1\}$ are uniformly random and independent except that $b = \bigoplus_{i=1}^n b_i$.
2. $E_1(b) = (b_1, \ldots, b_n)$ is sampled as follows. First sample $b_i' \in \{0, 1\}$ as in $E_0(b \oplus 1)$. Then, if $\sum_i b_i' - 0$, let $(b_1, \ldots, b_n) = (b_1', \ldots, b_n')$. Otherwise, pick a uniformly random $j \in \{1, \ldots, n\}$ for which $b_j' = 1$ and then let $b_j = 0$ and let $b_i = b_i'$ for $i \neq j$.

The event $\sum_i b_i' = 0$ happens with negligible probability, so we can analyze under the assumption that this does not happen. In that case the bits b_n and b_n' can be computed as $b_n = b \oplus \bigoplus_{i=1}^{n-1} b_i$ respectively $b_n' = b \oplus \bigoplus_{i=1}^{n-1} b_i'$. So, one can distinguish $D_0(b) = (b_1, \ldots, b_{n-1})$ and $D_1(b) = (b_1', \ldots, b_{n-1}')$ with the same advantage as one can distinguish $E_0(b)$ and $E_1(b)$. The distribution $D_0(b)$ consists of $n-1$ uniformly random bits. The distribution $D_1(b)$ consists of $n-1$ uniformly random bits, where we flipped a random occurence of 1 to 0. For $b \in \{0, 1\}^{n-1}$, let $\#_1(b) = \sum_{i=1}^{n-1} b_i$ be the number of 1's in the vector and let $\#_0(b) = n-1-\#_1(b)$ be the number of 0's. By the symmetry of the distributions, it is easy to see that one can distinguish $\#_1(D_0(b))$ and $\#_1(D_1(b))$ with the same advantage as one can distinguish $D_0(b)$ and $D_1(b)$. Since $\#_1(D_0(b))$ is binomially distributed with expectation $\frac{n-1}{2}$ and $\#_1(D_1(b)) = \#_1(D_0(b)) - 1$, it follows that an optimal distinguisher for $\#_1(D_0(b))$ and $\#_1(D_1(b))$ is to guess 0 if $\#_1(D) \geq \frac{n-2}{2}$ and guess 1 otherwise, as this is a maximum likelyhood distinguisher. The advantage of this distinguisher is

$$
\begin{aligned}
\mathrm{Adv} &= \frac{1}{2}\left| \Pr\left[\#_1(D_0(b)) \geq \frac{n-2}{2}\right] - \Pr\left[\#_1(D_1(b)) \geq \frac{n-2}{2}\right] \right| \\
&= \frac{1}{2}\left| \Pr\left[\#_1(D_0(b)) \geq \frac{n-2}{2}\right] - \Pr\left[\#_1(D_0(b)) \geq \frac{n-2}{2}+1\right] \right| \\
&= \frac{1}{2}\Pr\left[\#_1(D_0(b)) \in \left[\frac{n-2}{2}, \frac{n-2}{2}+1\right)\right] .
\end{aligned}
$$

From $\#_1(D_0(b)) = (n-1) - \#_0(D_0(b))$, we get that $2\#_1(D_0(b)) = \#_1(D_0(b)) + (n-1) - \#_0(D_0(b))$, so $\#_1(D_0(b)) = \frac{n-1}{2} + \frac{1}{2}(\#_1(D_0(b)) - \#_0(D_0(b)))$, and it follows that

$$
\begin{aligned}
\mathrm{Adv} &= \frac{1}{2}\Pr\left[\frac{1}{2}(\#_1(D_0(b)) - \#_0(D_0(b))) \in \left[-\frac{1}{2}, \frac{1}{2}\right)\right] \\
&= \frac{1}{2}\Pr\left[\frac{1}{2}\#_1(D_0(b)) - \frac{1}{2}\#_0(D_0(b)) \in \left[0, \frac{1}{2}\right)\right] .
\end{aligned}
$$

The last equality follows from n being odd. Consider then Lemma 4, with $p = q = \frac{1}{2}$ and $N_s = s - 1$. The variable X_s in the premise then has exactly the same distribution as $\frac{1}{2}\#_1(D_0(b)) - \frac{1}{2}\#_0(D_0(b))$ when $s = n$. Plugging $p = q = \frac{1}{2}$ and $N_s = n - 1$ into Lemma 4 we get that $\Pr\left[\frac{1}{2}\#_1(D_0(b)) - \frac{1}{2}\#_0(D_0(b)) \in [0, \frac{1}{2})\right] \leq \frac{2}{\sqrt{s-1}}$. $\qquad\square$

4.3   Poly-Bi-Deniability

We show that a multi-distributional bi-deniable scheme implies a poly-bi-deniable scheme. From a scheme $(\mathsf{G}, \mathsf{G_F}, \mathsf{E}, \mathsf{E_F}, \mathsf{D}, \mathsf{F_S}, \mathsf{F_R})$ we produce a scheme $(\mathsf{G'}, \mathsf{E'}, \mathsf{D'}, \mathsf{F_S'}, \mathsf{F_R'})$ which encrypts a single bit.

Key generation $\mathsf{G'}(1^\kappa)$: For $i = 1, \ldots, n^2$ sample random bits $a_i \in \{0,1\}$ and then sample $(pk_i, sk_i) \leftarrow \mathsf{G}^{a_i}(1^\kappa)$, where $\mathsf{G}^0 = \mathsf{G}$ and $\mathsf{G}^1 = \mathsf{G_F}$. Sample the a_i's independently with $\Pr[a_i = 0] = 1/n$. Output $(PK, SK) = ((pk_i)_{i=1}^{n^2}, (sk_i, a_i)_{i=1}^{n^2})$.

Encryption $\mathsf{E'}_{PK}(b)$: Parse PK as $(pk_i)_{i=1}^{n^2}$. For $i = 1, \ldots, n^2$
1. Sample uniformly random $b_i \in_R \{0,1\}$ and $m_i \in_R \{0,1\}^\kappa$ such that $b = \bigoplus_{i=1}^{n^2} b_i$.
2. Compute $c_i \leftarrow \mathsf{E}^{b_i}_{pk_i}(m_i', r_i)$, where $m_i' = b_i m_i$ ($0m_i = 0^\kappa$ and $1m_i = m_i$), $\mathsf{E}^0 = \mathsf{E}$ and $\mathsf{E}^1 = \mathsf{E_F}$.

Output $C = (c_i)_{i=1}^{n^2}$.

Decryption $\mathsf{D}'_{SK}(C)$: Parse SK as $(sk_i, a_i)_{i=1}^{n^2}$ and C as $(c_i)_{i=1}^{n^2}$. For $i = 1, \ldots, n^2$, compute $m'_i = \mathsf{D}_{sk}(c_i)$ and let $b'_i = 1$ if $m'_i \neq 0$ and $b'_i = 0$ if $m'_i = 0$. Output $b = \bigoplus_{i=1}^{n^2} b'_i$.

Fake (sender) $\mathsf{F_S}'(PK, b, (r_i, m_i, b_i)_{i=1}^{n^2}, b')$: If $b = b'$ output $(r_i, m_i, b_i)_{i=1}^{n^2}$. Otherwise parse PK as $(pk_i)_{i=1}^{n^2}$. Let $m' = \min\{m'_i = b_i m_i | i \in \{1, \ldots, n^2\} \wedge m'_i \neq 0^\kappa\}$ and pick the unique (ewnp.) index k for which $m'_k = b_k m_k = m'$ (notice this implies $b_k = 1$). I.e., k is the index of the c_i containing the small-est non-zero plaintext. The minimum is taken according to lexicographic or-der. Then let $r'_k = \mathsf{F_S}(pk_k, m'_k, r_k, 0^\kappa)$, $m'_k = m_k$ and $b'_k = 0$. For all $j \neq k$, let $r'_j = r_j$, $m'_j = m_j$ and $b'_j = b_j$. Output $(r_j, m_j, b_j)_{j=1}^{n^2}$.

Fake (receiver) $\mathsf{F_R}'(SK, C, b')$: If $\mathsf{D}'_{SK}(C) = b'$ output SK. Otherwise parse SK as $(sk_i, a_i)_{i=1}^{n^2}$ and C as $(c_i)_{i=1}^{n^2}$ and compute $m'_i = \mathsf{D}_{sk}(c_i)$. Let $m' = \min\{m'_i | i \in \{1, \ldots, n^2\} \wedge m'_i \neq 0^\kappa\}$ and pick the unique (ewnp.) index k for which $m'_k = m'$. I.e., k is the index of the c_i containing the smallest non-zero plaintext. The minimum is taken according to lexicographic order. If $a_k = 0$, then give up. If $a_k = 1$, then let $sk'_k = \mathsf{F_R}(sk_k, c_k, 0^\kappa)$ and $a'_k = 0$. For all $j \neq k$, let $sk'_j = sk_j$ and $a'_j = a_j$. Output $SK' = (sk'_j, a'_j)_{j=1}^{n^2}$.

Theorem 4. *If* $(\mathsf{G}, \mathsf{G_F}, \mathsf{E}, \mathsf{E_F}, \mathsf{D}, \mathsf{F_S}, \mathsf{F_R})$ *is multi-distributional bi-deniable, then* $(\mathsf{G}', \mathsf{E}', \mathsf{D}', \mathsf{F_S}', \mathsf{F_R}')$ *is* $O(n^{-1/2})$-*bi-deniable.*

Proof. Correctness follows by observing that $b'_i = b_i$ unless one of the uniformly random κ-bit messages m_i happens to be 0^κ, which is a negligible event. Semantic security is obvious. As for bi-deniability, by a hybrid argument similar to that in the proofs of Thm. 2 and Thm. 3, distinguishing the honest and faking game comes down to distinguishing the following two random encodings of a bit b.

1. $E_0(b) = (b_1, \ldots, b_{n^2}, a_1, \ldots, a_{n^2})$, where the $b_i \in \{0,1\}$ are sampled uni-formly at random except that $\bigoplus_{i=1}^{n^2} b_i = b$ and the $a_i \in \{0,1\}$ are sampled such that $\Pr[a_i = 0] = 1/n$.

2. $E_1(b) = (b_1, \ldots, b_{n^2}, a_1, \ldots, a_{n^2})$ is sampled as follows. First sample $b'_i, a'_i \in \{0,1\}$ as in $E_0(b \oplus 1)$. Then, if $\sum_i b'_i = 0$, let $(b_1, \ldots, b_{n^2}) = (b'_1, \ldots, b'_{n^2})$. Otherwise, pick a uniformly random $k \in \{1, \ldots, n^2\}$ for which $b'_k = 1$ and then let $b_k = 0$ and let $b_k = b'_k$ for $i \neq k$. If $a'_k = 1$ let $a_k = 0$ and let $a_i = a'_i$ for $i \neq k$.

It happens that $a'_k = 0$ with probability $1/n$, so by adding $1/n$ to the bound in the end, we can analyse under the assumption that $a'_k = 1$. In that case we can describe $E_1(b)$ as above, except that we pick k uniformly at random among the i's for which $b'_i = 1$ and $a'_i = 1$. Then we set $b_k = 0$ and $a_k = 0$ and set $b_{i \neq k} = b'_i$ and $a_{i \neq k} = a'_i$.

Given a vector $\boldsymbol{v} = (b_1, \ldots, b_{n^2}, a_1, \ldots, a_{n^2})$, we let $\#_{00}(\boldsymbol{v})$ be the number of i's for which $b_i = a_i = 0$ and we let $\#_{11}(\boldsymbol{v})$ be the number of i's for which $b_i = a_i = 1$. For simplicity we assume that b is uniformly random, such that $b_1, \ldots, b_{n^2}$ is uniform in $\{0,1\}^{n^2}$. Deriving the same bound for fixed $b = 0$ and $b = 1$ is straight-forward. Let $p = \frac{1}{2n}$ be the probability that $a_i = 0$ and $b_i = 0$. Let $q = \frac{n-1}{2n}$ be the probability that $a_i = 1$ and $b_i = 1$. The expected value of $\#_{00}(E_0(b))$ is pn^2. The expected value of $\#_{11}(E_0(b))$ is qn^2, and $\#_{00}(E_1(b)) = \#_{00}(E_0(b)) + 1$ and $\#_{11}(E_1(b)) = \#_{11}(E_0(b)) - 1$. From this it can be derived as in the proof of Thm. 3 that the maximum likelihood distinguisher for $E_0(b)$ and $E_1(b)$ guesses 0 if $q\#_{00} - p\#_{11} > 0$ and that its advantage is $\frac{1}{2}\Pr\left[q\#_{00}(E_0(b)) - p\#_{11}(E_1(b)) \in [0, \frac{1}{2})\right]$. Using Lemma 4 as in the proof of Thm. 3, with $s = n$, $N_s = s^2$ and the p and q defined above, it follows that

$$\Pr\left[q\#_{00}(E_0(b)) - p\#_{11}(E_1(b)) \in [0, \frac{1}{2})\right] \leq \frac{1}{\sqrt{s}}\left(\sqrt{2} + \frac{1}{\sqrt{\pi}}\right).$$

The theorem then follows from $\sqrt{2} + \frac{1}{\sqrt{\pi}} \leq 2$.   □

Acknowledgements. The authors would like to thank Adam O'Neill, Chris Peikert and Brent Waters for sharing with us an early copy of [OPW11].

References

[Bea97]   Beaver, D.: Plug and Play Encryption. In: Kaliski Jr., B.S. (ed.) CRYPTO 1997. LNCS, vol. 1294, pp. 75–89. Springer, Heidelberg (1997)

[CDNO97]   Canetti, R., Dwork, C., Naor, M., Ostrovsky, R.: Deniable Encryption. In: Kaliski Jr., B.S. (ed.) CRYPTO 1997. LNCS, vol. 1294, pp. 90–104. Springer, Heidelberg (1997)

[CDSMW09]   Choi, S.G., Dachman-Soled, D., Malkin, T., Wee, H.: Improved Non-Committing Encryption with Applications to Adaptively Secure Protocols. In: Matsui, M. (ed.) ASIACRYPT 2009. LNCS, vol. 5912, pp. 287–302. Springer, Heidelberg (2009)

[CFGN96]   Canetti, R., Feige, U., Goldreich, O., Naor, M.: Adaptively secure multi-party computation. In: STOC, pp. 639–648 (1996)

[DF11]   Dürmuth, M., Freeman, D.M.: Deniable Encryption with Negligible Detection Probability: An Interactive Construction. In: Paterson, K.G. (ed.) EUROCRYPT 2011. LNCS, vol. 6632, pp. 610–626. Springer, Heidelberg (2011)

[DN00]   Damgård, I.B., Nielsen, J.B.: Improved Non-Committing Encryption Schemes Based on a General Complexity Assumption. In: Bellare, M. (ed.) CRYPTO 2000. LNCS, vol. 1880, pp. 432–450. Springer, Heidelberg (2000)

[GWZ09]   Garay, J.A., Wichs, D., Zhou, H.-S.: Somewhat Non-Committing Encryption and Efficient Adaptively Secure Oblivious Transfer. In: Halevi, S. (ed.) CRYPTO 2009. LNCS, vol. 5677, pp. 505–523. Springer, Heidelberg (2009)

[KO04]   Katz, J., Ostrovsky, R.: Round-Optimal Secure Two-Party Computation. In: Franklin, M. (ed.) CRYPTO 2004. LNCS, vol. 3152, pp. 335–354. Springer, Heidelberg (2004)

[KS10]   Korolev, V.Y., Shevtsova, I.G.: On the upper bound for the absolute constant in the Berry–Esseen inequality. Theory of Probability and its Applications 54(4), 638–658 (2010)

[Nie02]   Nielsen, J.B.: Separating Random Oracle Proofs from Complexity Theoretic Proofs: The Non-Committing Encryption Case. In: Yung, M. (ed.) CRYPTO 2002. LNCS, vol. 2442, pp. 111–126. Springer, Heidelberg (2002)

[OPW11]   O'Neill, A., Peikert, C., Waters, B.: Bi-Deniable Public-Key Encryption. In: Rogaway, P. (ed.) CRYPTO 2011. LNCS, vol. 6841, pp. 525–542. Springer, Heidelberg (2011)

Bridging Broadcast Encryption and Group Key Agreement

Qianhong Wu[1,2], Bo Qin[1,3], Lei Zhang[4],
Josep Domingo-Ferrer[1], and Oriol Farràs[1,5]

[1] Universitat Rovira i Virgili, Department of Computer Engineering and
Mathematics, UNESCO Chair in Data Privacy, Tarragona, Catalonia
{qianhong.wu,bo.qin,josep.domingo,oriol.farras}@urv.cat
[2] Key Lab. of Aerospace Information Security and Trusted Computing, Ministry of
Education School of Computer, Wuhan University, China
[3] Dept. of Maths, School of Science, Xi'an University of Technology, China
[4] Software Engineering Institute, East China Normal University, Shanghai, China
leizhang@sei.ecnu.edu.cn
[5] Department of Computer Science, Ben Gurion University, Be'er-Sheva, Israel

Abstract. Broadcast encryption (BE) schemes allow a sender to securely broadcast to any subset of members but requires a trusted party to distribute decryption keys. Group key agreement (GKA) protocols enable a group of members to negotiate a common encryption key via open networks so that only the members can decrypt the ciphertexts encrypted under the shared encryption key, but a sender cannot exclude any particular member from decrypting the ciphertexts. In this paper, we bridge these two notions with a hybrid primitive referred to as contributory broadcast encryption (CBE). In this new primitive, a group of members negotiate a common public encryption key while each member holds a decryption key. A sender seeing the public group encryption key can limit the decryption to a subset of members of his choice. Following this model, we propose a CBE scheme with short ciphertexts. The scheme is proven to be fully collusion-resistant under the decision n-Bilinear Diffie-Hellman Exponentiation (BDHE) assumption in the standard model. We also illustrate a variant in which the communication and computation complexity is sub-linear with the group size. Of independent interest, we present a new BE scheme that is aggregatable. The aggregatability property is shown to be useful to construct advanced protocols.

Keywords: Broadcast encryption; Group key agreement; Contributory broadcast encryption; Provable Security.

1 Introduction

With the fast advance and pervasive deployment of the communication technologies, there is an increasing demand of versatile cryptographic primitives to protect modern communication and computation platforms. These new platforms, including instant-messaging tools, collaborative computing, mobile *ad hoc*

D.H. Lee and X. Wang (Eds.): ASIACRYPT 2011, LNCS 7073, pp. 143–160, 2011.

networks and social networks, allow exchanging data within *any subset* of their users. These new information technologies provide potential opportunities for organizations and individuals. For instance, the users of a social network may wish to share their private photos/videos with their friends; scientists from different places may want to collaborate in a research project by means of an insecure third-party platform.

These new applications call for cryptographic primitives allowing a sender to securely encrypt to any subset of the users of the services without relying on a fully trusted dealer. Broadcast encryption (BE) [15] is a well-studied primitive intended for secure group-oriented communications. It allows a sender to securely broadcast to any subset of the group members. Nevertheless, its security heavily relies on a trusted key server to generate and distribute secret decryption keys for the members; both the sender and the receivers must fully trust the key server who can read all communications to any subset of the group members.

Group key agreement (GKA) [20] is another well-established primitive to secure group-oriented communications. A conventional GKA protocol allows a group of members to establish a common secret key via open networks. However, whenever a sender wants to broadcast to a group, he must first join the group and run a GKA protocol to share a secret key with the intended members. To overcome this limitation, Wu *et al.* recently introduced asymmetric GKA [32] in which only a common group public key is negotiated and each group member holds a different decryption key. However, neither conventional symmetric GKA nor newly-introduced asymmetric GKA allows *the sender* to exclude any particular member *on demand*[1]. Hence, it is essential to find more flexible cryptographic primitives allowing dynamic broadcasts without a fully trusted dealer.

1.1   Our Contributions

In this paper we present the Contributory Broadcast Encryption (CBE) primitive, which is a hybrid of GKA and BE. The new cryptographic primitive is motivated by the emerging communication and computation platforms. In CBE, a group of members contribute to the public group encryption key, and a sender can securely broadcast to any subset of the group members chosen in an *ad hoc* way. Specifically, our main contributions can be summarized as follows.

First, we present a model of CBE and formalize its security definitions. CBE incorporates the underlying ideas of GKA and BE. In the set-up stage of a CBE scheme, a group of members interact via open networks to negotiate a common encryption key while each member holds a different secret decryption key. Using the common encryption key, anyone can encrypt any message to any subset of the group members and only the intended receivers can decrypt. Unlike GKA, CBE allows the sender to exclude some members from reading the ciphertexts.

---

[1] Dynamic GKA equipped with a leave sub-protocol allows a sender to exclude some members from decrypting ciphertexts. In this case, the sender has to negotiate with the remaining members for their agreement to run the leave sub-protocol. The sender cannot exclude any member on his own demand.

Compared to BE, CBE does not need a fully trusted third party to set up the system. We formalize collusion resistance by defining an attacker who can adaptively corrupt some members during the set-up stage and can also query the decryption keys of the group members after the system is set up. Even if the attacker fully controls all members outside the intended receivers, she cannot extract useful information from the ciphertext. A trivial CBE scheme can be constructed by concurrently encrypting to each member with her/his regular public key. Unfortunately, the trivial solution incurs a heavy encryption cost and produces linear-size ciphertexts. The challenge is to design CBE schemes with efficient encryption and short ciphertexts.

Second, we present the notion of aggregatable broadcast encryption (ABE) and construct a concrete ABE scheme. The construction is based on the newly introduced aggregatable signature-based broadcast (ASBB) primitive [32]. Our ABE construction is tightly proven to be fully collusion-resistant under the decision BDHE assumption, and offers short ciphertexts and efficient encryption. Further, the proposed ABE scheme is equipped with aggregatability, which means that different instances of the ABE scheme can be aggregated into a new instance. We observe that the BE schemes in the literature are not aggregatable. However, the aggregatability of ABE schemes seems very useful to design advanced protocols, as illustrated in the construction of our CBE scheme.

Finally, we construct an efficient CBE scheme with our ABE scheme as a building block. The CBE construction is proven to be semi-adaptively secure under the decision BDHE assumption in the standard model. Only one round is required to establish the public group encryption key and set up the CBE system. After the system set-up, the storage cost of both the sender and the group members is $O(n)$, where n is the number of group members participating in the set-up stage. However, the online complexity (which dominates the practicality of a CBE scheme) is very low. Indeed, at the sender's side, the encryption needs only $O(1)$ exponentiations and generates $O(1)$-size ciphertexts; and at the receivers' side, the decryption requires only $O(1)$ exponentiations and $O(1)$ bilinear map operations. We also illustrate a trade-off between the set-up complexity and the online performance. After the trade-off, the variant has $O(n^{2/3})$ complexity in communication, computation and storage. This is comparable to up-to-date regular BE schemes which have $O(n^{1/2})$ complexity in the same performance metrics, but our scheme does not require a trusted key dealer. As a versatile GKA scheme, our CBE does not require additional rounds to enable a *new sender* to broadcast to the group members or to let a sender *revoke* any subset of group members. These features are desirable for applications in which the sender and the group members may change frequently.

1.2 Related Work

Considerable efforts have been devoted to protect group communications. Among them, the most prominent notions are key agreement and broadcast encryption. Since the inception of the Diffie-Hellman protocol [14] in 1976, a number of proposals have addressed key agreement protocols for multiple parties. The schemes due to Ingemarsson *et al.* [20] and Steiner *et al.* [29] are designed for n parties and require $O(n)$ rounds. Tree key structures have been further proposed and reduced the number of rounds to $O(\log n)$ [23, 24, 27]. A multi-round GKA protocol poses a synchronism requirement on group members and it needs all group members to simultaneously stay online to complete the protocol. Several proposals (*e.g.*, [8, 18, 30]) have been motivated to optimize round complexity in GKA protocols. Burmester and Desmedt [12] proposed a two-round n-party GKA protocol for n parties. The Joux protocol [21] is one-round and only applicable to three parties. The work of Boneh and Silverberg [5] shows that a one-round $(n+1)$-party GKA protocol can be constructed from n-linear pairings. However, it remains unknown whether there exist n-linear pairings for $n > 2$.

Dynamic GKA protocols provide extra mechanisms to cope with member changes. Bresson *et al.* [9, 10] extended the protocol in [11] to dynamic GKA protocols which allow members to leave and join the group. The number of rounds in `set-up/join` algorithms of their protocols [9, 10] is linear with the group size, but the number of rounds in the `leave` algorithm is constant. The theoretical analysis [28] proves that, for any tree-based group key agreement scheme, the lower bound of the worst-case cost is $O(\log n)$ rounds for a member to join or leave. Without relying on a tree-based structure, Kim *et al.* [22] proposed a two-round dynamic GKA protocol. Recently, Abdalla *et al.* [1] presented a two-round dynamic GKA protocol in which only one round is required to cope with the change of members if they are in the initial group. Observing that existing GKA protocols cannot handle sender changes efficiently, Wu *et al.* presented the notion of asymmetric GKA [32] to support sender changes and their instantiated protocol allows anyone to securely broadcast to the group members.

BE is another well-established cryptographic primitive developed for secure group communications. BE schemes in the literature can be classified into two categories, *i.e.*, symmetric-key BE and public-key BE. In the symmetric-key setting, only the trusted center generates all the secret keys and broadcasts messages to users. Hence, only the key generation center can be the broadcaster or the sender. Fiat and Naor [15] first formalized broadcast encryption in the symmetric-key setting and proposed a systematic BE method. Similarly to the GKA setting, tree-based key structures were subsequently proposed to improve efficiency in symmetric-key BE systems [19, 31]. The state of the art along this research line is presented in [13].

Public-key BE schemes are more flexible in practice. In this setting, in addition to the secret keys for each user, the trusted center also generates a public key for all the users so that any one can play the role of a broadcaster or sender. Naor and Pinkas presented in [25] the first public-key BE scheme in which up to a threshold of users can be revoked. If more than this threshold of users are revoked, the scheme will be insecure and hence not fully collusion-resistant. Subsequently, by exploiting newly developed bilinear pairing technologies, a fully collusion-resistant public-key BE scheme was presented in [3] which has $O(\sqrt{n})$ complexity in key size, ciphertext size and computation cost. A recent scheme [26] slightly reduces the size of the key and the ciphertexts, although it still has sublinear complexity. The schemes presented in [4, 6, 17] strengthen the security concept of public-key BE schemes. However, as to performance, the sub-linear barrier $O(\sqrt{n})$ has not yet been broken.

Although both GKA and BE are used to secure group communications, they have very different features as they were initially developed for different types of group-oriented applications. First, GKA can be applied to *ad hoc* groups where there is no fully trusted party while BE is usually deployed to secure group communications where a fully trusted third party is available. Second, the encryption key in GKA protocols is usually established by group members in a contributory way, regardless of conventional symmetric GKAs or newly-introduced asymmetric GKAs. On the contrary, the encryption key in BE schemes is usually generated by a centralized key server. Third, the secret decryption key in GKA protocols is computed by each member with public inputs from other members and his/her own private inputs. Contrary to GKA protocols, the decryption key of each member in BE schemes is assigned by the dealer, which implies that the dealer can read all communications to any subset of the group members and n secure unicast channels have to be established before a BE scheme is set up. Finally, in a GKA protocol group members need to interact to update their keys if the membership changes, which implies that a sender cannot exclude some members from reading the ciphertexts. Unlike GKA, BE supports a much more flexible revocation mechanism. It allows a sender to choose the intended receivers on demand to read the ciphertexts. This revocation mechanism does not require cooperation among group members or extra interactions between the dealer and the group members. For the newly-emerging applications, the contributory feature of GKA protocols is desirable but GKA protocols do not allow a sender to exclude receivers from reading specific ciphertexts on demand; the flexible revocation mechanism of BE schemes is desirable but BE schemes heavily relies on a fully trusted authority that is hard to implement in the motivated scenarios. These observations inspire us to investigate more versatile cryptographic primitives to bridge the gap.

1.3 Paper Organization

The rest of the paper is organized as follows. In Section 2, we model CBE and define its security. In Section 3, we present a collusion-resistant regular public-key BE scheme with aggregatability. Efficient CBE schemes are realized in Section 4, and Section 5 concludes the paper.

2 Modeling Contributory Broadcast Encryption

We begin by formalizing the CBE notion bridging the GKA and BE primitives. In CBE, a group of members first jointly establish a public encryption key, then a sender can freely select which subset of the group members can decrypt the ciphertext. Our definition incorporates the up-to-date definitions of GKA [32] protocols and BE [3] schemes. Since the negotiated public key is usually employed to transmit session keys, we define a CBE scheme as a key encapsulation mechanism (KEM). Knowing this public encryption key, anyone can send a session key ξ to any subset of the initial group members. Only the intended receivers can extract ξ. Even if all the outsiders including group members not in the intended subset collude, they receive no information about ξ.

2.1 Syntax

We first define the algorithms that compose a CBE scheme. Let $\lambda \in \mathbb{N}$ denote the security parameter. Suppose that a group of members $\{\mathcal{U}_1, \cdots, \mathcal{U}_n\}$ wants to jointly establish a CBE system, where n is a positive integer and each member $\mathcal{U}_i$ is indexed by i for $1 \leq i \leq n$. We focus on bridging BE and GKA and we assume that the communications between members are authenticated, but we do not further elaborate on the authentication of the group members. Formally, a CBE scheme is a tuple $\mathcal{CBE} =$(ParaGen, CBSetup, CBEncrypt, CBDecrypt) of polynomial-time algorithms defined as follows.

ParaGen(1^λ). This algorithm is used to generate global parameters. It takes as input a security parameter λ and it outputs the system parameters, including the group size n.

CBSetup($\mathcal{U}_1(x_1), \cdots, \mathcal{U}_n(x_n)$). This interactive algorithm is jointly run by members $\mathcal{U}_1, \cdots, \mathcal{U}_n$ to set up a BE scheme. Each member $\mathcal{U}_i$ takes private input x_i (and her/his random coins representing the member's random inner state information). The communications between members go through public but authenticated channels. The algorithm will either abort or successfully terminate. If it terminates successfully, each user $\mathcal{U}_i$ outputs a decryption key dk_i securely kept by the user and a common group encryption key gek shared by all group members. The group encryption gek is publicly accessible. If the algorithm aborts, it outputs NULL. Here, we leave the input system parameters implicitly. We denote this procedure by $(\mathcal{U}_1(dk_1), \cdots, \mathcal{U}_n(dk_n); gek) \leftarrow$CBSetup($\mathcal{U}_1(x_1), \cdots, \mathcal{U}_n(x_n)$).

CBEncrypt($\mathbb{R}, gek$). This group encryption algorithm is run by a sender who is assumed to know the public group encryption key. The sender may or may not be a group member. The algorithm takes as inputs a receiver set $\mathbb{R} \subseteq \{1, \cdots, n\}$ and the public group encryption key gek, and it outputs a pair $\langle c, \xi \rangle$, where c is the ciphertext and ξ is the secret session key in a key space $\mathbb{K}$. Then $(c, \mathbb{R})$ is sent to the receivers.

CBDecrypt($\mathbb{R}, j, dk_j, c$). This decryption algorithm is run by each intended receiver. It takes as inputs the receiver set $\mathbb{R}$, an index $j \in \mathbb{R}$, the receiver's decryption key dk_j, a ciphertext c, and it outputs the secret session key ξ.

2.2 Security Definitions

The correctness of a CBE scheme means that if all members and the sender follow the scheme honestly, then the members in the receiver set can always correctly decrypt. Formally, the *correctness* of a CBE scheme is defined as follows.

Definition 1 (Correctness). *A CBE scheme is correct if for any parameter* $\lambda \in \mathbb{N}$ *and any element* ξ *in the session key space,* $(\mathcal{U}_1(dk_1), \cdots, \mathcal{U}_n(dk_n); gek)$ $\leftarrow$ CBSetup$(\mathcal{U}_1(x_1), \cdots, \mathcal{U}_n(x_n))$, *and* $(c, \xi) \leftarrow$CBEncrypt$(\mathbb{R}, gek)$, *it holds that* CBDecrypt$(\mathbb{R}, j, dk_j, c) = \xi$ *for any* $j \in \mathbb{R}$.

We next define the secrecy of a CBE scheme. In the above, to achieve better practicality, a CBE scheme is modeled as a KEM in which a sender sends a (short) secret session key to the intended receivers and simultaneously, (long) messages can be encrypted using a secure symmetric encryption algorithm with the session key. Hence, we define the secrecy of a CBE scheme by the indistinguishability of the encrypted session key from a random element in the session key space. Since there exist standard conversions (*e.g.*, [16]) from secure KEM against chosen-plaintext attacks (CPA) to secure encryption against adaptively chosen-ciphertext attacks (CCA2), it is sufficient to only define the CPA secrecy of CBE schemes. However, noting that CBE is designed for distributed applications where the users are likely to be corrupted, we include full collusion resistance into our secrecy definition.

The fully collusion-resistant secrecy of a CBE scheme is defined by the following secrecy game between a challenger $\mathcal{CH}$ and an attacker $\mathcal{A}$. The secrecy game is defined as follows.

Initial. The challenger $\mathcal{CH}$ runs ParaGen with a security parameter λ and obtains the system parameters. The system parameters are given to the attacker $\mathcal{A}$.

Queries. The attacker $\mathcal{A}$ can make the following queries to challenger $\mathcal{CH}$.

    Execute. The attacker $\mathcal{A}$ uses the identities of n members $\mathcal{U}_1, \cdots, \mathcal{U}_n$ to query the challenger $\mathcal{CH}$. The challenger runs CBSetup$(\mathcal{U}_1(x_1), \cdots, \mathcal{U}_n(x_n))$ on behalf of the n members, and responds with the group encryption key gek and the transcripts of CBSetup to the attacker $\mathcal{A}$.

    Corrupt. The attacker $\mathcal{A}$ sends i to the Corrupt oracle maintained by the challenger $\mathcal{CH}$, where $i \in \{1, \cdots, n\}$. The challenger $\mathcal{CH}$ returns the private input and inner random coins of $\mathcal{U}_i$ during the execution of CBSetup.

    Reveal. The attacker $\mathcal{A}$ sends i to the Reveal oracle maintained by the challenger $\mathcal{CH}$, where $i \in \{1, \cdots, n\}$. The challenger $\mathcal{CH}$ responds with dk_i, which is the decryption key of $\mathcal{U}_i$ after execution of CBSetup.

Challenge. At any point, the attacker $\mathcal{A}$ can choose a target set $\mathbb{R}^* \subseteq \{1, \cdots, n\}$ to attack, with a constraint that the indices in $\mathbb{R}^*$ have never been queried to the Corrupt oracle or the Reveal oracle. Receiving $\mathbb{R}^*$, the challenger $\mathcal{CH}$ randomly selects $\rho \in \{0, 1\}$ and responds with a challenge ciphertext c^*, where c^* is obtained from $(c^*, \xi) \leftarrow$CBEncrypt$(\mathbb{R}, gek)$ if $\rho = 1$, else if $\rho = 0$, c^* is randomly sampled from the image space of CBEncrypt.

Output. Finally, $\mathcal{A}$ outputs a bit ρ', its guess of ρ. The adversary wins if $\rho' = \rho$.

We define $\mathcal{A}$'s advantage $Adv_{CBE,\mathcal{A}}^{secrecy-fc}$ in winning the above fully collusion-resistant secrecy game as

$$Adv_{CBE,\mathcal{A}}^{secrecy-fc} = |\Pr[\rho = \rho'] - 1/2|.$$

Definition 2. *An n-party CBE scheme has adaptive (τ, n, ϵ)-secrecy against a full-collusion attack if there is no adversary $\mathcal{A}$ which runs in time at most τ and has advantage $Adv_{CBE,\mathcal{A}}^{secrecy-fc}$ at least ϵ in the above secrecy game. An n-party CBE scheme has semi-adaptive (τ, n, ϵ)-secrecy against a full-collusion attack if, for any attacker $\mathcal{A}'$ running in time τ, $\mathcal{A}'$'s advantage $Adv_{CBE,\mathcal{A}'}^{secrecy-fc}$ is less than ϵ in the above secrecy game, with extra constraints that $\mathcal{A}'$ (1) must commit to a set of indices $\tilde{\mathbb{R}} \subseteq \{1, \cdots, n\}$ before the Queries stage, (2) can only query Corrupt and Reveal with $i \notin \tilde{\mathbb{R}}$ and (3) can only choose $\mathbb{R}^* \subseteq \tilde{\mathbb{R}}$ to query $\mathcal{CH}$ in the Challenge stage.*

The above definition captures the full collusion resistance since the attacker is allowed to access the Corrupt and Reveal oracles. The Corrupt oracle is used to model an attacker who compromises some members during the set-up stage to establish the group encryption key. The Corrupt oracle is used to capture the decryption key leakage after the CBE system has been established. This difference can be used to differentiate the secrecy against attacks during the set-up stage from the secrecy against attacks after a CBE system is deployed.

2.3   Remarks on Complexity Bounds of CBE and BE Schemes

Before concrete CBE schemes are constructed, it is meaningful to examine the complexity bound of a CBE scheme for the purpose of guiding the design of CBE schemes.

A CBE scheme consists of an offline stage (consisting of ParaGen and CBSetup) to establish the group encryption key and an online stage enabling a sender to securely encrypt to intended receivers. Since CBE allows to revoke members, the members do not need to reassemble for a new run of the CBSetup procedure until some new members join. This implies that the practicality of a CBE scheme critically depends on the overheads of the CBEncrypt and CBDecrypt procedures for online encryption of session keys and decryption of ciphertexts. Hence, special efforts should be devoted to improve this online performance.

It is easy to see that there exists a trivial construction of CBE schemes. A group of n members independently generate public/secret key pairs in a standard public-key cryptosystem. The public group encryption key is a concatenation of each member's public key, and each member's decryption key is his/her secret key. To broadcast to a subset of the members, a sender first encrypts the session key using each member's public key and obtains the CBE ciphertext by concatenating the generated n ciphertexts in the underlying public-key cryptosytems. This trivial CBE has $n\tau_{PKE}$ online encryption cost, $n\ell_{PKC}$-size ciphertext, where ℓ_{PKC} is the binary length of the ciphertext in the standard public-key cryptosystem, and τ_{PKE} is the time to perform a standard public-key encryption operation. Hence, the upper bound of online complexity of a CBE scheme is $O(n)$.

We next analyze whether there exist CBE schemes with online complexity less than $O(n)$. From the definition of CBEncrypt, a sender has to read the indices in $\mathbb{R} \subseteq \{1, \cdots, n\}$ and perform some operations involving each index. This implies that the CBEncrypt procedure has a cost $|\mathbb{R}|\tau_{CEO}$, where $|\mathbb{R}| = n$ in the worst case and τ_{CEO} is the time to perform a basic cryptographic encryption operation involving each index. Also, the sender needs to send $(c, \mathbb{R})$ to the receivers. This requires $\ell_c + n$ bits, where ℓ_c is the binary size of the CBE ciphertext. The analysis shows that the lower bound of the online complexity of a CBE scheme is also $O(n)$.

From the above analysis, it would seem that no better than a trivial CBE can be done. However, a closer look shows this is not the case. First, a well-designed CBE can be more efficient than a trivial CBE if $\tau_{\text{CEO}} \ll \tau_{\text{PKE}}$ and the performance difference can be further amplified by the factor n. Second, ℓ_{PKC} is usually hundreds to thousands, thus a trivial CBE may consume hundreds to thousands times more bits than an elegantly-developed CBE if ℓ_c is independent of the group size n. Hence, the efforts to achieve non-trivial CBE schemes are meaningful in practice.

To highlight this point, we further look at regular public-key BE schemes. The definitions of encryption and decryption in our CBE are exactly the same as those of standard public-key BE schemes [3]. Hence, the above online complexity bounds also apply to regular BE systems. Furthermore, by slightly modifying the above trivial CBE, one can also obtain a trivial public-key BE scheme. To strictly follow the public-key BE definition, one just needs to let a trusted key dealer generate the public/secret key pairs for all members. The rest is the same as the trivial CBE. This implies that a trivial public-key BE scheme has exactly the same asymptotical complexity as the trivial one. However, as discussed above, it is still meaningful to construct non-trivial public-key BE schemes. Indeed, this work has attracted a lot of attention and numerous efforts (*e.g.*, [3, 4, 6, 26, 17]) have been devoted to reduce the ℓ_c size and the τ_{CEO} complexity. We do a parallel work in the CBE setting.

3   An Aggregatable BE Scheme

Previously, aggregatability was mainly considered in the signature setting [7] and exploited to reduce the signature verification time and the storage overhead when numerous signatures need to be verified and stored. In [32], Wu *et al.* first presented the ASBB notion and considered aggregatability in the static BE setting. In this section, we integrate aggregatability into dynamic BE schemes and instantiate an aggregatable BE (ABE) scheme.

3.1   Review of Aggregatable Signature-Based Broadcast

Our ABE scheme is based on the ASBB primitive [32]. An ASBB scheme consists of the algorithms *ParaGen, KeyGen, Sign, Verify, Encrypt* and *Decrypt*. *ParaGen* takes as input a security parameter λ and outputs the public parameters π. *KeyGen* takes input π and outputs a public/secret key pair (pk, sk). *Sign* takes as input the key pair (pk, sk) and a string s, and outputs a signature $\sigma(s)$. *Verify* takes as input the public key pk and the signature $\sigma(s)$ of the string s, and outputs 0 or 1. *Encrypt* takes as input a public key pk and a plaintext m, and outputs a ciphertext c. *Decrypt* takes as input the public key pk, a valid string-signature $(s, \sigma(s))$ and a ciphertext c, and outputs the plaintext m.

An ASBB scheme has a key-homomorphic property. This property states that, for any two public/secret key pairs (pk_1, sk_1) and (pk_2, sk_2) generated by running $KeyGen(\pi)$, two signatures $\sigma_1 = Sign(pk_1, sk_1, s)$, $\sigma_2 = Sign(pk_2, sk_2, s)$ on any message string s with respect to the two public keys, it holds that $Verify(pk_1 \otimes pk_2, s, \sigma_1 \odot \sigma_2) = 1$, where $\otimes : \Gamma \times \Gamma \to \Gamma$ and $\odot : \Omega \times \Omega \to \Omega$ are two efficient operations in the public key space Γ and the signature space Ω, respectively. Clearly, from the key-homomorphic property, we have that $Decrypt(pk_1 \otimes pk_2, s, \sigma_1 \odot \sigma_2, c) = m$ for any plaintext m and the corresponding ciphertext $c = Encrypt(pk_1 \otimes pk_2, m)$.

Furthermore, an ASBB scheme has an interesting property referred to as aggregatability. Assume that an adversary $\mathcal{A}$ knows $(\pi, pk_1, \cdots, pk_n)$, where π is the system parameters, and $pk_1, \cdots, pk_n$ are n different public keys generated by independently invoking $KeyGen$ of the ASBB scheme. For n public binary strings $s_1, \cdots, s_n \in \{0,1\}^*$, the adversary $\mathcal{A}$ is provided with valid signatures $\sigma_i(s_j)$ under pk_i for $1 \leq i, j \leq n$ and $i \neq j$. Due to the key-homomorphic property, $pk = pk_1 \otimes \cdots \otimes pk_n$ forms the public key of the aggregated ASBB instance. Aggregatability states that the new ASBB instance related to the aggregated public key pk is secure against any polynomial-time adversary $\mathcal{A}$. Wu *et al.*'s ASBB scheme [32] is briefly reviewed next.

- **ParaGen**(π). Let `PairGen` be an algorithm that, on input a security parameter 1^λ, outputs a tuple $\Upsilon = (p, \mathbb{G}, \mathbb{G}_T, e)$, where $\mathbb{G}$ and $\mathbb{G}_T$ have the same prime order p, and $e : \mathbb{G} \times \mathbb{G} \to \mathbb{G}_T$ is an efficient non-degenerate bilinear map such that $e(g, g) \neq 1$ for any generator g of $\mathbb{G}$, and for all $u, v \in \mathbb{Z}$, it holds that $e(g^u, g^v) = e(g, g)^{uv}$. Let $\Upsilon = (p, \mathbb{G}, \mathbb{G}_T, e) \leftarrow$ `PairGen`(1^λ), and g be a generator of $\mathbb{G}$, and $H : \{0,1\}^* \to \mathbb{G}$ be a cryptographic hash function. The system parameters are $\pi = (\Upsilon, g, H)$.
- **KeyGen**(π). Select at random $r \in \mathbb{Z}_p^*$, $X \in \mathbb{G} \setminus \{1\}$. Compute $R = g^{-r}$, $A = e(X, g)$. Output a public key $pk = (R, A)$ and its associating secret key $sk = (r, X)$.
- **Sign**(pk, sk, s). Take as inputs public key $pk = (R, A)$, secret key $sk = (r, X)$ and a string $s \in \{0,1\}^*$, and output a signature $\sigma = XH(s)^r$ on s.
- **Verify**(pk, s, σ). Take as inputs public key $pk = (R, A)$, a message-signature pair (s, σ), and output 1 if $e(\sigma, g)e(H(s), R) = A$ holds; else output 0.
- **Encryption**(pk, ξ). Given public key $pk = (R, A)$, for a plaintext $\xi \in \mathbb{G}_T$, randomly select $t \in \mathbb{Z}_p^*$ and compute $c_1 = g^t$, $c_2 = R^t$, $c_3 = \xi A^t$. Output $c = (c_1, c_2, c_3)$.
- **Decryption**(pk, s, σ, c). Given public key $pk = (R, A)$ and ciphertext $c = (c_1, c_2, c_3)$, anyone with a valid message-signature pair (s, σ) can extract $\xi = \frac{c_3}{e(\sigma, c_1)e(H(s), c_2)}$.

In the ASBB scheme, every signature under the public key can be used as a decryption key to decrypt ciphertexts generated with the same public key. This feature allows ASBB to be used as static broadcast schemes.

3.2   An Aggregatable BE Scheme Based on ASBB

We construct a BE scheme from the the ASBB scheme [32] and show the resulting BE scheme preserves aggregatability as that of the underlying ASBB scheme. The construction is conceptually simple. Assume that the j-th user holds decryption keys[2] corresponding to the indices $\{0, ..., n\} \setminus \{j\}$. An encrypter knows which public key he should use. For instance, if the encrypter doesn't want to revoke anybody, he encrypts using pk_0. If he wants to exclude i from decrypting, he encrypts using pk_i. If he wants to exclude i and j from decrypting, he encrypts by using an *aggregated* public key $pk_i \otimes pk_j$. In the same way, more users can be excluded from decrypting. With the parameters in the above setting, the proposal is realized as follows.

  - BSetup(n, N): The dealer randomly chooses $X_i \in \mathbb{G}, r_i \in \mathbb{Z}_p^*$ and computes $R_i = g^{-r_i}, A_i = e(X_i, g)$. The BE public key is $PK = ((R_0, A_0), \cdots, (R_n, A_n))$ and the BE secret key is $sk = ((r_0, X_0), \cdots, (r_n, X_n))$.
  - BKeyGen(j, SK): For $j = 1, \cdots, n$, the private key of the user j is $d_j = (\sigma_{0,j}, \cdots, \sigma_{j-1,j}, \sigma_{j+1,j}, \cdots, \sigma_{n,j}) : \sigma_{i,j} = X_i H(ID_j)^{r_i}$.
  - BEncryption($\mathbb{R}, PK$): Set $\overline{\mathbb{R}} = \{0, 1, \cdots, n\} \setminus \mathbb{R}$. Randomly pick t in $\mathbb{Z}_p$ and compute $c = (c_1, c_2) : c_1 = g^t, c_2 = (\prod_{i \in \overline{\mathbb{R}}} R_i)^t$. Set the session key $\xi = (\prod_{i \in \overline{\mathbb{R}}} A_i)^t$. Output (c, ξ) and send $(\mathbb{R}, c)$ to receivers.
  - BDecryption($\mathbb{R}, j, d_j, c, PK$): If $j \in \mathbb{R}$, the receiver j extracts ξ from c with private key d_j by computing $e(\prod_{i \in \overline{\mathbb{R}}} \sigma_{i,j}, c_1) e(H(ID_j), c_2) = \xi$.

The correctness of the BE scheme above follows from direct verification of the following equations
$$e(\prod_{i \in \overline{\mathbb{R}}} \sigma_{i,j}, c_1) e(H(ID_j), c_2) = e(\prod_{i \in \overline{\mathbb{R}}} X_i H(ID_j)^{r_i}, g^t) e(H(ID_j), \prod_{i \in \overline{\mathbb{R}}} g^{-r_i t})$$
$$= e(\prod_{i \in \overline{\mathbb{R}}} X_i, g)^t = (\prod_{i \in \overline{\mathbb{R}}} A_i)^t = \xi.$$

The security of our BE scheme relies on the decision n-BDHE assumption which was shown to be sound by Boneh *et al.* [2] in the generic group model.

Definition 3 (Decision n-BDHE Assumption). *Let $\mathbb{G}$ be a bilinear group of prime order p as defined above, g a generator of $\mathbb{G}$, and $h = g^t$ for some unknown $t \in \mathbb{Z}_p$. Denote $\overrightarrow{y}_{g,\alpha,n} = (g_1, \cdots, g_n, g_{n+2}, \cdots, g_{2n}) \in \mathbb{G}^{2n-1}$, where $g_i = g^{\alpha^i}$ for some unknown $\alpha \in \mathbb{Z}_p$. We say that an algorithm $\mathcal{B}$ that outputs $b \in \{0, 1\}$ has advantage ε in solving the decision n-BDHE assumption if $|\Pr[\mathcal{B}(g, h, \overrightarrow{y}_{g,\alpha,n}, e(g_{n+1}, h)) = 0] - \Pr[\mathcal{B}(g, h, \overrightarrow{y}_{g,\alpha,n}, Z) = 0)]| \geq \varepsilon$, where the probability is over the random choice of g in $\mathbb{G}$, the random choice $t, \alpha \in \mathbb{Z}_p$, the random choice of $Z \in \mathbb{G}_T$, and the random bits consumed by $\mathcal{B}$. We say that the decision (τ, ε, n)-BDHE assumption holds in $\mathbb{G}$ if no τ-time algorithm has advantage at least ε in solving the decision n-BDHE assumption.*

According to the BE security definition in [17], our scheme is fully collision-resistant under the Decision BDHE assumption. The proof is given in the full

---

[2] Here, user j's i-th decryption key corresponding to index $i \in \{0, ..., n\} \setminus \{j\}$ is a signature $\sigma_{i,j} = \sigma_i(ID_j)$ on user j's identity ID_j verifiable under the public key pk_i.

version of the paper [33]. One can further apply the generic Gentry-Waters transformation [17] to convert our semi-adaptive BE schemes into an adaptively secure one. The cost is to double the size of the public keys and the ciphertexts.

Theorem 1. *The proposed BE scheme for dynamic groups has full collusion resistance against semi-adaptive attacks in the random oracle model if the decision n-BDHE assumption holds. More formally, if there exists a semi-adaptive attacker $\mathcal{A}$ breaking our scheme with advantage ϵ in time τ, then there exists an algorithm $\mathcal{B}$ breaking the n-BDHE assumption with advantage ϵ in time $\tau' = \tau + \mathcal{O}((q_H + n^2)\tau_{Exp})$, where q_H is the number of queries to the random oracle from $\mathcal{A}$, and τ_{Exp} is the time to compute an exponentiation in $\mathbb{G}$ or $\mathbb{G}_T$.*

One may observe that, in the above BE scheme, if we replace $H(ID_j)$ with a random element h_j in $\mathbb{G}$, we obtain a semi-adaptive BE scheme with short ciphertexts in the standard model. In this case, to simulate h_j in the security proof, we just need to set $h_j = g^{\alpha^j} g^{v_j}$ for a randomly chosen value $v_j \in Z_p$, where g^{α^j} is obtained from the decision n-BDHE assumption.

3.3   Useful Properties

Our BE scheme inherits the key-homomorphic property of the underlying ASBB scheme. Consider the system parameters defined above. Let $PK_1 = (R_{0,1}, A_{0,1})$, $\cdots, (R_{n,1}, A_{n,1}))$ and $PK_2 = ((R_{0,2}, A_{0,2}), \cdots, (R_{n,2}, A_{n,2}))$ be the respective public keys of two random instances of the above BE scheme, and for $j = 1, \cdots, n$, let $d_{j,1} = (\sigma_{0,j,1}, \cdots, \sigma_{j-1,j,1}, \sigma_{j+1,j,1}, \cdots, \sigma_{n,i,1}) \in \mathbb{G}^n$ and $d_{j,2} = (\sigma_{0,j,2}, \cdots, \sigma_{j-1,j,2}, \sigma_{j+1,j,2}, \cdots, \sigma_{n,j,2}) \in \mathbb{G}^n$ be the respective decryption keys corresponding to index j under PK_1 and PK_2. Define $PK = PK_1 \circledast PK_2 = ((R_{0,1}R_{0,2}, A_{0,1}A_{0,2}), \cdots, (R_{n,1}R_{n,2}, A_{n,1}A_{n,2}))$ and define $dk_j = d_{j,1} \boxdot d_{j,2} = (\sigma_{0,j,1}\sigma_{0,j,2}, \cdots, \sigma_{j-1,j,1}\sigma_{j-1,j,2}, \sigma_{j+1,j,1}\sigma_{j+1,j,2}, \cdots, \sigma_{n,j,1}\sigma_{n,j,2})$. Then PK is the public key of a new instance of the above BE scheme and dk_j is the new decryption key corresponding to the index j. This fact can be directly verified.

Our BE scheme also preserves the aggregatability of the underlying ASBB scheme. Roughly speaking, a BE scheme is aggregatable if n instances of the BE scheme can be aggregated into a new BE instance secure against an attacker accessing some decryption keys of each instance, provided that the i-th decryption key corresponding to the i-th instance is unknown to the attacker for $i = 1, \cdots, n$. More formally, this property can be defined as follows.

Definition 4 (Aggregatability). *Consider the following game between an adversary $\mathcal{A}$ and a challenger $\mathcal{CH}$:*

- **Setup:** *$\mathcal{A}$ initializes the game with an integer n. $\mathcal{CH}$ replies with $(\pi, PK_1, \cdots, PK_n)$ which are the system parameters and the n independent public keys of the BE scheme.*
- **Corruption:** *For $1 \leq i, j \leq n$, where $i \neq j$, the adversary $\mathcal{A}$ is allowed to know the decryption keys $dk_{j,i}$ corresponding to index j with respect to the public key PK_i.*

- **Challenge:** $\mathcal{CH}$ and $\mathcal{A}$ run a standard Ind-CPA game under the aggregated public key $PK = PK_1 \circledast \cdots \circledast PK_n$. $\mathcal{A}$ wins if $\mathcal{A}$ outputs a correct guess bit. Denote $\mathcal{A}$'s advantage by $Adv_{\mathcal{A}} = |\Pr[win] - \frac{1}{2}|$.

A BE scheme is said to be (τ, ε, n)-aggregatable if no τ-time algorithm $\mathcal{A}$ has advantage $Adv_{\mathcal{A}} \geq \varepsilon$ in the above aggregatability game.

Theorem 2. *If there exists an attacker $\mathcal{A}$ who wins the aggregatability game with advantage ϵ in time τ, then there exists an algorithm $\mathcal{B}$ breaking the n-BDHE assumption with advantage ϵ in time $\tau' = \tau + \mathcal{O}((n^3)\tau_{Exp})$.*

For the proof of the previous theorem, we refer to Theorem 3 where we prove a stronger property in the sense that the attacker is additionally allowed to know the internal randomness used to compute $dk_{j,i}$ corresponding some PK_i for $1 \leq i, j \leq n$ where $i \neq j$.

4   Proposed CBE Scheme

In this section, we propose a CBE based on the above aggregatable BE scheme. The basic construction has short ciphertexts and long protocol transcripts. Then we show an efficient trade-off between ciphertexts and protocol transcripts.

4.1   High-Level Description

Our basic idea is to introduce the revocation mechanism of a regular BE scheme into the asymmetric GKA scheme [32]. To this end, each member acts as the dealer of the aggregatable BE scheme above. The k-th user publishes PK_k and $d_{j,k}$, where $d_{j,k}$ is the decryption key of PK_k corresponding to the index $j \in \{1, \cdots, n\} \setminus \{k\}$. Then the negotiated public key is $PK = PK_0 \circledast \cdots \circledast PK_n$. Each member j can compute the decryption key $dk_j = dk_{j,j} \boxdot_{k=1, k \neq j}^{n} dk_{j,k}$. Observe that $dk_{j,j}$ has never been published. Due to the key homomorphism of the BE scheme above, dk_j is a valid decryption key corresponding to PK. Hence, anyone knowing PK can encrypt to any subset of the members and the intended receivers can decrypt.

To guarantee the security of the resulting CBE scheme, we also need to show that *only* the intended receivers can decrypt. This is ensured by the fact that the underlying BE scheme is aggregatable. Indeed, although the Gentry-Waters BE scheme [17] is key-homomorphic, an analog of our CBE scheme using the Gentry-Waters BE scheme as a building block is shown to be insecure in [33], because the Gentry-Waters BE scheme is not aggregatable. We note that a *static* PKBE scheme without a dealer can be trivially obtained from the ASGKA protocol in [32]. This is realized by letting each member to register his/her published string as her public key. Then anyone knowing the public keys of all members can send encrypted messages to the group and only the group members can decrypt the message. However, no revocation mechanism is provided. To exclude some members, one may be motivated to modify the above trivial construction by using the aggregation of the public keys of the intended receivers as the sub-group public key. Clearly, this will allow the intended receivers to decrypt ciphertexts generated with this sub-group public key. Unfortunately, anyone (not necessary to be a revoked member) knowing the receivers' public keys can also decrypt, as shown in [33].

4.2   The Proposal

Based on our aggregatable BE scheme, we implement a CBE scheme with short ciphertexts. Assume that the group size is at most n. Let $\Upsilon = (p, \mathbb{G}, \mathbb{G}_T, e) \leftarrow$ $\mathsf{PairGen}(1^\lambda)$, and $g, h_1, \cdots, h_n$ be independent generators of $\mathbb{G}$. The system parameters are $\pi = (\lambda, n, \Upsilon, g, h_1, \cdots, h_n)$.

- **Setup.** The set-up of a CBE system consists of the following three procedures:
  - Group Key Agreement Execution: For $1 \leq k \leq n$, member k does the following:
    Randomly choose $X_{i,k} \in \mathbb{G}, r_{i,k} \in \mathbb{Z}_p^*$;
    Compute $R_{i,k} = g^{-r_{i,k}}, A_{i,k} = e(X_{i,k}, g)$;
    Set $PK_k = ((R_{0,k}, A_{0,k}), \cdots, (R_{n,k}, A_{n,k}))$;
    For $1 \leq j \leq n$, $j \neq k$, compute $\sigma_{i,j,k} = X_{i,k} h_j^{r_{i,k}}$ for $0 \leq i \leq n, i \neq j$;
    Set $d_{j,k} = (\sigma_{0,j,k}, \cdots, \sigma_{j-1,j,k}, \sigma_{j+1,j,k}, \cdots, \sigma_{n,j,k})$;
    Publish $(PK_k, d_{1,k}, \cdots, d_{k-1,k}, d_{k+1,k}, \cdots, d_{n,k})$ and keep $d_{k,k}$ secret.
  - Group Encryption Key Derivation: The group encryption key is $PK = PK_0 \circledast \cdots \circledast PK_n = ((R_0, A_0), \cdots, (R_n, A_n))$, where $R_i = \prod_{k=1}^n R_{i,k}$, $A_i = \prod_{k=1}^n A_{i,k}$ for $i = 0, \cdots, n$. The group encryption key PK is publicly computable.
  - Member Decryption Key Derivation: For $0 \leq i \leq n$, $1 \leq j \leq n$ and $i \neq j$, member j can compute decryption key $d_j = (\sigma_{0,j}, \cdots, \sigma_{j-1,j}, \sigma_{j+1,j}, \cdots, \sigma_{n,j})$, where $\sigma_{i,j} = \sigma_{i,j,j} \prod_{k=1, k \neq j}^n \sigma_{i,j,k} = \prod_{k=1}^n \sigma_{i,j,k} = \prod_{k=1}^n X_{i,k} h_j^{r_{i,k}}$.
- **CBEncrypt.** Assume that a sender (not necessarily a group member) wants to send to receivers in $\mathbb{R} \subseteq \{1, \cdots, n\}$ a session key ξ. Set $\overline{\mathbb{R}} = \{0, 1, \cdots, n\} \setminus \mathbb{R}$. Randomly pick t in $\mathbb{Z}_p$ and compute the ciphertext $c = (c_1, c_2)$ where $c_1 = g^t, c_2 = (\prod_{i \in \overline{\mathbb{R}}} R_i)^t$. Output (c, ξ) where $\xi = (\prod_{i \in \overline{\mathbb{R}}} A_i)^t$. Send $(\mathbb{R}, c)$ to the receivers.
- **CBDecrypt.** If $j \in \mathbb{R}$, receiver j can extract ξ from the ciphertext c with decryption key d_j by computing $e(\prod_{i \in \overline{\mathbb{R}}} \sigma_{i,j}, c_1) e(h_j, c_2) = \xi$.

The correctness of the proposed CBE scheme is correct directly follows from the fact that the underlying BE scheme is correct and key-homomorphic. As to security, we have the following theorem, whose proof is given in [33].

Theorem 3. *The proposed CBE scheme has fully collusion-resistant secrecy against semi-adaptive attacks in the standard model if the decision n-BDHE assumption holds. More formally, if there exists a semi-adaptive attacker $\mathcal{A}$ breaking our scheme with advantage ϵ in time τ, then there exists an algorithm $\mathcal{B}$ breaking the n-BDHE assumption with advantage ϵ in time $\tau' = \tau + O((n^3)\tau_{Exp})$.*

4.3   Discussion

We first examine the online complexity our scheme which is critical for the practicality of a CBE scheme. We use the widely-adopted metrics [3, 4, 6, 26, 17] for regular BE schemes. After the CBSetup procedure, a sender needs to retrieve and store the group public key PK consisting of n elements in $\mathbb{G}$ and n elements in $\mathbb{G}_T$. This requires about $150n$ bytes to achieve the security level of an RSA-1024 cryptosystem. Note that in the motivated applications, the group size is usually not very large. Consider an initial group of 100 users. The group public key is about $15K$ bytes long and acceptable in practice. Moreover, for encryption, the sender needs only two exponentiations and the ciphertext merely contains two elements in $\mathbb{G}$. This is about n times more efficient than the trivial solution. At the receiver's side, in addition to the description of the bilinear pair which may be shared by many other security applications, a receiver needs to store n elements in $\mathbb{G}$ for decryption. The storage cost of a receiver is about $22n$ bytes. For decryption, a receiver needs to compute two single-base bilinear pairings (or one two-base bilinear pairing). The online costs on the sides of both the sender and the receivers are really low.

We next discuss the complexity of the CBSetup procedure to set up a CBE system. The overhead incurred by this procedure is $O(n^2)$. However, in most cases, this procedure needs to be run only once and this can be done offline before online transmission of secret session keys. For instance, in the social networks example, a number of friends exchange their CBSetup transcripts and establish a CBE system to secure their subsequent sharing of private picture/videos. Since CBE allows revoking members, the members do not need to reassemble for a new run of the CBSetup procedure until some new friends join. From our personal experience, the group lifetime usually lasts from weeks to months. These observations imply that our protocol is practical in the real world.

Furthermore, if the initial group is too large, an efficient trade-off can be employed [3] to balance the online and offline costs. Suppose that n is a cube, $i.e.$, $n = n_1^3$, and the initial group has n members. We divide the full group into n_1^2 subgroups, each of which has n_1 members. By applying our basic CBE to each subgroup, we obtain a CBE scheme with $O(n_1^2)$-size transcripts per member during the offline stage of group key establishment; a sender needs to do $O(n_1^2)$ encryption operations of the basic CBE scheme, which produces $O(n_1^2)$-size ciphertexts. Consequently, we obtain a CBE scheme with $O(n^{\frac{2}{3}})$ complexity. This is comparable to up-to-date public-key BE systems whose complexity is $O(n^{\frac{1}{2}})$. For a group of 1000 users, our dealer-free BE scheme is about 10 times more efficient than the trivial solution. It is about 3 times less efficient than a public-key BE scheme, but our CBE does not require a trusted key dealer. The cost of versatileness is acceptable.

One may notice a subtlety in the above trade-off. When the basic CBE scheme is applied to each subgroup, members in each subgroup will extract the same session key, but members in different subgroups will have different session keys. This is inconsistent with the CBE definition in which all members should extract the same session key, even if the members are in different subgroups. This can be trivially addressed as follows. The sender additionally selects a string from the session key space and encrypts it for each subgroup with the session keys shared by each subgroup. Then all members can extract the same resulting session key. This introduces an additional $O(n^{\frac{2}{3}})$-size ciphertext if there are $O(n^{\frac{2}{3}})$ subgroups, but it does not affect the asymptotical complexity of the scheme after a trade-off.

Finally, we assume that the communication channels between members are authenticated during the CBSetup stage to establish the group encryption key. In practice, these authenticated channels can be the pre-existing ones between members (*e.g.*, in instant-messaging system and cooperative scientific computation) or be established by personal interaction (*e.g.*, some *ad hoc* network applications). This is plausible since CBE is usually deployed for cooperative members who may be friends. Note that the CBSetup sub-protocol requires only one round. An alternative option to achieve authentication is to let a partially trusted third party certify each member's protocol transcript. The third party plays a role similar to a certification authority in the popular PKI setting, and cannot read the plaintexts encrypted to the members. This is different from regular BE systems where the fully trusted dealer can decrypt all communications to the members. For instance, in a social network application, the service provider can serve as the partially trusted third party. This is also plausible since this kind of applications usually require users to register for service. In this case, the CBSetup transcript of each member can be viewed as her public key.

5   Conclusions

In this paper, we formalized the CBE primitive, which bridges the GKA and BE notions. In CBE, anyone can send secret messages to any subset of the group members, and the system does not require a trusted key server. Neither the change of the sender nor the dynamic choice of the intended receivers require extra rounds to negotiate group encryption/decryption keys. Following the CBE model, we instantiated an efficient CBE scheme that is secure in the standard model. As a versatile cryptographic primitive, our novel CBE notion opens a new avenue to establish secure broadcast channels and can be expected to secure numerous emerging distributed computation applications.

Acknowledgments. The authors gratefully acknowledge the anonymous reviewers for their invaluable comments. The authors are partly supported by the EU 7FP through project "DwB", the Spanish Government through projects CTV-09-634, PTA2009-2738-E, TSI-020302-2010-153, PT-430000- 2010-31, TIN2009-11689, CONSOLIDER INGENIO 2010 "ARES" CSD2007-0004 and TSI2007-65406-C03-01, by the Government of Catalonia under grant SGR2009-1135, and by the NSF of China through projects 60970114, 60970115, 60970116, 61173154, 61003214, 61173192, 91018008, 61021004 and 11061130539. The authors also acknowledge support by the Fundamental Research Funds for the Central Universities of China to Project 3103004, and Shaanxi Provincial Education Department through Scientific Research Program 2010JK727. The fourth author is partially supported as an ICREA-Acadèmia researcher by the Catalan Government. The fifth author is partially supported by ISF grant 938/09. The authors are with the UNESCO Chair in Data Privacy, but this paper does not necessarily reflect the position of UNESCO nor does it commit that organization.

References

1. Abdalla, M., Chevalier, C., Manulis, M., Pointcheval, D.: Flexible Group Key Exchange with On-demand Computation of Subgroup Keys. In: Bernstein, D.J., Lange, T. (eds.) AFRICACRYPT 2010. LNCS, vol. 6055, pp. 351–368. Springer, Heidelberg (2010)

2. Boneh, D., Boyen, X., Goh, E.J.: Hierarchical Identity Based Encryption with Constant Size Ciphertext. In: Cramer, R. (ed.) EUROCRYPT 2005. LNCS, vol. 3494, pp. 440–456. Springer, Heidelberg (2005)

3. Boneh, D., Gentry, C., Waters, B.: Collusion Resistant Broadcast Encryption with Short Ciphertexts and Private Keys. In: Shoup, V. (ed.) CRYPTO 2005. LNCS, vol. 3621, pp. 258–275. Springer, Heidelberg (2005)

4. Boneh, D., Sahai, A., Waters, B.: Fully Collusion Resistant Traitor Tracing with Short Ciphertexts and Private Keys. In: Vaudenay, S. (ed.) EUROCRYPT 2006. LNCS, vol. 4004, pp. 573–592. Springer, Heidelberg (2006)

5. Boneh, D., Silverberg, A.: Applications of Multilinear Forms to Crytography. Contemporary Mathematics, vol. 324, pp. 71–90 (2003)

6. Boneh, D., Waters, B.: A Fully Collusion Resistant Broadcast, Trace, and Revoke System. In: ACM CCS 2006, pp. 211–220. ACM Press (2006)

7. Boneh, D., Gentry, C., Lynn, B., Shacham, H.: Aggregate and Verifiably Encrypted Signatures from Bilinear Maps. In: Biham, E. (ed.) EUROCRYPT 2003. LNCS, vol. 2656, pp. 416–432. Springer, Heidelberg (2003)

8. Boyd, C., González-Nieto, J.M.: Round-Optimal Contributory Conference Key Agreement. In: Desmedt, Y.G. (ed.) PKC 2003. LNCS, vol. 2567, pp. 161–174. Springer, Heidelberg (2002)

9. Bresson, E., Chevassut, O., Pointcheval, D.: Provably Authenticated Group Diffie-Hellman Key Exchange - The Dynamic Case. In: Boyd, C. (ed.) ASIACRYPT 2001. LNCS, vol. 2248, pp. 290–309. Springer, Heidelberg (2001)

10. Bresson, E., Chevassut, O., Pointcheval, D.: Dynamic Group Diffie-Hellman Key Exchange under Standard Assumptions. In: Knudsen, L.R. (ed.) EUROCRYPT 2002. LNCS, vol. 2332, pp. 321–336. Springer, Heidelberg (2002)

11. Bresson, E., Chevassut, O., Pointcheval, D., Quisquater, J.-J.: Provably Authenticated Group Diffie-Hellman Key Exchange. In: ACM CCS 2001, pp. 255–264. ACM Press (2001)

12. Burmester, M., Desmedt, Y.: A Secure and Efficient Conference Key Distribution System. In: De Santis, A. (ed.) EUROCRYPT 1994. LNCS, vol. 950, pp. 275–286. Springer, Heidelberg (1995)

13. Cheon, J.H., Jho, N.S., Kim, M.H., Yoo, E.S.: Skipping, Cascade, and Combined Chain Schemes for Broadcast Encryption. IEEE Transactions Information Theory 54(11), 5155–5171 (2008)

14. Diffie, W., Hellman, M.: New Directions in Cryptography. IEEE Transactions on Information Theory 22(6), 644–654 (1976)

15. Fiat, A., Naor, M.: Broadcast Encryption. In: Stinson, D.R. (ed.) CRYPTO 1993. LNCS, vol. 773, pp. 480–491. Springer, Heidelberg (1994)

16. Fujisaki, E., Okamoto, T.: Secure Integration of Asymmetric and Symmetric Encryption Schemes. In: Wiener, M. (ed.) CRYPTO 1999. LNCS, vol. 1666, pp. 537–554. Springer, Heidelberg (1999)
17. Gentry, C., Waters, B.: Adaptive Security in Broadcast Encryption Systems (with Short Ciphertexts). In: Joux, A. (ed.) EUROCRYPT 2009. LNCS, vol. 5479, pp. 171–188. Springer, Heidelberg (2009)
18. Gorantla, M.C., Boyd, C., González Nieto, J.M., Manulis, M.: Generic One Round Group Key Exchange in the Standard Model. In: Lee, D., Hong, S. (eds.) ICISC 2009. LNCS, vol. 5984, pp. 1–15. Springer, Heidelberg (2010)
19. Halevy, D., Shamir, A.: The LSD Broadcast Encryption Scheme. In: Yung, M. (ed.) CRYPTO 2002. LNCS, vol. 2442, pp. 47–60. Springer, Heidelberg (2002)
20. Ingemarsson, I., Tang, D.T., Wong, C.K.: A Conference Key Distribution System. IEEE Transactions on Information Theory 28(5), 714–720 (1982)
21. Joux, A.: A One Round Protocol for Tripartite Diffie-Hellman. J. of Cryptology 17, 263–276 (2004)
22. Kim, H.J., Lee, S.M., Lee, D.H.: Constant-Round Authenticated Group Key Exchange for Dynamic Groups. In: Lee, P.J. (ed.) ASIACRYPT 2004. LNCS, vol. 3329, pp. 245–259. Springer, Heidelberg (2004)
23. Kim, Y., Perrig, A., Tsudik, G.: Tree-Based Group Key Agreement. ACM Transactions on Information System Security 7(1), 60–96 (2004)
24. Mao, Y., Sun, Y., Wu, M., Liu, K.J.R.: JET: Dynamic Join-Exit-Tree Amortization and Scheduling for Contributory Key Management. IEEE/ACM Transactions on Networking 14(5), 1128–1140 (2006)
25. Naor, M., Pinkas, B.: Efficient Trace and Revoke Schemes. In: Frankel, Y. (ed.) FC 2000. LNCS, vol. 1962, pp. 1–20. Springer, Heidelberg (2001)
26. Park, J.H., Kim, H.J., Sung, M.H., Lee, D.H.: Public Key Broadcast Encryption Schemes With Shorter Transmissions. IEEE Transactions on Broadcasting 54(3), 401–411 (2008)
27. Sherman, A., McGrew, D.: Key Establishment in Large Dynamic Groups Using One-way Function Trees. IEEE Transactions on Software Engineering 29(5), 444–458 (2003)
28. Snoeyink, J., Suri, S., Varghese, G.: A Lower Bound for Multicast Key Distribution. In: INFOCOM 2001, pp. 422–431. IEEE Press (2001)
29. Steiner, M., Tsudik, G., Waidner, M.: Key Agreement in Dynamic Peer Groups. IEEE Transactions on Parallel and Distributed Systems 11(8), 769–780 (2000)
30. Tzeng, W.-G., Tzeng, Z.-J.: Round-Efficient Conference Key Agreement Protocols with Provable Security. In: Okamoto, T. (ed.) ASIACRYPT 2000. LNCS, vol. 1976, pp. 614–627. Springer, Heidelberg (2000)
31. Wong, C.K., Gouda, M., Lam, S.: Secure Group Communications Using Key Graphs. IEEE/ACM Transactions on Networking 8(1), 16–30 (2000)
32. Wu, Q., Mu, Y., Susilo, W., Qin, B., Domingo-Ferrer, J.: Asymmetric Group Key Agreement. In: Joux, A. (ed.) EUROCRYPT 2009. LNCS, vol. 5479, pp. 153–170. Springer, Heidelberg (2009)
33. Wu, Q., Qin, B., Zhang, L., Domingo-Ferrer, J., Farras, O.: Bridging Broadcast Encryption and Group Key Agreement (full version), http://eprint.iacr.org

On the Joint Security of Encryption and Signature, Revisited

Kenneth G. Paterson[1,*], Jacob C.N. Schuldt[2,**], Martijn Stam[3], and Susan Thomson[1,***]

[1] Royal Holloway, University of London
[2] Research Center for Information Security, AIST, Japan
[3] University of Bristol

Abstract. We revisit the topic of joint security for combined public key schemes, wherein a single keypair is used for both encryption and signature primitives in a secure manner. While breaking the principle of key separation, such schemes have attractive properties and are sometimes used in practice. We give a general construction for a combined public key scheme having joint security that uses IBE as a component and that works in the standard model. We provide a more efficient direct construction, also in the standard model.

1 Introduction

Key separation versus key reuse: The folklore principle of key separation dictates using different keys for different cryptographic operations. While this is well-motivated by real-world, security engineering concerns, there are still situations where it is desirable to use the same key for multiple operations [15]. In the context of public key cryptography, using the same keypair for both encryption and signature primitives can reduce storage requirements (for certificates as well as keys), reduce the cost of key certification and the time taken to verify certificates, and reduce the footprint of cryptographic code. These savings may be critical in embedded systems and low-end smart card applications. As a prime example, the globally-deployed EMV standard for authenticating credit and debit card transactions allows the same keypair to be reused for encryption and signatures for precisely these reasons [11].

However, this approach of reusing keys is not without its problems. For example, there is the issue that encryption and signature keypairs may have different lifetimes, or that the private keys may require different levels of protection [15]. Most importantly of all, there is the question of whether it is *secure* to use the same keypair in two (or more) different primitives – perhaps the two uses will interact with one another badly, in such a way as to undermine the security of

---

* This author is supported by EPSRC Leadership Fellowship EP/H005455/1.
** This author is supported by a JSPS Fellowship for Young Scientists.
*** This author is supported by the EPSRC through Leadership Fellowship EP/H005455/1.

D.H. Lee and X. Wang (Eds.): ASIACRYPT 2011, LNCS 7073, pp. 161–178, 2011.

one or both of the primitives. In the case of textbook RSA, it is obvious that using the same keypair for decryption and signing is dangerous, since the signing and decryption functions are so closely related in this case. Security issues may still arise even if some standardized padding is used prior to encryption and signing [20]. In Section 3 we will provide another example in the context of encryption and signature primitives, where the individual components are secure (according to the usual notions of security for encryption and signature) but become completely insecure as soon as they are used in combination with one another. At the protocol level, Kelsey, Schneier and Wagner [18] gave examples of protocols that are individually secure, but that interact badly when a keypair is shared between them.

The formal study of the security of key reuse was initiated by Haber and Pinkas [15]. They introduced the concept of a *combined public key scheme*. Here, an encryption scheme and signature scheme are combined: the existing algorithms to encrypt, decrypt, sign and verify are preserved, but the two key generation algorithms are modified to produce a single algorithm. This algorithm outputs two keypairs, one for the encryption scheme and one for the signature scheme, with the keypairs no longer necessarily being independent. Indeed, under certain conditions, the two keypairs may be identical, in which case the savings described above may be realised. In other cases, the keypairs are not identical but can have some shared components, leading to more modest savings. Haber and Pinkas also introduced the natural security model for combined public key schemes, where the adversary against the encryption part of the scheme is equipped with a signature oracle in addition to the usual decryption oracle, and where the adversary against the signature part of the scheme is given a decryption oracle in addition to the usual signature oracle. In this setting, we talk about the *joint security* of the combined scheme.

Setting a benchmark: As we shall see in Section 3, there is a trivial "Cartesian product" construction for a combined public key scheme with joint security. The construction uses arbitrary encryption and signature schemes as components, and the combined scheme's keypair is just a pair of vectors whose components are the public/private keys of the component schemes. Thus the Cartesian product construction merely formalises the principle of key separation. This construction, while extremely simple, provides a benchmark by which other constructions can be judged. For example, if the objective is to minimise the public key size in a combined scheme, then any construction should aim to have shorter keys than can be obtained by instantiating the Cartesian product construction with the best available encryption and signature schemes.

Re-evaluating Haber-Pinkas: In this respect, we note that, while Haber and Pinkas considered various well-known concrete schemes and conditions under which their keys could be partially shared, none of their examples having provable security in the standard model lead to *identical* keypairs for both signature and encryption. Indeed, while the approach of Haber and Pinkas can be made to work in the random oracle model by careful oracle programming and domain separation, their approach does not naturally extend to the standard model. More specifically, in their approach, to be able to simulate the signing oracle in the IND-CCA security game, the public key of the combined scheme cannot be exactly the same as the public key of the underlying encryption scheme (otherwise, successful simulation would lead to a signature forgery). This makes it hard to achieve full effective overlap between the public keys for signing and encryption. For the (standard model) schemes considered by Haber and Pinkas this results in the requirements that part of the public key be specific to the encryption scheme and that another part of it be specific to the signature scheme. Furthermore, at the time of publication of [15] only a few secure (IND-CCA2, resp. EUF-CMA) and efficient standard-model schemes were known. Consequently, no "compatible" signature and encryption schemes were identified in [15] for the standard model.

Combined schemes from trapdoor permutations: The special case of combined schemes built from trapdoor permutations was considered in [8, 21]. Here, both sets of authors considered the use of various message padding schemes in conjunction with an arbitrary trapdoor permutation to build combined public key schemes having joint security. Specifically, Coron et al. [8] considered the case of PSS-R encoding, while Komano and Ohta [21] considered the cases of OAEP+ and REACT encodings. All of the results in these two papers are in the random oracle model. In further related, but distinct, work, Dodis et al. [10] (see also [9]) considered the use of message padding schemes and trapdoor permutations to build signcryption schemes. Dodis et al. showed, again in the random oracle model, how to build efficient, secure signcryption schemes in which each user's keypair, specifying a permutation and its trapdoor, is used for both signing and encryption purposes.

1.1   Our Contribution

We focus on the problem of how to construct combined public key schemes which are jointly secure in the standard model, a problem for which, as we have explained above, there currently exist no fully satisfactory solutions. Naturally, for reasons of practical efficiency, we are interested in minimising the size of keys (both public and private), ciphertexts, and signatures in such schemes. The complexity of the various algorithms needed to implement the schemes will also be an important consideration.

As a warm-up, in Section 3, we give the simple Cartesian product construction, as well as a construction showing that the general problem is not vacuous (i.e. that there exist insecure combined schemes whose component schemes are secure when used in isolation).

We then present in Section 4 a construction for a combined public key scheme using an IBE scheme as a component. The trick here is to use the IBE scheme in the Naor transform and the CHK transform *simultaneously* to create a combined public key scheme that is jointly secure, under rather weak requirements on the starting IBE scheme (specifically, the IBE scheme needs to be OW-ID-CPA and IND-sID-CPA secure). This construction extends easily to the (hierarchical) identity-based setting. Instantiating this construction using standard model secure IBE schemes from the literature already yields rather efficient combined schemes. For example, using an asymmetric pairing version of Gentry's IBE scheme [14], we can achieve a combined scheme in which, at the 128-bit security level, the public key size is 1536 bits, the signature size is 768 bits and the ciphertext size is 2304 bits (plus the size of a signature and a verification key for a one-time signature scheme), with joint security being based on a q-type assumption. This is already competitive with schemes arising from the Cartesian product construction.

We then provide a more efficient direct construction for a combined scheme with joint security in Section 5. This construction is based on the signature scheme of Boneh and Boyen [4] and a KEM obtained by applying the techniques by Boyen, Mei and Waters [7] to the second IBE scheme of Boneh and Boyen in [3]. At the 128-bit security level, it enjoys public keys that consist of 1280 bits, signatures that are 768 bits and a ciphertext overhead of just 512 bits. The signatures can be shrunk at the cost of increasing the public key size.

The ideas of this paper also have applications for signcryption. We show in the full version [24] that a (tag-based) combined public key scheme can be used to construct a signcryption scheme, using the "sign-then-encrypt" construction of [23], that is secure in the strongest security model for signcryption (achieving insider confidentiality and insider unforgeability in the multi-user setting). Instantiating this construction with our concrete combined public key scheme effectively solves the challenge implicitly laid down by Dodis et al. in [9], to construct an efficient standard model signcryption scheme in which a single short keypair can securely be used for both sender and receiver functions. Furthermore, we are able to show that the signcryption scheme we obtain is jointly secure when used in combination with *both* its signature and encryption components. Thus we are able to obtain a triple of functionalities (signcryption, signature, encryption) which are jointly secure using only a single keypair.

1.2   Further Related Work

Further work on combined public key schemes in the random oracle model, for both the normal public key setting and the identity-based setting can be found in [27]. In particular, it is proved that the identity-based signature scheme of Hess [16] and Boneh and Franklin's identity-based encryption scheme [6] can be used safely together.

The topic of joint security of combined public key schemes is somewhat linked to the topic of cryptographic agility [1], which considers security when the same key (or key pair) is used simultaneously in multiple instantiations of the *same* cryptographic primitive. This contrasts with joint security, where we are concerned with security when the same key pair is used simultaneously in instantiations of *different* cryptographic primitives. The connections between these different but evidently related topics remain to be explored.

2   Preliminaries

In our constructions, we will make use of a number of standard primitives, including digital signatures, (tag-based) public key encryption, identity-based encryption (IBE), a data encapsulation mechanism (DEM), and an always second-preimage resistant hash function. We refer the reader to the full version [24] for the standard definitions and security notions for these primitives. In the following, we briefly recall the properties of bilinear pairings as well as define the computational assumptions which we will make use of to prove the security of our concrete constructions.

Bilinear pairings: Let $\mathbb{G}_1 = \langle g_1 \rangle$, $\mathbb{G}_2 = \langle g_2 \rangle$, $\mathbb{G}_T$ be groups of prime order p. A pairing is a map $e : \mathbb{G}_1 \times \mathbb{G}_2 \to \mathbb{G}_T$ that satisfies the following properties:

1. Bilinear: For all $a, b \in \mathbb{Z}$, $e(g_1^a, g_2^b) = e(g_1, g_2)^{ab}$.
2. Non-degenerate: $e(g_1, g_2) \neq 1$.
3. Computable: There is an efficient algorithm to compute the map e.

Note that we work exclusively in the setting of asymmetric pairings, whereas schemes are often presented in the naive setting of symmetric pairings $e : \mathbb{G} \times \mathbb{G} \to \mathbb{G}_T$. At higher security levels (128 bits and above), asymmetric pairings are far more efficient both in terms of computation and in terms of the size of group elements [13]. As a concrete example, using BN curves [2] and sextic twists, we can attain the 128-bit security level with elements of $\mathbb{G}_1$ being represented by 256 bits and elements of $\mathbb{G}_2$ needing 512 bits. By exploiting compression techniques [26], elements of $\mathbb{G}_T$ in this case can be represented using 1024 bits. For further details on parameter selection for pairings, see [12].

Strong Diffie-Hellman (SDH) assumption [4]: Let $\mathbb{G}_1$ and $\mathbb{G}_2$ be two cyclic groups of prime order p, respectively generated by g_1 and g_2. In the bilinear group pair $(\mathbb{G}_1, \mathbb{G}_2)$, the q-SDH problem is stated as follows:

Given as input a $(q + 3)$-tuple of elements
$$\left(g_1, g_1^x, g_2, g_2^x, g_2^{(x^2)}, \ldots, g_2^{(x^q)} \right) \in \mathbb{G}_1^2 \times \mathbb{G}_2^{q+1}$$
output a pair $\left(c, g_2^{1/(x+c)} \right) \in \mathbb{Z}_p \times \mathbb{G}_2$ for a freely chosen value $c \in \mathbb{Z}_p \backslash \{-x\}$.

An algorithm $\mathcal{A}$ solves the q-SDH problem in the bilinear group pair $(\mathbb{G}_1, \mathbb{G}_2)$ with advantage ϵ if

$$\Pr \left[\mathcal{A} \left(g_1, g_1^x, g_2, g_2^x, g_2^{(x^2)}, \ldots, g_2^{(x^q)} \right) = \left(c, g_2^{1/(x+c)} \right) \right] \geq \epsilon,$$

where the probability is over the random choice of generators $g_1 \in \mathbb{G}_1$ and $g_2 \in \mathbb{G}_2$, the random choice of $x \in \mathbb{Z}_p^*$, and the random bits consumed by $\mathcal{A}$. We say that the (t, q, ϵ)-SDH assumption holds in $(\mathbb{G}_1, \mathbb{G}_2)$ if no t-time algorithm has advantage at least ϵ in solving the q-SDH problem in $(\mathbb{G}_1, \mathbb{G}_2)$.

Decisional Bilinear Diffie-Hellman Inversion (DBDHI) assumption [3]: Let $\mathbb{G}_1$ and $\mathbb{G}_2$ be two cyclic groups of prime order p, respectively generated by g_1 and g_2. In the bilinear group pair $(\mathbb{G}_1, \mathbb{G}_2)$, the q-DBDHI problem is stated as follows:

Given as input a $(q + 4)$-tuple of elements
$$\left(g_1, g_1^x, g_2, g_2^x, g_2^{(x^2)}, \ldots, g_2^{(x^q)}, T \right) \in \mathbb{G}_1^2 \times \mathbb{G}_2^{q+1} \times \mathbb{G}_T$$
output 0 if $T = e(g_1, g_2)^{1/x}$ or 1 if T is a random element in $\mathbb{G}_T$.

An algorithm $\mathcal{A}$ solves the q-DBDHI problem in the bilinear group pair $(\mathbb{G}_1, \mathbb{G}_2)$ with advantage ϵ if

$$\left| \Pr\left[\mathcal{A}\left(g_1, g_1^x, g_2, g_2^x, g_2^{(x^2)}, \ldots, g_2^{(x^q)}, e(g_1, g_2)^{1/x} \right) = 0 \right] \right.$$

$$\left. - \Pr\left[\mathcal{A}\left(g_1, g_1^x, g_2, g_2^x, g_2^{(x^2)}, \ldots, g_2^{(x^q)}, T \right) = 0 \right] \right| \geq \epsilon,$$

where the probability is over the random choice of generators $g_1 \in \mathbb{G}_1$ and $g_2 \in \mathbb{G}_2$, the random choice of $x \in \mathbb{Z}_p^*$, the random choice of $T \in \mathbb{G}_T$, and the random bits consumed by $\mathcal{A}$. We say that the (t, q, ϵ)-DBDHI assumption holds in $(\mathbb{G}_1, \mathbb{G}_2)$ if no t-time algorithm has advantage at least ϵ in solving the q-DBDHI problem in $(\mathbb{G}_1, \mathbb{G}_2)$.

3   Combined Signature and Encryption Schemes

A combined signature and encryption scheme is a combination of a signature scheme and a public key encryption scheme that share a key generation algorithm and hence a keypair (pk, sk). It comprises a tuple of algorithms (KeyGen, Sign, Verify, Encrypt, Decrypt) such that (KeyGen, Sign, Verify) form a signature scheme and (KeyGen, Encrypt, Decrypt) form a PKE scheme. Since the signature and PKE schemes share a keypair the standard notions of EUF-CMA and IND-CCA security need to be extended to reflect an adversary's ability to request both signatures and decryptions under the challenge public key. When defining a security game against a component of the scheme the nature of any additional oracles depends on the required security of the other components. For example, if EUF-CMA security of the signature component of a combined signature and encryption scheme is required, then it is necessary to provide the adversary with unrestricted access to a signature oracle when proving IND-CCA security of the encryption component of the scheme. The security definitions given implicitly in [8], considering IND-CCA security of the encryption component and EUF-CMA security of the signature component, are stated formally here.

EUF-CMA security in the presence of a decryption oracle: Let (KeyGen, Sign, Verify, Encrypt, Decrypt) be a combined signature and encryption scheme. Existential unforgeability of the signature component under an adaptive chosen message attack in the presence of an additional decryption oracle is defined through the following game between a challenger and an adversary $\mathcal{A}$.

Setup: The challenger generates a keypair $(pk, sk) \leftarrow \mathsf{KeyGen}(1^k)$ and gives $\mathcal{A}$ the challenge public key pk.

Query phase: $\mathcal{A}$ requests signatures on messages m_i of its choice. The challenger responds to each signature query with a signature $\sigma_i \leftarrow \mathsf{Sign}(sk, m_i)$. $\mathcal{A}$ also requests decryptions of ciphertexts c_i of its choice. The challenger responds to each decryption query with a message $m \leftarrow \mathsf{Decrypt}(sk, c_i)$ or a failure symbol $\perp$.

Forgery: $\mathcal{A}$ outputs a message signature pair (σ, m) such that m was not submitted to the signing oracle, and wins the game if $\mathsf{Verify}(pk, \sigma, m) = 1$.

The advantage of an adversary $\mathcal{A}$ is the probability it wins the above game.

A forger $\mathcal{A}$ (t, q_d, q_s, ϵ)-breaks the signature component of a combined signature and encryption scheme if $\mathcal{A}$ runs in time at most t, makes at most q_d decryption queries and q_s signature queries and has advantage at least ϵ. The signature component of a combined signature and encryption scheme is said to be (t, q_d, q_s, ϵ)-EUF-CMA secure in the presence of a decryption oracle if no forger (t, q_d, q_s, ϵ)-breaks it.

IND-CCA security in the presence of a signing oracle: Let $(\mathsf{KeyGen}, \mathsf{Sign}, \mathsf{Verify}, \mathsf{Encrypt}, \mathsf{Decrypt})$ be a combined signature and encryption scheme. Indistinguishability of the encryption component under an adaptive chosen ciphertext attack in the presence of an additional signing oracle is defined through the following game between a challenger and an adversary $\mathcal{A}$.

Setup: The challenger generates a keypair $(pk, sk) \leftarrow \mathsf{Keyen}(1^k)$ and gives $\mathcal{A}$ the challenge public key pk.

Phase 1: $\mathcal{A}$ requests decryptions of ciphertexts c_i of its choice. The challenger responds to each decryption query with a message $m \leftarrow \mathsf{Decrypt}(sk, c_i)$ or a failure symbol $\perp$. $\mathcal{A}$ also requests signatures on messages m_i of its choice. The challenger responds to each signature query with a signature $\sigma_i \leftarrow \mathsf{Sign}(sk, m_i)$.

Challenge: $\mathcal{A}$ chooses two equal length messages m_0, m_1. The challenger chooses a random bit b, computes $c^* \leftarrow \mathsf{Encrypt}(pk, m_b)$, and passes c^* to the adversary.

Phase 2: As Phase 1 but with the restriction that $\mathcal{A}$ must not request the decryption of the challenge ciphertext c^*.

Guess: $\mathcal{A}$ outputs a guess b' for b.

The advantage of $\mathcal{A}$ is $\left| \Pr[b' = b] - \frac{1}{2} \right|$.

An adversary $\mathcal{A}$ (t, q_d, q_s, ϵ)-breaks the encryption component of a combined signature and encryption scheme if $\mathcal{A}$ runs in time at most t, makes at most q_d decryption queries and q_s signature queries and has advantage at least ϵ. The encryption component of a combined signature and encryption scheme is said to be (t, q_d, q_s, ϵ)-IND-CCA secure in the presence of a signing oracle if no adversary (t, q_d, q_s, ϵ)-breaks it.

Informally, we say that a combined scheme is *jointly secure* if it is both EUF-CMA secure in the presence of a decryption oracle and IND-CCA secure in the presence of a signing oracle.

3.1   A Cartesian Product Construction

A trivial way of obtaining a system satisfying the above security properties is to concatenate the keys of an encryption scheme and signature scheme, then use the appropriate component of the compound key for each operation. This gives a combined signature and encryption scheme where the signature and encryption operations are essentially independent. Consequently their respective security properties are retained in the presence of the additional oracle. This simple construction sets a benchmark in terms of key size and other performance measures that any bespoke construction should best in one or more metrics.

Formally, let $S = (S.\mathsf{KeyGen}, S.\mathsf{Sign}, S.\mathsf{Verify})$ be a signature scheme, and let $\mathcal{E} = (\mathcal{E}.\mathsf{KeyGen}, \mathcal{E}.\mathsf{Encrypt}, \mathcal{E}.\mathsf{Decrypt})$ be an encryption scheme. Then the Cartesian product combined signature and encryption scheme $\mathsf{CartCSE}(\mathcal{E}, S)$ is constructed as follows:

$\mathsf{CartCSE}(\mathcal{E}, S).\mathsf{KeyGen}(1^k)$: Run $S.\mathsf{KeyGen}(1^k)$ to get (pk_s, sk_s). Run $\mathcal{E}.\mathsf{KeyGen}(1^k)$ to get (pk_e, sk_e). Output the public key $pk = (pk_s, pk_e)$ and the private key $sk = (sk_s, sk_e)$.

$\mathsf{CartCSE}(\mathcal{E}, S).\mathsf{Sign}(sk, m)$: Output $S.\mathsf{Sign}(sk_s, m)$.

$\mathsf{CartCSE}(\mathcal{E}, S).\mathsf{Verify}(pk, \sigma, m)$: Output $S.\mathsf{Verify}(pk_s, \sigma, m)$.

$\mathsf{CartCSE}(\mathcal{E}, S).\mathsf{Encrypt}(pk, m)$: Output $\mathcal{E}.\mathsf{Encrypt}(pk_e, m)$.

$\mathsf{CartCSE}(\mathcal{E}, S).\mathsf{Decrypt}(sk, c)$: Output $\mathcal{E}.\mathsf{Decrypt}(sk_e, c)$.

We omit the straightforward proof that this scheme is jointly secure if S is EUF-CMA secure and $\mathcal{E}$ is IND-CCA secure.

3.2   An Insecure CSE Scheme whose Components are Secure

To show that the definitions are not trivially satisfied, we give a pathological example to show that a PKE scheme and a signature scheme that are individually secure may not be secure when used in combination. Let $S = (S.\mathsf{KeyGen}, S.\mathsf{Sign}, S.\mathsf{Verify})$ be an EUF-CMA secure signature scheme, and let $\mathcal{E} = (\mathcal{E}.\mathsf{KeyGen}, \mathcal{E}.\mathsf{Encrypt}, \mathcal{E}.\mathsf{Decrypt})$ be an IND-CCA secure encryption scheme. A combined signature and encryption scheme $\mathsf{BadCSE}(\mathcal{E}, S)$ can be constructed as follows.

$\mathsf{BadCSE}(\mathcal{E}, S).\mathsf{KeyGen}(1^k)$: Run $S.\mathsf{KeyGen}(1^k)$ to get (pk_s, sk_s). Run $\mathcal{E}.\mathsf{KeyGen}(1^k)$ to get (pk_e, sk_e). Output the public key $pk = (pk_s, pk_e)$ and the private key $sk = (sk_s, sk_e)$.

$\mathsf{BadCSE}(\mathcal{E}, S).\mathsf{Sign}(sk, m)$: Compute $\sigma' = S.\mathsf{Sign}(sk_s, m)$. Output $\sigma = \sigma' \| sk_e$.

$\mathsf{BadCSE}(\mathcal{E}, S).\mathsf{Verify}(pk, \sigma, m)$: Parse σ as $\sigma' \| sk_e$. Run $S.\mathsf{Verify}(pk_s, \sigma', m)$ and output the result.

$\mathsf{BadCSE}(\mathcal{E}, \mathcal{S}).\mathsf{Encrypt}(pk, m)$: Output $c = \mathcal{E}.\mathsf{Encrypt}(pk_e, m)$.

$\mathsf{BadCSE}(\mathcal{E}, \mathcal{S}).\mathsf{Decrypt}(sk, c)$: Run $\mathcal{E}.\mathsf{Decrypt}(sk_e, c)$. If this decryption is successful, output the decrypted message. Otherwise (if $\perp$ was returned), output sk_s.

From the security of the base schemes it is easy to see that the signature scheme given by the algorithms $\mathsf{BadCSE}(\mathcal{E}, \mathcal{S}).\mathsf{KeyGen}$, $\mathsf{BadCSE}(\mathcal{E}, \mathcal{S}).\mathsf{Sign}$, $\mathsf{BadCSE}(\mathcal{E}, \mathcal{S}).\mathsf{Verify}$ is EUF-CMA secure, and the PKE scheme with algorithms $\mathsf{BadCSE}(\mathcal{E}, \mathcal{S}).\mathsf{KeyGen}$, $\mathsf{BadCSE}(\mathcal{E}, \mathcal{S}).\mathsf{Encrypt}$, $\mathsf{BadCSE}(\mathcal{E}, \mathcal{S}).\mathsf{Decrypt}$ is IND-CCA secure. However when key generation is shared a single signature reveals the PKE scheme's private key, and the decryption of a badly formed ciphertext reveals the private key of the signature scheme. Thus $\mathsf{BadCSE}(\mathcal{E}, \mathcal{S})$ is totally insecure, even though its component schemes are secure.

4   A Generic Construction from IBE

We show how to build a combined signature and encryption scheme from an IBE scheme $\mathcal{I}$ with algorithms $\mathcal{I}.\mathsf{Setup}$, $\mathcal{I}.\mathsf{Extract}$, $\mathcal{I}.\mathsf{Encrypt}$, $\mathcal{I}.\mathsf{Decrypt}$. We make use of a one time strongly secure signature scheme $\mathcal{OT}$ with algorithms $\mathcal{OT}.\mathsf{KeyGen}$, $\mathcal{OT}.\mathsf{Sign}(sk, m)$, $\mathcal{OT}.\mathsf{Verify}(pk, \sigma, m)$. The construction is particularly simple: the signature scheme component is constructed through the Naor transform [6] and the PKE scheme component through the CHK transform [5]. Since in the Naor construction signatures are just private keys from the IBE scheme, and these private keys can be used to decrypt ciphertexts in the PKE scheme resulting from the CHK transform, we use a bit prefix in the identity space to provide domain separation between the signatures and private keys.

We assume $\mathcal{I}$ has message space $\mathcal{M}$, ciphertext space $\mathcal{C}$ and identity space $\{0, 1\}^{n+1}$, and that $\mathcal{OT}$ has public key space $\{0, 1\}^n$. Then the signature scheme component of $\mathsf{CSE}(\mathcal{I})$ has message space $\{0, 1\}^n$ but can be extended to messages of arbitrary length through the use of a collision resistant hash function $H : \{0, 1\}^* \to \{0, 1\}^n$. The PKE component of $\mathsf{CSE}(\mathcal{I})$ has message space $\mathcal{M}$. The algorithms of $\mathsf{CSE}(\mathcal{I})$ are shown in Figure 1. In the full version [24] we show how the construction can be extended to support a tag-based encryption component.

Theorem 1. *Let $\mathcal{I}$ be a (t', q, ϵ)-OW-ID-CPA secure IBE scheme. Then the signature component of $\mathsf{CSE}(\mathcal{I})$ is (t, q_d, q_s, ϵ)-EUF-CMA secure in the presence of a decryption oracle provided that*

$$q_s + q_d \le q \qquad and \qquad t \le t' - q_d(T_v + T_d) - T_d,$$

where T_v is the maximum time for a verification in $\mathcal{OT}$ and T_d is the maximum time for a decryption in $\mathcal{I}$.

Proof of Theorem 1. Suppose there exists a forger $\mathcal{F}$ that (t, q_d, q_s, ϵ) breaks the EUF-CMA security of the signature component of $\mathsf{CSE}(\mathcal{I})$ in the presence of a decryption oracle. We construct an algorithm $\mathcal{A}$ that interacts with the forger $\mathcal{F}$ to (t', q, ϵ)-OW-ID-CPA break the IBE scheme $\mathcal{I}$.

$\mathsf{CSE}(\mathcal{I}).\mathsf{KeyGen}(1^k)$:
$\qquad (mpk, msk) \leftarrow \mathcal{I}.\mathsf{Setup}(1^k)$
$\qquad (pk, sk) = (mpk, msk)$
$\qquad$ return (pk, sk)

$\mathsf{CSE}(\mathcal{I}).\mathsf{Sign}(sk, m)$:
$\qquad ID = 0||m$
$\qquad \sigma \leftarrow \mathcal{I}.\mathsf{Extract}(sk, ID)$
$\qquad$ return σ

$\mathsf{CSE}(\mathcal{I}).\mathsf{Verify}(pk, \sigma, m)$:
$\qquad ID = 0||m$
$\qquad x \leftarrow_R \mathcal{M}$
$\qquad c \leftarrow \mathcal{I}.\mathsf{Encrypt}(pk, ID, x)$
$\qquad$ if $\mathcal{I}.\mathsf{Decrypt}(pk, \sigma, c) = x$
$\qquad$ then return 1
$\qquad$ else return 0

$\mathsf{CSE}(\mathcal{I}).\mathsf{Encrypt}(pk, m)$:
$\qquad (vk, sk') \leftarrow \mathcal{OT}.\mathsf{KeyGen}$
$\qquad ID = 1||vk$
$\qquad c' \leftarrow \mathcal{I}.\mathsf{Encrypt}(pk, ID, m)$
$\qquad \sigma \leftarrow \mathcal{OT}.\mathsf{Sign}(sk', c')$
$\qquad$ return (vk, σ, c')

$\mathsf{CSE}(\mathcal{I}).\mathsf{Decrypt}(sk, c)$:
$\qquad$ Parse c as (vk, σ, c')
$\qquad$ if $\mathcal{OT}.\mathsf{Verify}(vk, \sigma, c') = 1$
$\qquad$ then $ID = 1||vk$
$\qquad\qquad sk_{ID} \leftarrow \mathcal{I}.\mathsf{Extract}(sk, ID)$
$\qquad\qquad$ return $\mathcal{I}.\mathsf{Decrypt}(pk, sk_{ID}, c')$
$\qquad$ else return $\perp$

Fig. 1. Generic construction from IBE

Setup: $\mathcal{A}$ is given a master public key mpk which it gives to $\mathcal{F}$ as the public key.

Signing queries: In response to a request for a signature on message m, $\mathcal{A}$ queries its extraction oracle for the identity $ID = 0||m$ to obtain sk_{ID} which it returns to $\mathcal{F}$ as the signature.

Decryption queries: In response to a decryption query for a ciphertext $c = (vk, \sigma, c')$, $\mathcal{A}$ verifies that σ is a valid signature on c' with verification key vk. If it is not a valid signature, $\mathcal{A}$ returns $\perp$. If the signature is valid, $\mathcal{A}$ queries its extraction oracle for the identity $ID = 1||vk$ to obtain sk_{ID} which it uses to decrypt c', returning the output of the decryption operation as the result of the decryption query.

Forgery: Eventually $\mathcal{F}$ will return a forgery (σ^*, m^*) on a message m^* for which a signing query was not made. At this point $\mathcal{A}$ outputs $ID^* = 0||m^*$ as the target identity. This is a valid choice; since a signing query was not made for message m^* an extraction query was not made for $ID = 0||m^*$.

Challenge: $\mathcal{A}$ receives a ciphertext c^*, which is the encryption of a random message m for identity ID^*. If σ^* is a valid signature for message m^* then σ^* is a valid decryption key for identity ID^*. This allows $\mathcal{A}$ to decrypt c^* using $sk_{ID^*} = \sigma^*$ to retrieve the message m which it subsequently outputs.

$\mathcal{A}$ succeeds precisely when $\mathcal{F}$ succeeds, so if $\mathcal{F}$ outputs a valid forgery with probability ϵ in time t then algorithm $\mathcal{A}$ succeeds in time at most $t + q_d(T_v + T_d) + T_d$ with the same probability ϵ.

Theorem 2. *Let $\mathcal{I}$ be an (t_i, q_i, ϵ_i)-IND-sID-CPA secure IBE scheme and let $\mathcal{OT}$ be a (t_s, ϵ_s)-strongly unforgeable one time signature scheme. Then the encryption component of $\mathsf{CSE}(\mathcal{I})$ is (t, q_d, q_s, ϵ)-IND-CCA secure in the presence of a signing oracle provided that*

$$\epsilon > \frac{1}{2}\epsilon_s + \epsilon_i, \quad q_s + q_d < q_i, \quad and \quad t < t_i - T_{kg} - T_{sig} - q_d(T_v + T_d),$$

where T_{kg}, T_{sig} and T_v are the maximum times for key generation, signing and verifying respectively in $\mathcal{OT}$, and T_d is the maximum decryption time in $\mathcal{I}$.

Proof of Theorem 2. The proof follows closely that of Theorem 1 in [5]. Let $\mathcal{D}$ be an adversary against the IND-CCA security of the encryption component of $\mathsf{CSE}(\mathcal{I})$ in the presence of a signing oracle running in time at most t and making at most q_s signature queries and q_d decryption queries. We use $\mathcal{D}$ to build an IND-sID-CPA adversary $\mathcal{B}$ against $\mathcal{I}$ as follows.

Setup: $\mathcal{B}$ runs $\mathcal{OT}$.KeyGen to obtain a keypair (vk^*, sk^*) then submits $ID^* = 1\|vk^*$ as the target identity. $\mathcal{B}$ is then given master public key mpk which it gives to $\mathcal{D}$ as the challenge public key.

Decryption queries: We partition the decryption queries into three possible cases and show how $\mathcal{B}$ responds to each case. Suppose the query is for ciphertext (vk, σ, c'), and let $\mathcal{OT}$.Verify$(vk, \sigma, c') = validity$.

Case 1: $vk = vk^*$
  > If $validity = 0$ then $\mathcal{B}$ responds to the decryption query with $\bot$. If $validity = 1$ then a forgery has been made against $\mathcal{OT}$, call this event **Forge**. If **Forge** occurs, $\mathcal{B}$ aborts and outputs a random bit b'.

Case 2: $vk \neq vk^*$ and $validity = 0$
  > $\mathcal{B}$ responds to the decryption query with $\bot$.

Case 3: $vk \neq vk^*$ and $validity = 1$
  > $\mathcal{B}$ queries the extraction oracle for identity $ID = 1\|vk$ to obtain sk_{ID}, then uses sk_{ID} to decrypt c', responding to the decryption query with the output of the decryption operation.

Signature queries: In response to a signature query for message m, $\mathcal{B}$ queries its extraction oracle for identity $ID = 0\|m$ to obtain sk_{ID} which it returns as the signature.

Challenge: Eventually $\mathcal{D}$ will output a pair of messages m_0, m_1. $\mathcal{B}$ forwards these messages and receives a challenge ciphertext c^*. $\mathcal{B}$ calls $\mathcal{OT}$.Sign(sk^*, c^*) to obtain σ^* and sends $C = (vk^*, \sigma^*, c^*)$ to $\mathcal{D}$. $\mathcal{D}$ may make more signature and decryption queries under the restriction that it must not submit to the decryption oracle its challenge ciphertext C. $\mathcal{D}$ then submits a guess b' which $\mathcal{B}$ outputs as its guess.

$\mathcal{B}$ represents a legal strategy for attacking $\mathcal{I}$, in particular $\mathcal{B}$ never requests the private key corresponding to the target identity ID^*. Provided **Forge** does not occur, $\mathcal{B}$ provides a perfect simulation for $\mathcal{D}$ so $\mathcal{B}$ succeeds with the same probability as $\mathcal{D}$. If Forge does occur then $\mathcal{B}$ outputs a random bit and succeeds with probability $\frac{1}{2}$. Letting $\mathrm{Pr}_{\mathrm{IBE}}^{\mathcal{B}}[\mathrm{Succ}]$ denote the probability of $\mathcal{B}$ outputting the correct bit in the IBE security game and $\mathrm{Pr}_{\mathrm{PKE}}^{\mathcal{D}}[\mathrm{Succ}]$ denote the probability of $\mathcal{D}$ outputting the correct bit in the PKE security game, it can be seen that

$$\left| \mathrm{Pr}_{\mathrm{PKE}}^{\mathcal{D}}[\mathrm{Succ} \wedge \overline{\mathrm{Forge}}] + \frac{1}{2}\mathrm{Pr}_{\mathrm{PKE}}^{\mathcal{D}}[\mathrm{Forge}] - \frac{1}{2} \right| = \left| \mathrm{Pr}_{\mathrm{IBE}}^{\mathcal{B}}[\mathrm{Succ}] - \frac{1}{2} \right|.$$

Since $\mathcal{I}$ is an (t_i, q_i, ϵ_i)-IND-sID-CPA secure IBE scheme, $\left| \mathrm{Pr}_{\mathrm{IBE}}^{\mathcal{B}}[\mathrm{Succ}] - \frac{1}{2} \right| < \epsilon_i$. The event Forge represents a signature forgery against $\mathcal{OT}$, so $\mathrm{Pr}_{\mathrm{PKE}}^{\mathcal{D}}[\mathrm{Forge}] < \epsilon_s$. It follows that

$$\begin{aligned}
\epsilon &= \left| \mathrm{Pr}_{\mathrm{PKE}}^{\mathcal{D}}[\mathrm{Succ}] - \frac{1}{2} \right| \\
&\leq \left| \mathrm{Pr}_{\mathrm{PKE}}^{\mathcal{D}}[\mathrm{Succ} \wedge \mathrm{Forge}] - \frac{1}{2}\mathrm{Pr}_{\mathrm{PKE}}^{\mathcal{D}}[\mathrm{Forge}] \right| + \\
&\qquad\qquad \left| \mathrm{Pr}_{\mathrm{PKE}}^{\mathcal{D}}[\mathrm{Succ} \wedge \overline{\mathrm{Forge}}] + \frac{1}{2}\mathrm{Pr}_{\mathrm{PKE}}^{\mathcal{D}}[\mathrm{Forge}] - \frac{1}{2} \right| \\
&\leq \frac{1}{2}\mathrm{Pr}_{\mathrm{PKE}}^{\mathcal{D}}[\mathrm{Forge}] + \left| \mathrm{Pr}_{\mathrm{PKE}}^{\mathcal{D}}[\mathrm{Succ} \wedge \overline{\mathrm{Forge}}] + \frac{1}{2}\mathrm{Pr}_{\mathrm{PKE}}^{\mathcal{D}}[\mathrm{Forge}] - \frac{1}{2} \right| \\
&= \frac{1}{2}\mathrm{Pr}_{\mathrm{PKE}}^{\mathcal{D}}[\mathrm{Forge}] + \left| \mathrm{Pr}_{\mathrm{IBE}}^{\mathcal{B}}[\mathrm{Succ}] - \frac{1}{2} \right| \\
&\leq \frac{1}{2}\epsilon_s + \epsilon_i.
\end{aligned}$$

The running time of $\mathcal{B}$ is at most $t + T_{kg} + q_d(T_v + T_d) + T_{sig}$, and it asks at most $q_s + q_d$ private key extraction queries, so the theorem holds.

IBE schemes meeting the standard model security requirements include those of Gentry [14] and Waters [28]. The latter results in a large public key ($n+3$ group elements), though this could be reduced in practice by generating most of the elements from a seed in a pseudo-random manner. We focus on the instantiation of our construction using Gentry's scheme. This scheme was originally presented in the setting of symmetric pairings. When we translate it to the asymmetric setting (see the full version for details) and apply our construction at the 128-bit security level using BN curves with sextic twists, we obtain a combined public key scheme in which the public key consists of two elements of $\mathbb{G}_1$ and two elements of $\mathbb{G}_2$, giving a public key size of 1536 bits. Ciphertexts encrypt elements of $\mathbb{G}_T$ and consist of an element of $\mathbb{G}_1$, two elements of $\mathbb{G}_T$, and a verification key and signature from $\mathcal{OT}$, so are 2304 bits plus the bit length of a verification key and signature in $\mathcal{OT}$. Signatures consist of an element of $\mathbb{Z}_p$ and an element of $\mathbb{G}_2$, so are 768 bits in size. Here we assume that descriptions of groups and pairings are domain parameters that are omitted from our key size calculations. The security of this scheme depends on an assumption closely related to the decisional q-augmented bilinear Diffie-Hellman exponent assumption.

This construction could be improved further using the Boneh-Katz [5] alternative to the CHK transform. We omit the details in favour of our next scheme.

5   A More Efficient Construction

The following scheme is based on the signature scheme by Boneh and Boyen [4] and a KEM obtained by applying the techniques by Boyen, Mei and Waters [7] to the second IBE scheme by Boneh and Boyen in [3]. The schemes make use of a bilinear pairing $e : \mathbb{G}_1 \times \mathbb{G}_2 \to \mathbb{G}_T$, where the groups are of order p, and the KEM furthermore makes use of an always second-preimage resistant (aSec-secure) hash function $\mathsf{H} : \mathbb{G}_1 \to \{0,1\}^{n-1}$ where $2^n < p$. To obtain a full encryption scheme, the KEM is combined with a DEM, and we assume for simplicity that the key space of the DEM is $\mathcal{K} = \mathbb{G}_T$. Where a binary string is treated as a member of $\mathbb{Z}_p$ it is implicitly converted in the natural manner. The signature scheme supports messages in $\{0,1\}^{n-1}$, but can be extended to support message in $\{0,1\}^*$ by using a collision resistant hash function, while the encryption scheme supports messages of arbitrary length due to the use of a DEM. Note that to minimize the public key size and ciphertext overhead in the scheme, the elements of the public key are placed in the group $\mathbb{G}_1$. However, this implies that signatures contain an element of the group $\mathbb{G}_2$, having larger bit representations of elements.

KeyGen(1^k): Choose random generators $g_1 \in \mathbb{G}_1$, $g_2 \in \mathbb{G}_2$ and random integers $x, y \in \mathbb{Z}_p^*$, and compute $X = g_1^x$ and $Y = g_1^y$. The public key is (g_1, g_2, X, Y) and the private key is (x, y).

Sign(sk, m): To sign a message $m \in \{0,1\}^{n-1}$ first prepend a zero to m to give $m' = 0||m \in \{0,1\}^n$. Choose random $r \in \mathbb{Z}_p$. If $x + ry + m' \equiv 0 \bmod p$ then select another $r \in \mathbb{Z}_p$. Compute $\sigma = g_2^{\frac{1}{x+m'+yr}} \in \mathbb{G}_2$. The signature is $(\sigma, r) \in \mathbb{G}_2 \times \mathbb{Z}_p$.

Verify(pk, σ, m): If $e(X \cdot g_1^{m'} \cdot Y^r, \sigma) = e(g_1, g_2)$, where $m' = 0||m$, then return 1, otherwise return 0.

Encrypt(pk, m): To encrypt a message $m \in \{0,1\}^*$, choose random $s \in \mathbb{Z}_p^*$ and compute $c_1 = Y^s$ and $h = \mathsf{H}(c_1)$. Prepend a 1 to h to give $h' = 1||h \in \{0,1\}^n$, and compute $c_2 = X^s \cdot g_1^{s \cdot h'}$. Lastly, compute the key $K = e(g_1, g_2)^s \in \mathbb{G}_T$ and encrypt the message m using the DEM i.e. $c_3 = \mathsf{DEnc}(K, m)$. The ciphertext is $c = (c_1, c_2, c_3)$.

Decrypt(sk, c): To decrypt a ciphertext $c = (c_1, c_2, c_3)$, first compute $h = \mathsf{H}(c_1)$ and prepend a 1 to h to get $h' = 1||h$. If $c_1^{(x+h')/y} \neq c_2$, output $\perp$. Otherwise, compute the key $K = e(c_1, g_2^{1/y}) \in \mathbb{G}_T$, and output the message $m = \mathsf{DDec}(K, c_3)$.

We note that the computational cost of encryption and signature verification can be reduced by adding the redundant element $v = e(g_1, g_2)$ to the public key, but that this will significantly increase the public key size.

Theorem 3. *Suppose the (t', q, ϵ')-SDH assumption holds in $(\mathbb{G}_1, \mathbb{G}_2)$. Then the above combined public key scheme is (t, q_d, q_s, ϵ)-EUF-CMA secure in the presence of a decryption oracle given that*

$$q_s \leq q, \quad \epsilon \geq 2\epsilon' + q_s/p \approx 2\epsilon' \quad and \quad t \leq t' - \Theta(q_d T_p + (q_d + q^2) T_e),$$

where T_p is the maximum time for evaluating a pairing and T_e is the maximum time for computing an exponentiation in $\mathbb{G}_1$, $\mathbb{G}_2$ and $\mathbb{Z}_p$.

Theorem 4. *Suppose that the hash function H is (t_h, ϵ_h)-aSec secure, that the $(t_{dhi}, q_{dhi}, \epsilon_{dhi})$-DBDHI assumption holds in the groups $\mathbb{G}_1, \mathbb{G}_2$, and that the DEM is $(t_{dem}, q_{dem}, \epsilon_{dem})$-IND-CCA secure. Then the combined public key scheme above is $(t, q_d, q_s, \epsilon,)$-IND-CCA secure in the presence of a signing oracle given that*

$$q_s \leq q_{dhi}, \quad q_d \leq q_{dem}, \quad \epsilon \geq \epsilon_h + \epsilon_{dhi} + \epsilon_{dem} + q_s/p, \quad and$$
$$t \leq t_{min} - \Theta(q_d T_p + (q_{dhi} + q_d) T_e),$$

where $t_{min} = \min(t_h, t_{dhi}, t_{dem})$, T_p is the maximum time for evaluating a pairing, and T_e is the maximum time for computing an exponentiation in $\mathbb{G}_1, \mathbb{G}_2$.

The proofs of Theorems 3 and 4 can be found in the full version [24].

The above scheme provides public keys consisting of three group elements of $\mathbb{G}_1$ and one group element of $\mathbb{G}_2$. If the scheme is instantiated using BN curves with sextic twists mentioned above, this translates into a public key size of 1280 bits for a 128 bit security level. Furthermore, assuming that the DEM is redundancy-free (which can be achieved if the DEM is a strong pseudorandom permutation [25]), the total ciphertext overhead is just two group elements of $\mathbb{G}_1$ which translates into 512 bits. Signatures consist of a single group element of $\mathbb{G}_2$ and an element of $\mathbb{Z}_p$, and will be 768 bits. Again, we assume that descriptions of groups and pairings are ignored in these calculations.

In the full version, we show how the construction can be extended to support tag-based encryption. This property is required to allow us to use the scheme to instantiate our combined signcryption, signature and encryption scheme (see the full version for details).

6   Comparison of Schemes

In this section, we provide a comparison of the schemes arising from our IBE-based construction, our more efficient construction in Section 5 and the Cartesian product construction. In our comparison we will limit ourselves to other discrete-log/pairing-based schemes since provably secure (standard model) lattice-based schemes with short public keys are still unavailable and factoring-based schemes do not scale very well (for 128-bit security, the modulus would need to be > 3000 bits which is not competitive). We will include group generators in public key size calculations as the required number depends on the scheme, but we allow sharing of generators between signature and encryption component in Cartesian product instantiations to improve these constructions. Note that it is possible to reduce the private key of any scheme to a single short random seed by making the following simple modification to the scheme: to generate a public/private keypair, pick a random seed, generate the randomness required by the key generation algorithm by applying a pseudorandom generator to the seed, and generate the public/private keypair using this randomness, but store only the seed as the private key. Whenever the original private key is needed, re-compute this by applying the pseudorandom generator to the seed and re-run the key generation algorithm with the resulting randomness. This observation essentially makes the difference in private key sizes irrelevant, and we will not include this aspect in our comparison. We consider several instantiations of the Cartesian product construction with standard model secure encryption and signature schemes and give the results in Figure 2.

We will focus on Cartesian product instantiations using the scheme by Boneh and Boyen [4] as a signature component. This scheme is among the most efficient signature schemes and additionally has a short public key. To reduce the public key size even further, we can remove the redundant element $v = e(g_1, g_2)$ and place as many elements as possible in the group $\mathbb{G}_1$ of the pairing. The latter implies that signatures will be elements of $\mathbb{G}_2 \times \mathbb{Z}_p$ which results in an increase in signature size. However, since the Cartesian product constructions should compete with the combined public key schemes in terms of public key size, this tradeoff is desirable. While other signature schemes could be considered, we were not able to find a scheme providing shorter public keys without a significant disadvantage elsewhere. For instance, hash-based signature schemes give extremely short public keys (the hash function description plus the root digest), but result in signatures with length logarithmic in the number of messages to be signed. The signature scheme by Hofheinz and Kiltz [17] has shorter signatures than the Boneh-Boyen scheme and a public key consisting of a few group elements plus a hash key, but here the hash key will be long to achieve provable programmability.

For the encryption component, a relevant option is a DEM combined with the KEM obtained by applying the techniques by Boyen, Mei and Waters [7] to the second IBE scheme of Boneh and Boyen in [3], which also forms the basis of our concrete scheme. Combined with the Boneh-Boyen signature scheme, and assuming the group generators in the two schemes are shared, this yields a very efficient instantiation of the Cartesian product construction in which public keys consist of five group elements of $\mathbb{G}_1$, one group element of $\mathbb{G}_2$ (and a key defining a target collision resistant hash function). This is larger by two elements of $\mathbb{G}_1$ than the public key in our concrete construction from Section 5, which translates to a difference of 512 bits. Note that signature size, ciphertext overhead and computation costs are the same for the Cartesian product scheme and our construction.

Another encryption scheme to consider is that of Kurosawa and Desmedt [22]. Instantiating the Cartesian product construction with this scheme and the

Signature Scheme	PKE Scheme	Public Key Size	Signature Size	Ciphertext Overhead
BB [4]	BB [3] + BMW [7]	1792	768	512
BB [4]	KD [22]	2048	768	640
BB [4]	Kiltz [19]	1792	768	512
CSE(Gentry)		1536	768	$1280 + |vk_{OT}| + |\sigma_{OT}|$
Scheme from Sec. 5		1280	768	512

Fig. 2. Comparison of schemes at the 128-bit security level

Boneh-Boyen signature scheme yields a scheme with a public key consisting of six elements of $\mathbb{G}_1$, one element of $\mathbb{G}_2$ (and a key defining a target collision resistant hash), assuming that the Kurosawa-Desmedt scheme is implemented in $\mathbb{G}_1$. Hence, the public key will be larger by three group elements of $\mathbb{G}_1$ compared to our concrete construction, which equates to a difference of 768 bits at the 128-bit security level. Signature size and signing and verification costs will be the same as in our construction, whereas the ciphertext overhead will be slightly larger (an extra 128 bits) due to the requirement that the symmetric encryption scheme used in the Kurosawa-Desmedt scheme is authenticated. However, decryption costs will be lower since no pairing computations are required.

Lastly, the encryption scheme of Kiltz [19] might be considered. Again, combining this with the Boneh-Boyen signature scheme, and assuming group generators are shared, will yield a Cartesian product scheme with public keys consisting of five elements of $\mathbb{G}_1$ and one element of $\mathbb{G}_2$. This is two group elements of $\mathbb{G}_1$ larger than the public key of our concrete construction, which equates to an increase of 512 bits at the 128-bit security level. Signature size and ciphertext overhead will be the same while decryption in the Cartesian product scheme will be more efficient, since no pairing computations are required.

In summary, our concrete construction of a combined public key scheme admits shorter public keys than any instantiation of the Cartesian product construction of Section 3.1 with known standard model secure encryption and signature schemes, and furthermore enjoys compact ciphertexts and signatures.

7   Conclusions and Future Research

We have revisited the topic of joint security for combined public key schemes, focussing on the construction of schemes in the standard model, an issue not fully addressed in prior work. We gave a general construction for combined public key schemes from weakly secure IBE, as well as a more efficient concrete construction based on pairings. Using BN curves, these can be efficiently instantiated at high security levels and have performance that is competitive with the best schemes arising from the Cartesian product construction. Our results fill the gap left open in the original work of Haber and Pinkas [15], of constructing standard-model-secure combined public key schemes in which the signature and encryption components share an identical keypair. An interesting open problem is to construct efficient combined public key schemes in the standard model not using pairings. For example, is it possible to obtain joint security in the discrete log or in the RSA setting, in the standard model?

Our work points the way to an interesting new research area in cryptography, which closely relates to and generalises the topic of cryptographic agility [1]. The general question can be posed as follows: under what conditions is it safe to use the same key (or key pair) across multiple instantiations of the *same* or *different* cryptographic primitives?

References

1. Acar, T., Belenkiy, M., Bellare, M., Cash, D.: Cryptographic Agility and Its Relation to Circular Encryption. In: Gilbert, H. (ed.) EUROCRYPT 2010. LNCS, vol. 6110, pp. 403–422. Springer, Heidelberg (2010)
2. Barreto, P.S.L.M., Naehrig, M.: Pairing-Friendly Elliptic Curves of Prime Order. In: Preneel, B., Tavares, S.E. (eds.) SAC 2005. LNCS, vol. 3897, pp. 319–331. Springer, Heidelberg (2006)
3. Boneh, D., Boyen, X.: Efficient selective identity-based encryption without random oracles. Journal of Cryptology (2011),
   http://www.springerlink.com/content/n63632331k4q4h11/
4. Boneh, D., Boyen, X.: Short signatures without random oracles and the SDH assumption in bilinear groups. J. Cryptology 21(2), 149–177 (2008)
5. Boneh, D., Canetti, R., Halevi, S., Katz, J.: Chosen-ciphertext security from identity-based encryption. SIAM J. Comput. 36(5), 1301–1328 (2007)
6. Boneh, D., Franklin, M.K.: Identity-based encryption from the Weil pairing. SIAM J. Comput. 32(3), 586–615 (2003)
7. Boyen, X., Mei, Q., Waters, B.: Direct chosen ciphertext security from identity-based techniques. In: ACM Conference on Computer and Communications Security, pp. 320–329. ACM (2005)
8. Coron, J.S., Joye, M., Naccache, D., Paillier, P.: Universal Padding Schemes for RSA. In: Yung, M. (ed.) CRYPTO 2002. LNCS, vol. 2442, pp. 226–241. Springer, Heidelberg (2002)
9. Dodis, Y., Freedman, M.J., Jarecki, S., Walfish, S.: Optimal signcryption from any trapdoor permutation. Cryptology ePrint Archive, Report 2004/020 (2004),
   http://eprint.iacr.org/
10. Dodis, Y., Freedman, M.J., Jarecki, S., Walfish, S.: Versatile padding schemes for joint signature and encryption. In: ACM Conference on Computer and Communications Security, pp. 344–353. ACM (2004)
11. EMV Specifications, Version 4.2, Books 1–4 (June 2008), http://www.emvco.com/
12. Freeman, D., Scott, M., Teske, E.: A taxonomy of pairing-friendly elliptic curves. J. Cryptology 23(2), 224–280 (2010)
13. Galbraith, S.D., Paterson, K.G., Smart, N.P.: Pairings for cryptographers. Discrete Applied Mathematics 156(16), 3113–3121 (2008)
14. Gentry, C.: Practical Identity-Based Encryption Without Random Oracles. In: Vaudenay, S. (ed.) EUROCRYPT 2006. LNCS, vol. 4004, pp. 445–464. Springer, Heidelberg (2006)
15. Haber, S., Pinkas, B.: Securely combining public-key cryptosystems. In: ACM Conference on Computer and Communications Security, pp. 215–224. ACM (2001)

16. Hess, F.: Efficient Identity Based Signature Schemes Based on Pairings. In: Nyberg, K., Heys, H.M. (eds.) SAC 2002. LNCS, vol. 2595, pp. 310–324. Springer, Heidelberg (2003)

17. Hofheinz, D., Kiltz, E.: Programmable Hash Functions and Their Applications. In: Wagner, D. (ed.) CRYPTO 2008. LNCS, vol. 5157, pp. 21–38. Springer, Heidelberg (2008)

18. Kelsey, J., Schneier, B., Wagner, D.: Protocol Interactions and the Chosen Protocol Attack. In: Christianson, B., Lomas, M. (eds.) Security Protocols 1997. LNCS, vol. 1361, pp. 91–104. Springer, Heidelberg (1998)

19. Kiltz, E.: Chosen-Ciphertext Secure Key-Encapsulation Based on Gap Hashed Diffie-Hellman. In: Okamoto, T., Wang, X. (eds.) PKC 2007. LNCS, vol. 4450, pp. 282–297. Springer, Heidelberg (2007)

20. Klíma, V., Rosa, T.: Further Results and Considerations on Side Channel Attacks on RSA. In: Kaliski Jr., B.S., Koç, Ç.K., Paar, C. (eds.) CHES 2002. LNCS, vol. 2523, pp. 244–259. Springer, Heidelberg (2003)

21. Komano, Y., Ohta, K.: Efficient Universal Padding Techniques for Multiplicative Trapdoor One-Way Permutation. In: Boneh, D. (ed.) CRYPTO 2003. LNCS, vol. 2729, pp. 366–382. Springer, Heidelberg (2003)

22. Kurosawa, K., Desmedt, Y.G.: A New Paradigm of Hybrid Encryption Scheme. In: Franklin, M. (ed.) CRYPTO 2004. LNCS, vol. 3152, pp. 426–442. Springer, Heidelberg (2004)

23. Matsuda, T., Matsuura, K., Schuldt, J.C.N.: Efficient Constructions of Signcryption Schemes and Signcryption Composability. In: Roy, B., Sendrier, N. (eds.) INDOCRYPT 2009. LNCS, vol. 5922, pp. 321–342. Springer, Heidelberg (2009)

24. Paterson, K.G., Schuldt, J.C., Stam, M., Thomson, S.: On the joint security of encryption and signature, revisited. Cryptology ePrint Archive, Report 2011/486 (2011)

25. Phan, D.H., Pointcheval, D.: About the Security of Ciphers (Semantic Security and Pseudo-Random Permutations). In: Handschuh, H., Hasan, M.A. (eds.) SAC 2004. LNCS, vol. 3357, pp. 182–197. Springer, Heidelberg (2004)

26. Rubin, K., Silverberg, A.: Compression in finite fields and torus-based cryptography. SIAM J. Comput. 37(5), 1401–1428 (2008)

27. Vasco, M.I.G., Hess, F., Steinwandt, R.: Combined (identity-based) public key schemes. Cryptology ePrint Archive, Report 2008/466 (2008), http://eprint.iacr.org/

28. Waters, B.: Efficient Identity-Based Encryption Without Random Oracles. In: Cramer, R. (ed.) EUROCRYPT 2005. LNCS, vol. 3494, pp. 114–127. Springer, Heidelberg (2005)

Polly Cracker, Revisited[*]

Martin R. Albrecht[1], Pooya Farshim[2], Jean-Charles Faugère[1],
and Ludovic Perret[1]

[1] INRIA, Paris-Rocquencourt Center, SALSA Project
UPMC Univ Paris 06, UMR 7606, LIP6, F-75005, Paris, France
CNRS, UMR 7606, LIP6, F-75005, Paris, France
[2] Department of Computer Science, Darmstadt University of Technology, Germany
malb@lip6.fr, farshim@cased.de, jean-charles.faugere@inria.fr,
ludovic.perret@lip6.fr

Abstract. We initiate the formal treatment of cryptographic constructions ("Polly Cracker") based on the hardness of computing remainders
modulo an ideal over multivariate polynomial rings. We start by formalising the relation between the ideal remainder problem and the problem of computing a Gröbner basis. We show both positive and negative
results. On the negative side, we define a symmetric Polly Cracker encryption scheme and prove that this scheme only achieves bounded CPA
security. Furthermore, we show that a large class of algebraic transformations cannot convert this scheme to a fully secure Polly-Cracker-style
scheme. On the positive side, we formalise noisy variants of the ideal
membership, ideal remainder, and Gröbner basis problems. These problems can be seen as natural generalisations of the LWE problem and the
approximate GCD problem over polynomial rings. We then show that
noisy encoding of messages results in a fully IND-CPA-secure somewhat
homomorphic encryption scheme. Our results provide a new family of
somewhat homomorphic encryption schemes based on new, but natural,
hard problems. Our results also imply that Regev's LWE-based public-
key encryption scheme is (somewhat) *multiplicatively* homomorphic for
appropriate choices of parameters.

Keywords: Polly Cracker, Gröbner bases, LWE, Noisy encoding,
Homomorphic encryption, Public-key encryption, Provable security.

1 Introduction

BACKGROUND. Homomorphic encryption [19] is a cryptographic primitive which
allows performing arbitrary computations over encrypted data. From an

---

[*] The work described in this paper has been supported by the Royal Society grant
JP090728 and by the Commission of the European Communities through the ICT
program under contract ICT-2007-216676 (ECRYPT-II). M. Albrecht, J-C. Faugère,
and L. Perret were also supported by the french ANR under the Computer Algebra and Cryptography (CAC) project (ANR-09-JCJCJ-0064-01) and the EXACTA
project (ANR-09-BLAN-0371-01). P. Farshim was funded in part by the US Army
Research laboratory, the UK Ministry of Defense and was accomplished under Agreement Number W911NF-06-3-0001. Due to space limitations this work is only an
extended abstract of the full work available in [1].

D.H. Lee and X. Wang (Eds.): ASIACRYPT 2011, LNCS 7073, pp. 179–196, 2011.
© International Association for Cryptologic Research 2011

algebraic perspective, this homomorphic feature can be seen as the ability to evaluate multivariate polynomials over ciphertexts. Hence, an instantiation of homomorphic encryption over multivariate polynomials is perhaps the most natural strategy.

Indeed, let $\mathcal{I} \subset P = \mathbb{F}[x_0, \ldots, x_{n-1}]$ be some ideal. We can encrypt a message $m \in P/\mathcal{I}$ as $c = f + m$ for f randomly chosen in $\mathcal{I}$. Decryption is performed by computing remainders modulo $\mathcal{I}$. From the definition of an ideal the homomorphic features of this scheme follow. The problem of computing remainders modulo an ideal was solved by Buchberger in [8], where he introduced the notion of Gröbner bases, and gave an algorithm for computing such bases.

In fact, all known doubly homomorphic schemes are based on variants of the ideal remainder problem over various rings. For example in [13] the ring $\langle p \rangle \in \mathbb{Z}$ for p an odd integer is considered. In [19] ideals in a number field play the same role (cf. [29]). One can even view Regev's LWE-based public-key encryption scheme [25] in this framework. Finally, we note that the construction displayed above is essentially Polly Cracker (PC) [17]. However, despite their simplicity, our confidence in PC-style schemes has been shaken as almost all such proposals have been broken [15]. In fact, it is a long standing open research challenge to propose a secure PC-style encryption scheme [5].

CONTRIBUTIONS & ORGANISATION. Our contributions can be summarised as follows: 1) we initiate the formal treatment of PC-style schemes and characterise their security; 2) we show the impossibility of converting such schemes to fully IND-CPA-secure schemes through a large class of transformations; 3) we introduce natural noisy variants of classical problems related to Gröbner bases which also generalise previously considered noisy problems; and 4) we present a new somewhat (and doubly) homomorphic encryption scheme based on a new class of computationally hard problems.

In more detail, after settling notation in Section 2, we formalise various problems from commutative algebra in the language of game-based security definitions in Section 3. In particular, we show that computing remainders modulo an ideal with overwhelming probability is equivalent to computing a Gröbner basis for zero-dimensional ideals. We then show that deciding ideal membership and computing ideal remainders are equivalent for certain choices of parameters. We then introduce a symmetric variant of Polly Cracker and characterise its security guarantees. We show that this scheme achieves *bounded* IND-CPA security, and that this level of security is the best that one can hope for: we give an attacker which breaks the cryptosystem once enough ciphertexts are obtained.

In Section 5, we show the security limitations of the constructed scheme are in some sense *intrinsic*. More precisely, we show that a large class of algebraic transformation cannot turn this scheme into a (fully) IND-CPA secure and additively homomorphic PC-style scheme.

To go beyond this limitation, we consider a constructions where the encoding of messages is randomised. To prove security for such schemes, we consider noisy variants of the ideal membership and related problems. These can be seen as natural generalisations of the (decisional) LWE and the approximate GCD problems over polynomial rings (Section 6). After formalising and justifying the hardness of the noisy assumptions in Section 7, we show that noisy encoding of messages can indeed be used to construct a fully IND-CPA-secure somewhat homomorphic scheme. This result also implies that Regev's LWE-based public-key scheme is *multiplicatively* homomorphic under appropriate choices of parameters. Our result, together with a standard symmetric-to-asymmetric conversion for homomorphic schemes, provides a positive answer to the long standing open problem proposed by Barkee et al. [5]. In addition, we provide a new family of somewhat homomorphic schemes which are based on new natural variants of well-studied hard problems. Due to space limitations, we discuss concrete parameter choices and include a reference implementation in the full version of the paper [1]. There, we also show how our scheme allows proxy re-encryption of ciphertexts. This re-encryption procedure can be seen as trading noise for degree in ciphertexts. That is, we can control the growth of the ciphertext size due to multiplication by tolerating more noise. We note that this technique was recently and independently developed in [7]. In [1], we also show that our scheme achieves a limited form of key-dependent message (KDM) security in the standard model, where the least significant bit of the constant term of the key is encrypted. We leave it as an open problem to adapt the techniques of [2] to achieve full KDM security for the Polly Cracker with noise scheme.

1.1 Related Work

Polly Cracker. In 1993, Barkee et al. wrote a paper [5] whose aim was to dispel the urban legend that "Gröbner bases are hard to compute". Another goal of this paper was to direct research towards *sparse* systems of multivariate equations. To do so, the authors proposed the most obvious dense Gröbner-based cryptosystem, namely an instantiation of the construction mentioned at the beginning of the introduction. In their scheme, the public key is a set of polynomials $\{f_0, \ldots, f_{m-1}\} \subset \mathcal{I}$ which is used to construct an element $f \in \mathcal{I}$. Encryption of messages $m \in P/\mathcal{I}$ are computed as $c = \sum h_i f_i + m = f + m$ for $f \in \mathcal{I}$. The private key is a Gröbner basis G which allows to compute $m = c \mod \mathcal{I} = c \mod G$. As highlighted in [5] this scheme can be broken using results from [12] (cf. Theorem 2). At about the same time, and independently from the work of Barkee et al., Fellows and Koblitz [17] proposed a framework for the design of public-key cryptosystems. The ideas in [17] were similar to Barkee et al.'s, but differed in some details. However, the main instantiation of such a system was the Polly Cracker cryptosystem. Subsequently, a variety of sparse PC-style schemes were proposed. The focus on sparse polynomials aimed to prevent the attack based on Theorem 2, yet almost all of these schemes were broken. We point the reader to [15] for a good survey of various constructions and attacks. Currently, the only PC-style scheme which is not broken is the scheme in [9]. This scheme is based on binomial ideals (which in turn are closely related to lattices). Not only can our constructions be seen as instantiations of Polly Cracker (with and without noisy encoding of messages), they also allow security proofs based on the hardness of computational problems related to (multivariate) polynomial ideals with respect to random systems.

Homomorphic Encryption. With respect to doubly (i.e., additively and multiplicatively) homomorphic schemes, a number of different hardness assumptions and constructions appeared in the literature. These include the Ideal Coset Problem of Gentry [19], the approximate GCD problem over the Integers of van Dijk et al. [13], the Polynomial Coset Problem as proposed by Smart and Vercauteren in [29], the Approximate Unique Shortest Vector Problem, the Subgroup Decision Problem, and the Differential Knapsack Vector Problem which appear in [23]. The main difference between our work and previous work is that we base the security of our somewhat homomorphic scheme on *new* computational problems related to ideals over multivariate polynomial rings. Furthermore, due to the versatility of Gröbner basis theory, our work can be seen as a generalisation of a number of known schemes and their underlying hardness assumptions. However, while our construction is doubly homomorphic and reasonably efficient for low multiplicative circuit depths, it is currently an open problem how to make it bootstrappable and hence turn it into a fully homomorphic scheme.

$\mathcal{MQ}$ *Cryptography.* Our work bears some connection with public-key cryptosystems based on the hardness of solving multivariate quadratic equations ($\mathcal{MQ}$). The difference is that our cryptographic constructions enjoy strong reductions to the known and hard problem of solving a *random* system of equations, whereas the bulk of work in $\mathcal{MQ}$ cryptography relies on heuristic security arguments [14]. In contrast, our work is more in the direction of research initiated by Berbain et al. [6] who proposed a stream cipher whose security was reduced to the difficulty of solving a system of random multivariate quadratic equations over $\mathbb{F}_2$. Note also that the concept of adding noise to a system of multivariate equations has been also proposed by Gouget and Patarin in [21] for the design of an authentication scheme. Our work, however, presents a more general and complete treatment of problems related to ideals over multivariate polynomials – both with and without noise – and aims to provide a formal basis to assess the security of cryptosystems based on such problems.

2   Preliminaries

NOTATION. We write $x \leftarrow y$ for assigning value y to a variable x, and $x \leftarrow_\$ X$ for sampling x from a set X uniformly at random. If A is a probabilistic algorithm we write $y \leftarrow_\$ A(x_1, \ldots, x_n)$ for the action of running A on inputs $x_1, \ldots, x_n$ with uniformly chosen random coins, and assigning the result to y. For a random variable X we denote by $[X]$ the support of X, i.e., the set of all values that X takes with non-zero probability. We use ppt for probabilistic polynomial-time. We call $\eta(\lambda)$ negligible if $|\eta(\lambda)| \in \lambda^{-\omega(1)}$.

COMMUTATIVE ALGEBRA NOTATION. In [1] we recall some basic definitions related to Gröbner bases. For a more detailed treatment we refer to, for instance, [10]. We consider a polynomial ring $P = \mathbb{F}[x_0, \ldots, x_{n-1}]$ over some finite field (typically $\mathbb{F}_q$), some monomial ordering on elements of P, and a set of polynomials $f_0, \ldots, f_{m-1}$. We denote by $\mathrm{M}(f)$ the set of all monomials appearing in $f \in P$. By $\mathrm{LM}(f)$ we denote the leading monomial appearing in $f \in P$ according to the chosen term ordering. We denote by $\mathrm{LC}(f)$ the coefficient $\in \mathbb{F}$ corresponding to $\mathrm{LM}(f)$ in f and set $\mathrm{LT}(f) = \mathrm{LC}(f) \cdot \mathrm{LM}(f)$. We denote by $P_{<d}$ the set of polynomials of degree $< d$ (and analogously for the $>, \leq, \geq,$ and $=$ relations). We define $P_{=0}$ as the underling field including $0 \in \mathbb{F}$. We define $P_{<0}$ as zero. Finally, we denote by $M_{<m}$ the set of all monomials $< m$ for some monomial m (and analogously for the $>, \leq, \geq,$ and $=$ relations). We assume the usual power product representation for elements of P.

3   Gröbner Basis and Ideal Membership Problems

Following [11], we define *a computational polynomial ring scheme*. This is a general framework allowing to discuss in a concrete way the different families of rings that may be used in cryptographic applications. More formally, a computational polynomial ring scheme $\mathcal{P}$ is a sequence of probability distribution of *polynomial ring descriptions* $(\mathbf{P}_\lambda)_{\lambda \in \mathbb{N}}$. A polynomial ring description P specifies various algorithms associated with P such as computing ring operations, sampling elements, testing membership, encoding of elements, ordering of monomials, etc. We assume each polynomial ring distribution is over $n = n(\lambda)$ variables, for some polynomial $n(\lambda)$, and is over a finite prime field of size $q(\lambda)$.

In this work we denote by $\mathsf{GBGen}(1^\lambda, P, d)$ an arbitrary ppt algorithm which outputs a reduced Gröbner basis G for some zero-dimensional ideal $\mathcal{I} \subset P$ such that every element of G is of degree at most d. Of particular interest to this paper is the Gröbner basis generation algorithm shown in Algorithm 1 called $\mathsf{GBGen}_{\mathsf{dense}}(\cdot)$. (Algorithm $\mathsf{ReduceGB}(\cdot)$ is given in [1].) We show in [1] that $\mathsf{GBGen}_{\mathsf{dense}}(\cdot)$ returns a Gröbner basis. Throughout the paper we assume an implicit dependency of various parameters associated with P on the security parameter. Thus, we drop λ to ease notation.

Algorithm 1: Algorithm $\mathsf{GBGen}_{\mathsf{dense}}(1^\lambda, P, d)$

---

1 **begin**
2 $\quad$ **if** $d = 0$ **then return** $\{0\}$;
3 $\quad$ **for** $0 \leq i < n$ **do**
4 $\quad\quad$ **for** $m_j \in M_{<x_i^d}$ **do**
5 $\quad\quad\quad$ $c_{ij} \leftarrow_\$ \mathbb{F}_q$; $g_i \leftarrow g_i + c_{ij}m_j$;
6 $\quad$ **return** $\mathsf{ReduceGB}(\{x_0^d + g_0, \ldots, x_{n-1}^d + g_{n-1}\})$;
7 **end**

We can now formally define the problem of computing a Gröbner basis.

Definition 1. *The Gröbner basis problem is defined through the game denoted* $\mathsf{GB}_{\mathcal{P}, \mathsf{GBGen}(\cdot), d, b, m}$ *as shown in Figure 1. The advantage of a ppt algorithm $\mathcal{A}$ in* *solving the* GB *problem is defined as the probability of winning the game (i.e., the game returning* T*). An adversary is legitimate if it calls the* **Sample** *procedure at most* $m = m(\lambda)$ *times.*

Initialize$(1^\lambda, \mathcal{P}, d)$:	**Sample**():	**Finalize**():
begin	begin	begin
$\quad P \leftarrow_\$ \mathbf{P}_\lambda$;	$\quad f \leftarrow_\$ P_{\leq b}$;	$\quad$ return $(G = G')$;
$\quad G \leftarrow_\$ \mathsf{GBGen}(1^\lambda, P, d)$;	$\quad f \leftarrow f - (f \mod G)$;	end
$\quad$ return $(1^\lambda, P)$;	$\quad$ return f;	
end	end	

Fig. 1. Game $\mathsf{GB}_{\mathcal{P}, \mathsf{GBGen}(\cdot), d, b, m}$

We show in [1] that **Sample** returns elements of degree b which are uniformly distributed in $\langle G \rangle$. We recall that given a Gröbner basis G of an ideal $\mathcal{I}$, $r = f \mod \mathcal{I} = f \mod G$ is the normal form of f with respect to the ideal $\mathcal{I}$. We sometimes drop the explicit reference to $\mathcal{I}$ when it is clear from the context which ideal we are referring to, and simply refer to r as the normal form of f. Computing normal forms is the ideal remainder problem which we formalise below.

Definition 2. *The ideal remainder problem is defined through the game shown in Figure 2:* $\mathsf{IR}_{\mathcal{P}, \mathsf{GBGen}(\cdot), d, b, m}$. *The advantage of a ppt algorithm* $\mathcal{A}$ *in solving the* IR *problem is defined as the probability of winning the game minus* $1/q^{\dim_{\mathbb{F}_q}(P/\langle G \rangle)}$. *An adversary is legitimate if it calls the* **Sample** *procedure at most* $m = m(\lambda)$ *times.*

Initialize$(1^\lambda, \mathcal{P}, d)$:	**Sample**():	**Challenge**():	**Finalize**(r'):
begin	begin	begin	begin
$\quad P \leftarrow_\$ \mathbf{P}_\lambda$;	$\quad f \leftarrow_\$ P_{\leq b}$;	$\quad f \leftarrow_\$ P_{\leq b}$;	$\quad r \leftarrow f \mod G$;
$\quad G \leftarrow_\$ \mathsf{GBGen}(1^\lambda, P, d)$;	$\quad f' \leftarrow (f \mod G)$;	$\quad$ return f;	$\quad$ return $r = r'$;
$\quad$ return $(1^\lambda, P)$;	$\quad$ return $f - f'$;	end	end
end	end		

Fig. 2. Game $\mathsf{IR}_{\mathcal{P}, \mathsf{GBGen}(\cdot), d, b, m}$

In Lemma 1 below we prove a weak form of equivalence between the above problems. That is, we require that the IR adversary returns the correct answer with an *overwhelming* probability. This is due to the restriction that **Sample** can only be called a bounded number of times, and thus one cannot amplify the success probability of the IR adversary through repetition. The weak statement is sufficient in our context.

Lemma 1. *If the* GB *problem is hard, then the* IR *problem is weakly hard (i.e., cannot be solved with overwhelming probability). Furthermore, if the* IR *problem is hard then so is the* GB *problem.*

The precise theorem statement and a proof is given in [1]. Informally, the reduction of the GB problem to the IR problem works as follows. Consider an arbitrary element g_i in the Gröbner basis G. We can write g_i as $m_i + \tilde{g}_i$ for some $\tilde{g}_i < g_i$ and $m_i = \mathrm{LM}(g_i)$. Now, assume the normal form of m_i is r_i and suppose that $r_i < m_i$. This implies that $m_i = \sum_{j=0}^{n-1} h_j g_j + r_i$ for some $h_i \in P$. Hence, we have $m_i - r_i \in \langle G \rangle$, an element $\in \langle G \rangle$ with leading monomial m_i. Repeat this process for all monomials up to and including degree d and accumulate the results $m_i - r_i$ in a list $\tilde{G}$. The list $\tilde{G}$ is a list of elements $\in \langle G \rangle$ with $\mathrm{LM}(\tilde{G}) \supseteq \mathrm{LM}(G)$ which implies $\tilde{G}$ is a Gröbner basis. We note that this is the core idea behind the FGLM algorithm [16].

The decisional variant of the IR problem is to decide whether the normal form of some element modulo an ideal is zero or not, i.e., whether this element is in the ideal or not. This is the ideal membership problem formalised below.

Definition 3. *The ideal membership problem is defined through the the game denoted* $\mathsf{IM}_{\mathcal{P},\mathsf{GBGen}(\cdot),d,b,m}$ *as shown in Figure 3. The advantage of a ppt algorithm $\mathcal{A}$ in solving* IM *is defined as twice the probability of winning the game minus 1. An adversary is legitimate if it calls the* **Sample** *procedure at most $m = m(\lambda)$ times.*

Initialize($1^\lambda, \mathcal{P}, d$):	Sample():	Challenge():	proc. Finalize(c'):
begin	begin	begin	begin
$\quad P \leftarrow_\$ \mathbf{P}_\lambda;$	$\quad f \leftarrow_\$ P_{\leq b};$	$\quad f \leftarrow_\$ P_{\leq b};$	$\quad \mid$ return $(c = c');$
$\quad G \leftarrow_\$ \mathsf{GBGen}(1^\lambda, P, d);$	$\quad f' \leftarrow f \mod G;$	$\quad$ if $c = 1$ then	end
$\quad c \leftarrow_\$ \{0,1\};$	$\quad$ return $f - f';$	$\quad\quad f \leftarrow f - (f \mod G);$	
$\quad \mid$ return $(1^\lambda, P);$	end	$\quad$ return $f;$	
end		end	

Fig. 3. Game $\mathsf{IM}_{\mathcal{P},\mathsf{GBGen}(\cdot),d,b,m}$

Clearly any adversary which can solve the IR problem can also solve the IM problem. However, if the search space of reminders modulo $\langle G \rangle$ is sufficiently small, i.e., when $q^{\dim_{\mathbb{F}_q}(P/\langle G \rangle)} = \mathrm{poly}(\lambda)$, and under similar assumptions as for Lemma 1, one can also perform the converse reduction. That is, one can solve the IR problem using an oracle for the IM problem. Lemma 2 below proves this equivalence for the special case of $\mathsf{GBGen}_{\mathrm{dense}}(\cdot)$. Once again, this is sufficient in our context. As before, for Lemma 2 to be meaningful we require that the IM adversary returns the correct answer with *overwhelming* probability.

Lemma 2. *If the* IR *problem is hard, then the* IM *problem is weakly hard for poly-sized* $q^{\dim_{\mathbb{F}_q}(P/\langle G \rangle)}$. *Furthermore, if the* IM *problem is hard, then the* IR *problem is also hard.*

Informally, the construction of an IR adversary from an IM adversary proceeds as follows. Let $\tilde{f}$ be the challenge polynomial. The attacker simply exhaustively searches all elements of the $\mathbb{F}_q$ vector space $P/\langle G \rangle$ until the right remainder r is found. This occurs if $f - r \in \langle G \rangle$ and can be then detected using an IM adversary. However, there is a technical difficulty here. In general, the attacker does not necessarily know the support of $P/\langle G \rangle$ and hence cannot know how to construct r. However, in our case we assume that $\mathsf{GBGen}(\cdot) = \mathsf{GBGen}_{\mathsf{dense}}(\cdot)$ and this difficulty does not arise. In a more general setting, we would have to discover $P/\langle G \rangle$ as well (cf. proof of Lemma 4). See [1] for the proof.

Complexity estimation about Gröbner basis computations [1], together with the above results, lead to the following hardness assumptions.

Definition 4. *Let $\mathcal{P}$ be such that $n(\lambda) = \Omega(\lambda)$. Assume $b - d > 0$, $b > 1$, and that $m(\lambda) = c \cdot n(\lambda)$ for a constant $c \geq 1$. Then the advantage of any ppt algorithm in solving the $\mathsf{GB/IR/IM}$ problem is negligible as function of λ.*

4   Symmetric Polly Cracker: Noise-Free Version

4.1   Homomorphic Symmetric Encryption

SYNTAX. A *homomorphic symmetric-key encryption scheme* (HSKE) is specified by four ppt algorithms: 1) $\mathsf{Gen}(1^\lambda)$ is the key generation algorithm and returns a key pair $(\mathsf{SK}, \mathsf{PK})$, a message space $\mathsf{MsgSp}(\mathsf{PK})$ and a function space $\mathsf{FunSp}(\mathsf{PK})$. 2) $\mathsf{Enc}(\mathsf{m}, \mathsf{SK})$ is the encryption algorithm and returns a ciphertext c. 3) $\mathsf{Eval}(\mathsf{c}_0, \ldots, \mathsf{c}_{t-1}, C, \mathsf{PK})$ is the evaluation algorithm and outputs a ciphertext $\mathsf{c}_{\mathsf{evl}}$. 4) $\mathsf{Dec}(\mathsf{c}_{\mathsf{evl}}, \mathsf{SK})$ is the deterministic decryption algorithm and returns either a message m or a special failure symbol $\perp$.

CORRECTNESS. An HSKE scheme is correct if for any $\lambda \in \mathbb{N}$, any $(\mathsf{SK}, \mathsf{PK}) \in [\mathsf{Gen}(1^\lambda)]$, any t messages $\mathsf{m}_i \in \mathsf{MsgSp}(\mathsf{PK})$, any $\mathsf{c} \in [\mathsf{Enc}(\mathsf{m}, \mathsf{SK})]$, any circuit $C \in \mathsf{FunSp}(\mathsf{PK})$, any t ciphertexts $\mathsf{c}_i \in [\mathsf{Enc}(\mathsf{m}_i, \mathsf{PK})]$, and any evaluated ciphertext $\mathsf{c}_{\mathsf{evl}} \in [\mathsf{Eval}(\mathsf{c}_0, \ldots, \mathsf{c}_{t-1}, C, \mathsf{PK})]$, we have that $\mathsf{Dec}(\mathsf{c}_{\mathsf{evl}}, \mathsf{SK}) = C(\mathsf{m}_0, \ldots, \mathsf{m}_{t-1})$. We do not necessarily require correctness over freshly created ciphertexts.

COMPACTNESS. An HSKE scheme is compact if there exists a fixed polynomial bound $\mathsf{B}(\cdot)$ so that for any key pair $(\mathsf{SK}, \mathsf{PK}) \in [\mathsf{Gen}(1^\lambda)]$, any circuit $C \in \mathsf{FunSp}(\mathsf{PK})$, any set of t messages $\mathsf{m}_i \in \mathsf{MsgSp}(\mathsf{PK})$, any ciphertext $\mathsf{c}_i \in [\mathsf{Enc}(\mathsf{m}_i, \mathsf{SK})]$, and any evaluated ciphertext $\mathsf{c}_{\mathsf{evl}} \in [\mathsf{Eval}(\mathsf{c}_0, \ldots, \mathsf{c}_{t-1}, C, \mathsf{PK})]$, the size of $\mathsf{c}_{\mathsf{evl}}$ is at most $\mathsf{B}(\lambda + |C(\mathsf{m}_0, \ldots, \mathsf{m}_{t-1})|)$ (independently of the size of C).

The syntax, correctness, and compactness of a homomorphic public-key encryption scheme is defined similarly.

4.2   The Scheme

In this section we formally define the (noise-free) symmetric Polly Cracker encryption scheme. We present a family of schemes parameterised not only by the underlying computational polynomial ring scheme $\mathcal{P}$, but also by a Gröbner basis generation algorithm, which itself depends on a degree bound d, and a second degree bound b. Our parameterised scheme, which we write as $\mathcal{SPC}_{\mathcal{P}, \mathsf{GBGen}(\cdot), d, b}$, is presented in Figure 4. The message space is $P/\mathcal{I}$.

Correctness of evaluation can be verified by a straight-forward calculation. This scheme is not compact since multiplications square the size of the ciphertext.

$\text{Gen}_{\mathcal{P},\text{GBGen}(\cdot),d,b}(1^\lambda)$:	$\text{Enc}(m,\text{SK})$:	$\text{Dec}(c,\text{SK})$:	$\text{Eval}(c_0,\dots,c_{t-1},C,\text{PK})$:
begin	begin	begin	begin
$P \leftarrow_{\$} \mathbf{P}_\lambda$;	$f \leftarrow_{\$} P_{\leq b}$;	$m \leftarrow c \mod G$;	apply the Add and Mult
$G \leftarrow_{\$} \text{GBGen}(1^\lambda, P, d)$;	$f' \leftarrow f \mod G$;	return m;	gates of C over P;
$\text{SK} \leftarrow (G, P, b)$;	$f \leftarrow f - f'$;	end	return the result;
$\text{PK} \leftarrow (P, b)$;	$c \leftarrow m + f$;		end
return (SK, PK);	return c;		
end	end		

Fig. 4. The (noise-free) Symmetric Polly Cracker scheme $\mathcal{SPC}_{\mathcal{P},\text{GBGen}(\cdot),d,b}$

4.3 Security

We will show that the above scheme only achieves a weak version of chosen-plaintext security, which allows access to a limited number of ciphertexts.

Definition 5. *The m-time* IND-BCPA *security of a (homomorphic) symmetric-key encryption scheme $\mathcal{SKE}$ is defined though a game* IND-BCPA$_{m,\mathcal{SKE}}$*, which is similar to* IND-CPA *except that the adversary can query its encryption and left-or-right oracles a total of at most $m = m(\lambda)$ times. We say $\mathcal{SKE}$ is m-*IND-BCPA *secure if the advantage of any ppt adversary $\mathcal{A}$, defined as twice the probability of wining the game minus 1 is negligible.*

Theorem 1. *The scheme in Figure 4 is m-*IND-BCPA *secure iff the* IM *problem is hard.*

See [1] for the proof. As a corollary, observe that when $m(\lambda) = \mathcal{O}(\lambda^b)$ one can construct an adversary which breaks the IND-BCPA$_{m,\mathcal{SKE}}$ security of $\mathcal{SPC}$ in polynomial time. Thus we can only hope to achieve security in the bounded model for this scheme.

5 Symmetric-to-Asymmetric Conversion

Our goal for the rest of the paper is to convert the above scheme to one which is both fully IND-CPA secure and somewhat homomorphic. Once we achieve this, it is possible to construct a public-key scheme using the homomorphic features of the symmetric scheme by applying various generic conversions. In the literature there are two prominent such conversions:

(A) Publish a set of encryptions of zero F_0 as part of the public key. To encrypt $m \in \{0,1\}$ compute $c = \sum_{f_i \in S} f_i + m$ where S is a sparse subset of F_0 [13].
(B) Publish two sets F_0 and F_1 of encryptions of zero and one as part of the public key. To encrypt $m \in \{0,1\}$ compute $c = \sum_{f_i \in S_0} f_i + \sum_{f_j \in S_1} f_j$, with S_0 and S_1 being sparse subsets of F_0 and F_1 respectively such that the parity of $|S_1|$ is m. Decryption checks whether $\text{Dec}(c, \text{SK})$ is even or odd [27].

The security of the above transformations rests upon the (computational) in-distinguishability of asymmetric ciphertexts from those produced directly using the symmetric encryption algorithm. As noted above, since $\mathcal{SPC}$ is not IND-CPA secure the above transformations cannot be used. However, one could envisage a larger class of transformations which might lead to a fully secure additively homomorphic SKE (or equivalently an additively homomorphic PKE) scheme. In this section we rule out a large class of such transformations. To this end, we consider PKE schemes which lie within the following design methodology.

1. The secret key is the Gröbner basis G of a zero-dimensional ideal $\mathcal{I} \subset P$. The decryption algorithm computes $c \mod \mathcal{I} = c \mod G$ (perhaps together with some post-processing such as a mod 2 operation). Thus, the message space is (essentially) $P/\mathcal{I}$. We assume that $P/\mathcal{I}$ is known.
2. The public key consists of elements $f_i \in P$. We assume that the remainder of these elements modulo the ideal $\mathcal{I}$, i.e., $r_i := f_i \mod \mathcal{I}$, are known.
3. A ciphertext is computed using ring operations. In other words, it can be expressed as $f = \sum_{i=0}^{N-1} h_i f_i + r$. Here f_i are as in the public key, h_i are some polynomials (possibility depending on f_i), and r is an encoding in $P/\mathcal{I}$ of the message.
4. The construction of the ciphertext does not encode knowledge of $\mathcal{I}$ beyond f_i. That is, we have $\left(\sum_{i=0}^{N-1} h_i f_i + r\right) \mod \mathcal{I} = \sum_{i=0}^{N-1} h_i r_i + r$. Hence we have that $\left(\sum_{i=0}^{N-1} h_i r_i + r\right) \in P/\mathcal{I}$ as an element of P.
5. The security of the scheme relies on the fact that elements f produced at step (3) are computationally indistinguishable from random elements in $P_{\leq b}$.

Condition 4 imposes some real restrictions on the set of allowed transformation, but strikes a reasonable balance between allowing a general statement without ruling out too large a class of conversions. It requires that the r_i and r do not encode any information about the secret key. We currently require this restriction on the "expressive power" of r_i and r so as to make a general impossibility statement. If r_i and r produce a non-zero element in $\mathcal{I}$ using some arbitrary algorithm $\mathcal{A}$, we are unable to prove anything about the transformation. Furthermore, it is plausible that for any given $\mathcal{A}$ a similar impossibility result can be obtained if the remaining conditions hold.

Note that the two transformations above are special linear cases of this methodology. For transformation (A) we have that $f_i \in \mathcal{I}$ (hence $r_i = 0$), $h_i \in \{0, 1\}$ and $r = m$. For transformation (B) we have $r_i = 0$ if $f_i \in F_0$, $r_i = 1$ if $f_i \in F_1$, $h_i \in \{0, 1\}$, and $r = 0$.

To show that any conversion of the above form cannot lead to an IND-CPA-secure public-key scheme, we will use the following theorem which was also used in [5] to discourage the use of Gröbner bases for public-key schemes.

Theorem 2 ([12]). *Let* $\mathcal{I} = \langle f_0, \ldots, f_{m-1} \rangle$ *be an ideal in the polynomial ring* $P = \mathbb{F}[x_0, \ldots, x_{n-1}], h$ *be such that* $\deg(h) \leq D,$ *and*

$$h - (h \mod \mathcal{I}) = \sum_{i=0}^{m-1} h_i f_i, \text{ where } h_i \in P \text{ and } \deg(h_i f_i) \leq D.$$

Let G *be the output of some Gröbner basis computation algorithm up to degree* D. *Then* $h \mod \mathcal{I}$ *can be computed by polynomial reduction of* h *via* G.

The main result of this section is a consequence of the above theorem. It essentially states that uniformly sampling elements of the ideal up to some degree is equivalent to compute a Gröbner basis for the ideal. Note that in itself Theorem 2 does not provide this result, since there is no assumption about the "quality" of h. Hence, to prove this result we first show that the above methodology implies sampling as in Theorem 2 but with uniformly random output. Theorem 2 then allows us to compute normal forms which (because of the randomness of h) allows the computation of a Gröbner basis by Lemma 1. The proof of Theorem 3 is given in [1].

Theorem 3. *Let $G = \{g_0, \ldots, g_{s-1}\}$ be the reduced Gröbner basis of the zero-dimensional ideal $\mathcal{I}$ in the polynomial ring $P = \mathbb{F}[x_0, \ldots, x_{n-1}]$ where each $\deg(g_i) \leq d$. Assume that $P/\mathcal{I}$ is known. Furthermore, let $F = \{f_0, \ldots, f_{N-1}\}$ be a set of polynomials with known $r_i := f_i \mod \mathcal{I}$. Let $\mathcal{A}$ be a ppt algorithm which given F produces elements $f = \sum h_i f_i + r$ with $\deg(f) \leq b$, $h_i \in P$, $b \leq B$, $\deg(h_i f_i) \leq B$, and $(f \mod \mathcal{I}) = \sum h_i r_i + r$. Suppose further that the outputs of $\mathcal{A}$ are computationally indistinguishable from random elements in $P_{\leq b}$. Then there exists an algorithm which computes a Gröbner basis for $\mathcal{I}$ from F in $\mathcal{O}(n^{3B})$ field operations.*

Therefore, if for some degree $b \geq d$ computationally uniform elements of $P_{\leq b}$ can be produced using the public key $f_0, \ldots, f_{N-1}$, there is an attacker which recovers the secret key $g_0, \ldots, g_{s-1}$ in essentially the same complexity. Hence, while conceptually simple and provably secure up to some bound, our symmetric Polly Cracker scheme $\mathcal{SPC}_{\mathcal{P}, \mathsf{GBGen}(\cdot), d, b}$ does not provide a valid building block for constructing a fully homomorphic public-key encryption scheme.

REMARK. Although the above impossibility result is presented for public-key encryption schemes, due to the equivalence result of [27], it also rules out the existence of additively homomorphic symmetric PC-style schemes with full IND-CPA security.

6   Gröbner Bases with Noise

In this section, we introduce noisy variants of the problems presented in Section 3. The goal is to lift the restriction on the number of samples that the adversary can obtain, and following a similar design methodology to Polly Cracker, construct an IND-CPA-secure scheme. That is, we consider problems which naturally arise if we consider noisy encoding of messages in $\mathcal{SPC}$. Similarly to [13,26] we expect a problem which is efficiently solvable in the noise-free setting to be hard in the noisy setting. We will justify this assumption in Section 6.1 by arguing that our construction can be seen as a generalisation of [13,26]. The games below will be parameterised by a noise distribution. The discrete Gaussian distribution – denoted for $\chi_{\alpha,q}$ for standard deviation αq and modulus q – is of particular interest to us (cf. [25]).

We now define a noisy variant of the Gröbner basis problem. The task here is still to compute a Gröbner basis for some ideal $\mathcal{I}$. However, we are now only given access to a noisy sample oracle which provides polynomials which are not necessarily in $\mathcal{I}$ but rather are "close" approximations to elements of $\mathcal{I}$. Here the term "close" is made precise using a noise distribution χ on $P/\mathcal{I}$.

Definition 6. *The Gröbner basis with noise problem is defined through the game* $\mathsf{GBN}_{\mathcal{P},\mathsf{GBGen}(\cdot),d,b,\chi}$ *as shown in Figure 5. The advantage of a ppt algorithm $\mathcal{A}$ in solving the* GBN *problem is the probability of winning the game.*

Initialize($1^\lambda, \mathcal{P}, d$):	Sample():	Finalize(G'):
begin	begin	begin
$P \leftarrow_\$ \mathbf{P}_\lambda$;	$f \leftarrow_\$ P_{\leq b}$;	| return $(G = G')$;
$G \leftarrow_\$ \mathsf{GBGen}(1^\lambda, P, d)$;	$e \leftarrow_\$ \chi$;	end
return $(1^\lambda, P)$;	$f \leftarrow f - (f \mod G) + e$;	
end	return f;	
	end	

Fig. 5. Game $\mathsf{GBN}_{\mathcal{P},\mathsf{GBGen}(\cdot),d,b,\chi}$

The essential difference between the noisy and noise-free versions of the GB problem is that by adding noise we have eliminated the restriction on the adversary to call the **Sample** oracle a bounded number of times. The choice of χ greatly influences the hardness of the GBN problem.

As in the noise-free setting, we can ask various questions about the ideal $\mathcal{I}$ spanned by G. One such example is solving the ideal remainder problem with access to noisy samples from $\mathcal{I}$.

Definition 7. *The ideal remainder with noise problem is defined through the game* $\mathsf{IRN}_{\mathcal{P},\mathsf{GBGen}(\cdot),d,b,\chi}$ *as shown in Figure 6. The advantage of a ppt algorithm $\mathcal{A}$ is defined as the probability of winning the game minus* $1/q(\lambda)^{\dim_{\mathbb{F}}(P/\langle G \rangle)}$.

Initialize($1^\lambda, \mathcal{P}, d$):	Sample():	Challenge():	Finalize(r'):
begin	begin	begin	begin
$P \leftarrow_\$ \mathbf{P}_\lambda$;	$f \leftarrow_\$ P_{\leq b}$;	$f \leftarrow_\$ P_{\leq b}$;	$r" = f \mod G$;
$G \leftarrow_\$ \mathsf{GBGen}(1^\lambda, P, d)$;	$e \leftarrow_\$ \chi$;	return f;	return $r' = r"$;
return $(1^\lambda, P)$;	$f \leftarrow f - (f \mod G) + e$;	end	end
end	return f;		
	end		

Fig. 6. Game $\mathsf{IRN}_{\mathcal{P},\mathsf{GBGen}(\cdot),d,b,\chi}$

In fact, the above two problems are equivalent as shown in the lemma below. Compared to the noise-free version, we no longer need the IM adversary to be overwhelmingly successful, as there are no restrictions on the number of calls that can be made to the **Sample** procedure. The proof is given in [1].

Lemma 3. *The* IRN *problem is hard iff the* GBN *problem is hard.*

Similarly to the noise-free setting, the ideal membership with noise (IMN) problem is the decisional variant of the IRN (and hence the GBN) problem. However, in the noisy setting we have the choice between a noisy and noise-free challenge polynomial. In the definition below noisy challenges are provided and the adversary wins the game if he can distinguish whether an element was sampled uniformly from $P_{\leq b}$ or from $\mathcal{I} + \chi$.

Definition 8. *The ideal membership with noise problem is defined through the game* $\mathsf{IMN}_{\mathcal{P},\mathsf{GBGen}(\cdot),d,b,\chi}$ *as shown in Figure 7. The advantage of a ppt algorithm* $\mathcal{A}$ *in solving the* IMN *problem is as twice the probability of winning the game minus* 1.

Initialize$(1^\lambda, \mathcal{P}, d)$:	Sample():	Challenge():	Finalize(c'):
begin	**begin**	**begin**	**begin**
$\quad P \leftarrow_\$ \mathbf{P}_\lambda;$	$\quad f \leftarrow_\$ P_{\leq b};$	$\quad f \leftarrow_\$ P_{\leq b};$	$\quad$ **return** $(c' = c);$
$\quad G \leftarrow_\$ \mathsf{GBGen}(1^\lambda, P, d);$	$\quad e \leftarrow_\$ \chi;$	$\quad$ **if** $c = 1$ **then**	**end**
$\quad c \leftarrow_\$ \{0, 1\};$	$\quad f' \leftarrow f \mod G;$	$\quad\quad e \leftarrow_\$ \chi;$	
$\quad$ **return** $(1^\lambda, P);$	$\quad f \leftarrow f - f' + e;$	$\quad\quad f \leftarrow f - (f \mod G) + e;$	
end	$\quad$ **return** $f;$	$\quad$ **return** $f;$	
	end	**end**	

Fig. 7. Game $\mathsf{IMN}_{\mathcal{P},\mathsf{GBGen}(\cdot),d,b,\chi}$

Our definition of the IMN problem can be seen as an instantiation of Gentry's ideal coset problem [18] since both problems require distinguishing uniformly chosen elements in $P_{\leq b}$ from those in $\mathcal{I} + \chi$. Our problem, however, assumes noisy samples since it is clear from Section 3 that otherwise the problem is easy.

Again, we would like to have a decision-to-search reduction; that is, we would like to have an equivalence between the IRN and IMN problems. This equivalence holds when the search space of remainders is polynomial in λ, namely when $q(\lambda)^{\dim_{\mathbb{F}_q}(\mathcal{P}(\lambda)/\mathsf{GBGen}(\cdot))} = \mathrm{poly}(\lambda)$. The intuition behind this reduction is that the adversary can exhaustively search the quotient ring and use the IMN oracle to verify his guess. Once again, a technical difficulty arises as the adversary does not know the search space $P/\mathcal{I}$ and thus has to discover it during the attack. Again, the IMN adversary provides an oracle to accomplish this. This is formalised in the lemma below whose proof is in [1].

Lemma 4. *The* IMN *problem is hard iff the* IRN *problem is hard for poly-sized* $q^{\dim_{\mathbb{F}_q}(P/\langle G \rangle)}$.

Hence GBN is equivalent to IRN and IRN is equivalent to IMN under some additional assumptions about the size $P/\mathcal{I}$. Finally, for $d = 1$ (but arbitrarily b) we show that if we can solve the GBN problem on average, then we can also solve it for worst-case instances. This is turn increases our confidence in hardness of the GBN problem. The proof of the follow lemma is given in [1].

Lemma 5. *If the* GBN *problem is worst-case hard, then it is also average-case hard.*

6.1 Hardness Assumptions and Justifications

Let us now investigate the hardness of the GBN, IRN, and IMN problems.

RELATION TO LWE. It is easy to see that GBN can be considered as a non-linear generalisation of LWE if $q = \mathrm{poly}(n)$ is a prime. In other words, we have equivalence between these problems when $b = d = 1$ in GBN. This is formalised below (proof is in [1]).

Lemma 6. *If the* LWE *problem is hard then the* GBN *problem is also hard for* $b = d = 1$.

In the noise-free setting we assume that solving systems of equations of degree greater than 1 is harder than solving those of degree 1. More generally, we assume that equations of degree $b > b'$ are harder to solve than those of degree b'. Intuitively, equations of degree b' can be seen as those of degree b where the coefficients of higher degree monomials are set to zero. However, formalising this intuition for an adversary which expects uniformly distributed equations of degree b seems futile since producing such equations is equivalent to solving the system by Theorem 3.

   In the noisy setting this equivalence (i.e., Theorem 3) between sampling and solving no longer holds. However, we still need to deal with the distribution of noise. One strategy to show that difficulty increases with the degree parameter b is to allow for an increase of the noise level in the samples. We formalise this below (a proof is given in [1]) .

Lemma 7. *If the* GBN *problem is hard for degree* $2b$ *with noise* $\chi_{\sqrt{N}\alpha^2 q, q}$, $N = \binom{n+b}{b}$, *then it is also hard for degree* b *with noise* $\chi_{\alpha, q}$.

RELATION TO THE APPROXIMATE GCD PROBLEM. The GBN problem for $n = 1$ is the approximate GCD problem over $\mathbb{F}_q[x]$. Contrary to the approximate GCD problem over the integers (cf. [13]), this problem has not yet received much attention, and hence it is unclear under which parameters it is hard. However, as we discuss in [1], the notion of a Gröbner basis can be extended to $\mathbb{Z}[x_0, \ldots, x_{n-1}]$, which in turn implies a version of the GBN problem over $\mathbb{Z}$. This can be seen as a direct generalisation of the approximate GCD problem in $\mathbb{Z}$.

THE CASE $q = 2$. Recall that if $b = d = 1$ we have an equivalence with the LWE problem (or the well-known problem of learning parity with noise (LPN) if $q = 2$). More generally, for $d = 1$ we can reduce Max-3SAT instances to GBN instances by translating each clause individually to a Boolean polynomial. However, in Max-3SAT the number of samples is bounded and hence this reduction only shows the hardness of GBN with a bounded number of samples. Still, the Gröbner basis returned by an arbitrary algorithm $\mathcal{A}$ solving GBN using a bounded number of samples will provide a solution to the Max-3SAT problem. Vice versa, we may convert a GBN instance for $d = 1$ to a Max-SAT instance (more precisely Partial Max-Sat) by running an ANF to CNF conversion algorithm [4].

KNOWN ATTACKS. Finally, we consider known attacks to understand the difficulty of the GBN problem. Recall that if $b = 1$ Lemma 6 states that we can solve the LWE problem if we can solve the GBN problem. The converse also applies. Indeed, for any $b \geq d$ and $d = 1$ the best known attack against the

GBN problem for $d = 1$ is to reduce it to the LWE problem, similarly to the linearisation technique used for solving non-linear systems of equations in the noise-free setting. Let $N = \binom{n+b}{b}$ be the number of monomials up to degree b. Let $\mathcal{M} : P \to \mathbb{F}_q^N$ be a function which maps polynomials in P to vectors in $\mathbb{F}_q^N$ by assigning the i-th component of the image vector the coefficient of the i-th monomial $\in M_{\leq b}$. Then, in order to reduce GBN with n variables and degree b to LWE with N variables, reply to each LWE **Sample** query by calling the GBN **Sample** oracle to retrieve f, compute $v = \mathcal{M}(f)$ and return (a, b) with $a = (v_{N-1}, \ldots, v_1)$ and $b = -v_0$. When the LWE adversary queries **Finalize** on s, query the GBN **Finalize** on $[x_0 - s_0, \ldots, x_{n-1} - s_{n-1}]$. Correctness follows from the correctness of linearisation in the noise-free setting [3]. Furthermore, the LWE problem in N variables and with respect to the discrete Gaussian noise distribution $\chi_{\alpha,q}$ is considered to be hard if $\alpha \geq 3/2 \cdot \max(\frac{1}{q}, 2^{-2\sqrt{N \log q \log d}})$ for an appropriate choice of δ which is the quality of the approximation for the shortest vector problem. With current lattice algorithms $\delta = 1.01$ is hard and 1.005 infeasible [24].

Perhaps the most interesting attack on LWE from the perspective of this work is that due to Arora and Ge [3] which reduces the problem of solving linear systems with noise to the problem of solving (structured) non-linear noise-free systems. We may apply this technique directly to GBN, i.e., without going to LWE first, and reduce it to GB with large b. However, it seems this approach does not improve the asymptotic complexity of the attack. Finally, certain conditions to rule out exhaustive search must be imposed.

Definition 9. *Let* $b, d \in \mathbb{N}$ *with* $b \geq d \geq 1$. *Let* $\mathcal{P}$ *be a polynomial ring distribution and* $\chi_{\alpha,q}$ *be the discrete Gaussian distribution. Suppose the parameters* n, α, *and* q *(all being a function of* λ*) satisfy the following set of conditions:* 1) $n \geq \sqrt[b]{\lambda}$; 2) $(\alpha q)^{nd^n} \approx 2^\lambda$ *so exhaustive search over the noise or the secret key space is ruled out;* 3) $\alpha q \geq 8$ *as suggested in [22]; and 4) for* $N := \binom{n+b}{b}$, *and* $\delta := 1.005$ *we have* $\alpha \geq 3/2 \cdot \max\{\frac{1}{q}, 2^{-2\sqrt{N \log q \log \delta}}\}$, *and hence the best known attacks against the* LWE *problem are ruled out [24,28]. Then the advantage of any ppt algorithm in solving the* GBN, IRN, *and* IMN *problems is negligible.*

7   Polly Cracker with Noise

In this section we present a fully IND-CPA-secure PC-style symmetric encryption scheme. Our parameterised scheme, $\mathcal{SPCN}_{\mathcal{P},\mathsf{GBGen}(\cdot),d,b,\chi}$, is shown in Figure 8. Here we represent elements in $\mathbb{F}_q$ as integers in the interval $(-\lfloor \frac{q}{2} \rfloor, \lfloor \frac{q}{2} \rfloor]$. This representation is also used in the definition of noise. All the computations are performed in the ring P as generated by Gen. Furthermore we assume that $\gcd(2, q) = 1$. This condition is needed for the correctness and the security of our scheme. The message space is $\mathbb{F}_2$ (although we remark that this can be generalised to other small fields). Correctness of evaluation up to overflows can be established by a straight-forward calculation.

PERMITTED CIRCUITS. Circuits composed of Add and Mul gates can be seen as multivariate Boolean polynomials in t variables over $\mathbb{F}_2$. We can consider the

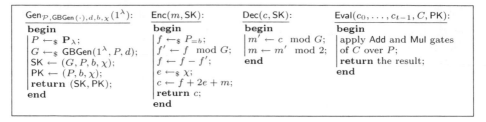

Fig. 8. The Symmetric Polly Cracker with Noise scheme $\mathcal{SPCN}_{\mathcal{P},\mathsf{GBGen}(\cdot),d,b,\chi}$

generalisation of this set of polynomials to $\mathbb{F}_q$ (i.e., the coefficients are in $\mathbb{F}_q$). In order to define the set of permitted circuits (which will be parameterised by $\alpha > 0$) we first embed the Boolean polynomials into the ring of polynomials over $\mathbb{Z}$. For $\chi_{\alpha,q}$ we have that the probability of the noise being larger than $k\alpha q$ is $< \exp(-k^2/2)$. We now say that a circuit is valid if for any $(s_0, \ldots, s_{t-1})$ with $s_i \leq t\alpha q$ we have that the outputs are less than q for some parameter t. This restriction ensures that no overflows occur when polynomials are evaluated over $\mathbb{F}_q$. In [1] we discuss how to set α and q in order to allow for evaluation of polynomials of some fixed degree μ and provide a Sage implementation [30].

COMPACTNESS. Additions do not increase the size of the ciphertext, but they do increase the size of the error by at most one bit. Multiplications square the size of the ciphertext and the bit-size of the the noise by approximately $\log(5e_0e_1)$ bits. In [1] we also provide a discussion on how to trade ciphertext size with noise, an avenue which is investigated independently in [7]. The theorem below, which is proven in [1], states the security properties of the above scheme.

Theorem 4. *If the* IMN *problem is hard, then the scheme in Figure 8 is secure.*

The above theorem together with the recent results in [27] which establish the equivalence of symmetric and asymmetric homomorphic encryption schemes leads to the first provably secure public-key encryption scheme from assumptions related to Gröbner bases for random systems. This provides a positive answer to the challenges raised by Barkee et al. [5] (and later also by Gentry [18]). We note here that the transformation – as briefly described in Section 5 – only use the additive features of the scheme and does not require full homomorphicity.

Acknowledgments. We would like to thank Carlos Cid for valuable feedback and discussions on this work. We would also like to thank Frederik Armknecht for helpful discussions on an earlier draft of this work.

References

1. Albrecht, M.R., Farshim, P., Faugère, J.-C., Perret, L.: Polly Cracker, revisited. Cryptology ePrint Archive, Report 2011/289 (2011)
2. Applebaum, B., Cash, D., Peikert, C., Sahai, A.: Fast Cryptographic Primitives and Circular-Secure Encryption Based on Hard Learning Problems. In: Halevi, S. (ed.) CRYPTO 2009. LNCS, vol. 5677, pp. 595–618. Springer, Heidelberg (2009)

3. Arora, S., Ge, R.: New Algorithms for Learning in Presence of Errors. In: Aceto, L., Henzinger, M., Sgall, J. (eds.) ICALP 2011, Part I. LNCS, vol. 6755, pp. 403–415. Springer, Heidelberg (2011)
4. Bard, G.V., Courtois, N.T., Jefferson, C.: Efficient methods for conversion and solution of sparse systems of low-degree multivariate polynomials over GF(2) via SAT-solvers. Cryptology ePrint Archive, Report 2007/024 (2007)
5. Barkee, B., Can, D.C., Ecks, J., Moriarty, T., Ree, R.F.: Why you cannot even hope to use Gröbner bases in public key cryptography: An open letter to a scientist who failed and a challenge to those who have not yet failed. J. of Symbolic Computations 18(6), 497–501 (1994)
6. Berbain, C., Gilbert, H., Patarin, J.: QUAD: A multivariate stream cipher with provable security. J. Symb. Comput. 44(12), 1703–1723 (2009)
7. Brakerski, Z., Vaikuntanathan, V.: Efficient fully homomorphic encryption from (standard) LWE. To appear in FOCS 2011 (2011)
8. Buchberger, B.: Ein Algorithmus zum Auffinden der Basiselemente des Restklassenrings nach einem nulldimensionalen Polynomideal. PhD thesis, Universität Innsbruck (1965)
9. Caboara, M., Caruso, F., Traverso, C.: Lattice Polly Cracker cryptosystems. Journal of Symbolic Computation 46, 534–549 (2011)
10. Cox, D., Little, J., O'Shea, D.: Ideals, Varieties, and Algorithms, 3rd edn. Springer, Heidelberg (2005)
11. Cramer, R., Shoup, V.: Design and analysis of practical public-key encryption schemes secure against adaptive chosen ciphertext attack. SIAM J. of Computing, 167–226 (2003)
12. Dickenstein, A., Fitchas, N., Giusti, M., Sessa, C.: The membership problem for unmixed polynomial ideals is solvable in single exponential time. Discrete Appl. Math. 33(1-3), 73–94 (1991)
13. van Dijk, M., Gentry, C., Halevi, S., Vaikuntanathan, V.: Fully Homomorphic Encryption Over the Integers. In: Gilbert, H. (ed.) EUROCRYPT 2010. LNCS, vol. 6110, pp. 24–43. Springer, Heidelberg (2010)
14. Ding, J., Yang, B.-Y.: Multivariate public key cryptography. In: Post-Quantum Cryptography, pp. 193–234. Springer, Heidelberg (2009)
15. dit Vehel, F.L., Marinari, M.G., Perret, L., Traverso, C.: A survey on Polly Cracker systems. In: Gröbner Bases. Coding and Cryptography, pp. 285–305. Springer, Heidelberg (2009)
16. Faugère, J.-C., Gianni, P.M., Lazard, D., Mora, T.: Efficient computation of zero-dimensional Gröbner bases by change of ordering. Journal of Symbolic Computation 16, 329–344 (1993)
17. Fellows, M., Koblitz, N.: Combinatorial cryptosystems galore! In: Finite Fields: Theory, Applications, and Algorithms. Contemporary Mathematics, vol. 168, pp. 51–61. AMS (1994)
18. Gentry, C.: A fully homomorphic encryption scheme. PhD thesis, Stanford University (2009)
19. Gentry, C.: Fully homomorphic encryption using ideal lattices. In: ACM Symposium on Theory of Computing, pp. 169–178 (2009)
20. Gentry, C., Halevi, S.: Implementing Gentry's Fully-Homomorphic Encryption Scheme. In: Paterson, K.G. (ed.) EUROCRYPT 2011. LNCS, vol. 6632, pp. 129–148. Springer, Heidelberg (2011)
21. Gouget, A., Patarin, J.: Probabilistic Multivariate Cryptography. In: Nguyên, P.Q. (ed.) VIETCRYPT 2006. LNCS, vol. 4341, pp. 1–18. Springer, Heidelberg (2006)

22. Lindner, R., Peikert, C.: Better Key Sizes (and Attacks) for LWE-Based Encryption. In: Kiayias, A. (ed.) CT-RSA 2011. LNCS, vol. 6558, pp. 319–339. Springer, Heidelberg (2011)
23. Melchor, C.A., Gaborit, P., Herranz, J.: Additively Homomorphic Encryption with d-Operand Multiplications. In: Rabin, T. (ed.) CRYPTO 2010. LNCS, vol. 6223, pp. 138–154. Springer, Heidelberg (2010)
24. Micciancio, D., Regev, O.: Lattice-based cryptography. In: Post-Quantum Cryptography, pp. 147–191. Springer, Heidelberg (2009)
25. Regev, O.: On lattices, learning with errors, random linear codes, and cryptography. Journal of the ACM 56, 34:1–34:40 (2009)
26. Regev, O.: The learning with errors problem. In: IEEE Conference on Computational Complexity 2010, pp. 191–204 (2010)
27. Rothblum, R.: Homomorphic Encryption: From Private-Key to Public-Key. In: Ishai, Y. (ed.) TCC 2011. LNCS, vol. 6597, pp. 219–234. Springer, Heidelberg (2011)
28. Rückert, M., Schneider, M.: Estimating the security of lattice-based cryptosystems. Cryptology ePrint Archive, Report 2010/137 (2010)
29. Smart, N.P., Vercauteren, F.: Fully Homomorphic Encryption with Relatively Small Key and Ciphertext Sizes. In: Nguyen, P.Q., Pointcheval, D. (eds.) PKC 2010. LNCS, vol. 6056, pp. 420–443. Springer, Heidelberg (2010)
30. Stein, W.A., et al.: Sage Mathematics Software. The Sage Development Team, Version 4.7.0 (2011), http://www.sagemath.org

Oblivious RAM with $O((\log N)^3)$ Worst-Case Cost

Elaine Shi[1,*], T.-H. Hubert Chan[2], Emil Stefanov[3,**], and Mingfei Li[2]

[1] UC Berkeley/PARC
[2] The University of Hong Kong
[3] UC Berkeley

Abstract. Oblivious RAM is a useful primitive that allows a client to hide its data access patterns from an untrusted server in storage outsourcing applications. Until recently, most prior works on Oblivious RAM aim to optimize its amortized cost, while suffering from linear or even higher worst-case cost. Such poor worst-case behavior renders these schemes impractical in realistic settings, since a data access request can occasionally be blocked waiting for an unreasonably large number of operations to complete.

This paper proposes novel Oblivious RAM constructions that achieves poly-logarithmic worst-case cost, while consuming constant client-side storage. To achieve the desired worst-case asymptotic performance, we propose a novel technique in which we organize the O-RAM storage into a binary tree over data buckets, while moving data blocks obliviously along tree edges.

1 Introduction

Oblivious RAM (or O-RAM for short) [5–7, 11, 12, 16] is a useful primitive for enabling privacy-preserving outsourced storage, where a client stores its data at a remote untrusted server. While standard encryption techniques allow the client to hide the contents of the data from the server, they do not guard the access patterns. As a result, the server can still learn sensitive information by examining the access patterns. For example, Pinkas and Reinman [12] gave an example in which a sequence of data access operations to specific locations (u_1, u_2, u_3) can indicate a certain stock trading transaction, and such financial information is often considered highly sensitive by organizations and individuals alike.

Oblivious RAM allows the client to completely hide its data access patterns from the untrusted server. It can be used in conjunction with encryption, to enable stronger privacy guarantees in outsourced storage applications. Not surprisingly, the client has to pay a certain cost in order to hide its access patterns from the server. Among all prior work in this space, the seminal constructions recently proposed by Goodrich and Mitzenmacher [7] achieve the best asymptotic performance in terms of amortized cost.

* This material is based upon work partially supported by the Air Force Office of Scientific Research under MURI Grant No. 22178970-4170 and No. FA9550-08-1-0352 Any opinions, findings, and conclusions or recommendations expressed in this material are those of the author(s) and do not necessarily reflect the views of the Air Force Office of Scientific Research.
** This material is based upon work partially supported by the National Science Foundation Graduate Research Fellowship under Grant No. DGE-0946797.

D.H. Lee and X. Wang (Eds.): ASIACRYPT 2011, LNCS 7073, pp. 197–214, 2011.

Table 1. Our contributions. The $\widetilde{O}$ notation hides poly log log N terms. The bounds for this paper hold with high probability $1 - \frac{1}{\text{poly}(N)}$, assuming that the total number of data access requests $M = \text{poly}(N)$, and that the block size $B \geq c \log N$ bits, for any constant $c > 1$. For a more precise statement of our bounds, please refer to Section 4. The BST bucket construction is due to an O-RAM construction by Damgård, Meldgaard, and Nielsen [4].

Scheme	Amortized Cost	Worst-case Cost	Client Storage	Server Storage
GO [6]	$O((\log N)^3)$	$O(N(\log N)^2)$	$O(1)$	$O(N \log N)$
WS [16]	$O((\log N)^2)$	$O(N \log N)$	$O(\sqrt{N})$	$O(N \log N)$
WSC [17]	$O(\log N \log \log N)$	$O(N \log \log N)$	$O(\sqrt{N})$	$O(N)$
PR [12]	$O((\log N)^2)$	$O(N \log N)$	$O(1)$	$O(N)$
GM [7]	$O((\log N)^2)$	$O(N \log N)$	$O(1)$	$O(N)$
	$O(\log N)$	$O(N)$	$O(\sqrt{N})$	$O(N)$
BMP [3]	$O(\sqrt{N})$	$O(\sqrt{N})$	$O(\sqrt{N})$	$O(N)$
SSS [15]	$O((\log N)^2)$	$O(\sqrt{N})$	$O(\sqrt{N})$	$O(N)$
This paper				
Trivial Bucket	$O((\log N)^3)$	$O((\log N)^3)$	$O(1)$	$O(N \log N)$
Square-Root Bucket	$\widetilde{O}((\log N)^{2.5})$	$\widetilde{O}((\log N)^3)$	$O(1)$	$O(N \log N)$
BST Bucket	$\widetilde{O}((\log N)^2)$	$\widetilde{O}((\log N)^3)$	$O(1)$	$\widetilde{O}(N \log N)$

Specifically, let N denote the maximum capacity of the O-RAM. Goodrich and Mitzenmacher show that with $O(1)$ client-side storage, one can achieve $O((\log N)^2)$ amortized cost, i.e., each oblivious data request translates into $O((\log N)^2)$ non-oblivious data access operations on average. Goodrich and Mitzenmacher also show that with $O(\sqrt{N})$ client-side storage, one can achieve $O(\log N)$ amortized cost [7].

O-RAM with sublinear worst-case cost. Until recently, most prior work on O-RAM optimizes for the amortized cost [6, 7, 12, 16], while not giving much consideration to the worst-case cost. Specifically, while achieving logarithmic or poly-logarithmic amortized cost, these constructions [6, 7, 12, 16] have a worst-case cost of $\Omega(N)$, due to the occasional reshuffling operations which can take up to $\Omega(N)$ time. Such $\Omega(N)$ worst-case behavior renders these schemes impractical in real-world applications; since every now and then, a data request can be blocked waiting for $\Omega(N)$ operations to complete. When this happens, the perceived waiting time for the user would be unacceptable.

The research community has only recently started to investigate O-RAMs with sublinear worst-case cost [3, 15]. Boneh, Mazieres, and Popa [3] proposed an O-RAM with $O(\sqrt{N})$ worst-case cost, however, at the expense of $O(\sqrt{N})$ (rather than poly-log) amortized cost. Stefanov, Shi, and Song [15] recently proposed an O-RAM with $O(\sqrt{N})$ worst-case cost , $O((\log N)^2)$ amortized cost, and $O(\sqrt{N})$ client-side storage.

1.1 Our Contributions

O-RAM with poly-log worst-case cost, and constant client-side storage. This paper proposes novel O-RAM constructions that achieve both poly-log amortized and worst-case cost, while consuming $O(1)$ client-side storage, and $O(N \log N)$ server-side storage. We offer two variants of our construction. The simpler variant (instantiated with the trivial bucket O-RAM) achieves $O((\log N)^3)$ amortized and worst-case cost. A slightly more sophisticated variant (instantiated with the Square-Root bucket O-RAM) achieves $\widetilde{O}((\log N)^{2.5})$ amortized cost, and $\widetilde{O}((\log N)^3)$ worst-case cost. We use the $\widetilde{O}$ notation to hide poly $\log \log$ terms from the asymptotic bounds.

These afore-mentioned bounds hold with very high probability (i.e., at least $1 - \frac{1}{\mathrm{poly}(N)}$), under realistic assumptions that the number of data requests $M = \mathrm{poly}(N)$, and that the block size $B \geq c \log N$ bits for any constant $c > 1$.

Novel binary-tree based technique. Most existing constructions [6, 7, 12, 16] are based on hierarchical solution initially proposed by Goldreich and Ostrovsky [6], and they suffer from $\Omega(N)$ worst-case cost due to the occasional reshuffling operation that can take up to $\Omega(N)$ time. Therefore, to reduce the worst-case cost, we wish to somehow spread the cost of reshuffling over time, so the worst-case cost can be amortized towards each O-RAM operation.

Unfortunately, due to certain technical constraints imposed by these constructions [6, 7, 12, 16], it does not seem possible to directly spread the cost of reshuffling over time. As a result, we propose a novel technique called the *binary-tree based construction* (Section 3). Basically, the server-side O-RAM storage is organized into a binary tree over small data buckets. Data blocks are evicted in an oblivious fashion along tree edges from the root bucket to the leaf buckets. While in spirit, the binary-tree based construction is trying to spread the reshuffling cost over time; in reality, its operational mechanisms bear little resemblance to prior schemes [7, 12, 16] based on Goldreich and Ostrovsky's original hierarchical solution [6]. Therefore, this represents an entirely new technique which has not been previously studied in the O-RAM literature.

While the basic binary-tree based construction achieves poly-logarithmic amortized and worst-case cost, it requires $\frac{N}{c}$ blocks of client-side storage for some constant $c > 1$. To reduce the client-side storage, we recursively apply our O-RAM construction over the index structure. Instead of storing the index structure on the client side, we store it in a separate and smaller O-RAM on the server side. We achieve $O(1)$ client-side storage through recursive application of our O-RAM construction over the index structure (Section 4).

Conceptual simplicity. Another notable characteristic of our constructions is their relative conceptual simplicity in comparison with most other existing constructions [6, 7, 12, 16]. In particular, the simpler variant of our construction (based on the trivial bucket O-RAM as described in Section 3) achieves $O((\log N)^3)$ amortized and worst-case cost while requiring no oblivious sorting or reshuffling, no hashing or Cuckoo hashing (or its oblivious simulation such as in the Goodrich-Mitzenmacher construction [7]). All O-RAM read and write operation behave uniformly in this simpler variant, and cost the same asymptotically.

1.2   Related Work

Oblivious RAM was first investigated by Goldreich and Ostrovsky [5, 6, 11] in the context of protecting software from piracy, and efficient simulation of programs on oblivious RAMs. Apart from proposing a seminal hierarchical solution with $O((\log N)^3)$ amortized cost, Goldreich and Ostrovsky [6] also demonstrate the following lower-bound: for an O-RAM of capacity N, the client has to pay an amortized cost of at least $\Omega(\log N)$. Recently, Beame and Machmouchi [2] improved the lower bound to $\Omega(\log N \log \log N)$.

Since the first investigation of Oblivious RAM by Goldreich and Ostrovsky [5, 6, 11], several constructions have been proposed subsequently [3, 7, 12, 15, 16]. Among these, the seminal constructions recently proposed by Goodrich and Mitzenmacher [7] achieve the best asymptotic performance in terms of amortized cost: with $O(1)$ client-side storage, their construction achieves $O((\log N)^2)$ amortized cost; and with $O(\sqrt{N})$ client-side storage, their construction achieves $O(\log N)$ amortized cost [7]. Pinkas and Reinman [12] also showed a similar result for the $O(1)$ client-side storage case; however, some researchers have pointed out a security flaw in their construction [7], which the authors of [12] have promised to fix in a future journal version.

For a fairly long time, almost all research in this space aimed to optimize the amortized cost, while neglecting the worst-case cost. Only very recently did the research community start to investigate O-RAM constructions with sublinear worst-case cost. As mentioned earlier, there have been two recent works [3, 15] aimed at achieving sublinear worst-case cost and making O-RAM practical. Boneh,Mazieres, and Popa [3] achieve $O(\sqrt{N})$ worst-case cost, however, at the expense of $O(\sqrt{N})$ amortized cost. Stefanov, Shi, and Song [15] recently proposed a novel O-RAM construction with $O(\sqrt{N})$ worst-case cost, $O((\log N)^2)$ amortized cost, and $O(\sqrt{N})$ client-side storage. Apart from this, Stefanov, Shi, and Song also offered another construction geared towards practical performance rather than asymptotics. This practical construction uses linear amount of client storage (with a very small constant), and achieves $O(\log N)$ amortized cost and $O(\sqrt{N})$ worst-case cost. Under realistic settings, it achieves $20 - 30X$ amortized cost, while storing $0.01\% - 0.3\%$ amount of total data at the client. To the best of our knowledge, this is the most practical scheme known to date.

We note that the hierarchical aspect of our binary-tree technique is partially inspired by the hierarchical solution originally proposed by Goldreich and Ostrovsky [6], and later adopted in many constructions [7, 12, 16]; while the eviction aspect is partially inspired by the background eviction idea originally proposed by Stefanov, Shi, and Song [15].

Our binary tree technique may also be superficially reminiscent of a construction by Damgård, Meldgaar, and Nielsen [4]. However, apart from that fact that both schemes rely on a binary tree, the internal mechanisms of our construction and the Damgård-Meldgaar-Nielsen construction are fundamentally different. Specifically, Damgård *et al.* primarily aim to avoid the need of random oracle or pseudo-random function, rather than improve worst-case cost. Their construction uses a binary search tree, and requires periodic reshuffling operations that can take $O(N \log N)$ time. In contrast, we use a binary tree (instead of a binary search tree), and we use a background eviction mechanism to circumvent the need for reshuffling.

Table 1 illustrates the asymptotic performance characteristics of various existing schemes, and positions our work in perspective of related work.

Concurrent/subsequent work. In concurrent/subsequent work, Goodrich *et al.* [8] and Kushilevitz *et al.* [10] also came up novel O-RAM constructions with poly-logarithmic overhead. Specifically, the construction by Goodrich *et al.* achieves $O((\log N)^2)$ worst-case cost with $O(1)$ memory; and and Kushilevitz *et al.* achieve $O(\frac{(\log N)^2}{\log \log N})$. Due to a larger constant in their asymptotic notations, in realistic scenarios, our scheme with the trivial bucket O-RAM is likely the most practical when the client-side storage is $O(1)$.

2   Preliminaries

Let N denote the O-RAM capacity, i.e., the maximum number of data blocks that an O-RAM can store. We assume that data is fetched and stored in atomic units called *blocks*. Let B denote the block size in terms of the number of bits. We assume that the block size $B \geq c \log N$, for some $c > 1$. Notice that this is true in almost all practical scenarios. We assume that each block has a global identifier $u \in \mathcal{U}$, where $\mathcal{U}$ denotes the universe of identifiers.

Throughout the paper, we use the asymptotic notation $\widetilde{O}(f(N))$ meaning $O(f(N)\text{poly} \log \log N)$ as a short-hand for hiding poly $\log \log N$ terms.

2.1   Defining O-RAM with Enriched Operations

The standard O-RAM adopted in prior work [5, 7, 12, 16] exports a Read and a Write interfaces. To hide whether the operation is a read or a write, either operation will generate both a read and a write to the O-RAM.

In this paper, we consider O-RAMs that support a few enriched operations. Therefore, we propose a modified O-RAM definition, exporting a ReadAndRemove primitive, and an Add primitive. We later show that given these two primitives, we can easily implement the standard O-RAM Read and Write operations. Moreover, given these two primitives, we can also support an enriched operation called Pop, which will be later needed in our constructions. Therefore, our modified O-RAM definition is more general than the standard O-RAM notion. The same modified O-RAM notion was adopted in the work by Stefanov, Shi, and Song [15].

Definition 1. *An Oblivious RAM (with enriched operations) is a suite of interactive protocols between a client and a server, comprising the following:*

ReadAndRemove(u): Given a private input $u \in \mathcal{U}$ which is a block identifier, the client performs an interactive protocol with the server to retrieve a block identified by u, and then remove it from the O-RAM. If u exists in the O-RAM, the content of the block data is returned to the client. Otherwise, $\bot$ is returned.

Add(u, data): The client is given private inputs $u \in \mathcal{U}$ and data $\in \{0, 1\}^B$, representing a block identifier and some data content respectively. *This operation must be immediately preceded by* ReadAndRemove(u) *such that block u no longer resides in the O-RAM.* The client then performs an interactive protocol with the server to write content data to the block identified by u, which is added to the O-RAM.

Definition 2 (Security definition). *Let* $y := ((\mathsf{op}_1, \mathsf{arg}_1), (\mathsf{op}_2, \mathsf{arg}_2), \ldots, (\mathsf{op}_M, \mathsf{arg}_M))$ *denote a data request sequence of length* M. *Each* op_i *denotes a* ReadAndRemove *or an* Add *operation. Moreover, if* op_i *is a* ReadAndRemove *operation, then* $\mathsf{arg}_i = \mathsf{u}_i$, *else if* op_i *is an* Add *operation, then* $\mathsf{arg}_i = (\mathsf{u}_i, \mathsf{data}_i)$, *where* u_i *denotes the identifier of the block being read or added, and* data_i *denotes the data content being written in the second case. Recall that if* op_i *is an* Add *operation with argument* $(\mathsf{u}_i, \mathsf{data}_i)$, *then* op_{i-1} *must be a* ReadAndRemove *operation with argument* $\mathsf{u}_{i-1} = \mathsf{u}_i$.

We use the notation $\mathsf{ops}(y)$ *to denote the sequence of operations associated with* y, *i.e.,* $\mathsf{ops}(y) := (\mathsf{op}_1, \mathsf{op}_2, \ldots, \mathsf{op}_M)$.

Let $A(y)$ *denote the (possibly randomized) sequence of accesses to the remote storage given the sequence of data requests* y. *An O-RAM construction is said to be secure if for any two data request sequences* y *and* z *such that* $|y| = |z|$, *and* $\mathsf{ops}(y) = \mathsf{ops}(z)$, *their access patterns* $A(y)$ *and* $A(z)$ *are computationally indistinguishable by anyone but the client.*

2.2   Relationship with the Standard O-RAM Definition

As mentioned earlier, our modified O-RAM notion is more general than the standard O-RAM notion, in the sense that given a modified O-RAM exporting ReadAndRemove and Add primitives, we can easily implement a standard O-RAM supporting Read and Write operations, as stated in the following observation.

Observation 1. *Given a modified O-RAM as defined above, we can construct a standard O-RAM, where a standard* Read(u) *operation is implemented by the operation* $\mathsf{data} \leftarrow$ ReadAndRemove(u) *followed by* Add(u, data), *and a standard* Write(u, data) *operation is implemented by the operation* $\mathsf{data}_0 \leftarrow$ ReadAndRemove(u) *followed by* Add(u, data) *operation.*

Most existing constructions [6, 7, 16] based on Goldreich and Ostrovsky's hierarchical solution [6] can be easily modified to support the ReadAndRemove and Add primitives.

2.3   Implementing Enriched Semantics

Implementing the Pop operation from the ReadAndRemove and Add primitives.
As mentioned earlier, our O-RAM storage is organized into a binary tree over buckets, where each bucket is a fully functional O-RAM by itself, referred to as a *bucket O-RAM*. For technical reasons which will become clear in Section 3, each bucket O-RAM needs to support not only the ReadAndRemove and Add operations (and hence the standard O-RAM Read and Write operations), but also a special-purpose operation called Pop().

The Pop() operation looks up a real data block and removes it from the O-RAM if one exists. Otherwise, it returns a dummy block $\perp$.

In our online full technical report [14], we present a constructive proof demonstrating that any O-RAM supporting the ReadAndRemove and Add primitives can be modified to support the Pop primitive as well; and the Pop operation costs asymptotically the same as the basic ReadAndRemove and Add primitives. We state this fact in the following lemma.

Lemma 1 (Additional Pop() **operation).** *Given any O-RAM construction of capacity* $3N$ *satisfying Definition 1, one can construct a new O-RAM of capacity* N *that not only provides a* ReadAndRemove(u) *and an* Add(u, data) *primitives (and hence, the standard* Read(u) *and* Write(u, data) *operations), but also provides a* Pop() *operation, where all operation preserve the asymptotic performance of the original O-RAM. Specifically, the* Pop() *operation selects an arbitrary block that currently exists in the O-RAM, reads it back and removes it from the O-RAM. If the O-RAM does not contain any real blocks, the* Pop *operation returns* $\perp$.

2.4 Encryption and Authentication

Similar to prior work in O-RAM [6, 7, 12, 16], we assume that all data blocks are encrypted using a semantically secure encryption scheme, so that two encryptions of the same plaintext cannot be linked. Furthermore, every time a data block is written back it is encrypted again using fresh randomness.

We also assume that the server does not tamper with or modify the data, since authentication and freshness can be achieved using standard techniques such as Message Authentication Codes (MAC), digital signatures, or authenticated data structures.

2.5 Two Simple O-RAM Constructions with Deterministic Guarantees

As mentioned earlier, our O-RAM storage is organized into a binary tree over small data buckets, where each bucket is a fully functional O-RAM by itself, referred to as a bucket O-RAM.

For technical reasons which will become clear in Section 3, we would like each bucket O-RAM to provide *deterministic* (as opposed to high probability) guarantees. Moreover, each bucket O-RAM needs to support *non-contiguous* block identifier space. We consider each block identifier $u \in \{0, 1\}^{\leq B}$, i.e., u can be an arbitrary string, as long as u can be described within one block. Furthermore, the set of block identifiers is unknown in advanced, but rather, determined dynamically during live operations of the bucket O-RAM. As long as the load of the bucket O-RAM never exceeds its capacity, the correct functioning of the bucket O-RAM should be guaranteed.

Below, we present the two candidate bucket O-RAMs constructions, called the trivial O-RAM and the Square-Root O-RAM respectively. They are modifications of the trivial O-RAM and the Square-Root O-RAM constructions originally proposed by Goldreich and Ostrovsky [6].

Trivial O-RAM. We can build a trivial O-RAM supporting non-contiguous block identifier space in the following way. Let N denote the O-RAM capacity. In the trivial O-RAM, the server side has a buffer storing N blocks, where each block is either a real block denoted (u, data), or a dummy block denoted $\perp$.

To perform a ReadAndRemove(u) operation, a client sequentially scans positions 0 through $N - 1$ in the server array: if the current block matches identifier u, the client remembers its content, and overwrites it with $\perp$; if the current block does not match identifier u, the client writes back the original block read.

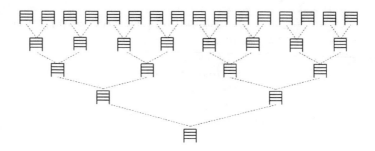

Fig. 1. Server-side storage hierarchy. The server-side O-RAM storage is organized into a binary tree over data buckets, where each bucket can hold up to $O(\log N)$ data blocks. A data block enters from the root bucket when written to the O-RAM, and then obliviously percolates down towards a random leaf over time, until the same block is accessed again.

To perform an $\mathsf{Add}(u, \mathsf{data})$ operation, a client sequentially scans positions 0 through $N - 1$ in the server buffer: the first time the client sees a dummy block, the client overwrites it with (u, data); otherwise, the client writes back the original block read.

As mentioned earlier, whenever blocks are written back to the server, they are re-encrypted in order to hide its contents from the server.

Clearly, the trivial O-RAM is secure, requires $O(N)$ amortized and worst-case cost, $O(N)$ server-side storage, and $O(1)$ client-side storage (since the client never downloads the entire array all at once, but performs the reads and updates in a streaming fashion).

Square-Root O-RAM [6]. Goldreich and Ostrovsky present a Square-Root O-RAM [6] which achieves $O(\sqrt{N} \log N)$ amortized cost, $O(N \log N)$ worst-case cost, $O(N)$ server-side storage, and $O(1)$ client-side storage. When using the deterministic AKS sorting network [1] to implement the reshuffling operation, the Square-Root O-RAM achieves deterministic (as opposed to high probability) guarantees. Although the original Square-Root O-RAM construction supports only contiguous block identifier space, it is not too difficult to modify it to support non-contiguous block identifier space, while preserving the same asymptotic performance. We defer the detailed description of this modified Square-Root O-RAM construction to our online full version [14].

3   Basic Construction

3.1   Overview of the Binary Tree Construction

We first describe a binary-tree based construction, which has two variants. The first variant makes use of the trivial bucket O-RAM and has amortized and worst case cost $O((\log N)^2)$; the second variant makes use of the Square-Root bucket O-RAM and has $\widetilde{O}((\log N)^{1.5})$ amortized cost, and $\widetilde{O}((\log N)^2)$ worst-case cost. Both variants require $\frac{N}{c}$ client-side storage, where $c > 1$ and we assume that the failure probability is $\frac{1}{\mathrm{poly}(N)}$ and the number of operations is $M = \mathrm{poly}(N)$, which is reasonable in practice (for instance $N = 10^6$ and $M = N^3 = 10^{18}$). Later, in Section 4, we describe how to apply our O-RAM construction recursively for the client-side storage, to achieve $O(1)$ client-side memory, while incurring a multiplicative factor of $O(\log N)$ to the amortized and worst-case costs.

As mentioned in Section 1, the motivation for the binary tree construction is to "in spirit" spread across time the reshuffling operations that commonly appear in existing constructions [5, 7, 12, 16]. However, since there is no trivial way to modify existing schemes to spread the reshuffling operation, we introduce a completely new technique based on the binary tree idea.

Server-side storage organization. In our construction, the server-side storage is organized into a binary tree of depth $D := \lceil \log_2 N \rceil$. For ease of explanation, let us assume that N is a power of 2 for the time being. In this way, there are exactly N leaf nodes in the tree.

Each node in the tree is a data bucket, which is a self-contained O-RAM of capacity $O(\log N)$, henceforth referred to as a *bucket O-RAM*. For technical reasons described later, each bucket O-RAM must have the following properties: (a) support non-contiguous identifier space, (b) support ReadAndRemove and Add primitives – from which we can also implement Read, Write, and Pop primitives as mentioned in Section 2, (c) has zero failure probability.[1]

There are two possible candidates for the bucket O-RAM, both of which are modifications of simple O-RAM constructions initially proposed by Goldreich and Ostrovsky [6], and described in more detail in Section 2.5.

1. **Trivial O-RAM.** Every operation is implemented by a sequential scan of all blocks in the server-side storage. For capacity L, the server-side storage is $O(L)$ and the cost of each operation (both amortized and worst-case) is $O(L)$.
2. **Square-Root O-RAM [6].** For capacity L, the Square-Root O-RAM achieves $O(L)$ server-side storage, $O(1)$ client-side storage, $O(\sqrt{L} \log L)$ amortized cost, and $O(L \log L)$ worst-case cost.

O-RAM operations. When data blocks are being written to the O-RAM, they are first added to the bucket at the root of the tree. As more data blocks are being added to a bucket, the bucket's load will increase. To avoid overflowing the capacity of a bucket O-RAM, data blocks residing in any non-leaf bucket are periodically evicted to its children buckets. More specifically, eviction is an oblivious protocol between the client and the server in which the client reads data blocks from selected buckets and writes each block to a child bucket.

Over time, each block will gradually percolate down a path in the tree towards a leaf bucket, until the block is read or written again. Whenever a block is being added to the root bucket, it will be logically assigned to a random leaf bucket, indexed by a string in $\{0,1\}^D$. Henceforth, this data block will gradually percolate down towards the designated leaf bucket, until the same data block is read or written again.

Suppose that at some point, a data block is currently logically assigned to leaf node $\ell \in \{0,1\}^D$. This means that a fresh copy of the data block exists somewhere along the path from the leaf node ℓ to the root. To find that data block, it suffices to search

---

[1] It would also be acceptable if a failure probability δ per operation would only incur a multiplicative factor of $O(\log \log \frac{1}{\delta})$ in the cost.

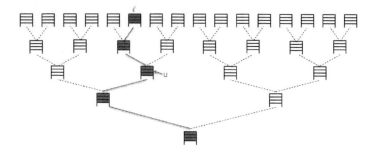

Fig. 2. Searching for a data block. A block u is logically associated with a leaf node ℓ at a given point time. To look up the block u, it suffices to search every bucket on the path from the leaf bucket ℓ to the root bucket (denoted by the shaded buckets in this figure). Every time a block is accessed, it will be logically assigned to a fresh random leaf node.

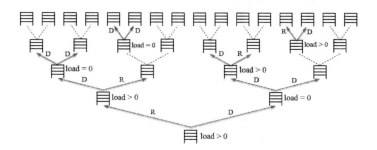

Fig. 3. Background evictions with eviction rate $\nu = 2$. Upon every data access operation, for each depth in the hierarchy, ν number of buckets are chosen randomly for eviction during which one data block (real or dummy) will be evicted to each of its children. If the bucket is loaded, then one real block and one dummy block are evicted. If the bucket is not loaded, two dummy blocks are evicted. In this figure, D denotes the eviction of a dummy block, and R denotes the eviction of a real block.

the data block in all buckets on the path from the designated leaf node to the root. We assume that when the data block is stored in a bucket, we store the tag ℓ along as well and we denote the block's contents by $(\mathsf{data}\|\ell)$.

Ensuring security. For security reasons, it is important to ensure the following:

- *Every time a block is accessed, its designated leaf node must be chosen independently at random.* This is necessary to ensure that two operations on the same data block are completely unlinkable.
- *The bucket sequence accessed during eviction process must reveal no information about the load of each bucket, or the data access sequence.* In our construction, the choice of which buckets to evict from is randomly selected, and independent from the load of the bucket, or the data access sequence. Furthermore, whenever a bucket is selected for eviction, we always write to both of its children – depending

on whether there are real blocks to evict, we would write a real or a dummy block to each of its children.

Client-side index. As each data block will be logically assigned to a random leaf node every time it is operated on, we need some data structure to remember where each block might be at any point of time. For this reason, the client stores a data structure of size $\frac{N \log N}{B}$ blocks, in which it records which leaf node is currently associated with each block. When $B \geq c \log N$, this index structure's size is a linear fraction of the capacity of the O-RAM. Therefore, in the basic scheme, we require $\frac{N}{c}$ client-side storage, where $c > 1$.

However, later in the recursive construction described in Section 4, we show how to apply our O-RAM construction recursively over the index structure to achieve $O(1)$ client-side storage.

A note about dummy blocks and dummy operations. To ensure the security of the O-RAM, in our construction, we often rely on *dummy blocks* and *dummy operations* to hide certain information from the untrusted server, such as whether a bucket is loaded, and where in the tree a block is headed.

For the purpose of this section, we adopt the following notion of dummy blocks and dummy operations. We will think of the dummy block as a regular but useless data block. We can dedicate a certain block identifier, e.g., $u = 0$ to serve as the dummy block. In this way, we simply deduct 1 from the O-RAM capacity, which does not affect the asymptotics. In our construction, every bucket may have a dummy block; while each real data block exists in at most one bucket.

Given the above notion of the dummy block, we can define a dummy O-RAM operation as a regular operation on the dedicated dummy block with $u = 0$. A dummy O-RAM operation serves no purpose other than ensuring the security of the O-RAM. Henceforth, with a slight abuse of notation, we use the symbol $\perp$ to denote a dummy data block or its identifier. We use the notations ReadAndRemove($\perp$), Add($\perp$), Read($\perp$) and Write($\perp$) to denote dummy O-RAM operations.

3.2   Detailed Construction

We define some notations in Table 2 which will be useful in the formal algorithm descriptions.

ReadAndRemove **operation.** The algorithm for performing a ReadAndRemove(u) operation is described in Figure 4. First, the client looks up its local index structure index to find out which leaf node ℓ the requested block u is associated with. We then generate a fresh random ℓ^* from $\{0, 1\}^D$ and overwrite index[u] $\leftarrow \ell^*$, i.e., block u is henceforth associated with a fresh random leaf node ℓ^*. Notice that this ensures no linkability between two operations on the same data block. In order to avoid extra index lookup for any following Add operation, ℓ^* is also stored in a global variable state.

Now, given that u is currently associated with leaf node ℓ, it means that a fresh copy of block u must reside in some bucket along the along the path from leaf ℓ to the root, denoted by $\mathcal{P}(\ell)$. If u is found in some bucket, we remove u from that bucket, and remember its the data content. Regardless of whether u has been found, we always

Table 2. Notations

D	$\lceil \log_2 N \rceil$
$u \in \{0, 1, \ldots, N-1\}$	global identifier of a block
index	client's index structure
$\text{index}[u] \in \{0, 1\}^D$	id of leaf node associated with block u, initially random
state	global variable to avoid unnecessary index lookup
root	root bucket of the binary tree
$\mathcal{P}(\ell)$	path from the leaf node ℓ to the root
$\text{Child}_b(\text{bucket})$, for $b \in \{0, 1\}$	the left or right child of a bucket
ν	eviction rate
$\text{UniformRandom}(S)$	Samples an element uniformly at random from the set S
$\text{UniformRandom}_\nu(S)$	Samples a subset of size ν uniformly at random from the set S
$\perp$	a dummy block or the identifier of a dummy block

continue our search all the way to the root. Note that to ensure obliviousness, it is important that the search does not abort prematurely even after finding block u. Finally, if the requested block u has been found, the ReadAndRemove algorithm returns its data contents; otherwise, the ReadAndRemove algorithm returns $\perp$.

Add operation. Also shown in Figure 4, the Add(u, data) operation reads the tag ℓ from state, which was just generated by the preceding ReadAndRemove(u) operation. The client writes the intended block $(u, \text{data}||\ell)$ to the root bucket.

Notice that here the client tags the data with ℓ, i.e., the id of the leaf node that block u would be logically associated with until the next operation on block u. The designated leaf node tag will become important when we recursively apply our O-RAM over the client's index structure, as described in Section 4. Specifically, the eviction algorithm will examine this designated leaf node tag to determine to which child node to evict this block. Observe that to preserve the desired asymptotics in the recursive construction, the eviction algorithm cannot afford to (recursively) look up the index structure to find the designated leaf node for a block. By tagging the data with its designated leaf, the eviction algorithm need not perform recursive lookups to the index structure.

Finally, at the end of every Add operation, the client invokes the background eviction process once. We now describe the background eviction algorithm.

Background evictions. Let ν denote the eviction rate. For the purpose of our asymptotic analysis, it suffices to let $\nu = 2$.

Whenever the background eviction algorithm is invoked, the client randomly selects ν buckets to evict at every depth of the tree.

If a bucket is selected for eviction, the client pops a block from the bucket O-RAM by calling the Pop operation (see Section 2.3 for how to implement the Pop operation given an O-RAM that supports ReadAndRemove and Write operations). If the bucket selected for eviction is loaded, then the Pop operation returns a real block and removes that block from the bucket O-RAM; otherwise, if the bucket is not loaded, the Pop operation returns a dummy block $\perp$.

Regardless of whether a real block or a dummy block is returned by the Pop operation, the client always performs a write to both children of the selected bucket:

ReadAndRemove(u):
1: $\ell^* \leftarrow$ UniformRandom($\{0, 1\}^D$)
2: $\ell \leftarrow$ index[u], index[u] $\leftarrow \ell^*$
3: state $\leftarrow \ell^*$ *//If an* Add *operation follows,* ℓ^* *will be used by* Add
4: data $\leftarrow \perp$
5: **for** each bucket on $\mathcal{P}(\ell)$ **do**   *//path from leaf ℓ to root*
6:    **if** $((\text{data}_0 || \ell_0) \leftarrow$ bucket.ReadAndRemove(u)$) \neq \perp$ **then**
7:       data $\leftarrow$ data$_0$ *//Notice that $\ell = \ell_0$*
8:    **end if**
9: **end for**
10: **return** data

Add(u, data):
1: $\ell \leftarrow$ state
2: root.Write(u, data$||\ell$)   *// Root bucket's O-RAM Write operation*
3: Call Evict(ν)
4: **return** data

Fig. 4. Algorithms for data access

1. If a dummy block is returned by Pop, the client simply performs a dummy write to
   both children buckets.
2. If a real block is returned, the client examines its designated leaf node tag to figure
   out the correct child node to evict this block to. Recall that this designated leaf node
   tag is added when the block is first written to the root bucket. (Note that although
   in the basic construction, the client can alternatively find out this information by
   looking up its local index structure; later in the recursive construction, the client
   will have to obtain this information through the designated leaf node tag.)
   Now, suppose that the block should be evicted to child $b \in \{0, 1\}$ of the selected
   bucket, the client then writes the block to child b, and writes a dummy block to
   child $1 - b$.

Regardless of which case, to ensure obliviousness, the two writes to the children nodes
must proceed in a predetermined order, e.g., first write a real or dummy block to child
0, and then write a real or dummy block to child 1.

3.3 Security Analysis

Theorem 1 (Security of Basic Construction). *Our Basic O-RAM Construction is secure in the sense of Definition 2, assuming that each bucket O-RAM is also secure.*

Proof. Observe that each bucket is itself a secure O-RAM. Hence, it suffices to show
that each type of operation induces independently the same distribution on the access
patterns of the buckets in the binary tree, regardless of the arguments.

For the ReadAndRemove(u) operation, the buckets along the path $\mathcal{P}(\ell)$ from the
root to the leaf indexed by $\ell =$ index(u) are accessed. Observe that ℓ is generated

```
Evict(ν):
 1: for d = 0 to D − 1 do
 2:     Let S denote the set of all buckets at depth d.
 3:     A ← UniformRandomν(S)
 4:     for each bucket ∈ A do
 5:         (u, data||ℓ) ← bucket.Pop()
 6:         b ← (d+1)-st bit of ℓ
 7:         blockb ← (u, data||ℓ),  block1−b ← ⊥
 8:         ∀b ∈ {0, 1} : Childb(bucket).Write(blockb)
 9:     end for
10: end for
```

Fig. 5. Background eviction algorithm with eviction rate ν

uniformly at random from $\{0, 1\}^D$. Hence, the distribution of buckets accessed is the buckets along the path to a random leaf. Moreover, each time ReadAndRemove(u) is called, a fresh random ℓ^* is generated to be stored in index(u) so that the next invocation of ReadAndRemove(u) will induce an independent random path of buckets.

For the Add(u, data) operation, the root bucket is always accessed. More buckets are accessed in the Evict subroutine. However, observe that the access pattern of the buckets are independent of the configuration of the data structure, namely two random buckets at each depth (other than the leaves) are chosen for eviction, followed by accesses to both child buckets.

3.4   Asymptotic Performance of the Basic Construction

We next analyze the server-side storage and the cost of each operation. If the capacity of each bucket is L, the server-side storage is $O(NL)$, because there are $O(N)$ buckets. If we use the trivial bucket O-RAM, each operation has cost $O(L \log N)$. If we use the Square-Root bucket O-RAM, each operation has amortized cost $O(\sqrt{L} \log L \log N)$ and worst case cost $O(L \log L \log N)$.

We prove the following lemma in Appendix A.

Lemma 2 (Each Bucket Has Small Load). *Let* $0 < \delta < \frac{1}{2^{2e}}$. *For a fixed time and a fixed bucket, the probability that the bucket has load more than* $\log_2 \frac{1}{\delta}$ *is at most* δ.

Applying Union Bound on Lemma 2 over all buckets and over all time steps, we have the following result.

Lemma 3 (Bucket Overflow). *Suppose* $0 < \delta < 1$ *and* $N, M \geq 10$. *Then, one can use bucket O-RAM with capacity* $O(\log \frac{MN}{\delta})$ *such that with probability at least* $1 - \delta$, *the Basic O-RAM Construction can support* M *operations without any bucket overflow.*

Lemma 3 gives an upper bound on the capacity of each bucket and from the above discussion, we have the following result.

Corollary 1. *The Basic O-RAM Construction can support* M *operations with failure probability at most* δ *using* $O(N \log \frac{MN}{\delta})$ *server-side storage and* $O(\frac{N \log N}{B})$ *client-side storage. The cost of each operation is as follows:*

Bucket O-RAM	Amortized	Worst-case
Trivial	$O(\log N \log \frac{MN}{\delta})$	$O(\log N \log \frac{MN}{\delta})$
Square-Root	$O(\log N \sqrt{\log \frac{MN}{\delta}} \log \log \frac{MN}{\delta})$	$O(\log N \log \frac{MN}{\delta} \log \log \frac{MN}{\delta})$

Specifically, if the number of data access requests $M = poly(N)$, then the basic construction with the trivial bucket O-RAM achieves $O((\log N)^2)$ amortized and worst-case cost; and the basic construction with the Square-Root bucket O-RAM achieves $\widetilde{O}((\log N))^{1.5}$ amortized cost, and $\widetilde{O}((\log N)^2)$ worst-case cost. Furthermore, no buckets will overflow with probability $1 - \frac{1}{poly(N)}$.

4 Recursive Construction and How to Achieve the Desired Asymptotics

The basic construction described in Section 3 achieves poly-logarithmic amortized and worst-case cost, but requires $\frac{N}{c}$ client-side storage, where $c = \frac{B}{\log N} > 1$.

In this section, we demonstrate how to recursively apply our O-RAM construction to the client's index structure to achieve $O(1)$ client-side storage, while incurring an $O(\log N)$ multiplicative factor in terms of the amortized and worst-case cost.

4.1 Recursive O-RAM Construction: $O(1)$ Client-Side Storage

Storing the index through recursion. In the basic construction, the client's index structure takes up at most $\frac{N \log N}{B} \leq \frac{N}{c}$ space, where $B \geq c \log N$. To achieve $O(1)$ client-side storage, we recursively apply our O-RAM over the index structure. Instead of storing the index structure on the client, we store the index structure in a separate O-RAM on the server side. At each step of the recursion, we effectively compress the O-RAM capacity by a factor of $c > 1$. Therefore, after $\log_c N$ levels of recursion, the index structure will be reduced to constant size.

To see how the recursion can be achieved, notice that Line 2 of the ReadAndRemove algorithm in Figure 4 can be replaced with a recursive O-RAM operation:

$$\text{O-RAM.Write}(\text{block\_id}(\text{index}[u]), \ell^*)$$

Here we have a slight abuse of notation, because in reality, the entry index[u] (stored sequentially according to u) resides in a larger block identified by block_id(index[u]), and one would have to first read that block, update the corresponding entry with ℓ^*, and then write the updated block back.

Theorem 2 (Recursive O-RAM Construction). *The Recursive O-RAM Construction can support M operations with failure probability at most δ using $O(N \log \frac{MN}{\delta})$ server-side storage and $O(1)$ client-side storage, and the cost of each operation is as follows:*

Bucket ORAM	Amortized	Worst-case
Trivial	$O(\log_c N \log N \log \frac{MN}{\delta})$	$O(\log_c N \log N \log \frac{MN}{\delta})$
Square-Root	$O(\log_c N \log N \sqrt{\log \frac{MN}{\delta}} \log \log \frac{MN}{\delta})$	$O(\log_c N \log N \log \frac{MN}{\delta} \log \log \frac{MN}{\delta})$

Specifically, if the number of data access requests $M = poly(N)$, then the recursive construction with the trivial bucket O-RAM achieves $O((\log N)^3)$ amortized and worst-case cost; and the recursive construction with the Square-Root bucket O-RAM achieves $\widetilde{O}((\log N))^{2.5}$ amortized cost, and $\widetilde{O}((\log N)^3)$ worst-case cost. Furthermore, no buckets will overflow with probability $1 - \frac{1}{poly(N)}$.

Proof. The $O(1)$ client-side storage is immediate, due to the fact that all client-side storage (including the state variable in Figure 4, and the shuffling buffer for the Square-Root bucket O-RAM) is transient state rather than persistent state, and therefore, all levels of recursion can share the same $O(1)$ client-side storage.

Observe that for each $j = 0, 1, \ldots, \lceil \log_c N \rceil$, the jth recursion produces a binary tree with $O(\frac{N}{c^j})$ buckets. Hence, there are totally $O(\sum_{j \geq 0} \frac{N}{c^j}) = O(N)$ buckets.

Recall that by Theorem 3, for each bucket and at the end of each operation, with probability at least η, the load of the bucket is at most $\log_2 \frac{1}{\eta}$. Since there are $O(N)$ buckets and M operations, we need to set $\eta = \Theta(\frac{\delta}{NM})$ to apply the Union Bound such that the overall failure probability (due to bucket overflow) is at most δ. It follows that the capacity of each bucket is $L = O(\log \frac{MN}{\delta})$. and hence the server-side storage is $O(NL) = O(N \log \frac{MN}{\delta})$.

Moreover, each operation on the Recursive O-RAM induces $O(\log \frac{N}{c^j})$ operations on the bucket O-RAMs in the jth binary tree. Hence, the total number of bucket O-RAM accesses is $Z = O(\sum_{j \geq 0} \log \frac{N}{c^j}) = O(\log_c N \log N)$.

If we use the trivial bucket O-RAM, each operation has cost $O(ZL)$.

If we use the Square-Root bucket O-RAM, the amortized cost is $O(Z\sqrt{L} \log L)$ and the worst-case cost is $O(ZL \log L)$, as required.

Remark 1. Observe that the BST O-RAM construction by Damgård, Meldgaard, and Nielsen [4] for capacity L has client storage $O(1)$, server storage $O(L \log L)$, amortized cost $O((\log L)^a)$ and worst-case cost $O((\log L)^b)$, where a and b are small integers. Hence, if we use the BST construction for out bucket O-RAM, the amortized cost of our binary scheme can be improved to $O(\log_c N \log N (\log \frac{MN}{\delta})^a) = \widetilde{O}((\log N)^2)$ and the worst-case cost to $O(\log_c N \log N \log \frac{MN}{\delta} (\log \log \frac{MN}{\delta})^b) = \widetilde{O}((\log N)^3)$, where $M = poly(N)$ and $\delta = \frac{1}{poly(N)}$, while the server storage cost is $\widetilde{O}(N \log N)$.

References

1. Ajtai, M., Komlós, J., Szemerédi, E.: Sorting in c log n parallel steps. Combinatorica 3, 1–19 (1983)
2. Beame, P., Machmouchi, W.: Making rams oblivious requires superlogarithmic overhead. Electronic Colloquium on Computational Complexity (ECCC) 17, 104 (2010)
3. Boneh, D., Mazieres, D., Popa, R.A.: Remote oblivious storage: Making oblivious ram practical (2011) (manuscript),
   http://dspace.mit.edu/bitstream/handle/1721.1/
   62006/MIT-CSAIL-TR-2011-018.pdf

4. Damgård, I., Meldgaard, S., Nielsen, J.B.: Perfectly Secure Oblivious RAM Without Random Oracles. In: Ishai, Y. (ed.) TCC 2011. LNCS, vol. 6597, pp. 144–163. Springer, Heidelberg (2011)
5. Goldreich, O.: Towards a theory of software protection and simulation by oblivious rams. In: STOC (1987)
6. Goldreich, O., Ostrovsky, R.: Software protection and simulation on oblivious rams. J. ACM (1996)
7. Goodrich, M.T., Mitzenmacher, M.: Mapreduce parallel cuckoo hashing and oblivious ram simulations. CoRR, abs/1007.1259 (2010)
8. Goodrich, M.T., Mitzenmacher, M., Ohrimenko, O., Tamassia, R.: Oblivious ram simulation with efficient worst-case access overhead. In: CCSW (2011)
9. Hsu, J., Burke, P.: Behavior of tandem buffers with geometric input and markovian output. IEEE Transactions on Communications 24, 358–361 (1976)
10. Kushilevitz, E., Lu, S., Ostrovsky, R.: On the (in)security of hash-based oblivious ram and a new balancing scheme, http://eprint.iacr.org/2011/327.pdf
11. Ostrovsky, R.: Efficient computation on oblivious rams. In: STOC (1990)
12. Pinkas, B., Reinman, T.: Oblivious RAM Revisited. In: Rabin, T. (ed.) CRYPTO 2010. LNCS, vol. 6223, pp. 502–519. Springer, Heidelberg (2010)
13. Raab, M., Steger, A.: "Balls into Bins" - A Simple and Tight Analysis. In: Rolim, J.D.P., Serna, M., Luby, M. (eds.) RANDOM 1998. LNCS, vol. 1518, pp. 159–170. Springer, Heidelberg (1998)
14. Shi, E., Chan, H., Stefanov, E., Li, M.: Oblivious ram with o((logn)³) worst-case cost. Online TR (2011), http://eprint.iacr.org/2011/407.pdf
15. Stefanov, E., Shi, E., Song, D.: Towards practical oblivious ram (2011) (manuscript)
16. Williams, P., Sion, R.: Usable PIR. In: NDSS (2008)
17. Williams, P., Sion, R., Carbunar, B.: Building castles out of mud: practical access pattern privacy and correctness on untrusted storage. In: CCS (2008)

Appendices

A   Bounding the Load of Each Bucket

In this section, we prove the following high probability statement for bounding the load in each bucket.

Theorem 3 (Each Bucket Has Small Load). *Let* $0 < \delta < \frac{1}{2^{2c}}$. *For a fixed time and a fixed bucket, the probability that the bucket has load more than* $\log_2 \frac{1}{\delta}$ *is at most* δ.

Recall that the number of levels is $L := \lceil \log_2 N \rceil$. We analyze the load according to the depth i of the bucket.

A.1   Bounding the Load for Levels 0 to $L - 1$ with Markov Process

Observe that in our scheme, when a block inside some bucket is accessed, the block is removed from the bucket. However, for the purpose of analysis, we assume that a block stays inside its bucket when it is accessed, i.e., a block can leave a bucket only when the bucket is chosen for eviction; moreover, since we are only concerned about the load of a bucket, for simplicity we also assume that the blocks arriving at a bucket are all distinct.

The load of a bucket in our scheme is always bounded above by the corresponding load in the modified process, which we analyze using a Markov process. If we assume that a bucket is initially empty, then its load will be stochastically dominated by the load under the stationary distribution.

Defining Markov Process $\mathcal{Q}(\alpha, \beta)$. Given $0 < \alpha \leq \beta \leq 1$, we describe a Markov process $\mathcal{Q}(\alpha, \beta)$ with non-negative integral states as follows. In order to illustrate the relationship between the Markov process and the load of a bucket, we define $\mathcal{Q}(\alpha, \beta)$ using the terminology related to the bucket. The state of the Markov process corresponds to the current load of a bucket. At any time step, the following happens independently of any past events in the specified order:

(a) With probability α, a block arrives at the bucket.
(b) If the load of the bucket is non-zero (maybe because a block has just arrived), then with probability β a block departs from the bucket.

Recall that when a block departs from a depth-i bucket, it arrives at one of the two depth-$(i + 1)$ child buckets uniformly at random.

Example. We immediately see that the root bucket is modeled by $\mathcal{Q}(1, 1)$ and a depth-1 bucket is modeled by $\mathcal{Q}(\frac{1}{2}, 1)$. Both cases are trivial because the load at the end of every time step is zero. One can see that at every time step a block arrives at one of the four depth-2 buckets uniformly at random and two out of the four buckets are chosen for eviction every step. Hence, each of the depth-2 buckets can be modeled by $\mathcal{Q}(\frac{1}{4}, \frac{1}{2})$. Using a classic queuing theory result by Hsu and Burke [9] we can show that at further depths, a block leaves a bucket with some fixed probability at every time step, so that independent arrivals are satisfied at the child buckets.

Corollary 2 (Load of an Internal Bucket). *For $2 \leq i < L$, under the stationary distribution, the probability that a depth-i bucket has load at least s is at most $\rho_i^s \leq \frac{1}{2^s}$; in particular, for $0 < \delta < 1$, with probability at least $1 - \delta$, its load is at most $\log_2 \frac{1}{\delta}$.*

Proof. The proof builds on top of a classic queuing theory result by Hsu and Burke [9]. Full proof is provide in our online technical report [14]. □

A.2 Bounding the Load of Level L with "Balls into Bins"

Observe that a block residing at a depth-L bucket traversed a random path from the root bucket to a random leaf bucket. Hence, given that a block is at depth L, the block is in one of the leaf buckets uniformly at random. Hence, to give an upper bound on the load of a leaf bucket at any single time step, we can imagine that each of the N blocks is placed independently in one of the leaf buckets uniformly at random. This can be analyzed by the well-known "Balls into Bins" process.

Corollary 3 (Load of a Leaf Bucket). *For each time step, for $0 < \delta < \frac{1}{2^{2c}}$, with probability at least $1 - \delta$, a leaf bucket has load at most $\log_2 \frac{1}{\delta}$.*

Proof. Using standard balls and bins analysis [13]. Full proof will be supplied in online technical report [14]. □

Noiseless Database Privacy

Raghav Bhaskar[1], Abhishek Bhowmick[2], Vipul Goyal[1],
Srivatsan Laxman[1], and Abhradeep Thakurta[3]

[1] Microsoft Research India
{rbhaskar,vipul,slaxman}@microsoft.com
[2] University of Texas, Austin
bhowmick@cs.utexas.edu
[3] Pennsylvania State University
azg161@cse.psu.edu

Abstract. Differential Privacy (DP) has emerged as a formal, flexible framework for privacy protection, with a guarantee that is agnostic to auxiliary information and that admits simple rules for composition. Benefits notwithstanding, a major drawback of DP is that it provides noisy[1] responses to queries, making it unsuitable for many applications. We propose a new notion called Noiseless Privacy that provides exact answers to queries, without adding any noise whatsoever. While the form of our guarantee is similar to DP, where the privacy comes from is very different, based on statistical assumptions on the data and on restrictions to the auxiliary information available to the adversary. We present a first set of results for Noiseless Privacy of arbitrary Boolean-function queries and of linear Real-function queries, when data are drawn independently, from nearly-uniform and Gaussian distributions respectively. We also derive simple rules for composition under models of dynamically changing data.

1 Introduction

Developing a mathematically sound notion of privacy is a difficult problem. Several definitions for database privacy have been proposed over the years, many of which were subsequently broken. For example, methods like k-anonymity [Swe02] and ℓ-diversity [MGKV06] are vulnerable to simple, practical attacks that can breach privacy of individual records [GKS08]. In 2006, Dwork *et al.* [DMNS06] made significant strides toward formal specification of privacy guarantees by introducing an information-theoretic notion called *Differential Privacy* (DP). For a detailed survey on DP see [Dwo08].

Definition 1 (ϵ-Differential Privacy [DMNS06]). *A randomized algorithm $\mathcal{A}$ is ϵ-differentially private if for all databases $T, T' \in \mathcal{D}^n$ differing in at most one record and all events $\mathcal{O} \subseteq Range(\mathcal{A})$, $\Pr[\mathcal{A}(T) \in \mathcal{O}] \leq e^\epsilon \Pr[\mathcal{A}(T') \in \mathcal{O}]$.*

---

[1] By noise we broadly refer to any external randomization introduced in the output by the privacy mechanism.

D.H. Lee and X. Wang (Eds.): ASIACRYPT 2011, LNCS 7073, pp. 215–232, 2011.

DP provides a flexible framework for privacy protection based on mechanisms that provide noisy responses to the database queries. The amount of noise introduced in the query-response is: 1) Independent of the actual data entries, 2) Based on the sensitivity of the query to "arbitrary" change of a small number of entries in the data, and 3) Agnostic to the auxiliary information available to the adversary. Their benefits notwithstanding, these properties of DP also result in high levels of noise in the DP output, oftentimes leading to unusable query responses [MKA+08]. Several applications, in fact, completely breakdown when even the slightest amount of noise is added to the output (For example, during a *financial audit*, noisy query-responses may reveal inconsistencies that may be wrongly interpreted as fraud). Besides, when transitioning from a noise-free regime, to incorporate privacy guarantees, the query-response mechanism must be re-programmed (to inject a calibrated amount of noise) and the mechanism *consuming* the DP output must be re-analyzed for its utility/effectiveness (since it must now operate on noisy, rather than exact, query-responses). Hence, the addition of noise to query-responses in the DP framework can be a major barrier to the adoption of DP in practice. Moreover, it is unclear if the DP guarantee (or for that matter, if *any* privacy guarantee) can provide meaningful privacy protection when the adversary has access to arbitrary auxiliary information. On the positive side, however, the structure of the DP guarantee makes it easy to derive simple rules of composition under multiple queries.

Noiseless Privacy: In this paper, we propose a new, also information-theoretic, notion of privacy called *Noiseless Privacy* that provides *exact* answers to database queries, without adding any noise whatsoever. While the form of our guarantee is similar to DP, where the privacy comes from is very different, and is based on: 1) A statistical (generative) model assumption for the database, 2) Restrictions on the kinds of auxiliary information available to the adversary. Both these assumptions are reasonable in many real-world settings; the former is, e.g., commonly used in machine learning, while the latter is natural when data is collected from a diverse network/collection of sources (e.g., from users of the world-wide web).

Consider an entry t_i in the database and two possible values a and b which it can take. Noiseless Privacy simply requires that the probability of the output (or the vector of outputs in-case of multiple queries) lying in a certain measurable set remains similar whether t_i takes value a or b. Here, the probability is taken over the choice of the database (coming from a certain distribution) and is conditioned on the auxiliary information (present with the adversary) about the database. See Definition 2 for formal details.

While the DP framework makes no assumptions about the data distribution or the auxiliary information available to the adversary, it requires the addition of external noise to query-responses. By contrast, in Noiseless Privacy, we study the privacy implications of providing noise-free responses to queries, but under assumptions governing the data distribution and limited auxiliary information.

At this point, we do not know how widely our privacy framework will be applicable in real systems. However, whenever privacy can be obtained in our framework (and our work shows there are significant non-trivial cases where Noiseless Privacy can be achieved) it comes for "free." Another practical benefit is that no changes are needed in the query-response or response-consumption mechanisms, only an analysis to "okay the system" to establish the necessary privacy guarantees is required. Moving forward, we believe that checking the feasibility of Noiseless Privacy is a useful first-step when designing privacy-preserving systems. Only when sufficient intrinsic entropy in the data cannot be established, do we need external noise-injection in the query-responses. This way, we would *pay for privacy only when strictly necessary.*

Our Results: In this work, we study certain types of boolean and real queries and show natural (and well understood) conditions under which Noiseless Privacy can be obtained with good parameters. We first focus on the (single) boolean query setting; i.e., the entries of the database as well as the query output have one bit of information each, with no auxiliary information available to the adversary. Our starting assumption is that each bit of the database is independently drawn from the uniform distribution (this assumption can be partially relaxed; see Section 3). We show that functions which are sufficiently "far" away from both 0-junta and 1-junta functions[2] satisfy Noiseless Privacy with "good" parameters. Note that functions which are close to either 0-junta or 1-junta do not represent an "aggregate statistic" of the database (which should depend on a large number of database entries). Hence, in real systems releasing some aggregate information about the database, we do expect such a condition to be naturally satisfied. Our proof of this theorem is rather intuitive and interestingly shows that these two (well understood) characteristics of the boolean functions are the only ones on which the privacy parameter depends. We extend our result to the case when the adversary has auxiliary information about some records in the database.

For functions over the reals with real outputs, we study two types of functions: (a) linear functions (i.e., where the output is a linear combination of the rows of the database), and, (b) sum of arbitrary functions of the database rows. These functions together cover a large class of aggregation functions that can support various data mining and machine learning tasks in the real-world. We show natural conditions on the database distribution for which Noiseless Privacy can be obtained with good parameters, even when the adversary has auxiliary information about some constant fraction of the dataset. We refer the reader to section 4.1 for more details.

Multiple Queries: The above results are for the case where the adversary is allowed to ask a single query, except for the case of linear real queries, where we have a result for multiple queries. In general, achieving composition in the Noiseless Privacy framework is tricky and privacy can completely breakdown even given a response to two different (carefully crafted) queries. The reason why such a composition is difficult to obtain in our setting is the lack of independence between the responses to the queries; the queries operate on the same database

---

[2] Roughly, an *i*-junta function is one which depends only upon *i* of the total input variables.

and might have complex interdependence on each other to enable an entry of the database to be deduced fully given the responses.

To break such interdependence in our setting, we introduce what we call the changing database model; we assume that between any two queries, a nontrivial fraction of the database has been "refreshed". The newly added entries (which may either replace some existing entries or be in addition to the existing entries) are independent of the old entries already present in the database. This helps us maintain some weak independence between different queries. We note that the setting of the changing database model is not unrealistic. Consider an organization that participates in a yearly industry-wide salary survey, where each organization submits relevant statistics about the salaries of its employees to some market research firms. A key requirement in such surveys is to maintain anonymity of its employees (and only give salary statistics based on the department, years of experience, etc.). A reasonable assumption in this setting is that a constant fraction of the employees will change every year (i.e., if the attrition rate of a firm is five percent, then roughly five percent of the entries can be expected to be refreshed every year). Apart from the above example, there are various other scenarios where the changing database model is realistic (i.e., when one is dealing with streaming data, data with a time window, etc.). Under such changing database model, we provide generalizations of our boolean as well as real query theorems to the case of multiple queries.

We also present other interesting results like obtaining Noiseless Privacy for symmetric boolean functions, "decomposable" functions, etc. In some cases, we in fact show positive results for Noiseless Privacy under multiple queries even in the *static database* model.

Future Work: Our works opens up an interesting direction for research in the area of database privacy. An obvious line to pursue is to expand the classes of functions and data distributions for which Noiseless Privacy can be achieved. Relaxing the independence assumption that our current results make on database records is another important topic. There is also scope to explore alternative ways of specifying the auxiliary information available to the adversary. In general, we believe that developing new techniques for analyzing statistical queries for Noiseless Privacy is an important direction of privacy research, that must go hand-in-hand with efforts toward new, more clever ways of adding smaller amounts of noise to achieve Differential Privacy.

Related Works: The line of works most related to ours is that of *query auditing* (see [KMN05] and [NMK+06]) where, given a database $T = \langle t_1, \cdots, t_n \rangle$ with real entries, a *query auditor* makes a decision as to whether or not a particular query can be answered. If the auditor decides to answer the query, then the answer is output without adding any noise. Since the decision of whether to answer a query can itself leak information about the database, the decision is randomized. This randomization can be viewed as injection of some form of noise into the query response. However, on the positive side, if a decision is made to answer the query, the answer never contains any noise, which is in harmony with the motivation of our present work. See our full version [BBG+11] for a more detailed comparison of our work to this and other related works.

2   Our Privacy Notion

In our present work, we investigate the possibility of guaranteeing privacy without adding any external noise. The main idea is to look for (and systematically categorize) query functions which under certain assumptions on the data generating distribution are inherently private (under our formal notion of privacy that we define shortly). Since, the output of the function itself is inherently private, there is no need to inject external noise. As a result the output of the function has no utility degradation. Formally, we define our new notion of privacy (called *Noiseless Privacy*) as follows:

Definition 2 (ϵ-Noiseless Privacy). *Let $\mathcal{D}$ be the domain from which the entries of the database are drawn. A deterministic query function $f : \mathcal{D}^n \to \mathcal{Y}$ is ϵ-noiseless private under a distribution D on $\mathcal{D}^n$ and some auxiliary information $\mathcal{A}ux$ (which the adversary might have), if for all measurable sets $\mathcal{O} \subseteq \mathcal{Y}$, for all $\ell \in [n]$ and for all $a, a' \in \mathcal{D}$,*

$$\Pr_{T \sim D}[f(T) \in \mathcal{O}|t_\ell = a, \mathcal{A}ux] \leq e^\epsilon \Pr_{T \sim D}[f(T) \in \mathcal{O}|t_\ell = a', \mathcal{A}ux]$$

where t_ℓ is the ℓ-th entry of the database T.

In comparison to Definition 1, the present definition differs at least in the following aspects, namely:

- unlike in Definition 1, it is possible for a non-trivial deterministic function f to satisfy Definition 2 with reasonable ϵ. For *e.g.*, XOR of all the bits of a boolean database (where each entry of the database is an unbiased random bit) satisfies Definition 2 with $\epsilon = 0$ where as Definition 1 is not satisfied for any finite ϵ.
- the privacy guarantee of Definition 2 is under a specific distribution D, where as Definition 1 is agnostic to any distributional assumption on the database.
- the privacy guarantee of Definition 2 is w.r.t. an auxiliary information $\mathcal{A}ux$ whereas differential privacy is oblivious to auxiliary information.

Intuitively, the above definition captures the change in adversary's belief about a particular output in the range of f in the presence or absence of a particular entry in the database. A comparable (and seemingly more direct) notion is to capture the change in adversary's belief about a particular entry before and after seeing the output. Formally,

Definition 3 (ϵ-Aposteriori Noiseless Privacy). *A deterministic query function $f : \mathcal{D}^n \to \mathcal{Y}$ is ϵ-Aposteriori Noiseless Private under a distribution D on $\mathcal{D}^n$ and some auxiliary information $\mathcal{A}ux$, if for all measurable sets $\mathcal{O} \subseteq \mathcal{Y}$, for all $\ell \in [n]$ and for all $a \in \mathcal{D}$,*

$$e^{-\epsilon} \leq \frac{\Pr_{T \sim D}[t_\ell = a|f(T) \in \mathcal{O}, \mathcal{A}ux]}{\Pr_{T \sim D}[t_\ell = a|\mathcal{A}ux]} \leq e^\epsilon$$

where t_ℓ is the ℓ-th entry of the database T.

The following fact shows that Definition 3 implies Definition 2 and vice versa with at most two times degradation in the privacy parameter ϵ. See the full version [BBG+11] for the proof.

Fact 1. *A query function f satisfies Definition 3 under a database generating distribution D and auxiliary information $\mathcal{A}ux$, if and only if it satisfies Definition 2 under the same distribution D and same auxiliary information $\mathcal{A}ux$. There is a possible deterioration of the privacy parameter ϵ by at most a factor of two in either direction.*

Hereafter, we will use Definition 2 as our defintion of Noiseless Privacy. We also introduce a relaxed notion of Noiseless Privacy called (ϵ, δ)-Noiseless Privacy, where with a small probability δ the ϵ-Noiseless Privacy does not hold. Here, the probability is taken over the choice of the database and the two possible values for the database entry in question. While for a strong privacy guarantee a negligible δ is desirable, a non-negligible δ may be tolerable in certain applications. The following definition captures this notion formally.

Definition 4 ((ϵ, δ)-Noiseless Privacy). *Let $f : \mathcal{D}^n \to \mathcal{Y}$ be a deterministic query function on a database of length n drawn from domain $\mathcal{D}$. Let D be a distribution on $\mathcal{D}^n$. Let $S_1 \subseteq \mathcal{Y}$ and $S_2 \subseteq \mathcal{D}$ be two sets such that for all $j \in [n]$, $\Pr_{T \sim D}[f(T) \in S_1] + \Pr_{T \sim D}[t_j \in S_2] \leq \delta$, where t_j is the j-th entry of T.*
The function f is said to be (ϵ, δ)-Noiseless Private under distribution D and some auxiliary information $\mathcal{A}ux$, if there exists S_1, S_2 as defined above such that, for all measurable sets $\mathcal{O} \subseteq \mathcal{Y} - S_1$, for all $a, a' \in \mathcal{D} - S_2$, and for all $\ell \in [n]$ the following holds:

$$\Pr_{T \sim D}[f(T) \in \mathcal{O}|t_\ell = a, \mathcal{A}ux] \leq e^\epsilon \Pr_{T \sim D}[f(T) \in \mathcal{O}|t_\ell = a', \mathcal{A}ux]$$

One kind of auxiliary information ($\mathcal{A}ux$) that we will consider is partial information about some subset of entries of the database (*i.e.* partial disclosure). But often, it is easier to analyze the privacy when $\mathcal{A}ux$ corresponds to a full disclosure (complete revelation) of a subset of entries rather than partial disclosure because it may be difficult to characterize the corresponding conditional probabilities. The following result shows that the privacy degradation when $\mathcal{A}ux$ corresponds to a partial disclosure of information about a subset of entries can never be worse than the privacy degradation under full disclosure of the same set of entries.

Theorem 1 (Auxiliary Information). *Consider a database T and a query function $f(\cdot)$ over T. Let $\mathcal{A}_p$ denote some partial information regarding some fixed (but typically unknown to the mechanism) subset $T' \subset T$. Let $\mathcal{A}_f$ denote the corresponding full information about the entries of T'. If $f(T)$ is (ϵ, δ)-Noiseless Private under (every possible value of) the auxiliary information $\mathcal{A}_f$ (full disclosure) provided to the adversary, then it is also (ϵ, δ)-Noiseless Private under auxiliary information $\mathcal{A}_p$ (partial disclosure).*

Sketch of the proof:

The partial information $\mathcal{A}_p$ induces a distribution over the space of possible full disclosures $\mathcal{A}_f$. Using the law of total probability, we can write

$$\Pr_{T \sim D}[f(T) \in \mathcal{O}|t_\ell = a, \mathcal{A}_p] = \int_{\mathcal{A}_f} \Pr_{T \sim D}[f(T) \in \mathcal{O}|t_\ell = a, \mathcal{A}_f] \, dF(\mathcal{A}_f|\mathcal{A}_p, t_\ell = a)$$

(1)

where $F(\mathcal{A}_f|\mathcal{A}_p, t_\ell = a)$ denotes the conditional distribution for $\mathcal{A}_f$ given $\mathcal{A}_p$ and $[t_\ell = a]$. Since $f(T)$ is (ϵ, δ)-Noiseless Private given $\mathcal{A}_f$, there exist appropriate sets S_1 and S_2 (see *Definition 4*) with $\Pr_{T \sim D}[f(T) \in S_1] + \Pr_{T \sim D}[t_j \in S_2] \leq \delta$ such that, for all measurable sets $\mathcal{O} \subseteq \mathcal{Y} - S_1$, for all $a, a' \in \mathcal{D} - S_2$, and for all $\ell \in [n]$ we have

$$\Pr_{T \sim D}[f(T) \in \mathcal{O}|t_\ell = a, \mathcal{A}_f] \leq e^\epsilon \Pr_{T \sim D}[f(T) \in \mathcal{O}|t_\ell = a', \mathcal{A}_f]$$

(2)

The conditional distribution on F given $\mathcal{A}_p$ and t_ℓ in (1) is in fact independent of t_ℓ (since we can only argue about the privacy of the ℓth entry of T if it has not been already disclosed *fully* in $\mathcal{A}_f$). Now, since $F(\mathcal{A}_f|\mathcal{A}_p, t_\ell = a) = F(\mathcal{A}_f|\mathcal{A}_p, t_\ell = a')$, we can integrate both sides of (2) with respect to the same distribution and obtain, for the same sets S_1 and S_2 as in (2):

$$\Pr_{T \sim D}[f(T) \in \mathcal{O}|t_\ell = a, \mathcal{A}_p] \leq e^\epsilon \Pr_{T \sim D}[f(T) \in \mathcal{O}|t_\ell = a', \mathcal{A}_p]$$

(3)

This completes the proof.

Composability. In many applications, privacy has to be achieved under multiple (partial) disclosures of the database. For instance, in database applications, several thousand user queries about the database entries are answered in a day. Thus, a general result which tells how the privacy guarantee changes (typically degrades) as more and more queries are answered is very useful and is referred to as *composability* of privacy under multiple queries. While in some scenarios (eg. streaming applications) the database can change in between queries (dynamic database), in other scenarios it remains the same (static database). Also, the queries can be of different types or multiple instances of the same type. As mentioned earlier, in Differential Privacy, the privacy guarantees degrade exponentially with the number of queries on a static database. The notion of Noiseless Privacy often fails to compose in the presence of multiple queries on a static database (an exception to this is given in Section 4.2). But we do present several composability results for multiple queries under dynamic databases.

Dynamic databases may arise in practical scenarios in several ways: (a) Growing database model: Here the database keeps growing with time, *e.g.* database of all registered cars. Thus, in-between subsequent releases of information, the database grows by some number k, (b) Streaming model: This is the more commonly encountered scenario, where the availability of limited memory/storage causes the replacement of some old data with new one. Thus, at the time of each query the database has some k new entries out of the total (fixed) n , and (c)

Random replacement model: A good generalization of the above two models, it replaces randomly chosen k entries from the database of size n with the new incoming entries.

In all the above models of dynamic databases, we assume that the number of new elements form a constant fraction of the database. In particular, if n is the current database size, then some $\rho n, (0 \le \rho \le 1)$ number of entries are old and the remaining $k = (1 - \rho)n$ entries are new. Our main result about composability of Noiseless Privacy holds for any query which has (ϵ, δ)-Noiseless Privacy under any auxiliary information about at most $\rho n, (0 \le \rho \le 1)$ elements of the database. Note that in the growing database model, the size of the largest database on which the query is made is assumed to be n and the maximum fraction of old entries is ρ.

Theorem 2 (Composition). *Consider a sequence of m queries, $f_i(\cdot)$, $i \in [m]$, over dynamically changing data, such that, the ith query operates on the subset T_i of data elements. For each $i \ge 2$, let T_i share no more than a constant fraction $\rho, (0 \le \rho \le 1)$ of elements with $\cup_{i' < i} T_{i'}$ (i.e., all except ρ fraction of the elements in the database are new). If every query $f_i(T_i)$, individually, is (ϵ_i, δ_i)-Noiseless Private under the release of auxiliary information about a constant fraction ρ of elements in T_i, then the sequence of queries is $(\sum_{i=1}^{m} \epsilon_i, \sum_{i=1}^{m} \delta_i)$-Noiseless Private over the entire data.*

Sketch of the proof:

To assess the privacy of the ℓth element t_ℓ, we write down the following probability:

$$\Pr_{T \sim D}[f_1(T_1) \in \mathcal{O}_1, \ldots, f_m(T_m) \in \mathcal{O}_m \mid t_\ell = a] = \Pr_{T \sim D}[f_1(T_1) \in \mathcal{O}_1 \mid t_\ell = a]$$

$$\times \prod_{i=2}^{m} \Pr_{T \sim D}[f_i(T_i) \in \mathcal{O}_i \mid f_1(T_1) \in \mathcal{O}_1, \ldots, f_{i-1}(T_{i-1}) \in \mathcal{O}_{i-1}, t_\ell = a] \quad (4)$$

Since T_i shares at most a constant fraction ρ of elements with $\cup_{i' < i} T_{i'}$, the sequence of query responses $\langle f_1(T_1), \ldots, f_{i-1}(T_{i-1}) \rangle$, can be thought of as revealing auxiliary (possibly partial) information about at most ρ fraction of elements in T_i. Under such auxiliary leakage, we are given that $f_i(T_i)$ is (ϵ_i, δ_i)-Noiseless Private, i.e., there exist appropriate sets S_1^i and S_2^i (see *Definition 4*) with $\Pr_{T \sim D}[f(T) \in S_1^i] + \Pr_{T \sim D}[t_j \in S_2^i] \le \delta_i$ such that, for all measurable sets $\mathcal{O} \subseteq \mathcal{Y} - S_1^i$, for all $a, a' \in \mathcal{D} - S_2^i$, we have

$$\Pr_{T \sim D}[f_i(T_i) \in \mathcal{O}_i \mid f_1(T_1) \in \mathcal{O}_1, \ldots, f_{i-1}(T_{i-1}) \in \mathcal{O}_{i-1}, t_\ell = a]$$

$$\le e^{\epsilon_i} \Pr_{T \sim D}[f_i(T_i) \in \mathcal{O}_i \mid f_1(T_1) \in \mathcal{O}_1, \ldots, f_{i-1}(T_{i-1}) \in \mathcal{O}_{i-1}, t_\ell = a'] \quad (5)$$

Setting $S_1 = \cup_i S_1^i$ and $S_2 = \cup_i S_2^i$, we have $\Pr_{T \sim D}[f(T) \in S_1] + \Pr_{T \sim D}[t_j \in S_2] \le \sum_{i=1}^{m} \delta_i$ and using (5) for each of the m terms in the RHS of (4) we get, for all measurable sets $\mathcal{O}_i \subseteq \mathcal{Y} - S_1$, for all $a, a' \in \mathcal{D} - S_2$,

$$\Pr_{T \sim D}[f_1(T_1) \in \mathcal{O}_1, \ldots, f_m(T_m) \in \mathcal{O}_m \mid t_\ell = a]$$

$$\leq e^{\sum_{i=1}^{m} \epsilon_i} \Pr_{T \sim D}[f_1(T_1) \in \mathcal{O}_1, \ldots, f_m(T_m) \in \mathcal{O}_m \mid t_\ell = a'] \qquad (6)$$

This completes the proof. See the full version [BBG+11] for other results under multiple queries.

3 Boolean Queries

In this section we study queries of the form $f : T \rightarrow \{0, 1\}$, *i.e.*, the query function f acts on a database $T \in \mathcal{D}^n$, where $\mathcal{D}$ is the domain from which the data entries are drawn.

3.1 The No Auxiliary Information Setting

We first study a simple and clean setting: the database entries are all drawn independently and the adversary has no auxiliary information about them. We discuss generalizations later on. Before we get into the details of privacy friendly functions under our setting, we need some of the terminologies from analysis of boolean functions literature.

Definition 5 (k-junta [KLM+09]). *A function $f : \{0, 1\}^n \rightarrow \{0, 1\}$ is said to be k-junta if it depends only on some subset of the n coordinates of size k .*

Definition 6 ($(1 - \tau)$-far from k-junta). *Let $\mathcal{F}$ be the class of all k-junta functions $f' : \{0, 1\}^n \rightarrow \{0, 1\}$ and let D be a distribution on $\{0, 1\}^n$. A function $f : \{0, 1\}^n \rightarrow \{0, 1\}$ is $(1 - \tau)$-far from k-junta under D if*

$$\max_{f' \in \mathcal{F}} | \Pr_{T \sim D}[f(T) = f'(T)] - \Pr_{T \sim D}[f(T) \neq f'(T)]| = \tau$$

It is easy to see that when D is a uniform distribution over n-bits, a k-junta is 0-far from the class of k-juntas and the parity function is 1-far from the class of all 1-juntas.

The theorem below is for the setting where the adversary has no auxiliary information about the database. Later on in this section, we show how to handle the case when the adversary may have a subset of the database entries.

Theorem 3. *Let D be an arbitrary distribution over $\{0, 1\}^n$ such that the marginal probability of the $i - th$ bit equaling 1 is p_i. Let $f : \{0, 1\}^n \rightarrow \{0, 1\}$ be a boolean function which is $(1 - \tau_1)$-far from 0-junta and $(1 - \tau_2)$-far from 1-junta under D. If $\frac{\tau_1 + \tau_2}{2} \leq \min_{i \in [n]} p_i$ and $\max_{i \in [n]} p_i \leq 1 - \frac{\tau_1 + \tau_2}{2}$, then f is $\left(\max_{i \in [n]} \max \left\{ \ln \frac{1 + (\tau_1 + \tau_2)/(2(1 - p_i))}{1 - (\tau_1 + \tau_2)/(2p_i)}, \ln \frac{1 + (\tau_1 + \tau_2)/(2p_i)}{1 - (\tau_1 + \tau_2)/(2(1 - p_i))} \right\} \right)$-Noiseless Private.*

Proof. Please refer to [BBG+11] for the proof.

Note that in the above theorem we do not assume independence among the entries in T. As a result we can handle databases with correlated entries. It is also worth mentioning here that all the other results in this section assume the entries in the database to be uncorrelated.

To get some more insight into the result let us consider $f(T)$ to be the XOR of all the bits of T. Let T be drawn from the uniform distribution. Then f is 1-far from both a 0-junta and a 1-junta. Hence, f is 0-Noiseless Private. Instead of the XOR, if we let f be the AND function, then we see that it is just $1 - \frac{1}{2^{n-1}}$-far from a 0-junta. The ratio in this case becomes ∞, which shows AND is not a very good function for providing ϵ-Noiseless Privacy for small ϵ. This is indeed the case because $\Pr_T[f(T) = 1 | t_i = 0] = 0$ for all i. However, we can capture functions like AND if we try to guarantee (ϵ, δ)-Noiseless Privacy. If we fix $\delta = \frac{1}{2^n}$ (which is basically the probability of the AND function yielding 1), we get $(0, \frac{1}{2^n})$-Noiseless Privacy for AND. This property is in fact not specific to AND. In fact one can easily guarantee (ϵ, δ)-Noiseless Privacy for any *symmetric boolean functions* (*i.e.*, the functions whose output does not change on any permutation of the input bits). We will discuss this result in a more general setting later.

3.2   Handling Auxiliary Information

We now study the setting where the adversary may have auxiliary information about a subset of the entries in the database. We study the privacy of the entries about whom the adversary has no auxiliary information.

Theorem 4. *Let D be the distribution over $\{0,1\}^n$ where the $i-th$ bit is chosen to be 1 independently with probability p_i. Let $f : \{0,1\}^n \to \{0,1\}$ be a boolean function which is $(1-2B)$-far away from $d+1$ junta, that is, for any function g that depends only on a subset S of $U = [n]$ of size $d+1$, $|Pr[f(U) = g(S)] - 1/2| < B$. Let T be a database drawn from D and let Γ be any adversarially chosen subset of variables that has been leaked with $|\Gamma| = d$. If $\frac{B}{\delta} < \min_{i \in [n]} p_i$ and if $\max_{i \in [n]} p_i \le 1 - \frac{B}{\delta}$, then function f is $(\max_{i \in [n] - \Gamma} \left(\max \left\{ \ln \left(\frac{1 + \frac{B}{\delta(1-p_i)}}{1 - \frac{B}{\delta p_i}} \right) \right., \right.$
$\left. \left. \ln \left(\frac{1 + \frac{B}{\delta p_i}}{1 - \frac{B}{\delta(1-p_i)}} \right) \right\} \right), 2\delta)$-Noiseless Private with respect to the bit $t_i \in T$, where $i \in [n] - \Gamma$.*

Proof. We analyze the ratio given that $\Gamma = t$ is such that $|Pr_R[f(R||t) = 0] - 1/2| < B/\delta$ and $|Pr_R[f(R||t) = t_i] - 1/2| < B/\delta$. This happens with probability at least $1 - \delta - \delta = 1 - 2\delta$. The proof is as follows. Here the notation $R||t$ refers to a database formed by combining R and t.

Lemma 1. *Let the underlying distribution be an arbitrary D where each bit is 1 independently with probability p_i. Under D, let f be far away from d junta, that is for any function g that depends only on a subset S (with $|S| = d$) of $U = [n]$, $|Pr_D[f(U) = g(S)] - 1/2| < A$. Let T be a database drawn from D and let Γ (with $|\Gamma| = d$) be any adversarial subset of entries of T that has been* *leaked. Then, with probability at least $1 - \delta$ over the choice of assignments t to Γ, $|Pr_R[f(R||t) = 0] - 1/2| < A/\delta$.*

Proof. Let $\Gamma \subset U = [n]$, $|\Gamma| = d$, be the set of indices leaked. Note that we use Γ to represent both the indices and the variables itself. Let $R = [n] - \Gamma$. We prove the lemma by contradiction. Suppose the claim is wrong. That is, with probability at least δ over Γ, $|Pr_R[f(R||t) = 0] - 1/2| > A/\delta$. Construct $g : \{0,1\}^d \to \{0,1\}$ as follows.

$$g(t) = \begin{cases} 0 \text{ if } Pr_R[f(R||t) = 0] \geq 1/2 \\ 1 \text{ otherwise} \end{cases}$$

Observe that g just depends on d variables. We shall now show predictability of f using g which contradicts farness from d junta. Let us evaluate $Pr[f(U) = g(\Gamma)]$. To that end, we partition the assignments t to T into three sets, S_1, S_2 and S_3. S_1 is the set of t such that $Pr_R[f(R||t) = 0] \geq 1/2 + A/\delta$, S_2 is the set of t such that $Pr_R[f(R||t) = 0] \leq 1/2 - A/\delta$ and S_3 is the set of remaining assignments. Now, from our assumption, we are given that $Pr[T \in S_1 \cup S_2] > \delta$. Also, it is easy to observe that for any t, $Pr_R[f(R||t) = g(t)] \geq 1/2$ by the choice of g. Now, we lower bound $Pr[f(U) = g(\Gamma)]$.

$$\begin{aligned} Pr[f(U) = g(\Gamma)] &= \mathbb{E}_\Gamma Pr_R[f(R||\Gamma) = g(\Gamma)] \\ &\geq Pr[\Gamma \in S_1](1/2 + A/\delta) \\ &\quad + Pr[\Gamma \in S_2](1/2 + A/\delta) + Pr[\Gamma \in S_3](1/2) \\ &\geq 1/2 + (A/\delta) Pr[\Gamma \in S_1 \cup S_2] \\ &\geq 1/2 + A \end{aligned}$$

This leads to a contradiction.

Lemma 2. *Let D be a distribution over $\{0,1\}^n$ where each bit is 1 independently with probability p_i. Under D, let f be far away from d junta, that is for any function g that depends only on a subset S (with $|S| = d$) of $U = [n]$, $|Pr_D[f(U) = g(S)] - 1/2| < B$. Let T be a database drawn from D and let Γ (with $|\Gamma| = d$) be any adversarial subset of entries of T that has been leaked. Then, with probability at least $1 - \delta$ over the choice of assignments t to Γ, $|Pr_R[f(R||t) = t_i] - 1/2| < B/\delta$, where t_i is the i-th entry of the database T.*

Proof. The proof of this lemma is identical to the previous proof. Please see [BBG+11] for the complete proof.

Following the proof structure of Theorem 3, let $N = Pr[f = 0|\Gamma = t, t_i = 0]$ and $D = Pr[f = 0|\Gamma = t, t_i = 1]$. Now,

$$(1 - p_i)N + p_i(1 - D) = 1/2 + B_i, \quad \text{where } |B_i| \leq B/\delta$$
$$(1 - p_i)N + p_i D = A, \quad \text{where } |A - 1/2| \leq B/\delta$$

We now use the argument from the proof of Theorem 3 to upper (lower) bound N/D. Since the bound holds with probability $1 - 2\delta$, we get $\max_{i \in [n]} p_i \leq 1 - \frac{B}{\delta}$;

hence f is $(\max_{i\in[n]-\Gamma}\left(\max\left\{\ln\left(\frac{1+\frac{B}{\delta(1-p_i)}}{1-\frac{B}{\delta p_i}}\right),\ln\left(\frac{1+\frac{B}{\delta p_i}}{1-\frac{B}{\delta(1-p_i)}}\right)\right\}\right),2\delta)$-Noiseless Private which again makes sense as long as $\frac{B}{\delta}<\min_{i\in[n]}p_i$ and $\max_{i\in[n]}p_i\leq 1-\frac{B}{\delta}$.

3.3 Handling Multiple Queries in Adversarial Refreshment Model

Unlike the static model, in this model we assume that every query is run on a database where some significant part of it is new. We focus on the following *adversarial replacement model*.

Definition 7 (d-Adversarial Refreshment Model). *Except for d adversarially chosen bits of the database T, the remaining bits are refreshed under the data generating distribution D before every query f_i.*

We demonstrate the composability of boolean to boolean queries (*i.e.*, $f:\{0,1\}^n\to\{0,1\}$) under this model.

By the reduction shown in Theorem 2, privacy under multiple queries follows from the privacy in single query under auxiliary information. We use Theorems 2 and 4 to obtain the following *composition* theorem for boolean functions.

Corollary 1. *Let f be far away from $d+1$ junta (with $d=O(n)$), that is for any function g that depends only on a subset S of $U=[n]$ of size $d+1$, $|Pr[f(U)=g(S)]-1/2|<B$. Let the database T be changed as per the d-Adversarial Refreshment Model and let $\hat{T}$ be the database formed by concatenating the new entries (in the d-Adversarial Refreshment Model) with the existing entries. Let the number of times that f has been queried is m. Under the conditions of Theorem 4, f is*

$$(m\max_{i\in[n]}\left(\max\left\{\ln\left(\frac{1+\frac{B}{\delta(1-p_i)}}{1-\frac{B}{\delta p_i}}\right),\ln\left(\frac{1+\frac{B}{\delta p_i}}{1-\frac{B}{\delta(1-p_i)}}\right)\right\}\right),2m\delta)\text{-Noiseless Private,}$$

where n is the size of the database $\hat{T}$ and p_i is the probability of the i-th bit of $\hat{T}$ being one.

Please refer to the full version of the paper [BBG+11] for results on the privacy of symmetric functions.

4   Real Queries

In this section, we study the privacy of functions which operate on databases with real entries and compute a real value as output. We view the database T as a collection of n random variables $\langle t_1, t_2, \ldots, t_n\rangle$ with the ith random variable representing the ith database item. First we analyze the privacy of a query that outputs the sum of functions of database rows, that is, $f_n(T)=\frac{1}{s_n}\sum_{i\in[n]}g_i(t_i)$, $s_n=\sum_{i\in[n]}\mathbb{E}[g_i^2(t_i)]$ in Section 4.1. We provide a set of assumptions about g_i, under which the response of a single such query can be provided with $(\frac{\ln n}{\sqrt[6]{n}},\frac{1}{\sqrt{n}})$-Noiseless Privacy guarantees in Theorem 5. While Theorem 5 is for an adversary that has no auxiliary information about the database,

Theorem 6 is for an adversary that may have auxiliary information about some constant fraction of the database. We note that this query function is important as many learning algorithms, including principal component analysis, k-means clustering and any algorithm in the *statistical query* framework can be captured by this type of query (see [BDMN05]). Next, in section 4.2, we study the case of simple linear queries of the form $f_n(T) = \sum_{i \in [n]} a_i t_i, a_i \in \mathbb{R}$ when t_i are drawn i.i.d. from a normal distribution. We show that we can allow upto $\sqrt[5]{n}$ query-responses (on a static database) while still providing (ϵ, δ)-Noiseless Privacy for any arbitrary ϵ and for δ negligible in n. Again, we give a theorem each for an adversary with no auxiliary information as well as for an adversary who may have auxiliary information about some constant fraction of the database. We present several results about the privacy of these two queries under the various changing databases models in section 4.3.

4.1   Sums of Functions of Database Rows

Let $T = \langle t_1, \cdots, t_n \rangle$ be a database where each $t_i \in \mathbb{R}$ is independently chosen and let $g_i : \mathbb{R} \to \mathbb{R}, \forall i \in [n]$ be a set of one-to-one real valued functions with the following properties: (i) $\forall i \in [n], \mathbb{E}[g_i(t_i)] = 0$, (ii) $\forall i \in [n], \mathbb{E}[g_i^2(t_i)] = O(1)$, (iii) $\forall i \in [n], \mathbb{E}[|g_i(t_i)|^3] = O(1)$, and (iv) The density function for $g_i(t_i), \forall i \in [n]$ exists and has a bounded derivative. We study the privacy of the following function on the database T: $Y_n = \frac{1}{s_n} \sum_{i=1}^{n} g_i(t_i)$ where $s_n^2 = \sum_{i=1}^{n} \mathbb{E}[g_i^2(t_i)]$. Using Hertz Theorem [Her69] (see [BBG+11]) we can derive the following uniform convergence result for the cdf of Y_n to the cdf of the standard normal.

Corollary 2 (Uniform Convergence of F_n to Φ). *Let F_n be the cdf of $Y_n = \frac{1}{s_n} \sum_{i=1}^{n} g_i(t_i)$ where $s_n^2 = \sum_{i=1}^{n} \mathbb{E}[g_i^2(t_i)]$ and let Φ denote the standard normal cdf. If $\mathbb{E}[g_i(t_i)] = 0$ and if $\mathbb{E}[g_i^2(t_i)], \mathbb{E}[|g_i(t_i)|^3] \sim O(1) \; \forall i \in [n]$, then Y_n converges in distribution uniformly to the standard normal random variable as follows: $|F_n(x) - \Phi(x)| \sim O\left(\frac{1}{\sqrt{n}}\right)$*

If the pdf f_n of Y_n exists and has a bounded derivative, we can further derive the convergence rate of the pdf f_n to the pdf ϕ of the standard normal random variable. This result about pdf convergence is required because we will need to calculate the conditional probabilities in our privacy definitions over all measurable sets $\mathcal{O}$ in the range of the query output (see Definitions 2 & 4). The result is presented in the following Lemma (Please refer to [BBG+11] for the proof).

Lemma 3 (Uniform Convergence of f_n to ϕ). *Let $f_n(\cdot)$ be the pdf of $Y_n = \frac{1}{s_n} \sum_{i=1}^{n} g_i(t_i)$ where $s_n^2 = \sum_{i=1}^{n} \mathbb{E}[g_i^2(t_i)]$ and let $\phi(\cdot)$ denote the standard normal pdf. If $\mathbb{E}[g_i(t_i)] = 0, \mathbb{E}[g_i^2(t_i)], \mathbb{E}[|g_i(t_i)|^3] \sim O(1) \; \forall i \in [n]$, and if $\forall i$, the densities of $g_i(t_i)$ exist and have bounded derivative then f_n converges uniformly to the standard normal pdf as follows: $|f_n(x) - \phi(x)| \sim O\left(\frac{1}{\sqrt[4]{n}}\right)$*

Theorem 5 (Privacy). *Let $T = \langle t_1, \cdots, t_n \rangle$ be a database where each $t_i \in \mathcal{D}$ is independently chosen. Let $g_i : \mathbb{R} \to \mathbb{R}, \forall i \in [n]$ be a set of one-to-one real valued* *functions and let* $Y_n = \frac{1}{s_n} \sum_{i=1}^{n} g_i(t_i)$, *where* $s_n^2 = \cdot \sum_{i=1}^{n} \mathbb{E}[g_i^2(t_i)]$ *and* $\forall i \in [n]$, $\mathbb{E}[g_i(t_i)] = 0$, $\mathbb{E}[g_i^2(t_i)]$, $\mathbb{E}[|g_i(t_i)|^3] \sim O(1)$ *and* $\forall i \in [n]$ *the density functions for* $g_i(t_i)$ *exist and have bounded derivative. Let the auxiliary information* $\mathcal{A}ux$ *be empty. Then,* Y_n *is* $\left(O\left(\frac{\ln n}{\sqrt[6]{n}} \right), \ O\left(\frac{1}{\sqrt{n}} \right) \right)$-*Noiseless Private.*

Sketch of the proof: Please see [BBG+11] for the full proof. To analyze the privacy of the ℓth entry in the database T, we consider the ratio $R = \text{pdf}(Y_n = a|t_\ell = \alpha)/\text{pdf}(Y_n = a|t_\ell = \beta)$. Setting $Z = \frac{1}{s_z} \sum_{i=1, i \neq \ell}^{n} g_i(t_i)$, where $s_z^2 = \sum_{i=1, i \neq \ell}^{n} \mathbb{E}[g_i^2(t_i)]$, we can rewrite this ratio as $R = \text{pdf}(Z = \frac{as_n - g_\ell(\alpha)}{s_z})/\text{pdf}(Z = \frac{as_n - g_\ell(\beta)}{s_z})$. Applying Lemma 3 to the convergence of the pdf of Z to ϕ, we can upper-bound R using a ratio of appropriate standard normal pdf evaluations. For suitable choice of parameters, this leads to $\ln R \sim O(\frac{\ln n}{\sqrt[6]{n}})$. Using Corollary 2, we can show that the probability of data corresponding to the unsuitable parameters is $O(\frac{1}{\sqrt{n}})$.

Theorem 6 (Privacy with auxiliary information). *Let* $T = \langle t_1, \cdots, t_n \rangle$ *be a database where each* $t_i \in \mathbb{R}$ *is independently chosen. Let* $g_i : \mathbb{R} \to \mathbb{R}, \forall i \in [n]$ *be a set of one-to-one real valued functions and let* $Y_n = \frac{1}{s_n} \sum_{i=1}^{n} g_i(t_i)$, *where* $s_n^2 = \cdot \sum_{i=1}^{n} \mathbb{E}[g_i^2(t_i)]$ *and* $\forall i \in [n]$, $\mathbb{E}[g_i(t_i)] = 0$, $\mathbb{E}[g_i^2(t_i)]$, $\mathbb{E}[|g_i(t_i)|^3] \sim O(1)$ *and* $\forall i \in [n]$ *the density functions for* $g_i(t_i)$ *exist and have bounded derivative. Let the auxiliary information* $\mathcal{A}ux$ *be any subset of* T *of size* ρn. *Then,* Y_n *is* $\left(O\left(\frac{\ln(n(1-\rho))}{\sqrt[6]{n(1-\rho)}} \right), \ O\left(\frac{1}{\sqrt{n(1-\rho)}} \right) \right)$-*Noiseless Private.*

Sketch of the proof: Please see [BBG+11] for the full proof. To analyze the privacy of the ℓth entry in the database T, we consider the ratio $R = \text{pdf}(Y_n = a|t_\ell = \alpha, \mathcal{A}ux)/\text{pdf}(Y_n = a|t_\ell = \beta, \mathcal{A}ux)$. Setting $Z = \frac{1}{s_z} \sum_{i \in [n] \setminus I(\mathcal{A}ux), i \neq \ell} g_i(t_i)$, where $s_z^2 = \sum_{i \in [n] \setminus I(\mathcal{A}ux), i \neq \ell} \mathbb{E}[g_i^2(t_i)]$, we can rewrite this ratio as $R = \text{pdf}(Z = \frac{z_0 - g_\ell(\alpha)}{s_z})/\text{pdf}(Z = \frac{z_0 - g_\ell(\beta)}{s_z})$, where $I(\mathcal{A}ux)$ is the index set of $\mathcal{A}ux$ and $z_0 = as_n - \sum_{j \in I(\mathcal{A}ux)} g_j(t_j)$. Thereafter, the proof is similar to the proof of Theorem 5 except that Z is now a sum of $n(1 - \rho)$ random variables instead of $n - 1$.

The above theorem and Theorem 1 together imply privacy of $Y_n = \frac{1}{s_n} \sum_{i=1}^{n} g_i(t_i)$ under any auxiliary information about a constant fraction of the database.

4.2   Privacy Analysis of $f_n^i(T) = \sum_{j \in [n]} a_{ij} t_j$

We consider a sequence of linear queries $f_n^i(T)$, $i = 1, 2, \ldots$ with constant and bounded coefficients for a static database T. For each $m = 1, 2, \ldots$, we ask if the set $\{f_n^i(T) : i = 1, \ldots, m\}$ of queries can have Noiseless Privacy guarantees.

Theorem 7 (Privacy). *Consider a database* $T = \langle t_1, \ldots, t_n \rangle$ *where each* t_j *is drawn i.i.d from* $\mathcal{N}(0, 1)$. *Let* $f_n^i(T) = \sum_{i \in [n]} a_{ij} t_j$, $i = 1, 2, \ldots$, *be a sequence of linear queries (over* T) *with constant coefficients* a_{ij}, $|a_{ij}| \leq 1$ *and at least two non-zero coefficients in each query. Assume the adversary does not have access* *to any auxiliary information. For every m, $1 \leq m \leq \sqrt[5]{n}$, the set of queries $\{f_n^1(T), \ldots, f_n^m(T)\}$ is $(\epsilon, negl(n))$-Noiseless Private for any constant ϵ, provided the following conditions hold: For all $i \in [m], \ell \in [n]$, $R(\ell, i) \leq 0.99 \sum_{j=1, j \neq \ell}^n a_{ij}^2$, where $R(\ell, i) = \sum_{k=1, k \neq i}^m |\sum_{j=1, j \neq \ell}^n a_{ij} a_{kj}|$.*

Sketch of the proof: Please refer to [BBG+11] for the complete proof. One can represent the sequence of queries and their corresponding answers via a system of linear equations $\boldsymbol{Y} = A\boldsymbol{T}$, where $\boldsymbol{Y}$ is the output vector and A (called the the *design matrix*) is a $m \times n$ matrix. Each row A^i of the matrix A represents the coefficients of the i-th query. Note that we cannot hope to allow more than n linearly independent *linear* queries. Because in that case the adversary can extract the entire database T from the query responses.

We will prove the privacy of the ℓth data item, t_ℓ for some $\ell \in [n]$. Let $Y_i = \sum_{j=1}^n a_{ij} t_j$, where t_j are sampled i.i.d. from $\mathcal{N}(0, 1)$. For any $\alpha, \beta \in \mathbb{R}$ and any $\boldsymbol{v} = (y_1, \cdots, y_m) \in \mathbb{R}^m$ the following ratio r needs to be bounded by e^ϵ to guarantee Noiseless Privacy: $r = \frac{\text{pdf}(Y_1 = y_1, \cdots, Y_m = y_m | t_\ell = \alpha)}{\text{pdf}(Y_1 = y_1, \cdots, Y_m = y_m | t_\ell = \beta)}$. If we define $Z_i = \sum_{j=1, j \neq \ell}^n a_{ij} t_j$ for $i \in [m]$, $r = \frac{\text{pdf}(Z_1 = y_1 - a_{1\ell}\alpha, \cdots, Z_m = y_m - a_{m\ell}\alpha)}{\text{pdf}(Z_1 = y_1 - a_{1\ell}\beta, \cdots, Z_m = y_m - a_{m\ell}\beta)}$.

Let $\tilde{A}$ denote the $m \times (n-1)$ matrix obtained by dropping ℓ^{th} column of A. We have $Z_i \sim \mathcal{N}(0, \sum_{j=1, j \neq \ell}^n a_{ij}^2)$ and the vector $\boldsymbol{Z} = (Z_1, \cdots, Z_m)$ follows the distribution $\mathcal{N}(0, \Sigma)$, where $\Sigma = \tilde{A}\tilde{A}^T$. The entries of Σ look like $\Sigma_{ik} = \sum_{j=1, j \neq \ell}^n a_{ij} a_{kj}$ and $dim(\Sigma) = m \times m$. The sum of absolute values of non-diagonal entries in the ith row of Σ is given by $R(\ell, i)$ and the ith diagonal entry is $\sum_{j=1, j \neq \ell}^n a_{ij}^2$ (denoted Σ_{ii}). By Gershgorin Circle Theorem (see [BBG+11]), the eigenvalues of Σ are lower-bounded by $\Sigma_{ii} - R(\ell, i)$ for some $i \in [m]$. The condition $R(\ell, i) \leq 0.99\Sigma_{ii}$ implies that every eigenvalue is at least $0.01 \times \sum_{j=1, j \neq \ell}^n a_{ij}^2$. Since at least two a_{ij}'s per query are strictly non-zero, Σ will have strictly positive eigenvalues, and since Σ is also real and symmetric, we know Σ is invertible. Hence, for a given vector $\boldsymbol{z} \in \mathbb{R}^m$, we can write $\text{pdf}(\boldsymbol{Z} = \boldsymbol{z}) = \frac{1}{(2\pi)^{m/2} |\Sigma|^{1/2}} exp(-\frac{1}{2} \boldsymbol{z}^T \Sigma^{-1} \boldsymbol{z})$. Then, for $\boldsymbol{z_\alpha} = \boldsymbol{y} - \alpha A_\ell$ and $\boldsymbol{z_\beta} = \boldsymbol{y} - \beta A_\ell$ where A_ℓ denotes the ℓth column of A, $r = exp\left(-\frac{1}{2}\left(\boldsymbol{z_\alpha}^T \Sigma^{-1} \boldsymbol{z_\alpha} - \boldsymbol{z_\beta}^T \Sigma^{-1} \boldsymbol{z_\beta}\right)\right)$ Let $\Sigma^{-1} = Q\Lambda Q^T$ be the eigen decomposition and let $\boldsymbol{z'_\alpha} = Q^T \boldsymbol{z_\alpha}$ and $\boldsymbol{z'_\beta} = Q^T \boldsymbol{z_\beta}$ under the eigen basis. Then, $r = exp\left(-\frac{1}{2}\sum_{i=1}^m \lambda_i \left((z'_{\alpha,i})^2 - (z'_{\beta,i})^2\right)\right)$, where $z'_{\alpha,i}$ is the i-th entry of $\boldsymbol{z'_\alpha}$, $z'_{\beta,i}$ is the i-th entry of $\boldsymbol{z'_\beta}$ and λ_i is the i-th eigen value of Σ^{-1}. Further it can be shown that,

$$r \leq exp\left(\frac{m\lambda_{\max}|\alpha - \beta|}{2} \sqrt{\sum_{i=1}^m (2y_i - a_{i\ell}(\alpha + \beta))^2} \sqrt{\sum_{i=1}^m a_{i\ell}^2}\right)$$

where $\lambda_{\max} = \arg\max_i \lambda_i$ and we have used the fact that L_1 norm $\leq \sqrt{m} L_2$ norm and that L_2 norms of $\boldsymbol{z'_\alpha}$ and $\boldsymbol{z'_\beta}$ are equal to L_2 norms of $\boldsymbol{z_\alpha}$ and $\boldsymbol{z_\beta}$ respectively. Thus, this ratio will be less than e^ϵ if:

$$\sqrt{\sum_{i=1}^m (2y_i - a_{i\ell}(\alpha + \beta))^2} \leq \frac{2\epsilon}{m|(\alpha - \beta)|\lambda_{\max}\|A_\ell\|} \tag{7}$$

For $i \in [m]$ let G_i denote the event $\left[|2y_i - a_{i\ell}(\alpha + \beta)| \leq \frac{2\epsilon}{m^{3/2}|(\alpha-\beta)|\lambda_{\max}\|A_\ell\|} \right]$. The conjunction of events represented by $G = \wedge_i G_i$ implies the inequality in (7). Then, in the last step of the proof, we show (see [BBG$^+$11]) that the probability of the event G^c (compliment of G) is negligible in n for any ϵ and $m \leq n^{\frac{1}{5}}$. The above theorem is also true if the expected value of the database entries is a non-zero constant. This is our next claim (see [BBG$^+$11] for the proof).

Claim 1. *If $Y = \sum_{i=1}^{n} a_i t_i$ is (ϵ, δ)-Noiseless Private for a database $T = \langle t_1, \cdots, t_n \rangle$ such that $\forall i, \mathbb{E}[t_i] = 0$, then $Y^* = \sum_{i=1}^{n} a_i t_i^*$, where $t_i^* = t_i + \mu_i$, is also (ϵ, δ)-Noiseless Private.*

The results of *Theorem 7* can be extended to the case when adversary has access to some auxiliary information, $\mathcal{Aux}$, provided that $\mathcal{Aux}$ only contains information about a constant fraction of entries, albeit with a stricter requirement on the coefficients of the queries ($0 < a_{ij} \leq 1$ instead of $|a_{ij}| \leq 1$).

Theorem 8 (Privacy with auxiliary information). *Consider a database $T = \langle t_1, \ldots, t_n \rangle$ where each t_j is drawn i.i.d from $\mathcal{N}(0, 1)$. Let $f_n^i(T) = \sum_{i \in [n]} a_{ij} t_j$, $i = 1, 2, \ldots,$ be a sequence of linear queries (over T) with constant coeficients a_{ij}, $0 < a_{ij} \leq 1$ and at least two non-zero coefficients in each query. Let $\mathcal{Aux}$ denote the auxiliary information that the adversary can access. If $\mathcal{Aux}$ only contains information about a constant fraction, ρ, of data entries in T, then, for every m, $1 \leq m \leq \sqrt[5]{n}$, the set of queries $\{f_n^1(T), \ldots, f_n^m(T)\}$ is $(\epsilon, negl(n))$-Noiseless Private for any constant ϵ, provided the following conditions hold: For all $i \in [m], \ell \in [n]$ and $(n - \rho n) \leq r \leq n$*

$$\min_{S_r} \sum_{j \in S_r} \left(0.99 a_{ij}^2 - \sum_{k=1, k \neq l}^{m} a_{ij} a_{kj} \right) \geq 0 \tag{8}$$

where S_r is the collection of all possible $(r-1)$-size subsets of $[n] \setminus \{\ell\}$. The test in (8) can be performed efficiently in $O(n \log n)$ time.

Sketch of the proof: We first give a proof for the case when the auxiliary information $\mathcal{Aux}$ is full disclosure of any r entries of the database. Thereafter, we use Theorem 1 to get privacy for the case when $\mathcal{Aux}$ is any partial information about at most r entries of the database. Fix a set $\widehat{I}$ of indices (out of $[n]$) that correspond to the elements in $\mathcal{Aux}$ (This set is known to the adversary, but not to the mechanism). Let $|\widehat{I}| = r$. The response Y_i to the ith query can be written as $Y_i = \widehat{Y}_i + \sum_{j \in \widehat{I}} a_{ij} t_j$, where $\widehat{Y}_i = \sum_{j \in [n] \setminus \widehat{I}} a_{ij} t_j$. Since the second term in the above summation is known to the adversary, the ratio R that we need to bound for Noiseless Privacy is given by

$$R = \frac{\text{pdf}(Y_1 = y_1, \ldots, Y_m = y_m \mid t_\ell = \alpha, \mathcal{Aux})}{\text{pdf}(Y_1 = y_1, \ldots, Y_m = y_m \mid t_\ell = \beta, \mathcal{Aux})} \tag{9}$$

$$= \frac{\text{pdf}(\widehat{Y}_i = y_i - \sum_{j \in \widehat{I}} a_{ij} t_j, \ i = 1, \ldots m \mid t_\ell = \alpha)}{\text{pdf}(\widehat{Y}_i = y_i - \sum_{j \in \widehat{I}} a_{ij} t_j, \ i = 1, \ldots, m \mid t_\ell = \beta)} \tag{10}$$

Applying *Theorem 7* to $\widehat{Y}_i$'s we get $(\epsilon, negl(n))$-Noiseless Privacy for any $m \leq \sqrt[5]{n}$, if $\forall i \in [m], \ell \in [n]$:

$$\sum_{j \in [n] \setminus \widehat{I}, j \neq \ell} 0.99a_{ij}^2 - \sum_{k=1, k \neq i}^{m} \left| \sum_{j \in [n] \setminus \widehat{I}, j \neq \ell} a_{ij}a_{kj} \right| \geq 0 \qquad (11)$$

Theorem 8 uses the stronger condition of $0 < a_{ij} \leq 1$ (compared to $|a_{ij}| \leq 1$ in *Theorem 7*). Hence, we can remove the mod signs and change order of summation to get the following equivalent test: For all $i \in [m], \ell \in [n]$,

$$\sum_{j \in [n] \setminus \widehat{I}, j \neq \ell} \left(0.99a_{ij}^2 - \sum_{k=1, k \neq i}^{m} a_{ij}a_{kj} \right) \geq 0 \qquad (12)$$

Since $\widehat{I}$ is not known to the mechanism, we need to perform this check for all $\widehat{I}$ and ensure that even the $\widehat{I}$ that minimizes the LHS above must be non-negative. This gives us the test of (8). We can first compute all entries inside the round braces of (12), and then sort and picking the first $(n - r)$ entries. This takes $O(n \log n)$ time. This completes the proof.

Finally, we point out that although *Theorem 8* requires $0 < a_{ij} \leq 1$, we can obtain a very similar result for the $|a_{ij}| \leq 1$ case as well. This is because (11) is true even for $|a_{ij}| \leq 1$. However, unlike for $0 < a_{ij} \leq 1$ (when (12) could be derived), testing (11) for all $\widehat{I}$ becomes combinatorial and inefficient.

4.3   Privacy under Multiple Queries on Changing Databases

Theorems 6 & 8 provide (ϵ, δ)-privacy guarantees under leakage of constant fraction of data as auxiliary information. From *Theorem 2*, this implies composition results under dynamically changing databases (e.g., if each query is (ϵ, δ)-Noiseless Private, composition of m such queries will be $(m\epsilon, m\delta)$-Noiseless Private). As discussed in Sec. 2, we get composition under growing, streaming and random replacement models. In addition, both the queries considered in this section are extendibile (see full version [BBG+11] for details) and thus, one can answer multiple *repeat* queries on a dynamic database (under growing data and streaming models) without degradation in privacy guarantee.

Acknowledgements. We thank Cynthia Dwork for suggesting the changing data model direction, among other useful comments. We also thank Adam Smith and Piyush Srivastava for many useful discussions and suggestions.

References

[BBG+11]   Bhaskar, R., Bhowmick, A., Goyal, V., Laxman, S., Thakurta, A.: Noiseless database privacy. Cryptology ePrint Archive, Report 2011/487 (2011), http://eprint.iacr.org/

[BDMN05]   Blum, A., Dwork, C., McSherry, F., Nissim, K.: Practical privacy: the sulq framework. In: PODS, pp. 128–138 (2005)

[DMNS06]   Dwork, C., McSherry, F., Nissim, K., Smith, A.: Calibrating Noise to Sensitivity in Private Data Analysis. In: Halevi, S., Rabin, T. (eds.) TCC 2006. LNCS, vol. 3876, pp. 265–284. Springer, Heidelberg (2006)

[Dwo08]   Dwork, C.: Differential Privacy: A Survey of Results. In: Agrawal, M., Du, D.-Z., Duan, Z., Li, A. (eds.) TAMC 2008. LNCS, vol. 4978, pp. 1–19. Springer, Heidelberg (2008)

[GKS08]   Ganta, S.R., Kasiviswanathan, S.P., Smith, A.: Composition attacks and auxiliary information in data privacy. In: KDD, pp. 265–273 (2008)

[Her69]   Hertz, E.S.: On convergence rates in the central limit theorem. Ann. Math. Statist. 40, 475–479 (1969)

[KLM+09]   Kolountzakis, M.N., Lipton, R.J., Markakis, E., Mehta, A., Vishnoi, N.K.: On the fourier spectrum of symmetric boolean functions. Combinatorica 29(3), 363–387 (2009)

[KMN05]   Kenthapadi, K., Mishra, N., Nissim, K.: Simulatable auditing. In: Proceedings of the Twenty-Fourth ACM SIGMOD-SIGACT-SIGART Symposium on Principles of Database Systems, PODS 2005, pp. 118–127. ACM, New York (2005)

[MGKV06]   Machanavajjhala, A., Gehrke, J., Kifer, D., Venkitasubramaniam, M.: l-diversity: Privacy beyond k-anonymity. In: ICDE, p. 24 (2006)

[MKA+08]   Machanavajjhala, A., Kifer, D., Abowd, J., Gehrke, J., Vilhuber, L.: Privacy: Theory meets practice on the map. In: ICDE 2008: Proceedings of the 2008 IEEE 24th International Conference on Data Engineering, pp. 277–286. IEEE Computer Society, Washington, DC (2008)

[NMK+06]   Nabar, S.U., Marthi, B., Kenthapadi, K., Mishra, N., Motwani, R.: Towards robustness in query auditing. In: VLDB, pp. 151–162 (2006)

[Swe02]   Sweeney, L.: k-anonymity: A model for protecting privacy. International Journal on Uncertainty, Fuzziness and Knowledge-based Systems 10(5), 557–570 (2002)

The Preimage Security of Double-Block-Length Compression Functions

Frederik Armknecht[1], Ewan Fleischmann[2], Matthias Krause[1],
Jooyoung Lee[3,*], Martijn Stam[4], and John Steinberger[5,**]

[1] Arbeitsgruppe Theoretische Informatik und Datensicherheit,
University of Mannheim, Germany
{armknecht,krause}@uni-mannheim.de

[2] Chair of Media Security, Bauhaus-University Weimar, Germany
ewan.fleischmann@uni-weimar.de

[3] Faculty of Mathematics and Statistics, Sejong University, Seoul, Korea
jlee05@sejong.ac.kr

[4] Dept. of Computer Science, University of Bristol, United Kingdom
m.stam@alumnus.tue.nl

[5] Institute of Theoretical Computer Science, Tsinghua University, Beijing, China
jpsteinb@gmail.com

Abstract. We present new techniques for deriving preimage resistance bounds for block cipher based double-block-length, double-call hash functions. We give improved bounds on the preimage security of the three "classical" double-block-length, double-call, block cipher-based compression functions, these being Abreast-DM, Tandem-DM and Hirose's scheme. For Hirose's scheme, we show that an adversary must make at least 2^{2n-5} block cipher queries to achieve chance 0.5 of inverting a randomly chosen point in the range. For Abreast-DM and Tandem-DM we show that at least 2^{2n-10} queries are necessary. These bounds improve upon the previous best bounds of $\Omega(2^n)$ queries, and are optimal up to a constant factor since the compression functions in question have range of size 2^{2n}.

Keywords: Hash Function, Preimage Resistance, Block Cipher, Beyond Birthday Bound, Foundations.

1 Introduction

Almost as soon as the idea of turning a block cipher into a hash function appeared [9], it became evident that, for typical block ciphers and security expectations, the hash function needs to output a digest that is considerably larger

* This research was supported by Basic Science Research Program through the National Research Foundation of Korea(NRF) funded by the Ministry of Education, Science and Technology(2011-0013560).

** Supported by the National Natural Science Foundation of China Grant 61033001, 61061130540, 61073174, by the National Basic Research Program of China Grant 2007CB807900, 2007CB807901 and by NSF grand CNS 0904380.

D.H. Lee and X. Wang (Eds.): ASIACRYPT 2011, LNCS 7073, pp. 233–251, 2011.

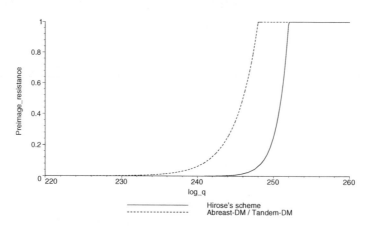

Fig. 1. Preimage bounds for the classical constructions

than the block cipher's block size. Consequently, many proposals of double-block-length, or more generally multi-block-length, hash functions have appeared in the literature. In this article we focus on a subclass of double-block-length constructions, where a $3n$-bit to $2n$-bit compression function makes two calls to a block cipher of $2n$-bit key and n-bit block.

Recently, for all three well-known members of this class—those being Tandem-DM [5], Abreast-DM [5] and Hirose's construction [4]—collision resistance has been successfully resolved [2,4,6,7]: for Abreast-DM and Hirose's scheme, $\Omega(2^n)$ queries to the underlying block cipher are needed to obtain a non-vanishing advantage in finding a collision. For Tandem-DM, $\Omega(2^{n-\log n})$ queries are needed, which is almost optimal ignoring log factors.

On the other hand, the corresponding situation for preimage resistance is far less satisfactory. Up to now, it has been an open problem to prove preimage resistance for values of q higher than 2^n for either Abreast-DM, Tandem-DM or Hirose. This is not to say that no dedicated preimage security proofs have appeared in the literature. For instance, Lee, Stam and Steinberger [7] provide a preimage resistance bound for Tandem-DM that is a lot closer to 2^n than a straightforward implication [10] of their collision bound would give. However, a "natural barrier" occurs once 2^n queries are reached: namely, a block cipher "loses randomness" after being queried $\Omega(2^n)$ times on the same key (for example, when $2^n - 1$ queries have been made to a block cipher under a given key, the answer to the last query under that key is deterministic). Going beyond the 2^n barrier seemed to require either a very technical probabilistic analysis, or some brand new idea. In this paper, we show a new idea which delivers tight bounds in a quite pain-free and non-technical fashion.

OUR CONTRIBUTION. In this paper, we prove that various compression functions that turn a block cipher of $2n$-bit key into a double-block-length hash function, have preimage resistance close to the optimal 2^{2n} in the ideal cipher model. Our analysis covers many practically relevant proposals, such as Abreast-DM, Hirose-DM and Tandem-DM. Bounds for the case $n = 128$ are depicted in Figure 1. At the heart of our result are so-called "super queries", a new technique to restrict the advantage of an adaptive preimage-finding adversary.

To build some intuition for our result, let us first consider the much easier problem of constructing a $3n$-bit to $2n$-bit compression function H based on two $3n$-bit to n-bit smaller underlying primitives f and f'. An obvious approach is simply to concatenate the outputs of f and f', that is let $H(B) = f(B)\|f'(B)$ for $B \in \{0,1\}^{3n}$. If f and f' are modeled as independent, ideal random functions, then it is not hard to see that H behaves ideally as well. In particular, it is preimage resistant up to 2^{2n} queries (to f and f').

When switching to a block cipher-based scenario, it is natural to replace f and f' in the construction above by E, resp. E', both run in Davies–Meyer mode. In other words, for block ciphers E and E' both with $2n$-bit keys and operating on n-bit blocks, define $H(A\|B) = (E_B(A) \oplus A)\|(E'_B(A) \oplus A)$ where $A \in \{0,1\}^n$ and $B \in \{0,1\}^{2n}$. While there is every reason to believe this construction maintains preimage resistance up to 2^{2n} queries, the standard proof technique against adaptive adversaries falls short significantly. Indeed, the usual argument goes that the i-th query an adversary makes to E using key K will return an answer uniform from a set of size at least $2^n - (i-1)$ and thus the probability of hitting a prespecified value is at most $1/(2^n - (i-1)) < 1/(2^n - q)$. Unfortunately, once q approaches 2^n, the denominator tends to zero (rendering the bound useless). As a result, one cannot hope to prove anything beyond 2^n queries using this method. This restriction holds even for a "typical" bound of type $q/(2^n - q)^2$.

When considering *non-adaptive* adversaries only, the situation is far less grim. Such adversaries need to commit to all queries in advance, which allows bounding the probability of each individual query hitting a prespecified value by 2^{-n}. While obviously there are dependencies (in the answers), these can safely be ignored when a union bound is later used to combine the various individual queries. Since the q offset has disappeared from the denominator, the typical bound $q/(2^n)^2$ *would* give the desired security.

Our solution, then, is to force an adaptive adversary to behave non-adaptively. As this might sound a bit cryptic, let us be more precise. Consider an adversary adaptively making queries to the block cipher, using the same key throughout. As soon as the number of queries *to this key* passes a certain threshold, we give the remaining queries to the block cipher using this very key *for free*. We will refer to this event as a *super query*. Since these free queries are all asked in one go, they can be dealt with non-adaptively, preempting the problems that occur (in standard proofs) due to adaptive queries. Nonetheless, for every super query we need to hand out a very large number of free queries, which can aid the adversary. Thus we need to limit the amount of super queries an adversary can make by setting the threshold that triggers a super query sufficiently high. In fact, we set the threshold at exactly half[1] the total number of queries that

---

[1] The "optimized" threshold turns out to be very near one half, but a bit less; we set the threshold at a half for simplicity in our proofs.

can be made under a given key (i.e., it is set at $2^n/2$ queries). This effectively doubles the adversary's query budget, since for every query the adversary makes it can get another one later "for free" (if it keeps on making queries under the same key), but such a doubling of the number of queries does not lead to an unacceptable deterioration of the security bound.

With this new technique in hand, we can prove in Section 3 that the construction H given above has indeed an asymptotically optimal preimage resistance bound. Afterwards, we revisit the proofs of preimage resistance of the three main double-block-length, double-call constructions: Hirose (Section 4), Abreast-DM (Section 5) and Tandem-DM (Section 6). An additional technical problem is that these compression functions each make two calls to the same block cipher, as opposed to using two calls to independent block ciphers. Ideally, to get a good bound, one would like to query the two calls necessary for a single compression function evaluation in conjunction (this would allow using the randomness of both calls simultaneously, potentially leading to a denominator 2^{2n} as desired for preimage resistance). For instance, in the context of collision resistance for Hirose-DM and Abreast-DM corresponding queries are grouped in cycles (of length 2 and 6, respectively) and all queries in a cycle are made simultaneously: if the adversary makes one query in a cycle, the remaining queries are handed out for free. Care has to be taken that these free queries and the free queries due to super queries do not reinforce each other to untenable levels.

For Hirose's scheme, there are no problems as the free queries introduced by a super query necessarily consist of full cycles only. The corresponding (upper) bound on the preimage finding advantage is $16q/2^{2n}$ which is as desired, up to a small factor. For Abreast-DM, however, the cyclic nature can no longer be exploited: any super query introduces many partial cycles, yet freely completing these might well trigger a new super query, etc.! Luckily, the original preimage proof for Tandem-DM [7] (which does not involve cycles) provides a way out of this conundrum. The downside however is that our preimage bound for Abreast-DM and Tandem-DM is slightly less tight than that for Hirose's scheme. Ignoring negligible terms, it grows roughly as $16\sqrt{q}/2^n$. Although this is faster than one might wish for (as can be seen in Figure 1), it does imply that $\Omega(2^{2n})$ queries are required to find a preimage with constant probability.

2   The Model

A block cipher is a function $E : \{0,1\}^m \times \{0,1\}^n \rightarrow \{0,1\}^n$ such that $E(K, \cdot)$ is a permutation of $\{0,1\}^n$ for each $K \in \{0,1\}^m$. We call m the *key size* and n the *block length* of the block cipher. It is customary to write $E_K(X)$ instead of $E(K, X)$ for $K \in \{0,1\}^m$, $X \in \{0,1\}^n$. The function $E_K^{-1}(\cdot)$ denotes the inverse of $E_K(\cdot)$ (as $E_K(\cdot)$ is a permutation). Henceforth, we will restrict to the case $m = 2n$ and we define $N = 2^n$.

A compression function H is block cipher-based if, in its execution, it has access to a block cipher. In this paper, we only discuss double-block-length, double-call constructions, meaning that H is a function from $3n$-bits to $2n$-bits making two calls to some underlying block cipher E. (This definition will become more concrete in the next sections.)

As our preimage security notion for H, we adopt everywhere preimage resistance in the information theoretic setting [10]. In this preimage resistance experiment, a computationally unbounded adversary with oracle access to a uniformly sampled block cipher $E : \{0,1\}^{2n} \times \{0,1\}^n \to \{0,1\}^n$ selects and announces a point $C \in \{0,1\}^{2n}$, before making queries to E. We allow the adversary to query both E and E^{-1}. After q queries to E, the *query history* of the attacker is the set of triples $\mathcal{Q} = \{(X_i, K_i, Y_i)\}_{i=1}^q$ such that $E_{K_i}(X_i) = Y_i$ and the attacker's i-th query is either $E_{K_i}(X_i)$ or $E_{K_i}^{-1}(Y_i)$ for $1 \le i \le q$. We say the attacker *succeeds* or *finds a preimage* if its query history $\mathcal{Q}$ contains the means of computing a preimage of C, in the sense that there exist values $B \in \{0,1\}^{3n}$, $K_1, K_2 \in \{0,1\}^{2n}$ and $X_1, X_2, Y_1, Y_2 \in \{0,1\}^n$ such that both (X_1, K_1, Y_1) and (X_2, K_2, Y_2) are in the query history $\mathcal{Q}$, $H(B) = C$ and the two queries used to evaluate $H(B)$ are precisely $E_{K_1}(X_1)$ and $E_{K_2}(X_2)$. In this case, we also say $\mathcal{Q}$ *contains a preimage* of C. We let $\mathsf{Preim}(\mathcal{Q})$ be the predicate that is true if and only if $\mathcal{Q}$ contains a preimage of C, where C is an elided-but-understood parameter of the predicate. We define

$$\mathbf{Adv}_H^{\mathrm{epre}}(q) = \max_A \Pr[\mathsf{Preim}(\mathcal{Q})]$$

where the maximum is taken over all adversaries making at most q queries, and where the probability is taken over the randomness of E as well as over the adversary's coins, if any.

For Tandem-DM, it turns out that the everywhere preimage resistance notion is slightly too strong, as there is one weak point (namely 0^{2n}) in the range, for which finding preimages is a bit easier. A simple adaptation of the everywhere preimage resistance definition is to disallow the adversary to choose $C = 0^{2n}$ as the target point [7]; we denote the corresponding advantage as

$$\mathbf{Adv}_H^{\mathrm{epre}\neq 0}(q) \ .$$

(We will still use the same predicate $\mathsf{Preim}(\mathcal{Q})$ though.)

A standard assumption made in ideal cipher proofs is that "the adversary never makes a query to which it already knows the answer". By this it is meant, for example, that one can assume the adversary never makes a query $E_K(X)$, obtaining an anwer Y, and then makes the query $E_K^{-1}(Y)$ (which will necessarily be answered by X). In the current context, where we consider adversaries making 2^n queries or more, this assumption should be more precisely restated as "the adversary never makes a query that will result in a triple (X, K, Y) which is already present in the query history". (This latter assumption can be made without loss of generality using the fact that $E_K(\cdot)$ is a permutation.) Indeed, if an adversary has made $2^n - 1$ queries under a key K, the result of the last query under that key is predetermined, and thus the adversary "already knows" the answer to this query. However, one should not forbid the adversary from making this query, since the query may be necessary to complete a preimage.

Our security proofs also use the notion of "free" queries. Formally, these can be modelled as queries which the adversary is "forced" to query (under certain conditions), but for which the adversary is not charged: they do not count towards the maximum of q queries which the adversary is allowed. However, these queries become part of the adversary's query history, just like other queries. In particular, the adversary is not allowed, later, to remake these queries "on its own" (due to the previously discussed assumption that the adversary never makes a query which it already owns). Observe that "free" queries are a common tool for analyzing the security of hash functions, e.g., see [2,3,6].

3   An Example Case

Before we apply the new technique of super queries to the analysis of three well-known constructions that compress $3n$ bits to $2n$ bits and that each call the same block cipher twice, we demonstrate our technique on the following simplest possible example. We consider the construction H_1, compressing $3n-1$ bits to $2n$ bits that makes two block cipher calls. Given a block cipher E of key length $m = 2n$ and block length n, an input block $X \in \{0,1\}^n$ and a key prefix $K \in \{0,1\}^{2n-1}$ we define

$$H_1(K, X) = (E_{K\|0}(X) \oplus X, E_{K\|1}(X) \oplus X)$$

where $\|$ denotes concatenation. If we consider the ideal cipher model, the two block cipher calls are independent. H_1 can be seen as a simple special case of a scenario where two different block ciphers are called and which is closely connected with the more general framework introduced by Özen and Stam [8,11] (with slightly different notation though).

Theorem 1. *Let* $H_1 : \{0,1\}^{3n-1} \to \{0,1\}^{2n}$ *be the block cipher-based compression function defined as above. Then*

$$\mathbf{Adv}_{H_1}^{\mathrm{epre}}(q) \leq 8q/N^2.$$

In particular, to achieve an advantage of $1/2$ the adversary has to make at least 2^{2n-4} queries.

Proof. Let $U\|V \in \{0,1\}^{2n}$ be the point to invert (chosen by the adversary before it makes any queries to E). We upper bound the probability that, in q queries, the adversary finds a point $A \in \{0,1\}^n$ and a key prefix $K \in \{0,1\}^{2n-1}$ such that $H_1(K\|A) = U\|V$. On top of the q queries the adversary wants to make, we give it several queries for free, to ensure that the elements $(X, K\|0, Y)$ and $(X, K\|1, Y')$ are always added to the query history as a pair. We call such a pair an "adjacent query pair" with respect to the key prefix $K \in \{0,1\}^{2n-1}$. The involved free queries are as follows.

Normal forward query. If the adversary queries $E_{K\|0}(X)$ (resp. $E_{K\|1}(X)$) for some key prefix $K \in \{0,1\}^{2n-1}$ and $X \in \{0,1\}^n$, we also give it for free $E_{K\|1}(X)$ (resp. $E_{K\|0}(X)$).

Normal inverse query. If the adversary queries $E_{K\|0}^{-1}(Y)$ (resp. $E_{K\|1}(Y')$) for some key prefix $K \in \{0,1\}^{2n-1}$ and receives answer X, we also give it for free $E_{K\|1}(X)$ (resp. $E_{K\|0}(X)$).

We now give further free queries to the adversary, in the fashion described next. After each adjacent query pair has been completed (namely, after the adversary has received the response to both its query and its associated free query, and after these have been placed in the query history), we check whether the key prefix used for the latest query is such that the (current) query history contains exactly $N/2$ adjacent query pairs with this key prefix. If so, we give *all* remaining adjacent query pairs under this key prefix for free to the adversary. There will be exactly $N/2$ such query pairs. We insert these $N/2$ free query pairs into the query history pair-by-pair (to maintain, mostly for conceptual simplicity, the adjacent pair structure of the query history). We note that, after these free queries have been inserted into the query history, the adversary cannot make any more queries under this key prefix, since the adversary is assumed never to make a query to which it knows the answer. When $N/2$ free query pairs are given to the adversary in the fashion just described, we say that a *super query* occurs. This can be summed up as follows:

Super query. When the query history contains $N/2$ adjacent query pairs all using the same key prefix $K \in \{0,1\}^{2n-1}$, all the remaining queries of the form $E_{K\|0}(\cdot)$ and $E_{K\|1}(\cdot)$ are given for free.

We say that an adjacent query pair $(X, K\|0, Y)$, $(X, K\|1, Y')$ is "winning", or "successful", if $X \oplus Y = U$ and $X \oplus Y' = V$. Thus the adversary obtains a preimage of $U\|V$ precisely if it obtains a winning adjacent query pair. This can occur in one of two ways: either the winning query pair is part of a super query, or not. We let $\mathsf{SuperQueryWin}(\mathcal{Q})$ denote the event that the adversary obtains a winning query pair that is part of a super query, and $\mathsf{NormalQueryWin}(\mathcal{Q})$ the event that the adversary obtains a winning query pair of normal queries. It thus suffices to upper bound

$$\Pr[\mathsf{SuperQueryWin}(\mathcal{Q})] + \Pr[\mathsf{NormalQueryWin}(\mathcal{Q})].$$

Here probabilities are taken (as usual) over the adversary's randomness (if any) and over the randomness of the ideal cipher.

We first upper bound $\Pr[\mathsf{NormalQueryWin}(\mathcal{Q})]$. Note that when the adversary makes, say, a forward query $E_{K\|0}(X)$, at most $N/2 - 1$ queries have been previously answered to the key $K\|0$ and at most $N/2 - 1$ queries have been previously answered to the key $K\|1$, since otherwise a super query for the key prefix K would have occurred. Thus the values $Y = E_{K\|0}(X)$ and $Y' = E_{K\|1}(X)$ come uniformly and independently at random from a set of size at least $N/2+1 \geq N/2$, and there is chance at most $(1/(N/2))^2 = 4/N^2$ that we obtain a winning pair of adjacent queries. The same is true if the adversary makes a forward query $E_{K\|1}(X)$, or an inverse query $E_{K\|0}^{-1}(Y)$, or an inverse query $E_{K\|1}^{-1}(Y')$. Since the adversary makes q queries in total, we therefore have

$$\Pr[\mathsf{NormalQueryWin}(\mathcal{Q})] \leq 4q/N^2. \tag{1}$$

We now bound $\Pr[\mathsf{SuperQueryWin}(\mathcal{Q})]$. Say a super query is about to occur on key prefix $K \in \{0,1\}^{2n-1}$, meaning that the value of $E_{K\|0}(\cdot)$ and $E_{K\|1}(\cdot)$ is already known on exactly $N/2$ points. Let us denote this set of points by $\mathcal{X}$, and let $\mathcal{Y} = E_{K\|0}(\mathcal{X})$ and $\mathcal{Y}' = E_{K\|1}(\mathcal{X})$. Further let $\mathcal{A} = \{0,1\}^n \backslash \mathcal{X}$, $\mathcal{B} = \{0,1\}^n \backslash \mathcal{Y}$, and $\mathcal{B}' = \{0,1\}^n \backslash \mathcal{Y}'$. Note that $|\mathcal{X}| = |\mathcal{Y}| = |\mathcal{Y}'| = |\mathcal{A}| = |\mathcal{B}| = |\mathcal{B}'| = N/2$.

Now let a point $A \in \mathcal{A}$ in the domain of the super query be arbitrarily fixed, and let us estimate the probability that point A induces a winning pair under E. If $A \oplus U \in \mathcal{Y}$ or if $A \oplus V \in \mathcal{Y}'$, this probability is zero. Consequently, let us suppose that $A \oplus U \in \mathcal{B}$ and $A \oplus V \in \mathcal{B}'$.

The probability (taken w.r.t. E) that $E_{K\|0}(A) = A \oplus U$ and $E_{K\|1}(A) = A \oplus V$ equals $\left(\frac{(N/2-1)!}{(N/2)!}\right)^2 = \left(\frac{1}{N/2}\right)^2$. Thus, by union bounding over A, we find that the probability of the super query producing a winning pair of adjacent queries is at most $N/2 \cdot \left(\frac{1}{N/2}\right)^2 = \frac{1}{N/2}$. We now observe that at most $q/(N/2)$ super queries can ever occur, since each super query requires a "setup" cost of $N/2$ queries. Thus

$$\Pr[\mathsf{SuperQueryWin}(\mathcal{Q})] \leq 4q/N^2. \tag{2}$$

Summing (1) and (2) completes the proof. □

4 Preimage Security Results for Hirose's Scheme

Hirose [4] introduced his $3n$-bit to $2n$-bit compression function making two calls to a block cipher of $2n$-bit key over 10 years after Abreast-DM and Tandem-DM (see the next Sections). Hirose's construction (Figure 2) is simpler than either of its predecessors and it uses a single keying schedule for the top and bottom block ciphers. Moreover, Hirose himself already proved birthday-type collision resistance for his construction in the ideal cipher model, thereby predating similar collision resistance analyses for Abreast-DM and Tandem-DM. Previously, Lee and Kwon [6] have shown that $\mathbf{Adv}^{\mathrm{epre}}_{\mathrm{Hir}}(q) \leq 2q/(N-2q)^2$, which becomes void once $q > N/2$. We improve upon this bound considerably.

Theorem 2. *Let* $\mathrm{Hir} : \{0,1\}^{3n} \to \{0,1\}^{2n}$ *be the block cipher-based compression function depicted in Figure 2. Then*

$$\mathbf{Adv}^{\mathrm{epre}}_{\mathrm{Hir}}(q) \leq 8q/N^2 + 8q/N(N-2).$$

In particular, $\mathbf{Adv}^{\mathrm{epre}}_{\mathrm{Hir}}(q)$ *is upper bounded by approximately* $16q/N^2$.

Proof. Let $U\|V \in \{0,1\}^{2n}$ be the point to invert (chosen by the adversary before it makes any queries to E). We upper bound the probability that, in q queries, the adversary finds a point $A\|L\|M \in \{0,1\}^{3n}$ such that $\mathrm{Hir}(A\|L\|M) = U\|V$.

When the adversary makes a *forward query* $E_{L\|M}(A)$ we give it for free, also, the answer to the query $E_{L\|M}(A \oplus c)$. Moreover when the adversary makes a *backward query* $E^{-1}_{L\|M}(R)$, resulting in an answer $A = E^{-1}_{L\|M}(R)$, we give it

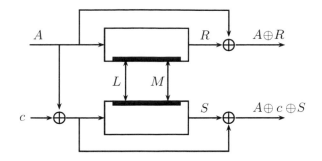

Fig. 2. Hirose's compression function. All wires carry n-bit values. The top and bottom block ciphers, which are the same block cipher, have $2n$-bit key and n-bit input/output. The wires A, L, M are the inputs to the compression function. The bottom left-hand wire is not an input; it carries an arbitrary nonzero constant c.

for free the answer to the forward query $E_{L\|M}(A \oplus c)$. Also, we assume that the adversary never makes a query to which it knows the answer (in the sense discussed in Section 2). Thus the elements of the adversary's query history $\mathcal{Q}$ can be paired into adjacent pairs of the form $(A, L\|M, R), (A \oplus c, L\|M, S)$. We call such a pair an "adjacent query pair". Furthermore, we define *super queries* analogously to the definition used in the proof of Theorem 1. More precisely, as soon as the (current) query history contains exactly $N/2$ queries with the same key, *all* remaining queries under this key are given for free to the adversary. (A minor difference with Theorem 1 is that it only takes $N/4$ queries to trigger a super query under a given key, instead of $N/2$.)

We say that an adjacent query pair $(A, L\|M, R), (A \oplus c, L\|M, S)$ is "winning", or "successful", if $A \oplus R = U$ and $A \oplus c \oplus S = V$, or if $A \oplus R = V$ and $A \oplus c \oplus S = U$. Thus the adversary obtains a preimage of $U\|V$ precisely if it obtains a winning adjacent query pair. This can occur in one of two ways: either the winning query pair is part of a super query, or not. We let $\mathsf{SuperQueryWin}(\mathcal{Q})$ denote the event that the adversary obtains a winning query pair that is part of a super query, and $\mathsf{NormalQueryWin}(\mathcal{Q})$ the event that the adversary obtains a winning query pair of normal queries. It thus suffices to upper bound

$$\Pr[\mathsf{SuperQueryWin}(\mathcal{Q})] + \Pr[\mathsf{NormalQueryWin}(\mathcal{Q})].$$

Here probabilities are taken (as usual) over the adversary's randomness (if any) and over the randomness of the ideal cipher.

We first upper bound $\Pr[\mathsf{NormalQueryWin}(\mathcal{Q})]$. Note that when the adversary makes, say, a forward query $E_{L\|M}(A)$, at most $N/2 - 2$ queries (counting free queries) have been previously answered with the key $L\|M$, since otherwise a super query for the key $L\|M$ would have occured. Thus the value $R = E_{L\|M}(A)$ comes uniformly at random from a set of size at least $N/2 + 2 \geq N/2$, and there is chance at most $2/(N/2) = 4/N$ that either $A \oplus R = U$ or $A \oplus R = V$ (this is also true if $U = V$). If, say, $A \oplus R = U$, there is further chance at most $1/(N/2) = 2/N$ that the free query $E_{L\|M}(A \oplus c)$ returns $A \oplus c \oplus V$, since the answer to the free query comes uniformly at random from a set of size at least $N/2+1 \leq N/2$. Other cases (e.g. when $A \oplus R = V$, and when the adversary makes a backward query $E_{L\|M}^{-1}(R)$) are similarly analyzed, showing that the adversary's chance of triggering the event $\mathsf{NormalQueryWin}(\mathcal{Q})$ at any given query is at most $(4/N)(2/N) = 8/N^2$. Since the adversary makes q queries total, we have

$$\Pr[\mathsf{NormalQueryWin}(\mathcal{Q})] \leq 8q/N^2. \tag{3}$$

We now bound $\Pr[\mathsf{SuperQueryWin}(\mathcal{Q})]$. Say a super query is about to occur on key $L\|M$, meaning that the value of $E_{L\|M}(\cdot)$ is already known on exactly $N/2$ points paired into $N/4$ query pairs. Let $A, A \oplus c$ be in the domain of the super query. (We say that a point $B \in \{0,1\}^n$ is "in the domain of the super query" if $E_{L\|M}(B)$ is not yet known, and will be queried as part of the super query; note that a point $A \in \{0,1\}^n$ is in the domain of the super query if and only if $A \oplus c$ is in the domain of the super query.) Then the probability that $E_{L\|M}(A) = U$ is either 0 if U is not in the range of the super query (meaning there is a normal query $E_{L\|M}(B) = U$ already present in the query history when the super query is made), or else is exactly $2/N$, since the value of $E_{L\|M}(A)$ returned by the super query is uniform at random in a set of size $N/2$. Thus, by a similar argument on V, the probability that $E_{L\|M}(A) \in \{U, V\}$ is at most $4/N$. Conditioning on the event $E_{L\|M}(A) \in \{U, V\}$, the probability that $E_{L\|M}(A \oplus c) \in \{U, V\}$ is at most $1/(N/2 - 1)$, since $E_{L\|M}(A \oplus c)$ is sampled uniformly at random from a set of size $N/2 - 1$, once the value $E_{L\|M}(A)$ is known. Thus the probability that the super query returns values such that the adjacent query pair $(A, L\|M, \cdot)$, $(A \oplus c, L\|M, \cdot)$ is winning is at most $4/N(N/2 - 1)$. But $A, A \oplus c$ were two arbitrary paired domain points; taking a union bound over the $N/4$ such pairs in the domain of the super query, we find that the probability of the super query producing a winning pair of adjacent queries is at most

$$(N/4) \cdot (4/N(N/2 - 1)) = 1/(N/2 - 1).$$

We now observe that at most $q/(N/4)$ super queries can ever occur, since each super query requires a "setup" cost of $N/4$ queries. Thus

$$\Pr[\mathsf{SuperQueryWin}(\mathcal{Q})] \leq 4q/N(N/2 - 1). \tag{4}$$

Summing (3) and (4) completes the proof.   □

5   Preimage Security Results for Abreast-DM

Abreast-DM, pictured in Figure 3, is one of the classical schemes for turning a $2n$-bit key block cipher into a $3n$-bit to $2n$-bit compression function. It was proposed by Lai and Massey in the same paper as Tandem-DM [5]. The collision resistance of Abreast-DM was independently resolved by Fleischmann, Gorski and Lucks [2] and Lee and Kwon [6], who both showed birthday-type collision resistance for Abreast-DM. Previously, Hirose [3] had given a collision resistance

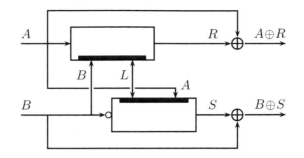

Fig. 3. The Abreast-DM compression function. The wires A, B, L are the inputs to the compression function. The empty circle at the left side of the bottom block cipher denotes bit complementation.

analysis for a general class of compression functions that included Abreast-DM as a special case, but under the assumption that the top and bottom block ciphers of the diagram be distinct. This assumption considerably simplifies the analysis (see also the later generalization by Özen and Stam [8]).

Previously, Lee and Kwon [6] have shown that $\mathbf{Adv}_{\mathrm{Abr}}^{\mathrm{epre}}(q) \leq 6q/(2^n - 6q)^2$. Although our bound for Abreast-DM (Theorem 3) is not as tight as our bound for Hirose's scheme (Theorem 2), it is clear from Corollary 1 below that our result significantly improves this bound.

Theorem 3. *Let* $\mathrm{Abr} : \{0,1\}^{3n} \rightarrow \{0,1\}^{2n}$ *be the block cipher-based compression function depicted in Figure 3. Let* $\alpha > 0$ *be an integer. Then*

$$\mathbf{Adv}_{\mathrm{Abr}}^{\mathrm{epre}}(q) \leq \frac{16\alpha}{N} + \frac{8q}{N^2(N-2)} + 2 \cdot \left(\frac{2eq}{\alpha N}\right)^\alpha + \frac{4q}{\alpha N}.$$

Proof. Let $U\|V$ be the point to invert, chosen by the adversary before any queries are made to E.

Unlike in the proof for Hirose's scheme, we do not give the adversary a free query after each query it makes. However, we still give the adversary "super queries" for free. More precisely, whenever the adversary has made $N/2$ queries under a given key $K\|L$, and after the $(N/2)$-th such query has been answered and placed in the query history, we give the remaining $N/2$ queries under the key $K\|L$ for free to the adversary, in any order. In this case, we say that a super query occurs; every query in the query history is either part of a super query, or not; in the latter case we call the query a "normal query". (Thus, in this theorem, normal queries are exactly the non-free queries.) Unlike in the proof of Theorem 2, there is no notion of an adjacent query pair. However, like in the proof of Theorem 2, we alert the reader to the fact that a "super query" consists of a set of $N/2$ queries, whereas a "normal query" is a single query.

We define an event $\mathsf{Lucky}(\mathcal{Q})$ on the query history; $\mathsf{Lucky}(\mathcal{Q})$ occurs if

$$|\{(X, K\|L, Y) \in \mathcal{Q} : X \oplus Y = U\}| > 2\alpha,$$

or if
$$|\{(X, K\|L, Y) \in \mathcal{Q} : X \oplus Y = V\}| > 2\alpha.$$

The adversary obtains a preimage of $U\|V$ precisely if it obtains queries of the form $(A, B\|L, R)$, $(\overline{B}, L\|A, S)$ such that $A \oplus R = U$ and $B \oplus S = V$, where $\overline{B}$ is bitwise complementation of B. It is easy to check that these two queries must be distinct, otherwise one obtains the contradiction $\overline{B} = A = L = B$. We call two such queries a "winning pair" of queries. Note, of course, that the queries in a winning pair need not be adjacent in the query history. We speak of the "first" and "second" query in a winning pair referring to the order in which they appear in the query history.

Let $\mathsf{WinNormal}(\mathcal{Q})$ be the event that the adversary obtains a winning pair in which the second query is a normal query. Let $\mathsf{WinSuper}_1(\mathcal{Q})$ be the event that the adversary obtains a winning pair in which the second query is part of a super query and the first is either normal or part of a super query, but is not part of the *same* super query as the second. Finally let $\mathsf{WinSuper}_2(\mathcal{Q})$ be the event that the adversary obtains a winning pair in which both queries of the pair are part of the same super query. It is then clear that if the adversary wins, one of the events
$$\mathsf{WinNormal}(\mathcal{Q}), \mathsf{WinSuper}_1(\mathcal{Q}) \text{ or } \mathsf{WinSuper}_2(\mathcal{Q})$$

occurs. In particular, thus, one of the four events
$$\mathsf{Lucky}(\mathcal{Q}), \mathsf{WinNormal}(\mathcal{Q}) \wedge \neg\mathsf{Lucky}(\mathcal{Q}), \mathsf{WinSuper}_1(\mathcal{Q}) \wedge \neg\mathsf{Lucky}(\mathcal{Q}),$$

$$\mathsf{WinSuper}_2(\mathcal{Q}) \wedge \neg\mathsf{Lucky}(\mathcal{Q})$$

must occur if the adversary wins. We upper bound the probability of each of these four events and sum the upper bounds in order to obtain an upper bound on the adversary's advantage.

We start by upper bounding $\Pr[\mathsf{Lucky}(\mathcal{Q})]$. For this we introduce two new events. Let $\mathcal{Q}_n$ be the restriction of $\mathcal{Q}$ to normal queries, and let $\mathcal{Q}_s$ be the restriction of $\mathcal{Q}$ to queries that are part of super queries. Let $\mathsf{Lucky}_n(\mathcal{Q})$ be the event that either
$$|\{(X, K\|L, Y) \in \mathcal{Q}_n : X \oplus Y = U\}| > \alpha,$$

or
$$|\{(X, K\|L, Y) \in \mathcal{Q}_n : X \oplus Y = V\}| > \alpha.$$

The event $\mathsf{Lucky}_s(\mathcal{Q})$ is likewise defined with respect to $\mathcal{Q}_s$. Obviously, $\mathsf{Lucky}(\mathcal{Q}) \implies \mathsf{Lucky}_n(\mathcal{Q}) \vee \mathsf{Lucky}_s(\mathcal{Q})$, so it suffices to upper bound $\mathsf{Lucky}_n(\mathcal{Q})$ and $\mathsf{Lucky}_s(\mathcal{Q})$ and to sum these upper bounds.

Since every answer to a normal query, forward or backward, comes at random from a set of size at least $N/2$, and since at most q normal queries are made, we have that
$$\Pr[\mathsf{Lucky}_n(\mathcal{Q})] \leq 2 \cdot \binom{q}{\alpha} \left(\frac{2}{N}\right)^\alpha \leq 2 \cdot \left(\frac{2eq}{\alpha N}\right)^\alpha.$$

To upper bound $\Pr[\mathsf{Lucky_s}(\mathcal{Q})]$, note that when a super query is made on key $K\|L$, the expected number of points $X \in \{0,1\}^n$ in the domain of the super query such that $X \oplus E_{K\|L}(X) = U$ is at most $(N/2) \cdot (2/N) = 1$, since for each individual such point the probability that $X \oplus E_{K\|L}(X) = U$ is either 0 (if $X \oplus U$ is not in the range of the super query) or $2/N$. Moreover there occur at most $q/(N/2) = 2q/N$ super queries, since it costs $N/2$ queries to setup a super query for a given key. Thus, the expectation of the random variable

$$|\{(X, K\|L, Y) \in \mathcal{Q}_s : X \oplus Y = U\}|,$$

taken over the coin tosses of the adversary and the randomness of E, is at most $2q/N \cdot 1 = 2q/N$. It then follows by Markov's inequality that the probability that

$$|\{(X, K\|L, Y) \in \mathcal{Q}_s : X \oplus Y = U\}| > \alpha$$

is at most $2q/\alpha N$. Then by a union bound and a symmetric argument (for $X \oplus Y = V$) , we obtain that $\Pr[\mathsf{Lucky_s}(\mathcal{Q})] \leq 4q/\alpha N$. Summing the upper bounds for $\Pr[\mathsf{Lucky_n}(\mathcal{Q})]$ and $\Pr[\mathsf{Lucky_s}(\mathcal{Q})]$, we thus obtain that

$$\Pr[\mathsf{Lucky}(\mathcal{Q})] \leq 2 \cdot \left(\frac{2eq}{\alpha N}\right)^{\alpha} + \frac{4q}{\alpha N}. \tag{5}$$

We now upper bound $\Pr[\mathsf{WinNormal}(\mathcal{Q}) \wedge \neg\mathsf{Lucky}(\mathcal{Q})]$. For this we use a "wish list" argument similar to that of [7]. As the adversary makes queries, we maintain two sequences $\mathcal{W}_T$ and $\mathcal{W}_B$ called *wish lists*. These are initially empty. For each query $(X, K\|L, Y)$ added to the query history (whether normal or part of a super query) we update the wish lists as follows:

1. If $X \oplus Y = U$ then $(\overline{K}, L\|X, K \oplus V)$ is added to $\mathcal{W}_B$.
2. If $X \oplus Y = V$ then $(L, \overline{X}\|K, L \oplus U)$ is added to $\mathcal{W}_T$.

We emphasize that $\mathcal{W}_B$ and $\mathcal{W}_T$ are sequences, not sets. The following properties are easy to check: (i) a query never "adds itself" to a wish list (namely, the queries inserted into the wish lists—if any—as a result of query $(X, K\|L, Y)$ being added to the query history, are distinct from $(X, K\|L, Y)$ itself); (ii) the elements of $\mathcal{W}_T$ are all distinct from one another, and the elements of $\mathcal{W}_B$ are all distinct from one another—namely, the same triple is never added twice to a wish list; (iii) the adversary obtains a winning pair precisely if a query is ever added to its query history that is already a member of one of its wish lists before the updating of the wish lists for that query (by property (i), however, we could equally well say "after the updating of the wish lists for that query"). Moreover, as long as $\neg\mathsf{Lucky}(\mathcal{Q})$ holds, the wish lists never exceed length 2α.

Let $E_{K\|L}(X)$ be a query made to E during the adversary's attack (either a normal query, or as part of a super query). If, at the moment when the query is being made, there is an element of the form $(X, K\|L, Y)$ in (at least) one of the wish lists for some $Y \in \{0,1\}^n$, then we say this wish list element is being "wished for" when the query $E_{K\|L}(X)$ is made. We similarly say the wish list element $(X, K\|L, Y)$ is being "wished for" if the query $E_{K\|L}^{-1}(Y)$ is made (note that in this case, the query $E^{-1}_{K\|L}(Y)$ is necessarily normal, since a super query is, by default, implemented by forward queries). We note, importantly, that any wish list element can only be wished for once, since $E_{K\|L}(\cdot)$ is a permutation.

Let $\mathsf{NormalWishGranted}_{\mathrm{T},i}$ be the event that a normal query $(X, K\|L, Y)$, when added to the query list, is equal to the i-th element of $\mathcal{W}_{\mathrm{T}}$ (presuming $\mathcal{W}_{\mathrm{T}}$ has length at least i when the query is added). Likewise define $\mathsf{NormalWishGranted}_{\mathrm{B},i}$ with respect to the list $\mathcal{W}_{\mathrm{B}}$. Then by the above remarks

$$\mathsf{WinNormal}(\mathcal{Q}) \wedge \neg\mathsf{Lucky}(\mathcal{Q}) \implies \bigvee_{i=1}^{2\alpha} \mathsf{NormalWishGranted}_{\mathrm{T},i} \vee$$
$$\bigvee_{i=1}^{2\alpha} \mathsf{NormalWishGranted}_{\mathrm{B},i}$$

so by a union bound

$$\Pr[\mathsf{WinNormal}(\mathcal{Q}) \wedge \neg\mathsf{Lucky}(\mathcal{Q})] \leq \sum_{i=1}^{2\alpha} \Pr[\mathsf{NormalWishGranted}_{\mathrm{T},i}] +$$
$$\sum_{i=1}^{2\alpha} \Pr[\mathsf{NormalWishGranted}_{\mathrm{B},i}].$$

Because each wish list element can only be wished for once and because a normal query is answered at random uniformly from a set of size at least $N/2$, we have

$$\Pr[\mathsf{NormalWishGranted}_{\mathrm{T},i}] \leq 2/N, \qquad \Pr[\mathsf{NormalWishGranted}_{\mathrm{B},i}] \leq 2/N$$

and therefore

$$\Pr[\mathsf{WinNormal}(\mathcal{Q}) \wedge \neg\mathsf{Lucky}(\mathcal{Q})] \leq 2 \cdot (4\alpha/N) = 8\alpha/N. \tag{6}$$

We now upper bound $\Pr[\mathsf{WinSuper}_1(\mathcal{Q}) \wedge \neg\mathsf{Lucky}(\mathcal{Q})]$. We keep the same definition of the wish lists $\mathcal{W}_{\mathrm{T}}$, $\mathcal{W}_{\mathrm{B}}$ as above. We let $\mathsf{SuperWishGranted}^1_{\mathrm{T},i}$ be the event that a query $(X, K\|L, Y)$ that is part of a super query is equal to the i-th element of $\mathcal{W}_{\mathrm{T}}$, where $\mathcal{W}_{\mathrm{T}}$ has length $\geq i$ before any of the super queries under key $K\|L$ have been made. The event $\mathsf{SuperWishGranted}^1_{\mathrm{B},i}$ is similarly defined. By the definition of $\mathsf{WinSuper}_1(\mathcal{Q})$ we have that

$$\Pr[\mathsf{WinSuper}_1(\mathcal{Q}) \wedge \neg\mathsf{Lucky}(\mathcal{Q})] \leq \sum_{i=1}^{2\alpha} \Pr[\mathsf{SuperWishGranted}^1_{\mathrm{T},i}] +$$
$$\sum_{i=1}^{2\alpha} \Pr[\mathsf{SuperWishGranted}^1_{\mathrm{B},i}].$$

Assume, for a given i, that the i-th element of $\mathcal{W}_{\mathrm{T}}$ (say) is $(X, K\|L, Y)$, and that a super query is about to be made for the key $K\|L$, and that X is in the domain of the super query. Then the probability that $E_{K\|L}(X) = Y$ is at most $2/N$ (more precisely, it is exactly $2/N$ unless Y is not in the super query's range, in which case it is 0). Thus, arguing similarly for the list $\mathcal{W}_B$, we obtain that

$$\Pr[\mathsf{SuperWishGranted}^1_{T,i}] \leq 2/N, \qquad \Pr[\mathsf{SuperWishGranted}^1_{B,i}] \leq 2/N.$$

Therefore

$$\Pr[\mathsf{WinSuper}_1(\mathcal{Q}) \wedge \neg\mathsf{Lucky}(\mathcal{Q})] \leq 8\alpha/N. \tag{7}$$

We finally bound $\Pr[\mathsf{WinSuper}_2(\mathcal{Q}) \wedge \neg\mathsf{Lucky}(\mathcal{Q})]$. In fact we upper bound the value $\Pr[\mathsf{WinSuper}_2(\mathcal{Q})]$, and we do not use a wish list argument. Note the event $\mathsf{WinSuper}_2(\mathcal{Q})$ can only occur when a super query is made on a key of the form $L\|L$, and then occurs only if both L and $\overline{L}$ are in the domain of the super query and if $E_{L\|L}(L) \oplus L = U$, $E_{L\|L}(\overline{L}) \oplus L = V$. It is easy to see that probability (when the super query is made) that these latter equalities hold is at most $(2/N)\cdot(1/(N/2-1))$. Since at most $q/(N/2)$ super queries are made, we therefore have

$$\Pr[\mathsf{WinSuper}_2(\mathcal{Q}) \wedge \neg\mathsf{Lucky}(\mathcal{Q})] \leq \Pr[\mathsf{WinSuper}_2(\mathcal{Q})] \leq 4q/N^2(N/2 - 1). \tag{8}$$

Finally, we obtain the theorem by summing (5), (6), (7) and (8).   □

Corollary 1. *We have*

$$\mathbf{Adv}^{\mathrm{epre}}_{\mathrm{Abr}}(2^{2n-10}) \leq 1/2 + o(1)$$

where the $o(1)$ term tends to 0 as $n \to \infty$.

Proof. By setting $\alpha = q^{1/2}/2$ (note that α is allowed to depend on q), the bound from Theorem 3 simplifies to

$$\frac{16q^{1/2}}{N} + \frac{8q}{N^2(N-2)} + 2 \cdot \left(\frac{4eq^{1/2}}{N}\right)^{q^{1/2}/2}$$

Suppose that $q = (cN)^2$ for some $0 < c < 1$, then this bound can be rewritten as

$$16c + \frac{8c^2}{N-2} + 2 \cdot (4ec)^{cN/2}.$$

For $4ec < 1$ this tends $16c$, so setting $c = 1/32$ gives us the claimed result.   □

6   Preimage Security Results for Tandem-DM

The Tandem-DM compression function, proposed by Lai and Massey in 1992 [5], is a $3n$-bit to $2n$-bit compression function based on two applications of a block cipher of $2n$-bit key and n-bit word length (Figure 4). The first (flawed) proof of collision security for Tandem-DM (by Fleischmann, Gorski and Lucks [1])

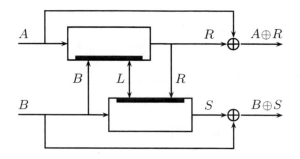

Fig. 4. The Tandem-DM compression function. The wires A, B, L are the inputs to the compression function.

did not appear until 2009. Later, Lee, Stam and Steinberger [7] gave a correct collision resistance analysis of Tandem-DM showing that indeed it has birthday-type collision security in the ideal cipher model (necessitating at least $2^{120.8}$ queries to break when the output length is $2n = 256$ bits). They also showed preimage resistance up to essentially 2^{128} queries (for $n = 128$), once $0^n \| 0^n$ is excluded as challenge digest. Our new bound is identical to the bound we gave for Abreast-DM, so in particular 2^{2n-10} queries are needed to obtain a preimage with probability ~ 0.5 (Corollary 2).

Theorem 4. *Let* $\mathrm{Tan} : \{0,1\}^{3n} \to \{0,1\}^{2n}$ *be the block cipher-based compression function depicted in Figure 4. Let* $\alpha > 0$ *be an integer. Then*

$$\mathbf{Adv}_{\mathrm{Tan}}^{\mathrm{epre} \neq 0}(q) \le \frac{16\alpha}{N} + \frac{8q}{N^2(N-2)} + 2 \cdot \left(\frac{2eq}{\alpha N}\right)^{\alpha} + \frac{4q}{\alpha N}.$$

Proof. Let $U \| V \neq 0^n \| 0^n$ be the point to invert, chosen by the adversary before making any queries to E.

We manage free queries exactly as for Abreast-DM; more precisely, when $N/2$ queries are made to E under a given key, we give the remaining $N/2$ queries under that key for free to the adversary, and this constitutes a "super query". No other free queries are given.

In the case of Tandem-DM, the adversary obtains a preimage of $U \| V$ precisely if it obtains queries of the form $(A, B \| L, R)$, $(B, L \| R, S)$ such that $A \oplus R = U$, $B \oplus S = V$. It is easy to see these two queries must be distinct, otherwise we would have $A = B = L = R = S$ and therefore $U \| V = 0^n \| 0^n$. We call two queries as above a "winning pair" of queries, where the two elements of a winning pair need not be adjacent in the query history (and could be in any order). We speak again of the "first" and "second" query in a winning pair referring to the order in which they appear in the query history.

We define the events $\mathsf{Lucky}(\mathcal{Q})$, $\mathsf{WinNormal}(\mathcal{Q})$, $\mathsf{WinSuper}_1(\mathcal{Q})$ and $\mathsf{WinSuper}_2(\mathcal{Q})$ as in the proof of Theorem 3 (but with respect, of course, to the new definition of "winning pair"). If the adversary wins, one of the events

$$\mathsf{Lucky}(\mathcal{Q}), \mathsf{WinNormal}(\mathcal{Q}) \wedge \neg\mathsf{Lucky}(\mathcal{Q}), \mathsf{WinSuper}_1(\mathcal{Q}) \wedge \neg\mathsf{Lucky}(\mathcal{Q}),$$

$$\mathsf{WinSuper}_2(\mathcal{Q}) \wedge \neg\mathsf{Lucky}(\mathcal{Q})$$

must occur. We upper bound the probability of each of these events separately. As in the case of Theorem 3, we have

$$\Pr[\mathsf{Lucky}(\mathcal{Q})] \leq 2 \cdot \left(\frac{2eq}{\alpha N}\right)^{\alpha} + \frac{4q}{\alpha N}. \tag{9}$$

To upper bound $\Pr[\mathsf{WinNormal}(\mathcal{Q}) \wedge \neg\mathsf{Lucky}(\mathcal{Q})]$, we again use wish lists. There are two wish lists, $\mathcal{W}_T$ and $\mathcal{W}_B$, which are initially empty and which are updated after each new query $(X, K\|L, Y)$ placed into the query history, according to the following rules:

1. If $X \oplus Y = U$ then $(K, L\|Y, K \oplus V)$ is added to $\mathcal{W}_B$.
2. If $X \oplus Y = V$ then $(L \oplus U, X\|K, L)$ is added to $\mathcal{W}_T$.

The same four properties from Theorem 3 are easy to check: (i) a query never "adds itself" to a wish list (this uses $U\|V \neq 0^n\|0^n$); (ii) the elements within each wish list are all distinct from one another; (iii) the adversary obtains a winning pair precisely if it obtains a query that is already in one of its wish lists (at the moment of insertion of that query into the query history). And by definition of $\mathsf{Lucky}(\mathcal{Q})$, the wish lists never exceed length 2α as long $\neg\mathsf{Lucky}(\mathcal{Q})$ holds.

Let $\mathsf{NormalWishGranted}_{T,i}$, $\mathsf{NormalWishGranted}_{B,i}$ be defined as in (the proof of) Theorem 3. Then, using exactly the same analysis as in the proof of Theorem 3, we have that

$$\Pr[\mathsf{NormalWishGranted}_{T,i}] \leq 2/N, \qquad \Pr[\mathsf{NormalWishGranted}_{B,i}] \leq 2/N$$

and that

$$\Pr[\mathsf{WinNormal}(\mathcal{Q}) \wedge \neg\mathsf{Lucky}(\mathcal{Q})] \leq 8\alpha/N. \tag{10}$$

Then also arguing word for word as in the proof of Theorem 3, we find that

$$\Pr[\mathsf{WinSuper}_1(\mathcal{Q}) \wedge \neg\mathsf{Lucky}(\mathcal{Q})] \leq 8\alpha/N. \tag{11}$$

We finally bound $\Pr[\mathsf{WinSuper}_2(\mathcal{Q}) \wedge \neg\mathsf{Lucky}(\mathcal{Q})]$. Note the event $\mathsf{WinSuper}_2(\mathcal{Q})$ can only occur when a super query occurs for a key of the form $L\|L$, and when that super query results in the triples $(U \oplus L, L\|L, L)$, $(L, L\|L, L \oplus V)$ being added to the query history. The probability that $E_{L\|L}(U \oplus L) = L$ is at most $2/N$, and, conditioned on the event that $E_{L\|L}(U \oplus L) = L$, the probability that $E_{L\|L}(L) = L \oplus V$ is at most $1/(N/2 - 1)$. Since at most $2q/N$ super queries occur, we thus find that

$$\Pr[\mathsf{WinSuper}_2(\mathcal{Q}) \wedge \neg\mathsf{Lucky}(\mathcal{Q})] \leq \Pr[\mathsf{WinSuper}_2(\mathcal{Q})] \leq 4q/N^2(N/2 - 1). \tag{12}$$

The theorem follows by summing (9), (10), (11) and (12).   □

As for Abreast-DM, we have the following corollary (with the same proof):

Corollary 2. *We have*

$$\mathbf{Adv}_{\mathrm{Tan}}^{\mathrm{epre}}(2^{2n-10}) \le 1/2 + o(1)$$

where the $o(1)$ term tends to 0 as $n \to \infty$.

7 Conclusion

In this work, we developed and applied new techniques for determining lower bounds with respect to preimage resistance. As opposed to existing techniques, statements on the security beyond the birthday bound are possible. We applied successfully these techniques to the three popular double-block-length, double-call, block cipher-based compression functions, these being Abreast-DM, Tandem-DM and Hirose's scheme.

Although these techniques allow for proving asymptotically optimal bounds, these bounds differ by constant factors from the best possible bound. This raises the question whether more accurate bounds can be derived, possibly revealing differences in the preimage resistance between the three constructions. A related question is the estimation of non-trivial upper bounds on the preimage resistance.

References

1. Fleischmann, E., Gorski, M., Lucks, S.: On the Security of TANDEM-DM. In: Dunkelman, O. (ed.) FSE 2009. LNCS, vol. 5665, pp. 84–103. Springer, Heidelberg (2009)
2. Fleischmann, E., Gorski, M., Lucks, S.: Security of Cyclic Double Block Length Hash Functions. In: Parker, M.G. (ed.) Cryptography and Coding 2009. LNCS, vol. 5921, pp. 153–175. Springer, Heidelberg (2009)
3. Hirose, S.: Provably Secure Double-Block-Length Hash Functions in a Black-Box Model. In: Park, C.-s., Chee, S. (eds.) ICISC 2004. LNCS, vol. 3506, pp. 330–342. Springer, Heidelberg (2005)
4. Hirose, S.: Some Plausible Constructions of Double-Block-Length Hash Functions. In: Robshaw, M.J.B. (ed.) FSE 2006. LNCS, vol. 4047, pp. 210–225. Springer, Heidelberg (2006)
5. Lai, X., Massey, J.L.: Hash Functions Based on Block Ciphers. In: Rueppel, R.A. (ed.) EUROCRYPT 1992. LNCS, vol. 658, pp. 55–70. Springer, Heidelberg (1993)
6. Lee, J., Kwon, D.: The security of Abreast-DM in the ideal cipher model, http://eprint.iacr.org/2009/225.pdf
7. Lee, J., Stam, M., Steinberger, J.: The Collision Security of Tandem-DM in the Ideal Cipher Model. In: Rogaway, P. (ed.) CRYPTO 2011. LNCS, vol. 6841, pp. 561–577. Springer, Heidelberg (2011)
8. Özen, O., Stam, M.: Another Glance at Double-Length Hashing. In: Parker, M.G. (ed.) Cryptography and Coding 2009. LNCS, vol. 5921, pp. 176–201. Springer, Heidelberg (2009)
9. Rabin, M.: Digitalized signatures. In: Foundations of Secure Computations, pp. 155–166. Academic Press (1978)

10. Rogaway, P., Shrimpton, T.: Cryptographic Hash-Function Basics: Definitions, Implications, and Separations for Preimage Resistance, Second-Preimage Resistance, and Collision-Resistance. In: Roy, B., Meier, W. (eds.) FSE 2004. LNCS, vol. 3017, pp. 371–388. Springer, Heidelberg (2004)
11. Stam, M.: Blockcipher-Based Hashing Revisited. In: Dunkelman, O. (ed.) FSE 2009. LNCS, vol. 5665, pp. 67–83. Springer, Heidelberg (2009)

Rebound Attack on JH42

María Naya-Plasencia[1,*], Deniz Toz[2,**], Kerem Varıcı[2,**]

[1] FHNW Windisch, Switzerland and University of Versailles, France
maria.naya-plasencia@prism.uvsq.fr
[2] Katholieke Universiteit Leuven, ESAT/COSIC and IBBT, Belgium
{deniz.toz,kerem.varici}@esat.kuleuven.be

Abstract. The hash function JH [20] is one of the five finalists of the NIST SHA-3 hash competition. It has been recently tweaked for the final by increasing its number of rounds from 35.5 to 42. The previously best known results on JH were semi-free-start near-collisions up to 22 rounds using multi-inbound rebound attacks. In this paper we provide a new differential path on 32 rounds. Using this path, we are able to build various semi-free-start internal-state near-collisions and the maximum number of rounds that we achieved is up to 37 rounds on 986 bits. Moreover, we build distinguishers in the full 42-round internal permutation. These are, to our knowledge, the first results faster than generic attack on the full internal permutation of JH42, the finalist version. These distinguishers also apply to the compression function.

Keywords: hash function, rebound attack, JH, cryptanalysis, SHA-3.

1 Introduction

A cryptographic hash function is a one way mathematical function that takes a message of arbitrary length as input and produces an output of fixed length, which is commonly called a fingerprint or message digest. Hash functions are fundamental components of many cryptographic applications such as digital signatures, authentication, key derivation, random number generation, etc. So, in terms of security any hash function should be preimage, second-preimage and collision resistant.

Most of the recent hash functions use either compression functions or internal permutations as building blocks in their design. In addition to the main properties mentioned above, some ideal properties should also be satisfied for the

---

* Supported by the National Competence Center in Research on Mobile Information and Communication Systems (NCCR-MICS), a center of the Swiss National Science Foundation under grant number 5005-67322 and by the French Agence Nationale de la Recherche through the SAPHIR2 project under Contract ANR-08-VERS-014.
** This work was sponsored by the Research Fund K.U.Leuven, by the IAP Programme P6/26 BCRYPT of the Belgian State (Belgian Science Policy) and by the European Commission through the ICT Programme under Contract ICT-2007-216676 (ECRYPT II). The information in this paper is provided as is, and no warranty is given or implied that the information is fit for any particular purpose. The user thereof uses the information at its sole risk and liability.

D.H. Lee and X. Wang (Eds.): ASIACRYPT 2011, LNCS 7073, pp. 252–269, 2011.

building blocks. This means that the algorithm should not have any structural weaknesses and should not be distinguishable from a random oracle. The absence of these properties on building blocks may not impact the security claims of the hash function immediately but it helps to point out the potential flaws in the design.

Since many of the hash standards [16,19] have been broken in recent years, the National Institute of Standards and Technology (NIST) announced a competition to replace the current standard SHA-2 with a new algorithm SHA-3. The hash function JH [20], designed by Hongjun Wu, is one of the five finalists of this competition. It is a very simple design and efficient in both software and hardware. JH supports four different hash sizes: 224, 256, 384 and 512-bit. It has been tweaked from the second round to the final round by increasing its number of rounds from 35.5 to 42. The new version is called JH42.

Related Work. We recall here the previously best known results on JH. A marginal preimage attack on the 512-bits hash function with a complexity in time and memory of 2^{507} was presented in [1]. Several multi-inbound rebound attacks were presented in [15], providing in particular a semi-free-start collision for 16 rounds with a complexity of 2^{190} in time and 2^{104} in memory and a semi-free-start near-collision for 22 rounds of compression function with a complexity of 2^{168} in time and 2^{143} in memory. In [12, Sec.4.1], improved complexities for these rebound attacks were provided: 2^{97} in time and memory for the 16 round semi-free-start collision and 2^{96} in time and memory for the 22 rounds semi-free-start near-collision for compression function.

Our Contributions. In this paper we apply, as in [15], a multi-inbound rebound attack, using 6 inbounds that cover rounds from 0 to 32. We first find partial solutions for the differential part of the path by using the ideas from [13]. Due to increased number of rounds compared with the previous attacks, the differential path will have several highly active peaks, instead of one as in [15]. This means that, while in the previous attacks finding the whole solution for the path could be easily done without contradicting any of the already fixed values from the inbounds, now finding the complete solution is the most expensive part. We propose here an algorithm that allows us to find whole solutions for rounds from 4 to 26 with an average complexity of 2^{64}. By repeating the algorithm, the attack can be started from round 0 and extended up to 37 rounds for building semi-free-start near-collisions on the internal state, since we have enough degrees of freedom. Based on the same differential characteristic, we also present distinguishers for 42 rounds of the internal permutation which is the first distinguisher on internal permutation faster than generic attack to the best of our knowledge. We summarize our main results in Table 1.

This paper is organized as follows: In Section 2, we give a brief description of the JH hash function, its properties and an overview of the rebound attack. In Section 3, we first describe the main idea of our attack and then give the semi-free internal near-collision results on the tweaked version JH42. Based on this

Table 1. Comparison of best attack results on JH (sfs: semi-free-start)

target	rounds	time comp.	memory comp.	attack type	generic comp.	sect.
hash function	16	2^{190}	2^{104}	sfs collision	2^{256}	[15]
hash function	16	$2^{96.1}$	$2^{96.1}$	sfs collision	2^{256}	[12]
comp. function	$19-22$	2^{168}	$2^{143.7}$	sfs near-collision	2^{236}	[15]
comp. function	$19-22$	$2^{95.6}$	$2^{95.6}$	sfs near-collision	2^{236}	[12]
comp. function	26	2^{112}	$2^{57.6}$	sfs near-collision	$2^{341.45}$	§3
comp. function	32	2^{304}	$2^{57.6}$	sfs near-collision	$2^{437.13}$	§3
comp. function	36	2^{352}	$2^{57.6}$	sfs near-collision	$2^{437.13}$	§3
comp. function	37	2^{352}	$2^{57.6}$	sfs near-collision	$2^{396.7}$	§3
internal perm.	42	2^{304}	$2^{57.6}$	distinguisher	2^{705}	§4
internal perm.	42	2^{352}	$2^{57.6}$	distinguisher	2^{762}	§4

results, we describe a distinguisher in Section 4 for the full internal permutation, that also applies to the full compression function. Finally, we conclude the paper and summarize our results in Section 5.

2   Preliminaries

2.1   The JH42 Hash Function

The hash function JH is an iterative hash function that accepts message blocks of 512 bits and produces a hash value of 224, 256, 384 and 512 bits. The message is padded to be a multiple of 512 bits. The bit '1' is appended to the end of the message, followed by $384 - 1 + (-l \mod 512)$ zero bits. Finally, a 128-bit block is appended which is the length of the message, l, represented in big endian form. Note that this scheme guarantees that at least 512 additional bits are padded.

In each iteration, the compression function F_d, given in Figure 1, is used to update the 2^{d+2} bits of the state H_{i-1} as follows:

$$H_i = F_d(H_{i-1}, M_i)$$

where H_{i-1} is the previous chaining value and M_i is the current message block. The compression function F_d is defined as follows:

$$F_d(H_{i-1}, M_i) = E_d(H_{i-1} \oplus (M_i || 0^{2^{d+1}})) \oplus (0^{2^{d+1}} || M_i)$$

Here, E_d is a permutation and is composed of an initial grouping of bits followed by $6(d-1)$ rounds, plus a final degrouping of bits. The grouping operation arranges bits in a way that the input to each S-Box has two bits from the message part and two bits from the chaining value. In each round, the input is divided into 2^d words and then each word passes through an S-Box. JH uses two 4-bit-to-4-bit S-Boxes ($S0$ and $S1$) and every round constant bit selects which S-Boxes are used. Then two consecutive words pass through the linear

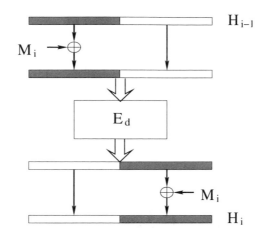

Fig. 1. The compression function F_d that transforms 2^{d+2} bits treated as 2^d words of four bits

transformation L, which is based on a $[4, 2, 3]$ Maximum Distance Separable (MDS) code over $GF(2^4)$. Finally all words are permuted by the permutation P_d. After the degrouping operation each bit returns to its original position.

The initial hash value H_0 is set depending on the message digest size. The first two bytes of H_{-1} are set as the message digest size, and the rest of the bytes of H_{-1} are set as zero. Then, $H_0 = F_d(H_{-1}, 0)$. Finally, the message digest is generated by truncating H_N where N is the number of blocks in the padded message, i.e, the last X bits of H_N are given as the message digest of JH-X where X = 224, 256, 384 and 512.

The official submitted version of JH42 has $d = 8$ and so the number of rounds is 42 and the size of the internal state is 1024 bits. Then, from now on, we will only consider E_8. For a more detailed information we refer to the specification of JH [20].

2.2   Properties of the Linear Transformation L

Since the linear transformation L implements a $[4, 2, 3]$ MDS code, any difference in one of the words of the input (output) will result in a difference in two words of the output (input). For a fixed L transformation, if one tries all possible 2^{16} pairs, the number of pairs satisfying the condition $2 \rightarrow 1$ or $1 \rightarrow 2$ is 3840, which gives a probability of $3840/65536 \approx 2^{-4.09}$. Note that, if the words are arranged in a way that they will be both active this probability increases to $3840/57600 \approx 2^{-3.91}$. For the latter case, if both words remain active ($2 \rightarrow 2$), the probability is $49920/57600 \approx 2^{-0.21}$.

2.3   Observations on the Compression Function

The grouping of bits at the beginning of the compression function assures that the input of every first layer S-Box is xor-ed with two message bits. Similarly, the output of each S-Box is xor-ed with two message bits. Therefore, for a random non-zero 4-bit difference, the probability that this difference is related to a message is $3/15 \approx 2^{-2.32}$.

The bit-slice implementation of F_d uses $d - 1$ different round functions. The main difference between these round functions is the permutation function. In each round permutation, the odd bits are swapped by $2^r \mod (d - 1)$ where r is the round number. Therefore, for the same input passing through multiple rounds, the output is identical to the output of the original round function for the $\alpha \cdot (d - 1)$-th round where α is any integer.

2.4   The Rebound Attack

The rebound attack was introduced by Mendel et al. [10]. The two main steps of the attack are called inbound phase and outbound phase. In the inbound phase, the available degrees of freedom are used to connect the middle rounds by using the match-in-the-middle technique and in the outbound phase connected truncated differentials are computed in both forward and backward direction.

This attack has been first used for the cryptanalysis of reduced versions of Whirlpool and Grøstl, and then extended to obtain distinguishers for the full Whirlpool compression function [6]. Later, linearized match-in-the-middle and start-from-the-middle techniques are introduced by Mendel et al. [9] to improve the rebound attack. Moreover, a sparse truncated differential path and state is used in the attack on LANE by Matusiewicz et al. [8] rather than using a full active state in the matching part of the attack. Then, these techniques were used to improve the results on AES-based algorithms in the following papers: [2,3,5,11,14,17,18].

3   Semi-free-start Internal Near-Collisions

In this section, we first present an outline for the rebound attack on reduced round versions of JH for all hash sizes. We use a differential characteristic that covers 32 rounds, and apply the start-from-the-middle technique by using six inbound phases with partially active states. We first describe how to solve the multi-inbound phase for the active bytes. Contrary to previous attacks on JH, we now have more fixed values from the inbound phases. So, in order to find a complete solution, we need to merge these fixed values without contradicting any of them. Therefore, we describe next how to match the passive bytes. Finally, we analyze the outbound part.

3.1   Matching the Active Bytes

Multi-inbound Phase. The multi-inbound phase of the attack covers 32 rounds and is composed of two parts. In the first part, we apply the start-from-the-middle-technique six times for rounds $0 - 4$, $4 - 10$, $10 - 16$, $16 - 20$, $20 - 26$

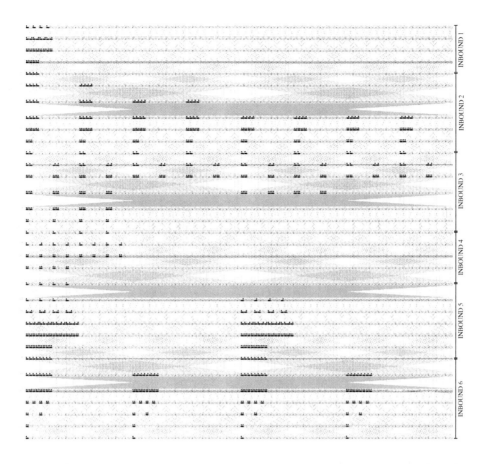

Fig. 2. Differential characteristic for 32 rounds of JH Compression Function (bit-slice representation)

and $26 - 32$. In the second part, we connect the resulting active bytes (hence the corresponding state values) by a match-in-the-middle step. The number of active S-Boxes in each of the sets is:

$$4 \leftarrow 8 \leftarrow 16 \rightarrow 8 \rightarrow 4 \tag{1}$$

$$4 \leftarrow 8 \leftarrow 16 \leftarrow 32 \leftarrow 64 \rightarrow 32 \rightarrow 16 \tag{2}$$

$$16 \leftarrow 32 \leftarrow 64 \rightarrow 32 \rightarrow 16 \rightarrow 8 \rightarrow 4 \tag{3}$$

$$4 \leftarrow 8 \leftarrow 16 \rightarrow 8 \rightarrow 4 \tag{4}$$

$$4 \leftarrow 8 \leftarrow 16 \leftarrow 32 \leftarrow 64 \rightarrow 32 \rightarrow 16 \tag{5}$$

$$16 \leftarrow 32 \leftarrow 64 \rightarrow 32 \rightarrow 16 \rightarrow 8 \rightarrow 4 \tag{6}$$

Here, the arrows represent the direction of the computations for the inbound phases and for a detailed sketch we refer to Figure 2. We start from the middle and then propagate outwards by computing the cross-product[1] of the sets and using the filtering conditions. For each inbound we try all possible 2^{16} pairs in Step 0. The number of sets, the bit length of the middle values (size) of each list, and the number of filtering conditions on words followed by the number of pairs in each set are given in Table 2. The complexities given in the Table 2 are not optimized yet, we will describe the improved complexities later in Section 3.1.

Merging Inbound Phases. The remaining pairs at inbound i are stored on list L_i. Connecting the six lists is performed in three steps as follows:

1. Whenever a pair is obtained from set 2, we check whether it exists in L_3 or not. If it does, another check is done for L_1. Since we have $2^{23.44}$ and $2^{83.96}$ elements in lists 1 and 3 respectively, $2^{83.96}$ pairs passing the second inbound phase, and 32-bit and 128-bit conditions for the matches, the expected number of remaining pairs is $2^{23.44} \cdot 2^{-32} \cdot (2^{83.96} \cdot 2^{-128} \cdot 2^{83.96}) = 2^{31.36}$. We store these these pairs in list A.

2. Similarly, whenever a pair is obtained from set 5, we check whether it exists in L_6 or not. If it does, another check is done for L_4. Since we have $2^{32.72}$ and $2^{83.96}$ elements in lists 4 and 6 respectively, 2^{80} pairs passing the fifth inbound phase, and 32-bit and 128-bit conditions for the matches, the expected number of remaining pairs is $2^{32.72} \cdot 2^{-32} \cdot (2^{83.96} \cdot 2^{-128} \cdot 2^{83.96}) = 2^{40.64}$. We store these pairs in list B.

3. Last step is merging these sets A and B. We have $2^{31.36}$ elements in A and $2^{40.64}$ elements in B and 32 bits of condition. Therefore the total expected number of remaining pairs is $2^{31.36} \cdot 2^{-32} \cdot 2^{40.64} = 2^{40}$.

Improving the complexity of finding a solution for the differential part. We have described how to obtain the existing 2^{40} solutions for the differential part. We are going to describe here a better way of doing the inbounds, as proposed in [12, Sec.4.1]. This new technique allows us to reduce the previous complexity from $2^{99.70}$ in time and $2^{83.96}$ in memory to $2^{69.6}$ in time and $2^{67.6}$ in memory. As in our further analysis we will just use one solution (and not 2^{40}) for the differential part, we will adapt the values being able to finally reduce the complexity of this part of the attack to $2^{59.6}$ in time and $2^{57.6}$ in memory. This memory is the memory bottleneck of all the analysis presented in this paper.

1. We consider the six inbounds as described in the previous section, with the difference that, for inbounds $2, 3, 5$ and 6 we will not perform the last step, but instead we obtain for each inbound $i \in \{2, 3, 5, 6\}$ two lists $L_{A,i}$ and $L_{B,i}$ as a result, each of size $2^{49.80}$ associated to half of the corresponding differential path. As mentioned before, we are only looking to find one solution

---

[1] cross-product is an operation on two arrays that results in another array whose elements are obtained by combining each element in the first array with every element in the second array.

Table 2. Overview of inbound phases of the attack on 32 rounds of JH

	Step	Size	Sets	Filtering Conditions	Pairs Remaining	Complexity Backwards	Complexity Forwards
Inbound 1	0	8	8	1	$2^{11.91}$	—	2^{16}
	1	16	4	2	2^{16}	$2^{23.91}$	—
	2	32	2	2	$2^{24.18}$	$2^{32.09}$	—
	3	64	1	4	$2^{32.72}$	$2^{48.46}$	—
	4	64	1	4^a	$2^{23.44}$		
Inbound 2	0	8	32	1	$2^{11.91}$	—	2^{16}
	1	16	16	2	2^{16}	$2^{23.91}$	—
	2	32	8	2	$2^{24.18}$	—	$2^{32.09}$
	3	64	4	4	$2^{32.72}$	$2^{48.46}$	—
	4	128	2	4	$2^{49.80}$	$2^{65.54}$	—
	5	256	1	4	$2^{83.96}$	$2^{99.70}$	—
Inbound 3	0	8	32	1	$2^{11.91}$	—	2^{16}
	1	16	16	2	2^{16}	$2^{23.91}$	—
	2	32	8	2	$2^{24.18}$	$2^{32.09}$	—
	3	64	4	4	$2^{32.72}$	—	$2^{48.46}$
	4	128	2	4	$2^{49.80}$	—	$2^{65.54}$
	5	256	1	4	$2^{83.96}$	—	$2^{99.70}$
Inbound 4	0	8	8	1	$2^{11.91}$	—	2^{16}
	1	16	4	2	2^{16}	$2^{23.91}$	—
	2	32	2	2	$2^{24.18}$	$2^{32.09}$	—
	3	64	1	4	$2^{32.72}$	$2^{48.46}$	—
Inbound 5	0	8	32	1	$2^{11.91}$	—	2^{16}
	1	16	16	2	2^{16}	$2^{23.91}$	—
	2	32	8	2	$2^{24.18}$	—	$2^{32.09}$
	3	64	4	4	$2^{32.72}$	$2^{48.26}$	—
	4	128	2	4	$2^{49.80}$	$2^{65.54}$	—
	5	256	1	4	$2^{83.96}$	$2^{99.70}$	—
Inbound 6	0	8	32	1	$2^{11.91}$	—	2^{16}
	1	16	16	2	2^{16}	$2^{23.91}$	—
	2	32	8	2	$2^{24.18}$	$2^{32.0}$	—
	3	64	4	4	$2^{32.72}$	—	$2^{48.46}$
	4	128	2	4	$2^{49.80}$	—	$2^{65.54}$
	5	256	1	4	$2^{83.96}$	—	$2^{99.70}$

^{a} Check whether the pairs satisfy the desired input difference

for the whole differential path. Then, instead of the $2^{49.80}$ existing solutions for each list, we can consider $2^{44.8}$ elements on each list.

2. First, we merge lists $L_{A,2}$ and $L_{A,3}$. We have 16-bit conditions on values and 16-bit conditions on differences. We obtain a new list $L_{A,23}$ of size $2^{44.8+44.8-32} = 2^{57.6}$. We do the same with $L_{B,2}$ and $L_{B,3}$ to obtain $L_{B,23}$.

Note that this list does not need to be stored, as we can perform the following step whenever an element is found.

3. In order to find a whole solution for the differential part of inbounds 2 and 3, one pair of elements from $L_{A,23}$ and from $L_{B,23}$ still needs to satisfy the following conditions: 32 bits from the parts $L_{A,2}$ and $L_{B,3}$, 32 bits from $L_{B,2}$ and $L_{A,3}$, 3.91×4 from the step 5 of inbound 2 that we have not yet verified and 3.91×4 from step 5 of inbound 3 that is not yet verified either. Therefore, we have 95.28-bit conditions in total to merge $L_{A,23}$ and $L_{B,23}$. For each element in $L_{B,23}$ we can check with constant cost if the corresponding element appears in $L_{A,23}$ (it can be done by a lookup in a table, representing the differential transitions of L and next by a lookup in the list $L_{A,23}$ to see if the wanted elements appear. See [13,12] and Figure 3 for more details). When we find a good pair, we store it in the list L_{23} that has a size of about $2^{19.92}$ elements satisfying the differential part of rounds from 4 to 16. The cost of this step is then $2^{57.6+1}$ in time and $2^{59.6}$ in memory.

4. Do the same with inbounds 5 and 6, to obtain list L_{56} of size $2^{19.92}$, with a cost of $2^{57.6+1}$ in time and $2^{57.6}$ in memory.

5. Merge the solutions obtained in the first inbound with the ones in L_{23}, obtaining a new set L_{123} of size $2^{19.92+23.44-32} = 2^{11.36}$.

6. Merge the solutions obtained from step 4 with list L_{56} obtaining a new one, L_{456} of size $2^{19.92+32.72-32} = 2^{20.64}$.

7. Finally, merging L_{123} and L_{456} gives $2^{11.36+20.64-32} = 1$ partial solution for the differential part of the path from round 0 to round 32.

The complexity of obtaining one partial solution for rounds from 0 to 32 is dominated by Steps $2 - 4$ of the algorithm. As a result, the complexity of matching the active bytes becomes $2^{59.6}$ in time and $2^{57.6}$ in memory.

3.2   Matching the Passive Bytes

In Figure 4, colored boxes denote the S-boxes whose values have already been fixed from the inbound phases. Note that, we have not treated the passive bits yet (i.e., found the remaining values that would complete the path). We will propose a way of finding 2^{32} solutions that verify the path from rounds 4 to 26 with time complexity 2^{96} and memory complexity $2^{51.58}$. This can be done in three steps as follows:

1. (Rounds 10 to 14): The sets of groups of 8 bits denoted by a, b, c, d, e, f in round 14 are independent of each other in this part of the path. In round 10, 32 bits are already fixed for each of these sets (groups of 4 bits denoted by A, B, C, D, E, F). By using all possible values of the remaining 96 passive bits (32 bits not fixed from A, B, C, D, E, F plus 64 from the remaining state at round 10), we can easily compute a list of 2^{96} elements with cost 2^{96} that satisfy the 32 bit conditions for each of the groups.

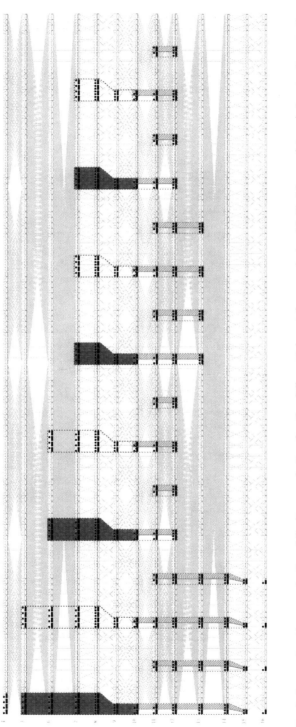

Fig. 3. Improving the complexity of finding a solution for the differential part for rounds 4 − 16: Red boxes (above) denote the sets $L_{A,2}$ (filled) and $L_{B,2}$, blue boxes (below) denote the sets $L_{A,3}$ and $L_{B,3}$ (filled)

2. (Rounds 14 to 20): In round 20, we have 256 bits (green S-boxes ■) whose values are fixed from the solutions of the second inbound phase. We can divide the state in round 19 (until the state in round 14) in 4 independent parts (m, n, o, p). In Figure 4, the fixed bits coming from round 20 are denoted by green lines and the ones of the first inbound phase are denoted in blue "■". Note that the three parts m, n, o are identical, while p is different since there are some differences and some additional fixed values in it.

   We fix the parts m and n to some values that satisfy all the conditions of the fixed bits in rounds 19 and 14. This can be done as follows: Similar to what we have done in step 1, we can divide the state of rounds $16 - 19$ (for each part separately) into four groups (x, y, z, u) such that they are independent of each other when computing forwards.

   In round 16, each group has 16 bits whose values have already been fixed and 48 bits of freedom. We see that each group affects only one fourth of the green lines (16 bits in total) in round 19. Therefore, there exist $2^{48-16} = 2^{32}$ possibilities for each group x, y, z, u but we just need one. This one can then be found with a cost of about 2^{16}.

3. (Merging): Each of the sets $L_a, \dots, L_f$ has 2^{96} possible values from step 1, and fixing m and n fixes 64 bits for each of them in round 14. This gives us in average $2^{96-64} = 2^{32}$ possible values for each set in the half of the state associated to o and p in round 14.

   For the part p we use the same idea explained in step 2. Group x is completely fixed due to the differential characteristic, and only the groups y, z, u have freedom, so there exists $(2^{32})^3 = 2^{96}$ possibilities. For each possibility, we compute the part of state in round 14 associated to p. We have 32 bits of condition for each of lists, and in average 2^{32} values are associated to each list. Thus, for each of the computed values, we will have only one remaining element that will determine the values at positions $a - f$ in the part o.

   Now, we have 2^{96} possible o values. The probability that a fixed value verifies the conditions of o in round 19 is $(2^{-4})^{16} = 2^{-64}$. Therefore, we obtain $2^{96-64} = 2^{32}$ solutions that verify the whole path from round 4 to round 26 with a complexity in time of 2^{96}.

Note that we do not need to store the lists $L_a, \dots, L_f$ of elements from round 14 each of size 2^{96} but we can instead store for each of them two lists of size 2^{48} corresponding to the upper and down halves of the corresponding groups in state 13. Then, when fixing a value of m and n we can check with a cost of 2^{32} which will be the list of 2^{32} values for o and p that we obtained in step 3.

Finally, we have obtained 2^{32} complete solutions for the path from 4 to 26 with a cost of 2^{96} in time, and $6 \cdot 2 \cdot 2^{48} \approx 2^{51.58}$ in memory.

Semi-free-start near-collisions up to 32 rounds: Up to now, we have found solutions for the passive bytes from rounds $4 - 26$. If we want a solution for the path from round 0 to round 26, we will have to repeat the previous procedure of matching the passive bytes 2^{16} times (as the probability of passing from round 0 to 4 is 2^{-48} and we have 2^{32} pairs). Then, we can find a solution for rounds $0 - 26$ with complexity 2^{112} in time. In order to extend this result to 32 rounds, we have to repeat the previous procedure 2^{192} times (since we have 64 and 128 bits of condition from rounds 26 and 27 respectively). Therefore, the complexity for finding a complete solutions for rounds from 0 to 32 is $2^{112} \cdot 2^{192} = 2^{304}$ in time.

Note that, we still have enough degrees of freedom. In step 1, we started with 768 bits (128×6 from the groups $a - f$) in round 14 and matched 192 bits (32×6 for $A - F$) in round 10. In Step 2, we have 48 bits in round 16 coming from the fourth inbound phase and we matched another 240 bits from the fifth inbound phase in round 19. So in total we have $768 - 192 - 48 - 240 = 288$ bits of degrees of freedom remaining.

3.3   Outbound Phase

The outbound phase of the attack is composed of 5 rounds in the forward direction. A detailed schema of this trail is shown in Figure 5 in appendix, and for the pairs that satisfy the inbound phase, we expect to see the following differential trail in the outbound phase:

$$\text{Inbound Phase} \;\rightarrow 4 \rightarrow 8 \rightarrow 16 \rightarrow 8 \rightarrow 4 \rightarrow 8$$

Semi-free-start near-collisions up to 37 rounds. For 32 rounds of the JH compression function, we obtain a semi-free-start near-collision for 1002 bits. We can simply increase the number of rounds by proceeding forwards in the outbound phase. Note that, we have an additional probability of $2^{-32} \times 2^{-16}$ coming from the eight filtering conditions in round 34 and the four filtering conditions in round 35. Thus, the complexity of the active part of the attack remains the same: $2^{59.6}$ in time and $2^{57.6}$ in memory. This is the case as one solution for the differential part is enough for the attack, as it will have different values at the bits with conditions in the outbound part when the passive part is modified. The complexity of the passive part becomes $2^{304} \cdot 2^{48} = 2^{352}$ in time and $2^{51.58}$ in memory.

The details can be seen in Table 3. We also take into account the colliding bits that we obtain at the output of the compression function after the final degrouping with the differences from the message.

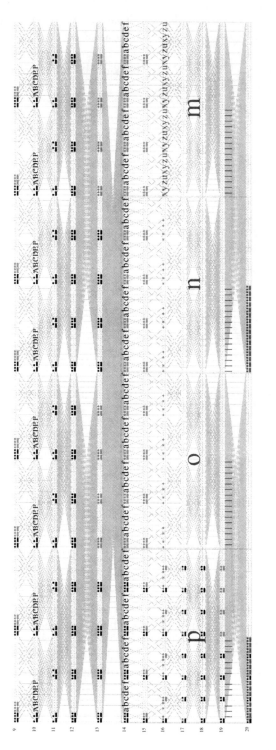

Fig. 4. Matching the passive bytes (bit-slice representation)

Table 3. Comparison of complexity of the generic attack for near-collisions and our results

#Rounds	# Colliding bits	Generic Attack Complexity	Our Results
23	892	$2^{230.51}$	$2^{59.6}$
24 − 26	762	$2^{99.18}$	$2^{59.6}$ [a]
26	960	$2^{341.45}$	2^{112}
27	896	$2^{236.06}$	2^{112}
32	1002	$2^{437.12}$	2^{304}
33	986	$2^{396.77}$	2^{304}
34	954	$2^{329.97}$	2^{304}
35	986	$2^{396.77}$	2^{336}
36	1002	$2^{437.12}$	2^{352}
37	986	$2^{396.77}$	2^{352}
38	928	$2^{284.45}$	2^{352}

[a] Obtained directly from the solutions of the active part, without need of matching the passive bits

4   Distinguishers on JH

Indifferentiability is considered to be a desirable property of any secure hash function design. Moreover, for many of the designs, the indifferentiability proofs for the mode of operation are based on the assumption that the underlying permutation (function) is ideal (i.e., random permutation). This is the case of the indifferentiability proof of JH [1], that supposes that E_d is a random permutation.

In this section, we present a distinguisher for E_8 showing that it is distinguishable form a random permutation. Using the differential path that we presented in the previous section, we can build the distinguishers on the full 42 rounds of the internal permutation E_8 with no additional complexity. As a result of our distinguisher, the proof from [1] does not apply to JH as the assumption of E_8 behaving like random does not hold. Next, we explain how these distinguishers on the internal permutation can be easily extended to distinguishers on the compression function.

There exists also a known trivial distinguisher on the construction of the compression function of JH: If the chaining value has a difference that can be cancelled by the message block, then the output will have a difference directly related to the one coming from the message block. This implies that both the message and the chaining values have differences. Contrary to the trivial one, our compression function distinguisher exploits the properties of the internal permutation and only needs differences in the message or in the chaining value.

4.1 Distinguishers on the Reduced Round Internal Permutation

Let us remark here briefly that if we find solutions for rounds 4 to 20, and then let them spread freely backward (difference in 64 bits) and forward (difference in 256 bits), we can obtain a distinguisher for 26 rounds with a much lower complexity: $2^{59.6}$ in time and $2^{57.6}$ in memory (the cost of the differential part). As in this paper the aim is reaching a higher number of rounds, we do not go further into the details.

4.2 Distinguishers on the Full Internal Permutation

In the previous sections we showed that a solution for 37 rounds can be obtained with a time complexity of 2^{352} in time and $2^{57.6}$ in memory. In Figure 5 from the appendix, we see how these active words diffuse to the state after 42 rounds with probability one. Therefore, before the degrouping operation we have 64 active and 192 passive words in the state. The number of active and passive bits still remain the same after the degrouping operation. It is important to remark that the positions of the active bits are fixed, also after the degrouping operation.

We can then build a distinguisher that will distinguish the 42-round permutation E_8 from a random permutation using this path. This distinguisher aims at finding a pair of input states (A, A') such that $E_8(A) \oplus E_8(A')$ collide in the 768 bits mentioned above. Let $A \oplus A' = \Delta_1$ correspond to the input difference of the differential path, then $|\Delta_1| = 8$ bits. Similarly, let $B = E_8(A)$ and $B' = E_8(A')$, then the output difference is $B \oplus B' = \Delta_2$ where $|\Delta_2| = 256$.

In the case of a random function, we calculate the complexity of such a distinguisher as follows: We fix the values of the passive bits in the input; but not the ones of the active bits. Then, we have $2^{|\Delta_1|}$ possibilities for the values from the active bits. We compute the output of E_8 for each one of these values and store them in a list. From this list we can obtain $\binom{2^{|\Delta_1|}}{2}$ pairs with the given input difference pattern. The probability of satisfying the desired output difference pattern is $2^{|\Delta_2|-1024}$ for each pair, so we repeat the procedure with a new value for the input passive bits until we find a solution. The time complexity of finding such an input pair will be:

$$\frac{2^{|\Delta_1|}}{2^{(|\Delta_1|-1)} \cdot (2^{|\Delta_1|} - 1) \cdot 2^{|\Delta_2|-1024}} = 2^{761}.$$

Instead, in our case the complexity of finding such an input pair is the complexity of finding a solution for the path, that is 2^{352} in time and $2^{57.6}$ in memory.

Another distinguisher of E_8 can be built if we consider the scenario where the differential path for rounds $0 - 4$ does not need to be verified, i.e., $|\Delta_1| = 64$. In this case, we consider that from round 4 to 0 we obtain the differences that propagate with probability one. Therefore, the matching of the passive part does not need to be repeated 2^{208} times but only 2^{160} (as we do not need 2^{48} extra repetitions for verifying rounds 0 to 4). The complexity of this distinguisher will then be 2^{304}, and provides a pair of inputs A and A' that produce an output with 768 colliding bits as the ones represented in Figure 5 from appendix. The complexity of such a generic distinguisher would be $\frac{2^{64}}{(2^{64}-1)\cdot2^{63-768}} = 2^{705}$, while in our case is 2^{304} in time and $2^{57.6}$ in memory.

4.3 Distinguishers on the Full Compression Function

We should emphasize that our distinguishers on E_8 can be easily converted to a distinguisher on the full compression function of JH42. We only need to xor this message difference to the output of E_8 as specified.

For our first distinguisher, the input difference is already arranged such that we only have difference in the message. These active bits coming from the message coincide with the active bits in the output at the xor operation. As a result, we have the same 768 passive bits. The same applies for our second distinguisher when we have differences only in the chaining value.

5 Conclusion

In this paper, we have presented semi-free-start internal near-collisions up to 37 rounds by using rebound attack techniques. We first obtained a 960-bit semi-free-start near-collision for 26 rounds of the JH compression function with a time complexity of 2^{112} and a memory complexity of $2^{57.6}$. We then extended this to 986-bit semi-free-start near-collision for 37 rounds by repeating the algorithm. Time complexity of the attack is increased to 2^{352} and the memory complexity remains the same. We also presented semi-free-start near-collision results for intermediate rounds $26 - 37$ in Table 3. Our findings are summarized in Table 1.

Even more, we have presented distinguishers on the full 42 rounds of the internal permutation E_8 of the tweaked SHA-3 finalist JH. The best distinguisher has a time complexity of 2^{304} in time and $2^{57.6}$ in memory and provides solutions for the differential path on the 42 rounds. Obtaining such a pair of inputs producing a same truncated differential in the output for a random function would cost 2^{705} in time. Our internal permutation distinguishers can easily be extended to compression function distinguishers with the same complexity.

Although our results do not present a threat to the security of the JH hash function, they invalidate the JH indifferentiability proof presented in [1].

References

1. Bhattacharyya, R., Mandal, A., Nandi, M.: Security Analysis of the Mode of JH Hash Function. In: Hong and Iwata [4], pp. 168–191
2. Burmester, M., Tsudik, G., Magliveras, S.S., Ilić, I. (eds.): ISC 2010. LNCS, vol. 6531. Springer, Heidelberg (2011)
3. Gilbert, H., Peyrin, T.: Super-sbox cryptanalysis: Improved attacks for aes-like permutations. In: Hong and Iwata [4], pp. 365–383
4. Hong, S., Iwata, T. (eds.): FSE 2010. LNCS, vol. 6147. Springer, Heidelberg (2010)

5. Ideguchi, K., Tischhauser, E., Preneel, B.: Improved collision attacks on the reduced-round grøstl hash function. In: Burmester, et al. (eds.) [2], pp. 1–16 (2010)
6. Lamberger, M., Mendel, F., Rechberger, C., Rijmen, V., Schläffer, M.: Rebound Distinguishers: Results on the Full Whirlpool Compression Function. In: Matsui [7], pp. 126–143
7. Matsui, M. (ed.): ASIACRYPT 2009. LNCS, vol. 5912. Springer, Heidelberg (2009)
8. Matusiewicz, K., Naya-Plasencia, M., Nikolic, I., Sasaki, Y., Schläffer, M.: Rebound Attack on the Full Lane Compression Function. In: Matsui [7], pp. 106–125
9. Mendel, F., Peyrin, T., Rechberger, C., Schläffer, M.: Improved Cryptanalysis of the Reduced Grøstl Compression Function, ECHO Permutation and AES Block Cipher. In: Jacobson Jr., M.J., Rijmen, V., Safavi-Naini, R. (eds.) SAC 2009. LNCS, vol. 5867, pp. 16–35. Springer, Heidelberg (2009)
10. Mendel, F., Rechberger, C., Schläffer, M., Thomsen, S.S.: The Rebound Attack: Cryptanalysis of Reduced Whirlpool and Grøstl. In: Dunkelman, O. (ed.) FSE 2009. LNCS, vol. 5665, pp. 260–276. Springer, Heidelberg (2009)
11. Mendel, F., Rechberger, C., Schläffer, M., Thomsen, S.S.: Rebound Attacks on the Reduced Grøstl Hash Function. In: Pieprzyk, J. (ed.) CT-RSA 2010. LNCS, vol. 5985, pp. 350–365. Springer, Heidelberg (2010)
12. Naya-Plasencia, M.: How to Improve Rebound Attacks. Cryptology ePrint Archive, Report 2010/607 (2010) (extended version), http://eprint.iacr.org/2010/607.pdf, http://eprint.iacr.org/
13. Naya-Plasencia, M.: How to Improve Rebound Attacks. In: Rogaway, P. (ed.) CRYPTO 2011. LNCS, vol. 6841, pp. 188–205. Springer, Heidelberg (2011)
14. Peyrin, T.: Improved Differential Attacks for Echo and Grøstl. In: Rabin, T. (ed.) CRYPTO 2010. LNCS, vol. 6223, pp. 370–392. Springer, Heidelberg (2010)
15. Rijmen, V., Toz, D., Varici, K.: Rebound Attack on Reduced-Round Versions of JH. In: Hong and Iwata [4], pp. 286–303
16. Rivest, R.L.: The MD5 Message-Digest Algorithm. RFC 1321 (1992), http://www.ietf.org/rfc/rfc1321.txt
17. Sasaki, Y., Li, Y., Wang, L., Sakiyama, K., Ohta, K.: Non-Full-Active Super-Sbox Analysis: Applications to ECHO and Grøstl. In: Abe, M. (ed.) ASIACRYPT 2010. LNCS, vol. 6477, pp. 38–55. Springer, Heidelberg (2010)
18. Schläffer, M.: Subspace Distinguisher for 5/8 Rounds of the ECHO-256 Hash Function. In: Biryukov, A., Gong, G., Stinson, D.R. (eds.) SAC 2010. LNCS, vol. 6544, pp. 369–387. Springer, Heidelberg (2011)
19. of Standards, N.I., Technology: FIPS 180-1:Secure Hash Standard (1995), http://csrc.nist.gov
20. Wu, H.: The Hash Function JH. Submission to NIST (2008), http://icsd.i2r.a-star.edu.sg/staff/hongjun/jh/jh_round2.pdf

A   Outbound Phase Figure

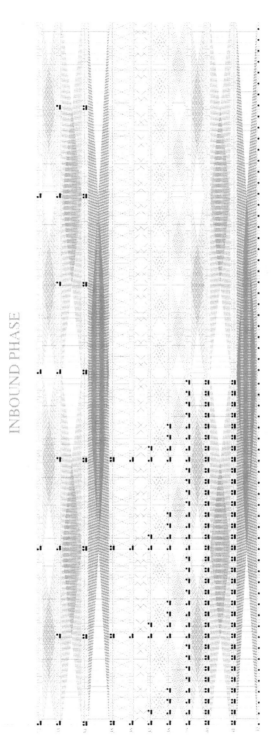

INBOUND PHASE

Fig. 5. Differential characteristic for the outbound phase of JH Compression Function (bit-slice representation)

Second-Order Differential Collisions for Reduced SHA-256

Alex Biryukov[1], Mario Lamberger[2], Florian Mendel[2], and Ivica Nikolić[1]

[1] University of Luxembourg, Luxembourg
[2] IAIK, Graz University of Technology, Austria

Abstract. In this work, we introduce a new non-random property for hash/compression functions using the theory of higher order differentials. Based on this, we show a second-order differential collision for the compression function of SHA-256 reduced to 47 out of 64 steps with practical complexity. We have implemented the attack and provide an example. Our results suggest that the security margin of SHA-256 is much lower than the security margin of most of the SHA-3 finalists in this setting. The techniques employed in this attack are based on a rectangle/boomerang approach and cover advanced search algorithms for good characteristics and message modification techniques. Our analysis also exposes flaws in all of the previously published related-key rectangle attacks on the SHACAL-2 block cipher, which is based on SHA-256. We provide valid rectangles for 48 steps of SHACAL-2.

Keywords: Hash functions, higher-order differentials, non-randomness, SHA-256, SHACAL-2.

1 Introduction

The significant advances in the field of hash function research that have been made in the recent years, had a formative influence on the landscape of hash functions. The analysis of MD5 and SHA-1 has convinced many cryptographers that these widely deployed hash functions can no longer be considered secure [39,40]. As a consequence, people are evaluating alternative hash functions in the SHA-3 initiative organized by NIST [29]. During this ongoing evaluation, not only the three classical security requirements (preimage resistance, 2nd preimage resistance and collision resistance) are considered. Researchers look at (semi-) free-start collisions, near-collisions, etc. Whenever a behavior different from the one expected of a 'random oracle' can be demonstrated for a new hash function, it is considered suspect, and so are the weaknesses that are demonstrated only for the compression function. In light of this, for four out of the five third round SHA-3 candidates the best attacks are in the framework of distinguishers: boomerang distinguisher for BLAKE [6], differential distinguisher for Grøstl [32], zero-sum distinguisher on Keccak [8] and rotational rebound distinguisher for Skein [17].

With the cryptographic community joining forces in the SHA-3 competition, the SHA-2 family gets considerably less attention. Apart from being marked

D.H. Lee and X. Wang (Eds.): ASIACRYPT 2011, LNCS 7073, pp. 270–287, 2011.
© International Association for Cryptologic Research 2011

as 'relying on the same design principle as SHA-1 and MD5', the best attack to date on SHA-256 is a collision attack for 24 out of 64 steps with practical complexity [13,33] and a preimage attack on 45 steps [18] having a complexity of $2^{255.5}$.

Higher-order differentials have been introduced by Lai in [21] and first applied to block ciphers by Knudsen in [20]. The application to stream ciphers was proposed by Dinur and Shamir in [10] and Vielhaber in [35]. First attempts to apply these strategies to hash functions were published in [2]. Recently, higher-order differential attacks have been applied to several hash functions submitted to the SHA-3 initiative organized by NIST such as BLAKE [6], Hamsi [7], Keccak [8], and Luffa [42].

In this work, we present a second-order differential collision for the SHA-256 compression function on 47 out of 64 steps having practical complexity. The attack is an application of higher-order differentials on hash functions. Table 3 shows the resulting example.

Since our attack technique resembles boomerang/rectangle attacks, known from the cryptanalysis of block ciphers, we use a strict criterion for checking that the switch in the middle does not contain any contradictions that can appear due to the independency assumption of the characteristics used in the rectangle. We show that all the previous related-key rectangle distinguishers for SHACAL-2 have a common flaw in the switch due to these assumptions and present a rectangle distinguisher for 48 steps that passes our check.

Our analysis shows that the compression functions exhibit non-random properties, though they do not lead to collision/preimage attacks on the hash functions. Nevertheless, the attacks give a clear indication that if we compare the security of SHA-256 to the security of the third round SHA-3 candidates, in the this setting, then SHA-256 has one of the lowest security margins.

2   Higher-Order Differential Collisions for Compression Functions

In this section, we give a high-level description of the attack. It is an application of higher-order differential cryptanalysis on hash functions. While a standard differential attack exploits the propagation of the difference between a pair of inputs to the corresponding output differences, a higher-order differential attack exploits the propagation of the difference between differences.

Higher-order differential cryptanalysis was introduced by Lai in [21] and subsequently applied by Knudsen in [20]. We recall the basic definitions that we will use in the subsequent sections.

Definition 1. *Let $(S, +)$ and $(T, +)$ be abelian groups. For a function $f \colon S \to T$, the derivative at a point $a_1 \in S$ is defined as*

$$\Delta_{(a_1)} f(y) = f(y + a_1) - f(y). \tag{1}$$

The i-th derivative of f at $(a_1, a_2, \ldots, a_i)$ is then recursively defined as

$$\Delta_{(a_1, \ldots, a_i)} f(y) = \Delta_{(a_i)}(\Delta_{(a_1, \ldots, a_{i-1})} f(y)). \tag{2}$$

Definition 2. *A one round differential of order i for a function $f : S \to T$ is an $(i + 1)$-tuple $(a_1, a_2, \ldots, a_i; b)$ such that*

$$\Delta_{(a_1, \ldots, a_i)} f(y) = b. \tag{3}$$

When applying differential cryptanalysis to a hash function, a collision for the hash function corresponds to a pair of inputs with output difference zero. Similarly, when using higher-order differentials we define a higher-order differential collision for a function as follows.

Definition 3. *An i-th-order differential collision for $f : S \to T$ is an i-tuple $(a_1, a_2, \ldots, a_i)$ together with a value y such that*

$$\Delta_{(a_1, \ldots, a_i)} f(y) = 0. \tag{4}$$

Note that the common definition of a collision for hash functions corresponds to a higher-order differential collision of order $i = 1$.

In this work, we concentrate on *second-order differential collisions, i.e.* $i = 2$:

$$f(y) - f(y + a_2) + f(y + a_1 + a_2) - f(y + a_1) = 0 \tag{5}$$

Further we assume that we have oracle access to a function $f : S \to T$ and measure the complexity in the number of queries to f, *i.e. query complexity*, while ignoring all other computations, memory accesses, etc. Additionally, we will restrict ourselves to functions f mapping to groups $(T, +)$ with $|T| = 2^n$ which are endowed with an additive operation.

Definition 4. *Let $f : S \to T$ be as above. A solution $(y, a_1, a_2) \in S^3$ to (5) is called trivial if the complexity of producing it is $O(1)$, otherwise it is called non-trivial.*

Lemma 1. *Let $f : S \to T$ be as above. Then, a trivial solution to (5) can be found if*

1. *f is linear, or*
2. *at least one of a_1, a_2 is zero, or*
3. *$a_1 = a_2$ and the group $(T, +)$ is of the form*

$$(T, +) \simeq (\mathbb{Z}_2, +)^{n-\ell} \oplus (\mathbb{Z}_{2^\ell}, +), \tag{6}$$

for small ℓ.

Proof. If f is a linear function, then (5) collapses and any choice of (y, a_1, a_2) is a valid solution. Under the assumption that f is drawn uniformly at random from all functions $f : S \to T$, and T is not as in (6), then the only trivial solution to equation (5) is when the inputs coincide, *i.e.* either $y = y + a_2$ and $y + a_1 + a_2 = y + a_1$ leading to the case where $a_2 = 0$, or $y = y + a_1$ and $y + a_1 + a_2 = y + a_2$ leading to $a_1 = 0$.

In the third case, equation (5) boils down to $2f(y) = 2f(y + a)$. In general this is a classical meet-in-the-middle problem, however if $(T, +)$ is as in (6), this equation holds with a probability $2^{-2(\ell-1)}$ for a random function f which leads to trivial solutions for small values of ℓ.

For all the other cases, the problem of finding a solution is an instance of the generalized birthday problem proposed by Wagner [37] and therefore the number of queries depends on n. ∎

We now want to lower bound the query complexity of producing a non-trivial differential collision of order 2.

Theorem 1. *For a function $f: S \rightarrow T$ with $|T| = 2^n$, the query complexity for producing a non-trivial differential collision of order 2 is $\Omega(2^{n/3})$.*

Proof. To find an input (y, a_1, a_2) such that (5) holds, one has to try around 2^n different tuples – otherwise the required value 0, may not appear. We can freely choose three input parameters, *i.e.* y, a_1, a_2, which then fix the remaining one. Therefore, (5) can be split into three parts (but not more!), and solved by generating three independent lists of values. Obviously, the number of queries is the lowest when these lists have equal size. Hence, to have a solution for (5), one has to choose $2^{n/3}$ values for each of y, a_1, a_2, and therefore the query complexity of a differential collision of order 2 for f is $\Omega(2^{n/3})$. ∎

Remark 1. We want to note that the actual complexity might be much higher in practice than this bound for the query complexity. We are not aware of any algorithm faster than $2^{n/2}$, since dividing (5) into three independent parts is not possible (one of the terms has all the inputs, and any substitution of variables leads to a similar case).

2.1 Second-Order Differential Collision for Block-Cipher-Based Compression Functions

In all of the following, we consider block ciphers $E : \{0,1\}^k \times \{0,1\}^n \rightarrow \{0,1\}^n$ where n denotes the block length and k is the key length. For our purposes, we will also need to endow $\{0,1\}^n$ with an additive group operation. It is however not important, in which way this is done. A natural way would be to simply use the XOR operation on $\{0,1\}^n$ or the identification $\{0,1\}^n \leftrightarrow \mathbb{Z}_{2^n}$ and define the addition of $a, b \in \{0,1\}^n$ by $a + b \bmod 2^n$. Alternatively, if we have an integer w dividing n, that is $n = \ell \cdot w$, we can use the bijection of $\{0,1\}^n$ and $\mathbb{Z}_{2^w}^{\ell}$ and define the addition as the word-wise modular addition, that is,

$$(\{0,1\}^n, +) := \underbrace{(\mathbb{Z}_{2^w}, +) \times \cdots \times (\mathbb{Z}_{2^w}, +)}_{\ell \text{ times}}. \tag{7}$$

The latter definition clearly aims very specifically at the SHA-2 design. However, the particular choice of the group law has no influence on our attack.

A well known construction to turn a block cipher into a compression function is the Davies-Meyer construction. The compression function call to produce the

i-th chaining value x_i from the i-th message block and the previous chaining value x_{i-1} has the form:

$$x_i = E(m_i, x_{i-1}) + x_{i-1} \tag{8}$$

When attacking block-cipher-based hash functions, the key is *not* a secret parameter so for the sake of readability, we will slightly restate the compression function computation (8) where we consider an input variable $y = (k||x) \in \{0,1\}^{k+n}$ so that a call to the block cipher can be written as $E(y)$. Then, the Davis-Meyer compression function looks like:

$$f(y) = E(y) + \tau_n(y), \tag{9}$$

where $\tau_n(y)$ represents the n least significant bits of y.

In an analogous manner, we can also write down the compression functions for the Matyas-Meyer-Oseas and the Miyaguchi-Preneel mode which are all covered by the following proposition.

Proposition 1. *For any block-cipher-based compression function which can be written in the form*

$$f(y) = E(y) + L(y), \tag{10}$$

where L is a linear function with respect to $+$, an i-th-order differential collision for the block cipher transfers to an i-th-order collision for the compression function for $i \geq 2$.

For the proof of Proposition 1, we will need following property of $\Delta_{(a_1,...,a_i)}f(y)$:

Proposition 2 (Lai [21]). *If $\deg(f)$ denotes the non-linear degree of a multivariate polynomial function f, then*

$$\deg(\Delta_{(a)}f(y)) \leq \deg(f(y)) - 1. \tag{11}$$

Proof (of Proposition 1). Let $\Delta_{(a_1,...,a_i)}E(y) = 0$ be an i-th-order differential collision for $E(y)$. Both the higher-order differential and the mode of operation for the compression function are defined with respect to the same additive operation on $\{0,1\}^n$. Thus, from (10) we get

$$\Delta_{(a_1,...,a_i)}(E(y) + L(y)) = \Delta_{(a_1,...,a_i)}E(y) + \Delta_{(a_1,...,a_i)}L(y),$$

so we see that all the terms vanish because the linear function $L(y)$ has degree one and so for $i \geq 2$ we end up with an i-th-order differential collision for the compression function because of Proposition 2. ∎

Hence, if we want to construct a second order collision for the compression function f it is sufficient to construct a second-order collision for the block cipher. The main idea of the attack is now to use two independent high probability differential characteristics – one in forward and one in backward direction – to construct a second-order differential collision for the block cipher E and hence due to Proposition 1, for the compression function.

Therefore, the underlying block cipher E is split into two subparts, $E = E_1 \circ E_0$. Furthermore, assume we are given two differentials for the two subparts, where one holds in the forward direction and one in the backward direction and we assume that both have high probability. This part of the strategy has been already applied in other cryptanalytic attacks, we refer to Section 2.2 for related work. We also want to stress, that due to our definition above, the following differentials are actually related-key differentials. We have

$$E_0^{-1}(y + \beta) - E_0^{-1}(y) = \alpha \tag{12}$$

and

$$E_1(y + \gamma) - E_1(y) = \delta \tag{13}$$

where the differential in E_0^{-1} holds with probability p_0 and in E_1 holds with probability p_1. Using these two differentials, we can now construct a second-order differential collision for the block cipher E. This can be summarized as follows (see also Figure 1).

1. Choose a random value for X and compute $X^* = X + \beta$, $Y = X + \gamma$, and $Y^* = X^* + \gamma$.
2. Compute backward from X, X^*, Y, Y^* using E_0^{-1} to obtain P, P^*, Q, Q^*.
3. Compute forward from X, X^*, Y, Y^* using E_1 to obtain C, C^*, D, D^*.
4. Check if $P^* - P = Q^* - Q$ and $D - C = D^* - C^*$ is fulfilled.

Due to (12) and (13),

$$P^* - P = Q^* - Q = \alpha, \quad \text{resp.} \quad D - C = D^* - C^* = \delta, \tag{14}$$

will hold with probability at least p_0^2 in the backward direction, resp. p_1^2 in the forward direction. Hence, assuming that the differentials are independent the attack succeeds with a probability of $p_0^2 \cdot p_1^2$. It has to be noted that this independence assumption is quite strong, *cf.* [28]. However, if this assumption holds, the expected number of solutions to (14) is 1, if we repeat the attack about $1/(p_0^2 \cdot p_1^2)$ times. As mentioned before, in our case, there is no secret key involved, so message modification techniques (*cf.* [40]) can be used to improve this complexity.

The crucial point now is that such a solution constitutes a second-order differential collision for the block cipher E. We can restate (14) as

$$Q^* - Q - P^* + P = 0 \tag{15}$$
$$E(Q^*) - E(P^*) - E(Q) + E(P) = 0 \tag{16}$$

If we set $\alpha := a_1$ and the difference $Q - P := a_2$ we can rewrite (16) as

$$E(P + a_1 + a_2) - E(P + a_1) - E(P + a_2) + E(P) = 0, \tag{17}$$

that is, we have found a second-order differential collision for the block cipher E. Because of Proposition 1 the same statement is true for the compression function.

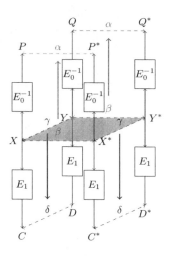

Fig. 1. Schematic view of the attack

2.2   Related Work

The attack presented in this paper stands in relation to previous results in the field of block cipher and hash function cryptanalysis. Figure 1 suggests that it stands between the *boomerang attack* and the *inside-out attack* which were both introduced by Wagner in [36] and also the *rectangle attack* by Biham *et al.* [3]. For the related-key setting, we refer to [4] (among others). We also want to refer to the *amplified boomerang attack* [16]. A previous application of the boomerang attack to block-cipher-based hash functions is due to Joux and Peyrin [15], who used the boomerang attack as a neutral bits tool. Another similar attack strategy for hash functions is the *rebound attack* introduced in [27]. Furthermore, the second-order differential related-key collisions for the block cipher used in Section 2.1 are called *differential q-multi-collisions* introduced by Biryukov *et al.* in [5] with $q = 2$. Recently, an attack framework similar to this was proposed in [6,22] and applied to HAVAL in [34].

3   Application to SHA-256

In the light of the breakthrough results of Wang *et al.* on the hash functions MD5 and SHA-1, the analysis of SHA-256 is of great interest. Moreover, SHA-2 is a reference point in terms of speed but also security for the SHA-3 candidates.

In the last few years several cryptanalytic results have been published for SHA-256. The security of SHA-256 against preimage attacks was first studied by Isobe and Shibutani in [14]. They presented a preimage attack on 24 steps. This was improved by Aoki *et al.* to 43 steps in [1] and later extended to 45 steps by Khovratovich *et al.* in [18]. All attacks are only slightly faster than the generic attack, which has a complexity of about 2^{256}. In [25], Mendel *et al.* studied the security of SHA-256 with respect to collision attacks. They presented the collision attack on SHA-256 reduced to 18 steps. After that these results have been improved by several researchers. In particular, Nikolić and Biryukov improved in [31] the collision techniques, leading to a collision attack for 23 steps of SHA-256. The best collision attacks so far are extensions of [31]. Indesteege *et al.* [13] and Sanadhya and Sarkar[33], both presented collision attacks for 24 steps. We want to note that in contrast to the preimage attacks all these attacks are of practical complexity. Furthermore, Indesteege *et al.* showed non-random properties for SHA-2 for up to 31 steps. At the rump session of Eurocrypt 2008, Yu and Wang announced that they had shown non-randomness for SHA-256 reduced to 39 steps [41]. In the same presentation they also provided a practical example for 33 steps. However, no details have been published to date. We are not aware of any attack on SHA-256 with practical complexity for more than 33 steps. In this section, we show how to construct a second-order differential collision for SHA-256 reduced to 47 (out of 64) steps, following the attack strategy described in the previous section. Since the complexity of the attack is quite low, only 2^{46} compression function evaluations, we implemented the attack. An example of a second-order differential collision for SHA-256 reduced to 47 steps is shown in Table 3.

3.1 Description of SHA-256

SHA-256 is an iterated hash function that processes 512-bit input message blocks and produces a 256-bit hash value. In the following, we briefly describe the hash function. It basically consists of two parts: the message expansion and the state update transformation. A detailed description of the hash function is given in [30].

Message Expansion. The message expansion of SHA-256 splits the 512-bit message block into 16 words M_i, $i = 0, \ldots, 15$, and expands them into 64 expanded message words W_i as follows:

$$W_i = \begin{cases} M_i & 0 \leq i < 16 \\ \sigma_1(W_{i-2}) + W_{i-7} + \sigma_0(W_{i-15}) + W_{i-16} & 16 \leq i < 64 \end{cases} . \qquad (18)$$

The functions $\sigma_0(X)$ and $\sigma_1(X)$ are given by

$$\begin{aligned} \sigma_0(X) &= (X \ggg 7) \oplus (X \ggg 18) \oplus (X \gg 3) \\ \sigma_1(X) &= (X \ggg 17) \oplus (X \ggg 19) \oplus (X \gg 10) \end{aligned} \qquad (19)$$

State Update Transformation. The state update transformation starts from a (fixed) initial value IV of eight 32-bit words and updates them in 64 steps. In each step one 32-bit word W_i is used to update the state variables $A_i, B_i, \ldots, H_i$ as follows:

$$\begin{aligned} T_1 &= H_i + \Sigma_1(E_i) + f_1(E_i, F_i, G_i) + K_i + W_i \ , \\ T_2 &= \Sigma_0(A_i) + f_0(A_i, B_i, C_i) \ , \\ A_{i+1} &= T_1 + T_2 \ , \quad B_{i+1} = A_i \ , \quad C_{i+1} = B_i \ , \quad D_{i+1} = C_i \ , \\ E_{i+1} &= D_i + T_1 \ , \quad F_{i+1} = E_i \ , \quad G_{i+1} = F_i \ , \quad H_{i+1} = G_i \ . \end{aligned} \qquad (20)$$

For the definition of the step constants K_i we refer to [30]. The bitwise Boolean functions f_1 and f_0 used in each step are defined as follows:

$$f_0(X, Y, Z) = X \wedge Y \oplus Y \wedge Z \oplus X \wedge Z$$
$$f_1(X, Y, Z) = X \wedge Y \oplus \neg X \wedge Z \tag{21}$$

The linear functions Σ_0 and Σ_1 are defined as follows:

$$\Sigma_0(X) = (X \ggg 2) \oplus (X \ggg 13) \oplus (X \ggg 22)$$
$$\Sigma_1(X) = (X \ggg 6) \oplus (X \ggg 11) \oplus (X \ggg 25) \tag{22}$$

After the last step of the state update transformation, the initial values are added to the output values of the last step (Davies-Meyer construction). The result is the final hash value or the initial value for the next message block.

3.2 Differential Characteristics

Finding the differential characteristics for both backward and forward direction is the most important and difficult part of the attack. Not only the differential characteristics need to be independent, but also they need to have high probability in order to result in a low attack complexity. As noted before, in general, the assumption on independent characteristics is quite strong, *cf.* [28].

We apply a particular approach to construct differential characteristics that are used to construct second-order differential collisions for reduced SHA-256. We run a full search for *sub-optimal* differential characteristics, *i.e.* characteristics with the following properties:

- use a linearized approximation of the attacked hash function, *i.e.* approximate all modular additions by the xor operation;
- approximate the Boolean functions f_0 and f_1 by the 0-function, except in the bits j, where either $\Delta A[j] = \Delta B[j] = \Delta C_i[j] = 1$ or $\Delta F[j] = \Delta G[j] = 1$ – in these bits approximate with 1. This requirement comes from the fact that if all three inputs to f_0 have a difference, then the output has a difference (with probability 1); a similar property holds for f_1. Note that it is possible to approximate some bits with either 0 or 1, however, this introduces a high branching leading to an infeasible search;
- the characteristic has a single bit difference in the message word at some step i ($i \leq 16$), followed by 15 message words without difference. When using such characteristic, 16 steps (the ones that follow i) can be passed with probability 1 – arguably, any characteristic that does not follow this strategy will have a low probability due to the fast diffusion of the difference coming from the message words. This type of characteristics was used to construct various related-key rectangle distinguishers for SHACAL-2 [11,19,23,24,38].

Once we have the set of sub-optimal characteristics, we try to combine them for the second-order differential collision scenario, *i.e.* try to check if the switch in the middle is possible. This is a very important requirement, as some of the characteristics cannot be combined, *i.e.* their combination leads to contradictions. Some of the conditions for the switch can be checked only by examining the differences in the characteristics, while other are checked by confirming experimentally the absence of contradictions in the switch.

Table 1. Differential characteristic for steps 1-22 using signed-bit-differences

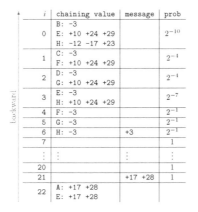

i	chaining value	message	prob
0	B: -3 E: +10 +24 +29 H: -12 -17 +23		2^{-10}
1	C: -3 F: +10 +24 +29		2^{-4}
2	D: -3 G: +10 +24 +29		2^{-4}
3	E: -3 H: +10 +24 +29		2^{-7}
4	F: -3		2^{-1}
5	G: -3		2^{-1}
6	H: -3	+3	2^{-1}
7			1
⋮	⋮	⋮	⋮
20			1
21		+17 +28	1
22	A: +17 +28 E: +17 +28		

In Table 1 and Table 2 the differential characteristics for both forward and backward direction are shown. Furthermore, the probabilities for each step of the differential characteristics are given. Note that for start we assume that the differential characteristic in the message expansion will hold with probability 1. To describe the differential characteristic we use signed-bit differences introduced by Wang *et al.* in the cryptanalysis of MD5 [40]. The advantage of using signed-bit differences is that there exists a unique mapping to both xor and modular differences. Another advantage is that the feedforward in SHA-256 is modular, hence no additional probability will be introduced for this operation.

3.3   Complexity of the Attack

Using the differential characteristics given in the previous section, we can construct a second-order differential collision for SHA-256 reduced to 47 out of 64 steps. The differential characteristic used in backward direction holds with probability 2^{-28} and the differential characteristic used in forward direction holds with probability 2^{-72}. Hence, assuming that the two differential characteristics are independent and using the most naive method, *i.e.* random trials, to fulfill all the conditions imposed by the differential characteristics would result in an attack complexity of $2^{2\cdot(72+28)} = 2^{200}$. This is too high for a practical attack on reduced SHA-256. However, the complexity can be significantly reduced by using message modification techniques. Moreover, some conditions at the end of the differential characteristics can be ignored which also improves the attack complexity.

Table 2. Differential characteristic for steps 23-47 using signed-bit-differences. Note that conditions imposed by the characteristic in steps 23-30 are fulfilled in a deterministic way using message modification techniques.

	i	chaining value	message	prob
	22	B: +3 +12 +14 +19 +23 +32 C: +25 E: -3 -7 -13 F: -12 -23 G: -25 H: -1 +3 +7 +14 +15 +24 +26 +28 -30	-25	2^{-22}
	23	C: +3 +12 +14 +19 +23 +32 D: +25 F: -3 -7 -13 G: -12 -23 H: -25		2^{-13}
	24	A: -25 D: +3 +12 +14 +19 +23 +32 G: -3 -7 -13 H: -12 -23		2^{-10}
forward	25	B: -25 E: +14 +19 +32 H: -3 -7 -13		2^{-7}
	26	C: -25 F: +14 +19 +32		2^{-4}
	27	D: -25 G: +14 +19 +32		2^{-4}
	28	E: -25 H: +14 +19 +32		2^{-4}
	29	F: -25		2^{-1}
	30	G: -25		2^{-1}
	31	H: -25	+25	1
	32			1
	⋮	⋮	⋮	⋮
	45			1
	46		-7 -18 -22	2^{-6}
	47	A: -7 -18 -22 E: -7 -18 -22		

(rightside annotation: *message modification*)

Ignoring conditions at the end. As was already observed in the cryptanalysis of SHA-1, conditions resulting from the modular addition in the last steps of the differential characteristic can be ignored [9,39]. The reason is that we do not care about carries in the last steps, since the modular difference will be the same. In the attack on SHA-256, we can ignore 6 conditions in step 46 in the characteristic used in forward direction and 3 conditions in step 1 in the characteristic used in backward direction. This improves the complexity of the attack by a factor of $2^{2 \cdot (3+6)} = 2^{18}$ resulting in a complexity of 2^{182}.

Impact of additional less probable characteristics. Even if all the message conditions for the two characteristics are already in place, there exist a number of differential characteristics which hold with the same or a slightly lower probability. Hence, it is advantageous to consider differentials. A similar effect has been exploited by Kelsey *et al.* in the amplified boomerang attack on block ciphers [16]. For hash functions, this has been systematically studied for SHA-1 in [26]. We achieve a significant speedup in the attack on SHA-256 by allowing these additional characteristics. For instance by changing the signs of the differences in chaining variable H_0, we get 2^3 additional differential characteristics for the backward direction which all hold with the same probability as the original differential characteristic given in Table 1. Similarly, we also get 2^3 additional differential characteristic by changing the signs of the differences in chaining variable H_3. This already improves the complexity of the attack by a factor of 2^6. Furthermore, if we do not block the input differences of f_1 and f_0 in step 1, we get 2^4 additional characteristics which again holds with the same probability. Thus, by allowing additional differential characteristics the complexity of the attack can be improved by a factor of 2^{10}, resulting in an attack complexity of 2^{172}. We want to stress, that in practice there exist many more additional differential characteristics that can be used and hence the attack complexity is much lower in practice.

Message modification. As already indicated in Section 2 message modification techniques can be used to significantly improve the complexity of the attack. The notion of message modification has been introduced by Wang *et al.* in the cryptanalysis of MD5 and other hash functions [40]. The main idea is to choose the message words and internal chaining variables in an attack on the hash function to fulfill the conditions imposed by the differential characteristic in a deterministic way.

Luckily, by using message modification techniques, we can fulfill all conditions imposed by the differential characteristic in steps 22-30 by choosing the expanded message words $W_{22}, \ldots, W_{30}$ accordingly. This improves the complexity of the attack by a factor of $2^{2 \cdot 66} = 2^{132}$ resulting in an attack complexity of 2^{40}.

Additional costs coming from the message expansion. So far we assumed that the differential characteristic in the message expansion of SHA-256 will hold with probability 1. However, since the message expansion of SHA-256 is not linear, this is not the case in practice. Indeed most of the conditions that have to be fulfilled to guaranty that the characteristic holds in the message expansion can be fulfilled by choosing the expanded message words and differences in steps 21-30 accordingly. Only the conditions for step 5 and step 6 imposed by the differential characteristic used in backward direction cannot be fulfilled deterministically (see Table 1). In step 6 we need that:

$$W_6^* - W_6 = 3 \tag{23}$$

Furthermore, to ensure that there will be no difference in W_5 we need that:

$$W_{21}^* - \sigma_0(W_6^*) - (W_{21} - \sigma_0(W_6)) = 0 \tag{24}$$

Since (23) will hold with a probability of 2^{-1} and (24) will hold with probability 2^{-2}, this adds an additional factor of $2^{2 \cdot 3} = 2^6$ to the attack complexity. Hence, the final complexity of the attack is 2^{46}. By Theorem 1, the complexity in the generic case is around 2^{85}.

Implementation. Even though the complexity of the attack was estimated to be about 2^{46}, we expected that the complexity will be lower in practice due to the impact of additional differential characteristics. This was confirmed by our implementation. In Table 3, an example of a second-order differential collision for 47 steps of SHA-256 is shown.

Table 3. Example of a second-order differential collision $f(y + a_1 + a_2) - f(y + a_1) - f(y + a_2) + f(y) = 0$ for 47 steps of the SHA-256 compression function

	89456784	4ef9daf6	0ab509f5	3fdf6c93	fe7afc67	b03ad81a	fd306df9	1d14cadd
y	daea3041	70f45fd7	4a03bf20	c13c961c	6a12c686	fc7be50c	7b060fc2	0ee1e276
	630c3c7e	734246a4	88401eb0	9aac88c1	4b6bca45	b777c1e6	5537cdb1	9b5bc93b

	00000000	00000000	00000000	00000000	00000000	00000000	00000004	00000000
a_1	00000000	00000000	00000000	00000000	00000000	00000000	00000000	00000000
	00000000	fffffffc	00000000	fffffffc	10800200	00000000	ff800000	803ef414

	2335e851	20f48326	69151911	f5cb76c2	b9d69e31	32685b9c	90cceff7	081ebbf7
a_2	967c8864	a43138d1	7e9a3eec	c39cf7d3	5914e008	8d0d3b73	e077c63f	d29db1b0
	742b8c01	92248811	a119f182	dd829be5	e3e1802e	21130e9f	1dacd7d3	8acf11fe

4   Applications to Related Primitives

The results presented in the previous section have a direct impact on the analysis of primitives similar to SHA-256. First of all, due to the similar design of SHA-256 and SHA-512 the attack extends in straight forward way to SHA-512. Second, our search for sub-optimal characteristics in SHA-256, can be used to find suitable characteristics for a related-key rectangle attack on the SHACAL-2 block cipher [12]. which is based on SHA-256. The block cipher proposed by Handschuh and Naccache in 2000 and was evaluated by NESSIE.

4.1   Application to SHA-512

The structure of SHA-512 is very similar to SHA-256 – only the size of the words is increased from 32 to 64 bits and the linear functions $\Sigma_0, \Sigma_1, \sigma_0, \sigma_1$ are redefined. Also the number of steps is increased from 64 to 80. Since the design of SHA-512 is very similar to SHA-256 the attack presented for SHA-256 extends in a straight forward way to SHA-512. Furthermore, due to the larger hash size of SHA-512 compared to SHA-256 also the complexity of the generic attack increases, *i.e.* it becomes around 2^{170}. Hence, the attack can be extended to more steps than it was the case for SHA-256 by adding steps at the beginning. Also, due to the larger word size and hence worse diffusion within the words adding steps in the middle becomes easier. Thus, we expect that several steps can be added in the middle as well. This is work in progress.

4.2   Application to SHACAL-2

In the past several related-key rectangle attacks have been published for the SHACAL-2 block cipher [11,19,23,24,38]. It is interesting to note that all of the published rectangles on SHACAL-2 contain a flaw in the analysis. This flaw is in the switch of the rectangle, since the used characteristics are not independent and the conditions cannot be satisfied simultaneously in both of the characteristics. In the rectangles in [24,38,11], in the the switch in the middle the following differences in bit 13 are defined: at the output of the backward characteristic $\Delta E[13] = 1$, $\Delta F[13] = \Delta G[13] = 0$; at the input of the forward characteristic $\Delta E[13] = 0, \Delta F[13] = 1, \Delta G[13] = 0$. At the first step of the forward characteristics it is assumed that the output difference of f_1 is zero. However, this is not possible for both of the characteristics. Since $\Delta F[13] = 1$, the value of $E[13]$ has to be 0. Then, in the second characteristic (on the other side of the rectangle), since the output difference $\Delta E[13]$ is 1, then this $E[13]$ will be 1, and therefore the output of f_1 in bit 13 will produce difference. A similar contradiction can be seen in [23]. First, since there is a difference in bit 13 in E_{25} coming from the upper trail, one needs the differences in F_{25} and G_{25} in bit 13 to be the same (have the same sign) in the lower trail (see Table 3), otherwise there will be a contradiction. In the next step we have $G_{26} = F_{25}$ and $H_{26} = G_{25}$ and hence the difference in bit 13 of G_{26} and H_{26} have the same sign. This leads now to a contradiction, since in the characteristic it is required that these two differences cancel out. However, since they have the same sign this is not possible and we get a contradiction. In [19], in the lower trail (Table 6) there are conditions on E_{24} in bits (2,14,15,24,25) to guarantee that the differences in G_{24} behave correctly, in particular the bit 24 of E_{24} has to be 1. But from the upper trail we get difference in W_{23} in bits 13,24, and 28, and hence E_{24} will have difference in bits 13,24,28. Therefore, E_{24} cannot take the value 1 (in these three bits) in both of the bottom characteristics. This can be fixed by allowing a carry from bit 13 to 24 to cancel the difference in bit 24, but then there will always be a difference in bit 14 and 15 which again leads to a contradiction.

Each of the published rectangle attack works for the whole key space. Further, we relax this requirement, *i.e.* we examine the security of the cipher in a weak-key class. These types of attacks are inline with the recent attacks on AES-256 [5]. We analyze a secret-key primitive, hence the message modification techniques presented in the previous section are not applicable and therefore the complexity of the attack is fully determined by the probability of the characteristics used in the rectangle. The probability of the characteristic in the key schedule not necessarily has to be 1 (it is a weak-key class), however this adds to the total complexity of the attack.

Our search for sub-optimal characteristics in SHA-256, can be used as well to find characteristics suitable for a related-key rectangle attack on SHACAL-2. Note that the search avoids using the above mentioned characteristics (with flaws), since it checks experimentally, that all the conditions on the switch can be satisfied.

We found a 48-step related-key rectangle distinguisher with two different characteristics, the first on 24 steps with 2^{-52}, and the second on 24 steps with 2^{-52} (see Table 4). The probability of the key schedule (message expansion) is $2^{-8.5}$. Therefore, the total probability of the rectangle is $2^{-216.5}$. Using some available techniques, e.g. the one presented in [24], we can add one step at the beginning, and two steps at the end of the rectangle, to obtain a key recovery attack on 51 steps of SHACAL-2.

Table 4. Differential characteristic using xor-differences for the rectangle distinguisher on 48 steps of SHACAL-2

i	chaining value	message	prob	i	chaining value	message	prob
0	C: 32 28 23 21 12 9 D: 2 F: 22 16 12 G: 32 21 12 H: 12 2		2^{-15}	24	C: 30 26 21 19 10 7 D: 32 F: 20 14 10 G: 30 19 10 H: 32 10		2^{-13}
1	A: 2 D: 32 28 23 21 12 9 G: 22 16 12 H: 32 21 12		2^{-11}	25	A: 32 D: 30 26 21 19 10 7 G: 20 14 10 H: 30 19 10		2^{-13}
2	B: 2 E: 28 23 9 H: 22 16 12		2^{-7}	26	B: 32 E: 26 21 7 H: 20 14 10		2^{-7}
3	C: 2 F: 28 23 9		2^{-4}	27	C: 32 F: 26 21 7		2^{-4}
4	D: 2 G: 28 23 9		2^{-4}	28	D: 32 G: 26 21 7		2^{-3}
5	E: 2 H: 28 23 9		2^{-4}	29	E: 32 H: 26 21 7		2^{-4}
6	F: 2		2^{-1}	30	F: 32		2^{-1}
7	G: 2		2^{-1}	31	G: 32		2^{-1}
8	H: 2	2	2^{-1}	32	H: 32	32	1
9			1	33			1
⋮	⋮	⋮	⋮	⋮	⋮	⋮	⋮
22			1	46			1
23		27 16	2^{-4}	47		29 25 14	2^{-6}
24	A: 27 16 E: 27 16			48	A: 29 25 14 E: 29 25 14		

5   Conclusions

In this work, we have shown an application of higher-order differential cryptanalysis on block-cipher-based hash functions. In our attack, we adapted several techniques known from block cipher cryptanalysis to hash functions. Applying these techniques to SHA-256 led to an attack for 47 (out of 64) steps of the compression function with practical complexity. The best known attack so far with practical complexity was for 33 steps. Since the structure of SHA-512 and SHA-256 is very similar, the attack transfers to SHA-512 in a straight forward way. Furthermore, due to the larger word size and output size, attacks for more steps may be expected. We also want to note that the attacks cannot be extended to the hash function to construct collisions or (second) preimages.

However, based on our results, a few conclusions can be deduced. First, SHA-256 has a low security margin against practical distinguishers. Its compression function seems to be weaker than those of the third round SHA-3 candidates, as none of them has practical distinguishers covering such a high percentage of the total number of steps.

Second, when applying boomerang/rectangle attacks to word oriented primitives, the switch in the middle has to be checked carefully – the flaws we have presented as well as our experiments indicate that only a very small percentage of characteristics (even with sparse input-output differences) can be combined.

Finally, the basic strategy described in this paper, *i.e.* linearize the compression function, search for sub-optimal characteristics and combine them in a boomerang/rectangle attack, can be used as a preliminary security analysis for hash functions in general.

Acknowledgements. The work in this paper has been supported in part by the Secure Information Technology Center - Austria (A-SIT), by the Austrian Science Fund (FWF), project P21936-N23 and by the European Commission under contract ICT-2007-216646 (ECRYPT II).

References

1. Aoki, K., Guo, J., Matusiewicz, K., Sasaki, Y., Wang, L.: Preimages for Step-Reduced SHA-2. In: Matsui, M. (ed.) ASIACRYPT 2009. LNCS, vol. 5912, pp. 578–597. Springer, Heidelberg (2009)
2. Aumasson, J.P., Dinur, I., Meier, W., Shamir, A.: Cube Testers and Key Recovery Attacks on Reduced-Round MD6 and Trivium. In: Dunkelman, O. (ed.) FSE 2009. LNCS, vol. 5665, pp. 1–22. Springer, Heidelberg (2009)
3. Biham, E., Dunkelman, O., Keller, N.: The Rectangle Attack - Rectangling the Serpent. In: Pfitzmann, B. (ed.) EUROCRYPT 2001. LNCS, vol. 2045, pp. 340–357. Springer, Heidelberg (2001)
4. Biham, E., Dunkelman, O., Keller, N.: Related-Key Boomerang and Rectangle Attacks. In: Cramer, R. (ed.) EUROCRYPT 2005. LNCS, vol. 3494, pp. 507–525. Springer, Heidelberg (2005)
5. Biryukov, A., Khovratovich, D., Nikolić, I.: Distinguisher and Related-Key Attack on the Full AES-256. In: Halevi, S. (ed.) CRYPTO 2009. LNCS, vol. 5677, pp. 231–249. Springer, Heidelberg (2009)
6. Biryukov, A., Nikolić, I., Roy, A.: Boomerang Attacks on BLAKE-32. In: Joux, A. (ed.) FSE 2011. LNCS, vol. 6733, pp. 218–237. Springer, Heidelberg (2011)
7. Boura, C., Canteaut, A.: Zero-Sum Distinguishers for Iterated Permutations and Application to KECCAK-f and Hamsi-256. In: Biryukov, A., Gong, G., Stinson, D.R. (eds.) SAC 2010. LNCS, vol. 6544, pp. 1–17. Springer, Heidelberg (2011)
8. Boura, C., Canteaut, A., De Cannière, C.: Higher-Order Differential Properties of KECCAK and *Luffa*. In: Joux, A. (ed.) FSE 2011. LNCS, vol. 6733, pp. 252–269. Springer, Heidelberg (2011)
9. De Cannière, C., Mendel, F., Rechberger, C.: Collisions for 70-Step SHA-1: On the Full Cost of Collision Search. In: Adams, C.M., Miri, A., Wiener, M.J. (eds.) SAC 2007. LNCS, vol. 4876, pp. 56–73. Springer, Heidelberg (2007)
10. Dinur, I., Shamir, A.: Cube Attacks on Tweakable Black Box Polynomials. In: Joux, A. (ed.) EUROCRYPT 2009. LNCS, vol. 5479, pp. 278–299. Springer, Heidelberg (2009)
11. Fleischmann, E., Gorski, M., Lucks, S.: Memoryless Related-Key Boomerang Attack on 39-Round SHACAL-2. In: Bao, F., Li, H., Wang, G. (eds.) ISPEC 2009. LNCS, vol. 5451, pp. 310–323. Springer, Heidelberg (2009)
12. Handschuh, H., Naccache, D.: SHACAL. Submitted as an NESSIE Candidate Algorithm (2000), http://www.cryptonessie.org
13. Indesteege, S., Mendel, F., Preneel, B., Rechberger, C.: Collisions and Other Non-random Properties for Step-Reduced SHA-256. In: Avanzi, R.M., Keliher, L., Sica, F. (eds.) SAC 2008. LNCS, vol. 5381, pp. 276–293. Springer, Heidelberg (2009)

14. Isobe, T., Shibutani, K.: Preimage Attacks on Reduced Tiger and SHA-2. In: Dunkelman, O. (ed.) FSE 2009. LNCS, vol. 5665, pp. 139–155. Springer, Heidelberg (2009)
15. Joux, A., Peyrin, T.: Hash Functions and the (Amplified) Boomerang Attack. In: Menezes, A. (ed.) CRYPTO 2007. LNCS, vol. 4622, pp. 244–263. Springer, Heidelberg (2007)
16. Kelsey, J., Kohno, T., Schneier, B.: Amplified Boomerang Attacks Against Reduced-Round MARS and Serpent. In: Schneier, B. (ed.) FSE 2000. LNCS, vol. 1978, pp. 75–93. Springer, Heidelberg (2001)
17. Khovratovich, D., Nikolić, I., Rechberger, C.: Rotational Rebound Attacks on Reduced Skein. In: Abe, M. (ed.) ASIACRYPT 2010. LNCS, vol. 6477, pp. 1–19. Springer, Heidelberg (2010)
18. Khovratovich, D., Rechberger, C., Savelieva, A.: Bicliques for Preimages: Attacks on Skein-512 and the SHA-2 family. Cryptology ePrint Archive, Report 2011/286 (2011)
19. Kim, J., Kim, G., Lee, S., Lim, J., Song, J.H.: Related-Key Attacks on Reduced Rounds of SHACAL-2. In: Canteaut, A., Viswanathan, K. (eds.) INDOCRYPT 2004. LNCS, vol. 3348, pp. 175–190. Springer, Heidelberg (2004)
20. Knudsen, L.R.: Truncated and Higher Order Differentials. In: Preneel, B. (ed.) FSE 1994. LNCS, vol. 1008, pp. 196–211. Springer, Heidelberg (1995)
21. Lai, X.: Higher Order Derivatives and Differential Cryptanalysis. In: Blahut, R.E., Costello Jr., D.J., Maurer, U., Mittelholzer, T. (eds.) Communications and Cryptography: Two Sides of One Tapestry, pp. 227–233. Kluwer Academic Publishers (1994)
22. Lamberger, M., Mendel, F.: Higher-Order Differential Attack on Reduced SHA-256. Cryptology ePrint Archive, Report 2011/037 (2011)
23. Lu, J., Kim, J.: Attacking 44 Rounds of the SHACAL-2 Block Cipher Using Related-Key Rectangle Cryptanalysis. IEICE Transactions 91-A(9), 2588–2596 (2008)
24. Lu, J., Kim, J., Keller, N., Dunkelman, O.: Related-Key Rectangle Attack on 42-Round SHACAL-2. In: Katsikas, S.K., López, J., Backes, M., Gritzalis, S., Preneel, B. (eds.) ISC 2006. LNCS, vol. 4176, pp. 85–100. Springer, Heidelberg (2006)
25. Mendel, F., Pramstaller, N., Rechberger, C., Rijmen, V.: Analysis of Step-Reduced SHA-256. In: Robshaw, M.J.B. (ed.) FSE 2006. LNCS, vol. 4047, pp. 126–143. Springer, Heidelberg (2006)
26. Mendel, F., Pramstaller, N., Rechberger, C., Rijmen, V.: The Impact of Carries on the Complexity of Collision Attacks on SHA-1. In: Robshaw, M.J.B. (ed.) FSE 2006. LNCS, vol. 4047, pp. 278–292. Springer, Heidelberg (2006)
27. Mendel, F., Rechberger, C., Schläffer, M., Thomsen, S.S.: The Rebound Attack: Cryptanalysis of Reduced Whirlpool and Grøstl. In: Dunkelman, O. (ed.) FSE 2009. LNCS, vol. 5665, pp. 260–276. Springer, Heidelberg (2009)
28. Murphy, S.: The Return of the Cryptographic Boomerang. IEEE Transactions on Information Theory 57(4), 2517–2521 (2011)
29. National Institute of Standards and Technology: Announcing Request for Candidate Algorithm Nominations for a New Cryptographic Hash Algorithm (SHA-3) Family. Federal Register 27(212), 62212–62220 (November 2007)
30. National Institute of Standards and Technology: FIPS PUB 180-3: Secure Hash Standard. Federal Information Processing Standards Publication 180-3, U.S. Department of Commerce (October 2008)
31. Nikolić, I., Biryukov, A.: Collisions for Step-Reduced SHA-256. In: Nyberg, K. (ed.) FSE 2008. LNCS, vol. 5086, pp. 1–15. Springer, Heidelberg (2008)

32. Peyrin, T.: Improved Differential Attacks for ECHO and Grøstl. In: Rabin, T. (ed.) CRYPTO 2010. LNCS, vol. 6223, pp. 370–392. Springer, Heidelberg (2010)
33. Sanadhya, S.K., Sarkar, P.: New Collision Attacks Against Up to 24-Step SHA-2. In: Chowdhury, D.R., Rijmen, V., Das, A. (eds.) INDOCRYPT 2008. LNCS, vol. 5365, pp. 91–103. Springer, Heidelberg (2008)
34. Sasaki, Y.: Boomerang Distinguishers on MD4-Based Hash Functions: First Practical Results on Full 5-Pass HAVAL. In: Miri, A., Vaudenay, S. (eds.) SAC. LNCS. Springer, Heidelberg (to appear, 2011)
35. Vielhaber, M.: Breaking ONE.FIVIUM by AIDA an Algebraic IV Differential Attack. Cryptology ePrint Archive, Report 2007/413 (2007)
36. Wagner, D.: The Boomerang Attack. In: Knudsen, L.R. (ed.) FSE 1999. LNCS, vol. 1636, pp. 156–170. Springer, Heidelberg (1999)
37. Wagner, D.: A Generalized Birthday Problem. In: Yung, M. (ed.) CRYPTO 2002. LNCS, vol. 2442, pp. 288–303. Springer, Heidelberg (2002)
38. Wang, G.: Related-Key Rectangle Attack on 43-Round SHACAL-2. In: Dawson, E., Wong, D.S. (eds.) ISPEC 2007. LNCS, vol. 4464, pp. 33–42. Springer, Heidelberg (2007)
39. Wang, X., Yin, Y.L., Yu, H.: Finding Collisions in the Full SHA-1. In: Shoup, V. (ed.) CRYPTO 2005. LNCS, vol. 3621, pp. 17–36. Springer, Heidelberg (2005)
40. Wang, X., Yu, H.: How to Break MD5 and Other Hash Functions. In: Cramer, R. (ed.) EUROCRYPT 2005. LNCS, vol. 3494, pp. 19–35. Springer, Heidelberg (2005)
41. Wang, X., Yu, H.: Non-randomness of 39-step SHA-256. Presented at Rump Session of EUROCRYPT (2008)
42. Watanabe, D., Hatano, Y., Yamada, T., Kaneko, T.: Higher Order Differential Attack on Step-Reduced Variants of *Luffa* v1. In: Hong, S., Iwata, T. (eds.) FSE 2010. LNCS, vol. 6147, pp. 270–285. Springer, Heidelberg (2010)

Finding SHA-2 Characteristics: Searching through a Minefield of Contradictions

Florian Mendel, Tomislav Nad, and Martin Schläffer

IAIK, Graz University of Technology, Austria
tomislav.nad@iaik.tugraz.at

Abstract. In this paper, we analyze the collision resistance of SHA-2 and provide the first results since the beginning of the NIST SHA-3 competition. We extend the previously best known semi-free-start collisions on SHA-256 from 24 to 32 (out of 64) steps and show a collision attack for 27 steps. All our attacks are practical and verified by colliding message pairs. We present the first automated tool for finding complex differential characteristics in SHA-2 and show that the techniques on SHA-1 cannot directly be applied to SHA-2. Due to the more complex structure of SHA-2 several new problems arise. Most importantly, a large amount of contradicting conditions occur which render most differential characteristics impossible. We show how to overcome these difficulties by including the search for conforming message pairs in the search for differential characteristics.

Keywords: hash functions, SHA-2, collision attack, differential characteristic, generalized conditions.

1 Introduction

Since the breakthrough results of Wang et al. [20,19], hash functions have been the target in many cryptanalytic attacks. These attack have especially shown that several well-known and commonly used algorithms such as MD5 and SHA-1 can no longer be considered to be secure. In fact, practical collisions have been shown for MD5 and collisions for SHA-1 can be constructed with a complexity of about 2^{63} [18]. For this reason, NIST has proposed the transition from SHA-1 to the SHA-2 family as a first solution. As a consequence, more and more companies and organizations are migrating to SHA-2. Hence, a detailed analysis of this hash function family is needed to get a good view on its security.

Although the design principles of SHA-2 are very similar to SHA-1, it is still unknown whether or how the attacks on MD5 and SHA-1 can be extended to SHA-2. Since 2008, no collision attacks have been published on SHA-2. One reason might be that the SHA-3 competition [9] initiated by NIST has attracted more attention by the cryptographic community. However, a more likely reason is the increased difficulty of extending previous collision attacks to more steps of SHA-2. In this work, we show that apart from a good attack strategy, advanced automated tools are essential to construct differential characteristics and to find confirming message pairs.

D.H. Lee and X. Wang (Eds.): ASIACRYPT 2011, LNCS 7073, pp. 288–307, 2011.

Related Work. In the past, several attempts have been made to apply the techniques known from the analysis of SHA-1 to SHA-2. The first known cryptanalysis of the SHA-2 family was published by Gilbert and Handschuh [3]. They have shown 9-step local collisions which hold with a probability of 2^{-66}. Hawkes et al. [6] have improved these results to get local collisions with a probability of 2^{-39} by considering modular differences.

In [8], Mendel et al. have analyzed how collision attacks can be applied to step reduced SHA-256. They have shown that the properties of the message expansion of SHA-256 prevent an efficient extension of the techniques of Chabaud and Joux [1] and Wang et al. [19]. Nevertheless, they presented a collision for 18 steps of SHA-256. In [12], Sanadhya and Sarkar have revisited the problem of obtaining a local collision for the SHA-2 family, and in [13] they have shown how to use one of these local collisions to construct another 18-step collision for SHA-256.

Finally, Nikolić and Biryukov [11] found a 9-step differential using modular differences which can be used to construct a practical collision for 21 steps and a semi-free-start collision for 23 steps of SHA-256. This was later extended to 22, 23 and 24 steps by Sanadhya and Sarkar in a series of papers [16,14,15]. The best known collision attack on SHA-256 so far was for 24 steps and has been found by Sanadhya and Sarkar [15], and Indesteege et al. [7].

All these results use rather simple differential characteristics which are constructed mostly manually or using basic cryptanalytic tools. However, the most efficient collision attacks on SHA-1 use more complex characteristics, especially in the first few steps of the attack. Constructing such complex characteristics is in general a difficult task. First, Wang et al. [19] have constructed such a characteristic for SHA-1 manually. Later, De Cannière and Rechberger [2] proposed a method to efficiently find such complex characteristics for SHA-1 in an automated way. Furthermore, also the best practical collision attack on SHA-1 (with the highest number of steps) is based on this approach [4].

Our Contribution. Currently, all collision attacks on SHA-2 are of practical complexity and based on the same basic idea: extending a local collision over 9 steps to more steps. As already mentioned in [7], this kind of attack is unlikely to be extended beyond 24 steps. In this work, we investigate new ideas to progress in the cryptanalysis of SHA-2. First, we extend the idea of finding local collisions to more than 9 steps by exploiting the nonlinearity of both the state update and message expansion.

To find such local collisions an automated tool to search for complex differential characteristics is needed. We start with the approach of De Cannière and Rechberger [2] on SHA-1. Unfortunately, their techniques cannot directly be applied to SHA-2. We have observed several problems in finding valid differential characteristics for SHA-2. In this work, we have identified these problems and show how to solve them efficiently. Most importantly, a very high number of contradicting conditions occurs which render most differential characteristics impossible.

To summarize, we present the first automatic tool to construct complex differential characteristics for reduced SHA-2. Applying our tool to SHA-256 results in practical examples of semi-free-start collisions for 32 and collisions for 27 out of 64 steps of SHA-256. The best semi-free-start collision and collision attack so far was on 24 steps of SHA-256.

Outline. The paper is structured as follows. In Section 2 we give a short description of SHA-256. In Section 3, we provide an overview of the general attack strategy and briefly mention which problems arise in the search for differential characteristics in SHA-2. In Section 4, we show how to efficiently propagate differences and conditions in SHA-2. Furthermore, we discuss why most differential characteristics are invalid and describe how to detect inconsistencies. In Section 5 we present our automated tool to construct complex differential characteristics and to find conforming message pairs in SHA-2. Finally, we conclude on our results in Section 6.

2   Description of SHA-256

SHA-256 is one of four hash functions defined in the Federal Information Processing Standard (FIPS-180-3) [10]. All four hash functions were designed by the National Security Agency (NSA) and issued by NIST in 2002. SHA-256 is an iterated cryptographic hash function with a hash output size of 256 bits, a message block size of 512 bits and using a word size of 32 bits. In the compression function of SHA-2, a state of eight chaining variables $A,\ldots,H$ is updated using 16 message words $M_0,\ldots,M_{15}$.

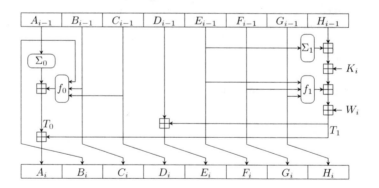

Fig. 1. The SHA-2 step update function

The compression function of SHA-256 consists of 64 identical step update functions which are illustrated in Fig.1 and given as follows:

$$
\begin{aligned}
T_0 &= \Sigma_0(A_{i-1}) + f_0(A_{i-1}, B_{i-1}, C_{i-1}) \\
T_1 &= \Sigma_1(E_{i-1}) + f_1(E_{i-1}, F_{i-1}, G_{i-1}) + H_{i-1} + K_i + W_i \\
A_i &= T_0 + T_1, \quad B_i = A_{i-1}, \quad C_i = B_{i-1}, \quad D_i = C_{i-1} \\
E_i &= D_{i-1} + T_1 \quad F_i = E_{i-1}, \quad G_i = F_{i-1}, \quad H_i = G_{i-1}
\end{aligned}
\tag{1}
$$

The Boolean functions f_0 (MAJ) and f_1 (IF) are given by

$$f_0(x, y, z) = (x \wedge y) \oplus (x \wedge z) \oplus (y \wedge z),$$
$$f_1(x, y, z) = (x \wedge y) \oplus (\neg x \wedge z).$$

The two $GF(2)$-linear functions Σ_0 and Σ_1 are defined as follows:

$$\Sigma_0(x) = x \ggg 2 \oplus x \ggg 13 \oplus x \ggg 22,$$
$$\Sigma_1(x) = x \ggg 6 \oplus x \ggg 11 \oplus x \ggg 25.$$

In the i-th step of the update function, a fixed constant K_i and the i-th word W_i of the expanded message are added to the state. The message expansion takes the 16 message words M_i as input and outputs 64 expanded message words W_i as follows:

$$W_i = \begin{cases} M_i & \text{for } 0 \leq i < 16 \\ \sigma_1(W_{i-2}) + W_{i-7} + \sigma_0(W_{i-15}) + W_{i-16} & \text{for } 16 \leq i < 64 \end{cases}$$

where the functions $\sigma_0(x)$ and $\sigma_1(x)$ are defined as follows:

$$\sigma_0(x) = x \ggg 7 \oplus x \ggg 18 \oplus x \gg 3,$$
$$\sigma_1(x) = x \ggg 17 \oplus x \ggg 19 \oplus x \gg 10.$$

3   Basic Attack Strategy

In this section, we give a brief overview of our attack strategy. We first describe how we generalize the approach of Nikolić and Biryukov [11] to find semi-free-start collisions on a higher number of steps. Due to this extension, differential characteristics cannot be constructed manually or semi-automatic anymore. Hence, we provide a fully automated tool to construct complex differential characteristics in SHA-2. Furthermore, we discuss why it is extremely difficult to find valid differential characteristics in SHA-2. In fact, we were not able to find a valid differential characteristic without including the search for a confirming message pair in the process. Therefore, the approach of first finding a valid differential characteristic and then, independently search for a conforming message pair does not apply very well to SHA-2. Hence, our attack strategy can be summarized as follows:

1. Determine a starting point for the search which results in an attack on a large number of steps. The resulting start characteristic should span over few steps and only some message words should contain differences.
2. Use an automated search tool to find a differential characteristic for the unrestricted intermediate steps including the message expansion.
3. Continue the search to find a conforming message pair. If no message pair can be found, adjust the differential characteristic accordingly.

Note that after step 2 it is not ensured that the differential characteristic is valid. If we cannot find a conforming message pair after a certain amount of time we go back to step 2 to adjust the differential characteristic.

3.1   Determining a Starting Point

By exploiting the nonlinearity of the step update function, Nikolić and Biryukov [11] found a 9-step differential characteristic for which it is not necessary to apply corrections (differences in the message words) in each step of the differential characteristic. The fact that not all (only 5 out of 9) message words contain differences helped to overcome several steps of the message expansion resulting in a collision and semi-free-start collision attack for 21 and 23 steps, respectively. Later this approach was extended to a collision attack on 24 steps [7,15]. However, as pointed out in [7] it is unlikely that this approach can be extended beyond 24 steps.

In our attack, we are using differential characteristics which span over $t \geq 9$ steps, which allows us to attack more steps of SHA-256. As in the attack of Nikolić and Biryukov we are interested in differential characteristics with differences in only a few message words. Then, large parts of the expanded message have no difference which in turn, results in an attack on more than 24 steps. Already by using a differential characteristic spanning over $t = 10$ steps (with differences in only 3 message words) we can construct a semi-free-start collision for 27 steps of SHA-256. This can be extended to 32 steps using a differential characteristic spanning over $t = 16$ steps with differences in 8 message words.

To construct these starting points, we first fix the value of t and consider only differential characteristics which may result in collisions on more than 24 steps. Then, we identify those message words which need to have differences such that the differential characteristic holds for the whole message expansion. Table 2 in Appendix A shows the used starting point for the attack on 32 steps. Note that we have further optimized the message difference slightly to keep it sparse, which reduces the search space for the automated tool.

3.2   Searching for Valid Differential Characteristics and Conforming Message Pairs in SHA-2

Once we have determined a good starting point we continue by constructing a valid differential characteristic for both the state update transformation and the message expansion. We have implemented an automated search tool for SHA-2 which is similar to the one proposed in [2] to construct complex characteristics for SHA-1. However, the increased complexity of SHA-2 compared to SHA-1 complicates a direct application of their approach. In the following, we briefly outline which problems occurred and how we have resolved them.

First of all, the larger state size, the combined update of two state variables, and the higher diffusion due to the Σ_i functions increases the complexity significantly. To limit these issues, we use an alternative description of SHA-2 where only two state variables are updated separately (see Section 4.1). Furthermore, we split up one SHA-2 step (including the nonlinear message expansion) into 9 less complex sub steps. This way, the propagation of differences can be implemented much more efficiently while losing only a small amount of information (see Section 4.3).

However, the main problem in SHA-2 is that it is difficult to determine whether a differential characteristic is valid, i.e. whether a conforming message pair exists. For example, a lot more conditions on two bits of the form $A_{i,j} = A_{i-1,j}$ occur in SHA-2, compared to SHA-1 for example. Furthermore, the orthogonal applications of the Σ_i and f_i functions results in cyclic conditions which contradict with a high probability (see Section 4.4). Additionally, more complex conditions on more bits occur. One reason for these additional conditions is that two state variables (A_i, E_i) are updated using a single message word (W_i). Unfortunately, it is not possible to determine all these conditions in general. However, we have implemented different tests to efficiently check for many contradictions (for more details, see Section 4.5).

Despite all these tests, we were not able to find a valid differential characteristic. At the end, even brute-forcing a single critical message word (a message word where most bits are already set) did not lead to a solution. Therefore, we have combined the search for differential characteristics with the search for a conforming message pair (see Section 5). During the message search, we first determine critical bits and backtrack if needed. This way complex hidden conditions are resolved at an earlier stage in the search. Furthermore, we correct impossible characteristics once they are detected.

4   Difference and Condition Propagation in SHA-2

We use generalized conditions to nonlinearly propagate differences and conditions in both the state update and message expansion of SHA-2. Generalized conditions are propagated in a bit sliced manner. Note that in the case of the SHA-2, one bit of A and E is updated using 15 input bits. Hence, to simplify the bit sliced step update, we use an alternative description of SHA-2.

4.1   Alternative Description of SHA-2

In the state update transformation of SHA-2, only two state variables are updated in each step, namely A_i and E_i. Therefore, we can redefine the state update such that only these two variables are involved. In this case, we get the following mapping between the original and new state variables:

A_i	B_i	C_i	D_i	E_i	F_i	G_i	H_i
A_i	A_{i-1}	A_{i-2}	A_{i-3}	E_i	E_{i-1}	E_{i-2}	E_{i-3}

Note that A_i is updated using an intermediate result of the step update of E_i (see Equation 1). Since this complicates the efficient bit sliced representation of the SHA-2 step update transformation we propose the following alternative description:

$$
\begin{aligned}
E_i &= E_{i-4} + \Sigma_1(E_{i-1}) + f_1(E_{i-1}, E_{i-2}, E_{i-3}) + A_{i-4} + K_i + W_i \\
A_i &= -A_{i-4} + \Sigma_0(A_{i-1}) + f_0(A_{i-1}, A_{i-2}, A_{i-3}) + E_i
\end{aligned}
\tag{2}
$$

In this case we get two SHA-1 like state update transformations, one for the left (A_i) and one for the right (E_i) side of the SHA-2 state update transformation. Note that in this description, the state variables $A_{-4}, \ldots, A_{-1}$ and $E_{-4}, \ldots, E_{-1}$ represent the chaining input or initial value of the compression function. The alternative description is also illustrated in Fig.2.

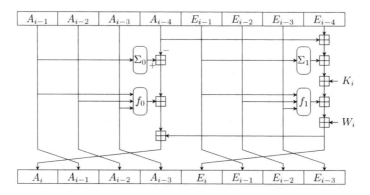

Fig. 2. Alternative description of the SHA-2 state update transformation

4.2   Generalized Conditions

Inspired by signed-bit differences [19], De Cannière and Rechberger introduced generalized conditions for differences, where all 16 possible conditions on a pair of bits are taken into account [2]. Table 1 lists all these possible conditions and introduces notations for the various cases.

Table 1. Notation for possible generalized conditions on a pair of bits [2]

(X_i, X_i^*)	$(0,0)$	$(1,0)$	$(0,1)$	$(1,1)$	(X_i, X_i^*)	$(0,0)$	$(1,0)$	$(0,1)$	$(1,1)$
?	✓	✓	✓	✓	3	✓	✓	-	-
-	✓	-	-	✓	5	✓	-	✓	-
x	-	✓	✓	-	7	✓	✓	✓	-
0	✓	-	-	-	A	-	✓	-	✓
u	-	✓	-	-	B	✓	✓	-	✓
n	-	-	✓	-	C	-	-	✓	✓
1	-	-	-	✓	D	✓	-	✓	✓
#	-	-	-	-	E	-	✓	✓	✓

Definition 1 (Generalized Conditions for Differences [2]). *Let $X \in \{0,1\}^n$ and $X^* \in \{0,1\}^n$, then the notation*

$$\nabla X = [c_{n-1}, \ldots, c_0],$$

where c_i denotes one of the conditions of Table 1 for the i-th bit, defines a subset of pairs $(X, X^*) \in \{0,1\}^n \times \{0,1\}^n$ that conforms to the specified conditions.

For example, all pairs of 8-bit words X and X^* that satisfy

$$\{(X, X^*) \in \{0,1\}^8 \times \{0,1\}^8 \mid X_7 \cdot X_7^* = 0, X_i = X_i^* \text{ for } 1 \leq i \leq 5, X_0 \neq X_0^*\},$$

can be conveniently written in the form

$$\nabla X = \text{[7?-----x]}.$$

4.3   Efficiently Implementing the Propagation of Generalized Conditions

We propagate generalized conditions similar as in the attack on SHA-1. However, the complexity of propagating generalized conditions increases exponentially with the number of input bits and additions. While there are only 6 input bits in the case of SHA-1 (excluding the carry), we have 9 input bits in the update of E_i and 8 input bits in the update of each of A_i and W_i in SHA-2.

To reduce the computational complexity of the propagation in SHA-2, we have further split the update of W_i, E_i and A_i into 3 sub steps. In more detail, we independently compute each output bit of the σ_i, Σ_i and f_i functions and then, compute the modular additions. This way, the number of input bits reduces to 3 for σ_i, Σ_i and f_i and we get at most 5 input bits for the modular additions. This split of functions reduces the computation complexity by a factor of about 100.

Furthermore, for the sub steps without modular addition we have precomputed the propagation of all generalized input conditions. For the modular additions we use a hash map to store already computed bit sliced results. In this case, the bit slice update of each sub step reduces to simple table or hash map lookups. Our experiments have shown a speedup of another factor 100 by caching already computed results. The drawback of this method is that we lose the relation between the sub steps compared to a combined propagation. Furthermore, due to memory restrictions we are not able to precompute or keep all possibilities for the modular additions.

4.4   Two-Bit Conditions

Apart from generalized conditions, additional conditions on more than a single bit are present in a differential characteristic. Especially, conditions on two bits are needed such that a differential path is valid. These two-bit conditions have already been used by Wang et al. in their attacks on the members of the MD4 family [17]. Such two-bit conditions occur mostly in the propagation of differences through the Boolean function. For example, if an input difference in A_{i-1} at bit position j should result in a zero output difference of $f_0(A_{i-1}, A_{i-2}, A_{i-3})$, the remaining two input bits should be equal. In this case, we get the two-bit condition $A_{i-2,j} = A_{i-3,j}$. Similar conditions occur not only in the f_i, σ_i and Σ_i functions but also in the modular additions.

Two-bit conditions are not covered by generalized conditions and thus, not shown in the characteristics given in [2]. However, two-bit conditions may lead to additional inconsistencies. For example, in two subsequent f_0 functions the following two contradicting conditions may occur:

$$(A_{i-2,j} = A_{i-3,j}) \wedge (A_{i-2,j} \neq A_{i-3,j}).$$

Since such contradicting conditions occur only rarely in SHA-1, simple additional checks are sufficient to verify whether a given differential characteristic is valid at the end of the search.

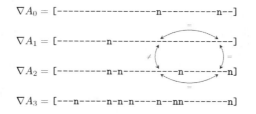

Fig. 3. Example of four cyclic and contradicting two-bit conditions. Such cases commonly occur in SHA-2 and are not covered by generalized conditions. For the two Σ_0 functions (XOR) we have twice $\Sigma_0(\mathbf{n}, \text{-}, \text{-}) = \mathbf{n}$ which results in the two equalities $A_{1,2} = A_{1,13}$ and $A_{2,2} = A_{2,13}$. For the f_0 function (MAJ) at bit position 2 we get $f_0(\text{-}, \text{-}, \mathbf{n}) = \mathbf{n}$ if and only if $A_{2,2} = A_{1,2}$, while for bit position 13 we get $f_0(\text{-}, \text{-}, \mathbf{n}) = \text{-}$ if and only if $A_{2,13} \neq A_{1,13}$. Note that in this example, all involved bits of E_i do not contain any difference.

This is not the case in SHA-2. Note that the nonlinear Boolean functions f_0 and f_1 update the same bit position of different words, while the linear Σ_i functions update different bit positions within the same word. Hence, more complex cyclic two-bit relations occur. A still simple example is given in Fig.3. In this case, 4 bits of two Σ_i and two Boolean functions are related in a cyclic form which results in a contradiction. We have observed that for a given differential characteristic even more complex relations with cycle lengths larger than 10 commonly occur. Of course already a single contradicting cycle results in an impossible differential characteristic.

4.5   Inconsistency Checks

To avoid inconsistent differential characteristics, we have evaluated a number of checks to detect contradictions as early and efficiently as possible. Note that a test which is able to detect many contradictions is usually also less efficient. However, also a simple test may detect a contradiction at a later point in the search. Due to the high number of complex conditions in SHA-2 and the difficulty to detect them we need to make a trade-off here.

Two-Bit Condition Check. Two-bit conditions are linear conditions in $GF(2)$ since such conditions can only be either equal ($A_{i,j} = A_{i-1,j}$) or non-equal ($A_{i,j} \neq A_{i-1,j}$). Contradictions in two-bit condition cycles can be efficiently detected by determining all two-bit conditions, setting up a linear system of equations and checking if the system can be solved using Gaussian elimination. Although a large number of contradictions are detected this way, most characteristics are still invalid after this check.

Complete Condition Check. A quite expensive test is to check for every bit restricted to '-' or 'x' whether both possible cases ('0' and '1', or 'n' and 'u') are indeed still valid. If both choices for a single bit are invalid we know that the whole characteristic is impossible. Of course these tests can be extended to other generalized conditions as well. However, it turned out to be more efficient to apply this check only rarely and only to specific conditions during the search. Furthermore, we have improved the speed of this complete test by applying it only to bits which are restricted by two-bit conditions.

Complete Condition Check on a Set of Bits. Since even the complete condition check is not able to detect many contradictions, we have analyzed different variants of setting all possibilities for all or selected combinations of 2, 3 or 4 bits. Such tests indeed detect more impossible characteristics but are very inefficient to compute and thus, cannot be used during the search for differential characteristics in SHA-2.

5   Searching for Differential Characteristics

In general, our search techniques can be divided into three parts: decision, deduction and backtracking. Note that the same separation is done in many other fields, like SAT solvers [5]. The first aspect of our search strategy is the decision, where we decide which bit is chosen and which condition is imposed at its position. In the deduction part we compute the propagation of the imposed condition and check for contradictions. If a contradiction occurs we need to backtrack and undo decisions, which is the third part of the search strategy. A basic search strategy to find differential characteristics has been described in [2] and works as follows.

Let U be the set of all '?' and 'x', then repeat the following until U is empty.
Decision
    1. Pick randomly a bit in U.
    2. Impose a '-' for a '?' or randomly a sign ('u' or 'n') for 'x'.
Deduction
    3. Compute the propagation.
    4. If a contradiction is detected start backtracking, else go to step 1.
Backtracking
    5. Jump back to an earlier state of the search and go to step 1.

We have applied this strategy to SHA-2 but could not find a valid differential characteristics. In any case at least one of the checks described in Section 4.5 failed. The reason for this is that conditions which are not covered by generalized or two-bit conditions appear much more often in SHA-2 than in SHA-1. Since more advanced checks are too expensive, we have developed a more sophisticated search strategy to find valid differential characteristics for SHA-2 as described in the next section.

5.1   Search Strategy

In our approach we already determine some message bits during the search for a differential characteristic. Generally speaking, we are combining the search for a conforming message pair with the search for a differential characteristic. In doing so we consider those bits much earlier, which are involved in many relations with other bits. This way, we can detect invalid characteristics at an early stage of the search. However, this should not be done too early to not restrict the message freedom too much. In addition, we are remembering critical bits during the search to improve the backtracking and speed-up the search process. In the following we describe the used search strategy in more detail.

In general we have two phases in our search strategy where different bits are chosen (guessed) and we switch between these two dynamically. In the following, we describe both phases in detail. Phase 1 can be described as follows.

Let U be the set of all '?' and 'x'. Repeat the following until U is empty:
Decision
1. Pick randomly a bit in U.
2. Impose a '-' for a '?' or randomly a sign ('u' or 'n') for 'x'.

Deduction
3. Compute the propagation as described in Section 4.3.
4. If a contradiction is detected start backtracking, else apply the additional checks of Section 4.5.
5. Continue with step 1 if all checks passed, if not start backtracking.

Backtracking
6. If the decision bit is 'x' try the second choice for the sign or if the decision bit is '?' impose a 'x'.
7. If still a contradiction occurs mark bit as critical.
8. Jump back until the critical bit ca be resolved.
9. Continue with step 1.

Note that, the additional checks in step 4 are optional and a trade-off between number of checks and speed has to be done. The additional steps in the back-tracking process improve the search speed significantly and prevent that critical bits result in a contradiction again.

Once phase 1 is finished (U is empty) we continue with phase 2 which can be summarized as follows.

Let U' be the set of all ' - ' with many two-bit conditions.
Repeat the following until U' is empty:

Decision

1. Pick randomly a bit in U'.
2. Impose randomly a '0' or '1'.

Deduction

3. Compute the propagation as described in Section 4.3.
4. If a contradiction is detected start backtracking, else apply additional checks from Section 4.5.
5. Continue with step 1 if all checks passed, if not start backtracking.

Backtracking

6. Try the second choice of the decision bit.
7. If still a contradiction occurs mark bit as critical.
8. Jump back until the critical bit can be resolved.
9. If necessary jump back to phase 1, otherwise continue with step 1.

Choosing a decision bit with many two-bit conditions ensures that bits which influence a lot of other bits are chosen first. Therefore, many other bits propagate by defining the value of a single bit. Furthermore, in step 7 and 8 of the backtracking we can also mark more than one bit as critical. We want to note that due to step 9, we actually switch quite often between both phases in our search.

Additionally, we restart the search from scratch after a certain amount of contradictions or iterations to terminate branches which appear to be stuck because of exploring a search space far from a solution.

5.2   Results

Using the start characteristic given in Table 2 and the search strategy described above, we can find a valid characteristic and confirming inputs which result in semi-free-collisions for 32 out of 64 steps of SHA-256. An example of a semi-free-start for 32 steps is shown in Table 4. The according differential characteristic and the set of conditions is given in Table 3 and Table 5. The find this example for 32 steps our tool was running a few days on a cluster with 32 nodes.

So far we have only considered semi-free-start collision attacks in which an attacker is allowed to choose the chaining value. However, in a collision attack on the hash function the chaining value is fixed, which makes an attack much more difficult. In order to construct a collision for step-reduced SHA-256, we are interested in differential characteristics with no differences in the first few message words. Then, the additional freedom in the first message words can be used to transform a semi-free-start collision into a real collision. Similar characteristics have also been used in the collision attacks on 24 steps of SHA-256 in [7].

By using a differential characteristic spanning over $t = 11$ steps with differences in only 5 expanded message words and with no differences in the first 7 message words (see Table 6) we are able to construct a collision for 27 steps of SHA-256. The colliding message pair is shown in Table 8 and the differential characteristic and the set of conditions is given in Table 7 and Table 9.

6   Conclusions and Future Work

In this paper, we have presented a collision for 27 and a semi-free-start collision for 32 steps of SHA-256 with practical complexity. This significantly improves upon the best previously published (semi-free-start) collision attacks on SHA-256 for up to 24 steps. We have extended and generalized existing approaches and developed a fully automatic tool to construct complex differential characteristics for SHA-2.

Our tool extends the techniques proposed by De Cannière and Rechberger to construct complex characteristics for SHA-1 using generalized conditions. The more complex structure of SHA-256 complicates a direct application of their approach. We have identified several problems and have shown how to overcome them. Most importantly, a high amount of found differential characteristics are invalid due to many contradicting conditions in SHA-2. We have resolved this problem by by identifying critical bits during the whole search process, and by combining the search for differential characteristics with the computation of conforming message pairs.

To summarize, the search for valid differential characteristics and conforming message pairs in SHA-2 is increasingly difficult and unpredictable, compared to more simple designs like MD5 and SHA-1. Nevertheless, we were able to construct a powerful tool to find practical examples for (semi-free-start) collisions in SHA-256 which can also be applied to other ARX based hash functions.

Acknowledgments. We would like to thank Vincent Rijmen, Christian Rechberger, Christophe De Cannière and the anonymous referees for useful comments and discussions. The work in this paper has been supported in part by the Secure Information Technology Center-Austria (A-SIT), by the European Commission under contract ICT-2007-216646 (ECRYPT II), by the Austrian Science Fund (FWF, project P21936) and the German Federal Office for Information Security (BSI).

References

1. Chabaud, F., Joux, A.: Differential Collisions in SHA-0. In: Krawczyk, H. (ed.) CRYPTO 1998. LNCS, vol. 1462, pp. 56–71. Springer, Heidelberg (1998)
2. De Cannière, C., Rechberger, C.: Finding SHA-1 Characteristics: General Results and Applications. In: Lai, X., Chen, K. (eds.) ASIACRYPT 2006. LNCS, vol. 4284, pp. 1–20. Springer, Heidelberg (2006)
3. Gilbert, H., Handschuh, H.: Security Analysis of SHA-256 and Sisters. In: Matsui, M., Zuccherato, R.J. (eds.) SAC 2003. LNCS, vol. 3006, pp. 175–193. Springer, Heidelberg (2004)
4. Grechnikov, E.: Collisions for 72-step and 73-step SHA-1: Improvements in the Method of Characteristics. Cryptology ePrint Archive, Report 2010/413 (2010)
5. Gu, J., Purdom, P.W., Franco, J., Wah, B.W.: Algorithms for the Satisfiability (SAT) Problem: A Survey. In: DIMACS Series in Discrete Mathematics and Theoretical Computer Science, pp. 19–152. American Mathematical Society (1996)

6. Hawkes, P., Paddon, M., Rose, G.G.: On Corrective Patterns for the SHA-2 Family. Cryptology ePrint Archive, Report 2004/207 (2004)

7. Indesteege, S., Mendel, F., Preneel, B., Rechberger, C.: Collisions and Other Non-random Properties for Step-Reduced SHA-256. In: Avanzi, R.M., Keliher, L., Sica, F. (eds.) SAC 2008. LNCS, vol. 5381, pp. 276–293. Springer, Heidelberg (2009)

8. Mendel, F., Pramstaller, N., Rechberger, C., Rijmen, V.: Analysis of Step-Reduced SHA-256. In: Robshaw, M.J.B. (ed.) FSE 2006. LNCS, vol. 4047, pp. 126–143. Springer, Heidelberg (2006)

9. National Institute of Standards and Technology: Cryptographic Hash Algorithm Competition (November 2007), http://csrc.nist.gov/groups/ST/hash/sha-3/index.html

10. National Institute of Standards and Technology: FIPS PUB 180-3: Secure Hash Standard. Federal Information Processing Standards Publication 180-3, U.S. Department of Commerce (October 2008), http://www.itl.nist.gov/fipspubs

11. Nikolić, I., Biryukov, A.: Collisions for Step-Reduced SHA-256. In: Nyberg, K. (ed.) FSE 2008. LNCS, vol. 5086, pp. 1–15. Springer, Heidelberg (2008)

12. Sanadhya, S.K., Sarkar, P.: New Local Collisions for the SHA-2 Hash Family. In: Nam, K.H., Rhee, G. (eds.) ICISC 2007. LNCS, vol. 4817, pp. 193–205. Springer, Heidelberg (2007)

13. Sanadhya, S.K., Sarkar, P.: Attacking Reduced Round SHA-256. In: Bellovin, S.M., Gennaro, R., Keromytis, A.D., Yung, M. (eds.) ACNS 2008. LNCS, vol. 5037, pp. 130–143. Springer, Heidelberg (2008)

14. Sanadhya, S.K., Sarkar, P.: Deterministic Constructions of 21-Step Collisions for the SHA-2 Hash Family. In: Wu, T.C., Lei, C.L., Rijmen, V., Lee, D.-T. (eds.) ISC 2008. LNCS, vol. 5222, pp. 244–259. Springer, Heidelberg (2008)

15. Sanadhya, S.K., Sarkar, P.: New Collision Attacks against Up to 24-Step SHA-2. In: Chowdhury, D.R., Rijmen, V., Das, A. (eds.) INDOCRYPT 2008. LNCS, vol. 5365, pp. 91–103. Springer, Heidelberg (2008)

16. Sanadhya, S.K., Sarkar, P.: Non-linear Reduced Round Attacks against SHA-2 Hash Family. In: Mu, Y., Susilo, W., Seberry, J. (eds.) ACISP 2008. LNCS, vol. 5107, pp. 254–266. Springer, Heidelberg (2008)

17. Wang, X., Lai, X., Feng, D., Chen, H., Yu, X.: Cryptanalysis of the Hash Functions MD4 and RIPEMD. In: Cramer, R. (ed.) EUROCRYPT 2005. LNCS, vol. 3494, pp. 1–18. Springer, Heidelberg (2005)

18. Wang, X., Yao, A., Yao, F.: New Collision Search for SHA-1. Presented at Rump Session of CRYPTO (2005)

19. Wang, X., Yin, Y.L., Yu, H.: Finding Collisions in the Full SHA-1. In: Shoup, V. (ed.) CRYPTO 2005. LNCS, vol. 3621, pp. 17–36. Springer, Heidelberg (2005)

20. Wang, X., Yu, H.: How to Break MD5 and Other Hash Functions. In: Cramer, R. (ed.) EUROCRYPT 2005. LNCS, vol. 3494, pp. 19–35. Springer, Heidelberg (2005)

A   Differential Characteristics and Conditions

Table 2. Starting point for a semi-free-start collision for 32 steps. Using the alternative description of SHA-2 (Section 4.1) and the notion of generalized conditions (Section 4.2).

i	∇A_i	∇E_i	∇W_i
-4	--------------------------------	--------------------------------	
-3	--------------------------------	--------------------------------	
-2	--------------------------------	--------------------------------	
-1	--------------------------------	--------------------------------	
0	--------------------------------	--------------------------------	--------------------------------
1	--------------------------------	--------------------------------	--------------------------------
2	-----------------------------x--	-----------------------------x--	-----------------------------x--
3	????????????????????????????????	????????????????????????????????	????????????????????????????????
4	????????????????????????????????	????????????????????????????????	????????????????????????????????
5	????????????????????????????????	????????????????????????????????	????????????????????????????????
6	????????????????????????????????	????????????????????????????????	????????????????????????????????
7	????????????????----------------	????????????????????????????????	????????????????????????????????
8	?????????????????---------------	????????????????????????????????	??????????????????x-------------
9	??????????????????x-------------	????????????????????????????????	--------------------------------
10	--------------------------------	????????????????????????????????	--------------------------------
11	--------------------------------	???????????????????-------------	--------------------------------
12	--------------------------------	?????????????????---------------	--------------------------------
13	--------------------------------	???????????????????x------------	--------------------------------
14	--------------------------------	--------------------------------	--------------------------------
15	--------------------------------	--------------------------------	--------------------------------
16	--------------------------------	--------------------------------	--------------------------------
17	--------------------------------	--------------------------------	----x---------x----------------
18	--------------------------------	--------------------------------	--------------------------------
19	--------------------------------	--------------------------------	--------------------------------
	. . .	. . .	. . .
30	--------------------------------	--------------------------------	--------------------------------
31	--------------------------------	--------------------------------	--------------------------------

Table 3. Characteristic for a semi-free-start collision for 32 steps of SHA-256

i	∇A_i	∇E_i	∇W_i
-4	-------------------------------	-------------------------------	
-3	-------------------------------	-------------------------------	
-2	-------------------------------	-------------------------------	
-1	-------------------------------	-------------------------------	
0	-------------------------------	----------------------------0--	-------------------------------
1	-----------------------1------	--0-0---1-1-1----1-0-0-------011-	-------------------------------
2	----------------------0--u--	--1-1-1000-1--11101101---1--1u0-	----------------------------u--
3	10n10nnn1n0n-11n1u01u11000uu0n0n	-1n1n10un0un101-n1n1n0110un0u0n0	uu-un-----un---n-u-uu-n---u--un-
4	-----n----------0---------0-1---	-0n0n1nuuun0-1u1unnnuu011n000nn1	1n---1u--uu1u-uu------nn--0n----
5	----------------n---------1-----	0u1nn1n-1010-00001u0101-11101110	01-1-un0-1-1n-nn1u1n0-0un-0-n--n
6	----------n--u---------u---n--	00u01un0000000n111u00100101uu11u	n----nnuu-n-nu---n--n----------
7	-------------------------------	-n10u000u1un0101nn10n00001n000u1	1n0001un10u0nnn-01n01u10000unnnn
8	-------------------------------	-10-1n0-0--1-01-0-1-0----n011-10	----u--------unnn---------0-----
9	---u---------u------------------	-0--u00-1-01-1--1---1----n1---0-	-------------------------------
10	-------------------------------	---nunn------n--n--------u-u-u--	-------------------------------
11	-------------------------------	---0-10----100--0-------1-0-0--	-------------------------------
12	-------------------------------	---0011----011--1--------0-1-1--	-------------------------------
13	-------------------------------	---un------unnnn----------------	-------------------------------
14	-------------------------------	---00-----00000----------------	-------------------------------
15	-------------------------------	---11-----11111----------------	-------------------------------
16	-------------------------------	-------------------------------	-------------------------------
17	-------------------------------	-------------------------------	----n-------n------
18	-------------------------------	-------------------------------	-------------------------------
19	-------------------------------	-------------------------------	-------------------------------
20	-------------------------------	-------------------------------	-------------------------------
21	-------------------------------	-------------------------------	-------------------------------
22	-------------------------------	-------------------------------	-------------------------------
23	-------------------------------	-------------------------------	-------------------------------
24	-------------------------------	-------------------------------	-------------------------------
25	-------------------------------	-------------------------------	-------------------------------
26	-------------------------------	-------------------------------	-------------------------------
27	-------------------------------	-------------------------------	-------------------------------
28	-------------------------------	-------------------------------	-------------------------------
29	-------------------------------	-------------------------------	-------------------------------
30	-------------------------------	-------------------------------	-------------------------------
31	-------------------------------	-------------------------------	-------------------------------

Table 4. Semi-free-start collision for 32 steps of SHA-256

h_0	764d264f 268a3366 285fecb1 4c389b22 75cd568d f5c8f99b 6e7a3cc3 1b4ea134
h_0^*	764d264f 268a3366 285fecb1 4c389b22 75cd568d f5c8f99b 6e7a3cc3 1b4ea134
Δh_0	00000000 00000000 00000000 00000000 00000000 00000000 00000000 00000000
m	52a600a8 2c3b8434 ea92dfcf d4eaf9ad b77fe08d 7c50e542 69c783a6 86a14e10
	baf88b0b 12665efb ce7c3a31 3030f09d 9bd52eb8 7549997e fa976e0d 86ebacbc
m^*	52a600a8 2c3b8434 ea92dfcb 0cdba38b f514e39d 7a5bb4cb ee6bcba6 c58f6a0f
	b2f78b0b 12665efb ce7c3a31 3030f09d 9bd52eb8 7549997e fa976e0d 86ebacbc
Δm	00000000 00000000 00000004 d8315a26 426b0310 060b5189 87ac4800 432e241f
	080f0000 00000000 00000000 00000000 00000000 00000000 00000000 00000000
h_1	d0b41ffa e1f519a2 e3cad2ed a19d5795 906ac05f c995f6c8 cf309f95 9fb9ca57
h_1^*	d0b41ffa e1f519a2 e3cad2ed a19d5795 906ac05f c995f6c8 cf309f95 9fb9ca57
Δh_1	00000000 00000000 00000000 00000000 00000000 00000000 00000000 00000000

Table 5. Set of conditions for the semi-free-start collision for 32 steps

i	set of conditions	
0	$E_{0,2} = 0$	1
1	$A_{1,5} = 1$, $A_{1,2} \neq A_{0,2}$, $A_{1,0} = E_{1,0}$, $E_{1,1} = 1$, $E_{1,2} = 1$, $E_{1,3} = 0$, $E_{1,11} = 0$, $E_{1,13} = 0$, $E_{1,15} = 1$, $E_{1,20} = 1$, $E_{1,23} = 1$, $E_{1,27} = 0$, $E_{1,29} = 0$	13
2	$A_{2,2} = 1$, $A_{2,5} = 0$, $A_{2,0} = A_{1,0}$, $A_{2,4} \neq A_{1,4}$, $A_{2,11} = A_{1,11}$, $A_{2,14} \neq A_{1,14}$, $A_{2,16} = A_{1,16}$, $A_{2,20} = A_{1,20}$, $A_{2,22} = A_{1,22}$, $A_{2,24} \neq A_{1,24}$, $A_{2,25} \neq A_{1,25}$, $A_{2,26} \neq A_{1,26}$, $A_{2,29} = A_{1,29}$, $A_{2,23} = A_{2,11}$, $A_{2,22} = A_{2,13}$, $A_{2,25} \neq A_{2,14}$, $E_{2,1} = 0$, $E_{2,2} = 1$, $E_{2,3} = 1$, $E_{2,6} = 1$, $E_{2,10} = 1$, $E_{2,11} = 0$, $E_{2,12} = 1$, $E_{2,13} = 1$, $E_{2,14} = 0$, $E_{2,15} = 1$, $E_{2,16} = 1$, $E_{2,17} = 1$, $E_{2,20} = 1$, $E_{2,22} = 0$, $E_{2,23} = 0$, $E_{2,24} = 0$, $E_{2,25} = 1$, $E_{2,27} = 1$, $E_{2,29} = 1$, $E_{2,5} \neq E_{1,5}$, $E_{2,21} = E_{2,7}$, $W_{2,2} = 1$, $W_{2,30} \neq W_{2,13}$, $W_{2,23} \neq W_{2,19}$	40
3	$A_{3,0} = 0$, $A_{3,1} = 0$, $A_{3,2} = 0$, $A_{3,3} = 0$, $A_{3,4} = 1$, $A_{3,5} = 1$, $A_{3,6} = 0$, $A_{3,7} = 0$, $A_{3,8} = 0$, $A_{3,9} = 1$, $A_{3,10} = 1$, $A_{3,11} = 1$, $A_{3,12} = 1$, $A_{3,13} = 0$, $A_{3,14} = 1$, $A_{3,16} = 1$, $A_{3,17} = 1$, $A_{3,18} = 1$, $A_{3,20} = 0$, $A_{3,21} = 0$, $A_{3,22} = 0$, $A_{3,23} = 1$, $A_{3,24} = 0$, $A_{3,25} = 0$, $A_{3,26} = 0$, $A_{3,27} = 0$, $A_{3,28} = 1$, $A_{3,29} = 0$, $A_{3,30} = 0$, $A_{3,31} = 1$, $E_{3,0} = 0$, $E_{3,1} = 0$, $E_{3,2} = 0$, $E_{3,3} = 1$, $E_{3,4} = 0$, $E_{3,5} = 0$, $E_{3,6} = 1$, $E_{3,7} = 0$, $E_{3,8} = 1$, $E_{3,9} = 1$, $E_{3,10} = 0$, $E_{3,11} = 0$, $E_{3,12} = 1$, $E_{3,13} = 0$, $E_{3,14} = 1$, $E_{3,15} = 0$, $E_{3,17} = 1$, $E_{3,18} = 0$, $E_{3,19} = 1$, $E_{3,20} = 0$, $E_{3,21} = 1$, $E_{3,22} = 0$, $E_{3,23} = 0$, $E_{3,24} = 1$, $E_{3,25} = 0$, $E_{3,26} = 1$, $E_{3,27} = 0$, $E_{3,28} = 1$, $E_{3,29} = 0$, $E_{3,30} = 1$, $W_{3,1} = 0$, $W_{3,2} = 1$, $W_{3,5} = 1$, $W_{3,9} = 0$, $W_{3,11} = 1$, $W_{3,12} = 1$, $W_{3,14} = 1$, $W_{3,16} = 0$, $W_{3,20} = 0$, $W_{3,21} = 1$, $W_{3,27} = 0$, $W_{3,28} = 0$, $W_{3,30} = 1$, $W_{3,31} = 1$, $W_{3,17} = W_{3,0}$, $W_{3,24} \neq W_{3,3}$, $W_{3,25} = W_{3,4}$, $W_{3,10} = W_{3,6}$, $W_{3,23} \neq W_{3,6}$, $W_{3,22} = W_{3,7}$, $W_{3,24} \neq W_{3,7}$, $W_{3,23} = W_{3,8}$, $W_{3,25} = W_{3,10}$, $W_{3,17} = W_{3,13}$, $W_{3,24} \neq W_{3,13}$, $W_{3,19} = W_{3,15}$, $W_{3,26} = W_{3,15}$, $W_{3,22} \neq W_{3,18}$, $W_{3,29} = W_{3,18}$, $W_{3,23} = W_{3,19}$, $W_{3,26} = W_{3,22}$	92
4	$A_{4,3} = 1$, $A_{4,5} = 0$, $A_{4,15} = 0$, $A_{4,26} = 0$, $A_{4,0} = A_{2,0}$, $A_{4,4} \neq A_{2,4}$, $A_{4,11} = A_{2,11}$, $A_{4,14} \neq A_{2,14}$, $A_{4,16} \neq A_{2,16}$, $A_{4,20} = A_{2,20}$, $A_{4,22} = A_{2,22}$, $A_{4,24} \neq A_{2,24}$, $A_{4,29} \neq A_{2,29}$, $A_{4,17} = A_{4,6}$, $E_{4,1} = 0$, $E_{4,2} = 0$, $E_{4,4} = 0$, $E_{4,6} = 0$, $E_{4,7} = 1$, $E_{4,8} = 1$, $E_{4,9} = 0$, $E_{4,10} = 1$, $E_{4,11} = 1$, $E_{4,12} = 0$, $E_{4,13} = 0$, $E_{4,14} = 0$, $E_{4,15} = 1$, $E_{4,16} = 1$, $E_{4,17} = 1$, $E_{4,18} = 1$, $E_{4,20} = 0$, $E_{4,21} = 0$, $E_{4,22} = 1$, $E_{4,23} = 1$, $E_{4,24} = 1$, $E_{4,25} = 0$, $E_{4,26} = 1$, $E_{4,27} = 0$, $E_{4,28} = 0$, $E_{4,29} = 0$, $E_{4,30} = 0$, $E_{4,19} \neq A_{3,19}$, $W_{4,4} = 0$, $W_{4,5} = 0$, $W_{4,8} = 0$, $W_{4,9} = 0$, $W_{4,16} = 1$, $W_{4,17} = 1$, $W_{4,19} = 1$, $W_{4,20} = 1$, $W_{4,21} = 1$, $W_{4,22} = 1$, $W_{4,25} = 1$, $W_{4,26} = 1$, $W_{4,30} = 0$, $W_{4,31} = 1$, $W_{4,18} \neq W_{4,1}$, $W_{4,13} = W_{4,2}$, $W_{4,23} \neq W_{4,2}$, $W_{4,10} = W_{4,6}$, $W_{4,11} \neq W_{4,7}$, $W_{4,23} = W_{4,12}$, $W_{4,27} = W_{4,12}$, $W_{4,28} = W_{4,13}$	67
5	$A_{5,5} = 1$, $A_{5,15} = 0$, $A_{5,0} \neq A_{4,0}$, $A_{5,2} \neq A_{4,2}$, $A_{5,4} \neq A_{4,4}$, $A_{5,6} \neq A_{4,6}$, $A_{5,11} = A_{4,11}$, $A_{5,14} = A_{4,14}$, $A_{5,16} \neq A_{4,16}$, $A_{5,18} = A_{4,18}$, $A_{5,20} = A_{4,20}$, $A_{5,22} = A_{4,22}$, $A_{5,24} = A_{4,24}$, $A_{5,25} = A_{4,25}$, $A_{5,29} \neq A_{4,29}$, $A_{5,26} = A_{5,3}$, $A_{5,24} = A_{5,4}$, $A_{5,27} \neq A_{5,6}$, $E_{5,0} = 0$, $E_{5,1} = 1$, $E_{5,2} = 1$, $E_{5,3} = 1$, $E_{5,4} = 1$, $E_{5,5} = 1$, $E_{5,6} = 1$, $E_{5,7} = 1$, $E_{5,9} = 1$, $E_{5,10} = 0$, $E_{5,11} = 1$, $E_{5,12} = 0$, $E_{5,13} = 1$, $E_{5,14} = 1$, $E_{5,15} = 0$, $E_{5,16} = 0$, $E_{5,17} = 0$, $E_{5,18} = 0$, $E_{5,20} = 0$, $E_{5,21} = 1$, $E_{5,22} = 0$, $E_{5,23} = 1$, $E_{5,25} = 0$, $E_{5,26} = 1$, $E_{5,27} = 0$, $E_{5,28} = 0$, $E_{5,29} = 1$, $E_{5,30} = 1$, $E_{5,31} = 0$, $E_{1,0} = A_{1,0}$, $W_{5,0} = 0$, $W_{5,3} = 0$, $W_{5,5} = 0$, $W_{5,7} = 0$, $W_{5,8} = 1$, $W_{5,9} = 0$, $W_{5,11} = 0$, $W_{5,12} = 0$, $W_{5,13} = 1$, $W_{5,14} = 1$, $W_{5,15} = 1$, $W_{5,16} = 0$, $W_{5,17} = 0$, $W_{5,19} = 0$, $W_{5,20} = 1$, $W_{5,22} = 1$, $W_{5,24} = 0$, $W_{5,25} = 0$, $W_{5,26} = 1$, $W_{5,28} = 1$, $W_{5,30} = 1$, $W_{5,31} = 0$, $W_{5,29} = W_{5,1}$, $W_{5,23} = W_{5,2}$, $W_{5,21} = W_{5,4}$, $W_{5,29} \neq W_{5,18}$	74
6	$A_{6,2} = 0$, $A_{6,6} = 1$, $A_{6,15} = 1$, $A_{6,18} = 0$, $A_{6,26} = A_{5,26}$, $A_{6,26} = A_{6,3}$, $A_{6,24} \neq A_{6,4}$, $A_{6,27} \neq A_{6,7}$, $A_{6,30} \neq A_{6,9}$, $A_{6,23} \neq A_{6,11}$, $A_{6,22} \neq A_{6,13}$, $A_{6,25} = A_{6,14}$, $A_{6,26} \neq A_{6,17}$, $E_{6,0} = 1$, $E_{6,1} = 1$, $E_{6,2} = 1$, $E_{6,3} = 1$, $E_{6,4} = 1$, $E_{6,5} = 1$, $E_{6,6} = 0$, $E_{6,7} = 1$, $E_{6,8} = 0$, $E_{6,9} = 0$, $E_{6,10} = 1$, $E_{6,11} = 0$, $E_{6,12} = 0$, $E_{6,13} = 1$, $E_{6,14} = 1$, $E_{6,15} = 1$, $E_{6,16} = 1$, $E_{6,17} = 0$, $E_{6,18} = 0$, $E_{6,19} = 0$, $E_{6,20} = 0$, $E_{6,21} = 0$, $E_{6,23} = 0$, $E_{6,24} = 0$, $E_{6,25} = 0$, $E_{6,26} = 1$, $E_{6,27} = 1$, $E_{6,28} = 0$, $E_{6,29} = 1$, $E_{6,30} = 0$, $E_{6,31} = 0$, $W_{6,11} = 0$, $W_{6,14} = 0$, $W_{6,18} = 1$, $W_{6,19} = 0$, $W_{6,21} = 0$, $W_{6,23} = 1$, $W_{6,24} = 1$, $W_{6,25} = 0$, $W_{6,26} = 0$, $W_{6,31} = 0$, $W_{6,17} \neq W_{6,0}$, $W_{6,28} = W_{6,0}$, $W_{6,22} = W_{6,1}$, $W_{6,7} \neq W_{6,3}$, $W_{6,20} = W_{6,3}$, $W_{6,8} \neq W_{6,4}$, $W_{6,22} = W_{6,5}$, $W_{6,10} = W_{6,6}$, $W_{6,27} \neq W_{6,6}$, $W_{6,22} = W_{6,7}$, $W_{6,28} = W_{6,7}$, $W_{6,12} \neq W_{6,8}$, $W_{6,29} = W_{6,8}$, $W_{6,13} \neq W_{6,9}$, $W_{6,30} = W_{6,9}$, $W_{6,27} = W_{6,10}$, $W_{6,30} = W_{6,15}$, $W_{6,20} \neq W_{6,16}$	73
7	$A_{7,2} = A_{5,2}$, $A_{7,6} \neq A_{5,6}$, $A_{7,18} = A_{5,18}$, $E_{7,0} = 1$, $E_{7,1} = 1$, $E_{7,2} = 0$, $E_{7,3} = 0$, $E_{7,4} = 0$, $E_{7,5} = 0$, $E_{7,6} = 1$, $E_{7,7} = 0$, $E_{7,8} = 0$, $E_{7,9} = 0$, $E_{7,10} = 0$, $E_{7,11} = 0$, $E_{7,12} = 0$, $E_{7,13} = 1$, $E_{7,14} = 0$, $E_{7,15} = 0$, $E_{7,16} = 1$, $E_{7,18} = 1$, $E_{7,19} = 0$, $E_{7,20} = 0$, $E_{7,21} = 1$, $E_{7,22} = 1$, $E_{7,23} = 1$, $E_{7,24} = 0$, $E_{7,25} = 0$, $E_{7,26} = 0$, $E_{7,27} = 1$, $E_{7,28} = 0$, $E_{7,29} = 1$, $E_{7,30} = 0$, $W_{7,0} = 0$, $W_{7,1} = 0$, $W_{7,2} = 0$, $W_{7,3} = 0$, $W_{7,4} = 1$, $W_{7,5} = 0$, $W_{7,6} = 0$, $W_{7,7} = 0$, $W_{7,8} = 0$, $W_{7,9} = 1$, $W_{7,10} = 1$, $W_{7,11} = 1$, $W_{7,12} = 0$, $W_{7,14} = 0$, $W_{7,15} = 0$, $W_{7,17} = 0$, $W_{7,19} = 0$, $W_{7,21} = 1$, $W_{7,22} = 0$, $W_{7,23} = 1$, $W_{7,24} = 0$, $W_{7,25} = 1$, $W_{7,26} = 1$, $W_{7,27} = 0$, $W_{7,28} = 0$, $W_{7,29} = 0$, $W_{7,30} = 0$, $W_{7,31} = 1$, $W_{7,16} \neq E_{3,16}$	66
8	$A_{8,2} = A_{7,2}$, $A_{8,6} \neq A_{7,6}$, $A_{8,15} \neq A_{7,15}$, $A_{8,16} = A_{7,16}$, $A_{8,18} = A_{7,18}$, $A_{8,27} \neq A_{7,27}$, $E_{8,0} = 0$, $E_{8,1} = 1$, $E_{8,3} = 1$, $E_{8,4} = 1$, $E_{8,5} = 0$, $E_{8,6} = 0$, $E_{8,11} = 0$, $E_{8,13} = 1$, $E_{8,15} = 0$, $E_{8,17} = 1$, $E_{8,18} = 0$, $E_{8,20} = 1$, $E_{8,23} = 0$, $E_{8,25} = 0$, $E_{8,26} = 0$, $E_{8,27} = 1$, $E_{8,29} = 0$, $E_{8,30} = 1$, $E_{8,12} = E_{8,7}$, $E_{8,21} \neq E_{8,8}$, $W_{8,5} = 0$, $W_{8,16} = 0$, $W_{8,17} = 0$, $W_{8,18} = 0$, $W_{8,19} = 1$, $W_{8,27} = 1$, $W_{8,0} \neq A_{4,0}$, $W_{8,1} = A_{4,1}$, $W_{8,4} \neq A_{4,4}$, $W_{8,21} = W_{8,0}$, $W_{8,22} = W_{8,1}$, $W_{8,23} \neq W_{8,2}$, $W_{8,7} \neq W_{8,3}$, $W_{8,8} \neq W_{8,4}$, $W_{8,23} \neq W_{8,6}$, $W_{8,31} \neq W_{8,10}$, $W_{8,28} = W_{8,13}$, $W_{8,29} = W_{8,14}$, $W_{8,15} = W_{8,31}$, $W_{8,31} = W_{8,20}$	46
9	$A_{9,16} = 1$, $A_{9,27} = 1$, $A_{9,25} = A_{9,5}$, $A_{9,15} = A_{9,6}$, $A_{9,18} \neq A_{9,7}$, $A_{9,28} = A_{9,7}$, $E_{9,1} = 0$, $E_{9,5} = 1$, $E_{9,6} = 0$, $E_{9,11} = 1$, $E_{9,15} = 1$, $E_{9,18} = 1$, $E_{9,20} = 1$, $E_{9,21} = 0$, $E_{9,23} = 1$, $E_{9,25} = 0$, $E_{9,26} = 0$, $E_{9,27} = 1$, $E_{9,30} = 0$, $E_{9,14} \neq E_{9,0}$, $E_{9,13} \neq E_{9,8}$, $E_{9,22} = E_{9,9}$	22
10	$A_{10,16} = A_{8,16}$, $A_{10,27} = A_{8,27}$, $E_{10,2} = 1$, $E_{10,4} = 1$, $E_{10,6} = 1$, $E_{10,15} = 0$, $E_{10,18} = 0$, $E_{10,25} = 0$, $E_{10,26} = 0$, $E_{10,27} = 1$, $E_{10,28} = 0$, $E_{10,13} \neq E_{10,0}$, $E_{10,14} \neq E_{10,0}$, $E_{10,23} = E_{10,5}$, $E_{10,20} = E_{10,7}$, $E_{10,21} \neq E_{10,8}$, $E_{10,22} = E_{10,9}$, $E_{10,23} = E_{10,10}$, $E_{10,29} = E_{10,10}$, $E_{10,30} \neq E_{10,12}$, $E_{10,31} \neq E_{10,13}$, $E_{10,29} \neq E_{10,16}$, $E_{10,22} \neq E_{10,17}$, $E_{10,24} = E_{10,19}$	25
11	$A_{11,16} = A_{10,16}$, $A_{11,27} = A_{10,27}$, $E_{11,2} = 0$, $E_{11,4} = 0$, $E_{11,6} = 1$, $E_{11,15} = 0$, $E_{11,18} = 0$, $E_{11,19} = 0$, $E_{11,20} = 1$, $E_{11,25} = 0$, $E_{11,26} = 1$, $E_{11,28} = 0$	12
12	$E_{12,2} = 1$, $E_{12,4} = 1$, $E_{12,6} = 0$, $E_{12,15} = 1$, $E_{12,18} = 1$, $E_{12,19} = 1$, $E_{12,20} = 0$, $E_{12,25} = 1$, $E_{12,26} = 1$, $E_{12,27} = 0$, $E_{12,28} = 0$, $E_{12,16} \neq E_{11,16}$	12
13	$E_{13,16} = 0$, $E_{13,17} = 0$, $E_{13,18} = 0$, $E_{13,19} = 0$, $E_{13,20} = 1$, $E_{13,27} = 0$, $E_{13,28} = 1$, $E_{13,13} \neq E_{13,0}$, $E_{13,14} = E_{13,0}$, $E_{13,6} = E_{13,1}$, $E_{13,14} \neq E_{13,1}$, $E_{13,15} \neq E_{13,1}$, $E_{13,15} = E_{13,2}$, $E_{13,29} = E_{13,2}$, $E_{13,21} \neq E_{13,3}$, $E_{13,30} \neq E_{13,3}$, $E_{13,22} \neq E_{13,4}$, $E_{13,31} \neq E_{13,4}$, $E_{13,23} = E_{13,5}$, $E_{13,24} \neq E_{13,6}$, $E_{13,25} = E_{13,7}$, $E_{13,13} \neq E_{13,8}$, $E_{13,30} = E_{13,11}$, $E_{13,31} \neq E_{13,12}$	25
14	$E_{14,16} = 0$, $E_{14,17} = 0$, $E_{14,18} = 0$, $E_{14,19} = 0$, $E_{14,20} = 0$, $E_{14,27} = 0$, $E_{14,28} = 0$	7
15	$E_{15,16} = 1$, $E_{15,17} = 1$, $E_{15,18} = 1$, $E_{15,19} = 1$, $E_{15,20} = 1$, $E_{15,27} = 1$, $E_{15,28} = 1$	7
17	$W_{17,16} = 0$, $W_{17,27} = 0$, $W_{17,14} \neq W_{4,15}$, $W_{17,4} = W_{17,2}$, $W_{17,29} \neq W_{17,20}$	5

Table 6. Starting point for a collision for 27 steps of SHA-256

i	∇A_i	∇E_i	∇W_i
-4	--------------------------------	--------------------------------	
-3	--------------------------------	--------------------------------	
-2	--------------------------------	--------------------------------	
-1	--------------------------------	--------------------------------	
0	--------------------------------	--------------------------------	--------------------------------
1	--------------------------------	--------------------------------	--------------------------------
2	--------------------------------	--------------------------------	--------------------------------
3	--------------------------------	--------------------------------	--------------------------------
4	--------------------------------	--------------------------------	--------------------------------
5	--------------------------------	--------------------------------	--------------------------------
6	--------------------------------	--------------------------------	--------------------------------
7	????????????????????????????????	????????????????????????????????	?????????????????????????????x??
8	????????????????????????????????	????????????????????????????????	????????????????????????????????
9	????????????????????????????????	????????????????????????????????	--------------------------------
10	--------------------------------	????????????????????????????????	--------------------------------
11	--------------------------------	????????????????????????????????	--------------------------------
12	--------------------------------	????????????????????????????????	????????????????????????????????
13	--------------------------------	????????????????????????????????	--------------------------------
14	--------------------------------	--------------------------------	--------------------------------
15	--------------------------------	--------------------------------	????????????????????????????????
16	--------------------------------	--------------------------------	--------------------------------
17	--------------------------------	--------------------------------	????????????????????????????????
18	--------------------------------	--------------------------------	--------------------------------
19	--------------------------------	--------------------------------	--------------------------------
20	--------------------------------	--------------------------------	--------------------------------
21	--------------------------------	--------------------------------	--------------------------------
22	--------------------------------	--------------------------------	--------------------------------
23	--------------------------------	--------------------------------	--------------------------------
24	--------------------------------	--------------------------------	--------------------------------
25	--------------------------------	--------------------------------	--------------------------------
26	--------------------------------	--------------------------------	--------------------------------

Table 7. Characteristic for a collision for 27 steps of SHA-256

i	∇A_i	∇E_i	∇W_i
-4	-------------------------------	--------------------------------	
-3	-------------------------------	--------------------------------	
-2	-------------------------------	--------------------------------	
-1	-------------------------------	--------------------------------	
0	-------------------------------	--------------------------------	--------------------------------
1	-------------------------------	--------------------------------	--------------------------------
2	-------------------------------	--------------------------------	--------------------------------
3	-------------------------------	--------------------------------	--------------------------------
4	-------------------------------	--------------------------------	--------------------------------
5	-------------------------------	---------------1--------1-----	--------------------------------
6	-------------------------------	-1--------0--0-10-1----0-0-----	--------------------------------
7	-------unn--u------n---nn-uuuu--	101-11---u10u1-0nuu-uuuu1n---n0-	00---1--un-0u-nuuuuu1-nu0n101n--
8	nnnnn-nnnn-------nuu----------	0n0n001001u-1u1n01un010n01n00110	-----u--n---n---------nn--------
9	----un--n--nu------nu-u--------	-1n1n1011u011100nn100u10-10000u-	--------------------------------
10	-------------------------------	u00000nuuu10uun01u00n00n110-u-u1	--------------------------------
11	-------------------------------	0n000uuuuu01010111n-uun01n000n01	--------------------------------
12	-------------------------------	01---1010u01u----111-010-0--110-	------110-u-------n0--u--n-n--nn
13	-------------------------------	01-10u1nunuuu---1110-1nn11---01-	--------------------------------
14	-------------------------------	-----1-01011----------00--------	--------------------------------
15	-------------------------------	-----1-001000---------11--------	0u1-nn-n-u-1u---11un0uu10u101u0-
16	-------------------------------	--------------------------------	--------------------------------
17	-------------------------------	--------------------------------	---0-1nnn---u-1-----10uu0-------
18	-------------------------------	--------------------------------	--------------------------------
19	-------------------------------	--------------------------------	--------------------------------
20	-------------------------------	--------------------------------	--------------------------------
21	-------------------------------	--------------------------------	--------------------------------
22	-------------------------------	--------------------------------	--------------------------------
23	-------------------------------	--------------------------------	--------------------------------
24	-------------------------------	--------------------------------	--------------------------------
25	-------------------------------	--------------------------------	--------------------------------
26	-------------------------------	--------------------------------	--------------------------------

Table 8. Collision for 27 steps of SHA-256

h_0	6a09e667 bb67ae85 3c6ef372 a54ff53a 510e527f 9b05688c 1f83d9ab 5be0cd19
m	725a0370 0daa9f1b 071d92df ec8282c1 7913134a bc2eb291 02d33a84 278dfd29 0c40f8ea d8bd68a0 0ce670c5 5ec7155d 9f6407a8 729fbfe8 aa7c7c08 607ae76d
m^*	725a0370 0daa9f1b 071d92df ec8282c1 7913134a bc2eb291 02d33a84 27460e6d 08c8fbea d8bd68a0 0ce670c5 5ec7155d 9f4425fb 729fbfe8 aa7c7c08 2d32d129
Δm	00000000 00000000 00000000 00000000 00000000 00000000 00000000 00cbf344 04880300 00000000 00000000 00000000 00202253 00000000 00000000 4d483644
h_1	5864015f 133494fa fa42bb35 94bc44f9 29eabb36 9e461e33 2eab27f8 106467c9

Table 9. Set of conditions for the collision for 27 steps

i	set of conditions	
5	$E_{5,6} = 1$, $E_{5,15} = 1$	2
6	$A_{6,2} = A_{5,2}$, $A_{6,3} \neq A_{5,3}$, $A_{6,7} = A_{5,7}$, $A_{6,8} \neq A_{5,8}$, $A_{6,19} \neq A_{5,19}$, $A_{6,22} \neq A_{5,22}$, $A_{6,23} = A_{5,23}$, $E_{6,6} = 0$, $E_{6,8} = 0$, $E_{6,13} = 1$, $E_{6,15} = 0$, $E_{6,16} = 1$, $E_{6,18} = 0$, $E_{6,21} = 0$, $E_{6,30} = 1$, $E_{6,2} = E_{5,2}$, $E_{6,9} \neq E_{5,9}$, $E_{6,14} = E_{5,14}$	18
7	$A_{7,2} = 1$, $A_{7,3} = 1$, $A_{7,4} = 1$, $A_{7,5} = 1$, $A_{7,7} = 0$, $A_{7,8} = 0$, $A_{7,12} = 0$, $A_{7,19} = 1$, $A_{7,22} = 0$, $A_{7,23} = 0$, $A_{7,24} = 1$, $A_{7,10} \neq A_{7,1}$, $A_{7,11} \neq A_{6,11}$, $A_{7,25} \neq A_{6,25}$, $A_{7,31} \neq A_{7,10}$, $A_{7,31} \neq A_{7,11}$, $A_{7,25} = A_{7,14}$, $A_{7,26} = A_{7,15}$, $A_{7,27} = A_{7,16}$, $A_{7,28} = A_{7,16}$, $A_{7,28} = A_{7,17}$, $A_{7,29} = A_{7,17}$, $A_{7,31} \neq A_{7,20}$, $E_{7,1} = 0$, $E_{7,2} = 0$, $E_{7,6} = 0$, $E_{7,7} = 1$, $E_{7,8} = 1$, $E_{7,9} = 1$, $E_{7,10} = 1$, $E_{7,11} = 1$, $E_{7,13} = 1$, $E_{7,14} = 1$, $E_{7,15} = 0$, $E_{7,16} = 0$, $E_{7,18} = 1$, $E_{7,19} = 1$, $E_{7,20} = 0$, $E_{7,21} = 1$, $E_{7,22} = 1$, $E_{7,26} = 1$, $E_{7,27} = 1$, $E_{7,29} = 1$, $E_{7,30} = 0$, $E_{7,31} = 1$, $E_{7,5} = E_{6,5}$, $E_{7,12} = E_{6,12}$, $E_{7,28} = E_{6,28}$, $E_{7,5} = E_{7,0}$, $E_{7,23} = E_{7,4}$, $E_{7,28} \neq E_{7,23}$, $W_{7,2} = 0$, $W_{7,3} = 1$, $W_{7,4} = 0$, $W_{7,5} = 1$, $W_{7,6} = 0$, $W_{7,7} = 0$, $W_{7,8} = 1$, $W_{7,9} = 0$, $W_{7,11} = 1$, $W_{7,12} = 1$, $W_{7,13} = 1$, $W_{7,14} = 1$, $W_{7,15} = 1$, $W_{7,16} = 1$, $W_{7,17} = 0$, $W_{7,19} = 1$, $W_{7,20} = 0$, $W_{7,22} = 0$, $W_{7,23} = 1$, $W_{7,26} = 1$, $W_{7,30} = 0$, $W_{7,31} = 0$, $W_{7,21} \neq W_{7,0}$, $W_{7,18} \neq W_{7,1}$, $W_{7,29} \neq W_{7,1}$, $W_{7,21} \neq W_{7,10}$, $W_{7,25} = W_{7,10}$, $W_{7,29} = W_{7,18}$, $W_{7,29} = W_{7,25}$	80
8	$A_{8,11} = 1$, $A_{8,12} = 1$, $A_{8,13} = 0$, $A_{8,22} = 0$, $A_{8,23} = 0$, $A_{8,24} = 0$, $A_{8,25} = 0$, $A_{8,27} = 0$, $A_{8,28} = 0$, $A_{8,29} = 0$, $A_{8,30} = 0$, $A_{8,31} = 0$, $A_{8,10} \neq W_{12,10}$, $A_{8,14} \neq W_{12,14}$, $A_{8,26} \neq W_{12,26}$, $A_{8,19} \neq A_{6,19}$, $A_{8,10} \neq A_{7,10}$, $A_{8,20} = A_{7,20}$, $A_{8,26} \neq A_{7,26}$, $A_{8,10} = A_{8,1}$, $A_{8,16} \neq A_{8,4}$, $A_{8,17} = A_{8,5}$, $A_{8,15} \neq A_{8,6}$, $A_{8,18} = A_{8,6}$, $A_{8,18} \neq A_{8,7}$, $A_{8,19} = A_{8,7}$, $A_{8,20} \neq A_{8,8}$, $E_{8,0} = 0$, $E_{8,1} = 1$, $E_{8,2} = 1$, $E_{8,3} = 0$, $E_{8,4} = 0$, $E_{8,5} = 0$, $E_{8,6} = 1$, $E_{8,7} = 0$, $E_{8,8} = 0$, $E_{8,9} = 0$, $E_{8,10} = 1$, $E_{8,11} = 0$, $E_{8,12} = 0$, $E_{8,13} = 1$, $E_{8,14} = 1$, $E_{8,15} = 0$, $E_{8,16} = 0$, $E_{8,17} = 1$, $E_{8,18} = 1$, $E_{8,19} = 1$, $E_{8,21} = 1$, $E_{8,22} = 1$, $E_{8,23} = 0$, $E_{8,24} = 0$, $E_{8,25} = 1$, $E_{8,26} = 0$, $E_{8,27} = 0$, $E_{8,28} = 0$, $E_{8,29} = 0$, $E_{8,30} = 0$, $E_{8,31} = 0$, $W_{8,8} = 0$, $W_{8,9} = 0$, $W_{8,19} = 0$, $W_{8,23} = 0$, $W_{8,26} = 1$, $W_{8,20} \neq W_{8,5}$, $W_{8,22} = W_{8,5}$, $W_{8,27} = W_{8,6}$, $W_{8,15} = W_{8,11}$, $W_{8,24} \neq W_{8,13}$, $W_{8,30} \neq W_{8,15}$, $W_{8,29} = W_{8,25}$	70
9	$A_{9,8} = 1$, $A_{9,10} = 1$, $A_{9,11} = 0$, $A_{9,19} = 1$, $A_{9,20} = 0$, $A_{9,23} = 0$, $A_{9,26} = 0$, $A_{9,27} = 1$, $A_{9,13} \neq A_{7,13}$, $A_{9,4} = A_{8,4}$, $A_{9,7} \neq A_{8,7}$, $A_{9,12} \neq A_{9,0}$, $A_{9,22} \neq A_{9,1}$, $A_{9,15} \neq A_{9,3}$, $A_{9,16} \neq A_{9,4}$, $A_{9,15} \neq A_{9,6}$, $A_{9,18} \neq A_{9,7}$, $A_{9,30} \neq A_{9,7}$, $A_{9,29} \neq A_{9,9}$, $A_{9,30} \neq A_{9,21}$, $A_{9,31} \neq A_{9,22}$, $E_{9,1} = 1$, $E_{9,2} = 0$, $E_{9,3} = 0$, $E_{9,4} = 0$, $E_{9,5} = 0$, $E_{9,6} = 1$, $E_{9,8} = 0$, $E_{9,9} = 1$, $E_{9,10} = 1$, $E_{9,11} = 0$, $E_{9,12} = 0$, $E_{9,13} = 1$, $E_{9,14} = 0$, $E_{9,15} = 0$, $E_{9,16} = 0$, $E_{9,17} = 0$, $E_{9,18} = 1$, $E_{9,19} = 1$, $E_{9,20} = 0$, $E_{9,21} = 0$, $E_{9,22} = 1$, $E_{9,23} = 1$, $E_{9,24} = 1$, $E_{9,25} = 0$, $E_{9,26} = 1$, $E_{9,27} = 0$, $E_{9,28} = 1$, $E_{9,29} = 0$, $E_{9,30} = 1$	50
10	$A_{10,8} \neq A_{8,8}$, $A_{10,10} \neq A_{8,10}$, $A_{10,19} \neq A_{8,19}$, $A_{10,20} = A_{8,20}$, $A_{10,26} \neq A_{8,26}$, $A_{10,12} = A_{9,12}$, $A_{10,13} = A_{9,13}$, $A_{10,22} = A_{9,22}$, $A_{10,24} \neq A_{9,24}$, $A_{10,25} \neq A_{9,25}$, $A_{10,28} = A_{9,28}$, $A_{10,29} = A_{9,29}$, $A_{10,30} \neq A_{9,30}$, $A_{10,31} \neq A_{9,31}$, $E_{10,0} = 1$, $E_{10,1} = 1$, $E_{10,3} = 1$, $E_{10,5} = 0$, $E_{10,6} = 1$, $E_{10,7} = 1$, $E_{10,8} = 0$, $E_{10,9} = 0$, $E_{10,10} = 0$, $E_{10,11} = 0$, $E_{10,12} = 0$, $E_{10,13} = 0$, $E_{10,14} = 1$, $E_{10,15} = 1$, $E_{10,16} = 0$, $E_{10,17} = 0$, $E_{10,18} = 1$, $E_{10,19} = 1$, $E_{10,20} = 0$, $E_{10,21} = 1$, $E_{10,22} = 1$, $E_{10,23} = 1$, $E_{10,24} = 1$, $E_{10,25} = 0$, $E_{10,26} = 0$, $E_{10,27} = 0$, $E_{10,28} = 0$, $E_{10,29} = 0$, $E_{10,30} = 0$, $E_{10,31} = 1$	44
11	$A_{11,8} = A_{10,8}$, $A_{11,10} = A_{10,10}$, $A_{11,11} \neq A_{10,11}$, $A_{11,19} \neq A_{10,19}$, $A_{11,20} = A_{10,20}$, $A_{11,23} = A_{10,23}$, $A_{11,26} \neq A_{10,26}$, $A_{11,27} \neq A_{10,27}$, $E_{11,0} = 1$, $E_{11,1} = 0$, $E_{11,2} = 0$, $E_{11,3} = 0$, $E_{11,4} = 0$, $E_{11,5} = 0$, $E_{11,6} = 0$, $E_{11,7} = 1$, $E_{11,8} = 0$, $E_{11,9} = 0$, $E_{11,10} = 1$, $E_{11,11} = 1$, $E_{11,13} = 0$, $E_{11,14} = 1$, $E_{11,15} = 1$, $E_{11,16} = 1$, $E_{11,17} = 0$, $E_{11,18} = 1$, $E_{11,19} = 0$, $E_{11,20} = 1$, $E_{11,21} = 0$, $E_{11,22} = 0$, $E_{11,23} = 1$, $E_{11,24} = 1$, $E_{11,25} = 1$, $E_{11,26} = 1$, $E_{11,27} = 0$, $E_{11,28} = 0$, $E_{11,29} = 0$, $E_{11,30} = 0$, $E_{11,31} = 0$	39
12	$E_{12,1} = 0$, $E_{12,2} = 1$, $E_{12,3} = 1$, $E_{12,6} = 0$, $E_{12,8} = 0$, $E_{12,9} = 1$, $E_{12,10} = 0$, $E_{12,12} = 1$, $E_{12,13} = 1$, $E_{12,14} = 1$, $E_{12,19} = 1$, $E_{12,20} = 1$, $E_{12,21} = 0$, $E_{12,22} = 1$, $E_{12,23} = 0$, $E_{12,24} = 1$, $E_{12,25} = 0$, $E_{12,26} = 1$, $E_{12,30} = 1$, $E_{12,31} = 0$, $E_{12,11} = W_{12,11}$, $E_{12,27} \neq W_{12,27}$, $E_{12,0} \neq A_{8,0}$, $E_{12,5} = E_{12,0}$, $W_{12,0} = 0$, $W_{12,1} = 0$, $W_{12,4} = 0$, $W_{12,6} = 0$, $W_{12,9} = 1$, $W_{12,12} = 0$, $W_{12,13} = 0$, $W_{12,21} = 1$, $W_{12,23} = 0$, $W_{12,24} = 1$, $W_{12,25} = 1$	35
13	$E_{13,1} = 1$, $E_{13,2} = 0$, $E_{13,6} = 1$, $E_{13,7} = 1$, $E_{13,8} = 0$, $E_{13,9} = 0$, $E_{13,10} = 1$, $E_{13,12} = 0$, $E_{13,13} = 1$, $E_{13,14} = 1$, $E_{13,15} = 1$, $E_{13,19} = 1$, $E_{13,20} = 1$, $E_{13,21} = 1$, $E_{13,22} = 0$, $E_{13,23} = 1$, $E_{13,24} = 0$, $E_{13,25} = 1$, $E_{13,26} = 1$, $E_{13,27} = 0$, $E_{13,28} = 1$, $E_{13,30} = 1$, $E_{13,31} = 0$, $E_{13,5} = E_{13,0}$, $E_{13,17} = E_{13,4}$, $E_{13,18} = E_{13,5}$, $E_{13,29} = E_{13,11}$	27
14	$E_{14,8} = 0$, $E_{14,9} = 0$, $E_{14,20} = 1$, $E_{14,21} = 1$, $E_{14,22} = 0$, $E_{14,23} = 1$, $E_{14,24} = 0$, $E_{14,26} = 1$	8
15	$E_{15,8} = 1$, $E_{15,9} = 1$, $E_{15,19} = 0$, $E_{15,20} = 0$, $E_{15,21} = 0$, $E_{15,22} = 1$, $E_{15,23} = 0$, $E_{15,24} = 0$, $E_{15,26} = 1$, $W_{15,1} = 0$, $W_{15,3} = 1$, $W_{15,4} = 0$, $W_{15,5} = 1$, $W_{15,6} = 1$, $W_{15,7} = 0$, $W_{15,8} = 1$, $W_{15,9} = 1$, $W_{15,10} = 1$, $W_{15,11} = 0$, $W_{15,12} = 0$, $W_{15,13} = 1$, $W_{15,14} = 1$, $W_{15,15} = 1$, $W_{15,19} = 1$, $W_{15,20} = 1$, $W_{15,22} = 1$, $W_{15,24} = 0$, $W_{15,26} = 0$, $W_{15,27} = 0$, $W_{15,29} = 1$, $W_{15,30} = 1$, $W_{15,31} = 0$, $W_{15,23} \neq W_{15,0}$, $W_{15,25} \neq W_{15,0}$, $W_{15,25} = W_{15,18}$, $W_{15,28} \neq W_{15,21}$	37
17	$W_{17,7} = 0$, $W_{17,8} = 1$, $W_{17,9} = 1$, $W_{17,10} = 0$, $W_{17,11} = 1$, $W_{17,17} = 1$, $W_{17,19} = 1$, $W_{17,23} = 0$, $W_{17,24} = 0$, $W_{17,25} = 0$, $W_{17,26} = 1$, $W_{17,28} = 0$, $W_{17,2} \neq W_{17,0}$, $W_{17,30} = W_{17,0}$, $W_{17,31} = W_{17,1}$, $W_{17,31} = W_{17,6}$, $W_{17,21} = W_{17,12}$, $W_{17,21} = W_{17,14}$, $W_{17,22} = W_{17,15}$, $W_{17,27} \neq W_{17,18}$	20

Cryptanalysis of ARMADILLO2*

Mohamed Ahmed Abdelraheem[1], Céline Blondeau[2], María Naya-Plasencia[3,**],
Marion Videau[4,5,***], and Erik Zenner[6,†]

[1] Technical University of Denmark, Department of Mathematics, Denmark
[2] INRIA, project-team SECRET, France
[3] FHNW, Windisch, Switzerland and University of Versailles, France
[4] Agence nationale de la sécurité des systèmes d'information, France
[5] Université Henri Poincaré-Nancy 1 / LORIA, France
[6] University of Applied Sciences Offenburg, Germany

Abstract. ARMADILLO2 is the recommended variant of a multi-purpose cryptographic primitive dedicated to hardware which has been
proposed by Badel et al. in [1]. In this paper, we describe a meet-in-
the-middle technique relying on the parallel matching algorithm that
allows us to invert the ARMADILLO2 function. This makes it possible
to perform a key recovery attack when used as a FIL-MAC. A variant
of this attack can also be applied to the stream cipher derived from
the PRNG mode. Finally we propose a (second) preimage attack when
used as a hash function. We have validated our attacks by implementing cryptanalysis on scaled variants. The experimental results match the
theoretical complexities.

In addition to these attacks, we present a generalization of the parallel
matching algorithm, which can be applied in a broader context than
attacking ARMADILLO2.

Keywords: ARMADILLO2, meet-in-the-middle, key recovery attack,
preimage attack, parallel matching algorithm.

1 Introduction

ARMADILLO is a multi-purpose cryptographic primitive dedicated to hardware which was proposed by Badel et al. in [1]. Two variants were presented:
ARMADILLO and ARMADILLO2, the latter being the recommended version.

---

* This work was partially supported by the European Commission through the ICT
programme under contract ICT-2007-216676 ECRYPT II.
** Supported by the National Competence Center in Research on Mobile Informa-
tion and Communication Systems (NCCR-MICS), a center of the Swiss National
Science Foundation under grant number 5005-67322 and by the French Agence
Nationale de la Recherche through the SAPHIR2 project under Contract ANR-
08-VERS-014.
*** Partially supported by the French Agence Nationale de la Recherche under Con-
tract ANR-06-SETI-013-RAPIDE.
† This work was produced while at the Technical University of Denmark.

D.H. Lee and X. Wang (Eds.): ASIACRYPT 2011, LNCS 7073, pp. 308–326, 2011.

In the following, the first variant will be denoted ARMADILLO1 to distinguish it from ARMADILLO2. Both variants comprise several versions, each one associated to a different set of parameters and to a different security level. For both primitives, several applications are proposed: fixed input-length MAC (FIL-MAC), pseudo-random number generator/pseudo-random function (PRNG/PRF), and hash function. In [6], authors present a polynomial attack on ARMADILLO1. Even if the design of ARMADILLO2 is similar to the design of the first version, authors of [6] claim that this attack can not be applied on ARMADILLO2.

The ARMADILLO family uses a parameterized internal permutation as a building block. This internal permutation is based on two bitwise permutations σ_0 and σ_1. In [1], these permutations are not specified, but some of the properties that they must satisfy are given.

In this paper we provide the first cryptanalysis of ARMADILLO2, the recommended variant. As the bitwise permutations σ_0 and σ_1 are not specified, we have performed our analysis under the reasonable assumption that they behave like random permutations. As a consequence, the results of this paper are independent of the choice for σ_0 and σ_1.

To perform our attack, we use a meet-in-the-middle approach and an evolved variant of the parallel matching algorithm introduced in [2] and generalized in [5,4]. Our method enables us to invert the building block of ARMADILLO2 for a chosen value of the public part of the input, when a part of the output is known. We can use this step to build key recovery attacks faster than exhaustive search on all versions of ARMADILLO2 used in the FIL-MAC application mode. Besides, we propose several trade-offs for the time and memory needed for these attacks. We also adapt the attack to recover the key when ARMADILLO2 is used as a stream cipher in the PRNG application mode. We further show how to build (second) preimage attacks faster than exhaustive search when using the hashing mode, and propose again several time-memory trade-offs. We have implemented the attacks on a scaled version of ARMADILLO2, and the experimental results confirm the theoretical predictions.

Organization of the paper. We briefly describe ARMADILLO2 in Section 2. In Section 3 we detail our technique for inverting its building block and we explain how to extend the parallel matching algorithm to the case of ARMADILLO2. In Section 4, we explain how to apply this technique to build a key recovery attack on the FIL-MAC application mode. We briefly show how to adapt this attack to the stream cipher scenario in Section 4.2. The (second) preimage attack on the hashing mode is presented in Section 5. In Section 6 we present the experimental results of the verification that we have done on a scaled version of the algorithm. Finally, in Section 7, we propose a general form of the parallel matching algorithm derived from our attacks which can hopefully be used in more general contexts.

2   Description of ARMADILLO2

The core of ARMADILLO is based on the so-called *data-dependent bit trans-positions* [3]. We recall the description of ARMADILLO2 given in [1] using the same notations.

2.1   Description

Let C be an initial vector of size c and U be a message block of size m. The size of the register $(C\|U)$ is $k = c + m$. The ARMADILLO2 function transforms the vector (C, U) into (V_c, V_t) as described in Figure 1:

$$\text{ARMADILLO2} \quad : \quad \mathbb{F}_2^c \times \mathbb{F}_2^m \to \mathbb{F}_2^c \times \mathbb{F}_2^m$$
$$(C, U) \quad \mapsto (V_c, V_t) = \text{ARMADILLO2}(C, U).$$

The function ARMADILLO2 relies on an internal bitwise parameterized permutation denoted by Q which is defined by a parameter A of size a and is applied to a vector B of size k:

$$Q \quad : \quad \mathbb{F}_2^a \times \mathbb{F}_2^k \to \mathbb{F}_2^k$$
$$(A, B) \quad \mapsto Q(A, B) = Q_A(B)$$

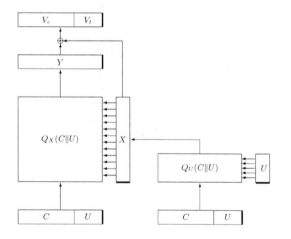

Fig. 1. ARMADILLO2

Let σ_0 and σ_1 be two fixed bitwise permutations of size k. In [1], the permutations are not defined but some criteria they should fulfil are given. As the attacks presented in this paper are valid for any bitwise permutations, we do not describe these properties. We just stress that in the following, when computing the complexities we assume that these permutations behave like random ones. We denote by γ a constant of size k defined by alternating 0's and 1's: $\gamma = 1010 \cdots 10$.

Using these notations, we can define Q which is used twice in the ARMADILLO2 function. Let A be a parameter and B be the internal state, the parameterized permutation Q (that we denote by Q_A when indicating the parameter is necessary) consists in $a = |A|$ simple steps. The i-th step of Q (reading A from its least significant bit to its most significant one) is defined by:

– an elementary bitwise permutation: $B \leftarrow \sigma_{A_i}(B)$, that is:
  • if the i-th bit of A equals 0 we apply σ_0 to the current state,
  • otherwise (if the i-th bit of A equals 1) we apply σ_1 to the current state,
– a constant addition (bitwise XOR) of γ: $B \leftarrow B \oplus \gamma$.

Using the definition of the permutation Q, we can describe the function ARMADILLO2. Let (C, U) be the input, then ARMADILLO2(C, U) is defined by:

– first compute $X \leftarrow Q_U(C\|U)$
– then compute $Y \leftarrow Q_X(C\|U)$
– finally compute $(V_c\|V_t) \leftarrow Y \oplus X$, the output is (V_c, V_t).

Actually c and m can take different values depending on the required security level. A summary of the sets of parameters for the different versions (A, B, C, D or E) proposed in [1] is given in Table 1.

Table 1. Sets of parameters for the different versions of ARMADILLO2

Version	k	c	m
A	128	80	48
B	192	128	64
C	240	160	80
D	288	192	96
E	384	256	128

2.2 A Multi-purpose Cryptographic Primitive

The general-purpose cryptographic function ARMADILLO2 can be used for three types of applications: FIL-MAC, hashing, and PRNG/PRF.

ARMADILLO2 in FIL-MAC mode. The secret key is C and the challenge, considered known by the attacker, is U. The response is V_t.

ARMADILLO2 in hashing mode. It uses a strengthened Merkle-Damgård construction, where V_c is the chaining value or the hash digest, and U is the message block.

ARMADILLO2 in PRNG/PRF mode. The output sequence is obtained by taking the first t bits of (V_c, V_t) after at least r iterations. For ARMADILLO2 the proposed values are $r = 1$ and $t = k$ (see [1, Sec. 6]). When used as a stream cipher, the secret key is C. The keystream is composed of k-bit frames indexed by U which is a public value.

3   Inverting the ARMADILLO2 Function

In [1] a sketch of a meet-in-the-middle (MITM) attack on ARMADILLO1, the first variant of the primitive, is given by the authors to prove *lower bounds* for the complexity and justify the choice of parameters. However, they do not develop further their analysis.

In this section we describe how to invert the ARMADILLO2 function when a part of the output (V_c, V_t) is known and U is chosen in the input $(C\|U)$. Inverting means that we recover C. The method we present can be performed for any arbitrary bitwise permutations σ_0 and σ_1. To conduct our analysis we suppose that they behave like random ones. Indeed, if the permutations σ_0 and σ_1 were not behaving like random ones, one could exploit their distributions to reduce the complexities of the attacks presented in this paper. Therefore, we are considering the worst case scenario for an attacker.

First, we describe the meet-in-the-middle technique we use. It provides two lists of partial states in the middle of the main permutation Q_X. To determine a list of possible values for C, we need to select a subset of the cartesian product of these two lists containing consistent couples of partial states. To build such a subset efficiently, we explain how to use an adaptation of the *parallel matching algorithm* presented in [2,5]. Then we present and apply the adapted algorithm and compute its time and memory complexities.

All cryptanalysis, we present, on the different applications of ARMADILLO2 relies on the technique for recovering C presented in this section.

3.1   The Meet-in-the-Middle Technique

Whatever mode ARMADILLO2 is embedded in, we use the following facts:

- We can choose the m-bit vector U, in the input vector $(C\|U)$.
- We know part of the output vector $(V_c\|V_t)$: the m-bit vector V_t in the FIL-MAC, the $(c+m)$-bit vector $(V_c\|V_t)$ in the PRNG/PRF and the c-bit vector V_c in the hash function.

We deal with two permutations: the pre-processing Q_U which is known as U is known and the main permutation Q_X which is unknown, and we exploit the three following equations:

- The permutation Q_U used in the pre-processing $X = Q_U(C\|U)$ is known. This implies that all the known bits in the input of the permutation can be traced to their corresponding positions in X. For instance, there are m coordinates of X whose values are determined by choosing U.
- The output of the main permutation $Y = (V_c\|V_t) \oplus X$ implies we know some bits of Y. The amount of known bits of Y is denoted by y and is depending on the mode we are focusing on through $(V_c\|V_t)$.
- In the sequel, we divide X in two parts: $X = (X_{\text{out}}\|X_{\text{in}})$. Then, the main permutation $Y = Q_X(C\|U)$ can be divided in two parts: $Q_{X_{\text{in}}}$ and $Q_{X_{\text{out}}}$ separated by a division line we call the *middle*, hence we perform the meet-in-the-middle technique between $Q_{X_{\text{in}}}$ and $Q_{X_{\text{out}}}^{-1}$.

As $(X_{\text{out}} \| X_{\text{in}}) = Q_U(C \| U)$, we denote by m_{in} (resp. m_{out}) the number of bits of U that are in X_{in} (resp. X_{out}). We have $m_{\text{out}} + m_{\text{in}} = m$. We denote by ℓ_{in} (resp. ℓ_{out}) the number of bits coming from C in X_{in} (resp. X_{out}). We have $\ell_{\text{out}} + \ell_{\text{in}} = c$. The meet-in-the-middle attack is done by guessing the ℓ_{in} unknown bits of X_{in} and the ℓ_{out} unknown bits of X_{out} independently.

First, consider the forward direction. We can trace the ℓ_{in} unknown bits of X_{in} back to C with Q_U^{-1}. Next, for each possible guess of X_{in}, we can trace the corresponding ℓ_{in} bits from C plus the m bits from U to their positions in the middle by computing $Q_{X_{\text{in}}}(C \| U)$. Then consider the backward direction, we can trace the y known bits of Y back to the middle for each possible guess of X_{out}, that is computing $Q_{X_{\text{out}}}^{-1}(Y)$. This way we can obtain two lists $\mathcal{L}_{\text{in}}$ and $\mathcal{L}_{\text{out}}$, of size $2^{\ell_{\text{in}}}$ and $2^{\ell_{\text{out}}}$ respectively, of elements that represent partially known states in the middle of Q_X.

To describe our meet-in-the-middle attack we represent the partial states in the middle of Q_X as ternary vectors with coordinate values from $\{0, 1, -\}$, where $-$ denotes a coordinate (or cell) whose value is unknown. We say that a cell is *active* if it contains 0 or 1 and *inactive* otherwise. The weight of a vector V, denoted by $\text{wt}(V)$, is the number of its active cells. Two partial states are a match if their colliding active cells have the same values.

The list $\mathcal{L}_{\text{in}}$ contains elements $Q_{X_{\text{in}}}(C \| U)$ whose weight is $x = \ell_{\text{in}} + m$. The list $\mathcal{L}_{\text{out}}$ contains elements $Q_{X_{\text{out}}}^{-1}(Y)$ whose weight is y. When taking one element from each list, the probability of finding a match will then depend on the number of collisions of active cells between these two elements.

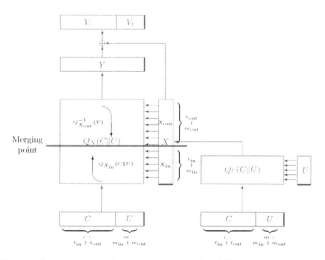

Fig. 2. Overview of the inversion of the ARMADILLO2 core function

Consider a vector A in $\{0, 1, -\}^k$ with weight a. We denote by $P_{[k,a,b]}(i)$ the probability over all the vectors $B \in \{0, 1, -\}^k$ with weight b of having i active cells at the same positions in A and B. This event corresponds to the situation where there are i active cells of B among the a active positions in A and the remaining $(b - i)$ active cells of B lie in the $(k - a)$ inactive positions in A. As the number of vectors of length k and weight b is $\binom{k}{b}$, we have:

$$P_{[k,a,b]}(i) = \frac{\binom{a}{i}\binom{k-a}{b-i}}{\binom{k}{b}} = \frac{\binom{b}{i}\binom{k-b}{a-i}}{\binom{k}{a}}.$$

Taking into account the probability of having active cells at the same positions in a pair of elements from $(\mathcal{L}_{in}, \mathcal{L}_{out})$ and the probability that these active cells do have the same value, we can compute the *expected probability* of finding a match for a pair of elements, that we will denote $2^{-N_{coll}}$. We have:

$$2^{-N_{coll}} = \sum_{i=0}^{y} 2^{-i} P_{[k,x,y]}(i).$$

This means that there will be a possible match with a probability of $2^{-N_{coll}}$. In total we will find $2^{\ell_{in}+\ell_{out}-N_{coll}}$ pairs of elements that pass this test. Each pair of elements defines a whole C value. Next, we just have to check which of these values is the correct one.

The big question now is that of the cost of checking which elements of the two lists $\mathcal{L}_{in}$ and $\mathcal{L}_{out}$ pass the test. The ternary alphabet of the elements and the changing positions of the active cells make it impossible to apply the approach of traditional MITM attacks — having an ordered list $\mathcal{L}_{in}$ and checking for each element in the list $\mathcal{L}_{out}$ if a match exists with cost 1 per element. Even more, a priori, for each element in $\mathcal{L}_{in}$ we would have to try if it matches each of the elements from $\mathcal{L}_{out}$ independently, which would yield the complexity of exhaustive search.

For solving this problem we adapt the algorithm described in [5, Sec. 2.3] as *parallel matching* to the case of ARMADILLO2. A generalized version of the algorithm is exposed in Section 7 with detailed complexity calculations and the link to our application case.

3.2   ARMADILLO2 Matching Problem: Matching Non-random Elements

Recently, new algorithms have been proposed in [5] to solve the problem of *merging* several lists of big sizes with respect to a given relation t that can be verified by tuples of elements. These new algorithms take advantage of the special structures that can be exhibited by t to reduce the complexity of solving this problem. As stated in [5], the problem of merging several lists can be reduced to the problem of merging two lists. Hereafter, we recall the reduced **Problem 1** proposed in [5] that we are interested in.

Problem 1 ([5]). *Let L_1 and L_2 be 2 lists of binary vectors of size 2^{ℓ_1} and 2^{ℓ_2} respectively. We denote by $\boldsymbol{x}$ a vector of L_1 and by $\boldsymbol{y}$ a vector of L_2.*

We assume that vectors $\boldsymbol{x}$ and $\boldsymbol{y}$ can be decomposed into z groups of s bits, i.e. $\boldsymbol{x}, \boldsymbol{y} \in (\{0,1\}^s)^z$ and $\boldsymbol{x} = (x_1, \ldots, x_z)$ (resp. $\boldsymbol{y} = (y_1, \ldots, y_z)$). The vectors in L_1 and L_2 are drawn uniformly and independently at random from $\{0,1\}^{sz}$.

Let t be a Boolean function, $t : \{0,1\}^{sz} \times \{0,1\}^{sz} \rightarrow \{0,1\}$ such that there exist some functions $t_j : \{0,1\}^s \times \{0,1\}^s \rightarrow \{0,1\}$ which verify:

$$t(\boldsymbol{x}, \boldsymbol{y}) = 1 \quad \Longleftrightarrow \quad \forall j, \; 1 \leq j \leq z, \quad t_j(x_j, y_j) = 1.$$

Problem 1 consists in computing the set $\mathcal{L}_{sol}$ of all 2-tuples $(\boldsymbol{x}, \boldsymbol{y})$ of $(L_1 \times L_2)$ verifying $t(\boldsymbol{x}, \boldsymbol{y}) = 1$. This operation is called merging the lists L_1 and L_2 with respect to t.

One of the algorithms proposed in [5] to solve **Problem 1** is the *parallel matching* algorithm, which is the one that provides the best time complexity when the number of possible associated elements to one element is bigger than the size of the other list, *i.e.*, when we can associate by t more than $|L_2|$ elements to an element from L_1 as well as more than $|L_1|$ elements to an element from L_2.

In our case, the lists $\mathcal{L}_{in}$ and $\mathcal{L}_{out}$ correspond to the lists L_1 and L_2 to merge but the application of this algorithm differs in two aspects. The first one is the alphabet, which is not binary anymore but ternary. The second aspect is the distribution of vectors in the lists. In **Problem 1**, the elements are drawn uniformly and independently at random while in our case the distribution is ruled by the MITM technique we use. For instance, all the elements of $\mathcal{L}_{in}$ have the same weight x and all the elements of $\mathcal{L}_{out}$ have the same weight y, which is far from the uniform case.

The function t is the *association rule* we use to select suitable vectors from $\mathcal{L}_{in}$ and $\mathcal{L}_{out}$. We say that two elements are *associated* if their colliding active cells have the same values. We can now specify a new **Problem 1** adapted for ARMADILLO2:

ARMADILLO2 Problem 1. Let $\mathcal{L}_{in}$ and $\mathcal{L}_{out}$ be 2 lists of ternary vectors of size $2^{\ell_{in}}$ and $2^{\ell_{out}}$ respectively. We denote by $\boldsymbol{x}$ a vector of $\mathcal{L}_{in}$ and by $\boldsymbol{y}$ a vector of $\mathcal{L}_{out}$, with $\boldsymbol{x}, \boldsymbol{y} \in \{0, 1, -\}^k$

The lists $\mathcal{L}_{in}$ and $\mathcal{L}_{out}$ are obtained by the MITM technique described in Paragraph 3.1.
Let $t : \{0, 1, -\}^k \times \{0, 1, -\}^k \rightarrow \{0, 1\}$ be the function defined by $t = t_1 \cdot t_2 \cdots t_{k-1} \cdot t_k$ and:

$$\forall j, \; 1 \leq j \leq k, \quad t_j : \{0, 1, -\} \times \{0, 1, -\} \rightarrow \{0, 1\},$$

x_j	0	0	0	1	1	1	$-$	$-$	$-$
y_j	0	1	$-$	0	1	$-$	0	1	$-$
$t_j(x_j, y_j)$	1	0	1	0	1	1	1	1	1

We say that $\boldsymbol{x}$ and $\boldsymbol{y}$ are associated if $t(\boldsymbol{x}, \boldsymbol{y}) = 1$.
ARMADILLO2 Problem 1 consists in merging the lists $\mathcal{L}_{in}$ and $\mathcal{L}_{out}$ with respect to t.

We can now adapt the parallel matching algorithm to ARMADILLO2 **Problem 1**.

3.3   Applying the Parallel Matching Algorithm to ARMADILLO2

The principle of the parallel matching algorithm is to consider *in parallel* the possible matches for the α first cells and the next β cells in the lists $\mathcal{L}_{\text{in}}$ and $\mathcal{L}_{\text{out}}$. The underlying idea is to improve, when possible, the complexity to find all the elements that are a match for the $(\alpha + \beta)$ first cells. To have a match between a vector in $\mathcal{L}_{\text{in}}$ and a vector in $\mathcal{L}_{\text{out}}$, the vectors should satisfy:

- the vector in $\mathcal{L}_{\text{in}}$ has u of its x active cells among the $(\alpha + \beta)$ first cells;
- the vector in $\mathcal{L}_{\text{out}}$ has v of its y active cells among the $(\alpha + \beta)$ first cells;
- looking at the $(\alpha + \beta)$ first cells, both vectors should have the same value at the same active position.

As x and y are the number of known bits from $(C\|U)$ and from Y resp. (see Fig. 2), the matching probability on the first $(\alpha + \beta)$ cells is:

$$2^{-N_{\text{coll}}^{\alpha+\beta}} = \sum_{u=0}^{x} P_{[k,\alpha+\beta,x]}(u) \cdot \sum_{v=0}^{y} P_{[k,\alpha+\beta,y]}(v) \cdot \sum_{w=0}^{v} 2^{-w} P_{[\alpha+\beta,v,u]}(w).$$

This means that we will find $2^{c-N_{\text{coll}}^{\alpha+\beta}}$ partial solutions. For each pair passing the test we will have to check next if the remaining $k - \alpha - \beta$ cells are verified.

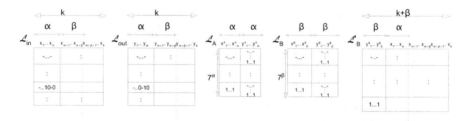

Fig. 3. Lists used in the parallel matching algorithm

In a pre-processing phase, we first need to build three lists, namely $\mathcal{L}_A$, $\mathcal{L}_B$, $\mathcal{L}_B'$, which are represented in Fig. 3.

List $\mathcal{L}_A$ contains all the elements of the form $(x_1^A \ldots x_\alpha^A, y_1^A \ldots y_\alpha^A)$ with $(x_1^A \ldots x_\alpha^A) \in \{0, 1, -\}^\alpha$ and $(y_1^A \ldots y_\alpha^A)$ being associated to $(x_1^A \ldots x_\alpha^A)$. The size of $\mathcal{L}_A$ is:

$$|\mathcal{L}_A| = \sum_{i=0}^{\alpha} \left(\binom{\alpha}{i} 2^i 3^{\alpha-i} 2^i \right) = 7^\alpha.$$

List $\mathcal{L}_B$ contains all the elements of the form $(x_1^B \ldots x_\beta^B, y_1^B \ldots y_\beta^B)$ with $(x_1^B \ldots x_\beta^B) \in \{0, 1, -\}^\beta$ and $(y_1^B, \ldots, y_\beta^B)$ being associated to $(x_1^B, \ldots, x_\beta^B)$. The size of $\mathcal{L}_B$ is:

$$|\mathcal{L}_B| = \sum_{i=0}^{\beta} \left(\binom{\beta}{i} 2^i 3^{\beta-i} 2^i \right) = 7^\beta.$$

List $\mathcal{L}'_B$ contains for each element $(x_1^B, \ldots, x_\beta^B, y_1^B, \ldots, y_\beta^B)$ in $\mathcal{L}_B$ all the elements $\boldsymbol{x}$ from $\mathcal{L}_{\text{in}}$ such that $(x_{\alpha+1} \ldots, x_{\alpha+\beta}) = (x_1^B, \ldots, x_\beta^B)$. Elements in $\mathcal{L}'_B$ are of the form $(y_1^B, \ldots, y_\beta^B, x_1, \ldots, x_k)$ indexed[1] by $(y_1^B \ldots, y_\beta^B, x_1, \ldots, x_\alpha)$. The probability for an element in $\mathcal{L}_{\text{in}}$ to have i active cells in its next β cells is $P_{[k,\beta,x]}(i)$. The size of $\mathcal{L}'_B$ is:

$$|\mathcal{L}'_B| = \sum_{i=0}^{\beta} \binom{\beta}{i} 2^i 3^{\beta-i} 2^i 2^{\ell_{\text{in}}} \frac{P_{[k,\beta,x]}(i)}{2^i \binom{\beta}{i}} = \sum_{i=0}^{\beta} 3^{\beta-i} 2^i 2^{\ell_{\text{in}}} P_{[k,\beta,x]}(i).$$

The cost of building $\mathcal{L}'_B$ is upper bounded by $(|\mathcal{L}'_B| + 3^\beta)$, where 3^β captures the cases where no element in $\mathcal{L}_{\text{in}}$ corresponds to elements in $\mathcal{L}_B$ and is normally negligible.

Next, we do the parallel matching. The probability for an element in $\mathcal{L}_{\text{out}}$ to have i active cells in its α first cells being $P_{[k,\alpha,y]}(i)$, for each element $(x_1^A \ldots x_\alpha^A, y_1^A \ldots y_\alpha^A)$ in $\mathcal{L}_A$ we consider the $2^{\ell_{\text{out}}} \frac{P_{[k,\alpha,y]}(i)}{2^i \binom{\alpha}{i}}$ elements $\boldsymbol{y}$ from $\mathcal{L}_{\text{out}}$ such that $(y_1, \ldots, y_\alpha) = (y_1^A, \ldots, y_\alpha^A)$. Then we check in $\mathcal{L}'_B$ if elements indexed by $(y_{\alpha+1} \ldots y_{\alpha+\beta}, x_1^A \ldots x_\alpha^A)$ exist. If this is the case, we check if each found pair of the form $(\boldsymbol{x}, \boldsymbol{y})$ verifies the remaining $(k - \alpha - \beta)$ cells. As we already noticed, we will find about $2^{c-N_{\text{coll}}^{\alpha+\beta}}$ partial solutions for which we will have to check whether or not they meet the remaining conditions.

The time complexity of this algorithm is:

$$\mathcal{O}\left(2^{c-N_{\text{coll}}^{\alpha+\beta}} + 7^\alpha + 7^\beta + \sum_{i=0}^{\beta} 3^{\beta-i} 2^i 2^{\ell_{\text{in}}} P_{[k,\beta,x]}(i) + \sum_{i=0}^{\alpha} 3^{\alpha-i} 2^i 2^{\ell_{\text{out}}} P_{[k,\alpha,y]}(i) \right).$$

The memory complexity is determined by $7^\alpha + 7^\beta + |\mathcal{L}'_B|$. We can notice that if

$$\sum_{i=0}^{\beta} 3^{\beta-i} 2^i 2^{\ell_{\text{in}}} P_{[k,\beta,x]}(i) > \sum_{i=0}^{\alpha} 3^{\alpha-i} 2^i 2^{\ell_{\text{out}}} P_{[k,\alpha,y]}(i),$$

we can exchange the roles of $\mathcal{L}_{\text{in}}$ and $\mathcal{L}_{\text{out}}$, so that the time complexity remains the same but the memory complexity will be reduced. The memory complexity is then:

$$\mathcal{O}\left(7^\alpha + 7^\beta + \min\left\{ \sum_{i=0}^{\beta} 3^{\beta-i} 2^i 2^{\ell_{\text{in}}} P_{[k,\beta,x]}(i), \sum_{i=0}^{\alpha} 3^{\alpha-i} 2^i 2^{\ell_{\text{out}}} P_{[k,\alpha,y]}(i) \right\} \right).$$

4   Meet in the Middle Key Recovery Attacks

4.1   Key Recovery Attack in the FIL-MAC Setting

In the FIL-MAC usage scenario, C is the secret key and U is the challenge. The response is the m-bit size vector V_t. In order to minimize the complexity of our attack, we want the number of known bits y from Y to be maximal. As

---

[1] We can use standard hash tables for storage and look up in constant time.

$Y = (V_c \| V_t) \oplus X$ and $X = Q_U(C\|U)$ it means that we are interested in having the maximum number of bits from U among the m less significant bits of X.

As we have m bits of freedom in U for choosing the permutation Q_U, we need the probability of having i known bits (from U) among the m first ones (of X), $P_{[k,m,m]}(i)$, to be bigger than 2^{-m}. Then to maximize the number of known bits in Y, we choose y as follows:

$$y = \max_{0 \leq i \leq m} \left\{ i : P_{[k,m,m]}(i) > 2^{-m} \right\}. \tag{1}$$

For instance for ARMADILLO2-A, we have $y = 38$ with a probability of $2^{-45.19} > 2^{-48}$.

Then, from now on, we assume that we know y among the m bits of the lower part of X and y bits at the same positions of Y.

Now, we can apply our meet-in-the-middle technique which allows us to recover the key. We have computed the optimal parameters for the different versions of ARMADILLO2, with different trade-offs — the generic attack has a complexity of 2^c. The results appear in Table 2.

For each version of ARMADILLO2 presented in Table 2, the first line corresponds to the ($\log_2$ of the) size of the lists $\mathcal{L}_{in}$ and $\mathcal{L}_{out}$ with the smallest time complexity. The second line corresponds to the best parameters when limiting the memory complexity to 2^{45}. In all cases, the complexity is determined by the parallel matching part of the attack. The data complexity of all the attacks is 1, that is, we only need one pair of plaintext/ciphertext to succeed.

Table 2. Complexities of the meet-in-the-middle key recovery attack on the FIL-MAC application

Version	c	m	ℓ_{out}	ℓ_{in}	α	β	$\log_2$(Time compl.)	$\log_2$(Mem. compl.)
ARMADILLO2-A	80	48	34	46	24	20	72.54	68.94
			18	62	16	9	75.05	45
ARMADILLO2-B	128	64	58	70	35	35	117.97	108.87
			38	90	2	16	125.15	45
ARMADILLO2-C	160	80	76	84	43	43	148.00	135.90
			35	125	4	16	156.63	45
ARMADILLO2-D	192	96	92	100	50	50	177.98	160.44
			29	163	11	12	187.86	45
ARMADILLO2-E	256	128	125	131	65	65	237.91	209.83
			29	227	11	13	251.55	45

4.2   Key Recovery Attack in the Stream Cipher Setting

As presented in [1], ARMADILLO2 can be used as a PRNG by taking the t first bits of (V_c, V_t) after at least r iterations. For ARMADILLO2, the authors state in [1, Sc. 6] that $r = 1$ and $t = k$ is a suitable parameter choice. If we want to use it as a stream cipher, the secret key is C. The keystream is composed of k-bit frames indexed by U which is a public value.

In this setting, we can perform an attack which is similar to the one on the FIL-MAC, but with different parameters. As we know more bits of the output of Q_X, $y = m + \ell_{out}$, complexities of the key recovery attack are lower.

In general, the best time complexity is obtained when $\ell_{in} = \ell_{out}$, as the number of known bits at each side is now $x = m + \ell_{in}$ in the input and $y = m + \ell_{out}$ in the output. In this context it also appears that the best time complexity occurs when $\alpha = \beta$. There might be a small difference between α and β when the leading term of the time complexity is $2^{c - N_{coll}^{\alpha+\beta}}$.

We present the best complexities we have computed for this attack in Table 3 — the generic attack has a complexity of 2^c. Other time-memory trade-offs would be possible. As in the previous section, we give as an example the best parameters when limiting the memory complexity to 2^{45}.

Table 3. Complexities of the meet-in-the-middle key recovery attack for the stream cipher with various trade-offs

Version	c	m	ℓ_{out}	ℓ_{in}	α	β	$\log_2$(Time compl.)	$\log_2$(Mem. compl.)
ARMADILLO2-A	80	48	40	40	19	19	65.23	62.91
			27	53	11	16	71.62	45
ARMADILLO2-B	128	64	64	64	31	32	104.71	101.75
			29	99	9	16	119.69	45
ARMADILLO2-C	160	80	80	80	39	40	130.53	127.49
			26	134	14	14	151.29	45
ARMADILLO2-D	192	96	96	96	47	48	156.35	153.23
			30	162	8	16	184.37	45
ARMADILLO2-E	256	128	128	128	64	64	207.96	205.93
			30	226	8	16	248.66	45

5   (Second) Preimage Attack on the Hashing Applications

We recall that the hash function built with ARMADILLO2 as a compression function follows a strengthened Merkle-Damgård construction, where the padding includes the message length. In this case C represents the input chaining value, U the message block and V_c the generated new chaining value and the hash digest. In [1] the authors state that (second) preimages are expected with a complexity of 2^c, the one of the generic attack. We show, in this section, how to build (second) preimage attacks with a smaller complexity.

5.1   Meet-in-the-Middle (Second) Preimage Attack

The principle of the attack is represented in Fig. 4. We first consider that the ARMADILLO2 function is invertible with a complexity of 2^q, given an output V_c and a message block. In the preimage attack, we choose and fix ℓ, the number of blocks of the preimage. In the second preimage attack, we can consider the length of the given message. Then, given a hash value h:

In the backward direction:
- We invert the insertion of the last block M_{pad} (padding). This step costs 2^q in a preimage scenario and 1 in a second preimage one. We get
$$\text{ARMADILLO2}^{-1}(h, M_{\text{pad}}) = S'.$$
- From state S', we can invert the compression function for 2^b different message blocks M_b with a cost 2^{b+q}, obtaining 2^b different intermediate states: $\text{ARMADILLO2}^{-1}(S', M_b) = S''$.

In the forward direction: From the initial chaining value, we insert 2^a messages of length $(\ell - 2)$ blocks, $\mathcal{M} = M_1 \| M_2 \| \ldots \| M_{\ell-2}$, obtaining 2^a intermediate states S. This can be done with a complexity of $\mathcal{O}((\ell - 2)2^a)$.

If we find a collision between one of the 2^a states S and one of the 2^b states S'', we have obtained a (second) preimage that is $\mathcal{M}\|M_b\|M_{\text{pad}}$.

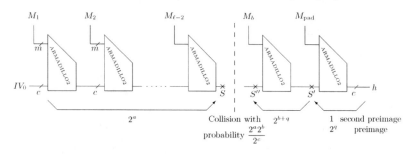

Fig. 4. Representation of the meet-in-the-middle (second) preimage attack

A collision occurs if $a + b \geq c$. The complexity of this attack is $2^a + 2^q + 2^{b+q}$ in time, where the middle term appears only in the case of a preimage attack and is negligible. The memory complexity is about 2^b (plus the memory needed for inverting the compression function). So if $2^q < 2^c$, we can find a and b so that $2^a + 2^{b+q} < 2^c$.

5.2   Inverting the Compression Function

In the previous section we showed that inverting the compression function for a chosen message block and for a given output can be done with a cost of $2^q < 2^c$. In this section we show how this complexity depends on the chosen message block, as the inversion can be seen as a key recovery similar to the one done in Section 4. In this case we know U (the message block) and V_c, and we want to find C. When inverting the function with the blocks M_b, we choose message blocks (U) that define permutations Q_U which put most of the m bits from U among the c most significant bits of X. This will result in better attacks, as the bits in Y known from U do not cost anything and this gives us more freedom when choosing the parameters ℓ_{in} and ℓ_{out}.

As before, we have 2^m possibilities for Q_U. We denote by n the number of bits of U in the c most significant bits of X. The number of message blocks (U) verifying this condition is:

$$N_{\text{block}}(n) = 2^m P_{[k,c,m]}(n).$$

In fact we are interested in the values of n which are the greatest possible (to lower the complexity) that still leaves enough message blocks to invert in order to obtain S''. It means that these values belong to a set $\{n_i\}$ such that:

$$\sum_{\{n_i\}} N_{\text{block}}(n_i) \geq 2^b.$$

As the output is V_c, the ℓ_{out} bits guessed from X are also known bits from the output of Q_X. The number of known bits of the output of Q_X is then defined by:

$$y = \min(c, \ell_{\text{out}} + n)$$

Compared to the key recovery attack, the number of known bits at the end of the permutation Q_X is significantly bigger, as we may know up to c bits, while in the previous case the maximal number for y was $y = \max_i \{i : P_{[k,m,m]}(i) > 2^{-m}\}$. To simplify the explanations, we concentrate on the case of ARMADILLO2-A, that can be directly adapted to any of the other versions. For $n = 48$ we have a probability $P_{[128,80,48]} = 2^{-44.171}$. This leaves $2^{48-44.171} = 2^{3.829}$ message blocks to invert which allow us to know $y = \min(80, \ell_{\text{out}} + 48)$ bits from the output of Q_X. As we need to invert 2^b message blocks, if b is bigger than 3.829, we have to consider next the message blocks with $n = 47$, that allow us to know $y = \min(80, \ell_{\text{out}} + 47)$ bits, and so on. For each n considered, the best time complexity (2^{q_n}) for inverting ARMADILLO2 might be different, but in practice, with at most two consecutive values of n we have enough message blocks for building the attack, and the complexity of inverting the compression function for these two different types of messages is very similar.

Table 4. Complexities for inverting the compression function

Version	c	m	ℓ_{out}	ℓ_{in}	n	$\log_2(N_{\text{block}}(n))$	α	β	$\log_2$(Time compl.)	$\log_2$(Mem. compl.)
ARMADILLO2-A 80 48			35	45	47	9.95	22	16	65.90	63.08
			35	45	48	3.83	22	16	65.90	63.08
			20	60	47	9.95	16	8	71.36	45
			27	53	48	3.83	11	16	71.62	45
ARMADILLO2-B 128 64			62	66	64	15.89	33	30	104.67	102.35
			33	95	64	15.89	6	16	120.41	45
ARMADILLO2-C 160 80			78	82	80	19.82	41	38	130.48	128.08
			26	134	80	19.82	11	16	152.24	45
ARMADILLO2-D 192 96			94	98	96	23.74	49	46	156.31	153.82
			30	162	96	23.74	8	16	184.37	45
ARMADILLO2-E 256 128			126	130	128	31.58	65	62	207.96	205.30
			34	222	128	31.58	5	16	249.47	45

For instance, in ARMADILLO2-A, we consider $n = 48, 47$, associated each to $2^{3.829}$ and $2^{9.96}$ possible message blocks respectively. The best time complexity for inverting the compression function in both cases is $2^{q_{48}} = 2^{q_{47}} = 2^{65.9}$, as we can see from Table 4. If we want to find the best parameters for a and b in the preimage attack, we can consider that $a + b = c$ and $2^b = 2^{b_{48}} + 2^{b_{47}}$, and we want that $2^a = 2^{b_{48}} 2^{65.9} + 2^{b_{47}} 2^{65.9} = 2^{65.9}(2^{b_{48}} + 2^{b_{47}})$, as the complexity of the attack is $\mathcal{O}(2^a + 2^{65.9}(2^{b_{48}} + 2^{b_{47}}))$. So if we choose the parameters correctly, the best time complexity will be $\mathcal{O}(2^{a+1})$.

In this particular case the time complexity for $n = 48$ and for $n = 47$ is the same, so finding the best b and a can be simplified by $b = \frac{c-q}{2}$ and $a = c - b$. We obtain $b = 7.275$, $a = 72.95$. We see that we do not have enough elements with $n = 48$ for inverting 2^b blocks, but we have enough with $n = 47$ alone. As the complexities are the same in both cases, we can just consider $b = b_{47}$. The best time complexity for the preimage attack that we can obtain is then $2^{73.95}$, with a memory complexity of $2^{63.08}$. Other trade-offs are possible by using other parameters for inverting the function, as shown in Table 5.

For the other versions of ARMADILLO2, the number of message blocks associated to $y = m$ is big enough for performing the 2^b inversions, so we do not consider other n's for computing the (second) preimage complexity. Then, $b = b_m = \frac{c - q_{\{n=m\}}}{2}$ and $a = c - b_m$.

Complexities for preimage attacks on the different versions of ARMADILLO2 are given in Table 5, where we can see two different complexities with different trade-offs for each version.

Table 5. Complexities of the (second) preimages attacks

Version	c	m	Best time		Time-memory trade-off	
			$\log_2$(Time compl.)	$\log_2$(Mem. compl.)	$\log_2$(Time compl.)	$\log_2$(Mem. compl.)
ARMADILLO2-A	80	48	73.95	63.08	76.81	45
ARMADILLO2-B	128	64	117.34	102.35	125.21	45
ARMADILLO2-C	160	80	146.24	128.08	157.12	45
ARMADILLO2-D	192	96	175.16	153.82	191.19	45
ARMADILLO2-E	256	128	232.98	205.30	253.74	45

6   Experimental Verifications

To verify the above theoretical results, we implemented the proposed key recovery attacks in the FIL-MAC and stream cipher settings against a scaled version of ARMADILLO2 that uses a 30-bit key and processes 18-bit messages, i.e. $c = 30$ and $m = 18$. We performed the attack 10 times for both the FIL-MAC and the PRNG settings where at each time we chose random permutations for both σ_0 and σ_1 and random messages U (in the FIL-MAC case U was chosen so that we got y bits from U among the m least significant bits of X).

As for each application the key is a 30-bit key, the generic attack requires a time complexity of 2^{30}. Using the parallel matching algorithm we decrease this complexity. Table 6 shows that the implementation results are very close to the theoretical estimates, confirming our analysis. We can also mention that we exchanged the role of $\mathcal{L}_{\text{in}}$ and $\mathcal{L}_{\text{out}}$ in our implementation of the attacks to minimize the memory needs.

Table 6. Key recovery attacks against a scaled version of ARMADILLO2 in the FIL-MAC and PRNG modes

		c	m	ℓ_{out}	ℓ_{in}	α	β	y	$\log_2(|\mathcal{L}'_B|)$	$\log_2(c - \frac{}{N_{\text{coll}}^{\alpha+\beta}})$	$\log_2(\text{Time compl.})$	$\log_2(\text{Mem. compl.})$
FIL-MAC	Impl.	30	18	12	18	8	6	14	23.477	27.537	27.874	24.066
	Theory	30	18	12	18	8	6	14	23.475	27.538	27.874	24.064
PRNG	Impl.	30	18	14	16	7	6	32	22.530	24.728	25.396	22.738
	Theory	30	18	14	16	7	6	32	22.530	24.735	25.401	22.738

7   Generalization of the Parallel Matching Algorithm

In Section 3, we managed to apply the parallel matching algorithm to invert the ARMADILLO2 function by modifying the merging **Problem 1** of [5].

When the number of possible associated elements to one element is bigger than the other list as it is the case for ARMADILLO2, we cannot apply a basic algorithm like the *instant matching* algorithm proposed in [5]. Instead, we can use either the *gradual matching* or the *parallel matching* algorithms also proposed in [5]. We are going to concentrate on the parallel matching algorithm which allows a significant reduction of the time complexity of solving **Problem 1**, while allowing several time-memory trade-offs.

We can state the generalized problem that also covers our attack on ARMADILLO2 and give the corresponding parallel matching algorithm. We believe that this more general problem will be useful for recognizing situations where the parallel matching can be applied, and solving them in an automatized way.

7.1   The Generalized Problem 1

As stated in [5], **Problem 1** for N lists can be reduced to 2 lists, therefore we will only consider the problem of merging 2 lists in the sequel.

Generalized Problem 1. *We are given 2 lists, L_1 and L_2 of size 2^{ℓ_1} and 2^{ℓ_2} respectively. We denote by $\boldsymbol{x}$ a vector of L_1 and by $\boldsymbol{y}$ a vector of L_2. Coordinates of $\boldsymbol{x}$ and $\boldsymbol{y}$ belong to a general alphabet $\mathcal{A}$.*

We assume that vectors $\boldsymbol{x}$ and $\boldsymbol{y}$ can be decomposed into z groups of s coordinates, i.e. $\boldsymbol{x}, \boldsymbol{y} \in (\mathcal{A}^s)^z$ and $\boldsymbol{x} = (x_1, \ldots, x_z)$ (resp. $\boldsymbol{y} = (y_1, \ldots, y_z)$).

We want to keep pairs of vectors verifying a given relation t: $t(\boldsymbol{x}, \boldsymbol{y}) = 1$. The relation t is group-wise, and is defined by $t : (\mathcal{A}^s)^z \times (\mathcal{A}^s)^z \rightarrow \{0, 1\}$ such that there exist some functions $t_j : \mathcal{A}^s \times \mathcal{A}^s \rightarrow \{0, 1\}$, verifying:

$$t(\boldsymbol{x}, \boldsymbol{y}) = 1 \iff \forall j, \ 1 \leq j \leq z, \quad t_j(x_j, y_j) = 1.$$

Generalized Problem 1 *consists in* merging *these* 2 *lists to obtain the set* $\mathcal{L}_{sol}$ *of all 2-tuples of* $(L_1 \times L_2)$ *verifying* $t(\boldsymbol{x}, \boldsymbol{y}) = 1$. *We say that* $\boldsymbol{x}$ *and* $\boldsymbol{y}$ *are associated in this case.*

In order to analyze the time and memory complexities of the attack we need to compute the size of $\mathcal{L}_{sol}$. This quantity depends on the probability that $t(\boldsymbol{x}, \boldsymbol{y}) = 1$. More precisely the complexities of the generalized parallel matching algorithm depends on the conditional probabilities: $\mathbf{Pr}_{y_j}[t_j(x_j, y_j) = 1 | x_j = a]$, $a \in \mathcal{A}^s$. We will denote these probabilities by $p_{j,a}$, $a \in \mathcal{A}^s$.

In [5] the elements of the lists L_1 and L_2 were binary (i.e. $\mathcal{A} = \{0, 1\}$) and random, and the probability of each t_j of being verified did not depend on the elements x_j or y_j. Let us consider as an example the case where $s = 1$ and t_j tests the equality of x_j and y_j. We have:

$$\forall j, \ 1 \le j \le z, \quad p_{j,0} = p_{j,1} = \frac{1}{2}.$$

In the case of the ARMADILLO2 cryptanalysis that we present in this paper, the alphabet is ternary (i.e. $\mathcal{A} = \{0, 1, -\}$) and the association rule (*see. AR-MADILLO2 Problem 1*) gives:

$$\forall j, \ 1 \le j \le z, \quad p_{j,0} = \frac{2}{3}, \ p_{j,1} = \frac{2}{3} \text{ and } p_{j,-} = 1$$

7.2   Generalized Parallel Matching Algorithm

First we need to build the three following lists:

List $\mathcal{L}_A$, of all the elements of the form $(x_1^A, \ldots, x_\alpha^A, y_1^A, \ldots, y_\alpha^A)$ with
$(x_1^A, \ldots, x_\alpha^A) \in (\mathcal{A}^s)^\alpha$ and $(y_1^A, \ldots, y_\alpha^A)$ being associated by t to $(x_1^A, \ldots, x_\alpha^A)$. The size of $\mathcal{L}_A$ is:

$$|\mathcal{L}_A| = \sum_{\mathbf{a} \in (\mathcal{A}^s)^\alpha} \prod_{j=1}^{\alpha} |\mathcal{A}|^s \, p_{j,\mathbf{a}_j}, \tag{2}$$

where $\mathbf{a}_j$ is the j-th coordinate of $\mathbf{a} \in (\mathcal{A}^s)^\alpha$.

List $\mathcal{L}_B$, of all the elements of the form $(x_1^B, \ldots, x_\beta^B, y_1^B, \ldots, y_\beta^B)$ with
$(x_1^B, \ldots, x_\beta^B) \in (\mathcal{A}^s)^\beta$ and $(y_1^B, \ldots, y_\beta^B)$ being associated by t to $(x_1^B, \ldots, x_\beta^B)$. The size of $\mathcal{L}_B$ is

$$|\mathcal{L}_B| = \sum_{\mathbf{b} \in (\mathcal{A}^s)^\beta} \prod_{j=1}^{\beta} |\mathcal{A}|^s \, p_{j,\mathbf{b}_j},$$

where $\mathbf{b}_j$ is the j-th coordinate of $\mathbf{b} \in (\mathcal{A}^s)^\beta$.

List $\mathcal{L}'_B$, containing for each element $(x_1^B, \ldots, x_\beta^B, y_1^B, \ldots, y_\beta^B)$ in $\mathcal{L}_B$ all the elements $\boldsymbol{x}$ from L_1 such that $(x_{\alpha+1} \ldots, x_{\alpha+\beta}) = (x_1^B, \ldots, x_\beta^B)$. Elements in $\mathcal{L}'_B$ are of the form $(y_1^B, \ldots, y_\beta^B, x_1, \ldots, x_z)$ indexed[2] by $(y_1^B \ldots, y_\beta^B, x_1, \ldots, x_\alpha)$.

---

[2] We can use standard hash tables for storage and look up in constant time.

If we denote by $P_{\mathbf{b},[\alpha+1,\alpha+\beta],L_1}$ the probability of having an element $\boldsymbol{x}$ from L_1 such that $(x_{\alpha+1}, \ldots, x_{\alpha+\beta}) = \mathbf{b}$, the size of $\mathcal{L}'_B$ is:

$$|\mathcal{L}'_B| = \sum_{\mathbf{b}\in(\mathcal{A}^s)^\beta} \left(\prod_{j=1}^\beta |\mathcal{A}|^s p_{j,\mathbf{b}_j} \right) 2^{\ell_1} P_{\mathbf{b},[\alpha+1,\alpha+\beta],L_1}.$$

The cost of building this list is upper-bounded by $(|\mathcal{L}'_B| + (|\mathcal{A}|)^\beta)$, where the second term captures the cases where no element in L_1 corresponds to elements in $\mathcal{L}_B$ and should be negligible.
In the case where

$$\sum_{\mathbf{a}\in(\mathcal{A}^s)^\alpha} \left(\prod_{j=1}^\alpha |\mathcal{A}|^s p_{j,\mathbf{a}_j} \right) 2^{\ell_2} P_{\mathbf{a},[\beta+1,\alpha+\beta],L_2} < \sum_{\mathbf{b}\in(\mathcal{A}^s)^\beta} \left(\prod_{j=1}^\beta |\mathcal{A}|^s p_{j,\mathbf{b}_j} \right) 2^{\ell_1} P_{\mathbf{b},[\alpha+1,\alpha+\beta],L_1}$$

we can swap L_1 and L_2, to reduce the memory complexity of the attack.

Next, we do the parallel matching. For each element $(x_1^A, \ldots, x_\alpha^A, y_1^A, \ldots, y_\alpha^A)$ in $\mathcal{L}_A$ we consider the $2^{\ell_2} P_{(y_1^A, \ldots, y_\alpha^A),[1,\alpha],L_2}$ elements $\boldsymbol{y}$ from L_2 such that $(y_1 \ldots y_\alpha)$ $= (y_1^A, \ldots, y_\alpha^A)$ and we check in $\mathcal{L}'_B$ if elements indexed by $(y_{\alpha+1} \ldots y_{\alpha+\beta}, x_1^A \ldots x_\alpha^A)$ exist. If this is the case, we check if each found pair of the form $(\boldsymbol{x}, \boldsymbol{y})$ verifies the remaining $(k - \alpha - \beta)$ cells. We denote by Ω the number of partial solutions for which we will have to check whether or not they meet the remaining conditions:

$$\Omega = 2^{\ell_1+\ell_2} \sum_{\mathbf{b}\in(\mathcal{A}^s)^{\alpha+\beta}} \left(\prod_{j=1}^{\alpha+\beta} p_{j,\mathbf{b}_j} \right) P_{\mathbf{b},[1,\alpha+\beta],L_1}$$

The time complexity of this algorithm is:

$$\mathcal{O}\left(\Omega + |\mathcal{L}_A| + |\mathcal{L}_B| + |\mathcal{L}'_B| + \sum_{\mathbf{a}\in(\mathcal{A}^s)^\alpha} \left(\prod_{j=1}^\alpha |\mathcal{A}|^s p_{j,\mathbf{a}_j} \right) 2^{\ell_2} P_{\mathbf{a},[\beta+1,\alpha+\beta],L_2} \right)$$

The memory complexity is determined by the size of the lists $\mathcal{L}_A$, $\mathcal{L}_B$ and $\mathcal{L}'_B$. Therefore the memory complexity is:

$$\sum_{\mathbf{a}\in(\mathcal{A}^s)^\alpha}\prod_{j=1}^\alpha |\mathcal{A}|^s p_{j,\mathbf{a}_j} + \sum_{\mathbf{b}\in(\mathcal{A}^s)^\beta}\prod_{j=1}^\beta |\mathcal{A}|^s p_{j,\mathbf{b}_j} + \sum_{\mathbf{b}\in(\mathcal{A}^s)^\beta} \left(\prod_{j=1}^\beta |\mathcal{A}|^s p_{j,\mathbf{b}_j} \right) 2^{\ell_1} P_{\mathbf{b},[\alpha+1,\alpha+\beta],L_1}$$

7.3   Link with Formulas in the Case of ARMADILLO

Using the previous formulas for the time and memory complexities, we can rediscover formulas of the time and memory complexities we have computed for ARMADILLO2 (see. Section 3.3). As these formulas depend essentially on the size of the different lists, we simply expose how to find the size of the list $|\mathcal{L}_A|$ using equation (2).

For ARMADILLO2, the probabilities $p_{j,a}$ are independent of the position j and $p_{j,a} = 2/3$ if and only if a is an active cell. Moreover, in this case, each cell is composed of one letter of the alphabet which means that $s = 1$. And we have:

$$|\mathcal{L}_A| = \sum_{\mathbf{a} \in (\mathcal{A}^s)^\alpha} \prod_{j=1}^\alpha |\mathcal{A}|^s \, p_{j,\mathbf{a}_j} = \sum_{\mathbf{a} \in \{0,1,-\}^\alpha} \prod_{j=1}^\alpha 3 \left(\frac{2}{3}\right)^{\mathrm{wt}(\mathbf{a})}$$

$$= \sum_{i=0}^\alpha \# \{\mathbf{a} : \mathrm{wt}(\mathbf{a}) = i\} \, 3^\alpha \left(\frac{2}{3}\right)^i = \sum_{i=0}^\alpha \binom{\alpha}{i} 2^i \left(\frac{2}{3}\right)^i 3^\alpha$$

The same method can be applied to find the size of the list $\mathcal{L}_B$ and $\mathcal{L}_B'$. Here we have $\Omega = 2^{c - N_{\mathrm{coll}}^{\alpha+\beta}}$.

8   Conclusion

In this paper, we have presented the first cryptanalysis of ARMADILLO2, the recommended variant of the ARMADILLO family. We propose a key recovery attack on all its versions for the FIL-MAC and the stream cipher mode, which works for any bitwise permutations σ_0 and σ_1. We give several time-memory trade-offs for its complexity. We also show how to build (second) preimage attacks when using the hashing mode.

Besides the results on ARMADILLO2, we have generalized the parallel matching algorithm presented in [5] for solving a wider **Problem 1** which includes the cases where the lists to merge do not have random elements. We believe that new types of meet-in-the-middle attacks might appear now given this algorithm that is cheaper than exhaustive search.

References

1. Badel, S., Dağtekin, N., Nakahara Jr., J., Ouafi, K., Reffé, N., Sepehrdad, P., Sušil, P., Vaudenay, S.: ARMADILLO: A Multi-purpose Cryptographic Primitive Dedicated to Hardware. In: Mangard, S., Standaert, F.-X. (eds.) CHES 2010. LNCS, vol. 6225, pp. 398–412. Springer, Heidelberg (2010)
2. Khovratovich, D., Naya-Plasencia, M., Röck, A., Schläffer, M.: Cryptanalysis of *Luffa* v2 Components. In: Biryukov, A., Gong, G., Stinson, D.R. (eds.) SAC 2010. LNCS, vol. 6544, pp. 388–409. Springer, Heidelberg (2011)
3. Moldovyan, A.A., Moldovyan, N.A.: A Cipher Based on Data-Dependent Permutations. Journal of Cryptology 15(1), 61–72 (2002)
4. Naya-Plasencia, M.: How to Improve Rebound Attacks. Tech. Rep. Report 2010/607, Cryptology ePrint Archive (2010), (extended version)
   http://eprint.iacr.org/2010/607.pdf
5. Naya-Plasencia, M.: How to Improve Rebound Attacks. In: Rogaway, P. (ed.) CRYPTO 2011. LNCS, vol. 6841, pp. 188–205. Springer, Heidelberg (2011)
6. Sepehrdad, P., Sušil, P., Vaudenay, S.: Fast Key Recovery Attack on ARMADILLO1 and Variants. In: Tenth Smart Card Research and Advanced Application Conference, CARDIS 2011. LNCS (to appear)

An Experimentally Verified Attack on Full Grain-128 Using Dedicated Reconfigurable Hardware

Itai Dinur[1], Tim Güneysu[2], Christof Paar[2],
Adi Shamir[1], and Ralf Zimmermann[2]

[1] Computer Science Department, The Weizmann Institute, Rehovot, Israel
[2] Horst Görtz Institute for IT Security, Ruhr-University Bochum, Germany

Abstract. In this paper we describe the first single-key attack which can recover the full key of the full version of Grain-128 for arbitrary keys by an algorithm which is significantly faster than exhaustive search (by a factor of about 2^{38}). It is based on a new version of a cube tester, which uses an improved choice of dynamic variables to eliminate the previously made assumption that ten particular key bits are zero. In addition, the new attack is much faster than the previous weak-key attack, and has a simpler key recovery process. Since it is extremely difficult to mathematically analyze the expected behavior of such attacks, we implemented it on RIVYERA, which is a new massively parallel reconfigurable hardware, and tested its main components for dozens of random keys. These tests experimentally verified the correctness and expected complexity of the attack, by finding a very significant bias in our new cube tester for about 7.5% of the keys we tested. This is the first time that the main components of a complex analytical attack are successfully realized against a full-size cipher with a special-purpose machine. Moreover, it is also the first attack that truly exploits the configurable nature of an FPGA-based cryptanalytical hardware.

Keywords: Grain-128, stream cipher, cryptanalysis, cube attacks, cube testers, RIVYERA, experimental verification.

1 Introduction

Grain-128 [3] is a 128-bit variant of the Grain scheme which was selected by the eSTREAM project in 2008 as one of the three recommended hardware-efficient stream ciphers. The only single-key attacks published so far on this scheme which were substantially faster than exhaustive search were either on a reduced number of rounds or on a specific class of weak keys which contains about one in a thousand keys. In this paper we describe the first attack which can be applied to the full scheme with arbitrary keys. It uses an improved cube distinguisher with new dynamic variables, which makes it possible to attack Grain-128 with no restriction on the key. Its main components were experimentally verified by running a 50-dimensional cube tester for 107 random keys and discovering a very

D.H. Lee and X. Wang (Eds.): ASIACRYPT 2011, LNCS 7073, pp. 327–343, 2011.

strong bias (of 50 zeroes out of 51 bits) in about 7.5% of these keys. For these keys, we expect the running time of our new attack to be about 2^{38} times faster than exhaustive search, using 2^{63} bits of memory. Our attack is thus both faster and more general than the best previous attack on Grain-128 [1], which was a weak-key attack on one in a thousand keys which was only 2^{15} times faster than exhaustive search. However, our attack does not seem to threaten the security of the original 80-bit Grain scheme.

In order to develop and experimentally verify the main components of the attack, we had to run thousands of summations over cubes of dimension 49 and 50 for dozens of randomly chosen keys, where each summation required the evaluation of 2^{49} or 2^{50} output bits of Grain-128 (running the time-consuming initialization phase of Grain-128 for about 2^{56} different key and IV values). This process is hardware-oriented, highly parallelizable, and well beyond the capabilities of a standard cluster of PC's. We thus decided to implement the attack on a new type of special purpose hardware consisting of 128 Spartan-3 FPGAs.

Special-purpose hardware, i. e., computing machines dedicated to cryptanalytical problems, have a long tradition in code-breaking, including attacks against the Enigma cipher during WWII [15]. Their use is promising if two conditions are fulfilled. First, the complexity of the cryptanalytical problem must be in the range of approximately $2^{50} \ldots 2^{64}$ operations. For problems with a lower complexity conventional computer clusters are typically sufficient, such as the linear cryptanalysis attack against DES [17] (which required 2^{43} DES evaluations), and more than 2^{64} operations are difficult to achieve with today's technology unless extremely large budgets are available. The second condition is that the computations involved are suited for customized hardware architectures, which is often the case in symmetric cryptanalysis. Both conditions are fulfilled for the building blocks of the Grain-128 attack described in this paper.

Even though it is widely speculated that government organizations have been using special-purpose hardware for a long time, there are only two confirmed reports about cryptanalytical machines in the open literature. In 1998, Deep Crack, an ASIC-based machine dedicated to brute-forcing DES, was introduced [16]. In 2006, COPACOBANA also allowed exhaustive key searches of DES, and in addition cryptanalysis of other ciphers [13]. However, in the latter case often only very small-scale versions of the ciphers are vulnerable. The paper at hand extends the previous work with respect to cryptanalysis with dedicated hardware in several ways. Our work is the first time that the main components of a complex analytical attack, i. e., not merely an exhaustive search, are successfully realized in a public way against a full-size cipher by using a special-purpose machine (previous attacks were either a simple exhaustive search sped up by a special-purpose hardware, or advanced attacks such as linear cryptanalysis which were realized in software on multiple workstations). Also, this is the first attack which makes use of the reconfigurable nature of the hardware. Our RIVYERA computer, consisting of 128 large FPGAs, is the most powerful cryptanalytical machine available outside government agencies (possessing more than four times as many logic resources as the COPACOBANA machine). This makes our attack an interesting case study about what type of cryptanalysis can be done with "university budgets" (as opposed to government budgets). As a final remark, it is worth noting that the same attack implemented on GPU clusters would require an extremely large number of graphic cards, which would not only require a very high budget but would consume considerably more electric energy to perform the same computations.

In the first part of this paper, we give the necessary background regarding Grain-128 and dynamic cube attacks and describe our new attack on Grain-128. In the second part of the paper, we present our FPGA implementation in detail.

2   Preliminaries

In this section we give a short description of Grain-128 [3], of cube testers (which were introduced in [2]), and of dynamic cube attacks (developed in [1]).

2.1   Description on Grain-128

The state of Grain-128 consists of a 128-bit LFSR and a 128-bit NFSR. The feedback functions of the LFSR and NFSR are respectively defined to be

$$s_{i+128} = s_i + s_{i+7} + s_{i+38} + s_{i+70} + s_{i+81} + s_{i+96}$$
$$b_{i+128} = s_i + b_i + b_{i+26} + b_{i+56} + b_{i+91} + b_{i+96} + b_{i+3}b_{i+67} + b_{i+11}b_{i+13} + b_{i+17}b_{i+18} +$$
$$b_{i+27}b_{i+59} + b_{i+40}b_{i+48} + b_{i+61}b_{i+65} + b_{i+68}b_{i+84}$$

The output function is defined as
$z_i = \sum_{j \in \mathcal{A}} b_{i+j} + h(x) + s_{i+93}$, where $\mathcal{A} = \{2, 15, 36, 45, 64, 73, 89\}$.
$h(x) = x_0x_1 + x_2x_3 + x_4x_5 + x_6x_7 + x_0x_4x_8$

where the variables x_0, x_1, x_2, x_3, x_4, x_5, x_6, x_7 and x_8 correspond to the tap positions b_{i+12}, s_{i+8}, s_{i+13}, s_{i+20}, b_{i+95}, s_{i+42}, s_{i+60}, s_{i+79} and s_{i+95} respectively.

Grain-128 is initialized with a 128-bit key that is loaded into the NFSR, and with a 96-bit IV that is loaded into the LFSR, while the remaining 32 LFSR bits are filled with 1's. The state is then clocked through 256 initialization rounds without producing an output, feeding the output back into the input of both registers.

2.2   Previous Results on Grain-128

All the previously published single-key attacks ([2], [5], [6], [7] and [8]) on Grain-128 which are substantially better than exhaustive search can only deal with simplified versions of the cryptosystem. In [9] a sliding property was used to speed-up exhaustive search by a factor of two. Related-key attacks on the full cipher were presented in [10]. However, the relevance of related-key attacks is disputed, and in this paper we concentrate on attacks in the single key model. The only significant known attack on the full version of Grain-128 in the single key model is given in [1], where dynamic cube attacks are used to break a particular subset of weak keys, which contains the 2^{-10} fraction of keys in which ten specific key bits are all zero. The attack is faster than exhaustive search in this weak key set by a factor of about 2^{15}. For the remaining 0.999 fraction of keys, there is no known attack which is significantly faster than exhaustive search.

2.3   Cube Testers

In almost any cryptographic scheme, each output bit can be described by a multivariate master polynomial $p(x_1, .., x_n, v_1, .., v_m)$ over GF(2) of secret variables x_i (key bits), and public variables v_j (plaintext bits in block ciphers and MACs, IV bits in stream ciphers). This polynomial is usually too large to write down or to manipulate in an explicit way, but its values can be evaluated by running the cryptographic algorithm as a black box. The cryptanalyst is allowed to tweak this master polynomial by assigning chosen values to the public variables (which result in multiple derived polynomials), but in single-key attacks he cannot modify the secret variables.

To simplify our notation, we ignore in the rest of this subsection the distinction between public and private variables. Given a multivariate master polynomial with n variables $p(x_1, .., x_n)$ over GF(2) in algebraic normal form (ANF), and a term t_I containing variables from an index subset I that are multiplied together, the polynomial can be written as the sum of terms which are supersets of I and terms that miss at least one variable from I:

$$p(x_1, .., x_n) \equiv t_I \cdot p_{S(I)} + q(x_1, .., x_n)$$

$p_{S(I)}$ is called the *superpoly* of I in p. Compared to p, the algebraic degree of the superpoly is reduced by at least the number of variables in t_I, and its number of terms is smaller.

Cube testers [2] are related to high order differential attacks [11]. The basic idea behind them is that the symbolic sum over GF(2) of all the derived polynomials obtained from the master polynomial by assigning all the possible 0/1 values to the subset of variables in the term t_I is exactly $p_{S(I)}$ which is the superpoly of t_I in $p(x_1, .., x_n)$. This simplified polynomial is more likely to exhibit non-random properties than the original polynomial P.

Cube testers work by evaluating superpolys of carefully selected terms t_I which are products of public variables, and trying to distinguish them from a random function. One of the natural properties that can be tested is balance: A random function is expected to contain as many zeroes as ones in its truth table. A superpoly that has a strongly unbalanced truth table can thus be used to distinguish the cryptosystem from a random polynomial by testing whether the sum of output values over an appropriate boolean cube evaluates as often to one as to zero (as a function of the public bits which are not summed over).

2.4   Dynamic Cube Attacks

Dynamic Cube Attacks exploit distinguishers obtained from cube testers to recover some secret key bits. This is reminiscent of the way that distinguishers are used in differential attacks to recover the last subkey in an iterated cryptosystem. In static cube testers (and other related attacks such as the original cube attack [18], and AIDA [19]), the values of all the public variables that are not summed over are fixed to a constant (usually zero), and thus they are called static variables. However, in dynamic cube attacks the values of some of the public variables that are not part of the cube are not fixed. Instead, each one of these variables (called dynamic variables) is assigned a function that depends on some of the cube public variables and on some private variables. Each such function is carefully chosen in order to simplify the resultant superpoly and thus to amplify the expected bias (or the non-randomness in general) of the cube tester.

The basic steps of the attack are briefly summarized below (for more details refer to [1], where the notion of dynamic cube attacks was introduced).

A preprocessing stage: We first choose some polynomials that we want to set to zero at all the vertices of the cube, and show how to nullify them by setting certain dynamic variables to appropriate expressions in terms of the other public and secret variables. To minimize the number of evaluations of the cryptosystem, we choose a big cube of dimension d and a set of subcubes to sum over during the online phase. We usually choose the subcubes of the highest dimension (namely d and $d-1$), which are the most likely to give a biased sum. We then determine a set of e expressions in the private variables that need to be guessed by the attacker in order to calculate the values of the dynamic variables during the cube summations.

Note that these steps have to be done only once for each cryptosystem, and the chosen parameters determine the running time and success probabilities of the actual attack, in the same way that finding a good differential property can improve the complexity of differential attacks on a cryptosystem.

The online phase of the attack has two parts:

Online Step 1

1. For each possible vector of values for the e secret expressions, sum modulo 2 the output bits over the subcubes chosen during preprocessing with the dynamic variables set accordingly, and obtain a list of sums (one bit per subcube).
2. Given the list of sums, calculate its score by measuring the non-randomness in the subcube sums. The output of this step is a sequence of lists sorted from the lowest score to the highest (in our notation the list with the lowest score has the largest bias, and is thus the most likely to be correct in our attack).

Given that the dimension of our big cube is d, the complexity of summing over all its subcubes is bounded by $d2^d$ (using the Moebius transform [12]). Assuming that we have to guess the values of e secret expressions in order to determine the values of the dynamic variables, the complexity of this step is bounded by $d2^{d+e}$ bit operations. Assuming that we have y dynamic variables, both the data and memory complexities are bounded by 2^{d+y} (since it is sufficient to obtain an output bit for every possible vertex of the cube and for every possible value of the dynamic variables).

Online Step 2. Given the sorted guess score list, we determine the most likely values for the secret expressions, for a subset of the secret expressions, or for the entire key. The specific details of this step vary according to the attack.

2.5   A Partial Simulation Phase

The complexity of executing online step 1 of the attack for a single key is $d2^{d+e}$ bit operations and 2^{d+y} cipher executions. In the case of Grain-128, these complexities are too high and thus we have to experimentally verify our attack with a simpler procedure. Our solution is to calculate the cube summations in online step 1 only for the correct guess of the e secret expressions. We then calculate the score of the correct guess and estimate its expected position g in the sorted list of score values by assuming that incorrect guesses will make the scheme behave as a random function. Consequently, if the cube sums for the correct guess detect a property that is satisfied by a random cipher with probability p, we estimate that the location of the correct guess in the sorted list will be $g \approx \max\{p \times 2^e, 1\}$ (as justified in [1]).

3   A New Approach for Attacking Grain-128

The starting point of our new attack on Grain-128 is the weak-key attack described in [1] and we repeat it here for the sake of completeness. Both our new attack and the attack described in [1] use only the first output bit of Grain-128 (with index $i = 257$). The output function of the cipher is a multivariate polynomial of degree 3 in the state, and its only term of degree 3 is $b_{i+12}b_{i+95}s_{i+95}$. Since this term is likely to contribute the most to the high degree terms in the output polynomial, we try to nullify it. Since b_{i+12} is the state bit that is calculated at the earliest stage of the initialization steps (compared to b_{i+95} and s_{i+95}), it should be the least complicated to nullify. However, after many initialization steps, the ANF of b_{i+12} becomes very complicated and it does not seem possible to nullify it in a direct way. Instead, the idea in [1] is to simplify (and not nullify) $b_{i+12}b_{i+95}s_{i+95}$, by nullifying b_{i-21} (which participated in the most significant terms of b_{i+12}, b_{i+95} and s_{i+95}). The ANF of the earlier b_{i-21} is much easier to analyze compared to the one of b_{i+12}, but it is still very complex. The solution adopted in [1] was to assume that 10 specific key bits are set to 0. This leads to a weak-key attack on Grain-128 which can only attack a particular fraction of 0.001 of the keys.

In order to attack a significant portion of all the possible keys, we use a different approach which nullifies state bits that are produced at an earlier stage of the encryption process. This approach weakens the resistance of the output of Grain-128 to cube testers, but in a more indirect way. In fact, the output function is a higher degree polynomial which can be more resistant to cube testers compared to [1]. This forces us to slightly increase the dimension d from 46 to 50. On the other hand, since we choose to nullify state bits that are produced at an earlier stage of the encryption process, their ANF is relatively simple and thus the number of secret expressions e that we need to guess is reduced from 61 to 39. Since the complexity of the attack is proportional to $d2^{d+e}$, the smaller value of e more than compensates for the slightly larger value of d. Our new strategy thus yields not only an attack which has a significant probability of success for all the keys rather than an attack on a particular subset of weak keys, but also a better improvement factor over exhaustive search (details are given at the end of this section).

In the new attack we decided to nullify b_{i-54}. This simplifies the ANF of the output function in two ways: It nullifies the ANF of the most significant term of b_{i-21} (the only term of degree 3), which has a large influence on the ANF of the output. In addition, setting b_{i-54} to zero nullifies the most significant terms of b_{i+62} and s_{i+62}, simplifying their ANF. This simplifies the ANF of the most significant terms of b_{i+95} and s_{i+95}, both participating in the most significant term of the output function. In addition to nullifying b_{i-54}, we nullify the most significant term of b_{i+12} (which has a large influence on the ANF of the output, as described in the first paragraph of this section), $b_{i-104}b_{i-21}s_{i-21}$, by nullifying b_{i-104}.

The parameter set we used for the new attack is given in table 1. Most of the dynamic variables are used in order to simplify the ANF of $b_{i-54} = b_{203}$ so that we can nullify it using one more dynamic variable with acceptable complexity. We now describe in detail how to perform the online phase of the attack, given this parameter set. Before executing these steps, one should take the following preparation steps in order to determine the list of e secret expressions in the key variables we have to guess during the actual attack.

1. Assign values to the dynamic variables given in table 1. This is a very simple process which is described in Appendix B of [1] (since the symbolic values of the dynamic variables contain hundreds of terms, we do not list them here, but rather refer to the process that calculates their values).
2. Given the symbolic form of a dynamic variable, look for all the terms which are combinations of variables from the big cube.
3. Rewrite the symbolic form as a sum of these terms, each one multiplied by an expression containing only secret variables.
4. Add the expressions of secret variables to the set of expressions that need to be guessed. Do not add expressions whose value can be deduced from the values of the expressions which are already in the set.

When we prepare the attack, we initially get 50 secret expressions. However, after removing 11 expressions which are dependent on the rest, the number of expressions that need to be guessed is reduced to 39. We are now ready to execute the online phase of the attack:

1. Obtain the first output bit produced by Grain-128 (after the full 256 initialization steps) with the fixed secret key and all the possible values of the variables of the big cube and the dynamic variables given in table 1 (the remaining public variables are set to zero). The dimension of the big cube is 50 and we have 13 dynamic variables and thus the total amount of data and memory required is $2^{50+13} = 2^{63}$ bits.

2. We have 2^{39} possible guesses for the secret expressions. Allocate a guess score array of 2^{39} entries (an entry per guess). For each possible value (guess) of the secret expressions:

    (a) Plug the values of these expressions into the dynamic variables (which thus become a function of the cube variables, but not the secret variables).

    (b) Our big cube in table 1 is of dimension 50. Allocate an array of 2^{50} bit entries. For each possible assignment to the cube variables:

        i. Calculate the values of the dynamic variables and obtain the corresponding output bit of Grain-128 from the data.

        ii. Copy the value of the output bit to the array entry whose index corresponds to the assignment of the cube variables.

    (c) Given the 2^{50}-bit array, sum over all the entry values that correspond to the 51 subcubes of the big cube which are of dimension 49 and 50. When summing over 49-dimensional cubes, keep the cube variable that is not summed over to zero. This step gives a list of 51 bits (subcube sums).

    (d) Given the 51 sums, calculate the score of the guess by measuring the fraction of bits which are equal to 1. Copy the score to the appropriate entry in the guess score array and continue to the next guess (item 2). If no more guesses remain go to the next step.

3. Sort the 2^{39} guess scores from the lowest score to the highest.

To justify item 2.c, we note that the largest biases are likely to be created by the largest cubes, and thus we only use cubes of dimension 50 and 49. To justify item 2.d, we note that the cube summations tend to yield sparse superpolys, which are all biased towards 0, and thus we can use the number of zeroes as a measure of non-randomness. The big cube in the parameter set is of dimension 50, which has 16 times more vertices than the cube used in [1] to attack the weak key set. The total complexity of algorithm above is about $50 \times 2^{50+39} < 2^{95}$ bit operations (it is dominated by item 2.c, which is performed once for each of the 2^{39} possible secret expression guesses).

Given the sorted guess array which is the output of online step 1, we are now ready to perform online step 2 of the attack (which recovers the secret key without going through the difficult step of solving the large system of polynomial equations). In order to optimize this step, we analyze the symbolic form of the secret expressions: Out of the 39 expressions (denoted by $s_1, s_2, ..., s_{39}$), 20 contain only a single key bit (denoted by $s_1, s_2, ..., s_{20}$). Moreover, 18 out of the remaining $39 - 20 = 19$ expressions (denoted by $s_{21}, s_{22}, ..., s_{38}$) are linear combinations of key bits, or can be made linear by fixing the values of 45 more key bits. Thus, we define the following few sets of linear expressions: Set 1 contains the 20 secret key bits $s_1, s_2, ..., s_{20}$. Set 2 contains the 45 key bits whose choice simplifies $s_{21}, s_{22}, ..., s_{38}$ into linear expressions. Set 3 contains the 18 linear expressions of $s_{21}, s_{22}, ..., s_{38}$ after plugging in the values of the $20 + 45 = 65$ key bits of the first two sets (note that the set itself depends on the values of the key bits in the first two sets). Altogether, the first three sets contain $20 + 45 + 18 = 85$ singletons or linear expressions. Set 4 contains $128 - 85 = 45$ linearly independent expressions which form a basis to the complementary subspace spanned by the first three sets. Note that given the 128 values of all the expressions contained in the 4 sets, it is easy to calculate the 128-bit key.

Our attack exploits the relatively simple form of 38 out of the 39 secret expressions in order to recover the key using basic linear algebra:

1. Consider the guesses from the lowest score to the highest. For each guess:
   (a) Obtain the value of the key bits of set 1, $s_1, s_2, ..., s_{20}$.
   (b) For each possible possible values of the 45 key bits of set 2:
       i. Plug in the (current) values of the key bits from sets 1 and 2 to the expressions of $s_{21}, s_{22}, ..., s_{38}$ and obtain set 3.
       ii. Obtain the values of the linear expressions of set 3 from the guess.
       iii. From the first 3 sets, obtain the 45 linear expressions of set 4 using Gaussian Elimination.
       iv. For all possible values of the 45 linear expressions of set 4 (iterated using Gray Coding to simplify the transitions between values):
           A. Given the values of the expressions of the 4 sets, derive the secret key.
           B. Run Grain-128 with the derived key and compare the result to a given (known) key stream. If there is equality, return the full key.

This algorithm contains 3 nested loops. The loop of item 1 is performed g times, where g is the expected position of the correct guess in the sorted guess array. The loop of item 1.b is performed 2^{45} times per guess. The loop of item 1.b.iv is performed 2^{45} per iteration of the previous loop. The loop of item 1.b contains linear algebra in item 1.b.iii whose complexity is clearly negligible compared to the inner loop of item 1.b.iv, which contains 2^{45} cipher evaluations. In the inner loop of step 1.b.iv (in item 1.b.iv.A) we need to derive the 128-bit key. In general, this is done by multiplying a 128×128 matrix with a 128-bit vector that corresponds to the values of the linear expressions. However, note that 65 key bits (of sets 1 and 2) are already known. Moreover, since we iterate the values of set 4 using Gray Coding (i. e., we flip the value of a single expression per iteration), we only need to perform the multiplication once and then calculate the difference from the previous iteration by adding a single vector to the previous value of the key. This optimization requires a few dozen bit operations, which is negligible compared to running Grain-128 in item 1.b.iv.B (which requires at least 1000 bit operation). Thus, the complexity of the exhaustive search per guess is about $2^{45+45} = 2^{90}$ cipher executions, which implies that the total complexity the algorithm is about $g \times 2^{90}$.

The attack is worse than exhaustive search if we have to try all the 2^{39} possible values of g, and thus it is crucial to provide strong experimental evidence that g is relatively small for a large fraction of keys. In order to estimate g, we executed the online part of the attack by calculating the score for the correct guess of the 39 expression values, and estimating how likely it is to get such a bias for incorrect guesses if we assume that they behave as random functions. We performed this simulation for 107 randomly chosen keys, out of which 8 gave a very significant bias in which at least 50 of the 51 cubes sums were zero. This is expected to occur in a random function with probability $p < 2^{-45}$, and thus we estimate that for about 7.5% of the keys, $g \approx \max\{2^{-45} \times 2^{39}, 1\} = 1$ and thus the correct guess of the 39 secret expressions will be the first in the sorted score list (additional keys among those we tested had smaller biases, and thus a larger g). The complexity of online step 2 of the attack is thus expected to be about 2^{90} cipher executions, which dominates the complexity of the attack (the complexity of online step 1 is about 2^{95} bit operations, which we estimate as $2^{95-10} = 2^{85}$ cipher executions). This gives an improvement factor of 2^{38} over the 2^{128} complexity of exhaustive search for a non-negligible fraction of keys, which is significantly better than the improvement factor of 2^{15} announced in [1] for the small subset of weak keys considered in that attack. We note that for most additional keys there is a continuous tradeoff between the fraction of keys that we can attack and the complexity of the attack on these keys.

Table 1. Parameter set for the attack on the full Grain-128, given output bit 257

Cube Indexes	{0,2,4,11,12,13,16,19,21,23,24,27,29,33,35,37,38,41,43,44,46, 47,49,52,53,54,55, 57,58,59,61,63,65,66,67,69,72,75,76,78,79,81,82,84,85,87,89,90,92,93}
Dynamic Variables	{31,3,5,6,8,9,10,15,7,25,42,83,1}
State Bits Nullified	{$b_{159}, b_{131}, b_{133}, b_{134}, b_{136}, b_{137}, b_{138}, b_{145}, s_{135}, b_{153}, b_{170}, b_{176}, b_{203}$}

4 Description of the Dedicated Hardware Used to Attack Grain-128

Cube attacks and testers are notoriously difficult to analyze mathematically. To test our attack experimentally and to verify its complexity, we had to try dozens of random keys, and thus to run thousands of cube summations of dimension 49 and 50 for multiple random keys. This is only marginally feasible on a large cluster of PCs, which are ill-suited for performing computations relying heavily on bit-permutations as needed for this kind of attack. We thus decided to experimentally verify our attack on dedicated reconfigurable hardware.

4.1 Architectural Considerations

We start with an evaluation of the online phase of the attack (for the correct guess of the 39 secret expression values) regarding possible optimizations in hardware. To get a better understanding of our implementation, we describe the basic work-flow in Figure 1: The software implementation of the attack uses a parameter set as input, e. g., the cube dimension, the cube itself, a base IV and the number of keys to attack. It selects a random key to attack and divides the big cube into smaller worker cubes and distributes them to worker threads running in parallel. Please note that for simplicity the figure shows only one worker. If 2^w workers are used, the iterations per worker are reduced from 2^d to 2^{d-w}.

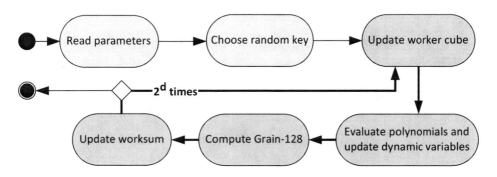

Fig. 1. Cube Attack — Program flow for cube dimension d

The darker nodes and the bold path show the steps of each independent thread: As each worker iterates over a distinct subset of the cube, it evaluates polynomials on the worker cube (dynamic variables) and updates the IV input to Grain-128. Using the generated IV and the random key, it computes the output of Grain-128 after the initialization phase. With this output, the thread updates an intermediate value — the worker sum — and starts the next iteration. In the end, the software combines all worker sums, evaluates the result and can chose a new random key to start again.

With a cube of dimension d, the attack on one key (for the correct guess of the 39 secret expression values) computes the first output bit of Grain-128 2^d times. Thus, in order to speed-up the attack, it is necessary to implement Grain-128 as efficiently as possible. The design of the stream cipher is highly suitable for hardware implementations: It consists mainly of two shift registers and some logic cells. As already proposed for cube testers on Grain-128 in [4], a fast and small FPGA implementation is a very good choice in comparison to a (bit-sliced) software implementation.

To create an independent worker on the FPGA, it is also required to implement the IV generation. To estimate the effort of building a full worker in hardware, we need to know how many dynamic inputs we have to consider: While dynamic modifications, e. g., iterating over arrays with dynamic step sizes, pose no problems in software, they can be very inefficient in hardware.

In order to compute the cipher, we need a key and an IV. The value of the key varies, as it is chosen at random. The IV is a 128 bit value, where each bit utilizes one of three functions: it is either a value given by the base IV provided by the parameter set, part of the (worker) cube or a dynamic variable. As the function of each bit is modified not only per parameter set, but also when assigning partial cubes to different workers, this input also varies. The first two functionalities are both restricted and can be realized by simple multiplexers in hardware. The dynamic variable on the other hand stores the result of a polynomial. As we have no set of pre-defined polynomials and they are derived at runtime, every possible combination of boolean functions over the worker cube (and thus over the complete 128 bits) must be realized. Even with tight restrictions like a maximum of terms per monomial and monomials per polynomial, it is impossible to provide the reconfigurable structure in hardware.

As a consequence, a fully dynamic approach leads to extremely large multiplexers and thus to very high area consumption on the FPGA, which is prohibitively slow. The completely opposite approach would be to utilize the complete area of an FPGA for massive parallel Grain-128 computations without additional logic. In this case, the communication between the host and the FPGA will be the bottleneck of the system and the parallel cores on the FPGA will idle.

For our attack, we use the RIVYERA special-purpose hardware cluster described in greater detail in Appendix A. For the following design decisions we remark that RIVYERA provides 128 powerful Spartan-3 FPGAs, which are tightly connected to an integrated server system powered by an Intel Core i7 920 with 8 logical CPU cores. This allows us to utilize dedicated hardware and use a multi-core architecture for the software part.

In order to implement the attack on the RIVYERA and benefit from its massive computing power, we propose the following implementation. Figure 2 shows the design of the modified attack. The software design is split into two parts: We use all but one core of the CPU to generate attack specific bitstreams, i. e., configuration files for the FPGAs, in parallel to prepare the computation on the FPGA cluster. Each of these generated designs configures the RIVYERA for a complete attack on one random key provided by the host PC. As soon as one bitstream was generated and waits in the queue, the remaining core programs all 128 FPGAs with it, starts the attack, waits for the computation to finish and stores the results.

In contrast to the first approach, which uses the generic structure realizable in software, we generate custom VHDL code containing constant settings and fixed boolean functions of the polynomials derived from the parameter set and the provided key. Building specific configuration files for each attack setup allows us to implement as many fully functional, independent, parallel workers as possible without the area consumption of complex control structures. In addition, only a single 7-bit parameter is necessary at runtime - to split the workspace between all 128 FPGAs - to start the computation and receive a d-bit return value. This efficiently circumvents all of the problems and overhead of a generic hardware design at the cost of rerunning the FPGA design flow for each parameter/key pair.

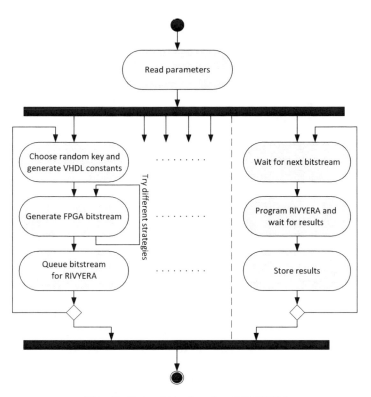

Fig. 2. Cube Attack using RIVYERA

Please note that in this approach the host software modifies a basic design by hard-coding conditions and adjusting internal bus and memory sizes for each attack. We optimized the basic layout as much as possible, but the different choices of polynomial functions lead to different combinatorial logic paths and routing decisions, which can change the critical path in hardware. As the clock frequency is linked to the critical path, we implemented different design strategies as well as multiple fall-back options to modify the clock frequency constraints in order to prevent parameter/key pairs from resulting in an invalid hardware configurations.

4.2 Hardware Implementation Results

In this section, we give a brief overview of the implementation and present results. As the total number of iterations for one attack (for the correct guess of the 39 secret expression values) is 2^d, the number of workers for an optimal setup has to be a power of two. Considering the area of a Spartan-3 5000 FPGA, we chose to implement a set of 2^4 independent workers per FPGA.

Figure 3 shows the top level overview. As mentioned before, creating an attack specific implementation allows us to strip down the communication interface and

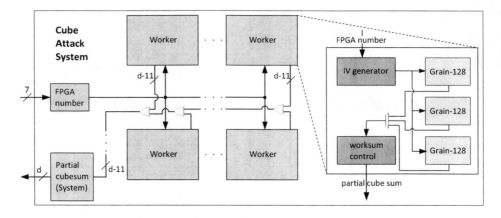

Fig. 3. FPGA Implementation of the online phase for cube dimension d

data paths to a minimum. This is very important as we cannot predict the impact of the (unknown) parameters and need to relax the design as much as possible.

Each of the workers consists of its own IV generator and controls three Grain-128 instances. The IV generator needs three clock cycles per IV and we need a corresponding number of Grain instances to process the output directly. As it is possible to run more than one initialization step per clock cycle in parallel, we had to find the most suitable time/area trade-off for the cipher implementation. Table 2 shows the synthesis results of our Grain implementation. In comparison, Aumasson et al. used 2^5 parallel steps, which is the maximum number of supported parallel steps without additional overhead, on the large Virtex-5 LX330 FPGA used in [4].

Table 2. Synthesis results of Grain-128 implementation on the Spartan-3 5000 FPGA with different numbers of parallel steps per clock cycle

Parallel Steps	2^0	2^1	2^2	2^3	2^4	2^5
Clock Cycles (Init)	256	128	64	32	16	8
Max. Frequency (MHz)	227	226	236	234	178	159
FPGA Resources (Slices)	165	170	197	239	311	418

The resulting attack system for the online phase — consisting of the software and the RIVYERA cluster — uses 16 workers per FPGA and 128 FPGAs on the cluster in parallel. This means that the number of Grain computations per worker is reduced to 2^{d-11}. The design ensures that each key can be attacked at the highest possible clock frequency, while it tries to keep the building time per configuration moderate.

Table 3 reflects the results of the generation process and the distribution of the configurations with respect to the different clock frequencies. It shows that the impact of the unknown parameters is predictable and that fallback strategies

Table 3. Results of the generation process for cubes of dimension 46, 47 and 50. The Duration is the time required for the RIVYERA cluster to complete the online phase. The Percentage row gives the percentage of configurations built with the given clock frequency out of the total number of configurations built with cubes of the same dimension.

Cube Dimension d	46			47	50	
Clock Frequency (MHz)	100	110	120	120	110	120
Configurations Built	1	7	8	6	60	93
Percentage	6.25	43.75	50	100	39.2	60.8
Online Phase Duration	17.2 min	15.6 min	14.3 min	28.6 min	4h 10 min	3h 49 min

are necessary. Please note that the new attack tries to generate configurations for multiple keys in parallel. This process — if several strategies are tried — may require more than 6 hours before the first configuration becomes available. Smaller cube dimensions, i.e., all cube dimensions lower than 48, result in very fast attacks and should be neglected, as the building time will exceed the duration of the attack in hardware. Further note that the duration of the attack increases exponentially in d, e.g., assuming 100 MHz as achievable for larger cube dimensions, $d = 53$ needs 1.5 days and $d = 54$ needs 3 days.

5   Conclusions

We presented the first attack on Grain-128 which is considerably faster than exhaustive search, and unlike previous attacks makes no assumptions on the secret key. While the full attack is infeasible, we can convincingly estimate its results by running a partial version in which all the e unknown secret expressions are set to their correct value. Due to its high complexity and hardware-oriented nature, the attack was developed and verified using a new type of dedicated hardware. Our experimental results show that for about 7.5% of the keys we get a huge improvement factor of 2^{38} over exhaustive search.

Acknowledgements. The authors thank Martin Ågren and the anonymous referees for their very helpful comments on this paper.

A   Design and Architecture of the RIVYERA Cluster

In this work we employ an enhanced version of the COPACOBANA special-purpose hardware cluster that was specifically designed for the task of crypt-analysis [13]. This enhanced cluster (also known as RIVYERA [14]) is populated with 128 Spartan-3 XC3S5000 FPGAs, each tightly coupled with 32MB memory. Each Spartan-3 XC3S5000 FPGA provides a sea of logic resources consisting of 33,280 slices and 104 BRAMs enabling the implementation even of complex functions in reconfigurable hardware. Eight FPGAs are soldered on individual card modules that are plugged into a backplane which implements a global systolic ring bus for high-performance communication. The internal ring bus is further connected via PCI Express to a host PC which is also installed in the same 19" housing of the cluster. Figure 4 provides an overview of the architecture of the RIVYERA special purpose cluster.

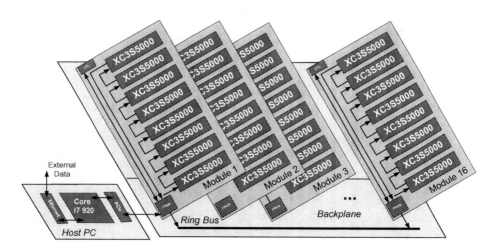

Fig. 4. Architecture of the RIVYERA cluster system

References

1. Dinur, I., Shamir, A.: Breaking Grain-128 with Dynamic Cube Attacks. In: Joux, A. (ed.) FSE 2011. LNCS, vol. 6733, pp. 167–187. Springer, Heidelberg (2011)
2. Aumasson, J.-P., Dinur, I., Meier, W., Shamir, A.: Cube Testers and Key Recovery Attacks on Reduced-Round MD6 and Trivium. In: Dunkelman, O. (ed.) FSE 2009. LNCS, vol. 5665, pp. 1–22. Springer, Heidelberg (2009)
3. Hell, M., Johansson, T., Maximov, A., Meier, W.: A Stream Cipher Proposal: Grain-128. In: IEEE International Symposium on Information Theory, ISIT 2006 (2006)
4. Aumasson, J.-P., Dinur, I., Henzen, L., Meier, W., Shamir, A.: Efficient FPGA Implementations of High-Dimensional Cube Testers on the Stream Cipher Grain-128. In: Workshop on Special-purpose Hardware for Attacking Cryptographic Systems – SHARCS 2009, September 9-10 (2009)
5. Knellwolf, S., Meier, W., Naya-Plasencia, M.: Conditional Differential Cryptanalysis of NLFSR-Based Cryptosystems. In: Abe, M. (ed.) ASIACRYPT 2010. LNCS, vol. 6477, pp. 130–145. Springer, Heidelberg (2010)
6. Englund, H., Johansson, T., Sönmez Turan, M.: A Framework for Chosen IV Statistical Analysis of Stream Ciphers. In: Srinathan, K., Rangan, C.P., Yung, M. (eds.) INDOCRYPT 2007. LNCS, vol. 4859, pp. 268–281. Springer, Heidelberg (2007)
7. Fischer, S., Khazaei, S., Meier, W.: Chosen IV Statistical Analysis for Key Recovery Attacks on Stream Ciphers. In: Vaudenay, S. (ed.) AFRICACRYPT 2008. LNCS, vol. 5023, pp. 236–245. Springer, Heidelberg (2008)

8. Stankovski, P.: Greedy Distinguishers and Nonrandomness Detectors. In: Gong, G., Gupta, K.C. (eds.) INDOCRYPT 2010. LNCS, vol. 6498, pp. 210–226. Springer, Heidelberg (2010)

9. De Cannière, C., Küçük, Ö., Preneel, B.: Analysis of Grain's Initialization Algorithm. In: Vaudenay, S. (ed.) AFRICACRYPT 2008. LNCS, vol. 5023, pp. 276–289. Springer, Heidelberg (2008)

10. Lee, Y., Jeong, K., Sung, J., Hong, S.: Related-Key Chosen IV Attacks on Grain-v1 and Grain-128. In: Mu, Y., Susilo, W., Seberry, J. (eds.) ACISP 2008. LNCS, vol. 5107, pp. 321–335. Springer, Heidelberg (2008)

11. Lai, X.: Higher Order Derivatives and Differential Cryptanalysis. In: "Symposium on Communication, Coding and Cryptography", in Honor of James L. Massey on the Occasion of his 60th Birthday, pp. 227–233 (1994)

12. Joux, A.: Algorithmic Cryptanalysis. Chapman & Hall, pp. 285–286

13. Güneysu, T., Kasper, T., Novotný, M., Paar, C., Rupp, A.: Cryptanalysis with COPACOBANA. IEEE Transactions on Computers 57(11), 1498–1513 (2008)

14. Güneysu, T., Pfeiffer, G., Paar, C., Schimmler, M.: Three Years of Evolution: Cryptanalysis with COPACOBANA. In: Workshop on Special-purpose Hardware for Attacking Cryptographic Systems – SHARCS 2009, September 9-10 (2009)

15. Budiansky, S.: Battle of Wits: the Complete Story of Codebreaking in World War II. Free Press (2000) ISBN: 9780684859323

16. Gilmore, J.: Cracking DES: Secrets of Encryption Research. Wiretap Politics & Chip Design. O'Reilly (July 1998)

17. Matsui, M.: The First Experimental Cryptanalysis of the Data Encryption Standard. In: Desmedt, Y.G. (ed.) CRYPTO 1994. LNCS, vol. 839, pp. 1–11. Springer, Heidelberg (1994)

18. Vielhaber, M.: Breaking ONE.FIVIUM by AIDA an Algebraic IV Differential Attack. Cryptology ePrint Archive, Report 2007/413 (2007)

19. Dinur, I., Shamir, A.: Cube Attacks on Tweakable Black Box Polynomials. In: Joux, A. (ed.) EUROCRYPT 2009. LNCS, vol. 5479, pp. 278–299. Springer, Heidelberg (2009)

Biclique Cryptanalysis of the Full AES*

Andrey Bogdanov[1,**], Dmitry Khovratovich[2], and Christian Rechberger[3,**]

[1] K.U. Leuven, Belgium
[2] Microsoft Research Redmond, USA
[3] ENS Paris and Chaire France Telecom, France

Abstract. Since Rijndael was chosen as the Advanced Encryption Standard (AES), improving upon 7-round attacks on the 128-bit key variant (out of 10 rounds) or upon 8-round attacks on the 192/256-bit key variants (out of 12/14 rounds) has been one of the most difficult challenges in the cryptanalysis of block ciphers for more than a decade. In this paper, we present the novel technique of block cipher cryptanalysis with bicliques, which leads to the following results:

- The first key recovery method for the full AES-128 with computational complexity $2^{126.1}$.
- The first key recovery method for the full AES-192 with computational complexity $2^{189.7}$.
- The first key recovery method for the full AES-256 with computational complexity $2^{254.4}$.
- Key recovery methods with lower complexity for the reduced-round versions of AES not considered before, including cryptanalysis of 8-round AES-128 with complexity $2^{124.9}$.
- Preimage search for compression functions based on the full AES versions faster than brute force.

In contrast to most shortcut attacks on AES variants, we *do not need to assume related-keys*. Most of our techniques only need a very small part of the codebook and have low memory requirements, and are practically verified to a large extent. As our cryptanalysis is of high computational complexity, it does not threaten the practical use of AES in any way.

Keywords: block ciphers, bicliques, AES, key recovery, preimage.

1 Introduction

Since the Advanced Encryption Standard competition finished in 2001, the world saw little progress in the cryptanalysis of block ciphers. In particular, the current standard AES is almost as secure as it was 10 years ago in the strongest and most practical model with a single unknown key. The former standard DES has not seen a major improvement since Matsui's seminal paper in 1993 [37].

In contrast, the area of hash function cryptanalysis is growing quickly, encouraged by the cryptanalysis of MD5 [48], of SHA-0 [6, 15] and SHA-1 [47],

---

* This is the proceedings version of the paper [12].
** The authors were visiting Microsoft Research Redmond while working on these results.

D.H. Lee and X. Wang (Eds.): ASIACRYPT 2011, LNCS 7073, pp. 344–371, 2011.

followed by a practical attack on protocols using MD5 [44, 45], preimage attacks on Tiger [28] and MD5 [43], etc. While differential cryptanalysis [7], a technique originally developed for block ciphers, was initially carried over to hash function analysis to enrich the cryptanalytic toolbox for hash functions, now cryptanalysts are looking for the opposite: a method of hash function analysis that would give new results on block ciphers. So far the most successful attempt is the analysis of AES with local collisions [8–11], but it is only applicable in the related-key model. In the latter model, an attacker works with plaintexts and ciphertexts that are produced under not only the unknown key, but also under other keys related to the first one in a way chosen by the adversary. Such a strong requirement is rarely practical and, thus, has not been considered to be a threat for the use of AES. Also, there has been no evidence that the local collision approach can facilitate an attack in the more practical and relevant single-key model.

State of the art for attacks on AES. AES with its wide-trail strategy was designed to withstand differential and linear cryptanalyses [18], so pure versions of these techniques have limited applications in attacks. With respect to AES, probably the most powerful single-key recovery methods designed so far are impossible differential cryptanalysis [5, 36] and Square attacks [17, 22]. Impossible differential cryptanalysis yielded the first attack on the 7-round AES-128 with non-marginal data complexity. The Square attack and its variations such as integral attack and multiset attack resulted in the cryptanalysis of round-reduced AES variants with lowest computational complexity to date, while the first attack on 8-round AES-192 with non-marginal data complexity has appeared only recently [22].

The situation is different in weaker attack models, where the related-key cryptanalysis was applied to the full versions of AES-192 and AES-256 [9], and the rebound attack demonstrated a non-random property in 8-round AES-128 [27, 33]. However, there is little evidence so far that carrying over these techniques to the most practical single-secret-key model is feasible. Note that no attack against the full AES-128 has been known even in the relate-key model or a hash mode.

Meet-in-the-middle attacks with bicliques. Meet-in-the-middle attacks on block ciphers have obtained less attention (see [13, 14, 16, 21, 24, 29, 49] for a list of the most interesting ones) than the differential, linear, impossible differential, and integral approaches. However, they are probably the most practical in terms of data complexity. A basic meet-in-the-middle attack requires only the information-theoretical minimum of plaintext-ciphertext pairs. The limited use of these attacks can be attributed to the requirement for large parts of the cipher to be independent of particular key bits. As this requirement is not met in AES and most AES candidates, the number of rounds broken with this technique is rather small [14, 21], which seems to prevent it from producing results on yet unbroken number of rounds in AES. We also mention that the collision attacks [19, 20] use some elements of the meet-in-the-middle framework.

In this paper we demonstrate that the meet-in-the-middle attacks on block ciphers have great potential if enhanced by a new concept called *bicliques*. The biclique concept was first introduced for hash cryptanalysis by Savelieva et al. [31]. It originates from the so-called splice-and-cut framework [1, 2, 28] in hash function cryptanalysis, more specifically its element called initial structure. The biclique approach led to the best preimage attacks on the SHA family of hash functions so far, including the attack on 50 rounds of SHA-512, and the first attack on a round-reduced Skein hash function [31]. We show how to carry over the concept of bicliques to block cipher cryptanalysis and get even more significant results, including the first key recovery for all versions of the full AES faster than brute force.

A biclique is characterized by its length (number of rounds covered) and dimension. The dimension is related to the cardinality of the biclique elements and is one of the factors that determines the advantage over brute force. The total cost of the key search with bicliques was two main contributors: firstly the cost of constructing the bicliques, and secondly the matching computations.

Two paradigms for key recovery with bicliques. Taking the biclique properties into account, we propose two different approaches, or paradigms, for key recovery. Suppose that the cipher admits the basic meet-in-the-middle attack on m (out of r) rounds. The first paradigm, the *long-biclique*, aims to construct a biclique for the remaining $r - m$ rounds. Though the dimension of the biclique decreases as r grows, small-dimension bicliques can be constructed with numerous tools and methods from differential cryptanalysis of block ciphers and hash functions: rebound attacks, trail backtracking, local collisions, etc. Also from an information-theoretic point of view, bicliques of dimension 1 are likely to exist in a cipher, regardless of the number of rounds. The computational bottleneck for this approach is usually the construction of the bicliques.

The second paradigm, the *independent-biclique*, aims to construct bicliques of higher dimensions for smaller $b < (r - m)$ number of rounds efficiently and cover the remaining rounds in a brute-force way with a new method of *matching with precomputations*. The construction of bicliques becomes much simpler with this approach, the computational bottleneck is hence the matching computation. Even though partial brute-force computations have been considered before for cryptanalytically improved preimage search methods for hash functions [1, 41], we show that its combination with biclique cryptanalysis allows for much larger savings of computations.

Results on AES. The biclique cryptanalysis successfully applies to all full versions of AES and compared to brute force provides a computational advantage of about a factor 3 to 5, depending on the version. Also, it yields advantages of up to a factor 15 for the key recovery of the AES versions with smaller but yet secure number of rounds. The largest factors are obtained in the independent-biclique paradigm and have success rate 1. We also provide complexities for finding compression function preimages for all full versions of AES when considered in hash modes. Our results on AES are summarized in Table 1 and 2, and an attempt to give an exhaustive overview with earlier results is given in Tables 4 and 5. The "full version" reference refers to [12].

Table 1. Biclique key recovery for AES

rounds	data	computations/succ.rate	memory	biclique length in rounds	reference
		AES-128 secret key recovery			
8	$2^{126.33}$	$2^{124.97}$	2^{102}	5	Full version
8	2^{127}	$2^{125.64}$	2^{32}	5	Full version
8	2^{88}	$2^{125.34}$	2^8	3	Sec. 6
10	2^{88}	$2^{126.18}$	2^8	3	Sec. 6
		AES-192 secret key recovery			
9	2^{80}	$2^{188.8}$	2^8	4	Full version
12	2^{80}	$2^{189.74}$	2^8	4	Full version
		AES-256 secret key recovery			
9	2^{120}	$2^{253.1}$	2^8	6	Sec. 7
9	2^{120}	$2^{251.92}$	2^8	4	Full version
14	2^{40}	$2^{254.42}$	2^8	4	Full version

Table 2. Biclique preimage search of AES in hash modes (compression function)

rounds	computations	succ.rate	memory	biclique length in rounds	reference
	AES-128 compression function preimage, Miyaguchi-Preneel				
10	$2^{125.83}$	0.632	2^8	3	Sec. 6
	AES-192 compression function preimage, Davies-Meyer				
12	$2^{125.71}$	0.632	2^8	4	Full version
	AES-256 compression function preimage, Davies-Meyer				
14	$2^{126.35}$	0.632	2^8	4	Full version

2   Biclique Cryptanalysis

Now we introduce the concept of biclique cryptanalysis applied to block ciphers. To make our approach clear for readers familiar with meet-in-the-middle attacks, we introduce most of the terminology while explaining how meet-in-the-middle works, and then proceed with bicliques.

2.1   Basic Meet-in-the-Middle Attack

An adversary chooses a partition of the key space into groups of keys of cardinality 2^{2d} each for some d. A key in a group is indexed as an element of a $2^d \times 2^d$ matrix: $K[i,j]$. The adversary selects an internal variable v in the data transform of the cipher such that

– as a function of a plaintext and a key, it is identical for all keys in a *row*:

$$P \xrightarrow[g_1]{K[i,\cdot]} v;$$

– as a function of a ciphertext and a key, it is identical for all keys in a *column*:

$$v \xleftarrow[g_2]{K[\cdot,j]} C,$$

where g_1 and g_2 form the cipher $E = g_2 \circ g_1$.

Given a plaintext-ciphertext pair (P,C) obtained under the secret key K_{secret}, an adversary computes 2^d possible values $\overrightarrow{v}$ and 2^d possible values $\overleftarrow{v}$ from the plaintext and from the ciphertext, respectively. A matching pair $\overrightarrow{v}_i = \overleftarrow{v}_j$ yields a key candidate $K[i,j]$. The expected number of key candidates depends on the bit size $|v|$ of v and is given by the formula $2^{2d-|v|}$. For $|v|$ close to d and larger, an attack has advantage of about 2^d over brute force search as it tests 2^{2d} keys with less than 2^d calls of the full cipher.

The basic meet-in-the-middle attack has clear limitations in block cipher cryptanalysis since an internal variable with the properties listed above can be found for a very small number of rounds only. We show how to bypass this obstacle with the concept of a *biclique*.

2.2   Bicliques

Now we introduce the notion of a biclique following [31]. Let f be a subcipher that maps an internal state S to the ciphertext C: $f_K(S) = C$. f connects 2^d internal states $\{S_j\}$ to 2^d ciphertexts $\{C_i\}$ with 2^{2d} keys $\{K[i,j]\}$:

$$\{K[i,j]\} = \begin{bmatrix} K[0,0] & K[0,1] & \dots & K[0,2^d-1] \\ \dots & & & \\ K[2^d-1,0] & K[2^d-1,1] & \dots & K[2^d-1,2^d-1] \end{bmatrix}.$$

The 3-tuple $[\{C_i\}, \{S_j\}, \{K[i,j]\}]$ is called a *d-dimensional biclique*, if

$$C_i = f_{K[i,j]}(S_j) \text{ for all } i,j \in \{0,\dots,2^d-1\}. \tag{1}$$

In other words, in a biclique, the key $K[i,j]$ maps the internal state S_j to the ciphertext C_i and vice versa. This is illustrated in Figure 1.

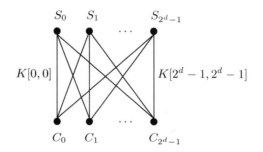

Fig. 1. *d*-dimensional biclique

2.3   The Flow of Biclique Cryptanalysis

Preparation. An adversary chooses a partition of the key space into groups of keys of cardinality 2^{2d} each for some d and considers the block cipher as a composition of two subciphers: $e = f \circ g$, where f follows g. A key in a group is indexed as an element of a $2^d \times 2^d$ matrix: $K[i,j]$.

Step 1. For each group of keys the adversary builds a structure of 2^d ciphertexts C_i and 2^d intermediate states S_j with respect to the group of keys $\{K[i,j]\}$ so that the partial decryption of C_i with $K[i,j]$ yields S_j. In other words, the structure satisfies the following condition:

$$\forall i,j : S_j \xrightarrow[f]{K[i,j]} C_i. \tag{2}$$

Step 2. The adversary asks the oracle to decrypt ciphertexts C_i with the secret key K_{secret} and obtains the 2^d plaintexts P_i:

$$C_i \xrightarrow[e^{-1}]{\text{decryption oracle}} P_i. \tag{3}$$

Step 3. If one of the tested keys $K[i,j]$ is the secret key K_{secret}, then it maps intermediate state S_j to the plaintext P_i. Therefore, the adversary checks if

$$\exists i,j : P_i \xrightarrow[g]{K[i,j]} S_j. \tag{4}$$

A valid pair proposes $K[i,j]$ as a key candidate.

3   New Tools and Techniques for Bicliques

In here we describe two approaches to construct bicliques, and propose a precomputation technique that speeds up the application of bicliques for key recovery. The exposition is largely independent of a cipher.

3.1   Bicliques from Independent Related-Key Differentials

A straightforward approach to find a d-dimensional biclique would be to fix 2^d states and 2^d ciphertexts, and derive a key for each pair to satisfy (2). This would require at least 2^{2d} key recovery attempts for f. A much more efficient way for the adversary is to choose the keys in advance and require them to conform to specific differentials as follows.

Let the key $K[0,0]$ map the intermediate state S_0 to the ciphertext C_0, and consider two sets of 2^d related-key differentials each over f with respect to the *base computation* $S_0 \xrightarrow[f]{K[0,0]} C_0$:

- Δ_i-**differentials.** A differential in the first set maps the input difference 0 to an output difference Δ_i under a key difference Δ_i^K:

$$0 \xmapsto[f]{\Delta_i^K} \Delta_i \text{ with } \Delta_0^K = 0 \text{ and } \Delta_0 = 0. \tag{5}$$

- ∇_j-**differentials.** A differential in the second set maps an input difference ∇_j to the output difference 0 under key difference ∇_j^K:

$$\nabla_j \xmapsto[f]{\nabla_j^K} 0 \text{ with } \nabla_0^K = 0 \text{ and } \nabla_0 = 0. \tag{6}$$

The tuple $(S_0, C_0, K[0,0])$ conforms to both sets of differentials by definition. If the trails of Δ_i-differentials do not share active nonlinear components (such as active S-boxes in AES) with the trails of ∇_j-differentials, then the tuple also conforms to 2^{2d} **combined (Δ_i, ∇_j)-differentials:**

$$\nabla_j \xmapsto[f]{\Delta_i^K \oplus \nabla_j^K} \Delta_i \text{ for } i, j \in \{0, \ldots, 2^d - 1\}, \tag{7}$$

which are obtained by formal xor of differentials (5) and (6) (and trails, if necessary). The proof follows from the fact that an active non-linear element in a trail of a combined differential is active in either Δ- or ∇-trail, hence its input still conforms to the corresponding trail by the assumption. A more formal and generic proof can be derived from the theory of boomerang attacks [46] and particularly from the concept of the S-box switch [9] and a sandwich attack [23]. Since Δ_i- and ∇_j-trails share no active non-linear elements, a boomerang based on them returns from the ciphertext with probability 1 as the quartet of states forms the boomerang rectangle at every step. In the special case where no nontrivial trail of one differential intersects with a nontrivial trail of the other differential, the differentials are completely independent and can be directly combined.

Substituting S_0, C_0, and $K[0,0]$ to the combined differentials (7), one obtains:

$$S_0 \oplus \nabla_j \xmapsto[f]{K[0,0] \oplus \Delta_i^K \oplus \nabla_j^K} C_0 \oplus \Delta_i. \tag{8}$$

Finally, we put

$$\begin{aligned} S_j &= S_0 \oplus \nabla_j, \\ C_i &= C_0 \oplus \Delta_i, \text{ and} \\ K[i,j] &= K[0,0] \oplus \Delta_i^K \oplus \nabla_j^K \end{aligned}$$

and get exactly the definition of a d-dimensional biclique (1). If $\Delta_i \neq \nabla_j$ for $i + j > 0$, then all keys $K[i,j]$ are different. The construction of a biclique is thus reduced to the computation of Δ_i and ∇_j, which requires no more than $2 \cdot 2^d$ computations of f.

The independency of the related-key differentials allows one to efficiently construct higher-dimensional bicliques and simplifies the partition of the key space. Though this approach turns out to be effective in the case of AES, the length of independent differentials (and hence a biclique) is limited by the diffusion properties of the cipher.

3.2 Bicliques from Interleaving Related-Key Differential Trails

The differential independency requirement appears to be a very strong requirement as it clearly limits the biclique length. An alternative way to construct a biclique is to consider interleaving differential trails. However, a primitive secure against differential cryptanalysis does not admit a long biclique of high dimension over itself, as such a biclique would consume too many degrees of freedom. For small dimensions, however, the biclique equations admit a rather simple differential representation, which allows a cryptanalyst to involve valuable tools from differential cryptanalysis of hash functions.

We outline here how bicliques of dimension 1 can be constructed in terms of differentials and differential trails with a procedure resembling the rebound attack [39]. We are also able to amortize the construction cost of a biclique by producing many more out of a single one. The construction algorithm is outlined as follows for a fixed key group $\{K[0,0], K[0,1], K[1,0], K[1,1]\}$, see also Figure 2:

- **Intermediate state T.** Choose an intermediate state T in subcipher f (over which the biclique is constructed). The position of T splits f into two parts : $f = f_2 \circ f_1$. f_1 maps S_j to T. f_2 maps T to C_i.
- **Δ- and ∇-trails.** Choose some truncated related-key differential trails: Δ-trails over f_1 and ∇-trails over f_2.
- **Inbound phase.** Guess the differences in the differential trails up to T. Get the values of T that satisfy the input and output differences over f.
- **Outbound phase.** Use the remaining degrees of freedom in the state to sustain difference propagation in trails.
- Output the states for the biclique.

We stress that the related-key trails are used in the single-key model.

Numerous optimizations of the outlined biclique construction algorithm are possible. For instance, it is not necessary to guess all differences in the trail, but only a part of them, and subsequently filter out the solutions. Instead of fixing the key group, it is also possible to fix only the difference between keys and derive actual values during the attack (the disadvantage of this approach is that key groups are generated online, and we have to take care of possible repetitions). It is also important to reduce an amortized cost of a biclique by producing new ones for other key group by some simple modification.

3.3 Matching with Precomputations

Here we describe the idea of matching with precomputations, which provides a significant computational advantage due to amortized computations. This is an efficient way to check Equation (4) in the procedure of biclique cryptanalysis.

First, the adversary computes and stores in memory $2 \cdot 2^d$ full computations

$$\text{for all } i \quad P_i \xrightarrow{\;K[i,0]\;} \overrightarrow{v} \quad \text{and} \quad \text{for all } j \quad \overleftarrow{v} \xleftarrow{\;K[0,j]\;} S_j$$

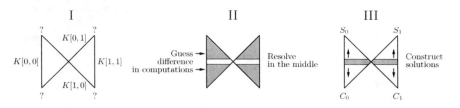

Fig. 2. Construction of a 1-dimensional biclique from dependent related-key differential trails: Guess difference between computations and derive states S_j and ciphertext C_i as conforming elements

up to some matching variable v, which can be a small part of the internal cipher state. Then for particular i, j he recomputes only those parts of the cipher that differ from the stored ones:

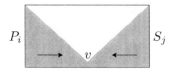

The amount of recalculation depends on the diffusion properties of both internal rounds and the key schedule of the cipher. The relatively slow diffusion in the AES key schedule allows the adversary to skip most recomputations of the key schedule operations.

4   Two Paradigms of Key Recovery

We have introduced different approaches to construct bicliques and to perform matching with precomputations. One may ask which approach is optimal and relevant. We have studied several block ciphers and hash functions, including different variants of AES, and it turns out that the optimal choice depends on a primitive, its diffusion properties, and features of the key schedule. This prepares the case to introduce two paradigms for key recovery, which differ both methodologically and in their use of tools.

To put our statement in context, let us consider the basic meet-in-the-middle attack (Section 2.1) and assume that it can be applied to m rounds of a primitive, while we are going to attack $r > m$ rounds.

4.1   Long-Biclique

Our first paradigm aims to construct a biclique over the remaining $(r - m)$ rounds so that the basic meet-in-the-middle attack can be applied with negligible modification. The first advantage of this approach is that theoretically we can get the same advantage as the basic attack if we manage to construct a biclique of appropriate dimension. If the dimension is inevitably small due to the diffusion, then we use the second advantage: the biclique construction methods based on differential cryptanalysis of block ciphers and hash functions.

The disadvantage of this paradigm is that the construction of bicliques over many rounds is very difficult. Therefore, we are limited in the total number of rounds that we can attack. Furthermore, the data complexity can be very large since we use all the degrees of freedom to construct a biclique and may have nothing left to impose restrictions on the plaintexts or ciphertexts.

Nevertheless, we expect this paradigm to benefit from the further development of differential cryptanalysis and the inside-out strategy and predict its applicability to many other ciphers.

Hence, to check (4) the adversary selects an internal variable $v \in V$ that can be computed as follows for each key group $\{K[i,j]\}$:

$$P \xrightarrow[\mathcal{E}_1]{K[i,\cdot]} v \xleftarrow[\mathcal{E}_2]{K[\cdot,j]} S. \tag{9}$$

Therefore, the computational complexity of matching is upper bounded by 2^d computations of the cipher.

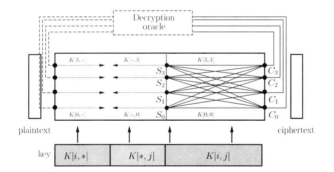

Fig. 3. Long-biclique attack with four states and four ciphertexts

Complexity of Key Recovery. Let us evaluate the full complexity of the long-biclique approach. Since the full key recovery is merely the application of Steps 1-3 2^{n-2d} times, we get the following equation:

$$C_{full} = 2^{n-2d} \left[C_{biclique} + C_{match} + C_{falsepos} \right],$$

where

- $C_{biclique}$ is the complexity of constructing a single biclique. Since the differential-based method is time-consuming, one has to amortize the construction cost by selecting a proper set of neutral bytes that do not affect the biclique equations.

- C_{match} is the complexity of the computation of the internal variable v 2^d times in each direction. It is upper bounded by 2^d calls of E.
- $C_{falsepos}$ is the complexity generated by false positives, which have to be matched on other variables. If we match on a single byte, the number of false positives is about 2^{2d-8}. Each requires only a few operations to re-check.

Generally, the complexity is dominated by C_{match} and hence has an advantage of at least 2^d over brute force. The memory complexity depends on the biclique construction procedure.

4.2   Independent-Biclique

Our second paradigm lets the attacker exploit the diffusion properties rather than differential properties, and does not aim to construct the longest biclique. In contrast, it proposes to construct shorter bicliques with high dimension by tools like independent related-key differentials (Section 3.1).

This approach has clear advantages. First, the data complexity can be made quite low. Since the biclique area is small, the attacker has more freedom to impose constraints on the ciphertext and hence restrict it to a particular set. Secondly, the attack gets a compact and small description, since the independent trails are generally short and self-explaining.

For further explanation, we recall the decomposition of the cipher:

$$E: \quad P \xrightarrow[\mathcal{E}_1]{} V \xrightarrow[\mathcal{E}_2]{} S \xrightarrow[\mathcal{E}_3]{} C,$$

In (4), the adversary detects the right key by computing an intermediate variable v in both directions:

$$P_i \xrightarrow[\mathcal{E}_1]{K[i,j]} \overrightarrow{v} \overset{?}{=} \overleftarrow{v} \xleftarrow[\mathcal{E}_2]{K[i,j]} S_j. \tag{10}$$

Since the meet-in-the-middle attack is no longer applicable to the $\mathcal{E}_2 \circ \mathcal{E}_1$, we apply the matching with precomputations (Section 3.3).

As with the long-biclique paradigm, 2^{2d} keys are tested using only 2^d intermediate cipher states. The precomputation of about 2^{d+1} matches allows for a significant complexity gain and is the major source of the computational advantage of our attacks on AES (Section 3.3). The advantage comes from the fact that in case of high dimension the basic computation has negligible cost, and the full complexity is determined by the amount of precomputation. By a careful choice of key groups, one is able to reduce the precomputation proportion to a very small factor, e.g. factor 1/15 in attacks on reduced-round versions of AES-256.

Complexity of Key Recovery. The full complexity of the independent biclique approach is evaluated as follows:

$$C_{full} = 2^{n-2d} \left[C_{biclique} + C_{precomp} + C_{recomp} + C_{falsepos} \right],$$

where

- $C_{precomp}$ is the complexity of the precomputation in Step 3. It is equivalent to less than 2^d runs of the subcipher g.
- C_{recomp} is the complexity of the recomputation of the internal variable v 2^{2d} times. It strongly depends on the diffusion properties of the cipher. For AES this value varies from $2^{2d-1.5}$ to 2^{2d-4}.

The biclique construction is quite cheap in this paradigm. The method in Section 3.1 enables construction of a biclique in only 2^{d+1} calls of subcipher f. Therefore, usually the full key recovery complexity will be dominated by $2^{n-2d} \cdot C_{recomp}$. However, it is dependent on the width of the matching variable and biclique dimension d too. We give more details for the case of AES in further sections. The memory complexity of the key recovery is upper-bounded by storing 2^d full computations of the cipher.

5   Description of AES

AES is a block cipher with 128-bit internal state and 128/192/256-bit key K (AES-128, AES-192, AES-256, respectively). The internal state is represented by a 4×4 byte matrix, and the key is represented by a $4 \times 4/4 \times 6/4 \times 8$ matrix.

The encryption works as follows. The plaintext is xored with the key, and then undergoes a sequence of 10/12/14 rounds. Each round consists of four transformations: nonlinear bytewise SubBytes, the byte permutation ShiftRows, linear transformation MixColumns, and the addition with a subkey AddRoundKey. MixColumns is omitted in the last round.

SubBytes is a nonlinear transformation operating on 8-bit S-boxes with maximum differential probability as low as 2^{-6} (for most cases 0 or 2^{-7}). The ShiftRows rotates bytes in row r by r positions to the left. The MixColumns is a linear transformation with branch number 5, i.e. in the column equation $(y_0, y_1, y_2, y_3) = MC(x_0, x_1, x_2, x_3)$ only 5 and more variables can be non-zero.

We address two internal states in each round as follows in AES-128: #1 is the state before SubBytes in round 1, #2 is the state after MixColumns in round 1, #3 is the state before SubBytes in round 2, ..., #19 is the state before SubBytes in round 10, #20 is the state after ShiftRows in round 10 (MixColumns is omitted in the last round). The states in the last round of AES-192 are addressed as #23 and #24, and of AES-256 as #27 and #28.

The subkeys come out of the key schedule procedure, which slightly differs for each version of AES. The key K is expanded to a sequence of keys $K^0, K^1, K^2, \ldots, K^{10}$, which form a 4×60 byte array. Then the 128-bit subkeys $\$0, \$1, \$2, \ldots, \$14$ come out of the sliding window with a 4-column step. The keys in the expanded key are formed as follows. First, $K^0 = K$. Then, column 0 of K^r is the column 0 of K^{r-1} xored with the nonlinear function (SK) of the last column of K^{r-1}. Subsequently, column i of K^r is the xor of column $i-1$ of K^r and of column i of K^{r-1}. In AES-256 column 3 undergoes SubBytes transformation while forming column 4.

Bytes within a state and a subkey are enumerated as follows

0	4	8	12
1	5	9	13
2	6	10	14
3	7	11	15

Byte i in state Q is addressed as Q_i.

6   Independent-Biclique: Key Recovery for the Full AES-128

In this section we describe a key recovery method on the full 10-round AES-128 using the independent-bilclique approach. The computational bottleneck will be the matching computation. See also Appendix A for an additional illustration.

Table 3. Parameters of the key recovery for the full AES-128

f		**Biclique**			
Rounds	Dimension	Δ^K bytes	∇^K bytes	Time	Memory
8-10	8	$\$8_8, \$8_{12}$	$\$8_1, \$8_9$	2^7	2^8

		Matching			
	g	Precomputation		Recomputation	
Rounds	v	Workload	Memory	SubBytes: forward	SubBytes: backward
1-7	$\#5_{12}$	$2^{8-\varepsilon}$	2^8	0.875	2.625

			Total complexity		
Memory	$C_{biclique}$	$C_{precomp}$	C_{recomp}	$C_{falsepos}$	C_{full}
2^8	2^7	2^7	$2^{14.14}$	2^8	$2^{126.18}$

6.1   Key Partitioning

For more clarity we define the key groups with respect to the subkey $8 of round 8 and enumerate the groups of keys by 2^{112} *base keys*. Since the AES-128 key schedule bijectively maps each key to $8, the enumeration is well-defined. The base keys $K[0,0]$ are all possible 2^{112} 16-byte values with two bytes fixed to 0 whereas the remaining 14 bytes run over all values:

The keys $\{K[i,j]\}$ in a group are enumerated by all possible byte differences i and j with respect to the base key $K[0,0]$:

This yields the partition of the round-8 subkey space, and hence the AES key space, into the 2^{112} groups of 2^{16} keys each.

6.2   3-Round Biclique of Dimension 8

We construct a 3-round biclique from combined related-key differentials as described in Section 3.1. The parameters of the key recovery are summarized in Table 3. The adversary fixes $C_0 = 0$ and derives $S_0 = f^{-1}_{K[0,0]}(C_0)$ (Figure 4, left). The Δ_i-differentials are based on the difference Δ_i^K in $8, and ∇_j-differentials are based on the difference ∇_j^K in $8:

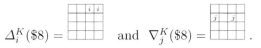

$$\Delta_i^K(\$8) = \qquad \text{and} \quad \nabla_j^K(\$8) = \qquad .$$

Both sets of differentials are depicted in Figure 4 in the truncated form. As they share no active S-boxes, the resulting combined differentials yield a biclique of dimension 8.

Since the Δ_i-differential affects only 12 bytes of the ciphertext, all the ciphertexts share the same values in bytes $C_{0,1,4,13}$. Furthermore, since $\Delta_i^K(\$10_{10}) = \Delta_i^K(\$10_{14})$, the ciphertext bytes C_{10} and C_{14} are also always equal. As a result, the data complexity does not exceed 2^{88}.

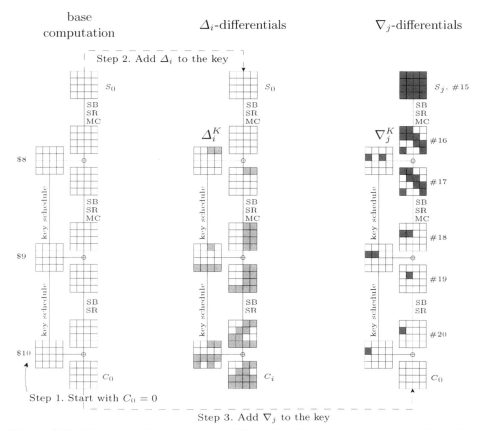

Fig. 4. AES-128 biclique from combined differentials: base computation as well as Δ_i- and ∇_j-differentials

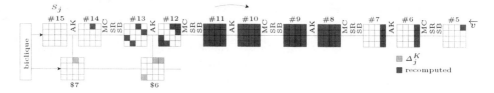

Fig. 5. Recomputation in the backward direction: AES-128

Forward computation. Now we figure out how the computation $P_i \xrightarrow{K[i,j]} \overrightarrow{v}$ differs from the stored one $P_i \xrightarrow{K[i,0]} \overrightarrow{v}_i$. Similarly, it is determined by the influence of the difference between keys $K[i,j]$ and $K[i,0]$, now applied to the plaintext. Thanks to the low diffusion of the AES key schedule and sparsity of the key difference in round 8, the whitening subkeys of $K[i,j]$ and $K[i,0]$ differ in 9 bytes only. The difference is no longer a linear function of j as it is in the computation of $\overleftarrow{v}$, but still requires only three s-boxes in the key schedule to recompute. The areas of internal states to be recomputed (with 13 S-boxes) are depicted in Figure 6.

6.3   Matching over 7 Rounds

Now we check whether the secret key K_{secret} belongs to the key group $\{K[i,j]\}$ according to Section 3.3. We make 2^{d+1} precomputations of v and store values as well as the intermediate states and subkeys in memory. Then we check (10) for every i, j by recomputing only those variables that differ from the ones stored in memory. Now we evaluate the amount of recomputation in both directions.

Backward direction. Let us figure out how the computation $\overleftarrow{v} \xleftarrow{K[i,j]} S_j$ differs from the stored one $\overleftarrow{v}_j \xleftarrow{K[0,j]} S_j$. It is determined by the influence of the difference between keys $K[i,j]$ and $K[0,j]$ (see the definition of the key group in Section 6.1). The difference in the subkey \$7 is non-zero in only one byte, so we have to recompute as few as four S-boxes in round 7 (state #13). The full area to be recomputed, which includes 41 S-boxes, is depicted in Figure 5. Note that the difference in the relevant subkeys is a linear function of i, and hence can be precomputed and stored.

Forward computation. Now we look at how the computation $P_i \xrightarrow{K[i,j]} \overrightarrow{v}$ differs from the stored one $P_i \xrightarrow{K[i,0]} \overrightarrow{v}_i$. Similarly, it is determined by the influence of the difference between keys $K[i,j]$ and $K[i,0]$, now applied to the plaintext. Thanks to the low diffusion of the AES key schedule and sparsity of the key difference in round 8, the whitening subkeys of $K[i,j]$ and $K[i,0]$ differ in 9 bytes only. The difference is no longer a linear function of j as it is involved into the computation of $\overleftarrow{v}$, but still requires only three S-boxes in the key schedule to recompute. This effect and the areas of internal states to be recomputed (with 13 S-boxes) are depicted in Figure 6.

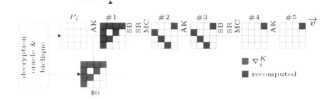

Fig. 6. Recomputation in the forward direction: AES-128

6.4   Complexities

Since only a portion of the round function is recomputed, one has to be highly accurate in evaluating the complexity C_{recomp}. A rough division of AES-128 into 10 rounds is not precise enough. For a more exact evaluation, we count the number of S-boxes in each SubBytes operation that we have to recompute, the number of active variables in MixColumns, the number of output variables that we need from MixColumns, and, finally, the number of S-boxes to recompute in the key schedule.

Altogether, we need an equivalent of 3.4375 SubBytes operations (i.e., 55 S-boxes), 2.3125 MixColumns operations, and a negligible amount of XORs in the key schedule. The number of SubBytes computations clearly is a larger summand. S-boxes are also the major contributor to the practical complexity of AES both in hardware and software. Therefore, if we aim for a single number that refers to the complexity, it makes sense to count the number of SubBytes operations that we need and compare it to that in the full cipher. The latter number is $10 + 2.5 = 12.5$ as we have to take the key schedule nonlinearity into account. As a result, C_{recomp} is equivalent to $2^{16} \cdot 3.4375/12.5 = 2^{14.14}$ runs of the full AES-128. The values $C_{biclique}$ and $C_{precomp}$ together do not exceed 2^8 calls of the full AES-128.

The full computational complexity amounts to about

$$2^{112} \left(2^7 + 2^7 + 2^{14.14} + 2^8 \right) = 2^{126.18}.$$

The memory requirement is upper-bounded by the storage of 2^8 full computations of g. Since the coverage of the key space by groups around base keys is complete, the success probability is 1.

This approach for 8-round AES-128 yields a key recovery with computational complexity about $2^{125.34}$, data complexity 2^{88}, memory complexity 2^8, and success probability 1. Similarly, preimage finding for the compression function of the full AES-128 in Miyaguchi-Preneel mode requires about $2^{125.83}$ computations, 2^8 memory, and has a success probability of about 0.6321.

7   Long-Biclique: 9-Round AES-256

Our attack is differential-based biclique attack (Section 3.2).

Step 1. A biclique of dimension 1 involves two states, two ciphertexts, and a group of four keys. The keys in the group are defined via the difference in subkeys:

$$K[0,1]: \qquad \$5(K[0,1]) \oplus \$5(K[0,0]) = \Delta K;$$
$$K[1,0]: \qquad \$6(K[1,0]) \oplus \$6(K[0,0]) = \nabla K;$$
$$K[1,1]: \qquad \$6(K[1,1]) \oplus \$6(K[0,1]) = \nabla K.$$

The differences ΔK and ∇K are defined columnwise:

$$\Delta K = (A, \overline{0}, \overline{0}, \overline{0}); \quad \nabla K = (B, B, \overline{0}, \overline{0}),$$

where

$$A = \text{MixColumns} \begin{pmatrix} 0 \\ 0 \\ 2 \\ 0 \end{pmatrix}; \quad B = \begin{pmatrix} 0 \\ 2 \\ 0\text{xb9} \\ 2 \end{pmatrix} = \text{MixColumns} \begin{pmatrix} 0\text{xd0} \\ 0\text{x69} \\ 0 \\ 0 \end{pmatrix}.$$

Let us note that the key relation in the next expanded key is still linear:

$$\$4(K[1,0]) \oplus \$4(K[0,0]) = \$4(K[1,1]) \oplus \$4(K[0,1]) = (B, \overline{0}, \overline{0}, \overline{0}).$$

Evidently, the groups do not intersect and cover the full key space. We split the 9-round AES-256 as follows:

- $\mathcal{E}_1$ is round 1.
- $\mathcal{E}_2$ is rounds 2-4.
- $\mathcal{E}_3$ is rounds 5-9.

Step 2. An illustration of steps 2(a) - 2(e) is given in Fig. 7.

Step 2 (a). The intermediate state T in $\mathcal{E}_3$ is the S-box layer in round 7. We construct truncated differential trails in rounds 5-6 based on the injection of ΔK after round 5 (Figure 7, left), and in rounds 7-9 based on the injection of ∇K before round 9 (Figure 7, right).

Step 2 (b). We guess the differences in the truncated trails up to T. We have four active S-boxes in round 6 and two active S-boxes in round 8. We also require Δ-trails to be equal. In total we make $2^{7 \cdot (4+2 \cdot 2)} = 2^{56}$ guesses.

Step 2 (c). For each S-box in round 7 that is active in both trails (eight in total) we take a quartet of values that conform to the input and output differences, being essentially the boomerang quartet for the S-box (one solution per S-box on average). For the remaining 8 S-boxes we take all possible values. Therefore, we have 2^{64} solutions for each guess in the inbound phase, or 2^{120} solutions in total.

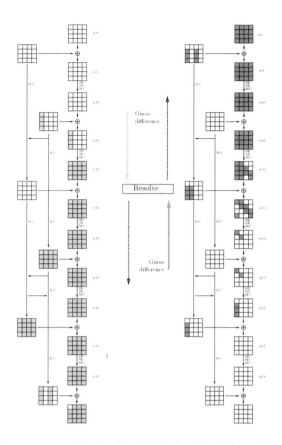

Fig. 7. Biclique construction in AES-256. Δ-trail (left) and ∇-trail (right).

Step 2 (d). Outbound phase: we filter out the solutions that do not conform to the differential trails in rounds 6 and 8. We have four active S-boxes in each Δ-trail, and two active S-boxes in each ∇-trail, hence 12 in total. Therefore, we get a 84-bit filter, and leave with 2^{36} bicliques.

Step 2 (e). Now we keep only the bicliques with byte $C_{0,0}$ equal to zero in both ciphertexts. This is a 16-bit filter, which reduces the number of bicliques to 2^{20}. We need only one.

Step 3-5. We ask for the decryption of two ciphertexts and get two plaintexts. The matching position (v) is the byte $\#3_{0,0}$. As demonstrated in Fig. 8, it is equal as a function of the plaintext for keys with difference ΔK (not affected by lightblue cells), and is also equal as a function of S for keys with difference ∇K (not affected by red cells). We compute v in both directions and check for the match.

Step 6. We can produce sufficiently many bicliques out of one to amortize the construction cost. Let us look at the subkey \$6 in the outbound phase. We can change its value to any of the 2^{96} specific values so that the active S-boxes in round 6 during the outbound phase are not affected. On the other hand, any change in bytes in rows 1,2,3 affects only those rows in the subkeys \$8 and \$9 and hence does not affect $C_{0,0}$. Therefore, we have $128 - 32 - 32 = 64$ neutral bits in \$6.

Similarly, we identify 9 bytes in \$7 that can be changed so that \$6, the active S-boxes in round 8, and the byte $C_{0,0}$ are unaffected. Those are bytes in the first three columns not on the main diagonal. Therefore, we have 72 neutral bits in \$7, and 136 neutral bits in total.

Complexity. A single biclique with $C_{0,0} = 0$ is constructed with complexity $2^{120-20} = 2^{100}$ and 2^8 memory needed for Step 2 (c). However, 136 neutral bits in the key reduce the amortized construction cost significantly. Let us compute the cost of constructing a new biclique according to Step 6. A change in a single byte in K^7 needs 5 S-boxes, 1 MC and several XORs recomputing for each ciphertext, which gives us the complexity of 10/16 AES rounds. This change also affects two bytes of K^5, so we have to recompute one half of round 5, with the resulting complexity of 1 AES round per biclique. The total amortized complexity is 1.625 AES rounds.

In the matching part we compute a single byte in two directions, thus spending 9/16 of a round in rounds 1-3, and full round 4, i.e. 3.125 full rounds per biclique. In total we need 4.75 AES rounds per biclique, i.e. $2^{-0.92}$ 9-round AES-256 calls. The complexity generated by false positives is at most 2^{-6} rounds per biclique. We need 2^{254} bicliques, so the total complexity is $2^{253.1}$.

The data complexity is 2^{120} since one ciphertext byte is always fixed. The success rate of the attack is 1, since we can generate many bicliques for each key group.

8   On Practical Verification

Especially for the type of cryptanalysis described in this paper where carrying out an attack in full is computationally infeasible, practical verification of attack details and steps is important in order to get confidence in it. To address this, we explicitly state the following:

- We verified all truncated differentials through AES-128/192/256 for all the attacks, including the independent bicliques.
- We constructed a real 6-round biclique for the 9-round AES-256 (Table 6). To make the algorithm in Section 7 practical, we fixed more key bytes than required. As a result, the construction cost for a single biclique dropped, but the amortized cost has increased.
- We verified that some difference guesses must be equal (like in the AES-256 attack) due to the branch number of MixColumns that results in a correlation of differences in the outbound phase.

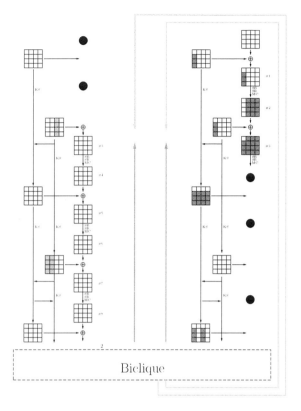

Fig. 8. Matching in AES-256. Byte $\#3_0$ can be computed in each direction.

9   Discussion and Conclusions

We propose the concept of bicliques for block cipher cryptanalysis and give various applications to AES, including a key recovery method for the full versions of AES-128, AES-192, and AES-256. Both the "long-biclique" and the "independent-biclique" approach we introduced feature conceptual novelties that we expect will find applications in other areas. For the "long-biclique" approach, it is the use of techniques from differential collision attacks on hash functions that forces two trails to be independent and hence allows to add more rounds at low amortized cost. For the "independent-biclique" approach, it is the *matching with precomputation* trick that allows to significantly reduce the cost of matching computations over more rounds in a MITM attack.

Using the latter approach on AES, we allow a small portion of the cipher to ie recomputed in every key test. The use of bicliques in combination with the technique of matching with precomputation, results in a surprisingly low recomputation in the innermost loop, varying from about 1/3 to approximately

1/5 of the cipher depending on the key size, while having data complexities of 2^{88}, 2^{80} and 2^{40} plaintext-ciphertext pairs, respectively. Arguably no known generic approach to key recovery allows for that gain. We notice that the data complexity of key recovery can be significantly reduced by sacrificing only a small factor of computational advantage.

To conclude, we discuss the properties of AES that allowed us to cover more rounds than in previous cryptanalysis, discuss the attained computational advantage, and list a number of problems to consider for future work.

9.1   What Properties of the AES Allowed to Obtain These New Results

Our approach heavily relies on the existence of high-probability related-key differentials over a part of the cipher. More specifically:

- The round transformation of AES is not designed to have strong resistance against several classes of attacks for a smaller number of rounds. The fact that our approach allows to split up the cipher into three parts exposes these properties even when considering the full cipher. Also, as already observed in [21, 42], the fact that the MixColumns transformation is omitted in the last round of AES helps to design attacks for more rounds.
- In the key schedule, we especially take advantage of the relatively slow backward diffusion. Whereas using key-schedule properties in related-key attacks is natural, there seem only a few examples in the literature where this is used in the arguably more relevant single-key setting. This includes the attack on the self-synchronized stream cipher Moustique [30], the lightweight block cipher KTANTAN [13], and recent improvements upon attacks on 8-rounds of AES-192 and AES-256 [22].

9.2   On the Computational Advantage of the Biclique Techniques

Most computational complexities in this paper are relatively close to those of generic attacks. In here we discuss why we think the complexity advantage is meaningful.

- Biclique cryptanalysis with the independent-biclique approach allows us to be very precise about the required computations. In all cases we arrive at computational complexities considerably lower than those of generic attacks.
- For long-biclique cryptanalysis, whenever it is difficult to be precise about certain parts of our estimates, we choose to be conservative, potentially resulting in an underestimate of the claimed improvement. Again, in all cases we arrive at a computational complexity that is considerably lower than that of generic attacks.
- Improved AES implementations (that may e.g. be used to speed-up brute force key search) will very likely also improve the biclique techniques we propose.
- To the best of our knowledge, there are no generic methods known that would speed-up key recovery given a part of the codebook.

9.3   Open Problems

There are a number of other settings this approach may be applied to. It will be interesting to study other block ciphers like the AES finalists or more recent proposals with respect to this class of attacks. A combination of the "long-biclique" and "independent-biclique" approaches may be a source for further improvements. Also, we may decide to drop the requirement of the biclique to be complete, i.e. instead of a complete bipartite graph consider a more general graph. There may be cases where different tradeoffs between success probability, complexity requirements, and even number of rounds are obtainable. Alternatively, this paper may inspire work on more generic attacks on block ciphers that try to take advantage of the fact that a small part of the codebook, or some memory, is available.

Acknowledgements. We thank Joan Daemen and Vincent Rijmen for their helpful feedback on the earlier versions of the paper. We also thank Pierre-Alain Fouque, Alexander Gotmanov, Gregor Leander, Søren Thomsen, and reviewers of ASIACRYPT 2011 for their comments. Part of this work was done while Andrey Bogdanov was visiting MSR Redmond and while Christian Rechberger was with K.U.Leuven and visiting MSR Redmond. This work was supported in part by the European Commission under contract ICT-2007-216646 ECRYPT NoE phase II and by the IAP Programme P6/26 BCRYPT of the Belgian State.

References

1. Aoki, K., Sasaki, Y.: Preimage Attacks on One-Block MD4, 63-Step MD5 and More. In: Avanzi, R.M., Keliher, L., Sica, F. (eds.) SAC 2008. LNCS, vol. 5381, pp. 103–119. Springer, Heidelberg (2009)
2. Aoki, K., Sasaki, Y.: Meet-in-the-Middle Preimage Attacks Against Reduced SHA-0 and SHA-1. In: Halevi, S. (ed.) CRYPTO 2009. LNCS, vol. 5677, pp. 70–89. Springer, Heidelberg (2009)
3. Bahrak, B., Aref, M.R.: A Novel Impossible Differential Cryptanalysis of AES. In: Lucks, S., Sadeghi, A.-R., Wolf, C. (eds.) WEWoRC 2007. LNCS, vol. 4945, pp. 152–156. Springer, Heidelberg (2008)
4. Bahrak, B., Aref, M.R.: Impossible differential attack on seven-round aes-128. IET Inf. Secur. 2(2), 28–32 (2008)
5. Biham, E., Biryukov, A., Shamir, A.: Miss in the Middle Attacks on IDEA and Khufu. In: Knudsen, L.R. (ed.) FSE 1999. LNCS, vol. 1636, pp. 124–138. Springer, Heidelberg (1999)
6. Biham, E., Chen, R., Joux, A., Carribault, P., Lemuet, C., Jalby, W.: Collisions of SHA-0 and Reduced SHA-1. In: Cramer, R. (ed.) EUROCRYPT 2005. LNCS, vol. 3494, pp. 36–57. Springer, Heidelberg (2005)
7. Biham, E., Shamir, A.: Differential Cryptanalysis of DES-like Cryptosystems. J. Cryptology 4(1), 3–72 (1991)
8. Biryukov, A., Dunkelman, O., Keller, N., Khovratovich, D., Shamir, A.: Key Recovery Attacks of Practical Complexity on AES-256 Variants with up to 10 Rounds. In: Gilbert, H. (ed.) EUROCRYPT 2010. LNCS, vol. 6110, pp. 299–319. Springer, Heidelberg (2010)

9. Biryukov, A., Khovratovich, D.: Related-Key Cryptanalysis of the Full AES-192 and AES-256. In: Matsui, M. (ed.) ASIACRYPT 2009. LNCS, vol. 5912, pp. 1–18. Springer, Heidelberg (2009)

10. Biryukov, A., Khovratovich, D., Nikolić, I.: Distinguisher and Related-Key Attack on the Full AES-256. In: Halevi, S. (ed.) CRYPTO 2009. LNCS, vol. 5677, pp. 231–249. Springer, Heidelberg (2009)

11. Biryukov, A., Nikolić, I.: Automatic Search for Related-Key Differential Characteristics in Byte-Oriented Block Ciphers: Application to AES, Camellia, Khazad and Others. In: Gilbert, H. (ed.) EUROCRYPT 2010. LNCS, vol. 6110, pp. 322–344. Springer, Heidelberg (2010)

12. Bogdanov, A., Khovratovich, D., Rechberger, C.: Biclique Cryptanalysis of the Full AES. Cryptology ePrint Archive, Report 2011/449 (2011), http://eprint.iacr.org/2011/449

13. Bogdanov, A., Rechberger, C.: A 3-Subset Meet-in-the-Middle Attack: Cryptanalysis of the Lightweight Block Cipher KTANTAN. In: Biryukov, A., Gong, G., Stinson, D.R. (eds.) SAC 2010. LNCS, vol. 6544, pp. 229–240. Springer, Heidelberg (2011)

14. Bouillaguet, C., Derbez, P., Fouque, P.-A.: Automatic Search of Attacks on Round-Reduced AES and Applications. In: Rogaway, P. (ed.) CRYPTO 2011. LNCS, vol. 6841, pp. 169–187. Springer, Heidelberg (2011)

15. Chabaud, F., Joux, A.: Differential Collisions in SHA-0. In: Krawczyk, H. (ed.) CRYPTO 1998. LNCS, vol. 1462, pp. 56–71. Springer, Heidelberg (1998)

16. Chaum, D., Evertse, J.-H.: Crytanalysis of DES with a Reduced Number of Rounds: Sequences of Linear Factors in Block Ciphers. In: Williams, H.C. (ed.) CRYPTO 1985. LNCS, vol. 218, pp. 192–211. Springer, Heidelberg (1986)

17. Daemen, J., Knudsen, L.R., Rijmen, V.: The Block Cipher SQUARE. In: Biham, E. (ed.) FSE 1997. LNCS, vol. 1267, pp. 149–165. Springer, Heidelberg (1997)

18. Daemen, J., Rijmen, V.: The Design of Rijndael: AES - The Advanced Encryption Standard. Springer, Heidelberg (2002)

19. Demirci, H., Selçuk, A.A.: A Meet-in-the-Middle Attack on 8-Round AES. In: Nyberg, K. (ed.) FSE 2008. LNCS, vol. 5086, pp. 116–126. Springer, Heidelberg (2008)

20. Demirci, H., Taşkın, İ., Çoban, M., Baysal, A.: Improved Meet-in-the-Middle Attacks on AES. In: Roy, B., Sendrier, N. (eds.) INDOCRYPT 2009. LNCS, vol. 5922, pp. 144–156. Springer, Heidelberg (2009)

21. Dunkelman, O., Keller, N.: The effects of the omission of last round's MixColumns on AES. Inf. Process. Lett. 110(8-9), 304–308 (2010)

22. Dunkelman, O., Keller, N., Shamir, A.: Improved Single-Key Attacks on 8-Round AES-192 and AES-256. In: Abe, M. (ed.) ASIACRYPT 2010. LNCS, vol. 6477, pp. 158–176. Springer, Heidelberg (2010)

23. Dunkelman, O., Keller, N., Shamir, A.: A Practical-Time Related-Key Attack on the KASUMI Cryptosystem Used in GSM and 3G Telephony. In: Rabin, T. (ed.) CRYPTO 2010. LNCS, vol. 6223, pp. 393–410. Springer, Heidelberg (2010)

24. Dunkelman, O., Sekar, G., Preneel, B.: Improved Meet-in-the-Middle Attacks on Reduced-Round DES. In: Srinathan, K., Rangan, C.P., Yung, M. (eds.) INDOCRYPT 2007. LNCS, vol. 4859, pp. 86–100. Springer, Heidelberg (2007)

25. Ferguson, N., Kelsey, J., Lucks, S., Schneier, B., Stay, M., Wagner, D., Whiting, D.: Improved Cryptanalysis of Rijndael. In: Schneier, B. (ed.) FSE 2000. LNCS, vol. 1978, pp. 213–230. Springer, Heidelberg (2001)

26. Gilbert, H., Minier, M.: A Collision Attack on 7 Rounds of Rijndael. In: AES Candidate Conference, pp. 230–241 (2000)

27. Gilbert, H., Peyrin, T.: Super-Sbox Cryptanalysis: Improved Attacks for AES-Like Permutations. In: Hong, S., Iwata, T. (eds.) FSE 2010. LNCS, vol. 6147, pp. 365–383. Springer, Heidelberg (2010)
28. Guo, J., Ling, S., Rechberger, C., Wang, H.: Advanced Meet-in-the-Middle Preimage Attacks: First Results on Full Tiger, and Improved Results on MD4 and SHA-2. In: Abe, M. (ed.) ASIACRYPT 2010. LNCS, vol. 6477, pp. 56–75. Springer, Heidelberg (2010)
29. Isobe, T.: A Single-Key Attack on the Full GOST Block Cipher. In: Joux, A. (ed.) FSE 2011. LNCS, vol. 6733, pp. 290–305. Springer, Heidelberg (2011)
30. Käsper, E., Rijmen, V., Bjørstad, T.E., Rechberger, C., Robshaw, M.J.B., Sekar, G.: Correlated Keystreams in MOUSTIQUE. In: Vaudenay, S. (ed.) AFRICACRYPT 2008. LNCS, vol. 5023, pp. 246–257. Springer, Heidelberg (2008)
31. Khovratovich, D., Rechberger, C., Savelieva, A.: Bicliques for preimages: attacks on Skein-512 and the SHA-2 family (2011), http://eprint.iacr.org/2011/286.pdf
32. Knudsen, L.R., Rijmen, V.: Known-Key Distinguishers for Some Block Ciphers. In: Kurosawa, K. (ed.) ASIACRYPT 2007. LNCS, vol. 4833, pp. 315–324. Springer, Heidelberg (2007)
33. Lamberger, M., Mendel, F., Rechberger, C., Rijmen, V., Schläffer, M.: Rebound Distinguishers: Results on the Full Whirlpool Compression Function. In: Matsui, M. (ed.) ASIACRYPT 2009. LNCS, vol. 5912, pp. 126–143. Springer, Heidelberg (2009)
34. Lu, J., Dunkelman, O., Keller, N., Kim, J.: New Impossible Differential Attacks on AES. In: Chowdhury, D.R., Rijmen, V., Das, A. (eds.) INDOCRYPT 2008. LNCS, vol. 5365, pp. 279–293. Springer, Heidelberg (2008)
35. Lucks, S.: Attacking seven rounds of Rijndael under 192-bit and 256-bit keys. In: AES Candidate Conference, pp. 215–229 (2000)
36. Mala, H., Dakhilalian, M., Rijmen, V., Modarres-Hashemi, M.: Improved Impossible Differential Cryptanalysis of 7-Round AES-128. In: Gong, G., Gupta, K.C. (eds.) INDOCRYPT 2010. LNCS, vol. 6498, pp. 282–291. Springer, Heidelberg (2010)
37. Matsui, M.: Linear Cryptanalysis Method for DES Cipher. In: Helleseth, T. (ed.) EUROCRYPT 1993. LNCS, vol. 765, pp. 386–397. Springer, Heidelberg (1994)
38. Mendel, F., Peyrin, T., Rechberger, C., Schläffer, M.: Improved Cryptanalysis of the Reduced Grøstl Compression Function, ECHO Permutation and AES Block Cipher. In: Jacobson Jr., M.J., Rijmen, V., Safavi-Naini, R. (eds.) SAC 2009. LNCS, vol. 5867, pp. 16–35. Springer, Heidelberg (2009)
39. Mendel, F., Rechberger, C., Schläffer, M., Thomsen, S.S.: The Rebound Attack: Cryptanalysis of Reduced Whirlpool and Grøstl. In: Dunkelman, O. (ed.) FSE 2009. LNCS, vol. 5665, pp. 260–276. Springer, Heidelberg (2009)
40. Phan, R.C.-W.: Impossible differential cryptanalysis of 7-round advanced encryption standard (AES). Inf. Process. Lett. 91(1), 33–38 (2004)
41. Rechberger, C.: Preimage Search for a Class of Block Cipher based Hash Functions with Less Computation (2008) (unpublished manuscript)
42. Sasaki, Y.: Meet-in-the-Middle Preimage Attacks on AES Hashing Modes and an Application to Whirlpool. In: Joux, A. (ed.) FSE 2011. LNCS, vol. 6733, pp. 378–396. Springer, Heidelberg (2011)
43. Sasaki, Y., Aoki, K.: Finding Preimages in Full MD5 Faster Than Exhaustive Search. In: Joux, A. (ed.) EUROCRYPT 2009. LNCS, vol. 5479, pp. 134–152. Springer, Heidelberg (2009)

44. Stevens, M., Lenstra, A.K., de Weger, B.: Chosen-Prefix Collisions for MD5 and Colliding X.509 Certificates for Different Identities. In: Naor, M. (ed.) EUROCRYPT 2007. LNCS, vol. 4515, pp. 1–22. Springer, Heidelberg (2007)
45. Stevens, M., Sotirov, A., Appelbaum, J., Lenstra, A.K., Molnar, D., Osvik, D.A., de Weger, B.: Short Chosen-Prefix Collisions for MD5 and the Creation of a Rogue CA Certificate. In: Halevi, S. (ed.) CRYPTO 2009. LNCS, vol. 5677, pp. 55–69. Springer, Heidelberg (2009)
46. Wagner, D.: The Boomerang Attack. In: Knudsen, L.R. (ed.) FSE 1999. LNCS, vol. 1636, pp. 156–170. Springer, Heidelberg (1999)
47. Wang, X., Yin, Y.L., Yu, H.: Finding Collisions in the Full SHA-1. In: Shoup, V. (ed.) CRYPTO 2005. LNCS, vol. 3621, pp. 17–36. Springer, Heidelberg (2005)
48. Wang, X., Yu, H.: How to Break MD5 and Other Hash Functions. In: Cramer, R. (ed.) EUROCRYPT 2005. LNCS, vol. 3494, pp. 19–35. Springer, Heidelberg (2005)
49. Wei, L., Rechberger, C., Guo, J., Wu, H., Wang, H., Ling, S.: Improved meet-in-the-middle cryptanalysis of KTANTAN. Cryptology ePrint Archive, Report 2011/201 (2011), http://eprint.iacr.org/
50. Zhang, W., Wu, W., Feng, D.: New Results on Impossible Differential Cryptanalysis of Reduced AES. In: Nam, K.-H., Rhee, G. (eds.) ICISC 2007. LNCS, vol. 4817, pp. 239–250. Springer, Heidelberg (2007)

A   Additional Illustration for the Case of Full AES-128

In Figure 9 we give an additional illustration of key recovery for the full AES-128 described in Section 6. It demonstrates biclique differentials, influence of key differences in matching, and the recomputations.

The influence of key differences in the matching part can be described as a truncated differential that starts with a zero difference in the plaintext (forward matching) or in the state (backward matching). Since both biclique and matching result from the same key differences, it is natural to depict the related differentials in the same computational flow (left and center schemes in Figure 9). We stress that the full 10-round picture does not represent a single differential trail, but it is rather a concatenation of trails in rounds 1–7 and 8–10, respectively.

The biclique differentials are depicted in pink (left, Δ-trail) and lightblue (center, ∇-trail) colors. The same for the matching: pink is the influence of ΔK on the backward computation, and lightblue is the influence of ∇K on the forward computation. The recomputation parts are derived as follows: formally overlap pink and blue schemes, then interleaving parts must be recomputed (darkgray cells). The lightgray cells are those excluded from recomputation since we do not match on the full state.

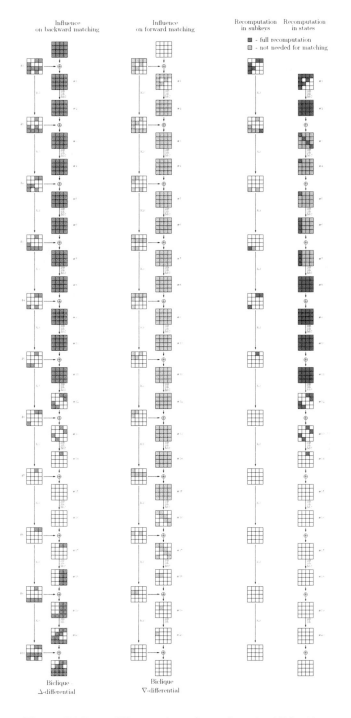

Fig. 9. Biclique differentials and matching in AES-128

Table 4. Summary of previous results on AES in the single-secret-key model for 7 or more rounds

rounds	data	workload	memory	method	reference
			AES-128		
7	$2^{127.997}$	2^{120}	2^{64}	Square	[25], 2000
7	2^{32}	$2^{128-\epsilon}$	2^{100}	Square-functional	[26], 2000
7	$2^{117.5}$	2^{123}	2^{109}	Impossible	[3], 2007
7	$2^{115.5}$	2^{119}	2^{45}	Impossible	[50], 2007
7	$2^{115.5}$	2^{119}	2^{109}	Impossible	[4], 2008
7	$2^{112.2}$	$2^{112} + 2^{117.2}$MA	2^{109}?	Impossible	[34] 2008
7	2^{80}	$2^{113}+2^{123}$ precomp.	2^{122}	MitM	[20], 2009
7	$2^{106.2}$	$2^{107.1} + 2^{117.2}$MA	$2^{94.2}$	Impossible	[36], 2010
7	2^{103}	2^{116}	2^{116}	Square-multiset	[22], 2010
			AES-192		
7	$2^{127.997}$	2^{120}	2^{64}	Square	[25], 2000
7	2^{36}	2^{155}	2^{32}	Square	[25], 2000
7	2^{32}	2^{182}	2^{32}	Square	[35], 2000
7	2^{32}	2^{140}	2^{84}	Square-functional	[26], 2000
7	2^{92}	2^{186}	2^{153}	Impossible	[40], 2004
7	$2^{115.5}$	2^{119}	2^{45}	Impossible	[50], 2007
7	2^{92}	2^{162}	2^{153}	Impossible	[50], 2007
7	$2^{91.2}$	$2^{139.2}$	2^{61}	Impossible	[34] 2008
7	$2^{113.8}$	$2^{118.8}$MA	$2^{89.2}$	Impossible	[34] 2008
7	2^{34+n}	$2^{74+n}+2^{208-n}$ precomp.	2^{206-n}	MitM	[19], 2008
7	2^{80}	$2^{113}+2^{123}$ precomp.	2^{122}	MitM	[20], 2009
7	2^{103}	2^{116}	2^{116}	Square-multiset	[22], 2010
8	$2^{127.997}$	2^{188}	2^{64}	Square	[25], 2000
8	2^{113}	2^{172}	2^{129}	Square-multiset	[22], 2010
			AES-256		
7	2^{36}	2^{172}	2^{32}	Square	[25], 2000
7	$2^{127.997}$	2^{120}	2^{64}	Square	[25], 2000
7	2^{32}	2^{200}	2^{32}	Square	[35], 2000
7	2^{32}	2^{184}	2^{140}	Square-functional	[26], 2000
7	$2^{92.5}$	$2^{250.5}$	2^{153}	Impossible	[40], 2004
7	$2^{115.5}$	2^{119}	2^{45}	Impossible	[50], 2007
7	$2^{113.8}$	$2^{118.8}$MA	$2^{89.2}$	Impossible	[34] 2008
7	2^{92}	2^{163}MA	2^{61}	Impossible	[34] 2008
7	2^{34+n}	$2^{74+n}+2^{208-n}$ precomp.	2^{206-n}	MitM	[19], 2008
7	2^{80}	$2^{113}+2^{123}$ precomp.	2^{122}	MitM	[20], 2009
8	$2^{127.997}$	2^{204}	2^{1044}	Square	[25], 2000
8	$2^{116.5}$	$2^{247.5}$	2^{45}	Impossible	[50], 2007
8	$2^{89.1}$	$2^{229.7}$MA	2^{97}	Impossible	[34] 2008
8	$2^{111.1}$	$2^{227.8}$MA	$2^{112.1}$	Impossible	[34] 2008
8	2^{34+n}	$2^{202+n}+2^{208-n}$ precomp.	2^{206-n}	MitM	[19], 2008
8	2^{80}	2^{241}	2^{123}	MitM	[20], 2009
8	2^{113}	2^{196}	2^{129}	Square-multiset	[22], 2010

Table 5. Summary of previous results on AES in hash-mode use, i.e. distinguishers in chosen and known-key models, or preimage or collision attacks

rounds	versions	type/mode	attack/gen.	memory	method	reference
7	all	known-key dist.	$2^{56}/2^{58}$?	–	Square	[32], 2007
7	all	chosen-key dist.	$2^{24}/2^{64}$	2^{16}	Rebound	[38], 2009
8	all	chosen-key dist.	$2^{48}/2^{64}$	2^{32}	Rebound	[27, 33],2009
14	256	chosen-key dist.	$2^{69}/2^{77}$	–	Boom-g	[10], 2009
6	all	collision/MMO+MP	$2^{56}/2^{64}$	2^{32}	Rebound	[33], 2009
7	all	near-coll./MMO	$2^{32}/2^{48}$	2^{32}	Rebound	[33], 2009
7	all	preimage/DM	$2^{120}/2^{128}$	2^{8}	Splice&Cut	[42], 2011
7	all	2nd-pre./MMO+MP	$2^{120}/2^{128}$	2^{8}	Splice&Cut	[42], 2011

Table 6. Example of a biclique for 9-round AES-256. S_i are states after MixColumns in round 5, C_i are ciphertexts.

S_0	S_1	C_0	C_1
40 8a ba 52	44 d2 66 7b	79 18 c0 8e	5d 08 b5 ac
30 4a 10 52	32 34 6e f7	67 ac 89 9e	e5 bd d3 54
34 b6 84 52	36 f4 b0 7a	2e 39 52 84	a0 ac d9 8a
b8 fe aa 52	b8 ba 71 3a	3c fd 40 26	09 6a 55 1e

$K[0,0]$: \$6, \$7	$K[0,1]$: \$6, \$7
7d 8a d8 a4 30 e8 0 0	7d 8a d8 a4 **34 ec 4 4**
12 a8 f9 31 5a 42 0 0	12 a8 f9 31 **58 40 2 2**
12 55 cd 0b 32 d6 0 0	12 55 cd 0b **30 d4 2 2**
58 66 d8 cf 54 f8 0 0	58 66 d8 cf **52 fe 6 6**

$K[1,0]$: \$6, \$7	$K[1,1]$: \$6, \$7
7d 8a d8 a4 30 e8 0 0	7d 8a d8 a4 **34 ec 4 4**
10 aa f9 31 5a 42 0 0	**10 aa** f9 31 **58 40 2 2**
ab ec cd 0b 32 d6 0 0	**ab ec** cd 0b **30 d4 2 2**
5a 64 d8 cf 54 f8 0 0	**5a 64** d8 cf **52 fe 6 6**

Tag Size *Does* Matter: Attacks and Proofs for the TLS Record Protocol

Kenneth G. Paterson[1], Thomas Ristenpart[2], and Thomas Shrimpton[3]

[1] Information Security Group, Royal Holloway, University of London, UK
[2] Dept. of Computer Sciences, University of Wisconsin–Madison, USA
[3] Dept. of Computer Science, Portland State University

Abstract. We analyze the security of the TLS Record Protocol, a MAC-then-Encode-then-Encrypt (MEE) scheme whose design targets confidentiality and integrity for application layer communications on the Internet. Our main results are twofold. First, we give a new distinguishing attack against TLS when variable length padding and short (truncated) MACs are used. This combination will arise when standardized TLS 1.2 extensions (RFC 6066) are implemented. Second, we show that when tags are longer, the TLS Record Protocol meets a new length-hiding authenticated encryption security notion that is stronger than IND-CCA.

1 Introduction

TLS is perhaps the Internet's most widely used security protocol. At its heart lies a sub-protocol for integrity-protecting and encrypting data, called the TLS Record Protocol. The current version of this protocol, TLS 1.2, is specified in [12], though earlier versions [10, 11] are still in widespread use. At a high level, the TLS Record Protocol makes use of a MAC-then-Encode-then-Encrypt (MEE) construction, where the "Encode" step takes care of any padding that might be needed prior to the encryption step. For reasons that will become clear, we focus on MEE when used with CBC mode.

In this case, TLS 1.2 works as follows to protect a message M whose bit-length $m = |M|$ must be a multiple of eight. Let n be the block size of the block cipher underlying CBC. Then, one chooses a fresh n-bit IV to use with CBC mode to encrypt the bit string $M \| T \| P \cdots_{p+1} P$. Here T is a τ-bit message authentication tag produced by running HMAC over M and some header information including a sequence number and $P \cdots_{p+1} P$ is the bit string formed by concatenating together $p + 1$ copies of the string P. The value P is the byte-encoding of the number p, which indicates the number of padding bytes. It is required that $\ell = m + \tau + 8(p+1)$ be a multiple of n. We refer to this scheme as MEE-TLS-CBC. A common instantiation uses AES and HMAC-SHA1, making $n = 128$ and $\tau = 160$.

Implementations can choose p in different ways. One is to use minimal-length padding by letting $p \geq 0$ to be the smallest possible value that results in ℓ being a multiple of n. Another is to use larger values of p in order to generate extra padding. GnuTLS [14], for example, randomly selects p from the set of possible

D.H. Lee and X. Wang (Eds.): ASIACRYPT 2011, LNCS 7073, pp. 372–389, 2011.

	MAC	Encoding	Security target
BN00 [3]	SUF-CMA	Concatenation	IND-CPA + PTXT
K01 [15]	SUF-CMA	Concatenation, tag fills one block	IND-CPA + CUF-CPA
K01*	SUF-CMA	Concatenation, tag fills one block	IND-CPA + CTXT
MT10 [18]	SUF-CMA	Any function	Secure channel
This work	PRF	TLS's padding, $m + \tau > n - 8$	LHAE

Fig. 1. Summary of positive results known about MEE under various assumptions about the MAC. The restriction on padding of our result involves the message length m, tag length τ, and block length n. Our attack shows the necessity of this restriction for security.

values. As indicated in the TLS specification, the intent is to combat traffic analysis attacks that exploit plaintext message lengths [16, 21, 23–26].

THIS PAPER. We provide the first analysis of the security of MEE-TLS-CBC as an authenticated encryption (AE) scheme. We start by strengthening traditional AE notions [3, 20] to cover the goal of hiding plaintext lengths that motivates the use of extra padding. Using our new length-hiding AE (LHAE) notion, we provide complementary negative and positive results about MEE-TLS-CBC for general m, τ, and n. When $m + \tau \leq n - 8$ and extra padding is used, we give an attack that allows a man-in-the-middle to readily distinguish between messages of different lengths. A variant of this attack rules out proving traditional AE security as well. On the other hand, we show that when $m + \tau > n - 8$ one provably achieves LHAE security. This positive result holds for a generalization of TLS encoding; it may be applicable in other settings where MEE is used with CBC.

In the current TLS standard [12], the allowed primitives are such that $n \leq 128$ and $\tau \geq 160$. Here the attack does not apply and our positive results provide strong evidence of security. More worrisome is the use of truncated MACs, where $\tau = 80$ and the attack would apply. Truncated MACs are used widely in other protocols (e.g., IPSec [17]) and are standardized as a TLS extension in RFC 6066 [13].

PRIOR WORK ON MEE. Before describing our results in a bit more detail, we briefly summarize the literature as it applies to MEE-TLS-CBC — see Figure 1. Bellare and Namprempre (BN00) [3] introduced two notions of integrity: integrity of plaintexts (PTXT) and of ciphertexts (CTXT). They showed that MEE with any invertible encoding step is IND-CPA and has integrity of plaintexts (PTXT) assuming the mac is strongly unforgeable (SUF-CMA), but argue that PTXT is insufficient for applications because one should target CTXT. Meeting both IND-CPA and CTXT is one of several equivalent formulations for AE security [20].

Krawczyk (K01) [15] analyzed a variant of MEE-TLS-CBC in which m must be a multiple of n, the tag length is $\tau = n$, and no padding is used. He showed that this variant —which does not arise in TLS— achieves a notion of integrity he calls CUF-CPA. This is weaker than CTXT, though a straightforward extension of

K01's techniques prove that this variant is both IND-CPA and CTXT secure; we list this as K01* in Figure 1. While we will build on the techniques underlying these results, the fact that they ignore padding makes them of limited direct relevance to TLS security. Indeed, as the attacks in [9, 22], discussed further below, indicate, the way padding is handled is crucial to the (in)security of MEE-TLS-CBC.

Maurer and Tackmann (MT10) [18] considered MEE with encoding steps being any function, thus restricting attention to minimum-length padding only. They provide a secure channel notion formalized within a new constructive cryptography framework, but the details of this framework (at the time of our writing) have not yet emerged, making comparison with our results for minimum-length padding TLS premature. Our approach uses a more traditional game-based treatment.

As it stands, none of the prior works analyze the AE security of the version of MEE-TLS-CBC used within the standard nor do they treat the length-hiding goal of extra padding.

LENGTH-HIDING ENCRYPTION. Our technical results begin by generalizing encryption to consider the length-hiding goal targeted by TLS. The explicitly stated intent is that applications should be able to hide the length of plaintexts up to some granularity. As mentioned above, the GnuTLS client [14] attempts to obfuscate plaintext length patterns by selecting the amount of padding for each message randomly. This means that for a given message length, the application may vary the amount of padding used. Standard-compliant decryption implementations must support ciphertexts including such extra padding.

This choice was perhaps prescient: attacks taking advantage of leaked plaintext lengths allow inferring web browsing habits [16, 21, 26] and voice-over-IP conversations [23–25]. Note that even when only minimal-length padding is used, MEE-TLS-CBC nevertheless seemingly should hide lengths that are padded to the same multiple of n. Given [16, 21, 23–26], MEE-TLS-CBC seems to have a small security advantage over MEE using OTP — the latter always leaks precise plaintext lengths. Traditional security notions that explicitly allow message lengths to leak (e.g., IND-CPA, IND-CCA) are too weak to surface this distinction.

To treat MEE-TLS-CBC in its full generality, then, we formalize *length-hiding encryption*. We extend the usual syntax of authenticated encryption scheme with associated data (AEAD) to allow the encryption algorithm to take an extra ciphertext-length parameter, in addition to the usual key, header, and message. This allows the user to indicate the desired length of ciphertext.

We correspondingly upgrade the traditional security notions, which do not capture length hiding, and introduce a *length-hiding authenticated encryption* (LHAE) security notion. Our all-in-one definition gives an attacker access to a left-or-right encryption oracle on pairs of chosen messages M_0, M_1 of arbitrary lengths and a *chosen* ciphertext-length. As usual, the attacker's job is to output its guess for a hidden bit b. The LHAE definition captures length hiding in settings where applications may adaptively vary padding per message (such as GnuTLS). Of course, a special case of our security notion is arrived at by restricting to same-length messages: this corresponds to a left-or-right indistinguishability variant of the all-in-one AE notion of Rogaway and Shrimpton [20]. Proving LHAE security therefore establishes AE security as a special case.

NEW ATTACKS AGAINST MEE-TLS-CBC. Our work brings to light interesting new attacks against TLS. Consider MEE-TLS-CBC with when $m + \tau \leq n - 8$. This means that a complete message M (of m bits), a tag, and at least one padding byte can fit into a single CBC block of size n. Then an attacker, given an encryption C of a message M that is created using longer-then-minimum padding, is able to create another encryption C' of the same message M; we call this a *decryption collision*.[1] This immediately violates the ciphertext integrity (CTXT) of MEE-TLS-CBC, thereby ruling out AE or LHAE security, and can easily be extended to build an IND-CCA distinguisher as well.

It may seem that this deficiency is not dangerous. After all it just shows that an attacker can generate a new ciphertext that decrypts to an already legitimately encrypted message, and this does not threaten the security of TLS as a secure channel protocol. Indeed, some formulations of channel security [6–8], including that of [18], explicitly exclude decryption collisions from being considered as an insecurity. Nevertheless, it rules out meeting the AE security notion targeted, and met, by other designs.

What's more, decryption collisions prove obviously damaging in the length-hiding setting. We will show that they can be used to allow an attacker to distinguish between encryptions of messages of *different* lengths, for example "YES" and "NO". This defeats the TLS design intention of hiding plaintext lengths at this level of granularity. The distinguishing attack would be simple to mount in practice by a man-in-the-middle.

TLS 1.2 (and older versions) specifies $n \in \{64, 128\}$ (DES, AES) and $\tau \geq 160$ (HMAC-SHAx), so this attack does not affect the security of TLS as specified in version 1.2. However 80-bit truncated MACs are explicitly defined for use in extensions to TLS 1.2 [13]. Our attack would therefore apply to TLS using CBC-AES with these truncated MACs and extra padding. We are unaware of any current implementations that are vulnerable, but this will change if, for example, GnuTLS implemented the TLS 1.2 truncated MAC extension.

LHAE SECURITY OF MEE-TLS-CBC. Now the good news. We complement our negative results by proving LHAE security for MEE-TLS-CBC exactly when the above attacks do not work: when $m + \tau > n - 8$ or no extra padding is used. The analysis is involved, as one may expect given the sharp divide between security and insecurity. Let us look at it from a high-level.

The natural starting point for our analysis is the K01* result for concatenation encoding, $\tau = n$, block-aligned tags, and no padding. Here one splits the task of proving authenticated encryption security into two key steps (leveraging

---

[1] The terminology from [6] would call this a replay. We reserve replay for the more traditional security goal of not accepting the same message twice, even if derived from the same ciphertext. Achieving replay resistance requires stateful decryption.

techniques from [20]): showing separately IND-CPA and CTXT security. IND-CPA security is immediate from the IND-CPA security of CBC mode. A general result gives that many-query CTXT is implied by q_d times the advantage of single-query CTXT where q_d is the number of decryption queries. So what remains is showing single-query CTXT. The K01* analysis applies the security of the block cipher as a strong PRP to move to a setting in which the adversary learns nothing about MAC tags from encryption queries and, moreover, for its single decryption query submits a ciphertext consisting of blocks that were output during encryption. The proof concludes via a case analysis partitioned according to which ciphertext blocks are used and how they relate to where tags were located within the encryption queries. The alignment of tags with block boundaries eases this analysis, but it is still relatively involved.

Several new difficulties arise in applying this approach to MEE-TLS-CBC. Foremost of these is that the case analysis becomes significantly more complex, as tags may (for example) span multiple blocks and variable-length padding is allowed. Also the K01* approach only provides a loose bound, approximately $2^{n/3}$, because it proves single-query CTXT and then uses a general hybrid argument to conclude multi-query CTXT. Finally, none of the general results apply to length-hiding encryption. The last issue is the easiest to handle, and in the full version we show that length-hiding IND-CPA and CTXT together imply LHAE. The other issues prove more troublesome. We therefore first simplify our task by introducing a new security notion that will enable further modularity.

COLLISION-RESISTANT DECRYPTION SECURITY. Recall that our attack above found decryption collisions: the adversary computed a new ciphertext that decrypts to a previously encrypted message. We formalize resistance to such attacks and call the resulting notion *collision-resistant decryption* (CRD). It turns out that CRD exactly characterizes the gap between CTXT and PTXT: we prove that a scheme is CTXT if and only if it is both PTXT and CRD.

With this new characterization of CTXT in hand, we proceed as follows. We show (in the full version) that MEE is length-hiding IND-CPA secure and PTXT secure. Both of these results follow straightforwardly from the techniques of [3]. Thus to show LHAE of MEE-TLS-CBC reduces to proving CRD security. Here we still have technical hurdles, including the fact that we must directly analyze multi-query CRD, deal with arbitrary tag locations and sizes, and account for variable length padding. What's more, we must observe precise requirements on tag and message lengths to avoid our attacks. To make this task slightly easier, we assume that the MAC is a secure PRF. While this is a stronger assumption than SUF-CMA, the MAC used by TLS is HMAC, which must be a good PRF in other parts of the TLS protocol.

STATEFUL LHAE. In fact the TLS record protocol uses both stateful encryption and stateful decryption, enabling replay resistance. We handle this, too. In the full version we formalize a stateful LHAE notion (generalizing a definition of [4]) and show that one can easily lift all our results to the stateful setting.

PRIOR VERSIONS OF TLS. We have concentrated on the TLS 1.2 standard, though all our results apply to TLS 1.1 as well. TLS 1.0 differs in two key ways, changing the applicability of our results. First, standard-compliant implementations of TLS 1.0 allowed an attacker to distinguish between decoding failures (arising from incorrectly formatted padding) and authentication failures (arising from MAC verification failures). It was shown in [9, 22] how this difference could be exploited to decrypt ciphertexts in the OpenSSL implementation of TLS. Consequently, the TLS 1.1 and 1.2 specifications mandate that implementations prevent such attacks by enforcing uniform error reporting (both in terms of timing and the actual message returned). Our positive results are in this uniform error reporting model and don't necessarily apply when non-uniform error reporting is in effect.

The second difference is that in TLS 1.0 CBC mode used chained IVs, meaning that the IV used to encrypt a message is set to the last ciphertext block from the previously sent ciphertext. As reported in [19], Rogaway and Dai found distinguishing attacks that exploit chained IVs, and so in TLS 1.1 and beyond, dedicated IVs are required. Our attacks and proof only apply when dedicated IVs are used as in TLS 1.1 and 1.2.

RECAP AND DISCUSSION. Putting together all our results, we see that the exact nature of encoding in MEE must be carefully considered when analysing protocols based upon it. Our attacks and positive results characterize the parameters under which MEE-TLS-CBC falls to (at least) distinguishing attacks and those under which we can have significantly better confidence in security via our proofs. To recap, tag size matters: too small and security fails, large enough and LHAE security can be proved.

We are in contact with those involved in TLS standardization, and hope that vulnerabilities in future versions can be avoided. There are several ways to protect TLS from these problems. For example, one could include the padding length in the MAC scope. Our attacks would no longer work and, in fact, one should be able to prove LHAE security. The best solution is to stop using using MEE-based encryption within TLS (and elsewhere). Instead, one could use Encrypt-then-MAC or one of the dedicated AE schemes. We note that our LHAE notion is interesting for these as well, allowing one to show, for example, that Encrypt-then-MAC achieves some degree of length hiding in the case where one uses CBC.

2   Notation, Syntax and Basic Security Notions

NOTATION. When $\mathcal{X}$ is a set, we write $X \leftarrow_\$ \mathcal{X}$ to mean that a element (named X) is uniformly sampled from $\mathcal{X}$. We overload the notation for probabilistic or stateful algorithms, writing $X \leftarrow_\$ M$ to mean that algorithm M runs and outputs value named X. The set $\{0,1\}^{\leq n}$ contains all bitstrings of length at most n bits, and as usual $\{0,1\}^*$ is the set of all finite length strings. When X and Y are strings, we write $X\|Y$ for their concatenation. When $X \in \{0,1\}^*$ we write $|X|$ for its length. For a tuple of strings $(X_1, X_2, \ldots, X_b)$ we define $|(X_1, X_2, \ldots, X_b)| = |X_1 \| X_2 \| \cdots \| X_b|$.

We often use the notation $M \Rightarrow x$ to denote the event (defined over some specified probability space) that at some algorithm M outputs value x.

An *adversary* A is a probabilistic algorithm that takes zero or more oracles, these denoted as superscripts.

FUNCTION FAMILIES, PRFS AND SPRPS. Fix sets $\mathcal{D}, \mathcal{R}$ and non-empty set $\mathcal{K}$. Let $F \colon \mathcal{K} \times \mathcal{D} \to \mathcal{R}$ be a mapping. For each $K \in \mathcal{K}$ we write $F_K(\cdot)$ for $F(K, \cdot)$ and thus think of F as a *function family* indexed by K. Let $\mathrm{Func}(\mathcal{D}, \mathcal{R})$ denote the set of all functions from $\mathcal{D}$ to $\mathcal{R}$. Let A be an adversary. We define
$$\mathbf{Adv}_F^{\mathrm{prf}}(A) = \Pr\left[K \leftarrow_\$ \mathcal{K}; A^{F_K(\cdot)} \Rightarrow 1 \right] - \Pr\left[f \leftarrow_\$ \mathrm{Func}(\mathcal{D}, \mathcal{R}) : A^{f(\cdot)} \Rightarrow 1 \right]$$
to be the *PRF*-advantage of A attacking F. We overload notation and write $\mathbf{Adv}_F^{\mathrm{prf}}(t, q, \mu)$ to mean the maximum of $\mathbf{Adv}_F^{\mathrm{prf}}(A)$ over all adversaries A that run in time t, ask q queries, these totalling μ bits in length.

Fix integers $k, n > 0$, and let $E \colon \{0,1\}^k \times \{0,1\}^n \to \{0,1\}^n$ be a function family. If for every $K \in \{0,1\}^k$ we have that $E_K(\cdot)$ is a permutation (bijective mapping), then E is a *blockcipher*, and we call n the *blocksize*. We write $\mathrm{Perm}(n)$ for the set of all permutations over $\{0,1\}^n$. We define $\mathbf{Adv}_E^{\mathrm{sprp}}(A) =$
$$\Pr\left[K \leftarrow_\$ \{0,1\}^k : A^{E_K(\cdot), E_K^{-1}(\cdot)} \Rightarrow 1 \right] - \Pr\left[\pi \leftarrow_\$ \mathrm{Perm}(n) : A^{\pi(\cdot), \pi^{-1}(\cdot)} \Rightarrow 1 \right]$$
to be the *strong PRP*-advantage of A attacking F. Again, we overload our notation and write $\mathbf{Adv}_E^{\mathrm{sprp}}(t, q_1, q_2)$ to mean the maximum of $\mathbf{Adv}_E^{\mathrm{sprp}}(A)$ over all adversaries A that run in time t, asking a total of q queries to its oracles.

ENCRYPTION SCHEMES AND MACS. An *encryption scheme* $\mathsf{SE} = (\mathsf{K}_{\mathsf{se}}, \mathsf{Enc}, \mathsf{Dec})$ is a triple of algorithms. The probabilistic algorithm K_{se} samples from a finite and non-empty set $\mathcal{K}_{\mathsf{se}}$. The *encryption algorithm* Enc and *decryption algorithm* Dec take an input $(K, \ell, H, M) \in \mathcal{K}_{\mathsf{se}} \times \mathbb{N} \times \{0,1\}^* \times \{0,1\}^*$ (the key, output length, associated data, and message or ciphertext) and outputs either a string or the distinguished output $\bot$. The encryption algorithm can be probabilistic while decryption is always deterministic. We assume there are sets $\mathcal{H} \subseteq \{0,1\}^*$ (the header space), $\mathcal{L} \subseteq \mathbb{N}$ (the requested length space), $\mathcal{M} \subseteq \{0,1\}^*$ (the message space) such that for all $K \in \mathcal{K}_{\mathsf{se}}$ it holds that $\Pr[\mathsf{Enc}_K(\ell, H, M) \in \{0,1\}^*] = 1$ if $(\ell, H, M) \in \mathcal{L} \times \mathcal{H} \times \mathcal{M}$ and $\Pr[\mathsf{Enc}_K(\ell, H, M) = \bot] = 1$ if $(\ell, H, M) \notin \mathcal{L} \times \mathcal{H} \times \mathcal{M}$. For correctness we require that for all $(K, \ell, H, M) \in \mathcal{K}_{\mathsf{se}} \times \mathcal{L}, \mathcal{H}, \mathcal{M})$ it holds that $\Pr[\mathsf{Dec}_K(H, \mathsf{Enc}_K(\ell, H, M)) = M] = 1$.

We further make a restriction that whether or not Enc returns $\bot$ does not vary with the message length (all other inputs kept equal). Formally, for all keys $(K, \ell, H) \in \mathcal{K}_{\mathsf{se}} \times \mathcal{L} \times \mathcal{H}$ and for all $M, M' \in \mathcal{M} \times \mathcal{M}$ such that $|M| = |M'|$ it holds that for all coins $\mathsf{Enc}_K(\ell, H, M) = \bot$ iff $\mathsf{Enc}_K(\ell, H, M') = \bot$.

Let us make a few comments on what this syntax captures. First, because ℓ is a parameter of encryption, the syntax supports encryption schemes that return variable-length ciphertexts of the same plaintext M. Second, for any fixed plaintext length m, either all $M \in \{0,1\}^m$ encrypt to valid ciphertexts, or none of them do. Third, if ℓ and M are such that encryption would return $\bot$ (e.g. because $\ell < |M|$, or the encryption algorithm does not support ciphertexts of length ℓ), then it does so always. Finally, since decryption does not take the length parameter ℓ, our correctness requirement implicitly demands that the length of the underlying plaintext can be inferred given (K, H, C) where $C = \mathsf{Enc}_K(\ell, H, M)$.

Let $E \colon \{0,1\}^k \times \{0,1\}^n \to \{0,1\}^n$ be a block cipher. Then the encryption scheme $\mathrm{CBC}[E]$ has message space $\{0,1\}^{n+}$ (all strings that are a multiple of n bits). Key generation K_{se} outputs a random $K \leftarrow_\$ \{0,1\}^k$. On input including a message $M = M_1 \| \cdots \| M_m \in \{0,1\}^{nm}$, encryption ignores any requested length or header inputs and returns the ciphertext $C_0 \| \ldots \| C_{m+1}$ where $C_0 \leftarrow_\$ \{0,1\}^n$ and $C_i \leftarrow E_K(C_{i-1} \oplus M_i)$ for $1 \le i \le m$.

Fix an integer $\tau > 0$. A *message authentication code* (MAC) is a function family $F \colon \mathcal{K}_{\mathrm{ma}} \times \mathcal{D} \to \{0,1\}^\tau$, where τ is the *tag length* of the MAC.

CONVENTIONS. The running time of algorithms (e.g. adversaries) is relative to some implicit underlying RAM model of computation. The running time of an adversary is assumed to include the time to execute the entire experiment in which it executes, including (for example) the time for its oracles to execute. Throughout we fix the convention that adversaries do not ask *pointless* queries: they do not query an oracle on a value outside of its domain, nor on values that are defined to cause a $\perp$ return value. Also, adversaries are assumed not to repeat queries to deterministic oracles. This convention is made without loss of generality.

3   MAC-Encode-Encrypt and the TLS Record Protocol

The TLS Record Protocol uses the MAC-then-encode-then-encrypt paradigm. The algorithm first applies a message authentication scheme to the message and header to derive a tag. The message and tag are then encoded into a bit string according to some encoding rules. Finally an encryption scheme is used to encrypt the result.

ENCODING SCHEMES. An *encoding scheme* $\mathsf{CODE} = (\mathsf{encode}, \mathsf{decode})$ is a pair of deterministic algorithms. The *encoding algorithm* encode takes an input $(\ell, M, T) \in \mathbb{N} \times \{0,1\}^* \times \{0,1\}^*$ (the output length, message, and tag) and returns a string of length ℓ or the distinguished symbol $\perp$. An encoding scheme is assume to have a fixed maximum allowable output length $\ell_{\max}$. If $\ell < |M| + |T|$ or $\ell > \ell_{\max}$ then encode returns $\perp$. The *decoding algorithm* decode takes an input in $\{0,1\}^*$ and returns an element of $\{0,1\}^* \times \{0,1\}^*$ or $(\perp, \perp)$. If either algorithm is called on an input outside of its specified domain, it returns an appropriate failure symbol. For correctness we require that, for all ℓ, M, and T such that $\mathsf{encode}(\ell, M, T) \ne \perp$ we have $\mathsf{decode}(\mathsf{encode}(\ell, M, T)) = (M, T)$.

THE MEE AEAD SCHEME. We define the MEE scheme that forms the basis for encryption in TLS, some modes of IPSec, and elsewhere. Fix some block size n. Let $\mathsf{SE} = (\overline{\mathsf{K}}, \overline{\mathsf{Enc}}, \overline{\mathsf{Dec}})$ be an encryption scheme with a message space $\{0,1\}^{n+}$ (all strings of length a multiple of n). We assume that, given inputs of an appropriate length, the algorithms $\overline{\mathsf{Enc}}, \overline{\mathsf{Dec}}$ are failure-free. Let $F \colon \mathcal{K}_{\mathrm{ma}} \times \{0,1\}^* \to \{0,1\}^\tau$ be a function. Let $\mathsf{CODE} = (\mathsf{encode}, \mathsf{decode})$ be an encoding

alg. K:	**alg. $\mathsf{Enc}_K(\ell, H, M)$:**	**alg. $\mathsf{Dec}_K(H, C)$:**
$K_{\mathrm{se}} \twoheadleftarrow \overline{\mathsf{K}}$	$(K_{\mathrm{ma}}, K_{\mathrm{se}}) \leftarrow K$	$(K_{\mathrm{ma}}, K_{\mathrm{se}}) \leftarrow K$
$K_{\mathrm{ma}} \twoheadleftarrow \mathcal{K}_{\mathrm{ma}}$	$T \twoheadleftarrow F_{K_{\mathrm{ma}}}(H, M)$	$X \leftarrow \overline{\mathsf{Dec}}_{K_{\mathrm{se}}}(C)$
Ret $(K_{\mathrm{ma}}, K_{\mathrm{se}})$	$X \leftarrow \mathsf{encode}(\ell - n, M, T)$	$(M, T) \leftarrow \mathsf{decode}(X)$
	If $X = \perp$ then Ret $\perp$	If $(M, T) = (\perp, \perp)$ then Ret $\perp$
	Ret $Y \twoheadleftarrow \overline{\mathsf{Enc}}_{K_{\mathrm{se}}}(X)$	If $F_{K_{\mathrm{ma}}}(H, M) \neq T$ then Ret $\perp$
		Ret M

Fig. 2. Algorithms for the MEE generic composition

scheme for which the outputs of encode all have bit lengths a positive multiple of n. Then $\mathsf{MEE}[F, \mathsf{CODE}, \mathsf{SE}] = (\mathsf{K}, \mathsf{Enc}, \mathsf{Dec})$ is defined as shown in Figure 2.

Notice that Enc takes as input a requested ciphertext length ℓ, as well as associated data H and message M. The inclusion of ℓ allows for variable length padding to be used, while the inclusion of H allows us to incorporate additional fields in the MAC scope, for example, TLS's sequence numbers and compression type and version fields. Notice that Dec can fail either because of a failure to properly decode the message X, or because of a failure to verify the MAC tag T. However, in our specification of the MEE scheme, these error events are *not* distinguishable. This prevents the attacks of [9, 22] and is in-line with the TLS specification [12]. In TLS, any such errors are *fatal*, leading to the destruction of the TLS connection and the disposal of the keys, meaning that an attacker can no longer interact with the protocol. In our description of MEE, these errors are non-fatal, allowing an attacker to continue to interact with the MEE scheme after an error has arisen. It is easy to see that security with non-fatal errors immediately implies security with fatal errors, since any adversary in the former case is more powerful than in the latter case. Thus any security results we prove about MEE will imply security for the more realistic version of MEE in which errors are fatal.

TLS ENCODING. Let $\mathsf{TLScode} = (\mathsf{TLSencode}, \mathsf{TLSdecode})$ be the encoding scheme defined in Figure 3. This scheme is parameterized by the integers ψ, n, and τ, representing the maximal padding length, a block length, and a tag length. Recall that we work with bits in our algorithmic descriptions and cryptographic analysis, rather than with bytes as in the TLS specification [12].

For TLS, ψ can be as large as 2048, since the longest padding pattern that is permitted consists of 256 copies of the byte value FF_x. However, an implementation may select a smaller value of ψ. Note that this scheme has a decoding algorithm permitting variable length padding of any length (not limited by ψ). This decoding algorithm checks every byte of padding to ensure that it is correct. It also allows the final message M (obtained after removing padding and parsing the resulting string into message M and MAC tag T) to be of zero length. Again, these choices are in accordance with the TLS specification [12].

GENERALIZING TLS ENCODING. For the purposes of our positive results, we will analyze a generalization of TLS encoding. An encoding scheme $\mathsf{CODE} =$

alg. TLSencode(ℓ, M, T):	**alg. TLSdecode(X):**
If $\ell \bmod n \neq 0$ then Ret $\perp$	If $|X| \bmod n \neq 0$ then Ret $(\perp, \perp)$
$p \leftarrow \ell - (|M| + |T|)$	$(X, P) \leftarrow \text{split}_{|X|-8,8}(X)$
If $p < 8$ then Ret $\perp$	$b \leftarrow \text{byte2int}(P)$
If $p > \psi$ then Ret $\perp$	$p \leftarrow 8 \cdot b$
If $p \bmod 8 \neq 0$ then Ret $\perp$	If $|X| - p - \tau < 0$ then Ret $(\perp, \perp)$
$P \leftarrow \text{int2byte}(p/8) - 1$	For $i = 1$ to b do
$X \leftarrow M \parallel T \parallel P \cdots_{P+1} P$	$\quad (X, P') \leftarrow \text{split}_{|X|-8,8}(X)$
Ret X	$\quad$ If $P \neq P'$ then Ret $(\perp, \perp)$
	$(M, T) \leftarrow \text{split}_{|X|-\tau,\tau}(X)$
	Ret (M, T)

Fig. 3. Algorithms for the TLS encoding scheme

(encode, decode) is *MEE sufficient* if it is parameterized by a block length n and tag length τ and has the following properties:

(1)   The output encode(ℓ, M, T) consists of a string $M \parallel T \parallel P \in \{0,1\}^{in}$ for some $i \geq 1$ and where $|P| = \ell - |M| - |T|$. The particular padding P is uniquely determined by $|P|$.

(2)   Algorithm decode(X) for $|X| = \ell$ returns (M, T) only if encode(ℓ, M, T) outputs $M \parallel T \parallel P$.

(3)   CODE yields *prefix-free padding*, which means that for any M, M' such that $|M| = |M'|$, for any T, T', the padding P returned by encode(ℓ, M, T) is not a prefix of the padding returned by encode(ℓ', M', T') for any $\ell \neq \ell'$.

One may be able to relax property (1) in various ways and still prove security, but we focus on this case for greatest simplicity (while still covering TLS encoding). Property (1) and the invertibility of encoding indicate that for any strings M, T and number ℓ for which encode(ℓ, M, T) does not output $\perp$, there is a single string P such that encode(ℓ, M, T) outputs $M \parallel T \parallel P$.

In the proof of our main technical result, Theorem 2, it will be useful to assume that one can extract from encode a routine called Pad that, on input $(|M|, \ell)$, simply returns the padding P from $M \parallel T \parallel P$. Similarly, it will be useful to assume that one can extract from decode: (1) a routine called Parse that, on input X, returns the appropriate triple M, T, P; and (2) a routine called PadCheck that, on input $(|M|, P, |X|)$, returns 1 if P is the correct padding, and 0 otherwise. It is easy to see that such routines can be extracted from TLSencode and TLSdecode.

For notational clarity and letting F be some function family that will be clear from context, we let MEE-GEN-CBC = MEE[F, CODE, CBC] be a mnemonic defining the scheme that uses a MEE-sufficient encoding scheme CODE with CBC. In particular, we let MEE-TLS-CBC = MEE[F, TLScode, CBC]. When we need to be explicit, we write CBC[E] to mean that CBC encryption is done over a function family E.

main LHAE$_{\mathsf{SE}}$:	procedure Enc(ℓ, H, M_0, M_1):	procedure Dec(H, C):
$K \xleftarrow{\$} \mathcal{K}_{\mathsf{se}}$	$C_0 \xleftarrow{\$} \mathsf{Enc}_K(\ell, H, M_0)$	If $b = 1 \wedge C \notin \mathcal{C}$ then
$b \xleftarrow{\$} \{0, 1\}$	$C_1 \xleftarrow{\$} \mathsf{Enc}_K(\ell, H, M_1)$	$\quad$ Ret $\mathsf{Dec}_K(H, C)$
$b' \leftarrow A^{\mathbf{Enc,Dec}}$	If $C_0 = \perp$ or $C_1 = \perp$ then	Ret $\perp$
Ret $(b' = b)$	$\quad$ Ret $\perp$	
	$\mathcal{C} \xleftarrow{\cup} C_b$; Ret C_b	

Fig. 4. Length-hiding AEAD security game

4   Length-Hiding Authenticated Encryption

Here we formalize security goals for the TLS Record Protocol, and establish some basic results about these goals. We target authenticated encryption security, which requires (informally) that an adversary cannot generate new, valid ciphertexts itself, nor learn partial information about encrypted plaintexts. Note that this implies traditional chosen-ciphertext attack security. One security aspect traditional AE security goals do not treat, however, is length hiding. As we saw in the previous section, the TLS standard includes the option for variable-length-padding so that applications can choose to hide exact message lengths. Even in the minimal-length-padding case some amount of length hiding could exist since one must pad to the next block boundary. Classical security goals, such as semantic security and the stronger AE notion mentioned above, explicitly leak message lengths. Thus one cannot use these to reason about the length-hiding capabilities of MEE-TLS-CBC. We therefore give a new security notion to capture length hiding under chosen-length attacks. It generalizes the randomized AEAD security notion given in [20].

LENGTH-HIDING AEAD SECURITY. Let $\mathsf{SE} = (\mathsf{K}_{\mathsf{se}}, \mathsf{Enc}, \mathsf{Dec})$ be an encryption scheme and let A be an adversary. Figure 4 details a security length-hiding authenticated-encryption game. We define the LHAE-advantage (of A) to be $\mathbf{Adv}_{\mathsf{SE}}^{\text{lh-ae}}(A) = 2 \cdot \Pr\left[\, \mathsf{LHAE}_{\mathsf{SE}}^A \Rightarrow \mathsf{true}\,\right] - 1$. Let LHAE1 (resp. LHAE0) be the LHAE game except with b set to one (resp. zero). Then a standard argument gives that $\mathbf{Adv}_{\mathsf{SE}}^{\text{lh-ae}}(A) = \Pr\left[\, \mathsf{LHAE1}_{\mathsf{SE}}^A \Rightarrow \mathsf{true}\,\right] - \Pr\left[\, \mathsf{LHAE0}_{\mathsf{SE}}^A \Rightarrow \mathsf{false}\,\right]$. We write $\mathbf{Adv}_{\mathsf{SE}}^{\text{lh-ae}}(q_e, \mu_e, q_d, \mu_d)$ to mean the maximum of $\mathbf{Adv}_{\mathsf{SE}}^{\text{lh-ae}}(A)$ taken over all adversaries A that run for t computational steps, asking at most q_e queries to its left oracle that result in ciphertexts of total length μ_e bits, and q_d queries to its right oracle that total μ_d bits in length. Restricting attention to adversaries A for which $q_d = \mu_d = 0$ yields a length-hiding version of the IND-CPA notion, which we denote by LH-IND-CPA. We let $\mathbf{Adv}_{\mathsf{SE}}^{\text{lh-ind-cpa}}(A) = \mathbf{Adv}_{\mathsf{SE}}^{\text{lh-ae}}(A)$ for A that make no decryption queries.

The LHAE notion captures chosen-length attacks along two dimensions. First, we allow $|M_0| \neq |M_1|$ unlike in previous formulations of encryption security. This captures that an attacker cannot distinguish between the encryptions of two chosen messages of arbitrary lengths. We only require that queried messages both encrypt to a ciphertext (not $\perp$). This restriction is necessary to avoid trivial wins in which an attacker abuses two tuples (ℓ, H, M_0) and (ℓ, H, M_1) for which only one is handled by encryption. Second, we allow the adversary to adaptively pick ℓ for each query. A weaker notion restricts attention to a specific ℓ for the entire experiment. Indeed, this fixed-ciphertext-length notion may be sufficient for some applications. Our attacks (in the next section) show insecurity against this weaker notion, and so, by extension, the LHAE notion. On the other hand, our proofs target the stronger notion, meaning when the proofs are applicable, length-hiding security is achieved even if applications dynamically change ciphertext lengths for a single key as done by GnuTLS [14] or if one implemented traffic morphing [27] using MEE.

5   Attacking TLS for Short Messages and Tags

Next we sketch attacks against the MEE scheme as used in TLS and as described in Section 3. In this section, for convenience, we work bytewise.

We give an attack that causes a decryption collision (recall: two valid ciphertexts that decrypt to the same plaintext). For concreteness, let $n = 128$ and $\tau = 80$. This would be the case for truncated MACs [13]. Now suppose the attacker can obtain a ciphertext $C = C_0||C_1||C_2$ for a message M with $|M| = 40$. Then the attacker computes a new ciphertext $C' = C_0' \parallel C_1$ where $C_0' = C_0 \oplus \text{0x00} \cdots_{14} \text{0x00 0x00 0x10}$, where $\text{0xab} \ldots_k \text{0xab}$ signifies a total of k copies of the byte value 0xab. The plaintext underlying the CBC mode ciphertext C is $M \parallel T \parallel \text{0x11} \cdots_{20} \text{0x11}$. It is easy to verify that the plaintext underlying C' is $M \parallel T \parallel \text{0x01}$, which is correctly formatted and, since it has the same message and tag as in C, will verify.

This attack can be extended to break **MEE-TLS-CBC** in the traditional IND-CCA sense. With parameters as before, suppose the attacker receives from its encryption oracle a 3-block encryption C of M_b, one of two 5-byte messages M_0, M_1. (The messages are the same length.) Then the attacker can modify C by truncation and bit flipping in the IV to produce a fresh ciphertext C' which is a valid encryption of M_b. At this point C' may be submitted to the decryption oracle and the returned plaintext will be M_b, allowing the attacker to win the IND-CCA game with probability 1. While this attack rules out MEE meeting IND-CCA security (for short messages and MACs), notice that it does not seem to translate into a mountable attack on TLS. This is because an attacker that intercepts C and sends C' instead will not see any difference in the behaviour of the TLS connection as compared to having just sent C. One may conclude from this that CTXT security, which is violated here, is overly strong and the abilty to find decryption collisions does not endanger security.

This intuition is wrong, and in fact what we'll see is that IND-CCA is in fact *too weak* to capture the problem that decryption collisions give rise to. Consider a client sending a short message, either "YES" or "NO" encoded as a 3-byte string or 2-byte string. Note these are of two *different* lengths, and so the IND-CCA security definition excludes such a pair from consideration. Let $M \in \{\text{YES}, \text{NO}\}$ denote the message the client encrypts, which is not known to the attacker. Assume the client uses extra padding (such as done by GnuTLS) to mask lengths; say the chosen extra padding during encryption was enough to fill up one extra block. The attacker intercepts the ciphertext $C = C_0 \parallel C_1 \parallel C_2$ generated by the client. It then generates a new ciphertext $C' = C'_0 \parallel C_1$ where

$$C'_0 = C_0 \oplus \text{0x00} \cdots_{12} \text{0x00 0x10 0x10 0x10 0x10} .$$

The attacker then forwards C' in place of C to the server and observes whether decryption succeeds (say, by seeing if the session is torn down). If decryption succeeds, the attacker knows that $M = \text{NO}$ and otherwise that $M = \text{YES}$. Why does this work? The plaintext for CBC underlying C is either $\text{NO}\|T_{no}\|\text{0x14}\cdots_{20}\text{0x14}$ or $\text{YES} \parallel T_{yes} \parallel \text{0x13}\cdots_{19}\text{0x13}$. If the former, then decrypting C' succeeds since the padding underlying C' is exactly $\text{0x04}\cdots_4\text{0x04}$. But in the latter case, the CBC decryption step applied to C' yields $\text{YES} \parallel T'_{yes} \parallel \text{0x03 0x03 0x03}$ where $T'_{yes} = T_{yes} \oplus \text{0x00}\cdots_9\text{0x00 0x10}$. Since the MAC tag is deterministic, it cannot be that this MAC verifies and so decryption fails.

This attack extends immediately to handle TLS's sequence numbers and associated data. It also extends to give LHAE attacks for a variety of pairs of message lengths, including combinations where one message is short (a few bytes) and the other is long (even up to 15 blocks in size). The example can be generalised to a variety of MAC sizes. Indeed, the attack still works in the extreme case where the MAC size is just 8 bits less than the block size[2], in which case one of the messages in the attack is of zero length, a length permitted in the TLS specification [12].

This distinguishing attack can be mounted in practice against TLS if an implementation uses sufficiently short MAC tags, such as those arising from the widespread use of truncated MACs (as done in IPsec and SSH). Fortunately TLS 1.2 does not support short enough MACs, but 80-bit truncated MACs are explicitly defined for use in extensions to TLS 1.2 [13]. In these extensions, then, we have a vulnerability: a man-in-the-middle attacker can violate TLS's confidentiality design goal.

6   The CRD Security Notion

We saw in the last section that MEE with TLS paddding is always LHAE insecure when $\tau + |M| \leq n - 8$ (where n is the underlying blockcipher length). Our goal in the rest of the paper is therefore to prove that when $\tau + |M| > n - 8$ the MEE scheme is LHAE secure for the generalized TLS encoding scheme described in Section 3. This will yield as a special case the first proof that the full TLS Record Protocol is secure for standard chosen-ciphertext attack models.

Consider first the non-length-hiding case. Then a natural approach is to target the two properties IND-CPA and ciphertext integrity (CTXT). Recall that CTXT [3] rules out the ability of an attacker to produce a valid ciphertext not before returned by an encryption oracle. A result of Rogaway and Shrimpton [20] states that satisfying both IND-CPA and CTXT is equivalent to AE security.

---

[2] This case is extreme because TLS is a byte-oriented protocol.

In the full versionwe state and prove a generalization of this equivalence for the length hiding setting and also show that MEE is length-hiding IND-CPA (LH-IND-CPA). The proofs are easy extensions of the proofs in the non-length-hiding setting.

The complexity of the analysis lies in showing CTXT. Consider the analysis by Krawczyk [15] for a restricted version of MEE with CBC that, unfortunately, does not cover any usage case of TLS. His proof shows that MEE is single-query CTXT in the case that $\tau = n$, and encoding is both injective and ensures that the tag fills exactly one plaintext block for the underlying encryption. These restrictions make a proof more manageable, in particular leading to a simpler final case analysis. In our setting, a direct CTXT analysis would require many more cases, these induced by the relaxation to variable length padding and the fact that tags may span multiple plaintext blocks. To ameliorate this complexity, we takemore modular approach to proving CTXT.

CRD SECURITY. We introduce a new notion of security for encryption schemes called *collision-resistant decryption* (CRD). This enables proofs of CTXT to be split into two self-contained parts and helps modularize our analysis further. Recall that plaintext integrity (PTXT) requires that an adversary not be able to construct a ciphertext that decrypts to a valid message that was not before queried to the encryption oracle. As mentioned above, CTXT rules out constructing any new ciphertext. As shown by Bellare and Namprempre [3], PTXT is a strictly weaker property than CTXT. We show that CRD is exactly the "gap" between the two properties. Informally, CRD security requires that an attacker cannot produce a new ciphertext that decrypts to a message previously queried to the encryption oracle. One can see, in fact, that the attacks of the previous section are, at their core, breaking MEE in the sense of CRD.

Let $\mathsf{SE} = (\mathsf{K_{se}}, \mathsf{Enc}, \mathsf{Dec})$ be an encryption scheme, and let A be an adversary. We define the collision-resistant decryption advantage of A as $\mathbf{Adv}_{\mathsf{SE}}^{\mathrm{crd}}(A) = \Pr\left[\mathrm{CRD}_{\mathsf{SE}}^{A} \Rightarrow \mathsf{true}\right]$ where the game $\mathrm{CRD}_{\mathsf{SE}}$ is defined in Figure 5. In the usual way, we write $\mathbf{Adv}_{\mathsf{SE}}^{\mathrm{crd}}(t, q_e, \mu_e, q_d, \mu_d)$ to mean the maximum of $\mathbf{Adv}_{\mathsf{SE}}^{\mathrm{crd}}(A)$ over all adversaries A that run for t computational steps, asking at most q_e queries to its encryption oracle that total at most μ_e bits in length and asking at most q_d queries to its test oracle that total at most μ_d bits in length.

Figure 5 also specifies the games $\mathrm{CTXT}_{\mathsf{SE}}$ and $\mathrm{PTXT}_{\mathsf{SE}}$. We similarly define $\mathbf{Adv}_{\mathsf{SE}}^{\mathrm{ctxt}}(A) = \Pr\left[\mathrm{CTXT}_{\mathsf{SE}}^{A} \Rightarrow \mathsf{true}\right]$ and $\mathbf{Adv}_{\mathsf{SE}}^{\mathrm{ptxt}}(A) = \Pr\left[\mathrm{PTXT}_{\mathsf{SE}}^{A} \Rightarrow \mathsf{true}\right]$. We also define $\mathbf{Adv}_{\mathsf{SE}}^{\mathrm{ctxt}}(t, q_e, \mu_e, q_d, \mu_d)$ and $\mathbf{Adv}_{\mathsf{SE}}^{\mathrm{ptxt}}(t, q_e, \mu_e, q_d, \mu_d)$ analogously.

The following theorem shows that the combination of PTXT and CRD security yields CTXT security. We omit the straightforward proof.

Theorem 1. (PTXT + CRD ⇒ CTXT) *Let* $\mathsf{SE} = (\mathsf{K_{se}}, \mathsf{Enc}, \mathsf{Dec})$ *be an encryption scheme. Then* $\mathbf{Adv}_{\mathsf{SE}}^{\mathrm{int\text{-}ctxt}}(t, q_e, \mu_e, q_d, \mu_d) \leq \mathbf{Adv}_{\mathsf{SE}}^{\mathrm{int\text{-}ptxt}}(t, q_e, \mu_e, q_d, \mu_d) + \mathbf{Adv}_{\mathsf{SE}}^{\mathrm{crd}}(t, q_e, \mu_e, q_d, \mu_d)$. □

Given Theorem 1 and our earlier remarks about LHAE being implied by LH-IND-CPA and CTXT, analyzing the LHAE security of any scheme can be separated into showing that LH-IND-CPA, PTXT and CRD are achieved. This

main $\mathrm{CTXT_{SE}}$:	**main** $\mathrm{PTXT_{SE}}$:	**main** $\mathrm{CRD_{SE}}$:
$K \leftarrow\!\!{}^{\$} \mathcal{K}_{se};\ \mathcal{S} \leftarrow \emptyset;\ i \leftarrow 0$	$K \leftarrow\!\!{}^{\$} \mathcal{K}_{se};\ \mathcal{S} \leftarrow \emptyset;\ i \leftarrow 0$	$K \leftarrow\!\!{}^{\$} \mathcal{K}_{se};\ \mathcal{S} \leftarrow \emptyset;\ i \leftarrow 0$
win $\leftarrow$ false	win $\leftarrow$ false	win $\leftarrow$ false
$(H^*, C^*) \leftarrow\!\!{}^{\$} A^{\mathbf{Enc},\mathbf{Test}}$	$(H^*, C^*) \leftarrow\!\!{}^{\$} A^{\mathbf{Enc},\mathbf{Test}}$	$(H^*, C^*) \leftarrow\!\!{}^{\$} A^{\mathbf{Enc},\mathbf{Test}}$
Ret win	Ret win	Ret win
procedure $\mathbf{Enc}(\ell, H, M)$:	**procedure** $\mathbf{Enc}(\ell, H, M)$:	**procedure** $\mathbf{Enc}(\ell, H, M)$:
$i \leftarrow i + 1$	$i \leftarrow i + 1$	$i \leftarrow i + 1$
$H_i \leftarrow H\ ;\ M_i \leftarrow M$	$H_i \leftarrow H\ ;\ M_i \leftarrow M$	$H_i \leftarrow H\ ;\ M_i \leftarrow M$
$C_i \leftarrow\!\!{}^{\$} \mathsf{Enc}_K(\ell, H_i, M_i)$	$C_i \leftarrow\!\!{}^{\$} \mathsf{Enc}_K(\ell, H_i, M_i)$	$C_i \leftarrow\!\!{}^{\$} \mathsf{Enc}_K(\ell, H_i, M_i)$
$\mathcal{S} \leftarrow \mathcal{S} \cup \{(H_i, C_i)\}$	$\mathcal{S} \leftarrow \mathcal{S} \cup \{(H_i, M_i)\}$	$\mathcal{S} \leftarrow \mathcal{S} \cup \{(H_i, C_i)\}$
Ret C_i	Ret C_i	Ret C_i
procedure $\mathbf{Test}(H^*, C^*)$:	**procedure** $\mathbf{Test}(H^*, C^*)$:	**procedure** $\mathbf{Test}(H^*, C^*)$:
$M^* \leftarrow \mathsf{Dec}_K(H^*, C^*)$	$M^* \leftarrow \mathsf{Dec}_K(H^*, C^*)$	$M^* \leftarrow \mathsf{Dec}_K(H^*, C^*)$
If $M^* \neq \bot \wedge (H^*, C^*) \notin \mathcal{S}$	If $M^* \neq \bot \wedge (H^*, M^*) \notin \mathcal{S}$	If $M^* \neq \bot \wedge (H^*, C^*) \notin \mathcal{S}$
then win $\leftarrow$ true	then win $\leftarrow$ true	$\wedge \exists i\ :\ (H^*, M^*)\ =$
Ret $(M^* \neq \bot)$	Ret $(M^* \neq \bot)$	(H_i, M_i)
		then win $\leftarrow$ true; Ret 1
		Ret 0

Fig. 5. The CTXT, PTXT, and CRD experiments. The set $\mathcal{S}$ and the counter i are global variables in each game.

modularity is particularly beneficial for the MEE construction, where showing LH-IND-CPA and PTXT is straightforward. We defer discussion of these results to the full version. Instead, we focus next on the most involved task: showing CRD security of MEE using CBC and TLS padding.

7   The CRD Security of **MEE-GEN-CBC**

In this section we give a formal security bound for MEE-GEN-CBC. In the following theorem we consider the case that $\tau \leq n$, where n is the blocksize of the blockcipher underlying CBC. In fact the bounds hold when $\tau > n$, too. Say that $\tau = n + n'$ for some $n' > 0$. Then we can reduce to the case considered by Theorem 2 by assuming that the adversary actually controls the first n' bits of T; essentially, they are treated as adversarially controlled message bits. Thus we can restrict our attention to the case that $\tau \leq n$, which simplifies our proof. Note that this does not significantly weaken our bound, since the dominating term is a function of n when $\tau > n$. We emphasize that, unlike prior proofs, we make no assumption about the position of the tag.

In what follows, let the *total plaintext length* of an encryption query (ℓ, H, M) in the CRD experiment be the total number of blocks that are consequently encrypted, i.e. the total number of blocks in $M \| T \| P$ where T is the tag and P is the padding.

Theorem 2. *Fix $n > 0$ and let $E \colon \{0,1\}^n \times \{0,1\}^n \to \{0,1\}^n$ be a blockcipher. Let $\mathsf{CODE} = (\mathsf{encode}, \mathsf{decode})$ be MEE sufficient with blocklength n and taglength $\tau \leq n$. Let $F \colon \mathcal{K} \times \{0,1\}^* \to \{0,1\}^\tau$ be a function family. Let $\mathsf{SE} = $ MEE-GEN-CBC, where CBC is over blockcipher E. Let A be a CRD-adversary that runs in time t; asks q_e encryption queries, the sum of whose total plaintext lengths is σ_e; and asks q_d Test queries, the sum of whose lengths is σ_d blocks. Let $\sigma = \sigma_e + \sigma_d$. Let $b_{\min}$ be the length (in bits) of the shortest message that A queries to its encryption oracle. Then, if $\tau + b_{\min} \geq n$, there exist adversaries B_1, B_2 such that*

$$\mathbf{Adv}^{\mathrm{crd}}_{\mathsf{SE}}(A) \leq \mathbf{Adv}^{\mathrm{prf}}_F(B_1) + \mathbf{Adv}^{\mathrm{sprp}}_E(B_2)$$
$$+ \frac{.5\sigma^2 + \sigma_e^2 + 2\sigma_d\alpha(\alpha+1)q_e + q_e q_d}{2^n} + \frac{q_e q_d}{2^\tau}$$

where where α is the number of distinct padding patterns. Here B_1 runs in time $t + \sigma\mathsf{Time}_E$ and asks at most $q + 1$ queries, and B_2 runs in time $t + O(\sigma)$ and asks at most σ queries. □

The proof can be found in the full version. We note that for TLS with full variable-length padding the parameter α is equal to 256.

Similarly, we can consider the case that minimal length padding is enforced by the encoding scheme. Equivalently, we can restrict to CRD adversaries that query ciphertext lengths ℓ that result in padding only to the closest blocklength. Let us call such adversaries *minimal-length padding respecting*. This case results in exactly the same bound. However for TLS with minimum-length padding the value of α changes to 16.

Corollary 1. *Let all quantities and objects be as in Theorem 2, except that A is a minimal-length padding respecting CRD-adversary. Then, if $\tau + b_{\min} \geq n$, there exist adversaries B_1, B_2*

$$\mathbf{Adv}^{\mathrm{crd}}_{\mathsf{SE}}(A) \leq \mathbf{Adv}^{\mathrm{prf}}_F(B_1) + \mathbf{Adv}^{\mathrm{sprp}}_E(B_2)$$
$$+ \frac{.5\sigma^2 + \sigma_e^2 + 2\sigma_d\alpha(\alpha+1)q_e + q_e q_d}{2^n} + \frac{q_e q_d}{2^\tau}$$

where where α is the number of distinct padding patterns. Here B_1 runs in time $t + \sigma\mathsf{Time}_E$ and asks at most $q + 1$ queries, and B_2 runs in time $t + O(\sigma)$ and asks at most σ queries. □

Acknowledgments. Kenny Paterson was supported by EPSRC Leadership Fellowship EP/H005455/1. Thomas Shrimpton was supported by NSF grants CNS 0627752 and CNS 0845610. Much of this work occurred while Ristenpart and Shrimpton visited RHUL, and they thank RHUL for their kind support.

References

1. Abadi, M., Rogaway, P.: Reconciling Two Views of Cryptography (The Computational Soundness of Formal Encryption). J. Cryptology 20(3), 395 (2007)

2. Bellare, M., Desai, A., Jokipii, E., Rogaway, P.: A concrete security treatment of symmetric encryption. In: Proceedings of 38th Annual Symposium on Foundations of Computer Science (FOCS 1997), pp. 394–403. IEEE (1997)
3. Bellare, M., Namprempre, C.: Authenticated Encryption: Relations among Notions and Analysis of the Generic Composition Paradigm. In: Okamoto, T. (ed.) ASIACRYPT 2000. LNCS, vol. 1976, pp. 531–545. Springer, Heidelberg (2000)
4. Bellare, M., Kohno, T., Namprempre, C.: Authenticated encrytion in SSH: Provably fixing the SSH binary packet protocol. In: ACM Conference on Computer and Communications Security, pp. 1–11 (2002)
5. Bellare, M., Rogaway, P.: The Security of Triple Encryption and a Framework for Code-Based Game-Playing Proofs. In: Vaudenay, S. (ed.) EUROCRYPT 2006. LNCS, vol. 4004, pp. 409–426. Springer, Heidelberg (2006)
6. Canetti, R., Krawczyk, H., Nielsen, J.B.: Relaxing Chosen Ciphertext Security of Encryption Schemes. In: Boneh, D. (ed.) CRYPTO 2003. LNCS, vol. 2729, pp. 565–582. Springer, Heidelberg (2003)
7. Canetti, R.: Universally Composable Security: A New Paradigm for Cryptographic Protocols. In: Foundations of Computer Science – FOCS (2001)
8. Canetti, R., Krawczyk, H.: Analysis of Key-Exchange Protocols and Their Use for Building Secure Channels. In: Pfitzmann, B. (ed.) EUROCRYPT 2001. LNCS, vol. 2045, pp. 453–474. Springer, Heidelberg (2001)
9. Canvel, B., Hiltgen, A.P., Vaudenay, S., Vuagnoux, M.: Password interception in a SSL/TLS Channel. In: Boneh, D. (ed.) CRYPTO 2003. LNCS, vol. 2729, pp. 583–599. Springer, Heidelberg (2003)
10. Dierks, T., Allen, C.: The TLS Protocol Version 1.0. RFC 2246 (January 1999), http://www.ietf.org/rfc/rfc2246.txt
11. Dierks, T., Rescorla, E.: The Transport Layer Security (TLS) Protocol Version 1.1. RFC 4346 (April 2006), http://www.ietf.org/rfc/rfc4346.txt
12. Dierks, T., Rescorla, E.: The Transport Layer Security (TLS) Protocol Version 1.2. RFC 5246 (August 2008), http://www.ietf.org/rfc/rfc5246.txt
13. Eastlake III, D.: Transport Layer Security (TLS) Extensions: Extension Definitions. RFC 6066 (January 2011), http://www.ietf.org/rfc/rfc6066.txt
14. GnuTLS Documentation (2011), http://www.gnu.org/software/gnutls/documentat.html
15. Krawczyk, H.: The Order of Encryption and Authentication for Protecting Communications (or: How Secure is SSL?). In: Kilian, J. (ed.) CRYPTO 2001. LNCS, vol. 2139, pp. 310–331. Springer, Heidelberg (2001)
16. Liberatore, M., Levine, B.: Inferring the source of encrypted HTTP connections. In: ACM Conference on Computer and Communications Security, pp. 255–263 (2006)
17. Manral, V.: Cryptographic Algorithm Implementation Requirements for Encapsulating Security Payload (ESP) and Authentication Header (AH). RFC 4835 (April 2007), http://www.ietf.org/rfc/rfc4835.txt
18. Maurer, U., Tackmann, B.: On the Soundness of Authenticate-then-Encrypt: Formalizing the Malleability of Symmetric Encryption. In: Proc. 2010 ACM Conference on Computer and Communications Security (CCS 2010), pp. 505–515. ACM (2010)
19. Möller, B.: Security of CBC Ciphersuites in SSL/TLS: Problems and Countermeasures, http://www.openssl.org/~bodo/tls-cbc.txt
20. Rogaway, P., Shrimpton, T.: A Provable-Security Treatment of the Key-Wrap Problem. In: Vaudenay, S. (ed.) EUROCRYPT 2006. LNCS, vol. 4004, pp. 373–390. Springer, Heidelberg (2006)

21. Sun, Q., Simon, D., Wang, Y., Russell, W., Padmanabhan, V., Qiu, L.: Statistical identification of encrypted web browsing traffic. In: IEEE Symposium on Security and Privacy, pp. 19–30 (2002)
22. Vaudenay, S.: Security Flaws Induced by CBC Padding - Applications to SSL, IPSEC, WTLS. In: Knudsen, L.R. (ed.) EUROCRYPT 2002. LNCS, vol. 2332, pp. 534–546. Springer, Heidelberg (2002)
23. White, A., Matthews, A., Snow, K., Monrose, F.: Phonotactic Reconstruction of Encrypted VoIP conversations: Hookt on fon-iks. In: IEEE Symposium on Security and Privacy (2011)
24. Wright, C., Ballard, L., Coull, S., Monrose, F., Masson, G.: Spot Me if You Can: Uncovering Spoken Phrases in Encrypted VoIP Conversations. In: IEEE Symposium on Security and Privacy, pp. 35–49 (2008)
25. Wright, C., Ballard, L., Coull, S., Monrose, F., Masson, G.: Uncovering Spoken Phrases in Encrypted Voice over IP Conversations. ACM Trans. Inf. Syst. Secur. 13(4) (2010)
26. Wright, C., Monrose, F., Masson, G.: On Inferring Application Protocol Behaviors in Encrypted Network Traffic. Journal of Machine Learning Research 6, 2745–2769 (2006)
27. Wright, C., Coull, S., Monrose, F.: Traffic Morphing: An Efficient Defense Against Statistical Traffic Analysis. In: Network and Distributed Security Symposium – NDSS (2009)

Resettable Cryptography in Constant Rounds – The Case of Zero Knowledge

Yi Deng[1,*], Dengguo Feng[2,**], Vipul Goyal[3], Dongdai Lin[2,***], Amit Sahai[4,†], and Moti Yung[5]

[1] NTU Singapore and SKLOIS, Institute of Software, CAS, China
[2] MSR India
[3] SKLOIS, Institute of Software, CAS, China
[4] UCLA
[5] Google Inc., USA

Abstract. A fundamental question in cryptography deals with understanding the role that randomness plays in cryptographic protocols and to what extent it is necessary. One particular line of works was initiated by Canetti, Goldreich, Goldwasser, and Micali (STOC 2000) who introduced the notion of resettable zero-knowledge, where the protocol must be zero-knowledge even if a cheating verifier can reset the prover and have several interactions in which the prover uses the same random tape. Soon afterwards, Barak, Goldreich, Goldwasser, and Lindell (FOCS 2001) studied the setting where the *verifier* uses a fixed random tape in multiple interactions. Subsequent to these works, a number of papers studied the notion of resettable protocols in the setting where *only one* of the participating parties uses a fixed random tape multiple times. The notion of resettable security has been studied in two main models: the plain model and the bare public key model (also introduced in the above paper by Canetti et. al.).

In a recent work, Deng, Goyal and Sahai (FOCS 2009) gave the first construction of a *simultaneous* resettable zero-knowledge protocol where both participants of the protocol can reuse a fixed random tape in any (polynomial) number of executions. Their construction however required $O(n^\epsilon)$ rounds of interaction between the prover and the verifier. Both in

---

 * Supported by the Singapore National Research Foundation under Research Grant NRF-CRP2-2007-03, the National Natural Science Foundation of China Under Grant NO.60803128, and the National 973 Program of China under Grant 2007CB311202.
 ** Supported by the National 973 Program of China under Grant 2007CB311202.
 *** Supported by the National 973 Program of China under Grant 2011CB302400 and the National Natural Science Foundation of China under Grant 60970152.
 † Supported in part from a DARPA/ONR PROCEED award, NSF grants 1118096, 1065276, 0916574 and 0830803, a Xerox Foundation Award, a Google Faculty Research Award, an equipment grant from Intel, and an Okawa Foundation Research Grant. This material is based upon work supported by the Defense Advanced Research Projects Agency through the U.S. O ffice of Naval Research under Contract N00014-11-1-0389. The views expressed are those of the author and do not reflect the official policy or position of the Department of Defense or the U.S. Government.

the plain as well as the BPK model, this construction remain the only known simultaneous resettable zero-knowledge protocols.

In this work, we study the question of round complexity of simultaneous resettable zero-knowledge in the BPK model. We present a *constant round* protocol in such a setting based on standard cryptographic assumptions. Our techniques are significantly different from the ones used by Deng, Goyal and Sahai.

1 Introduction

A fundamental question in cryptography deals with understanding the role that randomness plays in cryptographic protocols and to what extent it is necessary. Progress on this question was made relatively early with the result of Goldreich and Oren [GO94] showing that zero knowledge protocols cannot exist in the setting where the parties do not have access to any randomness resource at all. While this work showed that randomness cannot be completely eliminated, it simultaneously motivated several natural questions studying the "extent" to which randomness is necessary. A rich line of work deals with studying the usage of imperfect randomness in various settings (see [KLRZ08, DOPS04] and the references therein). Another line of work (and the one dealt with in this paper) studies whether all the random choices can be made "offline" and be fixed once and for all. In other words, is it possible to design cryptographic protocols where a party can reuse the same random tape in multiple (or even all) executions?

The question of reusing randomness in cryptographic protocols was first considered in the context of zero knowledge by Canetti, Goldreich, Goldwasser, and Micali [CGGM00] who proposed the notion of *resettable zero knowledge*. In resettable zero knowledge, the zero knowledge property is required to hold even if a malicious verifier can "reset" the prover to the initial state and start a new interaction *where the prover uses the same random tape*. Canetti et al. [CGGM00] proposed constructions of resettable zero knowledge protocols based on standard cryptographic assumptions. Barak, Goldreich, Goldwasser, and Lindell [BGGL01] showed how to construct zero knowledge protocols for opposite setting (where soundness is required to hold even if the verifier uses the same random tape in multiple executions), which following Micali and Reyzin [MR01b][1] they call resettably sound (rS) zero-knowledge. Barak et. al. also showed that any resettable sound zero-knowledge protocol must make use of non-black-box simulation techniques (introduced in a breakthrough work of Barak [Bar01]).

Subsequent to these two works, a number of papers have studied the notion of resettable security primarily in the setting where *only one* of the participating parties uses a fixed random tape multiple times. Protocols have been proposed in the so called *plain model* (cf. [CGGM00, BGGL01, BLV03, DL07a, GS09]). A larger body of literature studies resettable security in the so called *bare public key* (BPK) model. In the BPK model, a (possibly adversarial chosen) public

---

[1] Micali and Reyzin defined resettable soundness (and other soundness notions) in what is called the bare public key model.

key is selected and published by the verifier(s) before any protocol interaction starts [2]. Protocol for resettable security in the BPK model were studied in [CGGM00, MR01b, ZDLZ03, CPV04, DL07b, YZ07]. A more complete account of the related works is given in a later subsection.

In a recent work, Deng, Goyal and Sahai (FOCS 2009) gave the first construction of a *simultaneous* resettable zero-knowledge protocol where both participants of the protocol can reuse a fixed random tape in any (polynomial) number of executions. Their construction was in the plain model. The construction however required n^ϵ rounds of interaction between the prover and the verifier. Even in the BPK model, the DGS construction remains the best known simultaneous resettable zero-knowledge protocol. This motivates the following question:

"Does there exist a polylogarithmic (or even constant) round simultaneous resettable zero-knowledge protocol in the BPK model?"

Our Results. In this paper, we resolve the above question by constructing a constant round protocol for simultaneous resettable zero-knowledge in the BPK model. Our main theorem is as follows.

Theorem 1. If there exist trapdoor permutations and collision resistant hash function families, then there exist constant-round resettably-sound resettable ZK arguments for **NP** in the BPK model.

We leave open the question of round complexity of simultaneous resettable zero-knowledge in the plain model. Note that every resettable zero-knowledge protocol is also concurrent zero-knowledge [CGGM00]. Hence, a breakthrough will be required to construct a protocol in the plain model which matches the round complexity of the one in the BPK model given in our paper.

Our Techniques. The techniques used in our paper are quite different from the ones used in the DGS construction [DGS09]. Here we outline the main technical problem which is required to be resolved to obtain a constant round construction of simultaneous resettable zero-knowledge in the BPK model.

The source of large round complexity in the DGS construction is the usage of recursive rewinding strategies (cf. [RK99, KP01, PRS02]) which are coupled with a novel non-black-box simulation strategy. In the BPK model however, it is indeed possible to avoid recursive rewinding because of the existence of a "long term" trapdoor associated with the public key of the verifier (which the simulator can try to extract). At a high level, our protocol in the BPK would follow the following structure. The verifier would first prove knowledge of a long term trapdoor associated with the public key using a zero-knowledge protocol. The prover would then give a witness indistinguishable argument of knowledge (WIAOK) proving either $x \in L$ or that it "knows" such a trapdoor. Very roughly, now once the simulator extracts a long term trapdoor for a public key, it never needs to rewind a session with that public key (and the simulation can be done

---

[2] Such a model is quite different from having a "setup assumption" where one would assume, e.g., that a trusted party ensured that the public key was chosen correctly.

straight line). This would lead to a much simpler rewinding strategy avoiding large round complexity.

The key problem that arises while implementing the above approach in the simultaneous resettable setting is that obtaining a WIAOK protocol from the prover to the verifier is non-trivial and quite complex (since an adversarial verifier may rewind the prover to extract the witness). Instead, we would like to resort to using ZAPs [DN00] which are two round WI protocol (and hence already "secure" in the simultaneous resettable setting). Using a ZAP leads to the following problem. To arrive a contradiction in the proof of (resettable) soundness, the prover should be forced to prove a false statement about the trapdoor of the verifier (since we are not using an argument of *knowledge* protocol). This is turn means that the theorems the verifier proves about its long term trapdoor must also be false (this is important for the proof of resettable zero-knowledge to go through). However note that statements about the same public key (and the long term trapdoor) are being proven by the verifier in multiple sessions. To simulate its proof in all of those sessions, it seems that the verifier will need to use a (constant round) concurrent zero-knowledge protocol!

To overcome this problem, the verifier needs to be able to prove different statements in different sessions *with the same public key* such that some of them could be false while the others are true. This might suggest that the witness (containing the trapdoor) used by the verifier in each session is different. Yet we need that once we extract a trapdoor for any of these sessions, it should be a long term trapdoor which should enable the simulator to simulate every session with this public key (including even future sessions). Our protocol uses a careful technique to resolve this tension between "using sufficiently different witnesses in each session" and yet having "a common long term trapdoor binding them all". Our full protocol is described in Section 3.

Related Work. Subsequent to the works of Canetti et al. [CGGM00] and Barak et al. [BGGL01] described above, a number of works have investigated the problem of security against resetting attacks for zero-knowledge protocols in the plain model. Barak, Lindell, and Vadhan [BLV03] constructed the first constant-round public-coin argument that is *bounded* resettable zero-knowledge. Deng and Lin [DL07a] showed a zero-knowledge argument system that is bounded resettable zero-knowledge and satisfies a weak form of resettable soundness.

A larger body of work has investigated the same problems in a relaxed setting, called the "bare public key" (BPK) model, introduced by [CGGM00], which assumes that parties must register (arbitrarily chosen) public keys prior to any attack taking place. [CGGM00] presented a constant-round resettable zero-knowledge argument in the BPK model, the round complexity of which was improved by Micali and Reyzin [MR01b]. Micali and Reyzin [MR01b] also first investigated different notions of soundness in the BPK model, including the notion of resettable soundness. Di Crescenzo, Persiano, and Visconti [CPV04] described a resettable zero-knowledge protocol with concurrent soundness, and Deng and Lin [DL07b] improved the computational assumptions needed to obtain this result. Yung and Zhao [YZ07] also construct resettable zero-knowledge and concurrently sound arguments in the BPK model, using a general and efficient transformation. Micali and Reyzin [MR01a] also proposed a stronger variant of the BPK model for constructing bounded-secure protocols, and provided constant-round bounded resettable zero-knowledge arguments in this model; this result was strengthened by Zhao et al. [ZDLZ03] also in a bounded setting for resettable zero knowledge.

Goyal and Sahai [GS09] study the notion of general resettable two-party and multi-party computation and presented general feasibility results when only one of the parties may be reset. In this work, we restrict ourselves to the study of the zero-knowledge functionality.

Rest of this paper. We provide some basic definitions in section 2. In section 3, we construct a constant-round resettably-sound *concurrent* ZK arguments for NP in the BPK model. At last, we apply the transformation of Deng, Goyal and Sahai [DGS09] to the protocol constructed in section 3 to obtain our main result.

2 Definitions

Notation. We abbreviate probabilistic polynomial time as PPT. A function $f(n)$ is said to be negligible if for every polynomial $q(n)$ there exists an N such that for all $n \geq N$, $f(n) \leq 1/q(n)$. If L is a language in NP, we define the *associated relation* as the relation $R_L = \{(x, w) \mid x \in L; w$ is a witness for '$x \in L$'$\}$.

Interactive Arguments in the BPK Model. The bare public-key model (BPK model) assumes that:

- A public file F that is a collection of records, each containing a verifier's public key, is available to the prover.
- An (honest) prover P is an interactive polynomial-time algorithm that is given as inputs a secret parameter 1^n, a n-bit string $x \in L$, a witness w for $x \in L$, a public file F and a random tape r.
- An (honest) verifier V is an interactive polynomial-time algorithm that works in two stages. In stage one (key registration stage), on input a security parameter 1^n and a random tape r, V generates a key pair (pk, sk) and stores pk in the file F. In stage two (proof stage), on input sk, an n-bit string x and a random string ρ, V performs the interactive protocol with a prover, and outputs "accept x" or "reject x".

Definition 1 (Complete Interactive Arguments in the BPK Model). *We say that the protocol $< P, V >$ is complete for a language L in $\mathcal{NP}$, if for all n-bit string $x \in L$ and any witness w such that $(x, w) \in R_L$, the probability that V interacting with P on input w, outputs "reject x" is negligible in n.*

Malicious Resetting Provers in the BPK model. Let s be a positive polynomial and P^* be a PPT algorithm on input 1^n.

A resetting attack by a s-resetting malicious prover P^* in the BPK model is defined as the following process:

– Run the key generation stage of V on input 1^n and a random string r to obtain pk and sk. P^* obtains pk and V stores the corresponding sk.
– Choose $s(n)$ random string ρ_i, $1 \leq i \leq s(n)$, for V.
– P^* is allowed to choose an instance x and initiate any (polynomial) number of sessions with each verifier and interact with it in the second stage (proof stage) of the protocol. The i-th verifier uses input sk, ρ_i.

Definition 2 (Resettably sound arguments in the BPK model).
$< P, V >$ *satisfies* resettable soundness *for an NP language L in the BPK model if for all positive polynomial s, for all s-resetting malicious prover P^*, the probability that in an execution of resetting attack, P^* ever receives "accept x" for $x \notin L$ from any of these oracles is negligible in n*

Malicious Resetting/Concurrent Verifiers in the BPK model. A resetting attack by an (s, t)-resetting malicious PPT verifier V^*, for any two positive polynomials s and t, can be defined as the following process:

– In the key generation stage, on input 1^n, V^* receives s instances $x_1, ..., x_{s(n)} \in L$ of length n each, and, outputs an arbitrary public file F
– Choose $r_1, ..., r_{s(n)}$ for P uniformly at random.
– In proof stage, V^* starts in the final configuration of the key generation stage, is given oracle access to $s^3(n)$ provers, $P(x_i, w_i, pk_j, r_k, F)$, $1 \leq i, j, k \leq s(n)$.
– V^* finally outputs its entire view of the interaction (i.e., its random tape and the messages received from the provers). The total number of steps of V^* in both stages is at most $t(n)$.

The concurrent attack by V^* is defined in the same way except that we choose s^2 random tapes $r_{i,j}, 1 \leq i, j \leq s$, and V^* is allowed to interact with s^2 provers $P(x_i, w_i, pk_j, r_{i,j}, F)$ $(1 \leq i, j \leq s)$ concurrently. Note that here each random tape is used only once.

Definition 3 (Resettable zero-knowledge in the BPK model). $< P, V >$
is (non-black-box) resettable zero knowledge for an NP language L in the BPK model if for every pair of positive polynomials (s, t), for all (s, t)-resetting malicious verifier V^*, there exists a simulator S, given as input the description of V^*, such that for every $x_1, ..., x_{s(n)} \in L$, the following two distributions are computationally distinguishable:

   1. The output of V^* at the end of a resetting attack described above,

   2. The output of $S(V^*, x_1, ..., x_{s(n)})$.

Definition 4 (Concurrent zero-knowledge in the BPK model). $< P, V >$
is (non-black-box) concurrent zero-knowledge for an NP language L in the BPK model if for every pair of positive polynomials (s, t), for all (s, t)-concurrent malicious verifier V^*, there exists a simulator S, given as input the description of V^*, such that for every $x_1, ..., x_{s(n)} \in L$, the following two distributions are computationally distinguishable:

   1. The output of V^* at the end of a concurrent attack described above,

   2. The output of $S(V^*, x_1, ..., x_{s(n)})$.

3   Constructing Resettably-Sound *Concurrent* Zero Knowledge Arguments for NP in the BPK Model

As a first step towards obtaining a simultaneous resettable zero-knowledge protocol, we present a resettably-sound *concurrent* zero knowledge argument for an NP language in the BPK Model in this section. We will later show how to use a compiler described in [DGS09] to obtain our main theorem.

Let (G, E, D) be a semantically secure public-key encryption scheme, where G, E, and D denote key-generation algorithm, encryption algorithm, and decryption algorithm respectively. The commitment scheme Com is a statistically binding and computationally hiding commitment scheme. $Com(s, r)$ denotes the commitment to a string s using the random tape r. The protocol proceeds as follows.

The resettably-Sound Concurrent ZK Argument (P, V) in the BPK model

The key registration stage: V runs the key generation algorithm G of a semantically secure public key encryption scheme (G, E, D) twice independently, $(pk_0, sk_0) = G(1^n, r_0^k)$, $(pk_1, sk_1) = G(1^n, r_1^k)$, publishes (pk_0, pk_1) and stores r_b^k and sk_b for a random $b \in \{0, 1\}$.

The proof stage (main protocol):

Common input: x (supposedly in L) and verifier's public key (pk_0, pk_1).

P's private input: the witness w such that $(x, w) \in R_L$.

V's private input: the randomness r_b^k used in key generation for one of the public keys.

P's randomness: r_p.

V's randomness: r_v.

1. P sends a commitment $c = Com(e, r)$ to a random challenge e.
2. V Computes two ciphertexts of 0 under pk_0 and pk_1 independently, $c_0 = E(pk_0, 0, r_0)$, $c_1 = E(pk_1, 0, r_1)$; Send c_0, c_1 and the first message a of the 3-round WI proof of Hamiltonian Cycle for the following statement:
   (a) there exists r_b^k such that $(pk_b, sk_b) = G(1^n, r_b^k)$ (equivalently, "I know one of secret keys"); and,
   (b) there exist r_0 and r_1 such that $c_0 = E(pk_0, 0, r_0)$ and $c_1 = E(pk_1, 0, r_1)$ (i.e., both cipertexts are encryption of 0).
   The randomness used by V in this step as well as the rest of the protocol is generated by applying a pseudorandom function f_{r_v} to the first message c of the prover.
3. P sends e and executes the BGGL protocol in which P proves that either: 1) there exists r such that $c = Com(e, r)$, or, 2) $x \in L$.
4. V now responds to the challenge e by sending the final message z of the 3-round WI protocol of Hamiltonian Cycle.

5. P executes a ZAP in which P proves that either $x \in L$ or there exists r_d^k, $d \in \{0, 1\}$, such that $(pk_d, sk_d) = G(1^n, r_d^k)$ and $0 = D(sk_d, c_d)$ (i.e., one of the decryptions result to the message 0).

Remark 1. For simplicity of presentation, we view *com* and ZAPs as non-interactive protocol requiring only one message in each direction. However our construction can indeed use two round protocols for each in a straight-forward way.

Remark 2. Note that there is fine difference between the verifier and the prover in proving a ciphertext is an encryption of 0: the verifier uses the knowledge of randomness in encryption to prove the ciphertext is an encryption of 0, while the prover uses the knowledge of the secret key (more precisely, randomness that used to generate the public/secret key pair) to prove that one plaintext is actually 0. We stress that this difference is crucial for security proof. In the course of simulation, once our simulator extracts the randomness used for generating one of pk_0 and pk_1 (note that it does not need the randomness used in these encryptions by the verifier to execute a session), it can handle all sessions under the same public key (pk_0, pk_1). On the other hand, in the proof of soundness, the reduction algorithm, playing the role of verifier, needs only one of secret keys to execute a session, and this will enable it to use the power of cheating prover to either break the semantic security of the other public key scheme or break the WI property of the underlying 3-round WI protocol if such a cheating prover exists.

We now state the following theorem.

Theorem 2. The above protocol (P, V) is a resettably-sound concurrent zero knowledge argument.

The completeness is obvious. We will prove concurrent zero knowledge and resettable-soundness in next two subsections.

Hardness assumption. Note that the 2-round statistically-binding commitment scheme and semantically secure public key encryption scheme can be based on trapdoor permutations, which also imply the existence of ZAPs. In addition, we need to assume collision-resistant hash functions required for the resettably sound BGGL protocol (which makes use of non-black-box simulation techniques). Thus we can base the above resettably-sound concurrent ZK argument on the assumption of existence of trapdoor permutations and collision-resistant hash function families.

3.1 Proof of Concurrent Zero-Knowledge

Let V^* be an concurrent malicious verifier. Assume w.l.g. in real world, on input a fixed YES instance sequence $x_1, ..., x_{s(n)} \in L$ of length n each, V^* generates s public keys $F = ((pk_0^1, pk_1^1), ..., (pk_0^s, pk_1^s))$, and interacts with $s^2(n)$ incarnations of prover, $P(x_i, w_i, (pk_0^j, pk_1^j), r_{i,j}, F)$, $1 \le i, j \le s(n)$. We now construct a simulator S as required by definition 3.

S operates as follows. First, given a fixed YES instance sequence $x_1, ..., x_s \in L$ of length n each as input, S runs the key-generation phase of V^* to obtain the public file F.

In proof stage, the first task of S is to extract one r_b^k ($b \in \{0, 1\}$) for each public key pair (pk_0^j, pk_1^j) such that r_b^k is the randomness used for generating one public key pk_b^j. Note that once these r_b^k's are obtained, S is able to carry out all sessions successfully in a straight-line manner by decrypting one of two ciphertexts (and relying on the soundness of the WI protocol). We say a session under public key (pk_0^j, pk_1^j) is *solved* if S already extracted the corresponding randomness r_b^k; otherwise, we say it is *unsolved*.

The extraction is done in a sequential way. Once receiving an accepting execution of the 3-round WI protocol in an *unsolved* session under public key (pk_0^j, pk_1^j), S rewinds to the beginning of step 3, sends a random challenge e' and runs the simulator for BGGL protocol to prove that c is a commitment to e'. When another accepting execution of this subprotocol is obtained, S solved all sessions under this public key.

We would like to make the following remarks on the above extraction:

- The non-black-box simulator for the *standalone* BGGL protocol handles only a single session, but it runs in a concurrent setting. This means, during the execution of this subprotocol, many other sessions may appear. To deal with this issue, we have the following strategy. First observe that all the other sessions are being executed honestly by the simulator (and the current rewinding thread will be aborted if an unsolved session reaches its final prover message). Thus, we consider these sessions (and the part of the simulator handling these sessions) as part of the adversarial machine itself. Then our modified non-black-box simulator Sim will now simply act on this new machine (by using its code) instead of the original one.
- For the analysis of running time to go through, we use the Goldreich-Kahan technique to bound the running time of S.

The detailed description of S follows.

The Simulator S:

Input: the code of V^*, s YES instances $x_1, ..., x_s$.

1. select a random tape for V^*, and run the key-generation phase of V^* to obtain the public file $F = ((pk_0^1, pk_1^1), ..., (pk_0^s, pk_1^s))$.
2. Set $h \leftarrow (x_1, ..., x_s)$ and $\mathcal{S} \leftarrow \emptyset$.
3. Do the following:
   (a) Adopt the honest prover strategy until the final ZAP in every session, and extend h to include the transcript generated in this step. If V^* terminates during this step, return h; Otherwise, go to next step.
   (b) If a *solved* session reaches the final ZAP, use the relevant randomness and secret key to produce a prover message of the final ZAP, and extend h to include this message. If V^* terminates during this step, return h; Otherwise, go to next step.

(c) If an *unsolved* session reaches the end of of the underlying 3-round WI protocol, and the resulting transcript (a, e, z) so far is accepting, do the following:

- (**Estimation**) Suppose that the first two messages sent in the current session are $c, (c_0, c_1, a)$, and the corresponding public key is (pk_0^j, pk_1^j). Rewind P^* to the point (we call it **rewinding point**) where the verifier's message (c_0, c_1, a) was just sent, and repeat the following until it receives the accepting transcript (a, e, z) of the underlying 3 round WI argument n^2 times: send the honest challenge e and choose *independent randomness* to execute the underlying BGGL protocol honestly; when another unsolved session reaches the final ZAP, S aborts the current thread[3].
  We denote by X the total number of iterations (or threads) of this step.

- (**Extraction**) Rewind V^* to the above **rewinding point** again, and repeat the following until it obtains another accepting transcript (a, e', z') with $e \neq e'$ until the $X + 1^{st}$ iteration is reached. If all iterations fail, output "$\perp$".

  - For the current session, S send a new random challenge $e' \neq e$, and then runs the non-black-box simulator Sim to prove that c is a commitment to e', where Sim proceeds exactly the same as the simulator for the BGGL protocol (except for acting on the new adversarial machine as described earlier).
  - For any other solved session, S executes the strategy described in step b; if an unsolved session reaches the final ZAP, S aborts the current iteration.

(d) From the two accepting transcripts of the 3-round WI protocol (a, e, z) and (a, e', z'), compute the randomness r_b^k such that $(pk_b^j, sk_b^j) = G(1^n, r_b^k)$,[4] and update $\mathcal{S}$ to include r_b^k, and go to step 1. (Note that the above step 3(c) does not update history).

The concurrent zero knowledge property of our protocol follows from the following claims.

Claim 1 S runs in expected polynomial time.

Claim 2 The output h by S is indistinguishable from real interaction.

Proof of Claim 1. We first count the number of queries which the simulator makes to the adversary. Observe that the number of queries which S makes in a single solved session is a constant C. Suppose that for a specific session i, S enters step 3(c) with probability p_i, then we have for this session, the expected number of iterations in step 3(c) is at most $p_i \cdot (2n^2/p_i) < 2n^2$. Since V^* is only allowed to initiate s^2 sessions, the entire simulation of S will make an

---

[3] in this case, S cannot proceed further without knowledge of the relevant secret key.
[4] Note that we can also compute the randomness that were used in the two encryptions to 0, but we don't need it to carry out the final ZAP.

expected $s^2 \cdot C \cdot (2n^2 + 1)$ number of queries (which is polynomial). Since each query additionally requires only polynomial time, the overall running time of the simulator is expected polynomial.   □

Proof of Claim 2. We first prove the probability that S outputs ⊥ is negligible. Observe that S outputs ⊥ only if it fails to extract a relevant secret key.

Assume that for session i, S enters step 3(c) with probability p_i (taken over the random coins used in step 3 of the protocol; here prover proves that e is the correct challenge). We claim that in a single run of the **Extraction** in step 3(c), the probability that S obtains an accepting transcript of the 3-round WI protocol is at least $p_i - neg(n)$ for some negligible function neg (except for a negligible fraction of protocol prefixes, i.e., transcripts of steps 1 and 2), otherwise, we can use V^* to break either the computational-hiding property of the scheme Com or the zero knowledge property of the BGGL protocol.

Note that the Goldreich-Kahan technique [GK96] guarantees that, the estimation n^2/X of p_i is within a constant factor of p_i except with exponentially small probability, thus, we conclude that $X > n^2/(c \cdot p_i)$ holds for some constant c except with exponentially small probability.

Thus, the probability that S enters step 3(c) but doesn't extract out the randomness used in generation of some public key is

$$p_i(1 - p_i + neg)^X$$
$$\leq p_i(1 - p_i + neg)^{n^2/(c \cdot p_i)}$$

which is negligible.

Observe that the only difference between S and the honest prover is that they use different witness to carry out the final ZAP in each session. Now by the WI property of the ZAP, we conclude that h is indistinguishable from the real interaction between honest provers and V^*.   □

3.2   Proof of Resettable-Soundness

Assume that there is a PPT resetting P^* that can cheat an honest verifier V (and complete a protocol execution) on a NO instance x with noticeable probability p. We shall now consider the following 5 hybrid verifier strategies. We shall prove that in each hybrid, the probability of the verifier being able to cheat (in some session) is still noticeable. In the final hybrid, we note that the above cheating probability must be negligible by the soundness of the ZAP system (and thus arrive at a contradiction). We shall first describe the hybrid strategies and then argue that the probability of cheating remains negligible in each.

V_1: Follow the honest verifier strategy V, except that whenever V is instructed to applying the pseudorandom function specified by its random tape to generate randomness, V_1 uses truly random coins (while still making sure that for a given prover first message c, it always uses the same random coins).

V_2: Follow the strategy below.

1. In the key registration stage, V_2 acts exactly as V_1.
2. In the proof stage, V_2 first picks a session i at random.
   Suppose that the first prover message in session i is c, and that the public key is (pk_0, pk_1) and the secret key stored by V_2 is sk_b for some $b \in \{0, 1\}$.
3. For all sessions having a first prover message different than c, V_2 executes honest verifier's strategy throughout the entire interaction between P^* and V_2.
4. For all sessions having the first prover message c, V_2 executes honest verifier's strategy *until* when a session among them first completes an accepting proof via BGGL protocol for the correctness of challenge e, and then rewinds to the point where it received c for the first time, computes two encryptions of 0 under *both* public key pk_b and pk_{1-b} honestly again, produces a fake first massage a that can answer e successfully according to the 3-round WI protocol[5], and continue (without using the actual witness).

V_3: Follow the strategy of V_2 except that, in item 4 of V_2, computes an encryption of 0 under public key pk_b and an encryption of 1 under public key pk_{1-b} after extracting the challenge e and then rewinding (but produces the first message a in the same way as V_2),

V_4: Follow the strategy of V_3 except that, in *all sessions*, whenever V_3 needs to use r_b^k as partial witness to carry out the 3-round WI protocol, V_4 uses r_{1-b}^k.

V_5: Follow the strategy of V_4 except that, after rewinding, V_5 computes two encryptions of 1 under pk_0 and pk_1 respectively in those sessions having the first prover message c.

First, we have that P^* can cheat V_1 with probability negligibly close to p, due to the pseudorandomness of the pseudorandom function specified by the random tape of V.

We now prove that P^* can cheat V_2 in a session having the first prover message c with probability negligibly close to $p/poly$, where $poly$ is the total number of distinct first prover messages appeared in the whole interaction between P^* and V_2. Observe that for a randomly chosen first prover message c, P^* will cheat V_1 in a session having this first prover message with probability exactly $p/poly$, and that the only difference between the second run of V_2 and V_1 is the way in which the transcript (a, e, z) is produced. Since in the 3-round protocol for Hamiltonian Cycle, the simulated transcript (a, e, z) is computationally indistinguishable to a real one, we conclude that V_2 will accept with probability negligibly close to $p/poly$ in a session having the first prover message c.

We further claim that P^* can also cheat V_3 in a session having the first prover message c with probability negligibly close to $p/poly$. Notice that the only difference between V_2 and V_3 is, in their second run (after rewinding), V_2 encrypts to 0 under public key pk_{1-b}, while V_3 encrypts to 1 under public key

---

[5] In the 3-round WI protocol for Hamiltonian Cycle, given a challenge e, there exists a simple simulator that can produce an accepting transcript (a, e, z) efficiently.

pk_{1-b}. Notice also that in both their second runs, the message a is produced independently of these encryptions. Thus, if the aforementioned claim is false, we can construct an algorithm V_h to break the semantic security of the public key encryption scheme: V_h acts as V_2 except that, after rewinding, it obtains the ciphertext (that is supposed to be 0 or 1) under the public key pk_{1-b} from an external challenger, instead of computing this ciphertext itself; When P^* convinces V_h to accept in a session having the first prover message c, V_h outputs 0, otherwise, outputs 1. Observe that if the ciphertext obtained from encryption oracle is an encryption of 0, then V_h is identical to V_2; if this ciphertext is an encryption of 1, V_h is identical to V_3. Hence, in a session having the first message c, if there is a non-negligible gap between the probability that V_2 accepts and the probability that V_3 accepts, V_h breaks the semantic security of the underlying public key encryption scheme.

For strategies V_3 and V_4, we observe that the only difference between them is that they use different witnesses to carry out the 3-round WI protocol. Consider the following algorithm V_{wi}.

V_{wi}:
1. In the key registration stage, V_{wi} generates two public keys honestly, i.e., it computes $(pk_0, sk_0) = G(1^n, r_0^k)$, $(pk_1, sk_1) = G(1^n, r_1^k)$, publishes (pk_0, pk_1), chooses a random bit b and stores both r_0^k and r_1^k.
2. Like V_2, V_{wi} first picks a session i at random. Again, suppose that the first prover message in session i is c.
3. For all sessions having a first prover message different than c, when a session with a distinct first prover message c' was initiated for the first time, V_{wi} executes honest verifier's strategy to compute two encryptions of 0, $c_0 = E(pk_0, 0, r_0)$ and $c_1 = E(pk_1, 0, r_1)$, send (r_0^k, r_1^k, r_0, r_1) to an independent honest prover P_{wi} of the 3-round WI protocol, and forward the P_{wi}'s first message a' along with c_0, c_1 to P^*; Once a session with the first prover message c' first completes the correctness proof via BGGL protocol for the challenge e', V_{wi} sends e' to P_{wi} and forwards P_{wi}'s answer z' to P^*; in all sessions with c' as the first prover message, V_{wi} sends the same (a', c_0, c_1) to P^*, and if P^* reveals the same e' again and completes the correctness BGGL proof, V_{wi} answers with the same z'; Otherwise, V_{wi} outputs "failure".
4. When P^* sends c for the first time, V_{wi} acts the same as the above strategy: computes two encryptions of 0, sends all random tapes to an independent P_{wi} and forward P_{wi}'s first message a (and the two encryptions) to P^*. Once P^* repeats c, V_{wi} responds with the same a. When a session with the first prover message c first completes an accepting proof via BGGL protocol for the correctness of challenge e, it rewinds to the point where it received c for the first time, computes an encryptions of 0 under public key pk_b and an encryption of 1 under public key pk_{1-b}, produces a fake first massage a that can answer e successfully according to the 3-round WI protocol, and continue.

We first note that V_{wi} outputs "failure" only if P^* opens some commitment c' to two different values and gives two accepting proofs for both. Due to the statistically-binding property of the commitment scheme and resettable-soundness of the BGGL protocol, the probability that V_{wi} outputs "failure" is negligible. Note also that, each independent P_{wi} is run once (i.e., the 3-round WI protocol is executed in *concurrent* setting), and that if all these P_{wi}'s uses r_b^k (resp., r_{1-b}^k) as partial witness, then V_{wi} is identical to V_3 (resp., V_4). Note that the 3-round WI protocol is *concurrent* witness indistinguishable. Thus, we conclude that the probability that P^* cheats V_4 in a session with the first prover message c is negligibly close to $p/poly$.

Finally, notice that both V_4 and V_5 do not use the knowledge of the randomness r_b^k (used in generation the public/secret key pair (pk_b, sk_b)) to carry out any session in their entire interaction, and the only difference between them is that they encrypt different messages under pk_b in sessions having the first prover message c after rewinding. Similar to the analysis of V_2 and V_3, due to the semantic security of the public key encryption scheme (pk_b, sk_b), the probability that P^* cheats V_5 in a session with the first prover message c is negligibly close to $p/poly$. However, since both ciphertexts in these sessions are encryptions of 1, by the soundness of the ZAP system, P^* can cheat V_5 in any one of these sessions only with negligible probability. Thus we have p is negligible.

4   Simultaneous Resettable Zero-Knowledge Arguments for NP in the BPK model

In this section, we apply the transformation of [DGS09] to the resettably-sound concurrent ZK arguments presented in the last section, and obtain simultaneously resettable arguments for NP in the BPK model. This establishes theorem 1.

Given a resettably-sound concurrent ZK argument (P_{RC}, V_{RC}) for NP language L in the BPK model and a common input $x \in L$, the simultaneously resettable argument (P, V) for L proceeds as follows.

The key registration stage: V acts exactly the same as V_{RC} in the key registration stage.

The proof stage:

Common input: x (supposedly in L) and verifier's public key ver_k

P's randomness: (γ_p^1, γ_p^2)

V's randomness: (γ_v^1, γ_v^2)

1. P uses randomness γ_p^1 to generate a random string r_p (of appropriate length) and a first verifier message ρ_p of a ZAP system. P sends $C_p = Com(r_p)$ and ρ_p (where Com is a perfect binding commitment scheme).
2. V sets $(\tau_v^1, \tau_v^2) = f_{\gamma_v^1}(x, ver\_k, C_p)$. Using randomness τ_v^1, V generates the first verifier message ρ_v and compute a commitment $C_t = Com(0)$ to 0. V sends ρ_v and C_t.
3. V and P execute the BGGL protocol in which V uses random tape τ_v^2 and proves that C_t is a commitment to 0. In addition, in each verifier step in this subprotocol, P generates a ZAP proof along with each verifier message for the following OR statement:

(a) The current message is produced by an honest verifier of the BGGL protocol using random tape r_p, or,

(b) $x \in L$

4. V sets $(\tau_v^3, \tau_v^4) = f_{\gamma_v^2}(hist)$, where $hist$ is the history so far except those ZAP proofs. Using randomness τ_v^3, V sends a commitment $C_v = Com(\tau_v^3)$ to P. In the remaining steps, V uses randomness τ_v^4.

5. P sets $\tau_p = f_{\gamma_p^2}(hist)$. Using random tape τ_p, P and V execute (P_{RC}, V_{RC}) in which P proves $x \in L$, except that for every V_{RC}'s message, we have V give an additional ZAP proof for the following OR statements:

(a) the current message is produced by an honest verifier of (P_{RC}, V_{RC}) using random tape τ_v^3, or,

(b) C_t is a commitment to 1.

V accepts if and only if V_{RC} accepts the transcript of (P_{RC}, V_{RC}).

Remark. In [DGS09], the actual transformation of resettably-sound concurrent ZK argument into a resettably-sound resettable ZK argument takes two steps: 1) transform the resettably-sound concurrent ZK argument into a *hybrid* sound *hybrid* zero knowledge argument; 2) transform a hybrid sound hybrid zero knowledge protocol into a resettably-sound resettable zero knowledge protocol. The second step is done by simply having each party refresh their randomness via a pseudorandom function. Here for the sake of simplicity and keeping the proof short, we merge these two steps into a single transformation (and refer the reader to [DGS09] for a detailed formal presentation).

Theorem 3. The protocol (P, V) is a resettably-sound resettable zero knowledge.

Proof sketch. The proof of this theorem is similar in spirit to the one appeared in [DGS09]. Here we just give a proof outline.
The *completeness* is obvious.

Resettable-Soundness. For a given cheating prover P^* for (P, V) and a NO instance $x \notin L$, we can construct a series of hybrid verifiers to show the cheating probability is negligible just like the hybrid verifiers V_1, V_2, V_3, V_4 and V_5 we set up in the previous section. Whenever a hybrid verifier needs to rewind in some target sessions with a specific first prover message C_p, it always computes a commitment C_t to 1 in its first step, and then runs the simulator for the BGGL protocol to prove that C_t is a commitment to 0 in all sessions having the same first prover message C_t[6]; Whenever it produces a fake first message a of the underlying 3-round WI protocol in (P_{RC}, V_{RC}), it uses the witness for "C_t is a commitment to 1" to execute ZAP for the correctness of message a. Similar to the analysis presented in previous section, it is not hard to show that, if all building blocks are secure, the above protocol (P, V) is resettably-sound.

---

[6] Note that, all subexecutions of BGGL protocol in these sessions are actually *identical*, due to the resettable-soundness of ZAP and the instance x to be proven is a NO instance. This is why the simulator for BGGL protocol in the standalone setting works in this specific resettable setting.

Resettable ZK. Note that the BGGL protocol is resettably-sound, and hence for any malicious resetting verifier, if an execution of BGGL protocol in step 3 is accepting, the message C_t sent in step 2 is guaranteed to be a commitment to 0 (except with negligible probability). As a consequence, all verifier's messages sent in the subprotocol (P_{RC}, V_{RC}) are determined by the commitment C_v sent in step 4 and the session history of (P_{RC}, V_{RC}) due to the fact that ZAP is resettably-sound, that is, for a fixed session prefix until step 4, all subexecutions of (P_{RC}, V_{RC}) are *identical*. This observation enables us to adopt essentially the same simulation strategy of S which works for *concurrent* adversary and prove the property of resettable zero knowledge. Given a resetting verifier V^*, our simulator S' proceeds as follows. For all sessions, S' follows the honest prover strategy until step 4. When reaching the subprotocol (P_{RC}, V_{RC}), S' acts as the simulator S for (P_{RC}, V_{RC}). For those solved sessions, S' uses the relevant secret key as witness to carry out the final ZAP. When an unsolved session reaches the end of the 3-round WI protocol in (P_{RC}, V_{RC}), S' applies the extraction strategy of S to extract a secret key. We can perform a similar analysis and show that S' will run in expected polynomial time and its output is distinguishable from that in the real interaction.   □

Acknowledgement. Yi Deng would like to thank Ivan Visconti and Pino Persiano for helpful discussions. We thank anonymous referees for their constructive and valuable comments.

References

[Bar01]   Barak, B.: How to go beyond the black-box simulation barrier. In: FOCS, pp. 106–115 (2001)

[BGGL01]   Barak, B., Goldreich, O., Goldwasser, S., Lindell, Y.: Resettably-sound zero-knowledge and its applications. In: FOCS, pp. 116–125 (2001)

[BLV03]   Barak, B., Lindell, Y., Vadhan, S.P.: Lower bounds for non-black-box zero knowledge. In: FOCS, pp. 384–393 (2003)

[CGGM00]   Canetti, R., Goldreich, O., Goldwasser, S., Micali, S.: Resettable zero-knowledge (extended abstract). In: STOC, pp. 235–244 (2000)

[CPV04]   Di Crescenzo, G., Persiano, G., Visconti, I.: Constant-Round Resettable Zero Knowledge with Concurrent Soundness in the Bare Public-Key Model. In: Franklin, M. (ed.) CRYPTO 2004. LNCS, vol. 3152, pp. 237–253. Springer, Heidelberg (2004)

[DGS09]   Deng, Y., Goyal, V., Sahai, A.: Resolving the simultaneous resettability conjecture and a new non-black-box simulation strategy. In: FOCS, pp. 251–260. IEEE Computer Society (2009)

[DL07a]   Deng, Y., Lin, D.: Instance-Dependent Verifiable Random Functions and Their Application to Simultaneous Resettability. In: Naor, M. (ed.) EUROCRYPT 2007. LNCS, vol. 4515, pp. 148–168. Springer, Heidelberg (2007)

[DL07b]   Deng, Y., Lin, D.: Resettable Zero Knowledge with Concurrent Soundness in the Bare Public-Key Model Under Standard Assumption. In: Pei, D., Yung, M., Lin, D., Wu, C. (eds.) Inscrypt 2007. LNCS, vol. 4990, pp. 123–137. Springer, Heidelberg (2008)

[DN00]   Dwork, C., Naor, M.: Zaps and their applications. In: FOCS, pp. 283–293 (2000)
[DOPS04]   Dodis, Y., Ong, S.J., Prabhakaran, M., Sahai, A.: On the (im)possibility of cryptography with imperfect randomness. In: FOCS, pp. 196–205. IEEE Computer Society (2004)
[GK96]   Goldreich, O., Kahan, A.: How to construct constant-round zero-knowledge proof systems for np. J. Cryptology 9(3), 167–190 (1996)
[GO94]   Goldreich, O., Oren, Y.: Definitions and properties of zero-knowledge proof systems. J. Cryptology 7(1), 1–32 (1994)
[GS09]   Goyal, V., Sahai, A.: Resettably Secure Computation. In: Joux, A. (ed.) EUROCRYPT 2009. LNCS, vol. 5479, pp. 54–71. Springer, Heidelberg (2009)
[KLRZ08]   Kalai, Y.T., Li, X., Rao, A., Zuckerman, D.: Network extractor protocols. In: FOCS, pp. 654–663. IEEE Computer Society (2008)
[KP01]   Kilian, J., Petrank, E.: Concurrent and resettable zero-knowledge in poly-loalgorithm rounds. In: STOC, pp. 560–569 (2001)
[MR01a]   Micali, S., Reyzin, L.: Min-Round Resettable Zero-Knowledge in the Public-Key Model. In: Pfitzmann, B. (ed.) EUROCRYPT 2001. LNCS, vol. 2045, pp. 373–393. Springer, Heidelberg (2001)
[MR01b]   Micali, S., Reyzin, L.: Soundness in the Public-Key Model. In: Kilian, J. (ed.) CRYPTO 2001. LNCS, vol. 2139, pp. 542–565. Springer, Heidelberg (2001)
[PRS02]   Prabhakaran, M., Rosen, A., Sahai, A.: Concurrent zero knowledge with logarithmic round-complexity. In: FOCS, pp. 366–375 (2002)
[RK99]   Richardson, R., Kilian, J.: On the Concurrent Composition of Zero-Knowledge Proofs. In: Stern, J. (ed.) EUROCRYPT 1999. LNCS, vol. 1592, pp. 415–431. Springer, Heidelberg (1999)
[YZ07]   Yung, M., Zhao, Y.: Generic and Practical Resettable Zero-Knowledge in the Bare Public-Key Model. In: Naor, M. (ed.) EUROCRYPT 2007. LNCS, vol. 4515, pp. 129–147. Springer, Heidelberg (2007)
[ZDLZ03]   Zhao, Y., Deng, X., Lee, C.H., Zhu, H.: Resettable Zero-Knowledge in the Weak Public-Key Model. In: Biham, E. (ed.) EUROCRYPT 2003. LNCS, vol. 2656, pp. 123–139. Springer, Heidelberg (2003)

Two Provers in Isolation[*]

Claude Crépeau[1,**], Louis Salvail[2], Jean-Raymond Simard[3], and Alain Tapp[2]

[1] School of Computer Science, McGill University,
Montréal, QC, Canada
crepeau@cs.mcgill.ca
[2] Département d'Informatique et R.O., Université de Montréal,
Montréal, QC, Canada
{salvail,tappa}@iro.umontreal.ca
[3] GIRO inc., Montréal, QC, Canada
Jean-Raymond.Simard@GIRO.ca

Abstract. We revisit the Two-Prover Bit Commitment Scheme of BenOr, Goldwasser, Kilian and Wigderson [BGKW88]. First, we introduce Two-Prover Bit Commitment Schemes similar to theirs and demonstrate that although they are classically secure using their proof technique, we also show that if the provers are allowed to share quantum entanglement, they are able to successfully break the binding condition. Secondly, we translate this result in a purely classical setting and investigate the possibility of using this Bit Commitment scheme in applications. We observe that the security claim of [BGKW88] based on the assumption that the provers cannot communicate is not a sufficient criteria to obtain soundness. We develop a set of conditions, called *isolation*, that must be satisfied by any third party interacting with the provers to guarantee the binding property of the Bit Commitment.

1 Introduction

The notion of Multi-Prover Interactive Proofs was introduced by BenOr, Goldwasser, Kilian and Wigderson [BGKW88]. In the Two-Prover scenario, we have two provers, Peggy and Patty, that are allowed to share arbitrary information before the proof, but they become physically separated from each other during the execution of the proof, in order to prevent them from communicating. It was demonstrated by Babai, Fortnow, and Lund [BFL91] that Two-Prover Interactive Proofs (with a polynomial-time verifier) exist for all languages in NEXP-time. A fully parallel amalog was achieved by Lapidot and Shamir [LS97].

A quantum mechanical version of this scenario was considered by Kobayashi, Matsumoto, Yamakami and Yao [KM03, KMY03, Yao03]. To this day, it is still

---

[*] An earlier version of this work was presented under the title "Classical and Quantum Strategies for Two-Prover Bit Commitments", at QIP '06, *The 9th Workshop on Quantum Information Processing*, January 16-20, 2006, Paris.

[**] Supported in part by CIFAR, NSERC, MITACS, QuantumWorks and FQRNT's INTRIQ.

D.H. Lee and X. Wang (Eds.): ASIACRYPT 2011, LNCS 7073, pp. 407–430, 2011.

an open problem to establish the exact power of Multi-Prover Quantum Interactive Proofs. A rather vast litterature now exists on this topic (see [BHOP08], [CSUU07], [DLTW08], [IKM09], [IKPSY08], [KKMV08], [Weh06]). However, it is still not even clear whether two provers are as powerful as more-than-two provers.

The Two-Prover Zero-Knowledge Interactive Proofs of [BGKW88] rely on the construction of a Bit Commitment scheme, information theoretically secure under the assumption that the provers cannot communicate. We refer the reader to their paper to understand the application of this Bit Commitment scheme to the construction of Two-Prover Zero-Knowledge Proofs. We solely focus on their Bit Commitment scheme for the rest of our work. In this paper, we consider several important questions regarding Two-Prover Bit Commitment schemes. We do not limit our interest of Two-Prover Bit Commitment to the context of Zero-Knowledge proofs; as already discussed in [BGKW88] similar techniques lead them to a secure Oblivious Transfer under the same assumption. Given that *any* two-party computation may be achieved from Oblivious Transfer [Kil88], we consider the security of such Bit Commitment scheme in a very general context. We discuss at length the security in a very general composability situation.

In order to argue the security of their Bit Commitment scheme, the authors of [BGKW88] asserted the following assumption:

```
"there is no communication between the two provers while
  interacting with the verifier".
```

The current paper is concerned with the sufficiency of this assertion. We show is Section 3.2 that, although this assumption *must be made*, it is however considerably too weak, because we exhibit variations of the scheme that are equally binding classically but that are not at all binding if the provers were allowed to share entanglement. It is however a very well known fact that entanglement does not allow communication. Although it is true that they can cheat if they can communicate, it is also true that they can cheat without communicating. Therefore the assumption that the provers cannot communicate is too weak.

This observation can be turned into a purely classical argument by exhibiting a black-box two-party computation, that does not allow them to communicate, but that allows them to cheat the binding condition of the Bit Commitment scheme. This peculiar source of randomness may replace the entanglement used by the attack. Furthermore, the above assertion of BGKW can be interpreted as a prescription to the verifier that he should make sure not to help the provers to communicate while interacting with him. Again, this prescription would not prevent him from acting like the black-box we exhibit. Thus, a stronger prescription is mandatory in order to assert security.

We carefully define a notion of *isolation* by which the two provers may not communicate nor perform any non-local sampling beyond what is possible via quantum mechanics. We finally formalize a set of conditions that any third party involved in a Two-Prover Bit Commitment scheme may satisfy to make sure he does not break the assumption that the provers are in isolation. In particular, we make sure that if such a Bit Commitment scheme is used in another larger cryptographic protocol, its security properties will carry over to the larger context.

1.1 Related Work

The starting point of this research is clearly the Bit Commitment scheme introduced by BenOr, Goldwasser, Kilian and Wigderson [BGKW88]. The security of a Two-Prover Bit Commitment scheme against quantum adversaries has been considered in the past in the work of Brassard, Crépeau, Mayers and Salvail [BCMS98]. They showed that if such a Bit Commitment scheme is used in combination to the Quantum Oblivious Transfer protocol of [BCMS98] it is not sufficient to guarantee the security of the resulting QOT if the two provers can get back together at the end of the protocol. In the current work, we consider only the situation while the provers are isolated.

The research by Cleve, Høyer, Toner and Watrous [CHTW04] is the main inspiration of the current paper. They have established some relations between Two-Prover Interactive Proofs and so called "non-locality games". More precisely, they showed that certain languages have a classical Two-Prover Interactive Proof that looses soundness if the provers are allowed to share entanglement. Some of our results are very similar to this. However, our new contributions are numerous. While [CHTW04] focuses on languages, we focus on the tool known as Bit Commitment. This tool is used in many contexts other than proofs of membership to a language: proofs of knowledge, Oblivious Transfer, Zero-Knowledge proofs, general two-party computations. Moreover inspired by the observations of [CHTW04], we analyze the security of such Two-Prover tools in a completely classical situation. We conclude that proving security of such protocols is very subtle when used in combination with other such tools. We also argue that the claim of security of the protocols of [BGKW88] requires a lot more assumptions than the mere "no communication" assumption (even in the purely classical situation).

Despite the impossibility theorems of Mayers [May96] and Lo & Chau [LC97], the possibility of information theoretically secure Bit Commitment schemes in the Two-Prover model is *not* excluded in the classical and quantum models. Indeed, the computations sufficient to cheat the binding condition of a Quantum Bit Commitment scheme in the above "no-go" theorems cannot, in general, be performed by the two provers when they are isolated from each other. This is the reason why these theorems do not apply.

In a closely related piece of work, Kent [Ken05] showed how impossibility of communication, implemented through relativistic assumptions, may be used to obtain a Bit Commitment scheme similar to BGKW that can be constantly updated to avoid cheating. Kent proves the classical security of his scheme while remaining elusive about its quantum security. However, he claims security of one round (see [Ken05], Lemma 3, p. 329) of his protocol which is more or less the same as our Lemma 1. Unfortunately, his proof is incomplete as pointed out in our proof of the Lemma. But we clearly recongnized that he was first to address this question.

A very different set of results [BCU+06] relates non-locality boxes and two-party protocols such as Bit Commitment and Oblivious Transfer. These are only marginally connected to the current research. They showed how these cryptographic protocols may be securely implemented from those non-locality boxes. On the cotrary, we show how to break such protocols using non-locality boxes...

2   Preliminaries

2.1   Isolation

First let us define the condition imposed on the two provers: we use the word *isolation* to describe the relation between Peggy and Patty during the protocol. The intuitive meaning of this term is that Peggy and Patty cannot communicate with each other, since this condition is explicitly imposed by the Two-Prover model. However, we introduce this new terminology instead of the traditional "cannot communicate with one another" because we noticed that the meaning of "no-communication" is too weak and must be very clearly defined to produce valid security proofs. This *isolation* will be formally defined in Section 4. For now, the reader may follow his intuition and picture Peggy and Patty as restricted to compute their messages using only local variables.

2.2   Bit Commitment

The primitive known as "Bit Commitment" is a protocol in which a player Alice first sends some information to another player Bob, such that this information *binds* her to a particular bit value b. However, the information sent by Alice is not enough for Bob to learn b (b is *concealed*). At a later time, Alice sends the rest of the information to unveil the bit b, and she cannot change her mind to reveal $\bar{b}$ and convince Bob that this was the value to which she was committed in the first step. The following definitions will be used to characterize the security of a Bit Commitment scheme. Note that the function $\mu(n)$ always refers to a negligible function in n.

Definition 1. *A Bit Commitment scheme is* statistically concealing *if only a negligible amount of information on the committed bit can leak to the verifier before the unveiling stage.*

Definition 2. *A Bit Commitment scheme is* statistically binding *if, for $b \in \{0, 1\}$, the probability p_b that Alice successfully unveils for b satisfies*

$$p_0 + p_1 \leq 1 + \mu(n). \tag{1}$$

This binding condition was first proposed by Dumais, Mayers, and Salvail [DMS00], as a weaker substitute to the traditional definition $p_b \leq \mu(n)$ for either $b = 0$ or 1. This definition has been henceforward used to show security of many Bit Commitment schemes against quantum adversaries in various models, e.g. [DMS00, CLS01, DFSS05].

More recent definitions have been introduced since then ([DFRSS07]) that appear to be better characterization of Bit Commitment security in a quantum setting. However, we have not been able, so far, to find protocols that satisfy these definitions. This, we hope, will be part of future work in this area.

3   Two-Prover Bit Commitment scheme

For simplicity reasons, we replace the original scheme of [BGKW88] by a far simpler and compact version, which we call "simplified-BGKW" (or sBGKW as a short-hand). Still, we strongly recommend the reader to [BGKW88] for the details of the original construction. For an n-bit string r and a bit b, we define the n-bit string $b \cdot r := b \wedge r_1 || b \wedge r_2 || \ldots || b \wedge r_n$. The scheme is as follows:

Peggy and Patty agree on a uniform n-bit string w and a random bit d. They are then isolated from one another.

Protocol 31 (sBGKW - Commit to b)

  1: *Vic sends a random n-bit string r to Patty,*
  2: *Patty replies with $x := (d \cdot r) \oplus w$,*
  3: *Peggy announces $z := b \oplus d$.*

Protocol 32 (sBGKW - Unveil b)

  1: *Peggy announces bit b and the n-bit string w,*
  2: *Vic accepts iff $w = ((b \oplus z) \cdot r) \oplus x$.*

Note that at the unveiling stage, as in the original scheme it is not required that Peggy be the one announcing b. It is as good to let Vic deduce b: Vic computes $y := w \oplus x$, if $y = 0^n$ he sets $b := z$ and if $y = r$ he sets $b := \bar{z}$, and otherwise rejects. Indeed, Peggy may not even know b!

3.1   BGKW's Notion of Isolation

The assumption made in [BGKW88] is that Peggy and Patty are not allowed to communicate with each other. Based solely on that constraint, the following seems a "valid" security proof (it is more or less the same proof as in [BGKW88]).

Theorem 1. *Constraining the provers as in* [BGKW88], *the* sBGKW *protocol is secure classically.*

Proof. Vic does not know w, and w is uniformly distributed among all possible n-bit strings for both values of z. It follows that the two strings w and $r \oplus w$ have the exact same uniform distribution and are perfectly indistinguishable from one another. We can say the same for the pairs (z, w) and $(z, r \oplus w)$. Hence sBGKW is concealing.

Now suppose that Peggy and Patty would like to be able to unveil a certain instance of b both as 0 and as 1. To do so, Peggy would like to announce $\widehat{w}_b$ such that $\widehat{w}_b = (b \cdot r) \oplus x$. We note that this models the two possible dishonest behaviors for Peggy and Patty: honestly commit to $\bar{b}$ and try to change to b afterwards, and commit to nothing by sending some x and decide which b they want to unveil *only* at the unveiling stage. It follows that in both scenarios, a successful cheating strategy would allow to produce the two strings $\widehat{w}_0$ and $\widehat{w}_1$, such that $\{\widehat{w}_0, \widehat{w}_1\} = \{x, r \oplus x\}$. However, the string $\widehat{w}_0 \oplus \widehat{w}_1 = x \oplus r \oplus x = r$ is completely unknown to Peggy by the no-communication assumption. Therefore, even using unlimited computational power, her probability of issuing a valid pair $\widehat{w}_0, \widehat{w}_1$ is at most $1/2^n$. Hence sBGKW is binding.

Nevertheless, this result is incomplete[1]! Indeed, we show next how a correlated random variable can be used to invalidate the result of Theorem 1 while not violating the "no-communication" assumption. This suggest that the conventional wording "no-communication" is insufficient as it is not explicit enough to cover any kind of cheating mechanism Peggy and Patty can employ.

3.2   Cheating sBGKW with an NL-box

An **NL**-box, short-hand for "Non-Locality box" introduced by Popescu and Rohrlich [PR94, PR97], is a device with two inputs s and t, and two output bits u and v such that u and v are individually uniformly distributed and satisfy the relation $f(s,t) = u \oplus v$ for some function f. The pair (s, u) is on Peggy's side while the pair (t, v) is on Patty's side. Because u and v are individually uniformly distributed, no **NL**-box allow Peggy and Patty to communicate, in either direction. The **NL**-boxes are usually assumed as asynchronous devices, that is, feeding in the input s is sufficient to obtain u even if t has not been input yet, and likewise for t. Such a particular box, known as the **PR**-box, is defined for $f(s,t) = s \wedge t$, where s and t are binary inputs. It is known that two classical players can simulate the **PR**-box with success probability[2] at most 75% for all s, t, while quantum players sharing an entangled state can achieve a success probability of $\cos^2(\pi/8) \approx 85\%$ (consult [CHTW04] for details).

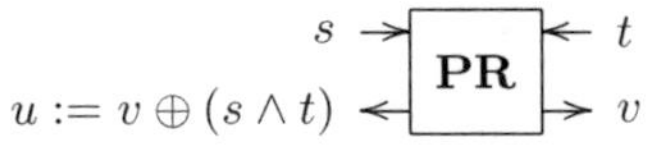

$$s \rightarrow \boxed{\textbf{PR}} \leftarrow t$$
$$u := v \oplus (s \wedge t) \leftarrow \phantom{\boxed{\textbf{PR}}} \rightarrow v$$

Fig. 1. the cheating **PR**-box

Let the two provers be given a black-box access to this **PR**-box. The following shows how this **PR**-box allows Peggy and Patty to unveil the bits committed

---

[1] The broad explanation is that we implicitly *assumed* the provers had only access to local variable. We'll see we need to guarantee this restriction for the proof to hold.

[2] This result is shown optimal by enumerating every possible classical strategies.

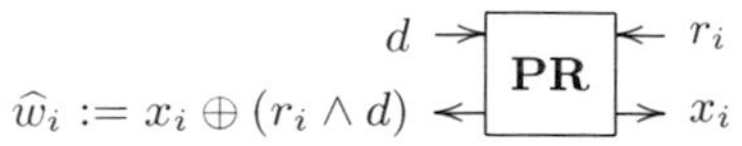

$$\widehat{w}_i := x_i \oplus (r_i \wedge d) \quad \boxed{\text{PR}} \quad \begin{array}{c} \leftarrow r_i \\ \rightarrow x_i \end{array}$$

Fig. 2. Using the **PR**-box

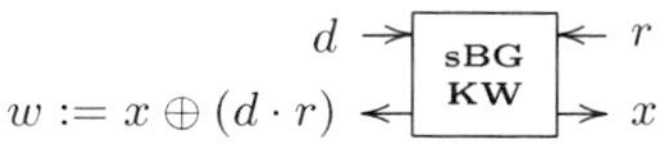

$$w := x \oplus (d \cdot r) \quad \boxed{\begin{array}{c}\text{sBG}\\\text{KW}\end{array}} \quad \begin{array}{c} \leftarrow r \\ \rightarrow x \end{array}$$

Fig. 3. The cheating **sBGKW**-box

through sBGKW in either way, at Peggy's will. For each position i, $1 \le i \le n$, Patty inputs in the **PR**-box the bit $s := r_i$ received from Vic and obtains output $x_i := u$ from the **PR**-box, which corresponds to the i-th bit of the commitment string. Patty sends x to Vic. Peggy discloses z a random bit to Vic. To unveil bit b, Peggy inputs $t := d := b \oplus z$ in the **PR**-box and obtains the output $\widehat{w}_i := v$ from the **PR**-box, which she sends to Vic together with b.

If $d = 0$ then $d \wedge r_i = 0$ and thus $\widehat{w}_i = x_i$ which is the right value she must disclose. If $d = 1$ then $d \wedge r_i = r_i$ and thus $\widehat{w}_i \oplus x_i = r_i$ or $\widehat{w}_i = x_i \oplus r_i$ which is again the right value she must disclose.

Indeed, we can view an arbitrary cheat on the sBGKW as a non-local computation between the provers as in Fig. 3. Essentially we have just demonstrated that an sBGKW-box can be emulated perfectly by perfect **PR**-boxes. However, a valid cheating strategy might not succeed 100% of the time, so an sBGKW-box that is correct 80% of the time, for instance, would be enough to break the binding property. It seems quite obvious, nevertheless, that a **PR**-box that is correct 80% of the time will not help implementing an sBGKW-box that is correct 80% of the time. For that matter, any **PR**-box that is correct a constant fraction $p < 1$ of the time will not help either...

It is not obvious that a **sBGKW**-box with error probability greater than zero is equivalent to the **PR**-box, but it would be very interesting to prove either way.

3.3 Quantumly Insecure - Two-Prover Bit Commitments

We exhibit an intermediate scheme to emphasize how shared entanglement can be used to cheat with probability almost one a classically "secure" Two-Prover Bit Commitment. The protocol is a weaker version of the sBGKW scheme, called wBGKW, where the acceptance criteria of the unveiling stage is loosen to tolerate some errors. A second protocol (available in Sub-Section 3.7) is also a modified version of the sBGKW scheme where the acceptance criteria is based on a game described later, called the Magic Square game.

A weaker acceptance criteria: the wBGKW scheme Consider a weaker acceptance criteria where the string $\widehat{w}$ sent by Peggy can differ in at most $n/5$ positions from what it should be. Formally the verifier Vic is to accept b if $d(\widehat{w}, ((b \oplus z) \cdot r) \oplus x) < n/5$, where $d(\cdot)$ is the binary Hamming distance. The interest of such a modification is that now a cheating quantum pair Peggy and Patty can use the non-local property of entanglement to approximate the **PR**-box and successfully cheat wBGKW, while, as we show next, the Bit Commitment is "secure" classically. To facilitate notation we add an index b to the string $\widehat{w}$, since $\widehat{w}$ is different whether we unveil zero or one. Also, define as B the random variable corresponding to the value they unveil.

Theorem 2. *For any classical strategy, the probability that it outputs a string $\widehat{w}_0$ when $B = 0$ and $\widehat{w}_1$ when $B = 1$ s.t. $E[d(\widehat{w}_b, ((b \oplus z) \cdot r) \oplus x)] < n/5$ for both values of b, is exponentially small in n.*

Proof (of Theorem 2).
  Wlog, we can assume the provers use a deterministic strategy that may produce such a $\widehat{w}_0$ when $B = 0$, and $\widehat{w}_1$ when $B = 1$, so they can in fact output *both* $\widehat{w}_0$ and $\widehat{w}_1$. Hence, Peggy can compute the string $\widehat{w}_0 \oplus \widehat{w}_1$. Recall that when $d(\widehat{w}_b, ((b \oplus z) \cdot r) \oplus x) = 0$ then $\widehat{w}_0 \oplus \widehat{w}_1 = r$. We want to determine the distance between $\widehat{w}_0 \oplus \widehat{w}_1$ and r in our situation. From the theorem's assumption, there exists a classical strategy that outputs $\widehat{w}_0$ *and* $\widehat{w}_1$ such that $E[d(\widehat{w}_b, ((b \oplus z) \cdot r) \oplus x)] < n/5$, for $b = 0, 1$. We easily obtain that for such a strategy, the expected distance from r is

$$E[d(\widehat{w}_0 \oplus \widehat{w}_1, r)] = E[d(\widehat{w}_0 \oplus \widehat{w}_1, x \oplus (x \oplus r))] \leq E[d(\widehat{w}_0, x)] + E[d(\widehat{w}_1, x \oplus r)] < 2n/5$$

by the triangular inequality. Using a standard Chernoff bound argument, and since r is absolutely unknown to Peggy, her probability of outputting a string $y = \widehat{w}_0 \oplus \widehat{w}_1$ such that $E[d(y, r)] < (1/2 - \epsilon) \cdot n$ is exponentially small in n for any $0 < \epsilon \leq 1/4$. Hence, because $1/4 < 2/5 < 1/2$, we conclude that such a strategy cannot exist except with exponentially small probability, and so unveiling *must* fail for one of the two possibilities.

Conversely, this scheme is almost totally insecure against quantum adversaries.

Theorem 3. *There exists a quantum strategy that successfully cheats the wBGKW scheme with probability $1 - \mu(n)$.*

Proof (of Theorem 3). We saw in Section 3.2 that the **PR**-box, taken as a black box, correctly produces the needed $\widehat{w}_b$ to unveil as b. Using the well-known result [e.g. [CHTW04]] that through entanglement, Peggy and Patty can optimally simulate the **PR**-box such that for each i taken independently, $1 \leq i \leq n$, the **PR**-box produces correlated outputs with probability $\cos^2(\pi/8) \approx 0.85$. Therefore, using the standard Chernoff bound, this independent quantum strategy yields that for both values of b,

$$E[d(\widehat{w}_b, ((b \oplus z) \cdot r) \oplus x)] = (1 - \cos^2(\pi/8)) \cdot n$$

with probability exponentially close to one. Having that $(1 - \cos^2(\pi/8)) \cdot n < 0.15 \cdot n < n/5$, we conclude that a pair of quantum provers defeats the binding condition of the scheme with probability $1 - \mu(n)$.

3.4   Discussion

The limitation of Theorem 1 (and Theorem 2) is that it claims that the following non-local computation, named **sBGKW2**-box (see Fig. 4) , is a communication

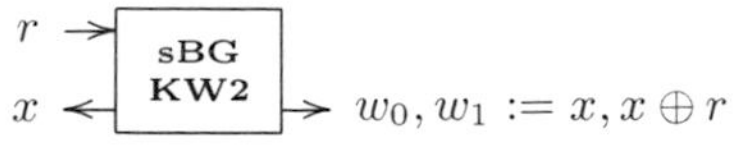

$$r \rightarrow \boxed{\begin{array}{c} \text{sBG} \\ \text{KW2} \end{array}}$$
$$x \leftarrow \quad \rightarrow w_0, w_1 := x, x \oplus r$$

Fig. 4. the cheating **sBGKW2**-box

device (which is obvious) assuming that any implementation of an **sBGKW**-box is sufficient to implement it (which is false, since the **sBGKW**-box is *not* a communication device, it is impossible to implement any communication device from it).

However, these proofs are not *wrong* either since it is impossible to accomplish the **sBGKW**-box without some sort of communication, which also works for the **sBGKW2**-box. In particular, it means that this proof is seriously context-dependent. In a context where Patty and Peggy have access to a third party that scrupulously monitors that they are not communicating with each other, the proof does not hold anymore because using the third party as a **sBGKW**-box is not excluded.

The bottom line here is that this proof is valid *solely* in a stand-alone security model. As soon as one starts composing such protocols, one has to, not only, monitor that the actions of the third party do not allow communication but also do not constitute any form of correlation between Patty and Peggy.

This demonstrates that certain non-local correlations are enough to cheat Two-Prover Bit Commitment schemes while they are not enough to communicate. Thus we have to define the prover's isolation in terms of these non-local correlations and not only in terms of communication. This is the purpose of Section 4.

3.5   A Non-Local Box to Cheat the Original BGKW Scheme

Similarly to the sBGKW scheme, we can define an analogous cheating box for the original BGKW scheme with two binary inputs s, t, and two uniformly generated ternary outputs x, y.

The original protocol goes as follows:
Peggy and Patty agree on a uniform n-trit string w. They are then isolated from one another.

Protocol 33 (BGKW - Commit to b)

  1: *Vic sends a random n-bit string r to Patty,*
  2: *Patty replies with x such that for all k, $x_k := \sigma_{r_k}(w_k) - b \bmod 3$.*

Protocol 34 (BGKW - Unveil b)

1: *Peggy announces bit b and the string w,*
2: *Vic accepts iff w is such that for all k, $b = \sigma_{r_k}(w_k) - x_k \mod 3$.*

Where the σ function of [BGKW88] can be re-written as the single expression: $\forall\, r \in \{0,1\}, w \in \{0,1,2\}$

$$\sigma_r(w) = (1+r)w \mod 3. \qquad (2)$$

So using (2), we want from the cheating **NL**-box that $u := (s+1)v - t \mod 3$ for each s, t, and uniformly chosen v. Because for any binary s, t we can easily define the inverse permutation over trits to be $v := (t+u)(s+1) \mod 3$, the following **PR3**-box does not allow to communicate since individually u and v are uniformly distributed.

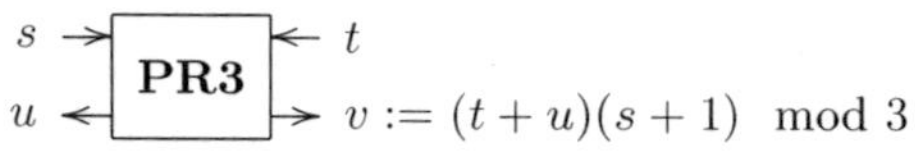

Fig. 5. A non-local box to cheat BGKW

It is not hard to verify that the **PR3**-box that implements this non-local computation from s, t is exactly the one needed to cheat the original BGKW scheme. As with the **PR**-box, for each round i, Peggy inputs in the box $s := r_i$ and obtains the trit $x_i := u$, which she sends to Vic. If Patty wants to unveil for b, she inputs $t := b$ in the **PR3**-box, which correctly outputs $\widehat{w}_i := v$. Clearly, they successfully cheat since

$$\forall\, i \quad (1+r_i)\widehat{w}_i - x_i \mod 3 = (1+r_i)(b+x_i)(1+r_i) - x_i \mod 3$$
$$= (1+r_i)^2(b+x_i) - x_i \mod 3$$
$$= (b+x_i) - x_i \mod 3$$
$$= b.$$

We can also demonstrate that the **PR3**-box is as powerful as the **PR**-box. It is straightforward to check that the outputs x' and y' depicted in Figure 6 are indeed the correct outputs to cheat the sBGKW scheme.

3.6   Magic Square Non-locality Game

A square is a 3×3 matrix whose entries are in $\{0,1\}$. A row is said to be *correct* if its parity is even, and a column is said to be *correct* if its parity is odd. We use the following definition of the Magic Square game (from [CHTW04]), which slightly differs from the original game due to Aravind [Ara02]. The verifier Vic picks at random a row or column, say column c_i, and a position x_j^i on c_i, $i, j \in \{1, 2, 3\}$.

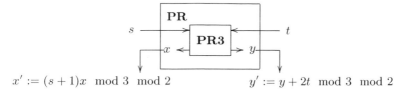

$$x' := (s+1)x \mod 3 \mod 2 \qquad\qquad y' := y + 2t \mod 3 \mod 2$$

Fig. 6. Reduction from the **PR**-box to the **PR3**-box.

He then asks the entries of column c_i to Peggy, and the value in position x_j^i to Patty. The two provers win if the parity of c_i is odd (more generally, if the row or column asked for is *correct*), and if the value returned by Patty matches the value at position x_j^i in Peggy's answer. The following defines the *validity* of a square.

Definition 3. *A (3 × 3) matrix S is valid for zero if all rows of S xor to 0, and S is valid for one when all columns of S xor to 1.*

For instance the following matrix S_0 is valid for zero while S_1 is valid for one:

$$S_0 = \begin{bmatrix} 0\,0\,0 \\ 0\,1\,1 \\ 1\,0\,1 \end{bmatrix}, \quad S_1 = \begin{bmatrix} 1\,0\,1 \\ 1\,1\,0 \\ 1\,0\,0 \end{bmatrix}. \tag{3}$$

Any classical strategy successfully wins this Magic Square game with probability at most $\left(\frac{17}{18}\right)$. Remarkably, there exists a quantum strategy that allows Peggy and Patty to successfully win this game *every* time, see [CHTW04, Ara02] for details.

3.7 Magic Square Bit Commitment

It is not hard to exploit the Magic Square game to build another Bit Commitment scheme. This scheme is particularly relevant in our study of Bit Commitments in the Two-Prover model as it is perfectly secure classically but can easily be cheated with probability one using a quantum strategy. The scheme is as follows:

Peggy and Patty agree on a random bit v and n random squares S_i such that S_i is valid for v. They are then isolated.

Protocol 35 (MSBC - Commit to b)

   1: *Peggy computes $x := v \oplus b$ and sends x to Vic.*
   2: *Vic picks a pair of random trits r_i and c_i and asks Peggy for $S_i(r_i, c_i)$.*

Protocol 36 (MSBC - Unveil b)

   1: *Peggy sends b to Vic,*
   2: *Vic asks Patty for row number r_i of S_i if $b = x$, or column number c_i of S_i if $b = \bar{x}$.*
   3: *Vic accepts b if, for each i, the row or column that should xor to b does, and if the entry returned by Peggy matches with Patty's answer. Vic rejects otherwise.*

Theorem 4. *Any classical strategy successfully cheats the binding property of the* MSBC *scheme with probability at most* $\left(\frac{8}{9}\right)^{n/6}$, *except with exponentially small probability.*

Proof (of Theorem 4).

Wlog, it is sufficient to consider deterministic strategies. Consider the strategy where only the entry $(2, 2)$ is used to make the square S_i correct for $\widehat{w}_i$. When $t_i = 0$ or 1, Peggy answers the line or column of S_i as is. However, when $t_i = 2$, she sets the entry $(2, 2)$ to the correct value such that a line xores to 0 or a column xores to 1. On query (y_i, z_i), Patty answers the entry (y_i, z_i) of S_i if $(y_i, z_i) \neq (2, 2)$, otherwise she answers 0. It is not hard to show that this strategy is optimal, since Peggy knows all the information (the S_i's, x, and r), and Patty knows nothing about x and r.

The problem for the provers is that whenever $b \cdot r_i = 1$, they succeed for at most only one of $b \in \{0, 1\}$. This is because the square S_i they share cannot be correct for both x_i and $\overline{x_i}$. Since r is uniformly distributed, by a Chernoff argument, r contains at least $n/3$ 1's. Thus, there is at least one of $b \in \{0, 1\}$ for which in at least $n/6$ challenges the provers will answer correctly with probability at most $8/9$ (the sum of the challenges where she succeeds with probability at most $8/9$ for 0, and those where she succeeds with probability at most $8/9$ for 1, adds up to $n/3$). Therefore, their probability of successfully cheating is at most $\left(\frac{8}{9}\right)^{n/6}$ for any classical strategy, except with exponentially small probability.

However, there exists a quantum strategy that allows Peggy and Patty to successfully break the binding condition with probability 1 by winning the Magic Square game *every* time.

Theorem 5. *There exists a quantum strategy that successfully cheats* MSBC *with probability 1.*

4   Defining and Checking Isolation

The existence of such an inputs-correlated[3] random variable, which does not allow communication but allows cheating of the sBGKW Two-Prover Bit Commitment scheme sheds some light on the limitations of the original assumption of [BGKW88].

Indeed, the assumption of [BGKW88] is necessary but not sufficient to guarantee the binding property of the Bit Commitment scheme. Among its weakness, we note that it does not *explicitly* force any cheating strategy to be repeatable. The **PR**-box not being a repeatable process[4] gives a first understanding why

---

[3] We emphasize that at least one of the "inputs" to the random variable needs to be obtained once the provers are isolated, otherwise such a random variable can be shared while the provers are together, and is thus useless to cheat the sBGKW scheme.

[4] The **PR**-box cannot be repeated to generate two *valid* strings $\widehat{w}_0$ and $\widehat{w}_1$.

we can still cheat the sBGKW scheme despite the result of Theorem 1, which implicitly assumed repeatability of the cheating strategy.

Clearly, to achieve the binding condition, a stronger assumption is needed. One could require that once the provers are isolated, there exists no mechanism by which they may sample a joint random variable which is dependent on the inputs they provide. We note that, among other things, this new condition excludes communication between the two provers, as desired. However, it excludes a lot more, such as shared entanglement! This is simply too strong; we need to be more subtle in the way we define this *"mechanism to sample a joint random variable"*.

It seems reasonable to believe that nature does not allow the existence of a **PR**-box (consult [CHTW04]). So why even ask for a stronger assumption than the no-communication assumption of [BGKW88]? Part of the answer is that Vic can play the role of the **PR**-box, or any other third party. In no circumstances can we ignore the fact that both Peggy and Patty individually talk to Vic. Definitely, we need to consider this aspect of the protocol with great care. For instance, consider the scenario where r is sent to Peggy but unveiling is not done immediately after committing, but rather once Vic and the two provers have been involved in other, unrelated, interactive protocols. It is perfectly conceivable that within those protocols, for each i, Peggy and Patty succeed in sending r_i and b to Vic, and then in a completely different context (or a moment of unawareness) Vic performs the required computation and output x_i and $\widehat{w}_i$, which are then sent respectively to Peggy and Patty. It is obvious that if such a computation, or any alike, can take place with enough probability then Peggy and Patty would succeed in cheating the sBGKW protocol!

More generally, we must not only consider Vic but any other third party, call it Ted, to which Peggy and Patty might have access to obtain correlated information. The previous situation highlights the fact that there is a whole class of functions with inputs coming from Peggy and Patty for which Ted must not send the outputs. Intuitively, each time Ted sends a message to either Peggy or Patty, he must ensure that the message does not outperform what Peggy and Patty can achieve using local variables in the sense of quantum mechanics. We propose two different approaches to formulate that statement as a criteria. The first considers the practical flavor of the problem, when Ted is working with instances of variables. The second approach is based on an information theoretic argument. At this point, we will not consider the scenario where the players can share quantum resources.

Let Peggy be identified by P_0 and Patty by P_1. The variable $D \in \{0,1\}$ is a reference to player P_D, and $T \in \{\emptyset, \{0\}, \{1\}, \{0,1\}\}$ is a tag appended to each message that indicates to Ted the player(s) that is (are) eligible for receiving this message, where $T = \{0,1\}$ means by both players and $T = \emptyset$ means by none of them. The message about to be sent from Ted to prover P_D is represented by $(m, T)_D$. We formalize Ted's behavior as follows.

Definition 4 (Practical criteria). *Ted is said to be a "secure third party" if* $\forall D \in \{0,1\}$, *Ted follows these points.*

1. *A message received from player P_D is tagged with $T := \{D\}$.*
2. *A message generated without involving any of the previous messages, e.g. picking a random string, is tagged with $T := \{0,1\}$.*
3. *A message obtained from a computation involving previous messages is tagged with the intersection of the tags of all the messages involved in that computation.*
4. *A message $(m,T)_D$ is sent to player P_D only if $D \in T$.*

Note: It is important that the communication pattern between Ted and the isolated provers be specified ahead of time, otherwise the traffic pattern (not only the message contents) may leak information.

We now explain why Ted will not send a message that allows P_0 and P_1 to communicate or establish non-local correlations. Let $(m,T)_D$ be the message Ted is about to send to player P_D. From the fourth point of Definition 4, Ted will send $(m,T)_D$ only if it is tagged $T = \{D\}$ or $\{0,1\}$. Looking at the message's tag assignment rule number 3, this happens only if there is absolutely no message tagged $\{1-D\}$ or $\emptyset$ used in the computation of $(m,T)_D$. Using an induction argument, it is not hard to see that this happens only when all the variables involved in the computation of $(m,T)_D$ are independent of the information of P_{1-D}, that is, they have been themselves generated using variables tagged $\{D\}$ or $\{0,1\}$. Thus, such a message $(m,T)_D$ is also independent of the information known only to P_{1-D}. Therefore, the messages sent by Ted do not let the two players communicate.

The case of non-locality is slightly more subtle, yet pretty straightforward. Recall that in a general non-local process, both players use a message each and receive a message uniformly distributed, from their point of view, such that the four messages satisfy a certain relation. The received message does not allow to communicate with the other player. Suppose P_{1-D} receives his message first. Since from his point of view, this message is uniformly distributed, Ted *can* in fact generate a uniformly distributed message, tag it with $T := \{0,1\}$ and send it to P_{1-D}. At this point, this behavior does not violate anything because non-locality has not been created yet. Then, Ted computes the message for P_D. Because this message needs to satisfy the relation that binds together the four messages, at least a message tagged with $T \neq \{D\}$ and one tagged with $T \neq \{1-D\}$ are used in its computation (it can be the same message), so the resulting message $(m,T)_D$ will be assigned a tag $T := \emptyset$ because the intersection does not contain $\{D\}$ nor $\{1-D\}$. This message $(m,\emptyset)_D$ is the one creating the non-local relation. However, from point 4 of Definition 4, since $D \notin \emptyset$, Ted will never send $(m,\emptyset)_D$.

As mentioned before the previous definition, we can alternatively formalize Ted's behavior in terms of entropy. The advantage of doing so is to enable analysis of existing protocols. To satisfy the above practical criteria, the wrapping protocol must be designed in a rather restricted way. To consider general protocols, we offer this alternate definition.

Let the message about to be sent from Ted to prover P_D be represented by the variable $(M,T)_D$. The set of variables $S_{D,T}$ represents all the variables (messages) with tag T sent by prover P_D to Ted, and the set of variables $R_{D,T}$ all the variables (messages) with tag T sent by Ted to prover P_D before $(M,T)_D$.

Definition 5 (Information based criteria). *Ted is said to be a "secure third party" if* $\forall D \in \{0, 1\}$, *Ted follows these points.*

1. *An information received from player P_D is tagged with $T := \{D\}$[5].*
2. *A variable M to be sent to P_D is tagged with the less restrictive tag $T \in \{\emptyset, \{D\}, \{0, 1\}\}$ that satisfies the following relation[6]. Note that the calligraphic tag $\mathcal{T}'$ stands for the tag $\{0, 1\}/(T \cap \{D\})$ and the calligraphic tag $\mathcal{T}''$ stands for the tag $\{D\} \cup (T \cap \{1 - D\})$.*

$$H((M, T)_D | S_{D,\{D\}}, R_{D,\{D\}}, R_{D,\{0,1\}}, S_{1-D,\mathcal{T}'}, R_{1-D,\mathcal{T}'}, R_{1-D,\{0,1\}})$$
$$= \quad H((M, T)_D | S_{D,\mathcal{T}''}, R_{D,\mathcal{T}''}, R_{D,\{0,1\}}, R_{1-D,\{0,1\}}) \quad (4)$$

3. *A variable $(M, T)_D$ is sent to player P_D only if $D \in T$.*

We warn the reader that the tags and players' variables D and $1 - D$ do not play any role in the computation of the entropies; they are only present to discriminate the variables and determine which ones to include in the conditional part of the entropies. Notice also that, contrary to Definition 4, a variable's tag is set only when Ted considers sending it to a player, except for incoming variables. This relaxation will turn out to be the key point to explain why this generalized definition is not stronger than local variables on the players' side.

The process of determining which tag to assign can be broken into two steps. We start with the empty tag $\emptyset$. The first step is to decide whether we can add $\{D\}$ to the tag, or not. Notice that the *right*-hand side of equation (4) is the same for $T \in \{\emptyset, \{D\}\}$. This results from the calligraphic tag $\mathcal{T}''$, which is equivalent to $\{D\}$ in this case. On the other hand, the calligraphic tag $\mathcal{T}'$ introduces the terms $S_{1-D,\{1-D\}}$ and $R_{1-D,\{1-D\}}$ in the *left*-hand side of equation (4) when $T = \{D\}$. Thus, if the result of this first step is that the tag is at least $\{D\}$, then it means that the message to be sent is independent of the private information held by P_{1-D}. However, if we find that the tag is not even $\{D\}$, then it means that the message to be sent has some dependencies with the private information of P_{1-D}, and therefore the message should not be sent.

If the first step terminates with a tag containing $\{D\}$, then we can move on to determine whether we can add $\{1 - D\}$ to the tag, or not. We note that $\mathcal{T}'$ won't change for $T \in \{\{D\}, \{0, 1\}\}$, so the *left*-hand side is invariant. However, the calligraphic tag $\mathcal{T}''$ will remove the terms $S_{D,\{D\}}$ and $R_{D,\{D\}}$ from the *right*-hand side if we consider the tag $T = \{0, 1\}$. Hence, if equation (4) is satisfied with $T = \{0, 1\}$, it means that the message to be sent is not only independent of the private information of P_{1-D} (from the first step), but also of the private information of P_D. It follows naturally that this message be eligible for distribution to both players.

---

[5] This implies that the sets $S_{D,\{0,1\}}$ and $S_{1-D,\{0,1\}}$ are always empty. Therefore we do not include them in equation (4), but a formal expression should include them in the conditional part on both sides of the equality.

[6] In order to write a clear equation, we had to specify to which player the message is intended. As a result, we did not include $\{1 - D\}$ in the set of possible tags. It turns out that the empty set tag is sufficient to cover both communication and correlation.

The interest of Definition 5 is that it is more flexible in the tag assignation than the practical Definition 4 (and thus more general). Indeed, whenever Ted deliberately randomizes a message with new [uniformly distributed] information, the information-based criteria concludes that there is no problem to send to P_D a message that would have been tagged with $T = \{1 - D\}$ or $\emptyset$ by the practical definition. The reason is that by randomizing completely all the [private] variables related to P_{1-D}, Ted is reducing the message he sends to P_D to what P_D can exactly achieve using local variables. That is to say, P_D already has (using local variables) a random view of P_{1-D}'s variables (and so of the global message), so there is no problem for Ted to first randomize P_{1-D}'s variables and then send this message to P_D. Note however that the variables used to randomize will never be sent to P_D since they now carry the sensible information. We give two examples of these particular cases in the Appendix A.

Henceforth, the Two-Prover model's assumption is based on this refined definition of isolation.

Definition 6. *We say that Peggy and Patty are* isolated *from one another if they cannot communicate with one another, and if they only have access as external resource to secure third parties.*

Using this new definition of isolation, we are now *guaranteed* that any strategy that Peggy and Patty try to perform through a third party can be achieved using *only* local variables on each side. Using this fact together with the general assumption that the cheating strategy is deterministic[7], it is straightforward to fix the proof of Theorem 1 by arguing that their classical strategy can be run on each copy of the information to output *both* $\widehat{w}_0$ and $\widehat{w}_1$.

5 Quantum Secure Bit Commitment in the Two-Prover Model

We now present the modified version of the sBGKW scheme, called the mBGKW scheme, and prove its security against quantum adversaries. Although the two schemes are almost identical, it turns out the proof against quantum provers is easier with the latter. The security of the sBGKW and BGKW schemes will follow as corollaries of mBGKW's security. The scheme is as follows:

Peggy and Patty agree on an n-bit string w. They are then isolated as in Definition 6.

Protocol 51 (mBGKW - Commit to b)

    1: *Vic sends two random n-bit strings r_0, r_1 to Peggy.*
    2: *Peggy replies with $x := r_b \oplus w$.*

---

[7] A probabilistic strategy can be made deterministic by fixing the randomness to the best sequence.

Protocol 52 (mBGKW - Unveil b)

   1: *Patty announces an n-bit string $\widehat{w}$*
   2: *Vic computes $r := \widehat{w} \oplus x$. He accepts iff $r \in \{r_0, r_1\}$ and deduces b from*
   $r = r_b$.

We want to show that the mBGKW scheme is secure against a quantum adversary. Clearly the commitment is concealing because Vic does not know w. This means that there exists w and w' such that $x = r_0 \oplus w = r_1 \oplus w'$, and Vic cannot determine which one has been used.

To prove that the binding property holds according to Definition 2, we again use the crucial observation that if Patty could simultaneously compute $(\widehat{w}_0, \widehat{w}_1)$, then she would learn $r_0 \oplus r_1 = \widehat{w}_0 \oplus \widehat{w}_1$. Let $p_\oplus := \Pr[\text{Patty determines } r_0 \oplus r_1]$. The next lemma relates $p_\oplus$ to $p_0 + p_1$ in the desired way. Notice however that because quantum information is involved this statement is much less straightforward than the classical analog: p_0 and p_1 still correspond to running the attack twice on the same data but an attacker cannot do both.

Lemma 1. *Assume Patty has probability p_b to unveil bit b successfully, for both values of b, and such that $p_0 + p_1 \geq 1 + \varepsilon$ for $\varepsilon > 0$. Then, Patty can guess $r_0 \oplus r_1$ with probability $p_\oplus \geq \varepsilon^2 / 4$.*

Proof (of Lemma 1).

   Assume without loss of generality that when the unveiling phase of mBGKW starts, Patty holds the pure state $|\psi\rangle \in \mathcal{H}^N$ of dimension $N \geq 2^n$. Note that we do not need to consider the whole bipartite state between Peggy and Patty since when the unveiling phase starts, Peggy does no longer play an active role in the protocol and no communication is allowed between the two; hence her system can be traced-out of the global Hilbert space. Moreover, by linearity, the proof also holds if $|\psi\rangle$ is replaced by a mixed state. Notice also that, from the new model's assumption, Peggy and Patty cannot do better using a third party than what they can achieve with entanglement.

   Generally speaking, Patty has two possible strategies depending upon the bit b she wants to unveil. When $B = 0$, she applies a unitary transform U_0 to $|\psi\rangle$ in order to get the state $|\psi_0\rangle := U_0|\psi\rangle$ that she measures in the computational basis $\{|w\rangle\langle w|\}_{w \in \{0,1\}^n}$ applied to the first n qubits of $|\psi_0\rangle$. When $B = 1$, she proceeds similarly with unitary transform U_1 allowing to prepare the state $|\psi_1\rangle := U_1|\psi\rangle$. She then measures $|\psi_1\rangle$ using the same measurement as for $B = 0$. All general measurement can be realized in this fashion, this is thus a general strategy for Patty. Notice that in the proof of Kent [Ken05], the use of unitary transformations U_0 and U_1 is obscured by the fact that he works with projective measurements. Notice also that the measurement on the first n qubits of $|\psi_b\rangle$ can alternatively be expressed by the measurement operators $\{|w\rangle\langle w| \otimes I_M\}_{w \in \{0,1\}^n}$ on the whole state $|\psi_b\rangle$, where I_M is the identity matrix on the system of dimension $M = N/2^n$.

From the values $r_0, r_1, x \in \{0,1\}^n$ announced by Vic and Peggy during the committing phase, we define $\widehat{w}_b := r_b \oplus x$ as the string Patty has to announce in order to open b with success. We have,

$$p_b = \langle \psi_b | \widehat{w}_b \rangle \langle \widehat{w}_b | \psi_b \rangle, \tag{5}$$

which by assumption satisfies

$$p_0 + p_1 \geq 1 + \varepsilon, \ \varepsilon > 0. \tag{6}$$

Notice that $\langle \psi_b | \widehat{w}_b \rangle$ is a generalized inner product[8] since $|\widehat{w}_b\rangle$ lives in a subspace of dimension 2^n in $\mathcal{H}^N$. Therefore when $\widehat{w}_b$ is obtained, there is some state left in $\mathcal{H}^N$ of dimension $N/2^n$ which we label as $|\widehat{v}_b\rangle$ (i.e. $|\psi_b\rangle$ has not been completely collapsed by the measurement). Thus, using (5) we can write $|\psi_b\rangle$ as

$$|\psi_b\rangle = \sqrt{p_b} |\widehat{w}_b\rangle |\widehat{v}_b\rangle + \sqrt{1 - p_b} |\widehat{w}_b^{\perp}\rangle, \tag{7}$$

where $\|\langle \widehat{v}_b | \langle \widehat{w}_b | \widehat{w}_b^{\perp} \rangle \|^2 = 0$. Note that the "state" $|\widehat{w}_b^{\perp}\rangle$ has not necessarily a physical signification. It is simply a mathematical tool that allows us to conveniently carry the statistics.

We want to determine a lower bound for the probability $p_{\oplus}$. One possible way for Patty to compute $r_0 \oplus r_1$ is to obtain $\widehat{w}_0$ and $\widehat{w}_1$ individually. Again, one possible way to do this is to use the following strategy:

1. Patty applies the strategy allowing to open $B = 0$ from $|\psi_0\rangle = U_0|\psi\rangle$ resulting in the state $|\tilde{\psi}_0\rangle$ after the measurement in the computational basis $\{|w\rangle\langle w|\}_{w \in \{0,1\}^n}$ has been performed on the first n qubits, and
2. Patty prepares $|\tilde{\psi}_1\rangle := U_1 U_0^{\dagger} |\tilde{\psi}_0\rangle$ before applying again the measurement in the computational basis $\{|w\rangle\langle w|\}_{w \in \{0,1\}^n}$ on the first n qubits.

Note that when preparing $|\tilde{\psi}_1\rangle$, we applied $U_0^{\dagger}$ before U_1. This is to put back the state $|\tilde{\psi}_0\rangle$ as close as possible as the original state $|\psi\rangle$. From (6) and for N big enough, the probability to measure $\widehat{w}_0$ in the first step is not too small and so, by applying the inverse of all the unitary transformations generated by U_0, the state $|\tilde{\psi}\rangle$ we get before applying U_1 is a good enough approximation of the original $|\psi\rangle$. Similarly we can say that the fidelity $F(|\tilde{\psi}\rangle, |\psi\rangle)$ is large enough. By invariance under unitary transformation, it follows that $|\tilde{\psi}_1\rangle$ approximates $|\psi_1\rangle$ with the same fidelity $F(|\tilde{\psi}\rangle, |\psi\rangle)$.

In the strategy described above, the probability to determine $r_0 \oplus r_1$ is

$$p_0 \cdot p_{\widehat{w}_1 | \widehat{w}_0} \ .$$

As we said earlier, this is only *one of the* possible strategies to determine $r_0 \oplus r_1$, thus

$$p_{\oplus} \geq p_0 \cdot p_{\widehat{w}_1 | \widehat{w}_0} \ .$$

---

[8] If $|w\rangle \in \mathcal{H}^M$ and $|\psi\rangle \in \mathcal{H}^N$ then for $|\psi\rangle^N = \sum_i \alpha_i |a_i\rangle^M \otimes |b_i\rangle^{N/M}$ we define $\langle w | \psi \rangle = \sum_i \alpha_i \langle w | a_i \rangle |b_i\rangle$.

Let us first find a lower bound on the probability $p_{\widehat{w}_1|\widehat{w}_0}$ to produce $\widehat{w}_1$ given that $\widehat{w}_0$ has already been produced after step 1. Since $\widehat{w}_0$ was obtained, the state $|\tilde{\psi}_0\rangle$ is equal to $|\widehat{w}_0\rangle|\widehat{v}_0\rangle$. We have,

$$|\tilde{\psi}_1\rangle = U_1 U_0^\dagger |\tilde{\psi}_0\rangle$$

$$= U_1 U_0^\dagger |\widehat{w}_0\rangle|\widehat{v}_0\rangle$$

$$= U_1 \left(U_0^\dagger \frac{|\psi_0\rangle}{\sqrt{p_0}} - U_0^\dagger \sqrt{\frac{1-p_0}{p_0}} |\widehat{w}_0^\perp\rangle \right) \tag{8}$$

$$= U_1 \frac{|\psi\rangle}{\sqrt{p_0}} - U_1 U_0^\dagger \sqrt{\frac{1-p_0}{p_0}} |\widehat{w}_0^\perp\rangle \tag{9}$$

$$= \frac{|\psi_1\rangle}{\sqrt{p_0}} - U_1 U_0^\dagger \sqrt{\frac{1-p_0}{p_0}} |\widehat{w}_0^\perp\rangle \tag{10}$$

$$= \frac{1}{\sqrt{p_0}} \left(\sqrt{p_1}|\widehat{w}_1\rangle|\widehat{v}_1\rangle + \sqrt{1-p_1}|\widehat{w}_1^\perp\rangle - U_1 U_0^\dagger \sqrt{1-p_0}|\widehat{w}_0^\perp\rangle \right), \tag{11}$$

where (8) follows from isolating $|\widehat{w}_0\rangle|\widehat{v}_0\rangle$ in (7), (9) and (10) are obtained by definition of U_0 and U_1 respectively, and (11) also follows from (7). At this point, Patty applies the measurement in the computational basis in order to obtain $\widehat{w}_1$. Since we are interested only in finding a lower bound, the probability to obtain $\widehat{w}_1$ is minimized when $U_1 U_0^\dagger |\widehat{w}_0^\perp\rangle = |\widehat{w}_1\rangle|\widehat{v}_1\rangle$. It easily follows that,

$$p_{\widehat{w}_1|\widehat{w}_0} = \langle\tilde{\psi}_1|\widehat{w}_1\rangle\langle\widehat{w}_1|\tilde{\psi}_1\rangle$$

$$\geq \frac{1}{p_0} \left(\sqrt{p_1} - \sqrt{1-p_0} \right)^2 \tag{12}$$

$$\geq \frac{1}{p_0} \left(\sqrt{p_1} - \sqrt{p_1-\varepsilon} \right)^2 \tag{13}$$

$$\geq \frac{\varepsilon^2}{4p_0}, \tag{14}$$

where (12) follows from (11), (13) is obtained from (6), and (14) follows from a Taylor expansion. Finally, (14) gives the desired result since

$$p_\oplus \geq p_0 \cdot p_{\widehat{w}_1|\widehat{w}_0} \geq \frac{\varepsilon^2}{4}.$$

Theorem 6. *If there exists an algorithm A that can cheat the mBGKW Bit Commitment scheme with probabilities $p_0 + p_1 > 1 + 2/\sqrt{2^n}$ then there exists an algorithm A' that can predict an unknown n-bit string $(r_0 \oplus r_1)$ with probabilities better than $1/2^n$, which is impossible.*

Proof (of Theorem 6). From the isolation assumption, we have

$$p_\oplus = \frac{1}{2^n}.$$

Using the result from Lemma 1,

$$\frac{1}{2^n} \geq \frac{\varepsilon^2}{4} \implies \varepsilon \leq \frac{1}{\sqrt{2^{n-2}}}. \tag{15}$$

It follows that the binding condition is satisfied: plugging (15) in Lemma 1, we get for any cheating strategies

$$p_0 + p_1 \leq 1 + \frac{1}{\sqrt{2^{n-2}}} \; .$$

Notice that the proof presented in Lemma 1 can easily be generalized to a whole class of Bit Commitment schemes with the properties that information unknown to Patty is sent to Peggy to commit, and an *exact* answer is needed from Patty to unveil successfully the committed bit. Theorem 6 therefore holds for a whole class of Bit Commitment schemes in the Two-Prover model.

Note that sBGKW is the same as mBGKW where $r_0 := 000...0$ is the all-zero string all the time. The statement and proof of Lemma 1 is equally valid for any fixed choice of either (but not both) r_0 or r_1 because the probability to predict $r_0 \oplus r_1$ remains exponentially small. Hence using only the model's assumption we get:

Corollary 1. *If there exists an algorithm A that can cheat the* sBGKW *Bit Commitment scheme with probabilities* $p_0 + p_1 > 1 + 2/\sqrt{2^n}$ *then there exists an algorithm A′ that can predict an unknown n-bit string r with probabilities better than* $1/2^n$, *which is impossible.*

However, as previously, this proof is valid *solely* in a stand-alone security model. As soon as one starts composing such protocols, this proof is not necessarily valid anymore.

6   Conclusion and Open Problems

This paper contained several results. It showed that Two-Prover Bit Commitment schemes may or not be secure quantumly when they are classically. It also considered for the first time ever the exact conditions that the provers and verifier must satisfy to obtain security proofs of such Bit Commitment schemes both classically and quantumly.

A natural question would be to determine if the binding condition of ALL Two-Prover Quantum Bit Commitment schemes can be broken by a non-local computation that does not allow to communicate. This would imply that the no-communication assumption is NEVER sufficient to asses security of such schemes. A hierarchy of non-local correlations may be imagined with higher up correlations simulating lower down correlations, but not the opposite. What is the Bit Commitment scheme that can be broken only by a very highest correlation ?

In our definition of Bit Commitment, we assessed that cheating meant $p_0 + p_1 > 1 + \epsilon$ for non-negligible ϵ. However, recently more precise binding conditions have been introduced [DFRSS07]. The results of this paper should be extended to suit this newer definition.

The last natural question that results from our work is to find the complexity class corresponding to Quantum Two-Prover Zero-Knowledge Interactive Proofs (and similarly for $k > 2$ provers). Remember that these questions are not even settled for Quantum Two-Prover Interactive Proofs alone. As soon as the verifier is also quantum it is not clear how Bit Commitments may be used to "encrypt" the verifier's computations, thus the classical methodologies fall apart.

Acknowledgements. We are thankful the anonymous referees (of several conferences) for their comments and numerous suggestions.

References

[Ara02]   Aravind, P.K.: Bell's theorem without inequalities and only two distant observers. Foundation of Physics Letters, 397–405 (2002)

[BCMS98]   Brassard, G., Crépeau, C., Mayers, D., Salvail, L.: Defeating classical bit commitment schemes with a quantum computer. ArXiv Quantum Physics e-prints (1998)

[BCU⁺06]   Buhrman, H., Christandl, M., Unger, F., Wehner, S., Winter, A.: Implications of superstrong nonlocality for cryptography. Proceedings of The Royal Society A 462(2071), 1919–1932 (2006)

[BFL91]   Babai, L., Fortnow, L., Lund, C.: Non-deterministic exponential time has two-prover interactive protocols. Computational Complexity 1, 3–40 (1991)

[BGKW88]   BenOr, M., Goldwasser, S., Kilian, J., Widgerson, A.: Multi-prover interactive proofs: how to remove intractability. In: STOC 1988: Proceedings of the Twentieth Annual ACM Symposium on Theory of Computing, pp. 113–131. ACM Press, New York (1988)

[BHOP08]   Ben-Or, M., Hassidim, A., Pilpel, H.: Quantum Multi Prover Interactive Proofs with Communicating Provers. In: 49th Annual IEEE Symposium on Foundations of Computer Science (FOCS 2008), pp. 467–476. IEEE Computer Society (2008)

[CHTW04]   Cleve, R., Hoyer, P., Toner, B., Watrous, J.: Consequences and limits of nonlocal strategies. In: CCC 2004: Proceedings of the 19th IEEE Annual Conference on Computational Complexity, pp. 236–249. IEEE Computer Society, Washington, DC, USA (2004)

[CLS01]   Crépeau, C., Légaré, F., Salvail, L.: How to Convert the Flavor of a Quantum Bit Commitment. In: Pfitzmann, B. (ed.) EUROCRYPT 2001. LNCS, vol. 2045, pp. 60–77. Springer, Heidelberg (2001)

[CSUU07]   Cleve, R., Slofstra, W., Unger, F., Upadhyay, S.: Perfect Parallel Repetition Theorem for Quantum XOR Proof Systems. In: CCC 2007: Proceedings of the 2007 IEEE 22nd Annual Conference on Computational Complexity, pp. 109–114. IEEE Computer Society, Los Alamitos (2007)

[DFSS05]   Damgård, I., Fehr, S., Salvail, L., Schaffner, C.: Cryptography in the bounded quantum-storage model. In: 46th Annual IEEE Symposium on Foundations of Computer Science (FOCS 2005), pp. 449–458. IEEE Computer Society (2005)

[DFRSS07]   Damgård, I., Fehr, S., Renner, R., Salvail, L., Schaffner, C.: A Tight High-Order Entropic Quantum Uncertainty Relation with Applications. In: Menezes, A. (ed.) CRYPTO 2007. LNCS, vol. 4622, pp. 360–378. Springer, Heidelberg (2007)

[DLTW08]   Doherty, A.C., Liang, Y.-C., Toner, B., Wehner, S.: The Quantum Moment Problem and Bounds on Entangled Multi-prover Games. In: CCC 2008: Proceedings of the 2008 IEEE 23rd Annual Conference on Computational Complexity, pp. 199–210. IEEE Computer Society, Washington, DC, USA (2008)

428   C. Crépeau et al.

[DMS00]   Dumais, P., Mayers, D., Salvail, L.: Perfectly concealing quantum bit com-
             mitment from any quantum one-way permutation, pp. 300–315 (2000)
[IKM09]   Ito, T., Kobayashi, H., Matsumoto, K.: Oracularization and Two-Prover
             One-Round Interactive Proofs against Nonlocal Strategies. In: CCC 2009:
             Proceedings of the 2009 IEEE 24th Annual Conference on Computational
             Complexity, pp. 217–228. IEEE Computer Society, Los Alamitos (2009)
[IKO03]   Ibaraki, T., Katoh, N., Ono, H. (eds.): ISAAC 2003. LNCS, vol. 2906.
             Springer, Heidelberg (2003)
[IKPSY08]   Ito, T., Kobayashi, H., Preda, D., Sun, X., Yao, A.C.-C.: Generalized
             Tsirelson Inequalities, Commuting-Operator Provers, and Multi-prover In-
             teractive Proof Systems. In: CCC 2008: Proceedings of the 2008 IEEE
             23rd Annual Conference on Computational Complexity, pp. 187–198. IEEE
             Computer Society, Washington, DC, USA (2008)
[Ken05]   Kent, A.: Secure classical bit commitment using fixed capacity communi-
             cation channels. J. Cryptology 18(4), 313–335 (2005)
[Kil88]   Kilian, J.: Founding cryptography on oblivious transfer. In: Proceedings
             of the Twentieth Annual ACM Symposium on Theory of Computing, pp.
             20–31 (1988)
[KKMV08]   Kempe, J., Kobayashi, H., Matsumoto, K., Vidick, T.: Using Entanglement
             in Quantum Multi-prover Interactive Proofs. In: CCC 2008: Proceedings
             of the 2008 IEEE 23rd Annual Conference on Computational Complexity,
             pp. 211–222. IEEE Computer Society, Washington, DC, USA (2008)
[KM03]   Kobayashi, H., Matsumoto, K.: Quantum multi-prover interactive proof
             systems with limited prior entanglement. J. Comput. Syst. Sci. 66(3), 429–
             450 (2003)
[KMY03]   Kobayashi, H., Matsumoto, K., Yamakami, T.: Quantum merlin-arthur
             proof systems: Are multiple merlins more helpful to arthur? In: Ibaraki, et
             al. (eds.) [IKO03], pp. 189–198 (2003)
[LC97]   Lo, H.-K., Chau, H.F.: Is quantum bit commitment really possible? Phys.
             Rev. Lett. 78(17), 3410–3413 (1997)
[LS97]   Lapidot, D., Shamir, A.: Fully parallelized multi-prover protocols for nexp-
             time. J. Comput. Syst. Sci. 54(2), 215–220 (1997)
[May96]   Mayers, D.: Unconditionally secure quantum bit commitment is impossible
             (November 1996)
[PR94]   Popescu, S., Rohrlich, D.: Nonlocality as an axiom. Foundations of
             Physics 24, 379 (1994)
[PR97]   Popescu, S., Rohrlich, D.: Causality and nonlocality as axioms for quantum
             mechanics. In: Symposium on Causality and Locality in Modern Physics
             and Astronomy (1997)
[Weh06]   Wehner, S.: Entanglement in Interactive Proof Systems with Binary An-
             swers. In: Durand, B., Thomas, W. (eds.) STACS 2006. LNCS, vol. 3884,
             pp. 162–171. Springer, Heidelberg (2006)
[Yao03]   Yao, A.C.-C.: Interactive proofs for quantum computation. In: Ibaraki, et
             al. (eds.) [IKO03], p. 1 (2003)

A   Isolation Examples

Example 1:
Let P_0 send to Ted a message represented by $(X, \{0\})_0$ (the variable X is tagged with $\{0\}$ and comes from P_0). Then Ted generates a uniform random variable $(W, T)_D$ (its tag and receiver have not been set yet) and produces the message $M = X \oplus W$ for P_1. Checking with equation (4) we see there is no problem setting M's tag to $\{1\}$, as

$$H((M, \{1\})_1 | (X, \{0\})_0) = H((W, T)_D) = H((M, \{1\})_1).$$

This is satisfied since $(W, T)_D$ is uniform and has never been sent. However, the practical definition would have assigned the tag $T := \{0\}$ since W's tag would have been $\{0, 1\}$ (by the second rule) and $\{0\} = \{0\} \cap \{0, 1\}$. Let Ted send $(M, \{1\})_1$. We now get that for *both* $D = 0$ and 1, if $T = \{D\}$ or $\{0, 1\}$ then the left-hand side of equation (4) for W is

$$H((W, T)_D | (X, \{0\})_0, (M, \{1\})_1) = 0,$$

and the right-hand side is respectively

$$H((W, \{0\})_0 | (X, \{0\})_0) = H((W, \{0\})_0) = 1,$$
$$H((W, \{1\})_1 | (M, \{1\})_1) = H((X, \{0\})_0) = 1,$$
$$H((W, \{0, 1\})_D) = 1.$$

Because equation (4) is not satisfied for both $T = \{D\}$ and $\{0, 1\}$, W's tag is set to $T := \emptyset$, and Ted should not send $(W, \emptyset)_D$ to neither of P_D, for $D = 0, 1$.

Example 2:
Similarly, we can send to P_1 a message M that would have been tagged $\emptyset$ by the practical definition. We take the **PR**-box relation as example. Suppose the variables $(X, \{0\})_0$ and $(Y, \{1\})_1$ have already been sent to Ted by the players (and tagged accordingly), and $(U, \{0, 1\})_0$ [9] has been sent by Ted to P_0. Let $(W, T)_D$ be a uniformly distributed random variable chosen by Ted, with $D \in \{0, 1\}$. Consider the following variable for P_1,

$$V = U \oplus (W \oplus X) \wedge T,$$

that is, we randomized the variable tagged $\{0\}$ (i.e. X) in the **PR**-box relation. In the practical definition, because W is chosen uniformly and independently of previous variables, the second rule would have assigned a tag $\{0, 1\}$ to it, and so V's tag would have been set to $\emptyset = \{0, 1\} \cap \{0, 1\} \cap \{0\} \cap \{1\}$. However, checking with equation (4), because W has not been sent yet, we get that there is no problem setting V's tag to $\{1\}$, as

$$H((V, \{1\})_1 | (Y, \{1\})_1, (X, \{0\})_0, (U, \{0, 1\})_0) = \frac{1}{2} = H((V, \{1\})_1 | (Y, \{1\})_1, (U, \{0, 1\})_0).$$

---

[9] It is straightforward to verify that this is the less restrictive tag.

So Ted would send this message $(V, \{1\})_1$ to P_1. Is this a problem? No, because the classical limitations of non-locality have not been violated yet! The reason is simple: by randomizing completely all the [private] variables related to P_0, Ted is reducing the message he sends to P_1 to what P_1 can exactly achieve using local variables. That is to say, P_1 already has a random view of P_0's variables, so there is no problem for Ted to first randomize P_0's variables and then send this message to P_1. If we make the calculations, we see that indeed, for the variable V sent, the relation

$$V = U \oplus X \wedge Y$$

holds with probability 75%, just as in the classical scenario, and no W will never let us beat that. Of course, as in the previous example, the variable $(W, T)_D$ used to randomize can never be disclosed to *any* of the two players, and equation (4) agrees with that (W's tag will be set to $T := \emptyset$ for both D).

Efficient Zero-Knowledge Arguments from Two-Tiered Homomorphic Commitments

Jens Groth*

University College London, UK
j.groth@ucl.ac.uk

Abstract. We construct practical and efficient zero-knowledge arguments with sublinear communication complexity. The arguments have perfect completeness, perfect special honest verifier zero-knowledge and computational soundness. Our zero-knowledge arguments rely on two-tiered homomorphic commitments for which pairing-based constructions already exist.

As a concrete application of our new zero-knowledge techniques, we look at the case of range proofs. To demonstrate a committed value belongs to a specific N-bit integer interval we only need to communicate $O(N^{\frac{1}{3}})$ group elements.

Keywords: Zero-knowledge arguments, sublinear communication, circuit satisfiability, range proofs, two-tiered homomorphic commitments.

1 Introduction

Zero-knowledge proofs introduced by Goldwasser, Micali and Rackoff [18] are fundamental building blocks in cryptography that are used in secure multi-party computation and numerous other protocols. Zero-knowledge proofs enable a prover to convince a verifier of the truth of a statement without leaking any other information. The central properties are captured in the notions of completeness, soundness and zero-knowledge.

Completeness: The prover can convince the verifier if the prover knows a witness testifying to the truth of the statement.

Soundness: A malicious prover cannot convince the verifier if the statement is false. We distinguish between computational soundness that protects against polynomial time cheating provers and statistical or perfect soundness where even an unbounded prover cannot convince the verifier of a false statement. We will call computationally sound proofs for *arguments*.

Zero-knowledge: A malicious verifier learns nothing except that the statement is true. We distinguish between computational zero-knowledge, where a polynomial time verifier learns nothing from the proof and statistical or perfect zero-knowledge, where even a verifier with unlimited resources learns nothing from the proof.

Recent works on zero-knowledge proofs [25] give us proofs with a communication complexity that grows linearly in the size of the statement to be proven and [25,26]

---

* Supported by EPSRC grant number EP/G013829/1.

D.H. Lee and X. Wang (Eds.): ASIACRYPT 2011, LNCS 7073, pp. 431–448, 2011.

also give us proofs where the communication complexity depends quasi-linearly on the witness-length. These works rely on standard assumptions; if one is willing to assume the existence of fully homomorphic encryption [15] the communication complexity can be reduced to the witness-length plus a small additive overhead [14,23].

For zero-knowledge *arguments* the communication complexity can be even lower. Kilian [27] gave a zero-knowledge argument for circuit satisfiability with polylogarithmic communication. His argument goes through the PCP-theorem [3,2,11] and uses a collision-free hash-function to build a hash-tree that includes the entire PCP though. Even with the best PCP constructions known to date [4] Kilian's argument has high computational complexity for practical parameters. Goldwasser, Kalai and Rothblum [17] improve that state of affairs by constructing arguments that have both low communication complexity and highly efficient verification.

A large body of research starting with Schnorr's identification protocols [32] deals with zero-knowledge proofs and arguments over prime order groups. A class of zero-knowledge proofs and arguments known as Σ-protocols [8] is often used in practical applications. Groth [22] also used prime order groups to develop practical sublinear size zero-knowledge arguments for statements relating to linear algebra over $\mathbb{Z}_p$ for large primes p.

One particular example of zero-knowledge arguments that has appeared in several applications, e.g., e-voting [10] and auctions [30] are range proofs. Here the prover holds a commitment to a value w and wants to convince the verifier that the value belongs to a specific integer interval $[A; B)$. Boudot [5], Lipmaa [29] and Groth [20] have given constant size zero-knowledge argument for interval membership based on the strong RSA assumption.

In prime order groups the best range proof technique known was for a long time to commit to the bits of the value and use OR-proofs [8] to show that the committed bits were 0 or 1. For N-bit integers this communicates $O(N)$ group elements. Camenisch, Chaabouni and Shelat [6] improved this in the bilinear group setting by giving a zero-knowledge range proof with communication complexity $O(\frac{N}{\log N})$. Chaabouni, Lipmaa and Shelat [7] improved this complexity with a factor 2.

Our contribution. We construct zero-knowledge arguments for circuit satisfiability and range proofs that have perfect completeness and perfect zero-knowledge. For simplicity our constructions are in the common reference string model, but typically the common reference string can be chosen by the verifier at the cost of one extra round in the beginning to get zero-knowledge arguments in the plain model; we refer to the remarks at end of Section 2.2 for further discussion.

The circuit satisfiability argument has communication complexity $O(N^{\frac{1}{3}})$ group elements when the circuit has N gates. The range proof has a size of $O(N^{\frac{1}{3}})$ group elements for N-bit intervals. The arguments have quasi-linear computational complexity for the prover and very efficient verification. An efficiency comparison of the arguments can be found in Tables 1 and 2.

In the tables we give the conservative estimate of $O(N \log^2 N)$ estimate for the prover's computation, but as we will discuss at the end of Section 3 it can often be reduced to $O(N \log N)$ using Fast Fourier Transform techniques. When comparing the range proofs, we are assuming a common reference string is available. This permits the

Table 1. Zero-knowledge arguments for satisfiability of circuits with N NAND-gates measured in group elements G, exponentiations E, and multiplications M

	Rounds	Comm.	Prover comp.	Verifier comp.	Assumption
Cramer et al. [8]	3	$O(N)$ G	$O(N)$ E	$O(N)$ E	Dlog
Groth [22]	5	$O(N^{\frac{1}{2}})$ G	$O(N \log^2 N)$ M	$O(N)$ M	DLog
This paper	7	$O(N^{\frac{1}{3}})$ G	$O(N \log^2 N)$ M	$O(N)$ M	DPair

Table 2. Range proofs in prime order groups measured in group elements G, exponentiations E, and multiplications M

	Rounds	Comm.	Prover comp.	Verifier comp.	Assumption
Camenisch et al. [6]	3	$O(\frac{N}{\log N})$ G	$O(\frac{N}{\log N})$ E	$O(\frac{N}{\log N})$ E	q-SDH
Chaabouni et al [7]	3	$O(\frac{N}{\log N})$ G	$O(\frac{N}{\log N})$ E	$O(\frac{N}{\log N})$ E	q-SDH
This paper	7	$O(N^{\frac{1}{3}})$ G	$O(N \log^2 N)$ M	$O(N^{\frac{1}{3}})$ M	DPair

incorporation of the initial messages in [6,7] into the common reference string such that their range proofs only use 3 rounds instead of 4 rounds.

Our zero-knowledge arguments can be instantiated in asymmetric bilinear groups where the computational double pairing assumption (Section 2.1) holds. In comparison, the range proofs [6,7] are based on the q-SDH assumption in bilinear groups.

Techniques. Our main technical contribution is the batch product argument that can be found in Section 3. Using homomorphic commitments to group elements [1,22] we can in combination with Pedersen commitments to multiple elements commit to N elements in $\mathbb{Z}_p$ using only $N^{\frac{1}{3}}$ group elements. Given $3N$ committed elements $u_i, v_i, w_i \in \mathbb{Z}_p$ we generalize techniques from [24,22] to develop a communication-efficient zero-knowledge argument for proving that the committed values all satisfy $u_i v_i = w_i$.

Since the commitments are homomorphic we can now do both additions and multiplications on the committed elements. This enables the prover to commit to the wires in a circuit and prove that they respect the NAND-gates.

For the range proof we commit to the bits $w_1, \ldots, w_N$ of the committed value. Using the batch product argument we can show with a communication complexity of $O(N^{\frac{1}{3}})$ group elements that the committed bits satify $w_i w_i = w_i$, which can only be true if $w_i \in \{0, 1\}$. Once we have the committed bits, we can then use the homomorphic properties of the commitment schemes to compute $w = \sum_{i=1}^{N} w_i 2^{i-1}$. This shows that w belongs to the range $[0; 2^N)$ and can be generalized to a range of the form $[A; B)$.

2 Preliminaries

We write $y = A(x; r)$ when the algorithm A on input x and randomness r, outputs y. We write $y \leftarrow A(x)$ for the process of picking randomness r at random and setting $y = A(x; r)$. We also write $y \leftarrow S$ for sampling y uniformly at random from the set S.

We give a security parameter λ written in unary as input to all parties in our protocols. Intuitively, the higher the security parameter the more secure the protocol. We say a function $f : \mathbb{N} \to [0, 1]$ is negligible if $f(\lambda) = O(\lambda^{-c})$ for every constant $c > 0$. We write $f \approx g$ when $|f(\lambda) - g(\lambda)|$ is negligible. We say f is overwhelming if $f \approx 1$.

2.1 Two-tiered Homomorphic Commitments

A commitment scheme allows Alice to compute and send a commitment to a secret message a. Later Alice may open the commitment and reveal to Bob that she committed to a. Commitments must be binding and hiding. Binding means that Alice cannot change her mind; a commitment can only be opened to one message a. Hiding means that Bob does not learn which message Alice committed to.

In the Pedersen commitment scheme [31] the public key contains the description of a group of prime order p and group elements g, h. A commitment to $a \in \mathbb{Z}_p$ is constructed by picking $r \leftarrow \mathbb{Z}_p$ and computing $c = g^a h^r$. This commitment scheme is very useful because it is homomorphic, i.e., the product of two commitments is $c \cdot c' = (g^a h^r)(g^b h^s) = g^{a+b} h^{r+s}$, which is a commitment to $a+b$. The Pedersen commitment can be generalized such that the public key contains $g_1, \ldots, g_n, h$ and a commitment to $(a_1, \ldots, a_n) \in \mathbb{Z}_p^n$ is computed as $h^r \prod_{k=1}^n g_k^{a_k}$.

Abe, Fuchsbauer, Groth, Haralambiev and Ohkubo [1,21] proposed commitment schemes for group elements. One of the commitment schemes uses a bilinear group with a pairing $e : \mathbb{G} \times \hat{\mathbb{G}} \to \mathbb{T}$. Here $\mathbb{G}, \hat{\mathbb{G}}, \mathbb{T}$ are cyclic groups of prime order p where we call $\mathbb{G}, \hat{\mathbb{G}}$ the base groups and $\mathbb{T}$ the target group. The pairing is efficiently computable, non-trivial and bilinear, i.e., for all x, y, a, b we have $e(x^a, y^b) = e(x, y)^{ab}$. The commitment scheme specifies non-trivial group elements $v, u_1, \ldots, u_m \in \hat{\mathbb{G}}$ and a commitment to $(c_1, \ldots, c_m) \in \mathbb{G}$ is computed by picking at random $t \in \mathbb{G}$ and computing $C = e(t, v) \prod_{j=1}^m e(c_j, u_j)$. The commitment scheme is computationally binding under the computational double pairing assumption, which states that given random $u, v \in \hat{\mathbb{G}}$ it is hard to find non-trivial $s, t \in \mathbb{G}$ such that $e(s, u) = e(t, v)$. The hardness of the computational double pairing assumption is implied by the decision Diffie-Hellman assumption in $\hat{\mathbb{G}}$ [1,21].[1] Furthermore, the bilinearity of the pairing means that the commitment scheme is homomorphic in the sense that

$$ C \cdot C' = \left(e(t, v) \prod_{j=1}^m e(c_j, u_j) \right) \left(e(t', v) \prod_{j=1}^m e(c'_j, u_j) \right) = e(tt', v) \prod_{j=1}^m e(c_j c'_j, u_j) $$

is a commitment to the entry-wise product of the messages.

Combining the two types of commitment schemes it is possible to commit to commitments. If we compute $c_j = h^{r_j} \prod_{k=1}^n g_k^{a_{jk}}$ and $C = e(t, v) \prod_{j=1}^m e(c_j, u_j)$ we have a single target group element that is a commitment to mn values $\{a_{jk}\}_{j=1,k=1}^{m,n}$. Since both commitment schemes are homomorphic the product of two commitments $C \cdot C'$ is a commitment to the sums of the messages $a_{jk} + a'_{jk}$. In our zero-knowledge arguments

---

[1] Galbraith, Paterson and Smart [12] classified bilinear groups into 3 types. The commitment scheme described above uses type II or type III bilinear groups. In a type I bilinear group we could instead use the decisional linear assumption based commitment scheme from [21].

the homomorphic and the length-reducing properties allow the prover to do computations on committed values in a verifiable manner and with little communication.

The commitment schemes described above provide an example of what we will call a two-tiered commitment scheme. With the Pedersen commitment scheme in mind we will for simplicity assume the randomness is drawn from $\mathbb{Z}_p$ but it would be easy to generalize to other randomizer spaces. Furthermore, in the example given above the Pedersen commitments are perfectly hiding and we can therefore use trivial randomness $t = 1$ in the commitments to Pedersen commitments. This observation is incorporated in the following definition of a two-tiered commitment scheme.

A two-tiered commitment scheme has three polynomial time algorithms $(\mathcal{K}, \mathrm{com}, \mathrm{com}^{(2)})$. $\mathcal{K}$ is a key generator that on security parameter λ and integers m, n returns a public key ck. The commitment key specifies cyclic groups $\mathbb{Z}_p$, $\mathbb{G}$ and $\mathbb{T}$ of prime order p. It also specifies how to efficiently compute $\mathrm{com}_{ck} : \mathbb{Z}_p^n \times \mathbb{Z}_p \to \mathbb{G}$ and $\mathrm{com}_{ck}^{(2)} : \mathbb{G}^m \to \mathbb{T}$.

Definition 1 (Homomorphic). *We say the two-tiered commitment scheme is homomorphic, when the maps com_{ck} and $\mathrm{com}_{ck}^{(2)}$ are $\mathbb{Z}_p$-linear.*

Definition 2 (Computationally binding). *The two-tiered commitment scheme $(\mathcal{K}, \mathrm{com}, \mathrm{com}^{(2)})$ is computationally binding if for all non-uniform polynomial time adversaries $\mathcal{A}$ and for all $m, n = \lambda^{O(1)}$*

$$\Pr\left[ck \leftarrow \mathcal{K}(1^\lambda, m, n); (\boldsymbol{a}, \boldsymbol{b}, r, s, \boldsymbol{c}, \boldsymbol{d}) \leftarrow \mathcal{A}(ck) : \ \boldsymbol{a} \neq \boldsymbol{b} \in \mathbb{Z}_p^n \ r, s \in \mathbb{Z}_p \ \boldsymbol{c} \neq \boldsymbol{d} \in \mathbb{G}^m\right.$$

$$\left. \mathrm{com}_{ck}(\boldsymbol{a}; r) = \mathrm{com}_{ck}(\boldsymbol{b}; s) \quad \text{or} \quad \mathrm{com}_{ck}^{(2)}(\boldsymbol{c}) = \mathrm{com}_{ck}^{(2)}(\boldsymbol{d})\right] \approx 0.$$

Definition 3 (Perfectly hiding). *The two-tiered commitment scheme $(\mathcal{K}, \mathrm{com}, \mathrm{com}^{(2)})$ is perfectly hiding if for all stateful adversaries $\mathcal{A}$ and all $m, n \in \lambda^{O(1)}$*

$$\Pr\left[ck \leftarrow \mathcal{K}(1^\lambda, m, n); \boldsymbol{a}_0, \boldsymbol{a}_1 \leftarrow \mathbb{Z}_p^n; b \leftarrow \{0, 1\}; c \leftarrow \mathrm{com}_{ck}(\boldsymbol{a}_b) : \mathcal{A}(ck, \boldsymbol{a}_0, \boldsymbol{a}_1, c) = b\right] = \frac{1}{2}.$$

The zero-knowledge arguments we describe will work over any two-tiered homomorphic commitment scheme with a large prime p. When giving concrete efficiency estimates we will assume we are using the bilinear group based scheme described earlier in this section. The public key for this commitment scheme consists of a description of a bilinear group $(p, \mathbb{G}, \hat{\mathbb{G}}, \mathbb{T}, e)$ and $m + n + 2$ group elements in $\mathbb{G}$ and $\hat{\mathbb{G}}$. We will be looking at statements of size N and the minimal communication complexity will be obtained when $m = O(N^{\frac{1}{3}})$ and $n = O(N^{\frac{1}{3}})$ giving a public key size of $O(N^{\frac{1}{3}})$ group elements.

2.2 Special Honest Verifier Zero-knowledge Arguments of Knowledge

We will for simplicity describe how our arguments work in the common reference string model and how to obtain zero-knowledge against honest-but-curious verifiers. Both of these restrictions can be removed at very small cost to get full zero-knowledge in the plain model as described in the remarks at the end.

Consider a triple of probabilistic polynomial time interactive algorithms $(\mathcal{K}, \mathcal{P}, \mathcal{V})$ called the common reference string generator, the prover and the verifier. The common reference string generator takes the security parameter λ as input in unary and some auxilliary input m, n that specifies the size of the statements and generates a common reference string. In the zero-knowledge arguments in this paper, the common reference string will contain the public key ck for a two-tiered commitment scheme.

Let R be a polynomial time decidable ternary relation. For a statement x we call w a witness if $(ck, x, w) \in R$. We define a corresponding common reference string dependent language L_{ck} consisting of statements x that have a witness w such that $(ck, x, w) \in R$. This is a natural generalization of NP-languages; when R ignores ck we have the standard notion of an NP-language.

We write $\mathrm{tr} \leftarrow \langle \mathcal{P}(s), \mathcal{V}(t) \rangle$ for the public transcript produced by $\mathcal{P}$ and $\mathcal{V}$ when interacting on inputs s and t. This transcript ends with $\mathcal{V}$ either accepting or rejecting. We sometimes shorten the notation by saying $\langle \mathcal{P}(s), \mathcal{V}(t) \rangle = b$, where $b = 0$ corresponds to $\mathcal{V}$ rejecting and $b = 1$ corresponds to $\mathcal{V}$ accepting.

Definition 4 (Argument). *The triple $(\mathcal{K}, \mathcal{P}, \mathcal{V})$ is an argument for relation R with perfect completeness if for all non-uniform polynomial time interactive adversaries $\mathcal{A}$ and all $m, n = \lambda^{O(1)}$ we have*

Perfect completeness:

$$\Pr\Big[ck \leftarrow \mathcal{K}(1^\lambda, m, n); (x, w) \leftarrow \mathcal{A}(ck) : (ck, x, w) \notin R \text{ or } \langle \mathcal{P}(ck, x, w), \mathcal{V}(ck, x)\rangle{=}1\Big]{=}1.$$

Computational soundness:

$$\Pr\Big[ck \leftarrow \mathcal{K}(1^\lambda, m, n); x \leftarrow \mathcal{A}(ck) : x \notin L_{ck} \text{ and } \langle \mathcal{A}, \mathcal{V}(ck, x)\rangle = 1\Big] \approx 0.$$

Definition 5 (Public coin argument). *An argument $(\mathcal{K}, \mathcal{P}, \mathcal{V})$ is public coin if the verifier's messages are chosen uniformly at random independently of the messages sent by the prover.*

We shall define an argument of knowledge through witness-extended emulation [19,28]. Informally, the definition says: given an adversary that produces an acceptable argument with probability ϵ, there exists an emulator that produces a similar argument with roughly the same probability ϵ and at the same time provides a witness.

Definition 6 (Witness-extended emulation). *We say the public coin argument $(\mathcal{K}, \mathcal{P}, \mathcal{V})$ has computational witness-extended emulation if for all deterministic polynomial time $\mathcal{P}^*$ there exists an expected polynomial time emulator $\mathcal{X}$ such that for all non-uniform polynomial time adversaries $\mathcal{A}$ and all $m, n = \lambda^{O(1)}$*

$$\Pr\Big[ck \leftarrow \mathcal{K}(1^\lambda, m, n); (x, s) \leftarrow \mathcal{A}(ck); \mathrm{tr} \leftarrow \langle \mathcal{P}^*(ck, x, s), \mathcal{V}(ck, x)\rangle : \mathcal{A}(\mathrm{tr}){=}1\Big]$$

$$\approx \Pr\Big[ck \leftarrow \mathcal{K}(1^\lambda, m, n); (x, s) \leftarrow \mathcal{A}(ck); (\mathrm{tr}, w) \leftarrow \mathcal{X}^{\langle \mathcal{P}^*(ck, x, s), \mathcal{V}(ck, x)\rangle}(ck, x) :$$

$$\mathcal{A}(\mathrm{tr}) = 1 \text{ and if } \mathrm{tr} \text{ is accepting then } (ck, x, w) \in R\Big],$$

where $\mathcal{X}$ has access to a transcript oracle $\langle \mathcal{P}^*(ck, x, s), \mathcal{V}(ck, x)\rangle$ that can be rewound to a particular round and run again with the verifier using fresh randomness.

We think of s as being the state of $\mathcal{P}^*$, including the randomness. Then we have an argument of knowledge in the sense that the emulator can extract a witness whenever $\mathcal{P}^*$ is able to make a convincing argument. This shows that the definition implies soundness. We remark that the verifier's randomness is part of the transcript and the prover is deterministic. So combining the emulated transcript with ck, x, s gives us the view of both the prover and the verifier and at the same time gives us the witness.

We define special honest verifier zero-knowledge (SHVZK) [8] for a public coin argument as the ability to simulate the transcript without access to the witness as long as the challenges are known in advance.

Definition 7 (Perfect special honest verifier zero-knowledge). *The public coin argument* $(\mathcal{K}, \mathcal{P}, \mathcal{V})$ *is a perfect special honest verifier zero-knowledge argument for* R *if there exists a probabilistic polynomial time simulator* S *such that for all non-uniform polynomial time adversaries* $\mathcal{A}$ *and all* $m, n = \lambda^{O(1)}$

$$\Pr\left[ck \leftarrow \mathcal{K}(1^{\lambda}, m, n); (x, w, \rho) \leftarrow \mathcal{A}(ck); \mathrm{tr} \leftarrow \langle \mathcal{P}(ck, x, w), \mathcal{V}(ck, x; \rho) \rangle : \right.$$

$$\left. (ck, x, w) \in R \text{ and } \mathcal{A}(\mathrm{tr}) = 1\right]$$

$$= \Pr\left[ck \leftarrow \mathcal{K}(1^{\lambda}, m, n); (x, w, \rho) \leftarrow \mathcal{A}(ck); \mathrm{tr} \leftarrow S(ck, x, \rho) : (ck, x, w) \in R \text{ and } \mathcal{A}(\mathrm{tr}) = 1\right].$$

The plain model. We will describe our arguments in the common reference string model where the prover and verifier have a trusted setup. If we want to work in the plain model we can add an initial round where the verifier picks the common reference string and sends it to the prover. Provided it can be verified that the verifier's initial message describes a valid common reference string this will still be perfect SHVZK because we do not rely on the simulator knowing any trapdoor information associated with the common reference string.

Full zero-knowledge. For simplicity, we focus on SHVZK arguments in this paper. There are very efficient standard techniques [9,13,19] to convert an SHVZK argument into a public-coin full zero-knowledge argument with a cheating verifier when a common reference string is available.

If we work in the plain model and let the verifier choose the common reference string, we can use coin-flipping techniques (for the full zero-knowledge property the coin-flips should be simulatable against a dishonest verifier) for the challenges to get private-coin[2] full zero-knowledge arguments against a cheating verifier. Challenges in our SHVZK arguments are very short so both in the case with and without a common reference string the overhead of getting full zero-knowledge is insignificant compared to the cost of the SHVZK arguments.

3   Batch Product Argument

We will now present our main technical contribution, which is a batch product argument for committed values $\{u_{ijk}, v_{ijk}, w_{ijk}\}_{i=1, j=1, k=1}^{M, m, n}$ satisfying $u_{ijk} v_{ijk} = w_{ijk}$. More

---

[2] Goldreich and Krawczyk [16] have shown that only languages in BPP have constant-round public-coin arguments.

precisely, the statement consists of commitments $C_{U_1}, C_{V_1}, C_{W_1}, \ldots, C_{U_M}, C_{V_M}, C_{W_M}$.
The prover argues knowledge of openings $u_{ijk}, r_{ij}, v_{ijk}, s_{ij}, w_{ijk}, t_{ij} \in \mathbb{Z}_p$ satisfying

$$
\begin{aligned}
c_{u_{ij}} &= \mathrm{com}_{ck}(u_{ij1}, \ldots, u_{ijn}; r_{ij}) & C_{U_i} &= \mathrm{com}_{ck}^{(2)}(c_{u_{i1}}, \ldots, c_{u_{im}}) \\
c_{v_{ij}} &= \mathrm{com}_{ck}(v_{ij1}, \ldots, v_{ijn}; s_{ij}) & C_{V_i} &= \mathrm{com}_{ck}^{(2)}(c_{v_{i1}}, \ldots, c_{v_{im}}) \\
c_{w_{ij}} &= \mathrm{com}_{ck}(w_{ij1}, \ldots, w_{ijn}; t_{ij}) & C_{W_i} &= \mathrm{com}_{ck}^{(2)}(c_{w_{i1}}, \ldots, c_{w_{im}}) \\
& \qquad\qquad u_{ijk} v_{ijk} = w_{ijk}.
\end{aligned}
$$

The argument will have communication complexity $O(M + m + n)$. In order to explain
the idea behind the argument let us first focus on soundness and for now postpone
the question of how to get SHVZK. In the argument, the prover will demonstrate that
she knows openings of $C_{U_i}, C_{V_i}, C_{W_i}$ to $c_{u_{ij}}, c_{v_{ij}}, c_{w_{ij}}$ and that she knows openings of
$c_{u_{ij}}, c_{v_{ij}}, c_{w_{ij}}$ using standard techniques. She will also know openings $a_\alpha, \rho_\alpha, b_\beta, \sigma_\beta \in$
$\mathbb{Z}_p$ of intermediate commitments $c_{a_\alpha} = \mathrm{com}_{ck}(a_\alpha; \rho_\alpha), c_{b_\beta} = \mathrm{com}_{ck}(b_\beta, \sigma_\beta)$ that she
sends during the argument and which will be specified later. The argument runs over
7 moves with the prover getting challenges $x, y, z \in \mathbb{Z}_p^*$ in round 2, 4 and 6. The
commitments c_{a_α} are sent in round 3 and the commitments c_{b_β} are sent in round 5.
This means a_α may depend on x but is independent of y and z, and b_β may depend on
both x and y but is independent of z.

The prover will demonstrate to the verifier that

$$
\sum_{i=1}^{M} \sum_{j=1}^{m} \sum_{k=1}^{n} (u_{ijk} v_{ijk} - w_{ijk}) x^{i(m+1)n+jn+k} = 0. \tag{1}
$$

Unless $u_{ijk} v_{ijk} = w_{ijk}$ for all choices of i, j, k this has negligible probability of
holding over a randomly chosen challenge $x \in \mathbb{Z}_p^*$. Our main obstacle is to build up
this polynomial and convince the verifier that the equality (1) holds true using only
$O(M + m + n)$ communication.

We carefully choose appropriate linear combinations of the commitments and by the
homomorphic property get corresponding linear combinations of the $u_{ijk}, v_{ijk}, w_{ijk}$
values such that the equality (1) emerges. During this process, we will also use exponentiations of some of the commitments to powers of x such that we get linear combinations of $u_{ijk} x^{i(m+1)n+jn+k}$ and $w_{ijk} x^{i(m+1)n+jn+k}$. Suppose for instance that the
prover after seeing x computes and opens

$$
\prod_{i=1}^{M} C_{U_i}^{x^{i(m+1)n}} = \mathrm{com}_{ck}^{(2)}(c_{u_1}, \ldots, c_{u_m}) \qquad \text{where} \qquad c_{u_j} = \prod_{i=1}^{M} c_{u_{ij}}^{x^{i(m+1)n}}
$$

$$
\prod_{i=1}^{M} C_{V_i} = \mathrm{com}_{ck}^{(2)}(c_{v_1}, \ldots, c_{v_m}) \qquad \text{where} \qquad c_{v_j} = \prod_{i=1}^{M} c_{v_{ij}}
$$

$$
\prod_{i=1}^{M} C_{W_i}^{x^{i(m+1)n}} = \mathrm{com}_{ck}^{(2)}(c_{w_1}, \ldots, c_{w_m}) \qquad \text{where} \qquad c_{w_j} = \prod_{i=1}^{M} c_{w_{ij}}^{x^{i(m+1)n}}
$$

and at the same time computes and opens

$$\prod_{j=1}^{m} c_{u_j}^{x^{jn}} = \mathrm{com}_{ck}(u_1,\dots,u_n;r) \qquad \text{where} \qquad u_k = \sum_{i=1}^{M}\sum_{j=1}^{m} u_{ijk}x^{i(m+1)n+jn}$$

$$\prod_{j=1}^{m} c_{v_j} = \mathrm{com}_{ck}(v_1,\dots,v_n;s) \qquad \text{where} \qquad v_k = \sum_{i=1}^{M}\sum_{j=1}^{m} v_{ijk}$$

$$\prod_{j=1}^{m} c_{w_j}^{x^{jn}} = \mathrm{com}_{ck}(w_1,\dots,w_n;t) \qquad \text{where} \qquad w_k = \sum_{i=1}^{M}\sum_{j=1}^{m} w_{ijk}x^{i(m+1)n+jn}$$

Using only $3m$ commitments and $3m+3$ elements in $\mathbb{Z}_p$ this tells the verifier

$$u_k x^k = \sum_{i=1}^{M}\sum_{j=1}^{m} u_{ijk}x^{i(m+1)n+jn+k} \qquad v_k = \sum_{i=1}^{M}\sum_{j=1}^{m} v_{ijk}$$

$$w_k x^k = \sum_{i=1}^{M}\sum_{j=1}^{m} w_{ijk}x^{i(m+1)n+jn+k}.$$

We now have that

$$\sum_{k=1}^{n}(u_k v_k - w_k)x^k$$

$$=\sum_{k=1}^{n}\left((\sum_{i=1}^{M}\sum_{j=1}^{m} u_{ijk}x^{i(m+1)n+jn+k})(\sum_{i'=1}^{M}\sum_{j'=1}^{m} v_{i'j'k}) - \sum_{i=1}^{M}\sum_{j=1}^{m} w_{ijk}x^{i(m+1)n+jn+k} \right)$$

contains the desired polynomial from (1) but there are some cross-terms corresponding to $i \neq i'$ or $j \neq j'$ so the polynomial given above may be non-zero.

We will choose the a_α and b_β values such that they cancel out the cross-terms. However, we have to be careful that there are only $O(M+m+n)$ of them and that they are feasible to compute. We will therefore use an interactive technique that will enable the verifier to pick a_α and b_α after seeing x. This introduces a second concern, namely to choose them in a way such that they do not affect the original equality we wish to get. We accomplish this by making sure that a_α and b_β are modified by factors y^α and z^β for $\alpha,\beta \neq 0$ while the desired equality does not contain any such factors. To make this happen we will modify the opening process of the commitments C_{U_i} and C_{V_i} described above to open

$$\prod_{i=1}^{M} C_{U_i}^{x^{i(m+1)n}y^i} = \mathrm{com}_{ck}^{(2)}(c_{u_1},\dots,c_{u_m}) \qquad \prod_{j=1}^{m} c_{u_j}^{x^{jn}z^j} = \mathrm{com}_{ck}(u_1,\dots,u_n;r)$$

$$\prod_{i=1}^{M} C_{V_i}^{y^{-i}} = \mathrm{com}_{ck}^{(2)}(c_{u_1},\dots,c_{u_m}) \qquad \prod_{j=1}^{m} c_{v_j}^{z^{-j}} = \mathrm{com}_{ck}(v_1,\dots,v_n;r)$$

This gives us

$$u_k x^k = \sum_{i=1}^{M} \sum_{j=1}^{m} u_{ijk} x^{i(m+1)n+jn+k} y^i z^j \qquad v_k = \sum_{i=1}^{M} \sum_{j=1}^{m} v_{ijk} y^{-i} z^{-j}.$$

We now have

$$\sum_{k=1}^{n} u_k x^k v_k = \sum_{k=1}^{n} \Big(\sum_{i=1}^{M} \sum_{j=1}^{m} u_{ijk} x^{i(m+1)n+jn+k} y^i z^j \Big) \Big(\sum_{i'=1}^{M} \sum_{j'=1}^{m} v_{i'j'k} y^{-i'} z^{-j'} \Big)$$

$$= \sum_{k=1}^{n} \sum_{i=1}^{M} \sum_{i'=1}^{M} \sum_{j=1}^{m} \sum_{j'=1}^{m} u_{ijk} x^{i(m+1)n+jn+k} v_{i'j'k} y^{i-i'} z^{j-j'}$$

By splitting the sum into three parts corresponding to the three cases $j = j', i = i'$ and $j = j', i \neq i'$ and $j \neq j'$ and subtracting the $w_k x^k$'s we get

$$\sum_{k=1}^{n} (u_k v_k - w_k) x^k = \sum_{k=1}^{n} \sum_{i=1}^{M} \sum_{j=1}^{m} (u_{ijk} v_{ijk} - w_{ijk}) x^{i(m+1)n+jn+k}$$

$$+ \sum_{k=1}^{n} \sum_{i=1}^{M} \sum_{\substack{i'=1 \\ i' \neq i}}^{M} \sum_{j=1}^{m} u_{ijk} x^{i(m+1)n+jn+k} v_{i'jk} y^{i-i'}$$

$$+ \sum_{k=1}^{n} \sum_{i=1}^{M} \sum_{i'=1}^{M} \sum_{j=1}^{m} \sum_{\substack{j'=1 \\ j' \neq j}}^{m} u_{ijk} x^{i(m+1)n+jn+k} v_{i'j'k} y^{i-i'} z^{j-j'} \qquad (2)$$

$$= \sum_{k=1}^{n} \sum_{i=1}^{M} \sum_{j=1}^{m} (u_{ijk} v_{ijk} - w_{ijk}) x^{i(m+1)n+jn+k}$$

$$+ \sum_{\substack{\alpha=-M \\ \alpha \neq 0}}^{M} \sum_{\substack{i=1,i'=1 \\ i-i'=\alpha}}^{M,M} \sum_{k=1}^{n} \sum_{j=1}^{m} u_{ijk} x^{i(m+1)n+jn+k} v_{i'jk} y^{\alpha}$$

$$+ \sum_{\substack{\beta=-m \\ \beta \neq 0}}^{m} \sum_{\substack{j=1,j'=1 \\ j-j'=\beta}}^{m,m} \sum_{k=1}^{n} \Big(\sum_{i=1}^{M} u_{ijk} x^{i(m+1)n+jn+k} y^i \Big) \Big(\sum_{i'=1}^{M} v_{i'j'k} y^{-i'} \Big) z^{\beta}$$

The prover will select

$$a_\alpha = \sum_{\substack{i=1,i'=1 \\ i-i'=\alpha}}^{M,M} \sum_{k=1}^{n} \sum_{j=1}^{m} u_{ijk} x^{i(m+1)n+jn+k} v_{i'jk}$$

$$b_\beta = \sum_{\substack{j=1,j'=1 \\ j-j'=\beta}}^{m,m} \sum_{k=1}^{n} \Big(\sum_{i=1}^{M} u_{ijk} x^{i(m+1)n+jn+k} y^i \Big) \Big(\sum_{i'=1}^{M} y^i v_{i'j'k} y^{-i'} \Big)$$

and send the commitments $\{c_{a_\alpha}\}_\alpha$ before seeing y and send $\{c_{b_\beta}\}_\beta$ before seeing z. She will reveal randomness $R \in \mathbb{Z}_p$ such that

$$\prod_{\substack{\alpha=-M \\ \alpha\neq0}}^{M} c_{a_\alpha}^{y^\alpha} \cdot \prod_{\substack{\beta=-m \\ \beta\neq0}}^{m} c_{b_\beta}^{z^\beta} = \mathrm{com}_{ck}(\sum_{k=1}^{n}(u_k v_k - w_k)x^k; R).$$

This corresponds to the values in the commitments satisfying

$$\sum_{\substack{\alpha=-M \\ \alpha\neq0}}^{M} a_\alpha y^\alpha + \sum_{\substack{\beta=-m \\ \beta\neq0}}^{m} b_\beta z^\beta = \sum_{k=1}^{n}(u_k v_k - w_k)x^k.$$

Keeping in mind the expansion of the right hand side (2) we get that with overwhelming probability over y, z this can only be true if equation (1) holds.

In order to make the protocol SHVZK we add some commitments and values such that $c_{u_j}, c_{v_j}, c_{w_j}$ and u_k, v_k, w_k cannot reveal anything about $u_{ijk}, v_{ijk}, w_{ijk}$. Furthermore, we add some d_k values and c_{d_k} commitments to cancel out new cross-terms arising from the added values. This gives us the full batch product argument below.

Common reference string: Two-tiered commitment key ck.
Statement: Commitments $C_{U_1}, C_{V_1}, C_{W_1} \ldots, C_{U_M}, C_{V_M}, C_{W_M} \in \mathbb{T}$.
Prover's witness: Values $u_{111}, v_{111}, w_{111}, \ldots, u_{Mmn}, v_{Mmn}, w_{Mmn} \in \mathbb{Z}_p$ and randomness $r_{11}, s_{11}, t_{11}, \ldots, r_{Mm}, s_{Mm}, t_{Mm} \in \mathbb{Z}_p$ such that for all $i \in \{1, \ldots, M\}, j \in \{1, \ldots, m\}, k \in \{1, \ldots, n\}$:

$$c_{u_{ij}} = \mathrm{com}_{ck}(u_{ij1}, \ldots, u_{ijn}; r_{ij}) \qquad C_{U_i} = \mathrm{com}_{ck}^{(2)}(c_{u_{i1}}, \ldots, c_{u_{im}})$$
$$c_{v_{ij}} = \mathrm{com}_{ck}(v_{ij1}, \ldots, v_{ijn}; s_{ij}) \qquad C_{V_i} = \mathrm{com}_{ck}^{(2)}(c_{v_{i1}}, \ldots, c_{v_{im}})$$
$$c_{w_{ij}} = \mathrm{com}_{ck}(w_{ij1}, \ldots, w_{ijn}; t_{ij}) \qquad C_{W_i} = \mathrm{com}_{ck}^{(2)}(c_{w_{i1}}, \ldots, c_{w_{im}})$$
$$u_{ijk} v_{ijk} = w_{ijk}.$$

1. $\mathcal{P} \to \mathcal{V}$: Pick $u_{00k}, v_{00k}, w_{00k} \leftarrow \mathbb{Z}_p$ and set $u_{0jk} = v_{0jk} = w_{0jk} = 0$ and $u_{i0k} = v_{i0k} = w_{i0k} = 0$ for $i \neq 0$ and $j \neq 0$. Pick $r_{00}, s_{00}, t_{00}, \tau_1, \ldots, \tau_n \leftarrow \mathbb{Z}_p$ and pick $r_{0j}, s_{0j}, t_{0j} \leftarrow \mathbb{Z}_p$. Compute for $j \in \{0, \ldots, m\}$ and $k \in \{1, \ldots, n\}$

$$c_{u_{0j}} = \mathrm{com}_{ck}(u_{0j1}, \ldots, u_{0jn}; r_{0j}) \qquad C_{U_0} = \mathrm{com}_{ck}^{(2)}(c_{u_{01}}, \ldots, c_{u_{0m}})$$
$$c_{v_{0j}} = \mathrm{com}_{ck}(v_{0j1}, \ldots, v_{0jn}; s_{0j}) \qquad C_{V_0} = \mathrm{com}_{ck}^{(2)}(c_{v_{01}}, \ldots, c_{v_{0m}})$$
$$c_{w_{0j}} = \mathrm{com}_{ck}(w_{0j1}, \ldots, w_{0jn}; t_{0j}) \qquad C_{W_0} = \mathrm{com}_{ck}^{(2)}(c_{w_{01}}, \ldots, c_{w_{0m}})$$
$$d_k = u_{00k} v_{00k} - w_{00k} \qquad c_{d_k} = \mathrm{com}_{ck}(d_k; \tau_k)$$

Send: $c_{u_{00}}, c_{v_{00}}, c_{w_{00}}, C_{U_0}, C_{V_0}, C_{W_0}, \{c_{d_k}\}_{k=1}^{n}$.
2. $\mathcal{P} \leftarrow \mathcal{V}$: $x \leftarrow \mathbb{Z}_p^*$.
3. $\mathcal{P} \to \mathcal{V}$: For $\alpha \in \{-M, \ldots, -1, 1, \ldots, M\}$ pick $\rho_\alpha \leftarrow \mathbb{Z}_p$ and compute

$$a_\alpha = \sum_{\substack{i=0,i'=0 \\ i-i'=\alpha}}^{M,M} \sum_{j=0}^{m} \sum_{k=1}^{n}(u_{ijk} x^{i(m+1)n+jn+k})v_{i'jk} \qquad c_{a_\alpha} = \mathrm{com}_{ck}(a_\alpha; \rho_\alpha).$$

Compute also for $j \in \{1, \ldots, m\}$

$$c_{u_j} = \prod_{i=0}^{M} c_{u_{ij}}^{x^{i(m+1)n} y^i} \qquad c_{v_j} = \prod_{i=0}^{M} c_{v_{ij}}^{y^{-i}} \qquad c_{w_j} = \prod_{i=0}^{M} c_{w_{ij}}^{x^{i(m+1)n}}.$$

Send: $\{c_{a_\alpha}\}_{\alpha \in \{-M, \ldots, -1, 1, \ldots, M\}}, \{c_{u_j}, c_{v_j}, c_{w_j}\}_{j=1}^{m}$.

4. $\mathcal{P} \leftarrow \mathcal{V}: y \leftarrow \mathbb{Z}_p^*$.

5. $\mathcal{P} \rightarrow \mathcal{V}:$ For $\beta \in \{-m, \ldots, -1, 1, \ldots, m\}$ pick $\sigma_\beta \leftarrow \mathbb{Z}_p$ and compute

$$b_\beta = \sum_{\substack{j=0, j'=0 \\ j-j'=\beta}}^{m,m} \sum_{k=1}^{n} \left(\sum_{i=0}^{M} u_{ijk} x^{i(m+1)n+jn+k} y^i \right) \left(\sum_{i'=0}^{M} v_{i'j'k} y^{-i'} \right)$$

Define $c_{b_\beta} = \operatorname{com}_{ck}(b_\beta; \sigma_\beta)$ and send: $\{c_{b_\beta}\}_{\beta \in \{-m, \ldots, -1, 1, \ldots, m\}}$.

6. $\mathcal{P} \leftarrow \mathcal{V}: z \leftarrow \mathbb{Z}_p^*$.

7. $\mathcal{P} \rightarrow \mathcal{V}:$ Compute for $k \in \{1, \ldots, n\}$

$$u_k = u_{00k} + \sum_{j=1}^{m} \sum_{i=0}^{M} u_{ijk} x^{i(m+1)n+jn} y^i z^j \qquad r = r_{00} + \sum_{j=1}^{m} \sum_{i=0}^{M} r_{ij} x^{i(m+1)n+jn} y^i z^j$$

$$v_k = v_{00k} + \sum_{j=1}^{m} \sum_{i=0}^{M} v_{ijk} y^{-i} z^{-j} \qquad s = s_{00} + \sum_{j=1}^{m} \sum_{i=0}^{M} s_{ij} y^{-i} z^{-j}$$

$$w_k = w_{00k} + \sum_{j=1}^{m} \sum_{i=0}^{M} w_{ijk} x^{i(m+1)n+jn} \qquad t = t_{00} + \sum_{j=1}^{m} \sum_{i=0}^{M} t_{ij} x^{i(m+1)n+jn}$$

$$R = \sum_{k=1}^{n} \tau_k x^k + \sum_{\substack{\alpha=-M \\ \alpha \neq 0}}^{M} \rho_\alpha y^\alpha + \sum_{\substack{\beta=-m \\ \beta \neq 0}}^{m} \sigma_\beta z^\beta$$

Send: $\{u_k, v_k, w_k\}_{k=1}^{n}, r, s, t, R$.

Verification: Accept the argument if the following holds

$$c_{u_{00}} \prod_{j=1}^{m} c_{u_j}^{x^{jn} z^j} = \operatorname{com}_{ck}(u_1, \ldots, u_n; r) \qquad \prod_{i=0}^{M} C_{U_i}^{x^{i(m+1)n} y^i} = \operatorname{com}_{ck}^{(2)}(c_{u_1}, \ldots, c_{u_m})$$

$$c_{v_{00}} \prod_{j=1}^{m} c_{v_j}^{z^{-j}} = \operatorname{com}_{ck}(v_1, \ldots, v_n; s) \qquad \prod_{i=0}^{M} C_{V_i}^{y^{-i}} = \operatorname{com}_{ck}^{(2)}(c_{v_1}, \ldots, c_{v_m})$$

$$c_{w_{00}} \prod_{j=1}^{m} c_{w_j}^{x^{jn}} = \operatorname{com}_{ck}(w_1, \ldots, w_n; t) \qquad \prod_{i=0}^{M} C_{W_i}^{x^{i(m+1)n}} = \operatorname{com}_{ck}^{(2)}(c_{w_1}, \ldots, c_{w_m})$$

$$\prod_{k=1}^{n} c_{d_k}^{x^k} \cdot \prod_{\substack{\alpha=-M \\ \alpha \neq 0}}^{M} c_{a_\alpha}^{y^\alpha} \cdot \prod_{\substack{\beta=-m \\ \beta \neq 0}}^{m} c_{b_\beta}^{z^\beta} = \operatorname{com}_{ck}(\sum_{k=1}^{n}(u_k v_k - w_k) x^k; R)$$

Theorem 1 (Full paper). *The argument given above has perfect completeness, perfect SHVZK and witness-extended emulation if the two-tiered commitment scheme is binding.*

Complexity. The communication complexity of the batch product argument is 3 elements in $\mathbb{T}$, $2M + 5m + n + 1$ elements in $\mathbb{G}$ and $3n + 7$ elements in $\mathbb{Z}_p$.

Let us estimate the computation complexity assuming that we use the two-tiered commitment scheme we described in Section 2.1 in an asymmetric bilinear group with base groups $\mathbb{G}, \hat{\mathbb{G}}$ and target group $\mathbb{T}$. The verifier's computation is $3m$ pairings and exponentiations in the target group $\mathbb{T}$ and $5M + 2m + 4n$ exponentiations in the base group $\mathbb{G}$. Using standard techniques for batch verification some of the equations can be combined in a randomized manner and we may also use multi-exponentiation techniques to reduce the complexity further to $O(\frac{M+m+n}{\log(M+m+n)})$ exponentiations.

A naïve implementation of the prover would require $3m$ pairings and $O(M+m+n)$ exponentiations and $O(N(M + m))$ multiplications in $\mathbb{Z}_p$, where $N - Mmn$. When M or m are large the latter complexity dominates.

We can use techniques for polynomial multiplication to reduce the prover's computation. Consider as an example the computation in round 3, where the prover computes

$$a_\alpha = \sum_{\substack{i=0,i'=0 \\ i-i'=\alpha}}^{M,M} \sum_{j=0}^{m} \sum_{k=1}^{n} (u_{ijk} x^{i(m+1)n+jn+k}) v_{i'jk}$$

for $\alpha = -M, \ldots, -1, 1, \ldots, M$. Define $\boldsymbol{u}_i = (u_{i01} x^{i(m+1)n+0n+1}, \ldots, u_{imn} x^{i(m+1)n+mn+n})$ and $\boldsymbol{v}_{i'} = (v_{i'01}, \ldots, v_{i'mn})$, which allows us to rewrite it as

$$a_\alpha = \sum_{\substack{i=0,i'=0 \\ i-i'=\alpha}}^{M,M} \boldsymbol{u}_i \boldsymbol{v}_{i'}^\top.$$

Observe that a_α is the $M + \alpha$'th coefficient of the polynomial

$$p(\omega) = \left(\sum_{i=0}^{M} \omega^i \boldsymbol{u}_i \right) \left(\sum_{i'=0}^{M} \omega^{M-i'} \boldsymbol{v}_{i'}^\top \right) \in \mathbb{Z}_p[\omega].$$

The degree of the polynomial is $2M$ so if we evaluate it in $2M + 1$ different points $\omega_1, \ldots, \omega_{2M+1} \in \mathbb{Z}_p$ we can use polynomial interpolation to recover the coefficients. The evaluation of $\sum_{j=0}^{M} \omega^i \boldsymbol{u}_i$ and $\sum_{i'=0}^{M} \omega^{M-i'} \boldsymbol{v}_{i'}^\top$ in $2M + 1$ different points can be done using $O(N \log^2 M)$ multiplications. If $2M|p - 1$ and M is a power of 2 we can pick $\omega_1, \ldots, \omega_{2M}$ as $2M$-roots of unity, i.e., $\omega_k^{2M} = 1$ and use the Fast Fourier Transform to reduce the cost further down to $O(N \log M)$ multiplications.[3] Similarly, we can compute $b_{-m}, \ldots, b_{-1}, b_1, \ldots, b_m$ using $O(N \log^2 m)$ multiplications or $O(N \log m)$ multiplications if $2m|p - 1$ and m is a power of 2.

Known values. Sometimes it will be useful to use publicly known values u_{ijk} in the argument. The trivial way to handle this is to use commitments $c_{u_{ij}} = \mathrm{com}_{ck}(u_{ij1}, \ldots, u_{ijn}; 0)$. Since they use trivial randomness, the verifier can check directly that $C_{U_1}, \ldots, C_{U_M}$ contain the correct values. A more careful inspection reveals

---

[3] It takes a while before the assymptotic behaviour kicks in, so for small M it may be better to use Toom-Cook related methods for computing the coefficients $a_{-M}, \ldots, a_M$.

that some efficiency savings can be made by abandoning the commitments $c_{u_{ij}}$ altogether. Since the u_{ijk} values are public we do not need to hide them, so the prover may choose $u_{0jk} = 0$. The verifier can now herself compute the resulting u_k values without using the commitments at all.

A similar analysis reveals that when w_{ijk} are known the prover does not need to communicate any C_{W_i} or c_{w_j} commitments since the verifier can compute w_k himself. In the special case where $w_{ijk} = 0$ this simplifies to fixing $w_k = 0$.

3.1   Inner Product Argument

A slight modification of the batch product argument allows the prover to demonstrate instead $\sum_{i=1}^{M} \sum_{j=1}^{m} \sum_{k=1}^{n} u_{ijk} v_{ijk} = \sum_{i=1}^{M} \sum_{j=1}^{m} \sum_{k=1}^{n} w_{ijk}$. The main observation is that we can fix $x = 1$ instead of letting the verifier choose it, in which case equation (1) gives us the desired equality.

The only issue in following this idea is the cross-terms arising from $u_{0jk}, v_{0jk}, w_{0jk}$. We therefore compute $C_{U_0}^x, C_{V_0}^x, C_{W_0}^x, c_{u_{00}}^x, c_{v_{00}}^x, c_{w_{00}}^x$ giving us commitments to $u_{0jk}x, v_{0jk}x, w_{0jk}x$. Since $x \in \mathbb{Z}_p^*$ these values will still ensure that $c_{u_j}, c_{v_j}, c_{w_j}, u_k, v_k, w_k$ do not leak any information about $u_{ijk}, v_{ijk}, w_{ijk}$. But since they are modified by a random factor x throughout the argument they will not interfere with the equation $\sum_{i=1}^{M} \sum_{j=1}^{m} \sum_{k=1}^{n} u_{ijk} v_{ijk} = \sum_{i=1}^{M} \sum_{j=1}^{m} \sum_{k=1}^{n} w_{ijk}$. To get perfect completeness, we use two commitments to d_1 and d_2 values to cancel out crossterms corresponding to x and x^2.

4   Arguments for Circuit Satisfiability

Using the batch product argument from Section 3 we can give a 7-move SHVZK argument for circuit satisfiability. Consider a boolean circuit consisting of $N - 1$ NAND-gates where the prover wants to convince the verifier that there is a satisfying assignment making the circuit output 1. If the output wire is w, we can add a new variable u and add a self-looping gate of the form $w = \neg(w \wedge u)$, which can only be satisfied if $w = 1$. The prover now has a circuit with N NAND-gates and no output and wants to demonstrate that there is an internally consistent assignment to the wires that respects all gates.

Let us without loss of generality consider a circuit with $N = Mmn$ NAND-gates for which the prover wants to demonstrate that there is a consistent assignment. The prover enumerates the two inputs and the output of each gate as $u_{ijk}, v_{ijk}, w_{ijk}$. The task is now to show that the committed values correspond to a satisfying assignment for the circuit.

The prover first shows that all the committed values are either 0 or 1 corresponding to truth values. This is done by using batch product arguments to show $u_{ijk} u_{ijk} = u_{ijk}, v_{ijk} v_{ijk} = v_{ijk}$ and $w_{ijk} w_{ijk} = w_{ijk}$, which can only be true if $u_{ijk}, v_{ijk}, w_{ijk} \in \{0, 1\}$.

The prover then uses the homomorphic property of the commitment scheme to compute commitments to $1 - w_{ijk}$. Using another batch product argument it can show $u_{ijk} v_{ijk} = 1 - w_{ijk}$, which means the committed values respect the NAND-gates.

Finally, using a technique from [22] it uses an inner product argument to show that all committed values u_{ijk}, v_{ijk} and w_{ijk} corresponding to the same wire x_ℓ are consistent with each other. We describe this technique in the full circuit satisfiability argument below.

Common reference string: Two-tiered commitment key ck.
Statement: $N = Mmn$ NAND-gates $x_{\ell_2} = \neg(x_{\ell_0} \wedge x_{\ell_1})$ over variables x_ℓ.
Prover's witness: An assigment to $\{x_\ell\}$ respecting all NAND-gates.
Argument: Label the inputs and outputs of the gates $\{u_{ijk}, v_{ijk}, w_{ijk}\}_{i=1,j=1,k=1}^{M,m,n}$.
Pick $r_{ij}, s_{ij}, t_{ij} \leftarrow \mathbb{Z}_p$ and compute the commitments

$$c_{u_{ij}} = \mathrm{com}_{ck}(u_{ij1}, \ldots, u_{ijn}; r_{ij}) \qquad C_{U_i} = \mathrm{com}_{ck}^{(2)}(c_{u_{i1}}, \ldots, c_{u_{im}})$$

$$c_{v_{ij}} = \mathrm{com}_{ck}(v_{ij1}, \ldots, v_{ijn}; s_{ij}) \qquad C_{V_i} = \mathrm{com}_{ck}^{(2)}(c_{v_{i1}}, \ldots, c_{v_{im}})$$

$$c_{w_{ij}} = \mathrm{com}_{ck}(w_{ij1}, \ldots, w_{ijn}; t_{ij}) \qquad C_{W_i} = \mathrm{com}_{ck}^{(2)}(c_{w_{i1}}, \ldots, c_{w_{im}})$$

Send $\{C_{U_i}, C_{V_i}, C_{W_i}\}_{i=1}^{M}$ to the verifier.
Engage in three batch product arguments with statements $\{C_{U_i}, C_{U_i}, C_{U_i}\}_{i=1}^{M}$, $\{C_{V_i}, C_{V_i}, C_{V_i}\}_{i=1}^{M}$ and $\{C_{W_i}, C_{W_i}, C_{W_i}\}_{i=1}^{M}$ in order to show that $u_{ijk}, v_{ijk}, w_{ijk} \in \{0, 1\}$.
Define $c_1 = \mathrm{com}_{ck}(1, \ldots, 1; 0)$ and $C_1 = \mathrm{com}_{ck}^{(2)}(c_1, \ldots, c_1)$. Engage in a batch product proof with statement $\{C_{U_1}, C_{V_1}, C_1 C_{W_i}^{-1}\}_{i=1}^{M}$ to show that the NAND-gates are respected.
There are $3N = 3Mmn$ committed values $u_{ijk}, v_{ijk}, w_{ijk}$. Let us rename them $\{b_i\}_{i=1}^{3N}$ and the corresponding commitments to $\{C_{B_i}\}_{i=1}^{3M}$. The same variable x_ℓ may appear n_ℓ times in the circuit as $b_{i_1}, \ldots, b_{i_{n_\ell}}$. Define π as the permutation in S_{3N} such that for each variable x_ℓ appearing n_ℓ times in the circuit the permutation makes a complete cycle $i_1 \rightarrow i_2 \rightarrow \ldots \rightarrow i_{n_\ell} \rightarrow i_1$ corresponding to those appearances.
The prover receives a challenge y from the verifier and defines $a_i = y^i - y^{\pi(i)}$. It uses the inner product argument[4] from Section 3.1 to demonstrate $\sum_{i=1}^{3N} a_i b_i = 0$. This shows that for random y

$$\sum_{i=1}^{3N} a_i b_i = \sum_{i=1}^{3N} (y^i - y^{\pi(i)}) b_i = \sum_{i=1}^{3N} y^i (b_i - b_{\pi^{-1}(i)}) = 0.$$

With overwhelming probability over y this shows $b_{\pi(i)} = b_i$ for all i thus proving that the values b_i and hence the values $u_{ijk}, v_{ijk}, w_{ijk}$ are consistent with the wires x_ℓ.
Verification: Verify the 4 batch product proofs and the inner product argument.

Theorem 2 (Full paper). *The argument for circuit satisfiability has perfect completeness, perfect SHVZK and witness-extended emulation.*

---

[4] The first round of the inner product argument can be run independently of y such that the total round complexity remains 7.

Arithmetic circuits. Using similar techniques as in the circuit satisfiability argument, we can also get an argument for the satisfiability of arithmetic circuits consisting of addition and multiplication gates over $\mathbb{Z}_p$. The prover commits to the values and uses the homomorphic property of the commitment scheme to show that addition gates are respected and the batch product argument to show that multiplication gates are respected. If there are publicly known constants (without loss of generality a multiple of mn) involved in the circuit, the prover commits to these using randomness 0 so the verifier can check directly that they are correct. As in the circuit satisfiability argument the prover also demonstrates that the committed values are consistent with the wiring of the arithmetic circuit. This gives an arithmetic circuit argument with communication complexity $O(M + m + n)$.

5   Range Arguments

As a concrete application of our batch product argument we will give a communication-efficient range proof. The prover has a commitment c and wants to convince the verifier that she knows an opening w, t such that $c = \text{com}_{ck}(w; t)$ and $w \in [A; B]$. Since the commitment is homomorphic, the problem can be simplified to demonstrating that she knows an opening of $c \cdot \text{com}_{ck}(-A; 0)$ in the range $[0; B - A]$. Let $N = \lfloor \log(B - A) \rfloor$. The prover can construct a commitment $c_{0/1} = \text{com}_{ck}(b; s)$ and show that it contains 0 or 1 using standard techniques. By showing that $c \cdot \text{com}_{ck}(-A; 0) \cdot c_{0/1}^{A-B+2^N}$ contains a value in the range $[0; 2^N)$ she convinces the verifier that $w \in [A; B]$.

We can therefore without loss of generality focus on demonstrating that a committed value w belongs to the interval $[0; 2^N)$. We will now give such a range argument that only communicates $O(N^{\frac{1}{3}})$ elements. The idea is that the prover will commit to the bit representation of w. Using a batch product argument the prover can demonstrate that the committed bits are 0 or 1. Furthermore, using techniques similar to the buildup of w_k in the batch product argument the prover will demonstrate that $w = \sum_{i=1}^{M} \sum_{j=1}^{m} \sum_{k=1}^{n} w_{ijk} 2^{imn+jn+k-1}$ using $O(M + m + n)$ communication. If $M = O(N^{\frac{1}{3}}), m = O(N^{\frac{1}{3}}), n = O(N^{\frac{1}{3}})$ the communication complexity is $O(N^{\frac{1}{3}})$ elements.

Common reference string: ck.
Statement: $c \in \mathbb{G}$.
Prover's witness: $w, t \in \mathbb{Z}_p$ such that $w \in [0; 2^N)$ and $c = \text{com}_{ck}(w; t)$.
Argument: Let $\{w_{ijk}\}_{i=1,j=1,k=1}^{M,m,n}$ be the bits of w. Pick $r_{ij} \leftarrow \mathbb{Z}_p$ and compute

$$c_{w_{ij}} = \text{com}_{ck}(w_{ij1}, \ldots, w_{ijn}; r_{ij}) \quad C_{W_i} = \text{com}_{ck}^{(2)}(c_{w_{i1}}, \ldots, c_{w_{im}}) \quad c_{w_j} = \prod_{i=1}^{M} c_{w_{ij}}^{2^{imn}}.$$

Pick $w_{01}, \ldots, w_{0n} \leftarrow \mathbb{Z}_p$ and $r_0, s_d \leftarrow \mathbb{Z}_p$ and compute $c_{w_0} = \text{com}_{ck}(w_{01}, \ldots, w_{0n}; r_0)$ and $c_d = \text{com}_{ck}(\sum_{k=1}^{n} w_{0k} 2^{k-1}; s_d)$.
Send $\{C_{W_i}\}_{i=1}^{M}, \{c_{w_j}\}_{j=0}^{m}$ and c_d to the verifier and get a challenge $x \leftarrow \mathbb{Z}_p^*$ back.
Compute

$$w_k = xw_{0k} + \sum_{i=1}^{M} \sum_{j=1}^{m} w_{ijk} 2^{imn+jn} \quad r = xr_0 + \sum_{i=1}^{M} \sum_{j=1}^{m} r_{ij} 2^{imn+jn} \quad s = s_d x + t$$

and send them to the verifier.

In parallel, engage in a batch product argument with statement$\{C_{W_i}, C_{W_i}, C_{W_i}\}_{i=1}^{M}$ to show that each w_{ijk} satisfies $w_{ijk}w_{ijk} = w_{ijk}$, which implies $w_{ijk} \in \{0,1\}$.

Verification: Verify that the batch product argument is valid and

$$\prod_{i=1}^{M} C_{W_i}^{2^{imn}} = \text{com}_{ck}^{(2)}(c_{w_1}, \ldots, c_{w_m}) \qquad c_{w_0}^x \prod_{j=1}^{m} c_{w_j}^{2^{jn}} = \text{com}_{ck}(w_1, \ldots, w_n; r)$$

$$c_d^x c = \text{com}_{ck}\left(\sum_{k=1}^{n} w_k 2^{k-1}; s\right).$$

Theorem 3 (Full paper). *The range argument given above has perfect completeness, perfect SHVZK and witness-extended emulation.*

References

1. Abe, M., Fuchsbauer, G., Groth, J., Haralambiev, K., Ohkubo, M.: Structure-Preserving Signatures and Commitments to Group Elements. In: Rabin, T. (ed.) CRYPTO 2010. LNCS, vol. 6223, pp. 209–236. Springer, Heidelberg (2010)
2. Arora, S., Lund, C., Motwani, R., Sudan, M., Szegedy, M.: Proof verification and the hardness of approximation problems. Journal of the ACM 45(3), 501–555 (1998)
3. Arora, S., Safra, S.: Probabilistic checking of proofs: a new characterization of NP. Journal of the ACM 45(1), 70–122 (1998)
4. Ben-Sasson, E., Goldreich, O., Harsha, P., Sudan, M., Vadhan, S.P.: Short PCPs verifiable in polylogarithmic time. In: IEEE Conference on Computational Complexity, pp. 120–134 (2005)
5. Boudot, F.: Efficient Proofs that a Committed Number Lies in an Interval. In: Preneel, B. (ed.) EUROCRYPT 2000. LNCS, vol. 1807, pp. 431–444. Springer, Heidelberg (2000)
6. Camenisch, J., Chaabouni, R., Shelat, A.: Efficient Protocols for Set Membership and Range Proofs. In: Pieprzyk, J. (ed.) ASIACRYPT 2008. LNCS, vol. 5350, pp. 234–252. Springer, Heidelberg (2008)
7. Chaabouni, R., Lipmaa, H., Shelat, A.: Additive Combinatorics and Discrete Logarithm Based Range Protocols. In: Steinfeld, R., Hawkes, P. (eds.) ACISP 2010. LNCS, vol. 6168, pp. 336–351. Springer, Heidelberg (2010)
8. Cramer, R., Damgård, I., Schoenmakers, B.: Proof of Partial Knowledge and Simplified Design of Witness Hiding Protocols. In: Desmedt, Y.G. (ed.) CRYPTO 1994. LNCS, vol. 839, pp. 174–187. Springer, Heidelberg (1994)
9. Damgård, I.: Efficient Concurrent Zero-Knowledge in the Auxiliary String Model. In: Preneel, B. (ed.) EUROCRYPT 2000. LNCS, vol. 1807, pp. 418–430. Springer, Heidelberg (2000)
10. Damgård, I., Jurik, M.J.: A Generalisation, a Simplification and Some Applications of Paillier's Probabilistic Public-Key System. In: Kim, K.-c. (ed.) PKC 2001. LNCS, vol. 1992, pp. 119–136. Springer, Heidelberg (2001)
11. Dinur, I.: The PCP theorem by gap amplification. Journal of the ACM 54(3) (2007)
12. Galbraith, S.D., Paterson, K.G., Smart, N.P.: Pairings for cryptographers. Discrete Applied Mathematics 156(16), 3113–3121 (2008)
13. Garay, J.A., MacKenzie, P.D., Yang, K.: Strengthening zero-knowledge protocols using signatures. Journal of Cryptology 19(2), 169–209 (2006)

14. Gentry, C.: A fully homomorphic encryption scheme. PhD thesis. Stanford University (2009)
15. Gentry, C.: Fully homomorphic encryption using ideal lattices. In: STOC, pp. 169–178 (2009)
16. Goldreich, O., Krawczyk, H.: On the composition of zero-knowledge proof systems. SIAM Journal of Computing 25(1), 169–192 (1996)
17. Goldwasser, S., Kalai, Y.T., Rothblum, G.N.: Delegating computation: interactive proofs for muggles. In: STOC, pp. 113–122 (2008)
18. Goldwasser, S., Micali, S., Rackoff, C.: The knowledge complexity of interactive proofs. SIAM Journal of Computing 18(1), 186–208 (1989)
19. Groth, J.: Honest verifier zero-knowledge arguments applied. Dissertation Series DS-04-3, BRICS, PhD thesis. xii+119 pp (2004)
20. Groth, J.: Non-interactive Zero-Knowledge Arguments for Voting. In: Ioannidis, J., Keromytis, A.D., Yung, M. (eds.) ACNS 2005. LNCS, vol. 3531, pp. 467–482. Springer, Heidelberg (2005)
21. Groth, J.: Homomorphic trapdoor commitments to group elements. Cryptology ePrint Archive, Report 2009/007 (2009)
22. Groth, J.: Linear Algebra with Sub-Linear Zero-Knowledge Arguments. In: Halevi, S. (ed.) CRYPTO 2009. LNCS, vol. 5677, pp. 192–208. Springer, Heidelberg (2009)
23. Groth, J.: Minimizing non-interactive zero-knowledge proofs using fully homomorphic encryption. Cryptology ePrint Archive, Report 2011/012 (2011)
24. Groth, J., Ishai, Y.: Sub-Linear Zero-Knowledge Argument for Correctness of a Shuffle. In: Smart, N.P. (ed.) EUROCRYPT 2008. LNCS, vol. 4965, pp. 379–396. Springer, Heidelberg (2008)
25. Ishai, Y., Kushilevitz, E., Ostrovsky, R., Sahai, A.: Zero-knowledge proofs from secure multiparty computation. SIAM Journal of Computing 39(3), 1121–1152 (2009)
26. Kalai, Y.T., Raz, R.: Interactive PCP. In: Aceto, L., Damgård, I., Goldberg, L.A., Halldórsson, M.M., Ingólfsdóttir, A., Walukiewicz, I. (eds.) ICALP 2008, Part II. LNCS, vol. 5126, pp. 536–547. Springer, Heidelberg (2008)
27. Kilian, J.: A note on efficient zero-knowledge proofs and arguments. In: STOC, pp. 723–732 (1992)
28. Lindell, Y.: Parallel coin-tossing and constant-round secure two-party computation. Journal of Cryptology 16(3), 143–184 (2003)
29. Lipmaa, H.: On Diophantine Complexity and Statistical Zero-Knowledge Arguments. In: Laih, C.-S. (ed.) ASIACRYPT 2003. LNCS, vol. 2894, pp. 398–415. Springer, Heidelberg (2003)
30. Lipmaa, H., Asokan, N., Niemi, V.: Secure Vickrey Auctions Without Threshold Trust. In: Blaze, M. (ed.) FC 2002. LNCS, vol. 2357, pp. 87–101. Springer, Heidelberg (2003)
31. Pedersen, T.P.: Non-Interactive and Information-Theoretic Secure Verifiable Secret Sharing. In: Feigenbaum, J. (ed.) CRYPTO 1991. LNCS, vol. 576, pp. 129–140. Springer, Heidelberg (1992)
32. Schnorr, C.-P.: Efficient signature generation by smart cards. Journal of Cryptology 4(3), 161–174 (1991)

A Framework for Practical Universally Composable Zero-Knowledge Protocols[*]

Jan Camenisch[1], Stephan Krenn[2], and Victor Shoup[3]

[1] IBM Research – Zurich, Rüschlikon, Switzerland
jca@zurich.ibm.com
[2] Bern University of Applied Sciences, Biel-Bienne, Switzerland, and
University of Fribourg, Switzerland
stephan.krenn@bfh.ch
[3] Department of Computer Science, New York University, USA
shoup@cs.nyu.edu

Abstract. Zero-knowledge proofs of knowledge (ZK-PoK) for discrete logarithms and related problems are indispensable for practical cryptographic protocols. Recently, Camenisch, Kiayias, and Yung provided a specification language (the *CKY-language*) for such protocols which allows for a modular design and protocol analysis: for every zero-knowledge proof specified in this language, protocol designers are ensured that there exists an efficient protocol which indeed proves the specified statement.

However, the protocols resulting from their compilation techniques only satisfy the classical notion of ZK-PoK, which is not retained are when they used as building blocks for higher-level applications or composed with other protocols. This problem can be tackled by moving to the Universal Composability (UC) framework, which guarantees retention of security when composing protocols in arbitrary ways. While there exist generic transformations from Σ-protocols to UC-secure protocols, these transformation are often too inefficient for practice.

In this paper we introduce a specification language akin to the CKY-language and a compiler such that the resulting protocols are UC-secure and efficient. To this end, we propose an extension of the UC-framework addressing the issue that UC-secure zero-knowledge proofs are by definition proofs *of knowledge*, and state a special composition theorem which allows one to use the weaker – but more efficient and often sufficient – notion of proofs *of membership* in the UC-framework. We believe that our contributions enable the design of practically efficient protocols that are UC-secure and thus themselves can be used as building blocks.

Keywords: UC-Framework, Protocol Design, Zero-Knowledge Proof.

1 Introduction

The probably most demanding task when designing a practical cryptographic protocol is to define its security properties and then to prove that it indeed

---

[*] This work was in part funded by the Swiss Hasler Foundation, and the EU FP7 grants 216483 and 216499, as well as by the NSF grant CNS-0716690.

D.H. Lee and X. Wang (Eds.): ASIACRYPT 2011, LNCS 7073, pp. 449–467, 2011.

satisfies them. For this security analysis it is often assumed that the "world" consists only of one instance of the protocol and only of the involved parties, rather than of many parties running many instances of the same protocol as well as other protocols at the same time. While this approach allows for a relatively simple analysis of protocols, it does not properly model reality and therefore provides little if any security guarantees. Also, this approach does not allow for a modular usage of the protocols, i.e., when a protocol is used as a building block for another protocol, the security analysis must be all done from scratch.

To address these problems, a number of frameworks have been proposed over the years, e.g., [1–3]. The so-called *Universal Composability* (UC) framework by Canetti [2] seems to be the most prevalent one. A fundamental result in this model is its very strong composition theorem: once a protocol is proved secure in this model, it can be used in arbitrary contexts retaining its security properties. This allows one to split a protocol into smaller subroutines so that the security of each subprotocol can be analyzed separately, making the security of the overall protocol much easier. In particular, each (sub-)protocol needs to be analyzed only once and for all and does not have to be repeated for each specific context.

This modularity and the high security guarantees suggest that protocols should always be designed and proven secure in the UC-framework. However, this is only the case for a small fraction of the proposed cryptographic schemes, such as oblivious transfer [4] and encryption- [5,6], and commitment schemes [7]. Furthermore, only very few UC-secure protocols are actually deployed in the real world, e.g., [8,9]. We believe that one main reason for this is the high computational overhead which is often required to achieve UC-security.

When designing practical cryptographic protocols, efficient *zero-knowledge proofs of knowledge* (ZK-PoK) for discrete logarithms and related problems have turned out to be indispensable. On a high level, these are two party protocols between a prover and a verifier which allow the former to convince the latter that it possesses some secret piece of information, without the verifier being able to learn anything about it. This allows protocol designers to enforce one party to assure other parties that its actions are consistent with its internal knowledge state. The shorthand notation for such proofs, introduced in [10], has been extensively used in the past and contributed to the wide employment of ZK-PoK in cryptographic design. This notation suggests using, e.g., $\mathsf{PK}\,[(\alpha) : y = g^{\alpha}]$ to denote a proof of the discrete logarithm $\alpha = \log_g y$, and it has appeared in many works sometimes with quite complex statements, e.g., [11–23]. This informal notion was recently formalized and refined by Camenisch, Kiayias and Yung who have provided a specification language (*CKY-language*) for such protocols [24]. The language allows for the modular design and analysis of cryptographic protocols: protocol designers just needs to specify the statement the ZK-PoK shall prove and, if the specification is in the CKY-language, they are ensured that the proof protocol exists and indeed proves the specified statement.

The realizations given by Camenisch et al. [24] are based on Σ-protocols and satisfy the classical notion of ZK-PoK but not that of UC zero-knowledge. On a high level, the problem here is that the classical notion only requires that a valid witness can be extracted from every convincing prover given rewindable access to that prover. However, in the UC-framework this has to be possible without rewinding. While generic transformations from Σ-protocols to UC-ZK protocols are known [25], they come along with a significant computational overhead, making the resulting protocols impracticable for real-world usage.

However, the security proofs of many cryptographic protocols only require the existence of a witness, and not that the prover actually knows it. Intuitively, this should be easier to achieve than proofs of knowledge. Yet, in the UC-framework zero-knowledge proofs are always proofs of *knowledge*. This is because otherwise the ideal functionality generally could not decide whether or not a given statement is true in polynomial time. In this paper we are aiming at closing the gap between high security guarantees and modularity on the one hand, and practical usability and efficiency of the resulting protocols on the other hand.

Our Contributions. We first present an exhaustive language and a compiler which allow protocol designers to efficiently and modularly specify and obtain UC-ZK protocols. We then give an extension of the UC-framework allowing protocol designers to also make usage of the more efficient proofs of *existence* (as opposed to proofs of *knowledge*), which we also incorporate into our language. Let us explain this in more detail in the next paragraphs.

A language for UC-ZK protocols. We provide an intuitive language for specifying ZK-PoK for discrete logarithms akin to the CKY-language [24] where the specification also allows one to assess the complexity of the specified protocol. We then provide a compiler which translates these specifications into concrete protocols. Even though this compiler is mainly based on existing techniques, it offers unified and unambiguous interfaces and semantics for the associated protocols for the first time. It thus enables protocol designers to treat specifications in our language as black-boxes, while having clearly defined security guarantees.

Proving existence rather than knowledge. In the UC-framework, all ZK proofs are necessarily proofs of *knowledge*. However, when designing higher-level protocols, it is often sufficient to prove that some computation was done correctly, but not to show that the secret quantities are actually known. To allow protocol designers to also make use of these more efficient protocols (which are not proofs *of knowledge* any more), we extend our language and provide the necessary framework to prove UC-security. Loosely speaking, we therefore formulate the *gullible ZK ideal functionality* $\mathcal{F}_{\text{gZK}}$, and provide a special composition theorem which allows protocol designers to use existence-proofs "as if they were ideal functionalities," if they are later instantiated as described in our compiler. Roughly, the theorem states that proving the correctness of a protocol using $\mathcal{F}_{\text{gZK}}$ in a slightly non-UC-compliant way is sufficient for the protocol where $\mathcal{F}_{\text{gZK}}$ is instantiated by the real-world protocol to be UC-secure in the standard sense.

Related Work. The UC-framework has first been introduced by Canetti [2]. The notion of Ω-protocols was introduced in [25, 26], and so far the most efficient UC-secure zero-knowledge proofs of knowledge have been proposed in [27]. Further, [28] analyzes UC-ZK in the presence of global setup [29]. The idea of committed proofs was first mentioned in [30]. We combine the techniques of [27, 30] to compile proof specifications in our language to real protocols. In particular this allows us to realize proofs *of existence*.

A language for specifying ZK-PoK for discrete logarithms was presented by Camenisch and Stadler [10] and later refined by Camenisch et al. [24], but neither of their realizations are UC-secure. Our notation is strongly inspired by theirs. In fact, our language has already turned out to be very useful to describe ZK-PoK in a companion paper [31], and in this paper we fulfill the promises given there.

Functionalities similar to $\mathcal{F}_{gZK}$ have already been used by Lindell [32, 33] and Pass and Rosen [34] in different contexts. That is, all this work is on two-party protocols which preserve their security guarantees under bounded-concurrent self-composition and not on full UC-security. Prabhakaran and Sahai [35, 36] also suggest generalizations of the UC-framework in which functionalities can be realized that cannot be realized in the plain UC-framework. Their work differs from ours in that they leave the standard model of polynomial time computation by granting the adversary access to some super-polynomially powerful oracle ("imaginary angel"), while our approach works in the standard computational model. Furthermore, they suggest generic solutions for ZK-PoK while we are aiming at practically efficient protocols. Finally, ideas similar to ours have also been suggested in unpublished work by Nielsen [37].

Roadmap. After introducing some notation, recapitulating fundamental theory and presenting two running examples in §2, we describe a basic language for specifying UC-secure ZK-PoK protocols in detail in §3. In §4, we show how proofs of existence rather than knowledge can be UC-realized, resulting in much more efficient protocols, and extend our language accordingly. In this section we further show how such specifications can be compiled to actual protocols. We give several extensions to our basic language in §5 and briefly conclude in §6.

2   Preliminaries

Let us introduce some notation first. By $s \in_R \mathcal{S}$ we denote the uniform random choice of some element s in set $\mathcal{S}$. The group of signed quadratic residues [38] for some modulus n is denoted by $\mathbb{SR}_n$. For two random ensembles, $\overset{s}{\approx}$ denotes statistical indistinguishability. Finally, two party protocols between parties P and V with common input y and private input w to P are written as $(\mathsf{P}(w), \mathsf{V})(y)$.

We assume that the reader is familiar with the notion of Σ- and Ω-protocols, and only give informal definitions here. A protocol $(\mathsf{P}(w), \mathsf{V})(y)$ is called a Σ-*protocol* [39], if it is an honest verifier ZK-PoK in the non-UC model, consisting of three messages being exchanged (a *commitment* t, a *challenge* $c \in_R \mathcal{C} = \{0, 1\}^k$, and a *response* r), such that the secret w can be computed from any two valid protocol transcripts with the same commitment but different challenges. A protocol is called an Ω-*protocol* [25], if it further takes a common reference string σ as additional input, such that when knowing a trapdoor to σ it is possible to compute the prover's secret input from any successful run of the protocol.

An Ω-protocol is said to be *f-extractable*, if it is not possible to compute w from any successful run, but only $f(w)$ for some function f. In particular, we will make use of two types of f-extractable protocols: one the one hand we will use $f(w_1, \ldots, w_n) = (w_1, \ldots, w_k)$ for some $k \leq n$, i.e., protocols which only allow to extract parts of the witness. On the other hand, we will have $f(w_1, \ldots, w_n) = (w_1, \ldots, w_n, \mathsf{A}(w_1, \ldots, w_n))$, i.e., functions f which in addition to all witnesses additionally output some further values depending on these witnesses. These constructions will allow for an efficiency speedup compared to using plain Ω-protocols, while often still ensuring appropriate security guarantees.

2.1   The UC-Model

We next briefly recapitulate the Universal Composability (UC) framework [2].

A *party* is a probabilistic polynomial time interactive Turing machine. Each party P is uniquely determined by a pair $(\mathtt{PID}_P, \mathtt{SID}_P)$, where $\mathtt{PID}_P$ and $\mathtt{SID}_P$ are its *party ID* and its *session ID*. Two parties share the same session ID if and only if they are participants of the same instance of a protocol. Party IDs are solely used to distinguish between participants of the same protocol instance. Following [31], we assume that session IDs are structured as pathnames. That is, for a protocol with session ID $\mathtt{SID}$, the session ID of any of its subprotocols is given by $\mathtt{SID/subsession}$, where $\mathtt{subsession}$ is a unique local identifier, containing the party IDs of all participating parties and shared public parameters.

The main concept of the UC framework is that of UC-emulation. Loosely speaking, a protocol ρ *UC-emulates* some protocol ϕ, if ρ does not affect the security of anything else than ϕ would have, no matter how many other instances of ρ or other protocols are executed concurrently. This implies that ρ can safely be used on behalf of ϕ without compromising security. The most interesting case is where ϕ is some *ideal functionality* $\mathcal{F}$, which can be thought of as an incorruptible trusted party that takes inputs from all parties, performs some local computations, and hands back outputs to the parties. Ideal functionalities can be seen as formal specifications of cryptographic tasks and are secure by definition. Now, if ρ UC-emulates $\mathcal{F}$, one can infer that ρ does not leak any other information to an adversary than $\mathcal{F}$ would have, and therefore securely realizes the given task in arbitrary contexts. For a more precise description see [2].

Protocols using an ideal functionality $\mathcal{F}$ as a subroutine are called $\mathcal{F}$-*hybrid*. If not stated otherwise, all protocols we are going to present are $\mathcal{F}_{\mathrm{ach}}$-hybrid protocols, where $\mathcal{F}_{\mathrm{ach}}$ is an ideal functionality realizing authenticated (but not necessarily private) channels. The functionality takes as input a message x from some a sender, and forwards it to a receiver. The adversary learns x, and, upon corruption of the sender, is allowed to change it before it is delivered.

The corruption model underlying our discussion is *adaptive corruptions with erasures*. This can be seen as a bit of a compromise: while only considering static corruptions would not properly reflect reality, assuming secure data erasures is necessary to obtain efficient protocols in this setting. However, even if implementing erasures might be difficult, it is not impossible.

The zero-knowledge functionality $\mathcal{F}_{\mathrm{ZK}}^{R,R'}$

1. Wait for an input ($\mathtt{prove}, y, w$) from P such that $(y, w) \in R$ if P is honest, or $(y, w) \in R'$ if P is corrupt. Send ($\mathtt{prove}, \ell(y)$) to $\mathcal{A}$. Further wait for a message $\mathtt{ready}$ from V, and send $\mathtt{ready}$ to $\mathcal{A}$.
2. Wait for a message $\mathtt{lock}$ from $\mathcal{A}$.
3. Upon receiving a message $\mathtt{done}$ from $\mathcal{A}$, send $\mathtt{done}$ to P. Further wait for an input $\mathtt{proof}$ from $\mathcal{A}$ and send ($\mathtt{proof}, y$) to V.

Corruption rules:

▷ If P gets corrupted after sending ($\mathtt{prove}, y, w$) and before Step 2, $\mathcal{A}$ is given (y, w) and is allowed to change this value to any value $(y', w') \in R'$ at any time before Step 2.

Fig. 1. The basic zero-knowledge functionality $\mathcal{F}_{\mathrm{ZK}}^{R,R'}$, parametrized by two binary relations R, R' such that $R' \supseteq R$ [31]

The Basic UC-ZK Ideal Functionality. In the following we discuss the basic ideal zero-knowledge functionality, which is formally specified in Figure 1. It is parametrized by two binary relations, R and R', which have the following meaning: the relation R specifies the set of inputs (y, w) the functionality accepts from an honest prover. For such inputs, the functionality informs the verifier that the prover knows a witness for y, while an adversary does not learn w. Yet, if the prover is corrupted, it is allowed to supply inputs from a binary relation $R' \supseteq R$, in which case the ZK property does not have to be satisfied any more.

The relation R might itself be parametrized by system parameters, specifying, e.g., the concrete groups being used. We will model all such parameters as public coin parameters, i.e., the environment might know the random coins being used to generate the system parameters. This is helpful if the same parameters are used in other protocols as well, e.g., to sign messages.

The functionality defined in Figure 1 differs from the standard one found in the literature in two ways. Firstly, we delay revealing the claimed statement y to V and $\mathcal{A}$ until the last possible moment, and only give $\ell(y)$ to the adversary in the first step, where ℓ is a leakage function, which roughly gives some information about the "size and shape" of y to $\mathcal{A}$ (to be precise, $\ell()$ is a parameter of $\mathcal{F}_{\mathrm{ZK}}$ as well which will be disregarded in the remaining discussion). This approach prevents the simulator from being over-committed in our constructions, and to the best of our knowledge $\mathcal{F}_{\mathrm{ZK}}^{R,R'}$ can safely be used instead of the standard UC-ZK functionality in any application. Secondly, we allow corrupt parties to supply witnesses from a larger set than honest parties. This relaxation stems from the *soundness gap* of most known efficient constructions for ZK-PoK for discrete logarithms in the non-UC case [40] (which are underlying the constructions for UC-ZK protocols): there, the verifier can only infer that the prover knows a witness w such that $(y, w) \in R'$, whereas an honest prover is ensured that for $(y, w) \in R$ the verifier cannot learn the secret. We further elaborate on this in §3.

The same formalization of the ZK functionality was also used in [31].

2.2   Running Examples

We next introduce two running examples, which we are going to use throughout the discussion to illustrate our techniques.

Example 2.1 (Running Example 1). Let be given an integer commitment $y \in \mathbb{SR}_n$ for some safe RSA modulus n. Let further be given two generators g, h of $\mathbb{SR}_n$. In this example, a prover is interested in proving knowledge of integers ω, ρ such that $y = g^\omega h^\rho$ and $\omega \geq 0$. □

Numerous practically relevant applications require such proof goals as basic building blocks for more complex protocols, e.g., [14, 16].

Example 2.2 (Running Example 2). Let be given a cyclic group $\mathcal{H}$ of prime order q, and two generators, g, h of $\mathcal{H}$. Let further be given a triple $(u_1, u_2, e) \in \mathcal{H}^3$, and let one be interested in proving that (u_1, u_2, e) is a valid encryption of $g^\alpha \in \mathcal{H}$ for some $\alpha \in \mathbb{Z}_q$ known to the prover under the semantically secure version of the Cramer-Shoup cryptosystem [30, 41]. That is, the task is to prove that (u_1, u_2, e) is of the form $(g^\rho, h^\rho, g^\alpha c^\rho)$ for a publicly known $c \in \mathcal{H}$. □

This example stems from [31], where such proofs are repeatedly needed in the context of credential-authenticated key-exchange and identification protocols.

3   A Language for Specifying UC-ZK Protocols

As shown in [42], any ideal functionality can be UC-realized given only functionalities realizing commitments and ZK proofs, respectively. This result suggests that ZK proofs are important building blocks of higher-level applications, and will thus often be deployed when UC-realizing cryptographic tasks.

Taking this as a motivation, we describe an intuitive language for specifying universally composable zero-knowledge protocols. The language is strongly inspired by the standard notation for describing ZK-PoK in the non-UC case which was introduced in [10]. We stress that similar to there, our notation does not only specify proof goals (i.e., what one wants to prove), but concrete protocols. Especially for our results given in §4, this unambiguity is important.

We start by describing a basic language, which allows one to specify arbitrary Boolean combinations of protocols proving knowledge of discrete logarithms (or representations) in arbitrary groups. In many cases the complexity of the resulting protocol can be inferred directly from the proof specification.

A protocol proving knowledge of integers $\omega_1, \ldots, \omega_n$ satisfying a *predicate* $\phi(\omega_1, \ldots, \omega_n)$ is denoted as follows:

$$\yen\, \omega_1 \in \mathcal{I}^*(m_{\omega_1}), \ldots, \omega_n \in \mathcal{I}^*(m_{\omega_n}) : \phi(\omega_1, \ldots, \omega_n). \tag{1}$$

Here, each witness ω_i belongs to some integer domain $\mathcal{I}^*(m_{\omega_i})$. The predicate $\phi(\omega_1, \ldots, \omega_n)$ is a Boolean formula containing ANDs ($\wedge$) and ORs ($\vee$), built from *atomic predicates* of the following form:

$$y = \prod_{i=1}^{u} g_i^{F_i(\omega_1, \ldots, \omega_n)}.$$

The g_i and y are elements of some commutative group, and the F_i are integer polynomials, i.e., $F_i \in \mathbb{Z}[X_1, \ldots, X_n]$. Similar to [10], we make the convention that values of which knowledge has to be proved are denoted by Greek letters, whereas all other quantities are assumed to be publicly known.

We next discuss the single components of our basic language in more detail.

Groups. Different atomic predicates may use different groups. Besides efficiently evaluable group operations we only require that group elements are efficiently recognizable, and that the group order does not contain any small prime divisors, where "small" can be seen as an implementation dependent parameter which typically will have $160 - 256$ bits. In particular, we do not make any intractability assumption for the groups.

We stress that the group of quadratic residues modulo a safe RSA modulus n (i.e., $n = pq$, where $p, q, \frac{p-1}{2}, \frac{q-1}{2}$ are prime, denoted by $\mathbb{QR}_n$) does not satisfy the above requirements, as group membership cannot be efficiently verified. We recommend using the group of *signed* quadratic residues instead [38].

Predicates. We allow predicates to be arbitrary combinations of atomic predicates by the Boolean connectives AND and OR. Also, witnesses may be reused across different atomic predicates.

Domains. We allow the secret values $\omega_1, \ldots, \omega_n$ to be arbitrary integers. However, for implementation issues, for each i an integer m_{ω_i} satisfying

$$\omega_i \in \mathcal{I}(m_{\omega_i}) := \{ l \in \mathbb{Z} : -m_{\omega_i} \le l \le m_{\omega_i} \}$$

is required. The value of m_{ω_i} can be chosen arbitrarily large, and is only needed for the protocols resulting from the construction in §4.1 to be statistically zero-knowledge for any $\omega_i \in \mathcal{I}(m_{\omega_i})$. They then guarantee that the prover knows witnesses in a larger interval, i.e., they prove knowledge of witnesses ω_i^* satisfying

$$\omega_i^* \in \mathcal{I}^*(m_{\omega_i}) := \{ l \in \mathbb{Z} : -tm_{\omega_i} \le l \le tm_{\omega_i} \},$$

where t is an implementation dependent parameter, which usually will have about $160-256$ bits and which is independent of the groups used in the predicate. In particular, $\mathcal{I}^*(m_{\omega_i})$ is thus uniquely defined even if ω_i is used across different atomic predicates. More precisely, we have $t \approx 2^{k+l} + 2^k - 1$, where 2^{-k} is the success probability of a malicious prover, and l is a security parameter controlling the tightness of the statistical ZK property of the protocol.

Formally, the gap between $\mathcal{I}(m_{\omega_i})$ and $\mathcal{I}^*(m_{\omega_i})$ is modeled by allowing corrupt provers to hand in values satisfying a relation $R' \supseteq R$ to the ideal functionality, whereas honest parties have to supply values in R, cf. §2.1.

As a special case, we allow to define $\mathcal{I}^*(m_{\omega_i}) = \mathcal{I}(m_{\omega_i}) = \mathbb{Z}_q$, if (i) the secret ω_i only occurs in atomic predicates for which the order of the group is known, and (ii) the integer q is a common multiple of all these group orders. This slightly increases the efficiency of the resulting protocols because of shorter exponents in the modular exponentiations in the protocol.

Induced Relation. Each proof specification spec of the form (1) induces two binary relations, $R = R(\text{spec}) \subseteq R'(\text{spec}) = R'$, and a protocol $\pi = \pi(\text{spec})$, cf. §4.1. The protocol π then UC-emulates $\mathcal{F}_{\text{ZK}}^{R,R'}$, i.e., it is zero-knowledge for $(y, w) \in R$, and guaranteed the verifier that the prover supplied $(y', w') \in R'$.

Let us now illustrate our basic language by means of our two running examples.

Example 3.1 (Running Example 1). We start by resolving the condition $\omega \geq 0$ into the form (1) by rewriting it to $\omega = \sum_{i=1}^{4} \chi_i^2$ [43]. Let ω be an element of $[-T, T]$, i.e., $m_\omega = T$. Then, clearly, we have that $m_{\chi_i} = \lfloor \sqrt{T} \rfloor$ for all i. Also, for y to be blinding, we can assume that $m_\rho = \lfloor n/4 \rfloor$.

The proof goal is thus given by:

$$\exists \, \rho \in \mathcal{I}^*(\lfloor n/4 \rfloor), \{\chi_i\}_{i=1}^{4} \in \mathcal{I}^*(\lfloor \sqrt{T} \rfloor) : y = g^{\chi_1^2 + \chi_2^2 + \chi_3^2 + \chi_4^2} h^\rho . \qquad \square$$

Example 3.2 (Running Example 2). In this case, all secret values are elements of $\mathbb{Z}_q$, where q is the order of $\mathcal{H}$. We therefore get the following proof specification:

$$\exists \, \alpha, \rho \in \mathbb{Z}_q : u_1 = g^\rho \wedge u_2 = h^\rho \wedge e = g^\alpha c^\rho .$$

In particular note that the requirement that $\text{ord}\,\mathcal{H}$ does not have small prime divisors is satisfied as q was assumed to be prime, cf. Example 2.1. $\qquad \square$

4 Proving Existence Rather Than Knowledge

Realizing ZK-PoK in the UC-framework is a computationally expensive task. On a high level this is because the simulator needs to be able to extract the secret witness without rewinding, and the most efficient currently known way to achieve this is to include Paillier encryptions of the witnesses into the proof. Now, in larger protocols, ZK-PoK are often only used to ensure that a computation was done correctly, and the simulators of these higher-level protocols do not make usage of the witnesses. For instance, in Example 2.2 proving the existence of ρ is sufficient to imply the required well-formedness of the ciphertext.

Thus, often a functionality realizing the following steps would be sufficient:

1. Wait for an input (prove, y, w) from P such that there is a $\tilde{w}$ satisfying $(y, \tilde{w}) \in R$ and $f(\tilde{w}) = w$, and send (prove, $\ell(y)$) to $\mathcal{A}$. Further wait for a message ready from V, and send ready to $\mathcal{A}$.
2. Wait for a message lock from $\mathcal{A}$.
3. Upon receiving a message done from $\mathcal{A}$, send done to P. Further wait for an input proof from $\mathcal{A}$ and send (proof, y) to V.

That is, one is aiming for a functionality which checks whether the prover knows some (partial) information $w = f(\tilde{w})$ for a full witness $\tilde{w}$, and informs the verifier if this is the case. However, the problem is that by definition any zero-knowledge proof in the UC-Framework is *always* a proof of knowledge. This is, because in general the existence of $\tilde{w}$ cannot be checked efficiently, and thus the witness has to be given as an input for the functionality to be able to check whether the statement is true. We now propose a framework that circumvents this problem and allows one to use proofs of existence in the UC-model.

We extend our basic language by the additional $\exists$-quantifier. For secrets quantified under $\exists$ (instead of $\forall$) only their existence (instead of their knowledge) is proved. A generalized specification of a proof goal now looks as follows:

$$\forall \left\{\omega_i \in \mathcal{I}^*(m_{\omega_i})\right\}_{i=1}^n : \exists \left\{\chi_j \in \mathcal{I}^*(m_{\chi_j})\right\}_{j=1}^m : \phi(\omega_1, \ldots, \omega_n, \chi_1, \ldots, \chi_m) \quad (2)$$

In the following we show how such specifications are compiled into protocols, and then describe the underlying theory and composition theorem which allow to use such specifications as modular building blocks in larger protocols.

4.1   Compiling Specifications to Protocols

Due to space limitations we here only give a brief overview about how protocol specifications are compiled into protocols. For a detailed description we refer to the full version of this paper [44].

▷ First, the proof specification is rewritten to a predicate which only contains atomic predicates having homogeneous linear relations in their exponents. This can be done by applying standard techniques [40, 43, 45–48].

▷ In a second step, the prover computes integer commitments η_i to all secret witnesses ω_i quantified by $\forall$.

▷ Next, using the technique proposed in [40], each conjunctive term in the specification (i.e., each subformula of ϕ not containing any OR connectives) is translated into a Σ-protocol which additionally proves that the witnesses being used are the same as in the η_i.

▷ Now, the different Σ-protocols are combined by the Boolean connectives as specified by the predicate ϕ [48, 49].

▷ As a fifth step, the Σ-protocol is transformed into an Ω-protocol [25, 26]. This is achieved by Paillier-encrypting the witnesses quantified by $\forall$ [50], and proving that the encrypted witnesses are the same as in the η_i.

▷ Using a simulation sound trapdoor commitment [27] and the committed-proof idea of [30], one finally obtains a protocol UC-emulating $\mathcal{F}_{\text{gZK}}^{R,R'}$.

Theorem 4.1. *Let* spec *be a proof specification of the form (1), and let* $R = R(\text{spec})$, $R' = R'(\text{spec})$, *and* $\pi = \pi(\text{spec})$. *Then* π *UC-realizes* $\mathcal{F}_{\text{ZK}}^{R,R'}$ *with respect to adaptive corruptions, assuming that securely erasing data is possible. If this is not the case, it still UC-realizes* $\mathcal{F}_{\text{ZK}}^{R,R'}$ *with respect to static corruptions.*

The proof of this theorem is a straightforward adaption of that in [27] and is omitted due to space limitations.

Let us discuss the potential speed-up and the semantical consequences coming along with the usage of the $\exists$-quantifier by means of our two running examples.

Example 4.2 (Running Example 1). For being able to see the speed-up, we first have to resolve the polynomial relation of Example 3.1. Using the technique from [43], we obtain the following equivalent proof specification:

$$\lambdabar \{\rho_i\}_{i=1}^4 \in \mathcal{I}^*(\lfloor n/4 \rfloor), \{\chi_i\}_{i=1}^4 \in \mathcal{I}^*(\lfloor \sqrt{T} \rfloor), \rho' \in \mathcal{I}^*((4\lfloor \sqrt{T} \rfloor + 1)\lfloor n/4 \rfloor) :$$
$$\bigwedge_{i=1}^4 y_i = g^{\chi_i} h^{\rho_i} \wedge y = y_1^{\chi_1} y_2^{\chi_2} y_3^{\chi_3} y_4^{\chi_4} h^{\rho'}$$

Keeping in mind that the χ_i, ρ_i and ρ' can be computed efficiently from ω, ρ using Lagrange's Four Square Theorem and the Rabin-Shallit algorithm [51], it is easy to see that this specification is semantically equivalent to the following:

$$\lambdabar \omega \in \mathcal{I}^*(T), \rho \in \mathcal{I}^*(\lfloor n/4 \rfloor) : \exists \{\rho_i\}_{i=1}^4 \in \mathcal{I}^*(\lfloor n/4 \rfloor), \{\chi_i\}_{i=1}^4 \in \mathcal{I}^*(\lfloor \sqrt{T} \rfloor),$$
$$\rho' \in \mathcal{I}^*((4\lfloor \sqrt{T} \rfloor + 1)\lfloor n/4 \rfloor) : y = g^{\omega} h^{\rho} \wedge \bigwedge_{i=1}^4 y_i = g^{\chi_i} h^{\rho_i} \wedge y = \prod_{i=1}^4 y_i^{\chi_i} h^{\rho'}$$

This rewriting yields a significant efficiency speedup, as only Paillier encryptions for 2 instead of 9 values are required. Overall, the prover (verifier) thus saves 14 (7) Paillier encryptions and evaluations of the integer commitment scheme.   □

In this example, changing from the $\lambdabar$- to the $\exists$-quantifier is a purely syntactical step, which increases the efficiency of the protocol. This can be seen by considering the underlying Ω-protocol as f-extractable, where $f(w) = (w, \mathsf{A}(w))$ and A is the algorithm of [51]. However, in general it is not possible to efficiently compute the witnesses quantified by $\exists$, and even their existence cannot be verified efficiently, as is illustrated by the following example.

Example 4.3 (Running Example 2). The following specification is sufficient for proving the required well-formedness of the ciphertext:

$$\lambdabar \alpha \in \mathbb{Z}_q : \exists \rho \in \mathbb{Z}_q : u_1 = g^{\rho} \wedge u_2 = h^{\rho} \wedge e = g^{\alpha} c^{\rho}.$$

This observation reduces the complexity of the prover's algorithms in the protocol by 2 Paillier encryptions and 2 evaluations of the integer commitment scheme (one each for their computation and their commitment in the Σ-protocol).   □

Here, the underlying Ω-protocol is f-extractable, where f is of the form $f(w_1, \ldots, w_n) = (w_1, \ldots, w_k)$ for $k < n$, such that the remaining w_i cannot be computed. This implies that in general it is not possible to construct an ideal functionality which captures the semantics of an expression such as (2), as it would have to run in probabilistic polynomial time by definition [2].

4.2   The Gullible ZK Functionality and a Composition Theorem

In the following we describe the theoretical framework which allows protocols designers to treat specifications containing values quantified by $\exists$ (almost) as if they were quantified by $\lambdabar$.

The gullible zero-knowledge functionality $\mathcal{F}_{\mathrm{gZK}}^{R,R'}$

1. Wait for an input $(\mathtt{prove}, y, (w, x))$ from P and send $(\mathtt{prove}, \ell(y))$ to $\mathcal{A}$. Further wait for a message $\mathtt{ready}$ from V, and send $\mathtt{ready}$ to $\mathcal{A}$.
2. Wait for a message $\mathtt{lock}$ from $\mathcal{A}$.
3. Upon receiving a message $\mathtt{done}$ from $\mathcal{A}$, send $\mathtt{done}$ to P. Further wait for an input $\mathtt{proof}$ from $\mathcal{A}$ and send $(\mathtt{proof}, y)$ to V.

Corruption rules:

▷ If P gets corrupted after sending $(\mathtt{prove}, y, (w, x))$ and before Step 2, $\mathcal{A}$ is given $(y, (w, \perp))$ and is allowed to change this value at any time before Step 2.

Fig. 2. The gullible zero-knowledge functionality $\mathcal{F}_{\mathrm{gZK}}^{R,R'}$ always informs the verifier that the proof was correct

The *gullible zero-knowledge functionality* $\mathcal{F}_{\mathrm{gZK}}^{R,R'}$ expects the prover to supply an image y and a pair (w, x) as inputs, and always informs the verifier that $(y, (w, x)) \in R'$, no matter whether this is the case or not, cf. Figure 2. For an honest prover, w will be the part of the witness for which knowledge has to be proved, whereas x is the part for which only existence has to be proved. Upon corruption of the prover, the adversary only learns y and w, but not x. This is to model the intuitive goal of proofs of existence appropriately.

Our special composition theorem guarantees that $\rho^{\pi/\mathcal{F}_{\mathrm{gZK}}^{R,R'}}$ UC-emulates some other protocol ϕ, if ρ UC-emulates ϕ with respect to a certain type of environments, called *nice environments*, which we define next. On a high level, these are environments which (almost) never ask the dummy adversary to send incorrect inputs to the gullible zero-knowledge functionality:

Definition 4.4. Let $\mathcal{A}^*$ be the dummy adversary attacking some $\mathcal{F}_{\mathrm{gZK}}^{R,R'}$-hybrid protocol ρ. We call an environment $\mathcal{Z}$ nice (with respect to ρ), if the statements it requires $\mathcal{A}^*$ to send to $\mathcal{F}_{\mathrm{gZK}}^{R,R'}$ acting as a prover are true with overwhelming probability. That is, with overwhelming probability $\mathcal{Z}$ asks $\mathcal{A}^*$ to send pairs $(y, (w, x))$ to $\mathcal{F}_{\mathrm{gZK}}^{R,R'}$, for which there is an $\tilde{w}$ satisfying $(y, \tilde{w}) \in R$ and $f(\tilde{w}) = w$.

Note that the value of x submitted by a nice environment is not restricted by this definition, but only w has to be a valid partial witness.

We now define UC-emulation with respect to nice environments:

Definition 4.5. Let ρ be an $\mathcal{F}_{\mathrm{gZK}}^{R,R'}$-hybrid protocol. We say that ρ UC-emulates a protocol ϕ with respect to the dummy adversary $\mathcal{A}^*$ and nice environments (w.r.t. ρ), if there is an efficient simulator $\mathcal{S}$ such that no nice environment can distinguish whether it is interacting with ρ and $\mathcal{A}^*$ or with ϕ and $\mathcal{S}$. That is, for every nice environment $\mathcal{Z}$ it holds that $\mathrm{EXEC}(\rho, \mathcal{A}^*, \mathcal{Z}) \approx \mathrm{EXEC}(\phi, \mathcal{S}, \mathcal{Z})$.

Here, $\mathrm{EXEC}(\rho, \mathcal{A}, \mathcal{Z})$ denotes the random variable given by the output of $\mathcal{Z}$ when interacting with ρ and $\mathcal{A}$, and analogously for $\mathrm{EXEC}(\phi, \mathcal{S}, \mathcal{Z})$.

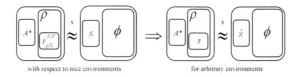

Fig. 3. Illustration of Theorem 4.6: for proving that $\rho^{\pi/\mathcal{F}_{\text{gZK}}^{R,R'}}$ UC-emulates ϕ, it is sufficient to show that ρ emulates ϕ for nice environments

Note that any non-nice environment could potentially distinguish between ρ and ϕ by just submitting a false statement, which will always be accepted by $\mathcal{F}_{\text{gZK}}^{R,R'}$. Informally, our special composition theorem now states that every non-nice environment can be detected if $\mathcal{F}_{\text{gZK}}^{R,R'}$ is instantiated by π as described in the previous section, and thus $\rho^{\pi/\mathcal{F}_{\text{gZK}}^{R,R'}}$ is secure against arbitrary environments. This allows a protocol designer to use ZK proofs of existence in a UC-compliant way, almost as if they were ZK-PoK. The theorem is illustrated in Figure 3.

Theorem 4.6. *Let* spec *be a proof specification of the form (2), and let* $R = R(\text{spec})$, $R' = R'(\text{spec})$, *and* $\pi = \pi(\text{spec})$. *Let further* ρ *be an* $\mathcal{F}_{\text{gZK}}^{R,R'}$-*hybrid protocol, such that* ρ *UC-emulates a protocol* ϕ *with respect to the dummy adversary and nice environments, and let* ρ, ϕ *be subroutine respecting. Then* $\rho^{\pi/\mathcal{F}_{\text{gZK}}^{R,R'}}$ *UC-emulates* ϕ *(in the standard sense) with respect to adaptive corruptions if securely erasing data is possible.*

Proof (Sketch). We omit a full proof here, and only give the underlying intuition. Let therefore S be the simulator for nice environments, which exists by assumption. We have to show that there exists an efficient simulator $\hat{S}$ such that for arbitrary environments $\mathcal{Z}$ we have that $\text{EXEC}(\rho^{\pi/\mathcal{F}_{\text{gZK}}^{R,R'}}, \mathcal{A}^*, \mathcal{Z}) \overset{s}{\approx} \text{EXEC}(\phi, \hat{S}, \mathcal{Z})$.

The idea is that $\hat{S}$ runs a copy of S and one of $\mathcal{A}^*$ internally, and all messages sent to or received from $\mathcal{Z}$ are routed through the simulated $\mathcal{A}^*$. In general, all communication is further forwarded to S, and $\hat{S}$ outputs whatever S does. The only exception is made when encountering a call to π between two parties, P and V. In this case $\hat{S}$ internally executes the protocol on the given inputs and behaves as follows (independent of the corruption state of the parties):

▷ If the run is successful, then with overwhelming probability the input was correct (i.e., $\mathcal{Z}$ "behaved nicely"), as the underlying Σ-protocol is an interactive proof system [52]. Thus, $\hat{S}$ proceeds like the simulator for Theorem 4.1, cf. [30] and [27], expect for the following difference: secret values quantified by $\exists$ are given to the attacker in the real protocol π, but not in the ideal functionality $\mathcal{F}_{\text{gZK}}$. This can be simulated because of the committed proof technique by choosing these secrets at random within their domains whenever necessary. Then, $\hat{S}$ computes the corresponding image y' and opens the commitment made in its first message accordingly. As in [27], this is possible because of the trapdoor property of the used commitment scheme. Note here that these values are deleted before sending out the final message, so the simulator never has to supply them after the adversary learned y.

▷ If however the run of π is not successful, the given input was incorrect. In this case, $\hat{\mathcal{S}}$ behaves as $\mathcal{S}$ in the case that no proof-message had been sent by the attacker.   □

The theorem can be applied as follows by a protocol designer: He first designs a high-level protocol using proofs of existence as if there was a corresponding ideal functionality. Then, in the security proof, he shows that the protocol using $\mathcal{F}_{\mathrm{gZK}}^{R,R'}$ UC-emulates a target functionality ϕ, where he may restrict himself to nice environments only. Finally, after instantiating $\mathcal{F}_{\mathrm{gZK}}^{R,R'}$ by $\pi(\mathtt{spec})$, he obtains a protocol emulating ϕ in the full UC-sense.

5   Enhancing the Basic Language

Even if the basic language presented in §3 allows one to describe almost arbitrary algebraic properties of and relations among the secret values, it might often be more convenient to declare them explicitly. Also, the requirement that all witnesses must be integers may seem overly restrictive.

To solve this problems, we next give some enhancements of our basic language. More precisely, we will first define a set of macros for specifying algebraic properties of the secret witnesses, and then give conditions under which knowledge of group elements can be proved instead of integers.

5.1   Using Macros to Specify Algebraic Properties of Witnesses

The language described in §3 does not allow to directly specify algebraic properties of the secrets or algebraic relations among them, and thus it becomes inconvenient to use for complex proof goals. We therefore extend the set of atomic predicates by so-called *macros*, which allow one to directly describe algebraic properties of the integer witnesses ω_i. In particular, we allow additional atomic predicates of the following forms, all of which can easily be translated into polynomial relations:

▷ $\omega \geq 0$. Such statements can easily be translated into statements of the above form by proving knowledge of integers $\chi_1, \ldots, \chi_4$ such that $\omega = \sum_{i=1}^{4} \chi_i^2$, see [43].

More generally, we also allow expressions of the form $\omega \in [a, b]$, where $a, b \in \mathbb{Z}$ are public. Such an expression is equivalent to $\omega - a \geq 0 \wedge b - \omega \geq 0$. If $b - a$ is even, this can be rewritten to the even more efficient proof goal $-(\omega - m)^2 \geq d^2$, where $m = \frac{a+b}{2}$ and $d = \frac{b-a}{2}$.

▷ $\gcd(\nu_1, \nu_2) = 1$, where each ν_1, ν_2 can be either public or private. As before, such expressions can be rewritten to a polynomial form by introducing additional integers α_1, α_2 and proving knowledge of $\alpha_1, \alpha_2, \nu_1, \nu_2$ such that $\alpha_1 \nu_1 + \alpha_2 \nu_2 = 1$.

▷ $\nu_1 | \nu_2$, where ν_1, ν_2 can be either public or private. By introducing an additional secret δ, such relations can be expressed in polynomial form as $\delta \nu_1 - \nu_2 = 0$.

Example 5.1 (Running Example 1). Using the first of our specific macros, a protocol for proving knowledge of a non-negative opening of the integer commitment y can be described as follows:

$$\exists \omega \in \mathcal{I}^*(T), \rho \in \mathcal{I}^*(\lfloor n/4 \rfloor) : y = g^\omega h^\rho \wedge \omega \geq 0 \qquad \square$$

Before moving to the next extension of our basic language, we point out that using macros impedes the possibility of estimating the computational costs of the protocol from its specification, which was a favorable property of our basic language. This can be seen by comparing Example 5.1 to Example 3.1: the seemingly simple macro $w \geq 0$ entails 5 atomic predicates, and 9 secret witnesses, and thus conceals very much of the computational costs of the resulting protocol.

As an important remark we note that every auxiliary variables χ_i, which has to be introduced when resolving any of these macros, can be quantified by $\exists$. This can easily be seen by noting that considering the resulting Ω-protocol as f-extractable for $f(w) = (w, \mathsf{A}(w))$, where A is the algorithm the honest prover used to compute the χ_i from w.

5.2   Proving Knowledge of Group Elements

Sometimes it is required to prove knowledge of *group elements* instead of integers, which is not possible in our basic language. For instance, one might be interested in proving possession of a digital signature on a given message, which, in the case of CL-signatures [53], essentially boils down to proving knowledge of a group element ω such that $e(\omega, z) = y$, where e is a bilinear map, and y, z are publicly known.

We thus also allow one to specify protocols proving knowledge of a preimage $\omega \in \mathcal{G}$ under some group homomorphism $\psi : \mathcal{G} \to \mathcal{H}$, if ψ satisfies two basic properties: (i) the finite group $\mathcal{G}$ comes along with a generator g and an upper bound B on its order, and (ii) the discrete logarithm problem is hard in $\mathcal{H}$. Then expressions of the following form, which, of course, can arbitrarily be combined with expressions of the basic language, may be used:

$$\exists \omega \in \mathcal{G} : y = \psi(\omega).$$

When compiling protocol specifications containing such expressions, one first has to perform the following steps, and then proceeds as in §4.1. The idea of the construction is to first blind the secret preimage ω using g, and then to prove knowledge of the blinding:

1. Set $m' = 2^l B$, where l is a security parameter.
2. Choose $\omega' \in_R \mathcal{I}(m')$, and set $u = g^{\omega'} \omega$, $y' = \psi(u) y^{-1}$, and $g' = \psi(g)$.
3. Rewrite the proof goal to $\exists \omega' \in \mathcal{I}^*(m') : y' = g'^{\omega'}$, and add u to commitment of the Σ-protocol.

6 Conclusion

We presented a framework enabling the use of efficient zero-knowledge protocols in the construction of UC-secure protocols. These protocols can be specified in a unified and unambiguous notation and then generated by a compiler. To make proving security of construction that make use proof of existence protocols easy, we provide a special composition theorem. By means of two running examples we illustrated that using proofs of *existence* (as opposed to proofs of *knowledge*) can significantly reduce the computational overhead required to achieve UC-security for many practical applications without affecting security.

We believe that by reducing the costs of UCZK protocols to a practically acceptable level in many cases our result can contribute to a wider employment of UC-secure protocols in the real world.

References

1. Backes, M., Pfitzmann, B., Waidner, M.: The reactive simulatability (RSIM) framework for asynchronous systems. Information and Computation 205(12), 1685–1720 (2007)
2. Canetti, R.: Universally Composable Security: A New Paradigm for Cryptographic Protocols. In: FOCS 2001, pp. 136–145. IEEE (2001), Revised version at http://eprint.iacr.org/2000/067
3. Pfitzmann, B., Waidner, M.: Composition and integrity preservation of secure reactive systems. In: ACM CCS 2000, pp. 245–254 (2000)
4. Peikert, C., Vaikuntanathan, V., Waters, B.: A Framework for Efficient and Composable Oblivious Transfer. In: Wagner, D. (ed.) CRYPTO 2008. LNCS, vol. 5157, pp. 554–571. Springer, Heidelberg (2008)
5. Küsters, R., Tuengerthal, M.: Universally Composable Symmetric Encryption. In: Computer Security Foundations Symposium – CSF 2009, pp. 293–307. IEEE (2009)
6. Laud, P., Ngo, L.: Threshold Homomorphic Encryption in the Universally Composable Cryptographic Library. In: Baek, J., Bao, F., Chen, K., Lai, X. (eds.) ProvSec 2008. LNCS, vol. 5324, pp. 298–312. Springer, Heidelberg (2008)
7. Lindell, Y.: Highly-Efficient Universally-Composable Commitments Based on the DDH Assumption. In: Paterson, K.G. (ed.) EUROCRYPT 2011. LNCS, vol. 6632, pp. 446–466. Springer, Heidelberg (2011)
8. Canetti, R., Krawczyk, H.: Security Analysis of IKE's Signature-Based Key-Exchange Protocol. In: Yung, M. (ed.) CRYPTO 2002. LNCS, vol. 2442, pp. 143–161. Springer, Heidelberg (2002)
9. Krawczyk, H.: HMQV: A High-Performance Secure Diffie-Hellman Protocol. In: Shoup, V. (ed.) CRYPTO 2005. LNCS, vol. 3621, pp. 546–566. Springer, Heidelberg (2005)
10. Camenisch, J., Stadler, M.: Efficient Group Signature Schemes for Large Groups (Extended Abstract). In: Kaliski Jr., B.S. (ed.) CRYPTO 1997. LNCS, vol. 1294, pp. 410–424. Springer, Heidelberg (1997)
11. Ateniese, G., Song, D.X., Tsudik, G.: Quasi-Efficient Revocation in Group Signatures. In: Blaze, M. (ed.) FC 2002. LNCS, vol. 2357, pp. 183–197. Springer, Heidelberg (2003)

12. Boudot, F.: Efficient Proofs that a Committed Number Lies in an Interval. In: Preneel, B. (ed.) EUROCRYPT 2000. LNCS, vol. 1807, pp. 431–444. Springer, Heidelberg (2000)
13. Bresson, E., Stern, J.: Efficient Revocation in Group Signatures. In: Kim, K.-c. (ed.) PKC 2001. LNCS, vol. 1992, pp. 190–206. Springer, Heidelberg (2001)
14. Brickell, E., Camenisch, J., Chen, L.: Direct Anonymous Attestation. In: ACM CCS 2004, pp. 132–145. ACM Press (2004)
15. Bussard, L., Roudier, Y., Molva, R.: Untraceable Secret Credentials: Trust Establishment with Privacy. In: PerCom Workshops, pp. 122–126. IEEE (2004)
16. Camenisch, J., Herreweghen, E.V.: Design and implementation of the idemix anonymous credential system. In: ACM CCS 2002, pp. 21–30. ACM Press (2002)
17. Camenisch, J., Shoup, V.: Practical Verifiable Encryption and Decryption of Discrete Logarithms. In: Boneh, D. (ed.) CRYPTO 2003. LNCS, vol. 2729, pp. 126–144. Springer, Heidelberg (2003)
18. Furukawa, J., Yonezawa, S.: Group Signatures with Separate and Distributed Authorities. In: Blundo, C., Cimato, S. (eds.) SCN 2004. LNCS, vol. 3352, pp. 77–90. Springer, Heidelberg (2005)
19. Nakanishi, T., Shiota, M., Sugiyama, Y.: An Efficient Online Electronic Cash with Unlinkable Exact Payments. In: Zhang, K., Zheng, Y. (eds.) ISC 2004. LNCS, vol. 3225, pp. 367–378. Springer, Heidelberg (2004)
20. Song, D.X.: Practical Forward Secure Group Signature Schemes. In: ACM CCS 2001, pp. 225–234. ACM press (2001)
21. Tang, C., Liu, Z., Wang, M.: A Verifiable Secret Sharing Scheme with Statistical zero-knowledge. Cryptology ePrint Archive, Report 2003/222 (2003), http://eprint.iacr.org/
22. Tsang, P.P., Wei, V.K.: Short Linkable Ring Signatures for E-Voting, E-Cash and Attestation. In: Deng, R.H., Bao, F., Pang, H., Zhou, J. (eds.) ISPEC 2005. LNCS, vol. 3439, pp. 48–60. Springer, Heidelberg (2005)
23. Tsang, P.P., Wei, V.K., Chan, T.K., Au, M.H., Liu, J.K., Wong, D.S.: Separable Linkable Threshold Ring Signatures. In: Canteaut, A., Viswanathan, K. (eds.) INDOCRYPT 2004. LNCS, vol. 3348, pp. 384–398. Springer, Heidelberg (2004)
24. Camenisch, J., Kiayias, A., Yung, M.: On the Portability of Generalized Schnorr Proofs. In: Joux, A. (ed.) EUROCRYPT 2009. LNCS, vol. 5479, pp. 425–442. Springer, Heidelberg (2009)
25. Garay, J.A., MacKenzie, P., Yang, K.: Strengthening Zero-Knowledge Protocols Using Signatures. Journal of Cryptology 19(2), 169–209 (2006)
26. Garay, J.A., MacKenzie, P., Yang, K.: Strengthening Zero-Knowledge Protocols Using Signatures. In: Biham, E. (ed.) EUROCRYPT 2003. LNCS, vol. 2656, pp. 177–194. Springer, Heidelberg (2003)
27. MacKenzie, P., Yang, K.: On Simulation-Sound Trapdoor Commitments. In: Cachin, C., Camenisch, J.L. (eds.) EUROCRYPT 2004. LNCS, vol. 3027, pp. 382–400. Springer, Heidelberg (2004)
28. Dodis, Y., Shoup, V., Walfish, S.: Efficient Constructions of Composable Commitments and Zero-Knowledge Proofs. In: Wagner, D. (ed.) CRYPTO 2008. LNCS, vol. 5157, pp. 515–535. Springer, Heidelberg (2008)
29. Canetti, R., Dodis, Y., Pass, R., Walfish, S.: Universally Composable Security with Global Setup. In: Vadhan, S.P. (ed.) TCC 2007. LNCS, vol. 4392, pp. 61–85. Springer, Heidelberg (2007)
30. Jarecki, S., Lysyanskaya, A.: Adaptively Secure Threshold Cryptography: Introducing Concurrency, Removing Erasures (Extended Abstract). In: Preneel, B. (ed.) EUROCRYPT 2000. LNCS, vol. 1807, pp. 221–242. Springer, Heidelberg (2000)

31. Camenisch, J., Casati, N., Groß, T., Shoup, V.: Credential Authenticated Identification and Key Exchange. In: Rabin, T. (ed.) CRYPTO 2010. LNCS, vol. 6223, pp. 255–276. Springer, Heidelberg (2010)

32. Lindell, Y.: Bounded-concurrent secure two-party computation without setup assumptions. In: STOC 2003, pp. 683–692. ACM Press (2003)

33. Lindell, Y.: Protocols for Bounded-Concurrent Secure Two-Party Computation in the Plain Model. Chicago Journal of Theoretical Computer Science 2006(1), 1–50 (2006)

34. Pass, R., Rosen, A.: Bounded-Concurrent Secure Two-Party Computation in a Constant Number of Rounds. In: FOCS 2003, pp. 404–413. IEEE (2003)

35. Prabhakaran, M., Sahai, A.: New notions of security: achieving universal composability without trusted setup. In: STOC 2004, pp. 242–251. ACM Press (2004)

36. Prabhakaran, M., Sahai, A.: Relaxing Environmental Security: Monitored Functionalities and Client-Server Computation. In: Kilian, J. (ed.) TCC 2005. LNCS, vol. 3378, pp. 104–127. Springer, Heidelberg (2005)

37. Nielsen, J.: Universally Composable Zero-Knowledge Proof of Membership. Technical report, University of Aarhus (2005)

38. Hofheinz, D., Kiltz, E.: The Group of Signed Quadratic Residues and Applications. In: Halevi, S. (ed.) CRYPTO 2009. LNCS, vol. 5677, pp. 637–653. Springer, Heidelberg (2009)

39. Cramer, R.: Modular Design of Secure yet Practical Cryptographic Protocols. PhD thesis. CWI and University of Amsterdam (1997)

40. Damgård, I., Fujisaki, E.: A Statistically-Hiding Integer Commitment Scheme Based on Groups with Hidden Order. In: Zheng, Y. (ed.) ASIACRYPT 2002. LNCS, vol. 2501, pp. 125–142. Springer, Heidelberg (2002)

41. Cramer, R., Shoup, V.: A Practical Public Key Cryptosystem Provably Secure against Adaptive Chosen Ciphertext Attack. In: Krawczyk, H. (ed.) CRYPTO 1998. LNCS, vol. 1462, pp. 13–25. Springer, Heidelberg (1998)

42. Canetti, R., Lindell, Y., Ostrovsky, R., Sahai, A.: Universally composable two-party and multi-party secure computation. In: STOC 2002, pp. 494–503. ACM Press (2002)

43. Lipmaa, H.: On Diophantine Complexity and Statistical Zero-Knowledge Arguments. In: Laih, C.-S. (ed.) ASIACRYPT 2003. LNCS, vol. 2894, pp. 398–415. Springer, Heidelberg (2003)

44. Camenisch, J., Krenn, S., Shoup, V.: A Framework for Practical Universally Composable Zero-Knowledge Protocols. Cryptology ePrint Archive, Report 2011/228 (2011), http://eprint.iacr.org/

45. Brands, S.: Rapid Demonstration of Linear Relations Connected by Boolean Operators. In: Fumy, W. (ed.) EUROCRYPT 1997. LNCS, vol. 1233, pp. 318–333. Springer, Heidelberg (1997)

46. Cramer, R., Damgård, I.: Zero-Knowledge Proofs for Finite Field Arithmetic or: Can Zero-Knowledge Be for Free? In: Krawczyk, H. (ed.) CRYPTO 1998. LNCS, vol. 1462, pp. 424–441. Springer, Heidelberg (1998)

47. Fujisaki, E., Okamoto, T.: Statistical Zero Knowledge Protocols to Prove Modular Polynomial Relations. In: Kaliski Jr., B.S. (ed.) CRYPTO 1997. LNCS, vol. 1294, pp. 16–30. Springer, Heidelberg (1997)

48. Smart, N.P. (ed.): Final Report on Unified Theoretical Framework of Efficient Zero-Knowledge Proofs of Knowledge. CACE Project Deliverable (2009), http://www.cace-project.eu

49. Cramer, R., Damgård, I., Schoenmakers, B.: Proof of Partial Knowledge and Simplified Design of Witness Hiding Protocols. In: Desmedt, Y.G. (ed.) CRYPTO 1994. LNCS, vol. 839, pp. 174–187. Springer, Heidelberg (1994)
50. Paillier, P.: Public-Key Cryptosystems Based on Composite Degree Residuosity Classes. In: Stern, J. (ed.) EUROCRYPT 1999. LNCS, vol. 1592, pp. 223–238. Springer, Heidelberg (1999)
51. Rabin, M., Shallit, J.: Randomized Algorithms in Number Theory. Communications in Pure and Applied Mathematics 39(S1), 239–256 (1986)
52. Goldwasser, S., Micali, S., Rackoff, C.: The knowledge complexity of interactive proof-systems. In: STOC 1985, pp. 291–304. ACM Press (1985)
53. Camenisch, J., Lysyanskaya, A.: Signature Schemes and Anonymous Credentials from Bilinear Maps. In: Franklin, M. (ed.) CRYPTO 2004. LNCS, vol. 3152, pp. 56–72. Springer, Heidelberg (2004)

Non-interactive and Re-usable
Universally Composable String Commitments
with Adaptive Security

Marc Fischlin[1], Benoît Libert[2], and Mark Manulis[1]

[1] TU Darmstadt & CASED, Germany
[2] Université catholique de Louvain, ICTEAM Institute, Belgium

Abstract. We present the first provably secure constructions of universally composable (UC) commitments (in pairing-friendly groups) that simultaneously combine the key properties of being *non-interactive*, supporting commitments to *strings* (instead of bits only), and offering *reusability of the common reference string* for multiple commitments. Our schemes are also *adaptively secure* assuming reliable erasures.

1   Introduction

UC-security. Cryptographic protocols being proven secure in the Universal Composability (UC) framework [6] bring several fundamental benefits compared to protocols for which only stand-alone proofs of security exist. A widely recognized advantage is that executions of UC-secure protocols remain secure in arbitrary, possibly malicious environments — essentially what one should expect from security protocols deployed in the real world. UC protocols do not receive much attention from practitioners, who in addition to security take many other factors into account such as efficiency and robustness, especially when it comes to protocols that require network communication. In this work we focus of universally composable commitment schemes [8] that are useful for various distributed applications.

UC commitments and their properties. In general commitment schemes are cryptographic protocols that proceed in two phases: In the *commit phase* the sender computes a commitment c to some message m and communicates c to the receiver; in the *open phase* the sender discloses the message m together with some proof d to provide assurance that m was indeed used in the commit phase. Typically, commitment schemes serve as building blocks in higher level applications, which is why striving for UC-security of these schemes is worthwhile. It is known that UC commitments imply key exchange and more general forms of secure two- and multi-party computation [9,12]. Unfortunately, security of commitment schemes under universal composition cannot be obtained without additional setup assumptions. A detailed explanation of the underlying simulation problem and work-around has been given in the seminal work by Canetti and Fischlin [8], who also showed that the UC-security of commitments

D.H. Lee and X. Wang (Eds.): ASIACRYPT 2011, LNCS 7073, pp. 468–485, 2011.

prevents their malleability, which is critical to many anticipated applications of these schemes. Since [8], one of the most basic and widely used setup assumptions is the Common Reference String (CRS) model, which is also used in our work. Note that alternative constructions of UC commitments appeal to stronger setup assumptions like random oracles [18] or hardware tokens [19]. In addition to setup assumptions prior work has identified several key properties, based on which UC commitment schemes are often compared. These properties (which we list below) may serve as "quality criteria" for UC commitments since they shed light on the security and potential practicality of the schemes.

EFFICIENCY. Several factors contribute to the overall efficiency of a UC commitment scheme. In particular, its *communication complexity* measures the total amount of bits (often in dependency on the security parameter) that are transmitted between the sender and the receiver during the both phases of the protocol. These costs also include the actual *commitment length*, *i.e.*, number of bits that receiver would have to store until the open phase. The *computational complexity* of a commitment schemes indicates the total amount of work performed by participants and is often given in form of costly public-key operations (e.g. modular exponentiations). Earlier UC commitments, e.g. [8,9], were *bit commitments* and required ℓ executions of the basic protocol to commit to an ℓ-bit string. This results usually in a commitment length of $\Omega(\ell \cdot \lambda)$, whereas the length should ideally be $O(\lambda)$ only.[1] Modern UC schemes, such as [13,12,5,22,20], are more efficient in that they can be used to commit to ℓ-bit strings directly without incurring an expansion factor proportional to the security parameter. Another efficiency indicator of UC commitments in the CRS model is the *length of the CRS*, which should ideally remain independent of the number of possible users. Note that this latter property is satisfied by many UC schemes today, e.g. [12,5,22,20].

CRS RE-USABILITY. UC commitments in the CRS model assume trusted generation of the CRS parameters. Of practical relevance is the question of whether these parameters are re-usable across polynomially many executions of the commitment protocol or whether they need to be set up for each new commit phase. Clearly, re-usability of CRS parameters is desirable in practice, where setting up these parameters anew for each commitment operation may not always be possible. Note that CRS re-usability is provided by many existing UC schemes, e.g. [13,12,5,20], though the CRS length in [13] is not constant.

INTERACTION. Another important property of a UC commitment scheme is whether its phases require interaction between the sender and the receiver. Ideally, UC commitment should be *non-interactive*, meaning that each phase should

---

[1] Due to the so-called extraction property of UC commitments [8] a commitment needs to somewhat contain the entire message, stipulating that the commitment itself is at least as large as the message. Hence, demanding a length $O(\lambda)$ usually requires $\ell \leq \lambda$.

contain at most one message sent by the sender towards the silent receiver. Such property is, for example, inherent to many regular (non-UC) commitments, e.g. [26]. Interactivity may increase the communication complexity by several factors, since in addition to the actual commitment length the amount of bits communicated during the interactive phases would have to be counted as well. For example, the two most recent interactive commitments by Lindell [20] have commitment lengths of only 4 resp. 6 group elements, while their total communication complexity amounts to 14 resp. 19 group elements (we remark that for concrete choices of parameters [20] still remains very efficient in this respect).

The actual advantage of non-interactive UC commitments from the practical point of view is resistance to denial of service attacks: Within an interactive phase (commit or open) parties maintain a state between the communication rounds. It is thus possible for an adversary (malicious sender/receiver or man-in-the-middle), by sending incorrectly formed messages during the interaction rounds, to lure parties into wasting their (computational) resources — something which does not happen in the non-interactive case. Note that, even if no adversary is present, interaction between the sender and the receiver may still be endangered by faults. Earlier UC bit commitments [8,9] were non-interactive. However, in the more desirable case of UC string commitments, the only known non-interactive scheme is due to Nishimaki *et al.* [22]. However, [22] does not allow CRS re-usability, which arguably diminishes the advantage gained through its non-interactivity. Other existing UC string commitments, e.g. [13,12,5,20], are all interactive, either in the commit or in the open phase.

UC commitments that have acceptably low computation and communication costs, allow CRS re-usability, and do not require any interaction between the sender and the receiver would already be ideal from the practical point of view. In addition to these properties there are further desirable properties which should also be assessed concerning their impact on their relevance in practice.

ADAPTIVE SECURITY. A typical question asked about UC-secure protocols is whether security is proven against static or adaptive adversaries. A *static* adversary can corrupt protocol participants at the outset of the protocol only. In case of UC commitments such corruptions would be allowed only prior to the execution of the commit phase, even before the CRS is generated. Since commitments always have two phases with the open phase being executed after the commit phase, it appears unrealistic to exclude corruptions between the two phases. Hence, *adaptive* UC-security, where the adversary can corrupt participants at any point in time, revealing all their secrets (including randomness being used), appears of higher practical relevance. We observe that some of known UC commitments are adaptively secure, e.g. [8,9,12,5,20].

SECURE ERASURES. Another property inherent to the UC-security of commitment schemes is whether they rely on the additional assumption that secrets can be securely erased. This assumption is often used in combination with adaptive security where secrets used in the commit phase that are no longer needed for the open phase are erased to allow simulation in case of later corruptions. Although secure erasures could be realized in practice, it is still desirable for a UC commitment scheme to avoid them. We observe that most adaptively secure UC commitments require secure erasures, the only exception (in addition to less efficient bit commitments from [8,9]) where adaptivity is achieved without erasures is the interactive string commitment by Damgård and Groth [12].

HARDNESS ASSUMPTIONS. Last but not least, in addition to an inevitable setup assumption (e.g. CRS) and possible reliance on secure erasures, UC-security of commitments is typically based on further hardness assumptions. These are either general assumptions such as existence of trapdoor permutations as in [8,9] or more concrete number-theoretic assumptions, which are more likely to give rise to efficient schemes. For example, UC commitments by Damgård and Nielsen [13] rely on p-subgroup [23] or Decision Composite Residuosity (DCR) assumption [25]. The DCR assumption has also been used in the UC commitments by Damgård and Groth [12] (together with Strong RSA (SRSA) assumption), by Camenisch and Shoup [5], and by Nishimaki *et al.* [22]. The recent UC commitments by Lindell [20] rely on the more established Decision Diffie-Hellman (DDH) assumption, which has also been used in one of the bit commitment schemes by Canetti and Fischlin [8] and in a particular instantiation of Nishimaki *et al.*'s scheme [22] with El-Gamal based matrix encryption of Peikert and Waters [27] (those communication complexity is asymptotically comparable to that of a bit commitment scheme though).

The current state of affairs is that *none* of the existing CRS-based UC-secure string commitment schemes fulfills all of the above mentioned "quality criteria".

1.1 Our Results and Comparison to Prior Work

Results. We propose the first UC-secure string commitment schemes in the (standard) CRS model with the so far unique combination of key properties: Our schemes have constant costs (*i.e.*, independent of the message length and the number of participants) for communication, computation, and CRS length. They offer re-usability of the CRS for polynomially many executions. Both schemes are completely non-interactive (*i.e.*, the commitment and opening phases both consist of a single message from the sender to the receiver). We prove their UC-security under adaptive corruptions (with erasures) using the well-known Decision Linear (DLIN) assumption [3]. As demonstrated in Table 1, such UC string commitments were not known to exist before. In particular, their ability to commit to strings with re-usable CRS in combination with non-interactivity and adaptive security seems so far unique.[2]

Our schemes are also the first UC-secure commitments designed for pairing-friendly groups. The main ingredients of our schemes are Groth-Sahai proofs [16]

---

[2] Zhu [30] claims to have a non-interactive, UC-secure string commitment without erasures for re-usable common reference strings; we were unable to verify the proof of the scheme, though. In fact, the encryption-based scheme does not seem to satisfy the usual equivocality property of such commitments.

Table 1. Comparison of UC commitment schemes in the CRS model

UC commitment schemes		comm. complexity in sec.par.(bits)	CRS re-usable	non-inter. phases	without erasures	adaptive security	hardness assumptions
CF01 (1)	[8]	$O(\ell \cdot \lambda)$	—	✓	✓	✓	TDP
CF01 (2)	[8]	$O(\ell \cdot \lambda)$	✓	✓	—	✓	CFP + CCA PKE
CF01 (3)	[8]	$O(\ell \cdot \lambda)$	✓	✓	✓	✓	DDH + UOWHF
CLOS02	[9]	$O(\ell \cdot \lambda)$	✓	✓	✓	✓	TDP
DN02 (1)	[13]	$18 \cdot \lambda$ (13824)	✓	—	✓	✓	p-subgroup
DN02 (2)	[13]	$24 \cdot \lambda$ (18432)	✓	—	✓	✓	DCR
DG03	[12]	$16 \cdot \lambda$ (12288)	✓	—	✓	✓	DCR + SRSA
CS03	[5]	$94 \cdot \lambda$ (72192)	✓	—	—	✓	DCR + CRHF
NFT09	[22]	$21 \cdot \lambda$ (16128)	—	✓	—	✓	DCR + sEUF-OT
NFT09	[22]	$O(\ell \cdot \lambda)$	—	✓	—	✓	DDH + sEUF-OT
Lin11 (1)	[20]	$14 \cdot \lambda$ (3584)	✓	—	✓	—	DDH + CRHF
Lin11 (2)	[20]	$19 \cdot \lambda$ (4864)	✓	—	—	✓	DDH + CRHF
Scheme I		$21 \cdot \lambda$ (5376)	✓	✓	—	✓	DLIN + CRHF
Scheme II		$40 \cdot \lambda$ (10240)	✓	✓	—	✓	DLIN + CRHF

Complexity costs: ℓ - length of committed messages, λ - security parameter,
(bits) - total number of communicated bits (based on λ)
In DN02, DG03, CS03, and DCR-based NFT09, λ is the length of the prime factor of N
(RSA modulus). We use $\lambda = 768$ bits.
In Lin11 λ is the length of the prime group order. We use $\lambda = 256$ bits.
In our schemes λ is the length of the prime group order of the input group.
We use $\lambda = 256$ bits (cf. [24] for parameter choice).
Hardness assumptions: TDP - trapdoor permutations, CFP - claw-free permutations,
UOWHF - universal one-way hash functions, CRHF - collision-resistant hash functions,
DDH - Decision Diffie-Hellman, DCR - Decision Composite Residuosity,
SRSA - Strong RSA, sEUF-OT - strongly unforgeable one-times signature,
DLIN - Decision Linear.

and Cramer-Shoup encryption (under DLIN assumption [3]). Although pairing operations are traditionally costlier in comparison to modular exponentiations in the RSA or Discrete Logarithm settings, constant costs incurred by our schemes seem still to be sufficient for practical purposes. As demonstrated in Table 1, the total communication costs of our schemes, when instantiated with appropriate security parameters, are lower than in all previous DCR-based constructions. For our first scheme, the costs are only slightly higher than for the recent (interactive) UC commitments by Lindell [20]. The entire communication complexity amounts to 21 group elements for our first scheme and 40 elements for our second scheme. Yet our schemes have opposite trade-offs regarding the two phases: Our first scheme outputs commitments containing only 5 group elements and transmits 16 elements in the open phase. In contrast, our second scheme requires 37 group elements to commit and only 3 elements to open.

Techniques. Our first scheme is inspired by the UC commitment scheme of Lindell [20], where the committer encrypts the message in the commit phase using the DDH-based Cramer-Shoup encryption scheme, and in the open phase, simply reveals the committed message and gives an interactive Sigma proof that the message is indeed the one encrypted in the ciphertext. Using non-interactive Groth-Sahai proofs we show that this interaction can be safely removed while preserving UC security and without losing much of the efficiency. We thus use the DLIN assumption instead of DDH. Observe that DLIN assumption is often referred to as a natural counterpart of the DDH assumption in bilinear groups where the latter does not hold. More surprisingly, when transforming Lindell's scheme, we also obtain security against adaptive corruptions essentially for free. That is, the basic scheme in [20] — which is the starting point for our first construction — is only secure against static corruptions. Lindell then provides additional means to derive a variant which withstands adaptive corruptions. In [20], there is no way to prove the basic scheme adaptively secure (even with reliable erasures) because the committer needs to store the randomness used to encrypt in order to give the interactive zero-knowledge proof in the opening phase, and thus cannot erase it after having committed. Having to present this randomness in case of adaptive corruption, however, inhibits the necessary equivocality property of commitments [8], the ability to adapt simulated commitments appropriately. In our case, the committer can compute the *non-interactive* proof already in the commitment phase and present it together with the message in the decommitment phase. By this, the committer only needs to store the proof and can erase any randomness from the commitment phase, buying us security against adaptive corruptions (with erasures).

At this point, we notice that Groth-Sahai proofs are widely used in many cryptographic constructions for reducing the amount of interaction. Interestingly, their applicability to the setting of UC commitments was not explored so far. We thus show that their techniques are powerful enough to allow construction of UC commitments with, up till now, unique properties. We demonstrate this not only with our first scheme, based on the Lindell's commitments (while using the DLIN assumption instead of DDH), but also with our second scheme, which builds upon Camenisch-Shoup commitments [5] with the difference that we work in a discrete logarithm setting instead of relying on the composite residuosity assumption as in [5].

We obtain our second scheme using pairing-based trapdoor commitments to group elements [10,15] in combination with Groth-Sahai proofs and DLIN-based Cramer-Shoup encryption. This scheme can be viewed as the UC secure non-interactive (pairing-based) version of the scheme from [5] with the following tweak: We use trapdoor commitments to group elements prior to the encryption scheme. Unlike [5], where a Pedersen commitment [26] to message M with randomness r is computed and followed by a verifiable encryption of (M, r), we trapdoor-commit to M (viewed as group element) and then encrypt only M. Yet, we can still extract an opening of the trapdoor commitment when the need arises in the security proof (due to the properties of Groth-Sahai commitments). The resulting scheme is somewhat more efficient in communication than if the full opening of the trapdoor commitment is encrypted as in the original construction [5]. We also notice that description of the UC commitment scheme in [5] was limited to the presentation of main ideas but a concrete specification and the eventual analysis of security were left open. With our pairing-based construction and the above mentioned tweak, we not only remove interaction in this scheme and significantly improve its communication complexity but essentially develop the initial ideas from [5] to a full-fledged specification and the corresponding proof of security.

Organization. We recall the basic building blocks that we need in Section 2. Section 3 then presents our non-interactive (adaptively) UC-secure string commitment scheme with re-usable CRS together with the detailed proof of security.

2   Preliminaries

2.1   Complexity Assumptions

In the paper, we use groups $(\mathbb{G}, \mathbb{G}_T)$ of prime order p with a generator $g \in \mathbb{G}$ and endowed with a mapping $e : \mathbb{G} \times \mathbb{G} \to \mathbb{G}_T$ such that $e(g^a, g^b) = e(g, g)^{ab}$ for all $a, b \in \mathbb{Z}_p$ and $e(g, h) \neq 1_{\mathbb{G}_T}$ whenever $g, h \neq 1_{\mathbb{G}}$. We occasionally consider the Cartesian product of groups as vector spaces where component-wise multiplication $(A, B, C) \cdot (X, Y, Z) = (AX, BY, CZ)$ is the vector addition and component-wise exponentiation $(A, B, C)^x = (A^x, B^x, C^x)$ is the scalar multiplication. In these groups, we rely on the following assumption.

Definition 1 ([3]). *The **Decision Linear Problem** (DLIN) in $\mathbb{G}$ consists in distinguishing the distribution $D_1 = \{(g, g^a, g^b, g^{ac}, g^{bd}, g^{c+d}) | a, b, c, d \xleftarrow{R} \mathbb{Z}_p^*\}$ from the distribution $D_2 = \{(g, g^a, g^b, g^{ac}, g^{bd}, g^z) | a, b, c, d, z \xleftarrow{R} \mathbb{Z}_p^*\}$.*

2.2   Groth-Sahai Proof Systems

In the following notations, for equal-dimension vectors $\boldsymbol{A}$ and $\boldsymbol{B}$ containing group elements, $\boldsymbol{A} \cdot \boldsymbol{B}$ stands for their component-wise product.

When based on the DLIN assumption, the Groth-Sahai (GS) proof systems [16] use a common reference string comprising vectors $\boldsymbol{g_1}, \boldsymbol{g_2}, \boldsymbol{g_3} \in \mathbb{G}^3$, where $\boldsymbol{g_1} = (g_1, 1, g)$, $\boldsymbol{g_2} = (1, g_2, g)$ for some $g_1, g_2 \in \mathbb{G}$. To commit to $X \in \mathbb{G}$, one sets $\boldsymbol{C} = (1, 1, X) \cdot \boldsymbol{g_1}^r \cdot \boldsymbol{g_2}^s \cdot \boldsymbol{g_3}^t$ with $r, s, t \xleftarrow{R} \mathbb{Z}_p^*$. When proofs should be perfectly sound, $\boldsymbol{g_3}$ is set as $\boldsymbol{g_3} = \boldsymbol{g_1}^{\xi_1} \cdot \boldsymbol{g_2}^{\xi_2}$ with $\xi_1, \xi_2 \xleftarrow{R} \mathbb{Z}_p^*$. Commitments $\boldsymbol{C} = (g_1^{r+\xi_1 t}, g_2^{s+\xi_2 t}, X \cdot g^{r+s+t(\xi_1+\xi_2)})$ are then Boneh-Boyen-Shacham (BBS) ciphertexts [3] that can be decrypted using $\alpha_1 = \log_g(g_1)$, $\alpha_2 = \log_g(g_2)$. In the witness indistinguishability (WI) setting, vectors $\boldsymbol{g_1}, \boldsymbol{g_2}, \boldsymbol{g_3}$ are linearly independent and $\boldsymbol{C}$ is a perfectly hiding commitment. Under the DLIN assumption, the two kinds of CRS are indistinguishable.

To commit to an exponent $x \in \mathbb{Z}_p$, one computes $\boldsymbol{C} = \boldsymbol{\varphi}^x \cdot \boldsymbol{g_1}^r \cdot \boldsymbol{g_2}^s$, with $r, s \xleftarrow{R} \mathbb{Z}_p^*$, using a CRS comprising vectors $\boldsymbol{\varphi}, \boldsymbol{g_1}, \boldsymbol{g_2}$. In the soundness setting $\boldsymbol{\varphi}, \boldsymbol{g_1}, \boldsymbol{g_2}$ are linearly independent vectors (typically, one chooses $\boldsymbol{\varphi} = \boldsymbol{g_3} \cdot (1, 1, g)$ where $\boldsymbol{g_3} = \boldsymbol{g_1}^{\xi_1} \cdot \boldsymbol{g_2}^{\xi_2}$) whereas, in the WI setting, choosing $\boldsymbol{\varphi} = \boldsymbol{g_1}^{\xi_1} \cdot \boldsymbol{g_2}^{\xi_2}$ gives a perfectly hiding commitment since $\boldsymbol{C}$ is always a BBS encryption of $1_{\mathbb{G}}$. On a perfectly sound CRS (where $\boldsymbol{g_3} = \boldsymbol{g_1}^{\xi_1} \cdot \boldsymbol{g_2}^{\xi_2}$ and $\boldsymbol{\varphi} = \boldsymbol{g_3} \cdot (1, 1, g)$), commitments to exponents are not fully extractable since the trapdoor (α_1, α_2) only allows recovering g^x from $\boldsymbol{C} = \boldsymbol{\varphi}^x \cdot \boldsymbol{g_1}^r \cdot \boldsymbol{g_2}^s$.   To prove that committed variables satisfy a set of relations, the Groth-Sahai techniques require one commitment per variable and one proof element (made of a constant number of group elements) per relation. Such proofs are available for pairing-product relations, which are of the type

$$\prod_{i=1}^{n} e(\mathcal{A}_i, \mathcal{X}_i) \cdot \prod_{i=1}^{n} \cdot \prod_{j=1}^{n} e(\mathcal{X}_i, \mathcal{X}_j)^{a_{ij}} = t_T, \tag{1}$$

for variables $\mathcal{X}_1, \ldots, \mathcal{X}_n \in \mathbb{G}$ and constants $t_T \in \mathbb{G}_T, \mathcal{A}_1, \ldots, \mathcal{A}_n \in \mathbb{G}, a_{ij} \in \mathbb{G}$, for $i, j \in \{1, \ldots, n\}$. Efficient proofs also exist for multi-exponentiation equations

$$\prod_{i=1}^{m} \mathcal{A}_i^{y_i} \cdot \prod_{j=1}^{n} \mathcal{X}_j^{b_j} \cdot \prod_{i=1}^{m} \cdot \prod_{j=1}^{n} \mathcal{X}_j^{y_i \gamma_{ij}} = T, \tag{2}$$

for variables $\mathcal{X}_1, \ldots, \mathcal{X}_n \in \mathbb{G}, y_1, \ldots, y_m \in \mathbb{Z}_p$ and constants $T, \mathcal{A}_1, \ldots, \mathcal{A}_m \in \mathbb{G}$, $b_1, \ldots, b_n \in \mathbb{Z}_p$ and $\gamma_{ij} \in \mathbb{G}$, for $i \in \{1, \ldots, m\}, j \in \{1, \ldots, n\}$.

Multi-exponentiation equations admit zero-knowledge proofs at no additional cost. On a simulated CRS (prepared for the WI setting), the trapdoor (ξ_1, ξ_2) makes it possible to simulate proofs without knowing witnesses, and simulated proofs are perfectly indistinguishable from real proofs. As for pairing-product equations, NIZK proofs are often possible (this is typically the case when the target element t_T has a special form) but usually come at some expense.

In both cases, proofs for quadratic equations (namely, when at least one of the coefficients a_{ij} and γ_{ij} is non-zero in (1) and (2), respectively) cost 9 group elements. Linear pairing-product equations (when $a_{ij} = 0$ for all i, j in (1)) take 3 group elements each. Linear multi-exponentiation equations of the type $\prod_{j=1}^{n} \mathcal{X}_j^{b_j} = T$ (resp. $\prod_{i=1}^{m} \mathcal{A}_i^{y_i} = T$) demand 3 (resp. 2) group elements.

2.3   Cramer-Shoup Encryption Based on DLIN Assumption

This section recalls a variant of the Cramer-Shoup encryption scheme [11] based on the DLIN assumption and suggested in [28,17]. The scheme offers IND-CCA2 security for encryption schemes with labels [29]. If we assume public generators g_1, g_2, g that are parts of public parameters (*i.e.*, a common reference string), the receiver's public key is made of

$$X_1 = g_1^{x_1} g^x \qquad X_3 = g_1^{x_3} g^y \qquad X_5 = g_1^{x_5} g^z$$
$$X_2 = g_2^{x_2} g^x \qquad X_4 = g_2^{x_4} g^y \qquad X_6 = g_2^{x_6} g^z.$$

To encrypt $m \in \mathbb{G}$ under the label L, the sender picks $r, s \xleftarrow{R} \mathbb{Z}_p^*$ and computes

$$\psi_{\mathsf{CS}} = (U_1, U_2, U_3, U_4, U_5) = \left(g_1^r, \; g_2^s, \; g^{r+s}, \; m \cdot X_5^r X_6^s, \; (X_1 X_3^\alpha)^r \cdot (X_2 X_4^\alpha)^s \right),$$

Functionality $\mathcal{F}_{\mathrm{MCOM}}$

$\mathcal{F}_{\mathrm{MCOM}}$ is parameterized by a message space $\mathcal{M}$ interacts with parties $P_1, \ldots, P_n$ and adversary $\mathcal{S}$ as follows.

- Upon receiving ($\mathtt{commit}, sid, cid, P_i, P_j, M$) from P_i, where $M \in \mathcal{M}$, record (sid, cid, P_i, P_j, M) and send a publicly delayed ($\mathtt{receipt}, sid, cid, P_i, P_j$) to P_j. Ignore any subsequent ($\mathtt{commit}, sid, cid, P_i, P_j, *$) messages.
- Upon receiving ($\mathtt{open}, sid, cid, P_i, P_j$) from P_i, if some tuple (cid, P_i, P_j, M) was previously recorded then send a publicly delayed ($\mathtt{open}, sid, cid, P_i, P_j, M$) to P_j. Otherwise halt.
- Upon receiving ($\mathtt{corrupt\text{-}committer}, sid, cid$) from the adversary, check if there is already an entry (sid, cid, P_i, P_j, M) and, if so, send M to the adversary. If the adversary provides some M' and ($\mathtt{receipt}, sid, cid, P_i, P_j$) has not yet been written on P_j's output tape, then change the record to (sid, cid, P_i, P_j, M').

Fig. 1. Functionality $\mathcal{F}_{\mathrm{MCOM}}$ for Multiple Commitments

where $\alpha = H(U_1, U_2, U_3, U_4, L) \in \mathbb{Z}_p$ is a collision-resistant[3] hash function. Given a pair (ψ_{CS}, L), the receiver computes α. If $U_5 \neq U_1^{x_1 + \alpha x_3} U_2^{x_2 + \alpha x_4} U_3^{x + \alpha y}$ then the receiver outputs $\bot$; otherwise he outputs $m = U_4/(U_1^{x_5} U_2^{x_6} U_3^z)$.

2.4   Ideal Functionality for Multiple Commitments

The ideal commitment functionality $\mathcal{F}_{\mathrm{MCOM}}$ described in Figure 1 is the one defined by Canetti and Fischlin [8] but, as in [18], we consider publicly delayed messages, where the message is delivered to the corresponding party only upon confirmation by the adversary (who sees the message first). Note that the functionality now takes another unique "commitment identifier" cid, which may be used if a sender commits to the same receiver multiple times within a session. We assume that the combination of sid, cid is globally unique.

3   Scheme I: A Tweak on Lindell's Scheme

Our first construction builds on Lindell's first interactive UC-secure commitment scheme from [5], which is only known to be secure against static corruptions in its original variant. We show how to utilize Groth-Sahai proofs so as to completely remove interaction, while still guaranteeing UC security (in the adaptive sense) and preserving all other valuable properties of the scheme.

---

[3] The security proofs of the original Cramer-Shoup encryption scheme [11] and its variants based on the DLIN assumption [17,28] only require a universal one-way hash function [21]. As mentioned in [4], for example, collision-resistance is needed when the scheme is extended so as to support labels.

CRS-Gen(λ): choose bilinear groups $(\mathbb{G}, \mathbb{G}_T)$ of order $p > 2^\lambda$, $g \xleftarrow{R} \mathbb{G}$ and $g_1 = g^{\alpha_1}$, $g_2 = g^{\alpha_2}$, with $\alpha_1, \alpha_2 \xleftarrow{R} \mathbb{Z}_p^*$. Define vectors $\boldsymbol{g_1} = (g_1, 1, g)$, $\boldsymbol{g_2} = (1, g_2, g)$ and $\boldsymbol{g_3} = \boldsymbol{g_1}^{\xi_1} \cdot \boldsymbol{g_2}^{\xi_2}$ with $\xi_1, \xi_2 \xleftarrow{R} \mathbb{Z}_p^*$, which form a Groth-Sahai CRS $\mathbf{g} = (\boldsymbol{g_1}, \boldsymbol{g_2}, \boldsymbol{g_3})$ for the perfect soundness setting. Then, choose a collision-resistant hash function $H : \{0,1\}^* \to \mathbb{Z}_p$ and generate a public key $\mathsf{pk} = (X_1, \ldots, X_6)$ for the linear Cramer-Shoup encryption scheme. The CRS consists of $\mathsf{crs} = \{\lambda, \mathbb{G}, \mathbb{G}_T, g, \mathbf{g}, H, \mathsf{pk}\}$.

Commit($\mathsf{crs}, M, sid, cid, P_i, P_j$): to commit to message $M \in \mathbb{G}$ for party P_j upon receiving a command ($\mathtt{commit}, sid, cid, P_i, P_j, M$), party P_i parses crs as $\{\lambda, \mathbb{G}, \mathbb{G}_T, g, \mathbf{g}, \mathbf{f}, \mathsf{pk}\}$, respectively, first fetches crs from $\mathcal{F}_{\mathrm{CRS}}$ if not done already, and then conducts the following steps.

1. Choose random exponents $r, s \xleftarrow{R} \mathbb{Z}_p$ and compute a Cramer-Shoup encryption $\psi_{\mathsf{CS}} = (U_1, U_2, U_3, U_4, U_5)$ of $M \in \mathbb{G}$ under the label $L = P_i || sid || cid$ and the public key $\mathsf{pk} \in \mathbb{G}^6$ as in Section 2.3.
2. Generate a NIZK proof $\pi_{val\text{-}enc}$ that $\psi_{\mathsf{CS}} = (U_1, U_2, U_3, U_4, U_5)$ is a valid encryption of $M \in \mathbb{G}$. This requires to commit to exponents r, s and prove that these exponents satisfy the multi-exponentiation equations

$$U_1 = g_1^r, \qquad\qquad U_2 = g_2^s, \qquad\qquad U_3 = g^{r+s}, \qquad (3)$$
$$U_4/M = X_5^r X_6^s, \qquad U_5 = (X_1 X_3^\alpha)^r \cdot (X_2 X_4^\alpha)^s$$

   (which only takes 5 times 2 elements as base elements are all public). Including commitments com_r and com_s to exponents r and s, the proof $\pi_{val\text{-}enc}$ demands 16 group elements overall.
3. P_i erases (r, s) after the generation of $\pi_{val\text{-}enc}$ but retains the state information $D_M = \pi_{val\text{-}enc}$.

The commitment $\sigma = \psi_{\mathsf{CS}}$ comprises 5 group elements. Upon receiving ($\mathtt{Com}, sid, cid, \sigma$) from P_i, party P_j verifies that $\sigma = \psi_{\mathsf{CS}}$ can be parsed as an element of $\mathbb{G}^5$. If yes, P_j outputs ($\mathtt{receipt}, sid, cid, P_i, P_j$). Otherwise, P_j ignores the message.

Open($\mathsf{crs}, M, D_M, sid, cid, P_i, P_j$): when receiving a command ($\mathtt{open}, sid, cid, P_i, P_j, M$), party P_i reveals M and his state information $D_M = \pi_{val\text{-}enc}$ to P_j.

Verify($\mathsf{crs}, (\mathtt{Com}, sid, cid, \sigma), M, D_M, sid, cid, P_i, P_j$): P_j verifies the proof $\pi_{val\text{-}enc}$ and ignores the opening if verification fails. If both proofs verify, P_j outputs ($\mathtt{open}, sid, cid, P_i, P_j, M$) iff cid has not been used with this committer previously. Otherwise, P_j also ignores the message.

Theorem 1. *The above commitment scheme securely realizes $\mathcal{F}_{\mathrm{MCOM}}$ in the CRS model against adaptive corruptions (assuming reliable erasure), provided that (i) the DLIN assumption holds in $\mathbb{G}$; (ii) H is collision-resistant.*

Proof. We construct an ideal-world adversary $\mathcal{S}$ that runs a black-box simulation of the real-world adversary $\mathcal{A}$ by simulating the protocol execution and relaying messages between $\mathcal{A}$ and the environment $\mathcal{Z}$. The ideal-world adversary $\mathcal{S}$ proceeds as follows in experiment IDEAL.

1. $\mathcal{S}$ sets up crs by choosing $\mathbf{g} = (\boldsymbol{g_1}, \boldsymbol{g_2}, \boldsymbol{g_3})$ as a Groth-Sahai CRS for the perfect WI setting (namely, $\boldsymbol{g_3} = \boldsymbol{g_1}^{\xi_1} \cdot \boldsymbol{g_2}^{\xi_2} \cdot (1, 1, g)^{-1}$ for some $\xi_1, \xi_2 \xleftarrow{R} \mathbb{Z}_p^*$). Also, $\mathcal{S}$ generates a public key pk $= (X_1, \ldots, X_6)$ as specified by the linear Cramer-Shoup encryption scheme.

2. When the environment $\mathcal{Z}$ requires some uncorrupted party P_i to commit to a message and send (Commit, sid, cid, P_i, P_j, M) to the functionality, the simulator $\mathcal{S}$ is notified that a commitment operation took place but does not know the committed message M. Therefore, $\mathcal{S}$ chooses a fake random message $R \xleftarrow{R} \mathbb{G}$ and computes a linear Cramer-Shoup encryption ψ_{CS} of $R \in \mathbb{G}$ using random exponents $r, s \xleftarrow{R} \mathbb{Z}_p$. The adversary $\mathcal{A}$ is then given (Com, sid, cid, σ) with $\sigma = \psi_{\text{CS}}$ and, when $\tilde{P}_j$ eventually obtains (Com, sid, cid, σ) and outputs (Receipt, sid, cid, P_i, P_j), the simulator $\mathcal{S}$ allows $\mathcal{F}_{\text{MCOM}}$ to proceed with the delivery of message (Commit, sid, cid, P_i, P_j) to P_j.

3. If $\mathcal{Z}$ requires some uncorrupted party P_i to open a previously generated commitment $\sigma = \psi_{\text{CS}}$ to some message $M \in \mathbb{G}$, $\mathcal{S}$ learns M from $\mathcal{F}_{\text{MCOM}}$ and, using the trapdoor $\xi_1, \xi_2 \in (\mathbb{Z}_p)^2$ of the simulated Groth-Sahai CRS, generates a simulated proof $\pi_{val\text{-}enc}$ that equations (3) are satisfied for the message M obtained from $\mathcal{F}_{\text{MCOM}}$. The internal state of $\tilde{P}_i$ is modified to be $D_M = \pi_{val\text{-}enc}$, which is also given to $\mathcal{A}$ as the real-world de-commitment. Before allowing $\mathcal{F}_{\text{MCOM}}$ to deliver the message (Open, sid, cid, P_i, P_j, M) to P_j, algorithm $\mathcal{S}$ waits for $\tilde{P}_j$ to acknowledge the opening in the simulation.

4. When the simulated adversary $\mathcal{A}$ delivers a commitment (Com, sid^*, cid^*, σ^*) for party $\tilde{P}_i$ to party $\tilde{P}_j$ and the latter still has not received a commitment with subsession ID cid^* from $\tilde{P}_i$, $\mathcal{S}$ proceeds as follows. If $\tilde{P}_i$ (and thus P_i as well) is uncorrupted, $\mathcal{S}$ notifies $\mathcal{F}_{\text{MCOM}}$ that the commitment (sid^*, cid^*) can be delivered. The Receipt message returned by $\mathcal{F}_{\text{MCOM}}$ is delivered to the dummy P_j as soon as the simulated $\tilde{P}_j$ outputs his own Receipt message. If $\tilde{P}_i$ is a corrupted party, then σ^* has to be extracted. Namely, $\mathcal{S}$ parses σ^* as ψ_{CS}^*. If $\psi_{\text{CS}}^* \notin \mathbb{G}^5$, $\mathcal{S}$ simply ignores the commitment. Otherwise, it uses the private key sk corresponding to pk to decrypt ψ_{CS}^*. If ψ_{CS}^* turns out to be an invalid Cramer-Shoup ciphertext, the commitment is ignored. Otherwise, $\mathcal{S}$ obtains the plaintext $M \in \mathbb{G}$ and sends (Commit, $sid^*, cid^*, P_i, P_j, M$) to $\mathcal{F}_{\text{MCOM}}$, which causes $\mathcal{F}_{\text{MCOM}}$ to prepare a Receipt message for P_j. The latter is delivered by $\mathcal{S}$ as soon as $\tilde{P}_j$ produces his own output.

5. If $\mathcal{A}$ gets a simulated corrupted party $\tilde{P}_i$ to correctly open a commitment (Com, sid^*, cid^*, σ^*) to message M^*, the ideal-world adversary $\mathcal{S}$ compares M^* to the message M that was previously extracted from σ^* and aborts if $M \neq M^*$. Otherwise, $\mathcal{S}$ sends (Open, sid, cid, P_i, P_j, M) on behalf of P_i to $\mathcal{F}_{\text{MCOM}}$. If $\mathcal{A}$ provides an incorrect opening, $\mathcal{S}$ simply ignores this opening.

6. If the simulated $\mathcal{A}$ decides to corrupt some party $\tilde{P}_i$, $\mathcal{S}$ corrupts the corresponding party P_i in the ideal world and obtains all his internal information. It also modifies all de-commitment information about the unopened commitments generated by $\tilde{P}_i$ so as to make it match the received de-commitment information of P_i. (Note that P_i is supposed to reliably delete the exponents

and to store only the group elements for decommitments.) This modified internal information is given to $\mathcal{A}$. For each commitment intended for P_j but for which P_j did not receive $(\texttt{Commit}, sid, cid, P_i, P_j)$, the newly corrupted $\tilde{P}_i$ is allowed to decide what the committed message will eventually be. A new message $M \in \mathbb{G}$ is thus supplied by $\mathcal{A}$ and $\mathcal{S}$ informs $\mathcal{F}_{\mathrm{MCOM}}$ that M supersedes the message chosen by P_i before his corruption.

To show that the output of the environment $\mathcal{Z}$ in the ideal world is indistinguishable from its output in the real world, we consider several hybrid experiments involving hybrid adversaries $\mathcal{S}_i$.

$\mathsf{HYB}^1_{\mathcal{S}_1, \mathcal{Z}}$: is identical to the real experiment with two differences. The first one is that the simulator $\mathcal{S}_1$ generates the CRS by choosing $\mathbf{g} = (\boldsymbol{g_1}, \boldsymbol{g_2}, \boldsymbol{g_3})$ for the WI setting (namely, $\boldsymbol{g_3}$ is chosen as $\boldsymbol{g_3} = \boldsymbol{g_1}^{\xi_1} \cdot \boldsymbol{g_2}^{\xi_2} \cdot (1, 1, g)^{-1}$) instead of the perfect soundness setting. The other difference is that honest parties generate commitments by computing ψ_{CS} as an encryption of a random group element $R \in \mathbb{G}$ instead of the real message M. The NIZK proof $\pi_{val\text{-}enc}$ is then simulated using the trapdoor (ξ_1, ξ_2) of the Groth-Sahai CRS $(\boldsymbol{g_1}, \boldsymbol{g_2}, \boldsymbol{g_3})$. Experiment $\mathsf{HYB}^1_{\mathcal{S}_1, \mathcal{Z}}$ proceeds almost identically to the ideal-world experiment: the only difference is that $\mathcal{S}_1$ does not extract messages that corrupted parties commit to and never has to abort.

We first observe that the output of the environment $\mathcal{Z}$ in $\mathsf{HYB}^1_{\mathcal{S}_1, \mathcal{Z}}$ is negligibly close to its output in the real experiment REAL if the linear Cramer-Shoup encryption scheme is IND-CPA and if the two types of Groth-Sahai reference strings are indistinguishable.

Claim. If the DLIN assumption holds in $\mathbb{G}$, the output of $\mathcal{Z}$ in REAL is negligibly different from its output in $\mathsf{HYB}^1_{\mathcal{S}_1, \mathcal{Z}}$.

Proof. The proof proceeds using two intermediate hybrid experiments HYB_0 and HYB'_0 between REAL and $\mathsf{HYB}^1_{\mathcal{S}_1, \mathcal{Z}}$. In HYB_0, the perfectly sound CRS $\mathbf{g} = (\boldsymbol{g_1}, \boldsymbol{g_2}, \boldsymbol{g_3})$, where $\boldsymbol{g_3} = \boldsymbol{g_1}^{\xi_1} \cdot \boldsymbol{g_2}^{\xi_2}$, is replaced by a fake CRS, where $\boldsymbol{g_3} = \boldsymbol{g_1}^{\xi_1} \cdot \boldsymbol{g_2}^{\xi_2} \cdot (1, 1, g)^{-1}$. It is clear that, under the DLIN assumption, this modification cannot affect $\mathcal{Z}$'s view.

Then, HYB'_0 is like HYB_0 with the difference that NIZK proofs $\pi_{val\text{-}enc}$ (which are generated when $\mathcal{S}_1$ has to open honestly generated commitments) are simulated using the trapdoor (ξ_1, ξ_2). Observe that proofs $\pi_{val\text{-}enc}$ are simulated proofs for true statements in HYB'_0. Since these proofs have the same distribution as real proofs on a fake CRS, $\mathcal{Z}$'s view is identical in HYB_0 and HYB'_0.

We now turn to the indistinguishability of HYB'_0 and $\mathsf{HYB}^1_{\mathcal{S}_1, \mathcal{Z}}$ and rely on the semantic security of the Cramer-Shoup cryptosystem, which is equivalent to the DLIN assumption. Namely, if there exist an environment $\mathcal{Z}$ and an adversary $\mathcal{A}$ for which the two experiments are distinguishable, there is an IND-CPA adversary $\mathcal{D}_{CPA}$ (in the sense of the left-or-right definition of [2]) against the linear Cramer-Shoup scheme. This adversary takes in an encryption key pk and proceeds as follows. (We merely provide a sketch here.) It uses a Groth-Sahai CRS $\mathbf{g} = (\boldsymbol{g_1}, \boldsymbol{g_2}, \boldsymbol{g_3})$ for the WI setting and the challenge Cramer-Shoup public key pk is used to complete the generation of crs. It then simulates adversary $\mathcal{A}$ with the left-or-right oracle and the simulation trapdoor (ξ_1, ξ_2) to simulate a NIZK proof. Algorithm $\mathcal{D}_{CPA}$ eventually outputs what the environment outputs. If the secret bit of the encryption oracle is $b = 0$, $\mathcal{D}_{CPA}$ is running experiment HYB_0' whereas, if $b = 1$, it is running $\mathsf{HYB}_{\mathcal{S}_1, \mathcal{Z}}^1$. The same argument as in [8, Theorem 8] shows that experiments REAL and $\mathsf{HYB}_{\mathcal{S}_1, \mathcal{Z}}^1$ are indistinguishable.   □

We observe that the only situation where experiments IDEAL and $\mathsf{HYB}_{\mathcal{S}_1, \mathcal{Z}}^1$ depart from each other is when, during the ideal experiment IDEAL, $\mathcal{S}$ gives a message M to $\mathcal{F}_{MCOM}$ when a corrupted party $\tilde{P}_i$ comes up with a commitment and, later on, $\tilde{P}_i$ opens that commitment to $M^* \neq M$. We are thus left with the task of bounding the probability of the latter event, which we call Fail, in IDEAL. To this end, we will actually rule out the possibility of such a mismatch in an experiment $\mathsf{IDEAL/GENUINE}$ where $\mathcal{A}$'s view is nearly the same as in the ideal experiment. We then argue that, if Fail occurs with non-negligible probability during IDEAL, the same holds in $\mathsf{IDEAL/GENUINE}$.

Experiment $\mathsf{IDEAL/GENUINE}$ is defined as being identical to IDEAL with two differences: (1) when honest parties generate commitments, the simulator $\mathcal{S}$ "magically" knows which message is being committed to and computes ψ_{CS} and the corresponding opening $\pi_{val-enc}$ according to the specification of the scheme; (2) $\mathcal{S}$ configures the Groth-Sahai CRS $\mathbf{g} = (\mathbf{g}_1, \mathbf{g}_2, \mathbf{g}_3)$ for the perfect soundness setting (namely, with $\mathbf{g}_3 = \mathbf{g}_1{}^{\xi_1} \cdot \mathbf{g}_2{}^{\xi_2}$, for some random $\xi_1, \xi_2 \in \mathbb{Z}_p$).

In $\mathsf{IDEAL/GENUINE}$, event Fail occurs if, on behalf of a corrupted player, the adversary $\mathcal{A}$ comes up with a commitment $\sigma^* = \psi_{CS}^*$ for which ψ_{CS}^* decrypts to M but $\mathcal{A}$ subsequently produces a convincing opening $\pi_{val-enc}^*$ proving that ψ_{CS}^* opens to $M^* \neq M$. As in IDEAL, $\mathcal{S}$ aborts if Fail occurs during $\mathsf{IDEAL/GENUINE}$. As will be argued later on, the probability of Fail is actually zero in $\mathsf{IDEAL/GENUINE}$.

Claim. If the DLIN assumption holds and if H is collision-resistant, the probability that event Fail occurs in IDEAL is negligibly close to its probability of occurring in experiment $\mathsf{IDEAL/GENUINE}$.

Proof. To prove the statement, we define experiments $\mathsf{IDEAL/GENUINE}^{(1)}$ and $\mathsf{IDEAL/GENUINE}^{(2)}$.

$\mathsf{IDEAL/GENUINE}^{(1)}$: is identical to IDEAL except that $\mathcal{S}$ knows which messages honest dummy parties commit to and computes ψ_{CS} as an encryption of the committed message M. On the other hand, NIZK proofs $\pi_{val-enc}$ are still simulated when these commitments have to be opened.

$\mathsf{IDEAL/GENUINE}^{(2)}$: is as $\mathsf{IDEAL/GENUINE}^{(1)}$ but, when the simulator $\mathcal{S}$ has to open honest parties' commitments, NIZK proofs $\pi_{val-enc}$ are calculated using the real witnesses instead of the simulation trapdoor (ξ_1, ξ_2).

$\mathsf{IDEAL/GENUINE}$: is the same as $\mathsf{IDEAL/GENUINE}^{(2)}$ with the difference that $\mathbf{g} = (\mathbf{g}_1, \mathbf{g}_2, \mathbf{g}_3)$ is defined to be a perfectly sound Groth-Sahai CRS.

Experiments IDEAL/GENUINE$^{(1)}$ and IDEAL/GENUINE$^{(2)}$ provide the adversary and $\mathcal{Z}$ with identical views since, in the WI setting, simulated proofs are distributed as real proofs. Also, it is straightforward that IDEAL/GENUINE and IDEAL/GENUINE$^{(2)}$ are indistinguishable under the DLIN assumption.

It remains to prove indistinguishability of IDEAL and IDEAL/GENUINE$^{(1)}$. To this end, we show that, if there exist an environment $\mathcal{Z}$ and an adversary $\mathcal{A}$ such that Fail occurs with noticeably different probabilities in the two experiments, there is a chosen-ciphertext adversary $\mathcal{D}_{CCA}$ against the linear Cramer-Shoup encryption scheme. Our adversary $\mathcal{D}_{CCA}$ takes as input a public key pk for the encryption scheme and is granted access to a decryption oracle. It then proceeds similar to $\mathcal{D}_{CPA}$ but this time uses its decryption oracle to extract messages from adversarial commitments (we omit a formal description here for space reasons). We observe that, if the challenger's bit is $b = 1$, $\mathcal{D}_{CCA}$ proceeds in such a way that $\mathcal{A}$'s view is exactly as in experiment IDEAL. If $b = 0$, $\mathcal{D}_{CCA}$ is running experiment IDEAL/GENUINE$^{(1)}$. Hence, as long as the linear Cramer-Shoup system is chosen-ciphertext secure, $\mathcal{D}_{CCA}$'s output probabilities in both experiments must be negligibly far apart.

In experiment IDEAL/GENUINE, it is easy to see that event Fail cannot occur whatsoever. Indeed, it would require the adversary to produce a valid proof for a false statement, which is precluded by the perfect soundness of Groth-Sahai proofs in the soundness setting. $\square$

4   Scheme II: A Tweak on the Camenisch-Shoup Scheme

4.1   Trapdoor Commitments to Group Elements

We need a trapdoor commitment scheme, suggested in [10], that allows committing to elements of a pairing-friendly group $\mathbb{G}$. To simplify our security analysis, we need commitments to consist of elements of the same group $\mathbb{G}$. We note that Groth's trapdoor commitment to group elements [15,1] could be used as well. However, our construction would then require to include NIZK proofs for pairing-product equations in each UC commitment, which would eventually result in longer commitment strings.

Such a trapdoor commitment can be obtained by modifying the opening phase of perfectly hiding Groth-Sahai commitments so as to enable trapdoor openings. This commitment uses a commitment key describing a prime order group $\mathbb{G}$ and $g \in \mathbb{G}$. The commitment key consists of vectors $(\boldsymbol{f_1}, \boldsymbol{f_2}, \boldsymbol{f_3})$ chosen as $\boldsymbol{f_1} = (f_1, 1, g)$, $\boldsymbol{f_2} = (1, f_2, g)$ and $\boldsymbol{f_3} = \boldsymbol{f_1}^{\chi_1} \cdot \boldsymbol{f_2}^{\chi_2} \cdot (1, 1, g)^{\chi_3}$, with $f_1, f_2 \xleftarrow{R} \mathbb{G}$, $\chi_1, \chi_2, \chi_3 \xleftarrow{R} \mathbb{Z}_p^*$. To commit to $X \in \mathbb{G}$, the sender picks $\theta_1, \theta_2, \theta_3 \xleftarrow{R} \mathbb{Z}_p^*$ and sets $\boldsymbol{C_X} = (1, 1, X) \cdot \boldsymbol{f_1}^{\theta_1} \cdot \boldsymbol{f_2}^{\theta_2} \cdot \boldsymbol{f_3}^{\theta_3}$, which, if $\boldsymbol{f_3}$ is parsed as $(f_{3,1}, f_{3,2}, f_{3,3})$, can be written $\boldsymbol{C_X} = (f_1^{\theta_1} \cdot f_{3,1}^{\theta_3}, f_2^{\theta_2} \cdot f_{3,2}^{\theta_3}, X \cdot g^{\theta_1 + \theta_2} \cdot f_{3,3}^{\theta_3})$. To open $\boldsymbol{C_X} = (C_1, C_2, C_3)$, the sender reveals $(D_1, D_2, D_3) = (g^{\theta_1}, g^{\theta_2}, g^{\theta_3})$ and X. The receiver is convinced that the committed value was X by checking that

$$\begin{cases} e(C_1, g) = e(f_1, D_1) \cdot e(f_{3,1}, D_3) \\ e(C_2, g) = e(f_2, D_2) \cdot e(f_{3,2}, D_3) \\ e(C_3, g) = e(X \cdot D_1 \cdot D_2, g) \cdot e(f_{3,3}, D_3). \end{cases}$$

If a sender can come up with distinct openings of C_X, we can easily construct a distinguisher for the DLIN assumption (and even break a computational assumption that implies DLIN), as noted in [10].

Using the trapdoor (χ_1, χ_2, χ_3), the sender can equivocate commitments when $\chi_3 \neq 0$. Given a commitment C_X and its opening $(X, (D_1, D_2, D_3))$, one can trapdoor open C_X to any other $X' \in \mathbb{G}$ (without knowing $\log_g(X')$) by computing $D'_1 = D_1 \cdot (X'/X)^{\chi_1/\chi_3}$, $D'_2 = D_2 \cdot (X'/X)^{\chi_2/\chi_3}$ and $D'_3 = (X/X')^{1/\chi_3} \cdot D_3$. The scheme is thus a trapdoor commitment whenever $\chi_3 \neq 0$. When $\chi_3 = 0$, the commitment is perfectly binding and even extractable with knowledge of discrete logarithms of the commitment key since X can be computed from (C_1, C_2, C_3) using $\beta_1 = \log_g(f_1)$, $\beta_2 = \log_g(f_2)$.

4.2   Construction

Our second construction builds upon the Camenisch-Shoup interactive UC-secure commitments [5]. The latter requires the committer to trapdoor-commit to the message m using some randomness r with the Pedersen trapdoor commitment [26] before encrypting m using a CCA2-secure encryption scheme supporting labels. In the committing phase, the sender then provides an interactive proof that the ciphertext ψ encrypts the plaintext which is committed to. To remove interaction from this construction, we use the Groth-Sahai techniques and combine them with the trapdoor commitment to group elements recalled in Section 4.1. The proof itself relies on a common reference string.

CRS-Gen(λ): choose bilinear groups $(\mathbb{G}, \mathbb{G}_T)$ of order $p > 2^\lambda$ with $g \xleftarrow{R} \mathbb{G}$ and compute $g_1 = g^{\alpha_1}$, $g_2 = g^{\alpha_2}$, $f_1 = g^{\beta_1}$, $f_2 = g^{\beta_2}$ with $\alpha_1, \alpha_2, \beta_1, \beta_2 \xleftarrow{R} \mathbb{Z}_p^*$. Define vectors $\boldsymbol{g_1} = (g_1, 1, g)$, $\boldsymbol{g_2} = (1, g_2, g)$ and $\boldsymbol{g_3} = \boldsymbol{g_1}^{\xi_1} \cdot \boldsymbol{g_2}^{\xi_2}$ with $\xi_1, \xi_2 \xleftarrow{R} \mathbb{Z}_p^*$, which form a Groth-Sahai CRS $\mathbf{g} = (\boldsymbol{g_1}, \boldsymbol{g_2}, \boldsymbol{g_3})$ for the perfect soundness setting. Then, define vectors $\boldsymbol{f_1} = (f_1, 1, g)$, $\boldsymbol{f_2} = (1, f_2, g)$ and $\boldsymbol{f_3} = \boldsymbol{f_1}^{\chi_1} \cdot \boldsymbol{f_2}^{\chi_2} \cdot (1, 1, g)^{\chi_3}$ with $\chi_1, \chi_2, \chi_3 \xleftarrow{R} \mathbb{Z}_p^*$, which form a public key $\mathbf{f} = (\boldsymbol{f_1}, \boldsymbol{f_2}, \boldsymbol{f_3})$ for the trapdoor commitment to group elements. Finally, choose a collision-resistant hash function $H : \{0,1\}^* \rightarrow \mathbb{Z}_p$ and generate a public key $\mathsf{pk} = (X_1, \ldots, X_6)$ for the linear Cramer-Shoup encryption scheme. The CRS consists of $\mathsf{crs} = \{\lambda, \mathbb{G}, \mathbb{G}_T, g, \mathbf{g}, \mathbf{f}, H, \mathsf{pk}\}$.

Commit$(\mathsf{crs}, M, sid, cid, P_i, P_j)$: to commit to message $M \in \mathbb{G}$ for party P_j upon receiving a command $(\mathtt{commit}, sid, cid, P_i, P_j, M)$, party P_i parses crs as $\{\lambda, \mathbb{G}, \mathbb{G}_T, g, \mathbf{g}, \mathbf{f}, \mathsf{pk}\}$, respectively, first fetches crs from $\mathcal{F}_{\mathrm{CRS}}$ if not done already, and then conducts the following steps.

1. Using vectors $\mathbf{f} = (\boldsymbol{f_1}, \boldsymbol{f_2}, \boldsymbol{f_3})$ as $\boldsymbol{f_1} = (f_1, 1, g)$, $\boldsymbol{f_2} = (1, f_2, g)$ and $\boldsymbol{f_3} = (f_{3,1}, f_{3,2}, f_{3,3})$, pick $\theta_1, \theta_2, \theta_3 \overset{R}{\leftarrow} \mathbb{Z}_p^*$ and compute a commitment to $M \in \mathbb{G}$ as

$$com_M = (c_{M,1}, c_{M,2}, c_{M,3}) = \left(f_1^{\theta_1} \cdot f_{3,1}^{\theta_3}, \ f_2^{\theta_2} \cdot f_{3,2}^{\theta_3}, \ M \cdot g^{\theta_1 + \theta_2} \cdot f_{3,3}^{\theta_3} \right).$$

2. Choose exponents $r, s \overset{R}{\leftarrow} \mathbb{Z}_p^*$ and compute a Cramer-Shoup encryption $\psi_{CS} = (U_1, U_2, U_3, U_4, U_5)$ of $M \in \mathbb{G}$ under the label $L = P_i \| sid \| cid$ and the public key $\mathsf{pk} \in \mathbb{G}^6$ as in Section 2.3.

3. Generate a NIZK proof $\pi_{val\text{-}enc}$ that $\psi_{CS} = (U_1, U_2, U_3, U_4, U_5)$ is a valid Cramer-Shoup encryption. This requires to commit to encryption exponents r, s and prove that these satisfy $U_1 = g_1^r$, $U_2 = g_2^s$, $U_3 = g^{r+s}$ and $U_5 = (X_1 X_3^\alpha)^r \cdot (X_2 X_4^\alpha)^s$ (which only takes 4 times 2 elements as base elements are all public). Including commitments com_r and com_s to exponents r and s, the proof $\pi_{val\text{-}enc}$ demands 14 group elements overall.

4. Generate a NIZK proof $\pi_{eq\text{-}com}$ that ψ_{CS} encrypts the same group element $M \in \mathbb{G}$ as the one that was committed to in com_M. In other words, prove that committed exponents $(r, s, \theta_1, \theta_2, \theta_3)$ satisfy

$$\left(\frac{U_1}{c_{M,1}}, \frac{U_2}{c_{M,2}}, \frac{U_4}{c_{M,3}} \right) = \left(g_1^r \cdot f_1^{-\theta_1} \cdot f_{3,1}^{-\theta_3}, \ g_2^s \cdot f_2^{-\theta_2} \cdot f_{3,2}^{-\theta_3}, \right.$$
$$\left. g^{-\theta_1-\theta_2} \cdot f_{3,3}^{-\theta_3} \cdot X_5^r \cdot X_6^s \right). \tag{4}$$

Commitments to r, s are already part of $\pi_{val\text{-}enc}$. Committing to $\theta_1, \theta_2, \theta_3$ takes 9 elements. Proving (4) requires 6 elements as each relation is linear. Hence, $\pi_{eq\text{-}com}$ requires 15 group elements and P_i erases $(r, s, \theta_1, \theta_2, \theta_3)$ after its generation but retains the information $D_M = (g^{\theta_1}, g^{\theta_2}, g^{\theta_3})$.

The entire commitment $\sigma = (com_M, \psi_{CS}, \pi_{val\text{-}enc}, \pi_{eq\text{-}com})$ takes 37 group elements. Upon receiving a commitment $(\mathsf{Com}, sid, cid, \sigma)$ from P_i, party P_j verifies the proofs $\pi_{val\text{-}enc}, \pi_{eq\text{-}com}$ in σ and, if correct, outputs $(\mathtt{receipt}, sid, cid, P_i, P_j)$; for invalid proofs P_j ignores the message.

Open$(crs, M, D_M, sid, cid, P_i, P_j)$**:** when receiving $(\mathtt{open}, sid, cid, P_i, P_j, M)$, P_i reveals M and $D_M = (D_1, D_2, D_3) = (g^{\theta_1}, g^{\theta_2}, g^{\theta_3})$ to P_j.

Verify$(crs, (\mathsf{Com}, sid, cid, \sigma), M, D_M, sid, cid, P_i, P_j)$**:** P_j verifies proofs $\pi_{val\text{-}enc}$, $\pi_{eq\text{-}com}$ (or recalls the previous check in the commitment phase) and ignores the opening if verification fails. If both proofs verify, P_j outputs $(\mathtt{open}, sid, cid, P_i, P_j, M)$ iff cid has not been used with this committer previously and the opening $D_M = (D_1, D_2, D_3)$ of com_M passes the verification test (as described in section 4.1). Otherwise, P_j also ignores the message.

4.3 Security

Theorem 2. *The above commitment scheme securely realizes $\mathcal{F}_{MCOM}$ in the CRS model against adaptive corruptions (assuming reliable erasure), provided that (i) the DLIN assumption holds in $\mathbb{G}$; (ii) the hash function H is collision-resistant. (The proof appears in the full version of the paper).*

5   Conclusion

In this paper we gave new constructions of efficient UC-secure commitment schemes in the CRS model, simultaneously supporting many useful properties: their commitment/opening phases are both non-interactive and they allow committing to strings rather than single bits while re-using the common reference string for an unbounded (but polynomial) number of commitments. Such UC secure commitments have not been known to exist so far. The only missing property, left as an open problem of our work, is to find new ways for eliminating the reliance on erasures (without introducing new assumptions, such as deployment of tamper-proof hardware that can be used in practice to avoid erasures, or using weaker adversary models that prevent adversarial access to ephemeral secrets).

Acknowledgments. Marc Fischlin was supported by grants Fi 940/2-1 and Fi 940/3-1 of the German Research Foundation (DFG). Benoît Libert acknowledges the Belgian Fund for Scientific Research (F.R.S.-F.N.R.S) for his "Chargé de recherches" fellowship and the BCRYPT Interuniversity Attraction Pole. Mark Manulis was supported by the DFG grant MA 4957. This work was also supported by CASED (www.cased.de).

References

1. Abe, M., Fuchsbauer, G., Groth, J., Haralambiev, K., Ohkubo, M.: Structure-Preserving Signatures and Commitments to Group Elements. In: Rabin, T. (ed.) CRYPTO 2010. LNCS, vol. 6223, pp. 209–236. Springer, Heidelberg (2010)
2. Bellare, M., Desai, A., Jokipii, E., Rogaway, P.: A Concrete Security Treatment of Symmetric Encryption. In: FOCS 1997, pp. 394–403 (1997)
3. Boneh, D., Boyen, X., Shacham, H.: Short Group Signatures. In: Franklin, M. (ed.) CRYPTO 2004. LNCS, vol. 3152, pp. 41–55. Springer, Heidelberg (2004)
4. Camenisch, J., Chandran, N., Shoup, V.: A Public Key Encryption Scheme Secure Against Key Dependent Chosen Plaintext and Adaptive Chosen Ciphertext Attacks. In: Joux, A. (ed.) EUROCRYPT 2009. LNCS, vol. 5479, pp. 351–368. Springer, Heidelberg (2009)
5. Camenisch, J., Shoup, V.: Practical Verifiable Encryption and Decryption of Discrete Logarithms. In: Boneh, D. (ed.) CRYPTO 2003. LNCS, vol. 2729, pp. 126–144. Springer, Heidelberg (2003)
6. Canetti, R.: Universally Composable Security: A New Paradigm for Cryptographic Protocols. In: FOCS 2001, pp. 136–145 (2001)
7. Canetti, R., Dodis, Y., Pass, R., Walfish, S.: Universally Composable Security with Global Setup. In: Vadhan, S.P. (ed.) TCC 2007. LNCS, vol. 4392, pp. 61–85. Springer, Heidelberg (2007)
8. Canetti, R., Fischlin, M.: Universally Composable Commitments. In: Kilian, J. (ed.) CRYPTO 2001. LNCS, vol. 2139, pp. 19–40. Springer, Heidelberg (2001)
9. Canetti, R., Lindell, Y., Ostrovsky, R., Sahai, A.: Universally composable two-party and multi-party secure computation. In: STOC 2002, pp. 494–503 (2002)
10. Cathalo, J., Libert, B., Yung, M.: Group Encryption: Non-Interactive Realization in the Standard Model. In: Matsui, M. (ed.) ASIACRYPT 2009. LNCS, vol. 5912, pp. 179–196. Springer, Heidelberg (2009)
11. Cramer, R., Shoup, V.: A Practical Public Key Cryptosystem Provably Secure Against Adaptive Chosen Ciphertext Attack. In: Krawczyk, H. (ed.) CRYPTO 1998. LNCS, vol. 1462, pp. 13–25. Springer, Heidelberg (1998)

12. Damgård, I., Groth, J.: Non-interactive and reusable non-malleable commitment schemes. In: STOC 2003, pp. 426–437 (2003)
13. Damgård, I., Nielsen, J.B.: Perfect Hiding and Perfect Binding Universally Composable Commitment Schemes with Constant Expansion Factor. In: Yung, M. (ed.) CRYPTO 2002. LNCS, vol. 2442, pp. 581–596. Springer, Heidelberg (2002)
14. Dolev, D., Dwork, C., Naor, M.: Non-malleable cryptography. In: STOC 1991, pp. 542–552. ACM Press (1991)
15. Groth, J.: Homomorphic trapdoor commitments to group elements. Cryptology ePrint Archive: Report 2009/007 (2009)
16. Groth, J., Sahai, A.: Efficient Non-Interactive Proof Systems for Bilinear Groups. In: Smart, N.P. (ed.) EUROCRYPT 2008. LNCS, vol. 4965, pp. 415–432. Springer, Heidelberg (2008)
17. Hofheinz, D., Kiltz, E.: Secure Hybrid Encryption from Weakened Key Encapsulation. In: Menezes, A. (ed.) CRYPTO 2007. LNCS, vol. 4622, pp. 553–571. Springer, Heidelberg (2007)
18. Hofheinz, D., Müller-Quade, J.: Universally Composable Commitments Using Random Oracles. In: Naor, M. (ed.) TCC 2004. LNCS, vol. 2951, pp. 58–76. Springer, Heidelberg (2004)
19. Katz, J.: Universally Composable Multi-party Computation Using Tamper-Proof Hardware. In: Naor, M. (ed.) EUROCRYPT 2007. LNCS, vol. 4515, pp. 115–128. Springer, Heidelberg (2007)
20. Lindell, Y.: Highly-Efficient Universally-Composable Commitments Based on the DDH Assumption. In: Paterson, K.G. (ed.) EUROCRYPT 2011. LNCS, vol. 6632, pp. 446–466. Springer, Heidelberg (2011)
21. Naor, M., Yung, M.: Universal one-way hash functions and their cryptographic applications. In: STOC 1989, pp. 33–43 (1989)
22. Nishimaki, R., Fujisaki, E., Tanaka, K.: Efficient Non-interactive Universally Composable String-Commitment Schemes. In: Pieprzyk, J., Zhang, F. (eds.) ProvSec 2009. LNCS, vol. 5848, pp. 3–18. Springer, Heidelberg (2009)
23. Okamoto, T., Uchiyama, S.: A New Public-Key Cryptosystem as Secure as Factoring. In: Nyberg, K. (ed.) EUROCRYPT 1998. LNCS, vol. 1403, pp. 308–318. Springer, Heidelberg (1998)
24. Page, D., Smart, N.P., Vercauteren, F.: A comparison of MNT curves and supersingular curves. Appl. Algebra Eng., Commun. Comput. 17(5), 379–392 (2006)
25. Paillier, P.: Public-Key Cryptosystems Based on Composite Degree Residuosity Classes. In: Stern, J. (ed.) EUROCRYPT 1999. LNCS, vol. 1592, pp. 223–238. Springer, Heidelberg (1999)
26. Pedersen, T.: Non-Interactive and Information-Theoretic Secure Verifiable Secret Sharing. In: Feigenbaum, J. (ed.) CRYPTO 1991. LNCS, vol. 576, pp. 129–140. Springer, Heidelberg (1992)
27. Peikert, C., Waters, B.: Lossy Trapdoor Functions and Their Applications. In: STOC 2008, pp. 187–196 (2008)
28. Shacham, H.: A Cramer-Shoup encryption scheme from the linear assumption and from progressively weaker linear variants. Cryptology ePrint Archive: Report 2007/074 (2007)
29. Shoup, V.: A proposal for the ISO standard for public-key encryption (version 2.1) (2001) (manuscript), http://shoup.net/
30. Zhu, H.: New Constructions for Reusable, Non-erasure and Universally Composable Commitments. In: Bao, F., Li, H., Wang, G. (eds.) ISPEC 2009. LNCS, vol. 5451, pp. 102–111. Springer, Heidelberg (2009)

Cryptography Secure against
Related-Key Attacks and Tampering

Mihir Bellare[1], David Cash[2], and Rachel Miller[3]

[1] Department of Computer Science & Engineering,
University of California San Diego
http://www.cs.ucsd.edu/users/mihir
[2] IBM T.J. Watson Research Center
http://www.cs.ucsd.edu/users/cdcash
[3] Department of Electrical Engineering and Computer Science, MIT
http://people.csail.mit.edu/rmiller/

Abstract. We show how to leverage the RKA (Related-Key Attack) security of blockciphers to provide RKA security for a suite of high-level primitives. This motivates a more general theoretical question, namely, when is it possible to transfer RKA security from a primitive P_1 to a primitive P_2? We provide both positive and negative answers. What emerges is a broad and high level picture of the way achievability of RKA security varies across primitives, showing, in particular, that some primitives resist "more" RKAs than others. A technical challenge was to achieve RKA security even for the practical classes of related-key deriving (RKD) functions underlying fault injection attacks that fail to satisfy the "claw-freeness" assumption made in previous works. We surmount this barrier for the first time based on the construction of PRGs that are not only RKA secure but satisfy a new notion of identity-collision-resistance.

1 Introduction

By fault injection [16,10] or other means, it is possible for an attacker to induce modifications in a hardware-stored key. When the attacker can subsequently observe the outcome of the cryptographic primitive under this modified key, we have a related-key attack (RKA) [5,19].

The key might be a signing key of a certificate authority or SSL server, a master key for an IBE system, or someone's decryption key. Once viewed merely as a way to study the security of blockciphers [9,27,5], RKAs emerge as real threats in practice and of interest for primitives beyond blockciphers.

It becomes of interest, accordingly, to achieve (provable) RKA security for popular high-level primitives. How can we do this?

PRACTICAL CONTRIBUTIONS. One approach to building RKA-secure high-level primitives is to do so directly, based, say, on standard number-theoretic assumptions. This, however, is likely to yield ad hoc results providing security against classes of attacks that are tied to the scheme algebra and may not reflect attacks in practice.

D.H. Lee and X. Wang (Eds.): ASIACRYPT 2011, LNCS 7073, pp. 486–503, 2011.

We take a different approach. RKA security is broadly accepted in practice as a requirement for blockciphers; in fact, AES was designed with the explicit goal of resisting RKAs. We currently have blockciphers whose resistance to RKAs is backed by fifteen years of cryptanalytic and design effort. We propose to leverage these efforts.

We will provide a general and systematic way to immunize any given instance of a high-level primitive against RKAs with the aid of an RKA-secure blockcipher, modeling the latter, for the purpose of proofs, as a RKA-secure PRF [5]. We will do this not only for symmetric primitives that are "close" to PRFs like symmetric encryption, but even for public-key encryption, signatures and identity-based encryption. Our methods are cheap, non-intrusive from the software perspective, and able to completely transfer all the RKA security of the blockcipher so that the high-level primitive resists attacks of the sort that arise in practice.

THEORETICAL CONTRIBUTIONS. The ability to transfer RKA security from PRFs to other primitives lead us to ask a broader theoretical question, namely, when is it possible to transfer RKA security from a primitive P_1 to a primitive P_2? We provide positive results across a diverse set of primitives, showing, for example, that RKA-secure IBE implies RKA-secure IND-CCA PKE. We also provide negative results showing, for example, that RKA-secure signatures do not imply RKA-secure PRFs.

All our results are expressed in a compact set-based framework. For any primitive P and class Φ of related-key deriving functions —functions the adversary is allowed to apply to the target key to get a related key— we define what it means for an instance of P to be Φ-RKA secure. We let $\mathbf{RKA}[P]$ be the set of all Φ such that there exists a Φ-RKA secure instance of primitive P. A transfer of RKA security from P_1 to P_2, expressed compactly as a set containment $\mathbf{RKA}[P_1] \subseteq \mathbf{RKA}[P_2]$, is a construction of a Φ-RKA secure instance of P_2 given both a normal-secure instance of P_2 and a Φ-RKA secure instance of P_1. Complementing this are non-containments of the form $\mathbf{RKA}[P_2] \not\subseteq \mathbf{RKA}[P_1]$, which show the existence of Φ such that there exists a Φ-RKA instance of P_2 yet *no* instance of P_1 can be Φ-RKA secure, indicating, in particular, that RKA security cannot be transferred from P_2 to P_1.

As Fig. 1 shows, we pick and then focus on a collection of central and representative cryptographic primitives. We then establish these containment and non-containment relations in a comprehensive and systematic way. What emerges is a broad and high level picture of the way achievability of RKA security varies across primitives, showing, in particular, that some primitives resist "more" RKAs than others.

We view these relations between $\mathbf{RKA}[P]$ sets as an analog of complexity theory, where we study relations between complexity classes in order to better understand the computational complexity of particular problems. Let us now look at all this more closely.

BACKGROUND. Related-key attacks were conceived in the context of blockciphers [9,27]. The first definitions were accordingly for PRFs [5]; for $F \colon \mathcal{K} \times \mathcal{D} \to \mathcal{R}$ they consider the game that picks a random challenge bit b and random target key $K \in \mathcal{K}$. For each $L \in \mathcal{K}$ the game picks a random function $G(L, \cdot): \mathcal{D} \to \mathcal{R}$, and next allows the adversary multiple queries to an oracle that given a pair (ϕ, x) with $\phi: \mathcal{K} \to \mathcal{K}$ and $x \in \mathcal{D}$ returns $F(\phi(K), x)$ if $b = 1$ and $G(\phi(K), x)$ if $b = 0$. They say that F is Φ-RKA secure, where Φ is a class of functions mapping $\mathcal{K}$ to $\mathcal{K}$, if the adversary has low advantage in predicting b when it is only allowed in its queries to use functions ϕ from Φ.

Let **RKA**[PRF] be the set of all Φ for which there exists a Φ-RKA secure PRF. Which Φ are in this set? All the evidence so far is that this question has no simple answer. Bellare and Kohno [5] gave natural examples of Φ not in **RKA**[PRF], showing the set is not universal. Membership of certain specific Φ in **RKA**[PRF] have been shown by explicit constructions of Φ-RKA PRFs, first under novel assumptions [28] and then under standard assumptions [3]. Beyond this we must rely on cryptanalysis. Modern blockciphers including AES are designed with the stated goal of RKA security. Accordingly we are willing to assume their Φ-RKA security —meaning that $\Phi \in$ **RKA**[PRF]— for whatever Φ cryptanalysts have been unable to find an attack.

BEYOND PRFS. Consideration of RKAs is now expanding to primitives beyond PRFs [20,2,22]. This is viewed partly as a natural extension of the questions on PRFs, and partly as motivated by the view of RKAs as a class of sidechannel attacks [19]. An RKA results when the attacker alters a hardware-stored key via tampering or fault injection [16,10] and subsequently observes the result of the evaluation of the primitive on the modified key. The concern that such attacks could be mounted on a signing key of a certificate authority or SSL server, a master key for an IBE system, or decryption keys of users makes achieving RKA security interesting for a wide range of high-level primitives.

DEFINITIONS. We focus on a small but representative set of primitives for which interesting variations in achievability of RKA security emerge. These are PRF (pseudorandom functions), Sig (Signatures), PKE-CCA (CCA-secure public-key encryption), SE-CCA (CCA-secure symmetric encryption), SE-CPA (CPA-secure symmetric encryption), IBE (identity-based encryption) and wPRF (weak PRFs [29]). We define what it means for an instance of P to be Φ-RKA secure for each P $\in$ {wPRF, IBE, Sig, SE-CCA, SE-CPA, PKE-CCA}. We follow the definitional paradigm of [5], but there are some delicate primitive-dependent choices that significantly affect the strength of the definitions and the challenge of achieving them (cf. Section 2). We let **RKA**[P] be the set of all Φ for which there exists a Φ-RKA secure instance of P. These sets are all non-trivial.

RELATIONS. We establish two kinds of relations between sets **RKA**[P$_1$] and **RKA**[P$_2$]:

- <u>Containment</u>: A proof that **RKA**[P$_1$] $\subseteq$ **RKA**[P$_2$], established by constructing a Φ-RKA secure instance of P$_2$ from a Φ-RKA secure instance of P$_1$, usually under the (minimal) additional assumption that one is given a normal-secure instance of P$_2$. Containments yield constructions of Φ-RKA secure instances of P$_2$.

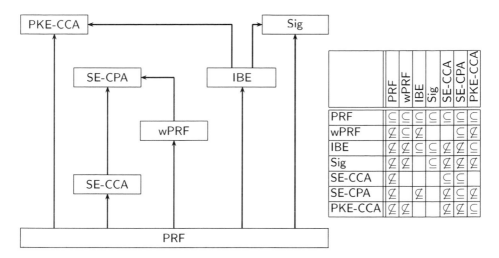

Fig. 1. Relations between RKA[P] classes. A containment $\mathbf{RKA}[\mathsf{P}_1] \subseteq \mathbf{RKA}[\mathsf{P}_2]$ is represented in the picture by an arrow $\mathsf{P}_1 \to \mathsf{P}_2$ and in the table by a "$\subseteq$" in the row P_1, column P_2 entry. A non-containment $\mathbf{RKA}[\mathsf{P}_1] \not\subseteq \mathbf{RKA}[\mathsf{P}_2]$ is represented in the table by a "$\not\subseteq$" in the row P_1, column P_2 entry. The picture does not show non-containments. The picture sometimes shows a redundant containment (for example the arrow $\mathsf{PRF} \to \mathsf{Sig}$ when there is already a path $\mathsf{PRF} \to \mathsf{IBE} \to \mathsf{Sig}$) because it corresponds to an interesting direct construction. A blank entry in the table means we do not know.

- <u>Non-containment:</u> A proof that $\mathbf{RKA}[\mathsf{P}_2] \not\subseteq \mathbf{RKA}[\mathsf{P}_1]$. Here we exhibit a particular Φ for which we (1) construct a Φ-RKA secure instance of P_1 under some reasonable assumption, and (2) show, via attack, that *any* instance of P_2 is Φ-RKA insecure.

We show that RKA-secure PRFs are powerful enablers of RKA-security: Given a Φ-RKA PRF and a normal-secure instance of P, we construct a Φ-RKA secure instance of P for all $\mathsf{P} \in \{\mathsf{wPRF}, \mathsf{IBE}, \mathsf{Sig}, \mathsf{SE\text{-}CCA}, \mathsf{SE\text{-}CPA}, \mathsf{PKE\text{-}CCA}\}$. This is represented by the string of containments in the first row of the table in Fig. 1. On the practical side, instantiating the PRF with a blockcipher yields a cheap way to immunize the other primitives against RKAs. On the theoretical side, instantiating the PRF with the construct of [3] yields Φ-RKA secure instances of the other primitives based on standard assumptions.

The separations shown in the first column of the table of Fig. 1, however, also show that RKA-PRFs are overkill: *all* the other primitives admit Φ-RKA secure instances for a Φ for which no Φ-RKA PRF exists. This leads one to ask whether there are alternative routes to RKA-secure constructions of beyond-PRF primitives.

We show that IBE is a particularly powerful starting point. We observe that Naor's transform preserves RKA-security, allowing us to turn a Φ-RKA secure IBE scheme into a Φ-RKA secure Sig scheme. Similarly, we show that the transform of Boneh, Canetti, Halevi and Katz (BCHK) [15] turns a Φ-RKA secure

IBE scheme into a Φ-RKA secure PKE-CCA scheme. What lends these transforms well to RKA-security is that they do not change the secret key. We also show that given a Φ-RKA secure wPRF we can build a Φ-RKA secure SE-CPA scheme. (A wPRF is like a PRF except that is only required to be secure on random inputs [29].) These results motivate finding new Φ-RKA secure IBE schemes and wPRFs.

As the table of Fig. 1 indicates, we show a number of other non-containments. Sig emerges as a very "RKA-resilient" primitive in the sense that it can be secure against strictly more RKAs than most other primitives. Some of the non-containments, such as **RKA**[PKE-CCA] $\not\subseteq$ **RKA**[SE-CPA] might seem odd; doesn't PKE always imply SE? What we are saying is that the trivial transformation of a PKE scheme to an SE one does not preserve RKA-security and, moreover, there are Φ for which *no transform exists* that can do this.

CLAWS OK. All previous constructions of Φ-RKA secure primitives [5,28,3,20,2,22,23] assume Φ is claw-free (distinct functions in ϕ disagree on all inputs) because it is hard to do the proofs otherwise, but the Φ underlying practical fault injection attacks are not claw-free, making it desirable to get constructions avoiding this assumption. For the first time, we are able to do this. In Section 2 we explain the technical difficulties and sketch our solution, which is based on the construction of a Φ-RKA PRG that has a novel property we call identity-collision-resistance (ICR), a variant of the collision-resistance property from [24].

RELATED WORK. The first theoretical treatment of RKAs was by Bellare and Kohno [5]; being inspired by blockciphers, the work addressed PRFs and PRPs. They showed examples of classes not in **RKA**[PRF], gave conditions on Φ for ideal ciphers to be Φ-RKA secure, and provided standard model constructs for some limited classes. Subsequently, constructions of Φ-RKA secure PRFs and PRPs for more interesting Φ were found, first under novel assumptions [28] and then under standard assumptions [3], and the results on ideal ciphers were extended in [1].

We are seeing growing interest in RKA security for primitives other than PRFs. Goldenberg and Liskov [20] study related-secret security of lower-level primitives, namely one-way functions, hardcore bits and pseudorandom generators. Applebaum, Harnik and Ishai [2] define RKA security for (randomized) symmetric encryption, gave several constructions achieving that definition for interesting Φ and then presented numerous applications. Connections with point obfuscation are made by Bitansky and Canetti [11].

Gennaro, Lysyanskaya, Malkin, Micali and Rabin [19] suggest that RKAs may arise by tampering. They show that one can achieve security when related keys are derived via arbitrary key modification, but assume an external trusted authority signs the original secret key and installs the signature on the device together with its own public key, the latter being "off limits" to the attacker. (Meaning, the related-key deriving functions may not modify them.) In our case, no such authority is assumed. The off-limit quantities are confined to pre-installed public parameters. No information that is a function of the parameters and the key is installed on the chip.

Ishai, Prabhakaran, Sahai and Wagner [25] are concerned with tampering of wires in the computation of a circuit while we are concerned with tampering with hardware-stored keys. Dziembowski, Pietrzak and Wichs [18] develop an information theoretic method for preventing tampering and show that a wide class of limited, but non-trivial, Φ can be achieved (unconditionally) for any so-called "interactive stateful system."

INDEPENDENT WORK. Interest in RKA security for higher-level primitives is evidenced by Goyal, O'Neill and Rao [22,23], who define correlated-input (CI) hash functions, show how to construct them from the q-DHI assumption based on Boneh-Boyen signatures [13,14] and the Dodis-Yampolskiy PRF [17], and apply this to get Φ-RKA secure signatures from q-DHI for a class Φ consisting of polynomials over a field of prime order. (They indicate their approach would also work for other primitives.) Their construction is similar to ours. Their definitions and results, unlike ours, are restricted to claw-free Φ. Also, we start from Φ-RKA-PRFs and thus get in-practice security for any class Φ for which blockciphers provide them, while they start from a number-theoretic assumption and get security for a specific class Φ, related to the scheme algebra. Their work and ours are concurrent and independent. (Ours was submitted to, and rejected from, Eurocrypt 2011, while theirs was submitted to, and accepted at, TCC 2011.)

Kalai, Kanukurthi and Sahai [26] provide encryption and signature schemes that protect against both tampering and leakage via the idea of key-updates that originated in forward-secure signatures [7]. They allow arbitrary tampering functions but only allow a bounded number of tampering queries within each time period. Their work and ours are again concurrent and independent.

2   Technical Approach

Before providing formal definitions, constructions and proofs of our many positive and negative results, we would like to illustrate one technical issue, namely the challenges created by Φ that are not claw-free and how we resolve them. For concreteness, our discussion is restricted to the design of Φ-RKA signatures based on Φ-RKA PRFs.

THE CLAW-FREENESS ASSUMPTION. All known constructions of Φ-RKA-secure primitives [5,28,3,20,2,22,23] are restricted to Φ that are *claw-free*. This means that any two distinct functions in Φ disagree on *all* inputs. This assumption is made for technical reasons; it seems hard to do simulations and proofs without it. Yet the assumption is undesirable, for many natural and practical classes of functions are *not* claw-free. For example, fault injection might be able to set a certain bit of the key to zero, and if Φ contains the corresponding function and the identity function then it is not claw-free. Any Φ that can set the key to a constant value is also not claw-free. Accordingly it is desirable to avoid this assumption. For the first time we are able to do so, via a new technical approach.

DEFINITIONS AND ISSUES. The degree to which claw-freeness is embedded in current approaches is made manifest by the fact that the very *definition* of Φ-RKA secure signatures of [22,23] assumes it and is unachievable without it. Let us take a closer look to see how.

The signature RKA-security game of [22,23] picks secret signing key sk and associated public verification key vk. It gives the adversary a signing oracle SIGN that takes m and $\phi \in \Phi$, and returns the signature of message m under key $\phi(sk)$. The adversary eventually outputs m, σ. Besides validity of m, σ under vk, winning requires that m be "new," meaning not "previously signed." The delicate question is, how do we define this? The choice of [22,23] is to disallow signing query id, m, where id is the identity function. But the adversary can easily define a function ϕ that is the identity on all but a negligible fraction of its inputs. A query ϕ, m is then valid since $\phi \neq$ id, but almost always returns the signature σ of m under sk, so the adversary can output m, σ and win. By assuming Φ is claw-free and contains id, [22,23] ensure that such a ϕ is not in Φ and the attack is ruled out.

Our altered definition of m being "new" is that there was no signing query ϕ, m with $\phi(sk) = sk$. This seems, indeed, the natural requirement, ruling out nothing more than that m was signed under sk.

We now have a much more general definition that is meaningful even for the non claw-free Φ that arise in practice, but it has a subtle feature that makes achieving it a challenge. Namely, *checking whether the adversary won apparently requires knowing sk* for we have to test whether or not $\phi(sk) = sk$. In the reduction proving security, we will be designing an adversary B attempting to distinguish "real" or "random" instances of some problem given an adversary A breaking the signature scheme; B will see if A won, declaring "real" if so and "random" otherwise. But B will be simulating A and will not know sk, so the difficulty is how it can test that A won.

OVERVIEW OF SOLUTION. We start from a Φ-RKA secure PRF $F: \mathcal{K} \times \mathcal{D} \rightarrow \mathcal{R}$ that has what we call a key fingerprint for the identity function. This is a relaxation of the notion of a key fingerprint of [3]. It consists of a vector $\mathbf{w}$ over $\mathcal{D}$ such that for all K and all $\phi \in \Phi$ with $\phi(K) \neq K$ there is some i such that $F(K, \mathbf{w}[i]) \neq F(\phi(K), \mathbf{w}[i])$. This allows statistical disambiguation of the original key K from other keys. Such fingerprints exist for the Φ-RKA PRFs of [3] and for blockciphers and are thus a mild assumption.

We now turn F into a PRG (Pseudorandom Generator) $\mathcal{G}$ that has two properties. First, it is Φ-RKA secure; this means the adversary has low advantage in determining the challenge bit b in the game that picks a random target key K and random function R, and then gives the adversary an oracle GEN that on input ϕ returns $\mathcal{G}(\phi(K))$ if $b = 1$ and $R(\phi(K))$ if $b = 0$. This is of course easily obtained from a Φ-RKA PRF. We call the new second property Φ-ICR (Identity-Collision-Resistant); this means that for a hidden key K, it is hard for the adversary to find $\phi \in \Phi$ such that $\phi(K) \neq K$ yet $\mathcal{G}(\phi(K)) = \mathcal{G}(K)$. At first it might seem this follows from Φ-RKA security but Proposition 2 shows it does not. However Proposition 3 shows how to build a PRG that is both Φ-RKA and Φ-ICR secure from a Φ-RKA PRF with an identity key fingerprint, without assuming Φ is claw-free.

We build our Φ-RKA secure signature scheme from this PRG $\mathcal{G}$ and a base (normal secure) signature scheme, as follows. The secret key of our new signature scheme is a key K for the PRG. The output of the PRG on input K, $\mathcal{G}(K)$, is used as randomness to run the key-generation algorithm $\mathcal{K}$ of the base signature scheme, yielding a public key pk which becomes the public key of our scheme, and the corresponding secret key which is discarded. (Recall the secret key of the new scheme is the PRG key K.) To sign a message m under K, run $\mathcal{G}$ on K to get coins for $\mathcal{K}$, run the latter with these coins to get pk, sk and finally sign m under sk with the base signature scheme. Verification is just as in the base signature scheme.

For the proof we must construct an adversary B breaking the Φ-RKA security of $\mathcal{G}$ given an adversary A breaking the Φ-RKA security of our signature scheme. B thinks of the key K underlying its game as the secret key for our signature scheme and then runs A. When A makes SIGN query ϕ, m, adversary B will call its GEN oracle on ϕ and use the result as coins for $\mathcal{K}$ to get a secret key under which it then signs m for A. Eventually A outputs a forgery attempt m, σ. The assumed security of the base signature scheme will make it unlikely that A's forgery is a winning one when GEN is underlain by a random function. So B would like to test if A's forgery was a winning one, outputting 1 if so and 0 otherwise, to win its game. The difficulty is that it cannot test this because, not knowing K, it cannot test whether or not A made a SIGN query ϕ, m with $\phi(K) = K$. The Φ-ICR property of $\mathcal{G}$ comes to the rescue, telling us that whether or not $\phi(K) = K$ may be determined by whether or not the outputs of $\mathcal{G}$ on these two inputs, which B *does* have, are the same.

This sketch still pushes under the rug several subtle details which are dealt with in the full proof of Theorem 5, to be found in the full version of this paper [4].

3   Preliminaries

NOTATION. For sets X, Y, Z let $\mathsf{Fun}(X, Y)$ be the set of all functions mapping X to Y, and let $\mathsf{FF}(X, Y, Z) = \mathsf{Fun}(X \times Y, Z)$. The empty string is denoted ε. If $\mathbf{v}$ is a vector then $|\mathbf{v}|$ denotes the number of its coordinates and $\mathbf{v}[i]$ denotes its i-th coordinate, meaning $\mathbf{v} = (\mathbf{v}[1], \dots, \mathbf{v}[|\mathbf{v}|])$. A (binary) string x is identified with a vector over $\{0, 1\}$ so that $|x|$ is its length and $x[i]$ is its i-th bit. If $a_1, \dots, a_n$ are strings then $a_1 \| \cdots \| a_n$ denotes their concatenation. If S is a set then $|S|$ denotes its size and $s \leftarrow_\$ S$ the operation of picking a random element of S and calling it s. We say that a real-valued function on the integers is *negligible* if it vanishes faster than the inverse of any polynomial.

ALGORITHMS. Unless otherwise indicated, an algorithm is PT (Polynomial Time) and may be randomized. An adversary is an algorithm. If A is an algorithm and $\mathbf{x}$ is a vector then $A(\mathbf{x})$ denotes the vector $(A(\mathbf{x}[1]), \dots, A(\mathbf{x}[|\mathbf{x}|]))$. By $y \leftarrow A(x_1, x_2, \dots; r)$ we denote the operation of running A on inputs $x_1, x_2, \dots$

and coins $r \in \{0, 1\}^*$. We denote by $y \leftarrow_{\$} A(x_1, x_2, \ldots)$ the operation of picking r at random and letting $y \leftarrow A(x_1, x_2, \ldots; r)$. We denote by $[A(x_1, x_2, \ldots)]$ the set of all possible outputs of A on inputs $x_1, x_2, \ldots$. We denote by $k \in \mathbb{N}$ the security parameter and by 1^k its unary encoding. It is assumed that the length of the output of any algorithm A depends only on the lengths of its inputs. In particular we can associate to single-input algorithm A its *output length* ℓ satisfying $|A(x)| = \ell(|x|)$ for all x. If A, B are algorithms then $A \| B$ denotes the algorithm that on any input x returns $A(x) \| B(x)$.

GAMES. Some of our definitions and proofs are expressed via code-based games [8]. Recall that such a game consists of an INITIALIZE procedure, procedures to respond to adversary oracle queries and a FINALIZE procedure. A game G is executed with an adversary A as follows. First, INITIALIZE executes on input 1^k and its output is the input to A. Then A executes, its oracle queries being answered by the corresponding procedures of G. When A terminates, its output becomes the input to the FINALIZE procedure. The output of the latter, denoted G^A, is called the output of the game. We let "$G^A \Rightarrow d$" denote the event that this game output takes value d. If FINALIZE is absent it is understood to be the identity function, so the game output is the adversary output. Boolean flags are assumed initialized to false.

4   Classes of RKDFs and RKA-PRFs

CLASSES OF RKDFs. In [5], a class Φ of related-key deriving functions (RKDFs) is a finite set of functions, all with the same domain and range. Our more general, asymptotic treatment requires extending this, in particular to allow the functions to depend on public parameters of the scheme. For us a *class* $\Phi = (\mathcal{P}, \mathcal{Q})$ of RKDFs, also called a RKA specification, is a pair of algorithms, the second deterministic. On input 1^k, parameter generation algorithm $\mathcal{P}$ produces parameters π. On input π, a key K and a description ϕ of an RKD function, the evaluation algorithm $\mathcal{Q}$ returns either a modified key or $\bot$. We require that for all ϕ, π, either $\mathcal{Q}(\pi, K, \phi) = \bot$ for all K or for no K. We let $\Phi_{\pi, \phi}(\cdot) = \mathcal{Q}(\pi, \cdot, \phi)$. We require that Φ always includes the identity function. (Formally, there is a special symbol id such that $\Phi_{\pi, \text{id}}(K) = K$ for all K, π. This is to ensure that Φ-RKA security always implies normal security.) We let ID be the class consisting of only the identity function, so that ID-RKA security will be normal security.

A scheme (regardless of the primitive) is a tuple $(\overline{\mathcal{P}}, \cdots)$ of algorithms, the first of which is a parameter generation algorithm that on input 1^k returns a string. If ℓ is the output length of $\mathcal{P}$, we say that $\Phi = (\mathcal{P}, \mathcal{Q})$ is *compatible* with the scheme if the string formed by the first $\ell(k)$ bits of the output of $\overline{\mathcal{P}}(1^k)$ is distributed identically to the output of $\mathcal{P}(1^k)$ for all $k \in \mathbb{N}$. This is done so that, in constructing one Φ-RKA primitive from another, we can extend the parameters of the constructed scheme beyond those of the original one without changing the class of RKDFs.

We say that $\Phi = (\mathcal{P}, \mathcal{Q})$ is *claw-free* if $\phi \neq \phi'$ implies $\mathcal{Q}(\pi, K, \phi) \neq \mathcal{Q}(\pi, K, \phi')$ (or both values are $\bot$) for all π, K. This property has been assumed almost

proc INITIALIZE // PRF	proc INITIALIZE // IDFP
$\pi \leftarrow_\$ \mathcal{P}(1^k)$; $K \leftarrow_\$ \mathcal{K}(\pi)$	$\pi \leftarrow_\$ \mathcal{P}(1^k)$
$b \leftarrow_\$ \{0,1\}$	$K \leftarrow_\$ \mathcal{K}(\pi)$
Return π	$\mathbf{w} \leftarrow_\$ \mathsf{IKfp}(\pi)$
	Return $\pi, \mathbf{w}$
proc FN(ϕ, x) // PRF	
$K' \leftarrow \Phi_{\pi,\phi}(K)$	proc FN(ϕ) // IDFP
If $K' = \bot$ then return $\bot$	$K' \leftarrow \Phi_{\pi,\phi}(K)$
If $b = 1$ then	If $(K' = \bot)$ then return $\bot$
$\quad T[K', x] \leftarrow \mathcal{F}(\pi, K', x)$	If $(K' \neq K)$ then
If $b = 0$ and $T[K', x] = \bot$ then	$\quad$ If $(\mathcal{F}(K', \mathbf{w}) = \mathcal{F}(K, \mathbf{w}))$ then
$\quad T[K', x] \leftarrow_\$ \mathsf{Rng}(\pi)$	$\quad\quad$ WIN $\leftarrow$ true
Return $T[K', x]$	Return $\mathcal{F}(K', \mathbf{w})$
proc FINALIZE(b') // PRF	proc FINALIZE() // IDFP
Return $(b = b')$	Return WIN

Fig. 2. Games defining Φ-RKA PRF security and Φ-IDFP security of function family $\mathcal{FF} = (\mathcal{P}, \mathcal{K}, \mathcal{F})$ having range $\mathsf{Rng}(\cdot)$

ubiquitously in previous work [5,28,20,3] because of the technical difficulties created by its absence, but its assumption is in fact quite restrictive since many natural classes do not have it. We are able to remove this assumption and provide constructs secure even for non-claw-free classes via new technical approaches. We let **CF** be the set of all Φ that are claw-free.

The class $\Phi^{\mathsf{const}} = (\mathcal{P}, \mathcal{Q}^{\mathsf{const}})$ of constant functions associated to class $\Phi = (\mathcal{P}, \mathcal{Q})$ is defined by $\Phi^{\mathsf{const}}_{\pi,a}(K) = a$ for all $K, a \in \{0,1\}^*$ and all π. The union $\Phi^1 \cup \Phi^2 = (\mathcal{P}, \mathcal{Q})$ of classes $\Phi^1 = (\mathcal{P}, \mathcal{Q}^1)$ and $\Phi^2 = (\mathcal{P}, \mathcal{Q}^2)$ is defined by having $\mathcal{Q}(\pi, K, \phi)$ parse ϕ as $i \parallel \phi^*$ for $i \in \{1,2\}$ and return $\mathcal{Q}^i(\pi, K, \phi^*)$.

DISCUSSION. In a non-asymptotic treatment, there is no formal line between "secure" and "insecure." This makes it unclear how to rigorously define the sets **RKA**[P]. Lead, accordingly, to pursue an asymptotic treatment, we introduce parameter dependence; this allows us to capture constructs in the literature [28,3] where RKDFs are defined over a group that is now parameter-dependent rather than fixed. (We note that even in the non-asymptotic case, a treatment like ours is needed to capture constructs in [28] relying on a RSA group defined by random primes. This issue is glossed over in [28].) A dividend of our treatment is a separation between an RKDF and its encoding, the latter being what an adversary actually queries, another issue glossed over in previous work.

FUNCTION FAMILIES. A function family $\mathcal{FF} = (\mathcal{P}, \mathcal{K}, \mathcal{F})$ consists of a parameter generator, a key generator, and an evaluator, the last deterministic. For each $k \in \mathbb{N}$ and $\pi \in [\mathcal{P}(1^k)]$, the scheme also defines PT decidable and sampleable sets $\mathsf{Dom}(\pi)$ and $\mathsf{Rng}(\pi)$ such that $\mathcal{F}(\pi, K, \cdot)$ maps elements of $\mathsf{Dom}(\pi)$ to $\mathsf{Rng}(\pi)$. We assume there are polynomials d, l, called the input and output lengths, respectively, such that $\mathsf{Dom}(\pi) \subseteq \{0,1\}^{d(k)}$ and $\mathsf{Rng}(\pi) \subseteq \{0,1\}^{l(k)}$. Unless otherwise indicated we assume $\mathsf{Rng}(\pi) = \{0,1\}^{l(k)}$ and $l(k) = \omega(\log(k))$ and $|\mathsf{Dom}(\pi)| \geq 2^k$ for all $\pi \in [\mathcal{P}(1^k)]$ and all $k \in \mathbb{N}$.

RKA-PRFs. Let $\mathcal{FF} = (\mathcal{P}, \mathcal{K}, \mathcal{F})$ be a function family as above. Game PRF of Fig. 2 is associated to $\mathcal{FF}$ and a RKA specification Φ that is compatible with $\mathcal{FF}$. Let $\mathbf{Adv}^{\text{prf-rka}}_{\mathcal{FF},A,\Phi}(k)$ equal $2\Pr[\text{PRF}^A \Rightarrow \text{true}] - 1$ when the game has input 1^k. We say $\mathcal{FF}$ is Φ-RKA secure if this advantage function is negligible.

IDENTITY KEY FINGERPRINTS. An identity key fingerprint function with vector length $v(\cdot)$ for $\mathcal{FF} = (\mathcal{P}, \mathcal{K}, \mathcal{F})$ is an algorithm IKfp that for every $\pi \in [\mathcal{P}(1^k)]$ and every $k \in \mathbb{N}$ returns, on input π, a $v(k)$-vector over $\text{Dom}(\pi)$ all of whose coordinates are distinct. Game IDFP of Fig. 2 is associated to $\mathcal{FF}$ and a RKA specification $\Phi = (\mathcal{P}, \mathcal{Q})$ that is compatible with $\mathcal{FF}$. Let $\mathbf{Adv}^{\text{idfp}}_{\mathcal{FF},A,\Phi}(k)$ equal $\Pr[\text{IDFP}^A \Rightarrow \text{true}]$ when the game has input 1^k. We say $\mathcal{FF}$ is Φ-IDFP secure if this advantage function is negligible.

The key fingerprint notion of [3] can be seen as allowing statistical disambiguation of any pair of keys. They showed that the Naor-Reingold PRF NR had such a fingerprint, but in general, it does not seem common. Interestingly, their own Φ-RKA PRFs, which build on NR, are not known to have such a fingerprint. Our relaxation can be seen as asking for computational disambiguation of the original key from other keys, and ends up being much easier to achieve. In particular, such fingerprints exist for the constructs of [3]. This is a consequence of something more general, namely that any Φ-RKA secure PRF with large enough range is Φ-IDFP secure if Φ is claw-free, using *any* point in the domain functioning as the fingerprint. This is formalized by Proposition 1 below, with a proof in [4]. Φ-IDFP security for the constructs of [3] follows as the Φ they use is claw-free.

Proposition 1. *Suppose Φ is claw-free and $\mathcal{FF}$ is a Φ-RKA secure PRF with associated domain $\text{Dom}(\cdot)$ and super-polynomial size range $\text{Rng}(\cdot)$. Let IKfp be any algorithm that on input π returns a 1-vector over $\text{Dom}(\pi)$. Then $\mathcal{FF}$ is Φ-IDFP secure.*

In practice Φ-IDFP security seems like a mild assumption even when Φ is not claw-free. A vector of a few, distinct domain points ought to be a suitable fingerprint for any practical blockcipher. This does not follow from a standard assumption on it such as PRF, but is consistent with properties assumed by cryptanalysts and can be proved in the ideal cipher model.

Φ-IDFP security of given Φ-RKA PRFs, even for non-claw-free Φ, will be important in the constructions underlying our containment results, and we make it a default assumption on a Φ-RKA PRF. The above shows that this is a mild and reasonable assumption.

RKA SETS. We say that an RKA specification $\Phi = (\mathcal{P}, \mathcal{Q})$ is achievable for the primitive PRF if there exists a Φ-RKA and Φ-IDFP secure PRF that is compatible with Φ. We let $\mathbf{RKA}[\text{PRF}]$ be the set of all Φ that are achievable for PRF.

WHAT CAN ATTACKS MODIFY? We view the system as a whole as having the following components: algorithms (code), parameters, public keys (if any) and secret keys. Of these, our convention is that only secret keys are subject to RKAs.

proc INITIALIZE // PRG
$\pi \leftarrow_\$ \mathcal{P}(1^k)$
$K \leftarrow_\$ \mathcal{K}(\pi) \; ; \; b \leftarrow_\$ \{0,1\}$
Return π

proc GEN(ϕ) // PRG
$K' \leftarrow \Phi_{\pi,\phi}(K)$
If $K' = \bot$ then return $\bot$
If $T[K'] = \bot$ then
    If $b = 1$ then $T[K'] \leftarrow \mathcal{G}(\pi, K')$
    Else $T[K'] \leftarrow_\$ \{0,1\}^{r(k)}$
Return $T[K']$

proc FINALIZE(b') // PRG
Return $(b = b')$

proc INITIALIZE // ICR
$\pi \leftarrow_\$ \mathcal{P}(1^k)$
$K \leftarrow_\$ \mathcal{K}(\pi) \; ; \; T_0 \leftarrow \mathcal{G}(\pi, K)$
Return π

proc GEN(ϕ) // ICR
$K' \leftarrow \Phi_{\pi,\phi}(K)$
If $K' = \bot$ then return $\bot$
$S \leftarrow \mathcal{G}(\pi, K')$
If $((S = T_0) \land (K \neq K'))$ then WIN $\leftarrow$ true
Return S

proc FINALIZE() // ICR
Return WIN

Fig. 3. Games defining Φ-RKA security and identity-collision-resistance for PRG $\mathcal{PRG} = (\mathcal{P}, \mathcal{K}, \mathcal{G}, r)$

This is not the only possible model, nor is it necessarily the most realistic if considering tampering attacks in practice, but it is a clear and interesting one with some justification. Parameters are systemwide, meaning fixed beforehand and independent of users, and may, in an implementation, be part of the algorithm code. Public keys are accompanied by certificates under a CA public key that is in the parameters, so if parameters are safe, so are public keys. This leaves secret keys as the main target. One consequence of this is that in a public key setting the attack is only on the holder of the secret key, meaning the signer for signatures and the receiver for encryption, while in the symmetric setting, both sender and receiver are under attack, making this setting more complicated.

We could consider attacks on public keys, but these are effectively attacks on parameters. Furthermore the only way for them to succeed is to modify the CA public key in the parameters in a rather special way, replacing it by some other key under which the attack produces signatures for the modified public key. "Natural" attacks caused by fault-injection are unlikely to do this, further supporting our convention of confining attacks to secret keys.

5   ICR PRGs: A Tool in Our Constructions

We will be using Φ-RKA PRFs to build Φ-RKA instances of many other primitives. An important technical difficulty will be to avoid assuming Φ is claw-free. A tool we introduce and use for this purpose is a Φ-RKA PRG satisfying a weak form of collision-resistance under RKA that we call Φ-ICR. In this section we define this primitive and show how to achieve it based on a Φ-RKA and Φ-IDFP secure PRF.

RKA PRGs. A PRG $\mathcal{PRG} = (\mathcal{P}, \mathcal{K}, \mathcal{G}, r)$ is specified by a parameter generation algorithm, a key generation algorithm, an evaluation algorithm and an output length $r(\cdot)$. Game PRG of Fig. 3 is associated to $\mathcal{PRG}$ and an RKA specification

Φ that is compatible with $\mathcal{PRG}$. Let $\mathbf{Adv}^{\mathrm{prg}}_{\mathcal{PRG},A,\Phi}(k) = 2\Pr[\mathrm{PRG}^A \Rightarrow \mathrm{true}] - 1$ when the game has input 1^k. We say $\mathcal{PRG}$ is Φ-RKA secure if this advantage function is negligible for all A.

We clarify that unlike a normal PRG [12], we don't require a Φ-RKA PRG to be length extending, meaning that outputs need not be longer than inputs. If one does want a length extending Φ-RKA PRG (we won't) one can get it by applying a normal-secure PRG to the output of a given Φ-RKA PRG.

ICR. We define and use a weak form of collision-resistance for PRGs which requires that the adversary be unable to find ϕ so that $\Phi_{\pi,\phi}(K) \neq K$ yet $\mathcal{G}(\Phi_{\pi,\phi}(K)) = \mathcal{G}(K)$. Game ICR of Fig. 3 is associated to $\mathcal{PRG}$ and a RKA specification Φ that is compatible with $\mathcal{PRG}$. Let $\mathbf{Adv}^{\mathrm{icr}}_{\mathcal{PRG},C,\Phi}(k)$ equal $2\Pr[\mathrm{ICR}^C \Rightarrow \mathrm{true}] - 1$ when the game has input 1^k. We say $\mathcal{PRG}$ is Φ-ICR (Identity-Collision-Resistant) secure if this advantage function is negligible.

DOES RKA SECURITY IMPLY ICR SECURITY? At first glance it would seem that if a PRG $\mathcal{PRG} = (\mathcal{P},\mathcal{K},\mathcal{G},r)$ is Φ-RKA secure then it is also Φ-ICR secure. Indeed, suppose an adversary has ϕ such that $\Phi_{\pi,\phi}(K) \neq K$ yet $\mathcal{G}(\Phi_{\pi,\phi}(K)) = \mathcal{G}(K)$. Let it query $R_0 \leftarrow \mathrm{GEN}(\mathrm{id})$ and $R_1 \leftarrow \mathrm{GEN}(\phi)$ and return 1 if $R_0 = R_1$ and 0 otherwise. In the real $(b = 1)$ case R_0, R_1 are equal but in the random $(b = 0)$ case they would appear very unlikely to be equal, so that that this strategy would appear to have high advantage in breaking the Φ-RKA security of $\mathcal{PRG}$. The catch is in our starting assumption, which made it appear that $\Phi_{\pi,\phi}(K) \neq K$ yet $\mathcal{G}(\Phi_{\pi,\phi}(K)) = \mathcal{G}(K)$ was an absolute fact, true both for $b = 0$ and $b = 1$. If $\Phi_{\pi,\phi}(K)$ and K are different in the real game but equal in the random game, the adversary sees an output collision in both cases and its advantage disappears. Can this actually happen? It can, and indeed the claim (that Φ-RKA security implies Φ-ICR security) is actually false:

Proposition 2. *Suppose there exists a normal-secure PRG $\overline{\mathcal{PRG}} = (\overline{\mathcal{P}},\overline{\mathcal{K}},\overline{\mathcal{G}},r)$ with $r(\cdot) = \omega(\log(\cdot))$. Then there exists a PRG $\mathcal{PRG} = (\overline{\mathcal{P}},\mathcal{K},\mathcal{G},r)$ and a class Φ such that $\mathcal{PRG}$ is Φ-RKA secure but $\mathcal{PRG}$ is not Φ-ICR secure.*

A proof is in [4]. Briefly, the constructed PRG $\mathcal{PRG}$ adds a redundant bit to the seed of $\overline{\mathcal{PRG}}$ so that seeds differing only in their first bits yield the same outputs, meaning create non-trivial collisions. But Φ is crafted so that that its members deviate from the identity function only in the real game, so that output collisions appear just as often in both cases but in the real game they are non-trivial while in the random game they are trivial.

CONSTRUCTION. We saw above that not all Φ-RKA PRGs are Φ-ICR secure. Our containments will rely crucially on ones that are. We obtain them from Φ-RKA PRFs that have key fingerprints for the identity function:

Proposition 3. *Let $\mathcal{FF} = (\mathcal{P},\mathcal{K},\mathcal{F})$ be a Φ-RKA PRF with output length l. Let IKfp be a Φ-IDFP secure identity key fingerprint function for $\mathcal{FF}$ with vector* proc INITIALIZE // Sig

$\pi \leftarrow_\$ \mathcal{P}(1^k)$; $M \leftarrow \emptyset$
$(vk, sk) \leftarrow_\$ \mathcal{K}(\pi)$
Return (π, vk)

proc SIGN(ϕ, m) // Sig

$sk' \leftarrow \Phi_{\pi,\phi}(sk)$
If $sk' = \bot$ then return $\bot$
If $sk' = sk$ then $M \leftarrow M \cup \{m\}$
Return $\sigma \leftarrow_\$ \mathcal{S}(\pi, sk', m)$

proc FINALIZE(m, σ) // Sig

Return $((\mathcal{V}(\pi, vk, m, \sigma) = 1) \wedge (m \notin M))$

proc FINALIZE(b') // IBE

Return $(b = b')$

proc INITIALIZE // IBE

$\pi \leftarrow_\$ \mathcal{P}(1^k)$; $(mpk, msk) \leftarrow_\$ \mathcal{M}(\pi)$
$b \leftarrow_\$ \{0, 1\}$; $id^* \leftarrow \bot$; $S \leftarrow \emptyset$
Return (π, mpk)

proc KD(ϕ, id) // IBE

$msk' \leftarrow \Phi_{\pi,\phi}(msk)$
If $msk' = \bot$ then return $\bot$
If $msk' = msk$ then $S \leftarrow S \cup \{id\}$
If $(msk' = msk) \wedge (id = id^*)$ then return $\bot$
Return $dk \leftarrow_\$ \mathcal{K}(\pi, mpk, msk', id)$

proc LR(id, m_0, m_1) // IBE

If $|m_0| \neq |m_1|$ then return $\bot$
$id^* \leftarrow id$; If $id^* \in S$ then return $\bot$
Return $C \leftarrow_\$ \mathcal{E}(\pi, mpk, id, m_b)$

proc FINALIZE(b') // IBE

Return $((b = b') \wedge (id^* \notin S))$

Fig. 4. Games defining Φ-RKA security for primitives Sig, IBE

length v. Let $r = lv$ and let $\overline{\mathcal{K}}$, on input $\pi \| \mathbf{w}$, return $\mathcal{K}(\pi)$. Define PRG $\mathcal{PRG} = (\mathcal{P} \| \mathsf{IKfp}, \overline{\mathcal{K}}, \mathcal{G}, r)$ via

$$\mathcal{G}(\pi \| \mathbf{w}, K) = \mathcal{F}(\pi, K, \mathbf{w}[1]) \| \cdots \| \mathcal{F}(\pi, K, \mathbf{w}[|\mathbf{w}|]) \ .$$

Then $\mathcal{PRG}$ is Φ-RKA secure and Φ-ICR secure.

6 Relations

We first present a containment and a non-containment related to Sig. Then we turn to IBE-related results. Other results can be found in [4].

SIGNATURES. A signature scheme $\mathcal{DS} = (\mathcal{P}, \mathcal{K}, \mathcal{S}, \mathcal{V})$ is specified as usual by its parameter generation, key generation, signing and verifying algorithms. Game Sig of Fig. 4 is associated to $\mathcal{DS}$ and an RKA specification Φ that is compatible with $\mathcal{DS}$. Let $\mathbf{Adv}_{\mathcal{DS},A,\Phi}^{\text{sig-rka}}(k) = \Pr[\text{Sig}^A \Rightarrow \text{true}]$ when the game has input 1^k. We say $\mathcal{DS}$ is Φ-RKA secure if this advantage function is negligible. Normal security of a signature scheme is recovered by considering Φ that contains only the identity function. One feature of the definition worth highlighting is the way we decide which messages are not legitimate forgeries. They are the ones signed with the real key sk, which means that oracle SIGN needs to check when a related key equals the real one and record the corresponding message, which is a source of challenges in reduction-based proofs.

ATTACKS. In [4] we present an attack, adapted from [6,19], that shows that there are some (quite simple) Φ such that *no* signature scheme is Φ-RKA secure, meaning $\Phi \notin \mathbf{RKA}[\text{Sig}]$. This indicates that the set $\mathbf{RKA}[\text{Sig}]$ is non-trivial. Similar attacks can be presented for other primitives.

FROM Φ-RKA PRGs TO Φ-RKA SIGNATURES. We will prove containments of the form **RKA**[PRF] $\subseteq$ **RKA**[P] by proving **RKA**[PRG] $\subseteq$ **RKA**[P] and exploiting the fact that **RKA**[PRF] $\subseteq$ **RKA**[PRG].

We start with a Φ-RKA PRG $\mathcal{PRG} = (\mathcal{P}, \mathcal{K}, \mathcal{G}, r)$ and a normal-secure signature scheme $\overline{\mathcal{DS}} = (\overline{\mathcal{P}}, \overline{\mathcal{K}}, \overline{\mathcal{S}}, \overline{\mathcal{V}})$ such that $r(\cdot)$ is the number of coins used by $\overline{\mathcal{K}}$. We now build another signature scheme $\mathcal{DS} = (\mathcal{P} \,\|\, \overline{\mathcal{P}}, \mathcal{K}', \mathcal{S}, \mathcal{V})$ as follows:

1. **Parameters:** Parameters for $\mathcal{DS}$ are the concatenation $\pi \,\|\, \overline{\pi}$ of independently generated parameters for $\mathcal{PRG}$ and $\overline{\mathcal{DS}}$.

2. **Keys:** Pick a random seed $K \leftarrow^{\$} \mathcal{K}(\pi)$ and let $(\overline{vk}, \overline{sk}) \leftarrow \overline{\mathcal{K}}(\overline{\pi}; \mathcal{G}(K))$ be the result of generating verifying and signing keys with coins $\mathcal{G}(K)$. The new signing key is K and the verifying key remains $\overline{vk}$. (Key $\overline{sk}$ is discarded.)

3. **Signing:** To sign message m with signing key K, recompute $(\overline{vk}, \overline{sk}) \leftarrow \overline{\mathcal{K}}(\overline{\pi}; \mathcal{G}(K))$ and then sign m under $\overline{\mathcal{S}}$ using $\overline{sk}$.

4. **Verifying:** Verify that σ is a base scheme signature of m under $\overline{vk}$ using $\overline{\mathcal{V}}$.

Signature scheme $\mathcal{DS}$ remains compatible with Φ since the parameters of $\mathcal{PRG}$ prefix those of $\mathcal{DS}$.

We want $\mathcal{DS}$ to inherit the Φ-RKA security of $\mathcal{PRG}$. In fact we will show more, namely that $\mathcal{DS}$ is $(\Phi \cup \Phi_c)$-RKA secure where Φ_c is the class of constant RKDFs associated to Φ. The intuition is deceptively simple. A signing query ϕ, m of an adversary A attacking $\mathcal{DS}$ results in a signature of m under what is effectively a fresh signing key, since it is generated using coins $\mathcal{G}(\phi(K))$ that are computationally independent of $\mathcal{G}(K)$ due to the assumed Φ-RKA security of the PRG. These can accordingly be simulated without access to K. On the other hand, signing queries in which ϕ is a constant function may be directly simulated. The first difficulty is that the adversary attacking the Φ-RKA security of $\mathcal{PRG}$ that we must build needs to know when A succeeds, and for this it needs to know when a derived seed equals the real one, and it is unclear how to do this without knowing the real seed. The second difficulty is that a queried constant might equal the key. We take an incremental approach to showing how these difficulties are resolved, beginning by assuming Φ is claw-free, which makes the first difficulty vanish:

Theorem 4. *Let signature scheme* $\mathcal{DS} = (\mathcal{P} \,\|\, \overline{\mathcal{P}}, \mathcal{K}', \mathcal{S}, \mathcal{V})$ *be constructed as above from* Φ-*RKA PRG* $\mathcal{PRG} = (\mathcal{P}, \mathcal{K}, \mathcal{G}, r)$ *and normal-secure signature scheme* $\overline{\mathcal{DS}} = (\overline{\mathcal{P}}, \overline{\mathcal{K}}, \overline{\mathcal{S}}, \overline{\mathcal{V}})$ *and assume* Φ *is claw-free. Then* $\mathcal{DS}$ *is* $(\Phi \cup \Phi_c)$-*RKA secure.*

A proof of Theorem 4 is in [4], and the intuition was discussed in Section 2. This result, however, is weaker than we would like, for, as we have already said, many interesting classes are not claw-free. Also, this result fails to prove **RKA**[PRF] $\subseteq$ **RKA**[Sig] since the first set may contain Φ that are not claw-free. To address this we show that the claw-freeness assumption on Φ can be replaced by the assumption that $\mathcal{PRG}$ is Φ-ICR secure:

Theorem 5. *Let signature scheme* $\mathcal{DS} = (\mathcal{P} \,\|\, \overline{\mathcal{P}}, \mathcal{K}', \mathcal{S}, \mathcal{V})$ *be constructed as above from* Φ-*RKA secure and* Φ-*ICR secure PRG* $\mathcal{PRG} = (\mathcal{P}, \mathcal{K}, \mathcal{G}, r)$ *and* normal-secure signature scheme $\overline{\mathcal{DS}} = (\overline{\mathcal{P}}, \overline{\mathcal{K}}, \overline{\mathcal{S}}, \overline{\mathcal{V}})$. Then $\mathcal{DS}$ is $(\Phi \cup \Phi_c)$-RKA secure.

A proof of Theorem 5 is in [4]. Proposition 3 says we can get the PRGs we want from Φ-RKA PRFs so Theorem 5 establishes the containment **RKA**[PRF] $\subseteq$ **RKA**[Sig]. (Theorem 4 only established **RKA**[PRF] $\cap$ **CF** $\subseteq$ **RKA**[Sig] $\cap$ **CF**.)

Our construction has the advantage that the verification process as well as the form of the signatures and public key are unchanged. This means it has minimal impact on software, making it easier to deploy than a totally new scheme. Signing in the scheme now involves evaluation of a Φ-RKA-PRG but this can be made cheap via an AES-based instantiation. However, signing also involves running the key-generation algorithm $\overline{\mathcal{K}}$ of the base scheme which might be expensive.

This construction also meets a stronger notion of Φ-RKA security where the adversary cannot even forge a signature relative to the public keys associated with the derived secret keys. We elaborate on this in [4].

Some base signature schemes lend themselves naturally and directly to immunization against RKAs via Φ-RKA PRFs. This is true for the binary-tree, one-time signature based scheme discussed in [21], where the secret key is already that of a PRF. If the latter is Φ-RKA secure we can show the signature scheme (unmodified) is too, and moreover also meets the strong version of the definition alluded to above. See [4].

SEPARATING Φ-RKA PRFS FROM Φ-RKA SIGNATURES. Having just shown that **RKA**[PRF] $\subseteq$ **RKA**[Sig] it is natural to ask whether the converse is true as well, meaning whether the sets are equal. The answer is no, so **RKA**[Sig] $\not\subseteq$ **RKA**[PRF]. The interpretation is that there exist Φ such that there exist Φ-RKA secure signatures, but there are *no* Φ-RKA PRFs. An example is when $\Phi = \Phi_c$ is the set of constant functions. Theorem 4 implies that there exists a Φ_c-RKA secure signature scheme by setting $\Phi = \emptyset$ in the theorem, so that $\mathcal{PRG}$ need only be a normal-secure PRG. But attacks from [5] show that no PRF can be Φ_c-RKA secure. Thus, this separation is quite easily obtained. In [4] we present others which are more interesting. This separation motivates finding other avenues to Φ-RKA signatures. Below we will show that IBE is one such avenue.

IBE. Our specification of an IBE scheme $\mathcal{IBE} = (\mathcal{P}, \mathcal{M}, \mathcal{K}, \mathcal{E}, \mathcal{D})$ adds a parameter generation algorithm $\mathcal{P}$ that given 1^k returns parameters π on which the masterkey generation algorithm $\mathcal{M}$ runs to produce the master public key mpk and master secret key msk. The rest is as usual except that algorithms get π as an additional input. Game IBE of Fig. 4 is associated to $\mathcal{IBE}$ and an RKA specification $\Phi = (\mathcal{P}, \mathcal{Q})$ that is compatible with $\mathcal{IBE}$. An adversary is allowed only one query to LR. Let $\mathbf{Adv}^{\text{ibe-rka}}_{\mathcal{IBE}, A, \Phi}(k)$ equal $2\Pr[\text{IBE}^A \Rightarrow \text{true}] - 1$ when the game has input 1^k. We say $\mathcal{IBE}$ is Φ-RKA secure if this advantage function is negligible. Here the feature of the definition worth remarking on is that the adversary loses if it ever issues a query to KD that contains the challenge identity *and* derives the same master secret key. In [4] we show (1) that the standard Naor transform preserves RKA security and thus **RKA**[IBE] $\subseteq$ **RKA**[Sig], and (2) that the BCHK transform [15] preserves RKA security and thus **RKA**[IBE] $\subseteq$ **RKA**[PKE-CCA].

OTHER RELATIONS. The remaining results and definitions from Fig. 1 are presented in [4].

Acknowledgments. We thank Susan Thomson, Martijn Stam, Pooya Farshim and the Asiacrypt 2011 reviewers for their comments and corrections. Mihir Bellare was supported in part by NSF grants CCF-0915675 and CNS-0904380. Work done while David Cash was at UCSD, supported in part by NSF grant CCF-0915675. Rachel Miller was supported in part by a DOD NDSEG Graduate Fellowship and NSF grant CCF-1018064.

References

1. Albrecht, M., Farshim, P., Paterson, K., Watson, G.: On cipher-dependent related-key attacks in the ideal-cipher model. Cryptology ePrint Archive, Report 2011/213 (2011), http://eprint.iacr.org/
2. Applebaum, B., Harnik, D., Ishai, Y.: Semantic security under related-key attacks and applications. In: Yao, A.C.-C. (ed.) ICS 2011. Tsinghua University Press (2011)
3. Bellare, M., Cash, D.: Pseudorandom Functions and Permutations Provably Secure Against Related-Key Attacks. In: Rabin, T. (ed.) CRYPTO 2010. LNCS, vol. 6223, pp. 666–684. Springer, Heidelberg (2010)
4. Bellare, M., Cash, D., Miller, R.: Cryptography secure against related-key attacks. Cryptology ePrint Archive, Report 2011/252, Full version of this paper (2011), http://eprint.iacr.org/2011/252
5. Bellare, M., Kohno, T.: A Theoretical Treatment of Related-Key Attacks: RKA-PRPs, RKA-PRFs, and Applications. In: Biham, E. (ed.) EUROCRYPT 2003. LNCS, vol. 2656, pp. 491–506. Springer, Heidelberg (2003)
6. Bellare, M., Kohno, T.: Hash Function Balance and its Impact on Birthday Attacks. In: Cachin, C., Camenisch, J. (eds.) EUROCRYPT 2004. LNCS, vol. 3027, pp. 401–418. Springer, Heidelberg (2004)
7. Bellare, M., Miner, S.K.: A Forward-Secure Digital Signature Scheme. In: Wiener, M.J. (ed.) CRYPTO 1999. LNCS, vol. 1666, pp. 431–448. Springer, Heidelberg (1999)
8. Bellare, M., Rogaway, P.: The Security of Triple Encryption and a Framework for Code-Based Game-Playing Proofs. In: Vaudenay, S. (ed.) EUROCRYPT 2006. LNCS, vol. 4004, pp. 409–426. Springer, Heidelberg (2006)
9. Biham, E.: New Types of Cryptanalytic Attacks Using Related Keys (Extended Abstract). In: Helleseth, T. (ed.) EUROCRYPT 1993. LNCS, vol. 765, pp. 398–409. Springer, Heidelberg (1994)
10. Biham, E., Shamir, A.: Differential Fault Analysis of Secret Key Cryptosystems. In: Kaliski Jr., B.S. (ed.) CRYPTO 1997. LNCS, vol. 1294, pp. 513–525. Springer, Heidelberg (1997)
11. Bitansky, N., Canetti, R.: On Strong Simulation and Composable Point Obfuscation. In: Rabin, T. (ed.) CRYPTO 2010. LNCS, vol. 6223, pp. 520–537. Springer, Heidelberg (2010)
12. Blum, M., Micali, S.: How to generate cryptographically strong sequences of pseudorandom bits. SIAM Journal on Computing 13(4), 850–864 (1984)
13. Boneh, D., Boyen, X.: Short Signatures Without Random Oracles. In: Cachin, C., Camenisch, J. (eds.) EUROCRYPT 2004. LNCS, vol. 3027, pp. 56–73. Springer, Heidelberg (2004)

14. Boneh, D., Boyen, X.: Short signatures without random oracles and the SDH assumption in bilinear groups. Journal of Cryptology 21(2), 149–177 (2008)
15. Boneh, D., Canetti, R., Halevi, S., Katz, J.: Chosen-ciphertext security from identity-based encryption. SIAM Journal on Computing 36(5), 915–942 (2006)
16. Boneh, D., DeMillo, R.A., Lipton, R.J.: On the Importance of Checking Cryptographic Protocols for Faults (Extended Abstract). In: Fumy, W. (ed.) EUROCRYPT 1997. LNCS, vol. 1233, pp. 37–51. Springer, Heidelberg (1997)
17. Dodis, Y., Yampolskiy, A.: A Verifiable Random Function with Short Proofs and Keys. In: Vaudenay, S. (ed.) PKC 2005. LNCS, vol. 3386, pp. 416–431. Springer, Heidelberg (2005)
18. Dziembowski, S., Pietrzak, K., Wichs, D.: Non-malleable codes. In: Yao, A.C.-C. (ed.) ICS 2010. Tsinghua University Press (2010)
19. Gennaro, R., Lysyanskaya, A., Malkin, T., Micali, S., Rabin, T.: Algorithmic Tamper-Proof (ATP) Security: Theoretical Foundations for Security against Hardware Tampering. In: Naor, M. (ed.) TCC 2004. LNCS, vol. 2951, pp. 258–277. Springer, Heidelberg (2004)
20. Goldenberg, D., Liskov, M.: On Related-Secret Pseudorandomness. In: Micciancio, D. (ed.) TCC 2010. LNCS, vol. 5978, pp. 255–272. Springer, Heidelberg (2010)
21. Goldreich, O.: Foundations of Cryptography: Basic Applications, vol. 2. Cambridge University Press, Cambridge (2004)
22. Goyal, V., O'Neill, A., Rao, V.: Correlated-Input Secure Hash Functions. In: Ishai, Y. (ed.) TCC 2011. LNCS, vol. 6597, pp. 182–200. Springer, Heidelberg (2011)
23. Goyal, V., O'Neill, A., Rao, V.: Correlated-input secure hash functions. Cryptology ePrint Archive, Report 2011/233, Full version of [22] (2011), http://eprint.iacr.org/
24. Halevi, S., Krawczyk, H.: Security under key-dependent inputs. In: Ning, P., di Vimercati, S.D.C., Syverson, P.F. (eds.) ACM CCS 2007, pp. 466–475. ACM Press (October 2007)
25. Ishai, Y., Prabhakaran, M., Sahai, A., Wagner, D.: Private Circuits II: Keeping Secrets in Tamperable Circuits. In: Vaudenay, S. (ed.) EUROCRYPT 2006. LNCS, vol. 4004, pp. 308–327. Springer, Heidelberg (2006)
26. Kalai, Y.T., Kanukurthi, B., Sahai, A.: Cryptography with Tamperable and Leaky Memory. In: Rogaway, P. (ed.) CRYPTO 2011. LNCS, vol. 6841, pp. 373–390. Springer, Heidelberg (2011)
27. Knudsen, L.R.: Cryptanalysis of LOKI91. In: Zheng, Y., Seberry, J. (eds.) AUSCRYPT 1992. LNCS, vol. 718, pp. 196–208. Springer, Heidelberg (1993)
28. Lucks, S.: Ciphers Secure against Related-Key Attacks. In: Roy, B.K., Meier, W. (eds.) FSE 2004. LNCS, vol. 3017, pp. 359–370. Springer, Heidelberg (2004)
29. Naor, M., Reingold, O.: Synthesizers and their application to the parallel construction of pseudo-random functions. J. Comput. Syst. Sci. 58(2), 336–375 (1999)

Counting Points on Genus 2 Curves
with Real Multiplication

Pierrick Gaudry[1], David Kohel[2], and Benjamin Smith[3]

[1] LORIA, CNRS / INRIA / Nancy Université,
Campus Scientifique, BP 239,
54500 Vandoeuvre lès Nancy, France
[2] Université de la Méditerranée,
Institut de Mathématiques de Luminy,
163, avenue de Luminy, Case 907,
13288 Marseille Cedex 9, France
[3] INRIA Saclay–Île-de-France,
Laboratoire d'Informatique de l'École polytechnique (LIX),
91128 Palaiseau Cedex, France

Abstract. We present an accelerated Schoof-type point-counting algorithm for curves of genus 2 equipped with an efficiently computable real multiplication endomorphism. Our new algorithm reduces the complexity of genus 2 point counting over a finite field $\mathbb{F}_q$ of large characteristic from $\tilde{O}(\log^8 q)$ to $\tilde{O}(\log^5 q)$. Using our algorithm we compute a 256-bit prime-order Jacobian, suitable for cryptographic applications, and also the order of a 1024-bit Jacobian.

1 Introduction

Cryptosystems based on curves of genus 2 offer per-bit security and efficiency comparable with elliptic curve cryptosystems. However, many of the computational problems related to creating secure instances of genus 2 cryptosystems are considerably more difficult than their elliptic curve analogues. Point counting—or, from a cryptographic point of view, computing the cardinality of a cryptographic group—offers a good example of this disparity, at least for curves defined over large prime fields. Indeed, while computing the order of a cryptographic-sized elliptic curve with the Schoof–Elkies–Atkin algorithm is now routine, computing the order of a comparable genus 2 Jacobian requires a significant computational effort [8,10].

In this article we describe a number of improvements to the classical Schoof–Pila algorithm for genus 2 curves with explicit and efficient real multiplication (RM). For explicit RM curves over $\mathbb{F}_p$, we reduce the complexity of Schoof–Pila from $\tilde{O}(\log^8 p)$ to $\tilde{O}(\log^5 p)$. We applied a first implementation of our algorithms to find prime-order Jacobians over 128-bit fields (comparable to prime-order elliptic curves over 256-bit fields, and therefore suitable for contemporary cryptographic applications). Going further, we were able to compute the order of an RM Jacobian over a 512-bit prime field, far beyond the cryptographic range.

D.H. Lee and X. Wang (Eds.): ASIACRYPT 2011, LNCS 7073, pp. 504–519, 2011.

(For comparison, the previous record computation in genus 2 was over a 128-bit field.)

While these RM curves are special, they are not "too special". Every ordinary genus 2 Jacobian over a finite field has RM; our special requirement is that this RM be known in advance and efficiently computable. The moduli of curves with RM by a fixed ring form 2-dimensional subvarieties (Humbert surfaces) in the 3-dimensional moduli space of all genus 2 curves. We can generate random curves with the specified RM by choosing random points on an explicit model of the corresponding Humbert surface [11]. In comparison with elliptic curves, for which the moduli space is one-dimensional, this still gives an additional degree of freedom in the random curve selection. To generate random curves with efficiently computable RM, we choose random curves from some known one and two-parameter families (see §4).

Curves with efficiently computable RM have an additional benefit in cryptography: the efficient endomorphism can be used to accelerate scalar multiplication on the Jacobian, yielding faster encryption and decryption [12,16,20]. The RM formulæ are also compatible with fast arithmetic based on theta functions [7].

2   Conventional Point Counting for Genus 2 Curves

Let $\mathcal{C}$ be a curve of genus 2 over a finite field $\mathbb{F}_q$ of odd characteristic, defined by an affine model $y^2 = f(x)$, where f is a squarefree polynomial of degree 5 or 6 over $\mathbb{F}_q$. Let $J_{\mathcal{C}}$ be the Jacobian of $\mathcal{C}$; we assume $J_{\mathcal{C}}$ is ordinary and absolutely simple. Points on $J_{\mathcal{C}}$ correspond to degree-0 divisor classes on $\mathcal{C}$; we use the Mumford representation for divisor classes together with the usual Cantor-style composition and reduction algorithms for divisor class arithmetic [6,3]. Multiplication by ℓ on $J_{\mathcal{C}}$ is denoted by $[\ell]$, and its kernel by $J_{\mathcal{C}}[\ell]$. More generally, if ϕ is an endomorphism of $J_{\mathcal{C}}$ then $J_{\mathcal{C}}[\phi] = \ker(\phi)$, and if S is a set of endomorphisms then $J_{\mathcal{C}}[S]$ denotes the intersection of $\ker(\phi)$ for ϕ in S.

2.1   The Characteristic Polynomial of Frobenius

We let π denote the Frobenius endomorphism of $J_{\mathcal{C}}$, with Rosati dual $\pi^{\dagger}$ (so $\pi\pi^{\dagger} = [q]$). The characteristic polynomial of π has the form

$$\chi(T) = T^4 - s_1 T^3 + (s_2 + 2q)\, T^2 - q s_1 T + q^2, \tag{1}$$

where s_1 and s_2 are integers (our s_2 is a translation of the standard definition). The polynomial $\chi(T)$ determines the cardinality of $J_{\mathcal{C}}(\mathbb{F}_{q^k})$ for all k: in particular, $\#J_{\mathcal{C}}(\mathbb{F}_q) = \chi(1)$. Determining $\chi(T)$ is called the *point counting problem*.

The polynomial $\chi(T)$ is a *Weil polynomial*: all of its complex roots lie on the circle $|z| = \sqrt{q}$. This implies the Weil bounds

$$|s_1| \leq 4\sqrt{q} \quad \text{and} \quad |s_2| \leq 4q.$$

The possible values of (s_1, s_2) do not fill the whole rectangle specified by the Weil bounds: Rück [18, Theorem 1.1] shows that s_1 and s_2 satisfy

$$s_1^2 - 4s_2 > 0 \quad \text{and} \quad s_2 + 4q > 2|s_1|\sqrt{q}.$$

The possible values of (s_1, s_2) therefore lie in the following domain:

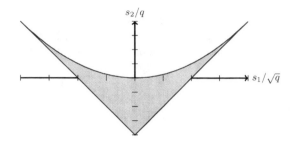

2.2   The Classical Schoof–Pila Algorithm for Genus 2 Curves

The objective of point counting is to compute $\chi(T)$, or equivalently the tuple of integers (s_1, s_2). When the characteristic of $\mathbb{F}_q$ is large, the conventional approach is to apply the Schoof–Pila algorithm as far as is practical, before passing to a baby-step giant-step algorithm if necessary (see §2.5).

The strategy of Schoof's algorithm and its generalizations is to compute the polynomials $\chi_\ell(T) = \chi(T) \bmod (\ell)$ for sufficiently many primes (or prime powers) ℓ to reconstruct $\chi(T)$ using the Chinese Remainder Theorem (CRT). Since $\chi_\ell(T)$ is the characteristic polynomial of π restricted to $J_C[\ell]$ (see [17, Proposition 2.1]), we have

$$\chi_\ell(\pi)(D) = 0 \text{ for all } D \text{ in } J_C[\ell].$$

Conversely, to compute $\chi_\ell(T)$ we let D be a generic element of $J_C[\ell]$ (as in §2.3 below), compute the three points

$$(\pi^2 + [\bar{q}])^2(D), \quad (\pi^2 + [\bar{q}])\pi(D), \text{ and } \pi^2(D),$$

and then search for the coefficients $(\bar{s}_1, \bar{s}_2)$ of $\chi_\ell(T)$ in $(\mathbb{Z}/\ell\mathbb{Z})^2$ for which the linear relation

$$(\pi^2 + [\bar{q}])^2(D) - [\bar{s}_1] (\pi^2 + [\bar{q}])\pi(D) + [\bar{s}_2] \pi^2(D) = 0 \tag{2}$$

holds. If the minimal polynomial of π on $J_C[\ell]$ is a proper divisor of $\chi_\ell(T)$—which occurs for at most a finite number of ℓ dividing $\mathrm{disc}(\chi)$—then the polynomial so determined is not unique, but $\chi_\ell(T)$ can be determined by deducing the correct multiplicities of its factors.

Once we have computed $\chi_\ell(T)$ for sufficiently many ℓ, we reconstruct $\chi(T)$ using the CRT. The Weil and Rück bounds together with a weak version of the prime number theorem tell us how many ℓ are required: Pila notes in [17, §1] that the set of $O(\log q)$ primes $\ell < 21 \log q$ will suffice. We analyse the complexity of the classical Schoof–Pila algorithm in §2.4.

2.3   Endomorphisms and Generic Kernel Elements

We now recall how to construct an effective version of a generic ℓ-torsion element. We present it in a slightly more general setting, so that we can use this ingredient in the subsequent RM-specific algorithm: we compute a generic element of the kernel of some endomorphism ϕ of J_C (the classical algorithm takes $\phi = \lfloor \ell \rfloor$).

Definition 1. *Fix an embedding $P \mapsto D_P$ of C in J_C. We say that an endomorphism ϕ of J_C is explicit if we can effectively compute polynomials d_0, d_1, d_2, e_0, e_1, and e_2 such that if $P = (x_P, y_P)$ is a generic point of C, then the Mumford representation of $\phi(D_P)$ is given by*

$$\phi(D_P) = \left(x^2 + \frac{d_1(x_P)}{d_2(x_P)} x + \frac{d_0(x_P)}{d_2(x_P)}, y - y_P \left(\frac{e_1(x_P)}{e_2(x_P)} x + \frac{e_0(x_P)}{e_2(x_P)} \right) \right). \quad (3)$$

The d_0, d_1, d_2, e_0, e_1, and e_2 are called the ϕ-division polynomials.

In the case $\phi = \lfloor \ell \rfloor$, the $\lfloor \ell \rfloor$-division polynomials are the ℓ-division polynomials of Cantor [4]. The ϕ-division polynomials depend on the choice of embedding $P \mapsto D_P$; we will make this choice explicit when computing the ϕ-division polynomials for each of our families in §4. Note that if ϕ is an explicit endomorphism, then we can use (3) (extending $\mathbb{Z}$-linearly) to evaluate $\phi(D)$ for general divisor classes D in J_C.

To compute a generic element of $J_C[\phi]$, we generalize the approach of [8] (which computes generic elements of $J_C[\ell]$). The resulting algorithm is essentially the same as in [8, §3] (except for the parasite computation step, which we omit) with ϕ-division polynomials replacing ℓ-division polynomials, so we will only briefly sketch it here.

Let $D = (x^2 + a_1 x + a_0, y - (b_1 x + b_0))$ be (the Mumford representation of) a generic point of J_C. We want to compute a triangular ideal I_ϕ in $\mathbb{F}_q[a_1, a_0, b_1, b_0]$ vanishing on the nonzero elements of $J_C[\phi]$. The element D equals $D_{(x_1,y_1)} + D_{(x_2,y_2)}$, where (x_1, y_1) and (x_2, y_2) are generic points of C. To find a triangular system of relations on the a_i and b_i such that D is in $J_C[\phi]$ we solve for x_1, y_1, x_2, and y_2 in

$$\phi(D_{(x_1,y_1)}) = -\phi(D_{(x_2,y_2)}),$$

using (3) and resultants computed with the evaluation–interpolation technique of [8, §3.1]. We then resymmetrize as in [8, §3.2] to express the result in terms of the a_i and b_i. We can now compute with a "generic" element $(x^2 + a_1 x + a_0, y - (b_1 x + b_0))$ of $J_C[\phi]$ by reducing the coefficients modulo I_ϕ after each operation.

Following the complexity analysis of [8, §3.5], we can compute a triangular representation for I_ϕ in $O(\delta^2 \mathsf{M}(\delta) \log \delta + \mathsf{M}(\delta^2) \log \delta)$ field operations, where δ is the maximum among the degrees of the ϕ-division polynomials, and $\mathsf{M}(d)$ is the number of operations required to multiply polynomials of degree d over $\mathbb{F}_q$. Using asymptotically fast multiplication algorithms, we can therefore compute I_ϕ in $\widetilde{O}(\delta^3)$ field operations. The degree of I_ϕ is in $O(\delta^2)$; with this triangular representation, each multiplication modulo I_ϕ costs $\widetilde{O}(\delta^2)$ field operations.

2.4 Complexity of Classical Schoof–Pila Point Counting

Proposition 1. *The complexity of the classical Schoof–Pila algorithm for a curve of genus 2 over $\mathbb{F}_q$ is in $O((\log q)^8)$.*

Proof. To determine $\chi(T)$, we need to compute $\chi_\ell(T)$ for $O(\log q)$ primes ℓ in $O(\log q)$. To compute $\chi_\ell(T)$, we must first compute the ℓ-division polynomials, which have degrees in $O(\ell^2)$. We then compute the kernel ideal I_ℓ; the total cost is in $\widetilde{O}(\ell^6)$ field operations, according to §2.3. The cost of checking (2) against a generic element of $J_\mathcal{C}[\ell]$ decomposes into the cost of computing Frobenius images of the generic element in $\widetilde{O}(\ell^4 \log q)$ and of finding the matching pair $(\bar{s}_1, \bar{s}_2)$ in $\widetilde{O}(\ell^5)$ field operations. So the total complexity for computing $\chi_\ell(T)$ is in $\widetilde{O}(\ell^4(\ell^2 + \log q))$ field operations. In terms of bit operations, for each ℓ bounded by $O(\log q)$, we compute $\chi_\ell(T)$ in time $\widetilde{O}((\log q)^7)$. The result follows from the addition of these costs for all the $O(\log q)$ different ℓ.   □

2.5   Baby-Step Giant-Step Algorithms

In practice, computing $\chi_\ell(T)$ with classical Schoof–Pila becomes impractical for large values of ℓ. The usual approach is to carry out the Schoof–Pila algorithm to the extent possible, obtaining congruences for s_1 and s_2 modulo some integer M, before completing the calculation using a generic group algorithm such as baby-step giant-step (BSGS). Our BSGS algorithm of choice is the low-memory parallelized variant of the Matsuo–Chao–Tsuji algorithm [9,13].

The Weil bounds imply that the search space of candidates for (s_1, s_2) is in $O(q^{3/2})$, so a pure BSGS approach finds (s_1, s_2) in time and space $\widetilde{O}(q^{3/4})$. However, when we apply BSGS after a partial Schoof–Pila computation, we have a congruence for (s_1, s_2) modulo M. If $M < 8\sqrt{q}$, then the size of the search space is reduced to $O(q^{3/2}/M^2)$, and the complexity for finding (s_1, s_2) is reduced to $\widetilde{O}(q^{3/4}/M)$. For larger M, the value of s_1 is fully determined; the problem is reduced to a one-dimensional search space of size $O(q/M)$, for which the complexity is $\widetilde{O}(\sqrt{q/M})$.

3   Point Counting in Genus 2 with Real Multiplication

By assumption, $J_\mathcal{C}$ is ordinary and simple, so $\chi(T)$ is an irreducible polynomial defining a quartic CM-field with real quadratic subfield $\mathbb{Q}(\sqrt{\Delta})$. We say that $J_\mathcal{C}$ (and $\mathcal{C}$) has *real multiplication* (RM) by $\mathbb{Q}(\sqrt{\Delta})$. For a randomly selected curve, Δ is in $O(q)$; but in the sequel we consider families of curves with RM by $\mathbb{Q}(\sqrt{\Delta})$ for small Δ ($= 5$ or 8), admitting an explicit (in the sense of Definition 1) endomorphism ϕ such that

$$\mathbb{Z}[\phi] = \mathbb{Q}(\sqrt{\Delta}) \cap \mathrm{End}(J_\mathcal{C}) \tag{4}$$

(that is, $\mathbb{Z}[\phi]$ is the full real subring of $\mathrm{End}(J_\mathcal{C})$), and

$$\mathrm{disc}\,(\mathbb{Z}[\phi]) = \Delta.$$

We presume that the trace $\mathrm{Tr}(\phi)$ and norm $\mathrm{N}(\phi)$, such that $\phi^2 - \mathrm{Tr}(\phi)\phi + \mathrm{N}(\phi) = 0$, are known. We also suppose that ϕ is *efficient*, in the following sense:

Definition 2. *We say that an explicit endomorphism ϕ is* efficiently computable *if evaluating ϕ at points of $J_C(\mathbb{F}_q)$ requires only $O(1)$ field operations (comparable to a few group operations in J_C). In practice, this means that the ϕ-division polynomials have small degree.*

The existence of an efficiently computable ϕ and knowledge of Δ allows us to make significant improvements to each stage of the Schoof–Pila algorithm. Briefly: in §3.2 we use ϕ to simplify the testing procedure for each ℓ; in §3.3 we show that when ℓ splits in $\mathbb{Z}[\phi]$, we can use ϕ to obtain a radical reduction in complexity for computing $\chi_\ell(T)$; and in §3.4 we show that knowing an effective ϕ allows us to use many fewer primes ℓ.

3.1   The RM Characteristic Polynomial

Let $\psi = \pi + \pi^\dagger$; we consider $\mathbb{Z}[\psi]$, a subring of the real quadratic subring of $\mathrm{End}(J_C)$. The characteristic polynomial of ψ is the *real Weil polynomial*

$$\xi(T) = T^2 - s_1 T + s_2; \tag{5}$$

the discriminant of $\mathbb{Z}[\psi]$ is $\Delta_0 = s_1^2 - 4s_2$. The analogue of Rück's bounds for (s_1, Δ_0) is

$$(|s_1| - 4\sqrt{q})^2 \geq \Delta_0 = s_1^2 - 4s_2 \geq 0. \tag{6}$$

Equation (4) implies that $\mathbb{Z}[\psi]$ is contained in $\mathbb{Z}[\phi]$, so there exist integers m and n such that

$$\psi = m + n\phi. \tag{7}$$

Both s_1 and s_2 are determined by m and n: we have

$$s_1 = \mathrm{Tr}(\psi) = 2m + n\mathrm{Tr}(\phi) \quad \text{and} \quad s_2 = \mathrm{N}(\psi) = (s_1^2 - n^2\Delta)/4. \tag{8}$$

In fact n is the conductor of $\mathbb{Z}[\psi]$ in $\mathbb{Z}[\phi]$ up to sign: $|n| = [\mathbb{Z}[\phi] : \mathbb{Z}[\psi]]$, and hence

$$\Delta_0 = \mathrm{disc}(\mathbb{Z}[\psi]) = s_1^2 - 4s_2 = n^2\Delta.$$

The square root of the bounds in (6) gives bounds on s_1 and n:

$$4\sqrt{q} - |s_1| \geq \sqrt{\Delta_0} = |n|\sqrt{\Delta} \geq 0;$$

In particular, $|s_1| \leq 4\sqrt{q}$ and $|n| \leq 4\sqrt{q/\Delta}$. Applying (8), we have the bounds

$$|m| \leq 2(|\mathrm{Tr}(\phi)| + \sqrt{\Delta})\sqrt{q/\Delta} \quad \text{and} \quad |n| \leq 4\sqrt{q/\Delta}. \tag{9}$$

Both $|m|$ and $|n|$ are in $O(\sqrt{q})$.

3.2   An Efficiently Computable RM Relation

We can use our efficiently computable endomorphism ϕ to replace the relation of (2) with a more efficiently computable alternative. Multiplying (7) through by π, we have

$$\psi\pi = \pi^2 + [q] = m\pi + n\phi\pi.$$

We can therefore compute $\bar{m} = m \bmod \ell$ and $\bar{n} = n \bmod \ell$ by letting D be a generic ℓ-torsion point, computing the three points

$$(\pi^2 + [\bar{q}])(D), \quad \pi(D), \quad \text{and} \quad \phi\pi(D),$$

and then searching for the $\bar{m}$ and $\bar{n}$ in $\mathbb{Z}/\ell\mathbb{Z}$ such that

$$(\pi^2 + [\bar{q}])(D) - [\bar{m}]\pi(D) - [\bar{n}]\phi\pi(D) = 0 \tag{10}$$

holds; we can find such an $\bar{m}$ and $\bar{n}$ in $O(\ell)$ group operations.

Solving (10) rather than (2) has several advantages. First, computing $(\pi^2 + [\bar{q}])(D)$, $\pi(D)$, and $\phi\pi(D)$ requires only two applications of Frobenius, instead of the four required to compute $(\pi^2 + [\bar{q}])^2(D)$, $(\pi^2 + [\bar{q}])\pi(D)$, and $\pi^2(D)$ (and Frobenius applications are costly in practice). Moreover, either s_2 needs to be determined in $O(q)$, or else the value of n in (2) leaves a sign ambiguity for each prime ℓ, because only $n^2 \bmod \ell$ can be deduced from $(\bar{s}_1, \bar{s}_2)$. In contrast, (10) determines n directly.

3.3   Exploiting Split Primes in $\mathbb{Q}(\sqrt{\Delta})$

Let $\mathbb{Z}[\phi] \subset \mathrm{End}(J_\mathcal{C})$ be an RM order in $\mathbb{Q}(\phi) \cong \mathbb{Q}(\sqrt{\Delta})$. Asymptotically, half of all primes ℓ split: $(\ell) = \mathfrak{p}_1\mathfrak{p}_2$ in $\mathbb{Z}[\phi]$, where $\mathfrak{p}_1 + \mathfrak{p}_2 = (1)$ (and this carries over to prime powers ℓ). This factorization gives a decomposition of the ℓ-torsion

$$J_\mathcal{C}[\ell] = J_\mathcal{C}[\mathfrak{p}_1] \oplus J_\mathcal{C}[\mathfrak{p}_2].$$

In particular, any ℓ-torsion point D can be uniquely expressed as a sum $D = D_1 + D_2$ where D_i is in $J_\mathcal{C}[\mathfrak{p}_i]$.

According to the Cohen–Lenstra heuristics [5], more than 75% of RM fields have class number 1; in each of the explicit RM families in §4, the order $\mathbb{Z}[\phi]$ has class number 1. All ideals are principal in such an order, so we may find a generator for each of the ideals $\mathfrak{p}_i$. The following lemma shows that we can find a generator which is not too large.

Lemma 1. *If $\mathfrak{p}$ is a principal ideal of norm ℓ in a real quadratic order $\mathbb{Z}[\phi]$, then there exists an effectively computable generator of $\mathfrak{p}$ with coefficients in $O(\sqrt{\ell})$.*

Proof. Let α be a generator of $\mathfrak{p}$, and ε a fundamental unit of $\mathbb{Z}[\phi]$. Let $\gamma \mapsto \gamma_1$ and $\gamma \mapsto \gamma_2$ be the two embeddings of $\mathbb{Z}[\phi]$ in $\mathbb{R}$, indexed so that $|\alpha_1| \geq |\alpha_2|$ and $|\varepsilon_1| > 1$ (replacing ε with ε^{-1} if necessary). Then $R = \log(|\varepsilon_1|)$ is the regulator of $\mathbb{Z}[\phi]$. Set $\beta = \varepsilon^{-k}\alpha$, where $k = [\log(|\alpha_1/\sqrt{\ell}|)/R]$; then $\beta = a + b\phi$ is a new generator for $\mathfrak{p}$ such that

$$-\frac{1}{2} \le \frac{\log(|\beta_i/\sqrt{\ell}|)}{R} \le \frac{1}{2}.$$

These bounds imply that $|\beta_1 + \beta_2| = |2a + b\mathrm{Tr}(\phi)|$ and $|\beta_1 - \beta_2| = |b\sqrt{\Delta}|$ are bounded by $2e^{R/2}\sqrt{\ell}$. Since $\mathrm{Tr}(\phi)$, Δ and R are fixed constants, $|a|$ and $|b|$ are in $O(\sqrt{\ell})$. The "effective" part of the result follows from classical algorithms for quadratic fields.   □

Lemma 2. *Let J_C be the Jacobian of a genus 2 curve over a finite field with an efficiently computable RM endomorphism ϕ. There exists an algorithm which, given a principal ideal $\mathfrak{p}$ of norm ℓ in $\mathbb{Z}[\phi]$, computes an explicit generator α of $\mathfrak{p}$ and the α-division polynomials in $O(\ell)$ field operations.*

Proof. By Lemma 1, we can compute a generator $\alpha = [a] + [b]\phi$ with a and b in $O(\sqrt{\ell})$. The $[a]$- and $[b]$-division polynomials have degrees in $O(\ell)$, and can be determined in $O(\ell)$ field operations. The division polynomials for the sum $\alpha = [a] + [b]\phi$ require one sum and one application of ϕ; and since ϕ is efficiently computable, this increases the division polynomial degrees and computing time by at most a constant factor.   □

We can now state the main theorem for RM point counting.

Theorem 1. *There exists an algorithm for the point counting problem in a family of genus 2 curves with efficiently computable RM of class number 1, whose complexity is in $\widetilde{O}((\log q)^5)$.*

Proof. Let J_C be a Jacobian in a family with efficiently computable RM by $\mathbb{Z}[\phi]$. Suppose that ℓ is prime, $(\ell) = \mathfrak{p}_1\mathfrak{p}_2$ in $\mathbb{Z}[\phi]$, and that the $\mathfrak{p}_i$ are principal. By Lemma 2 we can compute representative α-division polynomials for $\mathfrak{p}_1$ and $\mathfrak{p}_2$, and hence generic points D_i in $J_C[\mathfrak{p}_i]$, in time $\widetilde{O}(\ell)$.

We recall that (10) is the homomorphic image under π of the equation

$$\psi(D) - [\bar{m}](D) - [\bar{n}]\phi(D) = 0.$$

When applied to D_i in $J_C[\mathfrak{p}_i]$, the endomorphisms ψ and ϕ act as elements of $\mathbb{Z}[\phi]/\mathfrak{p}_i \cong \mathbb{Z}/\ell\mathbb{Z}$. Moreover $\bar{x}_i = \phi \bmod \mathfrak{p}_i$ is known, and it remains to determine $\bar{y}_i = \psi \bmod \mathfrak{p}_i$ by means of the discrete logarithm

$$\psi(D_i) = [\bar{y}_i](D_i) = [\bar{m} + \bar{n}\bar{x}_i](D_i)$$

in the cyclic group $\langle D_i \rangle \cong \mathbb{Z}/\ell\mathbb{Z}$. The application of π transports this discrete logarithm problem to that of solving for $\bar{y}_i$ in

$$D_i'' = [\bar{y}_i]D',$$

where $D_i' = \pi(D_i)$ and $D_i'' = (\pi^2 + [\bar{q}])(D_i)$. By the CRT, from $(\bar{y}_1, \bar{y}_2)$ in $(\mathbb{Z}/\ell\mathbb{Z})^2$ we recover $\bar{y}$ in $\mathbb{Z}[\phi]/(\ell)$, from which we solve for $(\bar{m}, \bar{n})$ in $(\mathbb{Z}/\ell\mathbb{Z})^2$ such that

$$\bar{y} = \bar{m} + \bar{n}\phi \in \mathbb{Z}[\phi]/(\ell).$$

The values of $(\bar{s}_1, \bar{s}_2)$ are then recovered from (8).

The ring $\mathbb{Z}[\phi]$ is fixed, so as $\log q$ goes to infinity we find that 50% of all primes ℓ split in $\mathbb{Z}[\phi]$ by the Chebotarev density theorem. It therefore suffices to consider split primes in $O(\log q)$. In comparison with the conventional algorithm presented in §2.2, we reduce from computation modulo the ideal for $J_\mathcal{C}[\ell]$ of degree in $O(\ell^4)$, to computation modulo the ideals for $J_\mathcal{C}[\mathfrak{p}_i]$ of degree in $O(\ell^2)$. This means a reduction from $\widetilde{O}(\ell^4(\ell^2 + \log q))$ to $\widetilde{O}(\ell^2(\ell + \log q))$ field operations for the determination of each $\chi_\ell(T)$, giving the stated reduction in total complexity from $\widetilde{O}((\log q)^8)$ to $\widetilde{O}((\log q)^5)$. □

Remark 1. Computing (m,n) instead of (s_1, s_2) allows us to reduce the number of primes ℓ to be considered by about a half, since by (9) their product needs to be in $O(\sqrt{q})$ instead of $O(q)$. While this changes only the constant in the asymptotic complexity of the algorithm, it yields a significant improvement in practice.

Remark 2. If the class number h of $\mathbb{Z}[\phi]$ is not 1, and if $(\ell) = \mathfrak{p}_1\mathfrak{p}_2$ where the $\mathfrak{p}_i$ are not principal, then we may apply Lemma 2 to a larger proportion of small ideals by using a small complementary ideal $(c) = \mathfrak{c}_1\mathfrak{c}_2$ such that the $\mathfrak{c}_i\mathfrak{p}_i$ are principal. Moreover, if $(\bar{m}, \bar{n})$ is known modulo c, this can be used to reduce the discrete log problem modulo ℓ. Again, since a fixed positive density $1/2h$ of primes are both split and principal, this does not affect the asymptotic complexity. We observe that the first discriminant with $h > 1$ is $\Delta = 65$, well beyond the current range for which an explicit RM construction is currently known.

3.4   Shrinking the BSGS Search Space

In the conventional Schoof–Pila context, we need to find s_1 in $O(\sqrt{q})$ and s_2 in $O(q)$. However, (7) and the effective form of (10) (valid for all D in $J_\mathcal{C}$) replace (s_1, s_2) with the tuple (m,n) of integers in $O(\sqrt{q})$. This reduces the search space size from $O(q^{3/2})$ to $O(q)$, so a BSGS strategy can find (m,n) (which determines (s_1, s_2)) in time and space $O(\sqrt{q})$, compared with $O(q^{3/4})$ when searching directly for (s_1, s_2).

As in the general case, if one knows (m,n) modulo an integer M, then the area of the search rectangle is reduced by a factor of M^2, so we find the tuple (m,n) in $O(\sqrt{q}/M)$ group operations. In contrast to the general case of §2.5, since m and n have the same order of magnitude, the speed-up is always by a factor of M.

4   Examples of Families of Curves with Explicit RM

We now exhibit some families of curves and efficient RM endomorphisms that can be used as sources of inputs to our algorithm.

4.1   Correspondences and Endomorphisms

To give a concrete representation for endomorphisms of $J_\mathcal{C}$, we use *correspondences*: that is, divisors on the surface $\mathcal{C} \times \mathcal{C}$. Suppose that $\mathcal{R}$ is a curve on $\mathcal{C} \times \mathcal{C}$, and let $\pi_1 : \mathcal{R} \to \mathcal{C}$ and $\pi_2 : \mathcal{R} \to \mathcal{C}$ be the restrictions to $\mathcal{R}$ of the natural projections from $\mathcal{C} \times \mathcal{C}$ onto its first and second factors. The pullback homomorphism $(\pi_1)^* : \mathrm{Pic}(\mathcal{C}) \to \mathrm{Pic}(\mathcal{R})$ is defined by

$$(\pi_1)^* \left(\Big[\sum_{P \in \mathcal{C}(\overline{\mathbb{F}}_q)} n_P P \Big] \right) = \Big[\sum_{P \in \mathcal{C}(\overline{\mathbb{F}}_q)} n_P \sum_{Q \in \pi_1^{-1}(P)} Q \Big],$$

where the preimages Q are counted with the appropriate multiplicities (we can always choose divisor class representatives so that each $\pi^{-1}(P)$ is zero-dimensional). The pushforward homomorphism $(\pi_2)_* : \mathrm{Pic}(\mathcal{R}) \to \mathrm{Pic}(\mathcal{C})$ is defined by

$$(\pi_2)_* \left(\Big[\sum_{Q \in \mathcal{R}(\overline{\mathbb{F}}_q)} n_Q Q \Big] \right) = \Big[\sum_{Q \in \mathcal{R}(\overline{\mathbb{F}}_q)} n_Q \pi_2(Q) \Big].$$

Note that $(\pi_1)^*$ maps $\mathrm{Pic}^n(\mathcal{C})$ into $\mathrm{Pic}^{(n \deg \pi_1)}(\mathcal{R})$ and $(\pi_2)_*$ maps $\mathrm{Pic}^n(\mathcal{R})$ into $\mathrm{Pic}^n(\mathcal{C})$ for all n. Hence $(\pi_2)_* \circ (\pi_1)^*$ maps $\mathrm{Pic}^0(\mathcal{C})$ into $\mathrm{Pic}^0(\mathcal{C})$, so we have an *induced endomorphism*

$$\phi = (\pi_2)_* \circ (\pi_1)^* : J_\mathcal{C} \to J_\mathcal{C}.$$

We write x_1, y_1 and x_2, y_2 for the coordinates on the first and second factors of $\mathcal{C} \times \mathcal{C}$, respectively (so $\pi_i(x_1, y_1, x_2, y_2) = (x_i, y_i)$). In our examples, the correspondence $\mathcal{R}$ will be defined by two equations:

$$\mathcal{R} = V(A(x_1, x_2), B(x_1, y_1, x_2, y_2)).$$

On the level of divisors, the image of a generic point $P = (x_P, y_P)$ of $\mathcal{C}$ (that is, a generic prime divisor) under the endomorphism ϕ is given by

$$\phi : (x_P, y_P) \longmapsto V(A(x_P, x), B(x_P, y_P, x, y)).$$

Using the relations $y_P^2 = f(x_P)$ and $y^2 = f(x)$ (and the fact that correspondences cut out by principal ideals induce the zero homomorphism), we can easily replace A and B with Cantor-reducible generators to derive the Mumford representation of $\phi(P)$, and thus the ϕ-division polynomials.

4.2   A 1-Dimensional Family with RM by $\mathbb{Z}[(1 + \sqrt{5})/2]$

Let t be a free parameter, and suppose that q is not a power of 5. Let $\mathcal{C}_T$ be the family of curves of genus 2 over $\mathbb{F}_q$ considered by Tautz, Top, and Verberkmoes in [21, Example 3.5], defined by

$$\mathcal{C}_T : y^2 = x^5 - 5x^3 + 5x + t.$$

Let $\tau_5 = \zeta_5 + \zeta_5^{-1}$, where ζ_5 is a 5th root of unity in $\overline{\mathbb{F}}_q$. Let ϕ_T be the endomorphism induced by the (constant) family of correspondences

$$\mathcal{R}_T = V\left(x_1^2 + x_2^2 - \tau_5 x_1 x_2 + \tau_5^2 - 4, y_1 - y_2\right) \subset \mathcal{C}_T \times \mathcal{C}_T.$$

(Note that $\mathcal{R}_T$ and ϕ_T are defined over $\mathbb{F}_q(\tau_5)$, which is equal to $\mathbb{F}_q$ if and only if $q \not\equiv \pm 2 \bmod 5$.) The family $\mathcal{C}_T$ has a unique point P_∞ at infinity, which we can use to define an embedding of $\mathcal{C}_T$ in $J_{\mathcal{C}_T}$ by

$$P = (x_P, y_P) \longmapsto D_P := [(P) - (P_\infty)] \leftrightarrow (x - x_P, y - y_P).$$

The ϕ_T-division polynomials with respect to this embedding are

$$d_2 = 1, \quad d_1 = -\tau_5 x, \quad d_0 = x^2 + \tau_5^2 - 4, \quad e_2 = 1, \quad e_1 = 0, \quad e_0 = 1.$$

Proposition 2. *The minimal polynomial of ϕ_T is $T^2 + T - 1$: that is, ϕ_T acts as multiplication by $-(1 + \sqrt{5})/2$ on $J_{\mathcal{C}_T}$. A prime ℓ splits into two principal ideals in $\mathbb{Z}[\phi_T]$ if and only if $\ell \equiv \pm 1 \bmod 5$.*

Proof. The first claim is proven in [21, §3.5]. More directly, if P and Q are generic points of $\mathcal{C}_T$, then on the level of divisors we find

$$(\phi_T^2 + \phi_T)((P) - (Q)) = (P) - (Q) + \mathrm{div}\left(\frac{y - y(P)}{y - y(Q)}\right).$$

Hence $\mathbb{Z}[\phi_T]$ is isomorphic to the ring of integers of $\mathbb{Q}(\sqrt{5})$. The primes ℓ splitting in $\mathbb{Q}(\sqrt{5})$ are precisely those congruent to ± 1 modulo 5; and since $\mathbb{Q}(\sqrt{5})$ has class number 1, the primes over ℓ are principal. $\square$

The Igusa invariants of $\mathcal{C}_T$, viewed as a point in weighted projective space, are $(140 : 550 : 20(32t^2 - 3) : 25(896t^2 - 3109) : 64(t^2 - 4)^2)$; in particular, $\mathcal{C}_T$ has a one-dimensional image in the moduli space of curves of genus 2. The Jacobian of the curve with the same defining equation over $\mathbb{Q}(t)$ is absolutely simple (cf. [12, Remark 15]).

4.3   A 2-Dimensional Family with RM by $\mathbb{Z}[(1 + \sqrt{5})/2]$

Let s and t be free parameters. Consider the family of genus 2 curves defined by $\mathcal{C}_H : y^2 = F_H(x)$, where

$$F_H(x) = sx^5 - (2s + t)x^4 + (s^2 + 3s + 2t - 1)x^3 - (3s + t - 3)x^2 + (s - 3)x + 1.$$

This family is essentially due to Humbert; it is equal to the family of Mestre [14, §2.1] with $(U, T) = (s, t)$, and the family of Wilson [22, Proposition 3.4.1] with $(A, B) = (s, -t-3s+3)$. The family has a full 2-dimensional image in the moduli space of genus 2 curves.

Let $\mathcal{R}_H$ be the family of correspondences on $\mathcal{C}_H \times \mathcal{C}_H$ defined by

$$\mathcal{R}_H = V\left(x_1^2 x_2^2 + s(s - 1)x_1 x_2 - s^2(x_1 - x_2) + s^2, y_1 - y_2\right);$$

let ϕ_H be the induced endomorphism. There is a unique point P_∞ at infinity on $\mathcal{C}_H$, which we can use to define an embedding of $\mathcal{C}_H$ in $J_{\mathcal{C}_H}$ by

$$P = (x_P, y_P) \longmapsto D_P := [(P) - (P_\infty)] \leftrightarrow (x - x_P, y - y_P).$$

The ϕ_H-division polynomials with respect to this embedding are

$$d_2 = x^2, \quad d_1 = (s^2 - s)x + s^2, \quad d_0 = -s^2 x + s^2, \quad e_2 = 1, \quad e_1 = 0, \quad e_0 = 1.$$

Proposition 3. *The minimal polynomial of ϕ_H is $T^2 + T - 1$: that is, ϕ_H acts as multipliction by $-(1 + \sqrt{5})/2$ on J_{C_H}. A prime ℓ splits into two principal ideals in $\mathbb{Z}[\phi_H]$ if and only if $\ell \equiv \pm 1 \bmod 5$.*

Proof. The first assertion is [14, Proposition 2] with $n = 5$; the rest of the proof is exactly as for Proposition 2.   □

4.4   A 2-Dimensional Family with RM by $\mathbb{Z}[\sqrt{2}]$

For an example with $\Delta = 8$, we present a twisted and reparametrized version of a construction due to Mestre [15]. Let s and t be free parameters, let $v(s)$ and $n(s)$ be the rational functions

$$v = v(s) := \frac{s^2 + 2}{s^2 - 2} \quad \text{and} \quad n = n(s) := \frac{4s(s^4 + 4)}{(s^2 - 2)^3},$$

and let $\mathcal{C}_M$ be the family of curves defined by

$$\mathcal{C}_M : y^2 = F_M(x) := (vx - 1)(x - v)(x^4 - tx^2 + vt - 1).$$

The family of correspondences on $\mathcal{C}_M \times \mathcal{C}_M$ defined by

$$\mathcal{R}_M = V \begin{pmatrix} x_1^2 x_2^2 - v^2(x_1^2 + x_2^2) + 1, \\ y_1 y_2 - n(x_1^2 + x_2^2 - t)(x_1 x_2 - v(x_1 + x_2) + 1) \end{pmatrix}$$

induces an endomorphism ϕ_M of $J_{\mathcal{C}_M}$.

The family $\mathcal{C}_M$ has two points at infinity, P_∞^+ and P_∞^-, which are generically only defined over a quadratic extension of $\mathbb{F}_q(s, t)$. Let $D_\infty = (P_\infty^+) + (P_\infty^-)$ denote the divisor at infinity. We can use the rational Weierstrass point $P_v = (v, 0)$ on $\mathcal{C}_M$ to define an embedding of $\mathcal{C}_M$ in $J_{\mathcal{C}_M}$ by

$$P = (x_P, y_P) \longmapsto D_P := [(P) + (P_v) - D_\infty]$$
$$\leftrightarrow \left((x - x_P)(x - v), y - \frac{y_P(x - v)}{x_P - v}(x - v) \right)$$

(appropriate composition and reduction algorithms for divisor class arithmetic on genus 2 curves with an even-degree model, such as $\mathcal{C}_M$, appear in [6]). The ϕ_M-division polynomials with respect to this embedding are

$$\begin{aligned}
d_2 &= x^2 - v^2, & e_2 &= (x^2 - v^2)F_M(x), \\
d_1 &= 0, & e_1 &= n(x - v)(x^4 - tx^2 + tv^2 - 1), \\
d_0 &= -v^2 x^2 + 1, & e_0 &= n(vx - 1)(x^4 - tx^2 + tv^2 - 1).
\end{aligned}$$

Proposition 4. *The minimal polynomial of ϕ_M is $T^2 - 2$: that is, ϕ_M acts as multiplication by $\sqrt{2}$ on $J_{\mathcal{C}_M}$. A prime ℓ splits into two principal ideals in $\mathbb{Z}[\phi_M]$ if and only if $\ell \equiv \pm 1 \bmod 8$.*

Proof. Let P and Q be generic points of $\mathcal{C}_M$. An elementary but lengthy calculation shows that on the level of divisors

$$\phi_M^2((P) - (Q)) = 2(P) - 2(Q) + \text{div}\left(\frac{x + x(P)}{x + x(Q)}\right),$$

so $\phi_M^2([D]) = 2[D]$ for all $[D]$ in $\text{Pic}^0(\mathcal{C}_M)$. Hence $\phi_M^2 = [2]$, and $\mathbb{Z}[\phi_M]$ is isomorphic to the maximal order of $\mathbb{Q}(\sqrt{2})$. The primes ℓ splitting in $\mathbb{Q}(\sqrt{2})$ are precisely those congruent to ± 1 modulo 8; further, $\mathbb{Q}(\sqrt{2})$ has class number 1, so the primes over ℓ are principal. □

Remark 3. As noted above, this construction is a twisted reparametrization of a family of isogenies described by Mestre in [15, §2.1]. Let a_1 and a_2 be the roots of $T^2 - tT + v^2 t - 1$ in $\overline{\mathbb{F}_q(v, t)}$. Mestre's curves C' and C are equal (over $\mathbb{F}_q(v, a_1, a_2)$) to our $\mathcal{C}_M$ and its quadratic twist by $A = 2(v^2 - 1)(v^2 + 1)^2 = (2n)^2$, respectively. We may specialize the proofs in [15] to show that $\mathcal{C}_M$ has a two-dimensional image in the moduli space of curves of genus 2, and that the Jacobian of the curve with the same defining equation over $\mathbb{Q}(s, t)$ is absolutely simple. Constructions of curves with RM by $\mathbb{Z}[\sqrt{2}]$ are further investigated in Bending's thesis [1].

Remark 4. Our algorithms should be readily adaptable to work with Kummer surfaces instead of Jacobians. In the notation of [7], the Kummers with parameters (a, b, c, d) satisfying $b^2 = a^2 - c^2 - d^2$ have RM by $\mathbb{Z}[\sqrt{2}]$, which can be made explicit as follows: the doubling algorithm decomposes into two identical steps, since $(A : B : C : D) = (a : b : c : d)$, and the components after one step are the coordinates of a Kummer point. This step therefore defines an efficiently computable endomorphism which squares to give multiplication by 2.

5   Numerical Experiments

We implemented our algorithm in C++ using the NTL library [19]. For non-critical steps, including computations in quadratic fields, we used Magma [2] for simplicity. With this implementation, determining $\chi(T)$ for a curve over a 128-bit prime field takes approximately 3 hours on one core of a Core2 processor at 2.83 GHz. This provides a proof of concept rather than an optimized implementation.

5.1   Cryptographic Curve Generation

When looking for cryptographic curves we used an early-abort strategy, switching to another curve as soon as either the order of the Jacobian or its twist cannot be prime. Using our adapted Schoof algorithm, we can guarantee that the group orders are not divisible by any prime that splits in the real field up to the CRT bound used. In fact, any prime that divides the group order of a curve having RM by the maximal order of $\mathbb{Q}(\sqrt{\Delta})$ must either be a split (or ramified) prime, or divide it with multiplicity 2. As a consequence, the early abort strategy works much better than in the classical Schoof algorithm, because it suffices to test half the number of primes up to our CRT bound.

We ran a search for a secure curve over a prime field of 128 bits, using a CRT bound of 131. Our series of computations frequently aborted early, and resulted in 245 curves for which $\chi(T)$ was fully determined, and for which neither the group order nor its twist was divisible by a prime less than 131. Together with the twists this provided 490 group orders, of which 27 were prime and therefore suitable for cryptographic use. We give here the data for one of these curves, that was furthermore twist-secure: the order of both the Jacobian and its twist are prime.

Let $q = 2^{128} + 573$, and let $\mathcal{C}/\mathbb{F}_q$ be the curve in the family $\mathcal{C}_\mathrm{T}$ of §4.2 specialized at $t = 75146620714142230387068843744286456025$. The characteristic polynomial $\chi(T)$ is determined by

$$s_1 = -26279773936397091867,$$
$$s_2 = -90827064182152428161138708787412643439,$$

giving prime group orders

$$\begin{aligned}
\#J_\mathcal{C}(\mathbb{F}_q) = {} & 11579208923731619543251352868591229880899 \\
& 5809621534164533135283195301868637471,
\end{aligned}$$
$$\begin{aligned}
\#J_{\mathcal{C}'}(\mathbb{F}_q) = {} & 11579208923731619541462844133146351767866 \\
& 5082003185737080136570606628937951 7451,
\end{aligned}$$

where $\mathcal{C}'$ denotes the quadratic twist of $\mathcal{C}$. Correctness of the orders is easily verified on random points in the Jacobians.

5.2   A Kilobit Jacobian

Let q be the prime $2^{512} + 1273$, and consider the curve over $\mathbb{F}_q$ from the family $\mathcal{C}_\mathrm{T}$ of §4.2 specialized at

$$\begin{aligned}
t = {} & 29085666333787272437998261129919801749774533003680957762232 \\
& 56986807375270272014471477919882845604269700820270816721532 \\
& 434975921085316560590832659122351278.
\end{aligned}$$

This value of t was randomly chosen, and carries no special structure. We computed the values of the pair $(s_1 \bmod \ell, n \bmod \ell)$ for this curve for each split prime ℓ up to 419; this was enough to uniquely determine the true value of (s_1, n) using the CRT. The numerical data for the curve follows:

$\Delta = 5$

$$\begin{aligned}
s_1 = {} & -10535684568225216385772683270554282199378670073368228748 \\
& 7810402851346035223080
\end{aligned}$$

$$\begin{aligned}
n = {} & -37786020778198256317368570028183842800473749792142072230 \\
& 9935490010350932 88492
\end{aligned}$$

$$\begin{aligned}
s_2 = {} & (s_1^2 - n^2\Delta)/4 \\
= {} & 990287025215436155679872249605061232893936642355960654938 \\
& 0080457770522333483406246939864255464288289545517520763844 \\
& 28888704295617466043679591527916629020
\end{aligned}$$

The order of the Jacobian is therefore

$$N = (1+q)^2 - s_1(1+q) + s_2$$
$$= 17976931348623159077293051907890247336179769789423065727343008115773267580550237573705948956144184541720417180780929444962763452801227364805323818926258902074851818089888868757737237328920325315884646393462965754493894524803468668112345681706310648544084486938739666585942218663644225871268417790010511900552 0.$$

The total runtime for this computation was about 80 days on a single core of a Core 2 clocked at 2.83 GHz. In practice, we use the inherent parallelism of the algorithm, running one prime ℓ on each available core.

We did not compute the characteristic polynomial modulo small prime powers (as in [10]), nor did we use BSGS to deduce the result from partial modular information as in §3.4 (indeed, we were more interested in measuring the behaviour of our algorithm for large values of ℓ). These improvements with an exponential-complexity nature bring much less than in the classical point counting algorithms, since they have to be balanced with a polynomial-time algorithm with a lower degree. For this example, we estimate that BSGS and small prime powers could have saved a factor of about 2 in the total runtime.

References

1. Bending, P.R.: Curves of genus 2 with $\sqrt{2}$ multiplication. Ph. D. thesis. University of Oxford (1998)
2. Bosma, W., Cannon, J., Playoust, C.: The Magma algebra system. I. The user language. J. Symbolic Comput. 24, 235–265 (1997)
3. Cantor, D.G.: Computing in the Jacobian of a hyperelliptic curve. Math. Comp. 48(177), 95–101 (1987)
4. Cantor, D.G.: On the analogue of the division polynomials for hyperelliptic curves. J. Reine Angew. Math. 447, 91–145 (1994)
5. Cohen, H., Lenstra Jr., H.W.: Heuristics on class groups of number fields. In: Number Theory, Noordwijkerhout 1983. Lecture Notes in Math., vol. 1068, pp. 33–62 (1984)
6. Galbraith, S.D., Harrison, M.C., Mireles Morales, D.J.: Efficient Hyperelliptic Arithmetic Using Balanced Representation for Divisors. In: van der Poorten, A.J., Stein, A. (eds.) ANTS-VIII 2008. LNCS, vol. 5011, pp. 342–356. Springer, Heidelberg (2008)
7. Gaudry, P.: Fast genus 2 arithmetic based on Theta functions. J. Math. Crypt. 1, 243–265 (2007)
8. Gaudry, P., Schost, É.: Construction of Secure Random Curves of Genus 2 Over Prime Fields. In: Cachin, C., Camenisch, J.L. (eds.) EUROCRYPT 2004. LNCS, vol. 3027, pp. 239–256. Springer, Heidelberg (2004)
9. Gaudry, P., Schost, É.: A Low-Memory Parallel Version of Matsuo, Chao, and Tsujii's Algorithm. In: Buell, D.A. (ed.) ANTS 2004. LNCS, vol. 3076, pp. 208–222. Springer, Heidelberg (2004)

10. Gaudry, P., Schost, É.: Genus 2 point counting over prime fields. Preprint (2010), http://hal.inria.fr/inria-00542650

11. Gruenewald, D.: Computing Humbert surfaces and applications. In: Arithmetic, Geometry, Cryptography and Coding Theory 2009. Contemp. Math., vol. 521, pp. 59–69 (2010)

12. Kohel, D.R., Smith, B.A.: Efficiently Computable Endomorphisms for Hyperelliptic Curves. In: Hess, F., Pauli, S., Pohst, M. (eds.) ANTS 2006. LNCS, vol. 4076, pp. 495–509. Springer, Heidelberg (2006)

13. Matsuo, K., Chao, J., Tsujii, S.: An Improved Baby Step Giant Step Algorithm for Point Counting of Hyperelliptic Curves over Finite Fields. In: Fieker, C., Kohel, D.R. (eds.) ANTS 2002. LNCS, vol. 2369, pp. 461–474. Springer, Heidelberg (2002)

14. Mestre, J.-F.: Familles de courbes hyperelliptiques à multiplications réelles'. In: Arithmetic algebraic geometry. Texel (1989); Progr. Math., vol. 89. Birkhäuser, Boston (1991)

15. Mestre, J.-F.: Couples de jacobiennes isogènes de courbes hyperelliptiques de genre arbitraire. Preprint, arXiv math.AG/0902.3470 v1 (2009)

16. Park, Y.-H., Jeong, S., Lim, J.: Speeding Up Point Multiplication on Hyperelliptic Curves with Efficiently-Computable Endomorphisms. In: Knudsen, L.R. (ed.) EUROCRYPT 2002. LNCS, vol. 2332, pp. 197–208. Springer, Heidelberg (2002)

17. Pila, J.: Frobenius maps of abelian varieties and finding roots of unity in finite fields. Math. Comp. 55(192), 745–763 (1990)

18. Rück, H.-G.: Abelian surfaces and jacobian varieties over finite fields. Compositio Math. 76(3), 351–366 (1990)

19. Shoup, V.: NTL: A Library for doing Number Theory, http://www.shoup.net/ntl/

20. Takashima, K.: A new type of fast endomorphisms on Jacobians of hyperelliptic curves and their cryptographic application. IEICE Trans. Fundamentals E89-A(1), 124–133 (2006)

21. Tautz, W., Top, J., Verberkmoes, A.: Explicit hyperelliptic curves with real multiplication and permutation polynomials. Canad. J. Math. 43(5), 1055–1064 (1991)

22. Wilson, J.: Curves of genus 2 with real multiplication by a square root of 5. Ph.D. thesis, University of Oxford (1998)

On the Efficiency of Bit Commitment Reductions

Samuel Ranellucci[1], Alain Tapp[1], Severin Winkler[2], and Jürg Wullschleger[1,3]

[1] DIRO, Université de Montréal, Quebec, Canada
[2] Institute of Theoretical Computer Science, ETH Zurich, Switzerland
[3] McGill University, Quebec, Canada

Abstract. Two fundamental building blocks of secure two-party computation are *oblivious transfer* and *bit commitment*. While there exist unconditionally secure implementations of oblivious transfer from noisy correlations or channels that achieve constant rates, similar constructions are not known for bit commitment.

In this paper, we show that any protocol that implements n instances of bit commitment with an error of at most 2^{-k} needs at least $\Omega(kn)$ instances of a given resource such as oblivious transfer or a noisy channel. This implies in particular that it is impossible to achieve a constant rate.

We then show that it is possible to circumvent the above lower bound by restricting the way in which the bit commitments can be opened. We present a protocol that achieves a constant rate in the special case where only a constant number of instances can be opened, which is optimal. Our protocol implements these restricted bit commitments from string commitments and is universally composable. The protocol provides significant speed-up over individual commitments in situations where restricted commitments are sufficient.

Keywords: secure two-party computation, bit commitment, string commitment, oblivious transfer, noisy channel, information theory.

1 Introduction

Commitment schemes [4] are one of the basic building blocks of two-party computation [42]. Commitments can be used in *coin-flipping* [4], *zero-knowledge proofs* [21,20], *zero-knowledge arguments* [7] or as a tool in general two-party computation protocols to prevent malicious players from actively cheating (see for example [14]).

A commitment scheme has two phases. In the *commit* phase, the sender has to decide on a value b. After the commit phase the value b is fixed and cannot be changed, while the receiver still does not get any information about its value. At a later time, the players may execute the second phase, called the *open* phase, where the bit b is revealed to the receiver. The scheme is called a *bit commitment* if b is only one bit, and it is called a *string commitment* if b is a longer bit string.

Bit commitments can be implemented from a wide variety of information-theoretic primitives [11,16,38,41]. There are protocols which implement a single

D.H. Lee and X. Wang (Eds.): ASIACRYPT 2011, LNCS 7073, pp. 520–537, 2011.

string commitment from noisy channels at a constant rate, meaning that the size of the string grows linearly with the number of instances of noisy channels used, which is essentially optimal [38]. Protocols which implement individual bit commitments at a constant rate, however, are not known. In [30] it has been shown that in any perfectly correct and perfectly hiding non-interactive bit commitment scheme from distributed randomness with a security of 2^{-k}, the size of the randomness given to the players must be at least $\Omega(k)$.

Another primitive that is of fundamental importance in two-party computation is oblivious transfer (OT) [36,32,19]. Oblivious transfer can be implemented from noisy channels [10,12,11,13], cryptogates [28] and weak variants of noisy channels [16,15,40,41]. While all these protocols require $\Omega(k)$ instances of a given primitive to implement a single OT with a security of 2^{-k}, it has been shown in [23,26,25,24] that there are more efficient protocols if many OTs are implemented at once. In the semi-honest model and in some cases also in the malicious model, it is possible to implement OT *at a constant rate*, which means n instances of OT can be implemented from just $O(n)$ instances of the given primitive, if n is big enough compared to the security parameter. It is, therefore, possible to achieve the lower bound for such reductions [17,2,39,37] up to a constant factor. In the following we address the question whether such efficient protocols also exist in the case of bit commitment.

1.1 Contribution

We show that — in contrast to implementations of OT — no constant rate reduction of bit commitment to distributed randomness can exist. More precisely, in Theorem 1 we show that if a protocol implements n bit commitments with a security of at least 2^{-k} from distributed randomness, then the mutual information between the sender's and the receiver's randomness must be almost kn or larger. Our proof is built on the insight that any such protocol must reveal at least k bits of information about the receiver's randomness for each committed bit that is opened. This implies that we need at least $\Omega(kn)$ instances of oblivious transfer or noisy channels to implement n bit commitments. Thus, executing for each bit commitment a protocol that uses $O(k)$ instances is optimal. In combination with the lower bound from [38], this bound can be generalized to string commitments: any protocol that implements n string commitments of length ℓ needs at least $\Omega(n(\ell + k))$ bits of distributed randomness.

However, in many applications of bit commitments the full strength of the commitment scheme is not required. For example in the famous zero-knowledge protocol of [20], it is only required that a constant number of committed bits can be opened. We show that restricting the ways in which the bit commitments can be opened enables us to implement more efficient schemes that circumvent our impossibility result.[1] We introduce a new concept that we call *bit commitments with restricted openings*. It allows a sender to commit to N bits, from which he

---

[1] Note that for the specific case of zero-knowledge proofs other, more efficient, techniques are known [29].

may open up to $r < N$ one by one. After that, he may only open all the remaining bits at once. Our protocol uses so-called *cover-free families*, and implements bit commitments with restricted openings from string commitments. Together with a simple construction of a cover-free family from [18], our results imply that for any prime power q, we can implement $N = q^2$ bit commitments from which r can be opened from $(r + 1)q$ string commitments of length q. (See Corollary 4 for the more general statement.) Together with the protocol from [38], we get a constant-rate bit commitment protocol from noisy channels, for any constant r. As bit commitments with restricted openings are strictly stronger than a string commitment, this is optimal. Together with another construction of a cover-free family from [6], it is possible to implement $N = 2^{\Omega(n/r^2)}$ bit commitments from n string commitments. We prove our protocol secure in the Universal Composability model (UC) [8].

We will prove our lower bounds for independent bit commitments in Section 2. In Section 3, we introduce commitments with restricted openings and give reductions to string commitments. Note that Section 3 can be read without reading Section 2.

1.2   Notation

In the following, the probability distribution of a random variable X is denoted by $P_X(x)$. The joint distribution $P_{XY}(x, y)$ defines a conditional distribution $P_{X|Y}(x, y) = P_{XY}(x, y)/P_Y(y)$ for all y with $P_Y(y) > 0$. The *statistical distance* between the distributions P_X and $P_{X'}$ over the domain $\mathcal{X}$ is defined as

$$\delta(P_X, P_{X'}) := \max_D | \Pr[D(X) = 1] - \Pr[D(X') = 1] | \,,$$

where we maximize over all (inefficient) distinguishers $D : \mathcal{X} \to \{0, 1\}$. We use the notation $[n]$ for the set $\{1, \ldots, n\}$. For a sequence $x = (x_1, \ldots, x_n)$ and $t \in [n]$, we denote by x^t the subsequence $(x_1, \ldots, x_t)$.

1.3   Information Theory

We will use the following tools from information theory in our proofs. We assume that the reader is familiar with the basic concepts of information theory, and refer to [9,22] for more details. The *conditional Shannon entropy* of X given Y is defined as[2]

$$\mathrm{H}(X \mid Y) := - \sum_{x,y} P_{XY}(x, y) \log P_{X|Y}(x, y) \,.$$

We use the notation

$$h(p) = -p \log(p) - (1 - p) \log(1 - p)$$

---

[2] All logarithms are binary, and we use the convention that $0 \cdot \log 0 = 0$.

for the binary entropy function, i.e., $h(p)$ is the entropy of the Bernoulli distribution[3] with parameter p. The *mutual information* of X and Y given Z is defined as

$$I(X;Y \mid Z) = H(X \mid Z) - H(X \mid YZ) .$$

The mutual information satisfies the following chain rule

$$I(X_1 \ldots X_n; Y) = \sum_{i=1}^{n} I(X_i; Y \mid X_1 \ldots X_{i-1}).$$

The Kullback-Leibler divergence or relative entropy of two distributions P_X and Q_X on $\mathcal{X}$ is defined as

$$D(P_X \parallel Q_X) = \sum_{x \in \mathcal{X}} P_X(x) \log \frac{P_X(x)}{Q_X(x)} .$$

The conditional divergence of two distributions P_{XY} and Q_{XY} on $\mathcal{X} \times \mathcal{Y}$ is defined as

$$D(P_{Y|X} \parallel Q_{Y|X}) = \sum_{x \in \mathcal{X}} P_X(x) D(P_{Y|X=x} \parallel Q_{Y|X=x}) .$$

The binary divergence of two probabilities p and q is defined as the divergence of the Bernoulli distributions with parameters p and q, i.e.,

$$d(p \parallel q) = p \log \frac{p}{q} + (1 - p) \log \frac{1 - p}{1 - q} .$$

The divergence (and hence also the conditional divergence) is always non-negative. Furthermore, we have the following chain rule

$$D(P_{XY} \parallel Q_{XY}) = D(P_X \parallel Q_X) + D(P_{Y|X} \parallel Q_{Y|X}) . \tag{1}$$

This implies

$$D(P_X P_{Y|X} \parallel P_X Q_{Y|X}) = D(P_{Y|X} \parallel Q_{Y|X}). \tag{2}$$

Let Q_X and P_X be two distributions over the inputs to the same channel $P_{Y|X}$. Then the divergence between the outputs $P_Y = \sum_x P_X P_{Y|X}$ and $Q_Y = \sum_x Q_X P_{Y|X}$ of the channel is not greater than the divergence between the inputs, i.e., the divergence satisfies the data-processing inequality

$$D(P_X \parallel Q_X) \geq D(P_Y \parallel Q_Y) . \tag{3}$$

Furthermore, for random variables X, Y and Z distributed according to P_{XYZ}

$$I(X;Y \mid Z) = D(P_{X|YZ} \parallel P_{X|Z}) . \tag{4}$$

Let $P_{X|Y=y} = P_{X|Y=y,Z=z}$ for all y, z (or $P_{Z|Y=y} = P_{Z|Y=y,X=x}$ for all y, z, which is equivalent). Then we say that X, Y and Z form a Markov-chain, denoted by $X \leftrightarrow Y \leftrightarrow Z$. If $W \leftrightarrow XZ \leftrightarrow Y$, then

$$I(X;Y \mid ZW) \leq I(X;Y \mid Z) . \tag{5}$$

---

[3] The Bernoulli distribution with parameter $p \in [0,1]$ takes on the value 1 with probability p and 0 otherwise.

2   Impossibility Results

2.1   Model and Security Definition

We will consider the following model: a trusted third party holds random variables (U, V) with a joint distribution P_{UV} and sends U to the sender and V to the receiver. The sender receives an input bit $b \in \{0, 1\}$. In the commit phase, the players exchange messages in several rounds. Let all the messages exchanged be M, which is a randomized function of (U, V, b). In the open phase, the sender sends b together with a value D_1 to the receiver. The receiver then sends a message E_1 to the receiver, who replies with a message D_2 and so on. Let $N := (D_1, E_1, D_2, E_2, \ldots, E_{t-1}, D_t)$ be the total communication in the open phase. (We assume that the number of rounds in the open phase is upper bounded by a constant t. By padding the protocol with empty rounds we can thus assume without loss of generality that the protocol uses t rounds in every execution.) Finally, the receiver accepts or rejects, which we model by a randomized function $F(b, V, M, N)$ that outputs 1 for accept and 0 for reject. Let the distribution in the honest setting be $P_{UVMN|B=b}$. We define three parameters that quantify the security for the sender and the receiver, respectively, and the correctness of the protocol.

- ε-correct: $\Pr[F(b, V, M, N) = 1] \geq 1 - \varepsilon$.
- β-hiding: $\delta(P_{VM|B=0}, P_{VM|B=1}) \leq \beta$.
- γ-binding: For any $b \in \{0, 1\}$ and for any malicious sender that is honest in the commit phase on input b and tries to open $1 - b$, we have $\Pr[F(1 - b, V, M, N') = 1] \leq \gamma$, where N' is the communication between the malicious sender and the honest receiver in the open phase.

Note that the above security conditions are not sufficient to prove the security of a protocol[4], but any sensible security definition for commitments implies these conditions. Since we only use the definition to prove the non-existence of certain protocols, this makes our result stronger.

2.2   Lower Bound for Multiple Bit Commitments

In the following we prove a lower bound on the mutual information between the randomness of the sender and the randomness of the receiver in any bit commitment protocol. First, we show the following technical lemma.

Lemma 1. *If a protocol that implements bit commitment from distributed randomness (U, V) is γ-binding, ε-correct and β-hiding, then for $b \in \{0, 1\}$*

$$\mathrm{d}(1 - \varepsilon \, \| \, \gamma + \beta) \leq \sum_{i=1}^{t} \mathrm{I}(D_i; V \mid MD^{i-1}E^{i-1}, B = b). \tag{6}$$

---

[4] To prove the security of a protocol one had to consider for example a malicious sender in the commit phase.

Proof. Let $b \in \{0,1\}$ and $\bar{b} := 1 - b$. Assume that the sender in the commit phase honestly commits to b. If she honestly opens b in the open phase, the communication can be modeled by a channel $P_{DE|VM}$ (that may depend on b) and the resulting distribution is

$$P_{DEVM|B=b} = P_{DE|VM} P_{VM|B=b} \, ,$$

We have omitted U as it does not play a role in the following arguments. The correctness property implies that an honest receiver accepts values drawn from this distribution with probability at least $1 - \varepsilon$. Let the sender commit to $\bar{b}$ and then try to open b by sampling her messages according to the distributions $P_{D_1|M}$ and $P_{D_i|MD^{i-1}E^{i-1}}$ for $2 \le i \le t$. (Note that the sender does not know V and, therefore, chooses her messages independently of V.) The communication in the opening phase can be modeled by a channel

$$Q_{DE|VM} := P_{D_1|M} P_{E_1|VMD_1} \cdots P_{D_t|MD^{t-1}E^{t-1}} \, .$$

The binding property implies that the receiver accepts values distributed according to $P_{VM|B=\bar{b}} Q_{DE|VM}$ with probability at most γ. $\delta(P_{VM|B=b}, P_{VM|B=\bar{b}}) \le \beta$ implies that

$$\delta(P_{VM|B=b} Q_{DE|VM}, P_{VM|B=\bar{b}} Q_{DE|VM}) \le \beta,$$

and hence values drawn from the distribution $P_{VM|B=b} Q_{DE|VM}$ are accepted with probability at most $\gamma + \beta$. Note that the bit indicating acceptance can also be modeled by a channel $P_{F|DEVM}$. Thus, we can apply the data-processing inequality (3) to bound $d(1 - \varepsilon \| \gamma + \beta)$. Using the chain rule (1) and the non-negativity of the relative entropy, we have (we omit conditioning on $B = b$ in the following)

$$
\begin{aligned}
d(1 - \varepsilon \| \gamma + \beta) &\le D(P_{VM} P_{DE|VM} \| P_{VM} Q_{DE|VM}) \\
&= D(P_{DE|VM} \| Q_{DE|VM}) \\
&= \sum_{i=1}^{t} D(P_{D_i|VMD^{i-1}E^{i-1}} \| P_{D_i|MD^{i-1}E^{i-1}}) \\
&\quad + \sum_{i=1}^{t-1} D(P_{E_i|VMD^iE^{i-1}} \| P_{E_i|VMD^iE^{i-1}}) \\
&= \sum_{i=1}^{t} D(P_{D_i|VMD^{i-1}E^{i-1}} \| P_{D_i|MD^{i-1}E^{i-1}}) \\
&= \sum_{i=1}^{t} I(D_i; V \mid MD^{i-1}E^{i-1})
\end{aligned}
$$

$\square$

The following lemma follows easily from Theorem 2.1 in [31]. We will use it to bound the right-hand side of (6) in the following.

Lemma 2. *Let $\varepsilon = \beta = \gamma = 2^{-k}$. Then, for $k \geq 3$, we have*

$$d(1 - \varepsilon \parallel \gamma + \beta) \geq (k - 2) \cdot \frac{2^{k-2} - 2}{2^{k-2} - 1}.$$

The following lemma generalizes the lower bounds on the size of the randomness for perfectly correct and perfectly hiding non-interactive schemes from [30] to arbitrary protocols. However, it also provides a more powerful result, namely a lower bound on the information that the communication in the open phase must reveal about the receiver's randomness V for any protocol that implements bit commitment from a shared distribution P_{UV}. The lower bound is essentially k if the error of the protocol is at most 2^{-k}. This stronger statement will allow us in the following to prove that there are no constant rate reductions of bit commitment to distributed randomness, the main result of this section.

Lemma 3. *Let $k \geq 3$. Then any 2^{-k}-secure bit commitment must have for $b \in \{0, 1\}$*

$$I(N; V \mid M, B = b) - I(N; V \mid UM, B = b)$$

$$= I(U; V \mid M, B = b) - I(U; V \mid MN, B = b) \geq (k - 2) \cdot \frac{2^{k-2} - 2}{2^{k-2} - 1}.$$

Proof. Again, we omit conditioning on $B = b$ in the following. Consider a protocol over t rounds in the open phase, i.e., the whole communication is $N = (D, E) = (D_1, E_1, \ldots, D_t)$. Since $D_i \leftrightarrow UMD^{i-1}E^{i-1} \leftrightarrow V$, we have $I(D_i; V \mid UMD^{i-1}E^{i-1}) = 0$. Hence,

$$I(NU; V \mid M) = I(U; V \mid M) + \sum_{i=1}^{t-1} I(E_i; V \mid UMD^i E^{i-1}).$$

Furthermore, from $E_i \leftrightarrow VMD^i E^{i-1} \leftrightarrow U$ and inequality (5) follows that for all i

$$I(E_i; V \mid MD^i E^{i-1}) \geq I(E_i; V \mid UMD^i E^{i-1}).$$

Hence, we have

$$I(N; V \mid M) = \sum_i I(E_i; V \mid MD^i E^{i-1}) + \sum_i I(D_i; V \mid MD^{i-1} E^{i-1})$$

$$\geq \sum_i I(E_i; V \mid UMD^i E^{i-1}) + \sum_i I(D_i; V \mid MD^{i-1} E^{i-1})$$

and

$$I(U; V \mid MN) = I(NU; V \mid M) - I(N; V \mid M)$$

$$= I(U; V \mid M) + \sum_i I(E_i; V \mid UMD^i E^{i-1}) - I(N; V \mid M)$$

$$\leq I(U; V \mid M) - \sum_i I(D_i; V \mid MD^{i-1} E^{i-1}).$$

The statement now follows from Lemma 1 and Lemma 2.   □

Next, we consider implementations of n individual bit commitments. The sender gets input $b^n = (b_1, \ldots, b_n)$ and commits to all bits at the same time, which results in the overall distribution

$$P_{UVM|B^n = b^n} = P_{UV} P_{M|UV,B^n = b^n} .$$

after the commit phase. To reveal the ith bit, the sender and the receiver interact resulting in the transcript N_i. The following theorem says that the mutual information between the sender's randomness U and the receiver's randomness V must be almost kn to implement n bit commitments with an error of at most 2^{-k}. The proof uses Lemma 3 to lower bound the information that the sender must reveal about V for every bit that he opens.

Theorem 1. *Let $k \geq 3$. Then any 2^{-k}-secure protocol that implements n bit commitments from randomness (U, V) must have for all $b^n \in \{0,1\}^n$*

$$I(U; V) \geq I(U; V \mid M, B = b^n) \geq n(k-2) \cdot \frac{2^{k-2} - 2}{2^{k-2} - 1} .$$

Proof. Let $\hat{i} \in [n]$. We first construct a commitment to a single bit, which will allow us to apply the bound from Lemma 3. This bit commitment is defined as follows: to commit to the bit b, the players execute the commit phase on input b^n, which is equal to the input bit b on position $\hat{i}$ and equal to the constant $\hat{b}^n \in \{0,1\}^n$ on all other positions. Additionally, (still as part of the commit phase), the sender opens the first $\hat{i} - 1$ commitments, which means that the messages $N^{\hat{i}-1}$ get exchanged. To open the commitment, the sender opens bit $\hat{i}$. This bit commitment scheme has at least the same security as the original commitment. Thus, Lemma 3 implies that (we omit conditioning on $B = b^n$ in the following)

$$I(U; V \mid MN^{\hat{i}}) \leq I(U; V \mid MN^{\hat{i}-1}) - (k-2) \cdot \frac{2^{k-2} - 2}{2^{k-2} - 1}. \tag{7}$$

Since this holds for all $\hat{i}$, we can apply (7) repeatedly to get

$$0 \leq I(U; V \mid MN^n)$$

$$\leq I(U; V \mid MN^{n-1}) - (k-2) \cdot \frac{2^{k-2} - 2}{2^{k-2} - 1}$$

$$\leq I(U; V \mid M) - n(k-2) \cdot \frac{2^{k-2} - 2}{2^{k-2} - 1}$$

By induction over all rounds of the commit protocol using (5) (see, for example, [37] for a detailed proof) it follows that

$$I(U; V \mid M) \leq I(U; V) .$$

This implies the statement. □

It is possible to securely implement 1-out-of-2 bit oblivious transfer $\left(\binom{2}{1}\text{-}\mathsf{OT}^1\right)$ from randomness distributed according to P_{UV} with $\mathrm{I}(U;V) = 1$ [3,1]. A binary symmetric noisy channel $((p)\text{-}\mathsf{BSNC})$ with crossover probability p can be implemented from randomness distributed according to P_{UV} with $\mathrm{I}(U;V) = 1 - h(p)$. Together with these reductions, Theorem 1 implies that (almost) kn instances of $\binom{2}{1}\text{-}\mathsf{OT}^1$ or $kn/(1 - h(p))$ instances of a $(p)\text{-}\mathsf{BSNC}$ are needed to implement n bit commitments with an error of at most 2^{-k}.

There exists a universally composable protocol[5] that implements bit commitment from $2k$ instances of $\binom{2}{1}\text{-}\mathsf{OT}^1$ with an error of at most 2^{-k}. Thus, n bit commitments can be implemented from $2n(k + \log(n))$ instances of $\binom{2}{1}\text{-}\mathsf{OT}^1$ with an error of at most $n \cdot 2^{-(k+\log(n))} = 2^{-k}$ using n parallel instances of this protocol. Theorem 1 shows that this is optimal up to a factor of 4 if $k \geq \log(n)$.

2.3   Lower Bounds for Multiple String Commitments

A *string commitment* is a generalization of bit commitment where the sender may commit to a bit-string of length $\ell \geq 1$. It is weaker than ℓ instances of bit commitment because the sender has to reveal all bits simultaneously. In [38] a lower bound on the conditional entropy of the sender's randomness U given the receiver's randomness V for any string commitment protocol from randomness (U, V) has been shown. This bound essentially says that $\mathrm{H}(U \mid V)$ must be greater than or equal to ℓ to implement a string commitment of length ℓ. The following lemma provides a similar bound for the security definition considered here. (The proof can be found in the full version of this paper [33].)

Lemma 4. *If any protocol implements an ℓ-bit string commitment from randomness (U, V) is ε-correct, $\beta-$hiding and γ-binding, then*

$$\mathrm{H}(U \mid V) \geq (1 - \varepsilon - \beta - \gamma)\ell - h(\beta) - h(\varepsilon + \gamma).$$

Together with the bound of Theorem 1, we obtain the following lower bound on the randomness of the sender in any bit commitment protocol.

Corollary 1. *Let $k \geq 3$. For any protocol that implements n individual ℓ-bit string commitments from randomness (U, V) with an error of at most 2^{-k}*

$$\mathrm{H}(U) \geq n(k + \ell - 2) \cdot \frac{2^{k-2} - 2}{2^{k-2} - 1} - 3 \cdot 2^{-k} \cdot n\ell - 3h(2^{-k}).$$

Proof. Using Lemma 4 and Theorem 1, we get

$$\mathrm{H}(U) = \mathrm{I}(U;V) + \mathrm{H}(U \mid V)$$

$$\geq n(k - 2) \cdot \frac{2^{k-2} - 2}{2^{k-2} - 1} + (1 - 3 \cdot 2^{-k})n\ell - h(2^{-k}) - h(2^{-k+1})$$

$$\geq n(k + \ell - 2) \cdot \frac{2^{k-2} - 2}{2^{k-2} - 1} - 3 \cdot 2^{-k} \cdot n\ell - 3h(2^{-k}).$$

$\square$

---

[5] See for example Claim 33 in the full version of [8].

In [5] it has been shown that any non-interactive perfectly hiding and perfectly correct bit commitment protocol from distributed randomness P_{UV} is at most $(2^{-H(V|U)})$-binding. This result implies stronger bounds than Theorem 1 and Lemma 4 for certain reductions. The following lemma provides a lower bound on the uncertainty of the sender about the receiver's randomness for any bit commitment protocol. This lower bound is essentially equal to k if the protocol is 2^{-k}-secure and implies, in particular, the result from [5].

Lemma 5. *If a protocol that implements bit commitment from randomness* (U, V) *is* γ-*binding,* ε-*correct and* β-*hiding, then*

$$\mathrm{d}(1 - \beta - \varepsilon \parallel \gamma) \leq \mathrm{H}(V \mid UM) \leq \mathrm{H}(V \mid U).$$

where M *is the whole communication in the commit phase. If* $\beta = \gamma = \varepsilon = 2^{-k}$, *then*

$$\mathrm{H}(V \mid U) \geq (k - 1) \cdot \frac{2^{k-1} - 4}{2^{k-1} - 1}. \tag{8}$$

Proof. We have $\delta(P_{VM|B=b}, P_{VM|B=\bar{b}}) \leq \beta$. This implies that the distribution $P_{U|VM,B=\bar{b}}P_{VM|B=b}$ is β-close to $P_{UVM|B=\bar{b}}$. Thus, when the sender honestly opens $\bar{b}$ starting from values distributed according to $P_{U|VM,B=\bar{b}}P_{VM|B=b}$, the receiver accepts the resulting values with probability at least $1 - \beta - \varepsilon$. We consider the following attack: the sender honestly commits to b, generates v' by applying $P_{V|UM,B=b}$ and then generates u by applying the channel $P_{U|VM,B=\bar{b}}$ to (v', m). When the sender now tries to open $\bar{b}$, the binding property guarantees that the receiver accepts the resulting values with probability at most γ. Thus, we can apply the data-processing inequality (3) to bound $\mathrm{d}(1 - \beta - \varepsilon \parallel \gamma)$. Let V' be a copy of V, i.e., a random variable with distribution $P_{VV'}(v, v) = P_V(v)$. Using the chain rule (2), we have

$$\mathrm{d}(1 - \beta - \varepsilon \parallel \gamma) \leq \mathrm{D}(P_{VV'|UM,B=b}P_{UM|B=b} \parallel P_{V|UM,B=b}P_{V|UM,B=b}P_{UM|B=b})$$
$$\leq \mathrm{D}(P_{VV'|UM,B=b} \parallel P_{V|UM,B=b}P_{V|UM,B=b})$$
$$= \mathrm{H}(V \mid UM, B = b)$$
$$\leq \mathrm{H}(V \mid U).$$

Using Lemma 2 this implies inequality (8).

$\square$

Consider a protocol that implements n bit commitment with security of 2^{-k} from n' instances of $\binom{2}{1}$-$\mathsf{OT}^{\ell'}$. Since $\binom{2}{1}$-$\mathsf{OT}^{\ell'}$ can be reduced to a shared distribution P_{UV} with $H(V|U) = 1$, Lemma 5 implies that $n' \geq (k-1) \cdot \frac{2^{k-1}-4}{2^{k-1}-1}$, i.e., one needs, independently of ℓ', almost k instances of OT.

Together with Theorem 1 and Lemma 4, this implies the following lower bound on the number of instances of OT needed to implement multiple string commitments, which demonstrates that all three lower bounds can be meaningful in this scenario.

Corollary 2. *Let $k \geq 3$. For any protocol that implements n individual ℓ-bit string commitments with an error of at most 2^{-k} from n' instances of $\binom{2}{1}$-$OT^{\ell'}$*

$$n' \geq \max\left(\frac{\ell n}{\ell'}(1 - 3 \cdot 2^{-k}) - \frac{3h(2^{-k})}{\ell'}, \frac{(k-2)n}{\ell'} \cdot \frac{2^{k-2} - 2}{2^{k-2} - 1}, (k-1)\frac{2^{k-1} - 4}{2^{k-1} - 1}\right).$$

3   Commitments with Restricted Openings

In this section, we will present a protocol that implements commitments with restricted openings from several instances of string commitment. We will use the Universal Composability model [8], and assume that the reader is familiar with it. In our proof, we will only consider static adversaries. For simplicity, we omit session IDs and players IDs.

String Commitment is a functionality that allows the sender to commit to a string of n bits, and to reveal the whole string later to the receiver. The receiver does not get to know the string before it is opened, and the sender cannot change the string once he has sent it.

Definition 1 (String-Commitment). *The functionality $\mathcal{F}_{SCOM}^n$ behaves as follows:*

- *Upon input (commit, b) with $b \in \{0,1\}^n$ from the sender: check that commit has not been sent yet. If so, send committed to the receiver and store b. Otherwise, ignore the message.*
- *Upon input openall from the sender: check if there has been a commit message before, and the commitment has not been opened yet. If so, send (openall, b) to the receiver and ignore the message otherwise.*

Note that given $\mathcal{F}_{SCOM}^n$, it is possible to commit to individual bits at different times: the sender simply commits to a random string $b' = (b'_1, \ldots, b'_n)$, and whenever he wants to commit to a bit b_i for $i \in [n]$, he sends $b_i \oplus b'_i$ to the receiver. On the other hand, it is not possible to open bits at different times using $\mathcal{F}_{SCOM}^n$.

Bit commitment is a string commitment of length 1, i.e., $\mathcal{F}_{BCOM} := \mathcal{F}_{SCOM}^1$. We denote n independent bit commitments by $(\mathcal{F}_{BCOM})^n$. Since $(\mathcal{F}_{BCOM})^n$ does allow bits to be opened at different times, it is strictly stronger than $\mathcal{F}_{SCOM}^n$. However, as we have seen in the last section, $(\mathcal{F}_{BCOM})^n$ is also quite expensive to implement in terms of resources needed. Therefore, we define a primitive that is somewhere between these two: *commitments with restricted openings* allow a sender to commit to n bits, but then he may only open r individual bits of his choice one by one. To open more than r bits, he has to open the remaining bits all at once.

Definition 2 (Commitments with restricted openings). *The functionality* $\mathcal{F}_{RCOM}^{n,r}$ *behaves as follows:*

- *Upon input (*commit*, b) with* $b \in \{0,1\}^n$ *from the sender: check that* commit *has not been sent yet. If so, send* committed *to the receiver and store b. Otherwise, ignore the message.*
- *Upon input (*open,i*) with* $i \in [n]$ *from the sender: check that there has been a commit message before, and that i has not been opened yet. Also check that the number of opened values so far is smaller than r. If so, send (*open*, i, b_i) to the receiver and ignore the message otherwise.*
- *Upon input* openall *from the sender: check if there has been a commit message before, and no* openall *message has been received yet from the sender. If so, send (*openall*, b) to the receiver and ignore the message otherwise.*

For $r = 0$ and $r = n$, commitment with restricted openings are equivalent to string commitments and individual bit commitments, respectively: $\mathcal{F}_{SCOM}^n = \mathcal{F}_{RCOM}^{n,0}$ and $(\mathcal{F}_{BCOM})^n \equiv \mathcal{F}_{RCOM}^{n,n}$.

Our protocol makes use of *cover-free families* [27,18,35,6], which are a generalization of *Sperner sets* [34]. Cover-free families are also known as *superimposed codes* and require that no set is covered by the union of r other sets.

Definition 3. *Let* $\mathcal{X}$ *be a set of n elements and let* $\mathcal{B}$ *be a set of subsets of* $\mathcal{X}$, *then* $(\mathcal{X}, \mathcal{B})$ *is a r-cover-free family* $r-\mathrm{CFF}(\mathcal{X}, \mathcal{B})$ *if for any r sets* $B_{i_1}, \ldots B_{i_r} \in \mathcal{B}$, *and any other* $B \in \mathcal{B}$, *it holds that*

$$B \not\subseteq \bigcup_{j=1}^{r} B_{i_j} .$$

Example 1. All subsets of $[n]$ of size s form a cover-free family for $r = 1$, because there is no subset that completely covers any other subset.

Here is a simple example of a cover-free family for $r > 1$ given in [18].

Example 2 ([18]). Let q be a prime power, and $d, r \in \mathbb{N}$ such that $rd < q$. Let $\mathcal{X} = \mathcal{Y} \times GF(q)$, where $\mathcal{Y} \subseteq GF(q)$ and $|\mathcal{Y}| = rd + 1$. An element B in the family $\mathcal{B}$ is constructed from a polynomial $p(y) := a_0 + y \cdot a_1 + \ldots + y^d \cdot a_d$ of degree d where $a_i \in GF(q)$ by $B := \{(y, p(y)) : y \in \mathcal{Y}\}$. Two polynomials of degree d intersect at most d times. Therefore, any union of r elements $B_1, \ldots B_r$ intersects any other element B at most $rd < |\mathcal{Y}|$ times, and therefore cannot cover B. $(\mathcal{X}, \mathcal{B})$ is therefore a r-cover-free family with $|\mathcal{X}| = (rd + 1)q$ and $|\mathcal{B}| = q^{d+1}$.

We now give a protocol that implements $\mathcal{F}_{RCOM}^{N,r}$ from n instances of $\mathcal{F}_{SCOM}^N$ using a $r-\mathrm{CFF}(\mathcal{X}, \mathcal{B})$, where $\mathcal{X} = \{1, \ldots, n\}$ and $\mathcal{B} = \{B_1, B_2, ..., B_N\}$.

Protocol 1.

- When the sender receives (commit, b), he chooses n uniformly chosen strings $c_1, \ldots c_n \in \{0,1\}^N$, with the restriction that for all $i \in [N]$ we have

$$\bigoplus_{j \in B_i} c_{j,i} = b_i .$$

For $j \in [n]$, the sender sends (commit, c_j) to the jth instances of $\mathcal{F}_{\text{SCOM}}^N$. After that he ignores all messages (commit, b').
- When the receiver has received committed from all instances of $\mathcal{F}_{\text{SCOM}}^N$, he outputs committed.
- For the first r times when the sender receives (open, i), he sends (open, i) to the receiver and openall to all instances of $\mathcal{F}_{\text{SCOM}}^N$ in B_i, if they have not been opened yet. After that, he ignores all messages (open, i).
- For the first r times when the receiver receives (open, i) from the sender and (open, c_j) from all instances $\mathcal{F}_{\text{SCOM}}^N$ in B_i, he outputs (open, $\bigoplus_{j \in B_i} c_{j,i}$). After that, he ignores these messages.
- When the sender receives openall, he sends openall to the receiver and to all instances of $\mathcal{F}_{\text{SCOM}}^N$. After that, he ignores all openall messages.
- When the receiver receives openall from the sender and (open, c_j) from all instances of $\mathcal{F}_{\text{SCOM}}^N$, he outputs (openall, $(b'_1, \ldots, b'_N)$), where $b'_i := \bigoplus_{j \in B_i} c_{j,i}$. After that, he ignores all messages openall.

Theorem 2. *Given an* $r-\text{CFF}(\mathcal{X}, \mathcal{B})$ *where* $|\mathcal{X}| = n$ *and* $|\mathcal{B}| = N$, *Protocol 1 UC-implements* $\mathcal{F}_{\text{RCOM}}^{N,r}$ *from* n *instances of* $\mathcal{F}_{\text{SCOM}}^N$.

Proof. It is easy to verify that the protocol is correct if the two players are honest.

Corrupted sender. First, we consider the case where the comitter is corrupted. He may send messages (commit, c_j) or openall to the instances of $\mathcal{F}_{\text{SCOM}}^N$, and message (open, i) or openall to the receiver.

Our simulator simulates the adversary, and records all messages sent out by the adversary. After receiving all messages (commit, c_j) to the instances of $\mathcal{F}_{\text{SCOM}}^N$, he calculates $b_i := \bigoplus_{j \in B_i} c_{j,i}$ for all i and sends (commit, $(b_1, \ldots, b_N)$) to $\mathcal{F}_{\text{RCOM}}^{N,r}$. After receiving (open, i) and all messages openall sent to the instances of $\mathcal{F}_{\text{SCOM}}^N$ in B_i, he sends (open, i) to $\mathcal{F}_{\text{RCOM}}^{N,r}$. After receiving openall sent to the receiver and all instances $\mathcal{F}_{\text{SCOM}}^N$, he sends openall to $\mathcal{F}_{\text{RCOM}}^{N,r}$. It is not difficult to verify that our simulation is perfect, and we get REAL≡ IDEAL.

Corrupted receiver. Let the receiver be corrupted by the adversary. He receives committed and (open,c_j) messages from the instances of $\mathcal{F}_{\text{SCOM}}^N$, and messages (open, i) and openall from the sender.

Our simulator simulates the adversary, and interacts with $\mathcal{F}_{\text{RCOM}}^{N,r}$ and the adversary. After receiving the **committed** message from $\mathcal{F}_{\text{RCOM}}^{N,r}$, it sends **committed** from all $\mathcal{F}_{\text{SCOM}}^{N}$ to the adversary. After receiving message (open, i, b_i) from $\mathcal{F}_{\text{RCOM}}^{N,r}$, he first sends (open, i) to the adversary. Then for all instances of $\mathcal{F}_{\text{SCOM}}^{N}$ in B_i which have not been opened yet, he chooses strings c_j uniformly at random, with the restriction that $\oplus_{j \in B_i} c_{j,i} = b_i$, and sends (open, c_j) from the jth instance of $\mathcal{F}_{\text{SCOM}}^{N}$ to the adversary. After receiving message $(\text{openall}, b)$ from $\mathcal{F}_{\text{RCOM}}^{N,r}$, he first sends **openall** to the adversary. Then for all instances of $\mathcal{F}_{\text{SCOM}}^{N}$ which have not been opened yet, he chooses the strings c_j uniformly at random, with the restriction that $\oplus_{j \in B_i} c_{j,i} = b_i$, and sends (open, c_j) from the jth instance of $\mathcal{F}_{\text{SCOM}}^{N}$ to the adversary.

To show that this simulation in the ideal setting is identical to the real setting, we have to show that they are identical after each step. It is easy to see that this is the case before anything has been opened, and after **openall** has been executed.

$\mathcal{F}_{\text{RCOM}}^{N,r}$ allows the sender to open at most r values. Assume that $s \leq r$ have been opened so far. Since $\mathcal{B}$ is a $r - \text{CFF}(\mathcal{X}, \mathcal{B})$, there is at least one instance of $\mathcal{F}_{\text{SCOM}}^{N}$ in B_i for all the remaining $i \in [N]$ that has not been opened yet. Since the ith bit of that string is uniform and all the ith bits of the strings in B_i add up to b_i, the bits at the ith position of all the opened strings are uniform and independent of each other and of the bit b_i. Therefore, the simulated values c_j sent to the adversary have the same distribution in the real and in the ideal setting. The simulation is again perfect, and we get REAL$\equiv$ IDEAL.   □

Note that in each instance of $\mathcal{F}_{\text{SCOM}}^{N}$ in Protocol 1, only a subset of the bits are actually used. Since they are at fixed positions and both players know where they are, they can be removed without changing the properties of the protocol. If we use the cover-free family from Example 1, the length of the string commitments used can be reduced to Ns/n, and we get the following corollary.

Corollary 3. *For any* $n \geq s \geq 1$ *and* $N = \binom{n}{s}$ *there exists a protocol that UC-implements* $\mathcal{F}_{\text{RCOM}}^{N,1}$ *from* $\left(\mathcal{F}_{\text{SCOM}}^{Ns/n}\right)^{n}$.

The protocol is optimal in the length of the strings up to a factor s; otherwise it would be possible to implement a string commitment of length bigger than $n \cdot \ell$ from n instances of string commitment of length ℓ, which is not possible. Thus, we can build $N = n(n-1)/2$ bit commitments (choosing $s = 2$), from which one can be opened, from n string commitments of length $n - 1$. When choosing $s = n/2$, we obtain an exponential number of committed bits from n strings, since $N = \binom{n}{n/2} > 2^{n/2}$.

If we use the cover-free family of Example 2, then the size of the commitments can be reduced by a factor of q because we can let all the bit commitments which have different values a_0 but the same values $a_1, ..., a_d$ share the same position in the string commitments. We get the following corollary.

Corollary 4. *Let* q *be a prime power,* $d < q$ *and* $N := q^{d+1}$. *There exists a protocol that UC-implements* $\mathcal{F}_{\text{RCOM}}^{N,r}$ *from* $(rd + 1)q$ *instances of* $\mathcal{F}_{\text{SCOM}}^{N/q}$.

This is optimal in the length of the strings up to a factor $rd + 1$; otherwise it would again be possible to implement a string commitment of length bigger than $n \cdot \ell$ from n instances of string commitment of length ℓ, which is not possible. Choosing $d = 1$, we get $N = q^2$ and $n = (r + 1)q$. Thus, there exists a protocol that uses $(r + 1)q$ string commitments of length q and implements q^2 bit commitments from which r can be opened.

To obtain an exponential number of bit commitments from n string commitments, we can use Corollary 1 in [6] which gives an explicit construction of a $t-\mathrm{CFF}(\mathcal{X}, \mathcal{B})$ where $|\mathcal{X}| < 24t^2 \log(|\mathcal{B}| + 2)$. Hence, we get the following result.

Corollary 5. *There exists a protocol that from $\mathcal{F}_{\mathrm{RCOM}}^{N,r}$ from $24r^2 \log(N + 2)$ instances of $\mathcal{F}_{\mathrm{SCOM}}^{N}$.*

This is close to the optimal efficiency we can expect from Protocol 1, as it has been shown in Theorem 1.1 in [35] that $t-\mathrm{CFF}(\mathcal{X}, \mathcal{B})$ must have

$$|\mathcal{X}| \geq c \cdot \frac{t^2}{\log t} \log |\mathcal{B}| ,$$

for a constant c.

Our protocols can be generalized in a simple way as follows: let $\mathcal{F}_{\mathrm{RCOM}}^{N,r,c}$ be the same functionality as $\mathcal{F}_{\mathrm{RCOM}}^{N,r}$ except that every bit is replaced by a block of size c. The sender can open up to r blocks, or all N blocks at the same time. It is not difficult to see that if Protocol 1 implements $\mathcal{F}_{\mathrm{RCOM}}^{N,r}$ from n instances of $\mathcal{F}_{\mathrm{SCOM}}^{\ell}$, then it can be transformed into a protocol that implements $\mathcal{F}_{\mathrm{RCOM}}^{N,r,c}$ from n instances of $\mathcal{F}_{\mathrm{SCOM}}^{\ell c}$.

3.1   Commitments from Noisy Channels at a Constant Rate

From Corollary 4 with $d = 1$ in combination with the string commitment protocol presented in [38], we get the following corollary.

Corollary 6. *For any constant r, there exists a protocol that implements $\mathcal{F}_{\mathrm{RCOM}}^{n,r}$ using only $O(n)$ noisy channels.*

This is optimal up to a constant factor.

4   Conclusions

In this work we have shown a strong lower bound for reductions of multiple bit commitments to other information theoretic primitives, such as oblivious transfer or noisy channels. Our bound shows that every single bit commitment needs at least $\Omega(k)$ instances of the underlying primitive. This makes bit commitments often much more costly to implement than oblivious transfer, for example. It would be interesting to see whether these results can be generalized to other functionalities.

We have presented a protocol that implements bit commitments more efficiently, when the number of bits that can be opened is restricted. Our protocol implements commitments with restricted openings from string commitments. We think that for some resources more efficient protocols might be possible by implementing them directly, instead of using string commitments as a building block.

Acknowledgements. AT is supported by Canada NSERC. SW is supported from the Swiss National Science Foundation and an ETHIIRA grant of ETH's research commission. JW is supported by the Canada-France NSERC-ANR project FREQUENCY.

References

1. Beaver, D.: Precomputing Oblivious Transfer. In: Coppersmith, D. (ed.) CRYPTO 1995. LNCS, vol. 963, pp. 97–109. Springer, Heidelberg (1995)
2. Beimel, A., Malkin, T.: A Quantitative Approach to Reductions in Secure Computation. In: Naor, M. (ed.) TCC 2004. LNCS, vol. 2951, pp. 238–257. Springer, Heidelberg (2004)
3. Bennett, C.H., Brassard, G., Crépeau, C., Skubiszewska, M.-H.: Practical Quantum Oblivious Transfer. In: Feigenbaum, J. (ed.) CRYPTO 1991. LNCS, vol. 576, pp. 351–366. Springer, Heidelberg (1992)
4. Blum, M.: Coin flipping by telephone a protocol for solving impossible problems. SIGACT News 15(1), 23–27 (1983)
5. Blundo, C., Masucci, B., Stinson, D.R., Wei, R.: Constructions and bounds for unconditionally secure non-interactive commitment schemes. Des. Codes Cryptography 26, 97–110 (2002)
6. De Bonis, A., Vaccaro, U.: Constructions of generalized superimposed codes with applications to group testing and conflict resolution in multiple access channels. Theor. Comput. Sci. 306, 223–243 (2003)
7. Brassard, G., Chaum, D., Crépeau, C.: Minimum disclosure proofs of knowledge. J. Comput. Syst. Sci. 37, 156–189 (1988)
8. Canetti, R.: Universally composable security: A new paradigm for cryptographic protocols. In: Proceedings of the 42nd Annual IEEE Symposium on Foundations of Computer Science (FOCS 2001), pp. 136–145 (2001), Updated Version at http://eprint.iacr.org/2000/067
9. Cover, T.M., Thomas, J.A.: Elements of Information Theory. Wiley-Interscience, New York (1991)
10. Crépeau, C.: Equivalence Between Two Flavours of Oblivious Transfers. In: Pomerance, C. (ed.) CRYPTO 1987. LNCS, vol. 293, pp. 350–354. Springer, Heidelberg (1988)
11. Crépeau, C.: Efficient Cryptographic Protocols Based on Noisy Channels. In: Fumy, W. (ed.) EUROCRYPT 1997. LNCS, vol. 1233, pp. 306–317. Springer, Heidelberg (1997)
12. Crépeau, C., Kilian, J.: Achieving oblivious transfer using weakened security assumptions (extended abstract). In: Proceedings of the 29th Annual IEEE Symposium on Foundations of Computer Science (FOCS 1988), pp. 42–52 (1988)

13. Crépeau, C., Morozov, K., Wolf, S.: Efficient Unconditional Oblivious Transfer from Almost Any Noisy Channel. In: Blundo, C., Cimato, S. (eds.) SCN 2004. LNCS, vol. 3352, pp. 47–59. Springer, Heidelberg (2005)
14. Crépeau, C., van de Graaf, J., Tapp, A.: Committed Oblivious Transfer and Private Multi-Party Computation. In: Coppersmith, D. (ed.) CRYPTO 1995. LNCS, vol. 963, pp. 110–123. Springer, Heidelberg (1995)
15. Damgård, I., Fehr, S., Morozov, K., Salvail, L.: Unfair Noisy Channels and Oblivious Transfer. In: Naor, M. (ed.) TCC 2004. LNCS, vol. 2951, pp. 355–373. Springer, Heidelberg (2004)
16. Damgård, I., Kilian, J., Salvail, L.: On the (Im)Possibility of Basing Oblivious Transfer and Bit Commitment on Weakened Security Assumptions. In: Stern, J. (ed.) EUROCRYPT 1999. LNCS, vol. 1592, pp. 56–73. Springer, Heidelberg (1999)
17. Dodis, Y., Micali, S.: Lower Bounds for Oblivious Transfer Reductions. In: Stern, J. (ed.) EUROCRYPT 1999. LNCS, vol. 1592, pp. 42–55. Springer, Heidelberg (1999)
18. Erdős, P., Frankl, P., Füredi, Z.: Families of finite sets in which no set is covered by the union of r others. Israel Journal of Mathematics 51(1-2), 79–89 (1985)
19. Even, S., Goldreich, O., Lempel, A.: A randomized protocol for signing contracts. Commun. ACM 28(6), 637–647 (1985)
20. Goldreich, O., Micali, S., Wigderson, A.: Proofs that yield nothing but their validity or all languages in np have zero-knowledge proof systems. J. ACM 38(3), 690–728 (1991)
21. Goldwasser, S., Micali, S., Rackoff, C.: The knowledge complexity of interactive proof-systems. In: Proceedings of the 17th Annual ACM Symposium on Theory of Computing (STOC 1985), pp. 291–304. ACM Press (1985)
22. Han, T.S., Kobayashi, K.: Mathematics of Information and Coding. American Mathematical Society, Boston (2001)
23. Harnik, D., Ishai, Y., Kushilevitz, E., Nielsen, J.B.: OT-Combiners Via Secure Computation. In: Canetti, R. (ed.) TCC 2008. LNCS, vol. 4948, pp. 393–411. Springer, Heidelberg (2008)
24. Ishai, Y., Kushilevitz, E., Ostrovsky, R., Prabhakaran, M., Sahai, A., Wullschleger, J.: Constant-rate Oblivious Transfer from Noisy Channels. In: Rogaway, P. (ed.) CRYPTO 2011. LNCS, vol. 6841, pp. 667–684. Springer, Heidelberg (2011)
25. Ishai, Y., Kushilevitz, E., Ostrovsky, R., Sahai, A.: Extracting correlations. In: Proceedings of the 50th Annual IEEE Symposium on Foundations of Computer Science (FOCS 2009), pp. 261–270 (2009)
26. Ishai, Y., Prabhakaran, M., Sahai, A.: Founding Cryptography on Oblivious Transfer – Efficiently. In: Wagner, D. (ed.) CRYPTO 2008. LNCS, vol. 5157, pp. 572–591. Springer, Heidelberg (2008)
27. Kautz, W., Singleton, R.: Nonrandom binary superimposed codes. IEEE Trans. on Information Theory 10(4), 363–377 (1964)
28. Kilian, J.: More general completeness theorems for secure two-party computation. In: Proceedings of the 32nd Annual ACM Symposium on Theory of Computing (STOC 2000), pp. 316–324. ACM Press (2000)
29. Kilian, J., Micali, S., Ostrovsky, R.: Minimum resource zero-knowledge proofs. In: Proceedings of the 30th Annual IEEE Symposium on Foundations of Computer Science (FOCS 1989), pp. 474–479. IEEE (1989)
30. Nascimento, A., Otsuka, A., Imai, H., Müller-Quade, J.: Unconditionally Secure Homomorphic Pre-Distributed Commitments. In: Fossorier, M.P.C., Høholdt, T., Poli, A. (eds.) AAECC 2003. LNCS, vol. 2643, pp. 604–604. Springer, Heidelberg (2003)

31. Ordentlich, E., Weinberger, M.J.: A distribution dependent refinement of pinsker's inequality. IEEE Transactions on Information Theory 51(5), 1836–1840 (2005)
32. Rabin, M.O.: How to exchange secrets by oblivious transfer. Technical Report TR-81, Harvard Aiken Computation Laboratory (1981)
33. Ranellucci, S., Tapp, A., Winkler, S., Wullschleger, J.: On the efficiency of bit commitment reductions. Cryptology ePrint Archive, Report 2011/324 (2011)
34. Sperner, J.: Ein Satz über Untermengen einer endlichen Menge. Math. Z. 27, 544–548 (1928)
35. Stinson, D.R., Wei, R., Zhu, L.: Some new bounds for cover-free families. J. Combin. Theory A 90, 224–234 (1999)
36. Wiesner, S.: Conjugate coding. SIGACT News 15(1), 78–88 (1983)
37. Winkler, S., Wullschleger, J.: On the Efficiency of Classical and Quantum Oblivious Transfer Reductions. In: Rabin, T. (ed.) CRYPTO 2010. LNCS, vol. 6223, pp. 707–723. Springer, Heidelberg (2010)
38. Winter, A., Nascimento, A.C.A., Imai, H.: Commitment capacity of discrete memoryless channels. In: IMA Int. Conf., pp. 35–51 (2003)
39. Wolf, S., Wullschleger, J.: New Monotones and Lower Bounds in Unconditional Two-Party Computation. In: Shoup, V. (ed.) CRYPTO 2005. LNCS, vol. 3621, pp. 467–477. Springer, Heidelberg (2005)
40. Wullschleger, J.: Oblivious-Transfer Amplification. In: Naor, M. (ed.) EUROCRYPT 2007. LNCS, vol. 4515, pp. 555–572. Springer, Heidelberg (2007)
41. Wullschleger, J.: Oblivious Transfer from Weak Noisy Channels. In: Reingold, O. (ed.) TCC 2009. LNCS, vol. 5444, pp. 332–349. Springer, Heidelberg (2009)
42. Yao, A.C.: Protocols for secure computations. In: Proceedings of the 23rd Annual IEEE Symposium on Foundations of Computer Science (FOCS 1982), pp. 160–164 (1982)

Secure Communication in Multicast Graphs

Qiushi Yang* and Yvo Desmedt**

Department of Computer Science, University College London, UK
{q.yang,y.desmedt}@cs.ucl.ac.uk

Abstract. In this paper we solve the problem of secure communication in multicast graphs, which has been open for over a decade. At Eurocrypt '98, Franklin and Wright initiated the study of secure communication against a Byzantine adversary on multicast channels in a neighbor network setting. Their model requires node-disjoint and neighbor-disjoint paths between a sender and a receiver. This requirement is too strong and hence not necessary in the general multicast graph setting. The research to find the lower and upper bounds on network connectivity for secure communication in multicast graphs has been carried out ever since. However, up until this day, there is no tight bound found for any level of security.

We study this problem from a new direction, i.e., we find the necessary and sufficient conditions (tight lower and upper bounds) for secure communication in the general adversary model with adversary structures, and then apply the results to the threshold model. Our solution uses an extended characterization of the multicast graphs, which is based on our observation on the eavesdropping and separating activities of the Byzantine adversary.

Keywords: secure communication, reliable communication, multicast, privacy, reliability, adversary structure.

1 Introduction

In most communication networks, a *sender* S and a *receiver* R are connected by unreliable and distrusted channels. The distrust of the channels is because of the assumption that there exists an adversary who, with unbounded computational power, can control some nodes on these channels. The interplay of network connectivity and secure communication between S and R has been studied extensively (see, e.g., [2,3,6,4,13]).

Secure communication is based on the problem of *secure message transmission* (SMT) between S and R. The aim of SMT is to enable a message to be transmitted from S to R *privately* (i.e., the adversary does not learn the message) and *reliably* (i.e., R can output the message correctly). In particular, *reliable message transmission* (RMT) is essential for all transmission protocols, and hence it has

---

* Part of this work was done while funded by UCL PhD Studentship.
** Part of this work was done while funded by EPSRC EP/C538285/1, by BT as BT Chair of Information Security, and by RCIS, AIST, Japan.

D.H. Lee and X. Wang (Eds.): ASIACRYPT 2011, LNCS 7073, pp. 538–555, 2011.

been studied exclusively. Normally there are two different measures of security or reliability: *perfect* (i.e., zero probability that the protocol fails to be secure or reliable) and *almost perfect* (i.e., an arbitrarily small probability that the protocol fails to be secure or reliable) [7].

The traditional studies of RMT and SMT consider a *point-to-point* network setting, where a sending node can transmit a message to a receiving node through a channel they choose. In the *threshold model* (t-bounded), the adversary is able to control up to t nodes in a network graph. The result by Dolev et al. [6] shows that $n > 2t$ *node-disjoint* paths are required for RMT and SMT between S and R. In [7], Franklin and Wright showed that the connectivity for almost perfect security can be reduced by using *multicast* channels.

A *multicast* channel allows a sending node to transmit a message to multiple receiving nodes. The study of secure multicast was initiated by Franklin and Yung in [9]. They used hypergraphs to model multicast networks, and studied privacy against a passive adversary (eavesdropper). Goldreich et al. [10] also studied multicast networks, but their work is in the full information model, which is different to the partial broadcast model in which we are interested. At Eurocrypt '98, Franklin and Wright [7] (see also [8]) first studied a *Byzantine* (active) adversary on multicast channels in *neighbor networks* (defined in [9]), in which a message multicast by a node is received—simultaneously and privately— by all its neighbors, where a neighbor is a node that shares a common edge with the sending node.[1] They found that with some properties of the multicast channels, only $n > t$ node-disjoint paths are needed for *almost perfectly* RMT and SMT. However, their setting is based on a strong assumption, that is, all paths between S and R must be *neighbor-disjoint* (i.e., there do not exist two paths that have a common neighbor node). Indeed, such a strong assumption may not be necessary in general multicast networks, and hence they gave the following open problem:

> ... if these n disjoint paths do not have disjoint neighborhood, then an adversary may be able to foil our protocols with $t < n$ faults by using one fault to eavesdrop on two disjoint lines. An obvious direction of further research is to characterize secure communication fully in this more general (multicast graph) setting.

Wang and Desmedt [14] further investigated the problem of secure communication in a more general multicast graph setting. They conjectured that a general connectivity (weaker than $n > t$ neighbor-disjoint) is the upper bound for achieving perfect privacy and almost perfect reliability (see Section 6 for more details). In another study, Desmedt and Wang [4] (see also [15]) extended this result. By using examples, they showed that the previously conjectured connectivity of [14] is not necessary, and they also proposed a lower bound for SMT and conjectured its tightness. Since it is very difficult to apply the threshold model in general

---

[1] For example, in Fig 1(a) in Section 3, when a message is multicast by node 2, it will be simultaneously received by nodes 1, 3 and 4. A multicast channel does not allow node 2 to send a message to node 1 and 3 without node 4 receiving it.

multicast graphs, up until this day, there has been no result that gives the necessary and sufficient conditions for RMT and SMT in multicast graphs.

Our contributions. We completely solve the problem of secure communication in multicast graphs (neighbor network setting), which has been open and studied for over a decade. We view this problem from a new direction. That is, our solution is based on two basic ideas: (1) a general graph setting can be applied naturally in the *general adversary model* with *adversary structures* (see, e.g., [11,13,5,17]); (2) a threshold corresponds to a special adversary structure. Thus we study multicast graphs in the general adversary model, and then apply the results to the threshold model.

We found that the current adversary structure model is not enough to characterize multicast graphs. Therefore, in Section 3, we give an extended characterization of the multicast graphs, which is based on our observation on the *eavesdropping* and *separating* activities of the adversary on the multicast channels. This characterization gives a clearer view on how the message can be securely transmitted over multicast graphs.

With the new characterization, we give the necessary and sufficient conditions for RMT and SMT respectively in Section 4 and Section 5. Besides proving that our conditions imply the lower bounds on network connectivity, we also provide message transmission protocols to show that these bounds are tight.

Finally in Section 6, we use our results in the general adversary model to find the necessary and sufficient conditions for RMT and SMT in the threshold model. Also by analyzing the previous results, we show how our results explain all the examples and prove all the conjectures in the previous work. Our final result regarding the tight bounds on network connectivity for RMT and SMT in multicast graphs is presented at the end of this paper.

2   Model

We abstract away the concrete network structure and model a *multicast communication neighbor network* by an *undirected graph* $G(V, E)$, whose nodes are the parties in the network and edges are private and authenticated multicast channels. Let $S, R \in V$, the paths between S and R are not necessarily node-disjoint.[2]

Let $\mathbb{F}$ be a *sufficiently large* finite field, we assume that $\mathbb{M} \subseteq \mathbb{F}$ is the message space from which S chooses messages. Let A be a set, we use $|A|$ to denote the number of elements in A, and we write $a \in_R A$ to indicate that a is chosen from A with respect to uniform distribution.

In the *threshold model*, an adversary can control up to t nodes in a graph, and hence control up to t node-disjoint paths. In the *general adversary model*, an adversary is characterized by an adversary structure, which is defined as follows (see [12,11]): Given a party set P, an adversary structure $\mathcal{A}$ on P is a subset

---

[2] Throughout the paper we consider only the simple paths. A simple path is a path with no repeated nodes.

of 2^P such that for any $A \in 2^P$, if $A \in \mathcal{A}$ and $A \supseteq A'$, then $A' \in \mathcal{A}$. The adversary is able to choose one set $A \in \mathcal{A}$ to control. It is straightforward that the threshold model is a special case of the general adversary model, because a threshold t can be seen as a special adversary structure $\mathcal{A}$ such that any set $A \in 2^P$ that has t parties or less is in $\mathcal{A}$.

In this paper we consider a *Byzantine* adversary who can exhibit an *active* behavior. A Byzantine adversary has unlimited resources and computational power. Not only can the adversary read the traffic through the parties it controls, but it can also decide, whether to deny or to modify the message, whether to follow the protocol or not, etc.

We use the security model given by Franklin and Wright [7]. Let Π be an SMT protocol. S starts with a message m^S drawn from a message space $\mathbb{M}$. At the end of Π, R outputs a message m^R. For any execution of the protocol Π, let adv be the adversary's view of the entire protocol, i.e., the behavior of the faulty nodes, the initial state of the adversary, and the coin flips of the adversary during the execution. We write $adv(m, r)$ to denote the adversary's view when $m^S = m$ and when the coin flips of the adversary are r.

Privacy. Π is ϵ-private if, for any two messages $m_1, m_2 \in \mathbb{M}$ and any r, we have $\sum_c |\Pr[adv(m_1, r) = c] - \Pr[adv(m_2, r) = c]| \leq 2\epsilon$. The probabilities are taken over the coin flips of the honest parties, and the sum is over all possible values of the adversary's view.

Reliability. Π is δ-reliable if, with probability at least $1 - \delta$, R outputs $m^R = m^S$ at the end of the protocol. The probability is over the choice of m^S and the coin flips of all parties.

Security. Π is (ϵ, δ)-secure if it is ϵ-private and δ-reliable.

We say Π is perfectly secure (PSMT) if it is a $(0,0)$-SMT protocol. In this paper, we also discuss reliability (without requirement for privacy): δ-RMT, 0-RMT, and almost perfect security: (ϵ, δ)-SMT and $(0, \delta)$-SMT. Note that in the rest of the paper, ϵ and δ only appear when studying almost perfect security, thus we let $\epsilon > 0$ and $0 < \delta < \frac{1}{2}$.

We employ the authentication code $\mathrm{auth}(m; a, b) = am + b$ for information-theoretically secure authentication. An authentication key $(a, b) \in_R \mathbb{F}^2$ can be used to authenticate one message m without revealing any information about the key itself.

3 Characterization of Multicast Graphs

In this section we characterize multicast graphs based on the adversary structures. We give an extended characterization which is essential for obtaining the necessary and sufficient conditions in the multicast model. This should give a clearer insight to the problems we are dealing with.

We let P be the set of all paths between S and R in a given graph $G(V, E)$. The adversary chooses a set of nodes $A \in \mathcal{A}$ to control, where $\mathcal{A}$ is an adversary structure on $V \setminus \{S, R\}$. For each path $p \in P$, we define *eavesdropping* and *separating* as follows.

Definition 1. *We say that the adversary can* eavesdrop *on p if it cannot control any node on p but can control some neighbors of p.*[3] *Suppose that the adversary can eavesdrop on p and there is an element a to be transmitted between S and R on p. We say that the adversary can* completely eavesdrop *on p if, despite what protocol is executed, the adversary can learn a by eavesdropping.*

Definition 2. *We say that the adversary can* separate S *and* R *on p if it can control some nodes on p. Suppose that the adversary can separate S and R on p and there are k elements $(a_1, \ldots, a_k) \in \mathbb{F}^k$ to be transmitted on p. We let $(a_1^S, \ldots, a_k^S)$ and $(a_1^R, \ldots, a_k^R)$ be the views of S and R respectively on these k elements at the end of any protocol. We say that the adversary can* completely separate S *and* R *if, despite what protocol is executed and how large k is, there exists a strategy of the adversary that causes $\forall i \ (1 \leq i \leq k) : a_i^S \neq a_i^R$.*

Next we show two lemmas regarding the eavesdropping and separating activities of the adversary on a single path $p \in P$. We assume that the path p is placed in a *left-to-right* direction, with S at the left end and R at the right end.

Lemma 1. *The adversary can completely eavesdrop on a path $p \in P$ if and only if it can eavesdrop on two adjacent nodes[4] on p.*

Proof. We first prove the "if" direction. The privacy problem has been studied by Franklin and Yung in [9]. They showed that private communication on p is possible only if, by removing all the faulty nodes and the hyperedges on which the faulty nodes are, path p remains.[5] Evidently, this necessary condition for privacy is satisfied if and only if the adversary *cannot* eavesdrop on two adjacent nodes on p (See Example 1 following this proof). Thus if the adversary can eavesdrop on two adjacent nodes on p, then it can completely eavesdrop on p.

Next we prove the "only if" direction. We give the following protocol, which allows S to send an element a^S to R with perfect privacy, when the adversary cannot eavesdrop on two adjacent nodes on p. First we assume that including S and R, there are $k + 2$ nodes $v_0, \ldots, v_{k+1}$ on p. We let S be node v_0, R be node v_{k+1}, and $v_1, \ldots, v_k$ be the other k nodes from left to right.

Single Path Private Propagation Protocol

1. For each $1 \leq i \leq k + 1$, v_i initiates an element $a_i \in_R \mathbb{F}$ and multicasts it. Thus for each $0 \leq i \leq k$, v_i receives element a_{i+1} from its right side neighbor node v_{i+1}.

---

[3] Obviously, if the adversary *can* control some nodes on p, then it can learn everything passing through those controlled nodes. However, for the purpose of our observation, we do not consider this activity as "eavesdropping", instead, we characterize it as "separating", which we describe in Definition 2.

[4] Two nodes $u, v \in V$ are said to be *adjacent* to one another if there is an edge $\{u, v\} \in E$ between them.

[5] In the threshold model where any t nodes can be the faulty, such connectivity is called the *weak t_{hyper}-connectivity*. We discuss this connectivity in more detail in Section 6.

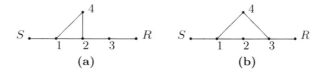

Fig. 1. Eavesdropping activities on a single path p

2. S sets $i := 1$ and multicasts $b_0 = a^S + a_1$. While $i \leq k$, v_i receives element b_{i-1} from its left side neighbor node v_{i-1}, v_i then multicasts $b_i = b_{i-1} - a_i + a_{i+1}$ and sets $i := i + 1$.
3. When $i = k + 1$, R receives element b_k from v_k, R then sets $a^R := b_k - a_{k+1}$.
   End.

Obviously, for each $0 \leq i \leq k$, the element that v_i multicasts is an encrypted ciphertext $b_i = a^S + a_{i+1}$. In order to decrypt a^S, the adversary needs to learn a pair (b_i, a_{i+1}) for some $0 \leq i \leq k$. Since b_i is multicast by v_i and a_{i+1} is multicast by v_{i+1}, the adversary who cannot eavesdrop on two adjacent nodes is not able to learn a^S by eavesdropping.   □

Single Path Eavesdropping Examples.

(a) If the adversary can eavesdrop on two adjacent nodes on path p, then the necessary condition of [9] is not satisfied. For example, in Fig 1(a), the faulty node is node 4 and the hyperedges are

$$(S, \{1\}), (1, \{S, 2, 4\}), (2, \{1, 3, 4\}), (3, \{2, R\}), (4, \{1, 2\}) \text{ and } (R, \{3\}).$$

By removing the hyperedges that node 4 is on, the remaining hyperedges are

$$(S, \{1\}), (3, \{2, R\}) \text{ and } (R, \{3\}).$$

Thus p does not remain because edge $\{1, 2\}$ is removed, and hence the condition of [9] is not satisfied.

(b) If the adversary cannot eavesdrop on two adjacent nodes on path p, then the necessary condition of [9] is satisfied. For example, in Fig 1(b), the faulty node is node 4 and the hyperedges are

$$(S, \{1\}), (1, \{S, 2, 4\}), (2, \{1, 3\}), (3, \{2, 4, R\}), (4, \{1, 3\}) \text{ and } (R, \{3\}).$$

By removing the hyperedges that node 4 is on, the remaining hyperedges are

$$(S, \{1\}), (2, \{1, 3\}) \text{ and } (R, \{3\}).$$

Thus p remains because all edges on p remain, and hence the condition of [9] is satisfied.

The different separating activities were observed by Franklin and Wright in [7], but here we extend their result and upgrade their protocol.

Lemma 2. (following [7]) *The adversary can completely separate S and R on a path $p \in P$ if and only if it can control two or more nodes on p.*

Proof. We refer the proof of the "if" direction to [8].

Next we prove the "only if" direction. We assume that including S and R, there are $k + 2$ nodes $v_0, \ldots, v_{k+1}$ on p. We let S be node v_0, R be node v_{k+1}, and $v_1, \ldots, v_k$ be the other k nodes from left to right. We show that with the following protocol, the adversary cannot completely separate S and R when k elements $(a_1, \ldots, a_k)$ are transmitted on p if the adversary can control no more than one node on p.

Single Path Distribution Protocol

1. For each $1 \leq i \leq k$, v_i initiates an element $a_i \in_R \mathbb{F}$ and multicasts it.
2. For each $1 \leq i \leq k$, the nodes on the left side of v_i execute an instance of the Single Path Private Propagation Protocol from v_{i-1} to S in which v_{i-1} sends a_i, and the nodes on the right side of v_i execute an instance of the Single Path Private Propagation Protocol from v_{i+1} to R in which v_{i+1} sends a_i.
3. At the end of the protocol, for each $1 \leq i \leq k$, S receives an element a_i^S and R receives an element a_i^R. If S (or R) receives nothing regarding element a_i for some $1 \leq i \leq k$, then S (or R) sets $a_i^S = 1$ (or $a_i^R = 1$). **End.**

Let v_e $(1 \leq e \leq k)$ be the only faulty node on p. It is straightforward that at the end of the protocol, $a_e^S = a_e^R$, even if v_e does not initiate and multicast any element (in this case $a_e^S = a_e^R = 1$). □

Next, we give the following two lemmas, which are trivial so we omit the proofs.

Lemma 3. *If the adversary can only control one node v on a path $p \in P$, then despite what protocol is executed on p, there exists a strategy of the adversary that causes the views of S and R to be different except for their views on the elements multicast by v.*

Lemma 4. *Given a node v on a path $p \in P$, if the adversary cannot separate S and R on p, completely eavesdrop on p, or control a neighbor of v, then during the execution of the Single Path Distribution Protocol on p, the adversary cannot learn the elements multicast by v.*

Having these lemmas, we now present an *extended characterization* $\zeta_\mathcal{A}$ of a multicast graph $G(V, E)$ given an adversary structure $\mathcal{A}$ on $V \setminus \{S, R\}$.

Definition 3. *Given a graph $G(V, E)$, let $\mathcal{A} = \{A_1, \ldots, A_z\}$ be an adversary structure on $V \setminus \{S, R\}$ and P be the set of all paths between S and R. An Extended Characterization of G given $\mathcal{A}$ is $\zeta_\mathcal{A} = \{\zeta_{A_1}, \ldots, \zeta_{A_z}\}$ where for each $1 \leq i \leq z$, we have $\zeta_{A_i} = (P_i^{(+)}, P_i^{(1)}, P_i^{(*)}, P_i)$ where*

- *$P_i^{(+)}$ is the set of all paths on each of which there are at least two nodes in A_i,*

- $P_i^{(1)}$ is the set of all paths on each of which there is exactly one node in A_i,
- $P_i^{(*)}$ is the set of all paths on each of which there is no node in A_i, but on each path in $P_i^{(*)}$, there are two adjacent nodes that both have neighbors in A_i, and
- $P_i = P_i^{(+)} \cup P_i^{(1)}$ is the set of all paths on each of which there is at least one node in A_i.

With the extended characterization $\zeta_{\mathcal{A}}$, we know that during the execution of any protocol, by choosing a set $A_i \in \mathcal{A}$ to control, the adversary can separate S and R on P_i, completely separate S and R on $P_i^{(+)}$ and completely eavesdrop on $P_i^{(*)}$.

Given any set $A_i \in \mathcal{A}$, we are particularly interested in the nodes of A_i on the paths of $P_i^{(1)}$. For each path $p \in P_i^{(1)}$, we use $A_i \sqcap p$ to denote the single node $v \in A_i$ that is on path p; i.e., $v = A_i \sqcap p$. Note that this notation is only used for the paths in $P_i^{(1)}$.

Definition 4. *Given a graph $G(V, E)$ and an adversary structure $\mathcal{A}$ on $V \setminus \{S, R\}$, we say that S and R are* highly $\mathcal{A}$-connected *if for any set $A_i \in \mathcal{A}$, we have $P_i \cup P_i^{(*)} \neq P$.*

Definition 5. *Given a graph $G(V, E)$ and an adversary structure $\mathcal{A}$ on $V \setminus \{S, R\}$, we say that S and R are* lowly 2$\mathcal{A}$-separated *if there exist two (not necessarily distinct) sets $A_1, A_2 \in \mathcal{A}$ such that*

(a) $P_1 \cup P_2 = P$, and

(b) $P_1^{(1)} = \emptyset$, or for each path $p \in P_1^{(1)}$, we have that $p \in P_2 \cup P_2^{(*)}$ or $A_1 \sqcap p$ has a neighbor in A_2, and

(c) $P_2^{(1)} = \emptyset$, or for each path $p \in P_2^{(1)}$, we have that $p \in P_1 \cup P_1^{(*)}$ or $A_2 \sqcap p$ has a neighbor in A_1.

We say that S and R are lowly 2$\mathcal{A}$-connected *if they are not lowly 2$\mathcal{A}$-separated.*

Lemma 5. *Given a graph $G(V, E)$ and an adversary structure $\mathcal{A}$ on $V \setminus \{S, R\}$, if S and R are lowly 2$\mathcal{A}$-connected, then for any set $A_i \in \mathcal{A}$, we have $P_i \neq P$.*

Proof. Assume there exits a set $A_i \in \mathcal{A}$ such that $P_i = P$, if we let both the sets A_1, A_2 of Definition 5 be A_i, then it is straightforward that S and R are lowly 2$\mathcal{A}$-separated. Thus we have a contradiction.   □

4   Reliable Communication

In this section, we discuss reliable communication with no requirement for privacy. We study almost perfect reliability (δ-RMT) in Section 4.1 and perfect reliability (0-RMT) in Section 4.2.

4.1   Almost Perfect Reliability

We give the necessary and sufficient condition for δ-RMT in multicast graphs.

Theorem 1. *Given a graph $G(V, E)$ and an adversary structure $\mathcal{A}$ on $V \setminus \{S, R\}$. The necessary and sufficient condition for δ-RMT from S to R is that S and R are lowly $2\mathcal{A}$-connected.*

Next, we use Lemma 7 and Lemma 8 to show the necessity and sufficiency of the condition respectively. Before we present these two lemmas, we first give the following Lemma 6, which is a key ingredient for proving the necessity.

Lemma 6. *If there exists two sets $A_1, A_2 \in \mathcal{A}$ such that $P_1^{(+)} \cup P_2^{(+)} = P$, and $\delta < \frac{1}{2}(1 - \frac{1}{|\mathbb{M}|})$, then δ-RMT from S to R is impossible.*

Proof. This lemma can be easily proven using a similar technique as that in [8, Theorem 5.1] and [5, Theorem 3]. See the full version of this paper [1].   □

Lemma 7. *The condition of Theorem 1 is necessary.*

Proof. It is straightforward that in order to achieve δ-reliability, it is necessary to have $P_i \neq P$ for any $A_i \in \mathcal{A}$; i.e., $P \setminus P_i \neq \emptyset$.

Next we prove the necessity of the condition by contradiction. We assume that S and R are lowly $2\mathcal{A}$-separated (i.e., there exist two sets $A_1, A_2 \in \mathcal{A}$ as they are in Definition 5) and there exists a δ-RMT protocol Π that transmits a message $m \in \mathbb{M}$ from S to R. Without loss of generality, we let $P_1 \cap P_2 = \emptyset$. Now if $P_1^{(1)} = \emptyset$ and $P_2^{(1)} = \emptyset$, then we have $P_1^{(+)} = P_1$ and $P_2^{(+)} = P_2$, and hence $P_1^{(+)} \cup P_2^{(+)} = P$ (following Definition 5(a)), thus due to Lemma 6, δ-RMT is impossible in the case. In the rest of our proof we let $P_1^{(1)} \neq \emptyset$ and/or $P_2^{(1)} \neq \emptyset$.

We make an observation on how protocol Π can achieve δ-reliability. Given a node v on a path $p \in P$, we use $(v \sim p)$ to denote the tuple of the elements that are multicast by v and received (in any way) by both S and R on p, and let $(v \sim p)^S$ and $(v \sim p)^R$ be the views of S and R respectively on $(v \sim p)$.

The strategy of the adversary is to choose an $e \in_R \{1, 2\}$ and control the set A_e. Let $d \in \{1, 2\}$ such that $d \neq e$, then R should be able to recover the actual message from the elements received on P_d. If, despite whether $e = 1$ or $e = 2$, $(v \sim p)^S \neq (v \sim p)^R$ for any v on any $p \in P_e$ (i.e., the views of S and R are completely different on P_e), then following Lemma 6, δ-RMT is impossible. Therefore, there must exist an $e \in \{1, 2\}$ such that $(v \sim p)^S = (v \sim p)^R$ is guaranteed for some v on some $p \in P_e$. We say that the tuple of elements $(v \sim p)$ where $p \in P_e$ such that $(v \sim p)^S = (v \sim p)^R$ *supports* the actual message. Following Lemma 2, the adversary can completely separate S and R on $P_e^{(+)}$ and cause $\forall (p \in P_e^{(+)}, v \text{ on } p) : (v \sim p)^S \neq (v \sim p)^R$. Following Lemma 3, for any path $p \in P_e^{(1)}$ (if $P_e^{(1)} \neq \emptyset$), $(v \sim p)^S = (v \sim p)^R$ can only be guaranteed if $v = A_e \sqcap p$. Therefore, there must exist an $e \in \{1, 2\}$ such that the actual message received on P_d is supported by some $((A_e \sqcap p) \sim p)$ where $p \in P_e^{(1)}$. Next, following Definition 5(b,c), for each path $p \in P_d^{(1)}$ (if $P_d^{(1)} \neq \emptyset$), we have case 1: $p \in P_e \cup P_e^{(*)}$, or case 2: $A_d \sqcap p$ has a neighbor in A_e. In case 1: $p \in P_e \cup P_e^{(*)}$, due to Lemma 1, there is no private transmission on path p whatsoever, so the adversary can learn $((A_d \sqcap p) \sim p)$. In case 2: $A_d \sqcap p$ has a neighbor in A_e, it is trivial that the adversary can learn $((A_d \sqcap p) \sim p)$.

To sum up, we can *conclude* that when the adversary chooses A_e to control, then the actual message, which can be recovered from the elements received on P_d, should be supported by some $((A_e \sqcap p) \sim p)$ where $p \in P_e^{(1)}$ (if $P_e^{(1)} \neq \emptyset$), and the adversary can learn $((A_d \sqcap p) \sim p)$ for each $p \in P_d^{(1)}$ (if $P_d^{(1)} \neq \emptyset$).

Now during the execution of the protocol Π, the adversary corrupts P_e and causes $(v \sim p)^S \neq (v \sim p)^R$ for all nodes v on all paths $p \in P_e$ except for $p \in P_e^{(1)}$ and $v = A_e \sqcap p$. This is possible due to Lemma 2 and Lemma 3. As we *concluded* above, the adversary can always learn $((A_d \sqcap p) \sim p)$ for each $p \in P_d^{(1)}$. Thus on P_e, the adversary simulates the protocol as S sent a message $m' \in \mathbb{M}$, and m' can be supported by $((A_d \sqcap p) \sim p)$, where $p \in P_d^{(1)}$.

Therefore, at the end of the protocol Π, despite whether $e = 1$ or $e = 2$, the view of R always consists of the following:

- on P_1, a message is recovered which can be supported by $((A_2 \sqcap p) \sim p)$ for any $p \in P_2^{(1)}$ (if $P_2^{(1)} \neq \emptyset$), but may not be supported by any other elements received on P_2;
- on P_2, a different message is recovered which can be supported by $((A_1 \sqcap p) \sim p)$ for any $p \in P_1^{(1)}$ (if $P_1^{(1)} \neq \emptyset$), but may not be supported by any other elements received on P_1.

Thus as we showed in Lemma 6, with probability $\delta \geq \frac{1}{2}(1 - \frac{1}{|\mathbb{M}|})$, R recovers the wrong message m'. We have a contradiction, which proves the necessity of the low 2$\mathcal{A}$-connectivity. $\qquad\square$

Let $P = \{p_1, \ldots, p_n\}$, we first generalize some of Franklin and Wright's protocols in multicast graphs.

Full Distribution Protocol

1. For each $1 \leq j \leq n$, the nodes on path p_j execute an instance of the Single Path Distribution Protocol for each node v_i on p_j to distribute an element $a_{i,j}$. The nodes not on p_j do not multicast anything.
2. At the end of the protocol, on each path p_j $(1 \leq j \leq n)$, S and R receive $a_{i,j}^S$ and $a_{i,j}^R$ respectively as the element initiated by node v_i on p_j.   **End.**

Private Propagation Protocol

1. For each $1 \leq j \leq n$, the nodes on path p_j execute an instance of the Single Path Private Propagation Protocol from S to R in which S sends an element a_j^S, and the nodes not on p_j do not multicast anything.
2. At the end of the protocol, on each path p_j $(1 \leq j \leq n)$, R receives a_j^R as the element that S initiated and propagated on p_j.   **End.**

Now we present the following protocol, which achieves δ-RMT for a message $m \in \mathbb{M}$ in a graph $G(V, E)$.

Reliable Transmission Protocol

1. The nodes of V execute an instance of the Full Distribution Protocol in which for each $1 \le j \le n$, the elements that node v_i on path p_j initiates are $(a_{i,j}, b_{i,j}) \in_R \mathbb{F}^2$. Let $(a_{i,j}^S, b_{i,j}^S)$ and $(a_{i,j}^R, b_{i,j}^R)$ be what S and R receive respectively regarding $(a_{i,j}, b_{i,j})$.

2. The nodes of V execute an instance of the Private Propagation Protocol from S to R in which S sends the same vector on all paths in P:

$$(m, \langle \operatorname{auth}(m; a_{i,j}^S, b_{i,j}^S) \rangle),$$

where $\langle \operatorname{auth}(m; a_{i,j}^S, b_{i,j}^S) \rangle$ is an ordered set of the authenticated m with *all* keys $(a_{i,j}^S, b_{i,j}^S)$ that S receives in Step 1. At the end of the instance, R receives a vector $(m_k, \langle u_{i,j,k} \rangle)$ on each path $p_k \in P$.

3. Given the vector $(m_k, \langle u_{i,j,k} \rangle)$ that R receives on p_k, if $\exists (i,j) : u_{i,j,k} = \operatorname{auth}(m_k; a_{i,j}^R, b_{i,j}^R)$, then we say that m_k is *qualified* on $(v_i \sim p_j)$. R finds an $A_f \in \mathcal{A}$ that satisfies the following three α-conditions:

   α-1 all vectors received on $P \setminus P_f$ are the same, say vector $(m_l, \langle u_{i,j,l} \rangle)$;

   α-2 $P_f^{(1)} = \emptyset$, or for each $p_j \in P_f^{(1)}$, m_l is qualified on $((A_f \sqcap p_j) \sim p_j)$;

   α-3 $P_f \cup P_f^{(*)} = P$, or for any vector $(m_k, \langle u_{i,j,k} \rangle)$ received on path $p_k \in P_f$ such that $m_k \ne m_l$, we have that m_k is *not* qualified on any $(v_i \sim p_j)$ where $p_j \in P \setminus (P_f \cup P_f^{(*)})$ and v_i does not have a neighbor in A_f.

   R then outputs the message m_l.   **End.**

Lemma 8. *The Reliable Transmission Protocol is a δ-RMT protocol under the condition of Theorem 1.*

Proof. It is straightforward that if the adversary cannot learn some $(a_{i,j}, b_{i,j})$ (initiated by v_i and multicast on p_j) but a *corrupted* m_k is qualified on $(v_i \sim p_j)$, then the Reliable Transmission Protocol fails. We use $\overline{RT}$ to denote the event when the above failure occurs and RT to denote the event otherwise. Let n be the total number of paths between S and R and y be the maximum number of nodes on any path, following the proof of [8, Theorem 3.4], the probability that the protocol fails is $\Pr[\overline{RT}] < \frac{yn^2}{|\mathbb{F}|}$. This probability is negligible in the security parameter (given $\mathbb{F}$ is sufficiently large). Next in our proof, we assume that the above failure does not happen. That is, *we analyze the protocol in the event RT*.

In the following, we first show that R can always find an $A_f \in \mathcal{A}$ that satisfies the three α-conditions, then we prove, by contradiction, that in the event RT, the message output by R is correct.

Now we show that there always exists an A_f that satisfies all three α-conditions, at least when the adversary chooses A_f to control so that P_f is *corrupted*. Since $P_f \ne P$ (following Lemma 5), we immediately have that condition α-1 is satisfied and m_l received on $P \setminus P_f$ is the actual message. If $P_f^{(1)} \ne \emptyset$, then as shown in the proof of Lemma 2, on each $p_j \in P_f^{(1)}$, S and R always have the same view on the key initiated by $A_f \sqcap p_j$. Thus it is clear that m_l is qualified on

$((A_f \sqcap p_j) \sim p_j)$, and hence condition α-2 is satisfied. If $P_f \cup P_f^{(*)} \neq P$, then the adversary cannot learn the key initiated by any node v_i which is on a path $p_j \in P \setminus (P_f \cup P_f^{(*)})$ if v_i does not have a neighbor in A_f. Thus without the above mentioned failure $\overline{RT}$, any faulty message $m_k \neq m_l$ cannot be qualified on such $(v_i \sim p_j)$, and hence condition α-3 is satisfied.

Next, using contradiction, we show that in the event RT, the message m_l that S outputs is the actual message. For contradiction, we assume that m_l is modified by the adversary who chooses a set $A_e \in \mathcal{A}$ to control, and all three α-conditions are satisfied. We now show that the three α-conditions imply the three properties of A_1, A_2 in Definition 5.

- From condition α-1, since all vectors received on $P \setminus P_f$ are modified, we have $P_e \cup P_f = P$ (i.e., corresponding to Definition 5(a)).

- Condition α-2 indicates that either $P_f^{(1)} = \emptyset$, or the adversary can learn the key initiated by node $A_f \sqcap p_j$ on any path $p_j \in P_f^{(1)}$ to make the faulty message m_l qualified on $((A_f \sqcap p_j) \sim p_j)$. Due to Lemma 4, this means that the adversary can separate S and R on p_j, completely eavesdrop on p_j or control a neighbor of $A_f \sqcap p_j$. Thus from condition α-2 we can conclude that $P_f^{(1)} = \emptyset$, or for each path $p_j \in P_f^{(1)}$, we have that $p_j \in P_e \cup P_e^{(*)}$ or $A_f \sqcap p_j$ has a neighbor in A_e (i.e., corresponding to Definition 5(c)).

- Finally, since $P_e \neq P$ and $P_e \cup P_f = P$, there exists at least one path $p_k \in P_f$ such that the message m_k received on p_k is the actual message. Due to condition α-3, there are two cases:

case 1 $P_f \cup P_f^{(*)} = P$, thus we have $P_e^{(1)} \subseteq P_f \cup P_f^{(*)} = P$;

case 2 The actual message m_k is *not* qualified on any $(v_i \sim p_j)$ where $p_j \in P \setminus (P_f \cup P_f^{(*)})$ and v_i does not have a neighbor in A_f. This implies that either $p_j \in P_e^{(+)}$, or $p_j \in P_e^{(1)}$ but any v_i on p_j that does not have a neighbor in A_f is not $A_e \sqcap p_j$ (because otherwise the actual message m_k should be qualified on $(v_i \sim p_j)$, due to the proof of Lemma 2). That is, if such $p_j \in P_e^{(1)}$ exists, then all the nodes on p_j that do not have a neighbor in A_f are not $A_e \sqcap p_j$. This implies that $A_e \sqcap p_j$ has a neighbor in A_f.

It is easy to conclude that in either case, $P_e^{(1)} = \emptyset$, or for each path $p_j \in P_e^{(1)}$, we have $p_j \in P_f \cup P_f^{(*)}$ or $A_e \sqcap p_j$ has a neighbor in A_f (i.e., corresponding to Definition 5(b)).

To sum up, A_e, A_f are as A_1, A_2 in Definition 5. This means S and R are lowly $2\mathcal{A}$-separated, which contradicts the condition of Theorem 1.

Therefore, at the end of the Reliable Transmission Protocol, R can recover $m_l = m$ with an arbitrarily small probability of failure (i.e., $\Pr[\overline{RT}] < \frac{yn^2}{|\mathbb{F}|}$). Thus the Reliable Transmission Protocol is a δ-RMT protocol.   □

4.2   Perfect Reliability

Here we study 0-RMT in multicast graphs. Similar to the result in [7], we show that the necessary and sufficient condition for 0-RMT in the multicast setting is the same as that in the point-to-point setting. The following theorem can be easily proven following some previous results in [8,5].

Theorem 2. *Given a graph $G(V, E)$ and an adversary structure $\mathcal{A}$ on $V \setminus \{S, R\}$. The necessary and sufficient condition for 0-RMT from S to R is that $P_i \cup P_j \neq P$ for any two sets $A_i, A_j \in \mathcal{A}$.*

Proof. See the full version of this paper [1].   □

5   Secure Communication

In this section we take the problem of achieving privacy into consideration. We study almost perfect security in Section 5.1; i.e., we discuss both (ϵ, δ)-SMT and $(0, \delta)$-SMT. In Section 5.2, we study $(0, 0)$-SMT that enables perfect security.

5.1   Almost Perfect Security

First we give the necessary and sufficient condition for (ϵ, δ)-SMT in multicast graphs. Unlike the setting in [7] in which the conditions for both δ-RMT and (ϵ, δ)-SMT are the same (i.e., $n > t$), in multicast graphs, (ϵ, δ)-SMT requires stronger connectivity than that for δ-RMT.

Theorem 3. *Given a graph $G(V, E)$ and an adversary structure $\mathcal{A}$ on $V \setminus \{S, R\}$. The necessary and sufficient condition for (ϵ, δ)-SMT from S to R is that S and R are highly $\mathcal{A}$-connected and lowly $2\mathcal{A}$-connected.*

Proof. We first prove the necessity of the condition. It is straightforward that the high $\mathcal{A}$-connectivity, i.e., $P_i \cup P_i^{(*)} \neq P$, is necessary for achieving ϵ-privacy, because otherwise there is no private transmission between S and R on any path in P. Moreover, as proven in Lemma 7, the low $2\mathcal{A}$-connectivity is necessary for achieving δ-reliability. Thus the condition is necessary for (ϵ, δ)-SMT.

Next we show that the condition is sufficient. Let $P = \{p_1, \ldots, p_n\}$, we give the following protocol (similar to [8,15]) for S to send a message $m \in \mathbb{M}$ to R.

Private Transmission Protocol

1. The nodes of V execute an instance of the Private Propagation Protocol from S to R in which for each $1 \leq j \leq n$, S sends a pair $(a_j^S, b_j^S) \in_R \mathbb{F}$ on path $p_j \in P$. At the end of the instance, R receives a pair (a_j^R, b_j^R) on each path $p_j \in P$.
2. R chooses an element $r^R \in_R \mathbb{F}$ and for each $1 \leq j \leq n$, computes $s_j^R = \mathrm{auth}(r^R; a_j^R, b_j^R)$. The nodes of V executes an instance of the Reliable Transmission Protocol from R to S in which R sends a vector $(r^R, s_1^R, \ldots, s_n^R)$. At the end of the instance, S outputs a vector $(r^S, s_1^S, \ldots, s_n^S)$.

3. S computes an index set $I = \{j | s_j^S = \text{auth}(r^S; a_j^S, b_j^S)\}$ and an encryption key $key = \sum_{j \in I} a_j^S$, and encrypts the message $c = m + key$. The nodes of V executes an instance of the Reliable Transmission Protocol from S to R in which S sends a vector (I, c). At the end of the instance, R outputs a vector (I', c').

4. R computes a decryption key $key' = \sum_{j \in I'} a_j^R$ and decrypts the message $m' = c' - key'$. **End.**

First we show that this protocol achieves ϵ-privacy. Suppose that the adversary chooses a set A_e to control. Since $P_e \cup P_e^{(*)} \neq P$, there exists a path $p_d \in P \setminus (P_e \cup P_e^{(*)})$. As shown in the proof of Lemma 1, the adversary cannot learn (a_d^S, b_d^S) in Step 1. Because $p_d \notin P_e$, we have $(a_d^R, b_d^R) = (a_d^S, b_d^S)$. Let RT denote the event that the instance of the Reliable Transmission Protocol in Step 2 succeeds and $\overline{RT}$ denote the event otherwise. In the event RT, $r^S = r^R$ and for each $1 \leq j \leq n$, we have $s_j^S = s_j^R$. This implies that $d \in I$. The adversary who cannot learn a_d^S by eavesdropping or by decoding s_d^R will not be able to compute key to decrypt m. That is, for any two messages $m_1, m_2 \in \mathbb{M}$ and any coin flips r, using the adversary's view adv, we have the following:

$$\sum_c | \Pr[adv(m_1, r) = c | RT] - \Pr[adv(m_2, r) = c | RT]| = 0 \qquad (1)$$

$$\sum_c | \Pr[adv(m_1, r) = c | \overline{RT}] - \Pr[adv(m_2, r) = c | \overline{RT}]| \leq |+1| + |-1| = 2 \quad (2)$$

Let $\Pr[\overline{RT}] = \epsilon$, which is arbitrarily small as we discussed in the proof of Lemma 8, by combining Eq. 1 and Eq. 2, we have the following:

$$\sum_c | \Pr[adv(m_1, r) = c] - \Pr[adv(m_2, r) = c]| \leq 0 \cdot \Pr[RT] + 2 \cdot \Pr[\overline{RT}] = 2\epsilon.$$

Thus the Private Transmission Protocol achieves ϵ-privacy.

Next we show that the protocol achieves δ-reliability. Let δ_1 be the probability that the instance of the Reliable Transmission Protocol in Step 2 fails and δ_2 be the probability that the instance in Step 3 fails. As we showed in the proof of Lemma 8, δ_1 and δ_2 are negligible in the security parameter. Let δ_3 be the probability that both the above mentioned instances succeed, but R outputs $m' \neq m$. This can only happen if there exists at least one $j \in I$ such that $a_j^S \neq a_j^R$. Since both reliable protocols succeed, the fact $j \in I$ implies $\text{auth}(r^R; a_j^S, b_j^S) = \text{auth}(r^R; a_j^R, b_j^R)$. That is,

$$a_j^S r^R + b_j^S = a_j^R r^R + b_j^R \Rightarrow r^R = \frac{b_j^R - b_j^S}{a_j^S - a_j^R} \in \mathbb{F}, \qquad (3)$$

where $a_j^S \neq a_j^R$. Since r^R is chosen with respect to the uniform distribution, if the adversary modifies (a_j^S, b_j^S) to (a_j^R, b_j^R) on path p_j in Step 1, then the probability that Eq. 3 is fulfilled is $\frac{1}{|\mathbb{F}|}$. Since the adversary can corrupt $|P_e|$ paths, it is straightforward that $\delta_3 = \frac{|P_e|}{|\mathbb{F}|} < \frac{n}{|\mathbb{F}|}$, which is much smaller than δ_1 and δ_2. Thus the final probability that the protocol fails to be reliable is

$$\delta = \delta_1 + (1 - \delta_1)\delta_2 + (1 - (\delta_1 + (1 - \delta_1)\delta_2))\delta_3 < \delta_1 + \delta_2 + \delta_3.$$

To sum up, the Private Transmission Protocol is an (ϵ, δ)-SMT protocol.   □

Note that the condition of Theorem 3 can be seen as it consists of two parts, with the high $\mathcal{A}$-connectivity enables private communication and the low $2\mathcal{A}$-connectivity enables δ-reliable communication. These two types of connectivity are *independent*. Indeed, with some examples in Section 6, we can show that they do not imply each other.

In [16], Yang and Desmedt proved that reducing the requirement for privacy does not weaken the minimal connectivity. In the following theorem, we show that the condition for (ϵ, δ)-SMT is also necessary and sufficient for $(0, \delta)$-SMT.

Theorem 4. *Given a graph $G(V, E)$ and an adversary structure $\mathcal{A}$ on $V \setminus \{S, R\}$. The necessary and sufficient condition for $(0, \delta)$-SMT from S to R is that S and R are highly $\mathcal{A}$-connected and lowly $2\mathcal{A}$-connected.*

Proof. It is straightforward that the condition is necessary. Next we show that the condition is sufficient by slightly amending the Private Transmission Protocol to the following protocol which achieves perfect privacy.

Perfectly Private Transmission Protocol

1. Same as Step 1 in the Private Transmission Protocol.
2. R chooses an element $r^R \in_R \mathbb{F}$ and for each $1 \leq j \leq n$, computes $s_j^R = \text{auth}(r^R; a_j^R, b_j^R)$. The nodes of V executes an instance of the Reliable Transmission Protocol from R to S in which R sends a vector $(r^R, s_1^R, \ldots, s_n^R)$. At the end of the instance, S distinguishes the following two cases:
   Case 1 If there exist two sets $A_{f_1}, A_{f_2} \in \mathcal{A}$ that satisfy all three α-conditions of the Reliable Transmission Protocol, and the two vectors (both regarding the vector $(r^R, s_1^R, \ldots, s_n^R)$) that S receives respectively on $P \setminus P_{f_1}$ and $P \setminus P_{f_2}$ are different, then S terminates the protocol.
   Case 2 Otherwise, S outputs a vector $(r^S, s_1^S, \ldots, s_n^S)$ and goes to Step 3.
3. Same as Step 3 in the Private Transmission Protocol.
4. Same as Step 4 in the Private Transmission Protocol.   **End.**

Now we show that this protocol achieves 0-privacy. Following the proof of Theorem 3, the privacy of the message transmission can only be breached in the event $\overline{RT}$. It is clear that the instance of the Reliable Transmission Protocol in Step 2 allows S to distinguish the events RT and $\overline{RT}$. As we showed in the proof of Lemma 8, in the event RT, only the correct vector can be output after the Reliable Transmission Protocol. This means if two different vectors can be output, then the event $\overline{RT}$ occurs. Thus in Step 2, Case 1 indicates $\overline{RT}$ and Case 2 indicates RT. In the event $\overline{RT}$, S terminates the protocol so the adversary learns nothing about the message. Thus the protocol achieves 0-privacy.[6] Next, using a similar proof as that for Theorem 3, we can prove that the Perfectly Private Transmission Protocol is also δ-reliable, which concludes the proof.   □

---

[6] A more formal proof is available in the full version of this paper [1].

5.2 Perfect Security

In [6], Dolev et al. showed that if σ is the maximum number of channels that a *listening* (passive) adversary can control and ρ is the maximum number of channels that a *disrupting* (active) adversary can control, then there must be at least $\max\{\sigma + \rho + 1, 2\rho + 1\}$ channels between S and R for PSMT (i.e., $(0,0)$-SMT). This setting can be generalized in our model as follows: given an adversary structure $\mathcal{A} = \{A_1, \ldots, A_z\}$, then $\{P_1 \cup P_1^{(*)}, \ldots, P_z \cup P_z^{(*)}\}$ consists of the subsets of paths a listening adversary can control and $\{P_1, \ldots, P_z\}$ consists of the subsets of paths a disrupting adversary can control. Thus we give the following theorem for $(0,0)$-SMT in multicast graphs.

Theorem 5. *Given a graph $G(V,E)$ and an adversary structure $\mathcal{A}$ on $V \setminus \{S, R\}$. The necessary and sufficient condition for (0,0)-SMT from S to R is that*
$$(P_i \cup P_i^{(*)}) \cup P_j \neq P \text{ for any } A_i, A_j \in \mathcal{A}.$$

Proof. See the full version of this paper [1]. □

6 Corresponding Threshold Model

In this section we use our results in the general adversary model to find the necessary and sufficient conditions for RMT and SMT in the threshold model. Because a threshold is a special case of an adversary structure, we re-define the threshold model in the adversary structure context.

Definition 6. *Given a graph $G(V,E)$, a threshold t is an adversary structure $\mathcal{A}^T \subseteq 2^{V \setminus \{S,R\}}$ such that $\forall (A \subseteq V \setminus \{S, R\}, |A| \leq t) : A \in \mathcal{A}^T$. Furthermore,*
- *we say that S and R are $t_{\zeta\text{-}private}$-connected if they are highly $\mathcal{A}^T$-connected;*
- *we say that S and R are $t_{\zeta\text{-}reliable}$-connected if they are lowly $2\mathcal{A}^T$-connected.*

It is easy to show that our results correspond to Franklin and Wright's [7] if the multicast graph only consists of n *node-disjoint* and *neighbor-disjoint* paths. For more details see the full version of this paper [1].

Next we discuss the connectivity in the general multicast graph setting with some previous results. In [4], Desmedt and Wang looked at four different types of connectivity. With slight changes, we show them in our model as follows.

- *t-connectivity.* For any $A \in \mathcal{A}^T$, after removing all nodes in A from G, there remains a path between S and R.
- *weak t_{hyper}-connectivity.* For any $A \in \mathcal{A}^T$, after removing from the hypergraph $H_G(V, E_H)$ all nodes in A and all hyperedges on each of which there is at least one node in A, there remains a path between S and R (see [9]).
- *$t_{neighbor}$-connectivity.* For any $A \in \mathcal{A}^T$, after removing all nodes in A and all their neighbors from G, there remains a path between S and R.
- *weak (n,t)-connectivity.* There are n node-disjoint paths $p_1, \ldots, p_n$ between S and R, and for any $A \in \mathcal{A}^T$, after removing all nodes in A and all their neighbors from G, there remains a path p_i $(1 \leq i \leq n)$ between S and R.

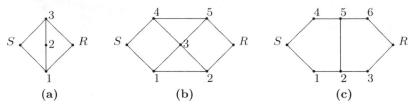

Fig. 2. Private and reliable connectivity

As we showed in the proof of Lemma 1, Franklin and Yung's weak t_{hyper}-connectivity [9] in a hypergraph H_G is essentially our $t_{\zeta-private}$-connectivity in a multicast graph G. Thus we use the $t_{\zeta-private}$-connectivity to replace the weak t_{hyper}-connectivity in the rest of the paper for a simpler presentation. Desmedt and Wang [4] showed that the following implications are strict:

$$\text{weak } (n,t)\text{-connectivity} \Rightarrow t_{neighbor}\text{-connectivity} \Rightarrow t_{\zeta-private}\text{-connectivity}$$
$$\Rightarrow t\text{-connectivity.}$$

In [14], Wang and Desmedt claimed that the weak (n,t)-connectivity is sufficient for $(0,\delta)$-SMT. Since weak (n,t)-connectivity $\Rightarrow t_{\zeta-private}$-connectivity, it is clear that 0-privacy can be achieved. However, δ-reliability is only achievable under their condition if weak (n,t)-connectivity $\Rightarrow t_{\zeta-reliable}$-connectivity. In [14], there is not a proper proof showing this implication. Thus their claim is only a conjecture. We leave this as an open problem.

Later study by Desmedt and Wang [4] showed that the conjectured upper bound, i.e., the weak (n,t)-connectivity, is not necessary for $(0,\delta)$-SMT, by showing an example, as Fig. 2(a), in which S and R are not weakly $(2,1)$-connected but $(0,\delta)$-SMT is possible. We observe that their protocol (appeared in [15]) is actually an (ϵ,δ)-SMT protocol but the claim is correct, because S and R are obviously $1_{\zeta-private}$-connected and $1_{\zeta-reliable}$-connected in Fig. 2(a). They also showed that the weak t_{hyper}-connectivity (i.e., the $t_{\zeta-private}$-connectivity) is the lower bound for $(0,\delta)$-SMT but not necessary for δ-RMT, as in Fig. 2(b) where S and R are not $1_{\zeta-private}$-connected but δ-RMT is possible. This claim is obvious under our condition because S and R are clearly $1_{\zeta-reliable}$-connected. Finally they conjectured that the weak t_{hyper}-connectivity (i.e., the $t_{\zeta-private}$-connectivity) is not sufficient for $(0,\delta)$-SMT, by asking whether $(0,\delta)$-SMT is possible in Fig. 2(c) such that S and R are $1_{\zeta-private}$-connected. Our condition proves their conjecture. Indeed, not only is $(0,\delta)$-SMT impossible in Fig. 2(c), but δ-RMT is also impossible, because S and R are not $1_{\zeta-reliable}$-connected. Therefore, our result explains all the examples and proves all the conjectures in the previous work.

Note that the examples of Fig. 2(b) and Fig. 2(c) also show that the $t_{\zeta-private}$-connectivity (or, the high $\mathcal{A}$-connectivity) and the $t_{\zeta-reliable}$-connectivity (or, the low $2\mathcal{A}$-connectivity) *do not imply each other*, because in Fig. 2(b), S and R are $1_{\zeta-reliable}$-connected but not $1_{\zeta-private}$-connected, and in Fig. 2(c), they are $1_{\zeta-private}$-connected but not $1_{\zeta-reliable}$-connected.

At the end, we present the following corollary as the final result of this paper.

Corollary 1. *Given a graph $G(V,E)$ and an adversary who can control up to t nodes in $V \setminus \{S, R\}$.*

- δ-RMT is possible if and only if S and R are $t_{\zeta\text{-reliable}}$-connected in G.
- 0-RMT is possible if and only if S and R are $2t$-connected in G.
- (ϵ, δ)-SMT or $(0, \delta)$-SMT is possible if and only if S and R are $t_{\zeta\text{-private}}$-connected and $t_{\zeta\text{-reliable}}$-connected in G.
- $(0,0)$-SMT is possible if and only if S and R are $(t_{\zeta\text{-private}} + t)$-connected in G. The $(t_{\zeta\text{-private}} + t)$-connectivity means that for any $A_i, A_j \in \mathcal{A}^T$, we have $(P_i \cup P_i^{(*)}) \cup P_j \neq P$.

References

1. The full version of this paper will be available on the authors' web pages
2. Ben-Or, M., Goldwasser, S., Wigderson, A.: Completeness theorems for non-cryptographic fault-tolerant distributed computing. In: Proc. ACM STOC 1988, pp. 1–10 (1988)
3. Chaum, D., Crépeau, C., Damgård, I.: Multiparty unconditionally secure protocols. In: Proc. ACM STOC 1988, pp. 11–19 (1988)
4. Desmedt, Y., Wang, Y.: Perfectly Secure Message Transmission Revisited. In: Knudsen, L.R. (ed.) EUROCRYPT 2002. LNCS, vol. 2332, pp. 502–517. Springer, Heidelberg (2002)
5. Desmedt, Y., Wang, Y., Burmester, M.: A Complete Characterization of Tolerable Adversary Structures for Secure Point-to-Point Transmissions without Feedback. In: Deng, X., Du, D.-Z. (eds.) ISAAC 2005. LNCS, vol. 3827, pp. 277–287. Springer, Heidelberg (2005)
6. Dolev, D., Dwork, C., Waarts, O., Yung, M.: Perfectly secure message transmission. J. ACM 40(1), 17–47 (1993)
7. Franklin, M.K., Wright, R.: Secure Communication in Minimal Connectivity Models. In: Nyberg, K. (ed.) EUROCRYPT 1998. LNCS, vol. 1403, pp. 346–360. Springer, Heidelberg (1998)
8. Franklin, M.K., Wright, R.: Secure communication in minimal connectivity models. J. Cryptology 13(1), 9–30 (2000)
9. Franklin, M.K., Yung, M.: Secure hypergraphs: Privacy from partial broadcast. In: Proc. ACM STOC 1995, pp. 36–44 (1995)
10. Goldreich, O., Goldwasser, S., Linial, N.: Fault-tolerant computation in the full information model. SIAM J. Comput. 27(2), 506–544 (1998)
11. Hirt, M., Maurer, U.M.: Player simulation and general adversary structures in perfect multiparty computation. J. Cryptology 13(1), 31–60 (2000)
12. Ito, M., Saito, A., Nishizeki, T.: Secret sharing schemes realizing general access structure. In: Proc. IEEE Globecom 1987, pp. 99–102 (1987)
13. Kumar, M., Goundan, P., Srinathan, K., Rangan, C.P.: On perfectly secure communication over arbitrary networks. In: Proc. ACM PODC 2002, pp. 293–202 (2002)
14. Wang, Y., Desmedt, Y.G.: Secure Communication in Broadcast Channels: The Answer to Franklin and Wright's Question. In: Stern, J. (ed.) EUROCRYPT 1999. LNCS, vol. 1592, pp. 446–458. Springer, Heidelberg (1999)
15. Wang, Y., Desmedt, Y.: Perfectly secure message transmission revisited. IEEE Transaction on Information Theory 54(6), 2582–2595 (2008)
16. Yang, Q., Desmedt, Y.: Cryptanalysis of Secure Message Transmission Protocols with Feedback. In: Kurosawa, K. (ed.) Information Theoretic Security. LNCS, vol. 5973, pp. 159–176. Springer, Heidelberg (2010)
17. Yang, Q., Desmedt, Y.: General Perfectly Secure Message Transmission using Linear Codes. In: Abe, M. (ed.) ASIACRYPT 2010. LNCS, vol. 6477, pp. 448–465. Springer, Heidelberg (2010)

Constant-Round Private Function Evaluation with Linear Complexity

Jonathan Katz[1,*] and Lior Malka[2,**]

[1] Dept. of Computer Science,
University of Maryland
jkatz@cs.umd.edu
[2] Intel
lior34@gmail.com

Abstract. We consider the problem of *private function evaluation* (PFE) in the two-party setting. Here, informally, one party holds an input x while the other holds a (circuit describing a) function f; the goal is for one (or both) of the parties to learn $f(x)$ while revealing nothing more to either party. In contrast to the usual setting of secure computation, where the function being computed is known to both parties, PFE is useful in settings where the function (i.e., algorithm) itself must remain secret, e.g., because it is proprietary or classified.

It is known that PFE can be reduced to standard secure computation by having the parties evaluate a *universal circuit*, and this is the approach taken in most prior work. Using a universal circuit, however, introduces additional overhead and results in a more complex implementation. We show here a completely new technique for PFE that avoids universal circuits, and results in constant-round protocols with communication/computational complexity *linear* in the size of the circuit computing f. This gives the first constant-round protocol for PFE with linear complexity (without using fully homomorphic encryption), even restricted to semi-honest adversaries.

1 Introduction

In the setting of two-party *private function evaluation* (PFE), a party P_1 holds an input x while another party P_2 holds a (circuit C_f describing a) function f; the goal is for one (or both) of the parties to learn the result $f(x)$ while not revealing to either party any information beyond this. (The parties do agree in advance on the *size* of the circuit being computed, as well as the input/output length. See Section 2.1 for further discussion.) PFE is useful when the function being computed must remain private, say because the function is classified, because revealing the function would lead to security vulnerabilities, or because the *implementation* of the function (e.g., the circuit C_f itself) is proprietary even if the function f is known [34, 6, 8, 9, 11–13, 19, 5, 33, 31, 3].

---

* Work supported by DARPA and NSF award #1111599.
** Work done while at the University of Maryland.

D.H. Lee and X. Wang (Eds.): ASIACRYPT 2011, LNCS 7073, pp. 556–571, 2011.

PFE stands in contrast to the standard setting of secure two-party computation [37, 14], where the parties hold inputs x and y, respectively, and wish to compute the result $f(x, y)$ for some *mutually known* function f using an agreed-upon circuit C_f for computing f. On the other hand, it is well known that the problem of PFE can be reduced to the problem of secure computation using *universal circuits*. In more detail, let U_n be some (fixed) universal circuit such that $U_n(x, C) = C(x)$ for every circuit C having at most n gates. (We implicitly assume here some fixed representation for circuits.) Then if $\mathcal{C}_n$ is the class of circuits having at most n gates, PFE for this class is solved by having the parties run a (standard) secure computation of U_n.

There are, however, drawbacks to using universal circuits to implement PFE. First is the resulting complexity: although PFE using universal circuits has been implemented [35], it is fair to say that it is more challenging, tedious, and error-prone to write code involving universal circuits than it is to implement secure computation "directly" using Yao's garbled circuit approach (as done, e.g., in [27, 26, 32, 16, 17]). Using universal circuits also impacts efficiency. Valiant [36] showed a construction of a universal circuit achieving (optimal) $|U_n| = O(n \log n)$; the construction is complex, however, and the constant terms (as well as the low-order terms) are significant. Kolesnikov and Schneider [23, 35] gave a simpler construction of universal circuits: they obtain the worse asymptotic bound $|U_n| = O(n \log^2 n)$, but their techniques are claimed to yield smaller universal circuits than Valiant's construction for "reasonable" values of n. (The exact improvement depends also on the number of inputs and outputs. We refer the reader to their work for a detailed comparison.) Even so, as secure two-party computation is used for ever-larger circuits (secure computation of circuits with up to 1 billion gates has been reported [17]), the overhead introduced by universal circuits becomes prohibitive. Indeed, the implementation of PFE by Kolesnikov and Schneider [23, 35] can handle circuits of only a few thousand gates [31].

Another approach to PFE is given by Abadi and Feigenbaum [1], who show a PFE protocol with complexity $O(n)$ but using $O(d)$ rounds, where d is (an upper bound on) the depth of the circuit being computed.

1.1 Contributions of Our Work

We show the first *constant-round* PFE protocols with *linear* complexity, without relying on fully homomorphic public-key encryption.[1] We begin by showing a protocol in the semi-honest setting; this illustrates our core techniques and represents what we consider to be our main contribution. (Semi-honest security was the focus of all prior work on PFE [34, 6, 8, 9, 11–13, 19, 5, 33, 31, 3].) Zero-knowledge proofs can be used in the standard way [15] to obtain security against malicious parties, still in constant rounds and with linear complexity; however, the resulting protocol is unlikely in practice to out-perform secure computation

---

[1] It is easy to construct constant-round, linear-complexity PFE from fully homomorphic encryption. But it is of theoretical interest to reduce the assumptions used, and of practical importance to avoid the overhead of fully homomorphic encryption.

of universal circuits using efficient protocols for the malicious setting (e.g., [24]). We sketch a more efficient construction for achieving security against a malicious P_1.

Our protocols rely on (singly) homomorphic public-key encryption as well as a symmetric-key encryption scheme secure against *linear* related-key attacks; see Definition 2. The former can be instantiated using various standard cryptosystems (e.g., [10, 30]); the latter can be instantiated in the random oracle model, or in a provable sense [2] based on the decisional Diffie-Hellman assumption.

In addition to the theoretical improvement, our approach should yield better performance in practice for PFE of large circuits and/or in certain settings. Specifically, although our protocol uses $O(n)$ public-key operations — in contrast to universal-circuit-based approaches that use $O(n \log n)$ or $O(n \log^2 n)$ symmetric-key operations[2] — our protocol has linear communication complexity, making it advantageous when network communication is expensive. Moreover, there are several ways our protocol can be improved (e.g., using elliptic-curve cryptography along with fast algorithms for performing multiple fixed-base exponentiations) to reduce its computational cost.

1.2 Overview of Our Techniques

Our main technical contribution, as noted above, is our idea for achieving PFE with linear complexity in the semi-honest setting; we describe this here. Our description is fairly detailed and we will refer to it in the formal description of our protocol later; it should also be possible to skim this section so as to obtain the main ideas. Our approach adapts Yao's garbled-circuit technique. At a very high level, our idea is to have P_1 generate a sequence of gates; P_2 then connects these gates together, using (singly) homomorphic encryption, in a manner that is oblivious to P_1, while still enabling P_1 to prepare a garbled circuit corresponding to the circuit C_f held by P_2. This idea of having one party connect gates of the circuit together is vaguely reminiscent of the "soldering" approach taken in [29]; our setting, however, is different than theirs (in [29] it was required that both parties know the circuit being computed), as is our implementation of the "soldering" step.

Say $x \in \{0,1\}^\ell$, and assume that f outputs a single bit and that C_f is known to contain exactly n NAND gates. (Neither of these assumptions is necessary, but we avoid complications for now.) It will be useful to distinguish between *outgoing wires* and *ingoing wires* of a circuit. Outgoing wires include the ℓ input wires of the circuit, along with the wire that exits each gate of the circuit; thus, in a circuit with ℓ inputs and n gates there are exactly $\ell + n$ outgoing wires. The ingoing wires are exactly the input wires to each gate of the circuit; thus, in a circuit with n two-input gates there are exactly $2n$ ingoing wires. A circuit is defined by specifying the output wires, and by giving a correspondence between

---

[2] This does not account for any oblivious transfers performed in the universal-circuit-based approaches. However the number of oblivious transfers scales linearly in the input length, not the circuit size.

outgoing wires and ingoing wires; e.g., specifying that outgoing wire i (which may be an input wire or a wire exiting some gate) connects to ingoing wires j, k, and ℓ. We stress that even though we speak of each internal gate as having only a single outgoing wire, we handle arbitrary fan-out since a single outgoing wire can be connected to several ingoing wires.

In our description below, we assume for concreteness that P_2 learns the output $f(x)$. However, it is trivial to modify our protocol (with no additional cost) so that only P_1 learns the output. See the remark at the end of this section.

The protocol begins by having P_1 generate and send a public key pk for a (singly) homomorphic encryption scheme Enc. Similar to Yao's garbled-circuit technique, P_1 then chooses $\ell + n$ pairs of random *keys* that will be assigned to each of the outgoing wires. Let s_i^b denote the key corresponding to bit b on wire i. Then P_1 sends

$$\left[\mathsf{Enc}_{pk}(s_1^0), \mathsf{Enc}_{pk}(s_1^1)\right], \ldots, \left[\mathsf{Enc}_{pk}(s_{\ell+n}^0), \mathsf{Enc}_{pk}(s_{\ell+n}^1)\right]$$

to P_2. (It will become clear from what follows that P_1 need not send the final encrypted pair $\left[\mathsf{Enc}_{pk}(s_{\ell+n}^0), \mathsf{Enc}_{pk}(s_{\ell+n}^1)\right]$. We include it above for clarity.)

P_2, in turn, *obliviously* defines keys for each of the $2n$ ingoing wires. P_2 sorts the gates of C_f topologically, so that if the outgoing wire from some gate i connects to an ingoing wire of some gate j then $i < j$. This defines a natural enumeration of the outgoing wires in the circuit: outgoing wires numbered from 1 to ℓ correspond to the input wires of the circuit, and outgoing wire $\ell + i$ (for $i \in \{1, \ldots, n\}$) corresponds to the wire exiting gate i. The output wire of the circuit corresponds to outgoing wire $\ell + n$. (Recall that here we assume f is boolean; in Section 3.1 we relax this.)

For each ingoing wire of the circuit, P_2 does as follows. Say the ingoing wire of some gate i is connected to outgoing wire j. Then P_2 chooses random a_i, b_i and defines the (encrypted) keys for this ingoing wire to be

$$\left[\mathsf{Enc}_{pk}(a_i \cdot s_j^0 + b_i), \ \mathsf{Enc}_{pk}(a_i \cdot s_j^1 + b_i)\right],$$

where the above is computed using the homomorphic properties of the encryption scheme. (In the above, the ciphertexts are re-randomized in the usual way.) Two observations are in order: first, the (unencrypted) keys $(r^0, r^1) \stackrel{\text{def}}{=} \left(a_i \cdot s_j^0 + b_i, \ a_i \cdot s_j^1 + b_i\right)$ are random and independent of j. Second, given s_j^b it is possible for P_2 to compute r^b (using a_i, b_i); without s_j^{1-b}, however, P_2 learns no information about r^{1-b}. (Recall we are in the semi-honest setting, so a_i, b_i are chosen at random.)

Expanding upon the above, say gate i of the circuit has its left ingoing wire connected to outgoing wire j and right ingoing wire connected to outgoing wire k. (As always, the outgoing wire from this gate is numbered $\ell + i$.) Then P_2 defines the encrypted "garbled gate"

$$\mathsf{encGG}_i = \begin{pmatrix} \left[\mathsf{Enc}_{pk}(a_i \cdot s_j^0 + b_i), \ \mathsf{Enc}_{pk}(a_i \cdot s_j^1 + b_i)\right] \\ \left[\mathsf{Enc}_{pk}(a_i' \cdot s_k^0 + b_i'), \ \mathsf{Enc}_{pk}(a_i' \cdot s_k^1 + b_i')\right] \\ \left[\mathsf{Enc}_{pk}(s_{\ell+i}^0), \ \mathsf{Enc}_{pk}(s_{\ell+i}^1)\right] \end{pmatrix},$$

where a_i, b_i, a'_i, b'_i are chosen uniformly at random. Finally, P_2 sends

$$\mathsf{encGG}_1, \ldots, \mathsf{encGG}_n$$

to P_1. (In fact P_2 need not transmit the final pair $[\mathsf{Enc}_{pk}(s^0_{\ell+i}), \mathsf{Enc}_{pk}(s^1_{\ell+i})]$ of each encrypted garbled gate, since P_1 knows it. We include it above for clarity.)

Upon receiving this message, P_1 decrypts each encGG to obtain, for each gate i, the three pairs of keys $([L^0_i, L^1_i], [R^0_i, R^1_i], [s^0_{\ell+i}, s^1_{\ell+i}])$. It then prepares a garbled version GG_i of this gate in the usual way: namely, it computes the four ciphertexts

$$C'_{b,c} \leftarrow \mathsf{sEnc}_{L^b_i} \left(\mathsf{sEnc}_{R^c_i} \left(s^{\mathrm{NAND}(b,c)}_{\ell+i} \right) \right), \quad b, c \in \{0, 1\}$$

(where sEnc denotes a symmetric-key encryption scheme), and sets GG_i to be the four ciphertexts $(C'_{0,0}, \ldots, C'_{1,1})$ in random permuted order. P_1 then sends $\mathsf{GG}_1, \ldots, \mathsf{GG}_n$ to P_2. In addition, P_1 sends the appropriate input-wire keys $s^{x_1}_1$, $\ldots, s^{x_\ell}_\ell$, as well as both output-wire keys $(s^0_{\ell+n}, s^1_{\ell+n})$.

P_2 now has enough information to compute the result, using a procedure analogous (but not identical) to what is done in a standard application of Yao's garbled-circuit methodology. P_2 begins knowing a key s_i for each outgoing wire $i \in \{1, \ldots, \ell\}$. (Recall these are the input wires of the circuit that correspond to P_1's input.) Inductively, P_2 can compute a key for every outgoing wire as follows: Consider the $(\ell + i)$th outgoing wire exiting from gate i, where the left ingoing wire to this gate is connected to outgoing wire $j < i$ and the right ingoing wire to this gate is connected to outgoing wire $k < i$. Assume P_2 has already determined keys s_j, s_k for outgoing wires j, k, respectively. P_2 computes keys $L_i = a_i s_j + b_i$ and $R_i = a'_i s_k + b'_i$ for the left and right *ingoing* wires to gate i. Then P_2 tries to decrypt each of the four ciphertexts in GG_i. With overwhelming probability, only one of these decryptions will be successful; the result of this successful decryption defines the key $s_{\ell+i}$ for outgoing wire $\ell + i$. Once P_2 has determined key $s_{\ell+n}$, it can check whether this corresponds to an output of '0' or '1' using the ordered pair $(s^0_{\ell+n}, s^1_{\ell+n})$ sent by P_1.

Further details, intuition for security of the above, proofs of security, and extensions to handle malicious behavior of P_1 are described in the sections that follow. A more efficient variant of the above protocol is described in Section 3.2.

Remark 1: It is trivial to modify the above protocol, at no additional cost, so that only P_1 learns the output (and P_2 learns nothing): first, change round 3 so that P_1 does not send the output-wire keys $(s^0_{\ell+n}, s^1_{\ell+n})$. Then when P_2 learns the final key $s_{\ell+n}$ it simply sends this key back to P_1, who can then check whether it is equal to $s^0_{\ell+n}$ or $s^1_{\ell+n}$.

1.3   Other Related Work

Several works have explored weaker variants of PFE. Paus et al. [31] consider *semi-private* function evaluation where the circuit *topology* (i.e., the connections between gates) is assumed to be known to both parties, but the boolean function computed by each gate can be hidden. Here we treat the more difficult case where *everything* about the circuit (except an upper bound on its size and the number of inputs/outputs) is hidden. Another direction has been to consider PFE for limited classes of functions: e.g., functions defined by low-depth circuits [34, 4], branching programs [19, 3], or polynomials [9, 28]. Here we handle functions defined by arbitrary (polynomial-size) circuits.

2 Definitions

Let k be the security parameter. A *distribution ensemble* $X = \{X(1^k, a)\}_{k \in \mathbb{N}, a \in \mathcal{D}}$ is an infinite sequence of random variables indexed by $k \in \mathbb{N}$ and $a \in \mathcal{D}$, for $\mathcal{D}$ some specified set. The two ensembles $X = \{X(1^k, a)\}_{k \in \mathbb{N}, a \in \mathcal{D}}$ and $Y = \{Y(1^k, a)\}_{k \in \mathbb{N}, a \in \mathcal{D}}$ are *computationally indistinguishable*, denoted $X \overset{c}{\equiv} Y$, if for every non-uniform polynomial-time algorithm D there exists a negligible function $\mu(\cdot)$ such that for every k and every $a \in \mathcal{D}$

$$\left| \Pr[D(X(1^k, a)) = 1] - \Pr[D(Y(1^k, a)) = 1] \right| \leq \mu(k).$$

2.1 Private Function Evaluation

Our definitions of security are standard, but we include them here for completeness. For simplicity, we treat the case where P_1 holds some value $x \in \{0, 1\}^\ell$ as input while P_2 holds a circuit C_f computing some deterministic function f; the goal of the protocol is for P_2 to learn $f(x)$. The definitions we provide here, as well as our protocols, extend easily to handle, e.g., additional input provided by P_2 (this can simply be incorporated into the circuit C_f), randomized functions f, or the case where P_1 receives output (see Remark 1 at the end of Section 1.2).

The problem of PFE is meaningless in practice if P_2 learns the output and f (resp., C_f) is allowed to be completely arbitrary: in that case P_2 could take $f(x) = x$ and learn P_1's entire input! It is thus reasonable to impose some restrictions on C_f. The most general formulation to assume that both parties fix some class $\mathcal{C}$ of circuits, and require that $C_f \in \mathcal{C}$; in that case we refer to the problem as $\mathcal{C}$-PFE. This encompasses both the case when P_1 knows some partial information about f (as in [31]), as well as the case where C_f is restricted in some way (e.g., to have low depth). In this work, we assume only that P_1 knows the input length ℓ, and upper bounds on the output length m and the number of gates n (i.e., $\mathcal{C}$ contains only circuits satisfying those constraints). Note that if $m \ll \ell$ then meaningful privacy of P_1's input is maintained regardless of what circuit $C_f \in \mathcal{C}$ is used by P_2.

There are two ways one could incorporate a security parameter k into the definition of the problem. The usual way, which we find less natural in our setting, is to allow the sizes of the inputs to grow and to set the security parameter equal to the input size(s). We prefer instead to treat the input domains (namely, $\{0, 1\}^\ell$ and some class of circuits $\mathcal{C}$) as fixed, and to treat k as an additional input.

A two-party protocol for $\mathcal{C}$-PFE is a protocol running in polynomial time and satisfying the following correctness requirement: if party P_1, holding input 1^k and x, and party P_2, holding input 1^k and $C_f \in \mathcal{C}$, run the protocol honestly, then (except with probability negligible in k) the output of P_2 is $C_f(x)$.

Security in the semi-honest case. In the semi-honest case we assume both parties follow the protocol honestly but may each try to learn some additional information from their (respective) view. Fix $\mathcal{C}$ and let Π be a protocol for $\mathcal{C}$-PFE. The *view* of the ith party during an execution of Π when the parties begin holding inputs x and C_f, respectively, and security parameter 1^k is denoted by $\text{VIEW}_i^\Pi(1^k, x, C_f)$. The view of P_i contains P_i's input and random tape, along with the sequence of messages received from the other party P_{3-i}.

When f is deterministic it suffices to consider the views of the parties in isolation, rather than their joint distribution [14, Sect. 7.2.2.1]. We thus have:

Definition 1. *Protocol Π is a* secure $\mathcal{C}$-PFE protocol for semi-honest adversaries *if there exist probabilistic polynomial-time simulators S_1, S_2 such that*

$$\left\{ S_1\left(1^k, x\right) \right\}_{k \in \mathbb{N}, \, x \in \{0,1\}^\ell, \, C_f \in \mathcal{C}} \overset{c}{\equiv} \left\{ \text{VIEW}_1^\Pi\left(1^k, x, C_f\right) \right\}_{k \in \mathbb{N}, \, x \in \{0,1\}^\ell, \, C_f \in \mathcal{C}}$$

$$\left\{ S_2\left(1^k, C_f, C_f(x)\right) \right\}_{k \in \mathbb{N}, \, x \in \{0,1\}^\ell, \, C_f \in \mathcal{C}} \overset{c}{\equiv} \left\{ \text{VIEW}_2^\Pi\left(1^k, x, C_f\right) \right\}_{k \in \mathbb{N}, \, x \in \{0,1\}^\ell, \, C_f \in \mathcal{C}}.$$

Security against malicious behavior. We refer to the full version of this paper [21] for a definition of security against malicious adversaries within the usual real/ideal framework [14].

2.2   Tools

We use a (singly) homomorphic public-key encryption scheme (Gen, Enc, Dec). The actual property we need is the ability to evaluate a *pairwise-independent function* on the plaintext space. If the plaintext space is a group $\mathbb{G}$ of prime order p, written additively, this can be achieved by mapping $a \in \mathbb{Z}_p$, $b \in \mathbb{G}$, and $\text{Enc}_{pk}(m)$ to a (random) encryption of $\text{Enc}_{pk}(am + b)$. Thus, e.g., standard El Gamal encryption [10] can be used (though $\mathbb{G}$ in that case is usually written multiplicatively). In fact, the plaintext space is not required to have prime order, as we only require "almost" pairwise-independence. In particular, Paillier encryption [30] could also be used.

We also use a symmetric-key encryption scheme (sEnc, sDec) whose key space is viewed as a group $\mathbb{G}(k)$ of order $p = p(k)$ that is, for simplicity, the same as the plaintext space of the public-key encryption scheme being used. (In practice, this can be achieved for any desired $\mathbb{G}$ by implementing encryption with key $g \in \mathbb{G}$ using AES with key SHA-1(g), truncated to 128 bits.) We impose the same requirements on (sEnc, sDec) as in [25]: namely, that it have *elusive* and *efficiently verifiable* range. (These properties are easily satisfied.) In addition, we require (sEnc, sDec) to satisfy a weak form of *related-key security* where, roughly, encryption remains secure even when performed using linearly related keys (where the linear relations are chosen at random). That is:

Definition 2. *Encryption scheme* (sEnc, sDec) *is* secure against linear related-key attacks *if the following is negligibly close (in k) to* $1/2$ *for all polynomials d and all* PPT *adversaries* $\mathcal{A}$:

$$\Pr\left[\begin{array}{l} s \leftarrow \mathbb{G}(k); c \leftarrow \{0,1\}; \\ a_1, \ldots a_d \leftarrow \mathbb{Z}_{p(k)} \quad : \mathcal{A}^{\mathsf{sEnc}^c_{a_1 s + b_1}(\cdot, \cdot), \ldots, \mathsf{sEnc}^c_{a_d s + b_d}(\cdot, \cdot)}(a_1, b_1, \ldots, a_d, b_d) = c \\ b_1, \ldots, b_d \leftarrow \mathbb{G}(k) \end{array}\right],$$

where $\mathsf{sEnc}^c_s(m_0, m_1) \stackrel{\text{def}}{=} \mathsf{sEnc}_s(m_c)$.

We remark that a weaker definition (where $\mathcal{A}$ queries each $\mathsf{sEnc}^c_{a_i s + b_i}(\cdot, \cdot)$ only on two inputs, chosen nonadaptively) suffices for our proof. It is easy to construct an encryption scheme satisfying the above definition using a (non-programmable) random oracle, and it would be surprising if standard encryption schemes based on AES could be shown not to satisfy the above definition. Moreover, recent work of Applebaum et al. [2] can be used to construct a scheme satisfying the above definition in a provable sense, based on the decisional Diffie-Hellman assumption.

3 A $\mathcal{C}$-PFE Protocol for Semi-honest Adversaries

3.1 Description of the Protocol

We now formally define our $\mathcal{C}$-PFE protocol for semi-honest adversaries. In our description here, we assume the reader is familiar with the protocol overview provided in Section 1.2.

We assume that all circuits in $\mathcal{C}$ are composed solely of NAND gates. This is for simplicity only, and our protocol can be easily modified to handle circuits over an arbitrary basis of 2-to-1 gates with only a small impact on the efficiency. Let n be an upper bound on the size of any circuit in $\mathcal{C}$, and let m be an upper bound on the number of outputs. By adjusting n appropriately, we may assume that every circuit in $\mathcal{C}$ has exactly m outputs (P_2 can always add "dummy" outputs that are fixed to some constant); that the output wires of the circuit do not connect to any other gates (this can be achieved by adding at most m gates to the circuit); and that every circuit in $\mathcal{C}$ contains exactly n gates (P_2 can add "dummy" gates whose output wires are connected to nothing). We make all these assumptions in what follows. We also assume that P_2 learns the output; however, it is trivial to modify the protocol so that P_1 learns the output; see Remark 1 in Section 1.2.

Recall from Section 1.2 that we distinguish between *outgoing wires* and *ingoing wires* of C_f. (Recall also that although each gate has only a single outgoing wire, we handle circuits with arbitrary fan-out since a single outgoing wire can be connected to several ingoing wires.) As in Section 1.2, party P_2 sorts the gates of C_f topologically and this defines an enumeration of the $N \stackrel{\text{def}}{=} \ell + n$ outgoing wires. The outgoing wires numbered from 1 to ℓ correspond to the ℓ input wires of the circuit, and outgoing wire $\ell + i$ (for $i \in \{1, \ldots, n\}$) corresponds to the output wire from gate i. The output wires of the circuit correspond to the m outgoing wires $N - m + 1, \ldots, N$.

We first define an algorithm encYao that prepares garbled gates as in Yao's protocol: encYao takes as input three pairs of keys and outputs four ciphertexts, and is defined as

$$\mathsf{encYao}\left([L^0, L^1], [R^0, R^1], [s^0, s^1]\right) \overset{\text{def}}{=} \left\{\mathsf{sEnc}_{L^b}\left(\mathsf{sEnc}_{R^c}\left(s^{\mathrm{NAND}(b,c)}\right)\right)\right\}_{b,c \in \{0,1\}},$$

where the four ciphertexts are in random permuted order. We analogously define an algorithm decYao that takes as input two keys (for each of two ingoing wires) and a garbled gate, and outputs a key for the outgoing wire; this algorithm, given keys L, R and four ciphertexts $\{C'_0, C'_1, C'_2, C'_3\}$, computes $\mathsf{sDec}_L(\mathsf{sDec}_R(C'_i))$ for all i and outputs the unique non-$\perp$ value that is obtained. (If more than one non-$\perp$ value results, this algorithm outputs $\perp$.)

Our protocol is described in Figure 1. Correctness holds with all but negligible probability, via an argument similar to the one in [25].

In our description of the protocol we aimed for clarity rather than efficiency, and several improvements are possible. For one, P_2 need not include $\left[\mathsf{Enc}_{pk}(s^0_{\ell+i}), \mathsf{Enc}_{pk}(s^1_{\ell+i})\right]$ as part of encGG_i since P_1 already knows these values. Furthermore, P_1 need not send

$$\left[\mathsf{Enc}_{pk}(s^0_{N-m+1}), \mathsf{Enc}_{pk}(s^1_{N-m+1})\right], \ldots, \left[\mathsf{Enc}_{pk}(s^0_N), \mathsf{Enc}_{pk}(s^1_N)\right]$$

in round 1 (since these outgoing wires do not connect to any ingoing wires). Moreover, P_1 can set $s^0_{N-m+1} = \cdots = s^0_N = 0$ and $s^1_{N-m+1} = \cdots = s^1_N = 1$ (and then there is no need to send the output-wires message in the third round); that is, for gates whose outgoing wires are the output of the circuit, P_1 can encrypt the wire value itself rather than encrypting a key that encodes the wire value.

Security against a semi-honest P_1 is easy to see. In fact, security in that case holds in a *statistical* sense. Indeed, with all but negligible probability it holds that $s^0_i \neq s^1_i$ for all $i \in \{1, \ldots, N\}$. Assuming this to be the case, the top two rows of each encGG_i sent by P_2 to P_1 in round 2 consist only of (random) encryptions of the four independent, uniform values

$$a_i \cdot s^0_j + b_i, \quad a_i \cdot s^1_j + b_i, \quad a'_i \cdot s^0_k + b'_i, \quad a'_i \cdot s^1_k + b'_i.$$

In particular, these values are independent of the interconnections between gates of C_f, and thus the view of P_1 is independent of the circuit held by P_2.

Security against a semi-honest P_2 holds *computationally*, assuming semantic security of the homomorphic encryption scheme and security against linear related-key attacks for the symmetric-key encryption scheme. Roughly, the initial encryptions sent to P_2 in round 1 do not reveal anything about the values s^0_i, s^1_i that P_1 assigns to each outgoing wire in the circuit. Thus, the information sent to P_2 in round 3 is essentially equivalent to the information sent to P_2 in a standard application of Yao's garbled-circuit methodology, with the only difference being that here ingoing wires and outgoing wires have different keys, and

Inputs: The security parameter is k. The input of P_1 is a value $x \in \{0,1\}^{\ell}$, and the input of P_2 is a circuit C_f with ℓ, n, m as described in the text.

Round 1 P_1 computes $(pk, sk) \leftarrow \mathsf{Gen}(1^k)$ and sends pk to P_2. In addition, P_1 chooses $N = \ell + n$ pairs of random keys s_i^0, s_i^1 for $i \in \{1, \ldots, N\}$. It then sends to P_2 the ciphertexts

$$\left[\mathsf{Enc}_{pk}(s_1^0), \mathsf{Enc}_{pk}(s_1^1) \right], \ldots, \left[\mathsf{Enc}_{pk}(s_N^0), \mathsf{Enc}_{pk}(s_N^1) \right].$$

Round 2 For each gate $i \in \{1, \ldots, n\}$ of C_f, with left ingoing wire connected to outgoing wire j, right ingoing wire connected to outgoing wire k, and outgoing wire $\ell + i$, party P_2 chooses a_i, b_i, a_i', b_i' uniformly (from the appropriate domains) and computes

$$\mathsf{encGG}_i = \left(\begin{array}{c} \left[\mathsf{Enc}_{pk}(a_i \cdot s_j^0 + b_i), \ \mathsf{Enc}_{pk}(a_i \cdot s_j^1 + b_i) \right] \\ \left[\mathsf{Enc}_{pk}(a_i' \cdot s_k^0 + b_i'), \ \mathsf{Enc}_{pk}(a_i' \cdot s_k^1 + b_i') \right] \\ \left[\mathsf{Enc}_{pk}(s_{\ell+i}^0), \ \mathsf{Enc}_{pk}(s_{\ell+i}^1) \right] \end{array} \right)$$

using the homomorphic properties of Enc. (In the above, each ciphertext is re-randomized.) Then P_2 sends $\mathsf{encGG}_1, \ldots, \mathsf{encGG}_n$ to P_1.

Round 3 For $i \in \{1, \ldots, n\}$, party P_1 decrypts encGG_i using sk to obtain the three pairs of keys $\mathsf{keys}_i \overset{\mathrm{def}}{=} ([L_i^0, L_i^1], [R_i^0, R_i^1], [s_{\ell+i}^0, s_{\ell+i}^1])$. It then computes $\mathsf{GG}_i \leftarrow \mathsf{encYao}(\mathsf{keys}_i)$, and sends $\mathsf{GG}_1, \ldots, \mathsf{GG}_n$ to P_2. Finally, P_1 sends

input-wires: $s_1^{x_1}, \ldots, s_{\ell}^{x_\ell}$; output-wires: $\left(s_{N-m+1}^0, s_{N-m+1}^1 \right), \ldots, \left(s_N^0, s_N^1 \right).$

Output determination Say P_1 sent input-wires: $s_1, \ldots, s_\ell$ to P_2 in the previous round. Then for all $i \in \{\ell + 1, \ldots, \ell + n\}$, party P_2 does: If the left ingoing wire of gate i is connected to outgoing wire $j < i$ and the right ingoing wire of gate i is connected to outgoing wire $k < i$, then (1) compute $L_i = a_i s_j + b_i$ and $R_i = a_i' s_k + b_i'$, and then (2) set $s_i = \mathsf{decYao}(L_i, R_i, \mathsf{GG}_i)$.

Once P_2 has computed $s_1, \ldots, s_{\ell+n}$, it sets the jth output bit o_j (for $j \in \{N - m + 1, \ldots, N\}$) to be the (unique) bit for which $s_j = s_j^{o_j}$.

Fig. 1. A $\mathcal{C}$-PFE protocol for semi-honest adversaries

P_2 must compute a key L_i on some ingoing wire by "translating" one of the keys s_j on the outgoing wire connected to that ingoing wire.

We have:

Theorem 1. *Assume the homomorphic encryption scheme is semantically secure, and the symmetric-key encryption scheme is secure against linear related-key attacks and has elusive and efficiently verifiable range. Then the protocol of Figure 1 is a secure $\mathcal{C}$-PFE protocol for semi-honest adversaries.*

Due to space limitations, a proof appears in the full version [21].

3.2   A More Efficient Variant

In this section we describe a more efficient variant of our protocol in which the wire labels are chosen in a coordinated fashion, as in [22]. Unfortunately, we are only able to prove security of the resulting protocol in the random oracle model; see further discussion at the end of this section.

We merely sketch the basic idea. Now, in round 1, P_1 chooses a global random shift r and $\ell + n$ outgoing-wire keys $\{s_i^0\}$; it then sets $s_i^1 = s_i^0 + r$ for all i. The first-round message from P_1 now contains pk and the $\ell + n$ ciphertexts $\mathsf{Enc}_{pk}(s_1^0), \ldots, \mathsf{Enc}_{pk}(s_{\ell+n}^0)$.

For each ingoing wire of the circuit, P_2 does as follows. Say this wire is connected to outgoing wire j. Then P_2 chooses random a and defines the (encrypted) 0-key for this ingoing wire to be (a re-randomization of) $\mathsf{Enc}_{pk}(s_j^0 + a)$, where this is computed using the homomorphic properties of the encryption scheme. Thus, if gate i of the circuit has its left ingoing wire connected to outgoing wire j and right ingoing wire connected to outgoing wire k, party P_2 defines the ith encrypted "garbled gate" via

$$\mathsf{encGG}_i = \begin{pmatrix} \mathsf{Enc}_{pk}(s_j^0 + a_i) \\ \mathsf{Enc}_{pk}(s_k^0 + a_i') \\ \mathsf{Enc}_{pk}(s_{\ell+i}^0) \end{pmatrix},$$

where a_i, a_i' are chosen uniformly at random. P_2 sends $\mathsf{encGG}_1, \ldots, \mathsf{encGG}_n$ to P_1.

Upon receiving this message, P_1 decrypts each encGG to obtain, for each gate i, the keys $\left(L_i^0, R_i^0, s_{\ell+i}^0,\right)$. It defines $L_i^1 = L_i^0 + r$ and $R_i^1 = R_i^0 + r$, and then prepares a garbled version GG_i of this gate as in the previous sections. P_2 can then compute the result as usual. The entire protocol is roughly twice as efficient as the original.

As we have mentioned, however, we are only able to prove security of this modified protocol in the (non-programmable) random oracle model. Although it may appear possible to prove security in the standard model if the symmetric-key encryption scheme satisfies a strong enough definition of security, we were not able to isolate any suitable definition. In particular, correlation robustness [18] does not appear to suffice, since there is a circularity when, e.g., keys $s, s+r, s', s'+r$ are used to encrypt keys s'' and $s'' + r$. (Some combination of correlation robustness and circular security appears necessary.) The same issue seems to be present in the works of [22, 29] as well.

4   Security for Malicious Adversaries

As noted in the Introduction, we can apply zero-knowledge proofs in the standard way [15] to obtain a protocol with linear complexity (and constant round complexity) that is secure against malicious adversaries. However, the resulting protocol is unlikely in practice to out-perform secure computation of universal circuits using efficient protocols for the malicious setting (e.g., [24]). Here, we sketch a more efficient construction that achieves security against a malicious P_1 only. As in the previous section, our goal here is not to optimize the efficiency of the resulting protocol but rather to illustrate the main ideas.

We continue to assume that P_2 learns the output, however Remark 1 of Section 1.2 applies here as well and so the protocol is easily modified so that only P_1 learns the output.

4.1 Protocol Modifications

We introduce the following changes to the protocol described in Section 3.1:

Proof of well-formedness of pk. We require P_1 to prove that the public key pk it sends in round 1 was output by the specified key-generation algorithm Gen. (This step is not necessary if it is possible to efficiently verify whether a given pk could have been output by Gen, as is the case with, e.g., El Gamal encryption.) We remark further that it suffices for the proof to be honest-verifier zero knowledge (since we only require security against a semi-honest P_2), and we do not require it to be a proof of knowledge.

The complexity of this step is independent of n.

Validity of outgoing-wire keys. Let $\left[C_1^0, C_1^1\right], \ldots, \left[C_N^0, C_N^1\right]$ denote the ciphertexts sent by P_1 in round 1. (Recall that it is supposed to be the case that $C_i^b = \mathsf{Enc}_{pk}(s_i^b)$.) We now require P_1 to prove that (1) each C_i^b is a well-formed ciphertext with respect to the public key pk (once again, this step is unnecessary if it is possible to efficiently verify validity of ciphertexts, as is the case with El Gamal encryption), and (2) for each i, the ciphertexts C_i^0, C_i^1 are encryptions of *distinct* values. If the encryption scheme is additively homomorphic, and we let s_i^0 (resp., s_i^1) denote the plaintext corresponding to C_i^0 (resp., C_i^1), then P_2 can compute $\mathsf{Enc}_{pk}(s_i^0 - s_i^1)$ and the latter step is equivalent to proving that this is not an encryption of 0. Once again, it suffices for these proofs to be honest-verifier zero knowledge and they are not required to be proofs of knowledge.

The complexity of this step is linear in n since the statement being proved can be written as a conjunction of n statements, each of size independent of n.

Correctness of garbled-circuit construction. We require P_1 to prove correctness of the garbled gates it sends to P_2 in the final round. This amounts to proving, for each $i \in \{1, \ldots, n\}$, that GG_i was correctly constructed from encGG_i. As before, it suffices for these proofs to be honest-verifier zero knowledge and they are not required to be proofs of knowledge.

The complexity of this step is linear in n since the statement being proved is a conjunction of n statements, each of which has size independent of n. We also note that by using an appropriate homomorphic encryption scheme and symmetric-key encryption scheme, these proofs can be made (reasonably) efficient using the techniques of Jarecki and Shmatikov [20] (who show efficient proofs for exactly this purpose, assuming a common reference string, using a variant of the Camenisch-Shoup encryption scheme [7]).

Correctness of input-wire and output-wire keys. Finally, P_1 is required to prove that the input-wire and output-wire keys it sends in the final round are correct. Let $\left[C_1^0, C_1^1\right], \ldots, \left[C_N^0, C_N^1\right]$ denote the ciphertexts sent by P_1 in round 1 (recall it is supposed to be the case that $C_i^b = \mathsf{Enc}_{pk}(s_i^b)$), and let

$$\text{input-wires: } s_1, \ldots, s_\ell \quad \text{and} \quad \text{output-wires: } \left(s_{N-m+1}^0, s_{N-m+1}^1\right), \ldots, \left(s_N^0, s_N^1\right)$$

be the values sent by P_1 in the last round. Then P_1 must prove that: (1) that for each index $i \in \{1, \ldots, \ell\}$, one of the ciphertexts C_i^0, C_i^1 is an encryption of the plaintext s_i, and (2) that for each index $i \in \{N - m + 1, \ldots, N\}$, the ciphertext C_i^0 (resp., C_i^1) is an encryption of s_i^0 (resp., s_i^1). It suffices for each of these proofs to be honest-verifier zero knowledge; the first set of proofs (proving correctness of the input-wire keys) must be proofs of knowledge to allow for input extraction. (Alternately, if the proof of well-formedness of the public key is a proof of knowledge then proofs of knowledge are not needed here.)

The complexity of this step is linear in $\ell + m$.

We remark that most of the above proofs can be implemented efficiently for any homomorphic encryption scheme. The main exception is the proof of correctness of the garbled-circuit construction; however, as noted above, there exists at least one specific homomorphic encryption scheme for which this step can be done reasonably efficiently [20]. A proof of the following appears in [21].

Theorem 2. *Under the same assumptions as in Theorem 1, the protocol of Figure 1 with the modifications described in the previous section is a secure C-PFE protocol for a malicious P_1.*

5   Conclusions and Future Work

We have shown the first constant-round protocol for PFE with complexity *linear* in the size of the circuit being computed (without relying on fully homomorphic encryption). Our results leave several interesting open questions:

- In addition to its theoretical importance, we believe our work is also of practical relevance: specifically, we expect that our approach to PFE will be both easier to implement and more efficient (for large circuits) than approaches relying on universal circuits. It remains to experimentally validate this claim.
- Our work leaves open the question of designing a fully secure protocol for PFE (i.e., PFE with security against a malicious P_1 and a malicious P_2) with linear complexity that would have better performance than what results from running a secure computation of universal circuits using efficient protocols for the malicious setting (e.g., [24]).
- It would also be interesting to further improve on the cryptographic assumptions needed for our results: e.g., to construct a protocol based on semantically secure symmetric-key encryption (without requiring related-key security), or to avoid the use of homomorphic public-key encryption.

The contents of this paper do not necessarily reflect the position or the policy of the US Government, and no official endorsement should be inferred.

References

1. Abadi, M., Feigenbaum, J.: Secure circuit evaluation. Journal of Cryptology 2(1), 1–12 (1990)
2. Applebaum, B., Harnik, D., Ishai, Y.: Semantic security under related-key attacks and applications. In: 2nd Symp. on Innovations in Computer Science, ICS (2011), http://eprint.iacr.org/2010/544
3. Barni, M., Failla, P., Kolesnikov, V., Lazzeretti, R., Sadeghi, A.-R., Schneider, T.: Secure Evaluation of Private Linear Branching Programs with Medical Applications. In: Backes, M., Ning, P. (eds.) ESORICS 2009. LNCS, vol. 5789, pp. 424–439. Springer, Heidelberg (2009)
4. Boneh, D., Goh, E.-J., Nissim, K.: Evaluating 2-DNF Formulas on Ciphertexts. In: Kilian, J. (ed.) TCC 2005. LNCS, vol. 3378, pp. 325–341. Springer, Heidelberg (2005)
5. Brickell, J., Porter, D.E., Shmatikov, V., Witchel, E.: Privacy-preserving remote diagnostics. In: 14th ACM Conf. on Computer and Communications Security (CCS), pp. 498–507. ACM Press (2007)
6. Cachin, C., Camenisch, J., Kilian, J., Müller, J.: One-Round Secure Computation and Secure Autonomous Mobile Agents. In: Welzl, E., Montanari, U., Rolim, J.D.P. (eds.) ICALP 2000. LNCS, vol. 1853, pp. 512–523. Springer, Heidelberg (2000)
7. Camenisch, J., Shoup, V.: Practical Verifiable Encryption and Decryption of Discrete Logarithms. In: Boneh, D. (ed.) CRYPTO 2003. LNCS, vol. 2729, pp. 126–144. Springer, Heidelberg (2003)
8. Canetti, R., Ishai, Y., Kumar, R., Reiter, M.K., Rubinfeld, R., Wright, R.N.: Selective private function evaluation with applications to private statistics. In: 20th Annual ACM Symposium on Principles of Distributed Computing (PODC), pp. 293–304. ACM Press (2001)
9. Chang, Y.-C., Lu, C.-J.: Oblivious Polynomial Evaluation and Oblivious Neural Learning. In: Boyd, C. (ed.) ASIACRYPT 2001. LNCS, vol. 2248, pp. 369–384. Springer, Heidelberg (2001)
10. El Gamal, T.: A public key cryptosystem and a signature scheme based on discrete logarithms. IEEE Trans. Information Theory 31, 469–472 (1985)
11. Frikken, K., Atallah, M., Li, J.: Hidden access control policies with hidden credentials. In: Proc. ACM Workshop on Privacy in the Electronic Society (WPES), p. 27. ACM (2004)
12. Frikken, K., Attallah, M., Zhang, C.: Privacy-preserving credit checking. In: ACM Conf. on Electronic Commerce (EC), pp. 147–154. ACM (2005)
13. Frikken, K.B., Li, J., Atallah, M.J.: Trust negotiation with hidden credentials, hidden policies, and policy cycles. In: Network and Distributed System Security Symposium (NDSS), pp. 157–172. The Internet Society (2006)
14. Goldreich, O.: Foundations of Cryptography. Basic Applications, vol. 2. Cambridge University Press, Cambridge (2004)
15. Goldreich, O., Micali, S., Wigderson, A.: How to play any mental game, or a completeness theorem for protocols with honest majority. In: 19th Annual ACM Symposium on Theory of Computing (STOC), pp. 218–229. ACM Press (1987)
16. Henecka, W., Kögl, S., Sadeghi, A.-R., Schneider, T., Wehrenberg, I.: TASTY: Tool for automating secure two-party computations. In: 17th ACM Conf. on Computer and Communications Security (CCS), pp. 451–462. ACM Press (2010)
17. Huang, Y., Evans, D., Katz, J., Malka, L.: Faster secure two-party computation using garbled circuits. In: 20th USENIX Security Symposium (2011)

18. Ishai, Y., Kilian, J., Nissim, K., Petrank, E.: Extending Oblivious Transfers Efficiently. In: Boneh, D. (ed.) CRYPTO 2003. LNCS, vol. 2729, pp. 145–161. Springer, Heidelberg (2003)
19. Ishai, Y., Paskin, A.: Evaluating Branching Programs on Encrypted Data. In: Vadhan, S.P. (ed.) TCC 2007. LNCS, vol. 4392, pp. 575–594. Springer, Heidelberg (2007)
20. Jarecki, S., Shmatikov, V.: Efficient Two-Party Secure Computation on Committed Inputs. In: Naor, M. (ed.) EUROCRYPT 2007. LNCS, vol. 4515, pp. 97–114. Springer, Heidelberg (2007)
21. Katz, J., Malka, L.: Constant-round private function evaluation with linear complexity, http://eprint.iacr.org/2010/528
22. Kolesnikov, V., Schneider, T.: Improved Garbled Circuit: Free XOR Gates and Applications. In: Aceto, L., Damgård, I., Goldberg, L.A., Halldórsson, M.M., Ingólfsdóttir, A., Walukiewicz, I. (eds.) ICALP 2008, Part II. LNCS, vol. 5126, pp. 486–498. Springer, Heidelberg (2008)
23. Kolesnikov, V., Schneider, T.: A Practical Universal Circuit Construction and Secure Evaluation of Private Functions. In: Tsudik, G. (ed.) FC 2008. LNCS, vol. 5143, pp. 83–97. Springer, Heidelberg (2008)
24. Lindell, Y., Pinkas, B.: An Efficient Protocol for Secure Two-Party Computation in the Presence of Malicious Adversaries. In: Naor, M. (ed.) EUROCRYPT 2007. LNCS, vol. 4515, pp. 52–78. Springer, Heidelberg (2007)
25. Lindell, Y., Pinkas, B.: A proof of security of Yao's protocol for two-party computation. Journal of Cryptology 22(2), 161–188 (2009)
26. Lindell, Y., Pinkas, B., Smart, N.P.: Implementing Two-Party Computation Efficiently with Security Against Malicious Adversaries. In: Ostrovsky, R., De Prisco, R., Visconti, I. (eds.) SCN 2008. LNCS, vol. 5229, pp. 2–20. Springer, Heidelberg (2008)
27. Malkhi, D., Nisan, N., Pinkas, B., Sella, Y.: Fairplay — a secure two-party computation system. In: Proc. 13th USENIX Security Symposium, pp. 287–302. USENIX Association (2004)
28. Naor, M., Pinkas, B.: Oblivious polynomial evaluation. SIAM Journal on Computing 35(5), 1254–1281 (2006)
29. Nielsen, J.B., Orlandi, C.: LEGO for Two-Party Secure Computation. In: Reingold, O. (ed.) TCC 2009. LNCS, vol. 5444, pp. 368–386. Springer, Heidelberg (2009)
30. Paillier, P.: Public-Key Cryptosystems Based on Composite Degree Residuosity Classes. In: Stern, J. (ed.) EUROCRYPT 1999. LNCS, vol. 1592, pp. 223–238. Springer, Heidelberg (1999)
31. Paus, A., Sadeghi, A.-R., Schneider, T.: Practical Secure Evaluation of Semi-Private Functions. In: Abdalla, M., Pointcheval, D., Fouque, P.-A., Vergnaud, D. (eds.) ACNS 2009. LNCS, vol. 5536, pp. 89–106. Springer, Heidelberg (2009)
32. Pinkas, B., Schneider, T., Smart, N.P., Williams, S.C.: Secure Two-Party Computation is Practical. In: Matsui, M. (ed.) ASIACRYPT 2009. LNCS, vol. 5912, pp. 250–267. Springer, Heidelberg (2009)
33. Sadeghi, A.-R., Schneider, T.: Generalized Universal Circuits for Secure Evaluation of Private Functions with Application to Data Classification. In: Lee, P.J., Cheon, J.H. (eds.) ICISC 2008. LNCS, vol. 5461, pp. 336–353. Springer, Heidelberg (2009)
34. Sander, T., Young, A., Yung, M.: Non-interactive cryptocomputing for NC^1. In: 40th Annual Symposium on Foundations of Computer Science (FOCS), pp. 554–567. IEEE (1999)

35. Schneider, T.: Practical secure function evaluation. Master's thesis, University Erlangen-Nürnberg (2008), http://thomasckneider.de/FairplayPF
36. Valiant, L.: Universal circuits. In: 8th Annual ACM Symposium on Theory of Computing (STOC), pp. 196–203. ACM Press (1976)
37. Yao, A.C.-C.: How to generate and exchange secrets. In: 27th Annual Symposium on Foundations of Computer Science (FOCS), pp. 162–167. IEEE (1986)

Constant-Rounds, Linear Multi-party Computation for Exponentiation and Modulo Reduction with Perfect Security[*]

Chao Ning[**] and Qiuliang Xu[***]

School of Computer Science and Technology, Shandong University,
Jinan, 250101, China
ncnfl@mail.sdu.edu.cn, xql@sdu.edu.cn

Abstract. Bit-decomposition is an important primitive in multi-party computation (MPC). With the help of bit-decomposition, we will be able to construct constant-rounds protocols for various MPC problems, such as *equality test, comparison, public modulo reduction* and *private exponentiation*, which are four main applications of bit-decomposition. However, when considering perfect security, bit-decomposition does *not* have a linear communication complexity; thus any protocols involving bit-decomposition inherit this inefficiency. Constructing protocols for MPC problems without relying on bit-decomposition is a meaningful work because this may provide us with perfectly secure protocols with linear communication complexity. It is already proved that *equality test, comparison* and *public modulo reduction* can be solved without involving bit-decomposition and the communication complexity can be reduced to linear. However, it remains an open problem whether *private exponentiation* could be done without relying on bit-decomposition. In this paper, maybe somewhat surprisingly, we show that it can. That is to say, we construct a *constant-rounds, linear, perfectly secure* protocol for private exponentiation *without* relying on bit-decomposition though it seems essential to this problem.

In a recent work, Ning and Xu proposed a generalization of bit-decomposi-tion and, as a simplification of their generalization, they also proposed a linear protocol for public modulo reduction. In this paper, we show that their generalization can be further generalized; more importantly, as a simplification of our further generalization, we propose a public modulo reduction protocol which is more efficient than theirs.

Keywords: Multi-party Computation, Perfectly Secure, Constant-Rounds, Linear, Exponentiation, Modulo Reduction, Bit-Decomposition.

---

[*] Supported by the National Natural Science Foundation of China Grant 60873232, 61173139, the National Basic Research Program of China Grant 2007CB807900, 2007CB807901, the National Natural Science Foundation of China Grant 61033001, 61061130540, 61073174.

[**] Chao Ning is now at ITCS, part of the IIIS at Tsinghua University, China.

[***] Corresponding author.

D.H. Lee and X. Wang (Eds.): ASIACRYPT 2011, LNCS 7073, pp. 572–589, 2011.

1   Introduction

Multi-party computation (MPC) is a powerful and interesting tool in cryptology. It allows a set of n mutually un-trusted parties to compute a predefined function f with their private information as inputs. After running the MPC protocol, the parties obtains only the predefined outputs but nothing else, and the privacy of their inputs is guaranteed. Although generic solutions for MPC (which can compute any function f) already exist [3,9], these solutions tend to be inefficient and thus not applicable for practical use. So, to fix this problem, we focus on constructing efficient protocols for specific functions.

Recently, in the work [6], Damgård *et al.* proposed a novel technique called *bit-decomposition* which can, in constant rounds, convert a polynomial sharing of secret x into the sharings of the bits of x. Bit-decomposition (which will often be referred to as **BD** hereafter for short) is a very useful tool for MPC. For example, after getting the sharings of the bits of some shared secrets using BD, we can securely perform Boolean operations on these secrets (such as computing the Hamming Weight, XOR, etc). Thus we can say that BD can be viewed as a "bridge" (in the world of MPC) connecting the arithmetic circuits and the Boolean circuits. What's more, with the help of BD, we can construct constant-rounds protocols for some very important basic problems in MPC, such as *equality test, comparison, public modulo reduction* and *private exponentiation*, which will be referred to as four main applications of BD. After getting the bit-wise sharings of the shared inputs to these problems (using BD), we will be able to use the *divide and conquer* technique to solve these problems.

However, a problem is, BD is relatively expensive in terms of round and communication complexities, and thus all the protocols relying on BD inherit this inefficiency. For example, the communication complexity of BD (with perfect security) is non-linear, thus all the protocols involving BD have a non-linear communication complexity. A feasible solution for this problem is to construct protocols for MPC problems without relying on BD. It is already proved that, three of the four main applications of BD, i.e. *equality test, comparison* and *public modulo reduction*, can be realized without relying on BD [11,12] and the main advantage is that the communication complexity can be reduced to linear (under the premise of ensuring perfect security). Thus a natural problem is whether a similar conclusion can be arrived at for another important application of BD: *private exponentiation*. This is generally believed to be impossible before (e.g. [11], Page 2; [14], Page 2), however, in this paper, we show that it can. What's more, we show an improvement of the public modulo reduction protocol (without BD) proposed in [12]. The details of our results are presented below. Here we'd like to argue that although these four applications of BD can be realized without involving BD, this does *not* mean BD is meaningless for these problems because all these protocols (without relying on BD) depend heavily on the ideas, techniques and sub-protocols of BD and thus can be viewed as an extension of the research on BD.

1.1   Our Results

First we introduce some necessary notations. In this paper, we concern mainly about MPC based on linear secret sharing schemes (LSSS). Assume that the underlying LSSS is built on field $\mathbb{Z}_p$ where p is a prime with bit-length l (i.e. $l = \lceil \log p \rceil$). For an element $x = (x_{l-1}, ..., x_1, x_0) \in \mathbb{Z}_p$, we use $[x]_p$ to denote "the sharing of x", and $[x]_B$ to denote "the bitwise sharing of x" (which will also be referred to as "the sharings of the bits of x" or "the shared base-2 form of x" in this paper), i.e. $[x]_B = ([x_{l-1}]_p, ..., [x_1]_p, [x_0]_p)$.

Our work is mainly about two basic problems in MPC: the *private exponentiation problem* and the *public modulo reduction problem*. We construct efficient protocols, which are *constant-rounds, linear and perfectly secure*, for these two problems. The details are presented below.

The *private exponentiation problem* can be formalized as:

$$[x^a \bmod p]_p \leftarrow \text{Private-Exponentiation}([x]_p, [a]_p)$$

where $x, a \in \mathbb{Z}_p$.

Hereafter we will refer to $[x^a \bmod p]_p$ as $[x^a]_p$ for simplicity. For solving this problem, it seems that we must involve BD to get the bitwise sharing of the exponent, i.e. $[a]_B$. This is exactly the case in the private exponentiation protocol in [6]. However, in this paper we show that this is not necessary. That is to say, the private exponentiation problem can also be solved without relying on BD and the communication complexity can also be reduced to linear (*in the input length l*). Compared with the private exponentiation protocol in [6] (denoted as Pri-Expo-BD($\cdot$) in this paper), our protocol (denoted as Pri-Expo$^+(\cdot)$) reaches lower round complexity and much lower communication complexity.

The *public modulo reduction problem* (which will be referred to as *Pub-MRP* for short) can be formalized as:

$$[x \bmod m]_p \leftarrow \text{Public-Modulo-Reduction}([x]_p, m)$$

where $x \in \mathbb{Z}_p$ and $m \in \{2, 3, ..., p-1\}$.

Our work on this problem can be viewed as an extension of [12], in which Ning and Xu proposed a generalization of BD and, as a simplification of their generalization, they proposed a linear protocol for Pub-MRP without involving BD (denoted as Pub-MR($\cdot$) in this paper). In this paper, we propose a further generalization of their generalization and, similarly and more importantly, as a simplification of our further generalization, we propose a protocol for Pub-MRP with improved efficiency (denoted as Pub-MR$^+(\cdot)$). Specifically, the round complexity of our Pub-MR$^+(\cdot)$ protocol is the same with Pub-MR($\cdot$) and, for relatively small m, the communication complexity is reduced by a factor of approximately 4.

We'd like to stress that all the protocols constructed in this paper are constant-rounds and perfectly secure. See Appendix A (Table 1) for an overview of our protocols. What's more, we strongly recommend the interested readers to read [13] which is the full version of this paper. Many of the details are omitted in the present paper due to space constraints.

1.2 Related Work

Both of the two problems considered in this paper, *exponentiation* and *modulo reduction*, are applications of bit-decomposition (BD). The problem of BD was first considered by Algesheimer *et al.* in [1], in which a partial solution was proposed. The first full solution for BD in the secret sharing setting was propose in [6] by Damgård *et al.* The main concern of this work is constant-rounds solution for BD and this is achieved by realizing various constant-rounds sub-protocols which are important building blocks for subsequent research including ours. What's more, as an application of BD, they also proposed a private exponentiation protocol which is the foundation of our work. Independently and concurrently, Schoenmakers and Tuyls [16] solved the problem of BD for MPC based on (Paillier) threshold homomorphic cryptosystems [4,7] and they concern mainly about efficient variations of BD for practical use. In the work [11], Nishide and Ohta proposed solutions for interval test, comparison and equality test of shared secrets without relying on the expensive BD protocol although it seems necessary. Their ideas and techniques play an important role in our work. Recently, Toft showed a novel technique that can reduce the communication complexity of BD to *almost linear* [18]. This is a very meaningful work and some key ideas of our work come from it. In a followup work, Reistad and Toft proposed a linear BD protocol [14], however, the security of this protocol is non-perfect.

As for the public modulo reduction problem (Pub-MRP), Guajardo *et al.* proposed a protocol for it in the threshold homomorphic setting without relying on BD [8]. Their protocol is very efficient (thus can be very useful for practical use) and is enlightening to this paper, however, they did not consider the general case (of Pub-MRP) where the inputs can be arbitrary size. In [12], Ning and Xu proposed a generalization of BD, and, as a simplification of their generalization, they proposed a linear protocol (without BD) for Pub-MRP which can deal with arbitrary inputs. Our work on Pub-MRP depends heavily on their work.

2 Preliminaries

In this section we introduce some important notations and known primitives.

2.1 Notations and Conventions

As mentioned above, the MPC considered in this paper is based on LSSS, such as Shamir's [15]. We denote the underlying field (of the LSSS) as $\mathbb{Z}_p$ where p is a prime with bit-length $l = \lceil \log p \rceil$. For a secret $x \in \mathbb{Z}_p$, we use $[x]_p$ to denote the sharing of x and $[x]_B = ([x_{l-1}]_p, ..., [x_1]_p, [x_0]_p)$ to denote the bitwise sharing of x. What's more, assume that there are n participants in the MPC protocol.

As in previous works, such as [6,11], we assume that the underlying LSSS allows to compute $[x + y \bmod p]_p$ from $[x]_p$ and $[y]_p$ without communication, and that it allows to compute $[xy \bmod p]_p$ from (public) $x \in \mathbb{Z}_p$ and $[y]_p$ without communication. We also assume that the LSSS allows to compute $[xy \bmod p]_p$

from $[x]_p$ and $[y]_p$ through communication among the parties and we call this procedure *secure multiplication* (or *multiplication* for simplicity). One invocation of this *multiplication* will be denoted as

$$[xy \bmod p]_p \leftarrow \text{Sec-Mult}([x]_p, [y]_p)$$

in which $[xy \bmod p]_p$ will be referred to as $[xy]_p$ for simplicity. Obviously, for MPC protocols, this multiplication protocol is a dominant factor of complexity as it involves communication. So, as in previous works, the round complexity of the (MPC) protocols is measured by the number of rounds of parallel invocations of multiplication (Sec-Mult($\cdot$)), and the communication complexity is measured by the number of invocations of multiplication. For example, if in all a protocol involves a multiplications in parallel and then another b multiplications in parallel, then we can say that the round complexity of this protocol is 2 and the communication complexity is $a + b$ multiplications. What's more, if a procedure does not involve any secure multiplication, then it can be viewed as free and will not count for complexity. For example, if we get $[x]_B$, then $[x]_p$ can be freely obtained by a linear combination since $x = \sum_{i=0}^{l-1} x_i \cdot 2^i$.

As in [11], when we write $[C]_p$, where C is a Boolean test, it means that $C \in \{0, 1\}$ and $C = 1$ iff C is true. For example, we use $[x \stackrel{?}{=} y]_p$ to denote the output of the equality test protocol, i.e. $(x \stackrel{?}{=} y) = 1$ iff $x = y$ holds.

Given $[c]_p$, we need a protocol to reveal c, which is denoted by $c \leftarrow \text{Reveal}([c]_p)$. Note that although this protocol involves communication, it does not count for (both round and communication) complexity because the communication it involves can be carried out through a public channel.

As in [17], we will often use the *conditional selection command* below:
$$[C]_p \leftarrow [b]_p \; ? \; [A]_p \; : \; [B]_p$$
in which $A, B, C \in \mathbb{Z}_p$ and $b \in \{0, 1\}$, and which means the following:
        If $b = 1$, then C is set to A; otherwise, C is set to B.
It is easy to see that this command can be realized by setting
$$[C]_p \leftarrow [b]_p([A]_p - [B]_p) + [B]_p$$
which costs only 1 round and 1 multiplication. We will frequently use this *conditional selection command* in this paper because it can make our protocols easier to be understood.

2.2   Known Primitives

We will now simply introduce some existing primitives which will be of importance later on. We refer the readers to [6,11,18] for detailed descriptions of these primitives.

- **Random Bit Protocol.** The **Random-Bit**($\cdot$) protocol has no input and it will output a shared uniformly random bit $[b]_p$ which is unknown to all parties. In the secret sharing setting, it takes only 2 rounds and 2 multiplications [6].

- **Bitwise Less-Than Protocol.** Given two bitwise shared inputs, $[x]_B$ and $[y]_B$, the **Bit-LessThan**$(\cdot)$ protocol can compute a shared bit $[x \overset{?}{<} y]_p$ which identifies whether $x < y$ holds. The complexity of this protocol can be referred to as 8 rounds and $14l$ multiplications when $l \geq 36$ holds which is often the case in practice [18,12].

- **Secure Inversion Protocol.** Given a shared *non-zero* secret $[x]_p$ as input, the secure inversion protocol **Sec-Inver**$(\cdot)$ will output $[x^{-1} \bmod p]_p$. This protocol will cost only 2 rounds and 2 multiplications [2,6,11].

- **Unbounded Fan-In Multiplication.** In this paper, we will often need to perform the unbounded fan-in secure multiplication [2,5], i.e. given l sharings $[A_0]_p, [A_1]_p, ..., [A_{l-1}]_p$ where $A_i \in \mathbb{Z}_p^*$ for $i \in \{0, 1, ..., l-1\}$, computing a sharing $[A]_p$ where $A = \prod_{i=0}^{l-1} A_i \mod p$. By the detailed analysis in [11], we get to know that this protocol, denoted as **Sec-Prod**$^*(\cdot)$ in this paper, can be realized in only 3 rounds and $5l$ multiplications.

- **Equal-Zero Test Protocol.** In [11], a linear protocol **Equ-Zero**$(\cdot)$ was proposed for testing whether a given secret $[x]_p$ is 0 or not, i.e. we have $[x \overset{?}{=} 0]_p \leftarrow \text{Equ-Zero}([x]_p)$. Obviously, this protocol can also be used to test "whether two shared secrets $[x]_p$ and $[y]_p$ are equal" because "$x = y$" $\Leftrightarrow$ "$(x - y) = 0$". The complexity of this protocol is 8 rounds and $81l$ multiplications.

- **Generation of Bitwise Shared Random Value.** This protocol, denoted by **Solved-Bits**$(\cdot)$, has no input and can output a bitwise shared random integer $[r]_B$ satisfying $r < p$. The complexity of this protocol can be referred to as 7 rounds and $56l$ multiplications when $l \geq 36$ [18].

- **Bit-Decomposition (BD).** In the secret sharing setting, the function of BD can be described as converting $[x]_p$ to $[x]_B$, i.e. we have $[x]_B \leftarrow \text{BD}([x]_p)$ [6,18]. To the best of our knowledge, currently the most efficient version of BD (*with perfect security*) was proposed in [18], whose complexity can be referred to as 23 rounds and $76l + 31l \log l$ multiplications when $l \geq 36$; in the text when analyzing the complexities of (exponentiation) protocols involving BD, we will refer to the complexity of BD as above. We note that [18] also proposed a BD protocol with *almost-linear* communication complexity (i.e. $O(l \log^* l)$ multiplications or even lower). This is of course a very meaningful work. However, inevitably the round complexity of this version of BD is relatively high and thus for obtaining (private exponentiation) protocols with close and comparable round complexities, (as well as for notational convenience) we do not referred to this BD protocol in detail in the text. (Although we focus mainly on the communication complexity of protocols, the round complexity should also be considered.) We also note that in [14], a linear BD is proposed, however, the security of this BD protocol is (at most) statistical; so in the text we will not refer to this BD protocol in detail neither, because we focus on protocols with perfect security.

3   Multi-party Computation for Private Exponentiation with BD

In [6], a constant-rounds private exponentiation protocol was constructed with the help of BD. This protocol is the foundation of our work and in our exponentiation protocol, we need to use the sub-protocols of it. So, in this section, we describe in detail this private exponentiation protocol *with* BD. We will first introduce two important sub-protocols of it, i.e. the *public exponentiation protocol* and the *bit exponentiation protocol*. All the protocols in this section are re-descriptions of the ones in [6] but with detailed analysis.

3.1   The Public Exponentiation Protocol

With a shared *non-zero* value $[x]_p$ (i.e. $x \in \mathbb{Z}_p^*$) and a public value $a \in \mathbb{Z}$ as inputs, the *public exponentiation protocol*, Pub-Expo($\cdot$), can compute $[x^a]_p$. The details are presented in Figure 1. Generally speaking, this protocol is a slightly improved version of the one in [6].

Protocol $[x^a]_p \leftarrow$ **Pub-Expo**$([x]_p, a)$

This protocol requires that $x \neq 0$.

1. Every party P_i ($i \in \{1, 2, ..., n\}$) picks a random integer $r_i \in \mathbb{Z}_p^*$ and computes r_i^{-a}. Then P_i shares r_i and r_i^{-a} between the parties, i.e. the parties get $[r_i]_p$ and $[r_i^{-a}]_p$.

2. The parties compute
$$[r]_p \leftarrow \text{Sec-Prod}^*([r_1]_p, [r_2]_p, ..., [r_n]_p)$$
$$[r^{-a}]_p \leftarrow \text{Sec-Prod}^*([r_1^{-a}]_p, [r_2^{-a}]_p, ..., [r_n^{-a}]_p)$$

3. $[xr]_p \leftarrow \text{Sec-Mult}([x]_p, [r]_p)$

4. $xr \leftarrow \text{Reveal}([xr]_p)$

5. **Return** $[x^a]_p = (xr)^a \cdot [r^{-a}]_p$

Fig. 1. The Public Exponentiation Protocol

As for the correctness, notice that in Step 4 we need to reveal the value of xr where r is non-zero, and it is easy to see that $xr = 0 \Leftrightarrow x = 0$. This is just why this protocol requires $x \neq 0$: if $x = 0$, then the parties will get to know this in this step. Also note that in this protocol the public exponent $(-a)$ is the additive inverse of a in the sense of mod $(p-1)$ *rather than* mod p.

Privacy is straightforward.

The complexity will be discussed in two cases: the semi-honest case and the malicious case. The difference between these two cases lies in Step 1 where every party P_i ($i \in \{1, 2, ..., n\}$) is required to distribute two sharings, $[r_i]_p$ and $[r_i^{-a}]_p$, between the parties. Below we will first analyze the complexity of this step. Before going on, recall the well-known fact that when considering the communication complexity of MPC protocols (in the LSSS setting), 1 invocation of the secure multiplication is equivalent to distributing n sharings between the n parties and thus the communication complexity of distributing 1 sharing (between the n parties) can be viewed as $\frac{1}{n}$ multiplications.

In the semi-honest case, all the n parties follow the protocol, so every party distributes 2 sharings between all the n parties, thus the complexity of Step 1 is 1 round and $\frac{2}{n} \cdot n = 2$ multiplications.

In the malicious case, as mentioned in [6], the complexity is much higher because we need to involve the *cut-and-choose* technique to make the protocol robust. Specifically, besides $[r_i]_p$ and $[r_i^{-a}]_p$, every party P_i ($i \in \{1, 2, ..., n\}$) is required to distribute another two sharings $[s_i]_p$ and $[s_i^{-a}]_p$. Then the parties involve the Random-Bit$(\cdot)$ protocol to jointly form a shared random bit $[b_i]_p$ and open it. Then they open $([s_i]_p, [s_i^{-a}]_p)$ or compute and open $([s_i r_i]_p, [s_i^{-a} r_i^{-a}]_p)$ according to the value of b_i and then verify that the first value is non-zero and that the second value is the $(-a)$'th power of the first. We call the above process *one instance of cut-and-choose*. For every party P_i ($i \in \{1, 2, ..., n\}$), to get a lower error probability, we can repeat the above process k (which satisfies $k \geq 1$ and which will be referred to as "*the security parameter for cut-and-choose*") times in parallel, leading to an error probability 2^{-k}. Then we can say that in all we need kn instances of cut-and-choose in parallel. As for the complexity of *one instance*, we notice the following facts: distributing $[s_i]_p$ and $[s_i^{-a}]_p$ between the parties involves $\frac{2}{n}$ multiplications; the generation of $[b_i]_p$ involves 2 rounds and 2 multiplications and can be scheduled in parallel with the process of distributing $[s_i]_p$ and $[s_i^{-a}]_p$; the computation of $([s_i r_i]_p, [s_i^{-a} r_i^{-a}]_p)$ involves 1 round and 2 multiplications and, obviously, on average we need only to compute $([s_i r_i]_p, [s_i^{-a} r_i^{-a}]_p)$ once every 2 instants of cut-and-choose because b_i is a uniformly random bit. So, on average, the complexity of one instance is (at most) $2 + 1 = 3$ rounds and $\frac{2}{n} + 2 + 2 \cdot \frac{1}{2} = \frac{2}{n} + 3$ multiplications. Recall that in all we need kn parallel instances of cut-and-choose. What's more, notice that the process of cut-and-choose can be scheduled in parallel with the process of distributing $[r_i]_p$ and $[r_i^{-a}]_p$. So, in the malicious case, the complexity of Step 1 is 3 rounds and $2 + kn \cdot (\frac{2}{n} + 3) = 2 + 2k + 3kn$ multiplications.

Then it is easy to see that, in the semi-honest case, the overall complexity of this Pub-Expo$(\cdot)$ protocol is $R_{pub} \triangleq R_{pub}^{s-h} = 1 + 3 + 1 = 5$ rounds and $C_{pub} \triangleq C_{pub}^{s-h} = 2 + 5n \cdot 2 + 1 = 10n + 3$ multiplications; in the malicious case, the overall complexity is $R_{pub} \triangleq R_{pub}^{mal} = 3 + 3 + 1 = 7$ rounds and $C_{pub} \triangleq C_{pub}^{mal} = (2 + 2k + 3kn) + 5n \cdot 2 + 1 = 3kn + 10n + 2k + 3$ multiplications[1]. Recall that n denotes the number of the parties and k is the security parameter for cut-and-choose. Hereafter, we will generally refer to the complexity of this protocol as R_{pub} rounds and C_{pub} multiplications. The values of R_{pub} and C_{pub}

---

[1] Thanks to one of the anonymous reviewers, we get to realize that (seen in isolation) we can combine Step 2 and Step 3 (in Figure 1 by viewing $[x]_p$ as one of the inputs of the Sec-Prod$^*()$ protocol for computing $[r]_p$) to save 1 round; this is of course a meaningful improvement for a "constant-rounds" protocol. However, considering the parallel invocations of this Pub-Expo() protocol (e.g. in the forthcoming protocol in Figure 5), we still separate these two steps (Step 2 and Step 3) when analyzing the complexity.

are determined by the adversaries considered; moreover, we can say that both R_{pub} and C_{pub} can be viewed as *constants* because they are independent from (the input length) l.

3.2   The Bit Exponentiation Protocol

With a shared *non-zero* value $[x]_p$ and a bitwise shared value $[a]_B = ([a_{l-1}]_p, ..., [a_1]_p, [a_0]_p)$ as inputs, the *bit exponentiation protocol*, Bit-Expo($\cdot$), can compute $[x^a]_p$. The details are seen in Figure 2.

Protocol $[x^a]_p \leftarrow$ **Bit-Expo**$([x]_p, [a]_B)$

This protocol requires that $x \neq 0$.

1. For $i = 0, 1, ..., l - 1$ in parallel: $[A_i]_p \leftarrow$ Pub-Expo($[x]_p, 2^i$)
2. For $i = 0, 1, ..., l - 1$ in parallel: $[B_i]_p \leftarrow [a_i]_p$? $[A_i]_p$: 1
3. **Return** $[x^a]_p \leftarrow$ Sec-Prod*$([B_{l-1}]_p, ..., [B_1]_p, [B_0]_p)$

Fig. 2. The Bit Exponentiation Protocol

Correctness and privacy is straightforward. The complexity of this protocol is $R_{pub} + 1 + 3 = R_{pub} + 4$ rounds and $C_{pub} \cdot l + l + 5l = (C_{pub} + 6)l$ multiplications.

3.3   The Private Exponentiation Protocol with BD

Here we come to the *private exponentiation protocol* relying on BD proposed in [6], which will be denoted by Pri-Expo-BD($\cdot$). Given two shared inputs $[x]_p$ and $[a]_p$, Pri-Expo-BD($\cdot$) will output $[x^a]_p$. This time, both x and a can be arbitrary values in $\mathbb{Z}_p$. See Figure 3 for the details.

Protocol $[x^a]_p \leftarrow$ **Pri-Expo-BD**$([x]_p, [a]_p)$

1. $[b]_p \leftarrow$ Equ-Zero($[x]_p$)
2. $[\tilde{x}]_p = [x]_p + [b]_p$
3. $[a]_B \leftarrow$ BD($[a]_p$)
4. $[\tilde{x}^a]_p \leftarrow$ Bit-Expo($[\tilde{x}]_p, [a]_B$)
5. **Return** $[x^a]_p = [\tilde{x}^a]_p - [b]_p$

Fig. 3. The Private Exponentiation Protocol *with* BD

As for the correctness, notice that $b = (x \overset{?}{=} 0)$ and that $[\tilde{x}]_p = [x]_p + [x \overset{?}{=} 0]_p$ is always non-zero and thus can be given to Bit-Expo($\cdot$) as the first input. What's more, it can be easily verified that $[x^a]_p = [\tilde{x}^a]_p - [x \overset{?}{=} 0]_p$ always holds no matter x is 0 or not. Using $\tilde{x}$ to substitute x to perform the protocol is in fact the "exception trick" proposed in [6] for handling the special case where $x = 0$.

Privacy follows readily from only using private sub-protocols.

The overall complexity of this protocol is $23 + (R_{pub} + 4) = R_{pub} + 27$ rounds and $81l + (76l + 31l \log l) + ((C_{pub} + 6)l) = 163l + C_{pub} \cdot l + 31l \log l$ multiplications. See [13] for the detailed complexity analysis.

4   Linear Multi-party Computation for Private Exponentiation

In this section, we propose a private exponentiation protocol with constant round complexity and linear communication complexity. Specifically, we will first show how to remove the invocation of BD to get a protocol with linear communication complexity. Then we will further improve this linear protocol to reduce the communication complexity considerably.

4.1   The Private Exponentiation Protocol without BD

See Figure 4 for our private exponentiation protocol *without* BD which will be denoted as Pri-Expo($\cdot$).

Protocol $[x^a]_p \leftarrow$ **Pri-Expo**($[x]_p, [a]_p$)
1. $[b]_p \leftarrow$ Equ-Zero($[x]_p$)
2. $[\tilde{x}]_p = [x]_p + [b]_p$
3. $[r]_B \leftarrow$ Solved-Bits() $\triangleright$ Recall that $[r]_B$ implies $[r]_p$.
4. $[c]_p = [a]_p + [r]_p$
5. $c \leftarrow$ Reveal($[c]_p$) $\triangleright$ $c = a + r \mod p$
6. $[C]_p \leftarrow$ Pub-Expo($[\tilde{x}]_p, c$) $\triangleright$ $C = \tilde{x}^c \mod p$
7. $[C']_p \leftarrow$ Sec-Mult($[C]_p, [\tilde{x}]_p$) $\triangleright$ $C' = C \cdot \tilde{x} = \tilde{x}^{c+1} = \tilde{x}^{c+1+\varphi(p)} = \tilde{x}^{c+p} \mod p$
8. $[f]_p \leftarrow$ Bit-LessThan($c, [r]_B$)
9. $[\tilde{C}]_p \leftarrow [f]_p \ ? \ [C']_p \ : \ [C]_p$ $\triangleright$ $\tilde{C} = \tilde{x}^{a+r} \mod p$
10. $[R]_p \leftarrow$ Bit-Expo($[\tilde{x}]_p, [r]_B$)
11. $[R^{-1}]_p \leftarrow$ Sec-Inver($[R]_p$) $\triangleright$ $R^{-1} = \tilde{x}^{-r} \mod p$
12. $[\tilde{x}^a]_p \leftarrow$ Sec-Mult($[\tilde{C}]_p, [R^{-1}]_p$)
13. **Return** $[x^a]_p = [\tilde{x}^a]_p - [b]_p$

Fig. 4. The Private Exponentiation Protocol *without* BD

Correctness: As for the correctness, similar to the Pri-Expo-BD($\cdot$) protocol (in Figure 3), we use the non-zero $[\tilde{x}]_p$ to substitute $[x]_p$ to perform the main process. The main idea of this protocol is as follows.

First we compute $[\tilde{C}]_p = [\tilde{x}^{a+r}]_p$. Notice that we have $c = a + r \mod p$ and there are two cases: no wrap-around mod p occurs or there is a wrap-around. In the former case, $a + r = c$ holds over the integers (or we can say "$a + r = c$ holds *unconditionally*") and then we have $c \geq r$ because $a \geq 0$; similarly, in the latter case, $a + r = c + p$ holds over the integers (i.e. $a + r = c + p$ holds *unconditionally*) and then we have $c < r$ because $c = r + (a - p)$ and $a < p$. So, for computing $[\tilde{x}^{a+r}]_p$, we can compute both of the two possible values of it, $[\tilde{x}^c]_p$ and $[\tilde{x}^{c+p}]_p$, and then *select* the correct one; this selection can be carried out by testing whether $c < r$ holds. What's more, when computing $[\tilde{x}^c]_p$ we need to involve the Pub-Expo$(\cdot)$ protocol; however, this is not necessary when computing $[\tilde{x}^{c+p}]_p$ because we have: $\tilde{x}^{c+p} = \tilde{x}^{c+p-(p-1)} = \tilde{x}^{c+1} = \tilde{x}^c \cdot \tilde{x} \mod p$.

Then, in the following steps, after getting $[R]_p = [\tilde{x}^r]_p$ using the Bit-Expo$(\cdot)$ protocol, we can obtain $[\tilde{x}^a]_p$ based on the simple fact $\tilde{x}^a = \tilde{x}^{a+r} \cdot (\tilde{x}^r)^{-1} \mod p$. Then finally $[x^a]_p$ can be easily obtained.

Privacy: Privacy is straightforward.

Complexity: As for the complexity, both in the semi-honest case and the malicious case, the complexity of this protocol (Pri-Expo$(\cdot)$) can be referred to as $8 + (R_{pub} + 6) + 1 = R_{pub} + 15$ rounds and $8l + 56l + C_{pub} + 1 + 14l + 1 + (C_{pub} + 6)l + 2 + 1 = 157l + C_{pub} \cdot l + C_{pub} + 5$ multiplications (See [13] for the detailed complexity analysis). Recall that both R_{pub} and C_{pub} can be viewed as *constants*, so this is a constant-rounds protocol with linear communication complexity. Compared with the (perfectly secure) Pri-Expo-BD$(\cdot)$ protocol proposed in [6] (whose complexity is $R_{pub} + 27$ rounds and $163l + C_{pub} \cdot l + 31l \log l$ multiplications), our protocol has a lower round complexity and a significantly lower communication complexity.

4.2   A Further Improvement

In this section, we make a further improvement of our Pri-Expo$(\cdot)$ protocol above by improving one of the sub-protocols of it, Bit-Expo$(\cdot)$, which is often the dominate factor of the communication complexity. The improved version of Pri-Expo$(\cdot)$ and Bit-Expo$(\cdot)$ will be denoted as Pri-Expo$^+(\cdot)$ and Bit-Expo$^+(\cdot)$ respectively. Generally speaking, by replacing the invocation of Bit-Expo$(\cdot)$ with Bit-Expo$^+(\cdot)$ in our Pri-Expo$(\cdot)$ protocol, we get our further improved private exponentiation protocol: Pri-Expo$^+(\cdot)$. The details are presented below.

In our Pri-Expo$(\cdot)$ protocol (in Figure 4), Bit-Expo$(\cdot)$ is a very important sub-protocol. Recall that the communication complexity of this sub-protocol is $(C_{pub} + 6)l$ multiplications; what's more, in the semi-honest case $C_{pub} = C_{pub}^{s-h} = 10n + 3$, and in the malicious case $C_{pub} = C_{pub}^{mal} = 3kn + 10n + 2k + 3$ (see Section 3.1). In many cases, Bit-Expo$(\cdot)$ is relatively expensive. For example, in the malicious case, if we set $n = 20$ and $k = 10$, then the communication complexity of Bit-Expo$(\cdot)$ will be $(C_{pub}^{mal} + 6)l = 829l$ multiplications; at the same time, the communication complexity of the (*whole*) Pri-Expo$(\cdot)$ protocol is $(C_{pub}^{mal} + 157)l + C_{pub}^{mal} + 5 = 980l + 828$ multiplications. So we can see that, in this case, Bit-Expo$(\cdot)$ is obviously a dominate factor of the communication complexity of Pri-Expo$(\cdot)$. So, reducing the communication complexity of Bit-Expo$(\cdot)$ is very meaningful.

The communication complexity of Bit-Expo$(\cdot)$ comes mainly from the l invocations of Pub-Expo$(\cdot)$ which is non-trivial (See Figure 2 and Figure 1). Here we show a technique that can reduce the number of invocations (of Pub-Expo($\cdot$)) to $2\sqrt{l}$ (with slight increase in round complexity) and thus reduce the communication complexity significantly. The main idea is presented below.

Consider the case that we want to compute x^a. We divide the given exponent $a = (a_{l-1}, ..., a_1, a_0)$ into s blocks, each of which contains t bits. Obviously, we have $s \cdot t = l$ and $1 \le s, t \le l$. We denote the i'th block of a as $a_i^{s \times t}$ for $i \in \{0, 1, ..., s-1\}$, and denote the j'th bit of the i'th block as $a_{i,j}^{s \times t}$ for $j \in \{0, 1, ..., t-1\}$. That is to say, we have

$$a = (a_{l-1}, ..., a_1, a_0) = \left(a_{s-1}^{s \times t}, ..., a_1^{s \times t}, a_0^{s \times t}\right)$$
$$= \left(\left(a_{s-1,t-1}^{s \times t}, ..., a_{s-1,1}^{s \times t}, a_{s-1,0}^{s \times t}\right), ..., \left(a_{1,t-1}^{s \times t}, ..., a_{1,1}^{s \times t}, a_{1,0}^{s \times t}\right), \left(a_{0,t-1}^{s \times t}, ..., a_{0,1}^{s \times t}, a_{0,0}^{s \times t}\right) \right)$$

Obviously, $a_i^{s \times t}$ can be viewed as the i'th digit of the base-2^t form of a. What's more, we have $a_{i,j}^{s \times t} = a_{i \cdot t + j}$. Now we have the following equations:

$$x^a = x^{\sum_{i=0}^{s-1} a_i^{s \times t} \cdot (2^t)^i} = \prod_{i=0}^{s-1} x^{a_i^{s \times t} \cdot (2^t)^i} = \prod_{i=0}^{s-1} \left(x^{a_i^{s \times t}}\right)^{(2^t)^i} = \prod_{i=0}^{s-1} \left(x^{\sum_{j=0}^{t-1} a_{i,j}^{s \times t} \cdot 2^j}\right)^{(2^t)^i}$$

$$= \prod_{i=0}^{s-1} \left(\prod_{j=0}^{t-1} x^{a_{i,j}^{s \times t} \cdot 2^j}\right)^{(2^t)^i} = \prod_{i=0}^{s-1} \left(\prod_{j=0}^{t-1} \left(x^{2^j}\right)^{a_{i,j}^{s \times t}}\right)^{(2^t)^i}$$

$$= \prod_{i=0}^{s-1} \left(\prod_{j=0}^{t-1} \left(a_{i,j}^{s \times t} \, ? \, x^{2^j} \, : \, 1\right)\right)^{(2^t)^i}$$

Based on the above facts, we propose our improved Bit-Expo($\cdot$) protocol, Bit-Expo$^+(\cdot)$, which is presented in Figure 5. Note that in Figure 5, for the convenience of the forthcoming discussions, the two variables, s and t, are not assigned. We will discuss how to assign them when analyzing the complexity of this protocol.

Protocol $[x^a]_p \leftarrow$ **Bit-Expo$^+([x]_p, [a]_B)$**

This protocol requires that $x \ne 0$.

1. For $j = 0, 1, ..., t-1$ in parallel: $[A_j]_p \leftarrow$ Pub-Expo($[x]_p, 2^j$)

2. For $i = 0, 1, ..., s-1$ in parallel do
    For $j = 0, 1, ..., t-1$ in parallel: $[B_{i,j}]_p \leftarrow [a_{i,j}^{s \times t}]_p \, ? \, [A_j]_p \, : \, 1$
    $[B_i]_p \leftarrow$ Sec-Prod$^*([B_{i,0}]_p, [B_{i,1}]_p, ..., [B_{i,t-1}]_p)$
    $[C_i]_p \leftarrow$ Pub-Expo$([B_i]_p, (2^t)^i)$
    End for

3. Return $[x^a]_p \leftarrow$ Sec-Prod$^*([C_0]_p, [C_1]_p, ..., [C_{s-1}]_p)$

Fig. 5. The *Improved* Bit Exponentiation Protocol

Correctness and privacy is straightforward. As for the complexity, notice that there are invocations of Pub-Expo($\cdot$) in both Step 1 and Step 2. One important point is, these two places of invocations can be scheduled *partially in parallel*. Specifically, when the invocations (of Pub-Expo($\cdot$)) in Step 1 are proceeding with the first two steps of Pub-Expo($\cdot$) (See Figure 1), the invocations in Step 2 can also proceed with them. That is to say, although these two places of invocations can *not* be scheduled (completely) in parallel, they will cost only 1 more round than one single invocation (note that Step 3 through Step 5 in Pub-Expo($\cdot$) (Figure 1) involve only 1 multiplication). So, the complexity of this protocol is $R_{pub} + 1 + 3 + 1 + 3 = R_{pub} + 8$ rounds and $C_{pub} \cdot t + t \cdot s + 5t \cdot s + C_{pub} \cdot s + 5s \leq C_{pub} \cdot (s + t) + 11l$ multiplications. (Recall that $s \cdot t = l$ and $1 \leq s \leq l$.)

It remains to assign concrete values to s and t. Note that we have "$s + t \geq 2\sqrt{s \cdot t} = 2\sqrt{l}$" and "$s + t = 2\sqrt{l}$ iff $s = t = \sqrt{l}$". So we should set $s = t = \sqrt{l}$, because in this case the communication complexity of this Bit-Expo$^+$($\cdot$) protocol will be the lowest, i.e. $C_{pub} \cdot 2\sqrt{l} + 11l$ multiplications. Then, if we replace the invocation of Bit-Expo($\cdot$) in our Pri-Expo($\cdot$) protocol (in Figure 4) with the Bit-Expo$^+$($\cdot$) protocol here, we will get an improved private exponentiation protocol (denoted as Pri-Expo$^+$($\cdot$)) whose complexity is $R_{pub} + 19$ rounds and $162l + C_{pub} \cdot 2\sqrt{l} + C_{pub} + 5$ multiplications. Compared with the Pri-Expo-BD($\cdot$) protocol in [6] (whose complexity is $R_{pub} + 27$ rounds and $163l + C_{pub} \cdot l + 31l \log l$ multiplications), our Pri-Expo$^+$($\cdot$) protocol reaches lower round complexity and much lower communication complexity. What's more, we can say that, the larger C_{pub} is (which implies larger n and k), the greater advantage our protocol has. For systems with relatively more participants, higher security requirements and longer input length (i.e. l), our protocol can be of overwhelming advantage. (See Appendix A (Table 1) for an overview.)

See [13] for some further discussions.

5   Further Generalization of BD and Improved Solution for Public Modulo Reduction

In this section, we propose a further generalization of BD and an improved solution for Pub-MRP. The work in this section depends *heavily* on the work in [12] which we'd strongly recommend the readers to read before going on.

Given a sharing of secret x, BD allows the parties to extract the shared *base-2 form* of x in constant rounds. In the work [12], Ning and Xu show us a generalization of BD which is named as "Base-m Digit-Decomposition" (or "Base-m Digit-Bit-Decomposition") and which can extract the shared (or bitwise shared) *base-m form* of x in constant rounds. We note that their generalization can be further generalized to a "*Hybrid-base Digit-Decomposition*" (or "*Hybrid-base Digit-Bit-Decomposition*") protocol which can extract the shared (or bitwise shared) *hybrid-base form* of x; here *hybrid-base* means the base of every digit can be different. For example, if we denote

"9 days 23 hours 59 minutes 59 seconds"

(which could be the "Time Left" before the submission deadline of this conference) as

$$x = \boxed{9\,|\,23\,|\,59\,|\,59}$$

then x can be used to represent the *total* seconds (left) and can be viewed as a hybrid-base integer with bases (from left to right) 10, 24, 60, 60. Here the left-most base (i.e. "10") can be set as we wish, but other bases are fixed.

Below we discuss the relationship between "the value of an integer" and "the bases" in another point of view. Specifically, we list 3 cases below.

1. **Getting the base-2 form of $x \in \mathbb{Z}_p$**
   In this case, we get $l = \lceil \log p \rceil$ bits $x_i \in \{0,1\}$ for $i \in \{0,1,...,l-1\}$ satisfying

$$x = \sum_{i=0}^{l-1} \left(x_i \cdot 2^i \right)$$

2. **Getting the base-m form of $x \in \mathbb{Z}_p$**
   Similarly, in this case, for the given base $m \geq 2$, we get $l^{(m)} = \lceil \log_m p \rceil$ digits $x_i^{(m)} \in \{0,1,...,m-1\}$ for $i \in \{0,1,...,l^{(m)}-1\}$ satisfying

$$x = \sum_{i=0}^{l^{(m)}-1} \left(x_i^{(m)} \cdot m^i \right)$$

3. **Getting the hybrid-base form of $x \in \mathbb{Z}_p$**
   Given an $l^{(M)}$ size "array of bases" $M[\,] = \left[m_{l^{(M)}-1}, ..., m_1, m_0 \right]$ satisfying $m_i \geq 2$ for $i \in \{0,1,...,l^{(M)}-1\}$ and $\prod_{i=0}^{l^{(M)}-2} m_i < p < \prod_{i=0}^{l^{(M)}-1} m_i$, we get $l^{(M)}$ digits $x_i^{(M)} \in \{0,1,...,m_i-1\}$ for $i \in \{0,1,...,l^{(M)}-1\}$ satisfying the following equation (*in which we set $m_{-1} = 1$*)

$$x = \sum_{i=0}^{l^{(M)}-1} \left(x_i^{(M)} \cdot \prod_{j=-1}^{i-1} m_j \right)$$

Here, we call $\left(x_{l^{(M)}-1}^{(M)}, ..., x_1^{(M)}, x_0^{(M)} \right)$ "*the hybrid-base form of x defined by $M[\,]$*".

It is easy to see that in the hybrid-base case (i.e. Case 3) if we set the "array of bases" $M[\,]$ to be $[m, ..., m, m]$ where $l^{(M)} = l^{(m)}$, then we will get the base-m case (i.e. Case 2); if we set $M[\,] = [2, ..., 2, 2]$ where $l^{(M)} = l$, we will get the base-2 case (i.e. Case 1).

Given a shared secret $[x]_p$ and an "array of bases" $M[\,]$, our "Hybrid-base Digit-Decomposition" (or "Hybrid-base Digit-Bit-Decomposition") protocol, whose asymptotic complexity is $O(1)$ rounds and $O(l^{(M)} \log l^{(M)} + l)$ [2] (or $O(l \log l)$) multiplications, can output the shared (or bitwise shared) hybrid-base

---

[2] This term was mistakenly written as $O(l^{(M)} \log l^{(M)})$ in the submission; thanks to one of the anonymous reviewers for pointing this out.

form of x defined by $M[\ ]$ (i.e. the sharings (or bitwise sharings) of all the digits $x_i^{(M)}$ for $i \in \{0, 1, ..., l^{(M)} - 1\}$) which will be referred to as $[x]_D^M$ (or $[x]_{D,B}^M$). That is to say, we have

$$[x]_D^M \leftarrow \text{Hybrid-Base-Digit-Decomposition}([x]_p, M[\ ]);$$

$$[x]_{D,B}^M \leftarrow \text{Hybrid-Base-Digit-Bit-Decomposition}([x]_p, M[\ ]).$$

The intuition behind our further generalization is similar to that of the generalization of BD in [12]. Specifically, as shown in [12], for getting the shared (or bitwise shared) base-m form of x, we need to randomize $[x]_p$ using a jointly generated random integer r whose bitwise shared base-m form is known to the parties; that is to say, the parties generate an array of bitwise shared base-m digits to form r; here, a *base-m digit* is in fact a non-negative integer less than m and the details of generating such a (bitwise shared) digit can be seen in [12] (the Random-Digit-Bit($\cdot$) protocol). Similarly, to obtain the shared (or bitwise shared) hybrid-base form of x (which is defined by $M[\ ]$), we should randomize $[x]_p$ using a (jointly generated random) integer r^+ whose bitwise shared hybrid-base form (*which is also defined by $M[\ ]$*) is known to the parties. This is the key idea of our further generalization and is also the key difference between the generalization in [12] and our further generalization.

More importantly, as a simplification of our "Hybrid-base Digit-Decomposition" protocol, we can get an *improved* public modulo reduction protocol (denoted as Pub-MR$^+(\cdot)$ here) which is more efficient than the one in [12] (denoted as Pub-MR($\cdot$)). Specifically, in [12], for solving Pub-MRP (i.e. computing $[x \bmod m]_p$ from $[x]_p$ and $m \in \{2, 3, ..., p - 1\}$), Ning and Xu view this problem as extracting only (the sharing of) the least significant base-m digit of x, and thus their modulo reduction protocol (i.e. Pub-MR($\cdot$)) can be viewed as a simplification of their "Base-m Digit-Decomposition" protocol (which extracts (the sharings of) all the base-m digits of x). In another point of view, we can say that they set $M[\ ] = [m, ..., m, m]$ and extract only (the sharing of) $x_0^{(M)}$ (see Case 3 above). This is of course correct because in this case we have $x = \sum_{i=0}^{l^{(M)}-1} \left(x_i^{(M)} \cdot m^i \right)$. However, this is not a must, and, enlightened by [11] and [8], we find that, by setting $M[\ ] = [2, ..., 2, 2, m]$ where $l^{(M)} = \left\lceil \log \left\lfloor \frac{p}{m} \right\rfloor \right\rceil + 1$, we can also solve Pub-MRP because in this case we have

$$x = \sum_{i=1}^{l^{(M)}-1} \left(x_i^{(M)} \cdot 2^{i-1} \cdot m \right) + x_0^{(M)}$$

and thus $x_0^{(M)} = (x \bmod m)$. That is to say, in the case where $M[\ ] = [2, ..., 2, 2, m]$, if we extract only (the sharing of) the least significant digit of x, which can be viewed as a simplification of our "Hybrid-base Digit-Decomposition", we can also get $[x \bmod m]_p$; this public modulo reduction protocol is just Pub-MR$^+(\cdot)$.

Comparison: Below we show the advantage of our Pub-MR$^+(\cdot)$ protocol over Pub-MR($\cdot$). Similar to the generalization and further generalization of BD, when computing $[x \bmod m]_p$ from $[x]_p$ and m, both Pub-MR($\cdot$) and our Pub-MR$^+(\cdot)$

need to use a jointly generated random integer to randomize $[x]_p$ [12]. The key difference between these two (modulo reduction) protocols lies just in the generation of this random integer. Specifically, in the Pub-MR$(\cdot)$ protocol, the random integer needed, denoted as r here, should be of a "hybrid-base" form defined by $M[\,] = [m, ..., m, m]$ where $l^{(M)} = l^{(m)}$; whereas in our Pub-MR$^+(\cdot)$ protocol, the random integer needed, denoted as r^+, should be of a hybrid-base form defined by $M[\,] = [2, ..., 2, 2, m]$ where $l^{(M)} = \lceil \log \lfloor \frac{p}{m} \rfloor \rceil + 1$. Then obviously, in Pub-MR$(\cdot)$ when generating r, we need to generate $l^{(M)} = l^{(m)}$ (bitwise shared) base-m digits, whereas in our Pub-MR$^+(\cdot)$ when generating r^+, we need only to generate 1 such base-m digit. This is just the advantage of our improved public modulo reduction protocol; reducing the demand for such base-m digits is very meaningful because the generation of them is a non-trivial work. Specifically, when m is a non-power of 2, roughly speaking the generation of 1 such digit will cost 8 rounds and $64L(m)$ multiplications where $L(m) \triangleq \lceil \log m \rceil$ denotes the bit-length of m [12].

Complexity: Finally, we conclude that, the complexity of our Pub-MR$^+(\cdot)$ protocol is 22 rounds and (about) $78l + 276L(m)$ multiplications. Compared with Pub-MR$(\cdot)$ (whose complexity is 22 rounds and (about) $326l + 28L(m)$ multiplications), we can see that, for relatively small m (thus $L(m)$ is very small), the communication complexity is reduced considerably. For example, in the case where $l = 256, m = 100$ (then $L(m) = \lceil \log 100 \rceil = 7$), the communication complexity is reduced by a factor of approximately 3.8.

6 Discussions

We note that using the ideas in [11], the round complexity of our private exponentiation protocols (as well as our public modulo reduction protocol) can be improved; the method is to use preprocessing, i.e. moving all the generation of (shared) random values (e.g. the invocations of Random-Bit$(\cdot)$) to the beginning of the whole protocol. In the analysis of the round complexity of our protocols, we simply ignore this for clarity.

An interesting point is that the communication complexity of our "Hybrid-base Digit-Decomposition" protocol and "Hybrid-base Digit-Bit-Decomposition" protocol can be reduced to "almost linear" using the techniques of [18]. Specifically, the only non-linear part of these two protocols is the computation of a *prefix*-∘ [6,12]; and the techniques proposed in [18], which is used to reduce the complexity of the only non-linear part of BD (computation of a *postfix-comparison*) to "almost linear", can also be used to reduce the complexity of the computation of this *prefix*-∘ to "almost linear".

7 Future Work

In our private exponentiation protocol, we need an important sub-protocol called *public exponentiation protocol* (i.e. Pub-Expo$(\cdot)$) for computing $[x^a \bmod p]_p$ from $[x]_p$ and a public a. A problem is, the communication complexity of this sub-protocol is relatively high and more importantly, the communication complexity depends on n and k (see Section 3.1 for the details). We leave it an open problem to construct more efficient protocols for this problem and protocols with communication complexity independent from n and k would be most welcome. What's more, in our private exponentiation protocol when involving Pub-Expo($\cdot$), the second input (i.e. the public a in Figure 1) is (almost) always a power of 2. So designing more efficient public exponentiation protocols for this special case is also meaningful.

Acknowledgments. We would like to thank the anonymous reviewers for their careful work and helpful comments.

References

1. Algesheimer, J., Camenisch, J.L., Shoup, V.: Efficient Computation Modulo A Shared Secret with Application to the Generation of Shared Safe-Prime Products. In: Yung, M. (ed.) CRYPTO 2002. LNCS, vol. 2442, pp. 417–432. Springer, Heidelberg (2002)
2. Bar-Ilan, J., Beaver, D.: Non-cryptographic fault-tolerant computing in a constant number of rounds of interaction. In: 8th ACM Symposium on Principles of Distributed Computing, pp. 201–209. ACM Press, New York (1989)
3. Ben-Or, M., Goldwasser, S., Wigderson, A.: Completeness Theorems for Noncryptographic Fault-Tolerant Distributed Computations. In: 20th Annual ACM Symposium on Theory of Computing, pp. 1–10. ACM Press, New York (1988)
4. Cramer, R., Damgård, I.B., Nielsen, J.B.: Multiparty Computation from Threshold Homomorphic Encryption. In: Pfitzmann, B. (ed.) EUROCRYPT 2001. LNCS, vol. 2045, pp. 280–300. Springer, Heidelberg (2001)
5. Chandra, A.K., Fortune, S., Lipton, R.J.: Unbounded Fan-In Circuits and Associative Functions. In: 15th Annual ACM Symposium on Theory of Computing, pp. 52–60. ACM Press, New York (1983)
6. Damgård, I.B., Fitzi, M., Kiltz, E., Nielsen, J.B., Toft, T.: Unconditionally Secure Constant-Rounds Multi-Party Computation for Equality, Comparison, Bits and Exponentiation. In: Halevi, S., Rabin, T. (eds.) TCC 2006. LNCS, vol. 3876, pp. 285–304. Springer, Heidelberg (2006)
7. Damgård, I.B., Nielsen, J.B.: Universally Composable Efficient Multiparty Computation from Threshold Homomorphic Encryption. In: Boneh, D. (ed.) CRYPTO 2003. LNCS, vol. 2729, pp. 247–264. Springer, Heidelberg (2003)
8. Guajardo, J., Mennink, B., Schoenmakers, B.: Modulo Reduction for Paillier Encryptions and Application to Secure Statistical Analysis. In: Sion, R. (ed.) FC 2010. LNCS, vol. 6052, pp. 375–382. Springer, Heidelberg (2010)
9. Goldreich, O., Micali, S., Wigderson, A.: How to Play Any Mental Game or A Complete Theorem for Protocols with Honest Majority. In: 19th Annual ACM Symposium on Theory of Computing, pp. 218–229. ACM Press, New York (1987)
10. Gennaro, R., Rabin, M.O., Rabin, T.: Simplified Vss and Fast-Track Multiparty Computations with Applications to Threshold Cryptography. In: 17th ACM Symposium on Principles of Distributed Computing, pp. 101–110. ACM Press, New York (1998)
11. Nishide, T., Ohta, K.: Multiparty Computation for Interval, Equality, and Comparison without Bit-Decomposition Protocol. In: Okamoto, T., Wang, X. (eds.) PKC 2007. LNCS, vol. 4450, pp. 343–360. Springer, Heidelberg (2007)

12. Ning, C., Xu, Q.: Multiparty Computation for Modulo Reduction without Bit-Decomposition and A Generalization to Bit-Decomposition. In: Abe, M. (ed.) ASI-ACRYPT 2010. LNCS, vol. 6477, pp. 483–500. Springer, Heidelberg (2010)
13. Ning, C., Xu, Q.: Constant-Rounds, Linear Multi-party Computation for Exponentiation and Modulo Reduction with Perfect Security. Cryptology ePrint Archive, Report 2011/069 (2011), http://eprint.iacr.org/2011/069
14. Reistad, T., Toft, T.: Linear, Constant-Rounds Bit-Decomposition. In: Lee, D., Hong, S. (eds.) ICISC 2009. LNCS, vol. 5984, pp. 245–257. Springer, Heidelberg (2010)
15. Shamir, A.: How to Share A Secret. Communications of the ACM 22(11), 612–613 (1979)
16. Schoenmakers, B., Tuyls, P.: Efficient Binary Conversion for Paillier Encrypted Values. In: Vaudenay, S. (ed.) EUROCRYPT 2006. LNCS, vol. 4004, pp. 522–537. Springer, Heidelberg (2006)
17. Toft, T.: Primitives and Applications for Multi-party Computation. PhD thesis, University of Aarhus (2007), http://www.daimi.au.dk/~ttoft/publications/dissertation.pdf
18. Toft, T.: Constant-Rounds, Almost-Linear Bit-Decomposition of Secret Shared Values. In: Fischlin, M. (ed.) CT-RSA 2009. LNCS, vol. 5473, pp. 357–371. Springer, Heidelberg (2009)

A   An Overview of the New Protocols

The details are presented in Table 1. Below are some notes.

As mentioned in Section 3.1, R_{pub} represents the round complexity of the public exponentiation protocol (i.e. Pub-Expo($\cdot$)) and C_{pub} represents the communication complexity (of Pub-Expo($\cdot$)), and the values of R_{pub} and C_{pub} are determined by the adversary considered. Specifically, in the semi-honest case, $R_{pub} = R_{pub}^{s-h} = 5$ and $C_{pub} = C_{pub}^{s-h} = 10n + 3$; in the malicious case, $R_{pub} = R_{pub}^{mal} = 7$ and $C_{pub} = C_{pub}^{mal} = 3kn + 10n + 2k + 3$, in which n denotes the number of the participants of the MPC protocol and k is the security parameter for cut-and-choose. Both R_{pub} and C_{pub} can be viewed as *constants* because they are independent from (the input length) l. What's more, as mentioned in Section 5, $L(m) = \lceil \log m \rceil$ represents the bit-length of m.

Table 1. Overview of The New Protocols

Protocol Description	Rounds	Multiplications
$[x^a]_p \leftarrow$ **Pub-Expo**$([x]_p, a)$	R_{pub}	C_{pub}
$[x^a]_p \leftarrow$ **Bit-Expo**$([x]_p, [a]_B)$	$R_{pub} + 4$	$C_{pub} \cdot l + 6l$
$[x^a]_p \leftarrow$ **Bit-Expo**$^+([x]_p, [a]_B)$	$R_{pub} + 8$	$C_{pub} \cdot 2\sqrt{l} + 11l$
$[x^a]_p \leftarrow$ **Pri-Expo-BD**$([x]_p, [a]_p)$	$R_{pub} + 27$	$163l + C_{pub} \cdot l + 31l \log l$
$[x^a]_p \leftarrow$ **Pri-Expo**$([x]_p, [a]_p)$	$R_{pub} + 15$	$157l + C_{pub} \cdot l + C_{pub} + 5$
$[x^a]_p \leftarrow$ **Pri-Expo**$^+([x]_p, [a]_p)$	$R_{pub} + 19$	$162l + C_{pub} \cdot 2\sqrt{l} + C_{pub} + 5$
$[x \bmod m]_p \leftarrow$ **Pub-MR**$([x]_p, m)$	22	$326l + 28L(m)$
$[x \bmod m]_p \leftarrow$ **Pub-MR**$^+([x]_p, m)$	22	$78l + 276L(m)$

Computational Verifiable Secret Sharing Revisited[*]

Michael Backes[1,2], Aniket Kate[1], and Arpita Patra[3,**]

[1] Max Planck Institute for Software Systems (MPI-SWS), Germany
{backes,aniket}@mpi-sws.org
[2] Saarland University, Germany
[3] Aarhus University, Denmark
arpita@cs.au.dk

Abstract. Verifiable secret sharing (VSS) is an important primitive in distributed cryptography that allows a dealer to share a secret among n parties in the presence of an adversary controlling at most t of them. In the *computational* setting, the feasibility of VSS schemes based on commitments was established over two decades ago. Interestingly, all known computational VSS schemes rely on the homomorphic nature of these commitments or achieve weaker guarantees. As homomorphism is not inherent to commitments or to the computational setting in general, a closer look at its utility to VSS is called for. In this work, we demonstrate that homomorphism of commitments is not a necessity for computational VSS in the synchronous or in the asynchronous communication model. We present new VSS schemes based only on the definitional properties of commitments that are almost as good as the existing VSS schemes based on homomorphic commitments. Importantly, they have significantly lower communication complexities than their (statistical or perfect) unconditional counterparts.

Further, in the synchronous communication model, we observe that a crucial interactive complexity measure of *round complexity* has never been formally studied for computational VSS. Interestingly, for the optimal resiliency conditions, the least possible round complexity in the known computational VSS schemes is identical to that in the (statistical or perfect) unconditional setting: three rounds. Considering the strength of the computational setting, this equivalence is certainly surprising. In this work, we show that three rounds are actually not mandatory for computational VSS. We present the first two-round VSS scheme for $n \geq 2t+1$ and lower-bound the result tightly by proving the impossibility of one-round computational VSS for $t \geq 2$ or $n \leq 3t$. We also include a new two-round VSS scheme using homomorphic commitments that has the same communication complexity as the well-known three-round Feldman and Pedersen VSS schemes.

Keywords: Verifiable Secret Sharing, Round Complexity, Commitments, Homomorphism.

---

[*] An extended version of this paper is available [1].
[**] Supported by Center for Research in the Foundations of Electronic Markets (CFEM), Denmark and Center for the Theory of Interactive Computation (CTIC).

D.H. Lee and X. Wang (Eds.): ASIACRYPT 2011, LNCS 7073, pp. 590–609, 2011.

1   Introduction

The notion of secret sharing was introduced independently by Shamir [30] and Blakley [2] in 1979. Since then, it has remained an important topic in cryptographic research. For integers n and t such that $n > t \geq 0$, an (n, t)-*secret sharing* scheme is a method used by a *dealer D* to share a secret s among a set of n parties (the *sharing* phase) in such a way that in the *reconstruction* phase any subset of $t + 1$ or more honest parties can compute the secret s, but subsets of size t or fewer cannot. Since in some secret sharing applications the dealer may benefit from behaving maliciously, parties also require a mechanism to confirm the correctness of the dealt values. To meet this requirement, Chor et al. [6] introduced the concept of *verifiable secret sharing* (VSS).

VSS has remained an important area of cryptographic research for the last two decades [3, 9–11, 13, 20, 21, 23, 26, 27]. In the literature, VSS schemes are categorized based on the adversarial computational power: computational VSS schemes and unconditional VSS schemes. In the former, the adversary is computationally bounded by a security parameter, while in the latter the adversary may possess unbounded computational power. Naturally, the computational VSS schemes are significantly more practical and efficient in terms of message and communication complexities as compared to the unconditional schemes. Thus, the majority of the recent research has been focussed on devising practical constructions for unconditional VSS. In this work, we revisit the concept of computational VSS [3, 9, 13, 26] to settle the round complexity of computational VSS based on minimal cryptographic assumptions (which is cryptographic commitment in our case) and to investigate the role of homomorphism of commitment schemes in the context of VSS.

Motivation and Contributions.   The major savings in the computational VSS schemes come from the use of cryptographic commitments. Interestingly, we find that all computational VSS schemes in the literature except [13, App. A] (which satisfies weaker conditions; see related work) require these commitments to be homomorphic. However, homomorphism is not inherent to cryptographic commitments; it is an additional property provided by discrete logarithm (DLog), Pedersen [27] and few other commitment schemes. As we elaborate later in the paper, commitments can be designed from general primitives such as one-way functions or collision-free hash functions; but, homomorphism may not be guaranteed in these constructions. Furthermore, relying on as little assumptions as possible without much loss in efficiency is always a general goal in cryptography. Therefore, computational VSS schemes based only on the definitional properties of commitments can be interesting to study.

In this paper, we show that homomorphism is not a necessity for VSS in both synchronous (known and bounded message delays) and asynchronous (unbounded message delays) communication model. While our VSS schemes (in both network settings) based on any commitment scheme are almost as good as the existing computational VSS protocols using homomorphic commitment schemes in terms of communication, they are considerably better than the unconditional VSS schemes.

In the synchronous communication model with a broadcast channel, Gennaro et al. [11] initiated the study of round complexity (number of rounds required to complete an execution) and proved a lower bound of three rounds during the sharing phase and one round during the reconstruction phase for unconditional VSS. The work was extended in [10, 20] with tight polynomial time constructions, and in [21, 23] by improving the bounds in a statistical scenario where the VSS properties are held *statistically* and can be violated with a negligible probability.

The round complexity of *computational* VSS has never being formally analyzed in the synchronous VSS literature. We observe that the round complexity of all known practical computational VSS protocols [9, 27] for the optimal resilience of $n \geq 2t + 1$ is the same as that of unconditional VSS schemes: three rounds in the sharing phase.[1] This similarity is surprising considering the usage of commitments in computational VSS. We analyze the round complexity of computational VSS with homomorphic and non-homomorphic commitments.

1. We show the impossibility of 1-round computational VSS protocol in the standard communication model under consideration; specifically, we prove that a computational VSS scheme with one round in the sharing phase is impossible for $t \geq 2$ or $n \leq 3t$. However, we find that there exists a special 1-round VSS construction for $t = 1$ and $n \geq 4$, when the dealer is one of the participants; we include the construction in the full version of the paper [1].
2. We then tighten our lower-bound result by providing a 2-round computational VSS scheme for $n \geq 2t + 1$ using any commitment scheme. Existing VSS schemes [9, 13, 27] based on homomorphic commitments require three rounds for $n \geq 2t+1$. Comparing with unconditional VSS schemes, we notice that the message (the number of messages transferred) and communication (the number of bits transferred) complexities of our scheme are at least a linear factor less. Also, our scheme is better in terms of round complexity or resilience bound as compared to all known unconditional VSS schemes.
   We then provide a VSS scheme for $n \geq 2t + 1$ using homomorphic commitments that has the same message and communication complexities but requires one less round of communication as compared to [9, 13, 27].

Organization. In the rest of this section, we review the related work. In Section 2, we describe our adversary model, and definitions of VSS and commitments. We present all our results for the synchronous model in Section 3 and those for the asynchronous model in Section 4. In Section 5, we discuss a few interesting open problems. Some of our proofs are shifted to the full version [1].

Related Work. For our work in the synchronous setting, we closely follow the network and adversary model of the best known VSS schemes: Feldman VSS [9] and Pedersen VSS [27]. These schemes are called *non-interactive* as

---

[1] Note that it is possible to reduce a round in sharing in [9, 27] but that asks for a sub-optimal resilience of $n \geq 3t + 1$. Further, with a much stronger assumption of non-interactive zero-knowledge (NIZK), it is possible to reduce the number of sharing rounds to one for $n \geq 2t + 1$ in the public key infrastructure [15].

they require unidirectional private links from the dealer to the parties; non-dealer parties speak only via the broadcast channel. Our protocol assumes nearly the same network model; however, in addition, we also allow parties to send messages to the dealer over the private channels. In practice, it is reasonable to assume that private links are bidirectional. Note that we do not need any private communication links between non-dealer parties.

It is also important to compare our results with unconditional VSS as we work towards reducing the cryptographic assumptions required for computational VSS. In unconditional or information theoretic settings, there are two different possibilities for the VSS properties; they can be held *perfectly* (i.e., error-free) or *statistically* with negligible error probability. Perfect VSS is possible if and only if $n \geq 3t+1$ [8], while statistical VSS is possible for $n \geq 2t+1$ [28], assuming a broadcast channel. Gennaro et al. [11] initiated the study of the round complexity of unconditional VSS, which was extended by Fitzi et al. [10] and Katz et al. [20]. They concentrate on unconditional VSS with perfect security and show that three rounds in the sharing phase are necessary and sufficient for $n \geq 3t+1$. In the statistical scenario, Patra et al. [23] show that $n \geq 3t+1$ is necessary and sufficient for 2-round statistical VSS. Recently, Kumaresan et al. [21] extended the result to prove that 3 rounds are enough for designing statistical VSS with $n \geq 2t+1$.

The round complexity is never studied formally for computational VSS. In the standard model that we follow, the best known computational VSS protocols [9, 13, 27] require two rounds; however, they work only for a suboptimal resilience of $n \geq 3t+1$. Although these schemes can also be adopted for $n \geq 2t+1$, they then ask for *three* rounds. In addition, the only known VSS scheme among these that does not mandate homomorphic commitments, [13, App. A], does not satisfies the generally required stronger commitment property described in Section 2.2. In this paper, we improve all the above results by showing that two rounds are necessary and sufficient for (stronger) VSS with $n \geq 2t+1$ using (homomorphic or non-homomorphic) cryptographic commitments. Note that it is also possible to achieve 1-round VSS in the presence of a public-key infrastructure (PKI) employing NIZK proofs [15]. However, NIZK proofs requires a common reference string or a random oracle. Furthermore, the scheme of [15] can only achieve computational secrecy, whereas our schemes can obtain unconditional or computational secrecy as required.

For our work in the asynchronous setting, we follow the standard model of Cachin et al. [3]. In the asynchronous setting, Cachin et al. [3], Zhou et al. [31], and more recently Schultz et al. [29] suggested computational VSS schemes. Of these, protocol by Cachin et al. is the most practical computational VSS protocol with $O(n^2)$ message complexity. However, all of these schemes rely on homomorphism of the commitment scheme. We avoid the use of homomorphism, while maintaining the message complexity of the VSS protocol by Cachin et al. [3]. Note that our protocol is significantly efficient in all aspects as compared to unconditional VSS schemes [4, 5, 24, 25] in the asynchronous setting.

2   Preliminaries

We work in the computational security setting, where κ denotes the security parameter of the system, in bits. We assume that the dealer's secret s lies over a finite field $\mathbb{F}_p$, where p is an κ bits long prime. Our polynomials for secret sharing belong to $\mathbb{F}_p[x]$ or $\mathbb{F}_p[x,y]$, and the indices for the parities are chosen from $\mathbb{Z}_p$. Without loss of generality, we assume these indices to be $\{1, \ldots, n\}$. A function $\epsilon(\cdot) : \mathbb{N} \to \mathbb{R}^+$ is called *negligible* if for all $c > 0$ there exists a κ_0 such that $\epsilon(\kappa) < 1/\kappa^c$ for all $\kappa > \kappa_0$. In the paper, $\epsilon(\cdot)$ denotes a negligible function.

2.1   Adversary Model

We consider a network of n parties $\mathcal{P} = \{P_1, P_2, \ldots, P_n\}$, where a distinguished party $D \in \mathcal{P}$ works as a dealer. Our adversary $\mathcal{A}$ is *t-bounded* and it can compromise and coordinate actions of up to t out of n parties. We also assume that the adversary is *adaptive*; it may corrupt any party at any instance during a protocol execution as long as the number of corruptions is bounded by t.

We work in the synchronous as well as the asynchronous settings , and postpone the discussions on communication setting to the respective sections (synchronous model in Section 3 and asynchronous model in Section 4).

2.2   VSS and Variants

We now present a definition of VSS [11]. A VSS protocol among n parties $\mathcal{P} = \{P_1, P_2, \ldots, P_n\}$ with a distinguished party $D \in \mathcal{P}$ consists of two phases: a *sharing* phase and a *reconstruction* phase.

Sharing. Initially, D holds an input s, referred to as the secret, and each party P_i may hold an independent random input r_i. At the end of the sharing phase, each honest party P_i holds a view v_i that may be required to reconstruct the dealer's secret later.

Reconstruction. In this phase, each party P_i publishes its entire view v_i from the sharing phase, and a reconstruction function $\texttt{Rec}(v_1; \ldots; v_n)$ is applied and is taken as the protocol's output.

We call an n-party VSS protocol, with t-bounded adversary $\mathcal{A}$, an (n, t)-VSS protocol if it satisfies the following conditions:

Secrecy. If D is honest then the adversary's view during the sharing phase reveals no information about s. More formally, the adversary's view is *identically distributed* for all different values of s.

Correctness. If D is honest then the honest parties output the secret s at the end of the reconstruction phase.

Commitment. If D is dishonest, then at the end of the sharing phase there exists a value $s^* \in \mathbb{F}_p \cup \{\perp\}$, such that at the end of the reconstruction phase all honest parties output s^*.

The sharing phase as well as the reconstruction phase may consist of several communication rounds. A VSS protocol is considered *efficient* if the total computation and communication performed by all the honest parties is polynomial in n and the security parameter κ. The optimal resiliency bound for VSS is $n \geq 2t + 1$ (in the presence of a broadcast channel) in the synchronous setting and $n \geq 3t + 1$ in the asynchronous setting.

Variants of VSS. A few variants of VSS have been introduced as required in secret sharing applications. We briefly describe those below.

1. In our VSS definition, we assume that secrecy is unconditional, while correctness and commitment are computational. We can have a variation where secrecy is computational, and correctness and commitment are unconditional in nature. This is easily possible as secrecy and correctness of a VSS scheme are derived respectively from the hiding and binding of the commitment scheme under use. Our lower bound results hold for this variation as well. However, for computationally secure VSS, we can prove security only against a *static* adversary that chooses t parties before a protocol execution starts.

2. In our VSS, the reconstruction may end with $\perp$. By fixing a default value in $\mathbb{F}_p$ (say 0) that will be output instead of $\perp$, it is possible to say that $s^* \in \mathbb{F}_p$. However, as suggested in [11, Sec. 2.1], there is even a stronger VSS definition possible. The stronger definition has exactly the same secrecy and correctness properties, but has a stronger commitment property:
   Strong Commitment. Even if D is dishonest, at the end of the sharing phase, each party locally outputs a share of a secret s^* chosen only from $\mathbb{F}_p$ such that shares from any $t + 1$ honest parties are consistent with s^*.
   For Shamir's secret sharing, this property means that at the end of the sharing phase, there exists a t-degree polynomial $f(x)$ such that a share s_i held by every honest party P_i is equal to $f(i)$. While our asynchronous protocol in Section 4.2 satisfies the basic VSS definition, our 2-round protocols in sections 3.2 and 3.4 satisfy the stronger definition. In the full version [1], we present an asynchronous protocol satisfying the stronger definition.

3. Another stronger variant of VSS considers dealer D to be an external party (i.e., $D \notin \mathcal{P}$) and allows the t-bounded adversary to corrupt the dealer and up to t additional parties in $\mathcal{P}$.
   Our lower bound results and all of our protocols except our one-round VSS protocol [1] work for this variant as well. We show that 1-round VSS with an external dealer is impossible even when $t = 1$ irrespective of the value of n and the number of rounds in the reconstruction phase.

We work on VSS as a standalone primitive in this paper. The required VSS properties, specially the commitment property, may change in some VSS application. We consider that to be an interesting future work and discuss in Section 5.

2.3   Commitment Schemes

A cryptographic commitment scheme is a two-phase cryptographic protocol between a *committer* and a *verifier*.

Commit Phase. Given a message m, a committer runs $[\mathcal{C}, (m, d)] = \mathsf{Commit}(m)$ and publishes $\mathcal{C}$ as a *commitment* that binds her to message m (*binding*) without revealing it (*hiding*). The function *may* output an opening value d.

Open Phase. The committer opens commitment $\mathcal{C}$ by revealing (m, d) to a verifier. The verifier can then check if the message is consistent with the commitment (i.e., $m \overset{?}{=} \mathsf{Open}(C, m, d)$).

We note that the commitment schemes also require a setup that generally involves choosing the cryptographic parameters. This can easily be included in the VSS setup and thus we do not consider it in detail.

A commitment scheme cannot be unconditional (perfect or statistical) binding and hiding at the same time. As a result, commitments come in two flavors: perfect (or statistical) binding but computational hiding commitments, and perfect (or statistical) hiding but computational binding commitments. There are many applications of commitments where they may never be opened or opened only after a while. In such scenarios, commitments of the second type are generally considered advantageous over the first type, since the committed values are hidden in information theoretic sense in the second type.

Perfect hiding but computational binding (under the DLog assumption) Pedersen commitment scheme [27] is the most commonly used commitment scheme in computational VSS. It has an interesting additive homomorphic property that a product of two commitments $\mathcal{C}_1$ and $\mathcal{C}_2$ (associated respectively with messages m_1 and m_2) commits to an addition of the committed messages $(m_1 + m_2)$. However, with its reliance on the DLog assumption, this commitment scheme will not be suitable once quantum computers arrive.

On the other hand, commitments of both types can be achieved from any one-way function (see [16] and references within). In this paper, we concentrate on the commitments of the second type, whose efficient constructions are possible from any claw-free permutation [14], any one-way permutation [22] or any collision-free hash function [17]. Along with being non-homomorphic, some of these commitment constructions are also interactive in the nature. We restrict ourselves to the non-interactive commitment constructions (e.g., [14] and [17]) as the interactive commitment constructions may increase the rounds complexity of our VSS schemes.

3   VSS in the Synchronous Network Model

Before presenting our results in the synchronous setting, we describe our synchronous communication model in detail.

3.1   Synchronous Communication Model

We closely follow the bounded-synchronous communication model in [9, 13, 27]. Here, the dealer is connected to every other party by a private, authenticated and bidirectional link. We do not require communication links between any two non-dealer parties in $\mathcal{P}$. We further assume that all parties have access to a common broadcast channel that allows a party to send a message to all other parties and every party is assured that all parties have received the same message in the same round.

In the synchronous model, the distributed protocols operate in a sequence of rounds. In each *round*, a party performs some local computation, sends messages (if any) to the dealer through the private and authenticated link, and broadcasts some information over the broadcast channel. By the end of the round, it also receives all messages sent or broadcast by the other parties in the same round.

Along with being adaptive and t-bounded, we allow the adversary to be *rushing*: in every round of communication it can wait to hear the messages of the honest parties before sending (or broadcasting) its own messages. By round complexity of VSS, we mean the number of rounds in the sharing phase only, since all of our protocols ask for single round during reconstruction.

3.2   2-Round VSS for $n \geq 2t + 1$ from Any Commitment

Here, we present a 2-round sharing and 1-round reconstruction VSS protocol for $n \geq 2t + 1$. Our 2-round VSS protocol allows any form of commitment. Feldman and Pedersen VSS schemes require three rounds for $n \geq 2t + 1$. The general structure of the sharing phase of their three round VSS schemes is: In the first (distribution) round, the dealer sends shares to parties and publishes a commitment on these shares. In the second round, parties may accuse (through broadcast) the dealer of sending inconsistent shares, which he resolves (through broadcast) in the third round. It is impossible to have distribution and accusation in the same round. Therefore, in order to reduce the number of rounds to two, the accusation and resolution rounds in VSS are collapsed into one round. To achieve this, the set of parties (in addition to dealer) performs some communication in the first round. We then employ a commitment-based modification of standard round-reduction technique from unconditional VSS protocols [11, Sect. 3.1]. It involves every party publicly committing to some randomness and sending that randomness to the dealer in the first round. The dealer uses this randomness as a blinding pad to broadcast the shares in the next round. Further, we use bivariate polynomial instead of univariate polynomials used in Feldman or Pedersen VSS. In the absence of homomorphism and without using bivariate polynomial, we do not know how the parties can check if the degree of a shared univariate polynomial is t without using expensive NIZK proofs.

Overview. In our 2-round protocol, dealer D chooses a t-degree symmetric bivariate polynomial $F(x, y)$ such that $F(0, 0) = s$, the secret that he wants to distribute. Note that all of our protocols in this paper work also with the asymmetric bivariate polynomials. However, for ease of understanding, we always use *symmetric* polynomials in our descriptions. Dealer D gives the univariate polynomial $f_i(x) = F(x, i)$ to every party P_i and publicly commits to evaluations $f_i(j)$ for $j \in [1, n]$. As already mentioned, we allow every party to communicate to D independently in the first round. Specifically, every party P_i sends n random

---

Protocol 2-Round-VSS$(D, \mathcal{P}, s)$: Sharing Phase (Two Rounds)

Round 1: Dealer D
- chooses a random symmetric bivariate polynomial $F(x, y)$ of degree-t such that $F(0, 0) = s$
- computes $[\mathsf{Com}_{ij}, (f_{ij}, r_{ij})] = \mathsf{Commit}(f_{ij})$ for $i, j \in [1, n]$ and $i \geq j$, where $f_{ij} = F(i, j)$
- assigns $\mathsf{Com}_{ij} = \mathsf{Com}_{ji}$ and $r_{ij} = r_{ji}$ for $i, j \in [1, n]$ and $i < j$
- sends (f_{ij}, r_{ij}) to P_i for $j \in [1, n]$ and broadcasts Com_{ij} for $i, j \in [1, n]$

Every other party P_i
- chooses two sets of n random values $(p_{i1}, \ldots, p_{in})$ and $(g_{i1}, \ldots, g_{in})$.
- computes $[\mathsf{PCom}_{ij}, (p_{ij}, q_{ij})] = \mathsf{Commit}(p_{ij})$ and $[\mathsf{GCom}_{ij}, (g_{ij}, h_{ij})] = \mathsf{Commit}(g_{ij})$ for $j \in [1, n]$.
- sends (p_{ij}, q_{ij}) and (g_{ij}, h_{ij}) for $j \in [1, n]$ to D, and broadcasts PCom_{ij} and GCom_{ij} for $j \in [1, n]$.

Round 2: Dealer D, for every party P_i,
- verifies if $p_{ij} \stackrel{?}{=} \mathsf{Open}(\mathsf{PCom}_{ij}, p_{ij}, q_{ij})$ and $g_{ij} \stackrel{?}{=} \mathsf{Open}(\mathsf{GCom}_{ij}, g_{ij}, h_{ij})$ for $j \in [1, n]$
- broadcasts $(\alpha_{ij}, \beta_{ij})$ for all $j \in [1, n]$ such that $\alpha_{ij} = f_{ij} + p_{ij}$ and $\beta_{ij} = r_{ij} + g_{ij}$ if the verification succeeds, and broadcasts (f_{ij}, r_{ij}) for all $j \in [1, n]$ otherwise.

Party P_i
- verifies if $\deg(f_i(x)) \stackrel{?}{=} t$ and $f_{ij} \stackrel{?}{=} \mathsf{Open}(\mathsf{Com}_{ij}, f_{ij}, r_{ij})$ for $j \in [1, n]$, where $f_i(x)$ is the polynomial defined by f_{ij}s for $j \in [1, n]$.
- broadcasts nothing if the verifications succeeds, and broadcasts (p_{ij}, q_{ij}) and (g_{ij}, h_{ij}) for $j \in [1, n]$ otherwise.

P_i is said to be **happy** if she broadcasts nothing, and considered **unhappy** otherwise.

Local Computation: Every party P_k
1. discards D and halts the execution of 2-Round-VSS, if D broadcasts
   - $\mathsf{Com}_{ij} \neq \mathsf{Com}_{ji}$ for some i and j
   - (f_{ij}, r_{ij}) such that $f_{ij} \neq \mathsf{Open}(\mathsf{Com}_{ij}, f_{ij}, r_{ij})$ for some i and j
   - f_{ij} for $j = [1, n]$ that define polynomial of degree $> t$ for some i
   - (f_{ij}, r_{ij}) and (f_{ji}, r_{ji}) for some i and j such that $(f_{ij} \neq f_{ji})$ or $(r_{ij} \neq r_{ji})$
   - $(\alpha_{ij}, \beta_{ij})$ and P_i broadcasts (p_{ij}, q_{ij}) and (g_{ij}, h_{ij}) such that $p_{ij} = \mathsf{Open}(\mathsf{PCom}_{ij}, p_{ij}, q_{ij}), g_{ij} = \mathsf{Open}(\mathsf{GCom}_{ij}, g_{ij}, h_{ij})$ for all j; and $(f'_{ij} \neq \mathsf{Open}(\mathsf{Com}_{ij}, f'_{ij}, r'_{ij})$ or $\deg(f'_i(x)) > t)$ where $f'_{ij} = \alpha_{ij} - p_{ij}, r'_{ij} = \beta_{ij} - g_{ij}$ and $f'_i(x)$ is the polynomial defined by f'_{ij}s for $j \in [1, n]$.
2. discards an **unhappy** party P_i, if she broadcasts p_{ij} and g_{ij} for $j \in [1, n]$ such that $p_{ij} \neq \mathsf{Open}(\mathsf{PCom}_{ij}, p_{ij}, q_{ij})$ or $g_{ij} \neq \mathsf{Open}(\mathsf{GCom}_{ij}, g_{ij}, h_{ij})$ for some j. Let $\mathcal{Q}$ be the set of non-discarded parties.
3. outputs (f_{kj}, r_{kj}) for $j \in [1, n]$ as received in round 1, if P_k is **happy** and in $\mathcal{Q}$. If she is **unhappy** and belongs to $\mathcal{Q}$ then she outputs (f_{kj}, r_{kj}) for $j \in [1, n]$ if they are broadcasted in round 2. Otherwise, P_k computes (f_{kj}, r_{kj}) for $j \in [1, n]$ as $f_{kj} = \alpha_{kj} - p_{kj}$ and $r_{kj} = \beta_{kj} - g_{kj}$.

Fig. 1. Sharing Phase of Protocol 2-Round-VSS$(D, \mathcal{P}, s)$ for $n \geq 2t + 1$

values privately to D and publicly commits them. At the end of the first round, every party checks the consistency of his received univariate polynomial with the commitments of D and D checks consistency of his received values with

---

Protocol 2-Round-VSS$(D, \mathcal{P}, s)$**: Reconstruction Phase (One Round)**

1. Each P_i in $\mathcal{Q}$ broadcasts (f'_{ij}, r'_{ij}) for $j \in [1, n]$

Local Computation: For every party P_k,
    1. Party $P_i \in \mathcal{Q}$ is said to be *confirmed* if $\deg(f'_i(x)) = t$ and $f'_{ij} = \mathsf{Open}(\mathsf{Com}_{ij}, f'_{ij}, r'_{ij})$ for $j \in [1, n]$, where $f'_i(x)$ is the polynomial defined by f'_{ij}'s for all $j \in [1, n]$.
    2. Consider $f'_i(x)$ polynomials of any $t + 1$ *confirmed* parties. Interpolate $F'(x, y)$ and output $s' = F'(0, 0)$.

Fig. 2. Reconstruction Phase of Protocol 2-Round-VSS$(D, \mathcal{P}, s)$ for $n \geq 2t + 1$

the corresponding commitments of the individual parties. The second round communication consists of only broadcasts. Any inconsistency between the public commitments and private values as well as the pairwise inconsistencies in the bivariate polynomial distribution (i.e, $f_i(j) \stackrel{?}{=} f_j(i)$) are sorted out in the second round. Note that there will be agreement among the parties at the end of local computation of sharing phase; i.e. every honest party knows if D is discarded, otherwise every honest party has identical copy of $\mathcal{Q}$, the set of parties allowed to participate in the reconstruction phase.

In the reconstruction phase, every party discloses their respective univariate polynomials. They are verified with respect to the public commitments and the consistent polynomials are used for the reconstruction of the bivariate polynomial and consequently the committed secret s. We present the protocol in Fig. 1 and Fig. 2. We prove that the 2-Round-VSS protocol satisfies the stronger variant of VSS defined in Section 2.2.

Theorem 1. *Protocol 2-Round-VSS is a VSS scheme for* $n \geq 2t + 1$.

Proof. We prove the secrecy, correctness and strong commitment properties of VSS to show that the above theorem holds.

Secrecy. The secrecy of the scheme follows from the unconditional hiding property of the underlying commitment function and the property of symmetric bivariate polynomial. D's public commitments Com_{ij}'s will be uniformly distributed given the unconditional hiding property of the underlying commitment function. Moreover, the α_{ij}, β_{ij} values for $j \in [1, n]$ corresponding to honest P_i's will be uniformly distributed. Now the secrecy of the constant term of the D's degree-t bivariate polynomial follows from the standard information-theoretic argument [27] against an adversary controlling at most t parties, i.e.,

$$\Pr[\mathcal{A} \text{ computes } s | \{V_i \text{ for any } t \text{ parties, Public Information}\}] = \Pr[\mathcal{A} \text{ computes } s],$$

where V_i represents all the information available at or computable by party P_i at the end of the sharing phase.

Correctness. If D is honest, then he will never be discarded. Moreover, all the honest parties will be **happy**. Now, correctness will follow if we show that a corrupted $P_i \in \mathcal{Q}$ is considered as *confirmed* only when she broadcasts correct polynomials in the reconstruction phase. Assume that corrupted P_i is considered to be *confirmed* even when she broadcasts f'_{ij} and r'_{ij} for $j \in [1, n]$, where these values are not equal to f_{ij} and r_{ij} (as given by D). We can then devise an algorithm to break the computational binding property of the commitment function using this adversary. Therefore, given that the commitment function achieves computational binding, all the *confirmed* parties disclose proper f_{ij} and r_{ij} for $j \in [1, n]$. Therefore, every honest party will correctly reconstruct $F(x, y)$ and consequently $s = F(0, 0)$.

Strong Commitment. We have to consider the case of a corrupted D. If D is discarded in the sharing phase, then every party may assume some default predefined value as D's secret. So we consider the case when D is not discarded.

Firstly, note that an honest party will never be discarded. Moreover at the end of sharing phase honest P_i will output n points (i.e. f_{ij}'s for all $j \in [1, n]$) on a degree-t polynomial $f_i(x)$ and n values r_{ij} such that for every honest P_j, it holds that $f_{ij} = f_{ji}$ and $r_{ij} = r_{ji}$. We show this by considering all the three cases for any pair of honest parties (P_i, P_j):

If P_i and P_j are happy, then we have $\mathsf{Com}_{ij} = \mathsf{Com}_{ji}$. Now P_i verified consistency of $(\mathsf{Com}_{ij}, f_{ij}, r_{ij})$, and P_j verified consistency of $(\mathsf{Com}_{ji}, f_{ji}, r_{ji})$. This implies the pair (f_{ij}, r_{ij}) is same as (f_{ji}, r_{ji}), unless corrupted D had broken the binding property of the commitment function.

If P_i is happy and P_j is unhappy, then $(\mathsf{Com}_{ij}, f_{ij}, r_{ij})$ is consistent and also $\mathsf{Com}_{ij} = \mathsf{Com}_{ji}$. For P_j, we have two cases: (1) D has broadcasted $f_j(k)$ and r_{jk} for $k \in [1, n]$; (2) D broadcasted α_{ik}, β_{ik} for $k \in [1, n]$ and P_j computed $f_{ik} = \alpha_{ik} - p_{ik}, r_{ik} = \beta_{ik} - g_{ik}$. However, in both the above cases, f_{ik} and r_{ik} are consistent with Com_{jk} for all $k \in [1, n]$ (for otherwise D would have been discarded). This also implies that tuple $(\mathsf{Com}_{ji}, f_{ji}, r_{ji})$ is consistent. Again unless corrupted D had broken the binding property of the commitment function, the pairs (f_{ij}, r_{ij}) and (f_{ji}, r_{ji}) are identical.

If P_i and P_j are unhappy, then D would have been discarded if the pairs (f_{ij}, r_{ij}) and (f_{ji}, r_{ji}) are not identical.

So unless corrupted D breaks the binding property of commitment function, the polynomials of the honest parties define symmetric bivariate polynomials, say $F(x, y)$. Now in the reconstruction phase, every honest party will be considered as *confirmed*. However, a corrupted party will be considered as *confirmed* if she broadcasts points on degree-t polynomial $f_i(x) = F(x, i)$ (assuming she does not break binding of commitment function). Let P_i broadcasts n points, say f'_{ij}'s, corresponding to $f'_i(x)$ that is different from $f_i(x)$. Then f_{ij} must be different from f'_{ij} at least for one j where P_j is honest. Then f'_{ij} will not be consistent with Com_{ij} and P_i will not be *confirmed*. Now it follows that the parties will reconstruct D's committed secret $s = F(0, 0)$ in the reconstruction phase.   □

The sharing phase of our 2-Round VSS protocol requires $O(n^2 \kappa)$ bits of broadcast and $O(n^2 \kappa)$ bits of private communication, while the reconstruction phase requires $O(n^2\kappa)$ bits of broadcast. This communication complexity is at least a linear factor lower than the unconditional VSS schemes for $n \geq 2t + 1$ [21]. On the other hand, it is also a linear factor higher than the communication complexity of 3-round Pedersen or Feldman VSS. This difference arises due to the use of bivariate polynomial in our protocol, which results from the lack of homomorphism in the commitment scheme under use. We suppose this increase in the communication complexity is a price paid for a reduction in the assumptions. In subsection 3.4, we present a more efficient VSS protocol using homomorphic commitments that has same communication complexity as Pedersen or Feldman VSS, but requires one less round of communication.

3.3 (Im)possibility Results for 1-Round VSS

Here, we prove the impossibility of 1-Round VSS except when $t = 1$ and $n \geq 4$, which lower-bounds computational VSS for $n \geq 2t + 1$ and any t to a round complexity of *two*. Our 2-round protocol presented in the previous section thus has an optimal round complexity. Our results hold irrespective of computational or unconditional nature of the secrecy property.

Theorem 2. *1-round VSS is impossible for $t > 1$ and $n \geq 4$, irrespective of the number of rounds in the reconstruction phase.*

Proof (Sketch). The proof of this theorem is very similar to the proof of Theorem 7 of [23]. We prove the theorem by contradiction. So we assume that 1-round VSS, say Π, with $t = 2$ exists. Without loss of generality, we assume D to be some party other than P_1. We then show that for any execution if party P_1 receives some particular piece of information from the dealer, then she will reconstruct a particular secret in the reconstruction phase irrespective of what $P_2, \ldots, P_n$ has received from the dealer. This of course allows us to show a breach of secrecy of Π, since P_1 could be the sole corrupted party and can distinguish the secret when he receives the particular information. We note that the proof does not make any assumption on the computational power of P_1 i.e. even a polynomial time P_1 can breach the secrecy. Since the proof strategy is very similar to the proof of Theorem 7 of [23], we skip the details here and present a detailed proof in the full version of the paper [1].

Theorem 3. *1-round VSS is impossible for $n \leq 3t$, irrespective of the number of rounds in the reconstruction phase.*

Proof (Sketch). This theorem is also proved by contradiction. In brief, we show that if such a scheme exists, then the the view of any t parties in the sharing phase must determine the secret. This further implies a breach of secrecy, since adversary $\mathcal{A}$ can corrupt and coordinate any t parties. A detailed proof appears in the full version of the paper [1].

In Theorem 3, we show that 1-round VSS is impossible for $n \leq 3t$, which implies the impossibility of 1-round VSS for $t = 1$ and $n \leq 3$. Further, in Theorem 2, we show that 1-round VSS is impossible for $t > 1$ and $n \geq 4$. Therefore, 1-round VSS, if possible, will work for $t = 1$ and $n \geq 4$. We present a 1-round protocol in support of the corollary in the full version of the paper.

VSS with an External Dealer. Here it can be shown that 1-round sharing VSS is impossible even in the presence of a single corruption apart from the dealer irrespective of the total number of parties and number of rounds in the reconstruction phase. Basically, we can follow the proof of Theorem 2 and arrive at the same contradiction while assuming $l = 1$ and the dealer is corrupted. Hence, we have the following theorem.

Theorem 4. *1-round VSS with external dealer is impossible for $t > 0$ irrespective of the number of parties and the number of rounds in reconstruction phase.*

3.4   An Efficient 2-round VSS Using Homomorphic Commitments

We now present a 2-round sharing, 1-round reconstruction VSS protocol for $n \geq 2t + 1$ using homomorphic commitments. It has the same message and communication complexities as that of Feldman and Pedersen VSS schemes, and requires one less round of interaction. The protocol is similar to our 2-round protocol in Section 3.2; however, we do not need bivariate polynomials here.

Without loss of generality, we use the Pedersen commitment scheme as a representative homomorphic commitment scheme. In the sharing phase, dealer D chooses two random degree-t polynomials $f(x)$ and $r(x)$ such that $f(0) = s$. Dealer D then sends $f_i = f(i)$ and $r_i = r(i)$ to each P_i over the private links and broadcasts commitments on the coefficients of $f(x)$ (using the coefficients of $r(x)$ as random strings). By the end of the second round, every honest party must hold the correct point on the committed polynomial. To ensure that every P_i sends two pairs (p_i, q_i) and (g_i, h_i) in $\mathbb{F}_p^2$ to dealer D and publicly commits p_i (using q_i as a random element) and g_i (using h_i as a random element). Broadcasts and local computations in the second round are very similar to 2-Round-VSS in Section 3.2. The protocol is presented in Fig. 3. Similar to 2-Round-VSS, we note that there will be agreement among the parties at the end of local computation of sharing phase on whether D is discarded or not. If D is not discarded, then every honest party will have identical copy of Q.

Theorem 5. *Protocol 2-Round-VSS-Hm is a VSS scheme for $n \geq 2t + 1$.*

The proof of the theorem closely follows from the proof of Theorem 1, and we include it in the full version of the paper.

The sharing phase requires $O(n\kappa)$ bits of communication over both the private links and the broadcast channel. The reconstruction phase requires $O(n\kappa)$ bits of communication over the broadcast channel.

4   VSS in the Asynchronous Communication Model

We now shift our focus to the asynchronous communication setting where VSS is possible for $n \geq 3t + 1$. As we discuss in the related work, all known computational VSS scheme [3, 29, 31] in the asynchronous communication setting rely on

---

Protocol 2-Round-VSS-Hm$(D, \mathcal{P}, s)$

Sharing Phase: Two Rounds

Round 1:

1. D selects two random polynomials $f(x)$ and $r(x)$ of degree-t, such that $f(0) = s$. Let $f(x) = a_0 + a_1 x + \ldots + a_t x^t$ and $r(x) = b_0 + b_1 x + \ldots + b_t x^t$.
2. For every $i \in [1, n]$, D sends $f_i = f(i)$ and $r_i = r(i)$ to P_i and broadcasts $\mathsf{Com}_i = \mathsf{Commit}(a_i, b_i)$ for $i = 0, \ldots, t$.
3. Every party P_i sends two pairs (p_i, q_i) and (g_i, h_i) in $\mathbb{F}_p^2$ to D and broadcasts commitments $\mathsf{PCom}_i = \mathsf{Commit}(p_i, q_i)$ and $\mathsf{GCom}_i = \mathsf{Commit}(g_i, h_i)$.

Round 2:

1. D checks if PCom_i and GCom_i are consistent with the received pairs (p_i, q_i) and (g_i, h_i). If they are not consistent, then D broadcasts (f_i, r_i); else he broadcasts $\alpha_i = f_i + p_i$ and $\beta_i = r_i + g_i$.
2. Party P_i checks if $\mathsf{Commit}(f_i, r_i) = \prod_{j=0}^{t} (\mathsf{Com}_i)^{i^j}$. If not, then P_i broadcasts pairs (p_i, q_i) and (g_i, h_i), else she broadcasts nothing. Party P_i is considered **happy** in the later case while she is **unhappy** in the former case.

Local Computation: Every party P_k

1. discards D and halts the execution of 2-Round-VSS-Hm, if D broadcasts
   (a) f_i, r_i for some i and $\mathsf{Commit}(f_i, r_i) \neq \prod_{j=0}^{t} (\mathsf{Com}_i)^{i^j}$.
   (b) α_i, β_i; and P_i broadcasts (p_i, q_i) and (g_i, h_i) such that $\mathsf{PCom}_i = \mathsf{Commit}(p_i, q_i)$ and $\mathsf{GCom}_i = \mathsf{Commit}(g_i, h_i)$; and $\mathsf{Commit}(f_i', r_i') \neq \prod_{j=0}^{t} (\mathsf{Com}_i)^{i^j}$ where $f_i' = \alpha_i - p_i$ and $r_i' = \beta_i - g_i$.
2. discards an **unhappy** party P_i if she broadcasts (p_i, q_i) and (g_i, h_i) such that $\mathsf{PCom}_i \neq \mathsf{Commit}(p_i, q_i)$ or $\mathsf{GCom}_i \neq \mathsf{Commit}(g_i, h_i)$. Let $\mathcal{Q}$ be the set of non-discarded parties.
3. outputs f_k, r_k as received from D in round 1, if P_k is in $\mathcal{Q}$ and **happy**. An unhappy P_k in $\mathcal{Q}$ outputs f_k, r_k if they are directly broadcasted by D in round 2. Else P_k computes f_k and r_k as $f_k = \alpha_k - p_k$ and $r_k = \beta_k - g_k$.

Reconstruction Phase: One Round

Round 1:

1. Each $P_i \in \mathcal{Q}$ broadcasts f_i' and r_i'.

Local Computation: For every party P_k,

1. Party $P_i \in \mathcal{Q}$ is said to be *confirmed* if $\mathsf{Commit}(f_i', r_i') = \prod_{j=0}^{t} (\mathsf{Com}_i)^{i^j}$.
2. Consider f_i' values of any $t + 1$ *confirmed* parties and interpolate $f'(x)$. Output $s' = f'(0)$.

Fig. 3. Protocol 2-Round-VSS-Hm for $n \geq 2t + 1$ with Homomorphic Commitments

homomorphism of commitments. In this section, we show that homomorphism is not necessary for computational VSS in the asynchronous communication setting. We build our protocol from asynchronous VSS of [3] as it is the only generic and efficient asynchronous VSS scheme known in the literature. Further, with its $O(n^2)$ messages complexity, it is extremely efficient in terms of the number of messages. We modify this scheme so that it satisfies the VSS properties when the underlying commitment need not be homomorphic. This protocol does not guarantee that every honest party receive his share of the secret. However, it guarantees that even a corrupted D can not commit to $\perp$ instead of a secret from

$\mathbb{F}_p$ (which is stronger than the basic definition given in section 2.2). We present another protocol in the full version that achieves the stronger definition where every party receives his share of the secret. Although this protocol increases the communication complexity by a linear factor in n, it is highly efficient in terms of communication when compared with the unconditional schemes [4, 5, 24, 25].

4.1   Asynchronous Communication Model

We follow the communication model of [3] and assume an asynchronous network of n parties $P_1, \ldots, P_n$ such that every pair of parties is connected by an authenticated and private communication link. We work against a t-bounded adaptive adversary that we defined in Section 2.1. In the asynchronous communication setting, we further assume that the adversary controls the network and may delay messages between any two honest parties. However, it cannot read or modify these messages as the links are private and authenticated, and it also has to eventually deliver all the messages by honest parties. In the asynchronous communication setting, a VSS scheme has to satisfy the liveness and agreement properties (also called as the termination conditions) along with the secrecy, correctness and commitment properties described in Section 2.2.

Liveness. If the dealer D is honest in the sharing phase, then all honest parties complete the sharing phase.

Agreement. If some honest party completes the sharing phase, then all honest parties complete the sharing phase eventually. If all honest parties subsequently start the reconstruction, then all honest parties will complete it.

4.2   VSS for $n \geq 3t + 1$ from Any Commitment

We observe that VSS of [3] heavily relies on homomorphism of the underlying commitment schemes and does not satisfy VSS properties if we replace the homomorphic commitments by non-homomorphic commitments (agreement property will not be satisfied). The incapability stems from the fact that verifying the following with respect to non-homomorphic commitment is not easy: given commitments on n values (associated with n indices), the underlying values define a degree-t polynomial. However, we find that with subtle enhancements to VSS of [3], one can obtain an asynchronous VSS protocol. In our enhanced protocol, a majority ($t + 1$ or more) of the honest parties receives proper share of the secret (t-degree univariate polynomial), while the remaining honest parties are assured that there are $t + 1$ or more honest parties that have received t-degree univariate polynomial and can complete the reconstruction phase. The message and communication complexities of our protocol are same as that of VSS of [3].

In our protocol, D chooses a symmetric bivariate polynomial $F(x, y)$ satisfying $F(0, 0) = s$. He then computes an $n \times n$ commitment matrix, Com such that $(i, j)^{th}$ entry in Com is the commitment on $F(i, j)$. Now D delivers $f_i(x) = F(x, i)$ and Com to every P_i. In the rest of the protocol the parties try to agree on Com and

---

Protocol AsynchVSS$(D, \mathcal{P}, s)$

Sharing Phase:
Code for D:

- Choose a random symmetric bivariate polynomial $F(x, y)$ of degree-t such that $F(0, 0) = s$.
- Compute $[\mathsf{Com}_{ij}, (f_{ij}, r_{ij})] = \mathsf{Commit}(f_{ij})$ for $i, j \in [1, n]$ and $i \geq j$, where $f_{ij} = F(i, j)$.
- Assign $\mathsf{Com}_{ij} = \mathsf{Com}_{ji}$ and $r_{ij} = r_{ji}$ for $i, j \in [1, n]$ and $i < j$. Let Com be the $n \times n$ matrix containing Com_{ij} for $j \in [1, n]$ in the ith row.
- Send $(\mathsf{send}, \mathsf{Com}, f_i(x), r_i(x))$ to P_i, where $f_i(x) = F(x, i)$, $r_i(x)$ is the degree-$(n-1)$ polynomial defined by the points $((1, r_{i1}), \ldots, (n, r_{in}))$.

Code for P_i:

- On receiving $(\mathsf{send}, \mathsf{Com}, f_i(x), r_i(x))$ from D, send $(\mathsf{echo}, \mathsf{Com})$ to every P_j if (a) Com is an $n \times n$ symmetric matrix and (b) $f_i(j) \stackrel{?}{=} \mathsf{Open}(\mathsf{Com}_{ij}, f_i(j), r_i(j))$.
- On receiving $(\mathsf{echo}, \mathsf{Com})$ from at least $2t + 1$ parties (possibly including itself) satisfying that Com received from P_j is same as received from D, send $(\mathsf{ready}, \mathsf{share\text{-}holder}, \mathsf{Com})$ to every P_j, if you have already sent out echo messages.
- If you have not sent out any ready signal before:
    1. on receiving ready messages from at least $t+1$ P_j's satisfying that Com received from P_j is same as received from D, send $(\mathsf{ready}, \mathsf{share\text{-}holder}, \mathsf{Com})$ to every P_j, if you have already sent out echo messages.
    2. on receiving $(\mathsf{ready}, \mathsf{share\text{-}holder}, \mathsf{Com})$ from at least $t + 1$ P_j's such that all the Com are same but do not match with the copy received from D, update your Com with this new matrix, delete everything else received from D and send $(\mathsf{ready}, \star, \mathsf{Com})$ to every P_j.
- On receiving ready signals from at least $2t+1$ parties such that all of them contain same Com as yours and at least $t + 1$ ready signals contain $\mathsf{share\text{-}holder}$, agree on Com and terminate.

Reconstruction Phase:
Code for P_i:

1. Send $(f_i(x), r_i(x))$ to every P_j if you had sent $(\mathsf{ready}, \mathsf{share\text{-}holder}, \mathsf{Com})$ in the sharing phase.
2. Wait for $t+1$ $(f_j(x), r_j(x))$ messages such that $f_j(x)$ is degree-t polynomial, $r_j(x)$ is degree-$(n-1)$ polynomial and $f_j(k) = \mathsf{Open}(\mathsf{Com}_{jk}, f_j(k), r_j(k))$ for all $k \in [1, n]$, interpolate $F(x, y)$ using those $t+1$ $f_j(x)$ polynomials, compute $s = F(0, 0)$ as the secret.

Fig. 4. Asynchronous VSS for $n \geq 3t + 1$ (optimal resilience)

check whether their polynomials are consistent with Com or not. We observe that the parties do not need to exchange and verify their common points on the bivariate polynomial, given that agreement on Com can be achieved. Because, the parties can now perform local consistency checking of their polynomial with Com. In our protocol, some honest parties may not receive polynomials consistent with Com, however, they still help to reach agreement on Com sensing that majority of the honest parties have received a common Com and also the polynomials received by them are consistent with Com. We describe the protocol in Fig. 4.

Lemma 1. *If an honest party P_i sends a* **ready** *message containing* Com *and a distinct honest party P_j sends a* **ready** *message containing* $\overline{\text{Com}}$, *then* Com $= \overline{\text{Com}}$.

Proof. We prove this by contradiction. Let there exists an honest pair (P_i, P_j) such that Com $\neq \overline{\text{Com}}$. The honest P_i communicates **ready** with Com if: (a) it receives (echo, Com) from at least $2t + 1$ parties OR b) it receives (ready, $\cdot$, Com) from at least $t + 1$ parties, where $\cdot$ can be either **share-holder** or $\star$. Similar reasons apply for P_j who sends $\overline{\text{Com}}$. If P_i and P_j send **ready** messages due to (a), then it implies that there is at least one honest party who communicates **echo** messages with Com as well as with $\overline{\text{Com}}$. This is impossible, since an honest party communicates **echo** with a unique matrix. For all other cases, we arrive at the contradiction that there is at least one honest party who sends **echo** with two different matrices or **ready** with two different matrices. We show this by considering the case when P_i sends **ready** due to (a) and P_j sends due to (b). The other cases will follow. P_j sends **ready** due to (b) implies that there is at least one honest party, say P_k who communicated **ready** with $\overline{\text{Com}}$ to P_j. Then by chain of arguments, we either get that honest P_i has sent **ready** with $\overline{\text{Com}}$ or get an honest party (possibly including P_i) who communicates **ready** with $\overline{\text{Com}}$ due to (a). In both cases, we arrive at contradiction, since no honest party can send **echo**/**ready** with two different matrices. Hence, we prove the lemma. □

Lemma 2. *If some honest party P_i has agreed on* Com, *then every honest party will eventually agree on* Com.

Proof. To prove the lemma, it is enough to prove the following: If some honest party P_i has received $2t + 1$ **ready** messages with Com such that at least $t + 1$ of them contain **share-holder**, then every honest party will eventually receive the same. If P_i receives **ready** messages as above, then there are at least $t + 1$ honest parties who send out **ready** messages with Com and at least one of the honest party's **ready** message must contain **share-holder**. An honest party sends out **ready** with **share-holder** in two cases: (a) She received at least $2t + 1$ **echo** message with Com and it has sent out **echo** with Com. Among these $2t + 1$ parties $t + 1$ are honest and they will eventually receive **ready** message from all the $t + 1$ honest parties who also sent the same to P_i (also by Lemma 2 if some honest party has sent a **ready** message with Com, then no other honest party will send **ready** with $\overline{\text{Com}}$). Hence these $t + 1$ honest parties will eventually send out **ready** with **share-holder**. Hence eventually every honest party will receive $2t + 1$ **ready** messages with Com such that at least $t + 1$ of them contain **share-holder**. (b) She received at least $(t + 1)$ **ready** messages with Com and she has sent out **echo** with Com. Among these $(t + 1)$, there is at least one honest party, say P_k. If P_k has sent **ready** with **share-holder**, then by recursive argument this case will boil down to case (a). However if P_k sends **ready** *without* **share-holder**, then he has received at least $t + 1$ **ready** massages with **share-holder** which ensures existence of another honest P_l who sent **ready** massage with **share-holder**. Now again by recursive argument, this case will boil down to case (a). □

Lemma 3. *If some honest party P_i has agreed on* Com, *then there is a set $\mathcal{H}$ of at least $t+1$ honest parties each holding degree-t polynomial $f_j(x)$ such that it is consistent with* Com *and there is a symmetric bivariate polynomial $F(x,y)$ such that $F(x,i) = f_i(x)$.*

Proof. If honest P_i has agreed on Com, then she has received $2t + 1$ ready messages with Com such that at least $t + 1$ of them contain share-holder. From the previous proof, eventually $t + 1$ honest parties (possibly including P_i) will eventually send out ready with share-holder. So there will be a set of at least $t+1$ honest parties who send out ready with share-holder. We claim that this set of honest parties, denoted by $\mathcal{H}$ will satisfy the conditions mentioned in the lemma statement. We notice that the honest parties in $\mathcal{H}$ never update Com and by previous lemma they eventually agree on the same. Also they send out echo well before sending out ready. This implies each honest party P_i in $\mathcal{H}$ ensures that her polynomial $f_i(x)$ (i.e. the points on it) are consistent with Com. Now we proceed to show that there is a symmetric bivariate polynomial $F(x,y)$ such that $F(x,i) = f_i(x)$. This can be shown by showing for every pair (P_i, P_j) from $\mathcal{H}$, $f_i(j) = f_j(i)$ holds good. This follows from the fact that P_i and P_j has same Com where they checked $\text{Com}_{ij} = \text{Com}_{ji}$ holds and then P_i and P_j individually ensured $f_i(j) \stackrel{?}{=} \text{Open}(\text{Com}_{ij}, f_i(j), r_i(j))$ and $f_j(i) \stackrel{?}{=} \text{Open}(\text{Com}_{ji}, f_j(i), r_j(i))$ respectively. If the above arguments do not hold then corrupted D has broken binding property of underlying commitment, as he knows how to open Com_{ij} in two different ways. □

Theorem 6. *Protocol* AsynchVSS *is an asynchronous VSS for $n \geq 3t + 1$.*

Proof. **Liveness.** If D is honest, then every honest party will eventually send out echo and then ready with share-holder. Since there are at least $2t + 1$ honest parties, every honest party will eventually agree on Com.
Agreement. Agreement follows from Lemma 2.
Correctness. Correctness follows from Lemma 2 and 3. Honest dealer case is easy to follow. For a corrupted dealer the unique secret determined in the sharing phase is nothing but the constant term of $F(x,y)$ defined by $\mathcal{H}$ in Lemma 3. In the reconstruction phase, all the parties will reconstruct D's secret using the polynomials sent by the honest parties in $\mathcal{H}$. Specifically, every honest party will definitely consider $f_j(x), r_j(x)$ sent by party P_j in $\mathcal{H}$. However, we will be done if we show that any wrong degree-t polynomial $\overline{f_j(x)}$ sent by a corrupted party P_j will never be considered (unless corrupted P_j breaks binding of commitment). This is ensured by the following check performed by an honest party before considering P_j's polynomial for the reconstruction of $F(x,y)$: $f_j(k) = \text{Open}(\text{Com}_{jk}, f_j(k), r_j(k))$ for all $k \in [1,n]$. This check ensures that $\overline{f_j(x)}$ must match with $f_j(x)$ at the $t + 1$ positions corresponding to $\mathcal{H}$. But then it implies $\overline{f_j(x)} = f_j(x)$.
Secrecy. Follows from the properties of bivariate polynomial and the hiding of underlying commitment scheme. □

5　Discussion and Future Work

In this paper, we considered computational VSS as a standalone primitive. Our VSS schemes may also be easily leveraged in applications such as asynchronous Byzantine agreement protocols [5]. However, other VSS applications such as proactive share renewal and share recovery schemes [3, 18] and distributed key generation [12, 19] heavily rely on homomorphism of the commitments. It represents an interesting open problem if we can do better than in the unconditional case (e.g., [7]) for these applications. Further, most of the threshold cryptographic protocols also rely on homomorphism to verify the correctness. It will be interesting to check the feasibility of these threshold protocols based our VSS schemes without using expensive zero-knowledge proofs.

Finally, our schemes based on the definitional properties of commitments are expensive (by a linear factor) in terms of communication complexity in comparison to the respective schemes employing homomorphic commitments. It is worthwhile to study whether this gap in communication complexity is inevitable.

Acknowledgements. We thank Jonathan Katz and our anonymous reviewers for their comments and suggestions on an earlier draft. We are also grateful to Ian Goldberg and Mehrdad Nojoumian for interesting initial discussions.

References

1. Backes, M., Kate, A., Patra, A.: Computational Verifiable Secret Sharing Revisited. Cryptology ePrint Archive, Report 2011/281 (2011)
2. Blakley, G.R.: Safeguarding Cryptographic Keys. In: The National Computer Conference, pp. 313–317 (1979)
3. Cachin, C., Kursawe, K., Lysyanskaya, A., Strobl, R.: Asynchronous Verifiable Secret Sharing and Proactive Cryptosystems. In: ACM CCS 2002, pp. 88–97 (2002)
4. Canetti, R.: Studies in Secure Multiparty Computation and Applications. Ph.D. thesis, The Weizmann Institute of Science (1996)
5. Canetti, R., Rabin, T.: Fast Asynchronous Byzantine Agreement with Optimal Resilience. In: ACM STOC 1993, pp. 42–51 (1993)
6. Chor, B., Goldwasser, S., Micali, S., Awerbuch, B.: Verifiable Secret Sharing and Achieving Simultaneity in the Presence of Faults. In: IEEE FOCS 1985, pp. 383–395 (1985)
7. D'Arco, P., Stinson, D.R.: On Unconditionally Secure Robust Distributed Key Distribution Centers. In: Zheng, Y. (ed.) ASIACRYPT 2002. LNCS, vol. 2501, pp. 346–363. Springer, Heidelberg (2002)
8. Dolev, D., Dwork, C., Waarts, O., Yung, M.: Perfectly secure message transmission. J. ACM 40(1), 17–47 (1993)
9. Feldman, P.: A Practical Scheme for Non-interactive Verifiable Secret Sharing. In: IEEE FOCS 1987, pp. 427–437 (1987)
10. Fitzi, M., Garay, J.A., Gollakota, S., Rangan, C.P., Srinathan, K.: Round-Optimal and Efficient Verifiable Secret Sharing. In: Halevi, S., Rabin, T. (eds.) TCC 2006. LNCS, vol. 3876, pp. 329–342. Springer, Heidelberg (2006)
11. Gennaro, R., Ishai, Y., Kushilevitz, E., Rabin, T.: The round complexity of verifiable secret sharing and secure multicast. In: ACM STOC 2001, pp. 580–589 (2001)

12. Gennaro, R., Jarecki, S., Krawczyk, H., Rabin, T.: Secure Distributed Key Generation for Discrete-Log Based Cryptosystems. J. of Cryptology 20(1), 51–83 (2007)
13. Gennaro, R., Rabin, M.O., Rabin, T.: Simplified VSS and Fact-Track Multiparty Computations with Applications to Threshold Cryptography. In: ACM PODC 1998, pp. 101–111 (1998)
14. Goldreich, O., Kahan, A.: How to Construct Constant-Round Zero-Knowledge Proof Systems for NP. J. Cryptology 9(3), 167–190 (1996)
15. Goldreich, O., Micali, S., Wigderson, A.: Proofs that yield nothing but their validity for all languages in np have zero-knowledge proof systems. J. ACM 38(3), 691–729 (1991)
16. Haitner, I., Reingold, O.: Statistically-hiding commitment from any one-way function. In: ACM STOC 2007, pp. 1–10 (2007)
17. Halevi, S., Micali, S.: Practical and Provably-Secure Commitment Schemes from Collision-Free Hashing. In: Koblitz, N. (ed.) CRYPTO 1996. LNCS, vol. 1109, pp. 201–215. Springer, Heidelberg (1996)
18. Herzberg, A., Jarecki, S., Krawczyk, H., Yung, M.: Proactive Secret Sharing or: How to Cope with Perpetual Leakage. In: Coppersmith, D. (ed.) CRYPTO 1995. LNCS, vol. 963, pp. 339–352. Springer, Heidelberg (1995)
19. Kate, A., Goldberg, I.: Distributed Key Generation for the Internet. In: Proc. Intl. Conf. on Distributed Computing Systems (ICDCS), pp. 119–128 (2009)
20. Katz, J., Koo, C.-Y., Kumaresan, R.: Improving the Round Complexity of VSS in Point-to-Point Networks. In: Aceto, L., Damgård, I., Goldberg, L.A., Halldórsson, M.M., Ingólfsdóttir, A., Walukiewicz, I. (eds.) ICALP 2008, Part II. LNCS, vol. 5126, pp. 499–510. Springer, Heidelberg (2008)
21. Kumaresan, R., Patra, A., Rangan, C.P.: The Round Complexity of Verifiable Secret Sharing: The Statistical Case. In: Abe, M. (ed.) ASIACRYPT 2010. LNCS, vol. 6477, pp. 431–447. Springer, Heidelberg (2010)
22. Naor, M., Ostrovsky, R., Venkatesan, R., Yung, M.: Perfect Zero-Knowledge Arguments for P Using Any One-Way Permutation. J. Cryptology 11(2), 87–108 (1998)
23. Patra, A., Choudhary, A., Rabin, T., Rangan, C.P.: The Round Complexity of Verifiable Secret Sharing Revisited. In: Halevi, S. (ed.) CRYPTO 2009. LNCS, vol. 5677, pp. 487–504. Springer, Heidelberg (2009)
24. Patra, A., Choudhary, A., Rangan, C.P.: Efficient Asynchronous Byzantine Agreement with Optimal Resilience. In: ACM PODC 2009, pp. 92–101 (2009)
25. Patra, A., Choudhary, A., Rangan, C.P.: Efficient Statistical Asynchronous Verifiable Secret Sharing with Optimal Resilience. In: Kurosawa, K. (ed.) Information Theoretic Security. LNCS, vol. 5973, pp. 74–92. Springer, Heidelberg (2010)
26. Pedersen, T.P.: A Threshold Cryptosystem Without a Trusted Party. In: Davies, D.W. (ed.) EUROCRYPT 1991. LNCS, vol. 547, pp. 522–526. Springer, Heidelberg (1991)
27. Pedersen, T.P.: Non-Interactive and Information-Theoretic Secure Verifiable Secret Sharing. In: Feigenbaum, J. (ed.) CRYPTO 1991. LNCS, vol. 576, pp. 129–140. Springer, Heidelberg (1992)
28. Rabin, T., Ben-Or, M.: Verifiable Secret Sharing and Multiparty Protocols with Honest Majority (Extended Abstract). In: ACM STOC 1989, pp. 73–85 (1989)
29. Schultz, D.A., Liskov, B., Liskov, M.: MPSS: Mobile Proactive Secret Sharing. ACM Trans. Inf. Syst. Secur. 13(4), 34 (2010)
30. Shamir, A.: How to Share a Secret. Commun. ACM 22(11), 612–613 (1979)
31. Zhou, L., Schneider, F.B., van Renesse, R.: APSS: Proactive Secret Sharing in Asynchronous Systems. ACM Trans. Inf. Syst. Secur. 8(3), 259–286 (2005)

Natural Generalizations of Threshold Secret Sharing

Oriol Farràs[1], Carles Padró[2], Chaoping Xing[2], and An Yang[2*]

[1] Universitat Rovira i Virgili, Tarragona, and Ben Gurion University, Be'er Sheva
[2] Nanyang Technological University, Singapore
oriol.farras@urv.cat, {carlespl,xingcp,yang0246}@ntu.edu.sg

Abstract. We present new families of access structures that, similarly to the multilevel and compartmented access structures introduced in previous works, are natural generalizations of threshold secret sharing. Namely, they admit an ideal linear secret sharing schemes over every large enough finite field, they can be described by a small number of parameters, and they have useful properties for the applications of secret sharing. The use of integer polymatroids makes it possible to find many new such families and it simplifies in great measure the proofs for the existence of ideal secret sharing schemes for them.

Keywords: Cryptography, secret sharing, ideal secret sharing schemes, multipartite secret sharing, integer polymatroids.

1 Introduction

The first proposed secret sharing schemes by Shamir [29] and by Blakley [6] have *threshold access structures*, that is, the qualified subsets are those having at least a certain number of participants. In addition, they are *ideal*, which means that every share has the same length as the secret. Moreover, as it was noticed by Bloom [7] and by Karnin, Greene and Hellman [19], they are *linear*, which implies that both the computation of the shares and the reconstruction of the secret can be performed by using basic linear algebra operations.

Even though there exists a linear secret sharing scheme for every access structure [4,18], the known general constructions are very inefficient because the length of the shares grows exponentially with the number of participants. Actually, the optimization of secret sharing schemes for general access structures has appeared to be an extremely difficult problem and not much is known about it. Readers are referred to [2] for a recent survey on this topic.

Nevertheless, this does not mean that efficient secret sharing schemes exist only for threshold access structures. Actually, the construction of ideal linear

* The work of the first author was supported by the Spanish Government through project CONSOLIDER INGENIO 2010 CSD2007-00004 "ARES" and by the Government of Catalonia through grant 2009 SGR 1135, and by ISF grant 938/09. The work of the other authors was supported by the Singapore National Research Foundation under Research Grant NRF-CRP2-2007-03.

D.H. Lee and X. Wang (Eds.): ASIACRYPT 2011, LNCS 7073, pp. 610–627, 2011.

secret sharing schemes for non-threshold access structures has attracted a lot of attention. This line of research was initiated by Kothari [20], who presented some ideas to construct ideal linear secret sharing schemes with hierarchical properties. Simmons [30] introduced the multilevel and compartmented access structures, and presented geometric constructions of ideal linear secret sharing schemes for some of them. Brickell [8] formalized the ideas in previous works [7,19,20,30] and introduced a powerful linear-algebraic method to construct ideal linear secret sharing schemes for non-threshold access structures. In addition, he used that method to construct such schemes for the families of access structures introduced by Simmons [30]. Tassa [31] and Tassa and Dyn [32] combined Brickell's [8] method with different kinds of polynomial interpolation to construct ideal linear secret sharing schemes for more general families of multilevel and compartmented access structures. Constructions for other interesting variants of compartmented access structures are given in [16,23]. All these families of access structures have some common features that are enumerated in the following.

1. They are natural and useful generalizations of threshold access structures. In the threshold case, all participants are equivalent, while the access structures in those families are *multipartite*, which means that the participants are divided into several parts and the participants in the same part play an equivalent role in the structure. In addition, they have some interesting properties for the applications of secret sharing. Some of them are useful for hierarchical organizations, while others can be used in situations requiring the agreement of several parties.

2. Similarly to the threshold ones, the access structures in those families admit a very compact description. Typically, they can be described by using a small number of parameters, at most linear on the number of parts.

3. They are *ideal access structures*, that is, they admit an ideal secret sharing scheme. Actually, every one of those access structures admits a *vector space secret sharing scheme*, that is, an ideal linear secret sharing scheme constructed by using the method proposed by Brickell [8]. Moreover, the only restriction on the fields over which these schemes are constructed is their size, and hence there is no required condition about their characteristic. Observe that this is also the case for threshold access structures, which admit vector space secret sharing schemes over every finite field with at least as many elements as the number of participants.

4. Even though the existence of ideal linear secret sharing schemes for those access structures has been proved, the known methods to construct such schemes are not efficient in general. This is an important difference to the threshold case, in which the construction proposed by Shamir [29] solves the problem. Tassa [31, Section 3.3] presented an efficient algorithm for the multilevel access structures. This is the only other family for which an efficient algorithm is known.

5. Determining over which fields those schemes can be constructed is another open problem. It is unsolved even for threshold access structures. In this case, it is equivalent to the problem considered in [1], and it is equivalent as well to

determine over which fields uniform matroids are representable [24, Problem 6.5.12, Conjecture 14.1.5], and also to determine the size of maximum arcs in projective spaces [27]. This is due to the well-known connection between threshold secret sharing and maximum distance separable codes [22]. Much less is known for the other families of multipartite access structures. Differently to the threshold case, there is a huge gap between the known lower and upper bounds on the minimum size of such fields.

Two questions naturally arise at this point. The first one is the search for new families of access structures with the properties above. The second one is to determine the existence of efficient methods to construct ideal linear secret sharing schemes for them, and to find better bounds on the minimum size of the fields over which such schemes can be found.

Another related line of work deals with the characterization of the ideal access structures in several families of multipartite access structures. The bipartite access structures [25] and the weighted threshold access structures [3] were the first families for which such a characterization was given. Some partial results about the tripartite case were presented in [10,16]. On the basis of the well known connection between ideal secret sharing schemes and matroids [9], Farràs, Martí-Farré and Padró [12] introduced integer polymatroids to study ideal multipartite secret sharing schemes. The power of this new mathematical tool was demonstrated in the same work by using it to characterize the ideal tripartite access structures. Subsequently, the use of integer polymatroids made it possible to characterize the ideal hierarchical access structures [14].

This work is devoted to the search for new families of ideal access structures that are among the most natural generalizations of threshold secret sharing, and to the efficiency analysis of the methods to construct ideal secret sharing schemes for them.

Our results strongly rely on the connection between integer polymatroids and ideal multipartite secret sharing presented in [12], which is summarized here in Theorem 2.2. The concepts, notation and related facts that are required to understand this result are recalled Section 2. Actually, the use of this tool provides important advantages in comparison to the techniques applied in previous constructions of ideal multipartite secret sharing schemes [8,16,23,25,30,31,32].

While no strong connection between all those families was previously known, a remarkable common feature is made apparent by identifying the integer polymatroids that are associated to those ideal multipartite access structures. Namely, they are Boolean polymatroids or basic transformations and combinations of Boolean polymatroids. This is of course a fundamental clue when trying to find new families of ideal access structures satisfying the aforementioned requirements.

By using other Boolean polymatroids, and by combining them in several different ways, we present a number of new families of ideal multipartite access structures. Specifically, we present in Section 4 several generalizations of the compartmented access structures introduced in [8,30,32]. Section 5 deals with some families of partially hierarchical access structures that can be defined from

Boolean polymatroids. For instance, we present a family of compartmented access structures in which every compartment has a hierarchy. Ideal (totally) hierarchical access structures, which were completely characterized in [14], are associated as well to a special class of Boolean polymatroids. Finally, we use another family of integer polymatroids, the uniform ones, to characterize in Section 6 the ideal members of another family of multipartite access structures: the ones that are invariant under every permutation of the parts.

All integer polymatroids that we use to find new families of ideal multipartite access structures can be defined by a small number of parameters, linear on the size of the ground set, and they are representable over every large enough finite field. Actually, these requirements are implied by the conditions we imposed on the access structures to be simple generalizations of threshold secret sharing. We analyze in Section 3 the basic integer polymatroids as well as the operations to modify and combine them that are used in our constructions. In particular, the result we prove in Proposition 3.4 is extremely useful.

We focus in this paper on a few examples that can be useful for the applications of secret sharing, but many other families can be described by using other integer polymatroids with those properties, and surely some other useful families will be found in future works.

Differently to the aforementioned previous works, our proofs that the structures in these new families are ideal are extremely concise. Of course, this is due to the use of integer polymatroids. In addition, some easily checkable necessary conditions that are derived from the results in [12] make it possible to prove that certain given multipartite access structures are not ideal. This simplifies as well the search for new families.

Even though the efficiency of the methods to construct actual ideal linear secret sharing schemes for those families of access structures has not been significantly improved by using the results from [12], they provide a unified framework in which the open problems related to that issue can be precisely stated. These open problems and some possible strategies to attack them are discussed in Section 7.

2   Preliminaries

2.1   Multipartite Access Structures and Their Geometric Representation

We introduce here some notation that will be used all through the paper. In addition, we present a very useful geometric representation of multipartite access structures that was introduced in [12,25].

We use $\mathbb{Z}_+$ to denote the set of the non-negative integers. For every $i, j \in \mathbb{Z}$ we write $[i, j] = \{i, i+1, \ldots, j\}$ if $i < j$, while $[i, i] = \{i\}$ and $[i, j] = \emptyset$ if $i > j$. Consider a finite set J. We notate J' for a set of the form $J' = J \cup \{p_0\}$ for some $p_0 \notin J$. Given two vectors $u = (u_i)_{i \in J}$ and $v = (v_i)_{i \in J}$ in $\mathbb{Z}^J$, we write $u \leq v$ if $u_i \leq v_i$ for every $i \in J$. The modulus $|u|$ of a vector $u \in \mathbb{Z}_+^J$ is defined by $|u| = \sum_{i \in J} u_i$. For every subset $X \subseteq J$, we notate $u(X) = (u_i)_{i \in X} \in \mathbb{Z}^X$.

The *support of* $u \in \mathbb{Z}^J$ is defined as $\mathrm{supp}(u) = \{i \in J : u_i \neq 0\}$. Finally, we consider the vectors $\mathbf{e}^i \in \mathbb{Z}^J$ such that $\mathbf{e}^i_j = 1$ if $j = i$ and $\mathbf{e}^i_j = 0$ otherwise.

For a finite set P, we notate $\mathcal{P}(P)$ for the power set of P, that is, the set of all subsets of P. A family $\Pi = (\Pi_i)_{i \in J}$ of subsets of P is called here a *partition of P* if $P = \bigcup_{i \in J} \Pi_i$ and $\Pi_i \cap \Pi_j = \emptyset$ whenever $i \neq j$. Observe that some of the parts may be empty. If $|J| = m$, we say that Π is an *m-partition* of P. For a partition Π of a set P, we consider the mapping $\Pi \colon \mathcal{P}(P) \to \mathbb{Z}^J_+$ defined by $\Pi(A) = (|A \cap \Pi_i|)_{i \in J}$. We write $\mathbf{P} = \Pi(\mathcal{P}(P)) = \{u \in \mathbb{Z}^J_+ : u \leq (|\Pi_i|)_{i \in J}\}$. For a partition Π of a set P, a Π*-permutation* is a permutation σ on P such that $\sigma(\Pi_i) = \Pi_i$ for every part Π_i of Π. An access structure on P is said to be Π*-partite* if every Π-permutation is an automorphism of it. If the number of parts in Π is m, such an access structure is called *m-partite*.

A multipartite access structure can be described in a compact way by taking into account that its members are determined by the number of elements they have in each part. If an access structure Γ on P is Π-partite, then $A \in \Gamma$ if and only if $\Pi(A) \in \Pi(\Gamma)$. That is, Γ is completely determined by the partition Π and set of vectors $\Pi(\Gamma) \subseteq \mathbf{P} \subseteq \mathbb{Z}^J_+$. Moreover, the set $\Pi(\Gamma) \subseteq \mathbf{P}$ is monotone increasing, that is, if $u \in \Pi(\Gamma)$ and $v \in \mathbf{P}$ are such that $u \leq v$, then $v \in \Pi(\Gamma)$. Therefore, $\Pi(\Gamma)$ is univocally determined by $\min \Pi(\Gamma)$, the family of its minimal vectors, that is, those representing the minimal qualified subsets of Γ. By an abuse of notation, we will use Γ to denote both a Π-partite access structure on P and the corresponding set $\Pi(\Gamma)$ of points in P, and the same applies to $\min \Gamma$.

2.2   Polymatroids and Matroids

A *polymatroid* $\mathcal{S}$ is a pair (J, h) formed by a finite set J, the *ground set*, and a *rank function* $h \colon \mathcal{P}(J) \to \mathbb{R}$ satisfying

1. $h(\emptyset) = 0$, and
2. h is *monotone increasing*: if $X \subseteq Y \subseteq J$, then $h(X) \leq h(Y)$, and
3. h is *submodular*: if $X, Y \subseteq J$, then $h(X \cup Y) + h(X \cap Y) \leq h(X) + h(Y)$.

If the rank function h is integer-valued, we say that $\mathcal{S}$ is an *integer polymatroid*. An integer polymatroid such that $h(X) \leq |X|$ for every $X \subseteq J$ is called a *matroid*. Readers that are unfamiliar with Matroid Theory are referred to the textbooks [24,33]. A detailed presentation about polymatroids can be found in [28, Chapter 44] or [17].

While matroids abstract some properties related to linear dependency of collections of vectors in a vector space, integer polymatroids do the same with collections of subspaces. Let V be a $\mathbb{K}$-vector space, and let $(V_i)_{i \in J}$ be a finite collection of subspaces of V. It is not difficult to check that the mapping $h \colon \mathcal{P}(J) \to \mathbb{Z}$ defined by $h(X) = \dim(\sum_{i \in X} V_i)$ is the rank function of an integer polymatroid. Integer polymatroids and, in particular, matroids that can be defined in this way are said to be $\mathbb{K}$*-representable*. Observe that, in a representable matroid, $\dim V_i \leq 1$ for every $i \in J$, and hence representations of matroids are considered as collections of vectors in a vector space.

Let $\mathcal{Z}$ be an integer polymatroid with ground set J. Consider the set $\mathcal{D}$ of the *integer independent vectors of* $\mathcal{Z}$, which is defined as

$$\mathcal{D} = \{u \in \mathbb{Z}_+^J \; : \; |u(X)| \le h(X) \text{ for every } X \subseteq J\}.$$

Integer polymatroids can be characterized by its *integer bases*, which are the maximal integer independent vectors. A nonempty subset $\mathcal{B} \subseteq \mathbb{Z}_+^J$ is the family of integer bases of an integer polymatroid if and only if it satisfies the following *exchange condition*.

- For every $u \in \mathcal{B}$ and $v \in \mathcal{B}$ with $u_i > v_i$, there exists $j \in J$ such that $u_j < v_j$ and $u - \mathbf{e}^i + \mathbf{e}^j \in \mathcal{B}$.

In particular, all bases have the same modulus. Every integer polymatroid is univocally determined by the family of its integer bases. Indeed, the rank function of $\mathcal{Z}$ is determined by $h(X) = \max\{|u(X)| \; : \; u \in \mathcal{B}\}$.

Since only integer polymatroids and integer vectors will be considered, we will omit the term "integer" most of the times when dealing with the integer independent vectors or the integer bases of an integer polymatroid.

If $\mathcal{D}$ is the family of independent vectors of an integer polymatroid $\mathcal{Z}$ on J, then, for every $X \subseteq J$, the set $\mathcal{D}|X = \{u(X) \; : \; u \in \mathcal{D}\} \subseteq \mathbb{Z}_+^X$ is the family of independent vectors of an integer polymatroid $\mathcal{Z}|X$ with ground set X. Clearly, the rank function $h|X$ of this polymatroid satisfies $(h|X)(Y) = h(Y)$ for every $Y \subseteq X$. Because of that, we will use the same symbol to denote both rank functions.

For an integer polymatroid $\mathcal{Z}$ and a subset $X \subseteq J$ of the ground set, we write $\mathcal{B}(\mathcal{Z}, X)$ to denote the family of the independent vectors $u \in \mathcal{D}$ such that $\text{supp}(u) \subseteq X$ and $|u| = h(X)$. Observe that there is a natural bijection between $\mathcal{B}(\mathcal{Z}, X)$ and the family of bases of the integer polymatroid $\mathcal{Z}|X$.

2.3   Integer Polymatroids and Multipartite Matroid Ports

The aim of this section is to summarize the results in [12] about ideal multipartite secret sharing schemes and their connection to integer polymatroids.

For a polymatroid $\mathcal{S}$ with ground set $J' = J \cup \{p_0\}$, the family $\Gamma_{p_0}(\mathcal{S}) = \{A \subseteq J \; : \; h(A \cup \{p_0\}) = h(A)\}$ of subsets of J is monotone increasing, and hence it is an access structure on J. If $\mathcal{S}$ is a matroid, then the access structure $\Gamma_{p_0}(\mathcal{S})$ is called the *port of the matroid* $\mathcal{S}$ *at the point* p_0. As a consequence of the results by Brickell [8] and by Brickell and Davenport [9], matroid ports play a very important role in secret sharing. Ports of $\mathbb{K}$-representable matroids are called $\mathbb{K}$-*vector space access structures*. Such an access structure admits an ideal scheme that is constructed according to the method given by Brickell [8]. In addition, Brickell and Davenport [9] proved that the access structure of every ideal secret sharing scheme is a matroid port. This result was generalized in [21] by proving that the access structure of a secret sharing scheme is a matroid port if the length of every share is less than $3/2$ times the length of the secret.

Definition 2.1. *Let* $\Pi = (\Pi_i)_{i \in J}$ *be a partition of a set P of participants. Consider an integer polymatroid $\mathcal{Z}'$ on J' with $h(\{p_0\}) = 1$ and $h(\{i\}) \leq |\Pi_i|$ for every $i \in J$, and take $\mathcal{Z} = \mathcal{Z}'|J$. We define a Π-partite access structure $\Gamma_{p_0}(\mathcal{Z}', \Pi)$ in the following way: a vector $u \in \mathbf{P}$ is in $\Gamma_{p_0}(\mathcal{Z}', \Pi)$ if and only if there exist a subset $X \in \Gamma_{p_0}(\mathcal{Z}')$ and a vector $v \in \mathcal{B}(\mathcal{Z}, X)$ such that $v \leq u$.*

The following theorem summarizes the results from [12] about the connection between ideal multipartite access structures and integer polymatroids. An access structure is said to be *connected* if all participants are in at least one minimal qualified subset.

Theorem 2.2 ([12]). *Let $\Pi = (\Pi_i)_{i \in J}$ be a partition of a set P. A Π-partite access structure Γ on P is a matroid port if and only if it is of the form $\Gamma_{p_0}(\mathcal{Z}', \Pi)$ for some integer polymatroid $\mathcal{Z}'$ on J' with $h(\{p_0\}) = 1$ and $h(\{i\}) \leq |\Pi_i|$ for every $i \in J$. In addition, if $\mathcal{Z}'$ is $\mathbb{K}$-representable, then $\Gamma_{p_0}(\mathcal{Z}', \Pi)$ is an $\mathbb{L}$-vector space access structure for every large enough finite extension $\mathbb{L}$ of $\mathbb{K}$. Moreover, if Γ is connected, the integer polymatroid $\mathcal{Z}'$ is univocally determined by Γ.*

3   Some Useful Integer Polymatroids

In order to find families of ideal multipartite access structures with the required properties, we need to find families of integer polymatroids that are representable over every large enough finite field and can be described in a compact way. To this end, we describe in the following two families of integer polymatroids, namely the Boolean and the uniform ones, and several operations to obtain new polymatroids from some given ones.

3.1   Operations on Polymatroids

We begin by presenting two operations on polymatroids: the sum and the truncation. The first one is a binary operation, while the second one is unitary.

The *sum $\mathcal{Z}_1 + \mathcal{Z}_2$ of two polymatroids* $\mathcal{Z}_1, \mathcal{Z}_2$ on the same ground set J and with rank functions h_1, h_2, respectively, is the polymatroid on J with rank function $h = h_1 + h_2$. If $\mathcal{Z}_1, \mathcal{Z}_2$ are $\mathbb{K}$-representable integer polymatroids, then their sum is $\mathbb{K}$-representable too. Clearly, if $\mathcal{Z}_1$ is represented by the vector subspaces $(V_i)_{i \in J}$ of V and $\mathcal{Z}_2$ is represented by the vector subspaces $(W_i)_{i \in J}$ of W, then the subspaces $(V_i \times W_i)_{i \in J}$ of $V \times W$ form a representation of the sum $\mathcal{Z}_1 + \mathcal{Z}_2$. If $\mathcal{D}_1, \mathcal{D}_2 \subseteq \mathbb{Z}_+^J$ are the sets of independent vectors of $\mathcal{Z}_1$ and $\mathcal{Z}_2$, respectively, then, as a consequence of [28, Theorem 44.6], the independent vectors of $\mathcal{Z}_1 + \mathcal{Z}_2$ are the ones in $\mathcal{D}_1 + \mathcal{D}_2 = \{u_1 + u_2 : u_1 \in \mathcal{D}_1, u_2 \in \mathcal{D}_2\}$. Therefore, the bases of $\mathcal{Z}_1 + \mathcal{Z}_2$ are the vectors in $\mathcal{B}_1 + \mathcal{B}_2$, where $\mathcal{B}_1, \mathcal{B}_2 \subseteq \mathbb{Z}_+^J$ are the families of the bases of those polymatroids.

For an integer polymatroid $\mathcal{Z}$ on J with rank function h and a positive integer t with $t \leq h(J)$, it is not difficult to prove that the map h' defined by $h'(X) = \min\{h(X), t\}$ is the rank function of an integer polymatroid on J, which is called the *t-truncation* of $\mathcal{Z}$. Observe that a vector $x \in \mathbb{Z}_+^J$ is a basis of the t-truncation of $\mathcal{Z}$ if and only if x is an independent vector of $\mathcal{Z}$ and $|x| = t$.

3.2   Boolean and Uniform Polymatroids

We introduce here two families of integer polymatroids.

The Boolean polymatroids form the first one. They are very simple integer polymatroids that are representable over every finite field. Consider a finite set B and a family $(B_i)_{i \in J}$ of subsets of B. Clearly, the map $h(X) = \left| \bigcup_{i \in X} B_i \right|$ for $X \subseteq J$ is the rank function of an integer polymatroid $\mathcal{Z}$ with ground set J. A *Boolean polymatroid* is an integer polymatroid that can be defined in this way. Boolean polymatroids are representable over every field $\mathbb{K}$. If $|B| = r$, we can assume that B is a basis of the vector space $V = \mathbb{K}^r$. For every $i \in J$, consider the vector subspace $V_i = \langle B_i \rangle$. Obviously, these subspaces form a $\mathbb{K}$-representation of $\mathcal{Z}$. The *modular polymatroids* are those having a *modular rank function*, that is, $h(X \cup Y) + h(X \cap Y) = h(X) + h(Y)$ for every $X, Y \subseteq J$. Every integer modular polymatroid is Boolean, and hence it is representable over every finite field. A Boolean polymatroid is modular if and only if the sets $(B_i)_{i \in J}$ are disjoint. Observe that the rank function of an integer modular polymatroid is of the form $h(X) = \sum_{i \in X} b_i$ for some vector $b \in \mathbb{Z}_+^J$. Actually, this vector is the only basis of such a polymatroid.

Proposition 3.1. *Every truncation of a Boolean polymatroid is representable over every large enough finite field.*

Proof. For a field $\mathbb{K}$ and a positive integer t, we consider the map $\psi_t \colon \mathbb{K} \to \mathbb{K}^t$ defined by $\psi_t(x) = (1, x, \ldots, x^{t-1})$. Observe that, for every t different field elements $x_1, \ldots, x_t \in \mathbb{K}$, the set of vectors $\{\psi_t(x_i) : i = 1, \ldots, t\}$ is linearly independent. Let $\mathcal{Z}$ be a Boolean polymatroid with ground set J, take $r = h(J)$, and consider a field $\mathbb{K}$ with $|\mathbb{K}| \geq r$. Take $B \subseteq \mathbb{K}$ with $|B| = r$ and a family $(B_i)_{i \in J}$ of subsets of B such that $h(X) = \left| \bigcup_{i \in X} B_i \right|$ for every $X \subseteq J$. For a positive integer $t \leq r$ and for every $i \in J$, consider the vector subspace $V_i \subseteq \mathbb{K}^t$ spanned by the vectors in $\{\psi_t(x) : x \in B_i\}$. Clearly, these subspaces form a $\mathbb{K}$-representation of the t-truncation of the Boolean polymatroid $\mathcal{Z}$.   □

The second family that is introduced in this section is the one of the uniform polymatroids. We say that a polymatroid $\mathcal{Z}$ with ground set J is *uniform* if every permutation on J is an automorphism of $\mathcal{Z}$. In this situation, the rank $h(X)$ of a set $X \subseteq J$ depends only on its cardinality, that is, there exist values $0 = h_0 \leq h_1 \leq \cdots \leq h_m$, where $m = |J|$, such that $h(X) = h_i$ for every $X \subseteq J$ with $|X| = i$. It is easy to see that such a sequence of values h_i defines a uniform polymatroid if and only if $h_i - h_{i-1} \geq h_{i+1} - h_i$ for every $i \in [1, m-1]$. Clearly, a uniform polymatroid is univocally determined by its *increment vector* $\delta = (\delta_1, \ldots, \delta_m)$, where $\delta_i = h_i - h_{i-1}$. Observe that $\delta \in \mathbb{R}^m$ is the increment vector of a uniform polymatroid if and only if $\delta_1 \geq \cdots \geq \delta_m \geq 0$. A uniform polymatroid is a matroid if and only if $\delta_i \in \{0, 1\}$ for every $i = 1, \ldots, m$. In this case, we obtain the *uniform matroid* $U_{r,m}$, where $r = \max\{i \in [1, m] : \delta_i = 1\}$. It is well known that $U_{r,m}$ is $\mathbb{K}$-representable whenever $|\mathbb{K}| \geq m$. Obviously, the sum of uniform polymatroids is a uniform polymatroid whose increment vector is obtained by summing up the corresponding increment vectors. The next result was proved in [13].

Proposition 3.2 ([13], Proposition 14). *Every uniform integer polymatroid is a sum of uniform matroids. In particular, every uniform integer polymatroid with ground set J is representable over every field $\mathbb{K}$ with $|\mathbb{K}| \geq |J|$.*

3.3 Multipartite Access Structures from Bases of Integer Polymatroids

We present in the following a consequence of Theorem 2.2 that is very useful in the search of new ideal multipartite access structures. Namely, we prove that a multipartite access structure is ideal if its minimal vectors coincide with the bases of a representable integer polymatroid. We need the following result, which is a consequence of [11, Proposition 2.3].

Proposition 3.3 ([11]). *Let $\mathcal{Z}$ be an integer polymatroid with ground set J and let Λ be an access structure on J. Then there exists an integer polymatroid $\mathcal{Z}'$ on J' with $h(\{p_0\}) = 1$ and $\mathcal{Z} = \mathcal{Z}'|J$ such that $\Lambda = \Gamma_{p_0}(\mathcal{Z}')$ if and only if the following conditions are satisfied.*

1. *If $X \subseteq Y \subseteq J$ and $X \notin \Lambda$ while $Y \in \Lambda$, then $h(X) \leq h(Y) - 1$.*
2. *If $X, Y \in \Lambda$ and $X \cap Y \notin \Lambda$, then $h(X \cup Y) + h(X \cap Y) \leq h(X) + h(Y) - 1$.*

Proposition 3.4. *Let $\mathcal{Z}$ be a $\mathbb{K}$-representable integer polymatroid on J and let Γ be a Π-partite access structure whose minimal vectors coincide with the bases of $\mathcal{Z}$. Then Γ is an $\mathbb{L}$-vector space access structure for every large enough finite extension $\mathbb{L}$ of $\mathbb{K}$.*

Proof. The access structure $\Lambda = \{X \subseteq J : h(X) = h(J)\}$ and the integer polymatroid $\mathcal{Z}$ satisfy the conditions in Proposition 3.3. Moreover, for this particular access structure, if $\mathcal{Z}$ is $\mathbb{K}$-representable, then the integer polymatroid $\mathcal{Z}'$ whose existence is given by Proposition 3.3 is $\mathbb{L}$-representable for every large enough finite algebraic extension $\mathbb{L}$ of $\mathbb{K}$. Indeed, consider a $\mathbb{K}$-vector space V and vector subspaces $(V_i)_{i \in J}$ forming a $\mathbb{K}$-representation of $\mathcal{Z}$. A representation of $\mathcal{Z}'$ is obtained by finding a vector $v_0 \in V$ such that $v_0 \notin \sum_{i \in X} V_i$ for every $X \subseteq J$ with $h(X) < h(J)$. Since $\sum_{i \in X} V_i \neq V$ if $h(X) < h(J)$, such a vector exists if $\mathbb{K}$ is large enough. Finally, it is not difficult to check that the minimal vectors of $\Gamma_{p_0}(\mathcal{Z}', \Pi)$ coincide with the bases of $\mathcal{Z}$. $\qquad\square$

4 Compartmented Access Structures

4.1 Compartmented Access Structures with Upper and Lower Bounds

Simmons [30] introduced compartmented access structures in opposition to the hierarchical ones. Basically, compartmented access structures can be seen as a modification of threshold access structures to be used in situations that require the agreement of several parties. In a compartmented structure, all minimal qualified subsets have the same size, but other requirements are added about the number of participants in every part, or the number of involved parts.

The first examples of compartmented access structures were introduced by Simmons [30]. Brickell [8] introduced a more general family, the so-called *compartmented access structures with lower bounds*, and showed how to construct ideal secret sharing schemes for them. These are the Π-partite access structures defined by $\min \Gamma = \{u \in \mathbf{P} : |u| = t \text{ and } u \geq a\}$ for some vector $a \in \mathbb{Z}_+^J$ and some positive integer t with $t \geq |a|$. The *compartmented access structures with upper bounds* are the Π-partite access structures with $\min \Gamma = \{u \in \mathbf{P} : |u| = t \text{ and } u \leq b\}$, where $b \in \mathbb{Z}_+^J$ and $t \in \mathbb{Z}_+$ are such that $b_i \leq t \leq |b|$ for every $i \in J$. They were introduced by Tassa and Dyn [32], who constructed ideal secret sharing schemes for them.

We introduce in the following a new family of compartmented access structures that generalize the previous ones. Namely, we prove that the compartmented access structures that are defined by imposing both upper and lower bounds on the number of participants in every part are ideal.

For a positive integer t and a pair of vectors $a, b \in \mathbb{Z}_+^J$ with $a \leq b \leq \Pi(P)$, and $|a| \leq t \leq |b|$, and $b_i \leq t$, consider the Π-partite access structure Γ defined by

$$\min \Gamma = \{x \in \mathbf{P} : |x| = t \text{ and } a \leq x \leq b\}. \tag{1}$$

The compartmented access structures with upper bounds and the ones with lower bounds correspond to the compartmented access structures defined above with $a = 0$ and with $b = \Pi(P)$, respectively. We prove in the following that the access structures (1) are ideal by checking that they are of the form $\Gamma_0(\mathcal{Z}', \Pi)$ for a certain family of representable integer polymatroids. Given a positive integer t and two vectors $a, b \in \mathbb{Z}_+^J$ with $a \leq b$ and $|a| \leq t \leq |b|$, consider the vector $c = b - a \in \mathbb{Z}_+^J$ and the integer $s = t - |a| \in \mathbb{Z}_+$. Let $\mathcal{Z}_1$ be the integer modular polymatroid defined by the vector a, and let $\mathcal{Z}_2$ be the s-truncation of the integer modular polymatroid defined by the vector c. Then the integer polymatroid $\mathcal{Z} = \mathcal{Z}_1 + \mathcal{Z}_2$ is representable over every large enough finite field. The family of bases of $\mathcal{Z}$ is $\mathcal{B} = \{x \in \mathbb{Z}_+^J : |x| = t \text{ and } a \leq x \leq b\}$. By Proposition 3.4, this proves that the compartmented access structures of the form (1) are vector space access structures over every large enough finite field.

4.2   Compartmented Compartments

We introduce next another family of compartmented access structures. In this case, instead of an upper bound for every compartment, we have upper bounds for groups of compartments. Take $J = [1, m] \times [1, n]$ and a partition $\Pi = (\Pi_{ij})_{(i,j) \in J}$ of the set P of participants. Take vectors $a \in \mathbb{Z}_+^J$ and $b \in \mathbb{Z}_+^m$, and an integer t with $|a| \leq t \leq |b|$ and $\sum_{j=1}^n a_{ij} \leq b_i \leq t$ for every $i \in [1, m]$. Consider the Π-partite access structure Γ defined by

$$\min \Gamma = \left\{ x \in \mathbf{P} : |x| = t, \text{ and } a \leq x, \text{ and } \sum_{j=1}^n x_{ij} \leq b_i \text{ for every } i \in [1, m] \right\}.$$

That is, the compartments are distributed into m groups and we have an upper bound for the number of participants in every group of compartments, while we have a lower bound for every compartment.

We prove next that these access structures admit a vector space secret sharing scheme over every large enough finite field. Consider the vector $c \in \mathbb{Z}_+^m$ defined by $c_i = b_i - \sum_{j=1}^n a_{ij}$ and the integer $s = t - |a| \in \mathbb{Z}_+$. Let $\mathcal{Z}_1$ be the integer modular polymatroid with ground set J defined by the vector a. Let $\mathcal{Z}_3$ the integer polymatroid with ground set J and family of bases

$$\mathcal{B}_3 = \left\{ x \in \mathbb{Z}_+^J \ : \ \sum_{j=1}^n x_{ij} = c_i \text{ for every } i \in [1, m] \right\},$$

and let $\mathcal{Z}_2$ be the s-truncation of $\mathcal{Z}_3$. Finally, take $\mathcal{Z} = \mathcal{Z}_1 + \mathcal{Z}_2$.

Lemma 4.1. *The minimal qualified sets of Γ coincide with the bases of $\mathcal{Z}$.*

Proof. Let $\mathcal{B}$ and $\mathcal{B}_2$ be the families of bases of $\mathcal{Z}$ and $\mathcal{Z}_2$, respectively. The bases of $\mathcal{Z}$ are precisely the vectors of the form $x = a + y$ with $y \in \mathcal{B}_2$. Observe that a vector $y \in \mathbb{Z}_+^J$ is in $\mathcal{B}_2$ if and only if $|y| = s$ and $\sum_{j=1}^n y_{ij} \le c_i$ for every $i \in [1, m]$. $\qquad\square$

Lemma 4.2. *The integer polymatroid $\mathcal{Z}$ is representable over every large enough finite field.*

Proof. We only have to prove that this holds for $\mathcal{Z}_2$. By Proposition 3.1, for every large enough finite field $\mathbb{K}$ there exist subspaces $(V_i)_{i \in [1,m]}$ of a $\mathbb{K}$-vector space V that form a representation of the s-truncation of the modular polymatroid with ground set $[1, m]$ defined by the vector c. Then the subspaces $(W_{ij})_{(i,j) \in J}$ of V with $W_{ij} = V_i$ for every $j \in [1, n]$ form a representation of $\mathcal{Z}_2$. $\qquad\square$

5   Ideal Partially Hierarchical Access Structures

5.1   Ideal Hierarchical Access Structures

For an access structure Γ on a set P, we say that a participant $p \in P$ is *hierarchically superior in Γ* to a participant $q \in P$, and we write $q \preceq p$, if $A \cup \{p\} \in \Gamma$ for every $A \subseteq P \smallsetminus \{p, q\}$ with $A \cup \{q\} \in \Gamma$. Two participants are *hierarchically equivalent* if $q \preceq p$ and $p \preceq q$. Observe that, if Γ is Π-partite, every pair of participants in the same part Π_i are hierarchically equivalent.

An access structure is *hierarchical* if every pair of participants are hierarchically comparable. In this situation, the hierarchical order $\preceq$ is a total order on Π. *Weighted threshold access structures*, which were introduced by Shamir [29] in his seminal work, are hierarchical, but they are not ideal in general. The ideal weighted threshold access structures were characterized by Beimel, Tassa and Weinreb [3]. Other examples of hierarchical access structures are the the multilevel access structures introduced by Simmons [30], which were proved to be ideal by Brickell [8], and the hierarchical threshold access structures presented by Tassa [31]. These were the only known families of ideal hierarchical access structures before the connection between integer polymatroids and ideal multipartite secret sharing presented in [12] made it possible to characterize the ideal hierarchical access structures [14]. Actually, all ideal hierarchical access structures are obtained from a special class of Boolean polymatroids [14] and, because of that, they are vector space access structures over every large enough finite field. Moreover, they admit a very compact description, as we see in the following.

Consider two sequences $\mathbf{a} = (a_0, \ldots, a_m)$ and $\mathbf{b} = (b_0, \ldots, b_m)$ of integer numbers such that $a_0 = a_1 = b_0 = 1$ and $a_i \le a_{i+1} \le b_i \le b_{i+1}$ for every $i \in [0, m-1]$. For $i \in [0, m]$, take the subsets $B_i = [a_i, b_i]$ of the set $B = [1, b_m]$ and consider the Boolean polymatroid $\mathcal{Z}' = \mathcal{Z}'(\mathbf{a}, \mathbf{b})$ with ground set $J' = [0, m]$ defined from them. It is proved in [14] (full version) that a vector $x \in \mathbf{P} \subseteq \mathbb{Z}_+^m$ is in the Π-partite access structure $\Gamma = \Gamma_0(\mathcal{Z}', \Pi)$ if and only if there exists $i_0 \in [1, m]$ such that $\sum_{j=1}^{i_0} x_j \ge b_{i_0}$, and $\sum_{j=1}^{i} x_j \ge a_{i+1} - 1$ for all $i \in [1, i_0 - 1]$. Therefore, the participants in Π_i are hierarchically superior to the participants in Π_j if $i \le j$, and hence every access structure of the form $\Gamma_0(\mathcal{Z}'(\mathbf{a}, \mathbf{b}), \Pi)$ is hierarchical. Moreover, every ideal hierarchical access structure is of this form or it can be obtained from a structure of this form by removing some participants [14].

In particular, if $a_i = 1$ for all $i \in [0, m]$ and $1 = b_0 \le b_1 < \cdots < b_m$, then $x \in \Gamma_0(\mathcal{Z}'(\mathbf{a}, \mathbf{b}), \Pi)$ if and only if $\sum_{j=1}^{i_0} x_j \ge b_{i_0}$ for some $i_0 \in [1, m]$. These are precisely the *multilevel access structures* introduced by Simmons [30], also called *disjunctive hierarchical threshold access structures* by other authors [31]. They were proved to be ideal by Brickell [8]. On the other hand, the *conjunctive hierarchical threshold access structures* for which Tassa [31] constructs ideal secret sharing schemes are obtained by considering $1 = a_0 = a_1 < \cdots < a_m$ and $1 = b_0 < b_1 = \cdots = b_m$. In this case, $x \in \Gamma_0(\mathcal{Z}'(\mathbf{a}, \mathbf{b}), \Pi)$ if and only if $\sum_{j=1}^{i} x_j \ge a_{i+1} - 1$ for all $i \in [1, m-1]$ and $\sum_{j=1}^{m} x_j \ge b_m$. Observe that, in an access structure in the first family, there may be qualified subsets involving only participants in the lowest level. This is not the case in any access structure in the second family, because every qualified subset must contain participants in the highest level.

By using the results in [14], we can find other ideal hierarchical access structures with more flexible properties. If we take, for instance, $\mathbf{a} = (1, 1, 1, 5, 5)$ and $\mathbf{b} = (1, 4, 6, 10, 12)$, every qualified subset in the hierarchical access structure $\Gamma_0(\mathcal{Z}'(\mathbf{a}, \mathbf{b}), \Pi)$ must contain participants in the first two levels, but some of them do not have any participant in the first level.

5.2   Partial Hierarchies from Boolean Polymatroids

Moreover, by considering other Boolean polymatroids, we can find other families of ideal access structures satisfying some given *partial hierarchy*, that is, Π-partite access structures in which the hierarchical relation $\preceq$ on Π is a partial order. We present next an example of such a family of ideal *partially hierarchical access structures*. Consider a family of subsets $(B_i)_{i \in [0,m]}$ of a finite set B satisfying:

- $|B_0| = 1$ and $B_0 \subseteq B_1$, while $B_0 \cap B_i = \emptyset$ if $i \in [2, m]$, and
- $B_1 \cap B_i \neq \emptyset$ for every $i \in [2, m]$, and
- $B_i \cap B_j = \emptyset$ for every $i, j \in [2, m]$ with $i \neq j$.

Let $\mathcal{Z}'$ be the Boolean polymatroid with ground set $J' = [0, m]$ defined from this family of subsets, and consider the Π-partite access structure $\Gamma = \Gamma_0(\mathcal{Z}', \Pi)$. Take $t_1 = |B_1|$ and $t_i = |B_i \setminus B_1|$, and $s_i = |B_i \cap B_1|$ for $i \in [2, m]$. Then a vector $x \in \mathbf{P}$ is in the access structure Γ if and only if there exist a vector $u \in \mathbf{P}$ such that

- $u \leq x$,
- $1 \in \mathrm{supp}(u) = X$, $|u| = \sum_{i \in X} t_i$,
- for every $Y \subseteq X$, $|u(Y)| \leq \sum_{i \in Y} (t_i + s_i)$, where $s_1 = 0$.

Clearly, $q \preceq p$ if $p \in \Pi_1$ and $q \in \Pi_i$ for some $i \in [2, m]$. On the other hand, any two participants in two different parts Π_i, Π_j with $i, j \in [2, m]$ are not hierarchically related.

5.3   Compartmented Access Structures with Hierarchical Compartments

We can consider as well compartmented access structures with hierarchical compartments. Take $J = [1, m] \times [1, n]$ and a partition $\Pi = (\Pi_{ij})_{(i,j) \in J}$ of the set P of participants. Consider a finite set B and a family of subsets $(B_{ij})_{(i,j) \in J}$ such that $B_{in} \subseteq \cdots \subseteq B_{i2} \subseteq B_{i1}$ for every $i \in [1, m]$, and $B_{11} \cup \cdots \cup B_{m1} = B$, and $B_{i1} \cap B_{j1} = \emptyset$ if $i \neq j$. Let $\mathcal{Z}$ be the t-truncation of the Boolean polymatroid defined by this family of subsets. If Γ is a Π-partite access structure such that its minimal vectors coincide with the bases of $\mathcal{Z}$, then Γ is a vector space access structure over every large enough finite field. We now describe Γ. For $(i, j) \in J$, take $b_{ij} = |B_{ij}|$. Consider the vector $b = (b_{11}, \ldots, b_{m1}) \in \mathbb{Z}_+^m$. Of course, $|b| = |B|$. Suppose $b_{i1} \leq t \leq |b|$ for every $i \in [1, m]$. It is not difficult to check that a vector $x \in \mathbb{Z}_+^J$ is a basis of $\mathcal{Z}$, and hence a minimal vector of Γ, if and only if $|x| = t$ and $\sum_{k=j}^n x_{ik} \leq b_{ij}$ for every $(i, j) \in J$. Observe that Γ can be seen as a compartmented access structure with compartments $\Pi_i = \bigcup_{j=1}^n \Pi_{ij}$ for $i \in [1, m]$, because every minimal qualified subset has exactly t participants, and at most b_{i1} of them in compartment Π_i. In addition, we have a hierarchy within every compartment. Actually, $q \preceq p$ if $p \in \Pi_{ij}$ and $q \in \Pi_{ik}$ with $j \leq k$.

6   Ideal Uniform Multipartite Access Structures

Herranz and Sáez [16, Section 3.2] introduced a family of ideal multipartite access structures that can be seen as a variant of the compartmented ones. Specifically, given integers $1 \leq k \leq t$, consider the Π-partite access structure defined by

$$\Gamma = \{x \in \mathbf{P} : |x| \geq t \text{ and } |\mathrm{supp}(x)| \geq k\}. \tag{2}$$

It is proved in [16] that Γ is a vector space access structure over every large enough finite field. Observe that the parts in the partition $\Pi = (\Pi_i)_{i \in J}$ are symmetrical in Γ. That is, the minimal vectors of Γ are invariant under any permutation on J. In the following, we characterize all ideal multipartite access structures with this property. We prove that all of them are vector space access structures over every large enough finite field.

A Π-partite access structure Γ is said to be *uniform* if the set $\min \Gamma \subseteq \mathbb{Z}_+^J$ of its minimal vectors is symmetric, that is, if $u = (u_i)_{i \in J} \in \min \Gamma$, then $\sigma u = (u_{\sigma i})_{i \in J} \in \min \Gamma$ for every permutation σ on J. In this section, we characterize the uniform multipartite access structures that admit an ideal secret sharing scheme. Moreover, we prove that all such access structures are vector space access structures over every large enough finite field. This is done by using the uniform integer polymatroids described in Section 3.2 to construct a family of uniform multipartite access structures that admit a vector space secret sharing scheme over every large enough finite field. Then we prove in Theorem 6.2 that every ideal uniform multipartite access structure is a member of this family.

Let $\mathcal{Z}$ be a uniform integer polymatroid with increment vector δ on a ground set J with $|J| = m$. For $i \in [0, m]$, consider $h_i = \sum_{j=1}^i \delta_j$, the values of the rank function of $\mathcal{Z}$. Recall that the (k, m)-threshold access structure on J consists of all subsets of J with at least k elements.

Lemma 6.1. *For an integer $k \in [1, m]$, there exists an integer polymatroid $\mathcal{Z}'_k$ on $J' = J \cup \{p_0\}$ with $h(\{p_0\}) = 1$ and $\mathcal{Z} = \mathcal{Z}'_k | J$ such that $\Gamma_{p_0}(\mathcal{Z}'_k)$ is the (k, m)-threshold access structure on J if and only if $1 \le k \le m - 1$ and $\delta_k > \delta_{k+1}$, or $k = m$ and $\delta_m > 0$.*

Proof. If there exists a polymatroid $\mathcal{Z}'$ with the required properties, then the first condition in Proposition 3.3 implies that $h_{k-1} < h_k$, while $h_{k+1} + h_{k-1} < 2h_k$ if $1 \le k \le m-1$ by the second one. Therefore, our condition is necessary. We prove now sufficiency. Let Λ be the (k, m)-threshold access structure on J. Observe that $h_k > h_{k-1}$ because $\delta_k > 0$, and hence $h(X) < h(Y)$ if $X \subseteq Y \subseteq J$ and $X \notin \Lambda$ while $Y \in \Lambda$. Consider now two subsets $X, Y \in \Lambda$ such that $X \cap Y \notin \Lambda$. This implies in particular that $k < m$. Take $r_1 = |X| \ge k$, $r_2 = |Y| \ge k$, and $s = |X \cap Y| < k$. Then $h_{r_1+r_2-s} - h_{r_2} = \sum_{i=1}^{r_1-s} \delta_{r_2+i} < \sum_{i=1}^{r_1-s} \delta_{s+i} = h_{r_1} - h_s$. The inequality holds because $k = s + i_0$ for some $i_0 \in [1, r_1 - s]$, and hence $\delta_{s+i_0} > \delta_{r_2+i_0}$. Therefore, $h(X \cup Y) + h(X \cap Y) < h(X) + h(Y)$. By Proposition 3.3, this concludes the proof. □

Consider an integer $k \in [1, m]$ in the conditions of Lemma 6.1 and the corresponding integer polymatroid $\mathcal{Z}'_k$. For a partition $\Pi = (\Pi_i)_{i \in J}$ of a set P of participants, consider the Π-partite access structure $\Gamma = \Gamma_{p_0}(\mathcal{Z}'_k, \Pi)$. A vector $v \in \mathbf{P}$ is in Γ if and only if there exists a vector u with $0 \le u \le v$ such that

- $s = |\operatorname{supp}(u)| \ge k$ and $|u| = h_s$, and
- $|u(Y)| \le h_i$ for every $i \in [1, m]$ and for every $Y \subseteq J$ with $|Y| = i$.

As a consequence of the next lemma, $\Gamma = \Gamma_{p_0}(\mathcal{Z}'_k, \Pi)$ is a vector space access structure over every large enough finite field. Moreover, every ideal uniform multipartite access structure is of this form. Due to space limitations, we skip the proof of this result, which will be given in the full version of this paper.

Theorem 6.2. *Let $\Pi = (\Pi_i)_{i \in J}$ with $|J| = m$ be a partition of a set P of participants and let Γ be a uniform Π-partite access structure. Then Γ is ideal if and only if there exist a uniform integer polymatroid $\mathcal{Z}$ on J and an integer $k \in [1, m]$ in the conditions of Lemma 6.1 such that $\Gamma = \Gamma_{p_0}(\mathcal{Z}'_k, \Pi)$. In particular, every ideal uniform multipartite access structure is a vector space access structure over every large enough finite field.*

The uniform multipartite access structures of the form (2) were proved to be ideal in [16]. By using the previous characterization, we obtain a shorter proof for this fact. Consider the uniform integer polymatroid $\mathcal{Z}$ on J with increment vector δ defined by $\delta_1 = t - k + 1$, and $\delta_i = 1$ if $i \in [2, k]$, and $\delta_i = 0$ if $i \in [k + 1, m]$. Consider the integer polymatroid $\mathcal{Z}'_k$ whose existence is given by Lemma 6.1. We claim that every Π-partite access structure Γ of the form (2) is equal to $\Gamma(\mathcal{Z}'_k, \Pi)$. Indeed, a vector $v \in \mathbf{P}$ is in $\Gamma(\mathcal{Z}'_k, \Pi)$ if and only if there exists a vector u with $0 \leq u \leq v$ such that

- $s = |\operatorname{supp}(u)| \geq k$ and $|u| = h_s = t$, and
- $|u(Y)| \leq h_i$ for every $i \in [1, m]$ and for every $Y \subseteq J$ with $|Y| = i$.

Since $h_i = t - k + i$ for every $i \in [1, k]$, it is clear that every vector $u \in \mathbf{P}$ satisfying the first condition satisfies as well the second one.

7   Efficiency of the Constructions of Ideal Multipartite Secret Sharing Schemes

Several families of ideal multipartite access structures have been presented in the previous sections. We proved that every one of these structures admits a vector space secret sharing scheme over every large enough finite field. Our proofs are not constructive, but a general method to construct vector space secret sharing schemes for multipartite access structures that are associated to representable integer polymatroids was given in [12]. Unfortunately, this method is not efficient, and no general efficient method is known.

Some issues related to the efficiency of the constructions of ideal schemes for several particular families of multipartite access structures have been considered [8,5,15,31,32]. We describe in the following a unified framework, derived from the general results in [12], in which those open problems can be more precisely stated.

Take $J = [1, m]$ and $J' = [0, m]$, and let $(\Pi_i)_{i \in J}$ be a partition of the set P of participants, where $|\Pi_i| = n_i$ and $|P| = n$. Consider an integer polymatroid $\mathcal{Z}' = (J', h)$ with $k_i = h(\{i\}) \leq n_i$ for every $i \in J$ and $k_0 = h(\{0\}) = 1$, and take $k = h(J')$. Consider as well a finite field $\mathbb{K}$ and a $\mathbb{K}$-representation $(V_i)_{i \in J'}$ of $\mathcal{Z}'$. In this situation, one has to find a matrix $M = (M_0|M_1|\cdots|M_m)$ over $\mathbb{K}$ with the following properties:

1. M_i is a $k \times n_i$ matrix ($n_0 = 1$) whose columns are vectors in V_i.
2. If $u = (u_0, u_1, \dots, u_m)$ is a basis of $\mathcal{Z}'$, every $k \times k$ submatrix of M formed by u_i columns in every M_i is nonsingular.

As a consequence of the results in [12], every such a matrix M defines a vector space secret sharing scheme for the multipartite access structure $\Gamma_0(\mathcal{Z}', \Pi)$.

One of the unsolved questions is to determine the minimum size of the fields over which there exists a vector space secret sharing scheme for $\Gamma_0(\mathcal{Z}', \Pi)$. An upper bound can be derived from [12, Corollary 6.7]. Namely, such a matrix M exists if $|\mathbb{K}| > \binom{n+1}{k}$. The best known lower bounds on $|\mathbb{K}|$ are linear on the number of participants, and they can be derived from [1, Lemma 1.2] and other known results about arcs in projective spaces. Even though very large fields are required in general to find such a matrix by using the known methods, the number of bits to represent the elements in the base field is polynomial on the number of participants, and hence the computation of the shares and the the the reconstruction of the secret value can be efficiently performed in such a vector space secret sharing scheme.

Another open problem is the existence of efficient methods to construct a vector space secret sharing scheme for $\Gamma = \Gamma_{p_0}(\mathcal{Z}', \Pi)$, that is, the existence of polynomial-time algorithms to compute a matrix M with the properties above. One important drawback is that no efficient method is known to check whether a matrix M satisfying Property 1 satisfies as well Property 2. Moreover, this seems to be related to some problems about representability of matroids that have been proved to be co-NP-hard [26].

We discuss in the following some general construction methods that can be derived from the techniques introduced in previous works [8,5,15,25,31,32] for particular families of multipartite access structures.

The first method, which was used in [8,25] and other works, consists basically in constructing the matrix M column by column, checking at every step that all submatrices that must be nonsingular are so. Arbitrary vectors from the subspaces V_i can be selected at every step, but maybe a wiser procedure is to take vectors of some special form as, for instance, Vandermonde linear combinations of some basis of V_i. In any case, an exponential number of determinants have to be computed.

A probabilistic algorithm was proposed in [31,32] for multilevel and compartmented access structures. Namely, the vectors from the subspaces V_i are selected at random. This method applies as well to the general case and the success probability is at least $1 - \binom{n+1}{k} N |\mathbb{K}|^{-1}$, where $N = \sum_{i \in J} k_i n_i$. By using this method, a matrix M that, with high probability, defines a secret sharing scheme for the given access structure can be obtained in polynomial time. Nevertheless, no efficient methods to check the validity of the output matrix are known.

Finally, we survey two different methods proposed by Brickell [8] and by Tassa [31] for the hierarchical threshold access structures. Other related solutions appeared in [5,15] for very particular cases of hierarchical threshold access structures. To better understand these methods, let us consider first the case of the threshold access structures. If the field $|\mathbb{K}|$ is very large, $n + 1$ randomly chosen vectors from $\mathbb{K}^k$ will define with high probability an ideal (k, n)-threshold scheme. Nevertheless, no efficient algorithm to check the validity of the output is available. One can instead choose $n + 1$ vectors of the Vandermonde form, and in this case an ideal (k, n)-threshold scheme is obtained, and of course we can check its validity in polynomial time. The solutions proposed in those works are based on the same idea. Namely, the vectors from the subspaces V_i have to be of some special form such that a matrix with the required properties is obtained and, in addition, the validity of the output can be efficiently checked. The solution proposed by Brickell [8] is not efficient because it requires to compute a primitive element in an extension field whose extension degree increases with the number of participants. The one proposed by Tassa [31, Section 3.3], which works only for prime fields, provides a polynomial time algorithm to construct a vector space secret sharing scheme for every hierarchical threshold access structure. The existence of similar efficient methods for other families of multipartite access structures is an open problem.

References

1. Ball, S.: On large subsets of a finite vector space in which every subset of basis size is a basis. Manuscript, available at the author's webpage (2010)
2. Beimel, A.: Secret-Sharing Schemes: A Survey. In: Chee, Y.M., Guo, Z., Ling, S., Shao, F., Tang, Y., Wang, H., Xing, C. (eds.) IWCC 2011. LNCS, vol. 6639, pp. 11–46. Springer, Heidelberg (2011)
3. Beimel, A., Tassa, T., Weinreb, E.: Characterizing Ideal Weighted Threshold Secret Sharing. SIAM J. Discrete Math. 22, 360–397 (2008)
4. Benaloh, J., Leichter, J.: Generalized Secret Sharing and Monotone Functions. In: Goldwasser, S. (ed.) CRYPTO 1988. LNCS, vol. 403, pp. 27–35. Springer, Heidelberg (1990)
5. Beutelspacher, A., Wettl, F.: On 2-level secret sharing. Des. Codes Cryptogr. 3, 127–134 (1993)
6. Blakley, G.R.: Safeguarding cryptographic keys. In: AFIPS Conference Proceedings, vol. 48, pp. 313–317 (1979)
7. Bloom, J.R.: A note on Superfast Threshold Schemes. Preprint, Texas A&M. Univ., Dept. of Mathematics (1981)
8. Brickell, E.F.: Some ideal secret sharing schemes. J. Combin. Math. and Combin. Comput. 9, 105–113 (1989)
9. Brickell, E.F., Davenport, D.M.: On the classification of ideal secret sharing schemes. J. Cryptology 4, 123–134 (1991)
10. Collins, M.J.: A Note on Ideal Tripartite Access Structures. Cryptology ePrint Archive, Report 2002/193 (2002), http://eprint.iacr.org/2002/193
11. Csirmaz, L.: The size of a share must be large. J. Cryptology 10, 223–231 (1997)
12. Farràs, O., Martí-Farré, J., Padró, C.: Ideal Multipartite Secret Sharing Schemes. Cryptology (2011) (Online First)
13. Farràs, O., Metcalf-Burton, J.R., Padró, C., Vázquez, L.: On the Optimization of Bipartite Secret Sharing Schemes. Des. Codes Cryptogr. (2011) (Online First)
14. Farràs, O., Padró, C.: Ideal Hierarchical Secret Sharing Schemes. In: Micciancio, D. (ed.) TCC 2010. LNCS, vol. 5978, pp. 219–236. Springer, Heidelberg (2010)

15. Giuletti, M., Vincenti, R.: Three-level secret sharing schemes from the twisted cubic. Discrete Mathematics 310, 3236–3240 (2010)
16. Herranz, J., Sáez, G.: New Results on Multipartite Access Structures. IEEE Proceedings on Information Security 153, 153–162 (2006)
17. Herzog, J., Hibi, T.: Discrete polymatroids. J. Algebraic Combin. 16, 239–268 (2002)
18. Ito, M., Saito, A., Nishizeki, T.: Secret sharing scheme realizing any access structure. In: Proc. IEEE Globecom 1987, pp. 99–102 (1987)
19. Karnin, E.D., Greene, J.W., Hellman, M.E.: On secret sharing systems. IEEE Trans. Inform. Theory 29, 35–41 (1983)
20. Kothari, S.C.: Generalized Linear Threshold Scheme. In: Blakely, G.R., Chaum, D. (eds.) CRYPTO 1984. LNCS, vol. 196, pp. 231–241. Springer, Heidelberg (1985)
21. Martí-Farré, J., Padró, C.: On Secret Sharing Schemes, Matroids and Polymatroids. J. Math. Cryptol. 4, 95–120 (2010)
22. Massey, J.L.: Minimal codewords and secret sharing. In: Proceedings of the 6-th Joint Swedish-Russian Workshop on Information Theory, Molle, Sweden, pp. 269–279 (August 1993)
23. Ng, S.-L.: Ideal secret sharing schemes with multipartite access structures. In: IEEE Proc.-Commun., vol. 153, pp. 165–168 (2006)
24. Oxley, J.G.: Matroid theory. Oxford Science Publications. The Clarendon Press, Oxford University Press, New York (1992)
25. Padró, C., Sáez, G.: Secret sharing schemes with bipartite access structure. IEEE Trans. Inform. Theory 46, 2596–2604 (2000)
26. Rao B.V., R., Sarma M.N., J.: On the Complexity of Matroid Isomorphism Problems. In: Frid, A., Morozov, A., Rybalchenko, A., Wagner, K.W. (eds.) CSR 2009. LNCS, vol. 5675, pp. 286–298. Springer, Heidelberg (2009)
27. Segre, B.: Curve razionali normali e k-archi negli spazi finiti. Ann. Mat. Pura Appl. 39, 357–379 (1955)
28. Schrijver, A.: Combinatorial optimization. Polyhedra and efficiency. Springer, Berlin (2003)
29. Shamir, A.: How to share a secret. Commun. of the ACM 22, 612–613 (1979)
30. Simmons, G.J.: How to (Really) Share a Secret. In: Goldwasser, S. (ed.) CRYPTO 1988. LNCS, vol. 403, pp. 390–448. Springer, Heidelberg (1990)
31. Tassa, T.: Hierarchical Threshold Secret Sharing. J. Cryptology 20, 237–264 (2007)
32. Tassa, T., Dyn, N.: Multipartite Secret Sharing by Bivariate Interpolation. J. Cryptology 22, 227–258 (2009)
33. Welsh, D.J.A.: Matroid Theory. Academic Press, London (1976)

Separating Short Structure-Preserving Signatures from Non-interactive Assumptions

Masayuki Abe, Jens Groth*, and Miyako Ohkubo

NTT Information Sharing Platform Laboratories
NTT Corporation
abe.masayuki@lab.ntt.co.jp
University College London, U.K.
j.groth@ucl.ac.uk
Security Architecture Laboratory, NSRI
NICT
m.ohkubo@nict.go.jp

Abstract. Structure-preserving signatures are signatures whose public keys, messages, and signatures are all group elements in bilinear groups, and the verification is done by evaluating pairing product equations. It is known that any structure-preserving signature in the asymmetric bilinear group setting must include at least 3 group elements per signature and a matching construction exists.

In this paper, we prove that optimally short structure preserving signatures cannot have a security proof by an algebraic reduction that reduces existential unforgeability against adaptive chosen message attacks to any non-interactive assumptions. Towards this end, we present a handy characterization of signature schemes that implies the separation.

Keywords. Structure-Preserving Signatures, Algebraic Reduction, Meta-Reduction.

1 Introduction

1.1 Background

When messages, signatures, and verification keys are elements of bilinear groups and the signature verification is done by evaluating pairing product equations, a signature scheme is called *structure-preserving* [2]. A structure-preserving signature (SPS for short) blends well with the Groth-Sahai non-interactive proof system [24], and enables the construction of efficient cryptographic protocols such as round-optimal blind signatures [4,2], traceable signatures [1], group encryption [10], proxy signatures [2], and delegatable credential systems [17].

---

* Supported by EPSRC grant number EP/G013829/1.

D.H. Lee and X. Wang (Eds.): ASIACRYPT 2011, LNCS 7073, pp. 628–646, 2011.

The first SPS was presented in [23] as a feasibility result. A variation of the Camenisch-Lysyanskaya signature scheme [9] introduced in [22] is an SPS that is secure against random message attacks. Schemes in [10] and [16] are efficient when signing a single group element, but their signature size grows linearly in the size of the message. The scheme in [16] is called automorphic as the message space includes its own public key, which is a useful feature in many applications. [2] presented the first constant-size SPS whose signature consists of 7 group elements. Yet shorter signatures have been pursued since then, however, [3] proved that any secure SPS in asymmetric bilinear groups requires at least 3 group elements. They presented a scheme matching the lower bound.

The 3-element SPS in [3] is based on a strong interactive assumption. They also constructed a 4-element SPS with a restricted message space based on a non-interactive assumption. It has been left as an open problem to find an optimal SPS based on a non-interactive assumption.

1.2   Black-Box Separations

A fully black-box reduction from a primitive B to a cryptographic scheme A is an algorithm R such that for any instance f of B and for any adversary E against A, if E breaks A^f then $R^{f,E}$ breaks f. A black-box separation is to show the absence of such an algorithm R. While there are number of non-black-box techniques, e.g., [5], black-box separations are meaningful as a convincing indication of the hardness of finding a reduction and as a guide to find a way to get around it. For variations and more discussion we refer to [33].

Oracle separation and meta-reduction are widely used techniques in showing a separation. Oracle separation is useful in showing the difficulty of constructing a cryptographic scheme from a minimal primitive such as a one-way function. Since black-box reductions relativise, showing the existence of an oracle that is useful in breaking A but useless in breaking B implies absence of black-box reductions from B to A. Since the seminal work by Impagliazzo and Rudich [26], numerous results have been found using this approach. In most cases, primitives are simple cryptographic objects such as one-way functions, and the schemes in question are non-interactive ones such as collision-free hash function [34] or signature schemes [20,14,13]. A recent work in [27] addresses more involved interactive schemes, blind signatures, by extending this line of techniques.

In the Meta-reduction approach, initiated by [7,11], the proof of separation is done by constructing an algorithm, a so-called meta-reduction, that uses a reduction as a black-box and solves a targeted problem, which can be the same as or different from the primitive the reduction is supposed to break. The intuition is that if a reduction is successful, the reduction breaks the underlying primitive by itself without help from the adversary. Proofs for separation exploits strong properties of the target schemes and underlying primitives. [15] exploits the blindness property in constructing a meta-reduction separating three-move blind signatures from non-interactive assumptions. In [32] a class of protocols, constant-round sequentially witness-hiding special-sound protocols for unique witness relations, is separated from any standard assumptions. It includes some practically important protocols such as Schnorr identification schemes.

Separation is often considered for limited classes of reductions. [31] assumes a key-preserving property where the same RSA moduli are used in all oracle calls. Later in [29] an assumption so-called instance non-malleability is introduced to ease the limitation. A variation in prime-order groups appears in [28]. In [7,11,30,8,19], a class of algorithms called *algebraic reductions* is considered. In this class, yielding a new group element is limited so that it is possible to extract its representation for relevant bases. As claimed in [7], the class of algebraic reductions are not overly restrictive. In particular, for prime order groups, all known efficient reductions fall into this class to the best of our knowledge.

1.3   Our Contribution

This paper shows that no algebraic reduction falls short in proving existential unforgeability against adaptive chosen message attacks of 3-element SPS in type-III bilinear groups [18] based on *any non-interactive assumption*. This gives a partial justification for the existing 3-element schemes with interactive assumptions since algebraic algorithms, while covering all known reduction algorithms in prime order groups, are not powerful enough to prove the security of a 3-element SPS.

Our separation follows the meta-reduction paradigm. However, instead of showing a monolithic proof that constructs a meta-reduction from scratch, we present a handy characterization that separates a signature scheme from any non-interactive assumptions. It facilitates the proofs, in particular when the reductions are restricted to a class of algorithms where knowledge extraction is given for free. The intuition behind our characterization is that if the signature scheme in question forces a reduction algorithm to know some information, e.g., the signing-key itself, to simulate the signing oracle in the EUF-CMA game, and this information is so essential that the adversary wins the game by seeing it, then the reduction algorithm can break the assumption without help from the adversary. Given the characterization, we show that such *crucial information* exists in any 3-element SPS when the reduction algorithm is algebraic. This gives us our separation from non-interactive assumptions.

2   Preliminaries

2.1   Digital Signature Scheme

We consider signature schemes that works over a set of common parameters, say GK. Concretely, there is a generator of the common parameters and the key generation algorithm takes GK as input. Such an extended formulation is often used in practical cryptographic protocols where many users share the group for efficiency reasons.

Definition 1 (Digital Signature Scheme). *A digital signature scheme Sig is a set of efficient algorithms* $(\mathcal{C}, \mathcal{K}, \mathcal{S}, \mathcal{V})$. $\mathcal{C}$ *is the common-parameter generator that takes security parameter* 1^λ *as input and outputs a common parameter GK.* $\mathcal{K}$ *is the key generator that takes GK as input and outputs a signing-key SK and verification-key VK. The keys include GK and the public-key defines a message space* Msp. $\mathcal{S}$ *is the signature generation algorithm that computes a signature* Σ *for input message M by using signing key SK.* $\mathcal{V}$ *is the verification algorithm that takes VK, M, and* Σ *and outputs 1 or 0 that represent acceptance and rejection, respectively.*

A signature scheme must be correct, i.e., it is required that for any keys generated by $\mathcal{K}$ and for any message in Msp, it holds that $1 = \mathcal{V}(VK, M, \mathcal{S}(SK, M))$. It is assumed that there exists an efficiently computable function $TstVk$ that takes λ and VK as input and checks the validity of VK such that if $0 \leftarrow TstVk(1^\lambda, VK)$ then $\mathcal{V}(VK, *, *)$ always returns 0, and if $1 \leftarrow TstVk(1^\lambda, VK)$ then the message space Msp is well defined and it is efficiently and efficiently sampleable. A signature Σ is called invalid (with respect to VK and M), if $1 \neq \mathcal{V}(VK, M, \Sigma)$. Otherwise, it is called valid.

We use the standard notion of existential unforgeability against adaptive chosen message attacks (EUF-CMA) [21] formally defined as follows.

Definition 2 (EUF-CMA). *A signature scheme* $Sig = (\mathcal{C}, \mathcal{K}, \mathcal{S}, \mathcal{V})$ *is existentially unforgeable against adaptive chosen message attacks if, for any* $\mathcal{A} \in$ PPT, *the probability*

$$\Pr \left[\begin{array}{l} GK \leftarrow \mathcal{C}(1^\lambda), \\ (VK, SK) \leftarrow \mathcal{K}(GK), \\ (M^\star, \Sigma^\star) \leftarrow \mathcal{A}^{\mathcal{S}(SK, \cdot)}(VK) \end{array} : M^\star \notin Q \wedge 1 \leftarrow \mathcal{V}(VK, M^\star, \Sigma^\star) \right]$$

is negligible in λ. *Here,* $\mathcal{S}(SK, \cdot)$ *is a signing oracle that takes message M and returns signatures* $\Sigma \leftarrow \mathcal{S}(SK, M)$. Q *is the set of messages submitted to the signing oracle.*

2.2   Bilinear Groups

In this paper, let $\mathcal{G}$ be a generator of bilinear groups. It takes security parameter 1^λ as input and outputs $\Lambda := (p, \mathbb{G}_1, \mathbb{G}_2, \mathbb{G}_T, e)$ where

- p is a λ-bit prime,
- $\mathbb{G}_1, \mathbb{G}_2, \mathbb{G}_T$ are groups of prime order p with efficiently computable group operations, membership tests, and bilinear mapping $e : \mathbb{G}_1 \times \mathbb{G}_2 \to \mathbb{G}_T$,
- $\forall G \in \mathbb{G}_1 \setminus \{1\}, H \in \mathbb{G}_2 \setminus \{1\}, e(G, H)$ generates $\mathbb{G}_T$, and
- $\forall A \in \mathbb{G}_1, \forall B \in \mathbb{G}_2, \forall x, y \in \mathbb{Z} : e(A^x, B^y) = e(A, B)^{xy}$.

By generic operations, we mean the group operation, membership testing, and bilinear mapping over the groups in Λ. In Type-III groups [18], no efficient isomorphisms are provided for either directions between $\mathbb{G}_1$ and $\mathbb{G}_2$. Throughout this paper, group descriptions Λ always describe Type-III groups.

By $\mathbb{G}_*$, we denote either $\mathbb{G}_1$ or $\mathbb{G}_2$ in Λ. For a vector of group elements $\boldsymbol{A} := (A_1, \ldots, A_k) \in \mathbb{G}_*^k$ and a vector of scalar values $\boldsymbol{x} := (x_1, \ldots, x_k) \in \mathbb{Z}_p^k$, we define the notation $\boldsymbol{A^x} = \prod_{i=1}^{k} A_i^{x_i}$.

2.3   Structure Preserving Signatures

For a description of bilinear groups $\Lambda = (p, \mathbb{G}_1, \mathbb{G}_2, \mathbb{G}_T, e)$, an equation of the form

$$\prod_i \prod_j e(A_i, B_j)^{a_{ij}} = Z$$

for constants $a_{ij} \in \mathbb{Z}_p, Z \in \mathbb{G}_T$, and constants or variables $A_i \in \mathbb{G}_1$, $B_j \in \mathbb{G}_2$ is called a pairing product equation (PPE for short).

Definition 3 (Structure-Preserving Signatures). *A signature scheme* $(\mathcal{C}, \mathcal{K}, \mathcal{S}, \mathcal{V})$ *is called structure preserving with respect to bilinear group generator* $\mathcal{G}$ *if*

- *Common parameter GK consists of a group description* Λ. *Constants* a_{ij} *in* $\mathbb{Z}_p$ *are also included in GK if any,*
- *Verification-key VK includes* Λ *and group elements in* $\mathbb{G}_1$, $\mathbb{G}_2$, *and* $\mathbb{G}_T$,
- *Messages M consists of group elements in* $\mathbb{G}_1$ *and* $\mathbb{G}_2$,
- *Signature* Σ *consists of group elements in* $\mathbb{G}_1$ *and* $\mathbb{G}_2$, *and*
- *Verification* $\mathcal{V}$ *evaluates membership in* $\mathbb{G}_1$ *and* $\mathbb{G}_2$ *and PPEs.*

In a narrow sense, SPS might be limited to $Z = 1$ and *VK* excluding elements in $\mathbb{G}_T$ so that accompanying witness-indistinguishable Groth-Sahai proofs can have the zero-knowledge property.

2.4   Algebraic Algorithms

An algorithm is called algebraic with respect to a group if it takes a vector of elements $\boldsymbol{X}$ in the group and outputs a group element Y and there is a corresponding algorithm called an extractor that can output the representation of Y with respect to $\boldsymbol{X}$. For instance, if the algebraic algorithm $\mathcal{R}$ takes $A, B \in \mathbb{G}_*$ as input and outputs $C \in \mathbb{G}_*$, then $\mathcal{R}$'s extractor $\mathcal{E}$ outputs (a, b) such that $C = A^a B^b$.

In the following, we give a formal definition of the minimal case where an algorithm takes group elements from one group as input and outputs only one group element.

Definition 4 (Algebraic Algorithm). *Let* $\mathcal{R}$ *be a probabilistic polynomial time algorithm that takes* Λ, *a string* $aux \in \{0,1\}^*$, *and group elements* $\boldsymbol{X} \in \mathbb{G}_*^k$ *for some k and* $\mathbb{G}_*$ *in* Λ *as input and outputs a group element in* $\mathbb{G}_*$ *and a string* $ext \in \{0,1\}^*$. $\mathcal{R}$ *is called algebraic with respect to* $\mathcal{G}$ *if there exists* $\mathcal{E} \in \mathsf{PPT}$ *getting the same input as* $\mathcal{R}$ *including the same random coins such that for* *any $\Lambda \leftarrow \mathcal{G}(1^\lambda)$ and all polynomial size $\boldsymbol{X}$ and aux, the following probability is negligible in λ.*

$$\Pr\left[\begin{matrix}(Y, ext) \leftarrow \mathcal{R}(\Lambda, \boldsymbol{X}, aux\,;\,r), \\ (\boldsymbol{y}, ext) \leftarrow \mathcal{E}(\Lambda, \boldsymbol{X}, aux\,;\,r)\end{matrix} \;:\; Y \neq \boldsymbol{X}^{\boldsymbol{y}}\right].$$

Please note that unlike the case of the knowledge of exponent assumptions [12,25,6] that assumes the presence of $\mathcal{E}$ for *any malicious* $\mathcal{R}$, here we try to capture the limitation of current technology in building reduction algorithms. It is in fact easy to imagine an algorithm $\mathcal{R}$ that may not be algebraic as defined above; $\mathcal{R}$ takes a string from *aux* and directly translates it as a group element in $\mathbb{G}_*$. For such $\mathcal{R}$ there may not be an efficient extractor $\mathcal{E}$. However, a reduction algorithm that chooses Y in this way will typically not be more useful than one that chooses Y with a known discrete logarithm with respect to $\boldsymbol{X}$. Accordingly, we consider algorithms that compute on explicitly given group elements. We also stress that we are only interested in capturing the structure of Y with respect to the base $\boldsymbol{X}$. It is possible that *aux* contains additional group elements and that $\mathcal{R}$ returns group elements in *ext* for which we do not care to know a representation with respect to $\boldsymbol{X}$.

The above definition extends naturally to $\mathcal{A}$ that takes group elements from both groups and outputs multiple group elements at the same time. Furthermore, we note that algorithms that outputs no group elements can also be regarded as algebraic by taking the identity as default output for such algorithms so that extracting the representation is trivial. Trivial algorithms that output group elements taken from inputs intact are algebraic, too.

The notion is also extended to oracle algorithms. Let $(Y, ext)[\boldsymbol{X}', aux'] \leftarrow \bar{\mathcal{R}}^O(\Lambda, \boldsymbol{X}, aux)$ denote an execution of $\bar{\mathcal{R}}$ accessing to oracle O where $[\boldsymbol{X}', aux']$ denotes all inputs to $\bar{\mathcal{R}}$ given from (all invocations of) O. We say that oracle algorithm $\bar{\mathcal{R}}$ is algebraic if there exists an algebraic algorithm $\mathcal{R}$, and the computation by $\bar{\mathcal{R}}^O$ is equivalent to the following sequence of computation. First set $\boldsymbol{X}_0 := \boldsymbol{X}$ and $aux_0 := aux$. Run $(\boldsymbol{Y}_1, ext_1 || \omega_1) \leftarrow \mathcal{R}(\Lambda, \boldsymbol{X}_0, aux_0)$ and repeat

$$(\boldsymbol{X}'_i, aux'_i) \leftarrow O(\Lambda, \boldsymbol{Y}_i, ext_i),$$

$$\boldsymbol{X}_{i+1} := \boldsymbol{X}_i || \boldsymbol{X}'_i, \; aux_{i+1} := \omega_i || aux'_i$$

$$(\boldsymbol{Y}_{i+1}, ext_{i+1} || \omega_{i+1}) \leftarrow \mathcal{R}(\Lambda, \boldsymbol{X}_{i+1}, aux_{i+1}).$$

for $i = 1$ until state ω_{i+1} explicitly indicates termination and $\boldsymbol{Y}_{i+1}$ includes Y. The extractor for $\bar{\mathcal{R}}$ is to compute $(\boldsymbol{y}, ext) \leftarrow \mathcal{E}^O(\Lambda, \boldsymbol{X}, aux)$ that fulfills $Y = (\boldsymbol{X}'')^{\boldsymbol{y}}$ for $\boldsymbol{X}'' = \boldsymbol{X} \cup \boldsymbol{X}'$. Such extractor can be constructed in straightforward manner by using the extractor for $\mathcal{R}$.

By $\mathtt{Cls}_{alb}$ we denote the set of all algebraic algorithms with respect to $\mathcal{G}$.

2.5   Non-interactive Hardness Assumptions

Intuitively, an assumption states that there is no algorithm $\mathcal{A}$ that is better than any known (typically trivial) algorithm U, which, for example, selects its output uniformly from a proper domain. In fact, our formulation is so general that it can capture too strong assumptions that never hold and too weak ones that always hold. But it does not matter for our purpose since we are to show the impossibility to reduce the security of a signature scheme to such (extreme) assumptions.

Definition 5 (Non-interactive Hardness Assumptions). *A non-interactive problem consists of a triple of algorithms $P = (I, V, U)$ where $I \in$ PPT is an instance generator, which takes 1^λ and outputs a pair of an instance and a witness, (y, w), and V is a verification algorithm that takes y, w and an answer x, and outputs 1 or 0 that represents acceptance or rejection, respectively. A non-interactive hardness assumption for problem P is to assume that, for any $\mathcal{A} \in$ PPT, the following advantage function Adv is negligible in λ.*

$$Adv_{\mathcal{A}}(1^\lambda) = \Pr[(y, w) \leftarrow I(1^\lambda), x \leftarrow \mathcal{A}(y) : 1 = V(y, x, w)]$$

$$- \Pr[(y, w) \leftarrow I(1^\lambda), x \leftarrow U(y) : 1 = V(y, x, w)] \qquad (1)$$

In search problems, U is typically set to an algorithm that returns constant $\bot$ (or a random answer x when the domain is uniformly sampleable). In decision problems, U typically returns 1 or 0 randomly so that the latter probability is $1/2$.

As we are concerned with structure preserving signatures, we consider hard problems that are defined over bilinear groups as follows.

Definition 6 (Hard Problem over $\mathcal{G}$). *A non-interactive problem P over bilinear group generator $\mathcal{G}$ is a non-interactive problem such that*

- *instance generator I runs $\Lambda \leftarrow \mathcal{G}(1^\lambda)$, and output y includes Λ, and*
- *there exists $\mathcal{A}$ that solves P with access to an oracle that solves the discrete logarithm problem for the groups in Λ.*

By NIP, we denote all non-interactive problems. Similarly, $\text{NIP}_{\mathcal{G}}$ denotes NIP over $\mathcal{G}$. Throughout the paper, we simply say that algorithm $\mathcal{A}$ solves problem P if advantage $\text{Adv}_{\mathcal{A}}(1^\lambda)$ is not negligible.

2.6   Black-Box Reduction and Meta-Reduction

When algorithm $\mathcal{R}$ is given $\mathcal{A}$ as black-box, denoted by $\mathcal{R}^{\mathcal{A}}$, we mean that $\mathcal{R}$ and $\mathcal{A}$ are given the same security parameter and $\mathcal{A}$ is given access to arbitrary number of copies of $\mathcal{A}$ as oracles. Interaction between $\mathcal{R}$ and $\mathcal{A}$ can be done in interleaving manner. If $\mathcal{A}$ is a randomized algorithm, $\mathcal{A}$ has random coins inside and every copy uses the same randomness. The security parameter and the random coins are out of the control of $\mathcal{R}$.

For problem P and signature scheme Sig, $\mathcal{R}$ is a fully black-box reduction if, for any (even inefficient) successful forger $\mathcal{A}$ for Sig, $\mathcal{R}^{\mathcal{A}}$ is successful in solving P. By Sig $\Rightarrow_{\mathcal{R}} P$, we mean that R is a black-box reduction from Sig to P. A separation between Sig and P is to show that for Sig and P, there is no such $\mathcal{R}$

under hardness assumption for problem P'. (The problem P' can be the same as P to make the separation unconditional.) Note that $\mathcal{R}$ depends on Sig and P. To claim that a class of hardness assumption falls short of proving the security of any construction of a signature scheme in a class by any black-box reduction, one need to show the absence of $\mathcal{R}$ for every signature and assumption in the respective classes.

In the meta-reduction paradigm, a proof typically begin with constructing a magic adversary $\mathcal{A}$ that is inefficient (or given access to powerful oracle) but successful in breaking Sig so that $\mathcal{R}^{\mathcal{A}}$ works as expected. It then constructs meta-reduction $\mathcal{M}$ that $\mathcal{M}^{\mathcal{R}}$ solves P'. A major task of $\mathcal{M}$ is to efficiently emulate $\mathcal{A}$ by rewinding $\mathcal{R}$ and/or exploiting special properties of $\mathcal{R}$ and Sig. If $\mathcal{M}$ is successful in the emulation, $\mathcal{M}^{\mathcal{R}}$ can be seen as a polynomial-time algorithm that solves P', which contradicts the assumed hardness of P'.

3   Crucial Relation

If any algorithm that simulates signatures must "know" the secret key, the unforgeability of the signature scheme cannot be proven by black-box reduction to any non-interactive assumption. We extend this idea in such a way that it is not necessary to know the entire secret key but some *crucial information* is necessary to conduct the simulation and sufficient to forge a signature if leaked to the adversary. Informally, crucial information is a witness for a binary relation, $\Psi(\theta, \varpi)$, which we call *crucial relation* defined over signatures θ and some sensitive information ϖ. The relation requires three properties: every θ has exactly one ϖ (uniqueness), whenever an entity is successful in producing signatures, it is possible to extract ϖ from the entity (extractability), and ϖ is useful enough to yield a forgery (usefulness). A crucial relation is defined with respect to a class of algorithms, $\mathtt{Cls} \subseteq \mathsf{PPT}$ to which the entity that generates θ belongs.

Let us first prepare some notations used in the formal definition. For a public key VK, a sequence of messages $\boldsymbol{M} = \{M_1, \ldots, M_n\} \in \mathsf{Msp}^n$ and signatures $\boldsymbol{\Sigma} = \{\Sigma_1, \ldots, \Sigma_n\}$, define $\mathcal{V}(\theta)$ for $\theta := (VK, \boldsymbol{M}, \boldsymbol{\Sigma})$ by a function that returns $\prod_{i=1}^{n} \mathcal{V}(VK, M_i, \Sigma_i)$.

Definition 7 (Crucial Relation). *Let* $\mathsf{Sig} = (\mathcal{C}, \mathcal{K}, \mathcal{S}, \mathcal{V})$ *be a signature scheme. Let* $\varpi \in \{0,1\}^*$ *and* $\theta = (VK, \boldsymbol{M}, \boldsymbol{\Sigma}) \in \{0,1\}^*$. *A relation* $\Psi(\theta, \varpi)$ *is a crucial relation for* Sig *with respect to a class of algorithms* Cls *if the following properties are provided.*

- *(Uniqueness) For every* $\theta := (VK, \boldsymbol{M}, \boldsymbol{\Sigma})$ *such that* $1 = \mathcal{V}(\theta)$, *there exists exactly one (polynomial size)* ϖ *fulfilling* $1 = \Psi(\theta, \varpi)$.
- *(Extractability) For any* $\mathcal{R} \in$ Cls, *there exists* $\mathcal{E} \in \mathsf{PPT}$ *and* $n > 0$ *such that, for any* $VK \in \{0,1\}^*$ *such that* $1 \leftarrow TstVk(1^\lambda, VK)$, *and any arbitrary string* φ *in* $1^\lambda || \{0,1\}^*$, *probability*

$$
\Pr \begin{bmatrix} \boldsymbol{M} \leftarrow \mathsf{Msp}^n \\ \boldsymbol{\Sigma} \leftarrow \mathcal{R}(\varphi, \boldsymbol{M}) \\ \varpi \leftarrow \mathcal{E}(\varphi, \boldsymbol{M}) \\ \theta := (VK, \boldsymbol{M}, \boldsymbol{\Sigma}) \end{bmatrix} \quad : \quad 1 = \mathcal{V}(\theta) \wedge 1 \neq \Psi(\theta, \varpi) \end{bmatrix} \tag{2}
$$

is negligible in λ. The probability is taken over the choice of $\boldsymbol{M}$ and the randomness given to $\mathcal{R}$. The same randomness is given to $\mathcal{E}$.

- *(Usefulness)* There exists an algorithm $\mathcal{B} \in$ PPT such that, for any $\theta := (VK, \boldsymbol{M}, \boldsymbol{\Sigma})$ and ϖ that satisfies $\Psi(\theta, \varpi) = 1$, the following probability is not negligible in λ.

$$\Pr\left[(M, \Sigma) \leftarrow \mathcal{B}(\theta, \varpi) \ : \ M \notin \boldsymbol{M} \wedge 1 = \mathcal{V}(VK, M, \Sigma)\right]$$

Remarks:

- The intuition of extractability is that whenever φ is helpful for $\mathcal{R}$ in computing valid signatures, extractor $\mathcal{E}$ should be successful in extracting ϖ from φ. This must hold even for non-legitimate VK as long as it is functional with respect to the verification.
- For $\mathcal{R}$ that is successful only with negligible probability, $\mathcal{E}$ can be an empty algorithm. So we only need to care for successful $\mathcal{R}$ that yields valid signatures. In particular, conditioned that $1 = \mathcal{V}(\theta)$ happens with noticeable probability, the conditional provability that $1 = \Psi(\theta, \varpi)$ is overwhelming.
- There may be many φ that make $\mathcal{R}$ produce the same $\boldsymbol{\Sigma}$ from the same VK and $\boldsymbol{M}$. Whichever φ is given, $\mathcal{E}$ must output the same ϖ.

Let $\text{SIGCR}_{\text{Cls}}$ denote signature schemes that has a crucial relation for a class of algorithms, Cls. We require Cls be a class of algorithms in PPT that satisfies the following trivial composition. For any $\mathcal{A} \in$ Cls, the following $\mathcal{A}'$ is also in Cls. $\mathcal{A}'$ takes inputs, say aux_1 and $X_1, \ldots, X_n$, and runs $\mathcal{A}$ as $(aux_{i+1}, Y_{i+1}) \leftarrow \mathcal{A}(aux_i, X_i)$ for $i = 1, ..., n$. $\mathcal{A}$ then picks some Y_i whose index is in the list specified in aux_1. Obviously, algebraic algorithms are in such a class. The following proof is given for such Cls.

Theorem 8. *For any signature scheme Sig in $\text{SIGCR}_{\text{Cls}}$, for any non-interactive problem P in NIP, there is no $\mathcal{R} \in$ Cls such that $\text{Sig} \Rightarrow_{\mathcal{R}} P$ if pseudo-random functions exit.*

Proof. Let O be a deterministic oracle that takes θ as input and returns ϖ that $1 = \Psi(\theta, \varpi)$ if it exists (otherwise return $\bot$). Consider the following all-powerful adversary $\mathcal{A}$ attacking Sig with access to O. Let f be a pseudo-random function. Given VK as input, $\mathcal{A}$ selects a random key for f and checks if $1 \leftarrow TstVk(1^\lambda, VK)$ (if not, $\mathcal{A}$ halts). Then it chooses $\boldsymbol{M}$ randomly from Msp^n for some constant n by using pseudo-randomness generated by $f(VK)$. Let $\boldsymbol{M} \leftarrow \text{Msp}^f(VK)$ denote these steps. $\mathcal{A}$ then send $\boldsymbol{M}$ to the signing oracle (simulated by $\mathcal{R}$). After receiving n signatures, $\boldsymbol{\Sigma}$, $\mathcal{A}$ aborts if $\boldsymbol{\Sigma}$ contains an invalid signature. Otherwise, $\mathcal{A}$ calls O with input $\theta = (VK, \boldsymbol{M}, \boldsymbol{\Sigma})$ and obtains ϖ. It then executes $(M, \Sigma) \leftarrow \mathcal{B}(VK, \boldsymbol{M}, \boldsymbol{\Sigma}, \varpi)$ and outputs (M, Σ).

To verify that above $\mathcal{A}^O$ is indeed a successful forger, consider that $\mathcal{A}^O$ is given legitimate VK and signatures generated by $\mathcal{S}(SK, \boldsymbol{M})$. By correctness of Sig and the uniqueness property, ϖ indeed exist and is uniquely defined. So O returns ϖ. Then due to the usefulness property, the output from $\mathcal{B}$ satisfies the predicates with probability not negligible in λ. Thus $\mathcal{A}^O$ is a successful forger against Sig.

Suppose that there exists $\mathcal{R} \in \mathtt{Cls}$ that Sig $\Rightarrow_{\mathcal{R}} P$ holds. Since $\mathcal{R}$ is a fully black-box reduction it must be successful with the above $\mathcal{A}^O$. Namely, $\mathrm{Adv}_{\mathcal{R}^{\mathcal{A}^O}}(1^\lambda)$ as defined in Definition 5 is not negligible.

Without loss of generality, we assume that $\mathcal{A}$ outputs n messages as $\boldsymbol{M}$ at once. We also assume, without loss of generality, that when $\mathcal{R}$ outputs something for interaction it also outputs the internal state φ at that moment. Then $\mathcal{R}$ is restarted taking φ and some data from the interaction as input.

We construct meta-reduction $\mathcal{M}$ that $\mathcal{M}^{\mathcal{R}}$ solves P. $\mathcal{M}$ emulates $\mathcal{A}^O$ without any oracles. By a session, we mean the conversation between $\mathcal{R}$ and a copy of $\mathcal{A}$ initiated by $\mathcal{R}$ with input VK_i to $\mathcal{A}$. Every session is labelled by an index. Given $y \leftarrow I(1^\lambda)$, $\mathcal{M}$ sets $\varphi_0 := y$. Let $\mathsf{BADSIG}[i]$ be a flag that indicates the presence of an invalid signature in i-th session. It is initialized to zero. $\mathcal{M}$ runs $\mathcal{R}(\varphi_0)$ and do as follows.

- If $\mathcal{R}$ outputs (φ_i, VK_j) to invoke j-th copy of $\mathcal{A}$, $\mathcal{M}$ checks $TstVk(1^\lambda, VK_j)$ and halt the session if it is not 1. Otherwise, $\mathcal{M}$ selects $\boldsymbol{M}_j \leftarrow \mathsf{Msp}_j^n$ (if the same VK_j has been observed before, say in session k, $\mathcal{M}$ uses the same $\boldsymbol{M}_k$ instead), and resume $\mathcal{R}$ as $\mathcal{R}(\varphi_i\|\boldsymbol{M}_j)$. Here Msp_j is the message space associated to VK_j.
- If $\mathcal{R}$ outputs $(\varphi_i, \Sigma_{k,\ell})$ for existing session k, $\mathcal{M}$ checks if $1 = \mathcal{V}(VK_k, M_{k,\ell}, \boldsymbol{\Sigma}_{k,\ell})$. If not, $\mathcal{M}$ sets $\mathsf{BADSIG}[k]$ to 1. It then continues as follows.
  - If $\ell < n$, $\mathcal{M}$ continues by running $\mathcal{R}(\varphi_i)$.
  - If $\ell = n$ and $\mathsf{BADSIG}[k] = 0$, then $\mathcal{M}$ extracts ϖ_k for this session as follows. Let φ_i be the internal state that $\mathcal{R}$ outputs with VK_k. Let $\boldsymbol{M}_{k'}$ be the last message $\mathcal{R}$ is given before outputting $\Sigma_{k,n}$. Let $\varphi_i' := \varphi_i\|\{\boldsymbol{M}_{k+1}, \dots, \boldsymbol{M}_{k'}\}$. Let $\mathcal{R}'$ be an algorithm associated to $\mathcal{R}$ that computes $\boldsymbol{\Sigma}_k \leftarrow \mathcal{R}'(\varphi_i', \boldsymbol{M}_k)$. $\mathcal{R}'$ is a simple algorithm that parses φ_i' into $\varphi_i\|\{\boldsymbol{M}_{k+1}, \dots, \boldsymbol{M}_{k'}\}$, runs $\mathcal{R}(\varphi_i, \boldsymbol{M}_k)$, continue running $\mathcal{R}$ giving messages $\boldsymbol{M}_{k+1}, \dots, \boldsymbol{M}_{k'}$ as input, and collects signatures $\Sigma_{k,i}$ for $i = 1, \dots, n$, and finally outputs $\boldsymbol{\Sigma}_k$. As $\mathcal{R}$ is in $\mathtt{Cls}$, so is $\mathcal{R}'$ as assumed to $\mathtt{Cls}$. Due to the extractability property, there exists polynomial-time $\mathcal{E}$ that computes ϖ_k for $\theta_k := (VK_k, \boldsymbol{M}_k, \boldsymbol{\Sigma}_k)$. Thus, $\mathcal{M}$ runs $\mathcal{E}(\varphi_i', \boldsymbol{M}_k)$ and obtains ϖ_k. As $\mathcal{V}(\theta_k) = 1$ holds, $1 = \Psi(\theta_k, \varpi_k)$ holds except for negligible probability. $\mathcal{M}$ then invokes $(M^\star_k, \Sigma^\star_k) \leftarrow \mathcal{B}(\theta_k, \varpi_k)$ and runs $\mathcal{R}(\varphi_i\|(M^\star_k, \Sigma^\star_k))$ to continue.
- If $\mathcal{R}$ outputs x, then $\mathcal{M}$ outputs x and halts.

Let $\mathrm{Adv}_{\mathcal{M}^{\mathcal{R}}}^P(1^\lambda)$ be the advantage of the above $\mathcal{M}$ in solving P. We show that the difference $|\mathrm{Adv}_{\mathcal{R}^{\mathcal{A}^O}}^P(1^\lambda) - \mathrm{Adv}_{\mathcal{M}^{\mathcal{R}}}^P(1^\lambda)|$ is negligible. We start from $\mathcal{M}^{\mathcal{R}}$ and modifies $\mathcal{M}$ slightly at a time. First replace truly random choice $\boldsymbol{M}_j \leftarrow \mathsf{Msp}_j^n$ with pseudo-random one $\boldsymbol{M} \leftarrow \mathsf{Msp}^f(VK)$. Call this modified algorithm $\mathcal{M}'$. The loss of the advantage by this modification is negligible due to the indistinguishability of f. We prove that by constructing a distinguisher $\mathcal{D}$ for f as follows. $\mathcal{D}$ runs $(y, w) \leftarrow I(1^\lambda)$ and emulate $\mathcal{M}^{\mathcal{R}}(y)$ as it is except that whenever $\mathcal{M}$ chooses

M_k, $\mathcal{D}$ sends VK_k to the challenger and obtains a string and use it as random coins to generate M_k. It then returns M_k to $\mathcal{R}$. When $\mathcal{M}$ terminates with x, $\mathcal{D}$ outputs $V(y, x, w)$. Obviously, if the strings from the challenger are truly random, $\mathcal{D}$ emulates $\mathcal{M}$. If, on the other hand, they are the output of f, $\mathcal{D}$ emulates $\mathcal{M}'$. Since the advantage of $\mathcal{D}$, say $\mathrm{Adv}_{\mathcal{D}}^f(1^\lambda)$, is assumed negligible, we have $|\mathrm{Adv}_{\mathcal{M}}^P(1^\lambda) - \mathrm{Adv}_{\mathcal{M}'}^P(1^\lambda)| = \mathrm{Adv}_{\mathcal{D}}^f(1^\lambda) < \mathrm{negl}(\lambda)$.

Next replace extractor $\mathcal{E}$ with oracle O. Call this modified algorithm $\mathcal{M}''$. We show that the loss of advantage by moving from $\mathcal{M}'$ to $\mathcal{M}''$ is negligible. Let

$$\Pr\left[\begin{matrix} M \leftarrow \mathsf{Msp}_j^n \\ \varpi \leftarrow \mathcal{E} \end{matrix}\right] \tag{3}$$

denote the probability presented in (2). We replace Msp_j^n and $\mathcal{E}$ with Msp^f and O accordingly with trivial meaning. With this notation, the loss of advantage is upper bound by

$$|\mathrm{Adv}_{\mathcal{M}'}^P(1^\lambda) - \mathrm{Adv}_{\mathcal{M}''}^P(1^\lambda)| \le \left|\Pr\left[\begin{matrix} M \leftarrow \mathsf{Msp}^f \\ \varpi \leftarrow \mathcal{E} \end{matrix}\right] - \Pr\left[\begin{matrix} M \leftarrow \mathsf{Msp}^f \\ \varpi \leftarrow O \end{matrix}\right]\right|. \tag{4}$$

To evaluate the right hand of (4), first observe that

$$\left|\Pr\left[\begin{matrix} M \leftarrow \mathsf{Msp}_j^n \\ \varpi \leftarrow \mathcal{E} \end{matrix}\right] - \Pr\left[\begin{matrix} M \leftarrow \mathsf{Msp}^f \\ \varpi \leftarrow \mathcal{E} \end{matrix}\right]\right| \tag{5}$$

is negligible due to the indistinguishability of f. Also,

$$\left|\Pr\left[\begin{matrix} M \leftarrow \mathsf{Msp}_j^n \\ \varpi \leftarrow \mathcal{E} \end{matrix}\right] - \Pr\left[\begin{matrix} M \leftarrow \mathsf{Msp}_j^n \\ \varpi \leftarrow O \end{matrix}\right]\right| \tag{6}$$

is negligible due to the extractability property. Finally observe that

$$\left|\Pr\left[\begin{matrix} M \leftarrow \mathsf{Msp}_j^n \\ \varpi \leftarrow O \end{matrix}\right] - \Pr\left[\begin{matrix} M \leftarrow \mathsf{Msp}^f \\ \varpi \leftarrow O \end{matrix}\right]\right| \tag{7}$$

is zero because oracle O never causes $1 \ne \Psi(\theta, \varpi)$ if $1 = \mathcal{V}(\theta)$ due to the uniqueness condition. Thus both probabilities in (7) are zero. Since (5) to (7) are all negligible, we conclude that (4) is negligible, too.

Finally, observe that $\mathcal{M}''$ is identical to $\mathcal{A}^O$. Accordingly, $|\mathrm{Adv}_{\mathcal{R}^{\mathcal{A}^O}}^P(1^\lambda) - \mathrm{Adv}_{\mathcal{M}^{\mathcal{R}}}^P(1^\lambda)|$ is negligible. Since $\mathcal{R}$ and $\mathcal{E}$ belongs to $\mathtt{Cls} \subseteq \mathtt{PPT}$ and $\mathcal{M}$ only performs operations that can be done in polynomial-time, the total running time of $\mathcal{M}$ and $\mathcal{R}$ remains polynomial. Thus $\mathcal{M}^{\mathcal{R}}$ forms a polynomial-time algorithm that solves P, which contradicts to the assumed hardness of P. $\square$

4 Crucial Relation in Size-3 SPS

We consider the class of algebraic reductions that make oracle calls with keys formed over over the groups for which it is defined as algebraic. This constraint plays a role when we construct an extractor for crucial relation based on the extractor associated with the algebraic reduction. Since the extractor works only for the groups the algebraic reduction is defined, so does the resulting extractor for crucial relation. Since the crucial relation involves the verification keys, we require all keys to be generated over the same groups the extractor works for. We call such algorithms *group-preserving algebraic reductions*. This notion has been used before in the literature, e.g., [19] and the constraint also has some similarity to key-preservation [31] and instance non-malleability [29].

Theorem 9. *There exists no group-preserving algebraic reduction that reduces the existential unforgeability of an SPS scheme to hardness of any problem in* $\text{NIP}_\mathcal{G}$ *if signatures consist of three base group elements.*

We prove Theorem 9 actually by proving the following lemma. Then applying Theorem 8 completes the proof.

Lemma 10. *Any SPS scheme with signature size 3 has a crucial relation with respect to group-preserving algebraic algorithms.*

We begin by recalling the result from [3] that any SPS scheme whose verification consists of one pairing product equation, or whose signature consists only of $\mathbb{G}_1$ or $\mathbb{G}_2$ is not EUF-CMA. A signature scheme for signing multiple elements at once can always be used to sign a single element by setting the other group elements to 1. Without loss of generality, it therefore suffices to consider schemes whose message consists of a single group element and where the signature consists of 2 elements in one group and 1 element in the other. We will also consider, without loss of generality, the case where the verification consists of two pairing product equations. The result applies to schemes with more than two verification equations as well and the proofs can be adopted with superficial changes.

Case of $\Sigma \in \mathbb{G}_1^2 \times \mathbb{G}_2$.
In any SPS whose signature consists of 3 group elements, $(R, S, T) \in \mathbb{G}_1^2 \times \mathbb{G}_2$, the verification predicate includes at least two pairing product equations that can be reduced to the following general form.

$$e(R, U_1 T^{a_1}) e(S, U_2 T^{a_2}) e(M, U_3 T^{a_3}) e(U_0, T^{a_4}) = Z_1 \tag{8}$$

$$e(R, V_1 T^{b_1}) e(S, V_2 T^{b_2}) e(M, V_3 T^{b_3}) e(V_0, T^{b_4}) = Z_2 \tag{9}$$

The group elements except for M, R, S and T are taken from the public key, and the constants in $\mathbb{Z}_p$ are taken from the common parameters. For a message M and a signature (R, S, T), let $\varphi_r, \alpha_r, \varphi_s, \alpha_s,$ and t be

$$R = G^{\varphi_r} M^{\alpha_r}, \quad S = G^{\varphi_s} M^{\alpha_s}, \quad \text{and} \quad T = H^t. \tag{10}$$

We consider $\varphi_r, \alpha_r, \varphi_s, \alpha_s$ be variables that fulfill relations determined by (8), (9) and (10). Let f_1 and f_2 be

$$f_1 = \alpha_r m + \varphi_r - r, \quad \text{and} \quad f_2 = \alpha_s m + \varphi_s - s \tag{11}$$

where small-case letters, r, s, and m, represents the discrete-logs (to base G) of group elements denoted by corresponding large-case letters. (This convention is used throughout this paper.) By replacing R and S in (8) with those in (10) and taking the discrete-logs with respect to base $e(G, H)$, we can represent (8) as $f_3 m + f_4 = 0$ where

$$f_3 = \alpha_r (u_1 + a_1 t) + \alpha_s (u_2 + a_2 t) + (u_3 + a_3 t), \text{ and} \tag{12}$$

$$f_4 = \varphi_r (u_1 + a_1 t) + \varphi_s (u_2 + a_2 t) + u_0 a_4 t - z_1. \tag{13}$$

Similarly, (9) can be represented as $f_5 m + f_6 = 0$ where

$$f_5 = \alpha_r (v_1 + b_1 t) + \alpha_s (v_2 + b_2 t) + (v_3 + b_3 t), \text{ and} \tag{14}$$

$$f_6 = \varphi_r (v_1 + b_1 t) + \varphi_s (v_2 + b_2 t) + v_0 b_4 t - z_2. \tag{15}$$

Consider a system of equations $Q := \{f_1 = 0, \ldots, f_6 = 0\}$. Focus on a non-redundant part, e.g., $f_1 = f_2 = f_3 = f_5 = 0$ which is represented as

$$\begin{pmatrix} m & 0 & 1 & 0 \\ 0 & m & 0 & 1 \\ u_1 + a_1 t & u_2 + a_2 t & 0 & 0 \\ v_1 + b_1 t & v_2 + b_2 t & 0 & 0 \end{pmatrix} \cdot \begin{pmatrix} \alpha_r \\ \alpha_s \\ \varphi_r \\ \varphi_s \end{pmatrix} = \begin{pmatrix} r \\ s \\ -(u_3 + a_3 t) \\ -(v_3 + b_3 t) \end{pmatrix}. \tag{16}$$

Let K_t denote the leftmost matrix in (16). It has rank 4, and

$$\det(K_t) = (a_1 b_2 - a_2 b_1) t^2 + (a_1 v_2 + u_1 b_2 - u_2 b_1 - a_2 v_1) t + (u_1 v_2 - u_2 v_1). \tag{17}$$

If $\det(K_t) \neq 0$, there exists unique $(\alpha_r, \alpha_s, \varphi_r, \varphi_s)$ that fulfills Q. Note that Q is defined with respect to the public key and M and T.

CRUCIAL RELATION. Now we are ready to define a crucial relation as follows. For $VK = (GK, U_0, U_1, U_2, U_3, U_4, V_0, V_1, V_2, V_3, V_4)$ and $\theta = (VK, \boldsymbol{M}, \boldsymbol{\Sigma})$, let $\varpi = (\alpha_r, \alpha_s, G^{\varphi_r}, G^{\varphi_s}, H^t)$. Relation $\Psi(\theta, \varpi)$ returns 1 if there exists a valid (M, R, S, T) in θ such that

- $T = H^t$,
- $(\alpha_r, \alpha_s, \varphi_r, \varphi_s)$ determined by ϖ fulfills Q w.r.t. VK and M, and
- (M, R, S, T) is the first one in θ that $\det(K_t) \neq 0$.

Relation Ψ also returns 1 if $\det(K_t) = 0$ for all (M, R, S, T) in θ and $\varpi = \perp$. Note that the second condition implies $R = G^{\varphi_r} M^{\alpha_r}$, $S = G^{\varphi_s} M^{\alpha_s}$. Such ϖ is extractable, unique, and useful as shown below.

UNIQUENESS. The first (M, Σ) with $\det(K_t) \neq 0$ is unique in θ (assuming that signatures are stored in order) if it exists. Then, ϖ is uniquely determined for such (M, Σ) from relation (16). When there is no (M, σ) with $\det(K_t) \neq 0$ exists in θ, ϖ is also uniquely defined to $\perp$. Accordingly, for any θ, there is unique ϖ such that $\Psi(\theta, \varpi) = 1$.

USEFULNESS. Given ϖ that satisfies $\Psi(\theta, \varpi) = 1$, a valid signature for arbitrary message can be created as follows. We first consider the case where $\varpi = (\alpha_r, \alpha_s, G^{\varphi_r}, G^{\varphi_s}, H^t) \neq \perp$. Given ϖ and arbitrary message $M^\star$, compute $R^\star = (G^{\varphi_r})M^{\star \alpha_r}$, $S^\star = (G^{\varphi_s})M^{\star \alpha_s}$, $T^\star = (H^t)$. To see that $\Sigma^\star = (R^\star, S^\star, T^\star)$ is a valid signature for $M^\star$, observe that the first verification predicate (8) is

$$e(R^\star, U_1 T^{a_1}) \, e(S^\star, U_2 T^{a_2}) \, e(M^\star, U_3 T^{a_3}) \, e(U_0, U_4 T^{a_4})$$

$$= e(G^{\varphi_r} M^{\star \alpha_r}, H^{u_1 + a_1 t}) \, e(G^{\varphi_s} M^{\star \alpha_s}, H^{u_2 + a_2 t})$$

$$e(M^\star, H^{u_3 + a_3 t}) \, e(G^{u_0}, H^{u_4 + a_4 t})$$

$$= e(M^\star, H)^{f_3} \, e(G, H)^{f_4}.$$

It results in 1 since ϖ satisfies $f_3 = f_4 = 0$. The second predicate can be verified in the same way. Thus, by choosing fresh $M^\star$, $(R^\star, S^\star, T^\star)$ is a successful forgery.

We next consider the case of $\varpi = \perp$. It means that $\det(K_t) = 0$ holds for all M and (R, S, T) in θ. We then present a concrete attack as follows. First we consider the case where (17) is not a zero polynomial. Since (17) is quadratic in t, there are at most two Ts for which $\det(K_t) = 0$. Given θ including more than three signatures, such T must appear more than once. Given two signatures (M_1, R_1, S_1, T) and (M_2, R_2, S_2, T) in θ, the forger computes random linear combination of the signatures as $(M^\star, R^\star, S^\star) = (M_1^{\beta_1} M_2^{\beta_2}, R_1^{\beta_1} R_2^{\beta_2}, S_1^{\beta_1} S_2^{\beta_2})$ for randomly chosen β_1 and β_2 that satisfies $\beta_1 + \beta_2 = 1$. Then $(R^\star, S^\star, T)$ is a valid signature for $M^\star$ that is random and fresh with high probability. (The forger chooses messages that are not 1 to make sure $M_1 \neq 1$ or $M_2 \neq 1$ to get $M^\star$ uniform.) Next consider the case where (17) is a zero polynomial. Then we have $a_1 b_2 = a_2 b_1$ and $u_1 v_2 = u_2 v_1$. Let δ_1 and δ_2 be

$$\delta_1 := \frac{b_1}{a_1} = \frac{b_2}{a_2}, \quad \text{and} \quad \delta_2 := \frac{v_1}{u_1} = \frac{v_2}{u_2}, \tag{18}$$

which are defined to zero if any of a_1, a_2, u_1 or u_2 is zero. Then, from $f_3 = f_5 = 0$ in (12) and (14), we have

$$\left(\frac{u_2 + a_2 t}{u_1 + a_1 t} - \frac{v_2 + b_2 t}{v_1 + b_1 t} \right) \alpha_s + \left(\frac{u_3 + a_3 t}{u_1 + a_1 t} - \frac{v_3 + b_3 t}{v_1 + b_1 t} \right) = 0. \tag{19}$$

The coefficient of α_s in (19) is zero since $\det(K_t) = 0$. Thus we have

$$\frac{u_3 + a_3 t}{u_1 + a_1 t} - \frac{v_3 + b_3 t}{v_1 + b_1 t} = 0. \tag{20}$$

Since (20) holds for any t, we have

$$\frac{b_3}{a_3} = \frac{b_1}{a_1} = \delta_1, \quad \text{and} \quad \frac{v_3}{u_3} = \frac{v_1}{u_1} = \delta_2. \tag{21}$$

Similarly, from $f_4 = f_6 = 0$ in (13) and (15), we have

$$\frac{v_0 \, b_4}{u_0 \, a_4} = \frac{b_1}{a_1} = \delta_1, \quad \text{and} \quad \frac{z_2}{z_1} = \frac{v_1}{u_1} = \delta_2. \tag{22}$$

From (18), (21) and (22), the second verification predicate (9) is

$$1 = e(R^{a_1} S^{a_2} M^{a_3} U_0^{a_4}, T)^{\delta_1} \cdot \{e(R, U_1)\, e(S, U_2)\, e(M, U_3)\, Z_1^{-1}\}^{\delta_2},$$

and the first verification predicate (8) is

$$1 = e(R^{a_1} S^{a_2} M^{a_3} U_0^{a_4}, T) \cdot \{e(R, U_1)\, e(S, U_2)\, e(M, U_3)\, Z_1^{-1}\}.$$

If $\delta_1 = \delta_2$, the verification predicates are in a linear relation. Thus they shrink into one predicate and the scheme is insecure. If $\delta_1 \neq \delta_2$, the equations hold if and only if

$$e(R^{a_1} S^{a_2} M^{a_3} U_0^{a_4}, T) = 1, \quad \text{and} \quad e(R, U_1)\, e(S, U_2)\, e(M, U_3)\, Z_1^{-1} = 1.$$

The first equation implies either $R^{a_1} S^{a_2} M^{a_3} U_0^{a_4} = 1$ or $T = 1$. For such a case, the following attack succeeds. Request three or more signatures on randomly chosen messages. Then find two signatures (M_1, R_1, S_1, T_1) and (M_2, R_2, S_2, T_2) such that $T_1 = T_2 = 1$ or $T_1 \cdot T_2 \neq 1$. Then, linear combination of the two signatures yields a new valid signature. That is, let $(M^\star, R^\star, S^\star) = (M_1^{\beta_1} M_2^{\beta_2}, R_1^{\beta_1} R_2^{\beta_2}, S_1^{\beta_1} S_2^{\beta_2})$ for randomly chosen β_1 and β_2 that satisfies $\beta_1 + \beta_2 = 1$. Then $(M^\star, R^\star, S^\star, T_1)$ is a valid fresh signature. Keeping the condition on T_1 and T_2 in mind, inspection is not hard and omitted. This concludes that a successful forgery is possible even for the case of $\varpi = \perp$.

EXTRACTABILITY. Observe that, for any algebraic algorithm that obtains M as input and computes group element R, there exists an extractor that outputs α_r such that $R = (G^{\varphi_r}) M^{\alpha_r}$ where (G^{φ_r}) part is computed by multi-base exponentiation of group elements except for M. Similarly, the extractor outputs α_s such that $S = (G^{\varphi_s}) M^{\alpha_s}$. Thus $(\alpha_1, \alpha_2, \varphi_1, \varphi_2)$ determined uniquely from extracted $(\alpha_1, \alpha_2, G^{\varphi_1}, G^{\varphi_2}, H^t)$ fulfills f_1 and f_2. We then claim that $f_i = 0$ for $i = 3, \ldots, 6$ also hold except for negligible probability. Otherwise, the algorithm can be used to solve the discrete-logarithm problem between G and M. As we can manipulate all group elements given to the algorithm so that all their discrete-logarithms are known except for M, we can compute φ_r (and φ_s) from the extracted exponents. Suppose that, without loss of generality, $f_3 \neq 0$ happens for $M \neq 1$. Since $f_3 m + f_4 = 0$ for valid signature, $f_4 \neq 0$ happens, too. Thus equation $f_3 m + f_4 = 0$ with non-zero f_3 and f_4 determine m. For the case of $f_5 \neq 0$, use equation $f_5 m + f_6 = 0$ with non-zero f_5 and f_6 instead. Accordingly, the extracted $(\alpha_1, \alpha_2, G^{\varphi_1}, G^{\varphi_2}, H^t)$ fulfills Q_t with overwhelming probability assuming the hardness of the discrete-logarithm problem in $\mathbb{G}_1$.

Since we can extract $(\alpha_1, \alpha_2, G^{\varphi_1}, G^{\varphi_2}, H^t)$ for all M and (R, S, T) in θ, a question is how to find the first one with $\det(K_t) \neq 0$ if it exists. It is done as follows. Suppose that θ includes more than six valid signatures, say (R_i, S_i, T_i) for M_i for $i = 1, \ldots, q$. Given corresponding α_{ri} and α_{si} that satisfies $f_1 = 0$ and $f_2 = 0$ from (12) and (13), one can solve the equations to obtain $(u_1, u_2, u_3, v_1, v_2, v_3)$ and every t_i. Observe that, when (12) and (14) are to be zero, we can represent

α_{ri} and α_{si} by

$$\alpha_{ri} = \{(u_3 + a_3\, t_i)(v_2 + b_2\, t_i) - (v_3 + b_3\, t_i)(u_2 + a_2\, t_i)\}/\det(K_{t_i})\,,\ \text{and}$$

$$\alpha_{si} = \{(v_3 + b_3\, t_i)(u_1 + a_1\, t_i) - (u_3 + a_3\, t_i)(v_1 + b_1\, t_i)\}/\det(K_{t_i}).$$

If $\det(K_{t_i}) \neq 0$, pair $(\alpha_{ri}, \alpha_{si})$ is unique to t_i. By using the extracted $(u_1, u_2, u_3, v_1, v_2, v_3)$ and t_i in each signature, we can find the smallest index $i^* \in \{1, \ldots, q\}$ at which $\det(K_{t_{i^*}}) \neq 0$ with respect to $(M_{i^*}, \Sigma_{i^*}) \in \boldsymbol{M} \times \boldsymbol{\Sigma}$, and assign ϖ accordingly. If there is no such index, we set $\varpi = \perp$. The success probability of the extraction is overwhelming since the probability of the extractor for the algebraic algorithm is overwhelming conditioned that given signatures are valid.

Case of $\Sigma \in \mathbb{G}_1 \times \mathbb{G}_2^2$.
As well as the previous case, any SPS with signature $(R, S, T) \in \mathbb{G}_1 \times \mathbb{G}_2^2$ for message $M \in \mathbb{G}_1$ verifies at least two pairing product equations that can be reduced to the following form.

$$e(R, U_1\, T^{a_1} S^{b_1})\, e(M, U_2\, T^{a_2} S^{b_2})\, e(U_3, T^{a_3})\, e(U_4, S^{b_4}) = Z_1 \qquad (23)$$

$$e(R, V_1\, T^{c_1} S^{d_1})\, e(M, V_2\, T^{c_2} S^{d_2})\, e(V_3, T^{c_3})\, e(V_4, S^{d_4}) = Z_2 \qquad (24)$$

Let $R = G^{\varphi_r} M^{\alpha_r}$. As before, we consider the relation in the exponent with respect to base $e(G, H)$. Then (23) and (24) are transformed as follows.

$$\{\alpha_r(u_1 + a_1 t + b_1 s) + (u_2 + a_2 t + b_2 s)\}\, m$$
$$+\ \varphi_r(u_1 + a_1 t + b_1 s) + u_3 a_3 t + u_4 b_4 s = z_1\,,\ \text{and} \qquad (25)$$
$$\{\alpha_r(v_1 + c_1 t + d_1 s) + (v_2 + c_2 t + d_2 s)\}\, m$$
$$+\ \varphi_r(v_1 + c_1 t + d_1 s) + v_3 c_3 t + v_4 d_4 s = z_2. \qquad (26)$$

Consider a system of equations $Q := \{f_1 = 0, \ldots, f_5 = 0\}$ where f_i is defined as

$$f_1 = \alpha_r m + \varphi_r - r, \qquad (27)$$

$$f_2 = \alpha_r(u_1 + a_1 t + b_1 s) + (u_2 + a_2 t + b_2 s), \qquad (28)$$

$$f_3 = \varphi_r(u_1 + a_1 t + b_1 s) + u_3 a_3 t + u_4 b_4 s - z_1, \qquad (29)$$

$$f_4 = \alpha_r(v_1 + c_1 t + d_1 s) + (v_2 + c_2 t + d_2 s)\,, \ \text{and} \qquad (30)$$

$$f_5 = \varphi_r(v_1 + c_1 t + d_1 s) + v_3 c_3 t + v_4 d_4 s - z_2. \qquad (31)$$

Note that, with the above definition, (25) and (26) can be written as $f_2 m + f_3 = 0$ and $f_4 m + f_5 = 0$, respectively. Also note that if $u_1 + a_1 t + b_1 s \neq 0$ or $v_1 + c_1 t + d_1 s \neq 0$, then α_r is uniquely determined by Q.

CRUCIAL RELATION. For $VK = (GK, G, H, U_0, U_1, U_2, U_3, V_0, V_1, V_2, V_3)$ and $\theta = (VK, \boldsymbol{M}, \boldsymbol{\Sigma})$, let $\varpi = (\alpha_r, G^{\varphi_r}, H^s, H^t)$. Relation $\Psi(\theta, \varpi)$ returns 1 if,

- $\varpi = \perp$, and there exists (M, R, S, T) in θ for which $u_1 + a_1 t + b_1 s = 0$ and $v_1 + c_1 t + d_1 s = 0$ hold, or

for the first (M, R, S, T) in θ,

- $R = G^{\varphi_r} M^{\alpha_r}$, $S = H^s$, and $T = H^t$ hold, and
- (α_r, φ_r) determined by ϖ fulfills Q with respect to VK, M, S, and T.

In the following, we show that such ϖ is unique, useful and extractable.

UNIQUENESS. If θ includes a signature that causes $u_1 + a_1 t + b_1 s = 0$ and $v_1 + c_1 t + d_1 s = 0$, then ϖ must be $\perp$ to have $\Psi(\theta, \varpi) = 1$. If θ does not, then each element in ϖ is uniquely determined from the first (M, R, S, T) in θ.

USEFULNESS. Given $\varpi = (\alpha_r, G^{\varphi_r}, H^s, H^t)$, pick random $M^\star$ and compute $R^\star = G^{\varphi_r} M^{\star \alpha_r}$, and set $S^\star = H^s$ and $T^\star = H^t$. Then $(M^\star, R^\star, S, T)$ is a valid forgery. If $\varpi = \perp$ and $\Psi(\theta, \varpi) = 1$, we show that the scheme is insecure. Suppose that $(u_1 + a_1 t + b_1 s = 0 \wedge v_1 + c_1 t + d_1 s = 0)$ happens with respect to (M, R, S, T) in θ. From (28) and (30), we have $u_2 + a_2 t + b_2 s = 0$ and $v_2 + c_2 t + d_2 s = 0$. It results in $U_2 T^{a_2} S^{b_2} = V_2 T^{c_2} S^{d_2} = 1$ in (23) and (24). Thus, $(M^\star, R, S, T)$ is a valid forgery.

EXTRACTABILITY. Given (M, R, S, T), relation $(u_1 + a_1 t + b_1 s = 0 \wedge v_1 + c_1 t + d_1 s = 0)$ can be verified by testing $(U_1 T^{a_1} S^{b_1} = 1 \wedge V_1 T^{c_1} S^{d_1} = 1)$. If it happens for any signature in θ, set $\varpi = \perp$. Suppose, without loss of generality, $u_1 + a_1 t + b_1 s \neq 0$ holds. Let (M, R, S, T) be the first signature in θ. For any algebraic algorithm that outputs (R, S, T) for given M, there exists an extractor that outputs α_r such that $R = G^{\varphi_r} M^{\alpha_r}$ for some φ_r. As argued before, this α_r fulfills Q except for negligible probability if the discrete-logarithm problem in $\mathbb{G}_1$ is hard. Thus outputting $\varpi = (\alpha_r, G^{\varphi_r}, S, T)$ completes the extraction.

5   Conclusion and Open Problems

Some ideas are suggested to get around our impossibility result. The first is to resort to interactive assumptions as done for constructing 3-element scheme in [3]. The second would be to go beyond the group-preserving algebraic reduction. It however needs a number theoretic breakthrough to exploit an adversary that works for a group with different prime order. More exotic approach is to find a non-blackbox reduction that uses the adversary in non-blackbox manner. It also needs a breakthrough technique to exploit the code of the adversary to handle number-theoretic object like bilinear groups.

While this paper focused on particular type of bilinear groups due to its importance, it is of interest to see whether similar result is obtained in other settings. Since known 4-element schemes based on non-interactive assumptions only sign messages in either of the base groups but not both, it would be worth pursuing a 4-element scheme that signs group elements from both groups at the same time, or to show the impossibility.

Acknowledgements. The first author thanks Takahiro Matsuda and Yutaka Kawai for their valuable comments on an early draft of this paper. Thanks also to Fumitaka Hoshino for discussions on the generation of bilinear groups. We are grateful to reviewers in Asiacrypt'11 for their instructive comments.

References

1. Abe, M., Chow, S.S.M., Haralambiev, K., Ohkubo, M.: Double-Trapdoor Anonymous Tags for Traceable Signatures. In: Lopez, J., Tsudik, G. (eds.) ACNS 2011. LNCS, vol. 6715, pp. 183–200. Springer, Heidelberg (2011)
2. Abe, M., Fuchsbauer, G., Groth, J., Haralambiev, K., Ohkubo, M.: Structure-Preserving Signatures and Commitments to Group Elements. In: Rabin, T. (ed.) CRYPTO 2010. LNCS, vol. 6223, pp. 209–236. Springer, Heidelberg (2010)
3. Abe, M., Groth, J., Haralambiev, K., Ohkubo, M.: Optimal Structure-Preserving Signatures in Asymmetric Bilinear Groups. In: Rogaway, P. (ed.) CRYPTO 2011. LNCS, vol. 6841, pp. 649–666. Springer, Heidelberg (2011)
4. Abe, M., Ohkubo, M.: A Framework for Universally Composable Non-Committing Blind Signatures. In: Matsui, M. (ed.) ASIACRYPT 2009. LNCS, vol. 5912, pp. 435–450. Springer, Heidelberg (2009)
5. Barak, B.: How to go beyond the black-box simulation barrier. In: FOCS 2001, pp. 106–115 (2001)
6. Bellare, M., Palacio, A.: The Knowledge-of-Exponent Assumptions and 3-Round Zero-Knowledge Protocols. In: Franklin, M. (ed.) CRYPTO 2004. LNCS, vol. 3152, pp. 273–289. Springer, Heidelberg (2004)
7. Boneh, D., Venkatesan, R.: Breaking RSA May Not be Equivalent to Factoring. In: Nyberg, K. (ed.) EUROCRYPT 1998. LNCS, vol. 1403, pp. 59–71. Springer, Heidelberg (1998)
8. Bresson, E., Monnerat, J., Vergnaud, D.: Separation Results on the "One-More" Computational Problems. In: Malkin, T. (ed.) CT-RSA 2008. LNCS, vol. 4964, pp. 71–87. Springer, Heidelberg (2008)
9. Camenisch, J., Lysyanskaya, A.: Signature Schemes and Anonymous Credentials from Bilinear Maps. In: Franklin, M. (ed.) CRYPTO 2004. LNCS, vol. 3152, pp. 56–72. Springer, Heidelberg (2004)
10. Cathalo, J., Libert, B., Yung, M.: Group Encryption: Non-Interactive Realization in the Standard Model. In: Matsui, M. (ed.) ASIACRYPT 2009. LNCS, vol. 5912, pp. 179–196. Springer, Heidelberg (2009)
11. Coron, J.-S.: Optimal Security Proofs for PSS and Other Signature Schemes. In: Knudsen, L.R. (ed.) EUROCRYPT 2002. LNCS, vol. 2332, pp. 272–287. Springer, Heidelberg (2002)
12. Damgård, I.: Towards Practical Public Key Systems Secure against Chosen Ciphertext Attacks. In: Feigenbaum, J. (ed.) CRYPTO 1991. LNCS, vol. 576, pp. 445–456. Springer, Heidelberg (1992)
13. Dodis, Y., Haitner, I., Tentes, A.: On the (in)security of RSA signatures. ePrint 2011/087 (2011)
14. Dodis, Y., Oliveira, R., Pietrzak, K.: On the Generic Insecurity of the Full Domain Hash. In: Shoup, V. (ed.) CRYPTO 2005. LNCS, vol. 3621, pp. 449–466. Springer, Heidelberg (2005)
15. Fischlin, M., Schröder, D.: On the Impossibility of Three-Move Blind Signature Schemes. In: Gilbert, H. (ed.) EUROCRYPT 2010. LNCS, vol. 6110, pp. 197–215. Springer, Heidelberg (2010)

16. Fuchsbauer, G.: Automorphic signatures in bilinear groups. ePrint 2009/320 (2009)
17. Fuchsbauer, G.: Commuting Signatures and Verifiable Encryption. In: Paterson, K.G. (ed.) EUROCRYPT 2011. LNCS, vol. 6632, pp. 224–245. Springer, Heidelberg (2011)
18. Galbraith, S., Paterson, K., Smart, N.: Pairings for cryptographers. ePrint 2006/165 (2006)
19. Garg, S., Bhaskar, R., Lokam, S.V.: Improved Bounds on Security Reductions for Discrete Log Based Signatures. In: Wagner, D. (ed.) CRYPTO 2008. LNCS, vol. 5157, pp. 93–107. Springer, Heidelberg (2008)
20. Gennaro, R., Gertner, Y., Katz, J., Trevisan, L.: Bounds on the efficiency of generic cryptographic constructions. SIAM J. Comput. 35(1), 217–246 (2005)
21. Goldwasser, S., Micali, S., Rivest, R.: A digital signature scheme secure against adaptive chosen-message attacks. SIAM J. Comp. 17(2), 281–308 (1988)
22. Green, M., Hohenberger, S.: Universally Composable Adaptive Oblivious Transfer. In: Pieprzyk, J. (ed.) ASIACRYPT 2008. LNCS, vol. 5350, pp. 179–197. Springer, Heidelberg (2008)
23. Groth, J.: Simulation-Sound Nizk Proofs for a Practical Language and Constant Size Group Signatures. In: Lai, X., Chen, K. (eds.) ASIACRYPT 2006. LNCS, vol. 4284, pp. 444–459. Springer, Heidelberg (2006)
24. Groth, J., Sahai, A.: Efficient Non-Interactive Proof Systems for Bilinear Groups. In: Smart, N.P. (ed.) EUROCRYPT 2008. LNCS, vol. 4965, pp. 415–432. Springer, Heidelberg (2008)
25. Hada, S., Tanaka, T.: On the Existence of 3-Round Zero-Knowledge Protocols. In: Krawczyk, H. (ed.) CRYPTO 1998. LNCS, vol. 1462, pp. 369–408. Springer, Heidelberg (1998); Full version available from IACR e-print archive 1999/009
26. Impagliazzo, R., Rudich, S.: Limits on the provable consequences of one-way permutations. In: STOC 1989, pp. 44–61. ACM (1989)
27. Katz, J., Schröder, D., Yerukhimovich, A.: Impossibility of Blind Signatures from One-Way Permutations. In: Ishai, Y. (ed.) TCC 2011. LNCS, vol. 6597, pp. 615–629. Springer, Heidelberg (2011)
28. Malkin, T., Moriarty, R., Yakovenko, N.: Generalized Environmental Security from Number Theoretic Assumptions. In: Halevi, S., Rabin, T. (eds.) TCC 2006. LNCS, vol. 3876, pp. 343–359. Springer, Heidelberg (2006)
29. Paillier, P.: Impossibility Proofs for RSA Signatures in the Standard Model. In: Abe, M. (ed.) CT-RSA 2007. LNCS, vol. 4377, pp. 31–48. Springer, Heidelberg (2006)
30. Paillier, P., Vergnaud, D.: Discrete-Log-Based Signatures May Not Be Equivalent to Discrete Log. In: Roy, B. (ed.) ASIACRYPT 2005. LNCS, vol. 3788, pp. 1–20. Springer, Heidelberg (2005)
31. Paillier, P., Villar, J.L.: Trading One-Wayness against Chosen-Ciphertext Security in Factoring-Based Encryption. In: Lai, X., Chen, K. (eds.) ASIACRYPT 2006. LNCS, vol. 4284, pp. 252–266. Springer, Heidelberg (2006)
32. Pass, R.: Limits of provable security from standard assumptions. In: STOC 2011, pp. 109–118. ACM (2011)
33. Reingold, O., Trevisan, L., Vadhan, S.P.: Notions of Reducibility Between Cryptographic Primitives. In: Naor, M. (ed.) TCC 2004. LNCS, vol. 2951, pp. 1–20. Springer, Heidelberg (2004)
34. Simon, D.R.: Finding Collisions on a One-Way Street: Can Secure Hash Functions be Based on General Assumptions? In: Nyberg, K. (ed.) EUROCRYPT 1998. LNCS, vol. 1403, pp. 334–345. Springer, Heidelberg (1998)

Short Signatures from Weaker Assumptions

Dennis Hofheinz[1], Tibor Jager[2], and Eike Kiltz[2]

[1] Institut für Kryptographie und Sicherheit,
Karlsruhe Institute of Technology, Germany
Dennis.Hofheinz@kit.edu
[2] Horst-Görtz Institute for IT Security, Ruhr-University Bochum, Germany
{**tibor.jager,eike.kiltz**}**@rub.de**

Abstract. We provide constructions of $(m, 1)$-programmable hash functions (PHFs) for $m \geq 2$. Mimicking certain programmability properties of random oracles, PHFs can, e.g., be plugged into the generic constructions by Hofheinz and Kiltz (J. Cryptol. 2011) to yield digital signature schemes from the strong RSA and strong q-Diffie-Hellman assumptions. As another application of PHFs, we propose new and efficient constructions of digital signature schemes from weaker assumptions, i.e., from the (standard, non-strong) RSA and the (standard, non-strong) q-Diffie-Hellman assumptions.

The resulting signature schemes offer interesting tradeoffs between efficiency/signature length and the size of the public-keys. For example, our q-Diffie-Hellman signatures can be as short as 200 bits; the signing algorithm of our Strong RSA signature scheme can be as efficient as the one in RSA full domain hash; compared to previous constructions, our RSA signatures are shorter (by a factor of roughly 2) and we obtain a considerable efficiency improvement (by an even larger factor). All our constructions are in the standard model, i.e., without random oracles.

Keywords: digital signatures, RSA assumption, q-DH assumption, programmable hash functions.

1 Introduction

Digital Signatures are one of the most fundamental cryptographic primitives. They are used as a building block in numerous high-level cryptographic protocols. Practical signature schemes are known whose security is based on relatively mild intractability assumptions such as the RSA [6] or the (bilinear) Computational Diffie-Hellman (CDH) assumption [13]. However, their security can only be proved in the random oracle model [5] with all its limitations (e.g., [17,26]).

STANDARD MODEL SIGNATURES. Signature schemes in the standard model (i.e., without using random oracles) are often considerably less efficient or based on much stronger assumptions. While tree-based signature schemes can be built from any one-way function [48], these constructions are far from practical. On the other hand, "Hash-and-sign" signatures are considerably more efficient, but the

D.H. Lee and X. Wang (Eds.): ASIACRYPT 2011, LNCS 7073, pp. 647–666, 2011.
© International Association for Cryptologic Research 2011

most efficient of these schemes rely on specific "strong" number theoretic hardness assumptions which we call Strong q-assumptions.[1] In Strong q-assumptions, an adversary is provided with a polynomial number of random "solved instances" and has to compute a new solved instance *of its choice*. For example, the schemes in [23,29,28,36,38,50] are based on the Strong (or, Flexible) RSA assumption and the schemes in [11,38,50] are based on the Strong q-Diffie-Hellman assumption. Both assumptions are considerably stronger than their "non-strong" counterparts (i.e., the q-Diffie-Hellman and the RSA assumptions, respectively), in which an adversary has to solve a *given, fixed instance*. (See the full version of this paper [34] for a discussion of the exact difference between strong and non-strong assumptions.)

PROGRAMMABLE HASH FUNCTIONS. In order to mimic certain "programmability properties" of random oracles, Hofheinz and Kiltz [38] introduced the combinatorial concept of programmable hash functions (PHF). (See Section 3 for a formal definition.) Among a number of other applications, they used PHFs as a building block for efficient and short hash-and-sign signatures based on the Strong RSA and the Strong q-Diffie-Hellman assumptions. Concretely, signatures in the Strong RSA based HK signature scheme $\mathsf{Sig}_{\mathsf{RSA}}[\mathsf{H}]$ are of the form $\mathsf{sig}(M) = (\mathsf{H}(M)^{1/e} \bmod N, e)$, where $N = pq$ is a public RSA modulus, $\mathsf{H}(\cdot)$ is a $(m, 1)$-PHF, and e is a *short* prime (chosen at random during the signing process). A given HK signature (σ, e) is verified by checking if $\sigma^e = \mathsf{H}(M) \bmod N$. The efficiency of the HK signature scheme is dominated by the time needed to generate the prime e, which (as shown in [38]) depends on the parameter m of the PHF: the bigger m, the smaller e and consequently the more efficient is the signing process.[2] Over bilinear groups there exists a similar construction, $\mathsf{Sig}_{\mathsf{S}-q\text{-}\mathsf{DH}}[\mathsf{H}]$, whose security is based on the Strong q-DH assumption. The main disadvantages of HK signatures is that their security relies on Strong assumptions, i.e., on the Strong RSA (Strong q-DH) and not on the standard RSA (q-DH) assumption.

RSA SIGNATURES. As a step towards practical signatures from the (standard) RSA assumption, Hohenberger and Waters [40,39] proposed the first hash-and-sign signature scheme (HW signatures) whose security is based on the RSA assumption. HW signatures are computed as $\mathsf{sig}(M) = g^{1/\mathsf{P}(M)} \bmod N$, where $g \in \mathbb{Z}_N^*$ is a public element and $\mathsf{P}(M) = e_1 \cdot \ldots \cdot e_{|M|}$ is the product of $|M|$ distinct primes. Here each prime e_i is uniquely determined by the i-bit prefix $M_{|i}$ of the message M, and for each generation of e_i a number of primality tests have to be executed which is the dominant running time of signing (and verifying). The above signature scheme is only weakly secure under the RSA

---

[1] There are exceptions, e.g., by Waters [53] (CDH assumption in bilinear groups), Hohenberger and Waters [40], and the lattice-based schemes [18,14] (SIS assumption). However, these are not among the most efficient "Hash-and-sign"-type schemes.

[2] We stress that the PHF parameter m does *not* directly correspond to the number of signatures that can be created during the security reduction. Rather, m indicates how many collisions of (honestly generated) e-values we can handle in the reduction. Hence, the larger m is, the smaller e can be chosen.

assumption, and a chameleon hash has to be used to make it fully secure, thereby doubling the signature size to two elements from $\mathbb{Z}_N$ and adding $\approx$ 2kbit to the public-key size [39]. The main disadvantage of HW signatures is, however, the generation and testing of the $|M|$ primes $e_1, \ldots, e_{|M|}$ necessary to compute the hash function $\mathsf{P}(M)$. Concretely, for $k = 80$ bits security, HW signatures need to generate $|M| = 160$ random primes for the signing process.

1.1   Summary of Our Contributions

As the main technical contribution we propose several new constructions of $(m, 1)$-PHFs for any $m \geq 1$. In particular, we solve the open problem posed in [38] of constructing deterministic $(m, 1)$-PHFs for $m > 2$. Even though our main applications are digital signatures we remark that PHFs are a very general framework for designing and analyzing cryptographic protocols in the Diffie-Hellman and RSA setting. For example, in [38], it was shown that PHFs imply collision-resistant hash functions and lead to elegant and simple proofs of Waters' IBE and signature schemes [53] and its countless variants (e.g., [15,7]). More importantly, a large body of cryptographic protocols with security in the standard model are using — implicitly or explicitly — the partitioning trick that is formalized in PHFs. To mention only a few examples, this ranges from collision-resistant hashing [20,4], digital signature schemes [12,53] (also in various flavors [47,51,8]), chosen-ciphertext secure encryption [15,41,35,37,14], identity-based encryption [9,10,42,18,1], attribute-based encryption [49] to symmetric authentication [43]. We expect that our new PHF constructions can also be applied to some of the mentioned applications.

We also show how to use our new $(m, 1)$-PHFs for generic constructions of short yet efficient hash-and-sign signatures whose security is based on weaker hardness assumptions: the q-DH and the RSA assumption. Whereas our q-DH schemes $\mathsf{Sig}_{q\text{-DH}}[\mathsf{H}]$ are (to the best of our knowledge) the first hash-and-sign schemes from this assumption, our RSA schemes $\mathsf{Sig}_{\mathsf{RSA}}[\mathsf{H}]$ and $\mathsf{Sig}_{\mathsf{RSA}}[\mathsf{H}]$ are conceptually different from HW signatures and we obtain a considerable efficiency improvement. A large number of new signature schemes with different tradeoffs can be derived by combining the generic signature schemes with PRFs. An overview of the efficiency of some resulting schemes and a comparison with existing schemes from [23,29,11,38,40] is provided in Table 1. Our new schemes offer different tradeoffs between signature size, efficiency, and public-key size. The bigger the parameter m in the $(m, 1)$-PHF, the larger the public-key size, the shorter the signatures. To obtain extremely short and/or efficient signatures, the size of the public key can get quite large. Concretely, with a public-key of size 26mbit we obtain 200 bit signatures from the (Strong) q-DH assumption. These are the shortest knwon standard-model digital signatures in bilinear groups. Remarkably, $\mathsf{Sig}_{\mathsf{SRSA}}[\mathsf{H}_{\mathsf{cfs}}]$ which instatiates the Strong RSA signatures from [38] with our new $(m, 1)$-PHF $\mathsf{H}_{\mathsf{cfs}}$ for $m \geq 6$, results in a hash-and-sign signature scheme where the signing procedure is dominated by one single modular exponentiation. This is the first RSA-based signature scheme whose signing complexity is not dominated by generating random primes.[3] Hence signing is essentially as efficient as RSA full-domain-hash [6] with the drawback of a huge public-key.

While these short signatures are mostly of theoretical interest and contribute to the problem of determining concrete bounds on the size of standard-model signatures, we think that in certain applications even a large public-key is tolerable. In particular, our public key sizes are still comparable to the ones of recently proposed lattice-based signatures [46,30,18,14].

We note furthermore, that it is possible to apply efficiency improvements from [40] to our RSA-based schemes as well. This allows us to reduce the number of primality tests required for signing and verification sigificantly. More precisely, it is possible to transform each signature scheme requiring λ primality tests into a scheme which requires only λ/c primality tests, at the cost of loosing a factor of 2^{-c} in the security reduction. For example, $\mathsf{Sig}^*_{\mathsf{RSA}}[\mathsf{H}_{\mathsf{Weak}}]^\S$ with $m = 11$ and $c = 40$ is a RSA-based signature scheme which requires only a *single* primality test for signing and verification, at the cost of loosing a factor of 2^{-40} in the security reduction.

1.2   Details of Our Contributions

Our main technical contribution to obtain shorter signatures are several new constructions of $(m, 1)$-PHFs for $m \geq 2$ (cf. Table 2 in Section 3). Using cover-free sets, we construct a deterministic $(m, 1)$-PHF $\mathsf{H}_{\mathsf{cfs}}$ with public parameters of $O(km^2)$ group elements. This solves the problem from [38] of constructing deterministic $(m, 1)$-PHFs for $m > 2$. We remark that cover-free sets were already used in [25,33,22] to construct identity-based encryption schemes. Furthermore, we propose a randomized $(m, 1)$-PHF $\mathsf{H}_{\mathsf{rand}}$ with public parameters of $O(m^2)$ group elements and small randomness space. Finally, we construct a weakly secure deterministic $(m, 1)$-PHF $\mathsf{H}_{\mathsf{Weak}}$ with public parameters of m group elements. The latter PHF already appeared implicitly in the context of identity/attribute-based encryption [19,49] (generalizing [9]). Weakly secure PHFs only yield weakly secure signature schemes that need to be "upgraded" to fully secure schemes using a chameleon hash function.

RSA SIGNATURES. Our new RSA signatures $\mathsf{Sig}_{\mathsf{RSA}}[\mathsf{H}]$ are of the form

$$\mathsf{sig}(M) = (\mathsf{H}(M)^{1/\mathsf{P}(s)} \bmod N, s), \tag{1}$$

where s is a short random bitstring, $\mathsf{H}(\cdot)$ is a $(m, 1)$-PHF, and $\mathsf{P}(s) := e_1 \cdot \ldots \cdot e_{|s|}$ is the product of $|s|$ primes $e_1, \ldots, e_{|s|}$, where the ith prime is uniquely determined by the ith prefix $s_{|i}$ of the randomness s. (If the PHF H is probabilistic, sig additionally contains a small random bitstring r.) Our security proof is along the lines of [38], but

---

[3] Since the complexity of finding a random μ-bit prime with error 2^{-k} is $O(k\mu^4)$, we expect that for $\mu \approx 60$ (or, equivalently, using $\mathsf{H}_{\mathsf{cfs}}$ with $m \geq 6$) a full exponentiation modulo a 1024-bit integer become roughly as expensive as generating a random μ-bit prime.

Table 1. Signature sizes of different schemes. Rows with grey background indicate new results from this paper. The chosen parameters provide unforgeability with $k = 80$ bits of security after revealing maximally $q = 2^{30}$ signatures. RSA signatures are instantiated with a modulus of $|N| = 1024$ bits, Bilinear signatures in asymmetric pairings using a BN curve [3] with $\log p = 160$ bits. (In this, we actually ignore the *multiplicative* reduction loss between a forger and, e.g., an RSA adversary.) We assume that elements in $\mathbb{G}_1$ can be represented by $|\mathbb{G}_1| = 160$ bits, while an element $\mathbb{G}_2$ by $|\mathbb{G}_2| = 320$ bits. The description of the bilinear group/modulus N is not counted in the public key. We assume $2k = 160$-bit messages in order to provide $k=80$ bits of security (to sign longer messages, we can apply a collision-resistant hash function first). The efficiency column counts the dominant operations for signing. For Bilinear and RSA signatures this counts the number of modular exponentiations, for RSA signatures $k \times \mathsf{P}_\mu$ counts the number of random μ-bit primes that need to be generated to evaluate function $\mathsf{P}(\cdot)$. (For $\mu \gg 60$, $1 \times \mathsf{P}_\mu$ takes more time than $1 \times \mathsf{Exp}$.) *The RSA-based chameleon hash function from [39] (which builds upon [2]) was used (adding $1 \times |\mathbb{Z}_N|$ to signature size). §Security reduction loses an additional factor of 2^{40}.

Signature scheme		Assumption	Sig. Size	Efficiency	PK size
Waters [53]		CDH	320	$2 \times \mathsf{Exp}$	26k
Boneh-Boyen [11]		Strong q-DH	320	$1 \times \mathsf{Exp}$	640
$\mathsf{Sig}_{\mathsf{S}\text{-}q\text{-DH}}[\mathsf{H}_{\mathsf{Wat}}]$ [38]		Strong q-DH	230	$1 \times \mathsf{Exp}$	26k
$\mathsf{Sig}_{\mathsf{S}\text{-}q\text{-DH}}[\mathsf{H}_{\mathsf{cfs}}]$	$(m=8)$	Strong q-DH	200	$1 \times \mathsf{Exp}$	26m
$\mathsf{Sig}_{q\text{-DH}}[\mathsf{H}_{\mathsf{Wat}}, \mathsf{H}_{\mathsf{Wat}}]$	$(m=2)$	q-DH	230	$1 \times \mathsf{Exp}$	48k
$\mathsf{Sig}_{q\text{-DH}}[\mathsf{H}_{\mathsf{cfs}}, \mathsf{H}_{\mathsf{Wat}}]$	$(m=8)$	q-DH	200	$1 \times \mathsf{Exp}$	26m
Cramer-Shoup [23]		Strong RSA	2208	$1 \times \mathsf{P}_{160}$	3k
Gennaro et. al.* [29]		Strong RSA	2048	$1 \times \mathsf{P}_{160}$	3k
$\mathsf{Sig}_{\mathsf{SRSA}}[\mathsf{H}_{\mathsf{Wat}}]$ [38]		Strong RSA	1104	$1 \times \mathsf{P}_{80}$	128k
$\mathsf{Sig}_{\mathsf{SRSA}}[\mathsf{H}_{\mathsf{cfs}}]$	$(m=6)$	Strong RSA	1068	$\approx 1 \times \mathsf{Exp}$	94m
$\mathsf{Sig}^*_{\mathsf{SRSA}}[\mathsf{H}_{\mathsf{Weak}}]$	$(m=6)$	Strong RSA	2092	$\approx 2 \times \mathsf{Exp}$	9k
Hohenberger-Waters* [40]		RSA	2048	$160 \times \mathsf{P}_{1024}$	3k
$\mathsf{Sig}^*_{\mathsf{RSA}}[\mathsf{H}_{\mathsf{Weak}}]$	$(m=2)$	RSA	2048	$70 \times \mathsf{P}_{1024}$	5k
$\mathsf{Sig}^*_{\mathsf{RSA}}[\mathsf{H}_{\mathsf{Weak}}]$	$(m=4)$	RSA	2048	$50 \times \mathsf{P}_{1024}$	7k
$\mathsf{Sig}_{\mathsf{RSA}}[\mathsf{H}_{\mathsf{Wat}}]$	$(m=2)$	RSA	1094	$70 \times \mathsf{P}_{1024}$	128k
$\mathsf{Sig}_{\mathsf{RSA}}[\mathsf{H}_{\mathsf{rand}}]$	$(m=4)$	RSA	1214	$50 \times \mathsf{P}_{1024}$	32k
$\mathsf{Sig}_{\mathsf{RSA}}[\mathsf{H}_{\mathsf{cfs}}]$	$(m=4)$	RSA	1074	$50 \times \mathsf{P}_{1024}$	40m
$\mathsf{Sig}^*_{\mathsf{RSA}}[\mathsf{H}_{\mathsf{Weak}}]^\S$	$(m=11)$	RSA	2048	$1 \times \mathsf{P}_{1024}$	14k

using P enables a reduction to the RSA assumption (Theorem 7) in the standard model. The main conceptual novelty is that we apply P to the randomness s rather than the message M as in HW signatures. Because the values s are relatively small, our scheme is considerably more efficient than that of [40].

Concretely, the length of s is controlled by the PHF parameter m as $|s| = \log q + k/m$, where q is an upper bound on the number of signatures the scheme supports. (See the full version [34] for a formal argument.) For $k = 80$ bits security and $q = 2^{30}$ (as recommended in [6]) we can make use of our new constructions of $(m, 1)$-PHFs with $m \geq 2$. For example, with a $(4, 1)$-PHF, the bitstring s can be as small as 50 bits which leads to very small signatures. More importantly, since the function $P(s)$ only has to generate $|s|$ distinct primes $e_1, \ldots, e_{|s|}$ (compared to $|M| \gg |s|$ primes in HW signatures), the signing and verification algorithms are considerably faster. The drawback of our new signature scheme is that the system parameters of H grow with m.

BILINEAR SIGNATURES. Our new q-DH signatures $\mathsf{Sig}_{q\text{-DH}}[H]$ are of the form

$$\mathsf{sig}(M) = (H(M)^{1/\mathsf{d}(s)}, s), \tag{2}$$

where again s is a short random bitstring, H is a $(m, 1)$ programmable hash function, and $\mathsf{d}(\cdot)$ is a special (secret) function mapping bitstrings to $\mathbb{Z}_p$. Since $\mathsf{D}(s) := g^{\mathsf{d}(s)}$ can be computed publicly, verification is done by using the properties of the bilinear group. Security is proved under the q-DH assumption in the standard model. Similar to our RSA-based signatures the length of s is controlled by the PHF parameter m. For example, for $m = 8$ we obtain standard-model signatures of size $|\mathbb{G}| + |s| = 160 + 40 = 200$ bits. We have to refer to the full version [34] for details.

FULL-DOMAIN HASH SIGNATURES. We remark that full-domain hash signature schemes over a homomorphic domain (e.g., RSA-FDH [6] and BLS signatures [13]) instantiated with $(m, 1)$-PHFs provide efficient m-time signature schemes without random oracles. This nicely complements the impossibility results from [26] who show that without the homomorphic property this is not possible. We remark that an instantiation of RSA-FDH as a m-time signature scheme was independently observed in [24].

PROOF TECHNIQUES AND RELATED WORK. Our RSA-based signature scheme represents a combination of techniques from [38] and [40]. Namely, in the basic RSA-based signature scheme from [38], a signature is of the form $(H(M)^{1/s} \bmod N, s)$ for a prime s. The use of a *programmable* hash function H enables very efficient schemes, whose security however cannot be reduced to the standard (non-strong) RSA problem, since a forged signature $(H(M)^{1/s^*}, s^*)$ corresponds to an RSA inversion with adversarially chosen exponent s^*. On the other hand, the (basic, weakly secure) signature scheme from [40] is of the form $g^{1/P(M)} \bmod N$. The special structure of P (which maps a message M to the product of $|M|$ primes) makes it possible to prove security under the standard RSA assumption. However, since P is applied to messages (i.e., 160-bit strings), evaluation of P requires a large number of primality tests. We combine the best of both worlds with signatures of the form $(H(M)^{1/P(s)} \bmod N, s)$ for *short* (e.g., 40-bit) random strings s. In contrast to the scheme of [40], this directly yields a fully secure signature scheme, so we do not need a chameleon hash function.

In the security proof of our RSA signatures we distinguish between two types of forgers: type I forgers recycle a value from $\{s_1, \ldots, s_q\}$ for the forgery, where the s_i's are the random bitstrings used for the simulated signatures; type II forgers use a new value $s^* \notin \{s_1, \ldots, s_q\}$ for the forgery and therefore are more difficult to reduce to the RSA assumption. For the reduction of type II forgers to the RSA assumption we can use a clever "prefix-guessing" technique from [40] to embed the prime e from the RSA challenge in the function $P(\cdot)$ such that the product $P(s^*)$ contains e.[4] Similar to the proof of HK signatures [38], the reduction for Type I forgers makes use of the $(m, 1)$ programmability of $H(\cdot)$.

Strong q-DH signatures from [38] can actually be viewed as our q-DH signatures from (2) instantiated with the special function $d(s) = x + s$ (where x is part of the secret-key). In our scheme, the leverage to obtain security from q-DH is that the function $D(s) := g^{d(s)}$ acts as a $(\mathsf{poly}, 1)$-PHF. That is, $d(\cdot)$ can be setup such that (with non-negligible probability) $d(s_i) = x + a(s_i)$ for $a(s_i) \neq 0$ but $d(s^*) = x$, where $s_1, \ldots, s_q$ is the randomness used for the generated signatures and s^* is the randomness used for the forgery.

1.3   Open Problems

A number of interesting open problems remain. We ask how to construct (deterministic) $(m, 1)$-PHFs for $m \geq 1$ with smaller parameters than the ones from Table 2. Since the constructions of cover free sets are known to be optimal up to a log factor, a new method will be required. Furthermore, obtaining truely practical signatures from the RSA or factoring assumption is still an open problem. In particular, we ask for a construction of hash-and-sign (strong) RSA signatures that do not require the generation of primes at signing.

2   Preliminaries

For $k \in \mathbb{N}$, we write 1^k for the string of k ones, and $[k]$ for $\{1, \ldots, k\}$. Moreover, $|x|$ denotes the length of a bitstring x, while $|S|$ denotes the size of a set S. Further, $s \xleftarrow{\$} S$ denotes the sampling a uniformly random element s of S. For an algorithm $\mathcal{A}$, we write $z \xleftarrow{\$} \mathcal{A}(x, y, \ldots)$ to indicate that $\mathcal{A}$ is a (probabilistic) algorithm that outputs z on input $(x, y, \ldots)$.

2.1   Digital Signatures

A digital signature scheme $\mathsf{Sig} = (\mathsf{Gen}, \mathsf{Sign}, \mathsf{Vfy})$ consists of three algorithms. Key generation Gen generates a keypair $(pk, sk) \xleftarrow{\$} \mathsf{Gen}(1^k)$ for a secret signing key sk and a public verification key pk. The signing algorithm Sign inputs a message and the secret signing key, and returns a signature $\sigma \xleftarrow{\$} \mathsf{Sign}(sk, m)$ of the message. The verification algorithm Vfy takes a verification key and a message with corresponding signature as input, and returns $b \leftarrow \mathsf{Vfy}(pk, m, \sigma)$ where $b \in \{\texttt{accept}, \texttt{reject}\}$. We require the usual correctness properties.

---

[4] More precisely, when simulating a type II forger, the values $s_1, \ldots, s_q$ are known in advance to the simulator. Since $s^* \notin \{s_1, \ldots, s_q\}$ there is some prefix $s^*_{|i}$ of s^* that is different from all prefixes of $s_1, \ldots, s_q$. We can guess the smallest such prefix such that the simulator knows $s^*_{|i}$ from the forgery at the beginning. This knowledge can be used to embed e from the RSA challenge in the function $P(\cdot)$ such that the product $P(s^*)$ contains e.

Let us recall the *existential unforgeability against chosen message attacks* (EUF-CMA) security experiment [31], played between a challenger and a forger $\mathcal{F}$.

1. The challenger runs Gen to generate a keypair (pk, sk). The forger receives pk as input.
2. The forger may ask the challenger to sign a number of messages. To query the i-th signature, $\mathcal{F}$ submits a message m_i to the challenger. The challenger returns a signature σ_i under sk for this message.
3. The forger outputs a message m^* and signature σ^*.

$\mathcal{F}$ wins the game, if accept $\leftarrow$ Vfy(pk, m^*, σ^*), that is, σ^* is a valid signature for m^*, and $m^* \neq m_i$ for all i. We say that $\mathcal{F}$ (t, q, ϵ)-breaks the EUF-CMA security of Sig, if $\mathcal{F}$ runs in time t, makes at most q signing queries, and has success probability ϵ. We say that Sig is EUF-CMA secure, or Sig is *fully* secure, if ϵ is negligible for any probabilistic polynomial-time algorithm $\mathcal{F}$.

We also say, that a scheme is *weakly* secure, if it meets the above security definition, but the adversary can not choose the messages to be signed adaptively. Instead it has to commit to a list $m_1, \ldots, m_q$ before seeing the public key. There exist efficient generic techniques to convert a weakly secure signature scheme into a fully secure one, e.g., using chameleon hashes [44].

2.2   Prime Numbers, Factoring, and the RSA Assumption

For $x \in \mathbb{N}$ let $\pi(x)$ denote the number of primes between 0 and x. The following lemma is a direct consequence of Chebyshev's bounds on $\pi(x)$ (see [32], for instance).

Lemma 1. $\frac{x}{\log_2 x} < \pi(x) < \frac{2x}{\log_2 x}$

We say that a prime p is a *safe* prime, if $p = 2p'+1$ and p' is also prime. Let p and q be two randomly chosen $k/2$-bit safe primes, and let $N = pq$. Let $e \in \mathbb{Z}_{\phi(n)}$ be a random integer, relatively prime to $\phi(N)$. We say that an algorithm $\mathcal{A}$ (t, ϵ)-breaks the RSA assumption, if $\mathcal{A}$ runs in time t and

$$\Pr[y^{1/e} \xleftarrow{\$} \mathcal{A}(N, e, y)] \geq \epsilon.$$

We assume that there exists no algorithm that (t, ϵ)-breaks the RSA assumption with polynomial t and non-negligible ϵ.

We denote with QR_N the group of quadratic residues modulo N. The following lemma, which is due to Shamir [52], is useful for the security proof of the generic RSA-based signature scheme described in Section 4.

Lemma 2. *There is an efficient algorithm that, on input $y, z \in \mathbb{Z}_N$ and integers $e, f \in \mathbb{Z}$ such that $\gcd(e, f) = 1$ and $z^e \equiv y^f \bmod n$, computes $x \in \mathbb{Z}_N$ satisfying $x^e \equiv y \bmod N$.*

2.3   Generalized Birthday Bound

Although not explicitly stated, the following lemma is implicit in [36]. We will apply it several times in the security proofs for our generic signature schemes.

Lemma 3. *Let A be a set with $|A| = a$. Let $X_1, \ldots, X_q$ be q independent random variables, taking uniformly random values from A. Then the probability that there exist $m + 1$ pairwise distinct indices $i_1, \ldots, i_{m+1}$ such that $X_{i_1} = \cdots = X_{i_{m+1}}$ is upper bounded by $\frac{q^{m+1}}{a^m}$.*

3   Programmable Hash Functions

3.1   Definitions

Let $G = (\mathbb{G}_k)$ be a family of groups, indexed by security parameter $k \in \mathbb{N}$. We omit the subscript when the reference to the security parameter is clear, thus write $\mathbb{G}$ for $\mathbb{G}_k$.

A *group hash function* H over $\mathbb{G}$ with input length $l = l(k)$ consists of two efficient algorithms PHF.Gen and PHF.Eval. The probabilistic algorithm $\kappa \xleftarrow{\$} \mathsf{PHF.Gen}(1^k)$ generates a hash key κ for security parameter k. Algorithm PHF.Eval is a deterministic algorithm, taking as input a hash function key κ and $X \in \{0, 1\}^l$, and returning $\mathsf{PHF.Eval}(\kappa, X) \in \mathbb{G}$.

Definition 1. *We say that a group hash function $H = (\mathsf{PHF.Gen}, \mathsf{PHF.Eval})$ is (m, n, γ, δ)-programmable, if there is an efficient trapdoor generation algorithm PHF.TrapGen and an efficient trapdoor evaluation algorithm PHF.TrapEval with the following properties.*

1. *The probabilistic algorithm $(\kappa, \tau) \xleftarrow{\$} \mathsf{PHF.TrapGen}(1^k, g, h)$ takes as input group elements $g, h \in \mathbb{G}$, and produces a hash function key κ together with trapdoor information τ.*

2. *For all generators $g, h \in \mathbb{G}$, the keys κ, κ', where $\kappa \xleftarrow{\$} \mathsf{PHF.Gen}(1^k)$ and $\kappa' \xleftarrow{\$} \mathsf{PHF.TrapGen}(1^k, g, h)$, are statistically γ-close.*

3. *On input $X \in \{0, 1\}^l$ and trapdoor information τ, the deterministic trapdoor evaluation algorithm $(a_X, b_X) \leftarrow \mathsf{PHF.TrapEval}(\tau, X)$ produces $a_X, b_X \in \mathbb{Z}$ so that for all $X \in \{0, 1\}^l$,*

$$\mathsf{PHF.Eval}(\kappa, X) = g^{a_X} h^{b_X}$$

4. *For all $g, h \in \mathbb{G}$, all κ generated by $\kappa \xleftarrow{\$} \mathsf{PHF.TrapGen}(1^k, g, h)$, and all $X_1, \ldots, X_m \in \{0, 1\}^l$ and $Z_1, \ldots, Z_n \in \{0, 1\}^l$ such that $X_i \neq Z_j$ for all i, j, we have*

$$\Pr[a_{X_1} = \cdots = a_{X_m} = 0 \text{ and } a_{Z_1}, \ldots, a_{Z_n} \neq 0] \geq \delta,$$

where $(a_{X_i}, b_{X_i}) = \mathsf{PHF.TrapEval}(\tau, X_i)$, $(a_{Z_j}, b_{Z_j}) = \mathsf{PHF.TrapEval}(\tau, Z_j)$, and the probability is taken over the trapdoor τ produced along with κ.

We also say that H is (m, n)-programmable for short, if γ is negligible and δ is noticeable. If H is $(1, q)$-programmable for every polynomial $q = q(k)$, then we say that H is $(1, \mathsf{poly})$-programmable.

In settings in which the group order is hidden, we will use a refinement of the PHF definition:

Definition 2. *A group hash function* $H = (\mathsf{RPHF.Gen}, \mathsf{RPHF.Eval})$ *is evasively* (m, n, γ, δ)-*programmable, if it is* (m, n, γ, δ)-*programmable as in Definition 1, but with the strengthened requirement*

4'. *For all prime numbers* e *with* $2^l < e \leq |\mathbb{G}|$, *all* $g, h \in \mathbb{G}$, *and all* κ *generated by* $\kappa \xleftarrow{\$} \mathsf{PHF.TrapGen}(1^k, g, h)$, *and all* $X_1, \ldots, X_m \in \{0, 1\}^l$ *and* $Z_1, \ldots, Z_n \in \{0, 1\}^l$ *such that* $X_i \neq Z_j$ *for all* i, j, *we have*

$$\Pr[a_{X_1} = \cdots = a_{X_m} = 0 \text{ and } \gcd(a_{Z_1}, e) = \cdots = \gcd(a_{Z_n}, e) = 1] \geq \delta.$$

Here a_{X_i} *and* a_{Z_j} *denote the output of the trapdoor evaluation algorithm* $(a_{X_i}, b_{X_i}) = \mathsf{PHF.TrapEval}(\tau, X_i)$ *and* $(a_{Z_j}, b_{Z_j}) = \mathsf{PHF.TrapEval}(\tau, Z_j)$, *and the probability is taken over the trapdoor* τ *produced along with* κ.

Hofheinz and Kiltz [36] have also introduced the notion of *randomized* programmable hash functions. A *randomized* group hash function H with input length $l = l(k)$ and randomness space $R = (\mathcal{R}_k)$ consists of two efficient algorithms $\mathsf{RPHF.Gen}$ and $\mathsf{RPHF.Eval}$. Algorithm $\mathsf{RPHF.Gen}$ is probabilistic, and generates a hash key $\kappa \xleftarrow{\$} \mathsf{RPHF.Gen}(1^k)$ for security parameter k. The deterministic algorithm $\mathsf{RPHF.Eval}$ takes randomness $r \in \mathcal{R}_k$ and $X \in \{0, 1\}^l$ as input, and returns a group element $\mathsf{RPHF.Eval}(\kappa, X) \in \mathbb{G}$.

Definition 3. *Let* $H = (\mathsf{RPHF.Gen}, \mathsf{RPHF.Eval})$ *be a randomized group hash function. We say that* H *is* (m, n, γ, δ)-*programmable, if there are efficient algorithms* $\mathsf{RPHF.TrapGen}$, $\mathsf{RPHF.TrapEval}$, *and* $\mathsf{RPHF.TrapRand}$ *such that:*

1. *The probabilistic algorithm* $\mathsf{RPHF.TrapGen}(1^k, g, h)$ *takes as input group elements* $g, h \in \mathbb{G}$, *and produces a key* κ *and trapdoor* τ. *For all generators* $g, h \in \mathbb{G}$, *the keys* $\kappa \xleftarrow{\$} \mathsf{RPHF.Gen}(1^k)$ *and* $\kappa' \xleftarrow{\$} \mathsf{RPHF.TrapGen}(1^k, g, h)$ *are statistically* γ-*close.*

2. *The deterministic trapdoor evaluation algorithm takes as input* $X \in \{0, 1\}^l$ *and* $r \in \mathcal{R}_k$, *and produces two functions* $(a_X(\cdot), b_X(\cdot)) \leftarrow \mathsf{RPHF.TrapEval}(\tau, X, r)$ *such that for all* $X \in \{0, 1\}^l$,

$$\mathsf{RPHF.Eval}(\kappa, X, r) = g^{a_X(r)} h^{b_X(r)}.$$

3. *On input of trapdoor* τ, $X \in \{0, 1\}^l$, *and index* $i \in [m]$, *the* $\mathsf{RPHF.TrapRand}$ *algorithm produces* $r \leftarrow \mathsf{RPHF.TrapRand}(\tau, X, i)$ *with* $r \in \mathcal{R}_k$. *For all* $g, h \in \mathbb{G}$, *all* κ *generated by* $(\kappa, \tau) \xleftarrow{\$} \mathsf{PHF.TrapGen}(1^k, g, h)$, *all* $X_1, \ldots, X_m$, *and* $r_{X_i} = \mathsf{RPHF.TrapRand}(\tau, X_i, i)$, *we require that the* r_{X_i} *are independent and uniformly distributed random variables over* $\mathcal{R}_k$.

4. *For all* $g, h \in \mathbb{G}$ *and all* κ *generated by* $(\kappa, \tau) \xleftarrow{\$} \mathsf{PHF.TrapGen}(1^k, g, h)$, *all* $X_1, \ldots, X_m \in \{0, 1\}^l$ *and* $Z_1, \ldots, Z_n \in \{0, 1\}^l$ *such that* $X_i \neq Z_j$, *and for all* $\tilde{r}_1, \ldots, \tilde{r}_n \in \mathcal{R}_k$ *and* $r_{X_i} \leftarrow \mathsf{RPHF.TrapRand}(\tau, X_i, i)$, *we have*

$$\Pr[a_{X_1}(r_{X_1}) = \cdots = a_{X_m}(r_{X_m}) = 0 \text{ and } a_{Z_1}(\tilde{r}_1), \ldots, a_{Z_n}(\tilde{r}_n) \neq 0] \geq \delta,$$

where the a_{X_i} *and* a_{Z_j} *are the output of the trapdoor evaluation* $(a_{X_i}, b_{X_i}) = \mathsf{RPHF.TrapEval}(\tau, X_i, r_{X_i})$ *and* $(a_{Z_j}, b_{Z_j}) = \mathsf{PHF.TrapEval}(\tau, Z_j, \tilde{r}_j)$, *and the*

Table 2. Overview of our constructions of (randomized/weak) programmable hash functions. Rows with grey background are new constructions from this paper.

Name	Type	Param. (m,n)	Size of κ	Randomness
H_{Wat} [53,36] (§ 3.2)	PHF	$(1, poly)$ and $(2,1)$	$(l+1) \times \|\mathbb{G}\|$	—
H_{cfs} (§ 3.3)	PHF	$(m,1)$	$(16m^2l+1) \times \|\mathbb{G}\|$	—
H_{rand} (§ 3.4)	RPHF	$(m,1)$	$(2m^2+1) \times \|\mathbb{G}\|$	$\{0,1\}^l$
H_{Weak} (§ 3.5)	weak PHF	$(m,1)$	$(m+1) \times \|\mathbb{G}\|$	—

probability is taken over the trapdoor τ produced along with κ. Here X_i may depend on X_j and r_{X_j} for $j < i$, and the $Z_1, \ldots, Z_n$ may depend on all X_i and r_i.
Again we omit γ and δ, if γ is negligible and δ is noticeable. Randomized evasively programmable hash functions *are defined as in Definition 2.*

In the remainder of this Section we propose a number of new PHFs offering different trade-offs. Our results are summarized in Table 2.

3.2 Multi-generator Programmable Hash Function

The programmable hash function described in Definition 4 below was (implicitly) introduced in [53]. An explicit analysis can be found in [36].

Definition 4. *Let $G = (\mathbb{G}_k)$ be a group family, and $l = l(k)$ be a polynomial. Let $H_{Wat} = (\mathsf{PHF.Gen}, \mathsf{PHF.Eval})$ be defined as follows.*
- *$\mathsf{PHF.Gen}(1^k)$ returns $\kappa = (h_0, \ldots, h_l)$, where $h_i \xleftarrow{\$} \mathbb{G}_k$ for $i \in [l]$.*
- *On input $X = (x_1, \ldots, x_l) \in \{0,1\}^l$ and $\kappa = (h_0, \ldots, h_l)$, $\mathsf{PHF.Eval}(\kappa, X)$ returns*

$$\mathsf{PHF.Eval}(\kappa, X) = h_0 \prod_{i=1}^{l} h_i^{x_i}.$$

Theorem 1 (Theorem 3.6 of [36]). *For any fixed polynomial $q = q(k)$ and any group with known order, H_{Wat} is evasively $(1, q, 0, O(1/(q\sqrt{l})))$-programmable and $(2, 1, 0, O(1/l))$-programmable hash function.*

Although evasive programmability was not introduced in [36], it follows from their proof, since the values of a_{Z_j} that occur there are bounded in the sense $|a_{Z_j}| < 2^l$. We remark that Theorem 1 also carries over to groups of unknown order.

3.3 A New Deterministic Programmable Hash Function

Let S, T be sets. We say that S does not cover T, if $T \not\subseteq S$. Let d, m, s be integers, and let $F = (F_i)_{i \in [s]}$ be a family of s subsets of $[d]$. We say that F is m-cover free, if for any set I containing (up to) m indices $I = \{i_1, \ldots, i_m\} \subseteq [s]$, it holds that $F_j \not\subseteq \bigcup_{i \in I} F_i$ for any j which is *not* contained in I. In other words, if $|I| \leq m$, then the union $\bigcup_{i \in I} F_i$ is not covering F_j for all $j \in [s] \setminus I$. We say that F is w-uniform, if $|F_i| = w$ for all $i \in [s]$.

Lemma 4 ([27,45]). *There is a deterministic polynomial-time algorithm that, on input of integers $m, s = 2^l$, returns $d \in \mathbb{N}$ and set family $F = (F_i)_{i \in [s]}$ such that F is m-cover free over $[d]$ and w-uniform, where $d \leq 16m^2 l$ and $w = d/4m$.*

In the following we will associate $X \in \{0,1\}^l$ to a subset F_i, $i \in [s]$, by interpreting X as an integer in the range $[0, 2^l - 1]$, and setting $i = X + 1$. We will write F_X to denote the subset associated to X.

Definition 5. *Let $G = (\mathbb{G}_k)$ be a group family, and $l = l(k)$ and $m = m(k)$ be polynomials. Let $s = 2^l$, $d = 16m^2 l$, and $w = d/4m$. We define a hash function $\mathsf{H_{cfs}} = (\mathsf{PHF.Gen}, \mathsf{PHF.Eval})$ be as follows.*
- *$\mathsf{PHF.Gen}(1^k)$ returns $\kappa = (h_1, \ldots, h_d)$, where $h_i \xleftarrow{\$} \mathbb{G}_k$ for $1 \leq i \leq d$.*
- *Let $F_X \subseteq [d]$ be the subset associated to $X \in [0, 2^l - 1]$. On input X and $\kappa = (h_1, \ldots, h_d)$, $\mathsf{PHF.Eval}(\kappa, X)$ returns*

$$\mathsf{PHF.Eval}(\kappa, X) = \prod_{i \in F_X} h_i.$$

Theorem 2. *Let $\mathbb{G} = \mathbb{G}_k$ be a group of known order p. $\mathsf{H_{cfs}}$ is an evasively $(m, 1, \gamma, \delta)$-programmable hash function with $\gamma = 0$ and $\delta = 1/(16m^2 l)$.*

Proof. Consider the following algorithms.
- $\mathsf{PHF.TrapGen}(1^k, g, h)$ samples d uniformly random integers $b_1, \ldots, b_d \xleftarrow{\$} \mathbb{Z}_p$ and an index $t \xleftarrow{\$} [d]$. Then it sets $h_t = g h^{b_t}$, and $h_i = h^{b_i}$ for all $i \in [1, d]$ with $i \neq t$. $\mathsf{PHF.TrapGen}$ returns (κ, τ) with $\tau = (t, b_1, \ldots, b_d)$ and $\kappa = (h_1, \ldots, h_d)$.
- On input (τ, X), $\mathsf{PHF.TrapEval}$ sets $b_X = \sum_{i \in F_X} b_i$, and $a_X = 1$ if $t \in F_X$, and $a_X = 0$ if $t \notin F_X$, and returns (a_X, b_X).

$\mathsf{PHF.TrapGen}$ outputs a vector of independent and uniformly distributed group elements, thus we have $\gamma = 0$. Fix $X_1, \ldots, X_m, Z \in [0, 2^l - 1]$. Since F is a m-cover free set family, there must be an index t' such that $t' \notin \bigcup_{i=1}^{m} F_{X_i}$, but $t' \in F_Z$. Since t is picked uniformly random among $16m^2 l$ possibilities, we have $t = t'$, and thus $a_{X_i} = 0$ and $a_Z = 1$, with probability $\delta = 1/(16m^2 l)$. Finally, $a_Z = 1$ implies $\gcd(a_Z, e) = 1$ for all primes e, thus $\mathsf{H_{cfs}}$ is evasively programmable.

Theorem 2 can be generalized to groups of hidden order. The proof proceeds exactly like the proof of Theorem 2, except that we have to approximate the group order. E.g., for the group of quadratic residues QR_n, we can sample random exponents $b_i \xleftarrow{\$} \mathbb{Z}_{n^2}$. This way, we can sample nearly uniform ($1/\sqrt{n}$-close) group elements $h_i = h^{b_i}$, which yields the following theorem.

Theorem 3. *Let $\mathbb{G} = \mathsf{QR}_n$ be the group of quadratic residues modulo $n = pq$, where p and q are safe distinct primes. $\mathsf{H_{cfs}}$ is a $(m, 1, \gamma, \delta)$-evasively programmable hash function over $\mathbb{G}$ with $\gamma = d/\sqrt{n}$ and $\delta = 1/(16m^2 l)$.*

3.4 A Randomized Programmable Hash Function

In [38] a randomized $(2,1)$-PHF was described which we now generalize to a randomzied $(m,1)$-PRF, for any $m \geq 1$.

Definition 6. *Let $G = (\mathbb{G}_k)$ be a group family, and $m = m(k)$ be a polynomial. In the following, let $[X]_{2^l} \in \mathbb{Z}$ denote a canonical interpretation of a field element $X \in \mathbb{F}_{2^l}$ as an integer between 0 and $2^l - 1$. We assume that X and $[X]_{2^l}$ are efficiently computable from one another. Let $\mathsf{H}_{\mathsf{rand}} = (\mathsf{PHF.Gen}, \mathsf{PHF.Eval})$ be defined as follows.*

- *$\mathsf{RPHF.Gen}(1^k)$ returns a uniformly sampled $\kappa = (h_0, (h_{i,j})_{(i,j) \in [2m] \times [m]}) \in \mathbb{G}^{2m^2+1}$.*
- *$\mathsf{RPHF.Eval}(\kappa, X; r)$ parses $X, r \in \mathbb{F}_{2^l}$, and computes and returns*

$$\mathsf{RPHF.Eval}_{\kappa}(X; r) = h_0 \prod_{i,j=1}^{m} h_{i,j}^{([iX+r]_{2^l})^j}.$$

Theorem 4. *For any group $\mathbb{G}$ of known order, $\mathsf{H}_{\mathsf{rand}}$ is evasively $(m, 1, 0, 1/2)$-programmable. For the group $\mathbb{G} = \mathsf{QR}_N$ of quadratic residues modulo $N = pq$ for safe distinct primes p and q, the function $\mathsf{H}_{\mathsf{rand}}$ is evasively $(m, 1, (2m^2 + 1)/\sqrt{N}, 1/2)$-programmable.*

The proof is given in the full version of this paper [34].

3.5 A Weak Programmable Hash Function

Essentially, a *weak* programmable hash function is a programmable hash function according to Definition 1, except that the trapdoor generation algorithm receives a list $X_1, \ldots, X_m \in \{0,1\}^l$ as additional input. On the one hand this allows us to construct significantly more efficient deterministic programmable hash functions, while on the other hand our generic signatures schemes are only weakly secure when instantiated with weak programmable hash functions. Fully secure signature schemes can be obtained by applying a generic conversion from weak to full security, for instance using chameleon hashes [44] which can be constructed based on standard assumptions like discrete logarithms [44], RSA [2,21,39], or factoring [44].

Definition 7. *A group hash function is a weak (m, n, γ, δ)-programmable hash function, if there is a (probabilistic) algorithm $\mathsf{PHF.TrapGen}$ and a (deterministic) algorithm $\mathsf{PHF.TrapEval}$ such that:*

1. *$(\kappa, \tau) \xleftarrow{\$} \mathsf{PHF.TrapGen}(1^k, g, h, X_1, \ldots, X_m)$ takes as input group elements $g, h \in \mathbb{G}$ and $X_1, \ldots, X_m \in \{0,1\}^l$, and produces a hash function key κ together with trapdoor information τ.*

2.-4. Like in Definition 1.

As before, we may omit γ and δ, if γ is negligible and δ is noticeable. Weak evasively programmable hash functions are defined as in Definition 2.

Interestingly, there is a very simple way to construct a randomized programmable hash function according to Definition 3 from any weak programmable hash function. Let us now describe our instantiation of a weak (evasively) programmable hash function. This PHF already appeared implicitly in [19,49] and [9] for $m = 1$.

Definition 8. *Let $G = (\mathbb{G}_k)$ be a group family, and $l = l(k)$ and $m = m(k)$ be polynomials. Let $H_{\mathsf{Weak}} = (\mathsf{PHF.Gen}, \mathsf{PHF.Eval})$ be defined as follows.*
  - *$\mathsf{PHF.Gen}(1^k)$ returns $\kappa = (h_0, \ldots, h_m)$, where $h_i \xleftarrow{\$} \mathbb{G}_k$ for $i \in \{0, \ldots, m\}$.*
  - *On input $X \in \{0,1\}^l$ and $\kappa = (h_0, \ldots, h_m)$, $\mathsf{PHF.Eval}(\kappa, X)$ returns*

$$\mathsf{PHF.Eval}(\kappa, X) = \prod_{i=0}^{m} h_i^{(X^i)}.$$

Here we interpret the l-bit strings X_i, $i \in [m]$, as integers in the canonical way.

Theorem 5. *Let $\mathbb{G} = \mathbb{G}_k$ be a group of known order p. H_{Weak} is a weak evasively $(m, 1, \gamma, \delta)$-programmable hash function with $\gamma = 0$ and $\delta = 1$.*

Again we can generalize Theorem 5 to groups of hidden order. The proof proceeds exactly like the proof of Theorem 5, except that we have to approximate the group order. For the group of quadratic residues QR_n, we can sample the random exponents b_i from $\mathbb{Z}_{n^2}$ for $i \in [0, m]$, which yields the following theorem.

Theorem 6. *Let $\mathbb{G} = \mathsf{QR}_N$ be the group of quadratic residues modulo $N = pq$, where p and q are safe distinct primes. H_{Weak} is a $(m, 1, \gamma, \delta)$-programmable hash function over $\mathbb{G}$ with $\gamma = (m + 1)/\sqrt{N}$ and $\delta = 0$.*

4   Signatures from the RSA Problem

4.1   Construction

Let $l = l(k)$ and $\lambda = \lambda(k)$ be polynomials. Let $H = (\mathsf{PHF.Gen}, \mathsf{PHF.Eval})$ be group hash functions over $\mathbb{G} = \mathsf{QR}_N$ with input length l. We define the signature scheme $\mathsf{Sig}_{\mathsf{RSA}}[H] = (\mathsf{Gen}, \mathsf{Sign}, \mathsf{Vfy})$ as follows.

$\mathsf{Gen}(1^k)$: The key generation algorithm picks two large safe $k/2$-bit primes p and q, and sets $N = pq$. Then it generates a group hash function key $\kappa \xleftarrow{\$} \mathsf{PHF.Gen}(1^k)$ for the group QR_N. Finally it chooses a random key K for the pseudorandom function $\mathsf{PRF} : \{0,1\}^* \to \{0,1\}^r$ and picks $c \xleftarrow{\$} \{0,1\}^r$, where $r = \lceil \log N \rceil$. These values define a function F as

$$\mathsf{F}(z) = \mathsf{PRF}_K(\mu||z) \oplus c,$$

where μ, called the *resolving index* of z, denotes the smallest positive integer such that $\mathsf{PRF}_K(\mu||z) \oplus c$ is an odd prime. Here $\oplus$ denotes the bit-wise XOR operation, and we interpret the r-bit string returned by F as an integer in the obvious way. (The definition of F is the same as in [40]. It is possible to replace the PRF with an $2k^2$-wise independent hash function [16].) The public key is $pk = (n, \kappa, K, c)$, the secret key is $sk = (pk, p, q)$.

In the following we will write $\mathsf{H}(M)$ shorthand for $\mathsf{PHF.Eval}(\kappa, M)$, and define $\mathsf{P} : \{0,1\}^\lambda \to \mathbb{N}$ as $\mathsf{P}(s) = \prod_{i=1}^\lambda \mathsf{F}(s_{|i})$, where $s_{|i}$ is the i-th prefix of s, i.e., the bit string consisting of the first i bits of s. We also define $s_{|0} = \emptyset$, where $\emptyset$ is the empty string, for technical reasons.

$\mathsf{Sign}(sk, M)$: On input of secret key sk and message $M \in \{0,1\}^l$, the signing algorithm picks $s \xleftarrow{\$} \{0,1\}^\lambda$ uniformly random and computes

$$\sigma = \mathsf{H}(M)^{1/\mathsf{P}(s)} \bmod N,$$

where the inverse of $\mathsf{P}(s)$ is computed modulo the order $\phi(n) = (p-1)(q-1)$ of the multiplicative group $\mathbb{Z}_N^*$. The signature is $(\sigma, s) \in \mathbb{Z}_N \times \{0,1\}^\lambda$.

$\mathsf{Vfy}(pk, M, (\sigma, s))$: On input of pk, message M, and signature (σ, s), return accept if

$$\mathsf{H}(M) = \sigma^{\mathsf{P}(s)} \bmod N.$$

Otherwise return reject.

Correctness. If $\sigma = \mathsf{H}(M)^{1/\mathsf{P}(s)}$, then we have $\sigma^{\mathsf{P}(s)} = \mathsf{H}(M)^{\mathsf{P}(s)/\mathsf{P}(s)} = \mathsf{H}(M)$.

Theorem 7. *Let* PRF *be a* (ϵ'', t'')-secure pseudo-random function and H be a $(m, 1, \gamma, \delta)$-evasively programmable hash function. Suppose there exists a (t, q, ϵ)-forger $\mathcal{F}$ breaking the existential forgery under adaptive chosen message attacks of $\mathsf{Sig_{RSA}}[\mathsf{H}]$. Then there exists an adversary that (t', ϵ')-breaks the RSA assumption with $t' \approx t$ and ϵ is bounded by*

$$(q+1)\lambda \left(\frac{4r^2}{\delta} \left(\epsilon' + \frac{r}{l \cdot 2^{r-l-1}} \right) + 3\epsilon'' + \frac{r(q+1)^2\lambda^2 + 2r + 1}{2^r} + \gamma + \frac{1}{2^{r-l}} \right)$$
$$+ \frac{q^{m+1}}{2^{m\lambda}}$$

We only give a brief proof outline here, and refer to the full version [34] for details. As customary in proofs for similar signature schemes (e.g., [23,28,36]), we distinguish between *Type I* and *Type II* forgers. A Type I forger forges a signature of the form (M^*, σ^*, s^*) with $s^* = s_i$ for some $i \in [q]$. (That is, a Type I forger reuses some s_i from a signature query.) A *Type II* forger returns a signature with a fresh s^*.

It will be easiest to first describe how to treat a Type II forger $\mathcal{F}$. Recall that we need to put up a simulation that is able to generate q signatures $(M_i, \sigma_i, s_i)_{i \in [q]}$ for adversarially chosen messages M_i. To do this, we choose all

s_i in advance. We then prepare the PHF H using PHF.TrapGen, but relative to generators g and h for which we know $\mathsf{P}(s_i)$-th roots. (That is, we set $g := \hat{g}^E$ and $h = \hat{h}^E$ for $E := \prod_i \mathsf{P}(s_i)$.) This allows to generate signatures for $\mathcal{F}$; also, by the security of the PHF H, this change goes unnoticed by $\mathcal{F}$. However, each time $\mathcal{F}$ outputs a new signature, it essentially outputs a fresh root $g^{1/\mathsf{P}(s^*)}$ of g, from which we can derive a $\mathsf{P}(s^*)$-th root of $\hat{g}$. To construct an RSA adversary from this experiment, we have to embed an auxiliary given exponent e into the definition of P, such that $\hat{g}^{1/\mathsf{P}(s^*)}$ allows to derive $\hat{g}^{1/e}$. This can be done along the lines of the proof of the Hohenberger-Waters scheme [40]. Concretely, for *initially given* values s_i and e, we can set up P such that (a) e does not divide any $\mathsf{P}(s_i)$, but (b) for any other fixed s^*, the probability that e divides $\mathsf{P}(s^*)$ is significant. Note that in our scheme, the s_i are chosen by the signer, and thus our simulation can select them in advance. In contrast to that, the HW scheme uses the signed messages M_i as arguments to P, and thus their argument achieves only a weaker form of security in which the forger has to commit to all signature queries beforehand.

Now the proof for Type I forgers proceeds similarly, but with the additional complication that we have to prepare one or more signatures of the form $\mathsf{H}(M_i)^{1/\mathsf{P}(s_i)}$ for the same $s_i = s^*$ that $\mathcal{F}$ eventually uses in his forgery. We resolve this complication by relying on the PHF properties of H. Namely, we first choose all s_i and guess i (i.e., the index of the s_i with $s_i = s^*$). We then prepare H with generators g, h such that we know all $\mathsf{P}(s_j)$th roots of h (for all j), and all $\mathsf{P}(s_j)$th roots of g *for all* $s_j \neq s_i$. Our hope is that whenever $\mathcal{F}$ asks for the signature of some M_j with $s_j = s_i$, we have $\mathsf{H}(M_i) \in \langle h \rangle$, so we can compute $\mathsf{H}(M_j)^{1/\mathsf{P}(s_j)}$. At the same time, we hope that $\mathsf{H}(M^*) \notin \langle h \rangle$ has a nontrivial g-factor, so we can build an RSA adversary as for Type II forgers. The PHF property of H guarantees a significant probability that this works out, provided that there are no more than m indices j with $s_j = s_i$ (i.e., provided that there are no $(m + 1)$-collisions). However, using a birthday bound, we can reasonably upper bound the probability of $(m + 1)$-collisions.

In the full version [34] we also give a variant of our scheme which is slightly more efficient but only offers weak security. A weakly secure signature scheme can be updated to a fully secure one by using a (randomized) Chameleon Hash Function.

Efficiency. Given $\mathsf{P}(s)$ and $\phi(N)$, computing $\sigma = \mathsf{H}(M)^{1/\mathsf{P}(s)}$ can also be carried out by one single exponentiation. Since one single evaluation of $\mathsf{P}(\cdot)$ has to perform (expected) λr many primality tests (for r-bit primes), the dominant part of signing and verification is to compute $\mathsf{P}(s)$, for $s \in \{0, 1\}^\lambda$. Theorem 7 tells us that if H is a $(m, 1)$-PHF we can set $\lambda = \log q + k/m$, see the full version [34] for more details.

Hohenberger and Waters [40] proposed several ways to improve the efficiency of their RSA-based signature scheme. These improvements apply to our RSA-based schemes as well. We refer to the full version [34] for details.

References

1. Agrawal, S., Boneh, D., Boyen, X.: Efficient Lattice (H)IBE in the Standard Model. In: Gilbert, H. (ed.) EUROCRYPT 2010. LNCS, vol. 6110, pp. 553–572. Springer, Heidelberg (2010)
2. Ateniese, G., de Medeiros, B.: Identity-Based Chameleon Hash and Applications. In: Juels, A. (ed.) FC 2004. LNCS, vol. 3110, pp. 164–180. Springer, Heidelberg (2004)
3. Barreto, P.S.L.M., Naehrig, M.: Pairing-Friendly Elliptic Curves of Prime Order. In: Preneel, B., Tavares, S. (eds.) SAC 2005. LNCS, vol. 3897, pp. 319–331. Springer, Heidelberg (2006)
4. Bellare, M., Goldreich, O., Goldwasser, S.: Incremental Cryptography: The Case of Hashing and Signing. In: Desmedt, Y.G. (ed.) CRYPTO 1994. LNCS, vol. 839, pp. 216–233. Springer, Heidelberg (1994)
5. Bellare, M., Rogaway, P.: Random oracles are practical: A paradigm for designing efficient protocols. In: Ashby, V. (ed.) ACM CCS 1993: 1st Conference on Computer and Communications Security, pp. 62–73. ACM Press (November 1993)
6. Bellare, M., Rogaway, P.: The Exact Security of Digital Signatures - How to Sign with RSA and Rabin. In: Maurer, U.M. (ed.) EUROCRYPT 1996. LNCS, vol. 1070, pp. 399–416. Springer, Heidelberg (1996)
7. Bentahar, K., Farshim, P., Malone-Lee, J., Smart, N.P.: Generic constructions of identity-based and certificateless KEMs. Journal of Cryptology 21(2), 178–199 (2008)
8. Blazy, O., Fuchsbauer, G., Pointcheval, D., Vergnaud, D.: Signatures on Randomizable Ciphertexts. In: Catalano, D., Fazio, N., Gennaro, R., Nicolosi, A. (eds.) PKC 2011. LNCS, vol. 6571, pp. 403–422. Springer, Heidelberg (2011)
9. Boneh, D., Boyen, X.: Efficient Selective-ID Secure Identity-Based Encryption without Random Oracles. In: Cachin, C., Camenisch, J. (eds.) EUROCRYPT 2004. LNCS, vol. 3027, pp. 223–238. Springer, Heidelberg (2004)
10. Boneh, D., Boyen, X.: Secure Identity Based Encryption without Random Oracles. In: Franklin, M. (ed.) CRYPTO 2004. LNCS, vol. 3152, pp. 443–459. Springer, Heidelberg (2004)
11. Boneh, D., Boyen, X.: Short Signatures without Random Oracles. In: Cachin, C., Camenisch, J. (eds.) EUROCRYPT 2004. LNCS, vol. 3027, pp. 56–73. Springer, Heidelberg (2004)
12. Boneh, D., Boyen, X.: Short signatures without random oracles and the SDH assumption in bilinear groups. Journal of Cryptology 21(2), 149–177 (2008)
13. Boneh, D., Lynn, B., Shacham, H.: Short Signatures from the Weil Pairing. In: Boyd, C. (ed.) ASIACRYPT 2001. LNCS, vol. 2248, pp. 514–532. Springer, Heidelberg (2001)
14. Boyen, X.: Lattice Mixing and Vanishing Trapdoors: A Framework for Fully Secure Short Signatures and More. In: Nguyen, P.Q., Pointcheval, D. (eds.) PKC 2010. LNCS, vol. 6056, pp. 499–517. Springer, Heidelberg (2010)
15. Boyen, X., Mei, Q., Waters, B.: Direct chosen ciphertext security from identity-based techniques. In: Atluri, V., Meadows, C., Juels, A. (eds.) ACM CCS 2005: 12th Conference on Computer and Communications Security, pp. 320–329. ACM Press (November 2005)

16. Cachin, C., Micali, S., Stadler, M.: Computationally Private Information Retrieval with Polylogarithmic Communication. In: Stern, J. (ed.) EUROCRYPT 1999. LNCS, vol. 1592, pp. 402–414. Springer, Heidelberg (1999)
17. Canetti, R., Goldreich, O., Halevi, S.: The random oracle methodology, revisited (preliminary version). In: 30th Annual ACM Symposium on Theory of Computing, pp. 209–218. ACM Press (May 1998)
18. Cash, D., Hofheinz, D., Kiltz, E., Peikert, C.: Bonsai Trees, or How to Delegate a Lattice Basis. In: Gilbert, H. (ed.) EUROCRYPT 2010. LNCS, vol. 6110, pp. 523–552. Springer, Heidelberg (2010)
19. Chatterjee, S., Sarkar, P.: Generalization of the Selective-ID Security Model for HIBE Protocols. In: Yung, M., Dodis, Y., Kiayias, A., Malkin, T. (eds.) PKC 2006. LNCS, vol. 3958, pp. 241–256. Springer, Heidelberg (2006)
20. Chaum, D., Evertse, J.-H., van de Graaf, J.: An Improved Protocol for Demonstrating Possession of Discrete Logarithms and Some Generalizations. In: Price, W.L., Chaum, D. (eds.) EUROCRYPT 1987. LNCS, vol. 304, pp. 127–141. Springer, Heidelberg (1988)
21. Chevallier-Mames, B., Joye, M.: A Practical and Tightly Secure Signature Scheme without Hash Function. In: Abe, M. (ed.) CT-RSA 2007. LNCS, vol. 4377, pp. 339–356. Springer, Heidelberg (2006)
22. Cramer, R., Hanaoka, G., Hofheinz, D., Imai, H., Kiltz, E., Pass, R., Shelat, A., Vaikuntanathan, V.: Bounded CCA2-Secure Encryption. In: Kurosawa, K. (ed.) ASIACRYPT 2007. LNCS, vol. 4833, pp. 502–518. Springer, Heidelberg (2007)
23. Cramer, R., Shoup, V.: Signature schemes based on the strong RSA assumption. In: ACM CCS 1999: 6th Conference on Computer and Communications Security, pp. 46–51. ACM Press (November 1999)
24. Dodis, Y., Haitner, I., Tentes, A.: On the (in)security of rsa signatures. Cryptology ePrint Archive, Report 2011/087 (2011), http://eprint.iacr.org/
25. Dodis, Y., Katz, J., Xu, S., Yung, M.: Key-Insulated Public Key Cryptosystems. In: Knudsen, L.R. (ed.) EUROCRYPT 2002. LNCS, vol. 2332, pp. 65–82. Springer, Heidelberg (2002)
26. Dodis, Y., Oliveira, R., Pietrzak, K.: On the Generic Insecurity of the Full Domain Hash. In: Shoup, V. (ed.) CRYPTO 2005. LNCS, vol. 3621, pp. 449–466. Springer, Heidelberg (2005)
27. Erdös, P., Frankel, P., Furedi, Z.: Families of finite sets in which no set is covered by the union of r others. Israeli Journal of Mathematics 51, 79–89 (1985)
28. Fischlin, M.: The Cramer-Shoup Strong-RSASignature Scheme Revisited. In: Desmedt, Y.G. (ed.) PKC 2003. LNCS, vol. 2567, pp. 116–129. Springer, Heidelberg (2002)
29. Gennaro, R., Halevi, S., Rabin, T.: Secure Hash-and-Sign Signatures without the Random Oracle. In: Stern, J. (ed.) EUROCRYPT 1999. LNCS, vol. 1592, pp. 123–139. Springer, Heidelberg (1999)
30. Gentry, C., Peikert, C., Vaikuntanathan, V.: Trapdoors for hard lattices and new cryptographic constructions. In: Ladner, R.E., Dwork, C. (eds.) 40th Annual ACM Symposium on Theory of Computing, pp. 197–206. ACM Press (May 2008)
31. Goldwasser, S., Micali, S., Rivest, R.L.: A digital signature scheme secure against adaptive chosen-message attacks. SIAM Journal on Computing 17(2), 281–308 (1988)
32. Hardy, G.H., Wright, E.M.: An Introduction to the Theory of Numbers, 5th edn. Oxford University Press (1979)

33. Heng, S.-H., Kurosawa, K.: k-Resilient Identity-Based Encryption in the Standard Model. In: Okamoto, T. (ed.) CT-RSA 2004. LNCS, vol. 2964, pp. 67–80. Springer, Heidelberg (2004)

34. Hofheinz, D., Jager, T., Kiltz, E.: Short signatures from weaker assumptions. Cryptology ePrint Archive, Report 2011/296 (2011), http://eprint.iacr.org/

35. Hofheinz, D., Kiltz, E.: Secure Hybrid Encryption from Weakened Key Encapsulation. In: Menezes, A. (ed.) CRYPTO 2007. LNCS, vol. 4622, pp. 553–571. Springer, Heidelberg (2007)

36. Hofheinz, D., Kiltz, E.: Programmable Hash Functions and their Applications. In: Wagner, D. (ed.) CRYPTO 2008. LNCS, vol. 5157, pp. 21–38. Springer, Heidelberg (2008)

37. Hofheinz, D., Kiltz, E.: Practical Chosen Ciphertext Secure Encryption from Factoring. In: Joux, A. (ed.) EUROCRYPT 2009. LNCS, vol. 5479, pp. 313–332. Springer, Heidelberg (2009)

38. Hofheinz, D., Kiltz, E.: Programmable hash functions and their applications. Journal of Cryptology, 1–44 (2011)

39. Hohenberger, S., Waters, B.: Realizing Hash-and-Sign Signatures under Standard Assumptions. In: Joux, A. (ed.) EUROCRYPT 2009. LNCS, vol. 5479, pp. 333–350. Springer, Heidelberg (2009)

40. Hohenberger, S., Waters, B.: Short and Stateless Signatures from the RSA Assumption. In: Halevi, S. (ed.) CRYPTO 2009. LNCS, vol. 5677, pp. 654–670. Springer, Heidelberg (2009)

41. Kiltz, E.: Chosen-Ciphertext Security from Tag-Based Encryption. In: Halevi, S., Rabin, T. (eds.) TCC 2006. LNCS, vol. 3876, pp. 581–600. Springer, Heidelberg (2006)

42. Kiltz, E., Galindo, D.: Direct chosen-ciphertext secure identity-based key encapsulation without random oracles. Theor. Comput. Sci. 410(47-49), 5093–5111 (2009)

43. Kiltz, E., Pietrzak, K., Cash, D., Jain, A., Venturi, D.: Efficient Authentication from Hard Learning Problems. In: Paterson, K.G. (ed.) EUROCRYPT 2011. LNCS, vol. 6632, pp. 7–26. Springer, Heidelberg (2011)

44. Krawczyk, H., Rabin, T.: Chameleon signatures. In: ISOC Network and Distributed System Security Symposium – NDSS 2000. The Internet Society (February 2000)

45. Kumar, R., Rajagopalan, S., Sahai, A.: Coding Constructions for Blacklisting Problems without Computational Assumptions. In: Wiener, M.J. (ed.) CRYPTO 1999. LNCS, vol. 1666, pp. 609–623. Springer, Heidelberg (1999)

46. Lyubashevsky, V., Micciancio, D.: Asymptotically Efficient Lattice-Based Digital Signatures. In: Canetti, R. (ed.) TCC 2008. LNCS, vol. 4948, pp. 37–54. Springer, Heidelberg (2008)

47. Okamoto, T.: Efficient Blind and Partially Blind Signatures without Random Oracles. In: Halevi, S., Rabin, T. (eds.) TCC 2006. LNCS, vol. 3876, pp. 80–99. Springer, Heidelberg (2006)

48. Rompel, J.: One-way functions are necessary and sufficient for secure signatures. In: 22nd Annual ACM Symposium on Theory of Computing, pp. 387–394. ACM Press (May 1990)

49. Sahai, A., Waters, B.R.: Fuzzy Identity-Based Encryption. In: Cramer, R. (ed.) EUROCRYPT 2005. LNCS, vol. 3494, pp. 457–473. Springer, Heidelberg (2005)

50. Schäge, S.: Tight Proofs for Signature Schemes without Random Oracles. In: Paterson, K.G. (ed.) EUROCRYPT 2011. LNCS, vol. 6632, pp. 189–206. Springer, Heidelberg (2011)

51. Schäge, S., Schwenk, J.: A CDH-Based Ring Signature Scheme with Short Signatures and Public Keys. In: Sion, R. (ed.) FC 2010. LNCS, vol. 6052, pp. 129–142. Springer, Heidelberg (2010)
52. Shamir, A.: On the generation of cryptographically strong pseudorandom sequences. ACM Trans. Comput. Syst. 1(1), 38–44 (1983)
53. Waters, B.R.: Efficient Identity-Based Encryption Without Random Oracles. In: Cramer, R. (ed.) EUROCRYPT 2005. LNCS, vol. 3494, pp. 114–127. Springer, Heidelberg (2005)

Practical Key-Recovery for All Possible Parameters of SFLASH

Charles Bouillaguet[1], Pierre-Alain Fouque[1], and Gilles Macario-Rat[2]

[1] École normale supérieure,
45 rue d'Ulm, 75005 Paris, France
{charles.bouillaguet,pierre-alain.fouque}@ens.fr
[2] Orange Labs
38–40, rue du Général Leclerc, 92794 Issy les Moulineaux Cedex 9, France
Gilles.Macariorat@orange-ftgroup.com

Abstract. In this paper we present a new practical key-recovery attack on the SFLASH signature scheme. SFLASH is a derivative of the older C^* encryption and signature scheme that was broken in 1995 by Patarin. In SFLASH, the public key is truncated, and this simple countermeasure prevents Patarin's attack. The scheme is well-known for having been considered secure and selected in 2004 by the NESSIE project of the European Union to be standardized.

However, SFLASH was practically broken in 2007 by Dubois, Fouque, Stern and Shamir. Their attack breaks the original (and most relevant) parameters, but does not apply when more than half of the public key is truncated. It is therefore possible to choose parameters such that SFLASH is not broken by the existing attacks, although it is less efficient.

We show a key-recovery attack that breaks the full range of parameters in practice, as soon as the information-theoretically required amount of information is available from the public-key. The attack uses new cryptanalytic tools, most notably pencils of matrices and quadratic forms.

1 Introduction

Multivariate cryptography is a brand that encompasses the (mostly public-key) cryptographic schemes whose security relies on the difficulty of solving systems of multivariate polynomial equations over a finite field. Even when restricted to quadratic polynomials, and to the smallest possible finite field, the problem is well-known to be NP-complete, not to mention very difficult in practice. In that restricted setting, the problem is often called Multivariate Quadratic (MQ for short). Because this mathematical problem is well-known and has a simple statement, it was very tempting to design cryptographic schemes relying on its hardness. This has the added benefit that no quantum algorithm is known to break MQ faster than in the classical world, unlike most number-theoretic hard problem that would fall to Shor's algorithm [16].

Multivariate polynomials have been used in cryptography as early as in 1984, mostly with the purpose of designing RSA variants with faster decryption [11,12,5].

D.H. Lee and X. Wang (Eds.): ASIACRYPT 2011, LNCS 7073, pp. 667–685, 2011.

At about the same time, Matsumoto and Imai designed the first public-key scheme explicitly based on the hardness of MQ. In fact, they had several proposal, but only a single one (their "Scheme A") made it to the general crypto community, and was presented at Eurocrypt'88 [10] under the name C^*. It is very similar to RSA, as its only non-linear component is a power function over a finite field. However, unlike RSA this power function is an easy-to-invert bijection, therefore in C^* it is composed with two secret invertible linear maps that destroy its algebraic structure. We therefore see C^* as an attempt to obfuscate a power function in $\mathbb{F}_{q^n}$ by presenting it as a collection of n quadratic polynomials in n variables over $\mathbb{F}_q$.

Several years later, Patarin found a devastating attack against C^*, allowing to decrypt and to forge signatures in a few seconds [13]. He showed that there always are bilinear relations between the ciphertext and the plaintext, which can be easily discovered by the adversary. This allows for an efficient attack by substituting the ciphertext into the bilinear relations, which results in a system of linear equations whose solution is the plaintext.

The SFLASH signature scheme [14] is a derivative of the original C^* that was proposed in 2001 by Courtois, Goubin and Patarin. It is famous for having been selected in 2003 by the NESSIE European project to be proposed to the standardization bodies.

The idea behind SFLASH is to take the original C^* but to throw away a part of the output. The resulting trapdoor one-way function can no longer be used for encryption, but it can still be used for signatures. This is achieved by removing a part of the public key, which is the obfuscated description of the power function. The idea of removing some of the public polynomials has been originally suggested by Shamir [15], and was called the "Minus transform". The original C^* with the minus transform is thus often called C^{*-}. This countermeasure is very effective since it avoids the reconstruction of the bilinear relations and makes it much harder to compute Gröbner basis of the public key.

SFLASH has in turn been very badly broken in 2007 when Dubois, Fouque, Stern and Shamir found a practical forgery attack [4,3], and further broken in 2008 when Fouque, Macario-Rat and Stern found a practical key-recovery attack [6]. Both attacks are very practical, defeating the actual SFLASH parameters in minutes. They are essentially polynomial in the security parameter(s), so that there is no hope that increasing them may make the scheme simultaneously secure and usable.

However, both attacks only apply *as long as the number of removed polynomials is less than half of the total number*. There are therefore *unbroken* ranges of parameters, even though they are less practical than the original (defeated) proposal. For instance, let us consider the parameters $q = 128$ and $n = 257$. The original C^* public key would be made of 257 polynomial in 257 variables over $\mathbb{F}_{128}$. If we throw away 75% of the public key, we obtain a C^{*-} public-key with 64 multivariate quadratic polynomials in 257 variables, and the existing attacks do no apply. The signatures are 1799-bit long, and the public-key is 1.8Mbyte long. Forging a signature by exhaustive search requires 2^{448} trials, and computing a Gröbner basis should require even more arithmetic operations.

Our Contribution. We show that SFLASH/C^{*-} can be broken regardless of the fraction of the public that was thrown away, thus improving on the previous attacks. We present a practical key-only attack that recovers the secret-key and applies as soon as *three* polynomials from the public key are available. This happens to be the information-theoretic minimum quantity of data required to uniquely characterize the set of possible secret keys. The attack has been implemented and tested. It runs very efficiently, and breaks in practice all the meaningful ranges of parameters. For instance, the particular parameters mentioned in the previous paragraph can be broken in about 10 hours using a single computer.

SFLASH had already been thrown out of the league of possible alternatives to RSA of discrete-logarithm based schemes by the previous attacks. The contribution of this work is not only to further break SFLASH, but also to introduce new cryptanalytic techniques. To achieve our results, we make use of mathematical tools that were not previously used in multivariate cryptanalysis, such as pencils of matrices or quadratic forms, adjugate matrices, simultaneous diagonalization of quadratic forms, kernels of quadratic forms, etc. We expect that some of these tools might apply further to other schemes, in particular those sharing some features with SFLASH, notably HFE.

1.1   Organization of the Paper

In section 2, we present some mathematical background. Then, in section 3, we describe the C^* and SFLASH signature schemes. In section 4, we investigate in great detail the mathematical properties of C^* and find exploitable relations between the secret and public keys. Finally, we expose our key-recovery attack in section 5, and give experimental results.

2   Mathematical Background

Finite Fields. Let $\mathbb{K}$ the finite field with q elements, where q is a power of two, and $\mathbb{F}$ an extension of $\mathbb{K}$ of degree n. Recall that $\mathbb{F}$ is isomorphic to $\mathbb{K}^n$, so that we often identify the two spaces. The *trace* on $\mathbb{F}$ over $\mathbb{K}$ is the $\mathbb{K}$-linear map defined by $\mathrm{Tr}_{\mathbb{F}/\mathbb{K}}(x) = x + x^q + \ldots + x^{q^{n-1}}$. The *norm* on $\mathbb{F}$ over $\mathbb{K}$ is defined by $\mathrm{N}_{\mathbb{F}/\mathbb{K}}(x) = x \cdot x^q \ldots \cdot x^{q^{n-1}}$. Both $\mathrm{Tr}_{\mathbb{F}/\mathbb{K}}$ and $\mathrm{N}_{\mathbb{F}/\mathbb{K}}$ are functions from $\mathbb{F}$ to $\mathbb{K}$, and we simply denote them Tr and N since there is no confusion. The map $x \mapsto x^q$ is called the Frobenius map, and it is a field automorphism.

Lemma 1. *For any $\mathbb{K}$-linear mapping L on $\mathbb{F}$ over $\mathbb{K}$, there exists an element λ of $\mathbb{F}$ such that, for all x in $\mathbb{F}$, $L(x) = \mathrm{Tr}(\lambda x)$. Moreover, if $\mathrm{Tr}(\lambda x) = 0$ for all $x \in \mathbb{F}$, then $\lambda = 0$.*

Quadratic forms. A quadratic form over $\mathbb{K}$ is a degree 2 homogeneous polynomial:

$$Q(x_1, \ldots, x_n) = \sum_{1 \leq i \leq j \leq n} a_{ij} \cdot x_i x_j \quad \text{with } a_{ij} \in \mathbb{K}.$$

It is well-known that over fields of characteristic not two, a quadratic form Q is uniquely represented by its *polar form*, *i.e.*, the symmetric bilinear form defined by $\psi(Q) : (x, y) \mapsto 1/2 \cdot (Q(x + y) - Q(x) - Q(y))$, with the nice property that $Q(x) = \psi(Q)(x, x)$. Over fields of characteristic two, this is however no longer possible, because the division by two is not defined. In this paper, we will slightly abuse the usual definition, and we define the polar form of a quadratic form to be the symmetric bilinear form:

$$\psi(Q) : (x, y) \mapsto Q(x + y) - Q(x) - Q(y)$$

Given a basis $b_1, \ldots, b_n$ of $\mathbb{F}$, $\psi(Q)$ can be represented by a $n \times n$ symmetric matrix whose (i, j) coefficient is $\psi(Q)(b_i, b_j)$. By an abuse of notation, we will often identify $\psi(Q)$ with its matrix representation.

The Kernel of a Quadratic Form. The *kernel* of a quadratic form Q, also called the *radical* of Q is the vector space of elements $a \in \mathbb{F}$ such that for any $x \in \mathbb{F}$, $\psi(Q)(x, a) = 0$. It is easy to see that the kernel of a quadratic form is the kernel of the matrix $\psi(Q)$. What makes the kernel interesting is that in characteristic two, when n is odd, all quadratic forms have a non-trivial kernel.

Theorem 1 ([1]). *Let q be a power of two, and let Q be a quadratic form over $\mathbb{K}$. Then the rank of $\psi(Q)$ is even.*

Linear Algebra. We denote the characteristic polynomial of M by $\chi(M)$. A *minor* of M is simply the determinant of a submatrix of M. We will use in the following the *adjugate* matrix $\text{adj}(M)$ of a matrix M. We recall that it is the transpose of the comatrix, which is the matrix of the cofactors. A cofactor of M, $\text{cof}_{i,j}(M)$ is the determinant of the submatrix $M_{\bar{i}}^{\bar{j}}$, where in this notation we refer to the matrix M without the ith row and the jth column. We lastly recall two well-known results connecting a matrix M and its adjugate.

Theorem 2 (Cayley-Hamilton). *If $\chi(M) = X^n + c_{n-1}X^{n-1} + \cdots + c_1 X + c_0$ is the characteristic polynomial of M, then:*

$$M^n + c_{n-1}M^{n-1} + \cdots + c_1 M + c_0 \cdot I_n = 0$$
$$M^{n-1} + c_{n-1}M^{n-2} + \cdots + c_2 M + c_1 \cdot I_n = \text{adj}(-M)$$

It follows that $-M \cdot \text{adj}(-M) = \text{adj}(-M) \cdot -M = \det(-M) \cdot I_n$

Lemma 2. *The rank of $\text{adj}(M)$ can be deduced from the rank of M:*

- *if $\text{rank}(M) = n$, then $\text{rank}(\text{adj}(M)) = n$.*
- *if $\text{rank}(M) = n - 1$, $\text{rank}(\text{adj}(M)) = 1$.*
- *In all other cases, $\text{rank}(\text{adj}(M)) = 0$.*

3   The C* and SFLASH Signature Schemes

The basic idea underlying both C* and SFLASH is to hide an easily invertible function ϕ in the large finite field $\mathbb{F}$ using two secret invertible linear (or affine) maps S and T which mix together the n coordinates of ϕ over the small field $\mathbb{K}$, with $\mathbf{PK} = T \circ \phi \circ S$. The signature of a message y is a vector x such that $\mathbf{PK}(x) = y$. The legitimate signer easily computes x by successively inverting T, ϕ and then S.

Let π be the canonical isomorphism between $\mathbb{K}^n$ and $\mathbb{F}$, and let ϕ be defined by $\phi(X) = X^{1+q^\theta}$. Enforcing that $\gcd(1 + q^\theta, q^n - 1) = 1$ makes ϕ bijective. Because we may write $\phi(X) = X \cdot X^{q^\theta}$, we find that ϕ is in fact the product of two linear functions (recall that the Frobenius map and its iterates are linear). It follows that $\pi \circ \phi \circ \pi^{-1}$ is a quadratic bijection of $\mathbb{K}^n$, i.e., that if $x \in \mathbb{K}^n$, then $\pi \circ \phi \circ \pi^{-1}$ is a vector whose coordinates are quadratic forms in the coordinates of x. For the sake of lighter notations, we omit π in the sequel.

The secret key of the scheme is composed by the two invertible $n \times n$ matrices S and T with coefficients in $\mathbb{K}$. The exponent θ and π are public parameters. The public-key of the scheme is formed by the representation over $\mathbb{K}^n$ of $T \circ \phi \circ S$. More precisely, if T_i denotes the i-th line of T, then the public key of C* is the vector of n quadratic forms over $\mathbb{K}^n$:

$$\mathcal{P}_i(x_1, \ldots, x_n) = \mathrm{Tr}\Big(T_i \cdot \phi\left(S\left(x_1, \ldots, x_n\right)\right)\Big) \quad 1 \le i \le n$$

The public key of SFLASH is composed of the first r quadratic forms $\mathcal{P}_1, \ldots, \mathcal{P}_r$. Typical values of the parameter may be the ones defined for SFLASH V3: $q = 128, n = 67, r = 56$ and $\theta = 33$.

Although the public key is a vector of polynomials in $(\mathbb{K}[x_1, \ldots, x_n])^n$, it is more convenient to see them as functions from $\mathbb{F}$ to $\mathbb{K}$. We therefore write

$$\mathcal{P}_i(x) = \mathrm{Tr}\left(T_i \cdot S(x)^{1+q^\theta}\right).$$

Equivalent Secret Keys. Given a public-key, there are many possible corresponding secret keys (there are "equivalent" secret keys [18]). A key-recovery attack is expected to retrieve one possible secret key amongst those generating the targeted public-key. The existence of many equivalent secret keys gives some freedom to the attacker: we may be guaranteed that there is an equivalent secret key satisfying some interesting property.

Lemma 3. *If (S, T) is an SFLASH secret-key that generates the public key $\mathbf{PK}$, then for any integer $k > 1$ there is an equivalent secret key (S', T') in which $T_i' = (T_i/T_1)^{q^k}$ (seeing the vectors T_i as elements of $\mathbb{F}$).*

Proof. Because the function $x \in \mathbb{F} \mapsto a \cdot x$ is linear over $\mathbb{F}$, it can be represented by a matrix M_a over $\mathbb{K}^n$. The key idea is that multiplications "commute" with the internal power function:

$$\mathcal{P}_i(x) = \mathrm{Tr}\left(\frac{T_i}{a^{1+q^\theta}} \cdot [a \times (S \cdot x)]^{1+q^\theta}\right)$$

Now, we pick a such that $a^{1+q^\theta} = T_1$ (this is always possible because the power function is bijective). Thus, a possible equivalent secret key is such that $T_i' = T_i/T_1$, and $S' = M_a \cdot S$.

Next, it follows from the definition of the trace, and from the identity $x^{q^n} = x$ which holds over $\mathbb{F}$ that $\mathrm{Tr}\left(x^{q^k}\right) = \mathrm{Tr}(x)$. This shows that

$$\mathcal{P}_i(x) = \mathrm{Tr}\left(\left(\frac{T_i}{a^{1+q^\theta}}\right)^{q^k} \cdot \left([(a \times (S \cdot x)]^{q^k}\right)^{1+q^\theta}\right)$$

Thus, if F denotes the matrix representing the Frobenius, *i.e.*, the linear map $x \mapsto x^q$ in $\mathbb{F}$, then a possible equivalent secret key is such that $T_i' = (T_i/T_1)^{q^k}$, and $S' = F^k \cdot M_a \cdot S$. $\qquad\square$

4 Mathematical Properties of C*− Public Keys

The aim of this section is to exhibit relations involving the secret elements S and the T_i's on the one hand, and the public key on the other hand, in such a way that the secrets can be easily reconstructed given only a small number of public polynomials.

For this purpose, we consider two public polynomials $\mathcal{P}_i$ and $\mathcal{P}_j$, and we define the *pencil of quadratic forms* $\mathbf{P} = \lambda \mathcal{P}_i + \mu \mathcal{P}_j$, with λ, μ in $\mathbb{K}$. We also define the *pencil of vectors* $\mathbf{T} = \lambda T_i + \mu T_j$, and because the Trace is $\mathbb{K}$-linear we have:

$$\mathbf{P}(X) = \mathrm{Tr}\left(\mathbf{T} \cdot S(X)^{1+q^\theta}\right). \tag{1}$$

We are interested in the *kernel* of $\mathbf{P}$, which is by definition the set of vectors a such that for any x, $\psi(\mathbf{P})(a, x) = 0$. In fact, it is simply the kernel of the matrix representation of the polar form $\psi(\mathbf{P})$. We first relate the kernel of $\mathbf{P}$ to the components of the secret key in section 4.1, and then with the components of the public-key in section 4.2. This allows us, by "transitivity", to find exploitable relations between the public key and the secret elements in section 4.3.

In the sequel, we adopt the typographic convention that any quantity that depends implicitly on λ and μ is written in bold.

4.1 Relations between the Kernel and the Secret-Key

It is not very surprising that the kernel of $\mathbf{P}$ admits a relatively simple expression in terms of the components of the secret key.

Theorem 3. *Given that n is odd, and $\gcd(\theta, n) = 1$, we have:*

(i)   *The kernel of $\mathbf{P}$ is $\left\{x \in \mathbb{K}^n \mid \mathbf{T} \cdot S(x)^{1+q^\theta} \in \mathbb{K}\right\}$.*

(ii) *The matrix pencil $\psi(\mathbf{P})$ has rank $n - 1$.*

(iii) *When $(\lambda, \mu) \neq (0, 0)$, there exists a unique vector $\mathbf{a} \in \mathbb{K}^n$ in the kernel of $\mathbf{P}$ such that $\mathbf{P}(\mathbf{a}) = 1$.*

(iv) *There exists* $\delta \in \mathbb{N}$ *such that* $\mathbf{a} = S^{-1}\left(\mathbf{T}^{\delta}\right)$. *A possible value for* δ *is*

$$\delta = \left(\frac{q}{2} - 1\right) \cdot \sum_{i=0}^{n-1} q^i + \sum_{i=(n+1)/2}^{n-1} q^{2i\theta} \tag{2}$$

Proof. It is known that the polar forms of C^* polynomials have a special shape:

$$\psi(\mathbf{P})(x, y) = \mathrm{Tr}\left(\mathbf{T} \cdot \left[S(x) \cdot S(y)^{q^\theta} + S(x)^{q^\theta} \cdot S(y)\right]\right)$$

After some manipulations, by exploiting the linearity of the Frobenius, of the Trace, and the fact that they commute, we find when $x \neq 0$:

$$\psi(\mathbf{P})(x, y) = \mathrm{Tr}\left(\left[\mathbf{T} \cdot S(x)^{1+q^\theta} + \left(\mathbf{T} \cdot S(x)^{1+q^\theta}\right)^{q^\theta}\right] \cdot \left(\frac{S(y)}{S(x)}\right)^{q^\theta}\right)$$

Now, inside the trace, the first term of the product depends only on x, and the second member takes all possible values in $\mathbb{F}$ when y ranges across $\mathbb{F}$, because S and the Frobenius are bijective. Lemma 1 then tells us that if $x \neq 0$ belongs to the kernel of $\mathbf{P}$, then

$$\mathbf{T} \cdot S(w)^{1+q^\theta} + \left(\mathbf{T} \cdot S(w)^{1+q^\theta}\right)^{q^\theta} = 0$$

It remains to show that the solutions of the equation $X + X^{q^\theta} = 0$ in $\mathbb{F}$ are precisely the elements of $\mathbb{K}$. It is easy to check that any $x \in \mathbb{K}$ is a solution, because the fields are of characteristic two, which makes the equation equivalent to $X = X^{q^\theta}$. The other direction is not much more difficult: by induction we find that $X = X^{q^{i\theta}}$ for any $i \in \mathbb{N}$. Since over $\mathbb{F}$ we always have $x = x^{q^n}$, then when $i\theta$ is congruent to 1 modulo n, the equation implies $X + X^q = 0$, which shows that the solutions all lies in $\mathbb{K}$. This establishes point (i).

Let us prove point (ii). The polar form $\psi(\mathbf{P})$ cannot be of rank n, because it is a skew-symmetric matrix and n is odd (this is well-known for matrices over fields, and is extended to the case of matrices multivariate polynomial rings in lemma 8, appendix A). Now, we show that the rank of $\psi(\mathbf{P})$ is greater than $n-1$. If we specialize (λ, μ) to any value in $\mathbb{K}^2$ distinct from $(0,0)$, then by point (i) $\psi(\lambda \mathcal{P}_i + \mu \mathcal{P}_j)$, seen as a matrix with entries in $\mathbb{K}$, has a kernel of dimension 1. By the rank theorem (over $\mathbb{K}$), its rank is then $n - 1$. This shows that there is a non-zero minor of dimension $(n - 1)$. This minor (seen as a polynomial in λ and μ) cannot be the zero polynomial, otherwise it could become non-zero for a particular choice of λ and μ in $\mathbb{K}$, hence the rank of $\psi(\mathbf{P})$ (seen as a matrix with entries in $\mathbb{K}[\lambda, \mu]$ has rank exactly $n - 1$.

Point (iii) follows immediately from (i) and from the fact S, T and the power function are bijective. To establish point (iv), we need to find a suitable value δ such that $S(a) = \mathbf{T}^\delta$. By definition of a, we should have $\left(\mathbf{T}^\delta\right)^{1+q^\theta} \cdot \mathbf{T} = 1$, so that δ satisfies the equation $1 + \delta(1 + q^\theta) = 0$ modulo $(q^n - 1)$. Checking that the given value of δ is valid is technical and not very interesting, and we refer the reader to [8] for more details.   □

The fourth point of theorem 3 makes it possible to explicitly write down the expression of $\mathbf{a}$, the kernel vector introduced in the proposition, as a function of λ and μ. Let us set $d = (n-1)/2$, and let us introduce $P_N, P_S \in \mathbb{F}[\lambda, \mu]$:

$$p_N = \mathrm{N}(\mathbf{T}) = \prod_{i=0}^{n-1} \left(\lambda \cdot T_1{}^{q^i} + \mu \cdot T_2{}^{q^i} \right)$$

$$p_S = S^{-1} \left(\prod_{i=(n+1)/2}^{n-1} \left(\lambda \cdot T_1{}^{q^{2i\theta}} + \mu \cdot T_2{}^{q^{2i\theta}} \right) \right) \tag{3}$$

The idea is that p_N only depends on T, while p_S depends "linearly" on S. It is fairly obvious that p_N has total degree n while p_S has total degree d. Next, we claim that p_N in fact has coefficients in $\mathbb{K}$. A possible way to see this is that because it coincides with the Norm, it takes values in $\mathbb{K}$ when $\lambda, \mu \in \mathbb{K}$, and therefore it could be interpolated as a polynomial of $\mathbb{K}[\lambda, \mu]$.

We have carefully chosen p_N and p_S so that the vector $\mathbf{a}$ defined in point (iii) of proposition 3 is such that:

$$\mathbf{a} = (p_N)^{q/2-1} \cdot p_S.$$

This fact is an easy consequence of the fourth point of proposition 3. Note that because p_N has values in $\mathbb{K}$, then $p_S(\lambda, \mu)$ spans the kernel of $\mathbf{P}$, but unlike $\mathbf{a}$, p_S does not a priori satisfy the additional condition that $\mathbf{P}(p_S) = 1$. It follows, by definition of $\mathbf{a}$, because $\mathbf{P}(\lambda x) = \lambda^2 \mathbf{P}(x)$ when $\lambda \in \mathbb{K}$ and because $x^{-1} = x^{q-2}$ in $\mathbb{K}$, that:

$$\mathbf{P}\left((p_N)^{q/2-1} \cdot p_S \right) = \frac{\mathbf{P}(p_S)}{p_N} = 1$$

And we find that $p_N = \mathbf{P}(p_S)$. The two polynomials p_N and p_S play a crucial role in the sequel: we will show in section 5 that knowing them is sufficient to reconstruct the secret key in polynomial time. In addition, we will also show that they can be reconstructed in polynomial time from the public-key. However, doing this requires some more mathematical machinery.

4.2   Relations between the Kernel and the Public Key

The kernel of $\mathbf{P}$ can be computed using only publicly available information, since it only depends on the public polynomials. If the values of λ and μ were fixed, this could be achieved with standard linear algebra. More sophisticated computer algebra systems have functions that compute a basis of the kernel in terms of λ and μ. We remove the need for such sophisticated operations by explicitly giving the form of the kernel.

Theorem 4. *Let $\mathbf{P}$ be a pencil of two public polynomials, $B = \{b_1, \ldots, b_n\}$ a basis of $\mathbb{K}^n$. There exists a vector $\mathbf{k} = (\mathbf{k}_1, \ldots, \mathbf{k}_n)$ of degree-d homogeneous bivariate polynomials in $\mathbb{K}[\lambda, \mu]$, such that:*

i) *The adjugate matrix of the polar form of* $\mathbf{P}$ *can be expressed as the tensor product of* $\mathbf{k}$ *with itself:*

$$adj\left(\psi(\mathbf{P})\right) = (\mathbf{k}_i \cdot \mathbf{k}_j)_{1 \le i,j \le n}$$

ii) *the kernel of* $\psi(\mathbf{P})$ *is spanned by* $\sum_{i=1}^{n} \mathbf{k}_i \cdot b_i$

Proof. According to theorem 3, item *(ii)*, the matrix pencil $\psi(\mathbf{P})$ is of rank $n-1$, and lemma 2 states that in this case adj$(\psi(\mathbf{P}))$ has rank 1. We will now show that adj$(\psi(\mathbf{P}))$ is the square of some other matrix, but we first require a technical lemma.

Lemma 4. *Let* $\mathbf{P}$ *be an arbitrary pencil of quadratic forms. There exists a family of bivariate polynomials* $\mathbf{p}_0, \dots, \mathbf{p}_d \in \mathbb{K}[\lambda, \mu]$ *such that* $\mathbf{p}_i$ *is homogeneous of degree* i, *and the characteristic polynomial of the polar form of* $\mathbf{P}$ *is:*

$$\chi\left(\psi\left(\mathbf{P}\right)\right) = \sum_{i=0}^{d} \mathbf{p}_i^{\,2} \cdot X^{n-2i}.$$

The proof of lemma 4 is postponed to appendix A. It follows from lemma 4 and theorem 2 that:

$$adj(\psi(\mathbf{P})) = \sum_{i=0}^{d} \mathbf{p}_i^{\,2} \cdot \psi(\mathbf{P})^{2d-2i} = \left(\sum_{i=0}^{d} \mathbf{p}_i \cdot \psi(\mathbf{P})^{d-i}\right)^2.$$

We denote by R the natural square-root of adj$(\psi(\mathbf{P}))$ occurring on the right-hand side. It is a symmetric matrix pencil whose coefficients are bivariate polynomials of degree d in λ and μ. Let us consider the i-th diagonal term of adj$(\psi(\mathbf{P}))$. We find:

$$adj(\psi(\mathbf{P}))_{i,i} = \sum_{j=1}^{n} R_{i,j} \cdot R_{j,i} = \left(\sum_{j=1}^{n} R_{i,j}\right)^2.$$

Consequently, let us define $\mathbf{k}_i = \sum_{j=1}^{n} R_{i,j}$. The previous equation tells us that adj$(\psi(\mathbf{P}))_{i,i} = \mathbf{k}_i^{\,2}$ for all $1 \le i \le n$. This establishes point (i) for the diagonal of adj$(\psi(\mathbf{P}))$ only.

Let us now consider the other terms with $i \ne j$. Since adj$(\psi(\mathbf{P}))$ is of rank 1, we know that all the minors of dimension 2 of adj$(\psi(\mathbf{P}))$ obtained by keeping only the i-th row and the j-th column is null. This yields:

$$adj(\psi(\mathbf{P}))_{i,i} \cdot adj(\psi(\mathbf{P}))_{j,j} + \left(adj(\psi(\mathbf{P}))_{i,j}\right)^2 = 0$$

and consequently adj$(\psi(\mathbf{P}))_{i,j} = \mathbf{k}_i \cdot \mathbf{k}_j$ (when the field is of characteristic two, the square root always exists and is unique because the Frobenius map is bijective). This completes the proof of (i).

Let us now focus on point (ii). One of the $\mathbf{k}_i$'s at least is non-zero, because $\mathrm{adj}(\psi(\mathbf{P}))$ is not the null matrix. We therefore assume (without loss of generality) that $\mathbf{k}_1$ is non-zero, and we consider the matrix relation given by theorem 2:

$$\psi(\mathbf{P}) \cdot \mathrm{adj}(\psi(\mathbf{P})) = 0.$$

Looking at the first column of the product, we conclude that

$$\psi(\mathbf{P}) \cdot \left(\sum_{i=1}^{n} \mathbf{k}_1 \mathbf{k}_i \cdot b_i \right) = 0,$$

and because $\mathbf{k}_1$ is non-zero, we conclude that $\psi(\mathbf{P}) \cdot (\sum_{i=1}^{n} \mathbf{k}_i \cdot b_i) = 0$.   □

In light of theorem 4, it seems that we can derive from the public key a polynomial whose properties mimic those of p_S. Keeping the notations of the theorem, we define:

$$\widetilde{p}_S = \sum_{i=1}^{n} \mathbf{k}_i \cdot b_i, \qquad \widetilde{p}_N = \mathbf{P}\,(\widetilde{p}_S)$$

We deduce from theorem 4 that $\widetilde{p}_S$ has the same degree as p_S, and that like p_S, it spans the kernel of $\psi(\mathbf{P})$. We also need to find a polynomial $\widetilde{p}_N$ that would be an analogous of p_N and that could be derived from the public key. Note that it immediately follows from theorem 4 that $\widetilde{p}_S$ spans the kernel of $\mathbf{P}$.

4.3   Relations between the Secret-Key and the Public-Key

The last (but not least) step of our analysis is to show that the two polynomials p_N, p_S derived from the secret key in section 4.1 on the one hand, and the polynomials $\widetilde{p}_N, \widetilde{p}_S$ derived from the public key in section 4.2 are in general equal up to a constant multiplicative factor.

Theorem 5. *If T_2/T_1 is primitive over $\mathbb{F}$ (i.e., generates the multiplicative group of $\mathbb{F}$), then there exists a constant $\zeta \neq 0$ in $\mathbb{K}$ such that $\widetilde{p}_S = \zeta \cdot p_S$, and (accordingly) $\widetilde{p}_N = \zeta^2 \cdot p_N$.*

Proof. The first step of the proof is to show that $\widetilde{p}_N$ has degree n, just like p_N. The polynomials $\mathbf{k}_1, \ldots, \mathbf{k_n}$ defined in theorem 4 have coefficients in $\mathbb{K}$, and are homogeneous of degree d. We can therefore find a family $c_0, \ldots, c_d$ of coefficients in $\mathbb{F}$ such that:

$$\widetilde{p}_S = \sum_{i=1}^{n} \mathbf{k}_i \cdot b_i = \sum_{i=0}^{d} c_i \cdot \lambda^{d-i} \mu^i. \tag{4}$$

It turns out that this family enjoys a nice property: over the subspace of $\mathbb{K}^n$ that it spans, the pencil $\mathbf{P}$ is in fact a diagonal form (i.e., the two public polynomial it is made of are simultaneously diagonal).

Lemma 5. $\psi(\mathbf{P})(c_i, c_j) = 0$ *for any $0 \leq i, j \leq d$.*

Lemma 6. *For any family $\{r_i\}_{0 \leq i \leq d}$ of polynomials over $\mathbb{K}$, we have*

$$\mathbf{P}\left(\sum_{i=0}^{d} r_i \cdot c_i\right) = \sum_{i=0}^{d} r_i^2 \cdot \mathbf{P}(c_i).$$

The proofs are postponed to appendix B. Applying lemma 6 to (4), we get:

$$\widetilde{p}_N = \mathbf{P}\left(\widetilde{p}_S\right) = \mathbf{P}\left(\sum_{i=0}^{d} c_i \cdot \lambda^{d-i}\mu^i\right) = \sum_{i=0}^{d}\left(\lambda\mathcal{P}_1(c_i) + \mu\mathcal{P}_2(c_i)\right) \cdot \lambda^{2d-2i}\mu^{2i}$$

From there, it is easy to see that $\widetilde{p}_N$ has degree $2d + 1 = n$.

Now that it has been established that p_N and $\widetilde{p}_N$ have the same degree, we will use irreducibility properties of p_N to conclude the proof of theorem 5. We first claim that the univariate polynomial $p_N(\lambda, 1) \in \mathbb{K}[\lambda]$ is irreducible over $\mathbb{K}$. After a few manipulations we find

$$p_N(\lambda, 1) = \mathrm{N}(\lambda T_1 + T_2) = \mathrm{N}(T_1) \cdot \mathrm{N}(\lambda + T_2/T_1).$$

Thus T_2/T_1, which is primitive over $\mathbb{F}$, is a root of $p_N(\lambda, 1)$, and this polynomial is therefore irreducible over $\mathbb{K}$.

Lemma 7. *There exist $\zeta, \widetilde{\zeta}$ in $\mathbb{K}[\lambda, \mu]$ such that:*

$$\zeta \cdot p_S = \widetilde{\zeta} \cdot \widetilde{p}_S \quad and \quad \gcd\left(\zeta, \widetilde{\zeta}\right) = 1.$$

Proof. First, the rank of the two-column matrix $(p_S, \widetilde{p}_S)$ is one. If it was two, then this matrix could be extended to a $n \times n$ matrix M of rank n. We then find that the rank of $\psi(\mathbf{P}) \cdot M$ would be at most $n - 2$, since its two first columns are null, which contradicts the fact established earlier that $\psi(\mathbf{P})$ has rank $n - 1$.

There exist polynomials $\{\ell_i\}$ such that $p_S = \sum_{i=1}^{n} \ell_i \cdot b_i$. We now argue that there exists an index i_0 such that $\mathbf{k}_{i_0} \neq 0$ and $\ell_{i_0} \neq 0$. The reasoning is by contradiction: assume that for all i we have $\mathbf{k}_i \cdot \ell_i = 0$. Since $\widetilde{p}_S \neq 0$ and $p_S \neq 0$, there exist indices i, j such that $\mathbf{k}_i \neq 0$ and $\ell_j \neq 0$. By hypothesis, $\mathbf{k}_j = 0$ and $\ell_i = 0$. But then, we find that $\mathbf{k}_i \cdot \ell_j + \mathbf{k}_j \cdot \ell_i = \mathbf{k}_i \cdot \ell_j \neq 0$. Consequently, a minor of dimension two of $(p_S, \widetilde{p}_S)$ is non-zero, which contradict the fact that it is of rank one.

We can therefore assume without loss of generality that $\mathbf{k}_1 \neq 0$ and $\ell_1 \neq 0$. The linear combination $\mathbf{k}_1 \cdot p_S + \ell_1 \cdot \widetilde{p}_S$ is null since by construction its first coordinate is zero, and the other coordinates are minors of dimension 2 of $(p_S, \widetilde{p}_S)$ and are also null. We can now assert that the pair

$$\left(\frac{\mathbf{k}_1}{\gcd\left(\mathbf{k}_1, \ell_1\right)}, \frac{\ell_1}{\gcd\left(\mathbf{k}_1, \ell_1\right)}\right)$$

satisfies the requirements of the lemma. □

Let $\left(\zeta, \tilde{\zeta}\right)$ a pair of bivariate polynomials over $\mathbb{K}$ satisfying lemma 7. By applying $\mathbf{P}$, we get: $\zeta^2 \cdot p_N = \tilde{\zeta}^2 \cdot \tilde{p}_N$. Any irreducible factor of $\tilde{\zeta}$ must divide p_N since it does not divide ζ. But because p_N is irreducible, $\tilde{\zeta}$ is necessarily of degree 0. And ζ is also degree 0 because p_N and $\widetilde{p_N}$ have the same degree. This concludes the proof of theorem 5.   □

We conclude this section by giving one last important but somewhat technical result. The polynomial p_S is "designed" to reveal the image of S on the subspace of $\mathbb{K}^n$ spanned by its $d+1$ coefficients (seen as vectors of $\mathbb{K}^n$). It does actually matter whether these are linearly independent or not.

Theorem 6. *The coefficients of the polynomials p_S form an independent family if and only if $(T_2/T_1)^{q^\theta}$ is not a root of the polynomials $x + x^{q^{2i\theta}}$ for $1 \le i \le d$. In particular, if n is a prime number this condition is satisfied since by assumption T_1 and T_2 are independent.*

The proof is given in appendix C.

5   The Attack

We are now ready to leverage our in-depth investigation of the properties of C*, by presenting a practical key-recovery attack that does not require any signature. The global attack strategy is to compute the polynomials $\tilde{p}_N$ and $\tilde{p}_S$ defined in section 4.2. Then, theorem 5 tells us that with non-negligible probability, these are equal to the polynomials p_N and p_S defined in section 4.1, from which the secret-key can be efficiently recovered.

Reconstructing the Polynomials p_N and p_S. Given a pencil $\mathbf{P} = \lambda \mathcal{P}_i + \mu \mathcal{P}_j$ of polynomials from the public key, we first show how the polynomials $\tilde{p}_N$ and $\tilde{p}_S$ defined in section 4.2 can be determined. More precisely, we show how to build a function KERNEL-RECOVERY($\mathbf{P}$) that returns the two polynomials p_N and p_S described in section 4.1. Because $p_N = \mathbf{P}(p_S)$, we focus our attention on the non-obvious part consisting in recovering p_S. This can be achieved in two different ways. A first possibility is to follow the proof of theorem 4, which results in the following procedure:

1. Compute the characteristic polynomial ζ of $\psi(\mathbf{P})$ and factor it into

$$\zeta = X \cdot \left(\sum_{i=0}^{d} \mathbf{P}_i^{\,2} \cdot X^{d-i} \right)^2$$

2. Compute the matrix $\mathbf{R} = \sum_{i=0}^{d} \mathbf{P}_i \cdot \psi(\mathbf{P})^{d-i}$ and let $\mathbf{k}_i = \sum_{j=1}^{n} \mathbf{R}_{i,j}$.
3. Finally let p_S be equal to $\sum_{i=1}^{n} \mathbf{k}_i \cdot b_i$

Note that computing the characteristic polynomial can be achieved over any commutative ring using the division-free algorithm of Mahajan and Vinay [9]. Computing the factorization of the characteristic polynomial is a (classical) multivariate factorization problem. Both functionality are available in several computer algebra systems, including (but not limited to) MAGMA [2] and SAGE [17].

Alternatively, we may directly compute a basis of $\ker \psi(\mathbf{P})$ (which is a module over $\mathbb{K}[\lambda, \mu]$) using the *ad hoc* function present in some computer algebra systems. This function is for instance available in MAGMA, and seems to rely on Gröbner basis computations. It is apparently much faster than the previous option.

From Kernel to Secret-Key. Let us call (T', S') the equivalent key we try to forge. Thanks to lemma 3, we know that we may without loss of generality assume that $T_1' = 1$ and $T_2' = (T_2/T_1)^{q^i}$, for any $i > 0$. This shows that if $(p_N, p_S) = \text{KERNEL-RECOVERY}(\lambda \mathcal{P}_1 + \mu \mathcal{P}_2)$, then we may safely choose T_2' to be any root of $p_N(\lambda, 1)$ different from one. We then focus on equation (3):

$$p_S = S^{-1} \left(\prod_{i=(n+1)/2}^{n-1} \left(\lambda \cdot T_1^{q^{2i\theta}} + \mu \cdot T_2^{q^{2i\theta}} \right) \right)$$

Given the values of T_1' and T_2', we may explicitly evaluate the product on the right-hand side. Identifying both sides coefficient-wise then reveals the image of S' on the subspace of K^n spanned by the $d+1$ coefficients of the product. Theorem 6 tells us that this subspace is of dimension $d+1$ with non-negligible probability.

To complete the key-recovery of the secret element, we use a third polynomial from the public-key. We compute $(p_N', p_S') = \text{KERNEL-RECOVERY}(\lambda \mathcal{P}_1 + \mu \mathcal{P}_3)$. Only one of the roots of p_N' yields a valid choice for T_3', therefore we pick one at random, and we will try again with another one in case of failure in the subsequent steps. Knowledge of T_1' and T_3' allows to discover the image of S' on another subspace spanned by $d+1$ generators following the same procedure.

At this point, we have learned the image of S' on $n+1$ vectors, and we really hope that S' is completely revealed. If it is not the case, we may try again with $\mathcal{P}_4$ instead of $\mathcal{P}_3$. Once S' is known, finding the other T_i's can be done by straightforward linear algebra. If no solution exists for any of them, then our guess for T_3' was wrong.

5.1 Complexity

We implemented the whole key-recovery using the MAGMA computer algebra system. The code of the full attack is 120 lines long, and is available on the web page of the first author. We first applied the attack to SFLASH v2 and SFLASH v3, that were already broken (universal forgery) by Dubois, Fouque, Stern and Shamir [3], and further broken (key-recovery) by Fouque, Macario-Rat and Stern [6]. We then applied the attack to SFLASH instances that cannot be broken by the existing attack, because the number of polynomials in the public

Table 1. Experimental results

SFLASH version	q	n	#public polynomials	Signature size	Already broken ?	Attack time	KeyGen time
v2	128	37	26 (70%)	259 bits	[3,6]	7s	0.1s
v3	128	67	56 (83%)	469 bits	[3,6]	47s	0.6s
	256	131	56 (42%)	1048 bits	No	17min	5s
	65536	257	64 (25%)	4112 bits	No	$\approx$ 10h	141s
	2	331	80 (24%)	331 bits	No	105min	16s
	2	521	80 (24%)	521 bits	No	$\approx$ 11h	62s
	2	1031	128 (12%)	1031 bits	No	...	680s

key is less than $n/2$. We tried various combinations of field size and variable numbers, and found out that the attack works quite well in practice, as Table 1 shows. There are thus no longer any practically unbroken set of parameters for SFLASH.

References

1. Albert, A.A.: Symmetric and alternate matrices in an arbitrary field, i. Transactions of the American Mathematical Society 43(3), 386–436 (1938)
2. Bosma, W., Cannon, J.J., Playoust, C.: The Magma Algebra System I: The User Language. J. Symb. Comput. 24(3/4), 235–265 (1997)
3. Dubois, V., Fouque, P.A., Shamir, A., Stern, J.: Practical Cryptanalysis of SFLASH. In: Menezes, A. (ed.) CRYPTO 2007. LNCS, vol. 4622, pp. 1–12. Springer, Heidelberg (2007)
4. Dubois, V., Fouque, P.A., Stern, J.: Cryptanalysis of SFLASH with Slightly Modified Parameters. In: Naor, M. (ed.) EUROCRYPT 2007. LNCS, vol. 4515, pp. 264–275. Springer, Heidelberg (2007)
5. Fell, H.J., Diffie, W.: Analysis of a Public Key Approach Based on Polynomial Substitution. In: Williams, H.C. (ed.) CRYPTO 1985. LNCS, vol. 218, pp. 340–349. Springer, Heidelberg (1986)
6. Fouque, P.A., Macario-Rat, G., Stern, J.: Key Recovery on Hidden Monomial Multivariate Schemes. In: Smart, N.P. (ed.) EUROCRYPT 2008. LNCS, vol. 4965, pp. 19–30. Springer, Heidelberg (2008)
7. Lidl, R., Niederreiter, H.: Finite Fields. Cambridge University Press (2008)
8. Macario-Rat, G.: Cryptanalyse de schémas multivariés et résolution du problème Isomorphisme de Polynômes. PhD thesis, Université Paris Diderot — Paris 7 (June 2010)
9. Mahajan, M., Vinay, V.: Determinant: Combinatorics, algorithms, and complexity. Chicago J. Theor. Comput. Sci. 1997 (1997)
10. Matsumoto, T., Imai, H.: Public Quadratic Polynomial-Tuples for Efficient Signature-Verification and Message-Encryption. In: Günther, C.G. (ed.) EUROCRYPT 1988. LNCS, vol. 330, pp. 419–453. Springer, Heidelberg (1988)
11. Ong, H., Schnorr, C.P., Shamir, A.: An efficient signature scheme based on quadratic equations. In: STOC, pp. 208–216. ACM (1984)

12. Ong, H., Schnorr, C.P., Shamir, A.: Efficient Signature Schemes Based on Poly-
    nomial Equations. In: Blakely, G.R., Chaum, D. (eds.) CRYPTO 1984. LNCS,
    vol. 196, pp. 37–46. Springer, Heidelberg (1985)
13. Patarin, J.: Cryptanalysis of the Matsumoto and Imai Public Key Scheme of Euro-
    crypt '88. In: Coppersmith, D. (ed.) CRYPTO 1995. LNCS, vol. 963, pp. 248–261.
    Springer, Heidelberg (1995)
14. Patarin, J., Courtois, N., Goubin, L.: SFLASH, a Fast Multivariate Signature Al-
    gorithm (2003), http://eprint.iacr.org/
15. Shamir, A.: Efficient Signature Schemes Based on Birational Permutations. In:
    Stinson, D.R. (ed.) CRYPTO 1993. LNCS, vol. 773, pp. 1–12. Springer, Heidelberg
    (1994)
16. Shor, P.W.: Polynomial-time algorithms for prime factorization and discrete loga-
    rithms on a quantum computer. SIAM J. Comput. 26(5), 1484–1509 (1997)
17. Stein, W., et al.: Sage Mathematics Software (Version 4.6.2). The Sage Develop-
    ment Team (2011), http://www.sagemath.org
18. Wolf, C., Preneel, B.: Equivalent Keys in HFE, C*, and Variations. In: Dawson, E.,
    Vaudenay, S. (eds.) Mycrypt 2005. LNCS, vol. 3715, pp. 33–49. Springer, Heidelberg
    (2005)

A   Mathematical Results

Lemma 8. *Let* $\mathbf{P} = \lambda A + \mu B$ *be a matrix pencil over* $\mathbb{K}$, *symmetric and with null diagonal, of any dimension* n. *Its determinant is a bivariate form of degree* n. *If* n *is odd,* $\det(M) = 0$, *and if* n *is even, there exists a bivariate form* k *over* $\mathbb{K}$ *of degree* $n/2$ *such that* $\det(M) = k^2$.

Proof. We will prove this result using a recurrence in a 2 by 2 step. For $n = 1$, $M = (0)$ and $\det(M) = 0$. For $n = 2$, we have $M = \begin{pmatrix} 0 & k \\ k & 0 \end{pmatrix}$, where k is a bivariate form of degree 1 and $\det(M) = k^2$. Now, let $n \geq 3$ and assume the property is true for $n - 2$. We will show that it is also true for n. We compute the determinant of M by developing according to the first column. Since the $(1,1)$-coefficient of M is null, we have $\det(M) = \sum_{i=2}^{n} M_{i,1} \det(M_{\overline{i}}^{\overline{1}})$, where $M_{i,1}$ denote the coefficient $(i,1)$ of M and $\det(M_{\overline{i}}^{\overline{1}})$ the $(1,i)$ minor. We can see that in all these minors, the first row has never been removed and always the first column. We can now do a development according to the first row and using the multi-linearity of the determinant, we get $\det(M) = \sum_{i=2}^{n} \sum_{j=2}^{n} M_{i,1} M_{1,j} \det(M_{\overline{1,i}}^{\overline{1,j}})$, where $M_{\overline{1,i}}^{\overline{1,j}}$ denote the matrix M by removing the rows 1 and i and the columns 1 and j. Since M is symmetric, we can add together the terms (i,j) and (j,i) for $i \neq j$ and these terms vanish. The determinant that we compute is equal to $\det(M) = \sum_{i=2}^{n} M_{i,1}^2 \det(M_{\overline{1,i}}^{\overline{1,i}})$. Now we can use the recurrence assumption and if n is odd, $\det(M) = 0$ and if n is even, $\det(M) = \sum_{i=2}^{n} M_{i,1}^2 k_i^2 = (\sum_{i=2}^{n} M_{i,1} k_i)^2$, where the forms k_i, for $i = 2, \ldots, n$ are of degree $(n-2)/2$. Consequently, the degree of the form $\sum_{i=2}^{n} M_{i,1} k_i$ is $n/2$. $\qquad\square$

Lemma 9. *Let* $\mathbf{P}_{\lambda,\mu}$ *be an arbitrary pencil of quadratic forms. There exist a family of bivariate polynomials* $\{\mathbf{p}_i\}_{0 \le i \le d}$ *in* $\mathbb{K}[x, y]$ *such that* p_i *is of degree* i, *and the characteristic polynomial of the polar form of* $\mathbf{P}$ *is:*

$$\chi\left(\psi\left(\mathbf{P}_{\lambda,\mu}\right)\right) = \sum_{i=0}^{d} \mathbf{p}_i{}^2 \cdot X^{n-2i}.$$

Proof. The result follows from lemma 8. The coefficient of X^{n-i} in $\chi\left(\psi\left(\mathbf{P}_{\lambda,\mu}\right)\right)$ is the sum of all M minors obtained by choosing $n - i$ diagonal terms and removing the $(n - i)$ corresponding rows and columns. The minors obtained are of dimension i. $\square$

B   Simultaneous Diagonalization of Two Quadratic Forms

Lemma 10. $\psi(\mathbf{P})(c_i, c_j) = 0$ *for* $0 \le i, j \le d.$

Proof. Let (λ, μ) and (λ', μ') two pairs of variables in $\mathbb{K}^2$ such that $\lambda\mu' + \lambda'\mu \ne 0$. Because $\widetilde{p}_S(\lambda, \mu)$ and $\widetilde{p}_S(\lambda', \mu')$ are the kernels of $\lambda\psi\left(\mathcal{P}_1\right) + \mu\psi\left(\mathcal{P}_2\right)$ and $\lambda'\psi\left(\mathcal{P}_1\right) + \mu'\psi\left(\mathcal{P}_2\right)$ respectively, we find:

$$(\lambda\psi\left(\mathcal{P}_1\right) + \mu\psi\left(\mathcal{P}_2\right))(\widetilde{p}_S(\lambda, \mu), \widetilde{p}_S(\lambda', \mu')) = 0$$
$$(\lambda'\psi\left(\mathcal{P}_1\right) + \mu'\psi\left(\mathcal{P}_2\right))(\widetilde{p}_S(\lambda, \mu), \widetilde{p}_S(\lambda', \mu')) = 0$$

By linear combination, we have

$$(\lambda\mu' + \lambda'\mu)\psi\left(\mathcal{P}_1\right)(\widetilde{p}_S(\lambda, \mu), \widetilde{p}_S(\lambda', \mu')) = 0$$
$$(\lambda\mu' + \lambda'\mu)\psi\left(\mathcal{P}_2\right)(\widetilde{p}_S(\lambda, \mu), \widetilde{p}_S(\lambda', \mu')) = 0$$

and since $(\lambda\mu' + \lambda'\mu) \ne 0$,

$$\psi\left(\mathcal{P}_1\right)(\widetilde{p}_S(\lambda, \mu), \widetilde{p}_S(\lambda', \mu')) = 0$$
$$\psi\left(\mathcal{P}_2\right)(\widetilde{p}_S(\lambda, \mu), \widetilde{p}_S(\lambda', \mu')) = 0.$$

Finally, thanks to the linearity of $\psi\left(\mathcal{P}_1\right)$ and $\psi\left(\mathcal{P}_2\right)$, we get:

$$\sum_{i=0}^{d}\sum_{j=0}^{d} \psi\left(\mathcal{P}_1\right)(c_i, c_j) \cdot \lambda^{d-i}\mu^i\lambda'^{d-j}\mu'^j = 0$$

$$\sum_{i=0}^{d}\sum_{j=0}^{d} \psi\left(\mathcal{P}_2\right)(c_i, c_j) \cdot \lambda^{d-i}\mu^i\lambda'^{d-j}\mu'^j = 0. \square$$

Lemma 11. *For any family* $\{r_i\}_{0 \le i \le d}$ *of polynomials over* $\mathbb{K}$, *we have:*

$$\mathbf{P}\left(\sum_{i=0}^{d} r_i \cdot c_i\right) = \sum_{i=0}^{d} r_i{}^2 \cdot \mathbf{P}(c_i).$$

Proof. We will prove it by induction on the number of non null polynomials in the family. We have $\mathbf{P}(r_1 \cdot c_1) = r_1^2 \cdot P(c_1)$ since $\mathbf{P}$ is a (pencil of) quadratic form(s) whose coefficients are bivariate polynomials over $\mathbb{K}$. Let us assume that the result holds for $k-1$ polynomials. According to the definition of the polar form, we can write:

$$\mathbf{P}\left(\sum_{j=1}^{k} r_j \cdot c_j\right) =$$

$$\mathbf{P}(r_1 c_1) + \mathbf{P}\left(\sum_{j=2}^{k} r_j \cdot c_j\right) + \psi(\mathbf{P})\left(r_1 \cdot c_1, \sum_{j=2}^{k} r_j \cdot c_j\right) =$$

$$r_1^2 \cdot P(c_1) + \sum_{j=2}^{k} r_j^2 \cdot \mathbf{P}(c_j) + \sum_{j=2}^{k} r_1 \cdot c_j \cdot \psi(\mathbf{P})(c_1, c_j).$$

And lemma 10 allows to conclude.

C   Showing Independence of the Coefficients of a Polynomial

We concentrate on a simpler polynomial of the form $\prod_{i=0}^{d-1}(x + t^{q^i})$.

Definition 1. *Let $d \geq 1$ a positive integer. We call elementary symmetric polynomials of order d, the $d+1$ polynomials with d variables $\sigma_{i,d}$, $0 \leq i \leq d$ defined implicitly by:*

$$\prod_{i=1}^{d}(X + X_i) = \sum_{i=0}^{d} \sigma_{i,d}(X_1, \ldots, X_d) X^{d-i}.$$

We also recall the following lemma useful to prove that a family of elements in $\mathbb{F}$ is independent [7].

Lemma 12. *Let $A = \{\alpha_i\}_{0 \leq i \leq d}$ a family of elements of $\mathbb{F}$. The elements in A are independent if and only if the determinant of the matrix $(\alpha_i^{q^j})_{0 \leq i,j \leq d}$ is non null.*

Let t an element of $\mathbb{F}$. In a first step we try to find an equivalent condition to the fact that the coefficients of the polynomial $\prod_{i=0}^{d-1}(x + t^{q^i})$ are independent. These coefficients can be expressed using the elementary symmetric polynomials. They are equal to $\{\sigma_{i,d}(t, t^q, \ldots, t^{q^{d-1}})\}_{0 \leq i \leq d}$.

We describe some notations. We denote by $s_{i,d}$ and Δ_d the mapping over $\mathbb{F}$ defined by:

$$s_{i,d}(x) = \sigma_{i,d}(x, x^q, \ldots, x^{q^{d-1}}),$$

$$\Delta_d(x) = \det((s_{i,d}(x)^{q^j})_{0 \leq i,j \leq d}).$$

Using the above lemma, and these notations, we can say that the coefficients of the polynomial $\prod_{i=d+1}^{n}(x+t^{q^{2i\theta}})$ are independent if and only if $\Delta_d(t) \neq 0$. In the following, we try to compute some simple expression for Δ_d.

Lemma 13. *For d and i integers such that $0 \leq i \leq d$, the Frobenius mapping commute with the mappings $s_{i,d}$, i.e. for every $x \in \mathbb{F}$, $s_{i,d}(x^q) = s_{i,d}(x)^q$.*

Proof. The mappings $s_{i,d}(x)$ are by construction sums of elementary functions $x \mapsto x^{q^{j_1}+\cdots+q^{j_i}}$, $0 \leq j_1 < \ldots < j_i \leq d-1$. The Frobenius mapping is linear and commute with each of these monomials. $\square$

Lemma 14. *For d and i integers such that $1 \leq i \leq d$, we have:*

$$s_{i,d}(x) + s_{i,d}(x^q) = s_{i-1,d-1}(x^q)(x + x^{q^i}).$$

Proof. We have the following relations:

$$\prod_{i=0}^{d-1}(X + x^{q^i}) + \prod_{i=0}^{d-1}(X + x^{q^{i+1}}) =$$
$$(x + x^{q^i})\prod_{i=1}^{d-1}(X + x^{q^i}) = (x + x^{q^d})\prod_{i=0}^{d-2}(X + x^{q^{i+1}})$$

and

$$\prod_{i=0}^{d-1}(X + x^{q^i}) = \sum_{i=0}^{d} s_{i,d}(x)X^{d-i}$$

$$\prod_{i=0}^{d-1}(X + x^{q^{i+1}}) = \sum_{i=0}^{d} s_{i,d}(x^q)X^{d-i}$$

$$\prod_{i=0}^{d-2}(X + x^{q^i}) = \sum_{i=0}^{d} s_{i,d-1}(x)X^{d-1-i}.$$

We get the desired equality by considering the coefficient X^{d-i}. $\square$

Lemma 15. *For $d \geq 1$, we have:*

$$\Delta_d(x) = \Delta_{d-1}(x^q)(x + x^{q^d})^{1+q+\cdots+q^{d-1}}.$$

Proof. The function Δ_d is a determinant of dimension $d+1$. We can note that the first line is composed of $d+1$ times the value 1 since for $0 \leq j \leq d$, $s_{0,d}^{q^j} = 1^{q^j} = 1$. We do not change the value of the determinant by adding each column to its right neighbor. After this operation, the first line is composed of one time the value 1 and d times the value 0. After this addition and using lemma 14, the term $(i+1, j+1)$ is:

$$s_{i+1,d}(x)^{q^j} + s_{i+1,d}(x)^{q^{j+1}} = (s_{i,d}(x) + s_{i,d}(x^q))^{q^j}$$
$$= s_{i,d+1}(x^q)^{q^j}(x + x^{q^d})^{q^j},$$

which correspond to the term (i, j) of $\Delta_{d-1}(x^q)$ times $(x + x^{q^d})^{q^j}$. By developing the determinant using its first row, we recover $\Delta_{d-1}(x^q)$ times the factors of each column, that is $\prod_{j=0}^{d-1}(x + x^{q^d})^{q^j}$. $\square$

Theorem 7. *For $d \geq 1$,*

$$\Delta_d(x) = \prod_{i=1}^{d} (x + x^{q^i})^{q^{d-i}+\dots+q^{d-1}}.$$

Proof. By induction. Indeed, the formula is straightforward for $d = 1$ and

$$\Delta_1(x) = \det \begin{pmatrix} 1 & 1 \\ x & x^q \end{pmatrix} = x + x^q.$$

Assume that it is true for $d - 1$, one gets:

$$\Delta_{d-1}(x^q) = \prod_{i=1}^{d-1} (x^q + x^{q^{i+1}})^{q^{d-1-i}+\dots+q^{d-2}}$$

$$\Delta_{d-1}(x^q) = \prod_{i=1}^{d-1} (x + x^{q^i})^{q^{d-i}+\dots+q^{d-1}}.$$

Using the formula of lemma 15, we get the result.   □

The Leakage-Resilience Limit of a Computational Problem Is Equal to Its Unpredictability Entropy

Divesh Aggarwal and Ueli Maurer

Department of Computer Science
ETH Zurich
CH-8092 Zurich, Switzerland
{diveshta,maurer}@inf.ethz.ch

Abstract. A cryptographic assumption is the (unproven) mathematical statement that a certain computational problem (e.g. factoring integers) is computationally hard. The leakage-resilience limit of a cryptographic assumption, and hence of a computational search problem, is the maximal number of bits of information that can be leaked (adaptively) about an instance, without making the problem easy to solve. This implies security of the underlying scheme against arbitrary side channel attacks by a computationally unbounded adversary as long as the number of leaked bits of information is less than the leakage resilience limit.

The hardness of a computational problem is typically characterized by the running time of the fastest (known) algorithm for solving it. We propose to consider, as another natural complexity-theoretic quantity, the success probability of the best polynomial-time algorithm (which can be exponentially small). We refer to its negative logarithm as the *unpredictability entropy* of the problem (which is defined up to an additive logarithmic term).

A main result of the paper is that the leakage-resilience limit and the unpredictability entropy are equal. This demonstrates, for the first time, the practical relevance of studying polynomial-time algorithms even for problems believed to be hard, and even if the success probability is too small to be of practical interest. With this view, we look at the best probabilistic polynomial time algorithms for the learning with errors and lattice problems that have in recent years gained relevance in cryptography.

We also introduce the concept of *witness compression* for computational problems, namely the reduction of a problem to another problem for which the witnesses are shorter. The length of the smallest achievable witness for a problem also corresponds to the *non-adaptive* leakage-resilience limit, and it is also shown to be equal to the unpredictability entropy of the problem. The witness compression concept is also of independent theoretical interest. An example of an implication of our result is that 3-SAT for n variables can be witness compressed from n bits (the variable assignments) to $0.41n$ bits.

D.H. Lee and X. Wang (Eds.): ASIACRYPT 2011, LNCS 7073, pp. 686–701, 2011.

1   Introduction and Motivation

1.1   Leakage Resilience of Cryptographic Assumptions

There have been many recent works (e.g., [2,3,10,13,16,14,21,33,37,38,40,42,43], and the references therein) aimed at designing cryptographic schemes that are secure against a large class of side-channel attacks. Some of these look at side channel attacks where the adversary can obtain some function of the secret key. We look at an even more general class of side-channel attacks where the adversary can obtain a *bounded amount of arbitrary information*. We model this kind of attack by allowing the adversary a bounded number of queries to an infinitely powerful oracle $\mathcal{O}$ that can be asked arbitrary binary (YES/NO) questions. This oracle was considered by Maurer [37] to study the hardness of factoring N given queries to this oracle.

Goldwasser et al [24] raised a more general question regarding leakage which is also the question that we are concerned with: Which of the cryptographic assumptions (rather than cryptographic schemes) are secure in the presence of leakage of some bits of information?

1.2   Complexity Notions

In this section, we introduce three notions, *unpredictability entropy*, *oracle complexity*, and *witness compressibility*, whose relationship we study in this paper.

A well-studied and realistic approach in the study of the computational complexity of a computational problem is to look at probabilistic polynomial time (PPT) algorithms that solve the problem. We define the *unpredictability entropy* [31] of a problem (essentially) as $-\log_2 p$, where p is the maximum possible success probability of a PPT algorithm for solving the problem. A common understanding is that the study of probabilistic algorithms makes sense only if the probability of success is non-negligible. While there have been a few results like [6,7,9,12,17,18,19,23,44,48] that look at the class of one-sided error probabilistic polynomial time (OPP) algorithms for decision problems with negligible success probability p, these are studied with the viewpoint of improving the bound on the exact worst-case complexity of the problem by repeating the algorithm $O(1/p)$ times and hence amplifying the success probability to a non-negligible quantity. However, we argue that PPT algorithms are interesting even if the success probability p is negligible and even if there exist other exact algorithms that run in time much less than $O(1/p)$.

Maurer [37] considered a class of PPT algorithms for search problems given the oracle $\mathcal{O}$. If the algorithm is allowed as many binary queries to the oracle as is the length of the solution/witness, then there is a trivial algorithm that solves the problem. Thus, this class of algorithms is looked at with the goal of minimizing the number of queries. The minimum number of queries required by a PPT algorithm for solving this problem with overwhelming probability is the *oracle complexity* (which is the same as the leakage-resilience limit) of the problem. A motivation for looking at such an oracle, as pointed out by the author, is to determine whether the difficulty of a certain problem can be concentrated in a few difficult bits leading to a new complexity theoretic classification of problems. This question was answered in the affirmative for the integer factorization problem in [37] but it remains open for other computational problems. Consider, for instance, the problem of computing discrete logarithms modulo a prime q. One can see that the hardness of this problem and the integer factorization problem is closely related in the sense that almost all algorithms for solving the factoring problem have a variant that solves the discrete logarithm problem modulo a prime. A survey of this can be found in Chapter 3 of [28]. However, the hardness of the two problems seems to differ significantly in terms of the number of queries to $\mathcal{O}$ required in order to solve these problems in polynomial time. Factoring can be solved with a small number of queries but, to the best of our knowledge, there exists no algorithm that solves the discrete logarithm problem with a non-trivial number (i.e., substantially less than the solution size) of queries to $\mathcal{O}$. Thus, finding the oracle complexity seems to be an interesting research area in itself.

We introduce another related notion called the *witness compressibility* of a problem. This is the smallest size k such that there is a PPT reduction that reduces the witness size of a given instance to at most k with overwhelming probability. This quantity can be seen as the non-adaptive leakage-resilience limit of an assumption about the hardness of the problem. A problem is not resilient to k bits of non-adaptive leakage if and only if it is witness compressible up to k bits. [1]

Note that the three quantities, i.e., unpredictability entropy, oracle complexity, and witness compressibility can only be defined up to an additive logarithmic term (see Section 2.2).

1.3   Our Contributions

We show that for all search problems with an efficiently computable verification predicate, the following are equivalent.

(i) There exists a PPT algorithm that solves a problem S with success probability $\Theta(2^{-k})$.

(ii) There exists a PPT algorithm that makes at most k queries to $\mathcal{O}$ and solves the problem S with a constant success probability.

(iii) There exists a PPT reduction that reduces the witness size of a given instance of S to at most k with constant probability.

This implies that the three quantities, i.e., unpredictability entropy, oracle complexity, and witness compressibility are essentially equal.

From this result, we get an exact characterization of the leakage-resilience of a cryptographic assumption about the hardness of some computational problem S in terms of the best possible PPT algorithm for S. A cryptographic assumption is robust up to k bits of leakage if and only if there is an algorithm that solves

---

[1] Witness compression should not be confused with instance compression that has been studied in [29,20].

the corresponding problem with probability $\Theta(2^{-k})$. This provides motivation
for improving the success probability of PPT algorithms for various computational problems. With this goal in mind, we present in this paper the best PPT
algorithms for some problems relevant in cryptography - in particular for the
learning with errors and lattice problems that have recently gained substantial
importance in cryptography.

The results of this paper also raise some interesting questions in complexity theory. One question this paper draws attention to is the following: Which
problems have optimal witness size, or stated differently, which problems can
or cannot be efficiently reduced to problems with a smaller witness size? Combining the results of [44] with our result gives evidence that the witness size
of Circuit-SAT cannot be compressed under reasonable complexity theoretic assumptions. However, for instance if we look at the 3-SAT problem, which is also
an NP-complete problem, combining our results with Schöning's PPT algorithm
[48] that solves 3-SAT with probability $(4/3)^{-n}$, we conclude that the witness of
3-SAT can be compressed to a $\log_2 4/3$-fraction, i.e., about 41.5% of its original
size.

1.4   Organization of This Paper

In Section 2, we introduce the definitions of problems and complexity notions
mentioned in the introduction. In Section 3, we prove the witness compression
lemma and establish the equivalence of (i), (ii) and (iii) mentioned in Section
1.2. In Section 4 we give/mention the best known PPT algorithms for some
problems relevant in cryptography. In Section 5, we conclude and give a list of
open problems that emerge from the results of this paper.

2   Definitions

2.1   Computational Search Problems

A computational search problem S is characterized by an instance space $\mathcal{X}$, a
solution (or witness) space $\mathcal{W}$, and a (verification) predicate $V : \mathcal{X} \times \mathcal{W} \rightarrow \{0,1\}$. Each element of $\mathcal{X}$ and $\mathcal{W}$ is assumed to be represented as a bitstring.
In this paper, unless otherwise stated, we consider problems for which there is
a polynomial time algorithm that computes the predicate V. We call this set of
problems $\mathcal{PC}$.[2]

The instance space $\mathcal{X}$ can be partitioned into two sets: the set $\mathcal{X}_1$ and $\mathcal{X}_0$ of
instances for which there exists a witness and for which there exists no witness,
respectively, i.e.,

$$\mathcal{X}_1 := \{x \in \mathcal{X} \mid \exists\, w \in \mathcal{W},\ V(x,w) = 1\}, \text{ and}$$

$$\mathcal{X}_0 := \{x \in \mathcal{X} \mid \forall\, w \in \mathcal{W},\ V(x,w) = 0\}.$$

---

[2] The name of this class, $\mathcal{PC}$, is taken from [26].

The sets $\mathcal{X}_1$ and $\mathcal{X}_0$ are sometimes referred to as the set of YES instances and that of NO instances, respectively.

We define $\gamma^V : \mathcal{X}_1 \mapsto \mathbb{N}$ as the size of the smallest witness for a given x, i.e.,

$$\gamma^V(x) := \min_{w \in \mathcal{W}, V(x,w)=1} |w| \,.^3$$

A *search problem* is the problem of finding, for a given element $x \in \mathcal{X}_1$, a witness $w \in \mathcal{W}$ such that $V(x, w) = 1$. By $\mathcal{O}$, we denote the infinitely powerful oracle that can answer arbitrary binary questions. The oracle, and hence the language in which questions are asked can be defined freely, and hence need not be specified (it can be thought of as being universally quantified).

Let $p : \mathbb{N} \times \mathbb{N} \mapsto [0, 1]$ and $q : \mathbb{N} \times \mathbb{N} \mapsto \mathbb{N} \cup \{0\}$ be functions.

Definition 1. Let $\mathsf{S} = (\mathcal{X}, \mathcal{W}, V)$ be a search problem. An algorithm $\mathcal{F}$ is called a (p, q)-*solver* for S if for all $m, n \in \mathbb{N}$ and for all $x \in \mathcal{X}_1$ such that $|x| \leq m$ and $\gamma(x) \leq n$, $\mathcal{F}$ makes at most $q(m, n)$ queries to $\mathcal{O}$, and with probability at least $p(m, n)$, computes a $w \in \mathcal{W}$ such that $V(x, w) = 1$.

In the above definition, $\mathcal{F}$ is called *efficient* if it runs in time polynomial in the size of input.

2.2   Complexity Notions

Now, we introduce the notion of witness compressibility. A problem is k-witness compressible if there exists another predicate V' such that for any given instance of the problem, there exists a witness of length at most k with respect to V', and given this witness one can efficiently compute a witness with respect to V. More formally,

Definition 2. A search problem S defined by $\mathsf{S} = (\mathcal{X}, \mathcal{W}, V)$ is *(deterministic) k-witness compressible* if there exists a witness set $\mathcal{W}'$, a predicate $V' : \mathcal{X} \times \mathcal{W}' \mapsto \{0, 1\}$, and a polynomial time algorithm $T : \mathcal{X} \times \mathcal{W}' \mapsto \mathcal{W}$ such that for all $x \in \mathcal{X}_1$,

- $\gamma^{V'}(x) \leq k(|x|, \gamma^V(x))$.
- For all $w \in \mathcal{W}'$, $V'(x, w) = 1$ if and only if $V(x, T(x, w)) = 1$.

As has been often seen in complexity theory, the best known PPT algorithm/reduction is significantly faster than the best known deterministic polynomial time algorithm/reduction, e.g. primality testing. In fact sometimes the former exists but the latter eludes discovery. Thus it is reasonable to look at the following randomized version of the above definition.

Definition 3. A search problem S defined by $\mathsf{S} = (\mathcal{X}, \mathcal{W}, V)$ is *k-witness compressible* within ϵ if there exists a witness set $\mathcal{W}'$, an efficiently samplable random variable S that takes values from a set $\mathcal{S}$, a set of predicates $V'_S : \mathcal{X} \times \mathcal{W}' \mapsto \{0, 1\}$, and polynomial time algorithms $T_S : \mathcal{X} \times \mathcal{W}' \mapsto \mathcal{W}$ parametrized by S such that for all $x \in \mathcal{X}_1$,

---

[3] We omit the predicate V if it is clear from the context.

- $\Pr\left(\gamma^{V'_S}(x) \leq k\left(|x|, \gamma^V(x)\right)\right) \geq 1 - \epsilon(|x|, \gamma^V(x)).^4$
- For all $w \in \mathcal{W}', s \in \mathcal{S}$, $V'_s(x,w) = 1$ if and only if $V(x, T_s(x,w)) = 1$.

With these definitions in place we now define the three quantities that we show, in this paper, are (essentially) equal. Let $k = k(m,n)$ be some integer valued function.

Definition 4. A search problem S has *unpredictability entropy at most k* if there exists an efficient $(2^{-k}, 0)$-solver for S.

Definition 5. A search problem S has *oracle-complexity at most k* if there exists an efficient $(1 - \epsilon, k)$-solver for S for some negligible function $\epsilon(m,n).^5$

Definition 6. A search problem S *is k-witness compressible* if S is k-witness compressible within ϵ for some negligible function $\epsilon(m,n)$.

Note that in these definitions, $k(m,n)$ is unique only up to an additive term of $O(\log_2 m)$. Also note that we can have an alternative version of these definitions where, for instance, the unpredictability entropy is *equal* to $k(m,n)$ (again, up to an additive term of $O(\log_2 m)$) by saying that there exists an efficient $(2^{-k(m,n)}, 0)$-solver but no efficient $(2^{-k(m,n)+\omega(\log_2 m)}, 0)$-solver for S. However, it would be cumbersome to make these alternative definitions precise and so we avoid them.

3   Relations between Complexity Notions for Search Problems

3.1   Two Simple Results

In this section, we give two simple relations between complexity notions for search problems.

Lemma 1. *For any search problem S and any functions $p = p(m,n)$, $q = q(m,n)$, and $k = k(m,n) \leq q(m,n)$, if there exists an efficient (p,q)-solver for S, then there exists an efficient $(p \cdot 2^{-k}, q - k)$-solver for S.*

Proof. Let S be a search problem and let $\mathcal{F}$ be an efficient (p,q)-solver for S. Let $\mathcal{F}'$ be an algorithm that simulates $\mathcal{F}$ except that it guesses the answer to the last k oracle queries uniformly at random. Thus $\mathcal{F}'$ makes $q - k$ queries and guesses the answer to the k queries correctly with probability 2^{-k} and hence succeeds in solving S with probability at least $p \cdot 2^{-k}$.

It is folklore as observed by a number of papers, e.g., [40,3,4] that non-adaptive leakage-resilience is the same as adaptive leakage-resilience. This can be seen in our terminology by the following lemma.

---

[4] The witness length is at most k with probability at least $1 - \epsilon$, where k and ϵ are both functions of $|x|$ and $\gamma^V(x)$.

[5] The term negligible, like the term efficient, is in terms of the input size m. So, for any m large enough, and any n, and any polynomial $P(.)$, $\epsilon(m,n) < 1/P(m)$.

Lemma 2. *For any functions $k = k(m, n)$ and $\epsilon = \epsilon(m, n)$, every search problem is k-witness compressible within ϵ if and only if it has an efficient $(1 - \epsilon, k)$-solver.*

Proof. ($\Rightarrow$) The idea is that, with probability $1 - \epsilon$, the witness size of an instance is reduced to size k, and hence we can use k queries to $\mathcal{O}$ to obtain a witness for the resulting instance.

Let $\mathsf{S} = (\mathcal{X}, \mathcal{W}, V)$ be the search problem. There exists some $\mathcal{W}'$, $V'_S : \mathcal{X} \times \mathcal{W}' \mapsto \{0, 1\}$ and $T_S : \mathcal{X} \times \mathcal{W}' \mapsto \mathcal{W}$ as in Definition 3. We give a polynomial time algorithm $\mathcal{F}$ that is a $(1 - \epsilon, k)$-solver for S. On input $x \in \mathcal{X}$, $\mathcal{F}$ generates $S = s$ and then uses k queries to $\mathcal{O}$ to ask for w', the string formed from the last k bits of a smallest length witness $w \in \mathcal{W}'$ (if it exists) such that $V'_s(x, w) = 1$. Then the algorithm outputs $T_s(x, w')$.

Let $m = |x|$ and $n = \gamma^V(x)$. With probability at least $1 - \epsilon$, $S = s$ such that the conditions of Definition 3 hold. Thus, $w' = w$ since $\gamma^{V'_s}(x) \leq k$. Hence $V'_s(x, w') = 1$, which implies $V(x, T_s(x, w')) = 1$.

($\Leftarrow$) Let $\mathcal{F}$ be a $(1 - \epsilon, k)$-solver for S. Define $\mathcal{W}'$ as the set of all bitstrings and let S denote the random choices made by $\mathcal{F}$. Define $T_s(x, w)$ to be the output of $\mathcal{F}$ on input x, $S = s$ and the result of the oracle queries equal to w. Further, define $V'_s(x, w)$ as $V(x, T_s(x, w))$. This gives the desired result.

3.2   The Witness Compression Lemma

We state a few lemmas that we need in order to prove the main lemma of this section.

Lemma 3. *Let $Y_1, \ldots, Y_t$ be pairwise independent binary random variables where $\Pr(Y_i = 1) = p$ for $1 \leq i \leq t$. Then*

$$\Pr(\exists i \in \{1, \ldots, t\} \ : \ Y_i = 1) \geq \max(tp - \frac{t^2 p^2}{2}, 1 - \frac{1}{tp})$$

Proof. We give two ways to bound the term on the left. Using Bonferroni inequalities [15],

$$\Pr(\exists i \in \{1, \ldots, t\} \ : \ Y_i = 1) = \Pr(Y_1 = 1 \vee Y_2 = 1 \vee \cdots \vee Y_t = 1)$$

$$\geq \sum_{1 \leq i \leq t} \Pr(Y_i = 1) - \sum_{1 \leq i_1 < i_2 \leq t} \Pr(Y_{i_1} = 1 \wedge Y_{i_2} = 1)$$

$$= tp - \frac{t(t-1)}{2} p^2$$

$$\geq tp - \frac{t^2 p^2}{2} \ .$$

Now, let $Y = Y_1 + \cdots + Y_t$. The expected value of Y is $E(Y) = tp$ and the variance of Y is $Var(Y) = tp(1 - p)$. Thus,

$$\Pr\left(\exists i \in \{1,\dots,t\} \, : \, Y_i = 1\right) = 1 - \Pr(Y = 0)$$
$$\geq 1 - \Pr\left(|Y - E(Y)| \geq E(Y)\right)$$
$$\geq 1 - \frac{Var(Y)}{E(Y)^2}$$
$$= 1 - \frac{1-p}{tp} \geq 1 - \frac{1}{tp} \, ,$$

where the second last inequality follows from the Chebyshev's inequality.

Lemma 4. *Let $\mathbb{F}$ be a finite field of cardinality 2^ℓ, let ϕ be a bijection from $\mathbb{F}$ to $\{0,1\}^\ell$, and let $T \subset \{0,1\}^\ell$. Further, let $y_1, \dots, y_t$ be some fixed distinct elements of $\mathbb{F}$. Then, for randomly chosen $A, B \in_R \mathbb{F}$, the probability that $\phi\left(Ay_i + B\right) \in T$ for some $1 \leq i \leq t$ is at least $\max\left(\frac{t|T|}{2^\ell} - \frac{t^2|T|^2}{2(2^\ell)^2}, 1 - \frac{2^\ell}{t|T|}\right)$.*[6]

Proof. Define binary random variables $Y_1, \dots, Y_t$ such that $Y_i = 1$ if $\phi\left(Ay_i + B\right) \in T$. Thus,

$$\Pr(Y_i = 1) = \frac{|T|}{2^\ell} \, ,$$

and it can be easily seen that the Y_i's are pairwise independent random variables. Therefore, by Lemma 3, the probability that $\phi\left(Ay_i + B\right) \in T$ for some $1 \leq i \leq t$ is at least

$$\max\left(\frac{t|T|}{2^\ell} - \frac{t^2|T|^2}{2(2^\ell)^2}, 1 - \frac{2^\ell}{t|T|}\right).$$

Now, we state the main lemma of this section.

Lemma 5. [Witness Compression Lemma] *Let $k = k(m,n)$ and $k' = k'(m,n) \geq k(m,n)$ be any functions. Every search problem with an efficient $(2^{-k}, 0)$-solver is k'-witness compressible within $\frac{1}{2^{k'-k}}$.*

Proof. Let $\mathsf{S} = (\mathcal{X}, \mathcal{W}, V)$ be a search problem and let $\mathcal{F}$ be an efficient $(2^{-k}, 0)$-solver for S. For a given input instance $x \in \mathcal{X}_1$, let $R \in \{0,1\}^\ell$ denote the random choices made by $\mathcal{F}$. Then,

$$\Pr\left(V(x, \mathcal{F}(x, R)) = 1\right) \geq 2^{-k} \, . \tag{1}$$

We define the set $\mathcal{R}(x)$ as the set of r such that $\mathcal{F}$ is successful in finding a witness for x for this choice of r, i.e.,

$$\mathcal{R}(x) = \{r \in \{0,1\}^\ell \mid V(x, \mathcal{F}(x, r)) = 1\} \, .$$

From (1), it follows that $|\mathcal{R}(x)| \geq 2^{\ell - k}$ for all $x \in \mathcal{X}_1$.

Now, let $\mathbb{F}, \phi, A, B$ and $y_1, \dots, y_t$ be as in Lemma 4. Thus, by using the second bound from Lemma 4, with $t = 2^{k'}$, and $T = \mathcal{R}(x)$, we get

$$\Pr\left(\exists 1 \leq i \leq 2^{k'} : \phi\left(Ay_i + B\right) \in \mathcal{R}(x)\right) \geq 1 - \frac{1}{2^{k'-k}} \, .$$

---

[6] Note that the result of this lemma will hold for any pairwise independent random function from $\mathbb{F}$ to itself, instead of $Ay + B$.

Then, let $S = (A, B)$ be uniformly distributed over $\mathbb{F} \times \mathbb{F}$. Furthermore, define $\mathcal{W}' = \{0, 1\}^*$,

$$T_S(x, w) = \mathcal{F}(x, \phi(Ay_w + B)), \text{ and } V_S'(x, w) = V(x, T_S(x, w)) .$$

In the above argument, we can also use the first bound from Lemma 4 to show that the problem is k-witness compressible within $\frac{1}{2}$.

3.3   The Main Result

Combining the results of Lemma 1, 2, and 5, we get the following result:

Theorem 1. *For any search problem* S, *and for any functions* $k = k(m, n)$ *and* $c = c(m, n) = \omega(\log_2 m)$:

- *If* S *is* k-*witness compressibile, then* S *has oracle complexity at most* k.
- *If* S *has oracle complexity at most* k, *then* S *has unpredictability entropy at most* k.
- *If* S *has unpredictability entropy at most* k, *then* S *is* $k + c$-*witness compressibile.*

Note that the results of this section are useful only if $k(m, n) = \omega(\log_2 m)$ because otherwise the corresponding search problem is solvable in expected polynomial time. Thus, without loss of generality, we can assume $k(m, n) = \omega(\log_2 m)$ and then choosing $c(m, n)$ as any function asymptotically smaller than k but larger than $\log_2 m$ (e.g. $c = \sqrt{k}$), we get that the three quantities in Theorem 1 are essentially equivalent for functions in $k + o(k)$.

Remark 1: Theorem 1 implies that an assumption of the hardness of a search problem S is secure up to k bits of leakage of arbitrary information if and only if there is no PPT algorithm that succeeds in solving S with probability $\Theta(2^{-k})$. However, the hardness assumptions we consider are worst case assumptions and not average case assumptions, which are more relevant in practice. Note that this is not a disadvantage, since our result implies a corresponding result for average case assumptions, just by restricting the set of instances of the problem to those where the problem is successful with significant (though possibly exponentially small) probability.

Remark 2: A similar result as Theorem 1 can also be proved for decision problems (using essentially the same proofs) but for that we need to be more careful in defining the oracle complexity of a problem and also the success probability of a PPT algorithm and we do not do so in this version of the paper.

4   PPT Algorithms for Problems Relevant in Cryptography

In this section, we give the best PPT algorithms known for various search problems relevant in cryptography. We look in more detail at the learning with errors and lattice problems that have been of interest in cryptography in recent years.

4.1 Factoring and Discrete Logarithms

There is a sequence of results [47,11,30] that show that partial information about p and q is enough to factor the RSA modulus pq. The best result in this direction is the result in [37] that, under a conjecture, shows that there is a polynomial time algorithm that factors N given $\epsilon \log_2 N$ questions to $\mathcal{O}$ where ϵ is some arbitrary constant. Equivalently, there exists a PPT algorithm that factors N with probability $2^{-\epsilon \log_2 N}$.

Even though the problem of computing discrete logarithms modulo a prime is closely related to the problem of factoring integers, to the best of our knowledge, there exists no non-trivial PPT algorithm for solving discrete logarithms in $\mathbb{Z}_p$. The same holds for the Computational Diffie Hellman problem.

It would be interesting to come up with an algorithm for solving discrete logarithm modulo a prime p that runs in time polynomial in $\log_2 p$ and succeeds with probability better than the trivial $\frac{\mathrm{poly}(\log_2 p)}{p}$.

4.2 Lattices

Preliminaries An n-dimensional lattice is a discrete additive subgroup of $\mathbb{R}^n$. A set of linearly independent vectors that generates a lattice is called a basis and is denoted by $\mathbf{B} = \{\mathbf{b}_1, \ldots, \mathbf{b}_n\} \subset \mathbb{R}^n$. The lattice Λ generated by the basis $\mathbf{B}$ is

$$\Lambda = \mathcal{L}(\mathbf{B}) = \left\{ \mathbf{Bz} = \sum_{i=1}^{n} z_i \mathbf{b}_i \; : \; z \in \mathbb{Z}^n \right\}.$$

For any point $\mathbf{t} \in \mathbb{R}^n$, the distance of $\mathbf{t}$ to the closest point in the lattice is written as $\mathrm{dist}(\mathbf{t}, \Lambda)$.

The Gram Schmidt orthogonalization of $\mathbf{B}$, denoted as $\{\tilde{\mathbf{b}}_1, \ldots, \tilde{\mathbf{b}}_n\}$, is defined as

$$\tilde{\mathbf{b}}_i = \mathbf{b}_i - \sum_{j=1}^{i-1} \mu_{i,j} \tilde{\mathbf{b}}_j, \text{ where } \mu_{i,j} = \frac{\langle \mathbf{b}_i, \tilde{\mathbf{b}}_j \rangle}{\langle \tilde{\mathbf{b}}_j, \tilde{\mathbf{b}}_j \rangle}.$$

By $\lambda_1(\Lambda)$, we denote the length of the shortest non-zero vector of the lattice Λ. For this paper, the lengths are always assumed to be in the ℓ_2 norm. If the lattice is clear from the context, then we write it simply as λ_1. It is well known and can be shown easily that

$$\lambda_1 \geq \min_i \|\tilde{\mathbf{b}}_i\|.$$

Definition 7. A basis $\mathbf{B} = \{\mathbf{b}_1, \ldots, \mathbf{b}_n\}$ is a δ-LLL Reduced Basis [35] if the following holds:

- $\forall\, 1 \leq j < i \leq n,\ \mu_{i,j} \leq \frac{1}{2}$,
- $\forall\, 1 \leq i < n,\ \delta \|\tilde{\mathbf{b}}_i\|^2 \leq \|\mu_{i+1,i} \tilde{\mathbf{b}}_i + \tilde{\mathbf{b}}_{i+1}\|^2$.

We choose $\delta = \frac{3}{4}$ and then it can be easily seen (e.g., refer to [25]) from the above definition that for a δ-LLL reduced basis, $\forall\, 1 \leq i < n$, $\|\mathbf{b}_i\| \leq \sqrt{2}\|\mathbf{b}_{i+1}\|$. Since there is an efficient algorithm [35] to compute an LLL-reduced basis, we assume, unless otherwise stated, that the given basis is always LLL-reduced and hence satisfies the above mentioned properties.

Now, we define some problems over lattices that we are interested in for this paper.

Definition 8. The *shortest vector problem* is defined as follows: Given a basis $\mathbf{B}$ of an n-dimensional lattice $\Lambda = \mathcal{L}(\mathbf{B})$, it is required to find a vector $\mathbf{v} \in \Lambda$ such that $\|\mathbf{v}\| = \lambda_1$.

A decision variant, whose hardness many cryptographic schemes are based on, is the gap shortest vector problem defined as follows.

Definition 9. The *gap shortest vector problem* GapSVP_γ for some $\gamma = \gamma(n)$ is defined as follows: Given a basis $\mathbf{B}$ of an n-dimensional lattice $\Lambda = \mathcal{L}(\mathbf{B})$ and $d > 0$ such that $d \notin [\lambda_1/\gamma, \lambda_1)$, decide whether $d \geq \lambda_1$ or $d < \lambda_1/\gamma$.

Next we define the closest vector problem (CVP) and bounded distance decoding (BDD) which is a special case of the CVP.

Definition 10. The *closest vector problem* CVP is defined as follows: Given a basis $\mathbf{B}$ of an n-dimensional lattice $\Lambda = \mathcal{L}(\mathbf{B})$, and $\mathbf{t} \in \mathbb{R}^n$, find $\mathbf{v} \in \Lambda$ such that $\|\mathbf{v} - \mathbf{t}\| = \mathrm{dist}(\mathbf{t}, \Lambda)$.

Definition 11. The α-*bounded distance decoding problem* BDD_α for some $0 < \alpha = \alpha(n) < 1/2$ is defined as follows: Given a basis $\mathbf{B}$ of an n-dimensional lattice $\Lambda = \mathcal{L}(\mathbf{B})$, and $\mathbf{t} \in \mathbb{R}^n$ such that $\mathrm{dist}(\mathbf{t}, \Lambda) \leq \alpha\lambda_1$, find $\mathbf{v} \in \Lambda$ such that $\|\mathbf{v} - \mathbf{t}\| = \mathrm{dist}(\mathbf{t}, \Lambda)$.

Shortest Vector Problem In this section, we give a polynomial time algorithm that computes the shortest vector of a lattice with probability $\frac{1}{2^{(n+1)(n+2)/4}}$. This algorithm, of course, also solves the GapSVP problem.

Theorem 2. *There exists a polynomial algorithm that, given a basis $\mathbf{B}$ of a lattice $\Lambda = \mathcal{L}(\mathbf{B})$, finds the shortest vector of Λ with probability $\frac{1}{2^{(n+1)(n+2)/4}}$.*

Proof. Since an LLL-reduced basis can be computed efficiently, we assume without loss of generality that $\mathbf{B}$ is an LLL-reduced basis. Let the shortest vector $\mathbf{u}$ of the lattice be $\mathbf{u} = a_1\tilde{\mathbf{b}}_1 + a_2\tilde{\mathbf{b}}_2 + \cdots + a_n\tilde{\mathbf{b}}_n$. Since $\tilde{\mathbf{b}}_1 = \mathbf{b}_1$ is a lattice vector, therefore $\|\mathbf{u}\| \leq \|\tilde{\mathbf{b}}_1\|$. By the property of the LLL basis, $\|\tilde{\mathbf{b}}_1\| \leq 2^{(i-1)/2}\|\tilde{\mathbf{b}}_i\|$, which implies $\|\mathbf{u}\| \leq 2^{(i-1)/2}\|\tilde{\mathbf{b}}_i\|$. Thus, $|a_i| \leq 2^{(i-1)/2}$. The component a_i is determined by the coefficients of $\mathbf{b}_i, \ldots, \mathbf{b}_n$ in $\mathbf{u}$. Thus, given the coefficients of $\mathbf{b}_n, \ldots, \mathbf{b}_{i+1}$, the coefficient of $\mathbf{b}_i$ can be chosen correctly with probability $1/(2 \cdot 2^{(i-1)/2}) = 2^{-(i+1)/2}$. This gives a polynomial time algorithm that succeeds in finding the shortest vector with probability

$$\prod_{i=1}^n 2^{-(i+1)/2} = 2^{-(n+1)(n+2)/4} .$$

Closest Vector Problem In this section, we give a polynomial time algorithm that solves the closest vector problem with probability $\frac{1}{2^{n(n+1)/4}}$.

Theorem 3. *There exists a polynomial time algorithm that, given a basis* $\mathbf{B}$ *of an n-dimensional lattice* $\Lambda = \mathcal{L}(\mathbf{B})$, *and* $\mathbf{t} \in \mathbb{R}^n$, *finds* $\mathbf{v} \in \Lambda$ *such that* $\|\mathbf{v} - \mathbf{t}\| = dist(\mathbf{t}, \Lambda)$ *with probability* $\frac{1}{2^{n(n+1)/4}}$.

Proof. Let $\mathbf{t} = \rho_1 \tilde{\mathbf{b}}_1 + \ldots \rho_n \tilde{\mathbf{b}}_n$ and let the closest vector to $\mathbf{t}$ in the lattice be $\mathbf{u} = a_1 \mathbf{b}_1 + a_2 \mathbf{b}_2 + \cdots + a_n \mathbf{b}_n$. Babai's algorithm [5] returns a vector $\mathbf{x}$ such that $\|\mathbf{x} - \mathbf{t}\| \le \frac{1}{2} 2^{n/2} \|\tilde{\mathbf{b}}_n\|$. Thus $\|\mathbf{u} - \mathbf{t}\| \le \|\mathbf{x} - \mathbf{t}\| \le \frac{1}{2} 2^{n/2} \|\tilde{\mathbf{b}}_n\|$, which implies $|a_n - \rho_n| \le \frac{1}{2} 2^{n/2}$. Thus the algorithm proceeds as follows: Choose $\hat{a}_n$ uniformly at random from $(\rho_n - \frac{1}{2} 2^{n/2}, \rho_n + \frac{1}{2} 2^{n/2})$ and recursively compute the closest vector to $\mathbf{t} - \hat{a}_n \mathbf{b}_n$ in the lattice $\mathcal{L}(\mathbf{b}_1, \ldots, \mathbf{b}_{n-1})$. The probability that $(\hat{a}_1, \ldots, \hat{a}_n) = (a_1, \ldots, a_n)$ is

$$\prod_{i=1}^{n} 2^{-i/2} = 2^{-n(n+1)/4} .$$

Bounded Distance Decoding (BDD) Problem The algorithm given in the previous section, of course, also solves the BDD problem since BDD is a special case of the closest vector problem. However, there exists an algorithm for BDD_α with a larger success probability $\frac{1}{\alpha^n 2^{(n+1)(n+2)/4}}$ as given below.

Theorem 4. *There exists a polynomial time algorithm that, given a basis* $\mathbf{B}$ *of an n-dimensional lattice* $\Lambda = \mathcal{L}(\mathbf{B})$, *and* $\mathbf{t} \in \mathbb{R}^n$ *such that* $dist(\mathbf{t}, \Lambda) \le \alpha \lambda_1$ *for some* $0 < \alpha(n) < 1/2$, *finds* $\mathbf{v} \in \Lambda$ *such that* $\|\mathbf{v} - \mathbf{t}\| = dist(\mathbf{t}, \Lambda)$ *with probability* $\frac{1}{\alpha^n 2^{(n+1)(n+2)/4}}$.

Proof. Since an LLL-reduced basis can be computed efficiently, we assume without loss of generality that $\mathbf{B}$ is an LLL-reduced basis. Let $\mathbf{t} = t_1 \tilde{\mathbf{b}}_1 + t_2 \tilde{\mathbf{b}}_2 + \cdots + t_n \tilde{\mathbf{b}}_n$ and the closest vector $\mathbf{u}$ of the lattice be $\mathbf{u} = u_1 \tilde{\mathbf{b}}_1 + u_2 \tilde{\mathbf{b}}_2 + \cdots + u_n \tilde{\mathbf{b}}_n$. Since $\tilde{\mathbf{b}}_1 = \mathbf{b}_1$ is a lattice vector, therefore $\|\mathbf{u} - \mathbf{t}\| \le \alpha \|\tilde{\mathbf{b}}_1\|$. By the property of the LLL basis, $\|\tilde{\mathbf{b}}_1\| \le 2^{(i-1)/2} \|\tilde{\mathbf{b}}_i\|$, which implies $\|\mathbf{u} - \mathbf{t}\| \le 2^{(i-1)/2} \alpha \|\tilde{\mathbf{b}}_i\|$. Thus, $|u_i - t_i| \le \alpha 2^{(i-1)/2}$. The component u_i is determined by the coefficients of $\mathbf{b}_i, \ldots, \mathbf{b}_n$ in $\mathbf{u}$. Thus, given the coefficients of $\mathbf{b}_n, \ldots, \mathbf{b}_{i+1}$, the coefficient of $\mathbf{b}_i$ can be chosen correctly with probability $1/(2\alpha \cdot 2^{(i-1)/2}) = 2^{-(i+1)/2} \alpha^{-1}$. This gives a polynomial time algorithm that succeeds in finding the shortest vector with probability

$$\prod_{i=1}^{n} \frac{1}{2^{(i+1)/2} \alpha} = \frac{1}{\alpha^n 2^{(n+1)(n+2)/4}} .$$

4.3   Learning with Errors and Its Relation to Lattice Problems

In this section, we mention the best PPT algorithm for the learning with errors (LWE) problem and its relation to the lattice problems with respect to leakage. The proofs and other details are omitted.

Theorem 5. *For some function $\beta = \beta(n)$ such that $\beta q = \omega(\log_2 n)$, $\beta(n) = o(1/\log_2 n)$ and $m \geq n$, there is a polynomial time algorithm that solves search-$LWE_{n,q,m,\overline{\psi}_\beta}$ with probability $(\beta q \log_2 n)^{-n}$ for a constant fraction α of the inputs.*

Note that it is straightforward to interpret the above mentioned algorithms for LWE and lattice problems as PPT algorithms that succeed with constant probability given $-\log_2 p$ queries to $\mathcal{O}$ (or equivalently $-\log_2 p$ bits of leakage), where p is the success probability of the algorithm. We do not need the witness compression lemma to make this conclusion. The witness compression lemma however implies that if there is any PPT algorithm for any of these problems that succeeds with probability $p' < p$, then there is a PPT algorithm that makes $-\log_2 p'$ queries to $\mathcal{O}$ and succeeds with constant probability.

It is common practice to base the LWE-based schemes on the hardness of lattice based schemes. In the same spirit, by a careful inspection of the reduction of BDD to LWE from [46], we get the following result:

Theorem 6. *If there exists a PPT algorithm that solves search-LWE_{n,q,m,ψ_β} with probability p then there exists a PPT algorithm that solves $BDD_{\frac{\beta}{n}}$ with probability $cp^{\lceil n/\log_2 q \rceil}$) for some constant c.*

By Theorem 6, we can base the LWE assumption with leakage on the exponential hardness of the BDD assumption as follows.

Corollary 1. *If there exists no polynomial time algorithm that solves $BDD_{\frac{\beta}{n}}$ with probability $2^{-\delta n^2}$, then the search-LWE_{n,q,m,ψ_β} assumption is robust to $\delta n \log_2 q - o(\log_2 q)$ bits of leakage.*

5   Conclusions and Open Problems

We show that the unpredictability entropy of a problem is equal to its leakage-resilience limit. This provides motivation to look at PPT algorithms for problems relevant in cryptography with maximum possible success probability. A question that is wide open is to what extent can the success probability of PPT algorithms be improved for various problems like the discrete logarithm problem, search LWE problem or various lattice problems. Note that if we repeatedly run algorithms for lattice problems given in Section 5 to amplify the success probability to a non-negligible quantity, we get algorithms with running time $2^{O(n^2)}$, which is much worse than the best known algorithms that run in time $2^{O(n)}$ [1,39]. Due to this large gap, one might expect that it should be possible to improve the success probability of a PPT algorithm and this has eluded discovery because of lack of attention to this question.

The witness compression lemma implies that the best known PPT algorithms, for instance [6,7,9,12,17,18,19,23], immediately give a lower bound on the maximum witness compressibility of the corresponding problems.

The results of [44] give evidence that perhaps Circuit-SAT is not witness compressible to any non-trivial witness size. In fact the result of [44], which shows that there exists no non-trivial PPT algorithm for Circuit-SAT (and hence for all **NP** problems) under reasonable complexity assumptions, can be proved by proving a decision version of the witness compression lemma. If there exist non-trivial PPT algorithms for all **NP** problems, we can repeatedly apply the witness compression lemma until the witness size is reduced to a constant, thus resulting in a sub-exponential time algorithm for any **NP** problem, which is not believed to be possible. It is interesting to look at the question of which are the other problems that, like Circuit-SAT are not witness compressible. The discrete logarithm problem modulo a prime seems to be a candidate.

Another interesting research direction is to look at PPT-reductions, i.e., PPT algorithms for solving one "hard" problem given a PPT algorithm for solving another problem (with possibly negligible success probability). Consider, for instance, the reduction of [36] from GapSVP to BDD. This reduction was derived from the main idea of [41] in obtaining the first public key cryptosystem whose hardness was based on the GapSVP. This reduction does not seem to translate easily to the case of PPT algorithms, since given a BDD oracle that solves the problem with an exponentially small probability, it is not clear how to use it to solve the GapSVP problem. If such a reduction was possible, we could base the leakage-resilience of the search LWE assumption on the exponential hardness of the GapSVP problem.

References

1. Ajtai, M., Kumar, R., Sivakumar, D.: A sieve algorithm for the shortest lattice vector problem. In: STOC 2001, pp. 601–610 (2001)
2. Akavia, A., Goldwasser, S., Vaikuntanathan, V.: Simultaneous Hardcore Bits and Cryptography Against Memory Attacks. In: Reingold, O. (ed.) TCC 2009. LNCS, vol. 5444, pp. 474–495. Springer, Heidelberg (2009)
3. Alwen, J., Dodis, Y., Wichs, D.: Leakage-Resilient Public-Key Cryptography in the Bounded-Retrieval Model. In: Halevi, S. (ed.) CRYPTO 2009. LNCS, vol. 5677, pp. 36–54. Springer, Heidelberg (2009)
4. Alwen, J., Dodis, Y., Wichs, D.: Survey: Leakage Resilience and the Bounded Retrieval Model. In: Kurosawa, K. (ed.) Information Theoretic Security. LNCS, vol. 5973, pp. 1–18. Springer, Heidelberg (2010)
5. Babai, L.: On Lovász' Lattice Reduction and the Nearest Lattice Point Problem. Combinatorica 6(1), 1–13 (1986)
6. Beigel, R., Eppstein, D.: 3-coloring in Time $o(1.3446^n)$: A No-mis Algorithm. In: FOCS 1995, pp. 444–452 (1995)
7. Beigel, R.: Finding Maximum Independent Sets in Sparse and General Graphs. In: SODA 1999, pp. 856–857 (1999)
8. Blum, A., Kalai, A., Wasserman, H.: Noise-tolerant Learning, the Parity Problem, and the Statistical Query Model. Journal of the ACM 50(4), 506–519 (2003)
9. Byskov, J.: Algorithms for k-colouring and Finding Maximal Independent Sets. In: SODA 2003, pp. 456–457 (2003)
10. Canetti, R., Dodis, Y., Halevi, S., Kushilevitz, E., Sahai, A.: Exposure-Resilient Functions and All-or-Nothing Transforms. In: Preneel, B. (ed.) EUROCRYPT 2000. LNCS, vol. 1807, pp. 453–469. Springer, Heidelberg (2000)

11. Coppersmith, D.: Finding a Small Root of a Bivariate Integer Equation; Factoring with High Bits Known. In: Maurer, U.M. (ed.) EUROCRYPT 1996. LNCS, vol. 1070, pp. 178–189. Springer, Heidelberg (1996)
12. Dantsin, E., Goerdt, A., Hirsch, E., Kannan, R., Kleinberg, J., Papadimitriou, C., Raghavan, P., Schöning, U.: A Deterministic $(2 - 2/(k + 1))^n$ Algorithm for k-SAT Based on Local Search. Theoretical Computer Science 289(1), 69–83 (2002)
13. Dodis, Y., Goldwasser, S., Kalai, Y., Peikert, C., Vaikuntanathan, V.: Public-Key Encryption Schemes with Auxiliary Inputs. In: Micciancio, D. (ed.) TCC 2010. LNCS, vol. 5978, pp. 361–381. Springer, Heidelberg (2010)
14. Dodis, Y., Kalai, Y., Lovett, S.: On Cryptography with Auxiliary Input. In: STOC 2009, pp. 621–630 (2009)
15. Dohmen, K.: Improved Bonferroni Inequalities with Applications: Inequalities and Identities of Inclusion-Exclusion Type. Springer, Berlin (2003)
16. Dziembowski, S., Pietrzak, K.: Leakage-resilient Cryptography. In: FOCS 2008, pp. 293–302 (2008)
17. Eppstein, D.: Improved Algorithms for 3-coloring, 3-edge-coloring, and Constraint Satisfaction. In: SODA 2001, pp. 329–337 (2001)
18. Eppstein, D.: Small Maximal Independent Sets and Faster Exact Graph Coloring. Journal of Graph Algorithms and Applications 7, 131–140 (2003)
19. Fomin, F., Grandoni, F., Kratsch, D.: Measure and Conquer: A Simple $o(2^{0.288n})$ Independent Set Algorithm. In: SODA 2006, pp. 18–25 (2006)
20. Fortnow, L., Santhanam, R.: Infeasibility of instance compression and succinct PCPs for NP. Journal of Computer and System Sciences 77(1), 91–106 (2011)
21. Faust, S., Kiltz, E., Pietrzak, K., Rothblum, G.: Leakage-Resilient Signatures. In: Micciancio, D. (ed.) TCC 2010. LNCS, vol. 5978, pp. 343–360. Springer, Heidelberg (2010)
22. Goldreich, O., Goldwasser, S.: On the Limits of Nonapproximability of Lattice Problems. Journal of Computation and Systems Sciences 60(3), 540–563 (2000)
23. Gramm, J., Hirsch, E., Niedermeier, R., Rossmanith, P.: Worst Case Upper Bounds for Max-2-sat with an Application to Max-cut. Discrete Applied Mathematics 130(2), 139–155 (2003)
24. Goldwasser, S., Kalai, Y., Peikert, C., Vaikuntanathan, V.: Robustness of the Learning With Errors Assumption. In: ICS 2010. Tsinghua University Press, Beijing (2010)
25. Goldwasser, S., Micciancio, D.: Complexity of Lattice Problems: a Cryptographic Perspective. The Kluwer International Series in Engineering and Computer Science, vol. 671. Kluwer Academic Publishers, Boston
26. Goldreich, O.: Computational Complexity: A Conceptual Perspective. Cambridge University Press, NY
27. Gentry, C., Peikert, C., Vaikuntanathan, V.: Trapdoors for hard lattices and new cryptographic constructions. In: STOC 2008, pp. 197–206 (2008)
28. Gregg, J.: On Factoring Integers and Evaluating Discrete Logarithms. Bachelor's Thesis. Harvard College, Cambridge, Massachusetts
29. Harnik, D., Naor, M.: On the compressibility of NP instances and cryptographic applications. In: FOCS 2006, pp. 719–728 (2006)
30. Heninger, N., Shacham, H.: Reconstructing RSA Private Keys from Random Key Bits. In: Halevi, S. (ed.) CRYPTO 2009. LNCS, vol. 5677, pp. 1–17. Springer, Heidelberg (2009)
31. Hsiao, C., Lu, C., Reyzin, L.: Conditional Computational Entropy, or Toward Separating Pseudoentropy from Compressibility. In: Naor, M. (ed.) EUROCRYPT 2007. LNCS, vol. 4515, pp. 169–186. Springer, Heidelberg (2007)

32. Klein, P.: Finding the Closest Vector When it is Unusually Close. In: SODA 2000, pp. 937–941 (2000)
33. Katz, J., Vaikuntanathan, V.: Signature Schemes with Bounded Leakage Resilience. In: Matsui, M. (ed.) ASIACRYPT 2009. LNCS, vol. 5912, pp. 703–720. Springer, Heidelberg (2009)
34. Lenstra, H.: Factoring Integers with Elliptic Curves. Annals of Mathematics 126, 649–673 (1987)
35. Lenstra, A., Lenstra, H., Lovász, L.: Factoring Polynomials wth Rational Coefficients. Mathematische Annalen 261(4), 515–534 (1982)
36. Lyubashevsky, V., Micciancio, D.: On Bounded Distance Decoding, Unique Shortest Vectors, and the Minimum Distance Problem. In: Halevi, S. (ed.) CRYPTO 2009. LNCS, vol. 5677, pp. 577–594. Springer, Heidelberg (2009)
37. Maurer, U.: On the Oracle Complexity of Factoring Integers. Computational Complexity 5(4), 237–247 (1996)
38. Micali, S., Reyzin, L.: Physically Observable Cryptography. In: Naor, M. (ed.) TCC 2004. LNCS, vol. 2951, pp. 278–296. Springer, Heidelberg (2004)
39. Micciancio, D., Voulgaris, P.: A deterministic single exponential time algorithm for most lattice problems based on voronoi cell computations. In: STOC 2010, pp. 351–358 (2010)
40. Naor, M., Segev, G.: Public-key Cryptosystems Resilient to Key Leakage. In: Halevi, S. (ed.) CRYPTO 2009. LNCS, vol. 5677, pp. 18–35. Springer, Heidelberg (2009)
41. Peikert, C.: Public-Key Cryptosystems from the Worst-Case Shortest Vector Problem. In: STOC 2009 (2009)
42. Pietrzak, K.: A Leakage-Resilient Mode of Operation. In: Joux, A. (ed.) EUROCRYPT 2009. LNCS, vol. 5479, pp. 462–482. Springer, Heidelberg (2009)
43. Petit, C., Standaert, F., Pereira, O., Malkin, T., Yung, M.: A Block Cipher Based Pseudo Random Number Generator Secure Against Side-channel Key Recovery. In: ASIACCS 2008, pp. 56–65 (2008)
44. Paturi, R., Pudlák, P.: On the Complexity of Circuit Satisfiability. In: STOC 2010 (2010)
45. Paturi, R., Pudlák, P., Zane, F.: Satisfiability Coding Lemma. In: FOCS 1997, pp. 566–574 (1997)
46. Regev, O.: On Lattices, Learning with Errors, Random Linear Codes, and Cryptography. In: STOC 2005 (2005)
47. Rivest, R., Shamir, A.: Efficient Factoring Based on Partial Information. In: Pichler, F. (ed.) EUROCRYPT 1985. LNCS, vol. 219, pp. 31–34. Springer, Heidelberg (1986)
48. Schöning, U.: A Probabilistic Algorithm for k-SAT and Constraint Satisfaction Problems. In: FOCS 1999 (1999)
49. Sipser, M.: A Complexity Theoretic Approach to Randomness. In: STOC 1983, pp. 330–335 (1983)
50. Stockmeyer, L.: The Complexity of Approximate Counting. In: STOC 1983, pp. 118–126 (1983)
51. Valiant, L., Vazirani, V.: NP is as Easy as Detecting Unique Solutions. Theoretical Computer Science 47, 85–93 (1986)

Leakage-Resilient Cryptography from the Inner-Product Extractor

Stefan Dziembowski[1,*] and Sebastian Faust[2,**]

[1] University of Warsaw and Sapienza University of Rome
[2] Aarhus University

Abstract. We present a generic method to secure various widely-used cryptosystems against *arbitrary* side-channel leakage, as long as the leakage adheres three restrictions: first, it is bounded per observation but in total can be arbitrary large. Second, memory parts leak *independently*, and, third, the randomness that is used for certain operations comes from a simple (non-uniform) distribution.

As a fundamental building block, we construct a scheme to store a cryptographic secret such that it remains *information theoretically* hidden, even given arbitrary continuous leakage from the storage. To this end, we use a randomized encoding and develop a method to securely *refresh* these encodings even in the presence of leakage. We then show that our encoding scheme exhibits an efficient additive homomorphism which can be used to protect important cryptographic tasks such as identification, signing and encryption. More precisely, we propose *efficient* implementations of the Okamoto identification scheme, and of an ElGamal-based cryptosystem with security against continuous leakage, as long as the leakage adheres the above mentioned restrictions. We prove security of the Okamoto scheme under the DL assumption and *CCA2 security* of our encryption scheme under the DDH assumption.

1 Introduction

In the last years, a large body of work attempts to analyze the effectiveness of side-channel countermeasures in a mathematically rigorous way. These works propose a physical model incorporating a (mostly broad) class of side-channel attacks and design new cryptographic schemes that provably withstand them under certain assumptions about the physical hardware (see, e.g., [24,11,12,16,9,5,23] and many more). By now we have seen new constructions for many important cryptographic primitives such as digital signature and public key encryption

---

* The European Research Council has provided financial support to the first author of this paper under the European Community's Seventh Framework Programme (FP7/2007-2013) / ERC grant agreement no CNTM-207908.
** Sebastian Faust acknowledges support from the Danish National Research Foundation and The National Science Foundation of China (under the grant 61061130540) for the Sino-Danish Center for the Theory of Interactive Computation, within part of this work was performed. Part of this work was done while being at KU Leuven.

D.H. Lee and X. Wang (Eds.): ASIACRYPT 2011, LNCS 7073, pp. 702–721, 2011.

schemes that are provably secure against surprisingly broad classes of leakage attacks.

Unfortunately, most of these new constructions are rather complicated non-standard schemes, often relying on a heavy cryptographic machinery, which makes them less appealing for implementations on computationally limited devices. In this work, we take a different approach: instead of developing new cryptographic schemes, we ask the natural question whether standard, widely-used cryptosystems can be implemented *efficiently* such that they remain secure in the presence of continuous bounded leakage. We answer this question affirmatively, and show a *generic* way that "compiles" various common cryptosystems into schemes that remain secure against a broad class of leakage attacks.

Similar to earlier work, we make certain restrictions on the leakage. We follow the work of Dziembowski and Pietrzak [11], and allow the leakage to be arbitrary as long as the following two restrictions are satisfied:

1. **Bounded leakage:** the amount of leakage in each round is bounded to λ bits (but overall can be arbitrary large).
2. **Independent leakage:** the computation can be structured into rounds, where each such round leaks independently (we define the notion of a "round" below).

Formally, this is modeled by letting the adversary in each round choose a polynomial time computable leakage function f with range $\{0,1\}^\lambda$, and then giving her $f(\tau)$ where τ is all the data that has been accessed during the current round. In addition to these two restrictions, we require that our device has access to a source of correlated randomness generated in a *leak-free* way – e.g., computed by a simple leak free component. We elaborate in the following on our leakage restrictions.

ON THE BOUNDED LEAKAGE ASSUMPTION. Most recent work on leakage resilient cryptography requires that the leakage is bounded per observation to some fraction of the secret key. This models the observation that in practice many side-channel attacks only exploit a polylogarithmic amount of information, and typically require thousands of observations until the single key can be recovered. This is, for instance, the case for DPA-based attacks where the power consumption is modeled by a weighted sum of the computation's intermediate values. We would like to mention that all our results also remain true in the *entropy loss model*, i.e., we do not necessarily require that the leakage is bounded to λ bits, but rather only need that the min entropy of the state remains sufficiently high even after given the leakage.

ON INDEPENDENT LEAKAGES. In this paper, we assume that the memory of the device is divided into three parts L, R and C where (L, C) and (R, C) leak independently. To use the independent leakage assumption, we structure the computation into rounds, where each round only accesses either (L, C) or (R, C). Similar assumptions have been used in several works [24,11,27,21,12].

ON LEAK-FREE COMPONENTS. We require that devices that implement our schemes have access to a source of correlated randomness sampled in a leak-free way. Such a source can, for instance, be implemented by a probabilistic leak-free component that outputs the correlated randomness. Of course, the assumption of a leak-free component is a strong requirement on the hardware, but let us argue why in our particular case it still may be a feasible assumption. As in earlier works that made use of leak-free components [15,13,20,16], we require that our component leaks from its outputs, but the leakage function is oblivious to its internals. To be more concrete, in the simplest case our component $\mathcal{O}$ outputs two random vectors $A, B \leftarrow \mathbb{F}^n$ (with $\mathbb{F}$ being a finite field and n being a statistical security parameter) such that their inner product is 0, i.e., $\sum_i A_i \cdot B_i = 0$. We require that A gets stored on one part of the memory, while B gets stored on the other, thus, we require that A and B leak independently.

Our component $\mathcal{O}$ exhibits several properties that are beneficial for implementations. First, $\mathcal{O}$ is simple and small. It can be implemented in size linear in n, as one simply needs to sample uniformly at random vectors A and $(B_1, \ldots, B_{n-1})$ and computes the last element B_n such that $\sum_i A_i \cdot B_i = 0$.[1] Second, $\mathcal{O}$ is used in a very limited way, namely, it is needed only when the secret key gets refreshed (cf. Section 1.2 for further discussion on this). Finally, $\mathcal{O}$ does not take any inputs, and hence its computation is completely independent of the actual computation (e.g., encryption or signing) that is carried out by the device. This not only allows to test the component independently from the actual cryptoscheme that is implemented, but moreover makes it much harder to attack by side-channel analysis, as successful attacks usually require some choice (or at least knowledge) over the inputs.

1.1   Leakage Resilient Standard Cryptographic Schemes

While in the last years tremendous progress has been made in the design of new cryptographic schemes with built-in leakage resilience, two common criticisms are frequently brought up:

1. Cryptographic schemes are rarely used stand-alone, but more often are part of an industrial standard. Even if desirable, it is unlikely that in the near future these standards will be adjusted to include recent scientific progress.
2. Many of the current leakage resilient cryptoschemes are complicated, rely on non-standard complexity assumptions and are often rather inefficient.

In this work, we are interested in techniques that allow for efficient leakage resilient implementations of widely-used cryptographic schemes. Before we given an overview of our contributions in the next section, we discuss some related literature that considered a similar question.

LEAKAGE RESILIENT CIRCUIT COMPILERS. One fundamental question in leakage resilient cryptography is whether *any* computation can be implemented in a way that resists certain side-channel leakages. This question has been studied in a series of works [19,13,20,16] and dates back to the work of Ishai et al. [19]. In

---

[1] For simplicity, we assume that L_n is non-zero.

particular, the works of Juma and Vahlis [20] and Goldwasser and Rothblum [16] study the question whether any computation can be implemented in a way that withstands arbitrary polynomial-time computable leakages. As a building block they use a public-key encryption scheme and encrypt the entire computation of the circuit. More precisely, the approach of Juma and Vahlis makes use of fully homomorphic encryption, while Goldwasser and Rothblum generate for each Boolean wire of the circuit a new key pair and encrypt the current value on the wire using the corresponding key. We would like to emphasize that all circuit compilers (except for the one of Ishai et al.) require leak-free components. Notice also that the work of Goldwasser and Rothblum and Juma and Vahlis requires the independent leakage assumption.

LEAKAGE RESILIENT ELGAMAL. While circuit compilers allow to secure any (cryptographic) computation against leakage, they typically suffer from a large efficiency overhead. A recent work of Kiltz and Pietrzak [21] makes progress in this direction. The authors show that certain standard cryptographic schemes can be implemented *efficiently* in a leakage resilient way. The main weakness of this work is that the security proof is given in the generic group model.

1.2   Our Contribution

In this paper, we show a *generic* method to implement various standard cryptographic schemes that are provably secure in the above described leakage model. More precisely, we propose an efficient and simple implementation of the Okamoto authentication/signature scheme and of an ElGamal-based encryption scheme, and prove the security of our implementations under continuous leakage attacks. We also discuss why our techniques are fairly general and may find applications for the secure implementation of various other cryptographic schemes. As a fundamental tool, we introduce an information theoretically secure scheme to refresh an encoded secret in the presence of continuous leakage. We detail on our results below.

LEAKAGE RESILIENT REFRESHING OF ENCODED SECRETS. Recently, Davi et al. [8] introduced the notion of leakage resilient storage (LRS). An LRS encodes a secret S such that given partial knowledge about the encoding an adversary does not obtain any knowledge about the encoded secret S. One of their instantiations relies on the inner product two-source extractor introduced in the seminal work of Chor and Goldreich [7]. In this scheme the secret S is encoded as a pair $(L, R) \in \mathbb{F}^n \times \mathbb{F}^n$, where $\mathbb{F}$ is some finite field, and $\langle L, R \rangle := \sum_i L_i \cdot R_i = S$. Unfortunately, the construction of Davi et al. has one important weakness: it can trivially be broken if an adversary continuously leaks from the two parts L and R. The first contribution of this paper is to propose an efficient refreshing scheme for the inner product based encoding.

   This is achieved by dividing the memory of the device into three parts L, R and C, where initially (L, R) are chosen uniformly subject to the constraint that $\langle L, R \rangle = S$, and C is empty. Our refreshing scheme Refresh takes as input (L, R) and outputs a fresh encoding (L', R') of S. The computation of Refresh will be structured into several rounds, where in each round we only touch either (L, C) or (R, C), but never L and R at the same time. We will allow the adversary to adaptively leak a bounded amount of information from (L, C) and (R, C). In fact, this is the only assumption we make, i.e., we do not require that the rounds of the computation leak independently. Since in our protocol the third part C is only used to "communicate" information between L and R, we will usually describe our schemes in form of a 2-party protocol: one party, P_L, is controlling L, while the second party, P_R, holds R. The third part C is used to store messages that are exchanged between the parties. Hence, instead of saying that we allow the adversary to retrieve information from (L, C) and (R, C), we can say that the leakage functions take as inputs all variables that are in the view of P_L or P_R.

Our protocol for the refreshing uses the following basic idea. Suppose initially P_L holds L and P_R holds R with $\langle L, R \rangle = S$, then we proceed as follows:

1. P_L chooses a vector X that is orthogonal to L, i.e., $\langle L, X \rangle = 0$, and sends it over to P_R.
2. P_R computes $R' := R + X$ and chooses a vector Y that is orthogonal to R' and sends it over to P_L.
3. P_L computes $L' := L + Y$.

The output of the protocol is (L', R'). By simple linear algebra it follows that $\langle L, R \rangle = \langle L', R' \rangle = S$. One may hope that the above scheme achieves security in the presence of continuous leakage. Perhaps counterintuitive, we show in the full version of this paper that this simple protocol can be broken if the leakage function can be evaluated on (L, X, Y) and (R, X, Y). To avoid this attack, we introduce a method for P_L to send a random X to P_R in an "oblivious" way, i.e., without actually learning anything about X, besides the fact that X is orthogonal to L (and symmetrically a similar protocol for P_R sending Y to P_L). We propose an efficient protocol that achieves this property by making use of our source of correlated randomness $(A, B) \leftarrow \mathcal{O}$. Notice that even given access to such a distribution, the refreshing of an encoded secret is a non-trivial task, as, e.g., just computing $L' = L + A$ and $R' = R + B$ does not preserve the secret.

The protocol that we eventually construct in Figure 1 solves actually a more general problem: we will consider schemes for storing vectors $S \in \mathbb{F}^m$, and the encoding of a secret S will be a random pair (L, R) where L is a vector of length n and R is an $n \times m$-matrix (where $n \gg m$ is some parameter), and $S = L \cdot R$.

LEAKAGE RESILIENT AUTHENTICATION AND SIGNATURES. We then use our protocol for refreshing an encoded secret as a building block to efficiently implement standard authentication and signature schemes. More concretely, we show that under the DL assumption a simple implementation of the widely-used Okamoto authentication scheme is secure against impersonation attacks even if the prover's computation leaks continuously. Using the standard Fiat-Shamir heuristic, we can turn our protocol into a leakage resilient signature scheme.

At a high level, our transformation of the standard Okamoto scheme encodes the original secret keys with our inner product based encoding scheme. Then, we carry out the computation of the prover in "encoded form", and finally after each execution of the prover, we refresh the encoded secrets using our leakage resilient refreshing scheme. To carry out the computation of the prover in an encoded, we use the following two observations about the inner product based encoding:

1. it exhibits an additive homomorphism, i.e., if we encode two secrets S_1, S_2 as (L, Q) and (L, R), then $(L, Q + R)$ represents an encoding of $S_1 + S_2$. Moreover, if Q and R are stored on the same memory part, then this computation can be carried out in a leakage resilient way.
2. for two secrets S_1 and S_2 and two group generators g_1 and g_2, it allows to compute $g_1^{S_1} \cdot g_2^{S_2}$ in a leakage-resilient way. To illustrate this, suppose that S_1 is encoded by (L, Q) and S_2 is encoded by (L, R). A protocol to compute $g_1^{S_1} \cdot g_2^{S_2}$ proceeds then as follows. P_R computes the vector $A := g_1^Q g_2^R = \left(g_1^{Q_1} g_2^{R_1}, \ldots, g_1^{Q_n} g_2^{R_n} \right)$ and sends it over to P_L. Next, P_L computes the vector $B := A^L = (A_1^{L_1}, \ldots, A_n^{L_n})$ and finally it computes $g_1^{S_1} g_2^{S_2} = \prod_i B_i$.

Together with our scheme for refreshing the inner product encoding, these both basic components suffice to implement the standard Okamoto authentication scheme in a leakage resilient way (cf. Section 4).

LEAKAGE RESILIENT CCA2-SECURE ENCRYPTION. As a third contribution, we show that a simple and efficient variant of the ElGamal cryptosystem can be proven to be CCA2 secure in the RO model even if the computation from the decryption process leaks continuously. We would like to emphasize that we allow the leakage to *depend* on the target ciphertext. We achieve this by exploiting the independent leakage assumption and carry out the computation using the above described protocol for secure exponentiation. We would like to note that even though our scheme uses a simulation sound (SS) NIZK, our construction is rather efficient, as SS-NIZKs can be implemented efficiently via the Fiat-Shamir heuristic. Notice that the Fiat-Shamir heuristic is the only place where the random oracle assumption is used.

A GENERAL PARADIGM FOR LEAKAGE RESILIENT IMPLEMENTATIONS. We observe that our methods for implementing cryptographic schemes is fairly general. Indeed, the two main properties that we require are

1. The secret key of the cryptosystem is an element in a finite field, and the scheme computes only a linear function of the secret keys, and
2. The secret key is hidden information theoretically even given the transcript that an adversary obtains when interacting with the cryptosystem.

Various other cryptosystems satisfy these properties. For instance, we can use our techniques to construct a (rather inefficient) leakage resilient CCA2-secure encryption scheme that is provably secure in the *standard model*.

COMPARISON TO OTHER RELATED WORK We would like to mention that in a series of important recent works [9,5,23,22,4] *new* schemes for leakage resilient signing and encryption (CPA-secure) have been proposed. While these works have an obvious advantage over our work by considering a more powerful leakage model, we would like to point out that these schemes are non-standard, rather inefficient and rely on non-standard assumptions. Very recently, Dodis et al. [10] introduced a method for storing and refreshing a secret. Their construction does not require leak-free components, but is rather inefficient and relies on computational assumptions. Moreover, it is not clear if it can be used for other purposes such as implementing standard cryptosystems.

2   Preliminaries

For a natural number n the set $\{1, \ldots, n\}$ will be denoted by $[n]$. If X is a random variable then we write $x \leftarrow X$ for the value that the random variable takes when sampled according to the distribution of X. In this paper, we will slightly abuse notation and also denote by X the probability distribution on the range of the variable. V is a row vector, and we denote by V^{T} its transposition. We let $\mathbb{F}$ be a finite field and for $m, n \in \mathbb{N}$, let $\mathbb{F}^{m \times n}$ denote the set of $m \times n$-matrices over $\mathbb{F}$. Typically, we use M_i to denote the column vectors of the matrix M. For a matrix $M \in \mathbb{F}^{m \times n}$ and an m bit vector $V \in \mathbb{F}^m$ we denote by $V \cdot M$ the n-element vector that results from matrix multiplication of V and M. For a natural number n by (0^n) we will denote the vector $(0, \ldots, 0)$ of length n. We will often use the set of non-singular $m \times m$ matrices denoted by $\mathsf{NonSing}^{m \times m}(\mathbb{F}) \subset \mathbb{F}^{m \times m}$.

Let in the rest of this work n be the statistical and k be the computational security parameter. Let $\mathbb{G}$ be a group of prime order p such that $\log_2(p) \geq k$. We denote by $(p, \mathbb{G}) \leftarrow \mathsf{G}$ a group sampling algorithm. Let g be a generator of $\mathbb{G}$, then for a (column/row) vector $A \in \mathbb{Z}_p^n$ we denote by g^A the vector $C = (g^{A_1}, \ldots, g^{A_n})$. Furthermore, let C^B be the vector $(g^{A_1 B_1}, \ldots, g^{A_n B_n})$.

Let X_0, X_1 be random variables distributed over $\mathcal{X}$ and Y be a random variable over a set $\mathcal{Y}$, then we define the statistical distance between X_0 and X_1 as $\Delta(X_0; X_1) = \sum_{x \in \mathcal{X}} 1/2 |\Pr[X_0 = x] - \Pr[X_1 = x]|$. Moreover, let $\Delta(X_0; X_1 | Y) \stackrel{\text{def}}{=} \Delta((Y, X_0); (Y, X_1))$ be the statistical distance conditioned on Y.

2.1   Model of Leakage

In this work, we assume that the memory of a physical device is split into two parts, which leak *independently*. We model this in form of a *leakage game*, where the adversary can *adaptively* learn information from each part of the memory. More formally, let $L, R \in \{0, 1\}^s$ be the two parts of the memory, then for a parameter $\lambda \in \mathbb{N}$, we define a λ-*leakage game* played between an adaptive adversary $\mathcal{A}$ – called a λ-*limited adversary* – and a *leakage oracle* $\Omega(L, R)$ as follows. For some $t \in \mathbb{N}$, the adversary $\mathcal{A}$ can adaptively issue a sequence $\{(f_i, x_i)\}_{i=1}^t$ of requests to the oracle $\Omega(L, R)$, where $x_i \in \{L, R\}$ and $f_i : \{0, 1\}^s \rightarrow \{0, 1\}^{\lambda_i}$. For the ith query the oracle replies with $f_i(x_i)$. The only restriction is that in total the adversary does not learn more than λ bits from each L and R. In the following, let $Out(\mathcal{A}, \Omega(L, R))$ be the output of $\mathcal{A}$ at the end of this game. Without loss of generality, we assume that $Out(\mathcal{A}, \Omega(L, R)) := (f_1(x_1), \ldots, f_t(x_t))$.

LEAKAGE FROM COMPUTATION. So far, we discussed how to model leakage from the memory of a device, where the memory is split into two parts (L, R). If the physical device carries out "some computation" using its memory (L, R), then this computation leaks information to the adversary. We model this in form of a two-party protocol $\Pi = (P_L, P_R)$ executed between the parties P_L and P_R.

Initially, the party P_L holds L, while P_R holds R. The execution of Π with initial inputs L and R, denoted by $\Pi(L, R)$, proceeds in rounds. In each round one player is active and sends messages to the other one. These messages can depend on his input (i.e., his initial state), his local randomness, and the messages that he received in earlier rounds. Additionally, the *user* of the protocol (or the adversary – in case the user is malicious) may interact with the protocol, i.e., he may receive messages from the players and send messages to them. For simplicity, we assume that messages that are sent by the user to the protocol are delivered to both parties P_L and P_R. At the end of the protocol's execution, the players P_L and P_R (resp.) may output a value L' and R' (resp.). These outputs may be viewed as the *new* internal state of the protocol.

One natural way to describe the leakage of the computation (and memory) of such a protocol is to allow the adversary to adaptively pick at the beginning of each round a leakage function f and give $f(\mathsf{state})$ to the adversary. Here, state contains the initial state of the active party, its local randomness and the messages sent and received during this round. Indeed, we allow the adversary to learn such leakages. To ease description, we consider however a stronger model, and use the concept of a leakage game introduced earlier in this section. More precisely, for player $P_x \in \{P_L, P_R\}$, we denote the local randomness that is used by P_x as ρ_x, and all the messages that are received *or* sent (including the messages from the user of the protocol) by M_x. At any point in time, we allow the adversary $\mathcal{A}$ to play a λ-leakage game against the leakage oracle $\Omega((L, \rho_L, M_L); (R, \rho_R, M_R))$. A technical problem may arise if $\mathcal{A}$ asks for leakages *before* sending regular messages to the players. In such a case parts of M_x may be undefined, and for simplicity, we will set them to constant 0. For some initial state (L, R), we denote the output of $\mathcal{A}$ after this process with $\mathcal{A} \leftrightarrows (\Pi(L, R) \to (L', R'))$.

As we are interested in the continuous leakage setting, we will mostly consider an adversary that runs in many executions of $\mathcal{A} \leftrightarrows (\Pi(L, R) \to (L', R'))$. For the ith execution of the protocol $\Pi(L^{i-1}, R^{i-1})$, we will write

$$\mathcal{A} \leftrightarrows \left(\Pi(L^{i-1}, R^{i-1}) \to (L^i, R^i)\right),$$

where the current initial state of this round is (L^{i-1}, R^{i-1}) and the new state of P_L and P_R will be (L^i, R^i). After $\mathcal{A} \leftrightarrows \left(\Pi(L^{i-1}, R^{i-1}) \to (L^i, R^i)\right)$, we assume that the players P_L and P_R erase their current state except for their new state L^i and R^i, respectively. For the ith execution of $\mathcal{A} \leftrightarrows \left(\Pi(L^{i-1}, R^{i-1}) \to (L^i, R^i)\right)$, we let the adversary interact with the leakage oracle $\Omega((L^{i-1}, \rho_L^i, M_L^i); (R^{i-1}, \rho_R^i, M_R^i))$. If $\mathcal{A}$ is a λ-limited adversary, then we allow him to learn up to λ bits from the oracle in each such execution.

2.2   Leakage-Resilient Storage

A leakage-resilient storage (LRS) $\Phi = (\mathsf{Encode}, \mathsf{Decode})$ allows to store a secret in an "encoded form" such that even given leakage from the encoding no adversary learns information about the encoded values. A simple LRS for the independent leakage model can be based on two source extractors. More precisely, an LRS for the independent leakage model is defined for message space $\mathcal{M}$ and encoding space $\mathcal{L} \times \mathcal{R}$ as follows:

- $\mathsf{Encode} : \mathcal{M} \to \mathcal{L} \times \mathcal{R}$ is a probabilistic, efficiently computable function and
- $\mathsf{Decode} : \mathcal{L} \times \mathcal{R} \to \mathcal{M}$ is a deterministic, efficiently computable function such that for every $S \in \mathcal{M}$ we have $\mathsf{Decode}(\mathsf{Encode}(S)) = S$.

An LRS Φ is said to be (λ, ϵ)-secure, if for any $S, S' \in \mathcal{M}$ and any λ-limited adversary $\mathcal{A}$, we have

$$\Delta(\mathit{Out}(\mathcal{A}, \Omega(L, R)); \mathit{Out}(\mathcal{A}, \Omega(L', R'))) \leq \epsilon,$$

where $(L, R) := \mathsf{Encode}(S)$ and $(L', R') := \mathsf{Encode}(S')$.

   We consider a leakage-resilient storage scheme that allows to efficiently store elements $S \in \mathbb{F}^m$ for some $m \in \mathbb{N}$. Namely, we propose $\Phi_{\mathbb{F}}^{n,m} = (\mathsf{Encode}_{\mathbb{F}}^{n,m}, \mathsf{Decode}_{\mathbb{F}}^{n,m})$ defined as follows:

- $\mathsf{Encode}_{\mathbb{F}}^{n,m}(S)$ first selects $L \leftarrow \mathbb{F}^n \setminus \{(0^n)\}$ at random, and then samples $R \leftarrow \mathbb{F}^{n \times m}$ such that $L \cdot R = S$. It outputs (L, R).
- $\mathsf{Decode}_{\mathbb{F}}^{n,m}(L, R)$ outputs $L \cdot R$.

The following lemma shows that $\Phi_{\mathbb{F}}^{n,m}$ is a secure LRS. The proof uses the fact that an inner product over a finite field is a two-source extractor [7,28] and appears in the full version.

Lemma 1. *Let* $m, n \in \mathbb{N}$ *with* $m < n$ *and let* $\mathbb{F}$ *such that* $|\mathbb{F}| = \Omega(n)$. *For any* $1/2 > \delta > 0, \gamma > 0$ *the LRS* $\Phi_{\mathbb{F}}^{n,m}$ *as defined above is* (λ, ϵ)-*secure, with* $\lambda = (1/2 - \delta)n \log |\mathbb{F}| - \log \gamma^{-1}$ *and* $\epsilon = 2m(|\mathbb{F}|^{m+1/2-n\delta} + |\mathbb{F}^m|\gamma)$.

The following is an instantiation of Lemma 1 for concrete parameters.

Corollary 1. *Suppose* $|\mathbb{F}| = \Omega(n)$ *and* $m < n/20$. *Then, LRS* $\Phi_{\mathbb{F}}^{n,m}$ *is* $(0.3 \cdot |\mathbb{F}^n|, negl(n))$-*secure, for some negligible function negl.*

3   Leakage-Resilient Refreshing of LRS

For a secret S and a leakage resilient storage $\Phi = (\mathsf{Encode}, \mathsf{Decode})$ with message space $\mathcal{M}$, we develop a probabilistic protocol $(L', R') \leftarrow \mathsf{Refresh}(L, R)$ that securely refreshes $(L, R) \leftarrow \mathsf{Encode}(S)$, even when the adversary can *continuously* observe the computation from the refreshing process. The only additional assumption that we make is that the protocol has access to a simple leak-free source $\mathcal{O}$ of correlated randomness.

Initially, P_L holds L and P_R holds R. At any point during the execution of the protocol, the adversary can interact with a leakage oracle and learn information about the internal state of P_L and P_R. At the end the players output the "refreshed" encoding (L', R'), i.e., the new state of the protocol. Notice that the only way in which the adversary can "interact" with the protocol is via the leakage oracle.

For *correctness*, we require that $\mathsf{Decode}(L, R) = \mathsf{Decode}(L', R')$ Informally, for security, we require that no λ-limited adversary can learn any significant information about S (for some parameter $\lambda \in \mathbb{N}$). We will define the security of the refreshing protocol using an indistinguishability notion. Intuitively, the definition says that for any two secrets $S, S' \in \mathcal{M}$ the view (i.e., the leakage) resulting from the execution of the refreshing of secret S is statistically close to the view from the refreshing of secret S'. Before we formally define security of our refreshing, we consider the following experiment, which runs the refreshing protocol for ℓ rounds and lets the adversary play a leakage game in each round. For a protocol Π, an LRS Φ, a λ-bounded adversary $\mathcal{A}$, $\ell \in \mathbb{N}$ and $S \in \mathcal{M}$, we have $\mathsf{Exp}_{(\Pi, \Phi)}(\mathcal{A}, S, \ell)$:

1. For a secret S, we generate the initial encoding as $(L^0, R^0) \leftarrow \mathsf{Encode}(S)$.
2. For $i = 1$ to ℓ run $\mathcal{A}$ against the *ith round* of the refreshing protocol: $\mathcal{A} \leftrightarrows \left(\Pi(L^{i-1}, R^{i-1}) \to (L^i, R^i) \right)$.
3. Return whatever $\mathcal{A}$ outputs.

Wlog. we assume that $\mathcal{A}$ outputs just a single bit $b \in \{0, 1\}$. To simplify notation, we will sometimes omit to specify Φ in $\mathsf{Exp}_{(\Pi, \Phi)}(\mathcal{A}, S, \ell)$ explicitly. We are now ready to define security of a refreshing protocol.

Definition 1 (A $(\ell, \lambda, \epsilon)$-refreshing protocol). *For a LRS Φ = (Encode, Decode) with message space $\mathcal{M}$, a refreshing protocol (Refresh, Φ) is $(\ell, \lambda, \epsilon)$-secure, if for every λ-limited adversary $\mathcal{A}$ and any two secrets $S, S' \in \mathcal{M}$, we have that $\Delta(\mathsf{Exp}_{(\mathsf{Refresh}, \Phi)}(\mathcal{A}, S, \ell); \mathsf{Exp}_{(\mathsf{Refresh}, \Phi)}(\mathcal{A}, S', \ell)) \leq \epsilon$.*

In the rest of this section, we construct a secure refreshing protocol for the LRS scheme $\Phi_{\mathbb{F}}^{n,m} = (\mathsf{Encode}_{\mathbb{F}}^{n,m}, \mathsf{Decode}_{\mathbb{F}}^{n,m})$ from Section 2.2. Our protocol can refresh an encoding $(L, R) \leftarrow \mathsf{Encode}_{\mathbb{F}}^{n,m}(S)$ any polynomial number of times, and guarantees security for λ being a constant fraction of the length of L and R (cf. Theorem 1 and Corollary 2 for the concrete parameters). To ease notation, we often omit to specify $\Phi_{\mathbb{F}}^{n,m}$ when talking about the refreshing protocol $(\mathsf{Refresh}_{\mathbb{F}}^{n,m}, \Phi_{\mathbb{F}}^{n,m})$ and just write Refresh.

As outlined in the introduction, we assume that the players have access to a non-uniform source of randomness. More precisely, they may access an oracle $\mathcal{O}$ that samples pairs $(A, B) \in \mathbb{F}^n \times \mathsf{NonSing}^{n \times m}(\mathbb{F})$ such that $A \neq (0^n)$ and $A \cdot B = (0^m)$. In each iteration the players will sample the oracle twice: once for refreshing the share of P_R (denote the sampled pair by (A, B)), and once for refreshing the share of P_L (denote the sampled pair by $(\tilde{A}, \tilde{B})$). The protocol is depicted on Fig. 1. To understand the main idea behind the protocol, the reader may initially disregard the checks (in Steps 1 and 4) that L and R' have full

---

Protocol $(L', R') \leftarrow \mathsf{Refresh}_{\mathbb{F}}^{n,m}(L, R)$:

Input (L, R): $L \in \mathbb{F}^n$ is given to P_L and $R \in \mathbb{F}^{n \times m}$ is given to P_R.

Refreshing the share of P_R:

1. If L does not have a full rank then the players abort. Let $(A, B) \leftarrow \mathcal{O}$ and give A to P_L and B to P_R.
2. Player P_L generates a random non-singular matrix $M \in \mathbb{F}^{n \times n}$ such that $L \cdot M = A$ and sends it to P_R.
3. Player P_R sets $X := M \cdot B$ and $R' := R + X$.

Refreshing the share of P_L:

4. If R' does not have a full rank then the players abort. Let $(\tilde{A}, \tilde{B}) \leftarrow \mathcal{O}$ and give $\tilde{A}$ to P_L and $\tilde{B}$ to P_R.
5. Player P_R generates a random non-singular matrix $\tilde{M} \in \mathbb{F}^{n \times n}$ such that $\tilde{M} \cdot R' = \tilde{B}$ and sends it to P_L.
6. Player P_L sets $Y := \tilde{A} \cdot \tilde{M}$ and $L' := L + Y$.

Output: The players output (L', R').

The adversary plays a λ-leakage game against:
$$\Omega\Big((L, A, M, \tilde{A}, \tilde{M}) \; ; \; (R, B, M, \tilde{B}, \tilde{M})\Big)$$

Fig. 1. Protocol $\mathsf{Refresh}_{\mathbb{F}}^{n,m}$. The oracle $\mathcal{O}$ samples randomly pairs $(A, B) \in \mathbb{F}^n \times \mathsf{NonSing}^{n \times m}(\mathbb{F})$ such that $A \neq (0^n)$ and $A \cdot B = (0^m)$. The text in the frame describes the leakage game played by the adversary. Note that sampling the random matrices in Steps 2 and 5 can be done efficiently.

rank (these checks were introduced only to facilitate the proof and only occur with very small probability). The reader may also initially assume that $m = 1$ (the case of $m > 1$ is a simple generalization of the $m = 1$ case). The main idea of our protocol is that first the players generate the value $X \in \mathbb{F}^{n \times m}$ such that $L \cdot X = (0^m)$, and then in Steps 3 the player P_R sets $R' := R + X$ (note that, by simple linear algebra $L \cdot R' = L \cdot (R + X) = L \cdot R + L \cdot X = L \cdot R$). Symmetrically, later, the players generate $Y \in \mathbb{F}^n$ such that $Y \cdot R' = (0^m)$ and set (in Step 6) $L' = L + Y$. By a similar reasoning as before we have $L' \cdot R' = L \cdot R'(= L \cdot R)$. The above analysis gives us the correctness of our protocol.

Lemma 2 (Correctness of the refreshing). *Assuming that the players P_L and P_R did not abort, we have for any $S \in \mathbb{F}^m$: $\mathsf{Decode}_{\mathbb{F}}^{n,m}(\mathsf{Refresh}_{\mathbb{F}}^{n,m}(S)) = S$.*

We now state our main theorem which shows that the protocol $\mathsf{Refresh}_{\mathbb{F}}^{n,m}$ from Figure 1 satisfies Definition 1. In the full version of this paper, we show that our refreshing is secure even if the adversary has some (not necessarily short) auxiliary information about the encoding.

Theorem 1 (Security of $\mathsf{Refresh}_{\mathbb{F}}^{n,m}$). *Let $m/3 \leq n$, $n \geq 16$ and $\ell \in \mathbb{N}$. Let n, m and $\mathbb{F}$ be such that $\Phi_{\mathbb{F}}^{n,m}$ is (λ, ϵ)-secure (for some λ and ϵ). The protocol $\mathsf{Refresh}_{\mathbb{F}}^{n,m}$ is a $(\ell, \lambda/2 - 1, \epsilon')$-refreshing protocol for an LRS $\Phi_{\mathbb{F}}^{n,m}$ with $\epsilon' := 2\ell |\mathbb{F}|^m (3 |\mathbb{F}|^m \epsilon + m |\mathbb{F}|^{-n-1})$.*

For the proof of this theorem, we will need to show that any adversary $\mathcal{A}$ that interacts for ℓ iterations with the refreshing experiment $\mathsf{Exp}_{\mathsf{Refresh}}$ (as given in

Definition 1), will only gain a negligible (in n) amount of information about the encoded secret S. Notice that this in particular means that $\mathcal{A}$'s interaction with the leakage oracle given in the frame of Figure 1 will not provide the adversary with information on the encoded secret. More formally, we will show that for every $(\lambda/2 - 1)$-limited $\mathcal{A}$ and every S, S' we have:

$$\Delta(\mathsf{Exp}_{\mathsf{Refresh}}(\mathcal{A}, S, \ell); \mathsf{Exp}_{\mathsf{Refresh}}(\mathcal{A}, S', \ell)) \le 2\ell |\mathbb{F}|^m (3 |\mathbb{F}|^m \epsilon + m |\mathbb{F}|^{-n-1}). \quad (1)$$

This will be proven using the standard technique called the "hybrid argument" by creating a sequence of "hybrid distributions". We will show that the first distribution in this sequence is statistically very close to $\mathsf{Exp}_{\mathsf{Refresh}}(\mathcal{A}, S, \ell)$, while the latter is close to $\mathsf{Exp}_{\mathsf{Refresh}}(\mathcal{A}, S', \ell)$. Moreover, each two consecutive distributions in the sequence will be statistically close. Hence, by applying the triangle inequality multiple times, we will obtain that $\mathsf{Exp}_{\mathsf{Refresh}}(\mathcal{A}, S, \ell)$ and $\mathsf{Exp}_{\mathsf{Refresh}}(\mathcal{A}, S', \ell)$ are close. The proof of the theorem is deferred to the full version of this paper. Combining Theorem 1 with Corollary 1 we get the following.

Corollary 2. *Let $n \in \mathbb{N}$ be the security parameter. Suppose $|\mathbb{F}| = \Omega(n)$ and let $m = o(n)$. Then $\mathsf{Refresh}_{\mathbb{F}}^{n,m}$ is a $(\ell, 0.15 \cdot n \log(|\mathbb{F}|) - 1, negl(n))$-refreshing protocol for the LRS $\Phi_{\mathbb{F}}^{n,m}$, where ℓ is a polynomial in n and $negl(n)$ is some negligible function.*

4 Identification and Signature Schemes

In an identification scheme ID a prover attempts to prove its identity to a verifier. For a security parameter k, ID consists out of three PPT algorithms $\mathsf{ID} = (\mathsf{KeyGen}, \mathcal{P}, \mathcal{V})$:

- $(pk, sk) \leftarrow \mathsf{KeyGen}(1^k)$: It outputs the public parameters of the scheme and a valid key pair.
- $(\mathcal{P}(pk, sk), \mathcal{V}(pk))$: An interactive protocol in which $\mathcal{P}$ tries to convince $\mathcal{V}$ of its identity by using his secret key sk. The verifier $\mathcal{V}$ outputs either *accept* or *reject*.

We require that ID is *complete*. This means that an honest prover will always be accepted by the verifier. The standard security definition of an identification scheme ID considers a polynomial-time adversary $\mathcal{A}$ that inputs the public key pk and interacts with the prover $\mathcal{P}(pk, sk)$ playing the role of a verifier. Then, $\mathcal{A}$ tries to impersonate $\mathcal{P}(pk, sk)$ by engaging in an interaction with $\mathcal{V}(pk)$. We say that the scheme is *secure* if every polynomial-time adversary $\mathcal{A}$ impersonates the prover with only negligible probability.

We extend this standard security to incorporate leakage from the prover's computation. To this end, we let the adversary take the role of $\mathcal{V}$ in the execution of the protocol $(\mathcal{P}(pk, sk), \mathcal{V}(pk))$ and allow him to obtain leakage from the prover's execution. We denote a single execution of this process by $\mathcal{A} \leftrightarrows (\mathcal{P}(sk) \to sk')$, where sk' may be the updated key.

Definition 2 (Security against Leakage and Impersonation Attacks (ID-LEAK security)). *Let $k \in \mathbb{N}$ be the security parameter. An identification scheme* ID $= (\mathsf{KeyGen}, \mathcal{P}, \mathcal{V})$ *is $\lambda(k)$-ID-LEAK secure if for any PPT $\lambda(k)$-limited adversary $\mathcal{A}$ it holds that the experiment below outputs 1 with probability at most $negl(k)$:*

1. *The challenger samples $(pk, sk^0) \leftarrow \mathsf{KeyGen}(1^k)$ and gives pk to $\mathcal{A}$.*
2. *Repeat for $i = 0 \ldots poly(k)$ times: $\mathcal{A} \leftrightarrows \left(\mathcal{P}(sk^i) \to sk^{i+1} \right)$, where in each execution the adversary can interact with the honest prover and gets up to $\lambda(k)$ bits about the current secret state sk^i and the randomness that is used.*
3. *$\mathcal{A}$ impersonates the prover and interacts with $\mathcal{V}(pk)$. If $\mathcal{V}(pk)$ accepts, then output 1; otherwise output 0.*

Notice that the adversary is allowed to obtain λ bits of information for *each* execution of the identification protocol.

4.1   A Construction of a Leakage-Resilient Identification Protocol

Our construction is based on the standard Okamoto identification scheme [25]. Let g_1 and g_2 be two generators of $\mathbb{G}$ such that $\alpha = \log_{g_1}(g_2)$ is unknown. The secret key sk is equal to $(x_1, x_2) \leftarrow \mathbb{Z}_p^2$ and the public key pk is $g_1^{x_1} \cdot g_2^{x_2}$.

1. $\mathcal{P}$ chooses $(w_1, w_2) \leftarrow \mathbb{Z}_p^2$, computes $a := g_1^{w_1} g_2^{w_2}$, and sends a to $\mathcal{V}$.
2. $\mathcal{V}$ chooses $c \leftarrow \mathbb{Z}_p$ and sends it to $\mathcal{P}$.
3. $\mathcal{P}$ computes $z_1 := w_1 + cx_1$ and $z_2 := w_2 + cx_2$ and sends (z_1, z_2) to $\mathcal{V}$.
4. $\mathcal{V}$ accepts if and only if $g_1^{z_1} g_2^{z_2} \stackrel{?}{=} a \cdot pk^c$.

We next describe how to implement the Okamoto scheme such that it remains secure even if the computation of the prover is carried out on a leaky device. Verification is as in the standard Okamoto scheme, while the key generation and the computation of the prover is adjusted to protect against leakage attacks. More precisely, instead of using $(x_1, x_2) \in \mathbb{Z}_p^2$ as secret key, we store $(L, (R_1, R_2)) \leftarrow \mathsf{Encode}_{\mathbb{F}}^{n,2}(x_1, x_2)$ and implement the computation of the prover as a two-party protocol run between $P_{\mathsf{L}}(L)$ and $P_{\mathsf{R}}(R_1, R_2)$. To this end, we will use the fact that the Okamoto identification protocol *only* requires to compute a linear function of the encoded secret key. The protocol is given in Figure 2.

Finally, we will combine our identification protocol with our protocol for refreshing to construct an identification scheme $\mathsf{Oka} = (\mathsf{KeyGen}, \mathcal{P}, \mathcal{V}, \mathsf{Refresh}_{\mathbb{Z}_p}^{n,2})$ that is ID-LEAK secure. More precisely, in the ith execution of $(\mathcal{P}(pk, (L, R)), \mathcal{V}(pk))$ after Step 5 in Figure 2, we execute $(L^{i+1}, R^{i+1}) \leftarrow \mathsf{Refresh}_{\mathbb{Z}_p}^{n,2}(L^i, R^i)$ and set the prover's secret key for the next round to $sk^{i+1} := (L^{i+1}, R^{i+1})$. Notice that in such a case, we include into the leakage oracle from the figure the variables that are used by the refreshing and let the adversary interact in each round with the following leakage oracle:

$$\Omega \left((L^i, U, Z, A, M, \tilde{A}, \tilde{M}) \, ; \, (R^i, W, A, M, \tilde{A}, \tilde{M}) \right).$$

Key generation KeyGen(1^k):

Sample $(p, \mathbb{G}) \leftarrow \mathsf{G}(1^k)$, generators $g_1, g_2 \leftarrow \mathbb{G}$, $S = (x_1, x_2) \leftarrow \mathbb{Z}_p^2$ and $(L, R) \leftarrow \mathsf{Encode}_{\mathbb{Z}_p}^{n,2}(S)$. Set $sk = (L, R)$ and $pk = (p, g_1, g_2, h := g_1^{x_1} g_2^{x_2})$.

The identification protocol $(\mathcal{P}(pk, (L, R)), \mathcal{V}(pk))$

Input for prover (L, R): L is given to P_L and R is given to P_R.

Prover $\mathcal{P}(pk, (L, R))$:	Verifier $\mathcal{V}(pk)$:
1. P_R samples $(W_1, W_2) \leftarrow \mathbb{Z}_p^{2n}$, computes $U := g_1^{W_1} \odot g_2^{W_2}$ and sets $W := (W_1^\mathsf{T}, W_2^\mathsf{T})$. The vector U is sent to P_L ($\odot$ is component-wise multiplication of vectors).	
2. P_L computes $V = U^L$ and $a = \prod_i V_i$. The value a is sent to $\mathcal{V}$.	
	3. Senc $c \leftarrow \mathbb{Z}_p$ to $\mathcal{P}$.
4. P_R computes the $n \times 2$ matrix $Z := W + cR$ and sends it to P_L.	
5. P_L computes $(z_1, z_2) = L \cdot Z$. The values (z_1, z_2) are given to $\mathcal{V}$.	

At any time, the adversary can play a λ-leakage game against: $\Omega\left((L, U, Z)\,;\,(R, W)\right)$. We set $Z = 0$ for leakage queries that are asked *before* c is fixed.

6. Accept iff $g_1^{z_1} g_2^{z_2} = ah^c$.

Fig. 2. The key generation algorithm and the protocol $(\mathcal{P}(pk, (L, R)), \mathcal{V}(pk))$ for identification. $(\mathcal{P}(pk, (L, R)), \mathcal{V}(pk))$ is an interactive protocol between a prover $\mathcal{P}$ and a verifier $\mathcal{V}$.

It is easy to see that the above protocol satisfies the completeness property. This is due to the correctness of the refreshing protocol, and the fact that messages that are exchanged by the parties $\mathcal{P}$ and $\mathcal{V}$ in Figure 2 are as in the original Okamoto protocol. The security of our protocol Oka is proven in the following theorem.

Theorem 2. Oka $=$ (KeyGen, $\mathcal{P}, \mathcal{V}$, Refresh$_{\mathbb{Z}_p}^{n,2}$) *is* $((0.15 \cdot n - 3) \log p - 1)$-*ID-LEAK secure, if the DL assumption holds.*

The proof follows from the following three observations:

1. We first consider a single execution of the protocol $(\mathcal{P}(pk, (L, R)), \mathcal{V}(pk))$ from Figure 2 and prove a simple property in the information theoretic setting. Namely, we show that the there exists an (unbounded) simulator with access to a leakage oracle $\Omega(L^*, R^*)$ can simulate $\mathcal{A}(pk)$'s view in $\mathcal{A} \leftrightarrows (\mathcal{P}(L, R)) \rightarrow (L, R)$. In this step the analysis neglects the leakage from the refreshing process as we consider only a *single* run of the protocol.

2. We next consider the setting where unbounded $\mathcal{A}$ runs in many iterations of $\mathcal{A} \leftrightarrows \left(\mathcal{P}(L^i, R^i) \right) \to (L^{i+1}, R^{i+1}) \right)$, where we also take into account that the refreshing of (L^i, R^i) leaks information. We will combine our results from the last section with the simulator defined in 1 to show that *any* unbounded adversary will only learn a negligible amount of information about the secret key.

3. Finally, we will argue why this proves the ID-Leak security of our scheme. To this end, we rely on a recent result of Dodis et al. [2], which shows security of the original Okamoto scheme for keys sampled from a high average min-entropy source.

LEAKAGE RESILIENT SIGNATURES It is well known fact that the Okamoto identification protocol can be turned into a signature scheme using the Fiat-Shamir heuristic. Similarly, we can turn the scheme from Figure 2 into a leakage resilient signature scheme which can be proven secure against continuous leakage attacks in the random oracle model under the DL assumption.

5 Leakage Resilient Encryption

In this section, we construct an efficient encryption schemes that is secure against continuous leakage attacks. Our construction is based on a variant of the ElGamal cryptosystem and is proven secure against adaptive chosen message and leakage attacks (CCLA2) in the Random Oracle model.

5.1 Definitions

For security parameter k a public-key encryption scheme $\mathsf{PKE} = (\mathsf{KeyGen}, \mathsf{Encr}, \mathsf{Decr})$ consists of three PPT algorithms.

- $(pk, sk) \leftarrow \mathsf{KeyGen}(1^k)$: It outputs a valid public/secret key pair.
- $c \leftarrow \mathsf{Encr}(pk, m)$: That is, a probabilistic algorithm that on input some message m and the public key pk outputs a ciphertext $c = \mathsf{Encr}(pk, m)$.
- $m = \mathsf{Decr}(sk, c)$: The decryption algorithm takes as input the secret key sk and a ciphertext c such that for any m we have $m = \mathsf{Decr}(sk, \mathsf{Encr}(pk, m))$.

To define security we allow the adversary to query the decryption oracle on some chosen ciphertext c, and additionally allow him to obtain a bounded amount of leakage from the decryption process. This may be repeated many times, hence, eventually the adversary may learn a large amount of information. Formally, we define security against adaptive chosen ciphertext and leakage attacks (IND-CCLA2 security) as follows.

Definition 3 (Security against Chosen Ciphertext Leakage Attacks (CCLA2-secure)). *Let $k \in \mathbb{N}$ be the security parameter. A public-key encryption scheme $\mathsf{PKE} = (\mathsf{KeyGen}, \mathsf{Encr}, \mathsf{Decr})$ is $\lambda(k)$-IND-CCLA2 secure if for any PPT $\lambda(k)$-limited adversary $\mathcal{A}$ the probability that the experiment below outputs 1 is at most $1/2 + negl(k)$.*

1. *Sample* $b \leftarrow \{0,1\}$ *and* $(pk, sk) \leftarrow \mathsf{KeyGen}(1^k)$. *Give* pk *to* $\mathcal{A}$.
2. *Repeat until* $\mathcal{A}(1^k)$ *outputs* (m_0, m_1): $\mathcal{A}(1^k) \leftrightarrows (\mathsf{Decr}(sk, c) \to sk')$, *where for each decryption query* c *the adversary additionally retrieves up to* $\lambda(k)$ *bits about the current secret state* sk. *Set the key for the next round to* $sk := sk'$.
3. *The challenger computes* $c^* \leftarrow \mathsf{Encr}(pk, m_b)$ *and gives it to* $\mathcal{A}$.
4. *Repeat until* $\mathcal{A}(1^k)$ *outputs* b': $\mathcal{A}(1^k) \leftrightarrows (\mathsf{Decr}(sk, c) \to sk')$, *where for each decryption query* $c \neq c^*$ *the adversary additionally retrieves up to* $\lambda(k)$ *bits about the current secret state* sk. *Set the key for the next round to* $sk := sk'$.
5. *If* $b = b'$ *then output 1; otherwise output 0.*

The weaker notion of CCLA1-security can be obtained by omitting Step 4 in the experiment above.

5.2 Efficient IND-CCLA2-secure Encryption

An important tool of our encryption scheme is a simulation-sound (SS) NIZK. Informally, a NIZK proof system is said to be simulation sound, if any adversary has negligible advantage in breaking soundness (i.e., forging an accepting proof for an invalid statement), *even* after seeing a bounded number of proofs for (in)valid statements. We refer the reader to [3,29] for the formal definition of NIZKs and simulation soundness. SS-NIZKs can be instantiated in the common random string model using the Groth-Sahai proof system [18] and the techniques of [17]. Unfortunately, this results into an impractical scheme. In contrast, in the random oracle model using the Fiat-Shamir heuristic [14] simulation soundness can be achieved efficiently. In particular, it has been proven in [1] that the standard Chaum-Pedersen protocol [6] for proving equivalence of discrete logarithms can be turned into a SS-NIZK using the Fiat-Shamir heuristic. Let in the following $(\mathsf{Prov}, \mathsf{Ver})$ denote such a non-interactive proof system for proving the equivalence of discrete logarithms.

Our scheme can be viewed as a leakage-resilient implementation of the following simple variant of the ElGamal encryption scheme using the above simulation sound NIZK. Let g_1, g_2 be two generators of a prime order p group $\mathbb{G}$. Let $sk = (x_1, x_2) \in \mathbb{Z}_p^2$ be the secret key and $pk = (g_1, g_2, h = g_1^{x_1} \cdot g_2^{x_2})$ the public key. To encrypt a message $m \in \mathbb{G}$, pick uniformly $r \leftarrow \mathbb{Z}_p$ and compute $c = (u := g_1^r, v := g_2^r, w := h^r m, \pi)$, where $\pi := \mathsf{Prov}(u, v, r)$ is a NIZK proof of $\log_{g_1}(u) = \log_{g_2}(v)$. To decrypt $c = (u, v, w, \pi)$, verify the NIZK, and if it accepts, output $w \cdot (u^{-x_1} \cdot v^{-x_2})$.

It can easily be shown that this scheme achieves standard CCA2 security in the RO model. In this section, we will show how to *implement* this scheme such that it remains secure even if the decryption continuously leaks information. Similar to our transformation of the Okamoto scheme, we store the secret key (x_1, x_2) as $(L, R) \leftarrow \mathsf{Encode}_{\mathbb{F}}^{n,2}(x_1, x_2)$ and implement the computation of the decryption process as a two-party protocol run between $P_L(L)$ and $P_R(R)$. The protocol for key generation and decryption is given in Figure 3. Finally, we will combine the protocol from Figure 3 with our refreshing protocol from Section 3 to construct an encryption scheme $\mathsf{PKE} = (\mathsf{KeyGen}, \mathsf{Encr}, \mathsf{Decr}, \mathsf{Refresh}_{\mathbb{Z}_p}^{n,2})$ that is CCLA2 secure.

Key generation KeyGen(1^k):

Let $(p, \mathbb{G}) \leftarrow \mathsf{G}(1^k)$, $g_1, g_2 \leftarrow \mathbb{G}$, $S = (x_1, x_2) \leftarrow \mathbb{Z}_p^2$ and $(L, R) \leftarrow \mathsf{Encode}_{\mathbb{Z}_p}^{n,2}(S)$. Let $sk = (L, R)$ and $pk = (p, g_1, g_2, h := g_1^{x_1} g_2^{x_2})$.

Encryption Encr(pk, m) :

Sample $r \leftarrow \mathbb{Z}_p$ uniformly at random and compute $c = (u := g_1^r, v := g_2^r, w := h^r m)$. Run the NIZK prover $\mathsf{Prov}(u, v, r)$ to obtain a proof π for $\log_{g_1}(u) = \log_{g_2}(v)$. Return (c, π).

The protocol for decryption Decr(sk, c) :

Input for decryption $sk := (L, R)$: L is given to P_L and R is given to P_R.

Both parties obtain c and parse it as (u, v, w, π). If $\mathsf{Ver}(u, v, \pi) = reject$ then abort; otherwise proceed as follows:

1. P_R computes the vector $U := u^{R_1} \odot v^{R_2}$. U is sent to P_L ($\odot$ denotes component-wise multiplication of vectors).
2. P_L computes $V = U^{-L}$ and outputs $w \prod_i V_i$.

> Notice that we can omit the leakage from the verification of the NIZK as it only includes publicly known values. At any time, the adversary can play a λ-leakage game against: $\Omega\left((L, U)\,;\, R\right)$.

Fig. 3. Our public-key encryption scheme PKE

The security analysis follows the outline given in the last section. We first show that the leakage from a single decryption query can be simulated in a perfect way with just access to a leakage oracle $\Omega(L^*, R^*)$. For this simulation to go through, we require that an adversary can only observe leakage from operations that involve the secret key, *if* the decryption oracle is queried on a valid ciphertexts. We call a ciphertext *valid*, if $\log_{g_1}(u) = \log_{g_2}(v)$ holds. Notice that this is also the reason why we need NIZKs and cannot use the standard techniques to get CCA1/2 security based on hash proof systems. In the next step, we show that even when the adversary can continuously obtain leakage from the decryption, he will not be able to learn information about the encoded secret key. To this end, we will combine the scheme from Figure 3 with our refreshing protocol $\mathsf{Refresh}_{\mathbb{Z}_p}^{n,2}$. In the following theorem, we show IND-CCLA2 security of our scheme.

Theorem 3. PKE *is* $(0.15 \cdot n \log p - 1)$-*IND-CCLA2 secure in the random oracle model, if the DDH assumption holds.*

6 A General Paradigm for Leakage-Resilient Cryptographic Schemes

In the last sections, we proposed leakage-resilient implementations of standard cryptographic schemes. Namely, we showed how to implement the standard Okamoto identification scheme and a variant of the ElGamal encryption scheme such that they satisfy strong security guarantees even under continuous leakage attacks. The security proof of both schemes relied on very similar observations, namely:

1. The underlying cryptographic scheme (e.g., the Okamoto scheme or the El-Gamal variant) computes only a linear function of the secret key. Notice that in the examples of the last section the linear function was computed in the exponen. This is not a problem as long as the computation can be carried out efficiently. This was indeed the case for the schemes of the last sections.
2. The secret key is hidden information theoretically even given the protocol transcript that an adversary obtains when interacting with the underlying cryptographic scheme. In the protocols from the last section, for instance, the secret key (x_1, x_2) was information theoretically hidden even given the corresponding public key. Furthermore, for the Okamoto scheme this holds even given (a, z_1, z_2), which were sent by the prover to the verifier.

Various other cryptographic schemes satisfy the above properties, and hence can be made secure against continuous leakage attacks. For instance, the Pedersen commitment scheme [26], which is information-theoretically hiding and at the same time only requires to compute a linear function of its secrets.[2] Another example of the above paradigm is a variant of the linear Cramer-Shoup cryptosystem as presented in [30]. Notice that as in the encryption scheme from Section 5, this requires to use as a check for the validity of the ciphertexts a NIZK proof system. One can instantiate such a NIZK in the standard model using the Groth-Sahai proof system [18]. This gives us an efficient CCLA1-secure public-key encryption scheme in the standard model, and a rather inefficient CCLA2-secure scheme using the extensions of [17]. We suggest that many other standard cryptographic schemes can be proven secure following the ideas that were presented in this paper.

Acknowledgments. The authors are grateful to Francesco Davi, Yevgeniy Dodis, Krzysztof Pietrzak, Leonid Reyzin and Daniele Venturi for helpful discussions on the problem of leakage-resilient refreshing.

References

1. Abdalla, M., Boyen, X., Chevalier, C., Pointcheval, D.: Distributed Public-Key Cryptography from Weak Secrets. In: Jarecki, S., Tsudik, G. (eds.) PKC 2009. LNCS, vol. 5443, pp. 139–159. Springer, Heidelberg (2009)
2. Alwen, J., Dodis, Y., Wichs, D.: Leakage-Resilient Public-Key Cryptography in the Bounded-Retrieval Model. In: Halevi, S. (ed.) CRYPTO 2009. LNCS, vol. 5677, pp. 36–54. Springer, Heidelberg (2009)
3. Blum, M., Feldman, P., Micali, S.: Non-interactive zero-knowledge and its applications (extended abstract). In: STOC, pp. 103–112 (1988)
4. Boyle, E., Segev, G., Wichs, D.: Fully Leakage-Resilient Signatures. In: Paterson, K.G. (ed.) EUROCRYPT 2011. LNCS, vol. 6632, pp. 89–108. Springer, Heidelberg (2011)

---

[2] Notice that we computed a Pedersen commitment as part of the prover's protocol in our implementation of the Okamoto scheme.

5. Brakerski, Z., Kalai, Y.T., Katz, J., Vaikuntanathan, V.: Overcoming the hole in the bucket: Public-key cryptography resilient to continual memory leakage. In: FOCS, pp. 501–510 (2010)
6. Chaum, D., Pedersen, T.P.: Wallet Databases with Observers. In: Brickell, E.F. (ed.) CRYPTO 1992. LNCS, vol. 740, pp. 89–105. Springer, Heidelberg (1993)
7. Chor, B., Goldreich, O.: Unbiased bits from sources of weak randomness and probabilistic communication complexity. SIAM J. Comput. 17(2), 230–261 (1988)
8. Davì, F., Dziembowski, S., Venturi, D.: Leakage-Resilient Storage. In: Garay, J.A., De Prisco, R. (eds.) SCN 2010. LNCS, vol. 6280, pp. 121–137. Springer, Heidelberg (2010)
9. Dodis, Y., Haralambiev, K., López-Alt, A., Wichs, D.: Cryptography against continuous memory attacks. In: FOCS, pp. 511–520 (2010)
10. Dodis, Y., Lewko, A., Waters, B., Wichs, D.: How to store a secret on continually leaky devices. Accepted to FOCS 2011 (2011)
11. Dziembowski, S., Pietrzak, K.: Leakage-resilient cryptography. In: FOCS 2008: Proceedings of the 49th Annual IEEE Symposium on Foundations of Computer Science. IEEE Computer Society, Washington, DC, USA (2008)
12. Faust, S., Kiltz, E., Pietrzak, K., Rothblum, G.N.: Leakage-Resilient Signatures. In: Micciancio, D. (ed.) TCC 2010. LNCS, vol. 5978, pp. 343–360. Springer, Heidelberg (2010)
13. Faust, S., Rabin, T., Reyzin, L., Tromer, E., Vaikuntanathan, V.: Protecting Circuits from Leakage: the Computationally-Bounded and Noisy Cases. In: Gilbert, H. (ed.) EUROCRYPT 2010. LNCS, vol. 6110, pp. 135–156. Springer, Heidelberg (2010)
14. Fiat, A., Shamir, A.: How to Prove Yourself: Practical Solutions to Identification and Signature Problems. In: Odlyzko, A.M. (ed.) CRYPTO 1986. LNCS, vol. 263, pp. 186–194. Springer, Heidelberg (1987)
15. Goldwasser, S., Kalai, Y.T., Rothblum, G.N.: One-Time Programs. In: Wagner, D. (ed.) CRYPTO 2008. LNCS, vol. 5157, pp. 39–56. Springer, Heidelberg (2008)
16. Goldwasser, S., Rothblum, G.N.: Securing Computation against Continuous Leakage. In: Rabin, T. (ed.) CRYPTO 2010. LNCS, vol. 6223, pp. 59–79. Springer, Heidelberg (2010)
17. Groth, J.: Simulation-Sound NIZK Proofs for a Practical Language and Constant Size Group Signatures. In: Lai, X., Chen, K. (eds.) ASIACRYPT 2006. LNCS, vol. 4284, pp. 444–459. Springer, Heidelberg (2006)
18. Groth, J., Sahai, A.: Efficient Non-interactive Proof Systems for Bilinear Groups. In: Smart, N.P. (ed.) EUROCRYPT 2008. LNCS, vol. 4965, pp. 415–432. Springer, Heidelberg (2008)
19. Ishai, Y., Sahai, A., Wagner, D.: Private Circuits: Securing Hardware against Probing Attacks. In: Boneh, D. (ed.) CRYPTO 2003. LNCS, vol. 2729, pp. 463–481. Springer, Heidelberg (2003)
20. Juma, A., Vahlis, Y.: Protecting Cryptographic Keys against Continual Leakage. In: Rabin, T. (ed.) CRYPTO 2010. LNCS, vol. 6223, pp. 41–58. Springer, Heidelberg (2010)
21. Kiltz, E., Pietrzak, K.: Leakage Resilient ElGamal Encryption. In: Abe, M. (ed.) ASIACRYPT 2010. LNCS, vol. 6477, pp. 595–612. Springer, Heidelberg (2010)
22. Lewko, A., Rouselakis, Y., Waters, B.: Achieving Leakage Resilience through Dual System Encryption. In: Ishai, Y. (ed.) TCC 2011. LNCS, vol. 6597, pp. 70–88. Springer, Heidelberg (2011)
23. Lewko, A., Lewko, M., Waters, B.: How to leak on key updates. In: To Appear at STOC 2011 (2011)

24. Micali, S., Reyzin, L.: Physically Observable Cryptography (Extended Abstract). In: Naor, M. (ed.) TCC 2004. LNCS, vol. 2951, pp. 278–296. Springer, Heidelberg (2004)

25. Okamoto, T.: Provably Secure and Practical Identification Schemes and Corresponding Signature Schemes. In: Brickell, E.F. (ed.) CRYPTO 1992. LNCS, vol. 740, pp. 31–53. Springer, Heidelberg (1993)

26. Pedersen, T.P.: Non-interactive and Information-Theoretic Secure Verifiable Secret Sharing. In: Feigenbaum, J. (ed.) CRYPTO 1991. LNCS, vol. 576, pp. 129–140. Springer, Heidelberg (1992)

27. Pietrzak, K.: A Leakage-Resilient Mode of Operation. In: Joux, A. (ed.) EUROCRYPT 2009. LNCS, vol. 5479, pp. 462–482. Springer, Heidelberg (2009)

28. Rao, A.: An exposition of bourgain's 2-source extractor. Electronic Colloquium on Computational Complexity (ECCC) 14(034) (2007)

29. Sahai, A.: Non-malleable non-interactive zero knowledge and adaptive chosen-ciphertext security. In: FOCS, pp. 543–553 (1999)

30. Shacham, H.: A cramer-shoup encryption scheme from the linear assumption and from progressively weaker linear variants. Cryptology ePrint Archive, Report 2007/074 (2007), http://eprint.iacr.org/

Program Obfuscation with Leaky Hardware

Nir Bitansky[1,*], Ran Canetti[1,2,*], Shafi Goldwasser[3], Shai Halevi[4],
Yael Tauman Kalai[5], and Guy N. Rothblum[5,**]

[1] Tel Aviv University
[2] Boston University
[3] MIT and Weizmann Institute of Science
[4] IBM T.J. Watson Research Center
[5] Microsoft Research

Abstract. We consider general program obfuscation mechanisms using
"somewhat trusted" hardware devices, with the goal of minimizing the
usage of the hardware, its complexity, and the required trust. Specifically,
our solution has the following properties:

(i) The obfuscation remains secure even if all the hardware devices in
use are *leaky*. That is, the adversary can obtain the result of evaluating
any function on the local state of the device, as long as this function has
short output. In addition the adversary also controls the communication
between the devices.

(ii) The number of hardware devices used in an obfuscation and the
amount of work they perform are polynomial in the security parameter
independently of the obfuscated function's complexity.

(iii) A (*universal*) set of hardware components, owned by the user, is
initialized only once and from that point on can be used with multiple
"software-based" obfuscations sent by different vendors.

1 Introduction

Program obfuscation is the process of making a program unintelligible while preserving its functionality. (For example, we may want to publish an encryption
program that allows anyone to encrypt messages without giving away the secret
key.) The goal of general program obfuscation is to devise a generic transformation that can be used to obfuscate any arbitrary input program.

It is known from prior work that general program obfuscation is possible with
the help of a completely trusted hardware device (e.g., [7, 28, 19]). On the other
hand, Barak et al. proved that software-only general program obfuscation is impossible, even for a very weak notion of obfuscation [6]. In this work we consider

* Supported by the Check Point Institute for Information Security, a Marie Curie
reintegration grant and an ISF grant.
** Most of this work was done while the author was at the Department of Computer
Science at Princeton University and Supported by NSF Grant CCF-0832797 and by
a Computing Innovation Fellowship.

D.H. Lee and X. Wang (Eds.): ASIACRYPT 2011, LNCS 7073, pp. 722–739, 2011.

an intermediate setting, where we can use hardware devices but these devices are not completely trusted. Specifically, we consider using leaky hardware devices, where an adversary controlling the devices is able to learn some information about their secret state, but not all of it.

We observe that the impossibility result of Barak et al. implies that hardware-assisted obfuscation using a single leaky device is also impossible, even if the hardware device leaks only a single bit (but this bit can be an arbitrary function of the device's state). See Section 1.3. Consequently, we consider a model in which several hardware devices are used, where each device can be locally leaky but the adversary cannot obtain leakage from the global state of all the devices together. Importantly, in addition to the leakage from the separate devices, our model also gives the adversary full control over the communication between them.

The outline of our solution is as follows: Starting from any hardware-assisted obfuscation solution that uses a completely trusted device (e.g., [19, 25]), we first transform that device into a system that resists leakage in the Micali-Reyzin model of "only computation leaks" (OCL) [29] (or actually in a slightly augmented OCL model). In principle, this can be done using OCL-compilers from the literature [27, 24, 22] (but see discussion in Section 1.4 about properties of these compilers). The result is a system that emulates the functionality of the original trusted device; however, now the system is made of several components and can resist leakage from each of the components separately.

This still does not solve our problem since the system that we get from OCL-compilers only resists leakage if the different components can interact with each other over secret and authenticated channels (see discussion in Section 1.3). We therefore show how to realize secure communication channels over insecure networks in a leakage-resilient manner. This construction, which uses non-committing encryption [12] and information theoretic MACs (e.g., [33, 3]), is the main technical novelty in the current work. See Section 1.4.

The transformation above provides an adequate level of security, but it is not as efficient and flexible as one would want. For one thing, the OCL-compilers in the literature [27, 24, 22] produce systems with roughly as many components as there are gates in the underlying trusted hardware device. We show that using fully homomorphic encryption [31, 18] and universal arguments [4] we can get a system where the number of components depends only on the security parameter and is (almost) independent of the complexity of the trusted hardware device that we are emulating. See Section 1.1.

Another drawback of the solution above is that it requires a new set of hardware devices for every program that we want to obfuscate. Instead, we would like to have just one set of devices, which are initialized once and thereafter can be used to obfuscate many programs. We show how to achieve such a reusable obfuscation system using a simple trick based on CCA-secure encryption, see Section 1.2.

We now proceed to provide more details on the various components of our solution.

1.1   Minimally Hardware-Assisted Obfuscation

Forgetting for the moment about leakage-resilience, we begin by describing a hardware-assisted obfuscating mechanism where the amount of work done by the trusted hardware is (almost) independent of the complexity of the program being obfuscated. The basic idea is folklore: The obfuscator encrypts the program f using a fully homomorphic encryption scheme [31, 18], gives the encrypted program to the evaluator and installs the decryption key in the trusted hardware device. Then, the evaluator can evaluate the program homomorphically on inputs of its choice and ask the device to decrypt.

Of course, the above does not quite work as is, since the hardware device can be used for unrestricted decryption (so in particular it can be used to decrypt the function f itself). To solve this, we make the evaluator prove to the device that the ciphertext to be decrypted was indeed computed by applying the homomorphic evaluation procedure on the encrypted program and some input. Note that to this end we must add the encrypted program itself or a short hash of it to the device (so as to make "the encrypted program" a well-defined quantity). To keep the device from doing a lot of work, the proof should be verifiable much more efficiently than the computation itself, e.g., using the "universal arguments" of Barak and Goldreich [4]. We formalize this idea and show that this obfuscation scheme satisfies a strong notion of simulation based obfuscation. It can even be implemented using stateless hardware with no source of internal randomness (so it is secure against concurrent executions and reset attacks). See Section 2 for more details.

1.2   Obfuscation Using Universal Hardware Devices

A side-effect of the above solution is that the trusted hardware device must be specialized for the particular program that we want to protect (e.g., by hard-wiring in it a hash of the encrypted program), so that it has a well-defined assertion to verify before decryption. Instead, we would like the end user to use a single *universal* hardware device to run all the obfuscated programs that it receives (possibly from different vendors).

We obtain this goal using a surprisingly simple mechanism: The trusted hardware device is installed with a secret decryption key of a CCA-secure cryptosystem, whose public key is known to all vendors. Obfuscation is done as before, except that the homomorphic decryption key and the hash of the encrypted program are encrypted using the CCA-secure public key and appended to the obfuscation. This results in a universal (or "sendable") obfuscation, the device is only initialized once and then everyone can use it to obfuscate their programs. See more details in Section 3.

1.3   Dealing with Leaky Hardware

The more fundamental problem with the hardware-assisted obfuscation is that the hardware must be fully leak-free and can only provide security as long as it is accessed as a black box. This assumption is not true in many deployments, so we replace it by the weaker assumption that our hardware components are "honest-but-leaky". Namely, in our model an obfuscated program consists of software that is entirely in the clear, combined with some *leaky hardware components*. Our goal is therefore to design an obfuscator that transforms any circuit with secrets into a system of software and hardware components that achieves strong black-box obfuscation even if the components can leak.

We remark that the impossibility of universal obfuscation [6] implies that more than one hardware component is necessary. To see this, observe that if we had a single hardware component that resists (even one-bit) arbitrary leakage then we immediately get a no-hardware obfuscation in the sense of Barak et al. [6]: The obfuscated program consists of our software and a full description of the hardware component (including all the embedded secrets). This must be a good obfuscation since any predicate that we can evaluate on this description can be seen as a one-bit leakage function evaluated on the state of the hardware component. If the device was resilient to arbitrary one-bit leakage, it would mean that any such leakage/predicate can be computed by a simulator that only has black-box access to the function; hence, we have a proper obfuscator.

The model of leaky distributed systems. Given the impossibility result for a single leaky hardware component, we concentrate on solutions that use multiple components. Namely, we have (polynomially) many hardware components, all of which are leaky. The adversary in our model can freely choose the inputs to the hardware components and obtain leakage by repeatedly choosing one component at a time and evaluating an arbitrary (polynomial-size) leakage function on the current state and randomness of that component. We place no restriction on the order or the number of times that components can be chosen to leak, so long as the total rate of leakage from each component is not too high.

In more detail, we consider continual leakage, where the lifetime of the system is partitioned into time units and within each time unit we have some bound on the number of leakage bits that the adversary can ask for. The components are running a randomized *refresh* protocol at the end of each time unit and erase their previous state.[1] A unique feature of our model is that the adversary sees and has complete control over all the communication between these components (including the communication needed for the refresh protocol). We term our leakage model the *leaky distributed system* model (LDS), indeed this is just the standard model of a distributed system with adversarially controlled communication, when we add to it the fact that the individual parties are leaky.

We stress that this model seems realistic: the different components can be implemented by physically (and even geographically) separated machines, amply justifying the assumption on separate leakage. We also note that a similar (but somewhat weaker) model was suggested recently by Akavia et al. [1], in the context of leakage-resilient encryption.

---

[1] This is reminiscent to the proactive security literature [30, 13].

Only-computation-leaks vs. leaky distributed systems. Our leakage model shares some similarities to the "only computation leaks" (OCL) model, in that the adversary can get leakage from different parts of the global state separately but not from the entire global state at once. These two models are nonetheless fundamentally different, for two reasons. One difference is that in the OCL the different components "interact" directly by writing to and reading from memory, and communication is neither controlled by nor visible to the adversary. In the LDS model, on the other hand, the adversary sees and controls the entire communication. Another difference is that in the OCL model, the adversary can only get leakage from the components in the order in which they perform the computation, whereas in LDS model, it can get leakage in any order.

An intermediate model, that we use as a technical tool in this work, is where the adversary can get leakage from the components in any order (as in the LDS model), but the components communicate securely as in the OCL model. For lack of a better name, we call this intermediate model the OCL^+ model. Clearly, resilience to leakage in the model of leaky distributed systems is strictly harder than in the OCL or OCL^+ models and every solution secure in our model will automatically be secure also in the two weaker models.

1.4   From OCL^+ to LDS

We present a transformation that takes any circuit secure in the OCL^+ model and converts it into a system of components that maintains the functionality and is secure in the model of leaky distributed systems. Recently, Goldwasser-Rothblum [22] constructed a universal compiler, which transforms any circuit into one that is secure in the OCL^+ model. (Unlike previous compilers [17, 24, 27], the [22] compiler does not require a leak-free hardware component.) Combining the compiler with our transformation, we obtain a compiler that takes any circuit and produces a system of components with the same functionality that is secure in the LDS model. The number of components in the resulting system is essentially the size of the original circuit, assuming we use the underlying Goldwasser-Rothblum compiler. However, as we explain in Section 1.5 below, we can reduce the number of components to be *independent* of the circuit size, by first applying the hardware-assisted obfuscator from Section 1.1.

The main gap between the OCL^+ model and our model of leaky distributed systems, is that in the former, communication between the components is completely secure, whereas in the latter it is adversarially controlled. In the heart of our transformation stands an implementation of *leakage-tolerant communication channels* that bridges the above gap, based on the following tools:

Non-Committing Encryption. Our main technical observation is that secret communication in the face of leakage can be obtained very simply using non-committing encryption [12]. Recall that non-committing encryption is a (potentially interactive) encryption scheme such that a simulator can generate a fake transcript, which can later be "opened" as either an encryption of zero or as an encryption of one. This holds even when the simulator needs to generate the randomness of both the sender and the receiver. In our context, the distributed components use non-committing encryption to preserve the privacy of their messages. The observation is that non-committing encryption can be used to implement "leakage resilient channels", in the sense that any leakage query on the state of the communicating parties could be transformed into a leakage query on the underlying message alone (see Section 4).

Leakage-resilient MACs. In addition to secrecy, we also need to ensure authenticity of the communication between the components. We observe that this can be done easily using information-theoretic MAC schemes based on universal-hashing [33, 3]. Roughly, each pair of components will maintain rolling MAC keys that are only used $\Theta(1)$ times. To authenticate a message, they will use the MAC key sent with the prior message and will send a new MAC key to be used for the next message. (We use a short MAC key to authenticate a much longer message, so the additional bandwidth needed for sending future MAC keys is tolerable.) Since these MAC schemes offer information-theoretic security, it is very easy to prove that they can also tolerate bounded leakage. Authenticating the communication assures that secrecy is kept (e.g. the adversary cannot have a component encrypt a secret message under an unauthentic key) and also ensures that the components remain "synchronized" (see Section 4).

1.5   The End-Result: Obfuscation with Leaky Hardware

To obfuscate a program, we first apply the hardware-assisted obfuscator from Section 1.1, thus obtaining a universal hardware device, whose size and amount of computation (per input) depend only on the security parameter, and which can be used to evaluate obfuscated programs from various vendors. We next apply the Goldwasser-Rothblum compiler [22], together with our transformation from Section 1.4, to the code of the hardware device, resulting in a system of components that can still be used for obfuscation in exactly the same way (as the universal device), but is now guaranteed to remain secure even if the components are leaky and even if the communication between them is adversarially controlled.

To obfuscate a program f using this system, the obfuscator generates keys for the FHE scheme and encrypts f under these keys. In addition, it uses the public CCA2 key generated with the original universal device to encrypt the secret FHE key together with a hash of the encrypted program. The encrypted program and parameters are then sent to the user. Evaluating the obfuscated program consists of running the FHE evaluation procedure and then interacting with the system of components (in a universal argument) to decrypt the resulting ciphertext. The system verifies the proof in a leakage-resilient manner and returns the decrypted result.

We remark that our transformation from any circuit/device to a leaky system of components, as well as our transformation from circuit-specific obfuscation schemes to general-purpose ones, are generic and can be applied to any device-assisted obfuscation scheme, such as the schemes of [19, 25]. When doing so, the end result will inherit the properties of the underlying scheme. In particular, when instantiated with [19, 25], the amount of work performed by the devices is proportional to the size of the entire computation (the hardware used for each gate in the obfuscated circuit).

1.6   Related Work

Research on formal notions of obfuscation essentially started with the work of Barak et. al. [6], who proved that software-only obfuscation is impossible in general. This was followed by other negative results [20] and some positive results for obfuscating very simple classes of functions (e.g., point functions) [32, 11, 15]. The sweeping negative results for software-only obfuscation motivated researchers to consider relaxed notions where some interesting special cases can be obfuscated (e.g., [23, 26, 8]).

In contrast, the early works of Best [7], Kent [28] and Goldreich and Ostrovsky [19] addressed the software-protection problem using a physically shielded full-blown CPU. The work of Goyal *et. al.* [25] showed that the same can be achieved also with small stateless hardware tokens. These solutions only consider perfectly opaque hardware. Furthermore, in these works the amount of work performed by the secure hardware device during the evaluation of one input is proportional to the size of the entire computation.[2]

The work by Goldwasser *et. al.* [21] on one-time programs shows that programs can be obfuscated using very simple hardware devices that do very little work. However, their resulting obfuscated program can be run *only once*.

Our focus on obfuscation with *leaky* hardware follows a large corpus of recent works addressing leakage-resilience cryptography (see, e.g., [16, 2] and references within). In particular, our construction uses results of Goldwasser and Rothblum [24, 22], which show how to convert circuits into ones that are secure in *only computation leaks* model of Micali and Reyzin [29] (or even in the stronger OCL^+ model described above).

Our construction of leakage-tolerant secure channels and the relation between leakage-tolerance and adaptive security were further investigated and generalized in [10], who consider general *universally composable* leaky protocols.

Organization In Section 2 we construct a hardware-assisted obfuscation scheme where the amount of work done by the hardware is minimal (polynomial in the security parameter). In Section 3 we show how to transform any "circuit-specific" scheme, such as the one constructed in Section 2, to a "general-purpose" scheme where the same hardware device can be used for multiple obfuscated programs. In Section 4 we show how to transform any hardware-assisted obfuscation, such as the above, to a leakage-resilient scheme. The full details and proofs as well as some of the secondary results can be found in the full version of this paper [9].

---

[2] On the other hand, the solutions in [19, 25] can be based on one-way functions, while our solution requires stronger tools such as FHE and universal arguments.

2   Hardware Assisted Obfuscation

In this section we construct a hardware assisted obfuscation scheme. The basic model and definitions are presented in Section 2.1. An overview of the construction is presented in Section 2.2. The detailed construction and its analysis can be found in the full version of this paper[9].

2.1   The Model

In the setting of hardware assisted obfuscation, a circuit C (taken from a family $\mathcal{C}_n$ of poly-size circuits) is obfuscated in two stages. First, the PPT obfuscation algorithm $\mathcal{O}$ is applied to C, producing the "software part" of the obfuscation obf, together with (secret) parameters params for device initialization. At the second stage, the hardware device HW is initialized with params. The evaluator is given obf and black-box access to the initialized device $\mathsf{HW}_{\mathsf{params}}$. In our security definition, we consider a setting in which the adversary is given $t = \mathrm{poly}\,(n)$ independent obfuscations of t circuits, where obfuscation i consists of a corresponding device $\mathsf{HW}_{\mathsf{params}_i}$ and obfuscated data obf_i. In this model each obfuscated circuit may have its own specialized device.

Definition 2.1 (Circuit-specific hardware-assisted obfuscation (CSHO)). $(\mathcal{O}, \mathsf{HW}, \mathsf{Eval})$ *is a* CSHO *scheme for a circuit ensemble* $\mathcal{C} = \{\mathcal{C}_n\}$, *if it satisfies:*

- **Functional Correctness.** Eval *is a poly-time oracle aided TM , such that for any* $n \in \mathbb{N}$, $C \in \mathcal{C}_n$ *and input* v *for* C: $\mathsf{Eval}^{\mathsf{HW}_{\mathsf{params}}}\left(1^{|C|}, \mathsf{obf}, v\right) = C\,(v)$, *where* $(\mathsf{obf}, \mathsf{params}) \leftarrow \mathcal{O}\,(C)$.
- **Circuit-Independent Efficiency.** *The size of* $\mathsf{HW}_{\mathsf{params}}$ *is* $\mathrm{poly}(n)$, *independently of* $|C|$, *where* $(\mathsf{params}, \mathsf{obf}) \leftarrow \mathcal{O}(C)$. *Also, during each run of* $\mathsf{Eval}^{\mathsf{HW}_{\mathsf{params}}}\left(1^{|C|}, \mathsf{obf}, v\right)$ *on any input* v, *the total amount of work performed by* $\mathsf{HW}_{\mathsf{params}}$ *is* $\mathrm{poly}(n)$, *independently of* $|C|$.
- **Polynomial Slowdown.** $\mathcal{O}$ *is a* PPT *algorithm. In particular, there is a polynomial* q, *such that for any* $n \in \mathbb{N}$ *and* $C \in \mathcal{C}_n$, $|\mathsf{obf}| \leq q\,(|C|)$.
- t-**Composable Virtual Black Box (VBB).** *Any adversary, given* t *obfuscations, can be simulated, given oracle access to the corresponding circuits. That is, for any* PPT $\mathcal{A}$ *(with arbitrary output) there is a* PPT $\mathcal{S}$ *such that:*

$$\left\{\mathcal{A}^{\mathsf{HW}_1, \dots, \mathsf{HW}_t}\,(z, \mathsf{obf}_1, \dots, \mathsf{obf}_t)\right\} \approx_c \left\{\mathcal{S}^{C_1, \dots, C_t}\,(z, 1^n, |C_1|, \dots, |C_t|)\right\}\ ,$$

where $C_1 \dots C_t \in \mathcal{C}_n$, $z \in \{0,1\}^{\mathrm{poly}(n)}$ *is an arbitrary auxiliary input,* $\mathsf{HW}_i = \mathsf{HW}_{\mathsf{params}_i}$ *and* $(\mathsf{obf}_i, \mathsf{params}_i) \leftarrow \mathcal{O}\,(C_i)$.
We say that the scheme is **stand-alone VBB** *if it is 1-composable. We say that the scheme is* **composable** *if its* t-composable for any polynomial t.

While previous solutions [19, 25] satisfy the correctness and security requirements of Definition 2.1, they require that the total amount of work performed by the device for a single evaluation is proportional to $|C|$, the size of the entire circuit. Namely, they do not achieve circuit-independent efficiency. In this section we show that how to construct schemes which do achieve this feature, based on a different approach. The main result is given by Theorem 2.1.

Theorem 2.1. *Assuming fully homomorphic encryption, there exists a compos-able* CSHO *scheme for all polynomial size circuit ensembles* $\mathcal{C} = \{\mathcal{C}_n\}$.

2.2   The Construction

We next overview the main aspects of the constructions.

The main ideas. Informally, given a FHE scheme $\mathcal{E}$, we obfuscate a circuit C by sampling $(\mathsf{sk}, \mathsf{pk}) \leftarrow \mathsf{Gen}\,(1^n)$, encrypting $\hat{C} = \mathsf{Enc}_{\mathsf{pk}}\,(C)$ and creating a "proof-checking decryption device" $\mathsf{HW} = \mathsf{HW}_{\mathsf{sk}}$ which is meant to decrypt "proper evaluations". The obfuscation consists of $\mathsf{obf} = (\hat{C}, \mathsf{pk})$ and oracle access to HW. To evaluate the obfuscation on input v, compute $e = \mathsf{Eval}_{\mathsf{pk}}(\hat{C}, U_{s,v})$, where $U_{s,v}$ is a universal circuit that given a circuit C of size s outputs $C\,(v)$.[3] Then, "prove" to HW that indeed $e = \mathsf{Eval}_{\mathsf{pk}}(\hat{C}, U_{s,v})$. In case HW is "convinced", it decrypts $C\,(v) = \mathsf{Dec}_{\mathsf{sk}}\,(e)$ and returns the result to the evaluator. Intuitively, the semantic security of $\mathcal{E}$ and the soundness of the proof system in use should prevent the evaluator from learning anything about the original circuit C other than its input-output behavior.

We briefly point out the main technical issues that arise when applying the above approach and the way we deal with these issues.

- **Minimizing the device's workload.** Proving the validity of an evaluated ciphertext e w.r.t. an encrypted circuit $\hat{C}$ amounts to proving that a poly$(|C|)$-long computation was performed correctly. However, the running time of our device should be independent of $|C|$ and hence cannot process such a computation. In fact, it cannot even process the assertion itself as it includes the poly$(|C|)$-long encryption $\hat{C}$. To overcome this, we use *universal arguments* (UA's) that also have a *proof of knowledge* property [4] and *collision resistant hashing*. Specifically, the device only stores a (short) hash $h(\hat{C})$ and the evaluator proves it "knows" an encrypted circuit $\hat{C}'$ with the same hash and that the evaluated ciphertext is the result of applying $\mathsf{Eval}_{\mathsf{pk}}$ to $\hat{C}'$ and the universal circuit $U_{s,v}$ (corresponding to some input v).
- **Using a stateless device with no fresh randomness.** Our device can be implemented as a boolean circuit that need not maintain a state between evaluator calls nor generate fresh randomness; in particular, it should withstand concurrent proof attempts and "reset attacks" (as termed by [14]). To enable this, we use similar techniques to those in [5]. Informally, these techniques allow transforming the UA protocol we use to a "resettable" protocol, where the verifier's randomness is fixed to some *pseudo random function.* [4]

---

[3] Abusing notation, we denote by Eval both evaluation algorithms $\mathsf{Eval}^{\mathsf{HW}_{\mathsf{params}}}(\mathsf{obf}, v)$ and $\mathsf{Eval}_{\mathsf{pk}}$. To distinguish between the two, we always denote the evaluation algorithm of the FHE scheme by $\mathsf{Eval}_{\mathsf{pk}}$.

[4] The mentioned techniques essentially transform any public-coin constant-round protocol to a "resettable" one.

3   General-Purpose (Sendable) Obfuscation

In this section we show how to convert any *circuit-specific* obfuscation scheme, such as the one in Section 2, to a scheme which uses a single *universal* (general-purpose) hardware device. The basic model and definitions are presented in Section 3.1, the transformation is presented in Section 3.2 and analyzed in the full version of this paper [9].

3.1   The Model

In circuit-specific obfuscation, the obfuscator gives the user a device that depends on the obfuscated circuit C. More precisely, the "specifying parameters" params, produced by $\mathcal{O}(C)$, depend on C and are hardwired into the device before it is sent to the user. Thus, each device supports only a single obfuscated circuit.

We consider a more natural setting in which different parties can send obfuscations to each other online, without the need of exchanging devices per each obfuscation. Informally, in this setting we assume that a trusted manufacturer creates devices, where each device is associated with private and public parameters (prv, pub). The private parameters are hardwired into the device and are never revealed (they can be destroyed), while the public ones are published together with the "identity" of the device (e.g., on the manufacturer's web page www.obfuscationdevices.com). Any user, who wishes to send an obfuscation of a circuit C to another user who holds such a device, retrieves the corresponding public parameters and sends the required obfuscation.

Concretely, a general-purpose obfuscation scheme consists of two randomized algorithms (Gen, $\mathcal{O}$) and a device HW. First, Gen (1^n) generates private and public parameters (prv, pub) (independently of any circuit). Then, HW is initialized with prv and the initialized device $\mathsf{HW}_{\mathsf{prv}}$ is given to the user. The corresponding pub are published. Anyone in hold of pub can obfuscate a circuit C by computing obf $\leftarrow \mathcal{O}(C, \mathsf{pub})$ and sending obf to the user holding the device.

Definition 3.1 (General-purpose   hardware-assisted   obfuscation (GPHO)). $(\mathcal{O}, \mathsf{Gen}, \mathsf{HW}, \mathsf{Eval})$ *is a* GPHO *scheme for* $\mathcal{C} = \{\mathcal{C}_n\}$ *if it satisfies:*

- **Functional Correctness.** Eval *is a polynomial-time oracle aided TM, such that for any* $n \in \mathbb{N}$, $C \in \mathcal{C}_n$ *and input* v *for* C: $\mathsf{Eval}^{\mathsf{HW}_{\mathsf{prv}}}(1^{|C|}, \mathsf{obf}, v) = C(v)$, *where* $(\mathsf{prv}, \mathsf{pub}) \leftarrow \mathsf{Gen}(1^n)$ *and* $\mathsf{obf} \leftarrow \mathcal{O}(C, \mathsf{pub})$.
- **Circuit-Independent Efficiency.** *The size of* $\mathsf{HW}_{\mathsf{prv}}$ *is polynomial in* n, *independent of* $|C|$, *where* $(\mathsf{prv}, \mathsf{pub}) \leftarrow \mathsf{Gen}(1^n)$. *Moreover, during each run of* $\mathsf{Eval}^{\mathsf{HW}_{\mathsf{prv}}}(1^{|C|}, \mathsf{obf}, v)$ *on any input* v, *the total amount of work performed by* $\mathsf{HW}_{\mathsf{prv}}$ *is polynomial in* n, *independent of* $|C|$.
- **Polynomial Slowdown.** $\mathcal{O}$ *and* Gen *are* PPT *algorithms. In particular, there is a polynomial* q *such that for any* $n \in \mathbb{N}$, $C \in \mathcal{C}_n$, $|\mathsf{pub}, \mathsf{prv}| \leq q(n)$ *and* $|\mathsf{obf}| \leq q(|C|)$.
- **Virtual Black Box (VBB).** *For any* PPT *adversary* $\mathcal{A}$ *and polynomial* t *there is a* PPT *simulator* $\mathcal{S}$ *such that:*

$$\left\{\mathcal{A}^{\mathsf{HW}_{\mathsf{prv}}}(z, \mathsf{obf}_1, \ldots, \mathsf{obf}_t)\right\} \approx_c \left\{\mathcal{S}^{C_1, \ldots, C_t}(z, 1^n, |C_1|, \ldots |C_t|)\right\} \;,$$

where $C_1 \ldots C_t \in \mathcal{C}_n$, $z \in \{0,1\}^{\mathrm{poly}(n)}$ is an arbitrary auxiliary input (prv, pub) $\leftarrow$ Gen (1^n) and obf$_i \leftarrow \mathcal{O}(C_i, \mathsf{pub})$.

3.2   The Transformation

Essentially, we wish to avoid restricting the device to a specific circuit C (like hard-wiring $h(\hat{C})$ into the device as done in our circuit-specific scheme). Instead, we would like to have the user "initialize" his device with the required parameters params for each obfuscation he wishes to evaluate. However, params cannot be explicitly given to the evaluator as they contain sensitive information.

For this purpose, we simply use a CCA2 public key encryption scheme. That is, the obfuscator will generate params, but instead of hard-wiring them into the hardware device (which will make the device circuit-specific), he will encrypt params and send the resulting ciphertext to the user. The fact that the underlying encryption scheme is CCA2 secure implies that the user can neither gain any information about params nor change it to related parameters params′.

More formally, the new general-purpose device HW′ is manufactured together with a pair of CCA2 keys (prv, pub) = (sk, pk). The secret key sk is hardwired into the device (and destroyed), while pk is published. Each device call is appended with the CCA2 encryption of params. The device HW′ answers its calls by first decrypting the encrypted parameters params and then applying the device HW$_{\mathsf{params}}$ of the underlying circuit-specific scheme (e.g. the scheme in Section 2). In the full version [9] we present the detailed construction and show:

Theorem 3.1. *Given a CCA2 encryption scheme, any circuit-specific obfuscation scheme as in Definition 2.1 can be transformed to a general-purpose one as in Definition 3.1.*

Corollary 3.1 (of Theorems 2.1,3.1). *Assume that there exists a fully homomorphic encryption scheme and a CCA2 encryption scheme, then there exists a general-purpose obfuscation scheme.*

Remark 3.1. The above transformation would also work (as is) for schemes with no circuit-independent efficiency. The amount of work performed by the general-purpose device is essentially inherited from the underlying scheme (with the fixed overhead of CCA2 decryption). In particular, we can apply it to the scheme of [25] and get a general-purpose solution that is based solely on the existence of CCA2 schemes, but which makes poly($|C|$) device calls.

4   Obfuscation with Leaky Hardware

We now turn to the task of dealing with leaky hardware. As we explained in the introduction, if we allow arbitrary leakage functions (even with small output) then it is impossible to obfuscate using a *single* leaky hardware device. Hence, our goal is to show how to use many leaky hardware devices to achieve obfuscation.

We first show how to obfuscate any function f using leaky hardware devices, where the number of devices is proportional to the size of the circuit computing f. Then, when we apply this obfuscator to the function computed by the hardware device from Section 2 (or Section 3, respectively), to get circuit-specific (or general-purpose, respectively) obfuscation with leaky hardware devices, where the number of devices is polynomial in the security parameter, *independent* of the function being obfuscated.

4.1   An Overview

In what follows, we give an informal definition of obfuscation with leaky hardware and a high-level overview of our construction. The formal definitions and detailed construction are given in Sections 4.2 and 4.3. The security analysis can be found in the full version of this paper [9].

The leaky distributed system (LDS) model. In the LDS model a functionality f (with secrets) is implemented by a system of multiple hardware components $(\mathsf{HW}_1, \mathsf{HW}_2, \ldots, \mathsf{HW}_m)$. The components can maintain a state and generate fresh randomness. To evaluate the functionality f, an input v is given to HW_1 and the components communicate to jointly compute $f(v)$, which is eventually outputted by HW_m. The adversary (evaluator) in our model can freely choose the inputs to the computation and is given full control over the communication between the components. In addition, the adversary can choose one component at a time and evaluate a leakage function on its inner state and randomness.

We consider a continual leakage model, where the lifetime of each component HW_i is partitioned into time periods (that are set according to the inputs that HW_i receives). At the end of each time period, HW_i "refreshes" its inner state by applying an Update procedure (that erases the previous state). The Update procedures performed at different components are coordinated by exchange of messages. As the rest of the computation, the Update procedure is also exposed to leakage and the adversary controls the exchange of messages during the update.

We place no restriction on the order and timing of the adversary's interaction with the system. In particular, it can pass messages to any component at any time and get leakage on any component at any time (which can depend on previous leakage and messages).

Constructing secure leaky distributed systems (LDS). Our goal is to compile (or "obfuscate") any functionality, given by some circuit C (with hardwired secrets), into an LDS that *perfectly protects* C, as long as **the leakage from each HW_i in each time period is bounded**. In the terminology of obfuscation, the LDS should perform as a *virtual black-box*: The view of any adversary $\mathcal{A}$ attacking the LDS can be simulated by a simulator $\mathcal{S}$ which can only access C as a black-box. In particular, $\mathcal{S}$ should simulate on its own the communication between the components and all the leakage. We achieve this goal in two main steps:

1. We apply the Goldwasser-Rothblum compiler to the circuit C to get a circuit that is secure in the (augmented) *only computation leaks* (OCL^+) model.

2. Then, we provide a general transformation that takes any OCL^+-secure circuit and transforms it to a secure LDS.

Hence, our main goal is to show that an adversary in the LDS model can be simulated by an adversary in the OCL^+ model (that does not witness the communication between the modules). Then, by the OCL^+-security (implied by the GR compiler), we can deduce that simulation can be done only with black-box access to the underlying functionality.

In the heart of our transformation stands an implementation of *leakage tolerant communication channels*. We first explain the main ideas required to achieve secrecy and then explain how to get authenticity.

Leaky secret channels from non-committing encryption. In the OCL^+ model, the components can securely exchange messages. Still, the adversary might get some leakage on the contents of these messages as the (leaky) state of the components includes the messages at some point. The OCL^+ security guarantee implies, however, that a bounded amount of leakage does not compromise the security of the entire system.

To enhance OCL^+-security to LDS-security we implement the secure communication channels. As explained above, we assume for now that the adversary delivers all messages intact and deal only with secrecy. The standard solution for secret channels would be to encrypt all communication between the components; however, in the face of leakage this approach encounters the following difficulty: Consider a sender component HW_S in the LDS model that wishes to communicate a message M to a receiver component HW_R (using some encryption scheme). Note that the adversary can obtain arbitrary (bounded) leakage on the state of both $\mathsf{HW}_S, \mathsf{HW}_R$, including leakage on both the plaintext M and the randomness r_S, r_R used to encrypt/decrypt. Moreover, the leakage function can depend on the corresponding ciphers which were already sent. This implies that naively simulating the communication (by say encryptions of 0) won't work.

Our main technical observation is that the above obstacle can be overcome using non-committing encryption (NCE) [12]. NCE schemes (which can potentially be interactive) allow simulating a fake cipher (or transcript) c together with two optional random strings $(r_S^0, r_S^1), (r_R^0, r_R^1)$ for both the sender S and the receiver R. The simulated cipher can later be "opened" as an encryption of either 1 or 0 (using the suitable randomness).[5] This tool allows us to show that the view of an attacker $\mathcal{A}$ in the LDS model can be simulated by an attacker $\mathcal{A}'$ in the OCL^+ model, provided that the components communicate using NCE.

Specifically, for any single bit message, the OCL^+ adversary $\mathcal{A}'$ (which does not see any communication) will use the NCE to generate fake communication

---

[5] NCE was so far mainly used in the setting of multi-party-computation as a tool for dealing with adaptive corruptions. Indeed, leakage can be viewed as a restricted form of "honest but curious" corruption, where the adversary learns part of the state, whereas in full corruption, it learns the entire state. In both cases, the choice of leakage/corruption is done adaptively according to the view of the adversary so far. The relation between leakage-tolerant protocols and adaptively secure protocols is further generalized in [10].

with corresponding randomness $\bar{r} = (r_S^0, r_S^1), (r_R^0, r_R^1)$. Then, when the simulated $\mathcal{A}$ performs a leakage query L to be evaluated on both the plaintext b and the encryption's randomness, $\mathcal{A}'$ can translate it to a new leakage query L' which **will only be evaluated on the plaintext message**. The leakage function L' will have the simulated randomness $\bar{r}$ hardwired into it and will choose which randomness to use according to the plaintext b.

Leakage resilient MACs. To deal with adversaries that interfere with message delivery we use leakage-resilient c-time MAC schemes. Informally, each two components maintain rolling MAC keys that are used at most $c = O(1)$ times. After $c - 1$ times the components run the Update protocol to regain fresh MAC keys. The communication during the update is done using NCE as described above, while authentication is done using the c-th application of the previous key.

4.2   The LDS Model

Our leakage model postulates an adversary $\mathcal{A}$ that interacts with a system of distributed leaky hardware components. Each component maintains a state and is capable of producing fresh randomness. At the onset of the interaction, the components are pre-loaded with some secret state and thereafter they can receive messages, send messages and leak information to the attacker. In our model all the I/O of the components and their communication is done via the attacker $\mathcal{A}$.

Definition 4.1 (Single-input leakage). *In a distributed single-input λ-leakage attack a* PPT *adversary $\mathcal{A}$ interacts with hardware components* $(\mathsf{HW}_1, \ldots, \mathsf{HW}_m)$ *and can do the following (in any order, possibly in an interleaving manner):*

1. *Feed $\mathcal{O}(C)$ a single input of his choice.*
2. *Interact with each component, sending it messages and receiving the resulting outputs and replies. These devices are message-driven, so they are activated by receiving messages from the attacker, then they compute and send the result, then wait for more messages.*
3. *Adaptively send up to λ 1-bit leakage queries to each of the hardware components. Each leakage query is modeled as a poly-size Boolean circuit and is applied to the entire state of a single hardware device. Without loss of generality, we can think of the state of the device as it was in the last time that the device was activated, including all the randomness that the device generated in order to deal with the last activation.*

We denote the output of $\mathcal{A}$ in such attack by $\mathcal{A}[\lambda : \mathsf{HW}_1, \ldots, \mathsf{HW}_m]$.

Definition 4.2 (Continual leakage). *A continual λ-leakage attack is an attack where a* PPT *adversary $\mathcal{A}$ repeats a single-input λ-leakage attack poly many times, where between any two consecutive attacks the devices' secret state is updated by applying a* PPT *algorithm* Update *to the state of each HW_i separately. $\mathcal{A}$ obtains leakage during the* Update *procedure, where the leakage function takes as input both the current secret state of HW_i and the randomness used by* Update.

We denote by time period t at device HW_i the time period between the beginning of the $(t-1)st$ Update procedure and the end of the t-th Update procedure (note that these time periods are overlapping).[6] We allow the adversary $\mathcal{A}$ to leak at most λ bits from each HW_i during each (local) time period.

We denote the output of $\mathcal{A}$ in such attack by $\mathcal{A}[\lambda : \mathsf{HW}_1, \ldots, \mathsf{HW}_m : \text{Update}]$.

Below we consider an obfuscator $\mathcal{O}$ that takes as input a circuit C and outputs an "obfuscated" version of C that uses leaky hardware devices as above. Namely, we have $(\mathsf{HW}_1, \ldots, \mathsf{HW}_m) \leftarrow \mathcal{O}(C)$, where the HW_i's are the leaky hardware devices, initialized with the appropriate circuits.

Remark 4.1. In Definitions 2.1 and 3.1, the obfuscator $\mathcal{O}$ outputs a "software part" obf and parameters params for initializing the hardware. In the current setting, the obfuscation does not contain a software part. The simplified notation $(\mathsf{HW}_1, \ldots, \mathsf{HW}_m) \leftarrow \mathcal{O}(C)$, should be interpreted as sampling $\{\text{params}_i\} \leftarrow \mathcal{O}(C)$ (where params_i corresponds to the i-th sub-computation) and initializing the hardware devices $\{\mathsf{HW}_i\}$ accordingly.

Definition 4.3. *We say that $\mathcal{O}$ is an LDS-obfuscator with continual λ-leaky hardware if for any circuit C and $(\mathsf{HW}_1, \ldots, \mathsf{HW}_m) \leftarrow \mathcal{O}(C)$, the distributed system $(\mathsf{HW}_1, \ldots, \mathsf{HW}_m)$ maintains the functionality of C when all the messages between them are delivered intact and in addition we have the following:*

For any PPT attacker $\mathcal{A}$, executing a continual λ-bit leakage attack, there exists a PPT simulator $\mathcal{S}$, such that for any ensemble of poly-size circuits $\{\mathcal{C}_n\}$:

$$\{\mathcal{A}(z)[\lambda : \mathsf{HW}_1, \ldots, \mathsf{HW}_m : \text{Update}]\}_{\substack{n \in \mathbb{N}, C \in \mathcal{C}_n \\ z \in \{0,1\}^{\text{poly}(n)}}} \approx_c \left\{\mathcal{S}^C(z, 1^{|C|})\right\}_{\substack{n \in \mathbb{N}, C \in \mathcal{C}_n \\ z \in \{0,1\}^{\text{poly}(n)}}},$$

where $(\mathsf{HW}_1, \ldots, \mathsf{HW}_m) \leftarrow \mathcal{O}(C)$ and z is an arbitrary auxiliary input.

4.3   The Construction

We build our solution using a compiler $\mathcal{C}$ that is secure in the continual λ-OCL^+ model. Namely, $\mathcal{C}$ converts any circuit C into a collection of leaky sub-components $(sub_1, \ldots, sub_m)$ (that also have an update procedure, Update') that is secure long as the adversary can only get λ leakage from each component in each time unit and cannot see or influence the communication between them. In our model, however, the communication is under the control of the adversary. To secure the communication, we use non-committing encryption and c-time leakage resilient MACs (as described in the overview).

The construction. Given a circuit C, the obfuscator $\mathcal{O}$ does the following:

---

[6] Intuitively, time period t is the entire period where the t-th updated secret states can be leaked. During the t-th Update procedure, both the $(t-1)st$ secret state and the t-th secret state may leak, which is why the time periods are overlapping.

1. Apply the λ-OCL$^+$ compiler $\mathcal{C}$ to C and obtain a circuit $C' = (sub_1, \ldots, sub_m)$ and an Update$'$ procedure, such that (C', Update') is secure in the continual λ-OCL$^+$ model.

   We assume for simplicity that: (a) sub_1 is the input module, that takes as input the "original" input $x \in \{0,1\}^n$ and passes it to the relevant sub_j's. (b) sub_m generates the final output. (c) The exchanged messages between the modules are all of the same size $\ell = \ell(n)$.

2. Put each module sub_i in a separate hardware component HW_i.

3. For every two communicating modules $i, j \in [m]$, generate a random key $K_{i,j} \leftarrow \{0,1\}^t$ for a λ-leakage-resilient MAC scheme $(\mathsf{MAC}, \mathsf{Vrfy})$, with keys of length $t = \Theta(\lambda)$. For every $i \in [m]$, hard-wire in HW_i the set of keys $\{(j, K_{i,j})\}$, for every j such that sub_j and sub_i communicate.

4. For every $i \in \{1, \ldots, m-1\}$ and every $j \in \{2, \ldots, m\}$, whenever sub_i is supposed to send a message $\mathbf{M} = (M_1, \ldots, M_\ell)$ to sub_j, the corresponding hardware HW_i sends $\mathbf{M}$ to HW_j using a non-committing encryption scheme $(\mathsf{NCGen}, \mathsf{NCEnc}, \mathsf{NCDec})$. Moreover, all the communication in this process is authenticated using the MAC scheme $(\mathsf{MAC}, \mathsf{Vrfy})$. More specifically, the hardware devices HW_i and HW_j communicate as follows:

   (a) Hardware HW_j does the following:
       i. For each $k \in [\ell]$, sample a random $r_{G,k} \in \{0,1\}^{\mathrm{poly}(n)}$ and compute $(e_k, d_k) = \mathsf{NCGen}(1^n; r_{G,k})$. Henceforth, let $\mathbf{e} = (e_1, \ldots, e_\ell), \mathbf{d} = (d_1, \ldots, d_\ell)$.
       ii. Compute $\sigma_{\mathbf{e}} = \mathsf{MAC}(\mathbf{e}; K_{i,j})$.
       iii. Send $(\mathbf{e}, \sigma_{\mathbf{e}})$ to HW_i and keep $\mathbf{d}$ as part of the secret state.

   (b) Hardware HW_i does the following:
       i. Verify that $\mathsf{Vrfy}(\mathbf{e}, \sigma_{\mathbf{e}}; K_{i,j}) = 1$ and verify that $(\mathbf{e}, \sigma_{\mathbf{e}})$ was not already sent by HW_j during this time period. If this check fails then discard the message $\mathbf{e}$.
       ii. If the check passes, for each $k \in [\ell]$ choose a random $r_{E,k} \in \{0,1\}^{\mathrm{poly}(n)}$, compute $c_k = \mathsf{NCEnc}(M_k, e_k; r_{E,k})$. Henceforth, let $\mathbf{c} = (c_1, \ldots, c_\ell)$.
       iii. Compute $\sigma_{\mathbf{c}} = \mathsf{MAC}(\mathbf{c}; K_{i,j})$.
       iv. Send $(\mathbf{c}, \sigma_{\mathbf{c}})$ to HW_j.

   (c) Hardware HW_j does the following:
       i. Verify that $\mathsf{Vrfy}(\mathbf{c}, \sigma_{\mathbf{c}}; K_{i,j}) = 1$ and verify that $(\mathbf{c}, \sigma_{\mathbf{c}})$ wasn't already sent by HW_i. If this check fails then discard the message $\mathbf{c}$.
       ii. If the check passes, compute for each $k \in [\ell]$, $M_i = \mathsf{NCDec}(c_i, d_i)$.

   Once HW_j gets $\mathbf{M}$, it runs sub_j on input $\mathbf{M}$ (unless sub_j is waiting for additional inputs).

5. Finally, HW_m sends an output message (assuming sub_m is the sub-computation that generates the outputs).

6. For each HW_i, after each "valid" activation (i.e., after it did its share in a computation), HW_i erases all its computations and updates its secret state, using an update procedure Update, defined as follows.

   (a) Apply the Update$'$ procedure to update the state of sub_i.

   (b) Refresh the MAC keys by choosing new random MAC keys $K'_{i,j}$ for every $j > i$ such that HW_i and HW_j communicate. Then send $K'_{i,j}$ to HW_j.

    (c) Erase the previous MAC keys $K_{i,j}$.

    (d) **Communication:** All the communication within the update procedure is done as in step 4. Namely, for each message, repeat steps $4(a) - 4(c)$, where the MACs are w.r.t. the previous MAC key $K_{i,j}$.

Theorem 4.1. *Assuming the compiler* C *used in the above construction is secure in the* λ*-OCL$^+$ model. Then the above construction yields an LDS-obfuscator with continual* λ*-leaky hardware* $HW_1, \ldots, HW_m$.

The proof of Theorem 4.1 is given in the full version of this paper [9].

References

[1] Akavia, A., Goldwasser, S., Hazay, C.: Distributed Public Key Encryption Schemes (2010) (manuscript)

[2] Akavia, A., Goldwasser, S., Vaikuntanathan, V.: Simultaneous Hardcore Bits and Cryptography against Memory Attacks. In: Reingold, O. (ed.) TCC 2009. LNCS, vol. 5444, pp. 474–495. Springer, Heidelberg (2009)

[3] Atici, M., Stinson, D.R.: Universal Hashing and Multiple Authentication. In: Koblitz, N. (ed.) CRYPTO 1996. LNCS, vol. 1109, pp. 16–30. Springer, Heidelberg (1996)

[4] Barak, B., Goldreich, O.: Universal arguments and their applications. SIAM J. Comput. 38(5), 1661–1694 (2008)

[5] Barak, B., Goldreich, O., Goldwasser, S., Lindell, Y.: Resettably-sound zero-knowledge and its applications. In: FOCS, pp. 116–125 (2001)

[6] Barak, B., Goldreich, O., Impagliazzo, R., Rudich, S., Sahai, A., Vadhan, S.P., Yang, K.: On the (Im)possibility of Obfuscating Programs. In: Kilian, J. (ed.) CRYPTO 2001. LNCS, vol. 2139, pp. 1–18. Springer, Heidelberg (2001)

[7] Best, R.M.: Microprocessor for executing enciphered programs. US Patent 4168396 (1979)

[8] Bitansky, N., Canetti, R.: On Strong Simulation and Composable Point Obfuscation. In: Rabin, T. (ed.) CRYPTO 2010. LNCS, vol. 6223, pp. 520–537. Springer, Heidelberg (2010)

[9] Bitansky, N., Canetti, R., Goldwasser, S., Halevi, S., Rothblum, G.: Obfuscation with leaky hardware (2011), Long Version on http://eprint.iacr.org

[10] Bitansky, N., Canetti, R., Halevi, S.: Leakage tolerant interactive protocols (2011) (manuscript), http://eprint.iacr.org/2011/204

[11] Canetti, R., Dakdouk, R.R.: Obfuscating Point Functions with Multibit Output. In: Smart, N.P. (ed.) EUROCRYPT 2008. LNCS, vol. 4965, pp. 489–508. Springer, Heidelberg (2008)

[12] Canetti, R., Feige, U., Goldreich, O., Naor, M.: Adaptively Secure Multi-party Computation. In: 28th Annual ACM Symposium on the Theory of Computing - STOC 1996, Philadelphia, PA, pp. 639–648. ACM (May 1996)

[13] Canetti, R., Gennaro, R., Herzberg, A., Naor, D.: Proactive security: Long-term Protection against break-ins. CryptoBytes 3(1) (1997)

[14] Canetti, R., Goldreich, O., Goldwasser, S., Micali, S.: Resettable zero-knowledge (extended abstract). In: STOC, pp. 235–244 (2000)

[15] Canetti, R., Rothblum, G.N., Varia, M.: Obfuscation of Hyperplane Membership. In: Micciancio, D. (ed.) TCC 2010. LNCS, vol. 5978, pp. 72–89. Springer, Heidelberg (2010)

[16] Dziembowski, S., Pietrzak, K.: Leakage-resilient cryptography. In: 49th FOCS - 2008, pp. 293–302. IEEE Computer Society (2008)

[17] Faust, S., Rabin, T., Reyzin, L., Tromer, E., Vaikuntanathan, V.: Protecting Circuits from Leakage: the Computationally-Bounded and Noisy Cases. In: Gilbert, H. (ed.) EUROCRYPT 2010. LNCS, vol. 6110, pp. 135–156. Springer, Heidelberg (2010)

[18] Gentry, C.: Fully homomorphic encryption using ideal lattices. In: Proceedings of the 41st ACM Symposium on Theory of Computing – STOC 2009, pp. 169–178. ACM (2009)

[19] Goldreich, O., Ostrovsky, R.: Software protection and simulation on oblivious rams. J. ACM 43(3), 431–473 (1996)

[20] Goldwasser, S., Kalai, Y.T.: On the impossibility of obfuscation with auxiliary input. In: 46th FOCS, pp. 553–562. IEEE Computer Society (2005)

[21] Goldwasser, S., Kalai, Y.T., Rothblum, G.N.: One-Time Programs. In: Wagner, D. (ed.) CRYPTO 2008. LNCS, vol. 5157, pp. 39–56. Springer, Heidelberg (2008)

[22] Goldwasser, S., Rothblum, G.: Unconditionally securing general computation against continuous only-computation leakage (2011) (manuscript)

[23] Goldwasser, S., Rothblum, G.N.: On Best-Possible Obfuscation. In: Vadhan, S.P. (ed.) TCC 2007. LNCS, vol. 4392, pp. 194–213. Springer, Heidelberg (2007)

[24] Goldwasser, S., Rothblum, G.N.: Securing Computation against Continuous Leakage. In: Rabin, T. (ed.) CRYPTO 2010. LNCS, vol. 6223, pp. 59–79. Springer, Heidelberg (2010)

[25] Goyal, V., Ishai, Y., Sahai, A., Venkatesan, R., Wadia, A.: Founding Cryptography on Tamper-Proof Hardware Tokens. In: Micciancio, D. (ed.) TCC 2010. LNCS, vol. 5978, pp. 308–326. Springer, Heidelberg (2010)

[26] Hofheinz, D., Malone-Lee, J., Stam, M.: Obfuscation for Cryptographic Purposes. In: Vadhan, S.P. (ed.) TCC 2007. LNCS, vol. 4392, pp. 214–232. Springer, Heidelberg (2007)

[27] Juma, A., Vahlis, Y.: Protecting Cryptographic Keys against Continual Leakage. In: Rabin, T. (ed.) CRYPTO 2010. LNCS, vol. 6223, pp. 41–58. Springer, Heidelberg (2010)

[28] Kent, S.T.: Protecting externally supplied software in small computers. PhD thesis, Massachusetts Institute of Technology (1981)

[29] Micali, S., Reyzin, L.: Physically Observable Cryptography. In: Naor, M. (ed.) TCC 2004. LNCS, vol. 2951, pp. 278–296. Springer, Heidelberg (2004)

[30] Ostrovsky, R., Yung, M.: How to withstand mobile virus attacks. In: 10th Annual ACM Symposium on Principles of Distributed Computing, PODC 1991, pp. 51–59. ACM (1991)

[31] Rivest, R., Adleman, L., Dertouzos, M.: On data banks and privacy homomorphisms. In: Foundations of Secure Computation, pp. 169–177. Academic Press (1978)

[32] Wee, H.: On obfuscating point functions. In: STOC 2005, pp. 523–532 (2005)

[33] Wegman, M., Carter, L.: New hash functions and their use in authentication and set equality. J. of Computer and System Sciences 22, 265–279 (1981)

BiTR: Built-in Tamper Resilience

Seung Geol Choi[1], Aggelos Kiayias[2,*], and Tal Malkin[3,**]

[1] University of Maryland
sgchoi@cs.umd.edu
[2] University of Connecticut
aggelos@cse.uconn.edu
[3] Columbia University
tal@cs.columbia.edu

Abstract. The assumption of the availability of tamper-proof hardware tokens has been used extensively in the design of cryptographic primitives. For example, Katz (Eurocrypt 2007) suggests them as an alternative to other setup assumptions, towards achieving general UC-secure multi-party computation. On the other hand, a lot of recent research has focused on protecting security of various cryptographic primitives against physical attacks such as leakage and tampering.

In this paper we put forward the notion of Built-in Tamper Resilience (BiTR) for cryptographic protocols, capturing the idea that the protocol that is encapsulated in a hardware token is designed in such a way so that tampering gives no advantage to an adversary. Our definition is within the UC model, and can be viewed as unifying and extending several prior related works. We provide a composition theorem for BiTR security of protocols, impossibility results, as well as several BiTR constructions for specific cryptographic protocols or tampering function classes. In particular, we achieve general UC-secure computation based on a hardware token that may be susceptible to affine tampering attacks. We also prove that two existing identification and signature schemes (by Schnorr and Okamoto, respectively) are already BiTR against affine attacks (without requiring any modification or endcoding). We next observe that non-malleable codes can be used as state encodings to achieve the BiTR property, and show new positive results for *deterministic* non-malleable encodings for various classes of tampering functions.

1 Introduction

Security Against Physical Attacks. Traditionally, cryptographic schemes have been analyzed assuming that an adversary has only *black-box* access to the underlying functionality, and no way to manipulate the internal state. For example, traditional security definitions for encryption schemes address an adversary who is given the public key — but not the private key — and tries to guess something about the plaintext of a challenge ciphertext, by applying some black-box attack (e.g., CPA or CCA). In practical situations, however, an adversary can often do more. For example, when using small

* Supported in part by NSF grants 0447808, 0831304, and 0831306.
** Supported in part by NSF grants 0831094 and 0347839.

D.H. Lee and X. Wang (Eds.): ASIACRYPT 2011, LNCS 7073, pp. 740–758, 2011.
© International Association for Cryptologic Research 2011

portable devices such as smart-cards or mobile-phones, an adversary can take hold of the device and apply a battery of attacks. One class of attacks are those that try to recover information via side channels such as power consumption [29], electromagnetic radiation [38], and timing [11]. To address these attacks, starting with the work of [27,33] there has been a surge of recent research activity on leakage-resilient cryptographic schemes. For example, refer to [41,37,1,22,10,19,32,9,31] and the references therein. The present work addresses *tampering attacks*, where an adversary can modify the secret data by applying various physical attacks (c.f., [2,8,7,40,4]). Currently, there are only a few results in this area [23,26,21].

Hardware Tokens. As discussed above, cryptographic primitives have traditionally been assumed to be tamper (and leakage) proof. In the context of larger cryptographic protocols, there have been many works that (implicitly or explicitly) used secure hardware as a tool to achieve security goals that could not be achieved otherwise. The work most relevant to ours is that of Katz [28], who suggests to use *tamper-proof hardware tokens* to achieve UC-secure [12] commitments. This allows achieving general feasibility results for UC-secure well-formed multi-party computation, where the parties, without any other setup assumptions, send each other tamper-proof hardware tokens implementing specific two-party protocols. There were several follow-up works such as [34,16,18,25,30,24,20], all of which assume a token that is tamper proof.

Given the wide applicability of tamper-proof tokens on one hand, and the reality of tampering attacks on the other, we ask the following natural question:

Can we relax the tamper-proof assumption, and get security using tamperable *hardware tokens?*

Clearly, for the most general interpretation of this question, the answer is typically negative. For example, if the result of [28] was achievable with arbitrarily-tamperable hardware token, that would give general UC-secure protocols in the "plain" model, which is known to be impossible [13]. In this work we address the above question in settings where the class of possible tampering functions and the class of protocols we wish to put in a token and protect are restricted.

1.1 Our Contributions

BiTR Definition. We provide a definition of Built-in Tamper Resilience (BiTR) for two party cryptographic protocols, capturing the idea that the protocol can be encapsulated in a hardware token, whose state may be tamperable. Our definition is very general, compatible with the UC setting [12], and implies that any BiTR protocol can be used as a hardware token within larger UC-protocols. Our definition may be viewed as unifying and generalizing previous definitions [23,26,21] and bringing them to the UC setting.

BiTR is a property of a cryptographic protocol M, which roughly says the following. Any adversary that is able to apply tampering functions from the class T on a token running M, can be simulated by an adversary that has no tampering capability, independently of the environment in which the tokens may be deployed.

The strongest result one would ideally want is a general compiler that takes an arbitrary protocol and transforms it to an equivalent protocol that is BiTR against arbitrary tampering functions, without having to encode the state into a larger one, and without requiring any additional randomness.[1] Since such a strong result is clearly impossible, we provide several specific results that trade off these parameters (see below), as well as the following composition theorem.

BiTR Composition.   As BiTR is a protocol centric property, the natural question that arises is whether it is preserved under composition. A useful result for a general theory of BiTR cryptography would be a general composition theorem which allows combining a BiTR protocol calling a subroutine and a BiTR implementation of that subroutine into one overall BiTR protocol. To this end, we characterize BiTR composition of protocols by introducing the notion of modular-BiTR which captures the property of being BiTR in the context of a larger protocol. We then prove that the property of modular-BiTR is *necessary and sufficient* for construction of composite BiTR protocols. At the same time we also derive a negative result, namely that modular-BiTR protocols that preserve the BiTR property in any possible context (something we term universal-BiTR) are unattainable assuming the existence of one-way functions, at least for non-trivial protocols. These results thus settle the question of BiTR composability.

BiTR Constructions without State Encoding.   We describe results for BiTR primitives that require no state encodings. It may come as a surprise that it is possible to prove a cryptographic protocol BiTR without any encoding and thus without any validation of the secret protocol state whatsoever. This stems from the power of our definitional framework for BiTR and the fact that it is can be achieved for specially selected and designed protocols and classes of tampering functions. We define the class $\mathcal{T}_{\text{aff}} = \{f_{a,b} \mid a \in \mathbb{Z}_q^*, b \in \mathbb{Z}_q, f_{a,b}(v) := av + b \bmod q\}$. That is, the adversary may apply a modular affine function of his choice to tamper the state. Affine tampering is an interesting class to consider as it has as special cases multiplication (e.g., shifting — which may be the result of tampering shift-register based memory storage), or addition (which may be result of bit flipping tampering).

We prove three protocols BiTR with respect to this class, where the tamper resilience is really "built-in" in the sense that no modification of the protocol or encoding of the state are necessary. The first one is Schnorr's identification (two-round) protocol [39]. The second is Okamoto's signature scheme [35]. Both protocols are interesting on their own (e.g., previous work [23] focused mostly on signature schemes), but the latter is also useful for the third protocol we prove affine-BiTR, described next.

UC-Secure Computation from tamperable tokens.   Katz's approach [28] for building UC-secure computation using hardware tokens allows a natural generalization that involves a commitment scheme with a special property, we call a *dual-mode parameter generation (DPG)* — depending on the mode of the parameter, the commitment scheme is either statistically hiding or a trapdoor commitment. We then observe that any DPG-commitment is sufficient for providing UC-secure multi-party computation assuming tamper proof tokens. Following this track, we present a new DPG-commitment scheme

---

[1] If an encoding ψ of the state is required, it is desirable that it is deterministic (randomness may not be available in some systems or expensive to generate), and that it has as high rate as possible. Ideally, an existing scheme can be proven BiTR *as-is*, without any state encoding at all.

that is BiTR against affine tampering functions, that relies on discrete-log based primitives including the digital signature scheme of Okamoto [35]. Thus, we obtain UC-secure general computation using hardware tokens tamperable with affine functions.

BiTR Constructions with State Encoding. We next discuss how one can take advantage of state consistency checks to design BiTR protocols. We observe first that non-malleable codes, introduced by Dziembowski, Pietrzak and Wichs [21] can be used as an encoding for proving the BiTR property of protocols. This gives rise to the problem of constructing such codes. Existing constructions [21] utilize randomness in calculating the encoding; we provide new constructions for such encodings focusing on purely *deterministic constructions*. In fact, when the protocol uses no randomness (e.g., a deterministic signing algorithm) or a finite amount of randomness (e.g., a prover in the resettable zero-knowledge [14] setting), by using deterministic encodings the token may dispense with the need of random number generation.

Our design approach takes advantage of a generalization of non-malleable encodings (called δ-non-malleable), and we show how they can be constructible for any given set of tampering functions (as long as they exist). Although inefficient for general tampering functions, the construction becomes useful if each function in the class $\mathcal{T}$ works independently on small blocks (of logarithmic size). In this case, we show that a non-malleable code for the overall state can be constructed efficiently by first applying Reed-Solomon code to the overall state and then applying δ-non-malleable codes for small blocks to the resulting codeword. We stress that this construction is intended as a feasibility result.

1.2 Related Work

We briefly describe the most relevant previous works addressing protection against tampering. We note that none of these works had addressed tampering in the context of UC-secure protocols.

Gennaro et al. [23] considered a device with two separate components: one is tamper-proof yet readable (circuitry), and the other is tamperable yet read-proof (memory). They defined algorithmic tamper-proof (ATP) security and explored its possibility for signature and decryption devices. Their definition of ATP security was given only for the specific tasks of signature and encryption. In contrast, our definition is simulation based, independent of the correctness or security objectives of the protocol, and we consider general two-party protocols (and the implications in the UC framework [12,28]).

Ishai et al. [26] considered an adversary who can tamper with the wires of a circuit. They showed a general compiler that outputs a self-destructing circuit that withstands such a tampering adversary. Considering that memory corresponds to a subset of the wires associated with the state in their model, the model seems stronger than ours (as we consider only the state, not the computation circuit). However, the tampering attack they considered is very limited: it modifies a bounded subset of the wires between each invocation, which corresponds to tampering memory only partially.

Dziembowski et al. [21] introduced the notion of non-malleable codes and tamper simulatability to address similar concerns as the present work. A distinguishing feature of BiTR security from their approach is that BiTR is protocol-centric. As such, it allows arguing about tamper resilience by taking advantage of specific protocol design features that enable BiTR even without any encodings. Moreover, the positive results of [21] require the introduction of additional circuitry or a randomness device; this may be infeasible, uneconomical or even unsafe in practice — it could be introducing new pathways for attacks. In contrast, our positive results do not require state encodings or when they do, they do not rely on randomness.

Bellare and Kohno defined security against related key attacks (RKA) for block ciphers [6], and there has been follow-up work [5,3] (see also the references therein). Roughly speaking, RKA-security as it applies to PRFs and encryption is a strengthening of the security definition of the underlying primitive (be it indistinguishability from random functions or semantic security). RKA-security was only shown against tampering that included addition or multiplication (but not both simultaneously). In fact, RKA-security for PRFs as defined in [5] is different from BiTR when applied to PRFs. A BiTR PRF is not necessarily RKA-secure since the BiTR simulator is allowed to take some liberties that would violate key independence under tampering as required by RKA-security. We do not pursue these relationships further here formally as it is our intention to capture BiTR in a weakest possible sense and investigate how it captures naturally in a simulation-based fashion the concept of tamper resilience for any cryptographic primitive.

2   BiTR Definitions

BiTR Protocols. Katz [28] modeled usage of a tamper-proof hardware token as an ideal functionality $\mathcal{F}_{wrap}$ in the UC framework. Here, we slightly modify the functionality so that it is parameterized by an interactive Turing machine (ITM) M for a two-party protocol[2] (see Fig. 1). The modification does not change the essence of the wrapper functionality; it merely binds honest parties to the use of a specific embedded program. Corrupted parties may embed an arbitrary program in the token by invoking Forge. We also define a new functionality $\mathcal{F}_{twrap}$ similar to $\mathcal{F}_{wrap}$ but with tampering allowed. Let $\mathcal{T}$ be a collection of (randomized) functions. Let $\psi = (E, D)$ be an encoding scheme[3]. The essential difference between $\mathcal{F}_{twrap}$ and $\mathcal{F}_{wrap}$ is the ability of the adversary to tamper with the internal state of the hardware token — a function drawn from $\mathcal{T}$ is applied on the internal state of the hardware token. This (weaker) ideal functionality notion is fundamental for the definition of BiTR that comes next.

We define a security notion for a protocol M, called Built-in Tamper Resilience (BiTR), which essentially requires that $\mathcal{F}_{twrap}(M)$ is interchangeable with $\mathcal{F}_{wrap}(M)$. We adopt the notations in the UC framework given by Canetti [12].

Definition 1 (BiTR protocol). *The protocol M is $(\mathcal{T}, \psi)$-BiTR if for any PPT $\mathcal{A}$, there exists a PPT $\mathcal{S}$ such that for any non-uniform PPT $\mathcal{Z}$,*

$$\text{IDEAL}_{\mathcal{F}_{twrap}(M,\mathcal{T},\psi),\mathcal{A},\mathcal{Z}} \approx \text{IDEAL}_{\mathcal{F}_{wrap}(M),\mathcal{S},\mathcal{Z}} ,$$

where $\approx$ denotes computational indistinguishability.

---

[2] We will interchangeably use protocols and ITMs, if there is no confusion.

[3] We will sometimes omit ψ from $\mathcal{F}_{twrap}$ when it is obvious from the context.

$\mathcal{F}_{wrap}(M)$ is parameterized by a polynomial p and a security parameter k.

Create: Upon receiving $\langle$Create, $sid, P, P', msg\rangle$ from party P: Let $msg' = ($Initialize, $msg)$. Run $M(msg')$ for at most $p(k)$ steps. Let out be the response of M (set out to $\perp$ if M does not respond). Let s' be the updated state of M. Send $\langle$Initialized, $sid, P', out\rangle$ to P, and $\langle$Create, $sid, P, P', 1^{|s'|}\rangle$ to P' and the adversary. If there is no record $(P, P', *, *)$, then store (P, P', M, s').

Forge: Upon receiving $\langle$Forge, $sid, P, P', M', s\rangle$ from the adversary, if P is not corrupted, do nothing. Otherwise, send $\langle$Create, $sid, P, P', 1^{|s|}\rangle$ to P'. If there is no record $(P, P', *, *)$, then store (P, P', M', s).

Run: Upon receiving $\langle$Run, $sid, P, msg\rangle$ from party P', find a record (P, P', K, s). If there is no such record, do nothing. Otherwise, do:
1. Run $K(msg; s)$ for at most $p(k)$ steps. Let out be the response of K (set out to $\perp$ if K does not respond). Let s' be the updated state of K. Send (sid, P, out) to P'.
2. Update the record with (P, P', K, s').

$\mathcal{F}_{twrap}(M, \mathcal{T}, \psi)$ is also parameterized by p and k (and $\psi = (E, D)$ is an encoding scheme).

Create: As in $\mathcal{F}_{wrap}(M)$ with the only change that state s' is stored as $E(s')$ in memory.

Forge: As in $\mathcal{F}_{wrap}(M)$.

Run: Upon receiving $\langle$Run, $sid, P, msg\rangle$ from party P', find a record $(P, P', K, \tilde{s})$. If there is no such record, do nothing. Otherwise, do:
1. (Tampering) If P' is corrupted and a record $\langle sid, P, P', \tau\rangle$ exists, set $\tilde{s} = \tau(\tilde{s})$ and erase the record.
2. (Decoding) If P is corrupted, set $s = \tilde{s}$; otherwise, set $s = D(\tilde{s})$. If $s = \perp$, send $(sid, P, \perp)$ to P' and stop.
3. Run $K(msg; s)$ for at most $p(k)$ steps. Let out be the response of K (set out to $\perp$ if K does not respond). Let s' be the updated state of K. Send (sid, P, out) to P'.
4. (Encoding) If P is corrupted, set $\tilde{s} = s'$; otherwise set $\tilde{s} = E(s')$. Update the record with $(P, P', K, \tilde{s})$.

Tamper: Upon receiving $\langle$Tamper, $sid, P, P', \tau\rangle$ from the adversary $\mathcal{A}$, if P' is not corrupted or $\tau \notin \mathcal{T}$, do nothing. Otherwise make a record (sid, P, P', τ) (erasing any previous record of the same form).

Fig. 1. Ideal functionalities $\mathcal{F}_{wrap}(M)$ and $\mathcal{F}_{twrap}(M, \mathcal{T}, \psi)$

In case $\psi = ($id, id$)$ (i.e., identify functions), we simply write $\mathcal{T}$-*BiTR*. Note that this definition is given through the ideal model, which implies (by the standard UC theorem) that whenever a tamper-proof token wrapping M can be used, it can be replaced by a $\mathcal{T}$-tamperable token wrapping M.[4] As a trivial example, every protocol is $\{$id$\}$-BiTR.

We note that the above definition is intended to capture in the weakest possible sense the fact that a protocol is tamper resilient within an arbitrary environment. A feature of the definition is that there is no restriction in the way the simulator accesses the underlying primitive (as long as no tampering is allowed). This enables, e.g., a signature to be called BiTR even if simulating tampered signatures requires untampered signatures on different chosen messages, or even on a larger number of chosen messages. We believe

---

[4] One could also consider a definition that requires this in the context of a *specific* UC-protocol. We believe our stronger definition, which holds for *any* UC-protocol using a token with M, is the right definition for built-in tamper resilience.

that this is the correct requirement for the definition to capture that "if the underlying primitive is secure without tampering, it is secure also with tampering" (in the signature example, security is unforgeability against any polynomial time chosen message attack). Nonetheless, it can be arguably even better to achieve BiTR security through a "tighter" simulation, where the BiTR simulator is somehow restricted to behave in a manner that is closer to the way $\mathcal{A}$ operates (except for tampering of course) or possibly even more restricted. For instance, one may restrict the number of times the token is accessed by the simulator to be upper bounded by the number of times $\mathcal{A}$ accesses the token. In fact all our positive results do satisfy this desired additional tighter simulation property. Taking this logic even further, one may even require that once tampering occurs the BiTR simulator can complete the simulation without accessing the token at all — effectively suggesting that tampering trivializes the token and makes it entirely simulatable. We believe that the ability of BiTR to be readily extended to capture such more powerful scenarios highlights the robustness of our notion and, even though these scenarios are not further pursued here, the present work provides the right basis for such upcoming investigations.

2.1   Composition of BiTR ITMs

It is natural to ask if a modular design approach applies to BiTR protocols. To investigate this question we need first to consider how to define the BiTR property in a setting where protocols are allowed to call subroutines.

Consider an ITM M_2 and another ITM M_1 that calls M_2 as a subroutine. We denote by $(M_1; M_2)$ the compound ITM. The internal state of $(M_1; M_2)$ is represented by the concatenation of the two states $s_1 \| s_2$ where s_1 and s_2 are the states of M_1 and M_2 at a certain moment of the runtime respectively. Let $\mathcal{F}_{twrap}(M_1; M_2, \mathcal{T}_1 \times \mathcal{T}_2, \psi_1 \times \psi_2)$ denote an ideal functionality that permits tampering with functions from $\mathcal{T}_1$ for the state of M_1 and from $\mathcal{T}_2$ for the state of M_2 while the states are encoded with ψ_1 and ψ_2 respectively. We can also consider a sequence of ITMs that call each other successively $\overline{M} = (M_1; \dots; M_n)$. We next generalize the BiTR notion for an ITM M_i employed in the context of $\overline{M}$ in a straightforward manner.

Definition 2 (modular BiTR protocol). *Given* $\overline{M} = (M_1; \dots; M_n)$, $\overline{\mathcal{T}} = \mathcal{T}_1 \times \dots \times \mathcal{T}_n$, *and* $\overline{\psi} = \psi_1 \times \dots \times \psi_n$, *for some* $i \in [n]$, *we say that* M_i *is modular-*$(\mathcal{T}_i, \psi_i)$*-BiTR with respect to* $\overline{M}, \overline{\mathcal{T}}$ *and* $\overline{\psi}$ *if for any PPT* $\mathcal{A}$ *there exists a PPT* $\mathcal{S}$ *such that for any non-uniform PPT* $\mathcal{Z}$,

$$\text{IDEAL}_{\mathcal{F}_{twrap}(\overline{M}, \overline{\mathcal{T}_i}, \overline{\psi}), \mathcal{A}, \mathcal{Z}} \approx \text{IDEAL}_{\mathcal{F}_{twrap}(\overline{M}, \overline{\mathcal{T}_{i+1}}, \overline{\psi}), \mathcal{S}, \mathcal{Z}},$$

where $\overline{\mathcal{T}_i} = \{id\} \times \dots \times \{id\} \times \mathcal{T}_i \times \dots \times \mathcal{T}_n$.

Roughly speaking, this definition requires that whatever the adversary can do by tampering M_i with $\mathcal{T}_i$ (on the left-hand side) should be also done without (on the right-hand side) in the context of $\overline{M}, \overline{\mathcal{T}}, \overline{\psi}$. For simplicity, if $\overline{M}, \overline{\mathcal{T}}, \overline{\psi}$ are clear from the context, we will omit a reference to it and call an ITM M_i simply modular-$(\mathcal{T}_i, \psi_i)$-BiTR.

The composition theorem below confirms that each ITM being modular BiTR is a necessary and sufficient condition for the overall compound ITM being BiTR.

Theorem 1 (BiTR Composition Theorem). *Consider protocols $M_1, \ldots, M_n$ with $\overline{M} = (M_1, \ldots, M_n)$ and $\mathcal{T} = \mathcal{T}_1 \times \ldots \times \mathcal{T}_n$, and $\psi = \psi_1 \times \ldots \times \psi_n$. It holds that M_i is modular-$(\mathcal{T}_i, \psi_i)$-BiTR for $i = 1, \ldots, n$, with respect to $\overline{M}, \overline{\mathcal{T}}, \overline{\psi}$ if and only if $(M_1; \ldots; M_n)$ is $(\mathcal{T}, \psi)$-BiTR.*

A natural task that arises next is to understand the modular-BiTR notion.

Context Sensitivity of Modular-BiTR Security. The modular-BiTR definition is context-sensitive; an ITM may be modular BiTR in some contexts but not in others, in particular depending on the overall compound token $\overline{M}$. This naturally begs a question whether there is a modular-BiTR ITM that is insensitive to the context. In this way, akin to a universally composable protocol, a universally BiTR ITM could be used modularly together with any other ITM and still retain its BiTR property. To capture this we formalize universal-BiTR security below, as well as a weaker variant of it that is called *universal-BiTR parent* which applies only to ITMs used as the parent in a sequence of ITMs.

Definition 3 (universal BiTR). *If an ITM M is modular-$(\mathcal{T}, \psi)$-BiTR with respect to any possible $\overline{M}, \overline{\mathcal{T}}, \overline{\psi}$ then we call M universal-$(\mathcal{T}, \psi)$-BiTR. If M is modular-$(\mathcal{T}, \psi)$-BiTR whenever M is used as the parent ITM then we call it universal-$(\mathcal{T}, \psi)$-BiTR parent.*

Not very surprisingly (and in a parallel to the case of UC protocols) this property is very difficult to achieve. In fact, we show that if one-way functions exist then *non-trivial* universal-BiTR ITMs do not exist. We first define non-triviality: an ITM M will be called non-trivial if the set of its states can be partitioned into at least two sets S_0, S_1 and there exists a set of inputs A that produce distinct outputs depending when the ITM M is called and its internal state belongs to S_0 or S_1. We call the pair of sets *a state partition for* M and the set A *the distinguishing input-set*. Note that if an ITM is trivial then for any partition of the set of states S_0, S_1 and any set of inputs A, the calling of the ITM M on A produces identical output. This effectively means that the ITM M does not utilize its internal state at all and obviously is BiTR by default. Regarding non-trivial ITMs we next prove that they cannot be $(\mathcal{T}, \psi)$-BiTR for any tampering function τ that switches the state between the two sets S_0, S_1, i.e., $\tau(S_0) \subseteq S_1, \tau(S_1) \subseteq S_0$. We call such tampering function *state-switching* for the ITM M. If an encoding ψ is involved, we call τ state-switching for the encoding ψ. We are now ready to prove our negative result.

Theorem 2. *Assuming one-way functions exist, there is no non-trivial universal-$(\mathcal{T}, \psi)$-BiTR ITM M such that $\mathcal{T}$ contains a state-switching function for M and the encoding ψ.*

Roughly speaking, the theorem holds since a parent ITM M_1 calling M_2 can make the message exchanges between them quite non-malleable by outputting a signature on these messages. In this context, no non-trivial M_2 can be modular-BiTR, and thus M_2 is not universal-BiTR. We note that the above theorem is quite final for the case of universal BiTR ITMs. It leaves only the possibility of proving the universal-BiTR property for trivial ITMs (that by default satisfy the notion) or for sets of functions that are not state-switching, i.e., essentially they do not affect the output of M and therefore inconsequential. This state of affairs is not foreign to properties that are supposed to universally compose. Indeed, in the case of UC-security large classes of functionalities are not UC-realizable [15]. To counter this issue, in the UC-setting one may seek setup assumptions to alleviate this problem, but in our setting setup assumptions should be avoided. For this reason, proving the modular-BiTR property within a given context is preferable.

On the other hand, the universal-BiTR parent property turns out to be feasible, and thus this leaves a context insensitive property to be utilized for modular design of BiTR protocols. We in fact take advantage of this, and jumping ahead, the parent ITM in the compound ITM used to achieve general UC-secure MPC in Section 4 satisfies this property and can be composed with any child ITM.

3 Affine BiTR Protocols without State Encoding

In this section, we show two protocols (for identification and signatures, respectively) that are BiTR against certain tampering functions, without using any modification or encoding. Specifically, we consider a tampering adversary that can modify the state of the hardware with *affine functions*. Assuming the state of the hardware is represented by variables of $\mathbb{Z}_q$ for some prime q, the adversary can choose a tampering $f_{a,b}$ on a variable v, which will change v into $f_{a,b}(v) = av + b \bmod q$. Let $\mathcal{T}_{\mathsf{aff}} = \{f_{a,b} \mid a \in \mathbb{Z}_q^*, b \in \mathbb{Z}_q\}$ and $\mathcal{T}_{\mathsf{aff}}^2 = \mathcal{T}_{\mathsf{aff}} \times \mathcal{T}_{\mathsf{aff}}$.

Schnorr Identification [39]. The Schnorr identification is a two-round two-party protocol between a prover and a verifier. The common input is $y = g^x$, where g is a generator of a cyclic group of size q, and the prover's auxiliary input is $x \in \mathbb{Z}_q$. The protocol proceeds as follows:

1. The prover picks a random $t \in \mathbb{Z}_q$ and sends $z = g^t$ to the verifier.
2. The verifier picks a random $c \in \mathbb{Z}_q$ and sends c to the prover, which in turn computes $s = cx + t \bmod q$ and sends s to the verifier. The verifier checks if $zy^c = g^s$.

We consider an ITM M on the prover side wrapped as a hardware token. This ITM is BiTR against affine functions. To see why it is BiTR, suppose that the adversary tampers with the state changing x into $ax + b$ for some a and b. In the second round, the BiTR simulator — given c, from the adversary, that is supposed to go to $\mathcal{F}_{twrap}(M; \mathcal{T}_{\mathsf{aff}})$ — has to find out an appropriate c' going to $\mathcal{F}_{wrap}(M)$ such that the simulator, on receiving $s' = c'x + t$ from $\mathcal{F}_{wrap}(M)$, can output $c(ax + b) + t$ that would come from $\mathcal{F}_{twrap}(M; \mathcal{T}_{\mathsf{aff}})$. In summary, given (a, b, c, s'), but not x or t, the simulator has to generate a correct output by controlling c'. It can do so by choosing $c' = ac$ and outputting $s' + cb$. Note that $s' + cb = c(ax + b) + t$.

Signature Scheme due to Okamoto [35]. The digital signature scheme of Okamoto [35] was employed in the context of designing blind signatures. Here we show that it is BiTR against affine functions. We give a brief description next. Let $(\mathbb{G}_1, \mathbb{G}_2)$ be a bilinear group as follows: (1) $\mathbb{G}_1$ and $\mathbb{G}_2$ are two cyclic groups of prime order q possibly with $\mathbb{G}_1 = \mathbb{G}_2$; (2) h_1 and h_2 are generators of $\mathbb{G}_1$ and $\mathbb{G}_2$ respectively; (3)

M_{oka}: The description of $\mathbb{G}_1$, $\mathbb{G}_2$, g_2, u_2, v_2, and a collision-resistant hashing function H : $\{0, 1\}^n \to \mathbb{Z}_q^*$ are embedded in the program as a public parameter. The state is $x \in \mathbb{Z}_q$.

Initialization

- Upon receiving a message (Initialize), choose $x \in_R \mathbb{Z}_q$, and $g_2, u_2, v_2 \in_R G_2$ and output (g_2, w_2, u_2, v_2).

Message Handling

- Upon receiving a message (Sign, m), Choose random $r, s \in \mathbb{Z}_q^*$ such that $x + r \neq 0$ (mod q). Compute $\sigma = (g_1^{H(m)} u_1 v_1^s)^{1/(x+r)}$ and output (σ, r, s).

Fig. 2. Okamoto signature M_{oka}

ψ is an isomorphism from $\mathbb{G}_2$ to $\mathbb{G}_1$ such that $\psi(h_2) = h_1$; (4) e is a non-degenerate bilinear map $e : \mathbb{G}_1 \times \mathbb{G}_2 \to \mathbb{G}_T$ where $|\mathbb{G}_T| = p$, $\forall u \in \mathbb{G}_1$ $\forall v \in \mathbb{G}_2$ $\forall a, b \in \mathbb{Z}$: $e(u^a, u^b) = e(u, v)^{ab}$.

The signature scheme below is secure against a chosen message attack under the Strong Diffie-Hellman assumption [35].

- Key Generation: Randomly select generators $g_2, u_2, v_2 \in G_2$ and compute $g_1 = \psi(g_2)$, $u_1 = \psi(u_2)$, and $v_1 = \psi(v_2)$. Choose a random $x \in \mathbb{Z}_q^*$ and compute $w_2 = g_2^x$. Verification key is $(g_1, g_2, w_2, u_2, v_2)$. Signing key is x.
- Signature of a message $m \in \mathbb{Z}_q^*$: Choose random $r, s \in \mathbb{Z}_q^*$. The signature is (σ, r, s) where $\sigma = (g_1^m u_1 v_1^s)^{1/(x+r)}$ and $x + r \neq 0$ (mod q).
- Verification of (m, σ, r, s): Check that $m, r, s, \in \mathbb{Z}_q^*$, $\sigma \in \mathbb{G}_1$, $\sigma \neq 1$, and $e(\sigma, w_2 g_2^r) = e(g_1, g_2^m u_2 v_2^s)$.

The signature token is described in Fig. 2. Similarly to the ITM for Schnorr signature scheme, this token can be shown to be BiTR against affine functions.

Theorem 3. *ITM M_{oka} in Fig. 2 is $\mathcal{T}_{\mathsf{aff}}$-BiTR.*

4   UC Secure Computation from Tamperable Tokens

In this section we examine the problem of achieving UC-secure computation relying on tamperable (rather than tamper-proof) tokens. Our starting point is the result of Katz [28], obtaining a UC commitment scheme (and general UC-secure computation) in the $\mathcal{F}_{wrap}(M)$-hybrid for an ITM M, which unfortunately, is not BiTR. However, we managed to change M so that the modified ITM M' is BiTR against affine functions, thus obtaining a UC commitment in the $\mathcal{F}_{twrap}(M')$-hybrid. Along the way, we present a generalization of Katz's scheme for building commitment schemes we call commitments with dual-mode parameter generation.

4.1   Katz's Commitment Scheme and its Generalization.

Intuitively, the UC-secure commitment scheme given by Katz [28] uses the tamper-proof hardware token to give the simulator the advantage over the adversary to force the commitment scheme to become extractable (in case the sender is corrupted) or equivocal (in case the receiver is corrupted). In spirit, this idea can be traced to mixed commitment schemes introduced in [17], although the two results differ greatly in techniques.

We abstract the approach of [28] to build UC commitments in Fig. 3. The UC commitment scheme is based on a primitive that we call commitment with dual-mode parameter generation (DPG-commitment for short).

Commitment Phase:
1. Each of the sender and the receiver calls $\mathcal{F}_{wrap}(M)$ with a Create message.
2. Each party executes the procedure dual-mode parameter generation with the $\mathcal{F}_{wrap}(M)$. Let pS be the parameter the receiver obtained, and pR be one the sender obtained. The parameters pR and pS are exchanged.
3. The sender commits to a message m by sending $\langle C_1, C_2, \pi \rangle$, where C_1 is a commitment to m based on the parameter pS, C_2 is a statistically-binding commitment to m, and π is WI proof that (1) C_1 and C_2 commits to the same message, or (2) pR was generated in the extraction mode.

Opening Phase:
1. The sender reveals $\langle m, \pi' \rangle$, where m is the committed message, π' is WI proof that (1) C_2 commits to m, or (2) pR was generated in the extraction mode.

Fig. 3. A UC Commitment that uses a DPG-commitment scheme Π with protocol M in the $\mathcal{F}_{wrap}(M)$-hybrid model.

A DPG-commitment is a commitment scheme whose parameter is generated by an interactive protocol M that is wrapped in a hardware token. Formally we define the following:

Definition 4 (DPG-Commitment scheme). *A commitment scheme $\Pi = (Com, Decom)$ that is parameterized by p, has a dual mode parameter generation (DPG-commitment) if there are ITMs M and P that form a two party protocol $\langle P, M \rangle$ and have the following properties:*

- *(Normal mode) For any PPT P^*, with overwhelming probability, the output of $\langle P^*, M \rangle$ satisfies that if it is not $\perp$ then it contains a parameter p over which the commitment scheme Π is unconditionally hiding.*
- *(Extraction mode) For any M^* with the same I/O as M, there is a PPT S that returns (p, t) such that the commitment scheme Π with the parameter p is a trapdoor commitment scheme with trapdoor t and the parameter generated by S is computationally indistinguishable from the parameter generated by $\langle P, M^* \rangle$.*

It is worth noting that DPG-commitments are very different from the mixed commitments of [17]. For one thing, contrary to mixed commitments, DPG-commitments do not have equivocal parameters. Moreover, mixed commitments have parameters that with overwhelming probability become extractable based on a trapdoor hidden in the common reference string. In contrast, DPG-commitments become extractable due to the manipulation of the parameter generation protocol M (specifically the ability of the simulator to rewind it). Now using the same arguments as in [28] it is possible to show that the commitment scheme in figure 3 is a UC-commitment provided that the underlying scheme used for C_1 is a DPG-commitment. We briefly sketch the proof argument. When the sender is corrupted, the simulator has to extract the committed message. This can be done by making pS extractable. Then, given a commitment $\langle C_1, C_2, \pi \rangle$ from the adversary, the simulator can extract the message committed to from C_1 using the trapdoor of pS. When the receiver is corrupted, the simulator can make the commitment equivocal by causing pR to be extractable. Using the trapdoor for pR as witness, the simulator can generate a WI proofs π and π' with respect to the condition (2) and thus open the commitment to an arbitrary message.

We next briefly argue that the construction suggested in [28] amounts to a DPG-commitment scheme. The token operates over a multiplicative cyclic group of prime order. In the first round, a party generates a cyclic group and sends to the token the group description and random elements g and h of the group; then, the token sends back a Pedersen commitment $c = \mathsf{com}(g_1, g_2)$ to random g_1, g_2 [36].[5] In the second final round, the party sends a random h_1, h_2, and then the token opens the commitment c and outputs the signature on $(g, h, \hat{g}_1, \hat{g}_2)$ where $\hat{g}_1 = g_1 h_1$ and $\hat{g}_2 = g_2 h_2$. With parameter $(g, h, \hat{g}_1, \hat{g}_2)$, commitment C_1 to a bit b is defined as $(g^{r_1} h^{r_2}, \hat{g}_1^{r_1} \hat{g}_2^{r_2} g^b)$ for randomly-chosen $r_1, r_2 \in \mathbb{Z}_q$. It is well-known (and easy to check) that if the parameter is a Diffie-Hellman (DH) tuple and $r = \log_g \hat{g}_1 = \log_h \hat{g}_2$ is known, then b can be efficiently extracted from the commitment. On the other hand, if it is a random tuple, this commitment scheme is perfectly hiding. Extraction mode is achieved by rewinding the code of a malicious token M^*. Specifically for a given M^*, the simulator S proceeds by picking a random DH tuple $(g, h, \hat{g}_1 = g^t, \hat{g}_2 = h^t)$ and running M^* once to reach a successful termination and learn the values g_1, g_2. Subsequently, it rewinds M^* right before the second round and selects $h_1 = \hat{g}_1/g_1$ and $h_2 = \hat{g}_2/g_2$. This will result in the parameter produced by M^* to be equal to the DH tuple, i.e., a parameter that is extractable with trapdoor t.

4.2 UC-Secure Commitment Scheme from a Tamperable Token

It is easy to see that the following result holds using the BiTR security properties.

Corollary 4. *If an ITM M, achieving parameters for DPG-commitment scheme, is $\mathcal{T}$-BiTR, then there exists a UC-secure commitment scheme in the $\mathcal{F}_{twrap}(M, \mathcal{T})$-hybrid model.*

Therefore, if the token used in [28] is $\mathcal{T}_{\mathsf{aff}}$-BiTR, then we obtain a UC-secure commitment scheme in the $\mathcal{F}_{twrap}(M, \mathcal{T}_{\mathsf{aff}})$-hybrid model. Unfortunately, the token is not $\mathcal{T}_{\mathsf{aff}}$-BiTR. We explain the issue below. Recall that in the first round the token sends a commitment to g_1, g_2. Suppose that $g_1 = g^{r_1}$ and $g_2 = g^{r_2}$ and that the values r_1 and r_2 are stored as state in the token after the first round. Suppose in addition that by tampering with an affine function the adversary causes the state to become $(ar_1 + b, r_2)$ for some a and b. Then, in the second round, the simulator — given h_1 and h_2 from

---

[5] We use a slightly different notation compared to [28] to unify the presentation with our BiTR token that is shown later.

Let $\mathbb{G}$ be the cyclic multiplicative group of size q defined by a safe prime $p = 2q + 1$ and g be a generator of $\mathbb{G}$. The description of $\mathbb{G}$ is embedded in the program. The state is $(r_1, r_2, s_1, s_2) \in \mathbb{Z}_q^4$. It uses a signature ITM K as a subprotocol.

Initialization

- Upon receiving a message (Initialize), call K with (Initialize), sets the state to all 0s and output whatever K outputs.

Message Handling

- Upon receiving a message h_0: Check h_0 is a generator of $\mathbb{G}$. If the checking fails, output $\perp$. Otherwise, pick $r_i, s_i \in_R \mathbb{Z}_q$ and compute Pedersen commitments $\mathsf{com}_i = g^{s_i} h_0^{\mathcal{X}(g_i)}$ for $i = 1, 2$, where $g_i = g^{r_i}$ and $\mathcal{X}$ is defined as: $\mathcal{X}(\alpha) = \alpha$ if $\alpha > p/2$, $p - \alpha$ otherwise. Output $(\mathsf{com}_1, \mathsf{com}_2)$.
- Upon receiving a message (h, h_1, h_2, x_1, x_2): Check $h, h_1, h_2 \in \mathbb{G}$, $x_1, x_2 \in \mathbb{Z}_q^*$. If the checking fails, output $\perp$. Otherwise, let $g_i = g^{r_i}$ and compute $\hat{g}_i = g_i^{x_i} h_i$ for $i = 1, 2$. Call K with $(Sign, (P, P', p, g, h, \hat{g}_1, \hat{g}_2))$ to get a signature σ. Output $(g_1, g_2, s_1, s_2, \sigma)$. Pick $r_i, s_i \in_R \mathbb{Z}_q$ for $i = 1, 2$.

Fig. 4. Dual parameter generating ITM M_{dpg} that is universal-BiTR parent

the adversary — has to send $\mathcal{F}_{wrap}$ appropriate messages h_1' and h_2 so that it can manipulate the output from $\mathcal{F}_{wrap}$ as if the result is from $\mathcal{F}_{twrap}$. Here the signature on $(g, h, \hat{g}_1, \hat{g}_2)$ is a critical obstacle, since the simulator cannot modify it (otherwise, it violates unforgeability of signature schemes). This means that for simulation to be successful it should hold that $\hat{g}_1 = g^{ar_1+b}h_1 = g^{r_1}h_1'$, i.e., the simulator should select $h_1' = g^{(a-1)r_1+b}h_1$. Unfortunately, the simulator does not know r_1 when it is supposed to send h_1'.

By slightly changing the token above, however, we manage to obtain a DPG-achieving ITM M_{dpg} that is BiTR against affine tampering functions. Its description is given in Fig. 4. First, we show M_{dpg} achieves parameters for DPG-commitment. Roughly speaking, the protocol in the normal mode generates a random tuple $(g, h, \hat{g}_1, \hat{g}_2)$, by multiplying random numbers g_1 and g_2 (from M_{dpg}) and random numbers h_1 and h_2 (from the party). Therefore, the probability that the tuple $(g, h, \hat{g}_1, \hat{g}_2)$ is a DH tuple is negligible since $\hat{g}_1$ and $\hat{g}_2$ are uniformly distributed. In the extraction mode, however, the simulator emulating $\mathcal{F}_{wrap}$ can rewind the ITM to cause $(g, h, \hat{g}_1, \hat{g}_2)$ to be a DH tuple. Specifically, the simulator picks a random DH tuple $(g, h, \hat{g}_1, \hat{g}_2)$ and, after finding out the values g_1, g_2, rewinds the machine right before the second round and sends $h_i = \hat{g}_i / g_i^{x_i}$ for $i = 1, 2$. Under the DDH assumption, parameters from the normal mode and from the extraction mode are indistinguishable.

More importantly, M_{dpg} is BiTR against affine tampering functions. To achieve BiTR security, we introduce x_1 and x_2. As before, suppose that the state for g_1 is changed from r_1 to $ar_1 + b$. In the second round, the simulator — given h_1 and x_1 — has to send appropriate h_1' and x_1' to $\mathcal{F}_{wrap}$ such that $\hat{g}_1 = g^{(ar_1+b)x_1}h_1 = g^{r_1 x_1'}h_1'$. This means that $h_1' = g^z h_1$ where $z = (ar_1 x_1 + bx_1 - r_1 x_1')$. The good news is that although the simulator doesn't know r_1, it does know how to pick x_1' to satisfy the equation: $x_1' = ax_1$. The value h_1' can be computed subsequently from the above equation.

Theorem 5. *The ITM* M_{dpg} *in Fig. 4 is* T^4_{aff}*-BiTR.*

Furthermore, the way the ITM M_{dpg} uses a signature scheme is simple enough (it simply passes through whatever it receives from the signature token) and we can easily extend the above lemma to prove that M_{dpg} is universal BiTR parent. We also show that the ITM for the Okamoto signature scheme M_{oka} is modular-T_{aff}-BiTR when used with M_{dpg}.

Lemma 6. *ITM* M_{oka} *in Fig. 2 is modular-*T_{aff}*-BiTR with respect to* $(M_{dpg}; M_{oka})$.

Applying the composition theorem (Theorem 1) along with Theorem 5 and Lemma 6 to the above scheme, we obtain a BiTR token that gives a UC commitment based on corollary 4.

Corollary 7. $(M_{dpg}; M_{oka})$ *is* T^5_{aff}*-BiTR.*

5   BiTR Protocols against General Classes of Tampering Functions

5.1   BiTR Protocols from Non-malleable Codes

In this section we will see how the BiTR property can be derived by implementing an integrity check in the form of an encoding ψ. A useful tool for this objective is the notion of non-malleable codes [21]. A pair of procedures (E, D) is a non-malleable code with respect to tampering functions T, if there is an algorithm S that detects whether the state becomes invalid, given only the tampering function t. In particular S should satisfy the following property: for all $x \in \{0, 1\}^n$ and $t \in T$, if $x = D(t(E(x)))$ (i.e., x stays the same even after applying the tampering t), it holds that $S(t) = \top$ with overwhelming probability, while otherwise $S(t)$ is statistically (or computationally) close to $D(t(E(x)))$. By encoding the state of a protocol with a non-malleable code it is possible to show the following restatement of Theorem 6.1 of [21] under the BiTR security framework.

Theorem 8 ([21]). *Let* T *be a class of tampering functions over* $\{0, 1\}^m$ *and* (E, D, S) *be a non-malleable code with respect to* T, *where* $E : \{0, 1\}^n \to \{0, 1\}^m$, $D : \{0, 1\}^m \to \{0, 1\}^n$ *and* S *are efficient procedures. Let* M *be any ITM whose state is of length* n. *Then* M *is* (T, ψ)*-BiTR where* $\psi = (E, D)$.

The above theorem suggests the importance of the problem of constructing non-malleable codes for a given class of tampering functions T. Some positive answers to this difficult question are given in [21] for a class of tampering functions that operate on each one of the bits of the state independently; they also provide a general feasibility result for tampering families of bounded size (with an inefficient construction); an important characteristic of those solutions is relying on the randomness of the encoding. Here we show a different set of positive results by considering the case of *deterministic* non-malleable codes, i.e., the setting where (E, D) are both deterministic functions.

In our result we will utilize a relaxation of non-malleable codes: $(E, D, Predict)$ is called a δ-non-malleable code with distance ϵ if for any $x \in \{0, 1\}^n$ and $t \in T$, it holds that (i) $D(E(x)) = x$, (ii) the probability that $D(t(E(x)))$ is neither x nor

$\perp$ is at most δ,[6] and (iii) $Predict(\cdot)$ outputs either $\top$ or $\perp$, and $|\Pr[D(t(E(x))) = x] - \Pr[Predict(t) = \top]| \leq \epsilon$. It is easy to see that if ϵ, δ are negligible the resulting code is non-malleable: given that δ is negligible, property (ii) suggests that D will return either the correct value or fail, and thus in case it fails, $Predict(\cdot)$ will return $\perp$ with about the same probability due to (iii). We call δ the crossover threshold and ϵ the predictability distance.

5.2   Constructing Deterministic Non-malleable Codes

Inefficient Construction for Any $\mathcal{T}$. We now consider the problem of constructing a δ-non-malleable code $E : \{0,1\}^n \to \{0,1\}^m$ for a given class of tampering functions and parameters δ, ϵ. We will only consider the case when $\delta > \epsilon$ as the other case is not useful. We note that the construction is inefficient for large m and n, but it becomes efficient for logarithmic values of m, n. Following this we utilize it in the construction of deterministic non-malleable codes.

For a given $t \in \mathcal{T}$ consider the graph G that is defined with vertex set $V = \{0,1\}^m$ with each edge (u_1, u_2) having weight $w_t(u_1, u_2) = \Pr[t(u_1) = u_2]$.[7] Finding a good δ-non-malleable code amounts to finding a partition $S, \overline{S} = V \setminus S$ of G satisfying the following properties that for each $t \in \mathcal{T}$:

- For all $u, v \in S$, it holds that $w_t(u, v) \leq \delta$.
- Either (i) $\forall u \in S : \sum_{v \in \overline{S}} w_t(u, v) \geq 1 - \epsilon$ or (ii) $\forall u \in S : \sum_{v \in \overline{S}} w_t(u, v) \leq \epsilon$.

If S satisfies condition (i) (resp., condition (ii)) for a given $t \in \mathcal{T}$, we will say that S is a *repeller* (resp., an *attractor*) with respect to t.

We next provide a simple algorithm that is guaranteed to produce a code of non-zero rate if such exists. Consider all pairs of vertices $\{u_1, u_2\}$ and classify them according to whether they are repellers or attractors with parameters δ, ϵ. Note that testing whether these sets are repellers or attractors requires $O(|V|)$ steps. We perform the same for all tampering functions $t \in \mathcal{T}$ and then consider only those sets that appear in the list of all tampering functions. Finally, we improve the size of such a selected pair by moving vertices from $\overline{S}$ to S provided that the repeller or attractor property is maintained. We note that this approach will enable us to reach a local maximum code nevertheless it is not guaranteed to find an optimal code.

Assume now that the output of the above procedure is the set $\mathcal{C} \subseteq V = \{0,1\}^m$. We next set $n = \lfloor \log_2 |\mathcal{C}| \rfloor$ and consider $E : \{0,1\}^n \to \{0,1\}^m$ an arbitrary injection from $\{0,1\}^n$ to $\mathcal{C}$. The decoding D is defined as the inverse of E when restricted on $\mathcal{C}$, and $\perp$ everywhere else. We next define $Predict$ as follows. On input t, if $\mathcal{C}$ is an attractor, then output ok; otherwise output $\perp$ (i.e., for the case $\mathcal{C}$ is an repeller).

---

[6] The tampering t may change the codeword x into another valid codeword.

[7]   In the above description, we assumed the probabilities $\Pr[t(c) = u]$ are known. If they are not known, they can be estimated using standard techniques. In particular, to evaluate the probability of an event A, repeat k independent experiments of A and denote the success ratio of the k experiments as $\hat{p}$. Let X_i be the probability that the i-th execution of the event A is successful. The expected value of $Y = \sum_{i=1}^{k} X_i$ is $k \cdot p$. Using the Chernoff bound it follows that $|\hat{p} - p| \leq 1/N$ with probability $1 - \gamma$ provided that $k = \Omega(N^2 \ln(\gamma^{-1}))$.

The rate of the constructed code is n/m, while the time-complexity of construct-
ing $E, D, Predict(\cdot)$ is $2^{\mathcal{O}(n)}|\mathcal{T}|$. The size of the circuit evaluating each one of these
functions is respectively $2^n, 2^m, |\mathcal{T}|$.

Theorem 9. *Fix any class of functions $\mathcal{T}$. If there exists a code $(E, D, Predict)$ with
rate > 0 that is δ-non-malleable w.r.t. $\mathcal{T}$ and distance ϵ, then such a code is produced
by the above procedure.*

When does a deterministic non-malleable code exist? The basic idea of the construction above was to search for a one-sided set of codewords and use it to define the
non-malleable code. The necessity of one-sidedness is easy to see since if the property
fails, i.e., $\epsilon < q_{u,t} < 1 - \epsilon$ for some t and u, the requirement on $Predict$ cannot hold
in general since it cannot predict with high probability what would happen in the real
world after tampering a state that is encoded as u. We now provide two illustrative examples and discuss the existence (and rate) of a deterministic non-malleable encoding
for them.

Example 1: Set Functions. If $\mathcal{T}$ contains a function t that sets the i-th bit of $u \in \{0, 1\}^m$
to 0, it follows that the code $\mathcal{C}$ we construct must obey that either all codewords have
the i-th bit set to 0 or all of them have the bit set to 1. This means that the inclusion
of any bit setting function in $\mathcal{T}$ cuts the size of the code $|\mathcal{C}|$ by half. There is no non-
malleable code when the collection $\mathcal{T}$ contains Set functions for every bit position (this
is consistent with the impossibility result of [23] for algorithmic tamper proof security
when Set functions are allowed for tampering).

Example 2: Differential Fault Analysis [8]. Consider a single function t which flips
each 1-bit to a 0-bit with probability β. Consider a code $\mathcal{C} \subseteq \{0, 1\}^m$ for which it
holds that all codewords in $\mathcal{C}$ have Hamming distance at least r between each other
and $0^m \in \mathcal{C}$. Then it is easy to see that δ, the probability of crossover, is at most β^r.
Further, now suppose that t is applied to an arbitrary codeword u in $\mathcal{C}$ other than 0^m.
We observe that the number of 1's in u is at least r (otherwise it would have been too
close to 0^m). It follows that t will change some of these 1's to 0's, with probability at
least $1 - (1 - \beta)^r$. It follows that we can predict the effect of the application of t with
this probability when we restrict to codewords in $\mathcal{C} \setminus \{0^m\}$. In summary, any code $\mathcal{C}$
over $\{0, 1\}^m$ with minimum distance r that contains 0^m allows for a β^r-non-malleable
code with $(1 - \beta)^r$ for t using the code $\mathcal{C} \setminus \{0^m\}$.

We can extend the above to the case when a compositions of t are allowed. Note
that a sequence of a applications of t will flip each 1-bit to a 0-bit with probability
$\beta + (1 - \beta)\beta + \ldots + (1 - \beta)^{a-1}\beta = 1 - (1 - \beta)^a$. The encoding now has crossover
$(1 - (1 - \beta)^a)^r \leq e^{-(1-\beta)^a r}$. Thus, from $e^{-(1-\beta)^a r} \leq \delta$, we obtain $r \geq (1/(1 -
\beta))^a \ln(1/\delta)$, i.e., when β is bounded away from 1, the minimum distance of the code
grows exponentially with a.

Efficient Construction for Localized $\mathcal{T}$. Now, we show a simple way to use the
(inefficient) construction of the beginning of the section with constant rate and any
cross-over $\delta < 1/2$, to achieve an efficient construction with negligible cross-over (and
thus, BiTR security for any protocol M whose state is encoded with the resulting code),
when the class contains only functions that can be split into independent tampering

of local (i.e., logarithmically small) blocks. Here we consider a tampering class $\mathcal{T}$ of polynomial size. Roughly speaking, the construction is achieved first by applying a Reed-Solomon code to the overall state and then by applying the δ-non-malleable code to the resulting codeword in small blocks. Let $\mathcal{T}^\ell$ denote $\mathcal{T} \times \cdots \times \mathcal{T}$ (with ℓ repetitions).

Theorem 10. *Let k be a security parameter. Let $\mathcal{T}$ be a class of functions over $\{0,1\}^m$ with $m = \mathcal{O}(\log k)$ for which a δ-non-malleable code exists and is efficiently constructible with rate r. Then there is an efficiently constructible deterministic non-malleable code w.r.t. $\mathcal{T}^\ell$ for any rate less than $(1-\delta)r$ provided $\ell / \log \ell = \omega(\log k)$.*

Acknowledgement. We are grateful to Li-Yang Tan and Daniel Wichs for useful discussions regarding this work.

References

1. Alwen, J., Dodis, Y., Wichs, D.: Leakage-Resilient Public-Key Cryptography in the Bounded-Retrieval Model. In: Halevi, S. (ed.) CRYPTO 2009. LNCS, vol. 5677, pp. 36–54. Springer, Heidelberg (2009)
2. Anderson, R.J., Kuhn, M.G.: Tamper resistance – a cautionary note. In: The Second USENIX Workshop on Electronic Commerce Proceedings, Oakland, California, pp. 1–11 (18-21, 1996)
3. Applebaum, B., Harnik, D., Ishai, Y.: Semantic security under related-key attacks and applications. In: Innovations in Computer Science - ICS 2011, pp. 45–60 (2011)
4. Bar-El, H., Choukri, H., Naccache, D., Tunstall, M., Whelan, C.: The sorcerers apprentice guide to fault attacks. Cryptology ePrint Archive, Report 2004/100 (2004), http://eprint.iacr.org/
5. Bellare, M., Cash, D.: Pseudorandom Functions and Permutations Provably Secure against Related-Key Attacks. In: Rabin, T. (ed.) CRYPTO 2010. LNCS, vol. 6223, pp. 666–684. Springer, Heidelberg (2010)
6. Bellare, M., Kohno, T.: A Theoretical Treatment of Related-Key Attacks: Rka-prps, Rka-prfs, and Applications. In: Biham, E. (ed.) EUROCRYPT 2003. LNCS, vol. 2656, pp. 491–506. Springer, Heidelberg (2003)
7. Biham, E., Shamir, A.: Differential Fault Analysis of Secret Key Cryptosystems. In: Kaliski Jr., B.S. (ed.) CRYPTO 1997. LNCS, vol. 1294, pp. 513–525. Springer, Heidelberg (1997)
8. Boneh, D., DeMillo, R.A., Lipton, R.J.: On the importance of eliminating errors in cryptographic computations. J. Cryptology 14(2), 101–119 (2001)
9. Boyle, E., Segev, G., Wichs, D.: Fully Leakage-Resilient Signatures. In: Paterson, K.G. (ed.) EUROCRYPT 2011. LNCS, vol. 6632, pp. 89–108. Springer, Heidelberg (2011)
10. Brakerski, Z., Kalai, Y.T., Katz, J., Vaikuntanathan, V.: Overcoming the hole in the bucket: Public-key cryptography resilient to continual memory leakage. In: FOCS, pp. 501–510 (2010)
11. Brumley, D., Boneh, D.: Remote timing attacks are practical. Computer Networks 48(5), 701–716 (2005)
12. Canetti, R.: Universally composable security: A new paradigm for cryptographic protocols. In: FOCS, pp. 136–145 (2001)
13. Canetti, R., Fischlin, M.: Universally Composable Commitments. In: Kilian, J. (ed.) CRYPTO 2001. LNCS, vol. 2139, pp. 19–40. Springer, Heidelberg (2001)
14. Canetti, R., Goldreich, O., Goldwasser, S., Micali, S.: Resettable zero-knowledge (extended abstract). In: STOC, pp. 235–244 (2000)

15. Canetti, R., Kushilevitz, E., Lindell, Y.: On the Limitations of Universally Composable Two-Party Computation Without Set-Up Assumptions. In: Biham, E. (ed.) EUROCRYPT 2003. LNCS, vol. 2656, pp. 68–86. Springer, Heidelberg (2003)
16. Chandran, N., Goyal, V., Sahai, A.: New Constructions for UC Secure Computation Using Tamper-Proof Hardware. In: Smart, N.P. (ed.) EUROCRYPT 2008. LNCS, vol. 4965, pp. 545–562. Springer, Heidelberg (2008)
17. Damgård, I., Nielsen, J.B.: Perfect Hiding and Perfect Binding Universally Composable Commitment Schemes with Constant Expansion Factor. In: Yung, M. (ed.) CRYPTO 2002. LNCS, vol. 2442, pp. 581–596. Springer, Heidelberg (2002)
18. Damgård, I., Nielsen, J.B., Wichs, D.: Isolated Proofs of Knowledge and Isolated Zero Knowledge. In: Smart, N.P. (ed.) EUROCRYPT 2008. LNCS, vol. 4965, pp. 509–526. Springer, Heidelberg (2008)
19. Dodis, Y., Haralambiev, K., López-Alt, A., Wichs, D.: Cryptography against continuous memory attacks. In: FOCS, pp. 511–520 (2010)
20. Döttling, N., Kraschewski, D., Müller-Quade, J.: Unconditional and Composable Security Using a Single Stateful Tamper-Proof Hardware Token. In: Ishai, Y. (ed.) TCC 2011. LNCS, vol. 6597, pp. 164–181. Springer, Heidelberg (2011)
21. Dziembowski, S., Pietrzak, K., Wichs, D.: Non-malleable codes. In: ICS, pp. 434–452 (2010)
22. Faust, S., Kiltz, E., Pietrzak, K., Rothblum, G.N.: Leakage-Resilient Signatures. In: Micciancio, D. (ed.) TCC 2010. LNCS, vol. 5978, pp. 343–360. Springer, Heidelberg (2010)
23. Gennaro, R., Lysyanskaya, A., Malkin, T., Micali, S., Rabin, T.: Algorithmic Tamper-Proof (ATP) Security: Theoretical Foundations for Security against Hardware Tampering. In: Naor, M. (ed.) TCC 2004. LNCS, vol. 2951, pp. 258–277. Springer, Heidelberg (2004)
24. Goyal, V., Ishai, Y., Mahmoody, M., Sahai, A.: Interactive Locking, Zero-Knowledge PCPs, and Unconditional Cryptography. In: Rabin, T. (ed.) CRYPTO 2010. LNCS, vol. 6223, pp. 173–190. Springer, Heidelberg (2010)
25. Goyal, V., Ishai, Y., Sahai, A., Venkatesan, R., Wadia, A.: Founding Cryptography on Tamper-Proof Hardware Tokens. In: Micciancio, D. (ed.) TCC 2010. LNCS, vol. 5978, pp. 308–326. Springer, Heidelberg (2010)
26. Ishai, Y., Prabhakaran, M., Sahai, A., Wagner, D.: Private Circuits II: Keeping Secrets in Tamperable Circuits. In: Vaudenay, S. (ed.) EUROCRYPT 2006. LNCS, vol. 4004, pp. 308–327. Springer, Heidelberg (2006)
27. Ishai, Y., Sahai, A., Wagner, D.: Private Circuits: Securing Hardware against Probing Attacks. In: Boneh, D. (ed.) CRYPTO 2003. LNCS, vol. 2729, pp. 463–481. Springer, Heidelberg (2003)
28. Katz, J.: Universally Composable Multi-party Computation Using Tamper-Proof Hardware. In: Naor, M. (ed.) EUROCRYPT 2007. LNCS, vol. 4515, pp. 115–128. Springer, Heidelberg (2007)
29. Kocher, P.C., Jaffe, J., Jun, B.: Differential Power Analysis. In: Wiener, M. (ed.) CRYPTO 1999. LNCS, vol. 1666, pp. 388–397. Springer, Heidelberg (1999)
30. Kolesnikov, V.: Truly Efficient String Oblivious Transfer Using Resettable Tamper-Proof Tokens. In: Micciancio, D. (ed.) TCC 2010. LNCS, vol. 5978, pp. 327–342. Springer, Heidelberg (2010)
31. Lewko, A.B., Lewko, M., Waters, B.: How to leak on key updates. In: STOC, pp. 725–734 (2011)
32. Malkin, T., Teranishi, I., Vahlis, Y., Yung, M.: Signatures Resilient to Continual Leakage on Memory and Computation. In: Ishai, Y. (ed.) TCC 2011. LNCS, vol. 6597, pp. 89–106. Springer, Heidelberg (2011)
33. Micali, S., Reyzin, L.: Physically Observable Cryptography (Extended Abstract). In: Naor, M. (ed.) TCC 2004. LNCS, vol. 2951, pp. 278–296. Springer, Heidelberg (2004)

34. Moran, T., Segev, G.: David and Goliath Commitments: UC Computation for Asymmetric Parties Using Tamper-Proof Hardware. In: Smart, N.P. (ed.) EUROCRYPT 2008. LNCS, vol. 4965, pp. 527–544. Springer, Heidelberg (2008)
35. Okamoto, T.: Efficient Blind and Partially Blind Signatures Without Random Oracles. In: Halevi, S., Rabin, T. (eds.) TCC 2006. LNCS, vol. 3876, pp. 80–99. Springer, Heidelberg (2006)
36. Pedersen, T.P.: Non-interactive and Information-Theoretic Secure Verifiable Secret Sharing. In: Feigenbaum, J. (ed.) CRYPTO 1991. LNCS, vol. 576, pp. 129–140. Springer, Heidelberg (1992)
37. Pietrzak, K.: A Leakage-Resilient Mode of Operation. In: Joux, A. (ed.) EUROCRYPT 2009. LNCS, vol. 5479, pp. 462–482. Springer, Heidelberg (2009)
38. Quisquater, J.J., Samyde, D.: ElectroMagnetic Analysis (EMA): Measures and Counter-Measures for Smart Cards. In: Attali, S., Jensen, T. (eds.) E-smart 2001. LNCS, vol. 2140, pp. 200–210. Springer, Heidelberg (2001)
39. Schnorr, C.P.: Efficient signature generation by smart cards. J. Cryptology 4(3), 161–174 (1991)
40. Skorobogatov, S.P.: Semi-invasive attacks – A new approach to hardware security analysis. Tech. Rep. UCAM-CL-TR-630, University of Cambridge, Computer Laboratory (April 2005), http://www.cl.cam.ac.uk/techreports/UCAM-CL-TR-630.pdf
41. Standaert, F.X., Malkin, T.G., Yung, M.: A Unified Framework for the Analysis of Side-Channel Key Recovery Attacks. In: Joux, A. (ed.) EUROCRYPT 2009. LNCS, vol. 5479, pp. 443–461. Springer, Heidelberg (2009)

Author Index